MECHANICAL ENGINEERING

FE Review Manual

KAPLAN

ENGINEERING EDUCATION

Executive Director of Engineering Education: Brian S. Reitzel, PE

MECHANICAL ENGINEERING FE REVIEW MANUAL

Published by Kaplan Engineering Education

1-877-884-0828

www.kaplanengineering.com

Printed in the United States of America.

13 14 15 10 9 8 7 6 5 4 3 2 1

ISBN: 978-1-4277-4528-6 / 1-4277-4528-5
PPN: 3200-3319

CONTENTS

PERMISSIONS

"Rules of Professional Conduct," Chapter 6, reprinted by permission of NCEES
Source: Model Rules, National Council of Examiners for Engineering and Surveying, 2007. www.ncees.org

Figure 10.7 reprinted by permission of ASME.
Source: Moody, L., F, Transactions of the ASME, Volume 66: pp. 671–684. 1944.

The figure on page 732 courtesy of DuPont.
Source: Thermodynamic Properties of HFC-134a, DuPont Company.

Chapter 13 Exhibit 2 and Figures 13.4a-b, 13.5a-b, 13.6a, 13.9, 13.10, 13.18, 13.21, 13.22 and 13.23 reprinted with permission of John Wiley & Sons, Inc.
Source: Callister, William D., Jr. Materials Science and Engineering: An Introduction, 6/e. J. Wiley & Sons. 2003.

Figures 13.4c, 13.5c, and 13.6b reprinted by permission of the estate of William G. Moffatt.
Source: Moffatt, William G. The Structure and Property of Materials, Volume 1. J. Wiley & Sons. 1964.

Tables 13.4 and 13.5 reprinted by permission of McGraw-Hill Companies.
Source: Fontana, M., Corrosion Engineering. McGraw-Hill Companies.

Figure 13.13 used by permission of ASM International.
Source: Mason, Clyde W., Introductory Physical Metallurgy: p. 33. 1947.

Figure 13.26 used by permission of ASM International.
Source: Rinebolt, J.A., and W. J. Harris, Jr., "Effect of Alloying Elements on Notch Toughness of Pearlitic Steels." Transactions of ASM, Volume 43: pp. 1175–1201. 1951

CHAPTER AUTHORS

David R. Arterburn, PhD, New Mexico Institute of Mining and Technology

Gary R. Crossman, PE, Old Dominion University

Fidelis O. Eke, PhD, University of California, Davis

Brian Flinn, PhD, PE, University of Washington

James R. Hutchinson, PhD, University of California, Davis

Ray W. James, PhD, PE, Texas A&M University College Station

Lincoln D. Jones, PE, San Jose State University

Sharad Laxpati, PhD, PE, University of Illinois, Chicago

Robert F. Michel, PE, Old Dominion University

Donald G. Newnan, PhD, PE, San Jose State University

Charles E. Smith, PhD, Oregon State University

SAMPLE EXAM CONTRIBUTORS

Lisa M. Anneberg, PhD, PE, Lawrence Tech University

Russ Callen, PhD, PE, Georgia Institute of Technology

Bruce DeVantier, PhD, PE, Southern Illinois University

Candace Christy Hickey, PE, Consultant

Forrest G. Lowe, PhD, PE, University of Missouri, Kansas City

Ronald McPherson, PhD, PE, Consultant

Larry D. Peel, PhD, Texas A&M University Kingsville

Sally Steadman, PhD, PE, University of South Alabama

CHAPTER 1

Introduction

HOW TO USE THIS BOOK

Mechanical Fundamentals of Engineering Review Manual is designed to help you prepare for the Fundamentals of Engineering (FE) Mechanical CBT exam. The book covers the full breadth and depth of topics covered by the exam.

Each chapter of this book covers a major topic on the exam, reviewing important terms, equations, concepts, analysis methods, and typical problems. Solved examples are provided throughout each chapter to help you apply the concepts and to model problems you may see on the exam. After reviewing the topic, you can work the end of chapter problems to test your understanding. The problems are typical of what you will see on the exam, and complete solutions are provided so that you can check your work and further refine your solution methodology.

The following sections provide you with additional details on the process of becoming a licensed professional engineer and on what to expect at the exam.

BECOMING A PROFESSIONAL ENGINEER

To achieve registration as a Professional Engineer, there are four distinct steps: (1) education, (2) the Fundamentals of Engineering/Engineer-in-Training (FE/EIT) exam, (3) professional experience, and (4) the professional engineer (PE) exam. These steps are described in the following sections.

Education

Generally, no college degree is required to be eligible to take the FE/EIT exam. The exact rules vary, but all states allow engineering students to take the FE/EIT exam before they graduate, usually in their senior year. Some states, in fact, have no education requirement at all. One merely need apply and pay the application fee. Perhaps the best time to take the exam is immediately following completion of related coursework. For most engineering students, this will be the end of the senior year.

Fundamentals of Engineering/ Engineer-in-Training Examination

This six-hour, multiple-choice examination is known by a variety of names—Fundamentals of Engineering, Engineer-in-Training (EIT), and Intern Engineer—but no matter what it is called, the exam is the same in all states. It is prepared and graded by the National Council of Examiners for Engineering and Surveying (NCEES).

Experience

States that allow engineering seniors to take the FE/EIT exam have no experience requirement. These same states, however, generally will allow other applicants to substitute acceptable experience for coursework. Still other states may allow a candidate to take the FE/EIT exam without any education or experience requirements.

Typically, several years of acceptable experience is required before you can take the Professional Engineer exam—the duration varies by state, and you should check with your state licensing board for details.

Professional Engineer Examination

The second national exam is called Principles and Practice of Engineering by NCEES, but many refer to it as the Professional Engineer exam or PE exam. All states, plus Guam, the District of Columbia, and Puerto Rico, use the same NCEES exam. Review materials for this exam are found in other engineering license review books.

FUNDAMENTALS OF ENGINEERING/ENGINEER-IN-TRAINING EXAMINATION

Laws have been passed that regulate the practice of engineering in order to protect the public from incompetent practitioners. Beginning in 1907 the individual states began passing *title* acts regulating who could call themselves engineers and offer services to the public. As the laws were strengthened, the practice of engineering was limited to those who were registered engineers, or to those working under the supervision of a registered engineer. Originally the laws were limited to civil engineering, but over time they have evolved so that the titles, and sometimes the practice, of most branches of engineering are included.

There is no national licensure law; licensure is based on individual state laws and is administered by boards of registration in each state. You can find a list of contact information for and links to the various state boards of registration at the Kaplan AE Web site: *www.kaplanengineering.com*. This list also shows the exam registration deadline for each state.

Examination Development

Initially, the states wrote their own examinations, but beginning in 1966 NCEES took over the task for some of the states. Now the NCEES exams are used by all states. Thus it is easy for engineers who move from one state to another to achieve licensure in the new state. About 50,000 engineers take the FE/EIT exam annually. This represents about 65% of the engineers graduated in the United States each year.

The development of the FE/EIT exam is the responsibility of the NCEES Committee on Examination for Professional Engineers. The committee is composed of people from industry, consulting, and education, all of whom are subject-matter experts. The test is intended to evaluate an individual's understanding of mathematics, basic sciences, and engineering sciences obtained in an accredited bachelor degree of engineering. Every five years or so, NCEES conducts an engineering task analysis survey. People in education are surveyed periodically to ensure the FE/EIT exam specifications reflect what is being taught.

The exam questions are prepared by the NCEES committee members, subject matter experts, and other volunteers. All people participating must hold professional licensure. When the questions have been written, they are circulated for review in workshop meetings and by mail. You will see mostly metric units (SI) on the exam. Some problems are posed in U.S. customary units (USCS) because the topics typically are taught that way. All problems are four-way multiple choice.

Examination Structure

The FE/EIT exam is a six-hour, computer-based test (CBT) that contains 110 multiple-choice questions. The six-hour time includes a tutorial, break, and brief survey at the conclusion. If you are taking the FE/EIT as a graduation requirement, your school may compel you to take the exam that matches the engineering discipline in which you are obtaining your degree. Otherwise, you can choose the afternoon exam you wish to take. There are two approaches to deciding.

One approach is to take the Other Disciplines exam regardless of your engineering discipline.

The second approach is to take the afternoon exam that matches the discipline you majored in. Particularly if you have been out of college for several years, practicing this discipline in your daily work, you will be very familiar and comfortable with the topics. This may be to your advantage during your review time and in the pressure of the exam itself.

Examination Procedure

Before the morning four-hour session begins, the proctors pass out exam booklets and a scoring sheet to each examinee. Space is provided on each page of the examination booklet for scratchwork. The scratchwork will *not* be considered in the scoring. Proctors will also provide each examinee with a mechanical pencil with eraser for use in writing answers; this is the only writing instrument allowed. Do not bring your own lead or eraser. If you need an additional pencil during the exam, a proctor will supply one.

The examination is closed book. You may not bring any reference materials with you to the exam. To replace your own materials, NCEES has prepared a *Fundamentals of Engineering (FE) Supplied-Reference Handbook*. The handbook contains engineering, scientific, and mathematical formulas and tables for use in the examination. Examinees will receive the handbook from their state registration board prior to the examination. The *Fundamentals of Engineering Supplied-Reference Handbook* is also included in the exam materials distributed at the beginning of each four-hour exam period.

There are three versions (A, B, and C) of the exam. These have the major subjects presented in a different order to reduce the possibility of examinees copying from one another. The first subject on your exam, for example, might be fluid mechanics, while the exam of the person next to you may have computers as the first subject.

The afternoon session begins following a one-hour lunch break. The afternoon exam booklets will be distributed along with a scoring sheet. There will be 60 multiple-choice questions, each of which carries twice the grading weight of the morning exam questions.

If you answer all questions more than 15 minutes early, you may turn in the exam materials and leave. If you finish in the last 15 minutes, however, you must remain to the end of the exam period to ensure a quiet environment for all those still working, and to ensure an orderly collection of materials.

Examination-Taking Suggestions

Those familiar with the psychology of examinations have several suggestions for examinees:

1. There are really two skills that examinees can develop and sharpen. One is the skill of illustrating one's knowledge. The other is the skill of familiarization with examination structure and procedure. The first can be enhanced by a systematic review of the subject matter. The second, exam-taking skills, can be improved by practice with sample problems—that is, problems that are presented in the exam format with similar content and level of difficulty.

2. Examinees should answer every problem, even if it is necessary to guess. There is no penalty for guessing. The best approach to guessing is to try to eliminate one or two of the four alternatives. If this can be done, the chance of selecting a correct answer obviously improves from 1 in 4 to 1 in 2 or 3.

3. Plan ahead with a strategy and a time allocation. Compute how much time you will allow for each of the subject areas. You might allocate a little less time per problem for the areas in which you are most proficient, leaving a little more time in subjects that are more difficult for you. Your time plan should include a reserve block for especially difficult problems, for checking your scoring sheet, and finally for making last-minute guesses on problems you did not work. Your strategy might also include time allotments for two passes through the exam—the first to work all problems for which answers are obvious to you, the second to return to the more complex, time-consuming problems and the ones at which you might need to guess.

4. Read all four multiple-choice answer options before making a selection. All distractors (wrong answers) are designed to be plausible. Only one option will be the best answer.

5. Do not change an answer unless you are absolutely certain you have made a mistake. Your first reaction is likely to be correct.

6. If time permits, check your work.

7. Do not sit next to a friend, a window, or other potential distraction.

License Review Books

To prepare for the FE/EIT exam you need two review books.

1. The FE Review Manual (for example, *Civil Engineering FE Review Manual*) if you are going to take one of the six branch exams. If you plan to take the Other Disciplines exam, this book covers those topics, so no supplement book is needed.
2. *Fundamentals of Engineering (FE) Supplied-Reference Handbook.* At some point this NCEES-prepared book will be provided to applicants by their state registration board. You may want to obtain a copy sooner so you will have ample time to study it before the exam. Pay close attention to the *Fundamentals of Engineering Supplied-Reference Handbook* and the notation used in it, because it is the only book you will have at the exam.

Textbooks

If you still have your university textbooks, they can be useful in preparing for the exam, unless they are out of date. To a great extent the books will be like old friends with familiar notation. You probably need both textbooks and license review books for efficient study and review.

Examination Day Preparations

The exam day will be a stressful and tiring one. You should take steps to eliminate the possibility of unpleasant surprises. If at all possible, visit the examination site ahead of time to determine the following:

1. How much time should you allow for travel to the exam on that day? Plan to arrive about 15 minutes early. That way you will have ample time, but not too much time. Arriving too early, and mingling with others who are also anxious, can increase your anxiety and nervousness.
2. Where will you park?
3. How does the exam site look? Will you have ample workspace? Will it be overly bright (sunglasses), or cold (sweater), or noisy (earplugs)? Would a cushion make the chair more comfortable?
4. Where are the drinking fountain and lavatory facilities?
5. What about food? Most states do not allow food in the test room (exceptions for ADA). Should you take something along for energy in the exam? A light bag lunch during the break makes sense.

Items to Take to the Examination

Although you may not bring books to the exam, you should bring the following:

- *Calculator*—NCEES has implemented a more stringent policy regarding permitted calculators. For a list of permitted models, see the NCEES Web site *(www.ncees.org)*. You also need to determine whether your state permits pre-programmed calculators. Bring extra batteries for your calculator just in case, and many people feel that bringing a second calculator is also a very good idea.
- *Clock*—You must have a time plan and a clock or wristwatch.
- *Exam Assignment Paperwork*—Take along the letter assigning you to the exam at the specified location to prove that you are the registered person. Also bring something with your name and picture (driver's license or identification card).
- *Items Suggested by Your Advance Visit*—If you visit the exam site, it will probably suggest an item or two that you need to add to your list.
- *Clothes*—Plan to wear comfortable clothes. You probably will do better if you are slightly cool, so it is wise to wear layered clothing.

Special Medical Condition

If you have a medical situation that may require special accommodation, notify the licensing board well in advance of exam day.

Examination Results

Your state board will notify you as to whether you have passed or failed roughly two weeks after the exam. Candidates who do not pass the exam the first time may take it again. If you do not pass, you will receive a report listing the percentages of questions you answered correctly for each topic area. This information can help focus the review efforts of candidates who need to retake the exam.

The FE/EIT exam is challenging, but analysis of previous pass rates shows that the majority of candidates do pass it the first time. By reviewing appropriate concepts and practicing with exam-style problems, you can be in that majority. Good luck!

Errata

The authors and publisher of this book have been careful to avoid errors, employing technical reviewers, copyeditors, and proofreaders to ensure the material is as flawless as possible. Any known errata and corrections are posted on the product page at our Web site, *www.kaplanengineering.com*. If you believe you have discovered an inaccuracy, please notify Customer Service at *kaplanaeinfo@kaplan.com*.

CHAPTER 2

Mathematics

David R. Arterburn and Thyagarajan Srinivasan

OUTLINE

ALGEBRA

Factorials

Definition. The factorial of a non-negative integer, n, is defined as $n!$ $n! = n(n-1)(n-2)(n-3)$. . . and so forth.

For example, $6! = 6(5)(4)(3)(2)(1) = 720$.

Also, $1! = 1$ and $0! = 1$.

Factorials can be written as multiples of other factorials:

$$6! = 6(5!) = 6(5)(4!) = 6(5)(4)(3!) = 6(5)(4)(3)(2!) = 6(5)(4)(3)(2)(1!) = 720$$

Factorials can be multiplied and divided.

$$\frac{n!}{(n-1)!} = \frac{n(n-1)!}{(n-1)!} = n$$

Exponents

Definition. Any number defined as base, b, can be multiplied by itself x number of times, which is denoted as b^x.

For example, $b^4 = b(b)(b)(b)$.

Properties of Exponents

$$b^0 = 1; \quad b^1 = b; \quad b^{-1} = \frac{1}{b}; \quad b^{-x} = \frac{1}{b^x}$$

$$b^{x+y} = \left(b^x\right)\left(b^y\right); \quad b^{x-y} = \frac{b^x}{b^y}; \quad b^{x \times y} = \left(b^x\right)^y$$

$$b^{\frac{1}{2}} = \sqrt{b}; \quad b^{\frac{1}{x}} \sqrt[x]{b}; \quad b^{\frac{x}{y}} = \left(\sqrt[y]{b}\right)^x$$

Logarithms

Definition. If b is a finite positive number, other than 1, and $b^x = N$, then x is the logarithm of N to the base b, or $\log_b N = x$. If $\log_b N = x$, then $b^x = N$.

Properties of Logarithms

$$\log_b b = 1; \ \log_b 1 = 0; \ \log_b 0 = \begin{cases} +\infty, \text{ when } b \text{ lies between 0 and 1} \\ +\infty, \text{ when } b \text{ lies between 1 and } \infty \end{cases}$$

$$\log_b \left(M \circ N\right) = \log_b M + \log_b N \qquad \log_b \ M/N = \log_b \ M - \log_b N$$

$$\log_b N^p = p \log_b N \qquad \log_b \sqrt[r]{N^p} = \frac{p}{r} \log_b N$$

$$\log_b N = \log_a N \,/\, \log_a b; \quad \log_b b^N = N; \quad b^{\log_b N} = N$$

Systems of Logarithms

Common (Briggsian)—base 10.

Natural (Napierian or hyperbolic)—base 2.7183 (designated by e or ε).

The abbreviation of *common logarithm* is log, and the abbreviation of *natural logarithm* is ln.

Example 2.1

(i) Solve for a if $\log_a 10 = 0.25$ and (ii) find $\log\left(\frac{1}{x}\right)$, if $\log x = 0.3332$.

Solution

(i) If $\log_a N = x$, then $N = a^x$ or $a = N^{(1/x)}$

Here, $N = 10$ and $x = 0.25$; then $a = 10^{\frac{1}{0.25}} = 10{,}000$

(ii) Since $\log \frac{M}{N} = \log M - \log N$, $\log\left(\frac{1}{x}\right) = \log 1 - \log x = -0.3332$

Example 2.2

Solve the equation $\log x + \log (x - 3) - \log 4 = 0$.

Solution

Since $(\log M + \log N - \log P) = \log (\frac{MN}{P})$, $\log x + \log (x - 3) - \log 4$

$= \log \left[\frac{x(x-3)}{4}\right] = 0$

$\left[\frac{x(x-3)}{4}\right] = 10^0 = 1$; simplifying, $x^2 - 3x - 4 = 0$. Finding the roots,

$x = 4$ and -1; then $x = 4$ (-1 is not an answer because the log of a negative number is undefined).

The Solution of Algebraic Equations

Definition. A root, x, is any value such that $f(x) = 0$.

The Quadratic Equation

If $ax^2 + bx + c = 0$, then

$$x = \frac{-b \pm \sqrt{b^2 - 4ac}}{2a}$$

If $b^2 - 4ac > 0$, the two roots are real and unequal; if $b^2 - 4ac = 0$, the two roots are real and equal; if $b^2 - 4ac < 0$, the two roots are imaginary.

Example 2.3

Find the root(s) of the following equations (i) $x + 4 = 0$, (ii) $x^2 + 4x + 3 = 0$, (iii) $x^2 - 4x + 4 = 0$, (iv) $x^2 + 4 = 0$, and (v) $3x^3 + 3x^2 - 18x = 0$.

Solution

For the quadratic equations in (ii), (iii), and (iv), use either the equation or simply a scientific calculator to find the roots. The results yield:

(i) -4, (ii) -1 and -3 (real and distinct roots), (iii) 2 and 2 (real and equal roots),

(iv) $+i2$ and $-i2$ (complex roots; always occur in pairs called *conjugates*)

For (v), factoring, $3x(x^2 + x - 6) = 0$. Roots are 0, -3, and 2.

Example 2.4

Find the equation whose roots are 3 and -2.

Solution

If the roots are x_1, x_2, x_3, etc., the equation is $(x - x_1)(x - x_2)(x - x_3)\ldots = 0$; here, $(x - 3)(x - (-2)) = 0$; $(x - 3)(x + 2) = 0$; $x^2 - x - 6 = 0$.

Progressions

Arithmetic Progression

An arithmetic progression is $a, a + d, a + 2d, a + 3d, \ldots$, where d = common difference.

The nth term is $t_n = a + (n - 1)d$

The sum of n terms is $S_n = \frac{n}{2}[2a + (n-1)d] = \frac{n}{2}(a + t_n)$

Geometric Progression

A geometric progression is $a, ar, ar^2, ar^3, \ldots$, where r = common ratio.

The nth term is $t_n = ar^{n-1}$

The sum of n terms is $S_n = a\left(\frac{1-r^n}{1-r}\right)$

If $r^2 < 1$, S_n approaches a definite limit as n increases indefinitely, and

$$S_\infty = \frac{a}{1-r}$$

Example 2.5

Consider the arithmetic progression 1, 3, 5, 7, 9, 11, 13, . . . (i) Find the sum of the first seven terms and (ii) the 18th term of the progression.

Solution

(i) First term $a = 1$, number of terms $n = 7$, and the common difference $d = 2$

sum $S_n = \frac{n}{2}[2a + (n-1)d] = \frac{7}{2}[2 + 6(2)] = 49$

(ii) Number of terms $n = 18$, first term $a = 1$, and the common difference $d = 2$

The last term or the 18th term is $= a + (n - 1)\, d = 1 + (18 - 1)2 = 35$.

Example 2.6

Find the sum of the series 1, 0.5, 0.25, 0.125, 0.0625, . . .

Solution

This geometric series is convergent. First term $a = 1$ and the common ratio $r = 0.5$.

As the number of terms tend to infinity, sum $S = \frac{a}{1-r} = \frac{1}{1-0.5} = 2$

Example 2.7

Consider the geometric progression 2, 4, 8, 16, 32, 64, 128, . . . (i) Find the sum of the first seven terms and (ii) the 20th term of the series.

Solution

(i) First term $a = 2$, common ratio $r = 2$, and the number of terms $n = 7$

sum $S = a\frac{(1-r^n)}{(1-r)} = \frac{2(1-2^7)}{(1-2)} = 256$

(ii) The 20th term of the series is $= ar^{(n-1)} = 2(2)^{(20-1)} = 1{,}048{,}576$.

COMPLEX QUANTITIES

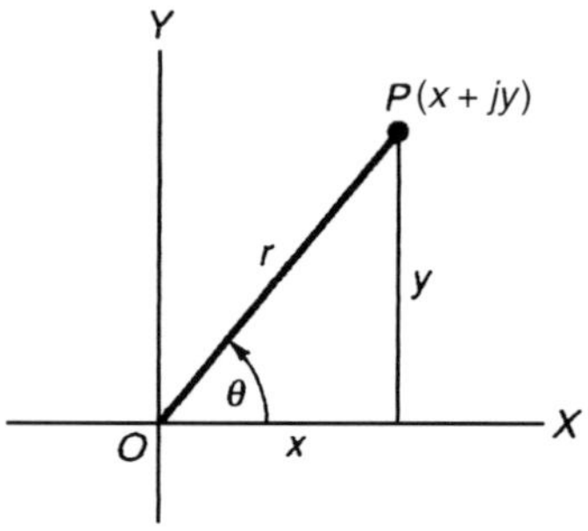

Figure 2.1

Definition and Representation of a Complex Quantity

If $z = x + jy$, where $j = \sqrt{-1}$ and x and y are real, z is called a complex quantity and is completely determined by x and y.

If $P(x, y)$ is a point in the plane (Figure 2.1), then the segment OP in magnitude and direction is said to represent the complex quantity $z = x + jy$.

If θ is the angle from OX to OP and r is the length of OP, then $z = x + jy = r(\cos\theta + j\sin\theta) = re^{j\theta}$, where $\theta = \tan^{-1} y/x$, $r = +\sqrt{x^2 + y^2}$ and e is the base of natural logarithms. The pair $x + jy$ and $x - jy$ are called complex conjugate quantities.

Properties of Complex Quantities

Let z, z_1, and z_2 represent complex quantities; then

Sum or difference: $z_1 \pm z_2 = (x_1 \pm x_2) + j(y_1 \pm y_2)$

Equation: If $z_1 = z_2$, then $x_1 = x_2$ and $y_1 = y_2$

Periodicity: $z = r(\cos\theta + j\sin\theta) = r[\cos(\theta + 2k\pi) + j\sin(\theta + 2k\pi)]$, or $z = re^{j\theta} = re^{j(\theta + 2k\pi)}$ and $e^{j2k\pi} = 1$, where k is any integer.

Exponential-trigonometric relations: $e^{jz} = \cos z + j\sin z$, $e^{-jz} = \cos z - j\sin z$,

$$\cos z = \frac{1}{2}\left(e^{jz} + e^{-jz}\right), \quad \sin z = \frac{1}{2j}\left(e^{jz} - e^{-jz}\right)$$

TRIGONOMETRY

Definition of an Angle

An angle is the amount of rotation (in a fixed plane) by which a straight line may be changed from one direction to any other direction. If the rotation is counterclockwise, the angle is said to be positive; if clockwise, negative.

Measure of an Angle

A degree is $\frac{1}{360}$ of the plane angle about a point, and a radian is the angle subtended at the center of a circle by an arc equal in length to the radius. One complete circle contains 180 degrees or 2π radians; 1 radian = $\pi/180$ degrees.

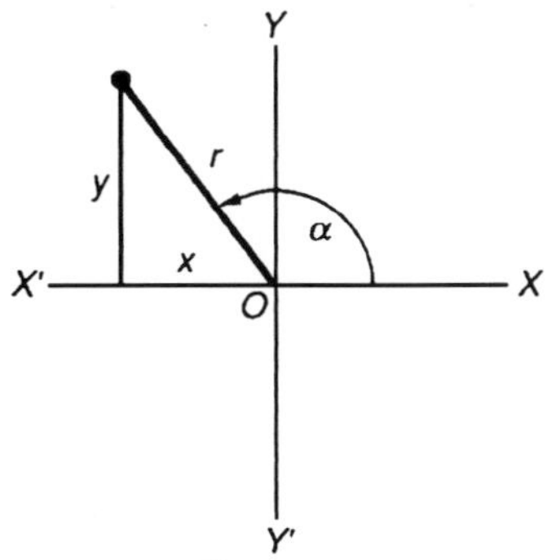

Figure 2.2

Trigonometric Functions of an Angle

sine (sin) $\alpha = y/r$ cosecant (csc) $\alpha = r/y$

cosine (cos) $\alpha = x/r$ secant (sec) $\alpha = r/x$

tangent (tan) $\alpha = y/x$ cotangent (cot) $\alpha = x/y$

The variable x is positive when measured along OX and negative along OX'. Similarly, y is positive when measured parallel to OY, and negative parallel to OY'.

$$\sin 0° = 0;\ \sin 90° = 1;\ \sin 180° = 0;\ \sin 270° = -1$$

$$\cos 0° = 1;\ \cos 90° = 0;\ \cos 180° = -1;\ \cos 270° = 0$$

Fundamental Relations among the Functions

$$\sin\alpha = \frac{1}{\csc\alpha};\quad \cos\alpha = \frac{1}{\sec\alpha};\quad \tan\alpha = \frac{1}{\cot\alpha} = \frac{\sin\alpha}{\cos\alpha}$$

$$\csc\alpha = \frac{1}{\sin\alpha};\quad \sec\alpha = \frac{1}{\cos\alpha};\quad \cot\alpha = \frac{1}{\tan\alpha} = \frac{\cos\alpha}{\sin\alpha}$$

$$\sin^2\alpha + \cos^2\alpha = 1;\quad \sec^2\alpha - \tan^2\alpha = 1;\quad \csc^2\alpha - \cot^2\alpha = 1$$

Functions of Multiple Angles

$$\sin 2\alpha = 2\sin\alpha\cos\alpha$$

$$\cos 2\alpha = 2\cos^2\alpha - 1 = 1 - 2\sin^2\alpha = \cos^2\alpha - \sin^2\alpha$$

$$\tan 2\alpha = (2\tan\alpha)/(1 - \tan^2\alpha)$$

$$\cot 2\alpha = (\cot^2\alpha - 1)/(2\cot\alpha)$$

Functions of Half Angles

$$\sin\frac{1}{2}\alpha = \sqrt{\frac{1-\cos\alpha}{2}};\quad \cos\frac{1}{2}\alpha = \sqrt{\frac{1+\cos\alpha}{2}}$$

$$\tan\frac{1}{2}\alpha = \frac{1-\cos\alpha}{\sin\alpha} = \frac{\sin\alpha}{1+\cos\alpha} = \sqrt{\frac{1-\cos\alpha}{1+\cos\alpha}}$$

Functions of Sum or Difference of Two Angles

$$\sin(\alpha \pm \beta) = \sin\alpha\cos\beta \pm \cos\alpha\sin\beta$$

$$\cos(\alpha \pm \beta) = \cos\alpha\cos\beta \mp \sin\alpha\sin\beta$$

$$\tan(\alpha \pm \beta) = \frac{\tan\alpha \pm \tan\beta}{1 \mp \tan\alpha\tan\beta}$$

Sums, Differences, and Products of Two Functions

$$\sin\alpha + \sin\beta = 2\sin\frac{1}{2}(\alpha+\beta)\cos\frac{1}{2}(\alpha-\beta)$$

$$\sin\alpha - \sin\beta = 2\cos\frac{1}{2}(\alpha+\beta)\sin\frac{1}{2}(\alpha-\beta)$$

$$\cos\alpha + \cos\beta = 2\cos\frac{1}{2}(\alpha+\beta)\cos\frac{1}{2}(\alpha-\beta)$$

$$\cos\alpha - \cos\beta = 2\sin\frac{1}{2}(\alpha+\beta)\sin\frac{1}{2}(\alpha-\beta)$$

$$\tan\alpha \pm \tan\beta = \frac{\sin(\alpha+\beta)}{\cos\alpha\cos\beta}$$

$$\sin^2\alpha - \sin^2\beta = \sin(\alpha+\beta)\sin(\alpha-\beta)$$

$$\cos^2\alpha - \cos^2\beta = \sin(\alpha+\beta)\sin(\alpha-\beta)$$

$$\cos^2\alpha - \sin^2\beta = \cos(\alpha+\beta)\cos(\alpha-\beta)$$

$$\sin\alpha\sin\beta = \frac{1}{2}\cos(\alpha-\beta) - \frac{1}{2}\cos(\alpha+\beta)$$

$$\cos\alpha\cos\beta = \frac{1}{2}\cos(\alpha-\beta) + \frac{1}{2}\cos(\alpha+\beta)$$

$$\sin\alpha\cos\beta = \frac{1}{2}\sin(\alpha+\beta) + \frac{1}{2}\sin(\alpha-\beta)$$

Example 2.8

Simplify the following expressions to trigonometric functions of angles less than 90 degrees. Note: There is more than one correct answer for each.

i) cos 370°
ii) sin 120°

Solution

Typical strategies for simplifying the functions are shown below:

i) cos 370° = cos (360° + 10°) = (cos 360°)(cos 10°) – (sin 360°)(sin 10°) = (1)(cos 10°) – (0)(sin 10°) = cos 10°

ii) sin 120° = sin (90° + 30°) = (sin 90°)(cos 30°) + (cos 90°)(sin 30°) = (1)(cos 30°) + (0)(sin 30°) = cos 30°

or

sin 120° = sin (180° – 60°) = (sin 180°)(cos 60°) – (cos 180°)(sin 60°) = (0)(cos 60°) – (–1)(sin 60°) = sin 60°

Properties of Plane Triangles

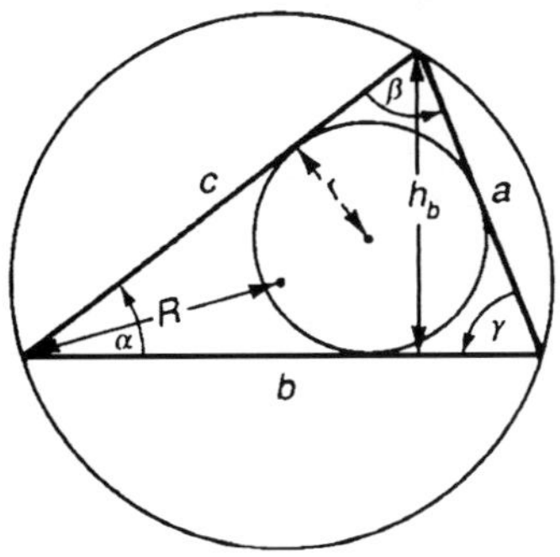

Figure 2.3

Notation. α, β, γ = angles; a, b, c = sides; A = area; h_b = altitude on b; $s = \frac{1}{2}(a+b+c)$; r = radius of inscribed circle; R = radius of circumscribed circle

$$\alpha + \beta + \gamma = 180° = \pi \text{ radians}$$

$$\frac{a}{\sin\alpha} = \frac{b}{\sin\beta} = \frac{c}{\sin\gamma}$$

$$\frac{a+b}{a-b} = \frac{\tan\frac{1}{2}(\alpha+\beta)}{\tan\frac{1}{2}(\alpha-\beta)}$$

$$a^2 = b^2 + c^2 - 2bc\cos\alpha \qquad a = b\cos\gamma + c\cos\beta$$

$$\cos\alpha = \frac{b^2+c^2-a^2}{2bc} \qquad \sin\alpha = \frac{2}{bc}\sqrt{s(s-a)(s-b)(s-c)}$$

$$\sin\frac{\alpha}{2}\sqrt{\frac{(s-b)(s-c)}{bc}} \qquad \cos\frac{\alpha}{2} = \sqrt{\frac{s(s-a)}{bc}}$$

$$\tan\frac{\alpha}{2} = \sqrt{\frac{(s-b)(s-c)}{s(s-a)}} = \frac{r}{s-a}$$

$$h_b = c\sin\alpha = a\sin\gamma = \frac{2}{b}\sqrt{s(s-a)(s-b)(s-c)}$$

$$r = \sqrt{\frac{(s-a)(s-b)(s-c)}{s}} = (s-a)\tan\frac{\alpha}{2}$$

$$R = \frac{a}{2\sin\alpha} = \frac{abc}{4A}$$

$$A = \frac{1}{2}bh_b = \frac{1}{2}ab\sin\gamma = \frac{a^2\sin\beta\sin\gamma}{2\sin\alpha} = \sqrt{s(s-a)(s-b)(s-c)} = rs$$

Example 2.9

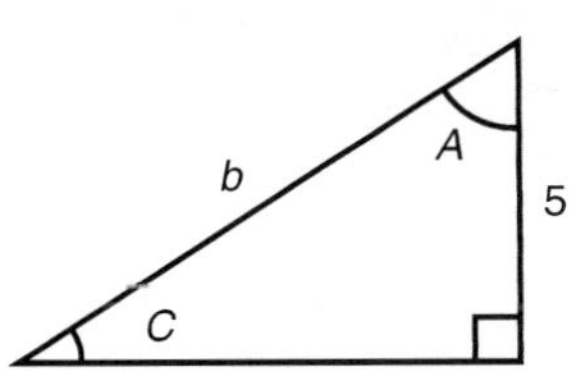

Figure 2.4

Find the side b and the angles A and C for the triangle in Figure 2.4.

Solution

$\tan(A) = 12/5$; then, Angle $A = \tan^{-1}(12/5) = 67.38°$

Since the sum $(A + C + 90°) = 180°$; $C = 90° - A = 22.62°$

Now, $\cos(A) = \frac{5}{b}$; then, $b = \frac{5}{\cos(A)} = 13$

Example 2.10

Find the side c and the angles A and B for the triangle in Figure 2.5.

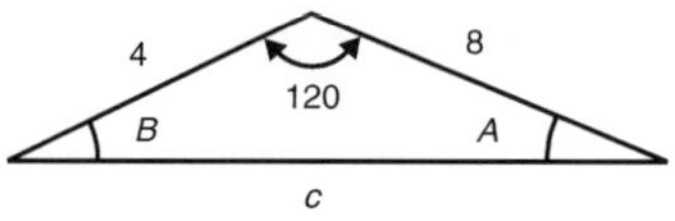

Figure 2.5

Solution

Since two sides and an included angle are given, use the law of cosines.

$c^2 = 4^2 + 8^2 - 2(4)(8)\cos 120$; solving $c = 10.583$

Now use law of sines to find the remaining angles.

$\frac{10.583}{\sin 120} = \frac{4}{\sin A} = \frac{8}{\sin B}$; solving, $A = 19.1°$ and $B = 40.89°$

(Check: sum of the angles = 180°)

Example 2.11

Simplify: (i) $(\sec^2\theta)(\sin^2\theta)$ (ii) $\sin(A + B) + \sin(A - B)$
(iii) $2\sin^2\theta + 1 + \cos^2\theta$

Solution

(i) $(1/\cos^2\theta)\sin^2\theta = \tan^2\theta$ (ii) $2\sin A\cos B$

(iii) $2\sin^2\theta + 1 + (2\cos^2\theta - 1) = 2(\sin^2\theta + \cos^2\theta) = 2$

GEOMETRY AND GEOMETRIC PROPERTIES (MENSURATION)

Notation. a, b, c, d, and s denote lengths, A denotes area, V denotes volume

Right Triangle

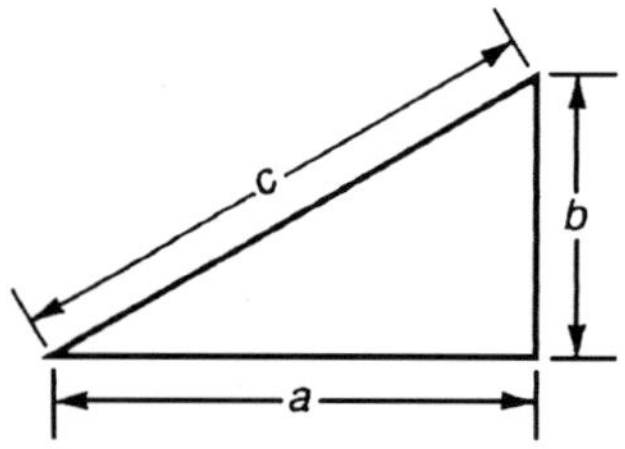

Figure 2.6

$$A = \frac{1}{2}ab$$

$$c = \sqrt{a^2 + b^2},\quad a = \sqrt{c^2 - b^2},\quad b = \sqrt{c^2 - a^2}$$

Oblique Triangle

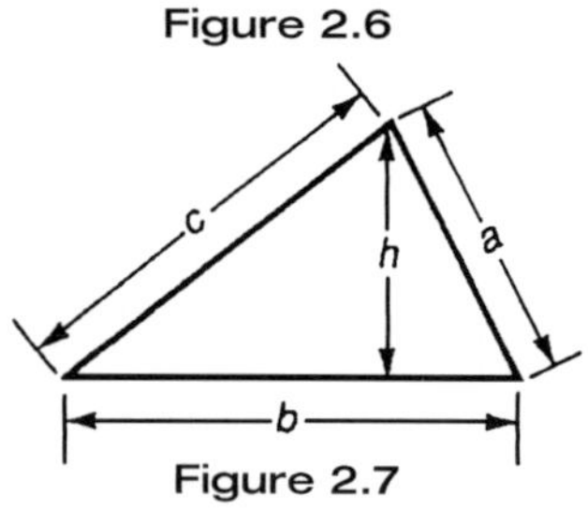

Figure 2.7

$$A = \frac{1}{2}bh$$

Equilateral Triangle

All sides are equal and all angles are 60°.

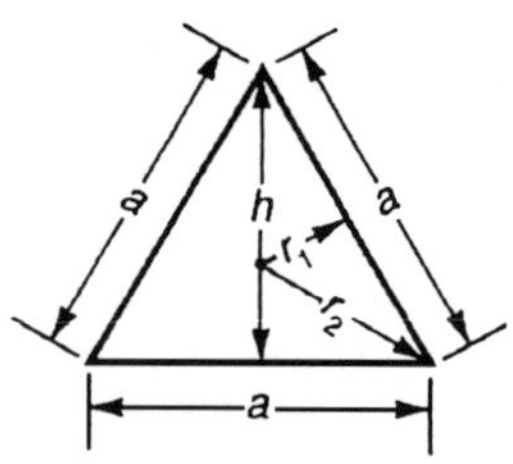

Figure 2.8

$$A = \frac{1}{2}ah = \frac{1}{4}a^2\sqrt{3},\qquad h = \frac{1}{2}a\sqrt{3},\qquad r_1 = \frac{a}{2\sqrt{3}},\qquad r_2 = \frac{a}{\sqrt{3}}$$

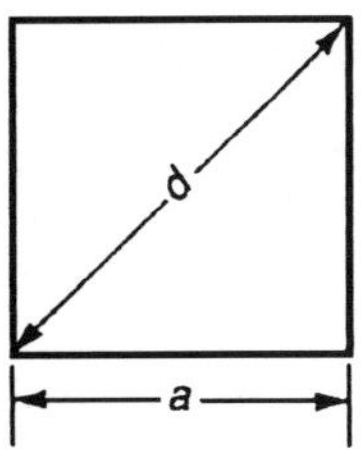

Figure 2.9

Square

All sides are equal, and all angles are 90°.

$$A = a^2, \qquad d = a\sqrt{2}$$

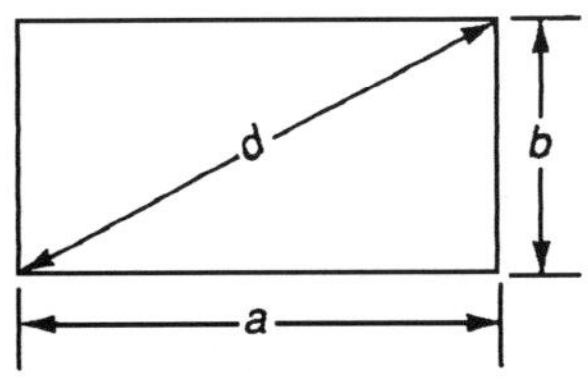

Figure 2.10

Rectangle

Opposite sides are equal and parallel, and all angles are 90°.

$$A = ab, \qquad d = \sqrt{a^2 + b^2}$$

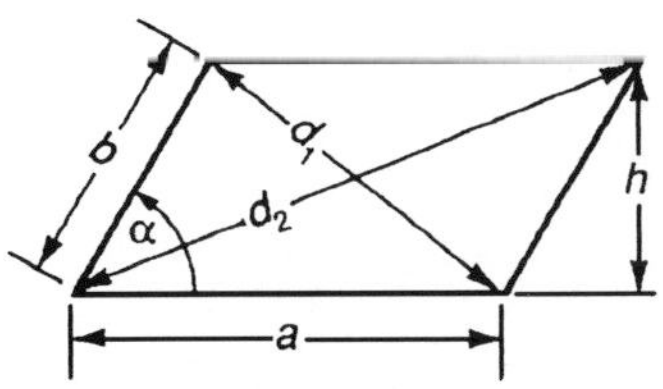

Figure 2.11

Parallelogram

Opposite sides are equal and parallel, and opposite angles are equal.

$$A = ah = ab \sin \alpha, \;\; d_1 = \sqrt{a^2 + b^2 - 2ab \cos \alpha}, \;\; d_2 = \sqrt{a^2 + b^2 + 2ab \cos \alpha}$$

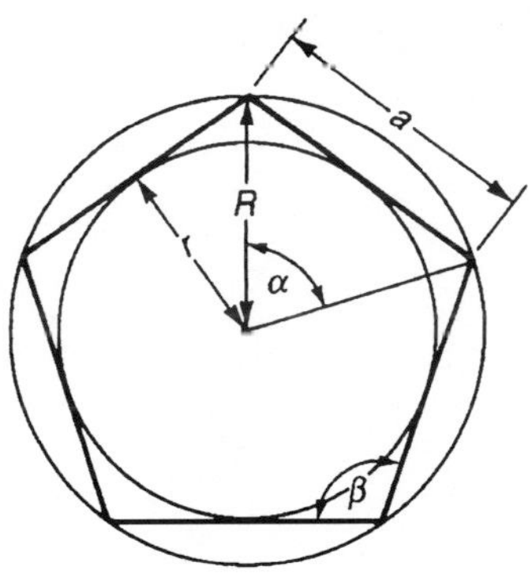

Figure 2.12

Regular Polygon of *n* Sides

All sides and all angles are equal.

$$\beta = \frac{n-2}{n} 180° = \frac{n-2}{n} \pi \text{ radians}, \qquad \alpha \frac{360°}{n} = \frac{2\pi}{n} \text{ radians}, \qquad A = \frac{nar}{2}$$

Circle

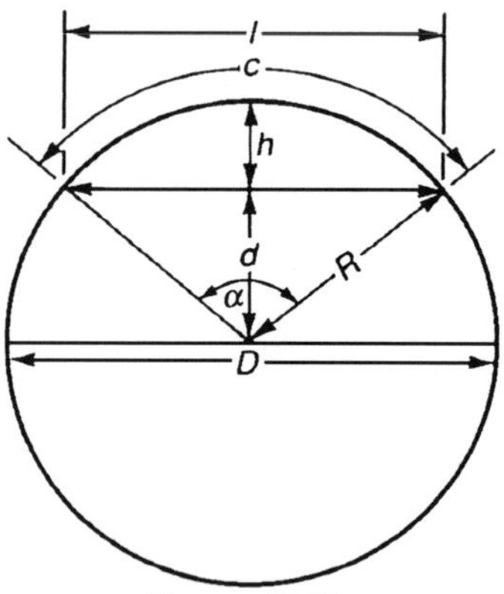

Figure 2.13

Notation. C = circumference, α = central angle in radians

$$C = \pi D = 2\pi R$$

$$c = R\alpha = \frac{1}{2} D\alpha = D \cos^{-1} \frac{d}{R} = D \tan^{-1} \frac{1}{2d}$$

$$l = 2\sqrt{R^2 - d^2} = 2R \sin \frac{\alpha}{2} = 2d \tan \frac{\alpha}{2} = 2d \tan \frac{c}{D}$$

$$d = \frac{1}{2}\sqrt{4R^2 - l^2} = \frac{1}{2}\sqrt{D^2 - l^2} = R \cos \frac{\alpha}{2}$$

$$h = R - d$$

$$\alpha = \frac{c}{R} = \frac{2c}{D} = 2 \cos^{-1} \frac{d}{R}$$

$$A_{(circle)} = \pi R^2 = \frac{1}{4}\pi D^2 = \frac{1}{2} RC = \frac{1}{4} DC$$

$$A_{(sector)} = \frac{1}{2} Rc = \frac{1}{2} R^2\alpha = \frac{1}{8} D^2\alpha$$

Ellipse

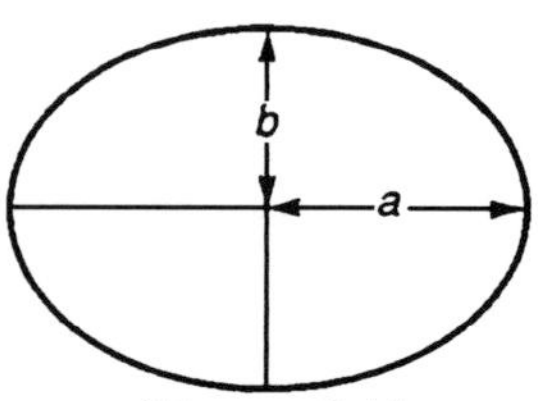

Figure 2.14

$$A = \pi ab$$

$$\text{Perimeter } (s) = \pi(a+b)\left[1 + \frac{1}{4}\left(\frac{a-b}{a+b}\right)^2 + \frac{1}{64}\left(\frac{a-b}{a+b}\right)^4 + \frac{1}{256}\left(\frac{a-b}{a+b}\right)^6 + \cdots\right]$$

$$\text{Perimeter } (s) \approx \pi \frac{a+b}{4}\left[3(1+\lambda) + \frac{1}{1-\lambda}\right], \quad \text{where } \lambda = \left[\frac{a-b}{2(a+b)}\right]^2$$

Parabola

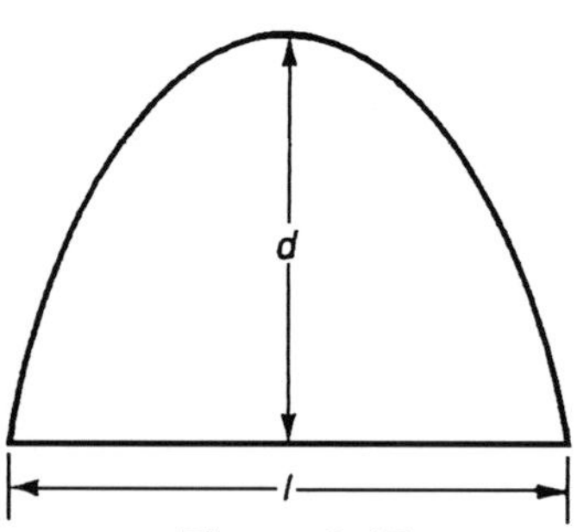

Figure 2.15

$$A = \frac{2}{3} ld$$

Cube

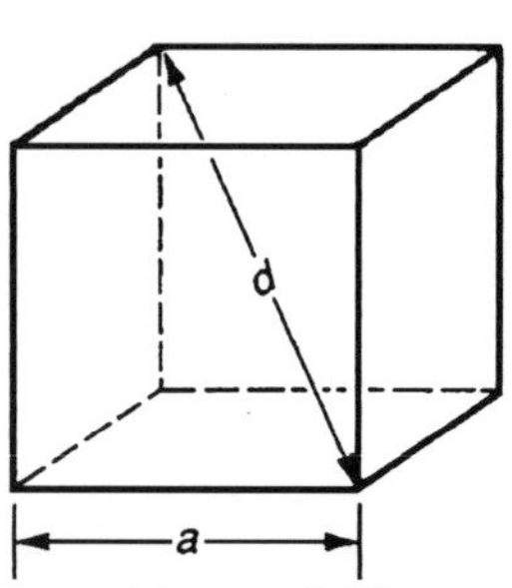

Figure 2.16

$$V = a^3 \qquad d = a\sqrt{3}$$

Total surface area $= 6a^2$

Prism or Cylinder

$V = $ (area of base) (altitude, h)

Lateral area = (perimeter of right section)(lateral edge, e)

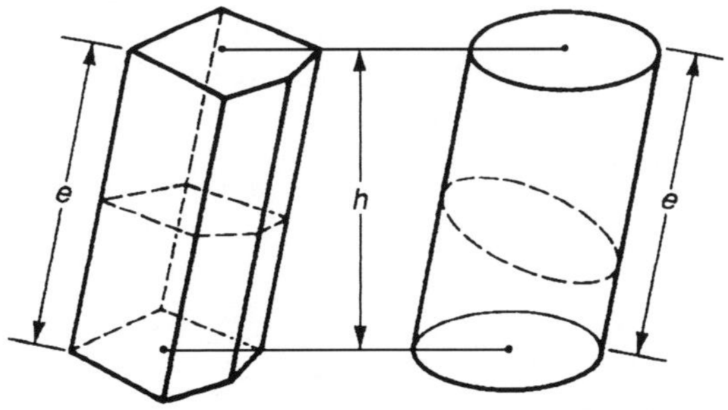

Figure 2.17

Pyramid or Cone

$$V = \frac{1}{3} \text{ (area of base) (altitude, } h\text{)}$$

$$\text{Lateral area of regular figure} = \frac{1}{2} \text{ (perimeter of base)(slant height, } s\text{)}$$

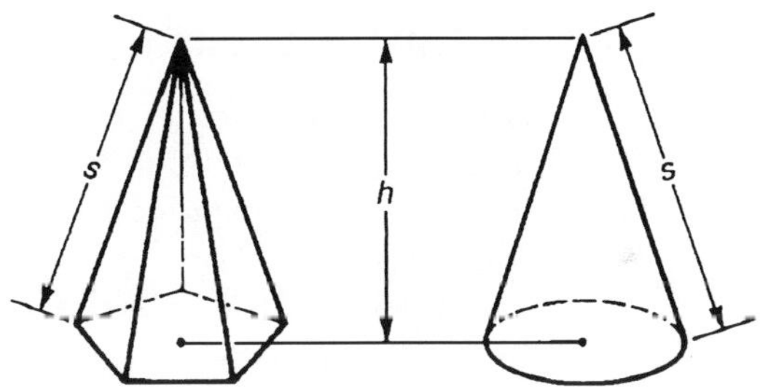

Figure 2.18

Sphere

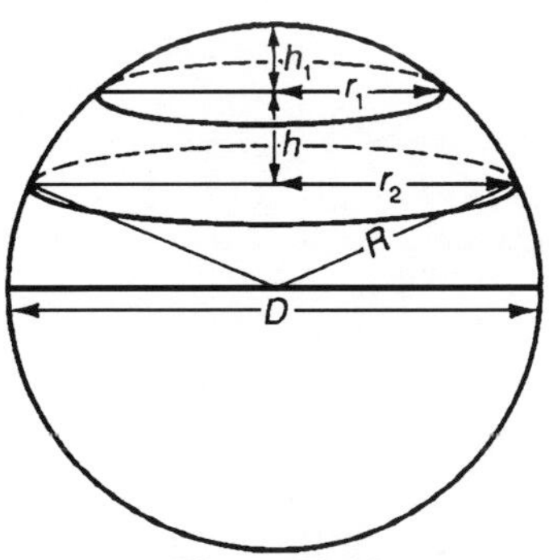

Figure 2.19

$$A_{(sphere)} = 4\pi R^2 = \pi D^2$$

$$A_{(zone)} = 2\pi Rh = \pi Dh$$

$$V_{(sphere)} = \frac{4}{3}\pi R^3 = \frac{1}{6}\pi D^3$$

$$V_{(spherical\ sector)} = \frac{2}{3}\pi R^2 h = \frac{1}{6}\pi D^2 h$$

PLANE ANALYTIC GEOMETRY

Rectangular Coordinates

Let two perpendicular lines, $X'X$ (x-axis) and $Y'Y$ (y-axis) meet at a point O (origin). The position of any point $P(x, y)$ is fixed by the distances x (abscissa) and y (ordinate) from $Y'Y$ and $X'X$, respectively, to P. Values of x are positive to the right and negative to the left of $Y'Y$; values of y are positive above and negative below $X'X$.

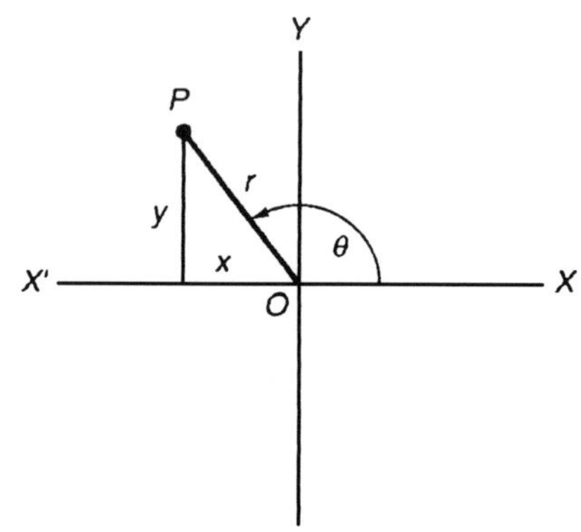

Figure 2.20

Polar Coordinates

Let O (origin or pole) be a point in the plane and OX (initial line) be any line through O. The position of any point $P\ (r, \theta)$ is fixed by the distance r (radius vector) from O to the point and the angle θ (vectorial angle) measured from OX to OP (Figure 2.20).

A value for r is positive and is measured along the terminal side of θ; a value for θ is positive when measured counterclockwise and negative when measured clockwise.

Relations Connecting Rectangular and Polar Coordinates

$$x = r\cos\theta, \quad y = r\sin\theta$$

$$r = \sqrt{x^2 + y^2}, \quad \theta = \tan^{-1}\frac{y}{x}, \quad \sin\theta = \frac{y}{\sqrt{x^2 + y^2}},$$

$$\cos\theta = \frac{x}{\sqrt{x^2 + y^2}}, \quad \tan\theta = \frac{y}{x}$$

Points and Slopes

Let $P_1(x_1, y_1)$ and $P_2(x_2, y_2)$ be any two points, and let α_1 be the angle from the x axis to P_1P_2, measured counterclockwise.

The length P_1P_2 is $d = \sqrt{(x_2 - x_1)^2 + (y_2 - y_1)^2}$.

The midpoint of P_1P_2 is $\left(\frac{x_1 + x_2}{2}, \frac{y_1 + y_2}{2}\right)$.

The point that divides P_1P_2 in the ratio $n_1{:}n_2$ is $\left(\frac{n_1x_2 + n_2x_1}{n_1 + n_2}, \frac{n_1y_2 + n_2y_1}{n_1 + n_2}\right)$.

The slope of P_1P_2 is $\tan\ \alpha = m = \frac{y_2 - y_1}{x_2 - x_1}$.

The angle between two lines of slopes m_1 and m_2 is $\beta = \tan^{-1}\frac{m_2 - m_1}{1 + m_1m_2}$.

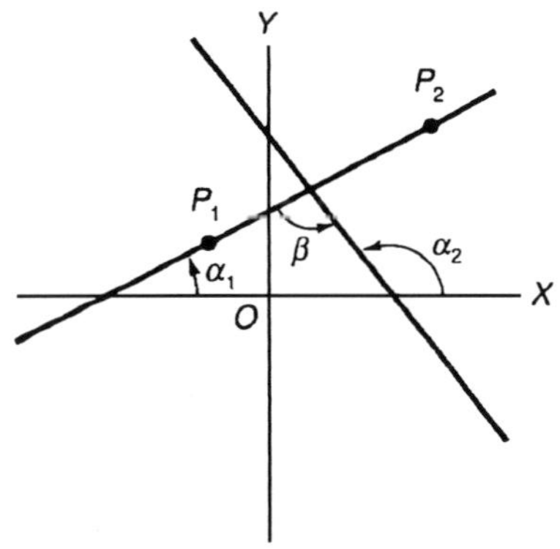

Figure 2.21

Two lines of slopes m_1 and m_2 are perpendicular if $m_2 = -\frac{1}{m_1}$.

Example **2.12**

Find the distance between the points (1, −2) and (−4, 2).

Solution

The distance is $d = \sqrt{(x_2 - x_1)^2 + (y_2 - y_1)^2} = \sqrt{[1-(-4)]^2 + [-2-2]^2}$
$= \sqrt{5^2 + 4^2} = 6.4$

Locus and Equation

The collection of all points that satisfy a given condition is called the **locus** of that condition; the condition expressed by means of the variable coordinates of any point on the locus is called the **equation of the locus**.

The locus may be represented by equations of three kinds: (1) a rectangular equation involves the rectangular coordinates (x, y); (2) a polar equation involves the polar coordinates (r, θ); and (3) parametric equations express x and y or r and θ in terms of a third independent variable called a parameter.

The following equations are generally given in the system in which they are most simply expressed; sometimes several forms of the equation in one or more systems are given.

Straight Line

$Ax + By + C = 0$ [$-A/B$ = slope]

$y = mx + b$ [m = slope, b = intercept on OY]

$y - y_1 = m(x - x_1)$ [m = slope, $P_1(x_1, y_1)$ is a known point on the line]

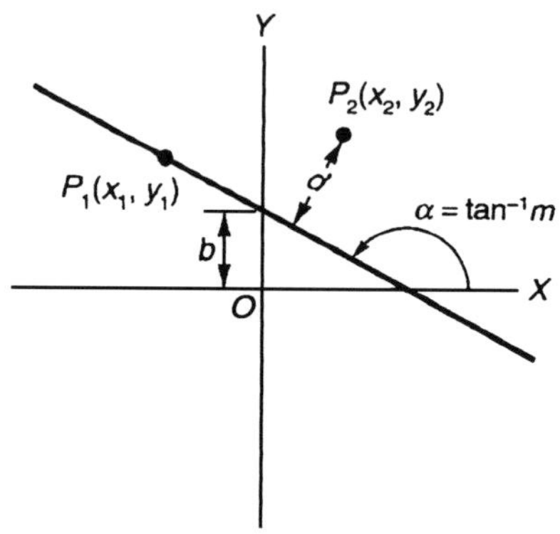

Figure 2.22

Example 2.13

Find the equation of the line passing through the points (2, 1) and (3, –3).

Solution

Slope can be found as $m = \dfrac{y_2 - y_1}{x_2 - x_1} = \dfrac{1-(-3)}{2-3} = -4.$

The point-slope form is $(y - 1) = -4(x - 2)$ or $(y + 3) = -4(x - 3)$.

Simplifying, either equation yields $y = -4x + 9$ or equivalently $4x + y - 9 = 0$.

Example 2.14

Find the equation of the straight line passing through the point (3, 1) and perpendicular to the line passing through the points (3, –2) and (–3, 7).

Solution

Slope of the line passing through (3, –2) and (–3, 7) is

$$m_1 = \frac{y_2 - y_1}{x_2 - x_1} = \frac{7-(-2)}{-3-3} = -\frac{3}{2}$$

Slope of the line passing through (3, 1) is $m_2 = -\dfrac{1}{m_1} = \dfrac{2}{3}$, since the two lines are perpendicular to each other.

Equation is $(y-1) = \dfrac{2}{3}(x - 3)$ or $2x - 3y - 3 = 0$.

Circle

The locus of a point at a constant distance (radius) from a fixed point C (center) is a circle.

$(x-h)^2+(y-k)^2=a^2$	$C(h, k)$, radius $=a$	
$r^2+b^2\pm 2\,br\cos(\theta-\beta)=a^2$	$C(b, \beta)$, radius $=a$	[Figure 2.23(a)]
$x^2+y^2=2ax$	$C(a, 0)$, radius $=a$	
$r=2a\cos\theta$	$C(a, 0)$, radius $=a$	[Figure 2.23(b)]
$x^2+y^2=2ay$	$C(0, a)$, *radius* $=a$	
$r=2a\sin\theta$	$C(0, a)$, *radius* $=a$	[Figure 2.23(c)]
$x^2+y^2=a^2$	$C(0, 0)$, radius $=a$	
$r=a$	$C(0, 0)$, radius $=a$	[Figure 2.23(d)]
$x=a\cos\phi,\ y=a\sin\phi$	$\phi=$ angle from OX to radius	

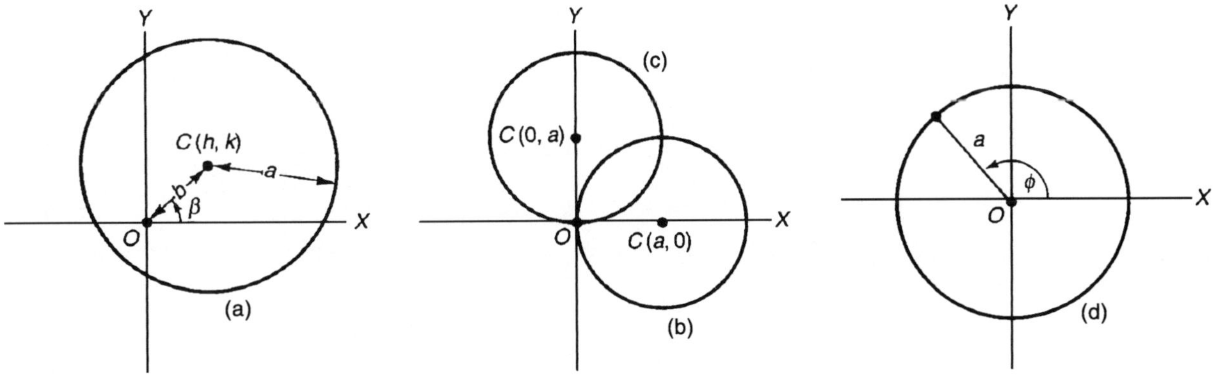

Figure 2.23

Example 2.15

Find the equation of the circle (i) with center at (0, 0) and radius 3, and (ii) with center at (1, 2) and radius 4.

Solution

Equation of a circle with center at (h, k) and radius r is $(x-h)^2+(y-k)^2=r^2$.

i) Here, $h=0$, $k=0$, and $r=3$; equation of the circle is $x^2+y^2=3^2$ or $x^2+y^2-9=0$
ii) Here, $h=1$, $k=2$, and $r=4$; equation of the circle is $(x-1)^2+(y-2)^2=4^2$
Simplifying, $x^2-2x+y^2-4y-11=0$

Conic

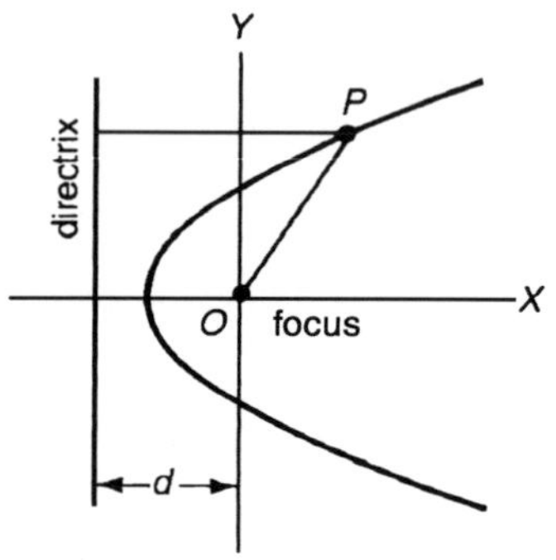

Figure 2.24

A **conic** is the locus of a point whose distance from a fixed point (focus) is in a constant ratio e, called the eccentricity, to its distance from a fixed straight line (directrix).

$$x^2 + y^2 = e^2(d + x)^2 \qquad d = \text{distance from focus to directrix}$$

$$r = \frac{de}{1 - e\cos\theta}$$

The conic is called a parabola when $e = 1$, an ellipse when $e < 1$, and a hyperbola when $e > 1$.

Example 2.16

Which conic section is represented by each of the following equations: (i) $x^2 + 4xy + 4y^2 + 2x = 10$ and (ii) $x^2 + y^2 - 2x - 4y - 11 = 0$.

Solution

The general equation of a conic section is $Ax^2 + 2Bxy + Cy^2 + 2Dx + 2Ey + F = 0$, where both A and C are not zeros. If $B^2 - AC > 0$, a *hyperbola* is defined; if $B^2 - AC = 0$, a *parabola* is defined; if $B^2 - AC < 0$, an *ellipse* is defined. (Note: If B is zero and A = C, a *circle* is defined.)

If A = B = C = 0, a *straight line* is defined.

If B = 0, A = C, a *circle* is defined with equation $x^2 + y^2 + 2ax + 2by + c = 0$.

Center is at (–a, –b) and radius = $\sqrt{a^2 + b^2 - c}$ provided $a^2 + b^2 - c > 0$.

(i) Here, A = 1, B = 2, and C = 4. Then, $B^2 - AC = (2)^2 - (1)(4) = 0$. The equation represents a parabola.

(ii) Here, A = 1, B = 0, C = 1. Because A = C and B = 0, the equation represents a circle.

[Note that the center is at (1, 2), and the radius is $\sqrt{(-1)^2 + (-2)^2 - (-11)} = 4$.]

Parabola

A parabola is a special case of a conic where $e = 1$.

$(y-k)^2 = a(x-h)$ Vertex (h, k), axis $\parallel OX$

$y^2 = ax$ Vertex $(0, 0)$, axis along OX [Figure 2.25(a)]

$(x-h)^2 = a(y-k)$ Vertex (h, k), axis $\parallel OY$

$x^2 = ay$ Vertex $(0, 0)$, axis along OY [Figure 2.25(b)]

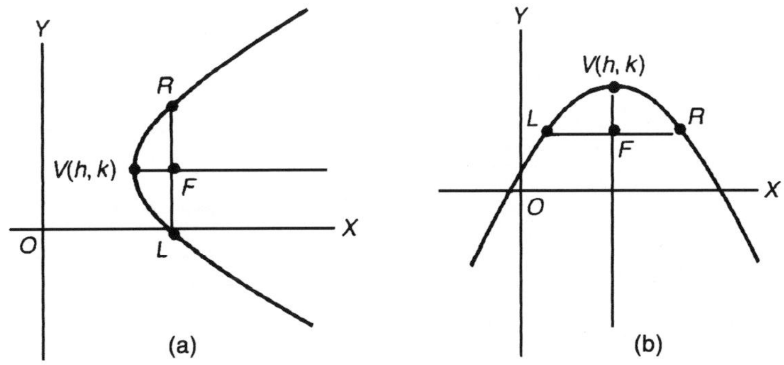

Figure 2.25

Distance from vertex to focus = $VF = \frac{1}{4}a$. Latus rectum = $LR = a$.

Example 2.17

Find the equation of a parabola (i) with center at (0, 0) and focus at (4, 0) and (ii) with center at (4, 2) and focus at (8, 2).

Solution

The equation of a parabola with center at (h, k) and focus at $(h + p/2, k)$ is given as

$(y - k)^2 = 2p(x - h)$

(i) $h = 0$; $k = 0$; $h + p/2 = 4$; then, $p = 8$ and the equation is $y^2 = 16x$.

(ii) $h = 4$; $k = 2$; $h + p/2 = 8$; then, $p = 8$ and the equation is $(y - 2)^2 = 16(x - 4)$.

Ellipse

This is a special case of a conic where $e < 1$.

$$\frac{(x-h)^2}{a^2}+\frac{(y-k)^2}{b^2}=1 \qquad \text{Center } (h, k), \text{ axes } \parallel OX, OY$$

$$\frac{x^2}{a^2}+\frac{y^2}{b^2}=1 \qquad \text{Center } (0, 0), \text{ axes along } OX, OY$$

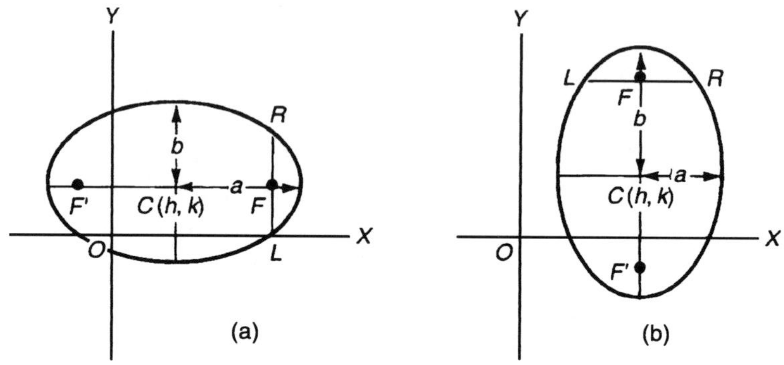

Figure 2.26

	$a > b$, Figure 2.26(a)	$b > a$, Figure 2.26(b)
Major axis	$2a$	$2b$
Minor axis	$2b$	$2a$
Distance from center to either focus	$\sqrt{a^2-b^2}$	$\sqrt{b^2-a^2}$
Latus rectum	$\frac{2b^2}{a}$	$\frac{2a^2}{b}$
Eccentricity, e	$\sqrt{\frac{a^2-b^2}{a}}$	$\sqrt{\frac{b^2-a^2}{b}}$
Sum of distances of any point P from the foci, $PF' + PF$	$2a$	$2b$

Example 2.18

Find the equation of an ellipse with center at origin, x-axis intercept (4, 0), and y-axis intercept (0, 2).

Solution

Equation of an ellipse with center at origin and x-axis intercept of $(a, 0)$ and y-axis intercept of $(0, b)$ is $\frac{x^2}{a^2}+\frac{y^2}{b^2}=1$. Here, $a = 4$ and $b = 2$; then, the equation is $\frac{x^2}{4^2}+\frac{y^2}{2^2}=1$.

Hyperbola

This is a special case of a conic where $e > 1$.

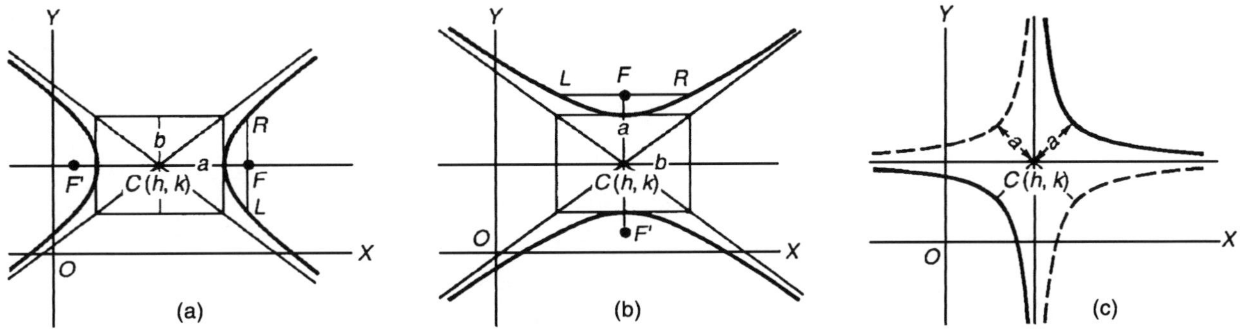

Figure 2.27

$$\frac{(x-h)^2}{a^2} - \frac{(y-k)^2}{b^2} = 1 \qquad C(h, k),\ \text{transverse axis} \parallel OX$$

$$\frac{x^2}{a^2} - \frac{y^2}{b^2} = 1 \qquad C(0,\ 0),\ \text{transverse axis along } OX$$

$$\frac{(y-k)^2}{a^2} - \frac{(x-h)^2}{b^2} = 1 \qquad C(h,\ k),\ \text{transverse axis} \parallel OY$$

$$\frac{y^2}{a^2} - \frac{x^2}{b^2} = 1 \qquad C(0, 0),\ \text{transverse along } OY$$

Transverse axis = $2a$; conjugate axis = $2b$

Distance from center to either focus = $\sqrt{a^2 + b^2}$

Latus rectum = $\dfrac{2b^2}{a}$

Eccentricity, $e = \dfrac{\sqrt{a^2 + b^2}}{a}$

Difference of distances of any point from the foci = $2a$.

The asymptotes are two lines through the center to which the branches of the hyperbola approach arbitrarily closely; their slopes are $\pm b/a$ [Figure 2.27(a)] or $\pm a/b$ [Figure 2.27(b)].

The rectangular (equilateral) hyperbola has $b = a$. The asymptotes are perpendicular to each other.

$$(x-h)(y-k) = \pm e = \sqrt{2} \qquad \text{Center } (h,k),\ \text{asymptotes} \parallel OX, OY$$

$$xy = \pm e = \sqrt{2} \qquad \text{Center } (0,0),\ \text{asymptotes along } OX, OY$$

The + sign gives the solid curves in Figure 2.27(c); the – sign gives the dotted curves in Figure 2.27(c).

Example 2.19

What is the equation of the hyperbola with center at origin, passing through (±2, 0), and an eccentricity of $\sqrt{10}$?

Solution

Equation of a hyperbola with center at origin, x-axis intercepts of (± a, 0), and eccentricity e is $\frac{x^2}{a^2} - \frac{y^2}{b^2} = 1$, where $b = a\sqrt{e^2 - 1}$. Here, $a = 2$; $e = \sqrt{10}$; then, $b = a\sqrt{e^2 - 1} = 2\sqrt{10 - 1} = 6$; equation is $\frac{x^2}{2^2} - \frac{y^2}{6^2} = 1$.

VECTORS

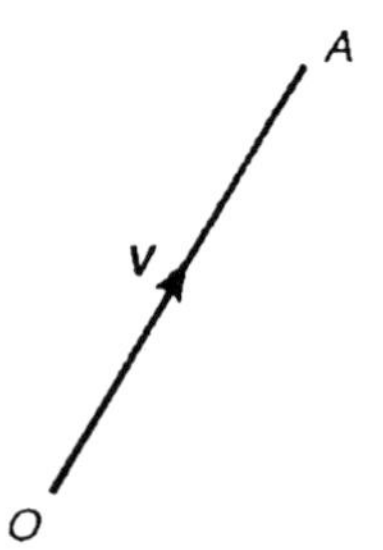

Figure 2.28

Definition and Graphical Representation of a Vector

A vector (**V**) is a quantity that is completely specified by magnitude *and* a direction. A scalar (s) is a quantity that is completely specified by a magnitude *only*.

The vector (**V**) may be represented geometrically by the segment $\overrightarrow{OA}$, the length of OA signifying the magnitude of **V** and the arrow carried by OA signifying the direction of **V**. The segment $\overrightarrow{AO}$ represents the vector –**V**.

Graphical Summation of Vectors

If $\mathbf{V}_1$ and $\mathbf{V}_2$ are two vectors, their graphical sum $\mathbf{V} = \mathbf{V}_1 + \mathbf{V}_2$ is formed by drawing the vector $\mathbf{V}_1 = \overrightarrow{OA}$, from any point O, and the vector $\mathbf{V}_2 = \overrightarrow{AB}$ from the end of $\mathbf{V}_1$ and joining O and B; then $\mathbf{V} = \overrightarrow{OB}$. Also, $\mathbf{V}_1 + \mathbf{V}_2 = \mathbf{V}_2 + \mathbf{V}_1$ and $\mathbf{V}_1 + \mathbf{V}_2 - \mathbf{V} = 0$ (Figure 2.29(a)).

Similarly, if $\mathbf{V}_1, \mathbf{V}_2, \mathbf{V}_3, \ldots, \mathbf{V}_n$ are any number of vectors drawn so that the initial point of one is the end point of the preceding one, then their graphical sum $\mathbf{V} = \mathbf{V}_1 + \mathbf{V}_2 + \ldots + \mathbf{V}_n$ is the vector joining the initial point of $\mathbf{V}_1$ with the end point of $\mathbf{V}_n$ (Figure 2.29(b)).

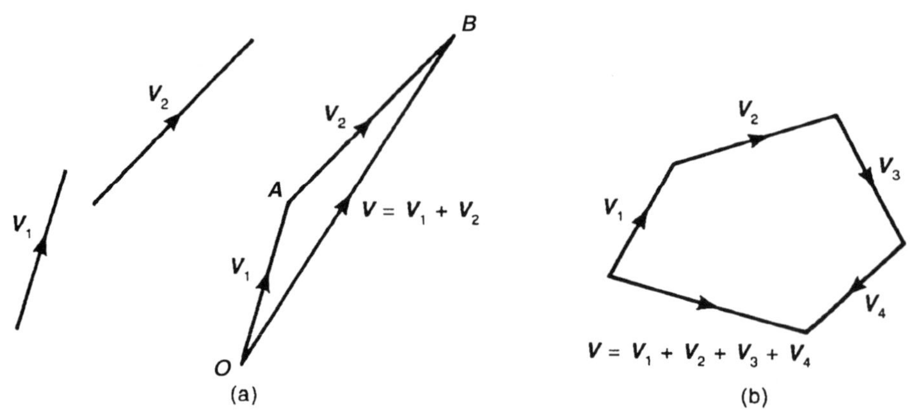

Figure 2.29

Analytic Representation of Vector Components

A vector **V** that is considered as lying in the *x*-*y* coordinate plane (Figure 2.30(a)) is completely determined by its horizontal and vertical components *x* and *y*. If **i** and **j** represent vectors of unit magnitude along *OX* and *OY*, respectively, and *a* and *b* are the magnitude of *x* and *y*, then **V** may be represented by $\mathbf{V} = a\mathbf{i} + b\mathbf{j}$, its magnitude by $|\mathbf{V}| = +\sqrt{a^2 + b^2}$, and its direction by $\alpha = \tan^{-1} b/a$.

A vector **V** in three-dimensional in space is completely determined by its components *x*, *y*, and *z* along three mutually perpendicular lines *OX*, *OY*, and *OZ*, directed as shown in Figure 2.30(b). If **i**, **j**, and **k** represent vectors of unit magnitude along *OX*, *OY*, *OZ*, respectively, and *a*, *b*, and *c* are the magnitudes of the components *x*, *y*, and *z*, respectively, then **V** may be represented by $\mathbf{V} = a\mathbf{i} + b\mathbf{j} + c\mathbf{k}$, its magnitude by, $|\mathbf{V}| = +\sqrt{a^2 + b^2 + c^2}$, and its direction by $\cos\alpha : \cos\beta : \cos\gamma = a : b : c$.

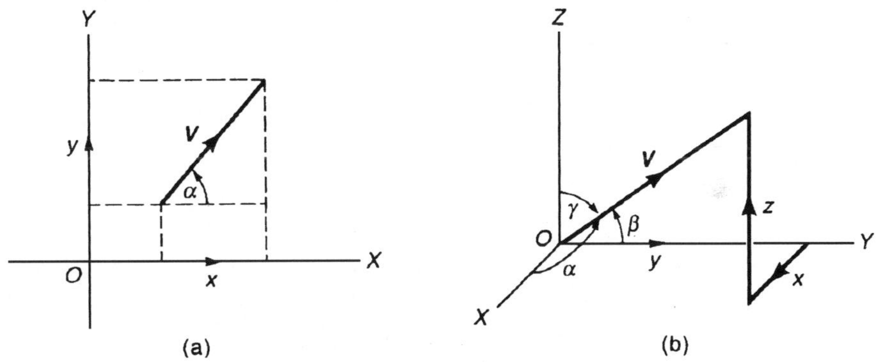

Figure 2.30

Properties of Vectors

$$\mathbf{V} = a\mathbf{i} + b\mathbf{j} \quad \text{or} \quad \mathbf{V} = a\mathbf{i} + b\mathbf{j} + c\mathbf{k}$$

Vector Sum V of any Number of Vectors, $\mathbf{V}_1$, $\mathbf{V}_2$, $\mathbf{V}_3$, ...

$$\mathbf{V} = \mathbf{V}_1 + \mathbf{V}_2 + \mathbf{V}_3 + \ldots = (a_1 + a_2 + a_3 + \ldots)\,\mathbf{i} + (b_1 + b_2 + b_3 + \ldots)\,\mathbf{j} + (c_1 + c_2 + c_3 + \ldots)\,\mathbf{k}$$

Product of a Vector V and a Scalar *s*

The product *s***V** has the same direction as **V**, and its magnitude is *s* times the magnitude of **V**.

$$s\mathbf{V} = (sa)\,\mathbf{i} + (sb)\,\mathbf{j} + (sc)\,\mathbf{k}$$

$$(s_1 + s_2)\,\mathbf{V} = s_1\mathbf{V} + s_2\mathbf{V} \qquad (\mathbf{V}_1 + \mathbf{V}_2)s = \mathbf{V}_1 s + \mathbf{V}_2 s$$

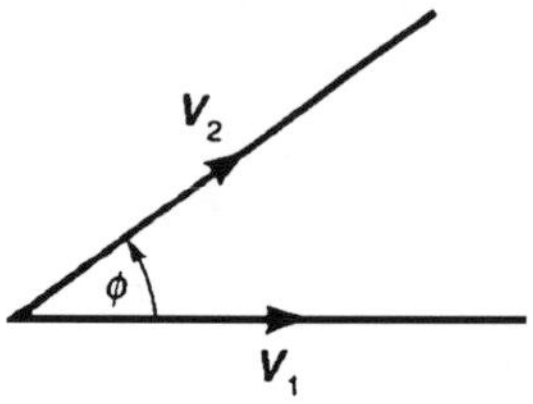

Figure 2.31

Scalar Product or Dot Product of Two Vectors: $\mathbf{V}_1 \bullet \mathbf{V}_2$

$$\mathbf{V}_1 \bullet \mathbf{V}_2 = |\mathbf{V}_1||\mathbf{V}_2|\cos\phi, \text{ where } \phi \text{ is the angle between } \mathbf{V}_1 \text{ and } \mathbf{V}_2$$

$$\mathbf{V}_1 \bullet \mathbf{V}_2 = \mathbf{V}_2 \bullet \mathbf{V}_1; \quad \mathbf{V}_1 \bullet \mathbf{V}_1 = |\mathbf{V}_1|^2; \quad (\mathbf{V}_1 + \mathbf{V}_2) \bullet \mathbf{V}_3 = \mathbf{V}_1 \bullet \mathbf{V}_3 + \mathbf{V}_2 \bullet \mathbf{V}_3$$

$$(\mathbf{V}_1 + \mathbf{V}_2) \bullet (\mathbf{V}_3 + \mathbf{V}_4) = \mathbf{V}_1 \bullet \mathbf{V}_3 + \mathbf{V}_1 \bullet \mathbf{V}_4 + \mathbf{V}_2 \bullet \mathbf{V}_3 + \mathbf{V}_2 \bullet \mathbf{V}_4$$

$$\mathbf{i} \bullet \mathbf{i} = \mathbf{j} \bullet \mathbf{j} = \mathbf{k} \bullet \mathbf{k} = 1; \quad \mathbf{i} \bullet \mathbf{j} = \mathbf{j} \bullet \mathbf{k} = \mathbf{k} \bullet \mathbf{i} = 0$$

In a plane, $\mathbf{V}_1 \bullet \mathbf{V}_2 = a_1 a_2 + b_1 b_2$; in space, $\mathbf{V}_1 \bullet \mathbf{V}_2 = a_1 a_2 + b_1 b_2 + c_1 c_2$.

The scalar product of two vectors $\mathbf{V}_1 \bullet \mathbf{V}_2$ is a scalar quantity and may physically represent the work done by a constant force of magnitude $|\mathbf{V}_1|$ on a particle moving through a distance $|\mathbf{V}_2|$, where ϕ is the angle between the direction of the force and the direction of motion.

Vector Product or Cross Product of Two Vectors: $\mathbf{V}_1 \times \mathbf{V}_2$

The vector product is $\mathbf{V}_1 \times \mathbf{V}_2 = \mathbf{l}\ |\mathbf{V}_1|\mathbf{V}_2|\sin\phi$, where ϕ is the angle from $\mathbf{V}_1$ to $\mathbf{V}_2$ and $\mathbf{l}$ is a unit vector perpendicular to the plane of the vectors $\mathbf{V}_1$ to $\mathbf{V}_2$ and so directed that a right-handed screw driven in the direction of $\mathbf{l}$ would carry $\mathbf{V}_1$ into $\mathbf{V}_2$.

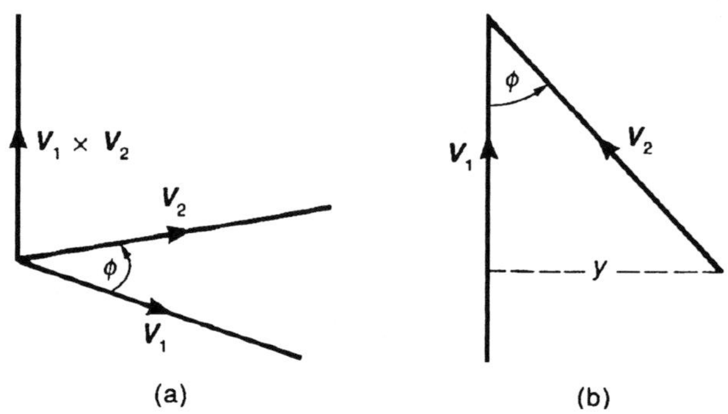

Figure 2.32

$$\mathbf{V}_1 \times \mathbf{V}_2 = -\mathbf{V}_2 \times \mathbf{V}_1; \quad \mathbf{V}_1 \times \mathbf{V}_1 = 0$$

$$(\mathbf{V}_1 + \mathbf{V}_2) \times \mathbf{V}_3 = \mathbf{V}_1 \times \mathbf{V}_3 + \mathbf{V}_2 \times \mathbf{V}_3$$

$$\mathbf{V}_1 \bullet (\mathbf{V}_2 \times \mathbf{V}_3) = \mathbf{V}_2 \bullet (\mathbf{V}_3 \times \mathbf{V}_1) = \mathbf{V}_3 \bullet (\mathbf{V}_1 \times \mathbf{V}_2)$$

$$\mathbf{i} \times \mathbf{i} = \mathbf{j} \times \mathbf{j} = \mathbf{k} \times \mathbf{k} = 0; \quad \mathbf{i} \times \mathbf{j} = \mathbf{k}; \quad \mathbf{j} \times \mathbf{k} = \mathbf{i}; \quad \mathbf{k} \times \mathbf{i} = \mathbf{j}$$

In the x-y plane, $\mathbf{V}_1 \times \mathbf{V}_2 = (a_1 b_2 - a_2 b_1)\mathbf{k}$.

In space, $\mathbf{V}_1 \times \mathbf{V}_2 = (b_2 c_3 - b_3 c_2)\,\mathbf{i} + (c_3 a_1 - c_1 a_3)\,\mathbf{j} + (a_1 b_2 - a_2 b_1)\,\mathbf{k}$.

The vector product of two vectors is a vector quantity and may physically represent the moment of a force $\mathbf{V}_1$ about a point O placed so that the moment arm is $y = |\mathbf{V}_2| \sin\phi$ (see Figure 2.32(b)).

Example 2.20

Vectors **A** and **B** are defined as: $\mathbf{A} = i - 2j + 3k$, $\mathbf{B} = 2i + j - 2k$

Find (i) **A** + **B**, (ii) **A** – **B**, (iii) 2**A**, (iv) |**A**|, (v) dot product **A**• **B**, and (vi) the cross product **A** × **B**.

Solution

(i) $\mathbf{A} + \mathbf{B} = 3i - j + k$ (ii) $\mathbf{A} - \mathbf{B} = -i - 3j + 5k$ (iii) $2\mathbf{A} = 2i - 4j + 6k$

(iv) $|\mathbf{A}| = \sqrt{1^2 + (-2)^2 + 3^2} = \sqrt{14}$

(v) Dot product $\mathbf{A} \bullet \mathbf{B} = (2) + (-2) + (-6) = -6$

(vi) Cross product $\mathbf{A} \times \mathbf{B} = \begin{vmatrix} i & j & k \\ 1 & -2 & 3 \\ 2 & 1 & -2 \end{vmatrix} = i + 8j + 5k$

LINEAR ALGEBRA

Matrix Operations

Matrices are rectangular arrays of real or complex numbers. Their great importance arises from the variety of operations that may be performed on them. Using the standard convention, the across-the-page lines are called **rows** and the up-and-down-the-page lines are **columns**. Entries in a matrix are **addressed** with double subscripts (always row first, then column). Thus the matrix

$$\mathbf{A} = \begin{bmatrix} 1 & 2 & 3 \\ 0 & 9 & -3 \end{bmatrix}$$

is 2 × 3, and the "9" is a_{22}. The "2" is a_{12} and the "0" is a_{21}. One also can refer to entries with square brackets, the "9" being $[A]_{22}$ and the "3" $[A]_{13}$.

If two matrices are the same size, they may be added: $[A + B]_{ij} = [A]_{ij} + [B]_{ij}$. Thus,

$$\begin{bmatrix} 1 & 2 & 3 \\ 0 & 9 & -3 \end{bmatrix} + \begin{bmatrix} 1 & 3 \\ 2 & 4 \end{bmatrix}$$

is not defined, but

$$\begin{bmatrix} 1 & 2 & 3 \\ 0 & 9 & -3 \end{bmatrix} + \begin{bmatrix} 1 & 3 & 5 \\ 2 & 4 & 6 \end{bmatrix} = \begin{bmatrix} 2 & 5 & 8 \\ 2 & 13 & 3 \end{bmatrix}$$

is proper.

Any matrix may be multiplied by a **scalar** (a number): $[c\mathbf{A}]_{ij} = c[\mathbf{A}]_{ij}$, so that

$$5\begin{bmatrix} 1 & 5 \\ 0 & 6 \end{bmatrix} = \begin{bmatrix} 5 & 25 \\ 0 & 30 \end{bmatrix}$$

The most peculiar matrix operation (and the most useful) is matrix multiplication. If **A** is $m \times n$ and **B** is $n \times p$, then **A** • **B** (or **AB**) is of size $m \times p$, and

$$[AB]_{ij} = \sum_{k=1}^{n} a_{jk} \bullet b_{kj}$$

The **dot product** (scalar product) of the ith row of **A** with the jth column of **B**, as in

$$\begin{bmatrix} 1 & 2 & 3 \\ 0 & 9 & -3 \end{bmatrix} \bullet \begin{bmatrix} 1 & 5 \\ 0 & 6 \end{bmatrix}$$

is not defined (owing to the mismatch of row and column lengths), but

$$\begin{bmatrix} 1 & 2 & 3 \\ 0 & 9 & -3 \end{bmatrix} \bullet \begin{bmatrix} 1 & 5 \\ 0 & 6 \\ 7 & 8 \end{bmatrix} = \begin{bmatrix} 1\,(1) + 2\,(0) + 3\,(7) & 1\,(5) + 2\,(6) + 3\,(8) \\ 0\,(1) + 9\,(0) - 3\,(7) & 0\,(5) + 9\,(6) - 3\,(8) \end{bmatrix} = \begin{bmatrix} 22 & 41 \\ -21 & 30 \end{bmatrix}$$

is correct.

A matrix with only one row or one column is called a vector, so a matrix times a vector is a vector (if defined). Thus **A** $(m \times n)$ • **X** $(n \times 1)$ = **Y**$(m \times 1)$, so a matrix can be thought of as an **operator** that takes vectors to vectors.

Another useful way of working with matrices is transposition: If **A** is $m \times n$, $\mathbf{A}^t$ is $n \times m$ and is the result of interchanging rows and columns. Hence

$$\begin{bmatrix} 1 & 2 & 3 \\ 0 & 9 & -3 \end{bmatrix}^t = \begin{bmatrix} 1 & 0 \\ 2 & 9 \\ 3 & -3 \end{bmatrix}$$

These various operations interact in the usual pleasant ways (and one decidedly unpleasant way); the standard convention is that all of the following combinations are defined:

$$\begin{gathered}
\mathbf{A} + \mathbf{B} = \mathbf{B} + \mathbf{A} \\
\mathbf{A} + (\mathbf{B} + \mathbf{C}) = (\mathbf{A} + \mathbf{B}) + \mathbf{C} \\
c(\mathbf{A} + \mathbf{B}) = c\mathbf{A} + c\mathbf{B} \\
(c + d)\,\mathbf{A} = c\mathbf{A} + d\mathbf{A} \\
(-1)\mathbf{A} + \mathbf{A} = (0\mathbf{A}) \\
\mathbf{A} \bullet \mathbf{B} \neq \mathbf{B} \bullet \mathbf{A} \text{ (in general)} \\
\mathbf{A} \bullet (\mathbf{B} \bullet \mathbf{C}) = (\mathbf{A} \bullet \mathbf{B}) \bullet \mathbf{C} \\
\mathbf{A} \bullet (\mathbf{B} + \mathbf{C}) = \mathbf{A} \bullet \mathbf{B} + \mathbf{A} \bullet \mathbf{C} \\
(\mathbf{A} + \mathbf{B}) \bullet \mathbf{C} = \mathbf{A} \bullet \mathbf{C} + \mathbf{B} \bullet \mathbf{C} \\
(\mathbf{A} + \mathbf{B})^t = \mathbf{A}^t + \mathbf{B}^t \\
(\mathbf{A} \bullet \mathbf{B})^t = \mathbf{B}^t \bullet \mathbf{A}^t
\end{gathered}$$

In addition, matrices **I**, which are $n \times n$ and whose entries are 1 on the diagonal $i = j$ and 0 elsewhere, are multiplicative identities: $\mathbf{A} \bullet \mathbf{I} = \mathbf{A}$ and $\mathbf{I} \bullet \mathbf{A} = \mathbf{A}$. Here the two **I** matrices may be different sizes; for example,

$$\begin{bmatrix} 1 & 2 & 3 \\ 0 & 9 & -3 \end{bmatrix} \bullet \begin{bmatrix} 1 & 0 & 0 \\ 0 & 1 & 0 \\ 0 & 0 & 1 \end{bmatrix} = \begin{bmatrix} 1 & 2 & 3 \\ 0 & 9 & -3 \end{bmatrix}$$

but

$$\begin{bmatrix} 1 & 0 \\ 0 & 1 \end{bmatrix} \bullet \begin{bmatrix} 1 & 2 & 3 \\ 0 & 9 & -3 \end{bmatrix} = \begin{bmatrix} 1 & 2 & 3 \\ 0 & 9 & -3 \end{bmatrix}$$

I is called the identity matrix, and the size is understood from context.

Example 2.21

Verify that the transpose of $\mathbf{A} + \mathbf{BC}$ is $\mathbf{C}^t\mathbf{B}^t + \mathbf{A}^t$ if

$$\mathbf{A} = \begin{bmatrix} 1 & 1 \\ 2 & 3 \end{bmatrix} \quad \mathbf{B} = \begin{bmatrix} 1 & 2 & 3 \\ 0 & 9 & -3 \end{bmatrix} \quad \mathbf{C} = \begin{bmatrix} 1 & 5 \\ 0 & 6 \\ 7 & 8 \end{bmatrix}$$

Solution

$$\mathbf{BC} \begin{bmatrix} 22 & 41 \\ -21 & 30 \end{bmatrix}, \text{ so } \mathbf{A} + \mathbf{BC} = \begin{bmatrix} 23 & 42 \\ -19 & 33 \end{bmatrix} \text{ and } [\mathbf{A} + \mathbf{BC}]^t = \begin{bmatrix} 23 & -19 \\ 42 & 33 \end{bmatrix}$$

On the other hand, $\mathbf{C}^t\mathbf{B}^t = \begin{bmatrix} 1 & 0 & 7 \\ 5 & 6 & 8 \end{bmatrix} \bullet \begin{bmatrix} 1 & 0 \\ 2 & 9 \\ 3 & -3 \end{bmatrix} = \begin{bmatrix} 22 & -21 \\ 41 & 30 \end{bmatrix}$ and

$$\mathbf{A}^t = \begin{bmatrix} 1 & 2 \\ 1 & 3 \end{bmatrix}, \text{ so } \mathbf{A}^t + \mathbf{C}^t\mathbf{B}^t = \mathbf{C}^t\mathbf{B}^t + \mathbf{A}^t = \begin{bmatrix} 23 & -19 \\ 42 & 33 \end{bmatrix}$$

Types of Matrices

Matrices are classified according to their appearance or the way they act. If **A** is square and $\mathbf{A}^t = \mathbf{A}$, then **A** is called symmetric. If $\mathbf{A}^t = -\mathbf{A}$, then it is skew-symmetric.

If **A** has complex entries, $\mathbf{A}^*$ then is called the Hermitian adjoint of **A**. If $\mathbf{A}^* = \bar{\mathbf{A}}^t$ (complex conjugate), then

$$\begin{bmatrix} 1+i & i \\ 3 & 4-i \end{bmatrix}^* = \begin{bmatrix} 1-i & -i \\ 3 & 4+i \end{bmatrix}^t = \begin{bmatrix} 1-i & 3 \\ -i & 4+i \end{bmatrix}$$

If $\mathbf{A} = \mathbf{A}^*$, then A is called Hermitian. If $\mathbf{A}^* = -\mathbf{A}$, the name is skew-Hermitian.

If **A** is square and $a_{ij} = 0$ unless $i = j$, **A** is called diagonal. If **A** is square and zero below the diagonal ($[\mathbf{A}]_{ij} = 0$ if $i > j$), **A** is called upper triangular. The transpose of such a matrix is called lower triangular.

If A is square and there is a matrix $\mathbf{A}^{-1}$ such that $\mathbf{A}^{-1} \bullet \mathbf{A} = \mathbf{A} \bullet \mathbf{A}^{-1} = \mathbf{I}$, **A** is nonsingular. Otherwise, it is singular. If **A** and **B** are both nonsingular $n \times n$ matrices, then **AB** is nonsingular and $(\mathbf{AB})^{-1} = \mathbf{B}^{-1}\mathbf{A}^{-1}$, because $(\mathbf{AB})(\mathbf{B}^{-1}\mathbf{A}^{-1}) = \mathbf{A}(\mathbf{B}\,\mathbf{B}^{-1})\mathbf{A}^{-1} = \mathbf{AIA}^{-1} = \mathbf{AA}^{-1} = \mathbf{I}$, as does $(\mathbf{B}^{-1}\mathbf{A}^{-1}) \bullet (\mathbf{AB})$.

If $\mathbf{A}'\mathbf{A} = \mathbf{AA}' = \mathbf{I}$ and **A** is real, it is called orthogonal (the reason will appear below). If $\mathbf{A}^*\mathbf{A} = \mathbf{AA}^* = \mathbf{I}$ (**A** complex), **A** is called unitary. If **A** commutes with $\mathbf{A}^*$, so that $\mathbf{AA}^* = \mathbf{A}^*\mathbf{A}$, then **A** is called normal.

Elementary Row and Column Operations

The most important tools used in dealing with matrices are the elementary operations: R for row, C for column. If **A** is given matrix, performing $R(i \leftrightarrow j)$ on **A** means interchanging Row i and Row j. $R_i(c)$ means multiplying Row i by the number c (except $c = 0$). $R_j + cR_i$ means multiply Row i by c and add this result into Row j $(i \neq j)$. Thus, if

$$\mathbf{A} = \begin{bmatrix} 1 & 2 & 3 \\ 4 & 5 & 6 \\ 7 & 8 & 0 \end{bmatrix}$$

then

$$R(2 \leftrightarrow 3)\,(\mathbf{A}) = \begin{bmatrix} 1 & 2 & 3 \\ 7 & 8 & 0 \\ 4 & 5 & 6 \end{bmatrix} \qquad C_1(2)(\mathbf{A}) = \begin{bmatrix} 2 & 2 & 3 \\ 8 & 5 & 6 \\ 14 & 8 & 0 \end{bmatrix}$$

$$R_1 - R_2(\mathbf{A}) = \begin{bmatrix} -3 & -3 & -3 \\ 4 & 5 & 6 \\ 7 & 8 & 0 \end{bmatrix}$$

These operations are used in reducing matrix problems to simpler ones.

Example 2.22

Solve $\mathbf{AX} = \mathbf{B}$ where

$$\mathbf{A} = \begin{bmatrix} 1 & 2 & 3 \\ 4 & 5 & 6 \\ 7 & 8 & 9 \end{bmatrix}, \quad \mathbf{X} = \begin{bmatrix} x \\ y \\ z \end{bmatrix}, \quad \mathbf{B} = \begin{bmatrix} 1 \\ 1 \\ 1 \end{bmatrix}$$

Solution

Form the "augmented" matrix

$$[\mathbf{A}|\mathbf{B}] = \begin{bmatrix} 1 & 2 & 3 & 1 \\ 4 & 5 & 6 & 1 \\ 7 & 8 & 9 & 1 \end{bmatrix}$$

and perform elementary row operations on this matrix until the solution is apparent:

$$\begin{bmatrix}1&2&3&1\\4&5&6&1\\7&8&9&1\end{bmatrix}\begin{matrix}R_2-4R_1\\R_3-7R_1\end{matrix}\begin{bmatrix}1&2&3&1\\0&-3&-6&-3\\7&-6&-12&-6\end{bmatrix}\begin{matrix}R_2\left(-\frac{1}{3}\right)\\R_3\left(-\frac{1}{6}\right)\end{matrix}\begin{bmatrix}1&2&3&1\\0&1&2&1\\0&1&2&1\end{bmatrix}$$

$$R_3-R_2\begin{bmatrix}1&2&3&1\\0&1&2&1\\0&0&0&0\end{bmatrix}$$

The answer is now apparent: $y + 2z = 1$ and $x + 2y + 3z = 1$, or, z arbitrary, $y = 1 - 2z$, $x = 1 - 2(1 - 2z) - 3z = -1 + z$. This system of equations has an infinite number of solutions.

Example 2.23

Solve the system of equations

$$\begin{aligned}x + y - z &= a\\2x - y + 3z &= 2\\3x + 2y + z &= 1\end{aligned}$$

for x, y, and z in terms of a.

Solution

Strip off the variables x, y, and z:

$$\begin{bmatrix}1&1&-1&a\\2&-1&3&2\\3&2&1&1\end{bmatrix}\begin{matrix}R_2-2R_1\\R_3-3R_1\end{matrix}\begin{bmatrix}1&1&-1&a\\0&-3&5&2-2a\\0&-1&4&1-3a\end{bmatrix}$$

$$\begin{matrix}R_2(2\leftrightarrow 3)\\R_2(-1)\end{matrix}\begin{bmatrix}1&1&-1&a\\0&1&-4&3a-1\\0&-3&5&2-2a\end{bmatrix}\begin{matrix}R_1-R_2\\R_3+3R_2\end{matrix}\begin{bmatrix}1&0&3&1-2a\\0&1&-4&3a-1\\0&0&-7&7a-1\end{bmatrix}$$

The solution is now clear:

$$z=\frac{7a-1}{-7}=-a+\frac{1}{7}$$

$$y=3a-1+4z=3a-1-4a+\frac{4}{7}=-a-\frac{3}{7}$$

$$x=1-2a-3z=1-2a+3a-\frac{3}{7}=a+\frac{4}{7}$$

Example 2.24

Find $\mathbf{A}^{-1}$ if

$$\mathbf{A} = \begin{bmatrix} 1 & 1 & -1 \\ 1 & 2 & 3 \\ 3 & 2 & 1 \end{bmatrix}$$

Solution

Since this amounts to solving $\mathbf{AX} = \mathbf{B}$ three times, with

$$\mathbf{B} = \begin{bmatrix} 1 \\ 0 \\ 0 \end{bmatrix} \quad \mathbf{B} = \begin{bmatrix} 0 \\ 1 \\ 0 \end{bmatrix} \quad \mathbf{B} = \begin{bmatrix} 0 \\ 0 \\ 1 \end{bmatrix}$$

form

$$[\mathbf{A}|\mathbf{I}] = \begin{bmatrix} 1 & 1 & -1 & 1 & 0 & 0 \\ 1 & 2 & 3 & 0 & 1 & 0 \\ 3 & 2 & 1 & 0 & 0 & 1 \end{bmatrix}$$

and perform row operations until a solution emerges.

$$[\mathbf{A}|\mathbf{I}] = \frac{R_2 - R_1}{R_3 - 3R_1} \begin{bmatrix} 1 & 1 & -1 & 1 & 0 & 0 \\ 0 & 1 & 4 & -1 & 1 & 0 \\ 0 & -1 & 4 & -3 & 0 & 1 \end{bmatrix}$$

$$\frac{R_1 - R_2}{R_3 + R_2} z \begin{bmatrix} 1 & 0 & -5 & 2 & -1 & 0 \\ 0 & 1 & 4 & -1 & 1 & 0 \\ 0 & 0 & 8 & -4 & 1 & 1 \end{bmatrix}$$

$$\begin{matrix} R_3\left(\frac{1}{8}\right) \\ \frac{R_2 - 4R_3}{R_1 + 5R_3} \end{matrix} \begin{bmatrix} 1 & 0 & 0 & -\frac{1}{2} & -\frac{3}{8} & \frac{5}{8} \\ 0 & 1 & 0 & 1 & \frac{1}{2} & -\frac{1}{2} \\ 0 & 0 & 1 & -\frac{1}{2} & \frac{1}{8} & \frac{1}{8} \end{bmatrix}$$

Thus,

$$\mathrm{A}^{-1} = \begin{bmatrix} -\frac{1}{2} & -\frac{3}{8} & \frac{5}{8} \\ 1 & \frac{1}{2} & -\frac{1}{2} \\ -\frac{1}{2} & \frac{1}{8} & \frac{1}{8} \end{bmatrix}$$

Example 2.25

Verify that $\mathbf{A}^{-1}\mathbf{A} = \mathbf{I}$ in Example 2.24.

Solution

$$8\mathbf{A}^{-1}\mathbf{A} = \begin{bmatrix} -4 & -3 & 5 \\ 8 & 4 & -4 \\ -4 & 1 & 1 \end{bmatrix} \begin{bmatrix} 1 & 1 & -1 \\ 1 & 2 & 3 \\ 3 & 2 & 1 \end{bmatrix}$$

$$= \begin{bmatrix} -4-3+15 & -4-6+10 & 4-9+5 \\ 8+4-12 & 8+8-8 & -8+12-4 \\ -4+1+3 & -4+2+2 & 4+3+1 \end{bmatrix} = 8\begin{bmatrix} 1 & 0 & 0 \\ 0 & 1 & 0 \\ 0 & 0 & 1 \end{bmatrix} = 8\mathbf{I}$$

Example 2.26

Describe the set of solutions of $\mathbf{AX} = \mathbf{B}$.

Solution

If $\mathbf{AX}_0 = \mathbf{B}$ is one solution, and $\mathbf{AY} = 0$, then $\mathbf{A}(\mathbf{X}_0 + \mathbf{Y})$ is a solution, so all solutions are of the form $\mathbf{X} = \mathbf{X}_0 + \mathbf{Y}$ where $\mathbf{AY} = 0$. Thus, if $\mathbf{N} = \{\mathbf{Y} : \mathbf{AY} = 0\}$ is the null space of $\mathbf{A}$, the set of solutions to $\mathbf{AX} = \mathbf{B}$ is $\mathbf{X}_0 + \mathbf{N} = \{\mathbf{X}_0 + \mathbf{Y} : \mathbf{Y} \in \mathbf{N}\}$.

Determinants

The determinant of a square matrix is a scalar representing the *volume* of the matrix in some sense. Matrices that are not square do not have determinants.

The determinant is frequently indicated by vertical lines, viz. $|A|$. It is a complicated formula, and one way to find it is by induction. The determinant of a 1×1 matrix is $|a| = a$. The determinant of a 2×2 matrix is

$$\begin{vmatrix} a & b \\ c & d \end{vmatrix} = ad - bc$$

The determinant of an $n \times n$ matrix is given in terms of n determinants, each of size $(n-1) \times (n-1)$. If $\mathbf{A}$ is $n \times n$ and $\mathbf{M}_{ij}$ is the matrix obtained by removing the ith row and the jth column from $\mathbf{A}$, then

$$|A| = \sum_{j=1}^{n} (-1)^{1+j} a_{1j} \, |M_{1j}|$$

Example 2.27

Find the determinant

$$\begin{vmatrix} 1 & 2 & 3 \\ 4 & 0 & 6 \\ 7 & 8 & 9 \end{vmatrix}$$

Solution

$$\begin{aligned} |A| &= (-1)^{1+1} a_{11} |M_{11}| + (-1)^{1+2} a_{12} |M_{12}| + (-1)^{1+3} a_{13} |M_{13}| \\ &= 1 \begin{vmatrix} 0 & 6 \\ 8 & 9 \end{vmatrix} - 2 \begin{vmatrix} 4 & 6 \\ 7 & 9 \end{vmatrix} + 3 \begin{vmatrix} 4 & 0 \\ 7 & 8 \end{vmatrix} \\ &= -48 - 2(36 - 42) + 3(32) = 60 \end{aligned}$$

Example 2.28

Find the determinant

$$\begin{vmatrix} 0 & 0 & 2 & 0 \\ 1 & 2 & 7 & 3 \\ 4 & 0 & 3 & 6 \\ 7 & 8 & -6 & 9 \end{vmatrix}$$

Solution

$$\begin{aligned} |A| &= a_{11} |M_{11}| - a_{12} |M_{12}| + a_{13} |M_{13}| - a_{14} |M_{14}| \\ &= 0 |M_{11}| - 0 \, |M_{12}| + 2 \, |M_{13}| - 0 \, |M_{14}| = 2\,(60) = 120 \end{aligned}$$

The last example provides a clue to the evaluation of large determinants, but the use of the first row of **A** in the definition of a determinant was arbitrary. For any row or column (fix i or j),

$$|A| = \sum_{j=1}^{n} (-1)^{1+j} a_{i1j} \, |M_{ij}|$$

The interaction of the determinant with elementary row or column operations is simple: Interchanging two rows changes the sign of the determinant; multiplying a row by a constant multiplies the determinant by that constant.

Example 2.29

Evaluate the determinant

$$\begin{vmatrix} 1 & 2 & 3 & 4 \\ 1 & 1 & 1 & 0 \\ 4 & 0 & 3 & 2 \\ 0 & 3 & 0 & 1 \end{vmatrix}$$

Solution

Choose a row or column with many zeroes and introduce still more:

$$|A| = C_2 - 3C_4\,|A| = \begin{vmatrix} 1 & -10 & 3 & 4 \\ 1 & 1 & 1 & 0 \\ 4 & -6 & 3 & 2 \\ 0 & 0 & 0 & 1 \end{vmatrix} = (-1)^{4+4} a_{44} \begin{vmatrix} 1 & -10 & 3 \\ 1 & 1 & 1 \\ 4 & -6 & 3 \end{vmatrix}$$

$$\underset{R_3 - 4R_1}{\overset{R_2 - R_1}{=}} \begin{vmatrix} 1 & -10 & 3 \\ 0 & 11 & -2 \\ 0 & 34 & -9 \end{vmatrix} = (-1)^{1+1} a_{11} \begin{vmatrix} 11 & -2 \\ 34 & -9 \end{vmatrix} = -99 + 68 = -31$$

Example 2.30

Find which values, if any, of the number c make **A** singular if

$$\mathbf{A} = \begin{vmatrix} 1 & 2 & c \\ 4 & 5 & 6 \\ 1 & 1 & 1 \end{vmatrix}$$

Solution

$|\mathbf{A}| = (-1)^2(5-6) + (-1)^3(2)(4-6) + (-1)^4 c(4-5) = -1 + 4 - c = 0.$ Hence **A** is singular for only one value of c, $c = 3$.

Cramer's Rule is a consequence of adj(A): If **A** is nonsingular, the ith component of the solution of $\mathbf{AX} = \mathbf{B}$ is $x_i = \dfrac{|A_i|}{|A|}$, where A_i is the result of replacing the ith column of **A** by **B**.

Example 2.31

Find x_2 in $\mathbf{AX} = \mathbf{B}$ by Cramer's Rule if

$$\mathbf{A} = \begin{bmatrix} 1 & 2 & 1 & 1 \\ 3 & 4 & 5 & -2 \\ 6 & 7 & 1 & 5 \\ -1 & 0 & 2 & 0 \end{bmatrix} \quad \text{and} \quad \mathbf{B} = \begin{bmatrix} 1 \\ 2 \\ 3 \\ 4 \end{bmatrix}$$

Solution

First,

$$|A| = \begin{vmatrix} 1 & 2 & 3 & 1 \\ 3 & 4 & 11 & -2 \\ 6 & 7 & 13 & 5 \\ -1 & 0 & 0 & 0 \end{vmatrix} = (-1)^{4+1}(-1) \begin{vmatrix} 2 & 3 & 1 \\ 4 & 11 & -2 \\ 7 & 13 & 5 \end{vmatrix}$$

$$= \begin{vmatrix} 0 & 0 & 1 \\ 8 & 17 & -2 \\ -3 & -2 & 5 \end{vmatrix} = (-1)^{1+3}(1) \begin{vmatrix} 8 & 17 \\ -3 & -2 \end{vmatrix} = -16 + 51 = 35$$

Next, the numerator of x_2 is

$$\begin{vmatrix} 1 & 1 & 1 & 1 \\ 3 & 2 & 5 & -2 \\ 6 & 3 & 1 & 5 \\ -1 & 4 & 2 & 0 \end{vmatrix} = \begin{vmatrix} 1 & 0 & 0 & 0 \\ 3 & -1 & 2 & -5 \\ 6 & -3 & -5 & -1 \\ -1 & 5 & 3 & 1 \end{vmatrix} = \begin{vmatrix} -1 & 2 & -5 \\ -3 & -5 & -1 \\ 5 & 3 & 1 \end{vmatrix} = \begin{vmatrix} -1 & 2 & -5 \\ 0 & -11 & 14 \\ 0 & 13 & -24 \end{vmatrix}$$

$$= -\begin{vmatrix} -11 & 14 \\ 13 & -24 \end{vmatrix} = -\begin{vmatrix} -11 & 14 \\ 2 & -10 \end{vmatrix} = -(110-28) = -82, x_2 = -\frac{82}{35}$$

NUMERICAL METHODS

This portion of numerical methods includes techniques of finding roots of polynomials by the Routh-Hurwitz criterion and Newton methods, Euler's techniques of numerical integration and the trapezoidal methods, and techniques of numerical solutions of differential equations.

Root Extraction

Routh-Hurwitz Method (without Actual Numerical Results)

Root extraction, even for simple roots (i.e., without imaginary parts), can become quite tedious. Before attempting to find roots, one should first ascertain whether they are really needed or whether just knowing the area of location of these roots will suffice. If all that is needed is knowing whether the roots are all in the left half-plane of the variable (such as is in the s-plane when using Laplace transforms—as is frequently the case in determining system stability in control systems), then one may use the Routh-Hurwitz criterion. This method is fast and easy even for higher-ordered equations. As an example, consider the following polynomial:

$$p_n(x) = \prod_{m=1}^{n}(x - x_m) = x^n + a_1 x^{n-1} + a_2 x^{n-2} + \cdots + a_{n-1} \quad \textbf{(2.1)}$$

Here, finding the roots, x_m, for $n > 3$ can become quite tedious without a computer; however, if one only needs to know if any of the roots have positive real parts, one can use the Routh-Hurwitz method. Here, an array is formed listing the coefficients of every other term starting with the highest power, n, on a line, followed by a line listing the coefficients of the terms left out of the first row. Following rows are constructed using Routh-Hurwitz techniques, and after completion of the array, one merely checks to see if all the signs are the same (unless there is a zero coefficient—then something else needs to be done) in the first column; if none, no roots will exist in the right half-plane. In case of zero coefficient, a simple technique is used; for details, see almost any text dealing with stability of control systems. A short example follows.

$F(s) = s^3 + 3s^2 + 10$ Array:	s^3	1	2	Where the s^1 term is formed as
	s^2	3	10	
$= (s + ?)(s + ?)(s + ?)$	s^1	$-\frac{4}{3}$	0	$(3 \times 2 - 10 \times 1)/3 = -\frac{4}{3}$. For details, refer to any text on control systems or numerical methods.
	s^0	10	0	

Here, there are two sign changes: one from 3 to $-\frac{4}{3}$, and one from $-\frac{4}{3}$ to 10. This means there will be two roots in the right half-plane of the s-plane, which yield an unstable system. This technique represents a great savings in time without having to factor the polynomial.

Newton's Method

The use of Newton's method of solving a polynomial and the use of iterative methods can greatly simplify a problem. This method utilizes synthetic division and is based upon the remainder theorem. This synthetic division requires estimating a root at the start, and, of course, the best estimate is the actual root. The root is the correct one when the remainder is zero. (There are several ways of estimating this root, including a slight modification of the Routh-Hurwitz criterion.)

If a $P_n(x)$ polynomial (see Equation (2.1)) is divided by an estimated factor $(x - x_1)$, the result is a reduced polynomial of degree $n-1$, $Q_{n-1}(x)$, plus a constant remainder of b_{n-1}. Thus, another way of describing Equation (2.1) is

$$P_n(x)/(x-x_1) = Q_{n-1}(x) + b_{n-1}/(x-x_1) \quad \text{or} \quad P_n(x) = (x-x_1)Q_{n-1}(x) + b_{n-1} \tag{2.2}$$

If one lets $x = x_1$, Equation (2.2) becomes

$$P_n(x = x_1) = (0)Q_{n-1}(x) + b_{n-1} = b_{n-1} \tag{2.3}$$

Equation 2.3 leads directly to the remainder theorem: "The remainder on division by $(x - x_1)$ is the value of the polynomial at $x = x_1$, $P_n(x_1)$."[1]

Newton's method (actually, the Newton-Raphson method) for finding the roots for an nth-order polynomial is an iterative process involving obtaining an estimated value of a root (leading to a simple computer program). The key to the process is getting the first estimate of a possible root. Without getting too involved, recall that the coefficient of x^{n-1} represents the sum of all of the roots and the last term represents the product of all n roots; then the first estimate can be "guessed" within a reasonable magnitude. After a first root is chosen, find the rate of change of the polynomial at the chosen value of the root to get the next, closer value of the root x_{n+1}. Thus the new root estimate is based on the last value chosen:

$$x_{n+1} = x_n - P_n(x_n)/P_n'(x_n), \quad \text{where } P_n'(x_n) = dP_n(x)/dx \text{ evaluated at } x = x_n \tag{2.4}$$

NUMERICAL INTEGRATION

Numerical integration routines are extremely useful in almost all simulation-type programs, design of digital filters, theory of z-transforms, and almost any problem solution involving differential equations. And because digital computers have essentially replaced analog computers (which were almost true integration devices), the techniques of approximating integration are well developed. Several of the techniques are briefly reviewed below.

1 Gerald & Wheatley, *Applied Numerical Analysis*, 3rd ed., Addison-Wesley, 1985.

Euler's Method

For a simple first-order differential equation, say $dx/dt + ax = af$, one could write the solution as a continuous integral or as an interval type one:

$$x(t) = \int^{t} [-ax(\tau) + af(\tau)]d\tau \quad \textbf{(2.5a)}$$

$$x(kT) = \int^{kT-T} [-ax + af]d\tau + \int_{kT-T}^{kT} [-ax + af]d\tau = x(kT - T) + A_{rect} \quad \textbf{(2.5b)}$$

Here, A_{rect} is the area of $(-ax + af)$ over the interval $(kT - T) < \tau < kT$. One now has a choice looking back over the rectangular area or looking forward. The rectangular width is, of course, T. For the forward-looking case, a first approximation for x_1 is[2]

$$\begin{aligned} x_1(kT) &= x_1(kT - T) + T[ax_1(kT - T) + af(kT - T)] \\ &= (1 - aT)x_1(kT - T) + aTf(kT - T) \end{aligned} \quad \textbf{(2.5c)}$$

Or, in general, for Euler's forward rectangle method, the integral may be approximated in its simplest form (using the notation $t_{k+1} - t_k$ for the width, instead of T, which is $kT–T$) as

$$\int_{t_k}^{t_{k+1}} x(\tau)d\tau \approx (t_{k+1} - t_k)x(t_k) \quad \textbf{(2.6)}$$

Trapezoidal Rule

This trapezoidal rule is based upon a straight-line approximation between the values of a function, $f(t)$, at t_0 and t_1. To find the area under the function, say a curve, is to evaluate the integral of the function between points a and b. The interval between these points is subdivided into subintervals; the area of each subinterval is approximated by a trapezoid between the end points. It will be necessary only to sum these individual trapezoids to get the whole area; by making the intervals all the same size, the solution will be simpler. For each interval of delta t (i.e., $t_{k+1} - t_k$), the area is then given by

$$\int_{t_k}^{t_{k+1}} x(\tau)\, d\tau \approx (1/2)(t_{k+1} - t_k)[x(t_{k+1}) + x(t_k)] \quad \textbf{(2.7)}$$

This equation gives good results if the delta t's are small, but it is for only one interval and is called the "local error." This error may be shown to be – (1/12) (delta t)$^3 f''(t = \xi_1)$, where ξ_1 is between t_0 and t_1. For a larger "global error" it may be shown that

$$\text{Global error} = -(1/12)(\text{delta } t)^3\, [f''(\xi_1) + f''(\xi_2) + \cdots + f''(\xi_n)] \quad \textbf{(2.8)}$$

Following through on Equation 2.8 allows one to predict the error for the trapezoidal integration. This technique is beyond the scope of this review or probably the examination; however, for those interested, please refer to pages 249–250 of the previously mentioned reference to Gerald & Wheatley.

---

[2] This method is as presented in Franklin & Powell, *Digital Control of Dynamic Systems*, Addison-Wesley, 1980, page 55.

NUMERICAL SOLUTIONS OF DIFFERENTIAL EQUATIONS

This solution will be based upon first-order ordinary differential equations. However, the method may be extended to higher-ordered equations by converting them to a matrix of first-ordered ones.

Integration routines produce values of system variables at specific points in time and update this information at each interval of delta time as T (delta $t = T = t_{k+1} - t_k$). Instead of a continuous function of time, $x(t)$, the variable x will be represented with discrete values such that $x(t)$ is represented by $x_0, x_1, x_2, \ldots, x_n$. Consider a simple differential equation as before as based upon Euler's method,

$$dx/dt + ax = f(t)$$

Now assume the delta time periods, T, are fixed (not all routines use fixed step sizes); then one writes the continuous equations as a difference equation where $dx/dt \approx (x_{k+1} - x_k)/T = -ax_k + f_k$ or, solving for the updated value, x_{k+1},

$$x_{k+1} = x_k - Tax_k + Tf_k \tag{2.9a}$$

For fixed increments by knowing the first value of $x_{k=0}$ (or the initial condition), one may calculate the solution for as many "next values" of x_{k+1} as desired for some value of T. The difference equation may be programmed in almost any high-level language on a digital computer; however, T must be small as compared to the shortest time constant of the equation (here, $1/a$).

The following equation—with the "f" term meaning "a function of" rather than as a "forcing function" term as used in Equation (2.5a)—is a more general form of Equation (2.9a). This equation is obtained by letting the notation x_{k+1} become $y\,[k+1\;\Delta t]$ and is written (perhaps somewhat more confusingly) as

$$y[(k+1)\Delta t] = y(k\Delta t) + \Delta t f[y(k\Delta t), k\Delta t)] \tag{2.9b}$$

Reduction of Differential Equation Order

To reduce the order of a linear time-dependent differential equation, the following technique is used. For example, assume a second-order equation: $x'' + ax' + bx = f(t)$. If we define $x = x_1$ and $x' = x_1{}' = x_2$, then

$$x_2' + ax_2 + bx_1 = f(t)$$
$$x_1' = x_2 \text{ (by definition)}$$
$$x_2' = -b_{x1} - ax_2 + f(t)$$

This technique can be extended to higher-order systems and, of course, be put into a matrix form (called the state variable form). And it can easily be set up as a matrix of first-order difference equations for solving digitally.

DIFFERENTIAL CALCULUS

Definition of a Function

Notation. A variable y is said to be a function of another variable x if, when x is given, y is determined. The symbols $f(x)$, $F(x)$, etc., represent various functions of x. The symbol $f(a)$ represents the value of $f(x)$ when $x = a$.

Definition of a Derivative

Let $y = f(x)$. If Δx is any increment (increase or decrease) given to x, and Δy is the corresponding increment in y, then the derivative of y with respect to x is the limit of the ratio of Δy to Δx as Δx approaches zero; that is,

$$\frac{dy}{dx} = \lim_{\Delta x \to 0} \frac{\Delta y}{\Delta x} = \lim_{\Delta x \to 0} \frac{f(x+\Delta x) - f(x)}{\Delta x} = f'(x)$$

Some Relations among Derivatives

If $x = f(y)$, then $\dfrac{dy}{dx} = 1 \div \dfrac{dx}{dy}$

If $x = f(t)$ and $y = F(t)$, then $\dfrac{dy}{dx} = \dfrac{dy}{dt} \div \dfrac{dx}{dt}$

If $y = f(u)$ and $u = F(x)$, then $\dfrac{dy}{dx} = \dfrac{dy}{du} \times \dfrac{du}{dx}$

If $x = f(t)$, then $f''(t) = \dfrac{d^2 x}{dt^2} = \dfrac{d}{dt}\left(\dfrac{dx}{dt}\right)$

Table of Derivatives

Functions of x are represented by u and v, and constants are represented by a, n, and e.

$$\frac{d}{dx}(x) = 1 \qquad \frac{d}{dx}(a) = 0$$

$$\frac{d}{dx}(u \pm v \pm \ldots) = \frac{du}{dx} \pm \frac{dv}{dx} \pm \ldots \qquad \frac{d}{dx}(au) = a\frac{du}{dx}$$

$$\frac{d}{dx}(uv) = u\frac{dv}{dx} + v\frac{du}{dx} \qquad \frac{d}{dx}\left(\frac{u}{v}\right) = \frac{v\dfrac{du}{dx} - u\dfrac{dv}{dx}}{v^2}$$

$$\frac{d}{dx}(u^n) = nu^{n-1}\frac{du}{dx} \qquad \frac{d}{dx}\log_a u = \frac{\log_a e}{u}\frac{du}{dx}$$

$$\frac{d}{dx}a^u = a^u \ln a \frac{du}{dx}$$

$$\frac{d}{dx}e^u = e^u \frac{du}{dx} \qquad \frac{d}{dx}u^v = vu^{v-1}\frac{du}{dx} + u^v \ln u \frac{dv}{dx}$$

$$\frac{d}{dx}\sin u = \cos u \frac{du}{dx} \qquad \frac{d}{dx}\cot u = -\csc^2 u \frac{du}{dx}$$

$$\frac{d}{dx}\cos u = -\sin u \frac{du}{dx} \qquad \frac{d}{dx}\sec u = \sec u \tan u \frac{du}{dx}$$

$$\frac{d}{dx}\tan u = \sec^2 u \frac{du}{dx} \qquad \frac{d}{dx}\csc u = -\csc u \cot u \frac{du}{dx}$$

$$\frac{d}{dx}\sin^{-1}u = \frac{1}{\sqrt{1-u^2}}\frac{du}{dx} \quad \text{where} \quad -\pi/2 \le \sin^{-1}u \ge \pi/2$$

$$\frac{d}{dx}\cos^{-1}u = -\frac{1}{\sqrt{1-u^2}}\frac{du}{dx} \quad \text{where} \quad 0 \le \cos^{-1}u \ge \pi$$

$$\frac{d}{dx}\tan^{-1}u = \frac{1}{1+u^2}\frac{du}{dx}$$

$$\frac{d}{dx}\cot^{-1}u = -\frac{1}{1+u^2}\frac{du}{dx}$$

$$\frac{d}{dx}\sec^{-1}u = \frac{1}{u\sqrt{u^2-1}}\frac{du}{dx} \quad \text{where} \quad 0 \le \sec^{-1}u \le \pi/2 \text{ and } -\pi \le \sec^{-1}u \le -\pi/2$$

$$\frac{d}{dx}\csc^{-1}u = -\frac{1}{u\sqrt{u^2-1}}\frac{du}{dx} \quad \text{where} \quad -\pi < \csc^{-1}u \le -\pi/2 \text{ and } 0 < \csc^{-1}u \le \pi/2$$

Example 2.32

Find the derivatives of the following functions with respect to x.

(i) $x^3 + 4e^{-2x}$, (ii) $x^2 \sin x$, (iii) $2\sin^2 x$, (iv) $(\frac{e^{-x}}{x})$

Solution

(i) $\frac{d}{dx}(x^3 + 4e^{-2x}) = \frac{d}{dx}(x^3) + \frac{d}{dx}(4e^{-2x}) = 3x^2 + 4(-2)e^{-2x} = 3x^2 - 8e^{-2x}$

(ii) $\frac{d}{dx}(x^2 \sin x) = x^2\frac{d}{dx}(\sin x) + (\sin x)\frac{d}{dx}(x^2) = x^2\cos x + 2x\sin x$

(iii) $\frac{d}{dx}(2\sin^2 x) = 2(2)(\sin^{2-1}x)\frac{d}{dx}(\sin x) = (4\sin x)\cos x$

(iv) $\frac{d}{dx}(\frac{e^{-x}}{x}) = \frac{x\frac{d}{dx}(e^{-x}) - e^{-x}\frac{d}{dx}(x)}{x^2} = \frac{x(-e^{-x}) - (e^{-x})1}{x^2} = \frac{-e^{-x}(x+1)}{x^2}$

Slope of a Curve: Tangent and Normal

The slope of the curve (slope of the tangent line to the curve) whose equation is $y = f(x)$ is

$$\text{Slope} = m = \tan\phi = \frac{dy}{dx} = f'(x)$$

$$\text{Slope at } x_1 \text{ is } m_1 = f'(x_1)$$

The equation of a tangent line at $P_1(x_1, y_1)$ is $y - y_1 = m_1(x - x_1)$. The equation of a normal at $P_1(x_1, y_1)$ is

$$y - y_1 = -\frac{1}{m_1}(x - x_1)$$

The angle β of the intersection of two curves whose slopes at a common point are m_1 and m_2 is

$$\beta = \tan^{-1}\frac{m_2 - m_1}{1 + m_1 m_2}$$

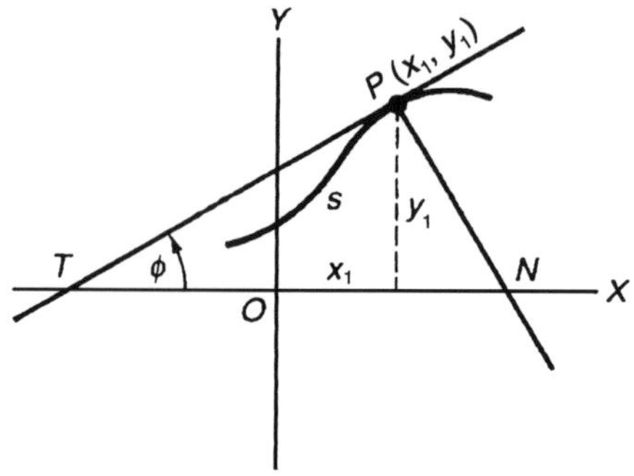

Figure 2.33

Example **2.33**

Find the slope of the curve $y = (4x^2 - 8)$ at $x = 1$.

Solution

Slope $\frac{dy}{dx} = \frac{d}{dx}(4x^2 - 8) = \frac{d}{dx}(4x^2) - \frac{d}{dx}(8) = 4\frac{d}{dx}(x^2) - 0 = 8x;$
at $x = 1$, slope = 8

Example **2.34**

Find the equation of the tangent to the curve $y = (x^2 - x - 4)$ at $(1, -4)$.

Solution

Slope of the curve $\frac{dy}{dx} = \frac{d}{dx}(x^2 - x - 4) = \frac{d}{dx}(x^2) - \frac{d}{dx}(x) - \frac{d}{dx}(4) = 2x - 1;$
at $(1, -4)$, slope $= 2(1) - 1 = 1$

Equation of the tangent is $(y + 4) = 1(x - 1)$ or $y = x - 5$.

Maximum and Minimum Values of a Function

The maximum or minimum value of a function $f(x)$ in an interval from $x = a$ to $x = b$ is the value of the function that is larger or smaller, respectively, than the values of the function in its immediate vicinity. Thus, the values of the function at M_1 and M_2 in Figure 2.34 are maxima, and its values at m_1 and m_2 are minima.

Test for a maximum at $x = x_1$: $f'(x_1) = 0$ or ∞, and $f''(x_1) < 0$

Test for minimum at $x = x_1$: $f'(x_1) = 0$ or ∞, and $f''(x_1) > 0$

If $f''(x_1) = 0$ or ∞, then for a maximum, $f'''(x_1) = 0$ or ∞ and $f^{IV}(x_1) < 0$; for a minimum, $f'''(x_1) = 0$ or ∞ and $f^{IV}(x_1) > 0$, and similarly if $f^{IV}(x_1) = 0$ or ∞, and so on, where f^{IV} represents the fourth derivative.

In a practical problem that suggests that the function $f(x)$ has a maximum or has a minimum in an interval from $x = a$ to $x = b$, simply equate $f'(x)$ to 0 and solve for the required value of x. To find the largest or smallest values of a function $f(x)$ in an interval from $x = a$ to $x = b$, find also the values $f(a)$ and $f(b)$. L and S may be the largest and smallest values, although they are not maximum or minimum values (see Figure 2.34).

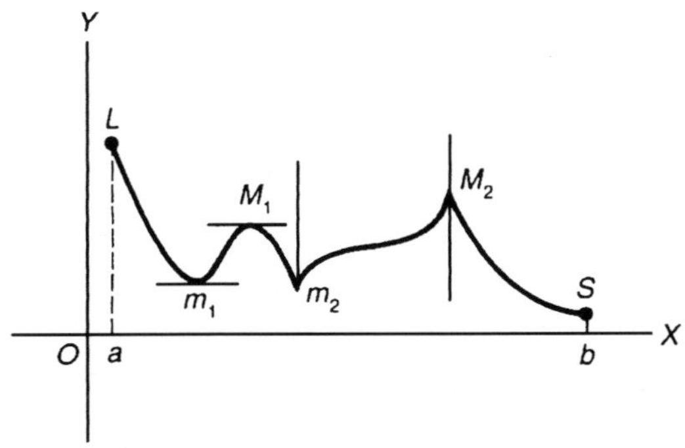

Figure 2.34

Points of Inflection of a Curve

Wherever $f''(x) < 0$, the curve is concave down.

Wherever $f''(x) > 0$, the curve is concave up.

The curve is said to have a point of inflection at $x = x_1$ if $f''(x_1) = 0$ or ∞, and the curve is concave up on one side of $x = x_1$ and concave down on the other (see points I_1 and I_2 in Figure 2.35).

Example 2.35

For the function $y = f(x) = (x^3 - 3x)$, find the maximum, minimum, and the point of inflection.

Solution

The derivative of $f(x)$, $f'(x) = 3x^2 - 3$.

Equating $f'(x)$ to 0, $3x^2 - 3 = 0$, which has roots at $x = 1$ and $- 1$.

The derivative of $f'(x)$, $f''(x) = 6x$. At $x = 1, f''(1) = 6$ and at $x = - 1, f''(-1) = - 6$.

Minimum occurs at $x = 1$, since $f''(1) > 0$ and $f''(1) = 0$. Value of minimum $= f(1) = - 2$.

Maximum occurs at $x = -1$, since $f'(- 1) < 0$ and $f'(- 1) = 0$. Value of maximum $= f(- 1) = 2$.

$f''(x) = 0$ at $x = 0$; $f''(x)$ changes sign as x passes through 0; then $x = 0$ is the inflection point (it is between 1 and –1).

Taylor and Maclaurin Series

In general, any $f(x)$ may be expanded into a **Taylor series**:

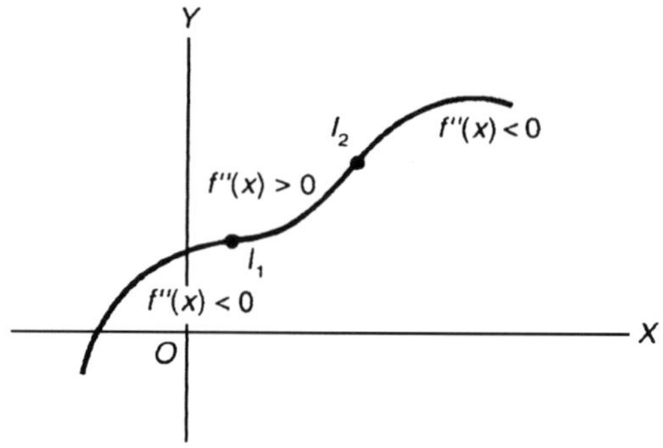

Figure 2.35

$$f(x) = f(a) + f'(a)\frac{x-a}{1} + f''(a)\frac{(x-a)^2}{2!} + f'''(a)\frac{(x-a)^3}{3!} + \ldots$$

$$+ f^{(n-1)}(a)\frac{(x-a)^{n-1}}{(n-1)!} + R_n$$

where a is any quantity whatever, so chosen that none of the expressions $f(a)$, $f'(a), f''(a), \ldots$ become infinite. If the series is to be used for the purpose of computing the approximate value of $f(x)$ for a given value of x, a should be chosen such that $(x - a)$ is numerically very small, and thus only a few terms of the series need be used. If $a = 0$, this series is called a Maclaurin series.

Example 2.36

Find the power series expansion of sin x about the point 0.

Solution

In the Taylor's series expansion, $f(x) = \sin(x)$ and $a = 0$. Derivatives of $f(x)$ are: $f'(x) = \cos(x), f''(x) = -\sin(x), f'''(x) = -\cos(x)$, and so on.

Substituting $f(x)$ and its derivatives at a in the expansion,

$$\sin(x) = \sin(0) + x\ \cos(0) + \frac{x^2(-\sin(0))}{2!} + \frac{x^3(-\cos(0))}{3!} + \ldots = x - \frac{x^3}{3!} + \frac{x^5}{5!} - \ldots$$

Evaluation of Indeterminate Forms

Let $f(x)$ and $F(x)$ be two functions of x, and let a be a value of x.

1. If $\dfrac{f(a)}{F(a)} = \dfrac{0}{0}$ or $\dfrac{\infty}{\infty}$, use $\dfrac{f'(a)}{F'(a)}$ for the value of this fraction.

   If $\dfrac{f'(a)}{F'(a)} = \dfrac{0}{0}$ or $\dfrac{\infty}{\infty}$, use $\dfrac{f''(a)}{F''(a)}$ for the value of this fraction, and so on.

2. If $f(a) \bullet F(a) = 0 \bullet \infty$ or if $f(a) - F(a) = \infty - \infty$, evaluate the expression by changing the product or difference to the form $\frac{0}{0}$ or $\frac{\infty}{\infty}$ and use the previous rule.

3. If $f(a)^{F(a)} = 0^0$ or ∞^0 or 1^∞, then form $e^{F(a) \bullet \ln f(a)}$, and the exponent, being of the form $0 \bullet \infty$, may be evaluated by rule 2.

Example 2.37

Find the following limits: (i) $\lim\limits_{x \to 0} \dfrac{\sin x}{x}$ (ii) $\lim\limits_{x \to 0} \dfrac{1 - \cos x}{x^2}$

Solution

(i) Since $\dfrac{\sin(0)}{0} = \dfrac{0}{0}$, we apply the limiting theorem:

$$\lim_{x \to 0} \frac{\sin x}{x} = \lim_{x \to 0} \frac{d(\sin x)}{d(x)} = \lim_{x \to 0} \frac{\cos x}{1} = 1$$

(ii) Since $\dfrac{1 - \cos(0)}{0^2} = \dfrac{0}{0}$, we apply the limiting theorem:

$$\lim_{x \to 0} \frac{1 - \cos x}{x^2} = \lim_{x \to 0} \frac{\sin x}{2x} = \lim_{x \to 0} \frac{\cos x}{2} = \frac{1}{2}$$

Example 2.38

The cubic $y = x^3 + x^2 - 3$ has one point of inflection. Where does it occur?

Solution

The answer requires knowing where y' changes sign. Now $y' = 3x^2 + 2x$ and $y'' = 6x + 2$, which is 0 where $x = -\frac{1}{3}$. Thus the only inflection point is at $x = -\frac{1}{3}$.

Example 2.39

The function of Example 2.38 has one local maximum and one local minimum. Where are they?

Solution

Setting $y' = 0$ ($3x^2 + 2x = 0$) yields $x = 0$ or $x = -\frac{2}{3}$. Since the second derivative is 2 at $x = 0$, this is the local minimum. At $x = -\frac{2}{3}$, $y'' = -2$, so $x = -\frac{2}{3}$ is the local maximum.

Differential of a Function

If $y = f(x)$ and Δx is an increment in x, then the differential of x equals the increment of x, or $dx = \Delta x$; and the differential of y is the derivative of y multiplied by the differential of x; thus

$$dy = \frac{dy}{dx}dx = \frac{df(x)}{dx}dx = f'(x)\,dx \quad \text{and} \quad \frac{dy}{dx} = dy \div dx$$

If $x = f_1(t)$ and $y = f_2(t)$, then $dx = f_1'(t)\,dt$, and $dy = f_2'(t)\,dt$.

Every derivative formula has a corresponding differential formula; thus, from the Table of Derivatives subsection, we have, for example,

$$d(uv) = u\,dv + v\,du; \quad d(\sin u) = \cos\, u\,du; \quad d(\tan^{-1} u) = \frac{du}{1+u^2}$$

Functions of Several Variables, Partial Derivatives, and Differentials

Let z be a function of two variables, $z = f(x, y)$; then its partial derivatives are

$$\frac{\partial z}{\partial x} = \frac{dz}{dx} \text{ when } y \text{ is kept constant} \qquad \frac{\partial z}{\partial y} = \frac{dz}{dy} \text{ when } x \text{ is kept constant}$$

Example 2.40

Find the partial derivatives of $f(x, y) = 4x^2y - 2y$ (i) with respect to x, and (ii) with respect to y.

Solution

(i) $$\frac{\partial f}{\partial x} = \frac{\partial}{\partial x}(4x^2y - 2y) = 4y\frac{\partial(x^2)}{\partial x} - 0 = (4y)(2x) = 8xy$$

(ii) $$\frac{\partial f}{\partial y} = \frac{\partial}{\partial y}(4x^2y - 2y) = 4x^2\frac{\partial(y)}{\partial y} - \frac{\partial(2y)}{\partial y} = 4x^2 - 2$$

Example 2.41

Two automobiles are approaching the origin. The first one is traveling from the left on the x-axis at 30 mph. The second is traveling from the top on the y-axis at 45 mph. How fast is the distance between them changing when the first is at (–5, 0) and the second is at (0, 10)? (Both coordinates are in miles.)

Solution

If $x(t)$ is taken as the position of the first auto at time t and $y(t)$ as the position of the second auto at time t, then the distance between them at time t is $s(t)=\sqrt{[x(t)]^2+[y(t)]^2}$. Using the chain rule,

$$s'(t)=\frac{ds}{dt}=\frac{1}{d_2 s(t)}\ \frac{d}{dt}\{[x(t)]^2+[y(t)]^2\}=\frac{1}{2s(t)}[2x(t)x'(t)+2y(t)y'(t)]$$

Now $x'(t) = 30$ and $y'(t) = -45$ for all t; and when $t = t_0$, $x(t_0) = -5$ and $y(t_0) = 10$.

Therefore,

$$s'(t_0)=\frac{-2(5)(30)-2(10)(45)}{2\sqrt{(-5)^2+(10)^2}}=\frac{-1200}{2\sqrt{125}}=\frac{-120}{\sqrt{5}}=24\sqrt{5}\approx 54$$

Thus, the two automobiles are "closing" at about 54 mph.

Example 2.42

How close do the two automobiles in Example 2.41 get?

Solution

One wants to minimize $s(t)$ in Example 2.41, so set $s'(t) = 0$. Thus,

$$\frac{xx'+yy'}{s}=0 \quad \text{or} \quad xx'+yy'=0$$

Since $x' = 30$ and $y' = -45$, $30x = 45y$. However, since $x'(t) = 30$, $x(t) = 30t + x_0$, and similarly $y(t) = -45t + y_0$. If one takes $t_0 = 0$ when the problem starts, $x_0 = -5$ and $y_0 = 10$, so $30(30t - 5) = 45(-45t + 10)$ gives time of minimum distance. Solving for t, factor out 75 from both sides to get $2(6t - 1) = 3(-9t + 2)$, or $39t = 8$.

Thus the minimum distance occurs at 8/39 of an hour after the initial conditions of Example 2.41. At this time $x = -5 + 240/39$ and $y = 10 - 360/39$, so $x = 45/39$ and $y = 30/39$. The minimum distance is

$$s\left(\frac{8}{39}\right)=\frac{\sqrt{(45)^2+(30)^2}}{39}=\frac{15}{39}\sqrt{9+4}=\frac{15\sqrt{13}}{39}\approx 1.4 \text{ miles}$$

Example 2.43

In Example 2.41, which reaches the origin car first, Car 1 or 2?

Solution

This is obvious if, in Example 2.42, one notices that x is positive and y is (still) positive. Alternatively, notice that Car 1 takes 5/30 of an hour to reach the origin and Car 2 takes 10/45 of an hour. The time 5/30 < 10/45, so Car 1 gets there first.

INTEGRAL CALCULUS

Definition of an Integral

The function $F(x)$ is said to be the integral of $f(x)$ if the derivative of $F(x)$ is $f(x)$, or if the differential of $F(x)$ is $f(x)\,dx$. In symbols,

$$F(x) = \int f(x)\,dx \quad \text{if} \quad \frac{dF(x)}{dx} = f(x), \qquad \text{or} \quad dF(x) = f(x)dx$$

In general, $\int f(x)\,dx = F(x) + C$, where C is an arbitrary constant.

Fundamental Theorems on Integrals

$$\int df(x) = f(x) + C$$

$$\int df(x)\,dx = f(x)\,dx$$

$$\int [f_1(x) \pm f_2(x) \pm \cdots]\,dx = \int f_1(x)\,dx \pm \int f_2(x)\,dx \pm \cdots$$

$\int af(x)\,dx = a\int f(x)\,dx$, where a is any constant

$\int u^n\,du = \frac{u^{n+1}}{n+1} + C \quad (n \neq -1)$, where u is any function of x

$\int \frac{du}{u} = \ln u + C$, where u is any function of x

$\int u\,dv = uv - \int v\,du$, where u and v are any functions of x

$$\int [u(x) \pm v(x)]\,dx = \int u(x)\,dx \pm \int v(x)\,dx$$

$$\int \frac{dx}{ax+b} = \frac{1}{a}\ln|ax+b|$$

$$\int \frac{dx}{\sqrt{x}} = 2\sqrt{x}$$

$$\int a^x dx = \frac{a^x}{\ln a}$$

$$\int \sin x \, dx = -\cos x$$

$$\int \sin^2 x \, dx = \frac{x}{2} - \frac{\sin 2x}{4}$$

$$\int x \sin x \, dx = \sin x - x \cos x$$

$$\int \cos x \, dx = \sin x$$

$$\int \cos^2 x \, dx = \frac{x}{2} + \frac{\sin 2x}{4}$$

$$\int x \cos x \, dx = \cos x + \sin x$$

$$\int \sin x \cos x \, dx = (\sin^2 x)/2$$

$$\int \tan x \, dx = -\ln|\cos x| = \ln|\sec x|$$

$$\int \tan^2 x \, dx = \tan x - x$$

$$\int \cot x \, dx = -\ln|\csc x| = \ln|\sin x|$$

$$\int \cot^2 x \, dx = -\cot x - x$$

$$\int e^{ax} dx = (1/a)e^{ax}$$

$$\int \ln x \, dx = x[\ln(x) - 1] \qquad (x > 0)$$

Example 2.44

Evaluate the following integrals:

(i) $\int (x^3 + 4)dx$ (ii) $\int \sin 3x dx$ (iii) $\int \sqrt{(1-x)}dx$

Solution

(i) $$\int (x^3 + 4)dx = \left(\frac{x^4}{4} + 4x\right) + \text{constant}$$

(ii) $\int \sin 3x dx$; let $t = 3x$; then $dt = 3dx$ and

$$\int \frac{1}{3} \sin t dt = -\frac{\cos t}{3} = \frac{-\cos 3x}{3} + \text{constant}$$

(iii) $\int \sqrt{(1-x)}dx$; let $t = 1 - x$; then $dt = -dx$;

$$\int -\sqrt{t} dt = -\frac{2}{3}t^{\frac{3}{2}} = -\frac{2}{3}(1-x)^{\frac{3}{2}} + \text{constant}$$

Moment of Inertia

Moment of inertia J of a mass m:

$$\text{About } OX: I_x = \int y^2 dm = \int r^2 \sin^2\theta \, dm$$

$$\text{About } OY: I_y = \int x^2 dm = \int r^2 \cos^2\theta \, dm$$

$$\text{About } O: J_0 = \int (x^2 + y^2)\, dm = \int r^2 \, dm$$

Center of Gravity

Coordinates $(\bar{x}, \bar{y})$ of the center of gravity of a mass m:

$$\bar{x} = \frac{\int x \, dm}{\int dm}, \qquad \bar{y} = \frac{\int y \, dm}{\int dm}$$

The center of gravity of the differential element of area may be taken at its midpoint. In the above equations, x and y are the coordinates of the center of gravity of the element.

Work

The work W done in moving a particle from $s = a$ to $s = b$ against a force whose component in the direction of motion is F_s is

$$dW = F_s \, ds, \qquad W = \int_a^b F_s \, ds$$

where F_s must be expressed as a function of s.

Example 2.45

Consider the function $y = x^2 + 1$ between $x = 0$ and $x = 2$. What is the area between the curve and the x-axis?

Solution

$$A = \int_0^2 (x^2 + 1)\, dx = \left(\frac{x^3}{3} + x\right)\Bigg|_0^2 = \frac{8}{3} + 2 = \frac{14}{3}$$

Example 2.46

Consider the area bounded by $x = y^2$, the x-axis, and the line $x = 4$. Find (1) the area; (2) the first moment of area with respect to the x-axis; (3) the first moment of area with respect to the y-axis; (4) the centroid; (5) the second moment of area with respect to the x-axis; (6) the second moment of area with respect to the y-axis; (7) the moment of inertia about the line $x = -2$; and (8) the moment of inertia about the line $y = 4$.

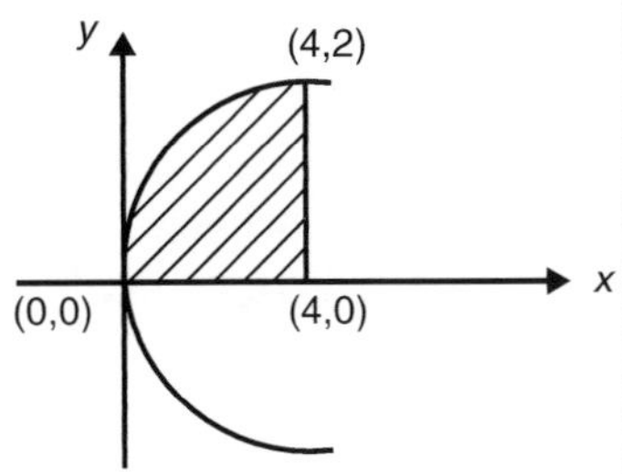

Exhibit 1

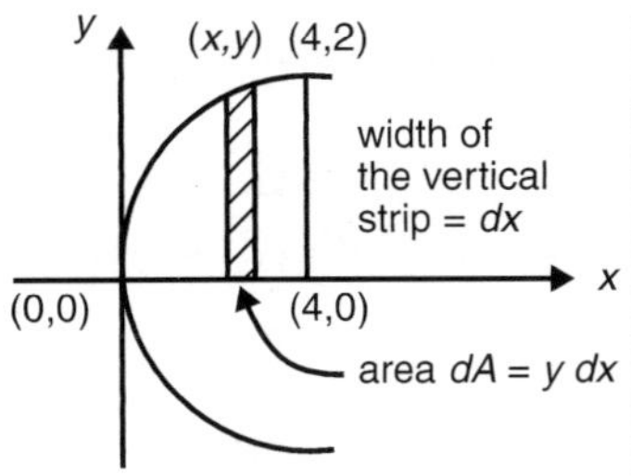

Exhibit 2

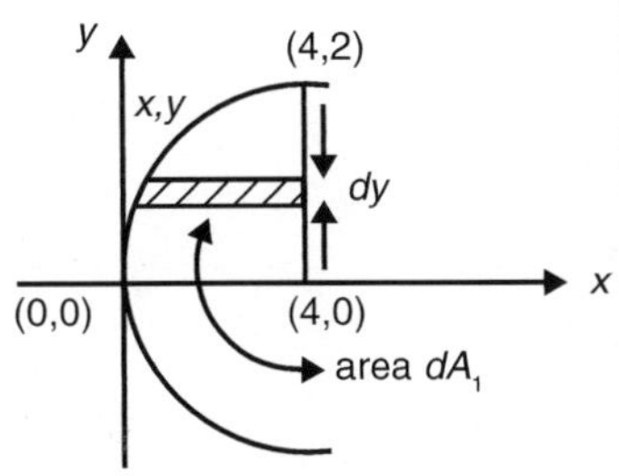

Exhibit 3

Solution

The area is shown shaded in Exhibit 1.

For the vertical strip shown in Exhibit 2, $dA = y\,dx = \sqrt{x}dx$.

(1) area $A = \int dA = \int ydx = \int_0^4 \sqrt{x}dx = \frac{16}{3}\text{cm}^2$

(3) first moment with respect to y-axis $M_y = \int xdA = \int_0^4 x\sqrt{x}dx = 12.8\text{ cm}^3$

(6) second moment with respect to y-axis

$$I_y = \int x^2 dA = \int_0^4 x^2\sqrt{x}dx = 36.57\text{ cm}^4$$

For the horizontal strip shown in Exhibit 3, $dA_1 = (4 - x)\,dy = (4 - y^2)\,dy$

$$\text{Note: area } A = \int_0^2 (4 - y^2)dy$$

(2) first moment with respect to x-axis $M_x = \int ydA_1 = \int_0^2 y(4 - y^2)dy = 4\text{ cm}^3$

(4) x coordinate of centroid $x_c = \frac{M_y}{A} = \frac{12.8}{5.33} = 2.4\text{ cm}$

y coordinate of the centroid $y_c = \frac{M_x}{A} = \frac{4}{5.33} = 0.75\text{ cm}$

(5) second moment with respect to the x-axis

$$I_x = \int y^2 dA_1 = \int_0^2 y^2(4 - y^2)dy = 4.27\text{ cm}^4$$

Now, using parallel-axis theorem,

(7) moment of inertia about the line $x = -2$ is $I_y + Ad^2 = 36.57 + (5.333)(2 + 2.4)^2 = 139.8\text{ cm}^4$

(8) moment of inertia about the line $y = 4$ is $I_x + Ad^2 = 4.27 + (5.333)(4 - 0.75)^2 = 60.6\text{ cm}^4$

DIFFERENTIAL EQUATIONS

Definitions

A **differential equation** is an equation involving differentials or derivatives.

The **order** of a differential equation is the order of the derivative of highest order that it contains.

The **degree** of a differential equation is the power to which the derivative of highest order in the equation is raised, that derivative entering the equation free from radicals.

The **solution** of a differential equation is the relation involving only the variables (but not their derivatives) and arbitrary constants, consistent with the given differential equation.

The most **general solution** of a differential equation of the *n*th order contains *n* arbitrary constants. If particular values are assigned to these arbitrary constants, the solution is called a particular solution.

Notation

Symbol or Abbreviation	Definition
M, N	Functions of x and y
X	Function of x alone or a constant
Y	Function of y alone or a constant
C, c	Arbitrary constants of integration
$a, b, k, 1, m, n$	Given constants

Equations of First Order and First Degree: *M dx* 1 *N dy* 5 0

Variables Separable: $X_1 Y_1\, dx + X_2 Y_2\, dy = 0$

Solution

$$\int \frac{X_1}{X_2} dx + \int \frac{Y_2}{Y_1} dy = 0$$

Linear Equation: $dy + (X_1 Y - X_2)\, dx = 0$

Solution

$$y = e^{-\int X_1 dx} \left(\int X_2 e^{\int X_1 dx} dx + C \right)$$

Second-Order Differential Equations

A second-order differential expression, $L(x, y, y', y'')$, is linear if

$$L(x, ay_1 + by_2, ay_1' + by_2', ay_1'' + by_2'') = aL(x, y_1, y_1', y_1'') + bL(x, y_2, y_2', y_2'')$$

or, if it has the form

$$L(x, y_1, y_1', y_1'') = f(x)y + g(x)y' + h(x)y''$$

A second-order linear differential equation is

$$L(x, y, y', y'') = F(x)$$

If $F(x) \equiv 0$, it is homogeneous; if $F(x)$ is nonzero, it is inhomogeneous.

Constant Coefficients

If $L = ay'' + by' + cy$ where a, b, and c are constants with $a \neq 0$, the first step is to solve the associated homogeneous equation $ay'' + by' + cy = 0$. By replacing y by 1, y' by r, and y'' by r^2, one obtains the characteristic equation $ar^2 + bc + c = 0$ with roots r_1 and r_2 obtained from factoring or from the quadratic formula. There are three cases to consider:

Case 1: $r_1 \neq r_2$, both real; $y = c_1 e^{r_1 x} + c_2 e^{r_2 x}$, where c_1 and c_2 are arbitrary constants.

Case 2: $r_1 = r_2$; $y = c_1 e^{r_1 x} + c_2 x e^{r_2 x}$.

Case 3: $r_1 = \alpha + j\beta$, $r_2 = \alpha - j\beta$, where α and β are real and $j^2 = -1$;
$y = d_1 e^{r_1 x} + d_2 e^{r_2 x} = e^{\alpha x}(c_1 \sin \beta x + c_2 \cos \beta x)$. In particular, if $\alpha = 0$,
$y = c_1 \sin \beta x + c_2 \cos \beta x$.

After finding the two solutions to the associated homogeneous equation (y_1 is the result of setting $c_1 = 1$ and $c_2 = 0$, whereas y_2 has $c_1 = 0$ and $c_2 = 1$), one proceeds in either of the following two ways.

Variation of Parameters

If $L(y) = F(x)$ in which the coefficient of y'' is 1, and $W(x) = y_1(x)y_2'(x) - y_1'(x)y_2(x)$ in which y_1 and y_2 are those solutions found above, and if

$$u_1' = \frac{-F(x)y_2(x)}{W(x)} \quad \text{and} \quad u_2' = \frac{F(x)y_1(x)}{W(x)}$$

one solution to the inhomogeneous equation is $y_p = u_1(x)y_1(x) + u_2(x)y_2(x)$.

Undetermined Coefficients

In this technique, one guesses y_p by using the following patterns.

One guesses that the solution may be of the same form as the $F(x)$ function but with coefficients to be determined. This method requires modification if $F(x)$ and $c_1y_1 + c_2y_2$ from the associated homogeneous equation interfere, but it is frequently easier than the integration required in the Variation of Parameters technique to construct u_1 and u_2 from their derivatives.

To guess, classify $F(x)$. If it is a polynomial of degree k, the guess will be a polynomial of degree k. However, if $r = 0$ occurs in the homogeneous equation, increase the degree of the polynomial by one. If $F(x)$ is a polynomial times e^{Ax}, so will be the guess. Once again, the degree may have to be increased by one. If $F(x)$ contains sines and cosines, so should the guess.

After making the guess, differentiate it twice and put it into the equation. The coefficients may be determined at this time.

When mixed forms of functions are present in $F(x)$—for example, $x^2 + 2 + 3 \sin 2x$—treat the terms $x^2 + 2$ and $3 \sin 2x$ independently. The principle of *superposition* then permits you to add the results.

Now, after finding the solution $c_1y_1 + c_2y_2$ to the associated homogeneous equation, and the particular solution y_p to the inhomogeneous equation, form the general solution $y = c_1y_1 + c_2y_2 + y_p$. If initial values are required, such as $y(1) = 2$ and $y'(1) = 3$, the final step is to determine the values of c_1 and c_2 that fit the initial conditions.

Example 2.47

Specify the order of each of the following differential equations and also identify each as linear/nonlinear and homogeneous/nonhomogeneous.

Note: $y' = \dfrac{dy}{dt}$; $y'' = \dfrac{d^2y}{dt^2}$

(i) $y'' + 4y' + 4y = 0$ (ii) $3y' + 2y = 0$ (iii) $y'' + 2y' + 2y = e^{-t}$

(iv) $2y' + y + 2 = 0$ (v) $y' + y^2 + 4 = 0$

Solution

(i) second order, linear, homogeneous

(ii) first order, linear, homogeneous

(iii) second order, linear, nonhomogeneous

(iv) first order, linear, nonhomogeneous

(v) first order, nonlinear, nonhomogeneous

Example 2.48

Find the homogeneous solutions for each of the following differential equations and specify whether it belongs to either overdamped, underdamped, or critically damped case.

(i) $y'' + 3y' + 2y = 0$ (ii) $y'' + 2y' + y = 4$ (iii) $y'' + 2y' + 2y = 0$

Solution

(i) Characteristic equation is $s^2 + 3s + 2 = 0$, and the characteristic roots are -1 and -2. Homogeneous solution is $y_h(t) = C_1e^{-1t} + C_2e^{-2t}$. The roots are real and distinct, so the function is overdamped.

(ii) Characteristic equation is $s^2 + 2s + 1 = 0$, and the characteristic roots are -1 and -1. Homogeneous solution is $y_h(t) = (C_1 + C_2t)e^{-1t}$. The roots are real and equal, so the function is critically damped.

(iii) Characteristic root is $s^2 + 2s + 2 = 0$ and the characteristic roots are $-1 + i1$ and $-1 - i1$. Homogeneous solution is $y_h(t) = e^{-1t}(C_1 \cos t + C_2 \sin t)$. As the roots are complex, the function is underdamped.

Example 2.49

Solve the differential equation y′ + 2y = 0 with initial condition y(0) = 3.

Solution

This is a first-order, linear, homogeneous differential equation with constant coefficient.

Characteristic equation is $s + 2 = 0$, and the root is –2; solution is $y(t) = Ce^{-2t}$.

To find C, $y(0) = C = 3$; then, $y(t) = 3e^{-2t}$.

Example 2.50

Solve the differential equation y″ + 2y′ + y = 4 with initial conditions y(0) = 0, y′(0) = 1.

Solution

Characteristic equation is $s^2 + 2s + 1 = 0$.

Characteristic roots are: –1 and –1 (roots are real and equal; the solution is critically damped).

Natural or homogeneous solution $y_h(t) = (C_1 + C_2t)e^{-t}$

Particular solution $y_p(t) = B$ due to forcing function $f(t) = 4$, a constant

Substituting $y_p(t)$ in the differential equation, $y_p'' + 2y_p' + y_p = 4$; $0 + 0 + B = 4$; or $B = 4$

Complete solution $y(t) = y_h(t) + y_p(t) = C_1e^{-t} + C_2te^{-t} + 4$

To find C_1 and C_2, use the initial conditions in $y(t)$ and $y'(t)$; $y'(t) = -C_1e^{-t} - C_2te^{-t} + C_2e^{-t}$

$y(0) = C_1 + B = 0$ and $y'(0) = -C_1 + C_2 = 1$; solving $C_1 = -4$; $C_2 = -3$

Complete solution is $y(t) = -4e^{-t} - 3te^{-t} + 4$.

Example 2.51

Find the solution of $y'' + 2y' = x^2 + 2 + 3\sin 2x$ subject to $y(0) = 1, y'(0) = 0$.

Solution

Begin with $y'' + 2y' = 0$. The characteristic equation is $r^2 + 2r = 0$, which has roots 0 and –2. Thus, the two solutions to the associated homogeneous equation are $y_1 = 1$ and $y_2 = e^{-2x}$. To use the method of Undetermined Coefficients, guess $(ax^2 + bx + c) \bullet x$ for the $x^2 + 2$ term (the x is needed because $y_1 = 1$). Differentiate twice and insert in the equation: $(6ax + 2b) + 2(3ax^2 + 2bx + c)$ should be the same as $x^2 + 2$. Thus, $6a = 1, 4b + 6a = 0$, and $2b + 2c = 2$. Consequently, $a = \frac{1}{6}$, $b = -\frac{1}{4}$, and $c = \frac{5}{4}$. Next, guess $c \sin 2x + d \cos 2x$ for the other term. Then $y'' + 2y' = -4c \sin 2x - 4d \cos 2x + 2(2c \cos 2x - 2d \sin 2x)$ should match $3 \sin 2x$, so $-4c - 4d = 3$ and $-4d + 4c = 0$. Thus $c = d = -\frac{3}{8}$.

Putting all this together, one has the general solution

$$y = A + Be^{-2x} + \frac{1}{6}x^3 - \frac{1}{4}x^2 + \frac{5}{4}x - \frac{3}{8}\sin 2x - \frac{3}{8}\cos 2x$$

Now to fit the initial conditions,

$$y(0) = 1 = A + B - \frac{3}{8} \text{ and } y'(0) = 0 = -2B + \frac{5}{4} - \frac{3}{4}$$

Hence, $B = \frac{1}{4}$ and A = $\frac{9}{8}$, so

$$y = \frac{9}{8} + \frac{1}{4}e^{-2x} + \frac{1}{6}x^3 - \frac{1}{4}x^2 + \frac{5}{4}x - \frac{3}{8}\sin 2x - \frac{3}{8}\cos 2x$$

Euler Equations

An equation of the form $x^2y'' + axy' + by = F(x)$, with a and b constants, may be solved as readily as the constant coefficient case. These are called **Euler equations.** Upon substituting $y = x^m$, one obtains the *indicial equation* $m(m - 1) + am + b = 0$. This quadratic equation has two roots, m_1 and m_2.

If $m_1 \neq m_2$, both real, then $y = c_1|x|^{m_1} + c_2|x|^{m_2}$.
If $m_1 = m_2$, then $y = |x|^{m_1}(c_1 + c_2 \ln|x|)$.

If $m_1 = p + jq$ and $m_2 = p - jq$, then $y = |x|^p[c_1\cos(q \ln|x|) + c_2\sin(q \ln|x|)]$.

Once y is determined, y_p for the inhomogeneous equation may be found by Variation of Parameters. The method of Undetermined Coefficients is not recommended for Euler equations.

Higher-order linear equations with constant coefficients or of Euler form may be solved analogously.

Laplace Transform

The Laplace transform is an operation that converts functions of x on the half-line $[0, \infty]$ into functions of p on some half-line (a, ∞). The damping power of e^{-xp} is the basis for this useful technique. If $f(x)$ is a piecewise continuous function on $[0, \infty]$ that does not grow too fast, $L(f)$ is the function of p defined by

$$L[f(p)] = \int_0^\infty e^{-xp} f(x)\,dx$$

for the values of p for which the integral converges. For example, if $f(x) \equiv 1$,

$$L(f) = L(1) = \int_0^\infty e^{-xp}\,dx = \frac{1}{p} \quad (\text{for } p > 0)$$

As a further example,

$$L(f) = L(e^{ax}) = \int_0^\infty e^{-xp}e^{ax}\,dx = \int_0^\infty e^{-x(p-a)}\,dx = \frac{1}{p-a} \quad (\text{for } p > a)$$

The basic connection between the Laplace transform and differential equations is the following result achieved by integration by parts:

$$L\left[y'(p)\right]=\int_0^{\infty} e^{-xp}y'(x)\,dx = y(x)e^{-xp}\Big|_0^{\infty}+p\int_0^{\infty} e^{-xp}y(x)\,dx=-y(0)+pL\left[y(p)\right]$$

Consequently, the solution to $ay''+by'+cy=F(x)$ may be obtained by transforming $L[ay''+by'+cy]=L(F)$, so $a[p^2L(y)-py(0)-y'(0)]+b[pL(y)-y(0)]+cL(y)=L(F)$. Solving for L(y),

$$L(y)=\frac{L(F)+apy(0)+ay'(0)+by(0)}{ap^2+bp+c}$$

If one were able to "invert" this result,

$$y=L^{-1}\left[\frac{L(F)+apy(0)+ay'(0)+by(0)}{ap^2+bp+c}\right]$$

the solution would appear, complete with initial values. The Laplace transform is invertible, and the process of finding $L^{-1}[f(p)]$ as a function of x is much like the process of integration.

Table 2.1 presents a tabulation of selected transforms, where the transform of $f(x)$ is called $F(p)$. Line 3 in Table 2.1 reveals that the operation of multiplying by x corresponds to the negative of the operation of differentiating with respect to p. Lines 2 and 8 have been previously discussed. The δ in Line 1 is a *pseudo-function* with great utility defined by $\delta(x)=0$ for all x except $x=0$, and $\int_0^{\infty}\delta(x)\,dx=1$, so $\delta(0)=+\infty$. The u in Line 7 is called the *Heaviside function*, and it is zero until $x-c>0$. Thus $u(x-c)f(x-c)$ is $f(x)$ shifted right to the point $x=c$. For example, if $f(x)=x$ and $c=1$, $u(x-1)f(x-1)$ has the graph shown in Figure 2.36, whereas f has the graph shown in Figure 2.37. The operation in the left column of Line 9 is a new way to multiply functions, called **convolution**.

Table 2.1 Selected transforms

$f(x)$	$F(p)$
1. δ	1
2. 1	p^{-1}
3. $xf(x)$	$\frac{-dF}{dp}$
4. $e^{ax}f(x)$	$F(p-a)$
5. sin ax	$\frac{a}{p^2+a^2}$
6. cos ax	$\frac{p}{p^2+a^2}$
7. $u(x-c)f(x-c)$	$e^{-cp}F(p)$
8. $f'(x)$	$pF(p)-f(0)$

(Continued)

(Continued)

9. $\int_0^x f(u)g(x-u)du$	$F(p) \bullet G(p)$
10. $f(x)$, if f is periodic, of period L	$\dfrac{\int_0^L e^{-px} f(x)dx}{1-e^{-pL}}$

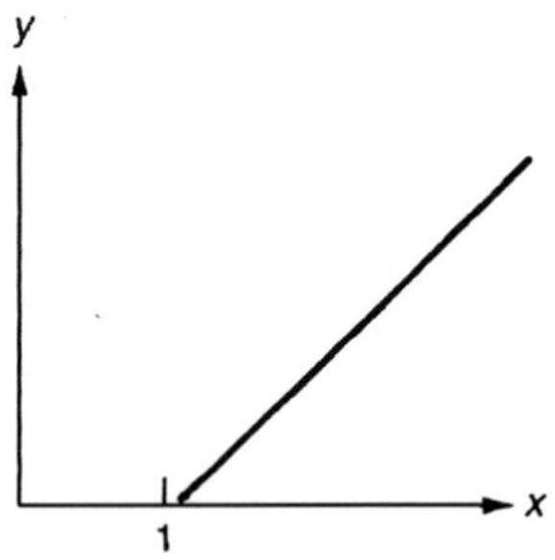

Figure 2.36

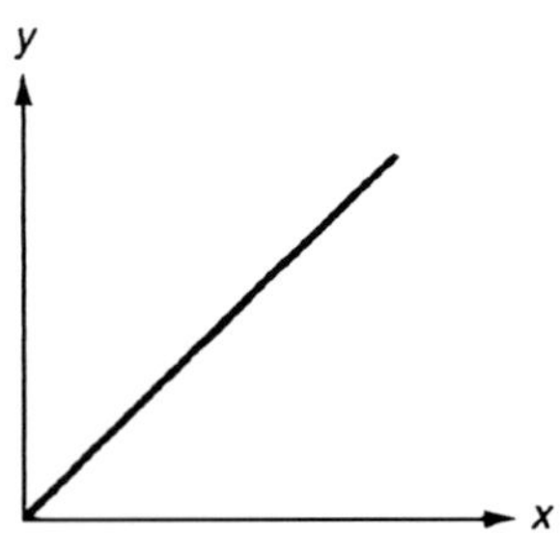

Figure 2.37

For an example of inversion of a transform, consider

$$F(p) = \frac{2p+3}{(p^2+1)(p-2)}$$

Begin by using partial fractions to write

$$F(p) = \frac{A}{p-2} + \frac{Bp+C}{p^2+1}$$

from which

$$F(p) = \frac{7/5}{p-2} + \frac{(-7/5)p - 4/5}{p^2+1}$$

Table 2.1 shows that $p/(p^2+1)$ is $L(\cos x)$, so

$$L^{-1}\left[\frac{(-7/5)p}{p^2+1}\right]$$

is $-(7/5)\cos x$, and similarly

$$L^{-1}\left(-\frac{4}{5} \bullet \frac{1}{p^2+1}\right)$$

is $-(4/5)\sin x$. Since $L(1) = 1/p$, $L(e^{2x})$ is $1/(p-2)$ by Line 4, and combining these three terms yields

$$f(x) = \frac{7}{5}e^{2x} - \frac{7}{5}\cos x - \frac{4}{5}\sin x$$

The *Heaviside* operation in Line 7 leads to an easy solution of differential equations whose right-hand side is not continuous. For example, consider the response of $y'' + y$ to a driving function $f(x)$ that is 1 for x between 0 and 2 and then becomes 0. Suppose $y(0) = 1$ and $y'(0) = 3$. By applying Line 8 in Table 2.1 twice,

$$L(y) = \frac{L(f) + p + 3}{p^2+1}$$

and since $f(x) = 1 - u(x-2)$,

$$L(f) = \frac{1}{p} - \frac{e^{-2p}}{p}$$

so

$$L(y) = \frac{p + 3 + (1 - e^{-2p})/p}{p^2+1}$$

The e^{-2p} portion represents delay, so write

$$L(y) = \frac{p^2+3p+1}{p(p^2+1)} - \frac{e^{-2p}}{p(p^2+1)} = \frac{1}{p} + \frac{3}{p^2+1} - e^{-2p}\left(\frac{1}{p} - \frac{p}{p^2+1}\right)$$

from which $y = 1 + 3\sin x - u(x-2) + u(x-2)\cos(x-2)$.

Example 2.52

Find the Laplace transform of each of the following functions:

(i) $2\,u(t) + e^{-3t}$ (ii) $2te^{-t}$ (iii) $e^{-t}\sin 4t$ (iv) $\sin 4t$ (v) $\cos 2t$

Solution

(i) $\frac{2}{s} + \frac{1}{s+3} = \frac{2s+7}{s(s+3)}$ (ii) $\frac{2}{(s+1)^2}$ (iii) $\frac{4}{(s+1)^2+4^2}$

(iv) $\frac{4}{s^2+4^2}$ (v) $\frac{s}{s^2+2^2}$

Example 2.53

Find an expression for $Y(s) = \pounds\, y(t)$ for each of the following differential equations.

(i) $\frac{dy}{dt} + 6\,y(t) = 4\,u(t);\ y(0) = 0$

(ii) $\frac{dy}{dt} + 2\,y(t) = 0;\ y(0) = -2$

(iii) $\frac{d^2y}{dt^2} + 2\frac{dy}{dt} + 3\,y(t) = \sin 2t$ with zero initial conditions

(iv) $\frac{d^2y}{dt^2} + 2\frac{dy}{dt} + y(t) = e^{-2t}$ with initial conditions $y(0) = -1$ and $\frac{dy(0)}{dt} = 1$

Solution

(i) $\left[sY(s) - 0\right] + 6Y(s) = \frac{4}{s};\ Y(s) = \frac{4}{s(s+6)}$

(ii) $\left[sY(s) - (-2)\right] + 2Y(s) = 0;\ Y(s) = -\frac{2}{s+2}$

(iii) $\left[s^2Y(s) - 0 - 0\right] + 2\left[sY(s) - 0\right] + 3Y(s) = \frac{2}{s^2+4};$

$$Y(s) = \frac{2}{(s^2+4)(s^2+2s+3)}$$

(iv) $\left[s^2Y(s)-(-s)-1\right]+2\left[sY(s)-(-1)\right]+Y(s)=\dfrac{1}{s+2}$

simplifying $Y(s)=\dfrac{(-s^2-3s-1)}{(s+2)(s^2+2s+1)}$

Example 2.54

Find the initial and final values of the functions whose Laplace transforms are given:

(i) $F(s)=\dfrac{2(s+1)}{s(s+4)(s+6)}$ (ii) $F(s)=\dfrac{4s}{s^2+2s+2}$

Solution

(i) Initial value is: $\lim_{t\to 0} f(t)=\lim_{s\to\infty} sF(s)=\lim_{s\to\infty} s\dfrac{2(s+1)}{s(s+4)(s+6)}=0$

Final value is: $\lim_{t\to\infty} f(t)=\lim_{s\to 0} sF(s)=\lim_{s\to 0} s\dfrac{2(s+1)}{s(s+4)(s+6)}=\dfrac{1}{12}$

(ii) Initial value is: $\lim_{t\to 0} f(t)=\lim_{s\to\infty} sF(s)=\lim_{s\to\infty} s\dfrac{4s}{s^2+2s+2}=4$

Final value is: $\lim_{t\to\infty} f(t)=\lim_{s\to 0} sF(s)=\lim_{s\to 0} s\dfrac{4s}{s^2+2s+2}=0$

FOURIER SERIES AND FOURIER TRANSFORM

Fourier Series

A periodic function $F(t)$ with period T can be expanded into Fourier series as

$$F(t)=a_0+\sum_{n=1}^{\infty}(a_n\cos n\omega_0 t+b_n\sin n\omega_0 t) \quad \textbf{(2.10)}$$

where $\omega_0=\dfrac{2\pi}{T}$ and the Fourier coefficients are defined as

$$a_0=\left(\frac{1}{T}\right)\int_0^T F(t)dt \quad \textbf{(2.11a)}$$

$$a_n=\left(\frac{2}{T}\right)\int_0^T F(t)\cos(n\omega_0 t)dt \quad \textbf{(2.11b)}$$

$$b_n=\left(\frac{2}{T}\right)\int_0^T F(t)\sin(nw_0 t)dt \quad \textbf{(2.111c)}$$

For a truncated series, the root mean square (RMS) value F_N is defined as

$$F_N^{\;2} = a_0^{\;2} + \left(\frac{1}{2}\right)\sum_{n=1}^{N}(a_n^{\;2} + b_n^{\;2}) \tag{2.12}$$

Example 2.55

Find the Fourier coefficients of the periodic waveform shown in Exhibit 4 and the RMS value of the truncated Fourier series including five harmonics.

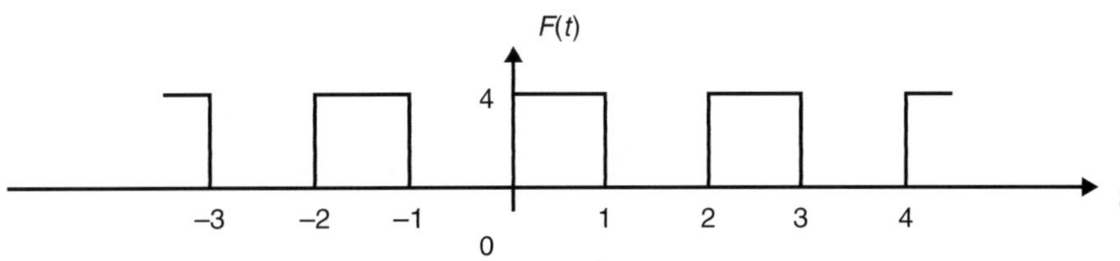

Exhibit 4

Solution

Period $T = 2\ s$; $w_0 = \dfrac{2\pi}{T} = \pi\ \ rad/s$

From Equation (2.1a) $a_0 = \dfrac{1}{2}\left[\displaystyle\int_0^1 4dt + \int_1^2 0dt\right] = 2$

From Equation (2.1b) $a_n = \dfrac{2}{2}\left[\displaystyle\int_0^1 4\cos n\omega_0 t dt + \int_1^2 0dt\right] = 0, \text{when } n \neq 0$

From Equation (2.1c) $b_n = \dfrac{2}{2}\displaystyle\int_0^1 4\sin n\omega_0 t dt + \int_1^2 0dt = \frac{4}{n\pi}(1 - \cos n\pi)$

$b_n = 0$ for all even n; $b_n = \dfrac{8}{n\pi}$ for all odd n

Fourier series is $F(t) = 2 + \dfrac{8}{\pi}\sin\ \omega_0 t + \dfrac{8}{3\pi}\sin\ 3\omega_0 t + \dfrac{8}{5\pi}\sin\ 5\omega_0 t + \ldots$

From Equation (2.2), for $N = 5$, $F_N^{\;2} = 2^2 + \dfrac{1}{2}\left[(\dfrac{8}{\pi})^2 + (\dfrac{8}{3\pi})^2 + (\dfrac{8}{5\pi})^2\right] = 7.732;$

$$F_N = 2.7807$$

Fourier Transform

The Fourier transform of a function $x(t)$ and its inverse relation are:

$$X(f) = \int_{-\infty}^{+\infty} x(t)\exp(-j2\pi ft)\,dt \tag{2.13a}$$

$$X(t) = \int_{-\infty}^{+\infty} x(f)\exp(j2\pi ft)\,dt \tag{2.13b}$$

Table 2.2 lists the Fourier transforms for a few commonly used functions in communication systems. The NCEES *Fundamentals of Engineering Supplied-Reference Handbook* includes a more complete table of Fourier transform pairs.

Table 2.2 Fourier Transform Pairs

$x(t)$		$X(f)$
Impulse at $t = 0$, $\delta(t)$	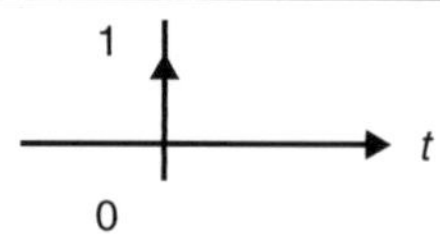	1
Unit step function, $u(t)$	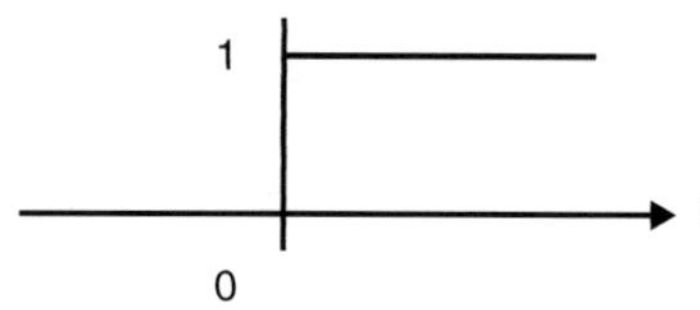	$\frac{1}{2}\delta(f)+\frac{1}{j2\pi f}$
Rectangular pulse, $\Pi(t/\tau)$	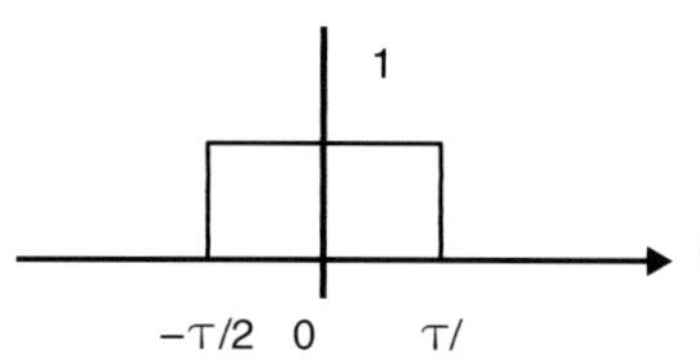	$\tau f \sin c(\tau f)$
Sinc function, sinc(at)	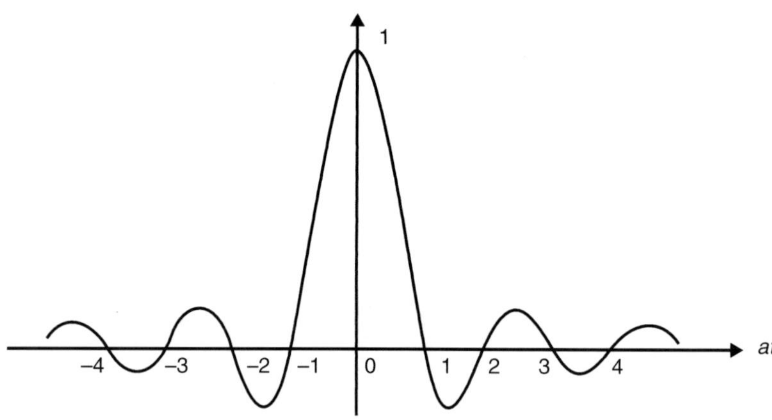	$\frac{1}{a}\Pi\left(\frac{f}{a}\right)$
Triangular pulse, $\Lambda(t/\tau)$	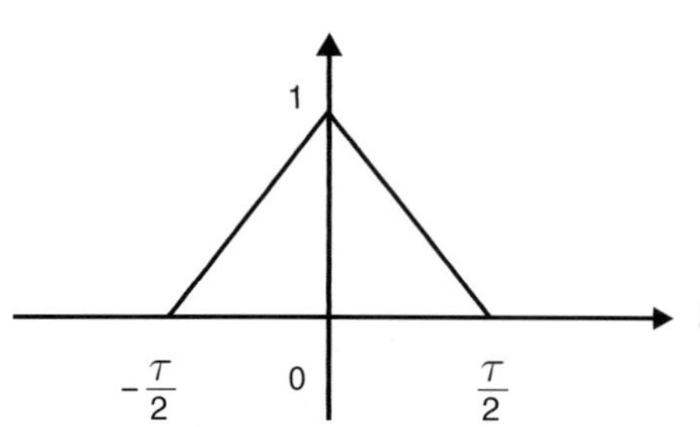	$\tau f \sin c^2(\tau f)$

Example 2.56

Find the Fourier transform of (i) 4 $\delta(t)$, (ii) 2 $u(t)$, (iii) $\Pi(t/2)$, and (iv) 6 $\cos(100\pi t)$.

Solution

(i) 4 (ii) $\delta(f) + 1/j\pi f$ (iii) $2f\,\text{sinc}(2f)$ (iv) $3[\delta(f-50) + \delta(f+50)]$

DIFFERENCE EQUATIONS AND Z-TRANSFORMS

Difference equations are used to model discrete systems; they are analogous to differential equations that describe continuous systems. The equations $y(k) - y(k-1) = 10$ and $y(k+1) - y(k) = 5$ are some examples of first order linear difference equations; $y(k) = y(k-1) + y(k-2)$ is an example of second order difference equation.

Example 2.57

Find the values of $y(1)$, $y(2)$, and $y(3)$ of the equation $y(k) - 1.01y(k-1) = -50$ with the initial condition $y(0) = 1000$.

Solution

$y(1) - 1.01\ y(0) = -50;\ y(1) = \$960$

$y(2) - 1.01\ y(1) = -50;\ y(2) = \$919.60$

$y(3) - 1.01\ y(2) = -50;\ y(3) = \$878.80$ and so on

Z-Transforms

Z-transform of a discrete sequence is defined as

$$F(z) = \sum_{k=0}^{\infty} f[k]z^{-k} \tag{2.14}$$

For example, if the discrete sequence is $f[k] = 0, 1, 4$, then its z-transform is $(F_z) = 0 + z^{-1} + 4z^{-2}$.

Example 2.58

Find the z-transform of the function $f(k) = 3\ u(k) + 2^k$ for $k \geq 0$.

Solution

Using the Z-transform table from the *Fundamentals of Engineering Supplied-Reference Handbook*, $F(z) =$

$$\frac{3}{1-z^{-1}} + \frac{1}{1-2z^{-1}}$$

Simplifying, $F(z) = \dfrac{3(1-2z^{-1}) + (1-z^{-1})}{(1-z^{-1})(1-2z^{-1})} = \dfrac{4-7z^{-1}}{(1-z^{-1})(1-2_z^{-1})}$

Multiplying by z^2, $F(z) = \dfrac{4z^2 - 7z}{z^2 - 3z + 2}$

PROBLEMS

2.1 The simplest value of $\frac{[(n+1)!]^2}{n!(n-1)!}$ is:

a. n^2
b. $n(n+1)$
c. $n+1$
d. $n(n+1)^2$

2.2 If $x^{3/4} = 8$, x equals:

a. 6
b. 9
c. –9
d. 16

2.3 If $\log_a 10 = 0.250$, $\log_{10} a$ equals:

a. 4
b. 0.50
c. 2
d. 0.25

2.4 If $\log_5 x = -1.8$, $\log_x 5$ is:

a. 0.35
b. 0.79
c. –0.56
d. undefined

2.5 If $\log x + \log (x - 10) - \log 2 = 1$, x is:

a. –1.708
b. 5.213
c. 7.824
d. 11.708

2.6 A right circular cone, cut parallel with the axis of symmetry, reveals a(n):

a. circle
b. hyperbola
c. eclipse
d. parabola

2.7 The expression $\frac{6!}{3!0!}$ is equal to:

a. ∞
b. 120
c. 2!
d. 0

2.8 To find the angles of a triangle, given only the lengths of the sides, one would use:

a. the law of cosines
b. the law of tangents
c. the law of sines
d. the inverse-square law

2.9 If $\sin\alpha = \dfrac{a}{\sqrt{a^2+b^2}}$, which of the following equations is true?

a. $\tan^{-1}\dfrac{b}{a} = \dfrac{\pi}{2} - \alpha$

b. $\tan^{-1}\dfrac{b}{a} = -\alpha$

c. $\cos^{-1}\dfrac{b}{\sqrt{a^2+b^2}} = \dfrac{\pi}{2} - \alpha$

d. $\cos^{-1}\dfrac{a}{\sqrt{a^2+b^2}} = \alpha$

2.10 The sine of 840° equals:
a. –cos 30°
b. –cos 60°
c. sin 30°
d. sin 60°

2.11 One root of $x^3 - 8x - 3 = 0$ is:
a. 2
b. 3
c. 4
d. 5

2.12 Roots of the equation $3x^3 - 3x^2 - 18x = 0$ are:
a. –2, 3
b. 0, – 2, 3
c. 2 + 1i, 2 – 1i
d. 0, 2, –3

2.13 The equation whose roots are –1 + i1 and –1 –i1 is given as:
a. $x^2 + 2x + 2 = 0$
b. $x^2 - 2x - 2 = 0$
c. $x^2 + 2 = 0$
d. $x^2 - 2 = 0$

2.14 Natural logarithms have a base of:
a. 3.1416
b. 2.171828
c. 10
d. 2.71828

2.15 $(5.743)^{1/30}$ equals:

a. 1.03
b. 1.04
c. 1.05
d. 1.06

2.16 The value of $\tan(A + B)$, where $\tan A = 1/3$ and $\tan B = 1/4$ (A and B are acute angles) is:

a. 7/12
b. 1/11
c. 7/11
d. 7/13

2.17 To cut a right circular cone in such a way as to reveal a parabola, it must be cut:

a. perpendicular to the axis of symmetry
b. at any acute angle to the axis of symmetry
c. at any obtuse angle to the axis of symmetry
d. none of these

2.18 The equation of the line perpendicular to $3y + 2x = 5$ and passing through $(-2, 5)$ is:

a. $2x = 3y$ c. $2y = 3x + 16$
b. $2y = 3x$ d. $3x = 2y + 8$

2.19 Equation of a line that has a slope of -2 and passes through $(2, 0)$ is:

a. $y = -2x$ c. $y = -2x + 4$
b. $y = 2x + 4$ d. $y = 2x - 4$

2.20 Equation of a line that intercepts the x-axis at $x = 4$ and the y-axis at $y = -6$ is:

a. $3x - 2y = 12$ c. $x + y = 6$
b. $2x - 3y = 12$ d. $x - y = 4$

2.21 The x-axis intercept and the y-axis intercept of the line $x + 3y + 9 = 0$ are:

a. 0 and 3 c. -3 and -9
b. -9 and -3 d. 9 and 0

2.22 The distance between the points $(1, 0, -2)$ and $(0, 2, 3)$ is:

a. 3.45 c. 6.71
b. 5.39 d. 7.48

2.23 The equation of a parabola with center at $(0, 0)$ and directrix at $x = -2$ is:

a. $y^2 = 4x$ c. $y^2 = 8x$
b. $y = 2x^2$ d. $y^2 = 2x$

2.24 The equation of the directrix of the parabola $y^2 = -4x$ is:

a. $y = 2$ c. $x + y = 0$
b. $x = 1$ d. $x = -2$

2.25 The equation of an ellipse with foci at $(\pm 2, 0)$ and directrix at $x = 6$ is:

a. $\frac{x^2}{12} + \frac{y^2}{8} = 1$

b. $\frac{x^2}{8} + \frac{y^2}{8} = 1$

c. $\frac{x^2}{12} + \frac{y^2}{12} = 1$

d. $\frac{x^2}{4} + \frac{y^2}{4} = 1$

2.26 The foci of the ellipse $\left(\frac{x}{3}\right)^2 + \left(\frac{y}{2}\right)^2 = 1$ are at:

a. $(\pm\sqrt{5}, 0)$
b. $(\pm\sqrt{2}, 0)$
c. $(0, \pm\sqrt{2})$
d. $(0, \pm\sqrt{5})$

2.27 The equation of a hyperbola with center at (0, 0), foci at $(\pm 4, 0)$, and eccentricity of 3 is:

a. $\frac{x^2}{16} - \frac{y^2}{16} = 1$

b. $\frac{x^2}{16} - \frac{y^2}{128} = 1$

c. $\frac{x^2}{64} - \frac{y^2}{48} = 1$

d. $\frac{x^2}{128} - \frac{y^2}{128} = 1$

2.28 The equation of a circle with center at (1, 2) and passing through the point (4, 6) is:

a. $x^2 + y^2 = 25$

b. $(x + 1)^2 + (y - 2)^2 = 25$

c. $x^2 + (y - 2)^2 = 25$

d. $(x - 1)^2 + (y - 2)^2 = 25$

2.29 The length of the tangent from (4, 8) to the circle $x^2 + (y - 1)^2 = 3^2$ is:

a. 3.81
b. 4.14
c. 5.66
d. 7.48

2.30 The conic section described by the equation $x^2 - 10xy + y^2 + x + y + 1 = 0$ is:

a. circle
b. parabola
c. hyperbola
d. ellipse

2.31 A triangle has sides of length 2, 3, and 4. The angle subtended by the sides of length 2 and 4 is:

a. 21.2° c. 46.6°
b. 35.0° d. 61.2°

2.32 Length a of one side of the triangle below is:

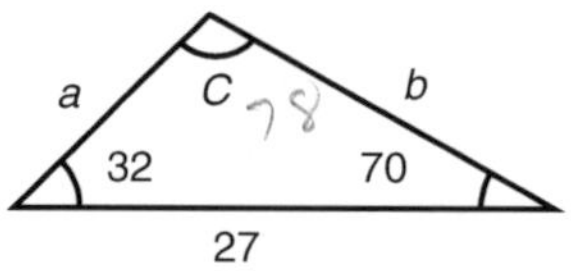

a. 25.9 c. 12.7
b. 19.1 d. 4.8

2.33 The relation $\sec\theta - (\sec\theta)(\sin^2\theta)$ can be simplified as:

a. $\sin\theta$ c. $\cot\theta$
b. $\tan\theta$ d. $\cos\theta$

2.34 If $\sin\theta = m$, $\cot\theta$ is:

a. $\dfrac{\sqrt{1-m^2}}{m}$

b. $\dfrac{m}{\sqrt{1-m^2}}$

c. $\sqrt{1-m^2}$

d. m

2.35 If vectors $\mathbf{A} = 3i - 6j + 2k$ and $\mathbf{B} = 10i + 4j - 6k$, their cross product $\mathbf{A} \times \mathbf{B}$ is:

a. $-12\mathbf{i} + 38\mathbf{j}$ c. $-12\mathbf{i} + 18\mathbf{j} + 24\mathbf{k}$
b. $12\mathbf{i} + 24\mathbf{j} + 36\mathbf{k}$ d. $28\mathbf{i} + 38\mathbf{j} + 72\mathbf{k}$

2.36 The sum of all integers from 10 to 50 (both inclusive) is:

a. 990 c. 1230
b. 1110 d. 1420

2.37 The 50th term of the series 10, 16, 22, 28, 34, 40. . . is:

a. 272 c. 428
b. 304 d. 584

2.38 Sum of the infinite series 4, 2, 1, 0.5, 0.25. . . is:

a. 8 c. 10,400
b. 128 d. ∞

2.39 If $\mathbf{A} = \begin{bmatrix} 1 & 2 & 3 \\ 1 & 2 & 9 \end{bmatrix}$ and $\mathbf{B} = \begin{bmatrix} 5 & 1 \\ 6 & 0 \\ 4 & 7 \end{bmatrix}$, the (2,1) entry of AB is:

a. 29 c. 33
b. 53 d. 64

2.40 The inverse of the matrix $\begin{bmatrix} 1 & 1 \\ 3 & 2 \end{bmatrix}$ is:

a. $\begin{bmatrix} 2 & -1 \\ -3 & 1 \end{bmatrix}$ b. $\begin{bmatrix} 2 & 3 \\ 1 & 1 \end{bmatrix}$ c. $\begin{bmatrix} 1 & 3 \\ 1 & 2 \end{bmatrix}$ d. $\begin{bmatrix} -2 & 1 \\ 3 & -1 \end{bmatrix}$

2.41 The determinant of the matrix $\begin{bmatrix} 1 & 2 & -1 \\ 3 & 0 & 2 \\ 2 & -2 & -1 \end{bmatrix}$ is:

a. 4 c. 24
b. 16 d. –16

2.42 In the system of equations

$$\begin{aligned} 3x_1 + 2x_2 - x_3 &= 5 \\ x_2 - x_3 &= 2 \\ x_1 + 2x_2 - 3x_3 &= -1 \end{aligned}$$

the value of x_2 = is:
a. 2 c. 4
b. –1 d. 6

2.43 What is the determinant of M?

$$M = \begin{bmatrix} 0 & 1 & 1 & 1 \\ 1 & 1 & 1 & 1 \\ 1 & 1 & 3 & 1 \\ 2 & 1 & 3 & 4 \end{bmatrix}$$

a. –6 c. 0
b. 6 d. 7

2.44 $\displaystyle\int_{\pi/2}^{\pi} \sin 2x \, dx =$

a. 2 c. 0
b. 1 d. –1

2.45 $\int_0^2 x^2\sqrt{1+x^3}\,dx =$

a. 52/9
b. 0
c. 52/3
d. 26/3

2.46 $\int_1^e x(\ln x)dx =$

a. $\frac{1}{2}e^2+1$
b. $\frac{1}{2}e^2-e+\frac{1}{2}$
c. $\frac{1}{4}e^2+\frac{1}{4}$
d. $\frac{1}{4}e^2-\frac{1}{2}e+\frac{1}{4}$

2.47 If the first derivative of the equation of a curve is constant, the curve is a:
a. circle
b. hyperbola
c. parabola
d. straight line

2.48 Which of the following is a characteristic of all trigonometric functions?
a. The values of all functions repeat themselves every 45 degrees.
b. All functions have units of length or angular measure.
c. The graphs of all functions are continuous.
d. All functions have dimensionless units.

2.49 For a given curve $y = f(x)$ that is continuous between $x = a$ and $x = b$, the average value of the curve between the ordinates at $x = a$ and $x = b$ is represented by:

a. $\dfrac{\int_a^b x^2\,dy}{b-a}$
b. $\dfrac{\int_a^b y^2\,dx}{b-a}$
c. $\dfrac{\int_a^b x\,dy}{a-b}$
d. $\dfrac{\int_a^b y\,dx}{b-a}$

2.50 If $y = \cos x$, $\dfrac{dy}{dx}$ is:

a. $\sin x$
b. $-\tan x \cos x$
c. $\dfrac{1}{\sec x}$
d. $\sec x \sin x$

2.51 The derivative of $\cos^3 5x$ is:

a. $3 \sin^2 5x$
b. $15 \sin^2 5x$
c. $\cos^2 5x \sin x$
d. $-15 \cos^2 5x \sin 5x$

2.52 The slope of the curve $y = 2x^3 - 3x$ at $x = 1$ is:

a. -1
b. 0
c. 3
d. 5

2.53 A stone is dropped from the top of a building at $t = 0$. The position of the stone is given by the equation $s(t) = 16\,t^2$ m. Acceleration of the stone (in m/s²) 2 seconds after it is dropped is:

a. 64 c. 24

b. 32 d. 16

2.54 Maximum value of the function $f(x) = x^3 - 5x - 4$ occurs at:

a. 0 c. –0.30

b. 1.29 d. –1.29

2.55 The partial derivative with respect to x of the function $xy^2 - 5y + 6$ is:

a. xy c. y^2

b. $2y$ d. $-5y$

2.56 The power series expansion of cos(x) about the point $x = 0$ is:

a. $1-\frac{x^2}{2!}+\frac{x^4}{4!}-\frac{x^6}{6!}+\ldots$ c. $1-\frac{x}{2!}+\frac{x}{4!}-\frac{x}{6!}+\ldots$

b. $1+\frac{x^2}{2!}+\frac{x^4}{4!}+\frac{x^6}{6!}+\ldots$ d. $1+\frac{x}{2!}+\frac{x}{4!}+\frac{x}{6!}+\ldots$

2.57 The value of $\lim\limits_{x\to 2}\frac{x^2-4}{x-2}$ is:

a. 0 c. 1

b. ∞ d. 4

2.58 If $A = \int_0^{\frac{\pi}{4}} \sin^2\theta\, d\theta,$ the value of A is:

a. 0.29 c. 1.75

b. 0.58 d. 3.14

2.59 If $x^3 + 3x^2y + y^3 = 4$ defines y implicitly, $dy/dx =$

a. $-\frac{x^2+2xy}{x^2+y^2}$ c. $-\frac{x^2+y^2}{x^2+2xy}$

b. $3x^2+3y^2$ d. $-\frac{x^2+2xy}{x^2+y^2}$

2.60 Estimate $\sqrt{34}$ using differentials. The answer is closest to:

a. $6+\frac{1}{6}$ c. 6

b. $6-\frac{1}{6}$ d. $6-\frac{1}{3}$

2.61 The only relative maximum of $f(x) = x^4 - \frac{4}{3}x^3 - 12x^2 + 1$ is:

a. -1 c. 0
b. 1 d. -1

2.62 The area between $y = x^2$ and $y = 2x + 3$ is:

a. 9 c. $6\frac{1}{3}$
b. 20 d. $10\frac{2}{3}$

2.63 The area enclosed by the curve $r = 2(\sin\theta + \cos\theta)$ is:

a. π c. 2π
b. $\frac{\pi}{2}$ d. $\pi\sqrt{2}$

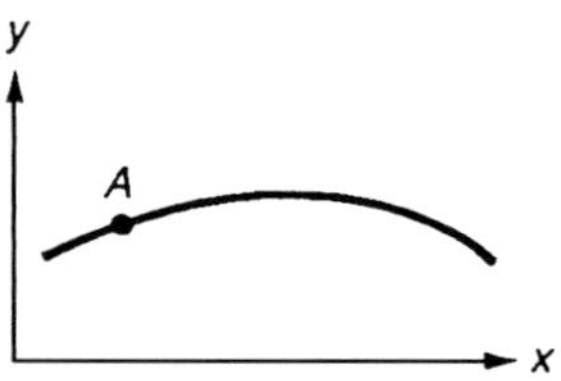

Exhibit 2.64

2.64 The curve in Exhibit 2.64 has the equation $y = f(x)$. At point A, what are the values of $\frac{dy}{dx}$ and $\frac{d^2y}{dx^2}$?

a. $\frac{dy}{dx} < 0, \frac{d^2y}{dx^2} < 0$ c. $\frac{dy}{dx} = 0, \frac{d^2y}{dx^2} = 0$
b. $\frac{dy}{dx} < 0, \frac{d^2y}{dx^2} > 0$ d. $\frac{dy}{dx} > 0, \frac{d^2y}{dx^2} < 0$

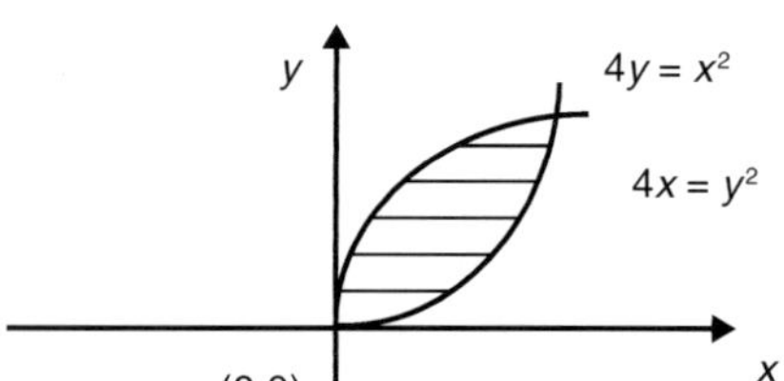

Exhibit 2.65

2.65 The area of the shaded region in Exhibit 2.65 is:

a. 1.37 c. 5.33
b. 3.82 d. 6.80

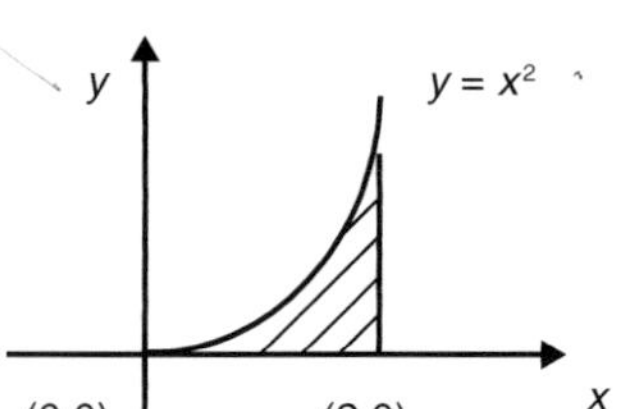

Exhibit 2.66

Refer to Exhibit 2.66 for problems 2.66 through 2.71.

2.66 Area of the shaded region is:

a. 1.10 c. 4.02
b. 2.67 d. 6.80

2.67 The first moment of the shaded area with respect to the x-axis is:

a. 1.20 c. 3.20
b. 2.30 d. 4.60

2.68 The first moment of the shaded area with respect to the y-axis is:

a. 0.85 c. 3.07
b. 1.90 d. 4.00

2.69 The centroid of the shaded area is:

a. (1.5, 1.2) c. (0, 1.2)
b. (1.5, 0) d. (1.0, 1.0)

2.70 The second moment of the shaded area with respect to the x-axis is:

a. 6.10 c. 12.11
b. 9.05 d. 18.32

2.71 The second moment of the shaded area with respect to the y-axis is:

a. 1.98 c. 4.63
b. 3.12 d. 6.40

2.72 If the characteristic roots of a differential equation are $-4 - i4$ and $-4 + i4$, the homogeneous solution is:

a. $C_1 \cos 4x + C_2 \sin 4x$ c. $C_1 e^{-i4x} + C_2 e^{i4x}$
b. $e^{-4x}(C_1 \cos 4x + C_2 \sin 4x)$ d. $C_1 \cos (4x + \theta)$

2.73 Characteristic roots of a differential equation are –2 and –2. If the forcing function is e^{-2x}, the particular solution, $y_p(x)$ is:

a. Ax^2 c. Axe^{-2x}
b. Ae^{-2x} d. Ax^2e^{2x}

2.74 The solution of the differential equation $y'' + 5y' + 6y = 2e^{-2x}$ with zero initial conditions is:

a. $y = -2e^{-2x} + 2e^{-3x} + 2xe^{-2x}$ c. $y = 2e^{-3x} + 2xe^{-2x}$
b. $y = x - 2e^{-2x} + 2e^{-3x}$ d. $y = -2e^{-2x} + 2xe^{-2x}$

2.75 $\lim\limits_{x \to 1} \dfrac{x^2 - 1}{x - 1} =$

a. 2 c. 0
b. ∞ d. 1

2.76 The solution to $xy' + 2y = e^{3x}$ is:

a. $y = e^{3x} - \dfrac{e^{3x}}{x} + \dfrac{c}{x}$

b. $y = \dfrac{xe^{3x} - 3e^{3x} + 3c}{3x^2}$

c. $y = xe^{3x} - 3e^{3x} + c$

d. $y + x = e^{3x} + c$

2.77 Solve $xy'' - 2(x+1)y' + (x+2)y = 0$.

a. $y = Ae^x + Bx^3e^x$
b. $y = Ae^x + Be^{2x}$
c. $y = A\sin(x+1) + B\cos(x+2)$
d. $y = Ae^x + Be^{-x}$

2.78 Solve $y'' + 4y = 8\sin x$.

a. $y = Ae^{2x} + Be^{-2x}$
b. $y = A\sin 2x + B\cos 2x$
c. $y = A\sin 2x + B\cos 2x + \sin x$
d. $y = A\sin 2x + B\cos 2x + \frac{8}{3}\sin x$

2.79 The family of trajectories orthogonal to the family $x^2 + y^2 = 2cy$ is:

a. $x - y = c$
b. $x^2 - y^2 = cx$
c. $x^2 + y^2 = c$
d. $x^2 + y^2 = 2cx$

2.80 The Laplace transform of the function $e^{-t}\cos(t)$ is:

a. $\dfrac{s}{s^2 + 1^2}$
b. $\dfrac{(s+1)}{(s+1)^2 + 1^2}$
c. $\dfrac{(s+1)}{s^2 + 1^2}$
d. $\dfrac{(s-1)}{(s-1)^2 + 1^2}$

2.81 For the differential equation $\dfrac{d^2y}{dt^2} + 2\dfrac{dy}{dt} + y(t) = \cos 2t$ with zero initial conditions, the Laplace transform $Y(s)$ of $y(t)$ is:

a. $\dfrac{s}{(s^2 + 2s + 1)}$
b. $\dfrac{2}{(s^2 + 2)(s^2 + 2s + 1)}$
c. $\dfrac{s}{(s^2 + 4)(s^2 + 2s + 1)}$
d. $\dfrac{s}{(s^2 + 4)(s^2 + 1)}$

2.82 The initial value of the function whose Laplace transform $F(s) = \dfrac{s+5}{s(s+1)(s+10)}$ is:

a. 2.0
b. 1.0
c. 0.5
d. 0

2.83 For the difference equation $y(k) = 2y(k-1) + 3y(k-2)$ with initial conditions $y(-1) = 1$; $y(-2) = 1$, the value of $y(1)$ is:

a. 1
b. 5
c. 13
d. 41

2.84 For the difference equation $y(k + 1) + y(k) = u(k)$ with $y(0) = 0$, find $Y(z)$, the z-transform of the function $y(k)$ is:

a. $\dfrac{1}{(z-1)(z+2)}$ c. $\dfrac{z}{(z+1)(z-2)}$

b. $\dfrac{z}{(z-1)(z+2)}$ d. $\dfrac{2}{(z+1)(z+2)}$

2.85 The initial value of the function whose Z-transform $F(z) = \dfrac{z+2}{z-4}$ is:

a. 0 c. 0.5
b. –0.5 d. 1.0

2.86 Find the Fourier coefficient b_3 of the waveform shown in Exhibit 2.86.

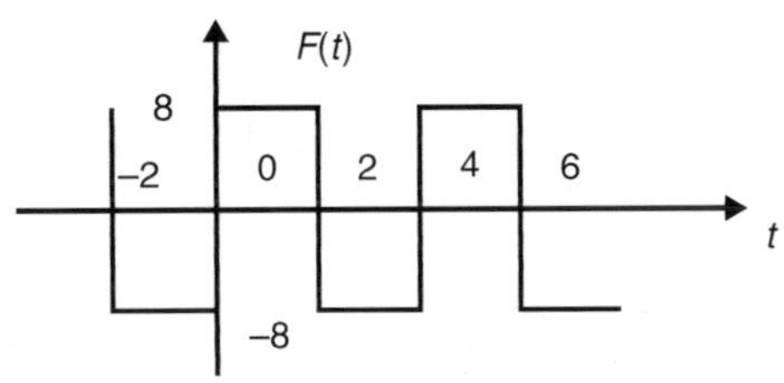

Exhibit 2.86

a. 1.17 c. 3.40
b. 2.23 d. 4.78

2.87 Find the Fourier transform, $F(f)$, of a triangular pulse of width 2.

a. $\text{sinc}(f)$ c. $\text{sinc}(2f)$
b. $\text{sinc}^2(f)$ d. $f\,\text{sinc}^2(f)$

2.88 Find the Fourier transform of $f(t) = \delta(t - 1) + \delta(t + 1)$.

a. $\exp(-j2\pi f)$ c. $2\cos(2\pi f)$
b. $\exp(j2\pi f)$ d. $j2\sin(2\pi f)$

2.89 Find $y(1)$ if $y(k) = 2y(k - 1) + 3y(k - 2)$ with $y(-1) = 1$; $y(-2) = 1$.

a. 1 c. 13
b. 5 d. 41

2.90 Find the final value of the function $f(k)$ whose Z-transform is $F(z) = \dfrac{2(z+1)}{(z-1)}$.

a. 1 c. 4
b. 2 d. ∞

SOLUTIONS

2.1 d. The value $(n+1)!$ may be written as $(n+1)(n)[(n-1)!]$. It may be written also as $n!(n+1)$. Hence the given expression may be written as follows:

$$\frac{\{(n+1)(n)[(n-1)!]\}\{n!(n+1)\}}{n!(n-1)!} = (n+1)^2 n$$

2.2 d. Raise both sides of the equation to the 4/3 power:

$$[x^{3/4}]^{4/3} = 8^{4/3}$$

$$x = \sqrt[3]{8^4} = \sqrt[3]{(2^3)^4} = 2^{\frac{3\bullet 4}{3}} = 2^4 = 16$$

2.3 a. $\log_a 10 = 0.250$ can be written as $10 = a^{0.250}$. Taking $\log_{10}$,

$$\log_{10} 10 = \log_{10} a^{0.250}$$

$$1 = 0.250 \log_{10} a$$

and

$$\log_{10} a = \frac{1}{0.250} = 4$$

2.4 c. Since $(\log_5 x)(\log_x 5)=1$, $\log_x 5 = \dfrac{1}{-1.8} = -0.556$

2.5 d. Since $\log(a) + \log(b) - \log(c) = \log \dfrac{ab}{c}$, the given equation simplifies to $\log \dfrac{x(x-10)}{2} = 1$. Equivalently, $\dfrac{x(x-10)}{2} = 10$ or $x^2 - 10x - 20 = 0$. The roots are 11.708 and –1.708. Since log is not defined for negative values, $x = 11.708$.

2.6 b.

2.7 b. $\dfrac{6!}{3!0!} = \dfrac{6(5)(4)(3)!}{3!(1)} = 120$

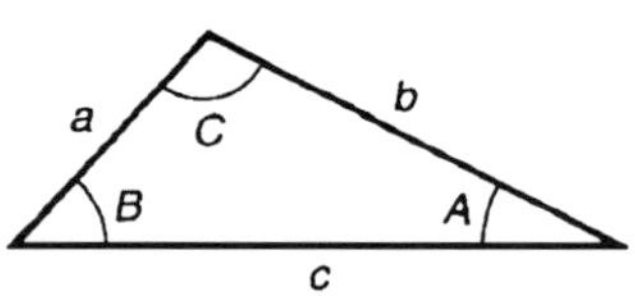

Exhibit 2.8

2.8 a. The law of cosines is $a^2 = b^2 + c^2 - 2bc \cos A$ for any plane triangle with angles A, B, C and sides a, b, c, respectively.

This law can be applied to solve for the angles, given three sides in a plane triangle (Exhibit 2.8).

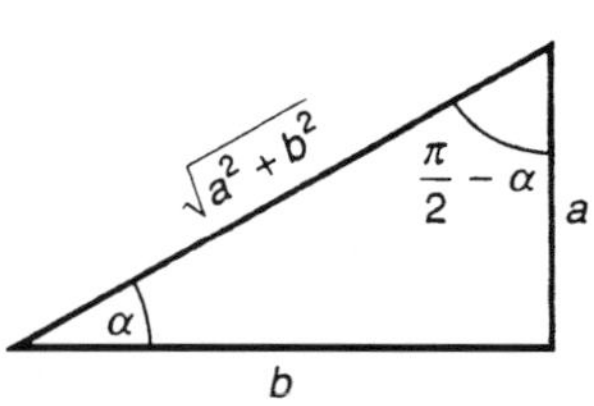

Exhibit 2.9

2.9 a. The triangle appears in Exhibit 2.9.

$$\tan\left(\frac{\pi}{2} - \alpha\right) = \frac{b}{a}$$

$$\tan^{-1}\frac{b}{a} = \frac{\pi}{2} - \alpha$$

2.10 d.

$$840° = 2(360) + 120 = 2(2\pi) \text{ rad} + 120°$$
$$\sin [2(2\pi) \text{ rad} + 120°] = \sin 120° = \sin 60°$$

2.11 b. The solution is obtained by seeing which of the five answers satisfies the equation.

x	$x^3 - 8x - 3$
2	–11
3	0
4	29
5	82
6	165

2.12 b. Any scientific calculator can be used to find the roots as 0, 3, and –2.

2.13 a. If x_1 and x_2 are the roots, the equation is $(x - x_1)(x - x_2) = 0$. Since the roots are $-1 + i$ and $-1 - i$, the equation is $(x - (-1 + i))(x - (-1 - i)) = 0$ or $(x + 1 - i)(x + 1 + i) = 0$ or $x^2 + 2x + 2 = 0$.

2.14 d. Common logarithms have base 10. Natural, or napierian, logarithms have base $e = 2.71828$.

2.15 d.

$$\log\ (5.743)^{1/30} = \frac{1}{30}\log 5.743 = \frac{1}{30}(0.7592) = 0.0253$$

The antilogarithm of 0.0253 is 1.06.

2.16 c.

$$\sin\ (A + B) = (\sin A \cos B) + (\cos A \sin B)$$

$$\cos\ (A + B) = (\cos A \cos B) - (\sin A \sin B)$$

$$\tan\ (A + B) = \frac{\sin(A + B)}{\cos(A + B)} = \frac{(\sin A \cos B) + (\cos A \sin B)}{(\cos A \cos B) - (\sin A \sin B)}$$

Dividing by $\cos A \cos B$,

$$\tan(A + B) = \frac{\dfrac{(\sin A\ \cos B)}{(\cos A\ \cos B)} + \dfrac{(\cos A\ \sin B)}{(\cos A\ \cos B)}}{\dfrac{(\cos A\ \cos B)}{(\cos A\ \cos B)} - \dfrac{(\sin A\ \sin B)}{(\cos A\ \cos B)}} = \frac{\tan A + \tan B}{1 - \tan A \tan B}$$

$$= \frac{\dfrac{1}{3} + \dfrac{1}{4}}{1 - \dfrac{1}{3} \times \dfrac{1}{4}} = \frac{\dfrac{4}{12} + \dfrac{3}{12}}{1 - \dfrac{1}{12}} = \frac{\dfrac{7}{12}}{\dfrac{11}{12}} = \frac{7}{11}$$

The problem could also be solved by determining angle A (whose tangent is 1/3) and angle B (whose tangent is 1/4). Then we could find the tangent of $(A + B)$.

$$\tan^{-1}\frac{1}{3}=18.435^\circ \quad \tan^{-1}\frac{1}{4}=14.036^\circ$$

$$\tan(18.435+14.036)^\circ=\tan(32.471^\circ)=0.6364=\frac{7}{11}$$

2.17 d. To reveal a parabola, a right circular cone must be cut parallel to an element of the cone and intersecting the axis of symmetry.

2.18 c. Rewriting the given line, $y=\frac{5}{3}-\frac{2}{3}x$. This line has slope $-\frac{2}{3}$, so a perpendicular line must have slope $\frac{3}{2}$. Using the point-slope form, $\frac{y-5}{x+3}=\frac{3}{2}$. Simplifying, $y = \frac{3}{2}x + 8$.

2.19 d. The equation of a straight line is $y = mx + b$ where the slope is m, (x, y) is any point on the line, and b is the y-axis intercept. Here, $m = -2$ and (2, 0) is a point. Substituting these values, $0 = -2(2) + b$ or $b = 4$. Then, the equation is $y = -2x + 4$ or $y + 2x = 4$.

2.20 a. (4,0) and (0, –6) are two points on the straight line. Then, the slope $m = \frac{y_2-y_1}{x_2-x_1}=\frac{0-(-6)}{4-0}=\frac{3}{2}$. Substituting one of the points, say, (0, –6) in the general equation $y = mx + b$; $-6 = 0 + b$. Then, $b = -6$ and the equation is $y = \left(\frac{3}{2}\right)x - 6$ or $3x - 2y = 12$.

2.21 b. For y-axis intercept, $x = 0$. Substituting this in the equation, $0 + 3y + 9 = 0$ or $y = -3$ is the y-axis intercept. For x-axis intercept, $y = 0$. Substituting this in the equation, $x + 0 + 9 = 0$ or $x = -9$ is the x-axis intercept.

2.22 b. Distance d between any two points (x_1, y_1, z_1) and (x_2, y_2, z_2) can be determined as $d^2 = (x_1 - x_2)^2 + (y_1 - y_2)^2 + (z_1 - z_2)^2$. Here, $d^2 = (1 - 0)^2 + (0 - 2)^2 + (-2 - 3)^2 = 30$; $d = 5.385$.

2.23 c. Directrix, $x = -p/2 = -2$ or $p = 4$. Equation of a parabola with center at origin is $y^2 = 2px$; as $p = 4$, $y^2 = 8x$.

2.24 b. Equation of a parabola with center at origin is $y^2 = 2px$. Since $y^2 = -4x$, $2p = -4$ or $p = -2$. Equation of a directrix is $x = (-p/2)$; since $p = 2$, the equation is $x = 1$.

2.25 a. Focus $ae = 2$ and directrix $\frac{a}{e} = 6$. Solving, $a^2 = 12$ and $e = 1/\sqrt{3}$. But $e = \sqrt{1-\frac{b^2}{a^2}}$; solving, $b^2 = 8$. The equation of an ellipse is

$$\frac{x^2}{a^2}+\frac{y^2}{b^2}=1 \text{ or } \frac{x^2}{12}+\frac{y^2}{8}=1.$$

2.26 **a.** The equation of an ellipse is $\frac{x^2}{a^2}+\frac{y^2}{b^2}=1$. Here, $a = 3$ and $b = 2$.

The eccentricity $e = \sqrt{1-\frac{b^2}{a^2}} = \sqrt{\frac{5}{9}}$, and the foci = $(\pm\, ae, 0) = (\pm\sqrt{5}, 0)$.

2.27 **b.** Focus $ae = 4$ and eccentricity $e = 3$. Solving, $a = 4/3$.

Also, $b = a\sqrt{e^2-1} = \frac{4\sqrt{8}}{3}$. Equation for a hyperbola is

$$\frac{x^2}{a^2}-\frac{y^2}{b^2}=1 \Rightarrow \frac{x^2}{16}-\frac{y^2}{128}=1.$$

2.28 **d.** Radius of a circle with center at (h, k) and passing through a point (x, y) is $r^2 = (x-h)^2 + (y-k)^2$. Here, $r^2 = (4-1)^2 + (6-2)^2 = 25$. Equation of a circle with center at (h, k) is $(x-h)^2 + (y-k)^2 = r^2$; here, $(x-1)^2 + (y-2)^2 = 25$.

2.29 **d.** Length of the tangent from any point (x', y') outside a circle with center at (h, k) and radius r is given as $t^2 = (x'-h)^2 + (y'-k)^2 - r^2$. Here, $(x', y') = (4, 8)$, center $(h, k) = (0, 1)$, and $r^2 = 3^2$. Then, $t^2 = (4-0)^2 + (8-1)^2 - 3^2 = 56$ or $t = 7.483$.

2.30 **c.** General equation of a conic section is $Ax^2 + 2Bxy + Cy^2 + 2Dx + 2Ey + F = 0$, where both A and C are not zeros. Here, A = 1; B = –5; C = 1; then, $(B^2 - AC) > 0 \Rightarrow$ hyperbola.

2.31 **c.** Since three sides of a triangle are given, use the law of cosines; $3^2 = 2^2 + 4^2 - 2(2)(4)\cos(\theta)$ where θ is the angle opposite to side 3. Solving, $\theta = 46.6°$.

2.32 **a.** Angle $C = 180 - (70 + 32) = 78°$. Using the law of sines,

$$\frac{a}{\sin 70}=\frac{b}{\sin 32}=\frac{27}{\sin 78}.$$ Solving, $a = 25.94$.

2.33 **d.** $(\sec\theta)(1-\sin^2\theta) = \sec\theta\cos^2\theta = \frac{1}{\cos\theta}\cos^2\theta = \cos\theta$

2.34 **a.** $\sin\theta=\frac{m}{1}=\frac{\text{opposite side}}{\text{hypotenuse}}$; then, adjacent side $=\sqrt{1-m^2}$

and $\cos\theta=\frac{\text{adjacent side}}{\text{hypotenuse}}=\frac{\sqrt{1-m^2}}{1}$; then, $\cot\theta=\frac{\cos\theta}{\sin\theta}=\frac{\sqrt{1-m^2}}{m}$

2.35 **d.** Cross product, $\mathbf{A}\times\mathbf{B} = \begin{vmatrix} i & j & k \\ 3 & -6 & 2 \\ 10 & 4 & -6 \end{vmatrix}$

Expanding, $i[(-6)(-6) - (2)(4)] - j[(3)(-6) - (2)(10)] + k[(3)(4) - (-6)(10)] = 28\mathbf{i} + 38\mathbf{j} + 72\mathbf{k}$

2.36 c. This is an arithmetic series; first term $a = 10$, last term $l = 50$, and common difference $d = 1$. The number of terms, n, can be calculated as, $l = a + (n - 1)d$; $50 = 10 + (n - 1)1$; solving, $n = 41$.

$$\text{sum } S = \frac{n(a+1)}{2} = \frac{41}{2}(10+50) = 1230$$

2.37 b. This is an arithmetic series; first term $a = 10$ and the common difference $d = 6$. Taking the number of terms n as 50, the nth term (last term) can be calculated as, $l = a + (n - 1)d$. In this case, $l = 10 + (50 - 1)6 = 304$.

2.38 a. This is a geometric series; first term $a = 4$ and common ratio $r = 0.5$. Since $r < 1$, the series is convergent and the sum as the number of terms n tend to infinity is $S = \dfrac{a}{1-r} = \dfrac{4}{1-0.5} = 8$.

2.39 b. To compute the (2,1) entry, take [1 2 9] • [5 6 4] = 5 + 12 + 36 = 53.

2.40 d. To invert a 2 × 2 matrix,

$$\begin{bmatrix} a & b \\ c & d \end{bmatrix}^{-1} = \frac{1}{ad-bc}\begin{bmatrix} d & -b \\ -c & a \end{bmatrix} = \frac{1}{2-3}\begin{bmatrix} 2 & -1 \\ -3 & 1 \end{bmatrix}$$

2.41 c. This 3 × 3 determinant can be computed quickly be expanding it in minors, especially around the second column:

$$\begin{vmatrix} 1 & 2 & -1 \\ 3 & 0 & 2 \\ 2 & -2 & 1 \end{vmatrix} = (-1)^{1+2}(2)\begin{vmatrix} 3 & 2 \\ 2 & -1 \end{vmatrix} + (-1)^{2+2}(0)\begin{vmatrix} 1 & -1 \\ 2 & -1 \end{vmatrix} + (-1)^{3+2}(-2)\begin{vmatrix} 1 & -1 \\ 3 & 2 \end{vmatrix}$$

$$= -2(-3-4) + 0 + 2(2+3) = 24$$

2.42 d. By Cramer's Rule,

$$x_2 = \frac{\text{Det}\begin{bmatrix} 3 & 5 & -1 \\ 0 & 2 & -1 \\ 1 & -1 & -3 \end{bmatrix}}{\text{Det}\begin{bmatrix} 3 & 2 & -1 \\ 0 & 1 & -1 \\ 1 & 2 & -3 \end{bmatrix}} = \frac{3\begin{vmatrix} 2 & -1 \\ -1 & -3 \end{vmatrix} + 1\begin{vmatrix} 5 & -1 \\ 2 & -1 \end{vmatrix}}{3\begin{vmatrix} 1 & -1 \\ 2 & -3 \end{vmatrix} + 1\begin{vmatrix} 2 & -1 \\ 1 & -1 \end{vmatrix}} = \frac{3(-7) + (-3) = -24}{3(-1) + (-1) = -4} = 6$$

2.43 **a.** To evaluate a 4×4 matrix, one must do some row or column operations and expand by minors:

$$\begin{bmatrix} 0 & 1 & 1 & 1 \\ 1 & 1 & 1 & 1 \\ 1 & 1 & 3 & 1 \\ 2 & 1 & 3 & 4 \end{bmatrix} \sim \begin{bmatrix} 0 & 1 & 1 & 1 \\ 1 & 1 & 1 & 1 \\ 0 & 0 & 2 & 0 \\ 0 & -1 & 1 & 2 \end{bmatrix}$$

Taking minors of column 1,

$$\text{Det}(M) = (1)(-1)^{2+1}\,\text{Det}\begin{bmatrix} 1 & 1 & 1 \\ 0 & 2 & 0 \\ -1 & 1 & 2 \end{bmatrix} = -(4+2) = -6$$

2.44 **d.**

$$\int_{\pi/2}^{\pi} \sin\, 2x\, dx = -\frac{1}{2}\cos\, 2x = -\frac{1}{2}\cos\, 2\pi + \frac{1}{2}\cos\,\pi = -\frac{1}{2} - \frac{1}{2} = -1$$

2.45 **a.** Let $u = 1 + x^3$, so $du = 3x^2\, dx$. The integral becomes $\frac{1}{3}\int_1^9 \sqrt{u}\, du$ $= \frac{2}{9}u^{3/2} = \frac{2}{9}(27-1)$.

2.46 **c.** Prepare to solve using $\int u\, dv = uv - \int v\, du$.

Let $u = \ln x$ and $dv = x\, dx$.

Therefore, $du = \frac{1}{x}dx$ and $v = \frac{1}{2}x^2$

$$\int_1^e x(\ln x)dx = \ln\, x\left(\frac{1}{2}x^2\right)\Big|_1^e - \int_1^e \frac{1}{2}x^2\left(\frac{1}{x}\right)dx$$

$$= \ln\, x\left(\frac{1}{2}x^2\right)\Big|_1^e - \frac{1}{4}x^2\Big|_1^e = \ln\, e\left(\frac{1}{2}e^2\right) - \ln 1\left(\frac{1}{2}1^2\right) - \left[\frac{1}{4}e^2 - \frac{1}{4}1^2\right]$$

$$= \frac{1}{4}e^2 + \frac{1}{4}$$

2.47 **d.** If $\frac{dy}{dx} = m$, $y = \int m\, dx = m\int dx = mx + b$, so $y = mx + b$ is a straight line.

2.48 **d.** All trigonometric functions are ratios of lengths, with the result that they are dimensionless.

2.49 d.

$$\text{Area} = \int_a^b y\,dx$$

$$\text{Average value} = \frac{\text{Area}}{\text{Base width}} = \frac{\int_a^b y\,dx}{b-a}$$

2.50 b. Since $\frac{dy}{dx} = -\sin x$ and $\tan x = \frac{\sin x}{\cos x}$, then $\sin x = \tan x \cos x$. Thus, the derivative is

$$\frac{dy}{dx} = -\tan x \cos x$$

2.51 d. Apply the chain rule. The "outside" function is u^3, so $y' = 3\cos^2 5x\,(\cos 5x)' = 3\cos^2 5x\,(-\sin 5x)\,(5)$.

252 c. Derivative of y, $y' = 6x^2 - 3$. Since the slope is the derivative, at $x = 1$ the slope is $y'|_{x=1} = 3$.

2.53 b. Position $s(t) = 16t^2$; then, velocity $v(t) = s'(t) = 32t$ m/s and acceleration $a(t) = s''(t) = 32$ m/s^2; at any time the acceleration is 32 m/s^2.

2.54 d. Derivative of y is $y' = 3x^2 - 5$. Equating y' to 0, $3x^2 - 5 = 0$ has roots at $x = -1.29, +1.29$.

Second derivative of y is $y'' = 6x$.

$y''|_{x=-1.29} < 0$; then, the maximum of y occurs at $x = -1.29$.

(Note: value of $y_{max} = (-1.29)^3 - 5(-1.29) - 4 = 0.30$

$y''|_{x=1.29} > 0$; then, the minimum of y occurs at $x = 1.29$; $y_{min} = (1.29)^3 - 5(1.29) - 4 = -8.30$.

At the inflection point $y'' = 0$: $y'' = 6x = 0$; then, $x = 0$. Also, y'' changes sign at $x = 0$.

2.55 c. y^2

2.56 a. $f(x) = \cos x$; $f'(x) = -\sin x$; $f''(x) = -\cos x$; $f''(x) = \sin x$; $f^{iv}(x) = \cos x$
Substituting these values in the Taylor's series expansion,

$$f(x) = \cos(0) + \frac{-\sin 0}{1!}x + \frac{-\cos 0}{2!}x^2 + \frac{\sin 0}{3!}x^3 + \frac{\cos 0}{4!}x^4$$

$$= 1 - \frac{x^2}{2!} + \frac{x^4}{4!} - \frac{x^6}{6!} + \ldots$$

2.57 d. Using L'Hopital's rule, $\lim_{x\to 2} \frac{x^2 - 4}{x - 2} = \lim_{x\to 2} \frac{2x}{1} = 4$.

2.58 a. Using the integral table, the integral =

$$2\left[\frac{\theta}{2}-\frac{\sin 2\theta}{4}\right]_0^{\pi/4}=\left[\left(\frac{\pi}{4}-\frac{1}{2}\right)-0\right]=0.285$$

2.59 a. Taking the derivative with respect to x,

$$3x^2+6xy+3x^2y'+3y^2y'=0,\ \text{so}\ y'=-\frac{3x^2+6xy}{3x^2+3y^2}$$

2.60 b. Since $\sqrt{36}=6$, take $x_0=36$ and $f(x)=\sqrt{x}$. In general,

$$\Delta y=f(x)-f(x_0)=f'(x_0)(x-x_0)$$

$$\sqrt{34}-\sqrt{36}=\frac{1}{2}\frac{1}{\sqrt{36}}(-2)=-\frac{1}{6}$$

2.61 b. Here,

$f'(x)=4x^3-4x^2-24x=x(4x^2-4x-24)=4x(x-3)(x+2)=0$, so possible extrema are at 0, 3, and –2. Since $f''(0)=-24$, it is the maximum (3 and –2 are minima). Since $f(0)=1$, the relative maximum is 1.

2.62 d. The line and the parabola intersect when $x^2=2x+3$, or $x^2-2x+1=4$, or $(x-1)^2=2^2$. The line is above the parabola, so

$$A=\int_{-1}^{3}\left(2x+3-x^2\right)dx=\left(x^2+3x-\frac{1}{3}x^3\right)\Bigg|_{-1}^{3}=9-\left(-\frac{5}{3}\right)$$

2.63 c. Multiply by r to obtain $x^2+y^2=2y+2x$, or $x^2-2x+y^2-2y=0$, or $(x-1)^2+(y-1)^2=2$ a circle centered at (1, 1) of radius $\sqrt{2}$. The area is $\pi(\sqrt{2})^2=2\pi$.

2.64 d. The first derivative $\frac{dy}{dx}$ is the slope of the curve. At point A the slope is positive. The second derivative $\frac{d^2y}{dx^2}$ gives the direction of bending. A negative value indicates the curve is concave downward.

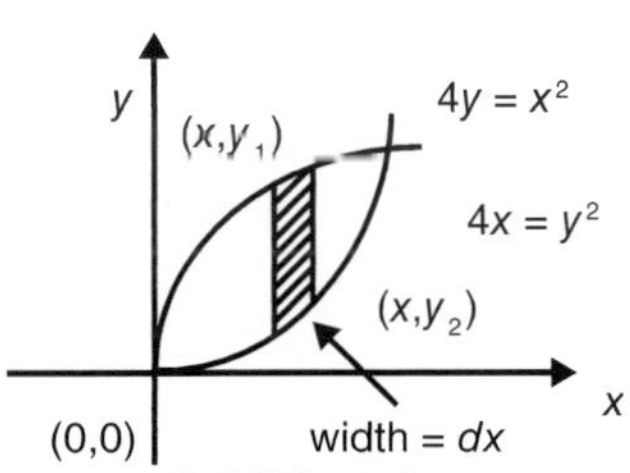

Exhibit 2.65a

2.65 c. Point of intersection of the curves is determined as $y^2=4x=4\sqrt{4y}$, $y=4$; $x=4$.

Area of the strip in Exhibit 2.65a, $dA=(y_1-y_2)dx=\left(\sqrt{4x}-\frac{x^2}{4}\right)dx$

$$\text{Area}=A=\int_{x=0}^{4}\left(\sqrt{4x}-\frac{x^2}{4}\right)dx=\frac{16}{3}$$

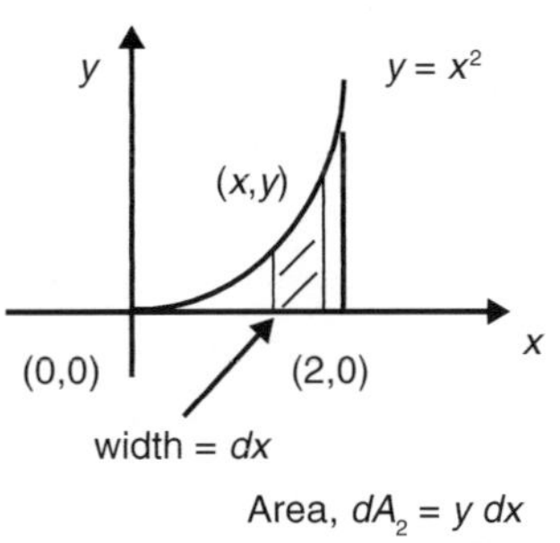

Exhibit 2.66a

2.66 b. In Exhibit 2.66a, area $dA_2=y dx$.

$$\text{Shaded area}=\int dA_2=\int_{x=0}^{2}y.dx=\int_{x=0}^{2}x^2dx=\frac{8}{3}$$

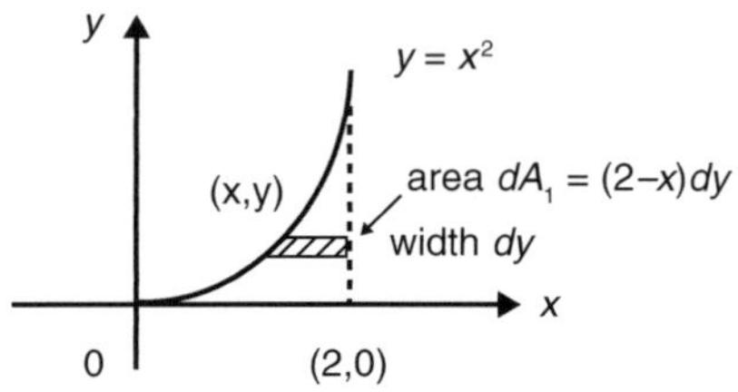

Exhibit 2.67

2.67 c. In Exhibit 2.67, area $dA_1 = (2 - x)dy$.

First moment with respect to x-axis of this area $dM_x = y\, dA_1$

First moment with respect to x-axis of shaded area is

$$M_x = \int dM_x = \int_{y=0}^{4} y(2-x)dy = \int_0^4 y(2-\sqrt{y})dy = 3.2$$

2.68 d. From Exhibit 2.66a, first moment with respect to y-axis of the shaded area dA_2 is $dM_y = x\, dA_2 = xy\, dx$; first moment with respect to y-axis of shaded area is

$$M_y = \int dM_y = \int_{x=0}^{2} x.y.dx = \int_0^2 x.x^2 dx = 4$$

2.69 a. x-coordinate $x_c = \dfrac{M_y}{area} = \dfrac{4}{2.67} = 1.5$; y-coordinate y_c

$$= \frac{M_x}{area} = \frac{3.2}{2.67} = 1.2$$

2.70 a. From Exhibit 2.67, moment of inertia with respect to x-axis of area dA_1 is $dMI_x = y^2 dA_1$. Moment of inertia with respect to x-axis of shaded area:

$$MI_x = \int dMI_x = \int y^2 dA_1 = \int_{y=0}^{4} y^2(2-x)dy = \int_0^4 y^2(2-\sqrt{y})dy$$

Integrating, $MI_x = 6.095$.

2.71 d. From Exhibit 2.66a, moment of inertia with respect to y-axis of area dA_2 is $dMI_y = x^2 dA_2$. Moment of inertia with respect to y-axis of shaded area:

$$MI_y = \int dMI_y = \int x^2 dA_2 = \int_{x=0}^{2} x^2.ydx = \int_{x=0}^{2} x^4 dx = 6.4$$

2.72 b. Homogeneous solution, $y_h(x) = e^{-4x}(C_1 \cos 4x + C_2 \sin 4x)$ or $e^{-4x}\, C_3 \cos(4x + \theta)$

2.73 d. Since the characteristic roots are –2 and –2, the particular solution is $y_p(x) = A\, x^2 e^{-2x}$.

2.74 **a.** The characteristic equation is $r^2 + 5r + 6 = 0$ and the characteristic roots are –2, –3. Then, the homogeneous solution is $y_h(x) = c_1e^{-2x} + c_2e^{-3x}$.

The particular solution due to e^{-2x} is $y_p(x) = Bxe^{-2x}$.

[As –2 is a characteristic root, e^{-2x} cannot be a particular solution.]

$y_p' = -2Be^{-2x}.x + Be^{-2x}$ and $y_p'' = 4Be^{-2x}.x - 2Be^{-2x} - 2Be^{-2x}$

$y_p'' + 5y_p' + 6y = 2e^{-2x}$ yields $B = 2$

$y(x) = 2e^{-2x}x + c_1e^{-2x} + c_2e^{-3x}$ $\quad y(0) = c_1 + c_2 = 0 \quad c_1 = -2$

$y'(x) = 2e^{-2x} - 4xe^{-2x} - 2c_1e^{-2x} - 3c_2 2e^{-3x}$ $\quad y'(0) = 2 - 2c_1 - 3c_2 = 0 \quad c_2 = 2$

The solution is $y = -2e^{-2x} + 2e^{-3x} + 2xe^{-2x}$.

2.75 **a.**

$$\lim_{x\to 1}\frac{x^2-1}{x-1} = \lim_{x\to 1}\frac{(x-1)(x+1)}{x-1} = \lim_{x\to 1}(x+1) = 2$$

2.76 **b.** This is linear equation, $y' + \frac{2}{x}y = \frac{1}{x}e^{3x}$. The integrating factor is $e^{\int \frac{2}{x}dx} = x^2$ so the equation becomes $d(x^2y) = xe^{3x}\,dx$. Integrating,

$x^2y = \frac{1}{3}xe^{3x} - \frac{1}{9}e^{3x} + c$, or $y = \frac{1}{3x}e^{3x} - \frac{1}{9x^2}e^{3x} + \frac{c}{x^2}$.

2.77 **a.** By inspection, $y_1 = e^x$ is one solution. Use reduction of order to obtain

$$\left(\frac{y_2}{y_1}\right)' = \frac{e^{\int \frac{-2(x+1)}{x}dx}}{y_1^2} = \frac{e^{2x+2\ln x}}{e^{2x}} = x^2$$

Hence $\frac{y_2}{y_1} = \frac{x^3}{3}$, so $y_2 = \frac{x^3}{3}y_1$. Suppressing the $\frac{1}{3}$, $y_2 = x^3e^x$.

2.78 **d.** The associated homogeneous equation, $y'' + 4y = 0$, has the solution $y_h = A\sin 2x + B\cos 2x$. Using the method of undetermined coefficients,

$$y_p = a\sin x + b\cos x$$
$$y_p'' = -a\sin x - b\cos x$$
$$y_p'' + 4y_p = (-a + 4a)\sin x + (-b + 4b)\cos x = 8\sin x$$
$$3a\sin x = 8\sin x$$
$$3b\cos x = 0\cos x$$

Thus, $b = 0$ and $a = \frac{8}{3}$.

2.79 d. Begin by eliminating c by solving for it from the derivative of the given equation: $2x + 2yy' = 2cy'$, $c = \dfrac{x + yy'}{y'}$,

$$x^2 + y^2 = 2\frac{x + yy'}{y'}y',\ x^2y' + y^2y' = 2xy + 2y^2y',\ \text{and}$$

$$y' = \frac{2xy}{x^2 - y^2}.$$

Now, the orthogonal family will have $y'_{\text{new}} = -\dfrac{1}{y'_{\text{old}}}$, so $y'_{\text{old}} = \dfrac{y^2 - x^2}{2xy}$.

Letting $u = \dfrac{y}{x}$, $xu' + u = \dfrac{u^2 - 1}{2u}$, $xu' = \dfrac{u^2 - 1 - 2u^2}{2u} = -\dfrac{1 + u^2}{2u}$,

and $\dfrac{2u\,du}{1 + u^2} = -\dfrac{dx}{x}$. Integrating, $\ln(1 + u^2) = -\ln|x| + c$, and

$$\ln\left\{\left[1 + \left(\frac{y}{x}\right)^2\right]|x|\right\} = c$$

$$\left|x + \frac{y^2}{x}\right| = e^c = c_1 > 0, \quad \text{or} \quad x + \frac{y^2}{x} = c_2\ (= \pm c_1)$$

Thus $x^2 + y^2 = c_2x$, or $x^2 + y^2 = 2cx$.

2.80 b. Using Laplace transform table, the transform is $\dfrac{(s + 1)}{(s + 1)^2 + 1^2}$.

2.81 c. Taking the Laplace transform of the differential equation and simplifying,

$$s^2Y(s) + 2sY(s) + Y(s) = \frac{s}{s^2 + 2^2} \Rightarrow Y(s) = \frac{s}{(s^2 + 4)(s^2 + 2s + 1)}$$

2.82 d. Initial value, $\lim\limits_{t \to 0} f(t) = \lim\limits_{s \to \infty} s\dfrac{(s + 5)}{s(s + 1)(s + 10)} = 0$

2.83 c. $y(k) = 2y(k - 1) + 3y(k - 2)$ with $y(-1) = 1$ and $y(-2) = 1$
For $k = 0$, $y(0) = 2y(-1) + 3y(-2) = 2(1) + 3(1) = 5$
For $k = 1$, $y(1) = 2y(0) + 3y(-1) = 2(5) + 3(1) = 13$

2.84 a. $y(k+1) + y(k) = u(k)$; taking Z-transform, $[z\,y(z) - 0] + y(z)$

$= \dfrac{1}{1 - z^{-1}}$. Simplifying, $y(z) = \dfrac{z}{(z - 1)(z + 2)}$.

2.85 d. Using the initial value theorem, the initial value,

$$\lim_{k \to 0} f(k) = \lim_{z \to \infty} \frac{z + 2}{z - 4} = 1$$

2.86 **c.** Period T = 4 s; $\omega_0 = \dfrac{2\pi}{T} = \dfrac{\pi}{2}$ rad/s;

$$b_3 = \frac{2}{4}\left[\int_0^2 8\sin\left(\frac{3\pi t}{2}\right)dt + \int_2^4 -8\sin\left(\frac{3\pi t}{2}\right)dt\right] = \frac{16}{3\pi}(1-\cos 3\pi) = 3.40$$

2.87 **d.** Since the pulse width is 2, $\tau = 1$ and from the Table, $X(f) = f\,\text{sinc}^2(f)$.

2.88 **c.** Using time shift theorem for Fourier transforms,

$X(f) = 1\exp(-j2\pi f) + 1\exp(+j2\pi f) = 2\cos(2\pi f)$

2.89 **c.** Substituting $k = 0$, $y(0) = 2y(-1) + 3y(-2) = 2 + 3 = 5$

Substituting $k = 1$, $y(1) = 2y(0) + 3y(-1) = 10 + 3 = 13$

2.90 **c.** $\text{Final value} = \lim_{k\to\infty} f(k) = \lim_{z\to 1}(1-z^{-1})F(z) = \lim_{z\to 1}(1-z^{-1})\dfrac{2(z+1)}{(z-1)} = 4$

CHAPTER 3

Probability and Statistics

David R. Arterburn, Wolfgang Baer, and Thyagarajan Srinivasan

OUTLINE

SETS AND SET OPERATIONS

A set is any well-defined list, collection, or class of objects. The objects in a set can be anything: numbers, letters, cards, people, and so on. They are called the **elements**, or **members**, of a set.

The name of a set is usually denoted by a capital letter, such as A, B, Y, Z. The elements of a set are usually denoted by small letters, such as a, b, y, z.

To specify that an element a is a member of a set B, we say "a is in B," which is written

$$a \in B$$

A set is called the **null**, or **empty**, set, denoted by ø, if it has no elements. We say the set A is a subset of the set B, written $A \subset B$, if all the members of A are also in B. The universal set, denoted by U, is the set that contains all the members of the subsets. The **complement** of a set A, denoted as A', consists of all the elements in U that are not in A.

Sets are defined in either of two ways: (1) by listing the members in a tabular form (for example, if A consists of the numbers 1, 3, 5, and 7, we write $A = \{1, 3, 5, 7\}$), or (2) by stating the properties that the members must satisfy in a set-builder form. For example, a set B containing all the odd numbers is written $B = \{x|x \text{ is odd}\}$, where x is an element of the set and the vertical line "|" is read "such that." The full notation is read "B is the set of numbers x such that x is odd."

Example 3.1

Define the set R of all outcomes of the roll of a six-sided die.

Solution

$R = \{1, 2, 3, 4, 5, 6\}$

Example 3.2

Let U be the set of integers. Define the complement of $B = \{x|x \text{ is odd}\}$.

Solution

$B' = \{x|x \text{ is even}\}$

Set Operations

The **union** of two sets A and B is defined as the set of all elements in either A or B and is traditionally written as $A \cup B$ or, alternatively, as A or B.

The **intersection** of two sets A and B is defined as the set of all elements in both A and B and is traditionally written $A \cap B$.

Set operations are well-behaved mathematically and follow these laws:

identity $\quad A \cup 0 = A \quad A \cup U = U$
$A \cap 0 = 0 \quad A \cap A = A$

complement $\quad A \cup A' = U \quad (A')' = A \quad A \cap A' = 0 \quad U' = 0$

commutative $\quad A \cup B = B \cup A \quad A \cap B = B \cap A$

associative $\quad (A \cup B) \cup C = A \cup (B \cup C)$
$(A \cap B) \cap C = A \cap (B \cap C)$

distributive $\quad A \cup (B \cap C) = (A \cup B) \cap (A \cup C)$
$A \cap (B \cup C) = (A \cap B) \cup (A \cap C)$

de Morgan's Law $\quad (A \cup B)' = A' \cap B'$
$(A \cap B)' = A' \cup B'$

For example, define the following sets:

$U = \{x \mid x \text{ is a person}\}$

$A = \{a \mid a \text{ is American}\}$

$F = \{f \mid f \text{ is French}\}$

$B = \{b \mid b \text{ is a person with dual French and American citizenship}\}$

Then the union of the sets A and F is the set of all people who are either American or French and is written $A \cup F$. The set of people who are not American is A'. The set of people who are not French is F'.

The set of people who are not French and not American is $A' \cap F'$. This is the same as $(A \cup F)'$, according to de Morgan's Law. In this example, de Morgan's Law means that all the people who are neither American nor French is the same as the set of all the people who are not French and not American.

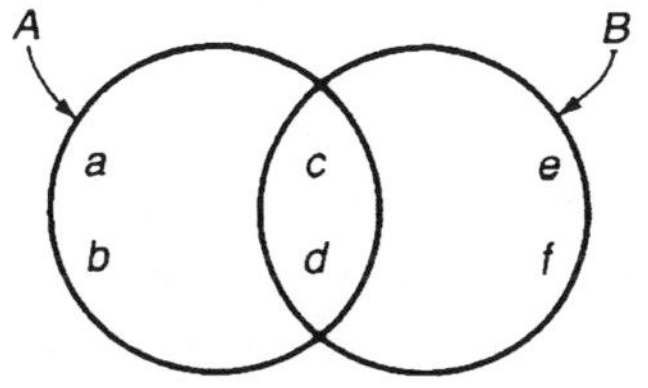

Figure 3.1

Venn Diagrams

A simple and intuitive way to represent these set relationships is known as the **Venn diagram**. Here we represent a set by a closed area and an element in the set by a point in the area. For example, let $A = \{a, b, c, d\}$ and $B = \{c, d, e, f\}$; the sets are represented in a **Venn diagram** as shown in Figure 3.1.

Relational properties such as the intersection of A and $B = \{c, d\}$ are intuitively understood as the overlap of two regions.

Example 3.3

Draw a Venn diagram of the relationship between the sets U, A, F, and B as defined in the previous section.

Solution

1. $A \subset U$. Americans are a subset of people (Exhibit 1).
2. $A \cup F$ is the set of Americans and French (Exhibit 2).
3. $B = A \cap F$ is the intersection of two sets; it represents Americans who are also French (Exhibit 3).

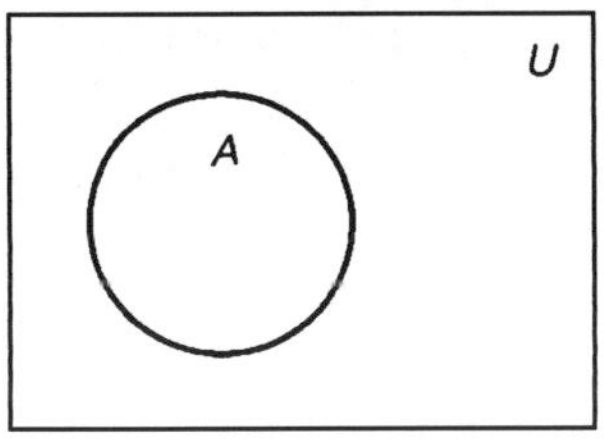

Exhibit 1

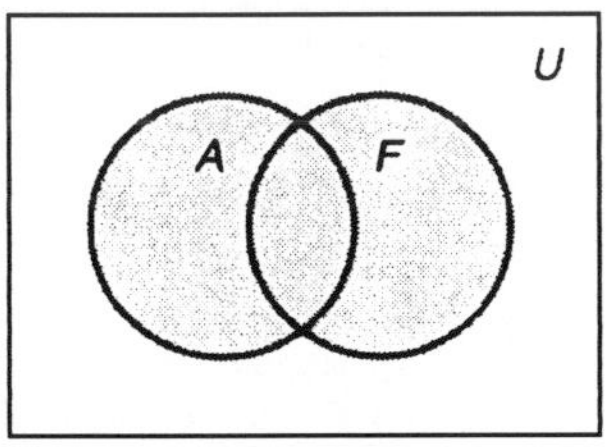

Exhibit 2

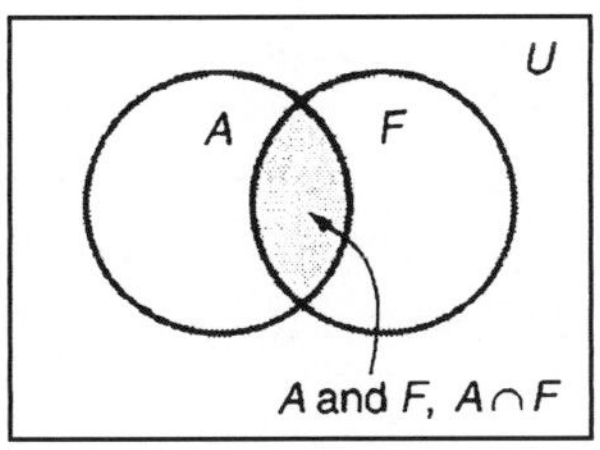

Exhibit 3

Product Sets

An ordered pair of elements a and b is denoted by (a, b). Two ordered pairs (a, b) and (c, d) are equal only if $a = c$ and $b = d$. For example, the ordered pairs (3, 4) and (4, 3) are different.

The product x of two sets A and B is denoted by $A \times B$ and consists of all ordered pairs of elements in A and B.

The two subsets of a product set can be treated like the intersection of two Cartesian coordinate axes. The subsets each act like a dimension, and the product set can be represented as a matrix where the elements of one subset are written along the row of the matrix and those of the other along the column.

For example, the set of results of a single coin toss is either heads (H) or tails (T) and is defined $R = \{H, T\}$. The result of two coin tosses is the product $R \times R$ and is defined as

$$R \times R = \{(H, H), (H, T), (T, H), (T, T)\}$$

COUNTING SETS

For discrete probability calculations it is important to count the number of elements in sets of possible outcomes. The primary method is simply to write down all the elements in a set and count them. For example, count the number of elements in the set S of all possible outcomes of a six-sided die throw. The set definition is $S = \{1, 2, 3, 4, 5, 6\}$. Simple counting gives six elements.

Most useful sets are large, and simple counting is too time-consuming. There are several methods for simplifying this task. One can use the product set concept introduced in the last section. If A is a set with n elements and B is a set with m elements, then the product set $A \times B$ has the arithmetic $n \times m$ number of elements. To simplify counting then, first count the sets making up the product set (usually containing a much smaller number of elements) and simply multiply these counts.

Example 3.4

Count the number of possible outcomes for tossing five coins.

Solution

The number of outcomes for a single toss defined by the set $R = \{H, T\}$ is 2. The result of five coin tosses is the product set $R \times R \times R \times R \times R$. The total number of possible outcomes is then the arithmetic product of the number of outcomes in each of the individual five tosses. This is $2 \times 2 \times 2 \times 2 \times 2 = 2^5 = 32$.

Permutations

If A is a set with n elements, a **permutation** of A is an ordered arrangement of A. Given the set $A = \{a, b, c\}$, the order a, b, c of the elements is one permutation. Any other order—for example, b, c, a—is another permutation.

The set B of all permutations of the set A is defined as the set of all arrangements of the three elements. These are

$$B = \{\{a, b, c\}, \{a, c, b\}, \{c, b, a\}, \{b, a, c\}, \{b, c, a\}, \{c, a, b\}\}$$

There are six permutations. This number also can be derived as follows. The number of ways an element can be chosen for the first space is three. Then there are two elements left. One of these can go in the second space. Then there is one element left. This must go in the third space. This gives the formula $3 \times 2 \times 1 = 6$. In general, the number of ways n distinct elements can be arranged is given by

$$n! = n \times (n-1) \times (n-2) \times \cdots 1$$

and is called the **factorial** of the number n. The factorial of 0 is 1 ($0! = 1$).

For example, count the number of ways a standard playing deck can be arranged. Since there are 52 distinct cards in a deck, there are 52! different arrangements, or permutations.

Now suppose we have the set of letters L in the word *obtuse* so that $L = \{o, b, t, u, s, e\}$. How many two-letter symbols could be made from this set? Notice that the letters are all distinct. We again count the number of ways the letters can be selected. For the first choice it is six; for the second choice it is five. The two selections are now complete. There are therefore $6 \times 5 = 30$ possibilities.

The general formula for the number of permutations, taking r items from a set of n, is given by

$$P(n, r) = n!/(n-r)!$$

Using this equation, one can express the previous example as $P(6, 2) = 6!/(6-2)! = 30$.

Example 3.5

A jeweler has nine different beads and a bracelet design that requires four beads. To find out which looks the best, he decides to try all the permutations. How many different bracelets will he have to try?

Solution

There are $n = 9$ beads. He selects $r = 4$ at a time. The order is important, because each arrangement of r beads on the bracelet makes a different bracelet. So the number of different bracelets is

$$P(9,4) = 9!/(9-4)! = 9 \times 8 \times 7 \times 6 = 3024$$

If the bracelet is a closed circle, there is no discernible difference when it is rotated. Then one observes four identical states for each unique bracelet. This is called ring permutation and is given by the formula

$$P_{\text{ring}}(n, r) = P(n, r)/r$$

There are only 3024/4 = 756 distinct ring bracelets the jeweler can make.

Example 3.6

(i) In how many ways can four people be asked to form a line of three people?
(ii) In how many ways can the letters of the word BEAUTY be arranged?
(iii) In how many ways can the letters of the word GOOD be arranged?

Solution

(i) $P(4,3) = \dfrac{4!}{(4-3)!} = 24$

(ii) $P(6,6) = \dfrac{6!}{(6-6)!} = 720$

(iii) $P(4;1,1,2) = \dfrac{4!}{1!1!2!} = 12$

Combinations

When the order of the set of r things that are selected from the set of n things does not matter, we talk about combinations.

Again consider the standard playing deck of 52 cards. How many hands of 5 cards can we get from a deck of 52 cards? Count the number of ways the hands can be drawn. The first draw can be any of the 52 cards. The second draw can only be one of the remaining 51 cards. The third draw can only be one of the remaining 50, the next is one of 49, the last one of 48. So the result is

$$52 \times 51 \times 50 \times 49 \times 48 = 52!/(52-5)!$$

This is the formula for permutations discussed in the last section. But the order in which we receive the cards is not important, so many of the hands are the same. In fact there are 5! similar arrangements of cards that make the same hand. The number of distinct hands is

$$(52 \times 51 \times 50 \times 49 \times 48)/(5 \times 4 \times 3 \times 2 \times 1) = 52!/[5! \times (52 - 5)!]$$

The general form r items taken from a set of n items when order is not important is written as the binomial coefficient $C(n, r)$, also written $\binom{n}{r}$, and is given by the formula

$$C(n, r) = \frac{n!}{r!(n-r)!}$$

Example 3.7

There are six skiers staying in a cabin with four bunks. How many combinations of people will be able to sleep in beds?

Solution

$C(6, 4) = 6!/[4! \times (6 - 4)!] = (6 \times 5 \times 4 \times 3 \times 2 \times 1)/[(4 \times 3 \times 2 \times 1) \times (2 \times 1)]$
$= 15$

PROBABILITY

Definitions

An **experiment**, or **trial**, is an action that can lead to a measurement.

Sampling is the act of taking a measurement. The **sample space** S is the set of all possible outcomes of an experiment (trial). An event e is one of the possible outcomes of the trial.

If an experiment can occur in n mutually exclusive and equally likely ways, and if m of these ways correspond to an event e, then the probability of the event is given by

$$P\{e\} = m/n$$

Example 3.8

A die is a cube of six faces designated as 1 through 6. The set of outcomes R of one die roll is defined as $R = \{1, 2, 3, 4, 5, 6\}$. If two dice are rolled, define trial, sample space, n, m, and the probability of rolling a seven when adding both dice together.

Solution

The trial is the rolling of two dice. The sample space is all possible outcomes of a two-dice roll, and the event is the outcome that the sum is 7.

The number of all possible outcomes, n, is the number of elements in the product set of the outcome of two dice when each is rolled independently. The product set is $R \times R$ and contains 36 elements.

The number of all possible ways, m, that the (7) event can occur is (1,6), (2,5), (3,4), (4,3), (5,2), and (6,1) for a total of six ways. The probability of rolling a 7 is $P\{7\} = \frac{6}{36} = \frac{1}{6}$.

Example 3.9

What is the probability of (i) a tail showing up when a fair coin is tossed, (ii) number 3 showing up when a fair die is tossed, and (iii) a red king is drawn from a deck of 52 cards.

Solution

(i) 1/2 (ii) 1/6 (iii) 2/52

General Character of Probability

The probability $P\{E\}$ of an event E is a real number in the range 0 through 1. Two theorems identify the range between which all probabilities are defined:

1. If ø is the null set, $P\{ø\} = 0$.
2. If S is the sample space, $P\{S\} = 1$.

The first states that the probability of an impossible event is zero, and the second states that, if an event is certain to occur, the probability is 1.

Complementary Probabilities

If E and E' are complementary events, $P\{E\} = 1 - P\{E'\}$. Complementary events are defined with respect to the sample space. The probability that an event E will happen is complementary to the probability that any of the other possible outcomes will happen.

Example 3.10

If the probability of throwing a 3 on a die is 1/6, what is the probability of not throwing a 3?

Solution

E is the probability of not throwing a 3, so $P\{E\} = 1 - P\{E'\} = 1 - \frac{1}{6} = \frac{5}{6}$.

Sometimes the complementary property of probabilities can be used to simplify calculations. This will happen when seeking the probability of an event that represents a larger fraction of the sample space than its complement.

Example 3.11

What is the probability $P\{E\}$ of getting at least one head in four coin tosses?

Solution

The complementary event $P\{E'\}$ to getting at least one head is getting no heads (or all tails) in four tosses. So the probability of getting at least one head is

$$P\{E\} = 1 - (0.5)^4 = 1 - 0.0625 = 0.9375$$

Joint Probability

The probability that a combination of events will occur is covered by joint probability rules. If E and F are two events, the joint probability is given by the rule

$$P\{E \cup F\} = P\{E\} + P\{F\} - P\{E \cap F\} \qquad \text{(Rule 1)}$$

A special case of the joint probability rule can be derived by considering two events, E and F, to be mutually exclusive. In this case the last term in Rule 1 is zero since $P\{E \cap F\} = P\{0\} = 0$. Thus, if E and F are mutually exclusive events,

$$P\{E \cup F\} = P\{E\} + P\{F\} \qquad \text{(Rule 2)}$$

Example 3.12

What is the probability of throwing a 7 or a 10 with two dice?

Solution

We will call the event of throwing a 7 A, and of throwing a 10 B. We know from previous examples that $P\{A\} = \frac{1}{6}$, and we can count outcomes to get $P\{B\} = \frac{1}{12}$. Applying the formula,

$$P\{A \cup B\} = P\{A\} + P\{B\} = \frac{1}{6} + \frac{1}{12} = \frac{1}{4}$$

If two events E and F are independent—that is, if they come from different sample spaces—then the probability that both will happen is given by the rule

$$P\{E \cap F\} = P\{E\} \times P\{F\} \qquad \text{(Rule 3)}$$

Example 3.13

What is the probability of throwing two heads in two coin tosses?

Solution

Call the throwing of one head E, the other F. The probability of throwing a single head is $P\{E\} = \frac{1}{2}$, and $P\{F\} = \frac{1}{2}$. The probability of throwing both heads is

$$P\{E \cap F\} = P\{E\} \times P\{F\} = \frac{1}{2} \times \frac{1}{2} = \frac{1}{4}$$

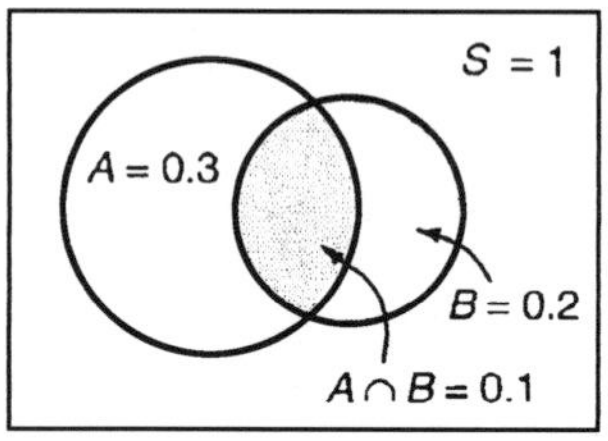

Figure 3.2
Venn diagram of joint probabilities

To visualize joint probabilities, we can use a Venn diagram showing two intersecting events, A and B, as shown in Figure 3.2. Let the normalized areas of each event represent the probability that the event will occur. For example, think of a random dart thrown at the Venn diagram: What are the chances of hitting one of the areas? Assume the areas correspond to probabilities and are given by $P\{S\} = 1$, $P\{A\} = 0.3$, $P\{B\} = 0.2$, and $P\{A \cap B\}$ is 0.6. The probability of hitting either area A or area B is calculated as the sum of the areas A and B minus the overlap area so it is not counted twice:

$$P\{A \cup B\} = 0.3 + 0.2 - .06 = .44$$

The result is also equal to the normalized area covered by A and B. The probability of hitting both A and B on one throw is simply the overlap area $P\{A \cap B\} = 0.1$.

If we throw two darts, the area S is used twice and represents two independent sample spaces. Hence Rule 3 applies.

Example 3.14

A die is tossed. Event A = {an odd number shows up}; event B = {a number > 4 shows up}.

(i) Find the probabilities, P(A) and P(B). (ii) What is the probability that either A or B or both occur?

Solution

(i) P(A) = P(1 or 3 or 5 showing up) = 3 (1/6) = 0.5
P(B) = P(5 or 6 showing up) = 2 (1/6) = 0.333

(ii) Event (A, B) = {5} and P(A, B) = 1/6;
then P(A + B) = P(A) + P(B) – P(AB) = 1/2 + 2/6 – 1/6 = 4/6
Check: Event (A + B) = {1,3,5,6}; then P(A + B) = 4/6

Conditional Probability

The conditional probability of an event E given an event F is denoted by $P\{E \mid F\}$ and is defined as

$$P\{E \mid F\} = P\{E \cap F\}/P\{F\} \qquad \text{for } P\{F\} \text{ not zero}$$

Example 3.15

Two six-sided dice, one red and one green, are tossed. What is the probability that the green die shows a 1, given that the sum of numbers on both dice is less than 4?

Solution

Let E be the event "green die shows 1" and let F be the event "sum of numbers shows less than four." Then

$$\begin{aligned} E &= \{(1,1), (1,2), (1,3), (1,4), (1,5), (1,6)\} \\ F &= \{(1,1), (1,2), (2,1)\} \\ E \cap F &= \{(1,1), (1,2)\} \\ P\{E \mid F\} &= P\{E \cap F\}/P\{F\} = (2/36)/(3/36) = 2/3 \end{aligned}$$

The generalized form of conditional probability is known as Bayes' theorem and is stated as follows: If $E_1, E_2, \ldots, E_n$ are n mutually exclusive events whose union is the sample space S, and E is any arbitrary event such that $P\{E\}$ is not zero, then

$$P\{E_k \mid E\} = \frac{P\{E_k\} \times P\{E \mid E_k\}}{\sum_{j=1}^{n} [P\{E_j\} \times P\{E \mid E_j\}]}$$

RANDOM VARIABLES

The method of random variables is a powerful concept. It casts the set-theory-based probability calculations of previous sections into a functional form and allows the application of standard mathematical tools to probability theory. It is often easy to solve fairly complex probability problems using random variables, although an approach different from the usual one is required.

A random variable, usually denoted by X, is a mapping of the sample space to some set of real numbers. The mapping transforms points of a sample space into points, or more accurately intervals, on the x-axis. The mapping, or random, variable X is called a discrete random variable if it assumes only a denumerable number of values on the x-axis. A random variable is called a continuous random variable if it assumes a continuum of values on the x-axis. The mapping is usually quite easy and intuitive for numerical events but provides no major advantage for nonnumerical discrete sample spaces, where counting remains the major tool.

Example 3.16

Cast the sample space of the outcomes of a roll of a die into random variable form.

Solution

The sample space is the set R defined by $R = \{1, 2, 3, 4, 5, 6\}$. These can easily by written along the x-axis as

$$R = \{1, 2, 3, 4, 5, 6\} \rightarrow 1 \mid 2 \mid 3 \mid 4 \mid 5 \mid 6 \mid x\text{-axis}$$

Probability Density Functions

A probability density function $f(x)$ is a mathematical rule that assigns a probability to the occurrence of the random variable x. Since the random variable is a mapping from trial outcomes, or events, to the numerical intervals on the x-axis, the probability that an event will occur is the area under the probability density function curve over the x interval defining the event.

For a continuous random variable the probability that an event E, mapped into an interval between x_1 and x_2, will occur is defined as

$$\int_{x_1}^{x_2} f(x)\,dx = P\{E\} \qquad \text{for } E \text{ mapped into } (x_1, x_2)$$

For a discrete case the formula is

$$\sum_{i=1}^{n} f(x_i) = P\{E\} \qquad \text{for } E \text{ containing } x_1, x_2, \ldots, x_n$$

It is assumed here that a step interval is associated with each value of x_i; therefore, the equivalent dx in the integral is 1 and is not required in the sum.

Example 3.17

The probability density function of a single six-sided die throw is shown graphically in Exhibit 4. The probability of throwing a 3 is given by the area under the curve over the interval assigned to the numeral 3, which is the step interval from 2.5 to 3.5.

Hence $P\{3\} = f(x) \times 1 = \frac{1}{6} \times 1 = \frac{1}{6}$.

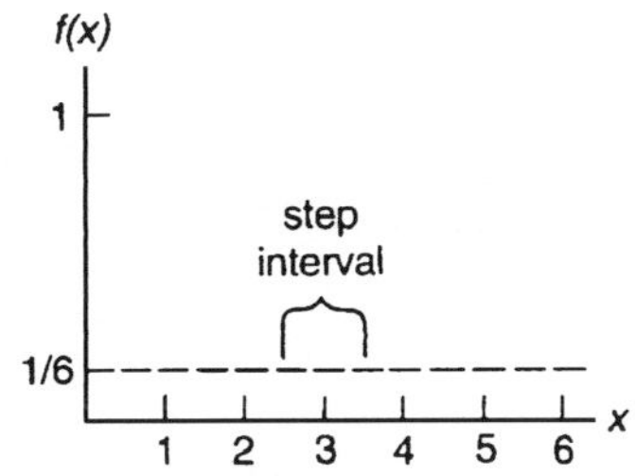

Exhibit 4

Example 3.18

A probability density function is defined as $f(x) = Ax^2$ for $-1 < x < 2$ and zero elsewhere. Find the value of A so that it is a valid density function.

Solution

For a valid density function, $\int_{-\infty}^{\infty} f(x)\,dx = 1$

Then, $\int_{-1}^{2} Ax^2\,dx = A\frac{9}{3} = 3A = 1$; then $A = \frac{1}{3}$

Properties of Probability Density Functions

The expected value $E\{X\}$ of a probability density function is also called the mean, and for a discrete case it is given by

$$E\{X\} = \sum x_i \times f(x_i) = u$$

The expected value of a continuous random variable is

$$E\{X\} = \int_{-\text{inf}}^{+\text{inf}} x \times f(x) \times dx = u$$

The expected value for a discrete random variable of a function $g(X)$ is given by

$$E\{g(X)\} = \sum g(x_i) \times f(x_i)$$

The expected value of a continuous random variable is

$$E\{g(X)\} = \int_{-\text{inf}}^{+\text{inf}} g(x) \times f(x) \times dx$$

Of special interest are the functions of the form

$$g(x) = (x - u)^r$$

These are the powers of the random variables around the mean. The expected values of these power functions are called the rth moments about the mean of the distribution, where r is the power. The second moment about the mean is also known as the variance and is calculated as follows:

$$V\{X\} = E\{(x-u)^2\} = E\{(x^2 - 2xu + u^2)\} = E\{x^2\} - E\{2xu\} + E\{u^2\}$$

Since u is a constant, the second term is $2u^2$ and the third term evaluates to u^2; therefore, the second moment about the mean becomes

$$V\{X\} = E\{x^2\} - u^2 = \sigma^2$$

The square root of the variance is signified by the Greek letter sigma and is called the **standard deviation**.

Example **3.19**

Calculate the mean and standard deviation of a single die throw.

Solution

This is a discrete function and can be calculated numerically by the discrete formulas given above. The mean, where $f(x_i) = \frac{1}{6}$ (all outcomes are equally likely), is given by

$$u = E\{X\} = \sum_{i=1}^{i=6} x_i \times f(x_i) = (1+2+3+4+5+6)/6 = \frac{21}{6} = 3.5$$

The standard deviation is given by

$$\sigma = \sqrt{V\{x\}} = \sqrt{E\{x^2\} - u^2} = \sqrt{[(1^2 + 2^2 + 3^2 + 4^2 + 5^2 + 6^2)/6] - 3.5^2} = 1.7$$

STATISTICAL TREATMENT OF DATA

Whether from the outcome of an experiment or trial, or simply the output of a number generator, we are constantly presented with numerical data. A statistical treatment of such data involves ordering, presentation, and analysis. The tools available for such treatment are generally applicable to a set of numbers and can be applied without much knowledge about the source of the data, although such knowledge is often necessary to make sensible use of the statistical results.

In its raw form, numerical data is simply a list of n numbers denoted by x_i, where $i = 1, 2, 3, \ldots, n$. There is no specific significance associated with the order implicit in the i numbers. They are names for the individuals in the list, although they are often associated with the order in which the raw data was recorded. For example, consider a box of 50 resistors. They are to be used in a sensitive circuit, and their resistances must be measured. The results of the 50 measurements are presented in the following table.

Table of Raw Measurements (Ω)				
101	105	110	115	82
86	91	96	117	112
109	103	89	97	98
101	104	99	95	97
85	90	94	112	107
103	94	98	106	98
114	112	108	101	99
93	96	99	104	90
109	106	101	93	92
104	99	109	100	107

Each number is named by the variable x_i, and there are $n = 50$ of them. The numbers range from 82 to 117.

Frequency Distribution

A systematic tool used in ordering data is the frequency distribution. The method requires counting the number of occurrences of raw numbers whose values fall within step intervals. The step intervals (or bins) are usually chosen to (1) be of constant size, (2) cover the range of numbers in the raw data, (3) be small enough in quantity to limit the amount of writing yet not have many empty steps, and (4) be sufficient in quantity so that significant information is not lost.

For example, the aforementioned raw data of measured resistances may be ordered in a frequency distribution table such as Table 3.1. Here the step interval is the event E of a random variable that can be mapped onto the x-axis. The set of eight events is the sample space. If we take a number randomly from the raw measurement set, the probability that it will be in bin 5 is

$$f(E_5) = P\{E_5\} = 10/50 = 0.2$$

Table 3.1 Frequency and cumulative frequency table

Event, E_i	Range, Ω	Frequency	Cumulative Frequency	Probability Density Function, $f(E_i)$
1	80–84	1	1	0.02
2	85–89	3	4	0.06
3	90–94	8	12	0.16
4	95–99	12	24	0.24
5	100–104	10	34	0.20
6	105–109	9	43	0.18
7	110–114	5	48	0.10
8	115–119	2	50	0.04

The last column in Table 3.1 is the probability density function of the distribution. The probability table can be plotted along the x-axis in several ways, as shown in Figures 3.3 through 3.5.

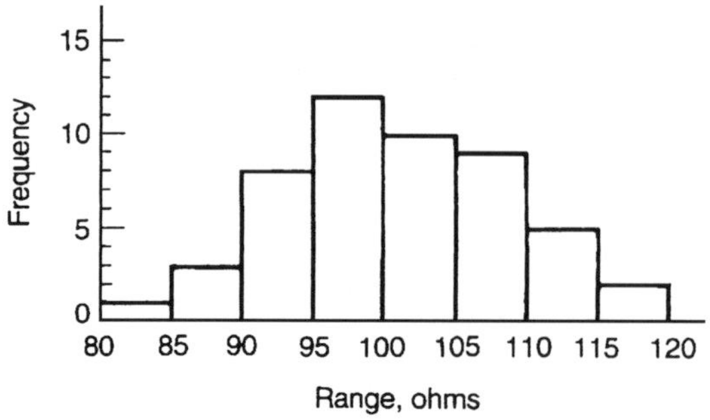

Figure 3.3 Histogram of resistance measurements

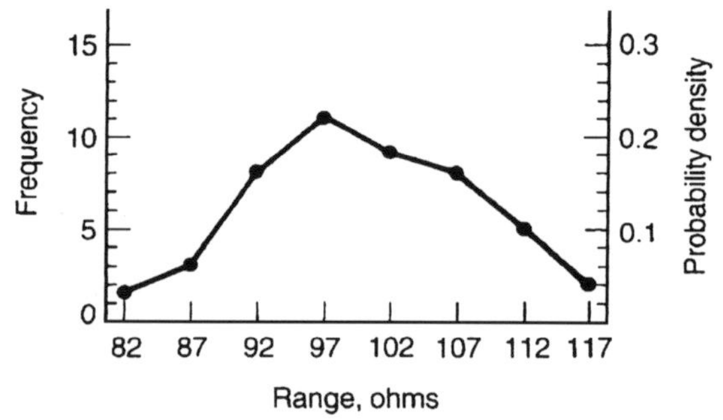

Figure 3.4 Frequency distribution and probability density plot

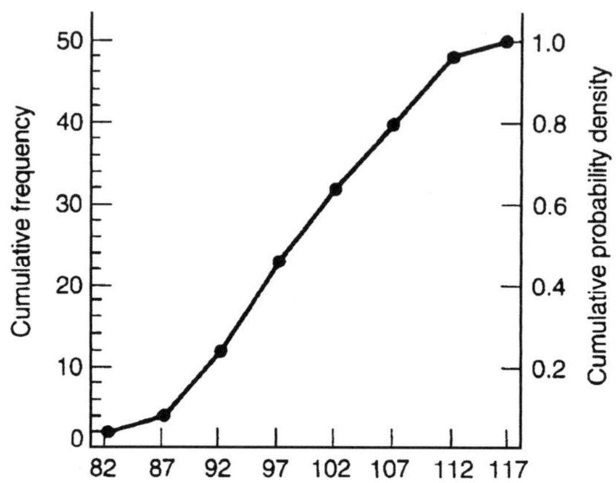

Figure 3.5 Cumulative frequency distribution and cumulative probability density

Standard Statistical Measures

There are several statistical quantities that can be calculated from a set of raw data and its distribution function. Some of the more important ones are listed here, together with the method of their calculation.

Mode The observed value that occurs most frequently; here the mode is bin 4 with a range of 95–99 Ω.

Median The point in the distribution that divides the number of observations such that half of the observations are above and half are below. The median is often the mean of the two middle values; here the median is 4.5 bins, 100 Ω.

Mean The arithmetic mean, or average, is calculated from raw data as

$$\mu = \frac{1}{n}\sum_{i=1}^{n} x_i = 100.6$$

It is calculated from the distribution function as

$$\mu = \sum_{i=1}^{m} b_i \times f(E_i) = 100.4$$

where b_i is the ith event value (for i = 1, b_i = 82) and m is the number of bins; $f(E_i)$ is the probability density function. (The two averages are not quite the same because of the information lost in assigning the step intervals.)

Standard deviation

(a) Computational form for the raw data:

$$\sigma = \sqrt{\frac{1}{n}\left[\left(\sum_{i=1}^{n} x_i^2\right) - n \times \mu^2\right]} = 8.08$$

(b) Computational form for the distribution function:

$$\sigma = \sqrt{\left\{\left[\sum_{i=1}^{m} b_i^2 \times f(E_i)\right] - \mu^2\right\}} = 8.02$$

Sample standard deviation

If the data set is a sample of a larger population, then the sample standard deviation is the best estimate of the standard deviation of the larger population.

The computational form for the raw data set is

$$\sigma = \sqrt{\frac{1}{n-1}\left[\left(\sum_{i=1}^{n} x_i^2\right) - n \times \mu^2\right]} = 8.166$$

Sample standard deviations and the use of $(n - 1)$ in the denominator are discussed in the section on sampling.

Skewness

This is a measure of the frequency distribution asymmetry and is approximately

skewness ≅ 3(mean – median)/(standard deviation)

Example 3.20

Two professors give the following scores to their students. What is the mode and arithmetic mean?

Frequency	1	3	6	11	13	10	2
Score	35	45	55	65	75	85	95

Solution

mode = 75; N = 1 + 3 + 6 + 11 + 13 + 10 + 2 = 46

weighted arithmetic mean = $\overline{X_w} = [35(1) + 45(3) + \ldots\ 95(2)]\,/\,46 = 70$

STANDARD DISTRIBUTION FUNCTIONS

In the previous section, we calculated several general properties of probability distribution functions.

To know the appropriate probability density function for an actual situation, two general methods are available:

1. The probability density function is actually calculated, as was done in the last section, by analyzing the physical mechanism by which experimental events and outcomes are generated and counting the number of ways an individual event occurs.
2. Recognition of an overall similarity between the present experiment and another for which the probability density function is already known permits the known behavior of the function to be applied to the new experiment. This work-saving method is by far the more popular one. Of course, to apply this method, it is necessary to have a repertoire of known probability functions and to understand the problem characteristics to which they apply.

This section lists several popular probability density functions and their characteristics.

Binomial Distribution

The binomial distribution applies when there is a set of discrete binary alternative outcomes. Deriving this distribution function helps one understand the class of problems to which it applies. For example, given a set of n events, each with a probability p of occurring, what is the probability that r of the events will occur and $(n - r)$ not occur?

The probability of one event occurring is p.

The probability of r events occurring is p^r.

The probability of $(n - r)$ events not occurring is $(1 - p)^{n-r}$.

The probability of exactly r events occurring and $(n - r)$ not occurring in a trial is given by the joint probability Rule 3:

$$P[r \cap (n-r)] = p^r \times (1-p)^{n-r}$$

However, there are many ways of choosing r occurrences out of n events. In fact, the number of different ways of choosing r items from a set of n items when order is not important is given by the binomial coefficient $C(n, r)$. The total probability of r occurrences from n trials, given an individual probability of occurrence as p, is thus given by

$$C(n,r) \times p^r \times (1-p)^{n-r} = f(r)$$

This is the **binomial probability density function**.

The mean of this density function is the first moment of the density function, or expected value, and is calculated as

$$E\{x\} = \sum_{r=0}^{n} r \times f(r) = \sum_{r=0}^{n} r \times \frac{n!}{(r)!(n-r)!} \times p^r \times (1-p)^{n-r}$$

This can be rewritten as

$$\sum_{r=1}^{n} \frac{n!}{(r-1)!(n-r)!} \times p^r \times (1-p)^{n-r}$$

We can now factor out the quantity $n \times p$ and let $r - 1 = y$. This can be rewritten as

$$n \times p \times \sum_{y=0}^{n-1} \frac{(n-1)!}{(y)!(n-1-y)!} \times p^y \times (1-p)^{n-1-y} = n \times p \times [p + (1-p)]^{n-1}$$

Since the sum is merely the expansion of a binomial raised to a power, and the number 1 raised to any power is 1, the mean is

$$\mu = n \times p$$

A similar calculation shows the variance is

$$\text{var} = n \times p \times (1-p)$$

The standard deviation is

$$\sigma = \sqrt{\text{var}} = \sqrt{n \times p \times (1-p)}$$

Example **3.21**

A truck carrying dairy products and eggs damages its suspension and 5% of the eggs break.

(i) What is the probability that a carton of 12 eggs will have exactly one broken egg?

(ii) What is the probability that one or more eggs in a carton will be broken?

Solution

(i) Since an egg is either broken or not broken, the binomial distribution applies. The probability p that an egg is broken is 0.05 and that one is not broken is $(1 - p) = 0.95$. From the equation for the binomial distribution, with $n = 12$ and $r = 1$,

$$p\{1\} = f(1) = C(12, 1) \times 0.05^1 \times 0.95^{11} = 12 \times 0.05 \times 0.57 = 0.34$$

(ii) The probability that one or more eggs will be broken can be calculated as the sum of each individual probability:

$$p\{x > 0\} = p\{1\} + p\{2\} + \cdots + p\{12\}$$

However, this requires 12 calculations. The problem can also be solved using the complementary rule:

$$p\{x > 0\} = 1 - p\{0\} = C(12, 0) \times 0.05^0 \times 0.95^{12} = 0.95^{12} = 0.54$$

Example 3.22

A biased coin is tossed. Find the probability that a head appears once in three trials.

P(Head) = p = 0.6

Solution

Here q = P(Head not occurring) or P(Tail) = 1 – 0.6 = 0.4.

Then P(1 Head) = C(3,1) $0.6^1\ 0.4^2$ = 0.2880.

Normal Distribution Function

The normal distribution, or Gaussian distribution, is widely used to represent the distribution of outcomes of experiments and measurements. It is popular because it can be derived from a few empirical assumptions about the errors presumed to cause the distribution of results about the mean. One assumption is that the error is the result of a combination of N elementary errors, each of magnitude e and equally likely to be positive or negative. The derivation then assumes $N \to \infty$ and $e \to 0$ in such a way as to leave the standard deviation constant. This error model is universal, since most experiments are analyzed to eliminate systematic errors. What remains is attributable to errors that are too small to explain systematically, so the normal probability distribution is evoked.

The form of the probability density and distribution functions for the **normal distribution** with a mean μ and variance σ^2 is given by

$$f(x) = \frac{e^{-(x-\mu)^2/2\sigma^2}}{\sigma\sqrt{2\pi}} \qquad -\infty < x < \infty$$

$$F(x) = \int_{-\infty}^{x} \frac{e^{-(x-\mu)^2/2\sigma^2}}{\sigma\sqrt{2\pi}}\, dt$$

The normal distribution is the typical bell-shaped curve shown in Figure 3.6. Here we see that the curve is symmetric about the mean μ. Its width and height are determined by the standard deviation σ. As σ increases, the curve becomes wider and lower.

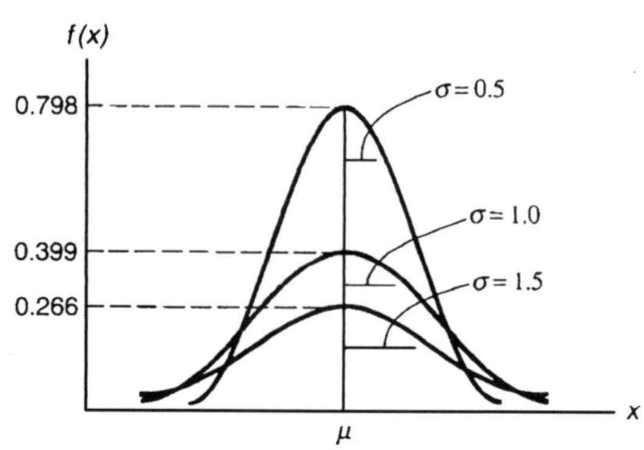

Figure 3.6 Normal distribution curve

Since this function is difficult to integrate, reference tables are used to calculate probabilities in a standard format; then the standard probabilities are converted to the actual variable required by the problem. The relation between the standard variable, z, and a typical problem variable, x, is

$$z = (x - \mu)/\sigma$$

Since μ and σ are constants, the standard probability at a value z is the same as the problem probability for the value at x.

The standard probability density function is

$$f(z) = \frac{1}{\sqrt{2\pi}} \times e^{-z^2/2}$$

Table 3.2 Standard probability table

z	$F(z)$	$f(z)$	z	$F(z)$	$f(z)$
0.0	0.5000	0.3989	2.0	0.9773	0.0540
0.1	0.5398	0.3970	2.1	0.9821	0.0440
0.2	0.5793	0.3910	2.2	0.9861	0.0355
0.3	0.6179	0.3814	2.3	0.9893	0.0283
0.4	0.6554	0.3683	2.4	0.9918	0.0224
0.5	0.6915	0.3521	2.5	0.9938	0.0175
0.6	0.7257	0.3332	2.6	0.9953	0.0136
0.7	0.7580	0.3123	2.7	0.9965	0.0104
0.8	0.7881	0.2897	2.8	0.9974	0.0079
0.9	0.8159	0.2661	2.9	0.9981	0.0060
1.0	0.8413	0.2420	3.0	0.9987	0.0044
1.1	0.8643	0.2179	3.1	0.9990	0.0033
1.2	0.8849	0.1942	3.2	0.9993	0.0024
1.3	0.9032	0.1714	3.3	0.9995	0.0017
1.4	0.9192	0.1497	3.4	0.9997	0.0012
1.5	0.9332	0.1295	3.5	0.9998	0.0009
1.6	0.9452	0.1109	3.6	0.9998	0.0006
1.7	0.9554	0.0940	3.7	0.9999	0.0004
1.8	0.9641	0.0790	3.8	0.9999	0.0003
1.9	0.9713	0.0656	3.9	1.0000	0.0002
			4.0	1.0000	0.0001

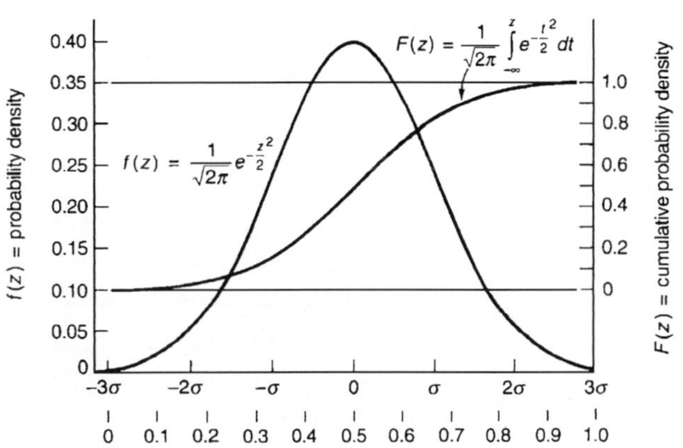

Figure 3.7 Normal error curve

The standard cumulative distribution function is

$$F(z) = \int_{-\infty}^{z} \frac{1}{\sqrt{2\pi}} \times e^{-t^2/2} \times dt$$

The standard probability function is shown graphically in Figure 3.7, and Table 3.2 shows the corresponding numerical values. The standard probability curve is symmetric about the origin and is given in terms of unit *sigma*. To use the table, remember that the function $F(z)$ is the area under the probability curve from minus infinity to the value z. The area under the curve up to $x = 0$ is therefore 0.5. Also, from symmetry,

$$F(-z) = 1 - F(z)$$

Example **3.23**

Find the probability that the standard variable z lies within (i) 1σ, (ii) 2σ, and (iii) 3σ of the mean.

Solution

(i) The probability is $P_1 = F(1.0) - F(-1.0)$. From the symmetry of F, $F(-1.0) = 1 - F(1.0)$, so

$$\begin{aligned} P_1 &= 2F(1.0) - 1 = 2(0.8413) - 1 \\ &= 0.6826 \end{aligned}$$

(ii) In this case, the probability is

$$\begin{aligned} P_2 &= F(2.0) - F(-2.0) \\ &= F(2.0) - [1 - F(2.0)] \\ &= 2F(2.0) - 1 = 2(0.9773) - 1 \\ &= 0.9546 \end{aligned}$$

(iii) In the same way,

$$\begin{aligned} P_3 &= 2F(3.0) - 1 \\ &= 2(0.9987) - 1 \\ &= 0.9974 \end{aligned}$$

Example **3.24**

A Gaussian random variable has a mean of 1830 and standard deviation of 460. Find the probability that the variable will be more than 2750.

Solution

$P(X > 2750) = 1 - P(X \leq 2750) = 1 - F[(2750 - 1830)/460] = 1 - F(2.0) = 1 - 0.9772 = 0.0228$

t-Distribution

The t-distribution is often used to test an assumption about a population mean when the parent population is known to be normally distributed but its standard deviation is unknown. In this case, the inferences made about the parent mean will depend upon the size of the samples being taken.

It is customary to describe the t-distribution in terms of the standard variable t and the number of degrees of freedom ν. The number of degrees of freedom is a measure of the number of independent observations in a sample that can be used to estimate the standard deviation of the parent population; the number of degrees of freedom ν is one less than the sample size ($\nu = n - 1$).

The density function of the t-distribution is given by

$$f(t) = \frac{\Gamma\left(\frac{\nu+1}{2}\right)}{\sqrt{\nu\pi}\,\Gamma\left(\frac{\nu}{2}\right)\left(1+t^2/\nu\right)^{(\nu+1)/2}}$$

and is provided in Table 3.3. The mean is $m = 0$, and the standard deviation is

$$\sigma = \sqrt{\frac{\nu}{\nu - 2}}$$

Table 3.3 t-Distribution; values of $t_{\alpha,\nu}$

Degrees of Freedom, ν	Area of the Tail				
	α = 0.10	α = 0.05	α = 0.025	α = 0.01	α = 0.005
1	3.078	6.314	12.706	31.821	63.657
2	1.886	2.920	4.303	6.965	9.925
3	1.638	2.353	3.182	4.541	5.841
4	1.533	2.132	2.776	3.747	4.604
5	1.476	2.015	2.571	3.365	4.032
6	1.440	1.943	2.447	3.143	3.707
7	1.415	1.895	2.365	2.998	3.499
8	1.397	1.860	2.306	2.896	3.355
9	1.383	1.833	2.262	2.821	3.250
10	1.372	1.812	2.228	2.764	3.169
11	1.363	1.796	2.201	2.718	3.106
12	1.356	1.782	2.179	2.681	3.055
13	1.350	1.771	2.160	2.650	3.012
14	1.345	1.761	2.145	2.624	2.977
15	1.341	1.753	2.131	2.602	2.947
16	1.337	1.746	2.120	2.583	2.921
17	1.333	1.740	2.110	2.567	2.898
18	1.330	1.734	2.101	2.552	2.878
19	1.328	1.729	2.093	2.539	2.861
20	1.325	1.725	2.086	2.528	2.845
21	1.323	1.721	2.080	2.518	2.831
22	1.321	1.717	2.074	2.508	2.819
23	1.319	1.714	2.069	2.500	2.807
24	1.318	1.711	2.064	2.492	2.797
25	1.316	1.708	2.060	2.485	2.787
26	1.315	1.706	2.056	2.479	2.779
27	1.314	1.703	2.052	2.473	2.771
28	1.313	1.701	2.048	2.467	2.763
29	1.311	1.699	2.045	2.462	2.756
inf.	1.282	1.645	1.960	2.326	2.576

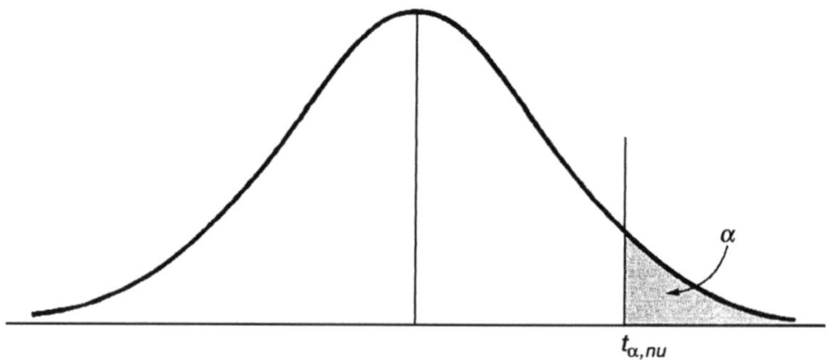

Figure 3.8

Probability questions involving the t-distribution can be answered by using the distribution function $t_{\alpha,\nu}$ shown in Figure 3.8. Table 3.3 gives the value of t as a function of the degrees of freedom v down the column and the area (α) of the tail across the top. The t-distribution is symmetric. As an example, the probability of t falling within ±3.0 when a sample size of 8 ($\nu = 7$) is selected is one minus twice the tail ($\alpha = 0.01$):

$$P\{-3.0 < t < 3.0\} = 1 - (2 \times 0.01) = 0.98$$

The t-distribution is a family of distributions that approaches the Gaussian distribution for large n.

X² DISTRIBUTION

In probability theory and statistics, the chi-square distribution (also chi-squared or x^2-distribution) is one of the most widely used theoretical probability distributions in inferential statistics (e.g., in statistical significance tests). It is useful because, under reasonable assumptions, easily calculated quantities can be proven to have distributions that approximate to the chi-square distribution if the null hypothesis is true.

The best known situations in which the chi-square distribution is used are the common chi-square tests for goodness of fit of an observed distribution to a theoretical one, and of the independence of two criteria of classification of qualitative data. Many other statistical tests also lead to a use of this distribution.

If $Z_1, Z_2, \ldots, Z_n$ are independent unit normal random variables, then

$$\chi^2 = Z_1^2 + Z_2^2 + \ldots + Z_n^2$$

is said to have a chi-square distribution with n degrees of freedom.

PROBLEMS

3.1 Define the set of all outcomes for the roll of two dice.

3.2 Draw a Venn diagram showing the following:

The universal set of all people in the United States as *U*

All the males as *M*

All the females as *F*

All the students of both sexes as *S*

All the students with grades above "B" as *A*

3.3 What is the probability of drawing a pair of aces in two cards when an ace has been drawn on the first card?

a. 1/13 c. 3/51
b. 1/26 d. 4/51

3.4 An auto manufacturer has three plants (A, B, C). Four out of 500 cars from Plant A must be recalled, 10 out of 800 from Plant B, and 10 out of 1000 from Plant C. Now a customer purchases a car from a dealer who gets 30% of his stock from Plant A, 40% from Plant B, and 30% from Plant C, and the car is recalled. What is the probability it was manufactured in Plant A?

a. 0.0008 c. 0.0125
b. 0.01 d. 0.2308

3.5 There are ten defectives per 1000 times of a product. What is the probability that there is one and only one defective in a random lot of 100?

a. 99×0.01^{99} c. 0.5
b. 0.01 d. 0.99^{99}

3.6 The probability that both stages of a two-stage missile will function correctly is 0.95. The probability that the first stage will function correctly is 0.98. What is the probability that the second stage will function correctly given that the first one does?

a. 0.99 c. 0.97
b. 0.98 d. 0.95

3.7 A standard deck of 52 playing cards is thoroughly shuffled. The probability that the first four cards dealt from the deck will be the four aces is closest to:

a. 2.0×10^{-1} c. 4.0×10^{-4}
b. 8.0×10^{-2} d. 4.0×10^{-6}

3.8 In statistics, the standard deviation measures:

a. a standard distance c. central tendency
b. a normal distance d. dispersion

3.9 There are three bins containing integrated circuits (ICs). One bin has two premium ICs, one has two regular ICs, and one has one premium IC and one regular IC. An IC is picked at random. It is found to be a premium IC. What is the probability that the remaining IC in that bin is also a premium IC?

a. $\frac{1}{5}$ c. $\frac{1}{3}$

b. $\frac{1}{4}$ d. $\frac{2}{3}$

3.10 How many teams of four can be formed from 35 people?

a. about 25,000 c. about 50,000

b. about 2,000,000 d. about 200,000

3.11 A bin contains 50 bolts, 10 of which are defective. If a worker grabs 5 bolts from the bin in one grab, what is the probability that no more than 2 of the 5 are bad?

a. about 0.5 c. about 0.90

b. about 0.75 d. about 0.95

3.12 How many three-letter codes may be formed from the English alphabet if no repetitions are allowed?

a. 26^3 c. $26 \times 25 \times 24$

b. 26/3 d. $26^3/3$

3.13 A widget has three parts, A, B, and C, with probabilities of 0.1, 0.2, and 0.25, respectively, of being defective. What is the probability that exactly one of these parts is defective?

a. 0.375 c. 0.95

b. 0.55 d. 0.005

3.14 If three students work on a certain math problem, student A has a probability of success of 0.5; student B, 0.4; and student C, 0.3. If they work independently, what is the probability that no one works the problem successfully?

a. 0.12 c. 0.32

b. 0.25 d. 0.21

3.15 A sample of 50 light bulbs is drawn from a large collection in which each bulb is good with a probability of 0.9. What is the approximate probability of having less than 3 bad bulbs in the 50?

a. 0.1 c. 0.3

b. 0.2 d. 0.4

3.16 The number of different 3-digit numbers that can be formed from the digits 1, 2, 3, 7, 8, 9 without reusing the digits is:

a. 10 c. 30

b. 20 d. 40

3.17 The number of different ways that a party of seven councilmen can be seated in a row is:

a. 1 c. 2080
b. 560 d. 5040

3.18 A student must answer six out of eight questions on an exam. The number of different ways in which he can do the exam is:

a. 8 c. 28
b. 18 d. 48

3.19 Repeat Problem 3.18 if the first two questions are mandatory.

a. 4 c. 12
b. 8 d. 15

3.20 A group of five women wishes to form a subcommittee consisting of two of them. The number of possible ways to do so is:

a. 5 c. 15
b. 10 d. 20

3.21 An integer has to be chosen from numbers between 1 and 100 (both inclusive). The probability of choosing a number divisible by 9 (with a remainder of 0) is:

a. 0 c. 0.11
b. 0.01 d. 0.91

3.22 Four fair coins are tossed. The probability of either one head or two heads showing up is:

a. 1/8 c. 4/8
b. 2/8 d. 5/8

3.23 Two identical bags contain ten apples and five oranges each. The probability of selecting an apple from the first bag and an orange from the second bag is:

a. 1/9 c. 3/9
b. 2/9 d. 4/9

3.24 A bag contains 5 red, 10 orange, 15 green, 20 violet, and 25 black cards. The probability that you will get a black card or a red card if you remove a card from the bag is:

a. 5/85 c. 25/85
b. 15/85 d. 35/85

3.25 Two bags each contain two orange balls, five white balls, and three red balls. The probability of selecting an orange ball from the first bag or a white ball from the other bag is:

a. 0 c. 0.6
b. 0.2 d. 1.0

3.26 A bag contains 100 balls numbered 1 to 100. One ball is drawn from the bag. What is the probability that the number on the ball will be even or greater than 72?

a. 0.64 c. 0.28
b. 0.50 d. 0.14

3.27 A circuit has two switches connected in series. For a signal to pass through, both switches must be closed. The probability that the first switch is closed is 0.95, and the probability that a signal passes through is 0.90. The probability that the second switch is closed is:

a. 0.8545 c. 0.9474
b. 0.9000 d. 0.9871

3.28 Four probability density distributions are shown in Exhibit 3.28. The only valid distribution is:

a. $f_1(x)$ c. $f_3(x)$
b. $f_2(x)$ d. $f_4(x)$

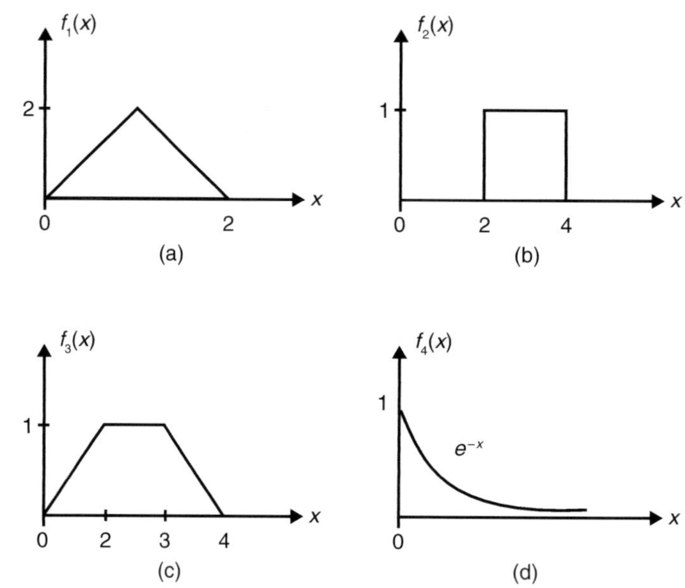

Exhibit 3.28

3.29 Four probability density distributions are shown in Exhibit 3.29. The only valid distribution is:

a. $f_1(x)$ c. $f_3(x)$
b. $f_2(x)$ d. $f_4(x)$

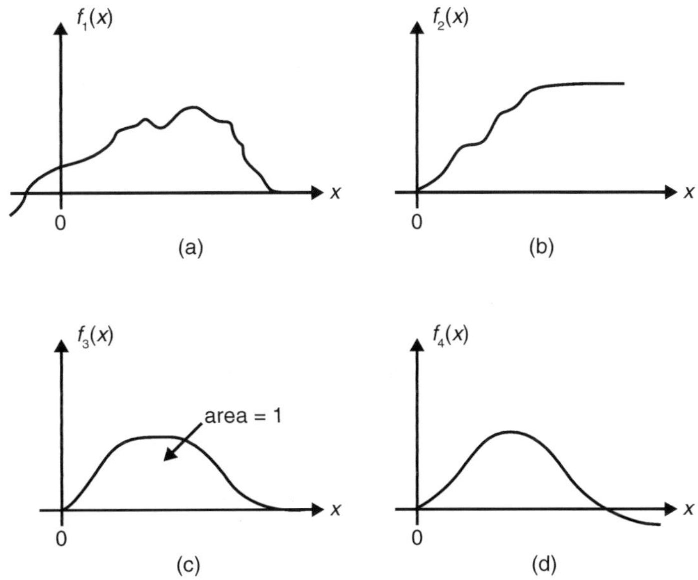

Exhibit 3.29

3.30 Four probability density distributions are shown in Exhibit 3.30. The only valid distribution is:

a. $f_1(x)$ c. $f_3(x)$
b. $f_2(x)$ d. $f_4(x)$

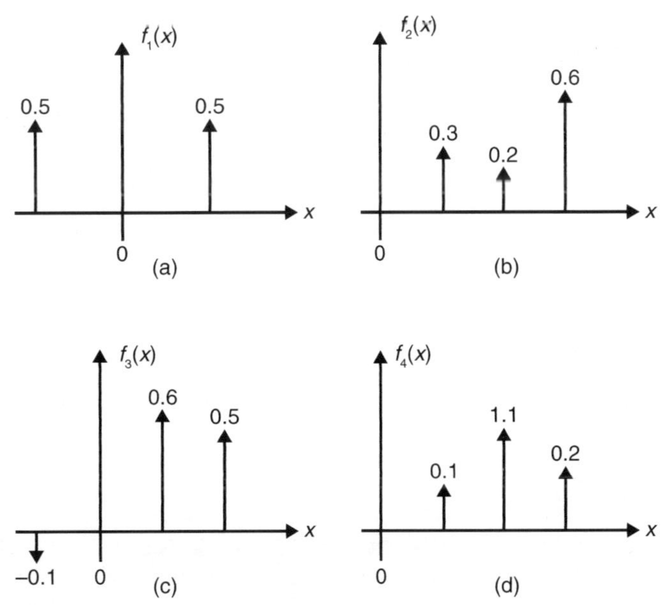

Exhibit 3.30

3.31 Four cumulative probability distribution functions are shown in Exhibit 3.31. The only valid distribution is:

a. $F_1(x)$ c. $F_3(x)$
b. $F_2(x)$ d. $F_4(x)$

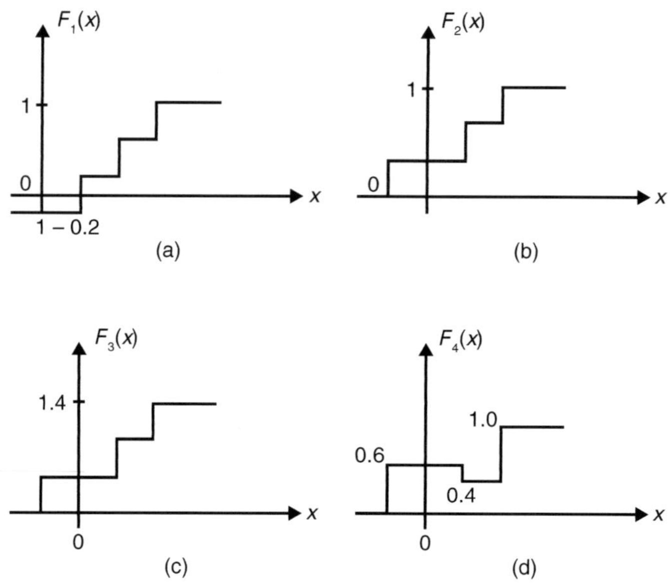

Exhibit 3.31

3.32 A coin is weighted so that heads is twice as likely to appear as tails. The probability that two heads occur in four tosses is:

a. 0.15 c. 0.45
b. 0.35 d. 0.75

3.33 It is given that 20% of all employees leave their jobs after one year. A company hired seven new employees. The probability that nobody will leave the company after one year is:

a. 0.1335 c. 0.3815
b. 0.2315 d. 0.6510

3.34 If four fair coins are tossed simultaneously, the probability that at least one head appears is:

a. 0.1335 c. 0.7815
b. 0.5635 d. 0.9375

3.35 For unit normal distribution, the probability that $(x > 3)$ is:

a. 0.0013 c. 0.1807
b. 0.0178 d. 0.5402

3.36 Scores in a particular game have a normal distribution with a mean of 30 and a standard deviation of 5. Contestants must score more than 26 to qualify for the finals. The probability of being disqualified in the qualifying round is:

a. 0.121 c. 0.304
b. 0.212 d. 0.540

3.37 The radial distance to the impact points for shells fired by a cannon is approximated by a normal Gaussian random variable with a mean of 2000 m and standard deviation of 40 m. When a target is located at 1980 m distance, the probability that shells will fall within ± 68 m of the target is:

a. 0.2341 c. 0.5847
b. 0.3248 d. 0.8710

3.38 The chance of a car being stolen from a residential area is 1 in 120. In one area there are five cars parked in front of the houses. The probability that none will be stolen is:

a. 0.0131 c. 0.5847
b. 0.3248 d. 0.9590

3.39 The standard deviation of the sequence 3, 4, 4, 5, 8, 8, 8, 10, 11, 15, 18, 20 is:

a. 5.36 c. 15.62
b. 9.35 d. 28.75

3.40 Weighted arithmetic mean of the following 50 data points is:

Frequency	3	8	18	12	9
Score	1.5	2.5	3.5	4.5	5.5

a. 1.56 c. 5.62
b. 3.82 d. 8.75

SOLUTIONS

3.1 We will write this in ordered pairs:

$$\begin{aligned}R = \{&(1,1),\\ &(1,2), (2,1),\\ &(1,3), (3,1), (2,2),\\ &(1,4), (4,1), (2,3), (3,2)\\ &(1,5), (5,1), (2,3), (3,2), (3,3),\\ &(1,6), (6,1), (5,2), (2,5), (3,4), (4,3),\\ &(2,6), (6,2), (5,3), (3,5), (4,4),\\ &(3,6), (6,3), (5,4), (4,5),\\ &(4,6), (6,4), (5,5),\\ &(5,6), (6,5),\\ &(6,6)\}\end{aligned}$$

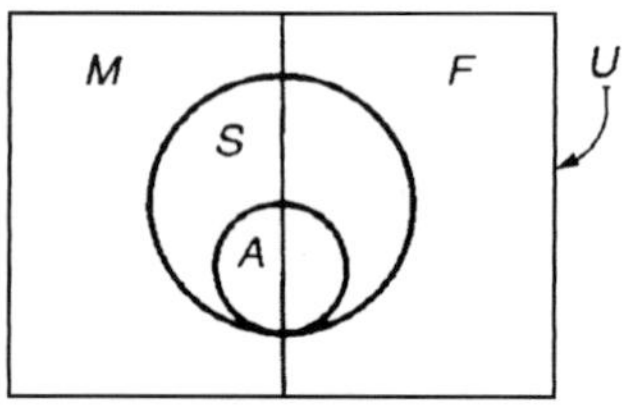

Exhibit 3.2

3.2 See Exhibit 3.2.

3.3 **c.** This is a conditional probability problem. Let B be "draw an ace," and let A be "draw a second ace": $P\{B\} = 4/52$ (1/13) and $P\{A\} = 3/51$. Then $P\{A|B\} = P\{A\} \times P\{B\}/P\{B\} = 3/51$.

3.4 **d.** This is a Bayes' theorem problem application because partitions are involved. The event E is a recall, with E_1 = Plant A, E_2 = Plant B, and E_3 = Plant C. The conditional probabilities of a recall from Plants E_1, E_2, and E_3 are

$$P(E \mid E_1) = 4/500 = 0.008$$
$$P(E \mid E_2) = 10/800 = 0.0125$$
$$P(E \mid E_3) = 10/1000 = 0.01$$

The probabilities that the dealer had a car from E_1, E_2, or E_3 are $P(E_1) = 0.3$, $P(E_2) = 0.4$, and $P(E_3) = 0.3$. Now applying Bayes' formula gives the probability that the recall was built in Plant A (E_1) as

$$P\{E_1|\text{recall}\} = \frac{P\{E_1\} \times P\{E|E_1\}}{P\{E_1\} \times P\{E \mid E_1\} + P\{E_2\} \times P\{E|E_2\} + P\{E_3\} \times P\{E \mid E_3\}}$$

$$= \frac{0.3 \times 0.008}{0.3 \times 0.008 + 0.4 \times 0.0125 + 0.3 \times 0.01} = 0.2308$$

3.5 **d.** The problem involves binomial probability. The probability that one item, selected at random, is defective is

$$p_{\text{defective}} = \frac{10}{1000} = 0.01$$

and the probability that one item is good (not defective) is

$$p_{\text{good}} = 1 - p_{\text{defective}} = 0.99$$

The probability that exactly one defective item will be found in a random sample of 100 items is given by the binomial $b(1, 100, 0.01)$, in which

No. defective, size sample, No. good

$$b(1, 100, 0.01) = C(100, 1)(0.01)^1(0.99)^{99}$$

$p_{defective}$, p_{good}

$C(n,r) = \binom{n}{r} = \frac{n!}{(n-r}$ is the number of combinations of n objects taken r at a time without concern for the order of arrangement.

$C(100, 1) = \frac{100!}{99!1!} = 100$, so $b(1, 100, 0.01) = 100(0.01)(0.99)^{99}$
$= 0.99^{99} = 0.3697$.

3.6 c. Here, $P(S_1) = 0.98$ and $P(S_2 \cap S_1) = 0.95$ are given. Hence the conditional probability $P(S_2 | S_1)$ is

$$P(S_2|S_1) = \frac{P(S_2 \cap S_1)}{P(S_1)} = \frac{0.95}{0.98} = 0.97$$

3.7 d. The probability of drawing an ace on the first card is 4/52. The probability that the second card is an ace is 3/51. The probability that the third card is an ace is 2/50, and probability for the fourth ace is 1/49. The probability that the first four cards will all be aces is

$$P = \frac{4}{52} \bullet \frac{3}{51} \bullet \frac{2}{50} \bullet \frac{1}{49} = 0.00\,0037 = 3.7 \times 10^{-6}$$

3.8 d.

3.9 d. Since the first IC that is picked is a premium IC, it was drawn from either bin 1 or bin 3. From the distribution of premium ICs, the probability that the premium IC came from bin 1 is $\frac{2}{3}$, and from bin 3 is $\frac{1}{3}$.

In bin 1, the probability that the remaining IC is a premium IC is 1; in bin 3, the probability is 0. Thus, the probability that the remaining IC is a premium IC is

$$\frac{2}{3}(1) + \frac{1}{3}(0) = \frac{2}{3}$$

An alternative solution using Bayes' theorem for conditional probability is

$$P(\text{bin 1} \mid \text{drew premium}) = \frac{P(\text{bin 1 and premium})}{P(\text{premium})}$$

$$= \frac{P(\text{premium} \mid \text{bin 1}) \bullet P(\text{bin 1})}{\sum_{i=1}^{3} P(\text{premium} \mid \text{bin 1}) P(\text{bin 1})}$$

$$= \frac{1\left(\frac{1}{3}\right)}{1\left(\frac{1}{3}\right) + 0\left(\frac{1}{3}\right) + \frac{1}{2}\left(\frac{1}{3}\right)} = \frac{2}{3}$$

3.10 c. The answer is the binomial coefficient

$$\binom{35}{4} = \frac{35 \bullet 34 \bullet 33 \bullet 32}{4 \bullet 3 \bullet 2 \bullet 1} = 35 \bullet 34 \bullet 11 \bullet 4 = 52,360$$

3.11 d. The total number of choices of 5 is $\binom{50}{5}$. Of these, $\binom{40}{5}$ have no bad bolts, $\binom{40}{4} \times \binom{10}{1}$ have one bad bolt, and $\binom{40}{3}\binom{10}{2}$ have two bad bolts. Thus,

$$\frac{\binom{40}{5} + \binom{40}{4}\binom{10}{1} + \binom{40}{3}\binom{10}{2}}{\binom{50}{5}}$$

$$= \frac{\frac{40 \bullet 39 \bullet 38 \bullet 37 \bullet 36}{5 \bullet 4 \bullet 3 \bullet 2} + \frac{40 \bullet 39 \bullet 38 \bullet 37}{4 \bullet 3 \bullet 2} \bullet 10 + \frac{40 \bullet 39 \bullet 38}{3 \bullet 2} \bullet \frac{10 \bullet 9}{2}}{\frac{50 \bullet 49 \bullet 48 \bullet 47 \bullet 46}{5 \bullet 4 \bullet 3 \bullet 2}}$$

$$= \frac{658,008 + 913,900 + 444,600}{2,118,760} = 0.9517$$

3.12 c. There are 26 choices for the first letter; 25 remain for the second, and 24 for the third.

3.13 a. The probability that only A is defective is

$$0.1 \times (1 - 0.2) \times (1 - 0.25) = 0.06$$

The probability that only B is defective is

$$(1 - 0.1) \times (0.2) \times (1 - 0.25) = 0.135$$

The probability that only C is defective is

$$(1 - 0.1) \times (1 - 0.2) \times (0.25) = 0.18$$

Now add to find the final probability, which is

$$0.06 + 0.135 + 0.18 = 0.375$$

3.14 d. Simply multiply the complementary probabilities $(1 - 0.5) \times (1 - 0.4) \times (1 - 0.3) = 0.21$.

3.15 a. Apply the binomial distribution. The probability of 0 bad is $(0.9)^{50}$; of 1 bad, $\binom{50}{1}(0.1)(0.9)^{49}$; and of 2 bad, $\binom{50}{1}(0.1)^2(0.9)^{48}$. Adding these, $(0.9)^{48}[(0.9)^2 + 5.0(0.9) + 1225(0.1)^2] = 0.112$.

3.16 b. This is the permutation of arranging 3 objects out of 6:

$$P(6,3) = \frac{6!}{(6-3)!} = 20$$

3.17 d. This is the permutation of arranging 7 persons out of 7:

$$P(7,7) = \frac{7!}{(7-7)!} = 5040$$

3.18 c. This is the selection (or combination) of 6 out of 8:

$$C(8,6) = \frac{8!}{(8-6)6!} = 28$$

3.19 d. Since two questions are mandatory, only four questions have to be selected out of six.

Then, $C(6,4) = \frac{6!}{(6-4)!4!} = 15$.

3.20 b. This is the selection (or combination) of 2 out of 5:

$$C(5,2) = \frac{5!}{3!2!} = 10$$

3.21 c. Since there are 11 integers that are exactly divisible by 9, probability = 11/100 = 0.11.

3.22 d. $P(1 \text{ head}) = \frac{C(4,1)}{2^4} = \frac{4}{16}$; $\quad P(2 \text{ heads}) = \frac{C(4,2)}{2^4} = \frac{6}{16}$

Since P(1 head AND 2 heads) is 0, P(1 head or 2 heads) = (4/16) + (6/16) = 5/8.

3.23 b. Let event A = (Apple from a bag) and event B = (Orange from the other bag).

Then, P(A) = 10/15 and P(B) = 5/15.

Since the events are independent, P(A AND B) = P(A) P(B) = (10/15) (5/15) = 2/9.

3.24 d. P(Black OR Red) = P(Black) + P(Red) – P(Black AND Red) = (5/85) + (30/85) – 0 = 35/85.

3.25 c. Let event A = (Orange from a bag) and event B = (White from the other bag).

P(A) = 2/10 = 0.2 and P(B) = 5/10 = 0.5

P(A OR B) = P(A) + P(B) – P(A AND B) = 0.2 + 0.5 – (0.2)(0.5) = 0.6

Note: P(A AND B) = P(A) P(B) as events A and B are independent.

3.26 a. Let event A = (number is even) and event B = (number > 72).

P(A) = 50/100 = 0.5 and P(B) = 28/100 = 0.28

Event (A and B) = (number is odd and > 72); P(A AND B) = 14/100 = 0.14

P(A OR B) = P(A) + P(B) – P(A AND B) = 0.50 + 0.28 – 0.14 = 0.64

3.27 c. P(both closed) = P(1 is closed) P(2 is closed)

0.90 = (0.95) P(2 is closed); then, P(2 is closed) = 0.90/0.95 = 0.9474

3.28 d. For a valid probability density function, $\int_{-\infty}^{\infty} f(x)\,dx = 1$, or the total area under the curve should be 1. For $f_1(x)$, the area is (1/2)(2)(2) = 2; for $f_2(x)$, it is 1(4 – 2) = 2; for $f_3(x)$, it is (1/2)(4 + 1) = 2.5. For $f_4(x)$, $\int_0^{\infty} e^{-x}\,dx = 1$. So, $f_4(x)$ is the only valid distribution.

3.29 c. For a valid probability distribution function, both $f(\infty)$ and $f(-\infty)$ should be zero and $f(x) \geq 0$ for all x. Also, $\int_{-\infty}^{\infty} f(x)\,dx = 1$. Then, $f_1(x)$ and $f_4(x)$ are not valid since $f(x) < 0$ for certain values of x. $f_2(x)$ is not valid because $f(\infty)$ is not 0. $f_3(x)$ is the valid function as it satisfies all the conditions.

3.30 a. The distribution is discrete, but the rules are similar to those of Problem 3.29. $f_1(x)$ is the only valid distribution. $f_2(x)$ is not valid since the sum of the densities (equivalent to integrating) is more than 1. $f_3(x)$ has a negative value. $f_4(x)$ has a value more than 1.

3.31 b. For a valid cumulative probability distribution function $F(x)$, the following rules apply: $F(-\infty) = 0$, $F(\infty) = 1$, $0 \leq F(x) \leq 1$, and $F(x_1) \leq F(x_2)$ if $x_1 < x_2$. For $F_1(x)$, $F(x)$ has a negative value; for $F_3(x)$, $F(\infty)$ is more than 1; for $F_4(x)$, the rule $F(x_1) \leq F(x_2)$ if $x_1 < x_2$ fails. Only $F_2(x)$ obeys all the rules.

3.32 a. Let p = P(head on the first toss); then, P(tail on the first toss) = $1 - p$

But, $p = 2(1 - p)$; solving, $p = 0.667$.

P(two heads in four tosses) = C(4,2)$(0.667)^2(1 - 0.667)^2$ = 0.1481

3.33 **a.** P(leaving the job) = 0.25; P(none will leave the job) = $C(7,0)(0.25)^0 (1 - 0.25)^7 = 0.1335$

3.34 **d.** P(head) = p = 0.5; P(at least one head) = 1 – P(no head) = 1 – $C(4,0)(0.5)^0(1 - 0.5)^4 = 0.9375$

3.35 **a.** Using the normal distribution table, P(X > 3) = 1 – F(3) = 0.0013.

3.36 **b.** $P\{X \le 26\} = F(26) = F\left(\frac{26-30}{5}\right) = F(-0.8) = 1 - F(0.8) = 1 - 0.7881 = 0.2119$

3.37 **d.** $P\{1980 - 68 < x \le 1980 + 68\} = F(2048) - F(1912)$

$$= F\left(\frac{2048-2000}{40}\right) - F\left(\frac{1912-2000}{40}\right) = F(1.20) - F(-2.2)$$

$$= 0.8849 - \{1 - 0.9861) = 0.8710$$

3.38 **d.** $C(5,0)(1/120)^0(119/120)^5 = 0.9590$

3.39 **a.** $\text{mean} = \overline{X} = \frac{\sum x}{n} = \frac{114}{12} = 9.5$

$$\text{variance, } \sigma^2 = (1/12)\left[(3-9.5)^2 + (4-9.5)^2 + \ldots\right] = 28.75$$

standard deviation $\sigma = 5.36$

3.40 **b.** $\frac{3(1.5) + 8(2.5) + \ldots + 9(5.5)}{3 + 8 + 18 + \ldots + 9} = 3.82$

CHAPTER 4

Computers

Donald G. Newnan, Sharad Laxpati, and Kenneth Olree

OUTLINE

INTRODUCTION

Current information for the FE exam indicates that approximately 7% of the *morning session* of the exam will contain questions related to computers, regardless of exam discipline. This means that there will be approximately eight or nine questions related to computers in the morning session for all FE examinees. There may be additional questions related to *computer systems* in the afternoon for examinees taking the *electrical* FE exam. All examinees should have basic familiarity with the following computer topics:

- Terminology (e.g., memory types, CPU, baud rates, Internet)
- Spreadsheets (e.g., addresses, interpretation, what if, copying formulas)
- Structured programming (e.g., assignment statements, loops and branches, function calls)

These three broad areas are specifically identified as topic areas addressed on the FE exam by NCEES. These and similar topics will be covered in this review chapter. More advanced computer topics, such as computer architecture, interfacing, microprocessors, and software design, which could be covered in the *afternoon section* of the *electrical* exam, should be reviewed using materials appropriate to that exam.

TERMINOLOGY

Types of Computers and Networks

There are many types of computers. The predecessors of all modern day computers were **mainframe** computers. Mainframe computers are large computers that typically fill an entire room, require air conditioning, and support many hundreds of users. Large and especially fast computers are known as **supercomputers**. Supercomputers are used to model very complex phenomena such as the operation of the human brain, air currents resulting from a fire, or nuclear explosions. Supercomputers are sometimes referred to as **Cray** computers because Cray Research, Inc. (founded by Seymour Cray) is a noted manufacturer of supercomputers. Mainframe and supercomputers tend to be expensive to buy and maintain.

Slightly smaller computers, which might still fill a room but support only 10 to 100 users, are known as **minicomputers**. Minicomputers are not in much use today, having been displaced by networks of smaller computers.

Smaller and moderately less powerful computers used for engineering and scientific calculations are often referred to as **workstations**. Workstations typically are designed for one user but can support multiple users. Computers that are less powerful than workstations and that are used in the home or office are typically known as **PCs** (personal computers). PCs derive their name from the IBM-PC, but other similarly used computers, such as Apple's Mac family of computers, also are typically referred to as PCs. Both workstations and PCs are **microcomputers**. Over time, the prefix "micro" has tended to be dropped from the term, and most people simply refer to microcomputers as computers. Similarly, in recent years, the distinction between workstations and PCs has blurred, and low-end workstations may also be referred to simply as PCs.

The computer chips that go into consumer devices, such as microwave ovens, washing machines, and cellular telephones, are known as **microprocessors**. If the chip has advanced systems on it such as analog-to-digital converters, timer systems, and significant memory, then it is known as a **microcontroller**.

To obtain large computing power at small cost, computers are sometimes networked together such that individual computers can process different parts of one problem. This is sometimes referred to as **cluster computing** or **grid computing**. Cluster computing typically takes place on computers located in close proximity to one another. A common cluster computing arrangement is a **Beowulf Cluster**. Grid computing typically uses computers more widely dispersed over the **Internet**, a worldwide network of computers for the purpose of sharing information. An arrangement in which computing services, such as network support, hardware and software upgrades, and data storage, take place over the Internet is known as **cloud computing** or **computing in the cloud**.

A lot of information available on the Internet can be viewed using an **Internet browser**, a computer program written to easily access information provided on the Internet using **http (hypertext transfer protocol)**, a standard method for publishing information on the Internet. A key feature of http is the use of a **URL**, or **universal resource locator**. A URL provides a standard format for providing a computer address of where information resides on the Internet. **Search engines**, such as Google, Yahoo, and Bing, can provide URLs for sought-after information.

Computer Communication

Computer communication has always been important. Typically a monitor and keyboard were used to access mainframe computers. The monitor and keyboard were known as a **terminal**. Today, a PC is often used as a terminal for large computers. If the terminal has very low processing power, it is known as a **thin client**. If the terminal has significant processing power, it is known as a **fat client** or **thick client**.

Computers communicate by sending **bits** of data. A bit is the simplest piece of information that can be stored in a computer. It is represented by a one or a zero. Representing information in only one of two possible states is known as **binary** representation.

Early computers communicated to terminals and other equipment at slow rates typically measured in hundreds or thousands of **baud**. The baud rate is the greatest number of state transitions per second in a communication system. In older technology, this is equivalent to the **bit rate**, the maximum number of bits that can be transmitted per second. However, modern communication systems may send more than one bit of information per state transition, so baud and bit rate should not be used interchangeably. The speed of **RS-232** communication, a common serial communication standard, has typically been expressed in baud. Its maximum data rate is about 20,000 baud.

Serial communication is a method of communication in which only one bit of information is sent at a time. Sometimes the bit that is sent is not received correctly. To prevent this type of error, serial communication systems often send a **parity bit**. A parity bit is an extra bit sent with a group of other bits (usually numbering seven or eight) such that the number of 1s or 0s sent is always even or odd. If the receiver detects that it has received the wrong number of even or odd bits, it can request retransmission of the data. Serial communication tends to be much slower than **parallel** communication in which many bits are sent simultaneously over separate physical lines.

RS-232 serial ports have been common on many PCs for many years; however, communication on PCs now typically takes place over **Ethernet, firewire**, or **USB**. Ethernet is a communication standard primarily used for **local area networks** (LANs), physically connected computers sharing a small geographic region. Ethernet now supports data speeds up to 100 billion bits per second. Firewire and USB standards support speeds up to about 5 billion bits per second and are primarily used to connect computers to other noncomputer devices, such as digital video cameras, external hard drives, and other personal electronic devices. **Wide area networks** (WANs) are used to link LANs and other computer networks together over far-reaching geographical areas.

Example 4.1

Which of the following communication protocols would be most suitable for a new data communication center's local area network (LAN)?

(i) RS-232 (ii) Ethernet (iii) firewire (iv) USB

Solution

The correct choice is (ii) Ethernet. All other options are considerably slower than Ethernet and more likely to be used to connect computers to other noncomputer devices rather than to other computers. As more information is being sent over networks, a new data communication center needs the fastest communication standard.

Computer Memory

Computer memory and communication speeds are typically expressed in bits (b) or **bytes** (B) and bits per second (bps) or bytes per second (B/s). A byte consists of eight bits.

As computer memory has increased in size, prefixes have been used to indicate larger units of computer memory. This has led to some confusion in specifying computer memory. Because computer memory is utilized in sizes consisting of powers of two, a **kilobyte (KB)** is not 1000 bytes, but is actually 1024 (2^{10}) bytes. The confusion exists because the prefix kilo means 1000 in SI units. The confusion is even more pronounced because in computer information storage (such as in hard drives, tape drives, and removable disks) and in other computer-related usage (such as in clock speeds, operations per second, or data transfer rates), the standard SI meaning of the prefix is *typically* used. Consequently, a **megabyte (MB)** is 1,048,576 (2^{20}) bytes if referring to computer memory and is 1,000,000 bytes if referring to the storage capacity of a removable disk. A **gigabyte (GB)** indicates 2^{30} bytes of computer memory or 10^9 bytes of disk storage capacity. Likewise, a **terabyte (TB)** consists of 2^{40} bytes if referring to computer memory or 10^{12} bytes if referring to the total amount of data to be transferred to a computer over the Internet.

If the base 2 usage of the prefix is meant, especially when using it in reference to something other than computer memory, then the prefixes should be altered to kibi (2^{10}), mebi (2^{20}), gibi (2^{30}), and tebi (2^{40}) to reduce confusion. The **kibibyte**, **mebibyte**, **gibibyte**, and **tebibyte** are then abbreviated **KiB**, **MiB**, **GiB**, and **TiB**, respectively. Thus, a 200 GiB hard drive has $200 \times 2^{30} = 214{,}748{,}364{,}800$ bytes of storage capacity, whereas a 200 GB hard drive has $200 \times 10^9 = 200{,}000{,}000{,}000$ bytes of storage capacity.

Example 4.2

A supercomputer has 200 TB of computer memory. How many bits does this represent?

Solution

Because the reference is to computer memory, the TB abbreviation indicates the nonstandard use of the prefix *tera* to mean 2^{40}. Additionally, the number of bits must be found from the number of bytes (8 bits = 1 byte).

$$C_aH_bO_cN_d + \left(\frac{4a + b - 2c + 3d}{4}\right)O_2 \rightarrow aCO_2 + \left(\frac{b - 3d}{2}\right)H_2O + dNH_3$$

Computer memory is typically referred to as belonging to **RAM** or **ROM**. For historical reasons, the name RAM was created to indicate *random access memory*, and ROM indicated *read-only memory*. However, the term *RAM* now indicates memory that is **volatile**, and *ROM* indicates memory that is **nonvolatile**. Volatile memory is memory in which the data is lost if power is removed, whereas nonvolatile memory retains its data even if power is removed. RAM is used to temporarily store data and program code while a computer is in operation. ROM is used to permanently store programs and other information that doesn't change (or at least doesn't change very often).

Although there are other types of RAM, the two most common types are **static RAM (SRAM)** and **dynamic RAM (DRAM)**. SRAM is faster than DRAM but is more expensive, takes up more space, and consumes more energy. It is typically used for a computer's **cache**. A cache is a separate memory structure that is used to rapidly access frequently used data or program code. Because of the disadvantages of SRAM, cache memory is typically much smaller than main memory. DRAM is typically used for main memory. DRAM is so named because the information is stored as a charge on a capacitor that gradually loses its charge. To keep from losing the data, the memory system must provide a periodic refresh charge. RAM chips are built on silicon wafers using complementary metal oxide semiconductor **(CMOS)** circuits. CMOS circuits are an especially energy-efficient means of storing data in RAM, especially if the transitions between the binary states occur infrequently.

Read-only memory (ROM) was initially just as the name states—memory that could only be read. The data or program code was placed into the memory chip during manufacture. This memory evolved into programmable ROM **(PROM)** memory, which was memory that could be programmed after manufacture but only programmed once. After a PROM was programmed, it could only be read and not written to again. This was followed by **EPROM** memory. EPROM stands for erasable programmable ROM. This memory type could be written to many times but only after erasing it via exposure to ultraviolet light through a quartz crystal window manufactured into the chip housing. **EEPROM** followed EPROM. EEPROM stands for electrically erasable PROM. Instead of needing ultraviolet light to erase the memory, this memory could be erased and reprogrammed by applying appropriate voltages to the chip.

Flash memory is a type of EEPROM and is commonly accessed via a USB port. This arrangement is referred to as a **pen drive**, **thumb drive**, **flash drive**, or **jump drive**. Sometimes hard drives and **CD** (compact disc) or **DVD** (digital video disc or digital versatile disc) drives are considered to be ROM, especially if the CD or DVD disc can only be written to once.

Data transmission is slower in ROM as compared to RAM. Data transmission to or from a hard drive, CD drive, or DVD drive is extremely slow as compared to accessing other memory types. Although considerably slower, these drives have substantially more memory capacity than internal RAM or ROM. Storage capacity for hard drives is now commonly measured in hundreds of GB or tens of TB.

Example **4.3**

Which of the following memory types would most likely be used for a computer's cache memory?

(i) SRAM (ii) DRAM (iii) EPROM (iv) FLASH

Solution

The correct answer is (i) SRAM because it is the fastest type of RAM. ROM types of memory would never be used for a cache.

Example **4.4**

To what do the first two letters in the name of the memory type known as EEPROM refer?

(i) Enhanced Emitter (ii) Energy Efficient (iii) Exception Event

(iv) Electrically Erasable

Solution

The correct answer is (iv) Electrically Erasable. EEPROM is a type of read only memory (ROM) that can be erased and reprogrammed by applying appropriate voltages to the memory chip.

Program Execution

Execution of computer programs to process data takes place in the **central processing unit (CPU)**. This is a specialized circuit that once consisted of vacuum tubes or mechanical relays but is now built on silicon as an **integrated circuit (IC)**. An IC is a complex electrical circuit consisting of many transistors to implement one or more functions on a single silicon wafer. CPUs have many substructures, such as **arithmetic and logic units (ALUs)**, registers (also commonly known as **accumulators**), **buses**, and **control units**.

- **ALUs** are responsible for implementing data manipulation via arithmetic instructions such as add, subtract, and multiply. They also implement logic instructions such as logical AND, OR, and XOR (exclusive or).
- **Registers** or **accumulators** are special memory locations within the CPU used for holding data and intermediate results that are currently being accessed and manipulated by the ALU. They are also used for keeping track of which instruction in memory is to be executed next and for keeping track of the CPU's current operational state.

- **Buses** are the electrically conductive paths on which code, data, control information, and memory locations (addresses) are sent and received.
- The **control unit** is the circuit that makes all the subsystems work together appropriately. The control unit ensures data is written to or read from the appropriate registers at the appropriate times. The control unit may also assist with getting the next instruction from memory and other similar tasks.

The four main functions of the CPU are to *fetch*, *decode*, *execute*, and *write back* program code and data. The speed at which a CPU can accomplish these functions is often measured in **MIPS**, **million instructions per second**, also sometimes referred to as **MOPS**, **million operations per second**.

A single integrated circuit having more than one CPU is referred to as a **multicore processor**. Multicore processors can increase processing speed by sharing operations between two or more CPUs. Splitting processing tasks so that they can be shared between processors is known as **parallel programming**. An arrangement in which a CPU accesses both program code and data from a single memory structure is known as the **von Neumann architecture**. The **Harvard architecture** is an arrangement in which the CPU accesses program code from a memory structure that is separate and distinct from another memory structure used for holding data.

Number Systems and Data Types

CPUs are designed to work with code and data as binary numbers (0s and 1s). A computer program represented in binary is often referred to as **machine code**. As the digits of a binary number are examined from right to left, each digit represents a higher consecutive integer power of two. For example, the binary number 1101 represents the decimal number 13 because $1 \times 2^3 + 1 \times 2^2 + 0 \times 2^1 + 1 \times 2^0 = 8 + 4 + 0 + 1 = 13$.

Example 4.5

The binary number 110 corresponds to what decimal (base 10) number?

Solution

$$1 \times 2^2 + 1 \times 2^1 + 0 \times 2^0 = 4 + 2 = 6$$

The conversion of a decimal number to a binary number can be achieved by the method of remainders as follows. A decimal integer is divided by 2, giving an integer quotient and a remainder. This process is repeated until the quotient becomes 0. The remainders (in the reverse order) form the binary number. The following example illustrates this process.

Example 4.6

Convert decimal number 43 to a binary number.

Solution

	Quotient		Remainder
43 ÷ 2 =	21	+	1
21 ÷ 2 =	10	+	1
10 ÷ 2 =	5	+	0
5 ÷ 2 =	2	+	1
2 ÷ 2 =	1	+	0
1 ÷ 2 =	0	+	1

Answer: $(43)_{10} = (101011)_2$

Although the conversion from decimal to binary is straightforward, binary numbers are typically difficult for humans to read and understand, especially as the number of digits increase. A small improvement in readability was made by introducing **hexadecimal** (or **hex**) numbers. Hex numbers are formed by taking four binary digits at a time and using them to represent a base 16 number. This is accomplished by using the letters A, B, C, D, E, and F to represent decimal numbers 10, 11, 12, 13, 14, and 15, respectively. Thus, the binary number 1101, hexadecimal number D, and decimal number 13 are all equivalent. The equivalent decimal number can be found for a hexadecimal number in a manner similar to that for a binary number.

Example 4.7

Find the equivalent decimal value for the hexadecimal number 25E6.

Solution

$2 \times 16^3 + 5 \times 16^2 + 14 \times 16^1 + 6 \times 16^0 = 8192 + 1280 + 224 + 6 = 9702$

Fractional binary values can be represented by including digits to the right of a radix point (commonly understood as the decimal point for decimal numbers). For instance, the binary number 1101.101 is equivalent to the decimal number 13.625 because $1101.101 = 1 \times 2^3 + 1 \times 2^2 + 0 \times 2^1 + 1 \times 2^0 + 1 \times 2^{-1} + 0 \times 2^{-2} + 1 \times 2^{-3}$, which is equivalent to $8 + 4 + 0 + 1 + 0.5 + 0 + 0.125 = 13.625$.

A similar procedure can be followed for hex numbers. To find equivalent binary or hex numbers from a decimal number, repeatedly divide the decimal number and all remainders by integer powers of 2 or 16, respectively. For instance, the decimal number 2672 is A70 in hex because 16^2 divides into 2672 ten times and 16^1 divides into the remainder seven times and 16^0 divides into the final remainder zero times.

Alternatively, conversion from decimal to hexadecimal may be carried out in a manner similar to that for decimal to binary conversion, with the divisor 2 (or multiplier in the case of fractions) replaced by 16.

Example 4.8

Convert the base 10 integer 458.75 to base 16 equivalent value.

	Quotient		Remainder	Hexadecimal
458 ÷ 16	28	+	10	A
28 ÷ 16	1	+	12	C
1 ÷ 16	0	+	1	1
	Integer		**Remainder**	**Hexadecimal**
.75 × 16	12		0	C
458.75 = 1CA.C				

If fractional values of a number are never needed, all of the available bits can be used to represent an **integer** data type. Because there must always be a finite number of bits used to represent a number, fractional numbers will often suffer from **round off**, or **rounding error**. For instance, if a binary number only had four digits available to represent a fractional value, any fraction that couldn't be exactly represented as an integer multiple of $1/2^4$ would be represented as the nearest integer multiple of $1/2^4$. This creates a rounding error of as much as $1/2^5 = 0.03125$. A limited number of digits for the fractional portion creates a limit to the precision of a binary number.

Similarly, a limited number of digits for the integer portion creates a limit to the range of a binary number. To help alleviate these limitations, the **floating point** data type was created. Floating point data permits the radix point to move (float) among the binary digits depending on whether greater range of the number is needed or whether increased precision is needed. It can be thought of as a type of scientific notation representation of numbers for computers. Common floating point number standards consist of **single precision**, **double precision**, and **quadruple precision**. These standards are able to accurately represent about 7, 16, and 34 significant figures, respectively. If a computation is attempted that cannot be represented by the computer, a **NaN** (Not a Number) data type might result. Examples of these types of computation might be trying to divide by zero or taking the square root of a negative number.

Character data types are often used to represent letters of the alphabet or the character representation of the numerals zero through nine or punctuation. Some other special symbols and control characters may also be represented by the character data type. Some of these include the symbol for a new line, carriage return, and bell sound. Common codes for representing character data include **ASCII (American Standard Code for Information Interchange), EBCDIC (Extended Binary Coded Decimal Information Code), and Unicode**. ASCII and EBCDIC are used for the English language alphabet and symbols. Unicode is an evolving standard used to represent alphabets and writing symbols of many different languages of the world. When character data is put together, as in forming a name, it is often referred to as a **string**.

Example 4.9

Which of the following would most likely be an example of a character string?

(i) 241.38
(ii) 2.4138e2
(iii) 241
(iv) Two hundred forty-one and thirty-eight hundredths

Solution

The correct answer is (iv) Two hundred forty-one and thirty-eight hundredths. Although all of the answers could be strings if the character representation of each of the digits were being displayed, only (iv) is the *most likely* example of a character string. This is because answers (i) and (ii) could also be examples of floating point data types, and answer (iii) could be an example of an integer data type. Only answer (iv) uses a significant number of alphabetic letters and dashes to guarantee it to be a character string.

Placing a number of data values in a continuous section of memory with one data value sequentially following the next is often referred to as a **data array**. This arrangement often makes processing of the data faster because the data can be easily loaded into the computer's cache memory for ready access. Data arrays may be one-dimensional arrays (as in a single column or row) or two-dimensional arrays in which there are multiple columns and rows of data (as in a matrix). Higher dimensional arrays are possible but not as common. A data array often uses a **pointer**, a variable that "points" to the location of the array, to provide access to the data within the array.

Computer Languages

To process data, computer programs are necessary. Originally, programs were written by physical switches or circuits that were opened or closed to represent the 0s or 1s of machine code. However, this was very difficult for humans to do without making mistakes, so **assembly languages** (or **assembler languages**) were developed. Assembly languages use short **mnemonics** to correspond to a specific instruction available to the CPU. For example, the instruction LDAA might stand for Load Accumulator A. Once the program is written in assembly language, a separate program known as an **assembler** converts the assembly code into machine code. The resulting file created is known as an **executable** because it can be directly loaded and executed on the computer.

Even though assembly languages were a vast improvement over writing programs in machine code, they were still difficult to use, especially as data processing tasks became more complicated. For this reason, **high-level languages** were created to make programming easier. Examples of high-level languages include Fortran, Basic, C, Java, Matlab, and Python. Similar to assemblers, **compilers** convert high-level languages into machine code so that executables are available to work directly on the computer. As long as a compiler is available for any given computer, high-level languages can usually be compiled to work on the computer with little or no modification to the original programming code. **Compiler errors** occur when the program is compiled if there are errors in the code that the compiler cannot process. A syntax error is a common error made by programmers. A **syntax error** occurs when the rules for how the language is to be used are not followed. An example of a syntax error is trying to use a variable before the variable is **declared** in a statement that names the variable and describes its data type.

Interpreted languages are also considered high-level languages. Interpreted languages do not need to be compiled and an executable is not generated. Interpreted languages are "interpreted" into machine code at runtime. This has some advantages in that variables and arrays can be created and altered "on the fly," but also has disadvantages in that the code executes more slowly, and an **interpreter** must exist on the machine in order for the interpreted language to run. Fortran and C are usually compiled into executables; whereas Matlab and Python are usually interpreted.

High-level languages use **functions** (which may also be known as **subroutines**, **procedures**, or **methods**) to break code into smaller reusable sections. Functions often need **parameters** (the datum or data) passed to them to execute properly. Parameters may be passed using **call-by-value** or **call-by-reference**. Call-by-value sends the data values directly to the function, whereas call-by-reference sends the function the memory location of where the data is stored. Functions may or may not return values to the program that originally called it.

Structured programs use three main structures to process data. These are **sequence**, **looping** (or **repetition**), and **decision** (or **choice**). Sequence is the process by which code executes only one instruction at a time, and each instruction executed follows in sequential order as determined by the programmer. Looping involves execution of an instruction or a sequential list of instructions repeatedly until some condition is satisfied. The determination for looping can be made prior to the loop beginning (pre-test) or after the loop has started execution (post-test). In structured programming, loops should only have one entry point and exit point. A decision is used to determine whether a given condition is true or false or what condition exists so that a program **branch** (a change in the flow of program execution) can take place.

As programs have become larger, **object oriented programming (OOP)** has become a major programming paradigm. Object oriented programming organizes programs around the data as objects rather than organizing around the tasks that must be performed on the data. Many believe OOP makes maintenance of programs easier, especially as the programming code becomes very large.

Operating Systems

A PC needs programs in order to interface with its own input and output devices, control those devices, and manage resources. These programs are known as **operating systems**. Some common operating systems include Unix, Linux, Mac OS, and Microsoft Windows. Operating system functions are generally accessed through an **API** (an **application program interface**). An API ensures that the operating system is in control of resources and that computer hardware is not *directly* controlled by the user or other programs. Operating systems permit other work productivity programs (**application programs**) to also run on the computer. Common application programs include programs for word processing, e-mail, spreadsheets, database management, presentation graphics, Internet browsers, and accounting programs. In addition to these, engineers often use application programs for **CAD (computer aided design)** and **CAM (computer aided manufacturing)**. Many application programs work with the user using a **GUI (graphical user interface)**. A GUI makes interacting with an application program (or operating system) easier by visually displaying options available and making those options easily selectable by a mouse or other input device.

Computer Security

Many operating systems and networking application programs implement security features to protect computer data and prevent unauthorized use of computers. A **password** is a sequence of letters, numbers, and other characters that must be provided before the operating system or networking program will permit access to computer data and programs. Typically, passwords should be changed often and consist of an unobvious sequence of character data involving both upper and lowercase letters, numbers, and other symbols. In some cases, passwords may be necessary to access computer data.

Encryption is also used to protect passwords and sensitive data. Encryption scrambles and encodes the data or password such that it is not readily seen or understandable without a key. A common method of providing encryption over the Internet is by using **SSL**, **secure socket layer**. The URL for Web sites using SSL begin with https rather than http.

Damage to data can occur due to a **computer virus**, **malware**, **spyware**, or other deliberate destruction by an unauthorized user. A computer virus is software written to cause damage to another computer's files or operation and to replicate itself so that it can be spread to other computers. **Antivirus software** is software that attempts to recognize computer viruses and inactivate them before they can infect a computer or replicate. Malware is any software written by unscrupulous people to cause damage to a computer's data, whether or not it is self-replicating. Spyware is software written to record or search for sensitive information on a computer to send to the spyware's programmer in order to steal information such as credit card numbers or computer passwords. People who write such programs are known as **crackers**, and sometimes **hackers**, although the term *hacker* is sometimes used to denote skilled programmers who don't necessarily have any malicious intent for the code they write.

Often a **firewall** is used to prevent unauthorized access to a computer over the Internet. A firewall carefully checks transmitted information over a network to make sure only authorized information is being conveyed. Some computers require **biometric data** to gain access to the computer. Biometric data is information about an individual that is uneasy to copy and is often used to provide a higher level of security for gaining computer access. Examples of biometric data would include a fingerprint or iris scan.

Some unauthorized people will try to gain access to computers by **spoofing**. This is an attempt to gain access by convincing the computer's security features that the unauthorized person is someone else or another computer that does have authorized access. Another commonly used method to try to gain unauthorized access to a computer is by **phishing**. Phishing is an attempt, usually via e-mail, to convince a user that the person sending the e-mail is someone who can be trusted and that the user should send confidential information such as passwords or banking information.

STRUCTURED PROGRAMMING

To solve a problem by using a computer program, an algorithm is usually developed. An **algorithm** is an ordered sequence of steps to take to arrive at a solution.

In the early days of computers and computer programs, there were no clear and concise rules to follow in implementing an algorithm in a computer program. Two of the earliest computer languages, assembly and BASIC, permitted the use of the *jump* and *goto* instructions, respectively. Although a computer could implement these instructions without any problems, human programmers overused them to their own detriment. Because no structured rules existed, programmers were left to implement the algorithm in whatever way they could to arrive at a solution. The resulting code solutions often were difficult to read and understand by people other than the original programmer. Such code also was prone to error because the programmer had not considered all the possible circumstances in which program execution could be altered by a goto statement. If the code ever needed to be upgraded, it was often difficult to follow the logical sequence and know what parts of the code should be changed and what should remain the same. Thus, program maintenance was difficult to achieve. Large blocks of code were often written to implement the algorithm all at once. This too made it difficult for others to keep track of all the variables that were being used, what the variables' current values were, and when program execution left the current sequence or jumped back in. As programming tasks became more complicated and larger code needed to be written, a real need for code that was easy to read and understand became apparent.

Structured programming is a concept generally credited to Edsger Dijkstra. He advocated much less dependence on the *goto* statement, even that it should be eliminated. Dijkstra articulated simple rules to make programming easier to read, understand, and maintain. To eliminate the rat's nest of code generated by overuse of the *goto* statement, structured programming specifies that there should be only one entry point and exit point within a loop, function call, or sequence of instructions. The entry or exit point should be determined by a decision statement that clearly indicates the condition or conditions under which the program execution could change (that is, under what conditions the program could *branch* to execute other instructions). Similarly, structured programming specifies that code should be modular. This means that large problems should be broken into smaller problems that can be solved first; the smaller solutions are then pieced together to solve the large problem. This top-down design approach means that function calls should be used to solve smaller portions of the problem. This keeps the code modular so that it is easy to see what small problem is being solved, the logical steps taken to solve it, and how it relates to solving the big problem without the confusion of solving too much at one time. Once the function is developed and well tested, it can be reused in other code, making new code development times shorter and facilitating code upgrades and maintenance. Indenting distinct chunks of code improves its readability and is another important contribution of structured programming. For example, the programmer might indent a section of instructions all pertaining to the same loop function.

In structured programming, only three main control structures are needed:

1. Statements for sequential execution of instructions, such as assignment statements
2. Decision statements, such as an *if-then-else*, for causing program branching
3. Repetition statements, such as the *while-do* or *for* looping commands

To also help develop algorithms and make programs easier to read and understand, programmers use flowcharts and pseudocode to illustrate the algorithm.

Example 4.10

Which of the following is considered to be good structured programming technique?

(i) Frequent use of the *goto* instruction

(ii) Frequent use of the *jump* instruction

(iii) An absence of function calls

(iv) Code that is broken into small sections

Solution

The correct answer is (iv) Code that is broken into small sections. This is the idea of modularity that makes the code easily readable and understandable as subsections that solve a small part of the larger problem rather than trying to solve the entire problem in one large block of code.

Flowcharts

An algorithmic flowchart is a pictorial representation of the step-by-step solution of a problem using standard symbols. Some of the commonly used shapes are shown in Figure 4.1. Consider the simple problem in Example 4.11.

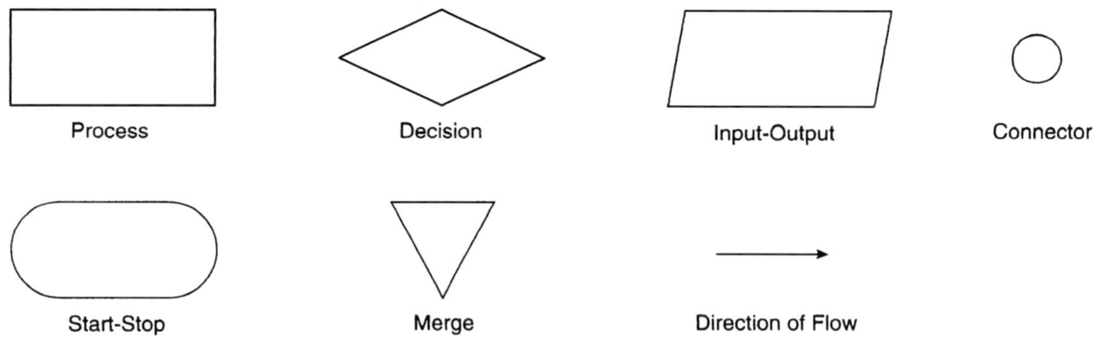

Figure 4.1 Flowchart Symbols

Example 4.11

A present sum of money (P) at an annual interest rate (I), if kept in a bank for N years, would amount to a future sum (F) at the end of that time according to the equation $F = P(1 + I)^N$. Prepare a flowchart for $P = \$100$, $I = 0.07$, and $N = 5$ years. Then compute and output the values of F for all values of N from 1 to 5.

Solution

Exhibit 1 shows a flowchart for this situation.

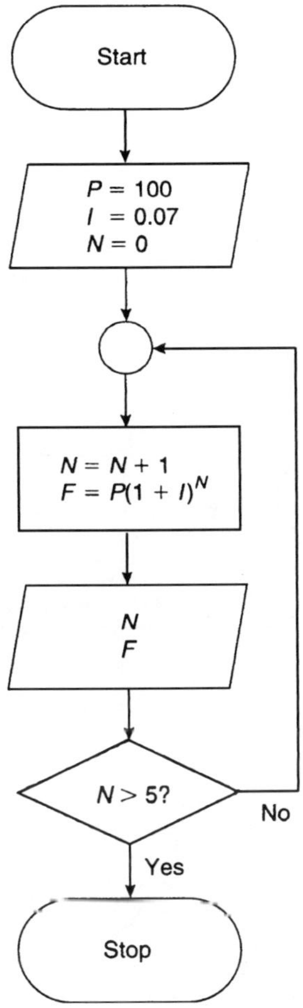

Exhibit 1 Algorithmic Flowchart

Example 4.12

Consider the flowchart in Exhibit 2.
The computation does which of the following?

(i) Inputs hours worked and hourly pay and outputs the weekly paycheck for 40 hours or less.

(ii) Inputs hours worked and hourly pay and outputs the weekly paycheck for hours worked including over 40 hours at premium pay.

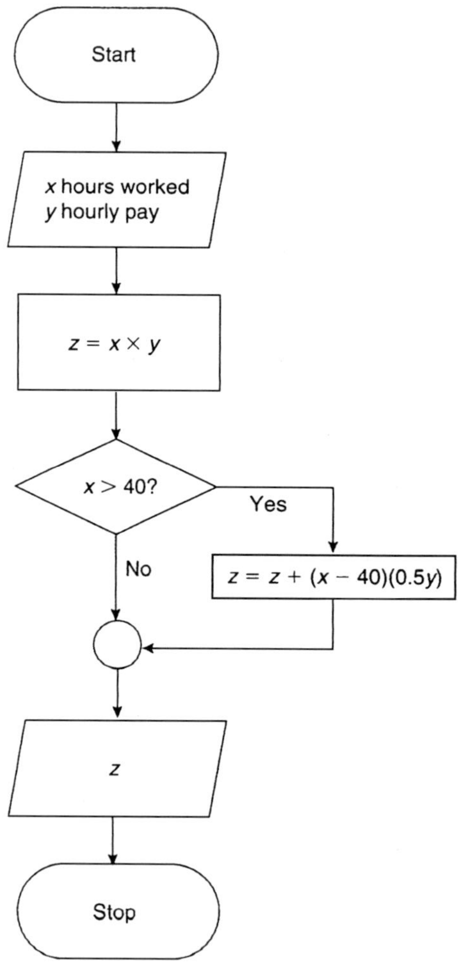

Exhibit 2

Solution

The answer is (ii).

Pseudocode

Pseudocode is an English-like language representation of computer programming. It is carefully organized to be more precise than a simple statement, but may lack the detailed precision of a flowchart or of the computer program itself.

Example 4.13

Prepare pseudocode for the computer problem described in Example 4.11.

Solution

```
INPUT P, I, and N = 0
DOWHILE N < 5
COMPUTE N = N + 1
   F = P (1 + I)^N
OUTPUT N, F
IF N > 5 THEN ENDDO
```

SPREADSHEETS

For today's engineers the ability to create and use spreadsheets is essential. Although several spreadsheet application programs exist, the most popular is Microsoft's Excel. The following explanations and examples are applicable to any spreadsheet package.

As noted in Figure 4.2, three types of information may be entered into a spreadsheet: text, values, and formulas. Text includes labels, headings, and explanatory text (columns A and B of Figure 4.2). Values are numbers, times, or dates (columns C, D, and E of Figure 4.2). Formulas combine operators and values in an algebraic expression (column F of Figure 4.2).

Figure 4.2 Example Spreadsheet

	A	B	C	D	E	F
1	**Last Name**	**First Name**	**Date**	**Age**	**Answers Correct**	**Percentage Correct**
2	Dempsey	Lou	4/17/2010	22	138	76.7
3	Johnson	Nathan	10/30/2010	26	105	58.3
4	Smith	Julie	10/30/2010	21	162	90.0

A **cell** is the intercept of a column and a row. Its location is based upon its column-row location; for example, B3 is the intercept of column B and row 3. Columns are labeled with letters across the top of the spreadsheet, and rows are labeled with numbers on the side. To change a cell entry, the cell must be highlighted using either an address or a pointer.

A group of cells may be called out by using a **range**. Cells A1, A2, A3, A4 could be called out using the range reference A1:A4 (or A1..A4). Similarly, the range A2, B2, C2, D2 could use the range reference A2:D2 (or A2..D2).

In order to call out a block of cells, a range callout might be A2:C4 (or A2..C4) and would reference the following cells:

A2 B2 C2
A3 B3 C3
A4 B4 C4

Formulas may include cell references, operators (such as +, –, *, /), and functions (such as SUM, AVERAGE). The formula SUM(A2:A6) or SUM(A2..A6) would be evaluated as equal to A2 + A3 + A4 + A5 + A6. Built-in formulas such as SUM or AVERAGE must be preceded by the @ symbol or an equals sign (=) in order to be recognized as a formula as opposed to text.

Relational References

Most spreadsheet references are relative to the cell's position. For example, if the content of cell A5 contains B4, then the value of A5 is the value of the cell up one and over one. The relational reference is most frequently used in tabulations, as in the following example for an inventory where cost times quantity equals value and the sum of the values yields the total inventory cost.

Inventory		Valuation	
	A	B	C
1 Item	Cost	Quantity	Value
2 box	5.2	2	10.4
3 tie	3.4	3	10.2
4 shoe	2.4	2	4.8
5 hat	1.0	1	1.0
6 Sum			26.4

In C2, the formula is A2*B2, in C3 A3*B3, and so forth. For the summation the function SUM is used; for example, SUM(C2:C5) in C6.

Instead of typing in each cell's formula, the formula can be copied from the first cell to all of the subsequent cells by first highlighting cell C2 then dragging the mouse to include cell C5. The first active cell, C2, would be displayed in the edit window. Typing the formula for C2 as = A2*B2 and holding the control key down and pressing the enter key will copy the relational formula to each of the highlighted cells. Since the call is relational, the formula in cell C3 is evaluated as A3*B3. In C4 the cell is evaluated A4*B4. Similarly, edit operations to include copy or fill operations simplify the duplication of relational formulas from previously filled-in cells.

Arithmetic Order of Operation

Operations in equations use the following sequence for precedence: exponentiation, multiplication, or division, followed by addition or subtraction. Parentheses in formulas supercede normal operator order.

Absolute References

Sometimes one must use a reference to a cell that should not be changed, such as a data variable. An absolute reference can be specified by inserting a dollar sign ($) before the column-row reference. If B2 is the data entry cell, then by using B2 as its reference in another cell, the call will always be evaluated to cell B2, regardless of which cell a formula might be copied to. Mixed reference can be made by using the dollar sign for only one of the elements of the reference. For example, the reference B$2 is a mixed reference in that the row does not change but the column remains a relational reference. Thus, a formula copied using this mixed reference would be relative to column B but unchanging in its reference to row 2.

The power of spreadsheets makes repetitive calculations very easy. Changing a value in one cell automatically updates every other cell that references it. Furthermore, program-like commands can be used in spreadsheets by executing a macro. Macros permit the execution of a large number of spreadsheet commands that have been previously recorded or programmed by the user.

All spreadsheet programs allow for changes in the appearance of the spreadsheet. Headings, borders, or type fonts are usual customizing tools. Spreadsheets also permit presentation of data as line graphs, bar graphs, pie charts, and other figures. This makes spreadsheets especially valuable for report generation and presentation.

Example **4.14**

Which of the following tasks is possible within a spreadsheet?

(i) Creating a graph
(ii) Creating reports
(iii) Analyzing data
(iv) All of these are possible.

Solution

The correct answer is (iv). All of these are possible within a spreadsheet. All of these tasks are easily done in a spreadsheet. There are very few things that can't be done, especially if the user knows how to use macros.

Example **4.15**

A spreadsheet has the values 5, 10, and 15 in cells A1, A2, and A3, respectively; and the formula \$A\$1 + A2 in cell B2. What value will display in cell B3 if the formula from cell B2 is copied to cell B3?

(i) 15
(ii) 20
(iii) 25
(iv) 30

Solution

The correct answer is (ii) 20. The \$ symbol represents absolute addressing within spreadsheets, and its absence indicates relative addressing. Thus, when the formula is copied from cell B2 to B3, the values added are the value 5 in cell A1 and the value 15 in cell A3. Thus, the formula in cell B3 will be \$A\$1 + A3, and the value displayed will be 20.

Example **4.16**

Which of the following would likely be most difficult to incorporate into a spreadsheet cell?

(i) A jpeg image
(ii) Text representing a name
(iii) Value representing currency
(iv) Cell shading and/or borders

Solution

The correct answer is (i) a jpeg image. Although it is quite easy to insert an image into a spreadsheet, inserting one into an individual cell is not possible. There are ways to make it seem as if an image is inserted into a cell, but this is not easily done. The other answer choices are all easily inserted or performed within a spreadsheet cell.

Example 4.17

A spreadsheet has the values 2, 4, and 6 in cells A1, A2, and A3, respectively. It also has values 3, 5, and 7 in cells B1, B2, and B3, respectively. What is displayed in cell D8 if it contains the formula @SUM(A1:B3)?

(i) 12
(ii) 15
(iii) 27
(iv) An error would result.

Solution

The correct answer is (iii) 27. Referencing cells in this manner provides the sum of all the values in the block. Where the formula is located (in this case, cell D8) is irrelevant unless the formula is copied to another cell. The format of the formula is correct in that it must start with the @ or = symbol, and the argument for the formula must either be specified with a colon (A1:B3) or two periods (A1..B3).

GLOSSARY OF COMPUTER TERMS

Accumulators	Registers that hold data, addresses, or instructions for further manipulations in the ALU
Address bus	Two-way parallel path that connects processors and memory containing addresses
AI	Artificial intelligence
Algorithm	A sequence of steps applied to a given data set that solves the intended problem
Alphanumeric data	Data containing the characters a, b, c, …, z, 0, 1, 2, …, 9
ALU	Arithmetic and logic unit
ASCII	American Standard Code for Information Interchange, 7 bit/character (Pronounced AS-key)
Asynchronous	Form of communications in which message data transfer is not synchronous, with the basic transfer rate requiring start/stop protocol
Baud rate	Number of state transitions that can be realized per second. Often equivalent to bits per second
BIOS	Basic input/output system
Bit	0 or 1
Buffer	Temporary storage device
Byte	8 bits
Cache memory	Fast look-ahead memory connecting processors with memory, offering faster access to often-used data
Channel	Logic path for signals or data
CISC	Complex instruction-set computing
Clock rate	Cycles per second
Control bus	Separate physical path for control and status information
Control unit	Ensures all subsystems within CPU work together appropriately
CPU	Central Processing Unit, the primary processor
Data buffer	Temporary storage of data
Data bus	Separate physical path dedicated for data
Digital	Discrete level or valued quantification, as opposed to analog or continuous valued
Duplex communication	Communications mode where data is transmitted in both directions at the same time
Dynamic memory	Storage that must be continually hardware-refreshed to retain valid information
EBCDIC	Extended Binary Coded Decimal Interchange Code—8 bits/character (pronounced EB-see-dick)
EPROM	Erasable programmable read-only memory
Expert systems	Programs with AI, which imitate the knowledge of a human expert
Floppy disk	Removable disk media in various sizes, $5^1/_4$", $3^1/_2$"
Flowchart	Graphical depiction of logic using shapes and lines
Gbyte (GB)	Gigabytes: 1,073,741,824 or 2^{30} bytes; a measurement of computer memory

Half-duplex communication	Two-way communications path in which only one direction operates at a time (transmit or receive)
Handshaking	Communications protocol to start/stop data transfer
Hard drive	Disk that has nonremovable media
Hardware	Physical elements of a system
Hexadecimal	Numbering system (base 16) that uses 0–9, A, B, …, F
Hierarchical database	Database organization containing hierarchy of indexes/keys to records
I/O	Input/output devices such as terminal, keyboard, mouse, printer
IR	Instruction register
Kbytes (KB)	Kilobytes: 1024 or 2^{10} bytes; a measurement of computer memory
LAN	Local area network
LIFO	Last in–first out
LSI	Large scale integration
Main memory	That memory seen by the CPU
Mbytes (MB)	Megabytes: 1,048, 576 or 2^{20} bytes; a measurement of computer memory
Memory	Generic term for random access storage
Microprocessor	Computer architecture with Central Processing Unit in one LSI chip
MODEM	Modulator-demodulator
MOS	Metal oxide semiconductor
Multiplexer	Device that switches several input sources, one at a time, to an output
Nibble	Four bits
Nonvolatile memory	As opposed to volatile memory, does not need power to retain its present state
OCR	Optical character recognition
OS	Operating system
OS memory	Memory dedicated to the OS, not usable for other functions
Parallel interface	A character (8-bit) or word (16-bit) interface with as many wires as bits in interface plus data clock wire.
Parity	Method for detecting errors in data: one extra bit carried with data, to make the sum of one bit in a data stream even or odd
PC	Program counter or personal computer
Peripheral devices	Input/output devices not contained in main processing hardware
Program	A sequence of computer instructions
PROM	Programmable read-only memory
Protocols	Established set of handshaking rules enabling communications
Pseudocode	An English-like way of representing programming control structures
RAM	Random access memory
Real time/Batch	Method of program execution: real-time implies immediate execution; batch mode is postponed until run on a group of related activities
RISC	Reduced instruction set computing
ROM	Read-only memory
Sequential storage	Memory (usually tape) accessed only in sequential order (n, $n + 1$, …)
Serial interface	Single data stream that encodes data by individual bit per clock period
Simplex communication	One-way communication
Static memory	Memory that does not require intermediate refresh cycles to retain state
Structured programming	Use of programming constructs, such as Do-While or If-Then-Else, to produce code that is logical and easy to follow
Synchronous	Communications mode in which data and clock are at same rate
Transmission speed	Rate at which data is moved, in baud (bits per second, bps)
Virtual memory	Addressable memory outside physical address bus limits through use of memory mapped pages
Volatile memory	Memory whose contents are lost when power is removed
Words	8, 16, or 32 bits
WYSIWYG	What you see is what you get
16-bit	Basic organization of data with 2 bytes per word
32-bit	Basic organization of data with 4 bytes per word
64-bit	Basic organization of data with 8 bytes per word

PROBLEMS

4.1 In spreadsheets, what is the easier way to write B1 + B2 + B3 + B4 + B5?

a. Sum (B1:B5) c. @B1..B5SUM
b. (B1..B5) Sum d. @SUMB2..B5

4.2 The address of the cell located at row 23 and column C is:

a. 23C c. C.23
b. C23 d. 23.C

4.3 Which of the following is *FALSE*?

a. Flowcharts use symbols to represent input/output, decision branches, process statements, and other operations.
b. Pseudocode is an English-like description of a program.
c. Pseudocode uses symbols to represent steps in a program.
d. Structured programming breaks a program into logical steps or calls to subprograms.

4.4 In pseudocode using DOWHILE, which of the following is *TRUE*?

a. DOWHILE is normally used for decision branching.
b. The DOWHILE test condition must be false to continue the loop.
c. The DOWHILE test condition tests at the beginning of the loop.
d. The DOWHILE test condition tests at the end of the loop.

4.5 A spreadsheet contains the following formulas in the cells:

	A	B	C
1		A1 +1	B1 +1
2	A1 ^2	B1^2	C1^2
3	Sum (A1:A2)	Sum (B1:B2)	Sum (C1:C2)

If 2 is placed in cell A1, what is the value in cell C3?

a. 12 c. 8
b. 20 d. 28

4.6 A spreadsheet contains the following:

	A	B	C	D
1		3	4	5
2	2	A$2		
3	4			
4	6			

If you copy the formula from B2 into D4, what is the equivalent formula in D4?

a. A$2 c. C4
b. C4 d. C$2

4.7 The hexadecimal number 2DB.A is most nearly equivalent to which decimal number?

a. 731.625
b. 731.10
c. 453.625
d. 341.10

4.8 The decimal number 1938.25 is most nearly equivalent to which hexadecimal number?

a. $(792.25)_{16}$
b. $(792.4)_{16}$
c. $(279.4)_{16}$
d. $(279.04)_{16}$

4.9 The type of office computer that would most likely be found at an engineer's desk is a:

a. mainframe
b. minicomputer
c. workstation
d. microcontroller

4.10 Which of the following is *NOT* considered an Internet search engine?

a. Bing
b. Google
c. Unix
d. Yahoo

4.11 Which term describes a worldwide network of computers for sharing information?

a. Beowulf cluster
b. Cloud computing
c. Internet
d. LAN

4.12 The baud rate of an old serial communication system is the same as the bit rate. If the baud rate is 9600, transmitting a 500 KB file would take:

a. 1 minute
b. 7 minutes
c. 2 minutes
d. 18 minutes

4.13 If the data rate for USB 3.0 is 4.8 billion bps, the file size that can be sent in 30 seconds is:

a. 1800 KB
b. 18 MB
c. 1.8 GB
d. 18 GB

4.14 Which of the following types of memory is considered to be nonvolatile?

a. Cache
b. DRAM
c. Flash
d. SRAM

4.15 Which of the following drives typically uses EEPROM memory?

a. CD
b. DVD
c. Hard drive
d. Thumb drive

4.16 Which of the following is *NOT* likely to be found as a subunit of a CPU?

a. ALU
b. Control unit
c. Registers
d. Transformer

4.17 Which of the following data types would give the *BEST* precision?

a. Floating point
b. Integer
c. Long integer
d. Unsigned integer

4.18 Which of the following is *NOT* a standard for encoding character data types?

a. ALPHA
b. ASCII
c. EBCDIC
d. Unicode

4.19 Which of the following programming language types would be the hardest for a human to code, read, and understand?

a. Assembly
b. High-level
c. Interpreted
d. Machine

4.20 Each of the following is another term for "function" *EXCEPT*:

a. declarative
b. method
c. procedure
d. subroutine

4.21 An application program is *MOST* likely to use which of the following to interact with the computer's operating system?

a. API
b. CAM
c. GUI
d. URL

4.22 Which of the following is *NOT* used as a safeguard for preventing unauthorized computer access?

a. Antivirus software
b. Biometrics
c. Firewall
d. Spoofing

4.23 A spreadsheet has the values 15, 18, and 32 in cells A1, A2, and A3, respectively. If cell B1 has the formula A1 + A2 in it, and cell B2 has the formula B1 + A3 in it, what will be displayed in cell B2?

a. 32
b. 55
c. 65
d. 98

4.24 If the formulas in cells B1 and B2 from Problem 4.23 are now copied to cells C1 and C2, respectively, what value will be displayed in cell C2?

a. 55
b. 65
c. 98
d. 130

4.25 If the conditions described in problems 4.23 and 4.24 exist, and if the value in cell A1 is changed from 15 to 10, which cells would show a change in values displayed?

a. All cell values would change.
b. Only cell A1would change.
c. Only cells A2 and A3 do not change.
d. Only row 1 would change.

4.26 The operation used within a spreadsheet to execute many previously recorded commands in a manner similar to running a program is a:

a. functional
b. macro
c. system tool
d. vector tool

4.27 In structured programming, how many possible exit points should there be for a loop structure?

a. One
b. Two
c. None
d. As many as desired

4.28 A change in the order of program execution is known as:

a. branching
b. inclusion
c. slippage
d. yielding

4.29 Giving a variable an initial value or modifying its value is known as:

a. an assignment
b. a disclosure
c. a fixture
d. a method

4.30 In a flowchart, an oval-shaped symbol represents:

a. a decision
b. initialization
c. input or output
d. program start or stop

4.31 Pseudocode is useful for developing an algorithm because:

a. it can be directly compiled into machine language
b. it has a very easy-to-read, English-like structure
c. it is highly structured, making mistakes unlikely
d. it requires the programmer to think about all parts of a problem simultaneously

SOLUTIONS

4.1 **a.** Sum (B1:B5) or @Sum (B1..B5)

4.2 **b.**

4.3 **c.** Pseudocode does not use symbols but uses English-like statements such as IF-THEN and DOWHILE.

4.4 **c.** IF-THEN is normally used for branching. The DOWHILE test condition must be true to continue branching, and the test is done at the beginning of the loop. The DOUNTIL test is done at the end of the loop.

4.5 **b.** Plugging 2 into cell A1 of the spreadsheet produces the following spreadsheet display:

	A	B	C
1	2	3	4
2	4	9	16
3	6	12	20

The value of C3 is 20.

4.6 **d.** The formula contains mixed references. The $ implies absolute row reference, whereas the column is relative. The result of any copy would eliminate any answer except for the absolute row 2 entry. The relative column reference A gets replaced by C. The cell contains C$2.

4.7 **a.**

$$(2DB)_{16} = 2 \times 16^2 + 13 \times 16^1 + 11 \times 16^0 = 512 + 208 + 11 = 731$$
$$(.A)_{16} = 10 \times 16^{-1} = .625$$
$$(2DB.A)_{16} = 731.625$$

4.8 **b.**

		Quotient		Remainder	Hexadecimal Digit
1938 ÷ 16	=	121	+	2	2
121 ÷ 16	=	7	+	9	9
7 ÷ 16	=	0	+	7	7

$$(1938)_{10} = (792)_{16}$$

		Integer		Fraction	Hexadecimal Digit
0.25 × 16	=	4	+	0.00	4

$$(.25)_{10} = (.4)_{16}$$
$$(1938.25)_{10} = (792.4)_{16}$$

4.9 **c.** The key terms are *office* and *desk* and *most likely*. Only a workstation or PC (microcomputer) would be most likely to be at an engineer's desk. A microcontroller could conceivably be at an engineer's desk, but it is only likely if the engineer works with embedded systems.

4.10 **c.** Unix is an operating system, not an Internet search engine.

4.11 **c.** The key term is *worldwide*. Only the Internet fits this description.

4.12 **b.** In this case, the problem explicitly states that the baud rate and the bit rate are the same. This means that the communication system can transmit 9600 *bits* per second. To find the total number of seconds needed to send the file, multiply the file size by 8 bits per byte and divide this result by the baud rate. Finally, convert the number of seconds to minutes by dividing by 60 seconds per minute.

$$\text{Total time} = \frac{\left(500\times 10^3\ \text{B}\right)\left(\dfrac{8\ \text{bits}}{1\ \text{B}}\right)}{\dfrac{9600\ \text{bits}}{\text{s}}} \times \frac{1\ \text{minute}}{60\ \text{s}} = 6.94\ \text{minutes}$$

Note that had the file been 500 KiB, the transmission would still have been about 7 minutes.

4.13 **d.** To find the file size (in bits), multiply the data transmission rate by the total time. Then, convert the answer to bytes by multiplying 1 B/8 bits. Finally, use the standard SI definitions for KB, MB, or GB to determine the file size as provided in the answer choices.

$$\text{File size} = \frac{4.8\times 10^9\ \text{b}}{\text{s}} \times 30\ \text{s} \times \frac{1\ \text{B}}{8\ \text{b}} \times \frac{1\ \text{GB}}{1\times 10^9\ \text{B}} = 18\ \text{GB}$$

4.14 **c.** All of the memory types are RAM except for Flash memory. Flash memory is a type of EEPROM, which is nonvolatile memory.

4.15 **d.** A thumb drive uses Flash memory, which is a type of EEPROM.

4.16 **d.** Transformers are used to increase or decrease alternating currents and voltages. They are not found in CPUs.

4.17 **a.** Answers b, c, and d are all forms of integer data types; they cannot represent fractional values. The floating point data type permits the radix point to "float," or move, so that all or most of the significant figures can be used to represent a fractional value. For instance, the integer data types have poor precision in representing π because they can only represent π as the integer 3. The single precision floating point data type can represent π as 3.141593.

4.18 **a.** ASCII, EBCDIC, and Unicode are all standards for encoding character data.

4.19 **d.** Machine language consists of 1s and 0s. It would be the hardest for a human to read and understand.

4.20 **a.** Methods, procedures, and subroutines are all synonymous terms for function.

4.21 **a.** An application program is most likely to use an API, an application program interface, to interact with the operating system.

4.22 **d.** Spoofing refers to a method of attempting to gain unauthorized computer access. It is not a preventative safeguard.

4.23 **c.** With values 15, 18, and 32 in cells A1, A2, and A3, respectively, and with the formulas A1 + A2 in cell B1 and B1 + A3 in cell B2, cell B1 will have 15 + 18 = 33 in it, and cell B2 will have 33 + 32 = 65. In this case, the absolute addressing is irrelevant to the final calculation. In other words, the answer would have been the same if cell B2 had contained the formula B1 + A3.

4.24 **d.** In this problem, the relative or absolute addressing scheme is important to the copying procedure. After copying, cell C1 will have the relative addressing formula B1 + B2. Cell C2 will have a mixture of relative addressing and absolute addressing. Cell C2 will contain C1 + A3. With these formulas, C1 will have the value 33 + 65 = 98, and C2 will have 98 + 32 = 130.

4.25 **c.** Given that the conditions in problems 4.23 and 4.24 exist, changing the value of A1 will change every cell that directly or indirectly references A1. Changing the value of A1 from 15 to 10 would give the following results: A1 = 10, A2 = 18, A3 = 32, B1 = 10 + 18 = 28, B2 = 28 + 32 = 60, C1 = 28 + 60 = 88, and C2 = 88 + 32 = 120. Thus, all cells change except for cells A2 and A3.

4.26 **b.** A macro is a sequence of spreadsheet commands that can be recorded and executed, as is done in traditional computer programming.

4.27 **a.** In structured programming, there should be only one exit point for each loop structure. Having more than one exit point makes debugging and maintaining code very difficult.

4.28 **a.** The term branch or branching indicates a change in the order of program execution.

4.29 **a.** An assignment statement assigns or changes a value within a variable.

4.30 **d.** An oval shape denotes program initiation (start) or termination (stop). Decisions are represented by diamonds. Inputs and outputs are represented by rhomboids. Processing steps (which might include an initialization procedure) are presented by rectangles.

4.31 **b.** Pseudocode is useful for developing an algorithm because of its natural, easy-to-read, English-like structure. It permits the programmer to capture the basic ideas and sequences of the algorithm on paper without having to worry about programming syntax and other rules.

CHAPTER 5

Ethics and Professional Practices

Ray W. James

OUTLINE

The topic of ethics and business practices represents about 7% of the morning FE/EIT exam. This is about the same fraction as represented by morning-section questions on subjects such as probability and statistics, material properties, strength of materials, and fluid mechanics, so it is clearly an important subject. Questions in the section on ethics and business practices of the exam may cover the following topics:

- Code of ethics (professional and technical societies)
- Agreements and contracts
- Ethical versus legal behavior
- Professional liability
- Public protection issues (for example, licensing boards)

MORALS, PERSONAL ETHICS, AND PROFESSIONAL ETHICS

To put professional ethics for engineers in perspective, it is helpful to distinguish it from morals and personal ethics. *Morals* are beliefs about right and wrong behaviors that are widely held by significant portions of a given culture. Obviously, morals will vary from culture to culture, and though some are common across different cultures, there seems to be no universal moral code.

Personal ethics are the beliefs that individuals hold that often are more restrictive than and sometimes contradictory to the morals of the culture. An example of personal ethics that might be more restrictive than morals might be the belief of an individual that alcohol should not be consumed in a culture that accepts the use of alcohol.

Professional ethics, on the other hand, is the formally adopted code of behavior by a group of professionals held out to society as that profession's pledge about how the profession will interact with society. Such rules or codes represent the agreed-on basis for a successful relationship between the profession and the society it serves.

The engineering profession has adopted several such codes of ethics, and different practitioners may adhere to or be bound by codes that vary by professional discipline but are similar in their basics. Codes adopted by the state boards of registration are typically codified into law and are legally binding for licensed engineers in the respective state. Codes adopted by professional societies are not legally binding but are voluntarily adhered to by members of those societies. The successful understanding of professional ethics for engineers requires an understanding of various codes; for purposes of examining registration applicants, the NCEES has adopted a "model code" that includes many canons common to most codes adopted nationwide.

CODES OF ETHICS

Codes of ethics are published by professional and technical societies and by licensing boards. Why are codes published and why are they important? These fundamental questions are at the heart of the definition of a "profession." Some important aspects of the definition of a profession might include skills and knowledge vital to society; extensive and intellectual education and training important for proper practice in the profession; an importance of autonomous action by practitioners; a recognition by society of these aspects, leading to a governmentally endorsed monopoly on the practice of the profession; and a reliance on published standards of ethical conduct, usually in the form of a code of ethics (Harris, et al., 2005). Such codes are published and followed to maintain a high standard of confidence in the profession by the public served by the practicing professionals, because without high standards of confidence, the ability of a profession to serve the public need may be seriously impaired.

The FE exam questions on ethics and business practices are based on the NCEES code of ethics, a concise body of model rules designed to guide state boards and practitioners as a model of good practice in the regulation of engineering. These rules do not bind any engineer, but the codes of ethics published by individual state boards and of professional societies will be very similar to these in principle.

NCEES Model Rules of Professional Conduct

A. Licensee's Obligation to Society

1. Licensees, in the performance of their services for clients, employers, and customers, shall be cognizant that their first and foremost responsibility is to the public welfare.
2. Licensees shall approve and seal only those design documents and surveys that conform to accepted engineering and surveying standards and safeguard the life, health, property, and welfare of the public.
3. Licensees shall notify their employer or client and such other authority as may be appropriate when their professional judgment is overruled under circumstances where the life, health, property, or welfare of the public is endangered.
4. Licensees shall be objective and truthful in professional reports, statements, or testimony. They shall include all relevant and pertinent information in such reports, statements, or testimony.
5. Licensees shall express a professional opinion publicly only when it is founded upon an adequate knowledge of the facts and a competent evaluation of the subject matter.
6. Licensees shall issue no statements, criticisms, or arguments on technical matters which are inspired or paid for by interested parties, unless they explicitly identify the interested parties on whose behalf they are speaking and reveal any interest they have in the matters.
7. Licensees shall not permit the use of their name or firm name by, nor associate in the business ventures with, any person or firm which is engaging in fraudulent or dishonest business or professional practices.
8. Licensees having knowledge of possible violations of any of these Rules of Professional Conduct shall provide the board with the information and assistance necessary to make the final determination of such violation. (Section 150, Disciplinary Action, NCEES Model Law)

B. Licensee's Obligation to Employer and Clients

1. Licensees shall undertake assignments only when qualified by education or experience in the specific technical fields of engineering or surveying involved.
2. Licensees shall not affix their signatures or seals to any plans or documents dealing with subject matter in which they lack competence, nor to any such plan or document not prepared under their direct control and personal supervision.
3. Licensees may accept assignments for coordination of an entire project, provided that each design segment is signed and sealed by the licensee responsible for preparation of that design segment.
4. Licensees shall not reveal facts, data, or information obtained in a professional capacity without the prior consent of the client or employer except as authorized or required by law. Licensees shall not solicit or accept

gratuities, directly or indirectly, from contractors, their agents, or other parties in connection with work for employers or clients.

5. Licensees shall make full prior disclosures to their employers or clients of potential conflicts of interest or other circumstances which could influence or appear to influence their judgment or the quality of their service.
6. Licensees shall not accept compensation, financial or otherwise, from more than one party for services pertaining to the same project, unless the circumstances are fully disclosed and agreed to by all interested parties.
7. Licensees shall not solicit or accept a professional contract from a governmental body on which a principal or officer of their organization serves as a member. Conversely, licensees serving as members, advisors, or employees of a government body or department, who are the principals or employees of a private concern, shall not participate in decisions with respect to professional services offered or provided by said concern to the governmental body which they serve. (Section 150, Disciplinary Action, NCEES Model Law)

C. Licensee's Obligation to Other Licensees

1. Licensees shall not falsify or permit misrepresentation of their, or their associates', academic or professional qualifications. They shall not misrepresent or exaggerate their degree of responsibility in prior assignments nor the complexity of said assignments. Presentations incident to the solicitation of employment or business shall not misrepresent pertinent facts concerning employers, employees, associates, joint ventures, or past accomplishments.
2. Licensees shall not offer, give, solicit, or receive, either directly or indirectly, any commission, or gift, or other valuable consideration in order to secure work, and shall not make any political contribution with the intent to influence the award of a contract by public authority.
3. Licensees shall not attempt to injure, maliciously or falsely, directly or indirectly, the professional reputation, prospects, practice, or employment of other licensees, nor indiscriminately criticize other licensees' work. (Section 150, Disciplinary Action, NCEES Model Law)

Many ethical questions arise in the formulation of business practices. Professionals should appreciate that expressions like "all is fair in business" and "let the buyer beware" can conflict with fundamental ideas about how a professional engineer should practice. The reputation of the profession, not only the individual professional, is critically important to the ability of all engineers to discharge their duty to protect the public health, safety, and welfare.

The *NCEES Model Rules* addressing a licensee's obligation to other licensees prohibit misrepresentation or exaggeration of academic or professional qualifications, experience, level of responsibility, prior projects, or any other pertinent facts that might be used by a potential client or employer to choose an engineer.

The *Model Rules* also prohibit gifts, commissions, or other valuable consideration to secure work. Political contributions intended to influence public authorities responsible for awarding contracts are also prohibited.

Often, these rules are misunderstood in the arena of foreign practice. Increasingly, engineering is practiced globally, and engineers must deal with foreign clients and foreign governmental officials, many times on foreign soil where laws and especially cultural practices vary greatly. In the United States, the federal Foreign

Corrupt Practices Act (FCPA) is a relatively recent recognition and regulation of this problem. Among other purposes, it provides clearer legal boundaries for U.S. engineers involved with international projects.

According to the FCPA, it is not illegal for a U.S. engineer to make petty extortion payments ("grease payments," "expediting payments," and "facilitating payments" are common expressions) to governmental officials when progress of otherwise legitimate projects is delayed by demands for such payments consistent with prevailing practice in that country. It is illegal, however, for U.S. engineers to give valuable gifts or payments to develop contracts for *new business*. In some cultures, reciprocal, expensive gift giving is an important part of business relationships, and the reciprocal nature of this practice can make it acceptable under the FCPA. Most commonly, when the engineer's responsibilities include interactions with foreign clients or partners, the engineer's corporate employer will publish detailed and conservative guidelines intended to guide the engineer in these ethical questions.

The practicing engineer should always be watchful of established and, especially, new business practices to be sure the practices are consistent with the codes of ethics he or she is following.

Example **5.1**

The *NCEES Model Rules of Professional Conduct* allow an engineer to do which one of the following?

(i) Accept money from contractors in connection with work for an employer or client

(ii) Compete with other engineers in seeking to provide professional services

(iii) Accept a professional contract from a governmental body even though a principal or officer of the engineer's firm serves as a member of the governmental body

(iv) Sign or seal all design segments of the project as the coordinator of an entire project

Solution

Although the other items are not allowed by the *Model Rules*, nowhere does it say that an engineer cannot compete with other engineers in seeking to provide professional services. But, of course, he or she should conduct business in an ethical manner. The correct answer is (ii).

AGREEMENTS AND CONTRACTS

One aspect of business practice is understanding the concepts and terminology of agreements and contracts.

Elements of a Contract

Contracts may be formed by two or more parties; that is, there must be a party to make an offer and a party to accept.

To be enforceable in a court of law, a contract must contain the following five essential elements:

1. There must be a mutual agreement.
2. The subject matter must be lawful.
3. There must be a valid consideration.
4. The parties must be legally competent.
5. To be a formal contract, the contract must comply with the provisions of the law with regard to form.

A *formal contract* depends on a particular form or mode of expression for legal efficacy. All other contracts are called *informal contracts* since they do not depend on mere formality for their legal existence.

Contract and Related Legal Terminology

Case law—the body of law created by courts interpreting statute law. Judges use precedents, the outcome of similar cases, to construct logically their decision in a given issue.

Changed or concealed conditions—in construction contracting, it is important to specify how changed or concealed conditions will be handled, usually by changes in the contract terms. For example, if an excavation project is slowed by a difficult soil pocket between soil corings, the excavation contractor may be able to support a claim for increased costs due to these unforeseen conditions. When the concealed conditions are such that they should have been foreseen, such claims are more difficult to support.

Common law—the body of rules of action and principles that derive their authority solely from usage and customs.

Damages for delays—in many contracts, completion time is an important concern, and contractual clauses addressing penalties for delays (or rewards for early completion) are often incorporated.

Equal or approved equivalent—terms used in specifications for materials to permit use of alternative but equal material when an original material is not available or an equivalent material can be obtained at lower cost. The engineer is responsible for approving the alternative material.

Equity—system of doctrines supplementing common and statute law, such as the Maxims of Equity.

Errors and omissions—term used to describe the kind of mistakes that can be made by engineers and architects leading to damage to the client. Often, this risk is protected by liability insurance policies.

Force account—a method of work by which the owner elects to do work with his or her own forces instead of employing a construction contractor. Under this method, the owner maintains direct supervision over the work, furnishes all materials and equipment, and employs workers on his or her own payroll.

Hold harmless—clauses are often included requiring one party to agree not to make a claim against the other and sometimes to cooperate in the defense of the other party if a claim is made by a third party.

Incorporate by reference—the act of making a document legally binding by referencing it within a contract, although it is not attached to or reproduced in the contract. This is done to eliminate unnecessary repetition.

Indemnify—to protect another person against loss or damage, as with an insurance policy.
Liquidated damages—a specific sum of money expressly stipulated as the amount of damages to be recovered by either party for a breach of the agreement by the other.
Mechanics' liens—legal mechanism by which unpaid contractors, suppliers, mechanics, or laborers are allowed to claim or repossess construction materials that have been delivered to the worksite in lieu of payment.
Plans—the drawings that show the physical characteristics of the work to be done. The plans and specifications form the guide and standards of performance that will be required.
Punitive damages—a sum of money used to punish the defendant in certain situations involving willful, wanton, malicious, or negligent torts.
Specifications—written instructions that accompany and supplement the plans. The specifications cover the quality of the materials, workmanship, and other technical requirements. The plans and specifications form the guide and standards of performance that will be required.
Statute law—acts or rules established by legislative action.
Statute of limitations—a time limit on claims resulting from design or construction errors, usually beginning with the date the work was performed, but in some cases beginning on the date the deficiency could first have been discovered.
Surety bond—bonds issued by a third party to guarantee the faithful performance of the contractor. Surety bonds are normally used in connection with competitive-bid contracts, namely, bid bonds, performance bonds, and payment bonds.
Workers' compensation—insurance protecting laborers and subcontractors in case of an on-the-job injury; it is often required of contractors.

ETHICAL VERSUS LEGAL BEHAVIOR

Engineers have a clear obligation to adhere to all laws and regulations in their work—what they do must be done legally. But the obligation goes beyond this. Unlike the world of business where cutthroat but legal practices are commonly condoned and frequently rewarded, engineers assume important obligations to the public and to the profession that restrict how they must practice and that often are much more stringent than law or regulation.

When you realize that restricting the practice of engineering to certain licensed professionals by the state is essentially a state-provided monopoly, you may begin to see why there is a difference. Competitive businesses compete in many ways to gain the kind of advantage in their field that engineers and other licensed professionals are given by the state.

Aggressive advertising is one example of a business practice that engineers avoid, even though it is not illegal or prohibited. Before 1978, it was common for professional societies to prohibit or narrowly restrict advertising by their practitioners; however, in 1978 the U.S. Supreme Court ruled such broad restrictions unconstitutional, allowing only reasonable restraints on advertising by professional societies. Since that time, engineering societies have adopted guidelines on advertising. Other professions have been less successful in regulating advertising. For example, the profusion of television advertising by lawyers, and the language of those advertisements, contrasts with the practice of engineering professionals where advertising is more commonly seen in technical journals or trade literature. Many believe the legal profession has suffered a loss of respect as a result, while the profession of engineering still is held in high regard by the public. It is in the interest of the engineering profession to avoid this kind of advertising, even though it is legal, because it can damage the reputation of the profession.

Another example of the importance of self-regulation is the engineer's responsibility to the environment. Although many laws and regulations restrict engineering practices that might damage the environment, there are still many legal ways to accomplish engineering projects that can have adverse environmental effects. Increasingly, codes of ethics are adding requirements for the engineer to consider the environment or the "sustainability" of proposed engineering projects. The engineer's ethical responsibility to work toward sustainable development may go beyond any legal requirements intended to prevent environmental damage.

Conflicts of Interest

A conflict of interest is any situation where the decision of an engineer can have some significant effect on his or her financial situation. It would be a clear conflict of interest for a designing engineer to specify exclusively some component that is only available from a supplier in which that engineer has a significant financial interest, when other components from other suppliers would serve equally well. Engineers must avoid even the *appearance* of a conflict of interest. This is critically important for the reputation of the profession, which the engineer is charged with protecting, in order for engineers to effectively serve the public interest.

An apparent conflict of interest is any situation that might appear to an outside observer to be an actual conflict of interest. For example, if the engineer in the case mentioned above had subsequently divested himself of all interest in the supplier, there is no longer an actual conflict of interest. However, to an outside observer with imperfect information, there might be the appearance of a conflict, resulting in the perception of unethical behavior in the public eye.

The usual remedy for conflicts of interest and apparent conflicts of interest is disclosure and, often, recusal. The engineer's interest must be disclosed in advance, generally to a supervisor, and recusal must at least be discussed. In many cases, recusal may not be necessary, but disclosure is vitally important. In every case, the public perception of the conflict must be considered, with the goal of protecting the reputation of the individuals and the profession.

PROFESSIONAL LIABILITY

Good engineering practice includes numerous checks and conservative principles of design to protect against blunders, but occasional errors and omissions can result in damage or injury. The engineer is responsible for such damage or injury, and it is good practice to carry errors and omissions insurance to provide appropriate compensation to any injured party, whether a client or a member of the public. Such insurance can be a significant cost in some fields of engineering, but it represents a cost of doing business that should be reflected in the fees charged. The most important factor in preventing errors and blunders is to provide adequate time for careful review of all steps in the project by knowledgeable senior licensed engineers. Frantic schedules and unrealistic deadlines can significantly increase the risk.

PUBLIC PROTECTION ISSUES

State boards in all 50 states and the District of Columbia are charged by their states with the responsibility for the licensing of engineers and the regulation of the practice of engineering to protect the health, safety, and welfare of the public. Licensed engineers in each state are legally bound by laws and regulations published by the respective state board. The boards are generally made up of engineers appointed by the state governor; sometimes nonengineering members also are appointed to make sure the public is adequately represented.

State boards commonly issue cease and desist letters to nonengineers who have firms or businesses with names that imply engineering services are being offered to the public or who may actually be offering engineering services without the required state license. These boards also regulate the practice of engineering by their registrants, often sanctioning registrants for inappropriate business practices or engineering design decisions. Many boards require continuing education by registrants for maintenance of proficiency. A weakness of many boards is in the area of discipline for incompetent practices, but this weakness is often offset by tort law whereby incompetent practitioners who cause damage or injury are commonly subject to significant legal damages.

REFERENCES

Harris, Charles E., Jr., Michael S. Pritchard, and Michael J. Rabins. *Engineering Ethics: Concepts and Cases*. Thompson Wadsworth, 2005.

National Council of Examiners for Engineering and Surveying. *Model Rules*, September 2006.

PROBLEMS

5.1 Jim is a PE working for an HVAC designer who often must specify compressors and other equipment for his many clients. He reports to Joan, the VP of engineering. Jim specifies compressors from several different manufacturers and suppliers based on the technical specifications and on his experience with those products in past projects. Joan's long-time friend Charlie, who has been working in technical sales of construction materials, takes a new job with one of the compressor suppliers that Jim deals with from time to time. Charlie calls on Joan, inviting her and any of her HVAC designers to lunch to discuss a new line of high-efficiency compressors; Joan invites Jim to come along. Jim should:

a. decline to attend the lunch, citing concerns about conflict of interest
b. agree to attend the lunch but insist on paying for his own meal
c. agree to attend the lunch and learn about the new line of compressors
d. report Joan to the state board and never specify compressors from that supplier again

5.2 Harry C. is an experienced geotechnical engineer who has many years' experience as a PE designing geotechnical projects and who is very familiar with the rules regarding the requirement for trench shoring and trench boxes to protect construction workers during excavations. During a vacation visit to a neighboring state, he observes a city sewer construction project with several workers in an unprotected deep trench, which, to Harry's experienced eye, is probably not safe without a trench box or shoring. Harry should:

a. remember that he is not licensed in the neighboring state and has no authority to interfere
b. approach the contractor's construction foreman and insist that work be halted until the safety of the trench is investigated
c. advise all the workers in the trench that they are in danger and encourage them to go on strike for safer conditions
d. contact the city engineer to report his concerns

5.3 Engineering student Travis is eagerly anticipating his graduation in three months and has interviewed with several firms for entry-level employment as an electrical engineer. He has received two offers to work for firms A and B in a nearby city, and after comparing the jobs, salaries, and benefits and discussing the choice with his faculty advisor, he telephones firm A whose offer is more appealing and advises them he will accept their job offer. Two weeks later he is contacted by firm C in a different city with a job offer that includes a salary more than 15% higher than the offer he has accepted plus a generous relocation allowance. Travis should:

a. decline the offer from firm C, explaining that he has already accepted a position
b. contact firm A and ask if he can reconsider his decision
c. contact firm A to give them a fair chance to match the offer from firm C
d. advise firm C that he can accept their offer if they will contact firm A to inform them of this change

5.4 EIT Jerry works for a small civil engineering firm that provides general civil engineering design services for several municipalities in the region. He has become concerned that his PE supervisor Eddie is not giving careful reviews to Jerry's work before sealing the drawings and approving them for construction. Jerry asks Eddie to review with him the design assumptions from Jerry's latest design, a steel fire exit staircase to be added to an elementary school building, because he has concerns about the appropriate design loadings. But before the design assumptions are reviewed, Jerry notices the drawings have been approved and released to the fabricator. Jerry should:

a. quit his job and find another employer
b. take a review course in live loadings for steel structures
c. in the future mark each drawing he prepares "Not Approved for Construction"
d. None of the above

5.5 Dr. Willis Hemmings, PE, is an engineering professor whose research in fire protection engineering is nationally recognized. He is retained as an expert witness for the defendant, a structural engineering design firm, in a lawsuit filed by a firefighter who was injured while fighting a fire in a steel structure that collapsed during the fire. The plaintiff's lawyer alleges that the original design of certain components of the fire protection system protecting the steel structural members was inadequate. Hemmings reviews the original design documents, which call for a protective coating that is slightly thinner than is required by the local building code. Hemmings testifies that even though the specified coating is thinner than required, he believes that the design was sound because the product used is applied by a new process that is probably more efficient and the thinner coating probably gave the same level of protection. He bases his testimony on his national reputation as an expert in this field. Such expert testimony is:

a. a commonly accepted method of certifying good engineering design in tort law
b. legal only when given by a licensed professional engineer like Hemmings
c. unethical because it contradicts accepted practice without supporting tests or other data
d. effective only because Hemmings is involved in cutting-edge research

5.6 Jackie is a young PE who works for a garden tool manufacturer that has produced about 100,000 shovels, rakes, and other garden implements annually for more than 20 years. The company recently won a contract to manufacture and supply 5000 folding entrenching tools of an existing design to a Central American military client. The vice president of marketing has been working to develop contracts with other military clients and asks Jackie to prepare a statement of qualifications (SOQ). Jackie is asked to describe the design group (consisting of two engineers, one EIT, one student intern, three CAD technicians, and one IT technician) as a "team of eight tool design engineers," and to describe the company as "experienced in the design, testing, and manufacturing of military equipment, with a recent production history of over two million entrenching tools and related hardware." Jackie should:

a. check the production records to be sure the figures cited are accurate
b. ask the vice president to sign off on the draft of the SOQ
c. object to describing the qualifications and experiences of her group in an exaggerated way
d. be sure to mention that she is a PE and list the states in which she is licensed

5.7 The Ford Motor Company paid millions of dollars to individuals injured and killed in crashes of the Ford Pinto, which had a fuel tank and filler system that sometimes ruptured in rear-end collisions, spilling gasoline and causing fires. While many considered the filler system design deficient because of this tendency, one important factor played a role in the lawsuits. An internal Ford memo was discovered that included the cost-benefit calculations Ford managers used in making the decision not to improve the tank/filler system design. This memo was significant because:

a. it is unethical to use the cost-benefit method for safety-related decisions
b. it is illegal to estimate the value of human life in cost-benefit calculations
c. state law requires estimates of the value of human life be at least $500,000 in such calculations
d. None of the above

5.8 Charles is tasked to write specifications for electric motors and pumps for a new sanitary sewage treatment plant his employer is designing for a municipal client. Charles is concerned that he doesn't have a very good knowledge about current pump design standards but is willing to learn. Charles's fiancée is an accountant employed by a pump distributor and offers to provide Charles with a binder of specifications for all the pumps her firm distributes. Charles should:

a. decline to accept the binder, citing concerns about conflict of interest
b. accept the binder but turn it over to his employer's technical librarian without reviewing it
c. accept the binder and study the materials to gain a better understanding of pump design and specifications
d. ask his fiancée if she knows an applications engineer at her firm who would draft specifications for him

5.9 Professor Martinez is a PE who teaches chemical engineering classes at a small engineering school. His student, Erica, recently graduated and took a job with WECHO, a small firm that provides chemicals and support to oil well drilling operations. WECHO has never employed an engineer and has hired Erica, partly on Prof. Martinez's strong recommendation, in hopes that she will one day become their chief engineer. After she has worked at WECHO for about two years, her supervisor Harry calls Prof. Martinez to explain that WECHO has been required to complete an environmental assessment before deploying a new surfactant, and the assessment must be sealed by a PE. Harry explains that Erica has done all the research to collect data and answer questions on the assessment, and everyone at WECHO agrees that she has done a superb job in completing the assessment, but it still requires the seal of a PE before submission. Harry asks Prof. Martinez if he can review Erica's work and seal the report, reminding him that he had given a glowing recommendation of Erica at the time WECHO hired her. Prof. Martinez should:

a. negotiate a consulting contract to allow him sufficient time and funding to review the report before sealing it
b. require Erica to first sign the report as an EIT and graduate engineer before reviewing it
c. require WECHO to purchase a bond against environmental damage before sealing the report
d. decline to review or seal the report, citing responsible charge issues

5.10 Jack Krompten, PE, is an experienced civil engineer working for a land development firm that has completed several successful residential subdivision developments in WoodAcres, a suburban bedroom community of a large, sprawling, and rapidly growing city. The WoodAcres city engineer, who also served half-time as the mayor, has retired, and the city council realizes that with rapid growth ahead it will be important to hire a new city engineer. They approach Krompton with an offer of half time city engineer, suggesting that he can keep his current job while discharging the responsibilities of the city engineer—primarily reviewing plans for future residential subdivision developments in WoodAcres. Krompton should:

a. recognize that by holding two jobs he is being paid by two parties for the same work
b. insist that he can only accept the offer if his present employer agrees to reduce his responsibilities to half-time
c. recognize that a 60-hour workweek schedule will take time away from his family
d. recognize that this arrangement will probably create a conflict of interest and refuse the offer

5.11 Willis is an aerospace engineering lab test engineer who works for a space systems contractor certifying components for spacecraft service. He is in charge of a team of technicians testing a new circuit breaker design made of lighter weight materials intended for service in unpressurized compartments in rockets and spacecraft. The new design has passed all tests except for some minor overheating during certain rare electrical load conditions. The lead technician notices that this overheating does not occur when a fan is used to cool the test apparatus and proposes to run the test with the fan to complete the certification process. He points out that the load conditions will only occur during thruster operation in space, which is a much colder environment. Willis should:

a. agree to the lead technician's suggestion, since he has many years of experience in testing and certification
b. agree to run the test as suggested but include a footnote explaining the use of the fan
c. insist on running the test as specified without the use of the fan
d. report the technician to the state board for falsifying test reports

5.12 Shamar is a registered PE mechanical engineer assigned as a project manager on a new transmission line project. He is tasked to build a project team to include several engineers and EITs that will be responsible for design and construction of 7.6 miles of high tension transmission lines consisting of steel towers and aluminum conductors in an existing right of way. He realizes that foundation design and soil mechanics will be an important technical area to his project, and he has never studied these subjects. He wonders if he is qualified to supervise such a project. He should:

a. meet with his supervisor to decline the assignment
b. decline the assignment and contact the state board to report that he is being asked to take responsibility for tasks he is not knowledgeable about
c. accept the assignment and check out an introductory soil mechanics textbook from the firm's technical library
d. accept the assignment and be sure his team includes licensed engineers with expertise in these areas

5.13 Julio is a design engineer working for a sheet metal fabricating firm. He is tasked with the design of a portable steel tank for compressed air to be mass produced and sold to consumers for pressurizing automobile tires. He designs a cylindrical tank to be manufactured by rolling sheet metal into a cylinder, closing with a longitudinal weld along the top, and welding on two elliptical heads. His design drawings are approved by his supervisor, Sonja, a licensed engineer, and by the vice president of manufacturing, but when the client reviews the designs, he asks the VP to change the design so that the longitudinal weld along the top is moved to the bottom where it will not be visible to improve the esthetics and marketability of the product. The VP agrees with this change. Julio learns of this change and objects, citing concerns about corrosion at the weld if it is on the bottom. Sonja forwards Julio's objection with a recommendation against the change to the VP, with a copy to the client, but the VP insists, saying esthetics is very important in this product. Julio should:

a. accept the fact that esthetics governs this aspect of the design
b. write a letter to the client stating his objections
c. put a clear disclaimer on the drawing indicating his objections
d. contact the state board to report that his recommendation has been overruled by the VP

5.14 William is a PE who designs industrial incineration systems. He is working on a system to incinerate toxic wastes, and his employer has developed advanced technology using higher temperatures and chemical-specific catalyst systems that minimize the risk to the environment, workers, and the public. A public hearing is scheduled to address questions of safety and environmental risk posed by the project, and William is briefed by the corporate VP for public affairs about how to handle questions from the public. He is told to buy a new suit, project an air of technical competence, point out that his firm is the industry leader with many successful projects around the world, and describe the proposed system as one with "zero risk" to the public. William should:

a. follow his instructions to the letter
b. insist that his old suit is adequate, because he refuses to appear more successful than he really is, but follow the other instructions
c. follow all instructions, except use the term "minimal risk" rather than "zero risk"
d. resign from his position and look for a different employer who won't ask him to face the public

5.15 Darlene is a metallurgical engineering EIT who works for a firm that manufactures automotive body panels. She has been tasked with improvements to the design of inner fender and trunk floor panels to reduce corrosion damage. After several weeks of study and comparison of alternatives, she submits a new trunk floor panel design utilizing a weldable stainless steel that will significantly reduce corrosion compared to the galvanized carbon steel alternatives she has been considering. The new panels will cost more, however, and after much study and debate, the VP of manufacturing rejects her design and approves an alternative made of a cheaper material. Darlene should:

a. accept the decision and work to finalize a workable design
b. resign from her position, since her employer has lost confidence in her
c. contact the state board to advise that her design decision has been overturned by a nonengineering manager
d. None of the above

5.16 Matt is a young PE who has just started his own consulting practice after six years of work with a small consulting firm providing structural engineering design services to architects. He has worked on steel and timber framed churches, prestressed and reinforced concrete parking structures, and many tilt-up strip center buildings. His expertise has been in the area of design of tilt-up concrete construction, where he has developed some innovative details regarding reinforcement at lifting points. His building designs, when constructed by experienced contractors, have reduced construction times and costs. Because of his expertise, he is approached by lawyer Marlene, who tells Matt that she represents a construction worker who is suing a project owner, contractor, and designer over a construction accident in which a tilt-up wall was dropped during construction, seriously injuring several workers. Marlene asks Matt to serve as an expert witness to assess the design and construction practices in the project and testify as to the causes of the accident. Marlene has taken the case on a contingency fee basis, in which she will earn 40% of any settlement, and she asks Matt if he would rather be paid by the hour for his study and testimony or instead accept 5% of any settlement, which she believes could be as high as $25 million. Matt should:

a. compare the 5% contingency fee with an expected fee based on his hourly rates, realizing that there is some chance he will earn nothing
b. be sure to have Marlene put the contingency fee arrangement in the form of a legal contract
c. decline the contingency fee arrangement and bill on an hourly basis
d. accept the contingency fee arrangement but donate the difference over his hourly rate to charity

5.17 Victor is a consulting engineer who is also in charge of a crew providing land surveying and subdivision design services to developers. He has been contracted to provide a survey of a 14-acre tract where a local developer is contemplating a subdivision, and he realizes that his crew had surveyed this same tract last month for another developer who has abandoned the project. He reprints the survey drawings, changing the title block for the new client. With respect to billing for the drawings, Victor should:

a. bill the new client the same as he billed the original client to be fair to both
b. bill the new client for half of the amount billed to the original client
c. bill the new client only for any work he did to change the drawings and reprint them
d. provide the new client with the drawings without any charge

5.18 Frank is a PE who works for ELEC, an electrical engineering design and construction firm. Frank's job is estimating construction costs, bidding construction projects, and supervision of design of electrical systems for buildings. Frank's bright EIT of five years, Linda, has just received her PE license and has resigned her position to open her own consulting business in a nearby community. Until she is replaced, Frank will have to also do all detail design of electrical systems for their projects. Frank receives a request for proposal (RFP) from a general contractor regarding design of electrical systems for a local independent school district. He realizes that his firm may be in competition with Linda for the engineering design, and knowing that Linda's salary was about half of his, he expects she may have a competitive advantage. Frank should:

a. ask Linda not to bid on this project
b. remind his contact with the general contractor that Linda has just left his firm, that she is inexperienced, and hint that she was sometimes slow to complete her design assignments
c. emphasize his 18 years of experience and subsequent design efficiency in his proposal
d. promote a CAD technician to a designer position so he can show a lower billing rate for engineering design hours

5.19 It is important to avoid the appearance of a conflict of interest because:

a. the engineer's judgment might be adversely affected
b. the engineer's client might suffer financial damages
c. the appearance of a conflict of interest is a misdemeanor
d. the appearance of a conflict of interest damages the reputation of the profession

5.20 The code of ethics published by the American Society of Civil Engineers is:

a. legally binding on all licensed engineers practicing civil engineering
b. adhered to voluntarily by members of the ASCE as a condition of membership
c. legally binding on all engineers with a degree in civil engineering
d. published only as a training guideline for young civil engineers

5.21 Which statement *MOST* accurately describes an engineer's responsibility to the environment?

a. The engineer has a legal obligation to make sure all development is sustainable.

b. The engineer has no obligation to the environment beyond protecting public health and safety.

c. The engineer has a moral obligation to consider the impact of his or her work on the environment.

d. The engineer's environmental responsibility is primarily governed by specific state laws.

SOLUTIONS

5.1 c. Jim can accept this invitation. We can assume there is no corporate policy prohibiting or restricting lunch invitations since VP Joan has accepted the invitation; therefore, there is no reason for Jim to decline the invitation. The opportunity to learn more about the new product is useful to him, his employer, and his clients; the cost of the lunch presumably would not be considered a "valuable" gift; and the lunch would not create either a conflict of interest or the appearance of a conflict to a reasonable person. If instead of lunch the offer involved a 10-day elk hunting trip or a vacation in the south of France, the solution would be very different because of the obvious "value" of the gift.

5.2 d. Doing nothing (a) is not an option if Harry really believes the trench represents a serious hazard to the workers. His code of ethics requires him to remember that his first and foremost responsibility is the public welfare, which includes the safety of the construction workers. Answers (b) and (c) are not the best way to proceed; his concerns should be reported to an engineer with some authority over the project. Since this is a city-contracted sewer improvement project, the city engineer will have project responsibility and will be the appropriate individual for Harry to take his concerns to.

5.3 a. Travis should decline the offer from firm C, explaining that he has already accepted an offer to work for firm A. While the *NCEES Model Rules* don't specifically address issues of personal integrity, it is clear that the engineer's obligation to employer and clients will not be satisfied by any decision that ignores Travis's verbal agreement to employment with firm A. Furthermore, such actions will tarnish his integrity in the eyes of firm A and by implication will harm the reputation and credibility of other students and the profession.

5.4 d. None of the first three solutions will address the concerns Jerry has raised about the safety of the particular project in question, so (d) is the correct solution. He should instead meet with Eddie to discuss the details of his design and make a determination if the design is completed safely. If it isn't, he will need to take further action to stop fabrication and construction while the design is reviewed and possibly modified to address any deficiencies. After this is done, he might want to consider all three of the other choices for his future. If he is thwarted in these responsibilities by Eddie, he should contact the state board with his concerns.

5.5 c. Hemmings cannot offer expert opinion that is contrary to accepted engineering practice without supporting that opinion with computer modeling, lab test results, or study of the literature. He can offer expert opinion that the design is not in line with accepted engineering practice without any supporting calculations, but he can't maintain that a substandard practice is acceptable without rational supporting evidence. Hemmings's credentials and experience may qualify him as an expert, but they do not relieve him of the requirement to base his professional opinion on facts.

5.6 **c.** Jackie should object to the request to exaggerate the size, qualifications, and experience level of her design group. The *Model Rules* require engineers to "be objective and truthful" in all professional matters, and the suggested exaggerations are clearly in conflict with this requirement.

5.7 **d.** It is not unethical or illegal to use the cost-benefit method, nor are certain values for human life prescribed by law; the answer is none of the above. The assumed values and calculations in the memo may have appeared callous or inflammatory to the juries in the resulting lawsuits, but they were not unethical or illegal. They may have been imprudent—an important lesson is that the public (jury) apparently objected to a design decision that increased the risk of a post-crash fire for such a small net benefit. Even though risk of death had been considered by the designers, the mode of death (burning to death in otherwise survivable crashes) was a factor in the strong reactions by the juries.

5.8 **c.** Studying products from a distributor is perfectly acceptable and can be a good way to gain a better understanding of pump equipment on the market. Accepting the binder does not represent a conflict of interest as implied by choices (a) and (b). Option (d) would be clearly setting himself up for an apparent conflict of interest.

5.9 **d.** Since Prof. Martinez has not been in responsible charge of the development of the assessment, he can't seal it, regardless of the level of review or the capabilities of the EIT who has done the work. Engineering students should be cautious in accepting a position as the sole engineering employee in a small firm; they are first encouraged to gain experience as an EIT under the guidance of an experienced PE and qualify for licensure as a PE before taking a position where licensure might be needed.

5.10 **d.** Krompton should recognize that the proposed arrangement will create a conflict of interest by placing him in charge of reviewing and approving plans from developers and potential developers that are in direct competition with his own employer. Any decision by him that might tend to make development less profitable for other developers could be advantageous to his employer by making their services more cost effective. Even if he were able to make all decisions rationally and without bias, the appearance of a conflict of interest would be very real and would cause a loss of credibility in the city engineer's actions and damage the reputation of the engineering profession.

5.11 **c.** Willis should insist on running the test as specified. The technician's suggestion to use a fan to "fudge" the test is technically indefensible, as well as unethical, in any case—the fan simulates convection cooling, which does not occur in the vacuum of space. Even if the technician had suggested a way to simulate an increased radiative heat flux, changes to a specified test procedure are not made casually. Much more study, documentation, and higher level approvals are involved.

5.12 **d.** Shamar should accept the assignment and select team members so that all areas of needed expertise are represented by licensed individuals who can seal appropriate portions of the plans. Option (a) is not recommended for an engineer's career advancement—he is expected to accept assignments of increasing responsibility; (b) is detrimental to his career—he will cause unnecessary concern with the board and with his supervisors; and (c) studying an introductory soil mechanics textbook may give him a better understanding of the problem but probably won't qualify him to seal foundation plans for the project described.

5.13 **d.** Julio is objecting because he knows water accumulates in compressed air storage tanks and causes corrosion, particularly at the bottom where the water will collect. Because of the metallurgy at the weld, corrosion is more aggressive at the weld site, and Julio considers this fundamentally a bad design if the weld is at the bottom. Since his recommendation (based on technical reasons and an increased risk to the public) is overruled under circumstances where the safety of the public is endangered, he is obligated (by the NCEES rules) to "notify his employer or client and other such authority." Having already notified his employer (through Sonja and the VP of manufacturing) and the client, all of whom except Sonja are part of the problem, his next logical step is to contact the state board with his concerns. This is rarely necessary; in most cases, the client and VP will be very interested in Julio's objection as it is based on public safety and will serve to reduce the company's own liability. However, if the employer and VP of manufacturing persist as described here, Julio must take additional action. It is not unreasonable to expect that Julio may face some sort of sanction from a management team that has put him in this position. He may even need a lawyer as this unpleasant situation deteriorates. Whistle-blowing should be considered the solution of last resort, as in this case.

5.14 **c.** William should avoid the use of the term "zero risk"; he knows that no project has zero risk. He should instead look for ways to quantify the risk that will be informative and meaningful to the public and try to convey the attitude of concern for minimizing the risk consistent with the potential public benefits of the project (increased employment and tax base). He should not do (a), and (b) does not address the problem of misinforming the public about the risk. There is no need to resign; to do so will not help his career or the project.

5.15 a. Darlene should accept the decision and work to make the chosen design successful and profitable for her employer. She does not need to resign—the business decision probably does not reflect a lack of confidence. It was made based on costs and profitability, not public safety, so she should not contact authorities to complain that a manager has overruled her. This kind of decision should be considered a "management decision" because it affects business and profits. Decisions that adversely affect public health, safety, or welfare should be considered "engineering decisions," and when these are overturned by nontechnical managers for other reasons, the engineer may be justified or even obligated to report this to authorities.

5.16 c. Matt should decline the contingency fee arrangement. While lawyers commonly work on contingency fee arrangements, engineers can't do this. Any contingency fee arrangement would put the engineer in a conflict of interest situation, where his engineering judgment can influence his income. In such a situation, his engineering judgment may not be sound or will at least appear to be conflicted to an outside observer.

5.17 c. Victor should be cautious in making sure that no additional surveying or resurveying is needed because of any changes to the tract. If additional fieldwork is not needed, he should bill the new client only for the work required to edit and reprint the drawings. He should not bill the new client the same as the original client, because that would be billing two clients for the same work, which is specifically prohibited by the *Model Rules*. Billing for half of the original amount is the same, just for an arbitrary amount. When two clients require the same survey simultaneously, it may make sense for Victor to facilitate a partnership in the project, but this is not always feasible when one client has already been billed for work done.

5.18 c. Option (c) is the only ethical and practical solution listed. Option (a) asking a competitor not to bid is not practical. Option (b) starting rumors about Linda's capabilities is clearly unethical. And option (d) is troublesome—experienced CAD technicians can do some aspects of design if closely supervised by a PE, but it isn't really necessary to make such a promotion just to show a lower billing rate. The billing rate for engineering design could be maintained at the same (competitive) level as when Linda was employed, even if the actual design work is done by Frank at twice the salary until a replacement for Linda can be hired. This in itself is not unethical, but if Frank does not budget sufficient time to do the design work in addition to his other work, it becomes a question of ethics. An engineer must allow sufficient time to do a professional job. The most practical and desirable solution is not listed—Frank should expedite hiring a qualified replacement for Linda, and he might have to consider declining the opportunity to bid on some projects until she is replaced.

5.19 **d.** Even the appearance of a conflict of interest can damage the reputation of the individuals involved and the profession as a whole, and such situations should be avoided. If there is no actual conflict of interest, the engineer's judgment will not be affected and the client will not suffer damages; nor is it criminal.

5.20 **b.** Codes of ethics published by professional societies are voluntarily adhered to by membership. They don't carry the weight of law but are much more than training guidelines—members who do not adhere to the society's code can be sanctioned by the society or forfeit their membership.

5.21 **c.** Many professional societies require the engineer to "consider" the impact on the environment. Some say he or she should consider whether the development is "sustainable." Most legal restrictions only require the engineer to prevent certain kinds of environmental damage and do not require sustainable development. There are, of course, specific state laws that must be followed, but many federal laws and regulations also apply. The engineer's responsibility is broader than laws and regulations in any event, making choice (c) the best answer.

CHAPTER 6

Engineering Economics

Donald G. Newnan

OUTLINE

This is a review of the field known as **engineering economics**, **engineering economy**, or **engineering economic analysis**. Since engineering economics is straightforward and logical, even people who have not had a formal course should be able to gain sufficient knowledge from this chapter to successfully solve most engineering economics problems.

There are 30 example problems throughout this review. These examples are an integral part of the review and should be examined as you come to them.

The field of engineering economics uses mathematical and economic techniques to systematically analyze situations that pose alternative courses of action. The initial step in engineering economics problems is to resolve a situation, or each possible alternative in a given situation, into its favorable and unfavorable consequences or factors. These are then measured in some common unit, usually money. Factors that cannot readily be equated to money are called **intangible** or **irreducible** factors. Such factors are considered in conjunction with the monetary analysis when making the final decision on proposed courses of action.

CASH FLOW

A cash flow table shows the *money consequences* of a situation and its timing. For example, a simple problem might be to list the year-by-year consequences of purchasing and owning a used car:

Year	Cash Flow	
Beginning of first year 0	–$4500	Car purchased now for $4500 cash. (The minus sign indicates a disbursement.)
End of year 1	−350	Maintenance costs are $350 per year.
End of year 2	−350	
End of year 3	−350	
End of year 4	−350 +2000	This car is sold at the end of the fourth year for $2000. (The plus sign represents the receipt of money.)

This same cash flow may be represented graphically, as shown in Figure 6.1. The upward arrow represents a receipt of money, and the downward arrows represent disbursements. The horizontal axis represents the passage of time.

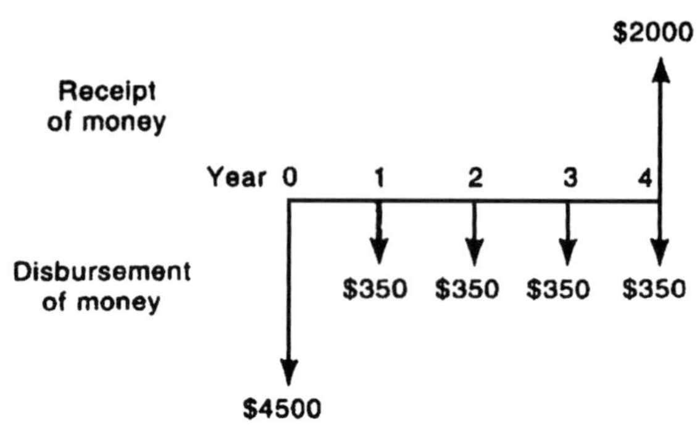

Figure 6.1 Cash flow diagram

Example 6.1

In January 2003, a firm purchased a used copier for $500. Repairs cost nothing in 2003 or 2004. Repairs were $85 in 2005, $130 in 2006, and $140 in 2007. The machine was sold in 2007 for $300. Complete the cash flow table.

Solution

Unless otherwise stated, the customary assumption is a beginning-of-year purchase, followed by end-of-year receipts or disbursements, and an end-of-year resale or salvage value. Thus the copier repairs and the copier sale are assumed to occur at the end of the year. Letting a minus sign represent a disbursement of money and a plus sign a receipt of money, we are able to set up the cash flow table:

Year	Cash Flow
Beginning of 2003	–$500
End of 2003	0
End of 2004	0
End of 2005	–85
End of 2006	–130
End of 2007	+160

Notice that at the end of 2007, the cash flow table shows +160, which is the net sum of –140 and +300. If we define year 0 as the beginning of 2003, the cash flow table becomes:

Year	Cash Flow
0	–$500
1	0
2	0
3	–85
4	–130
5	+160

From this cash flow table, the definitions of year 0 and year 1 become clear. Year 0 is defined as the *beginning* of year 1. Year 1 is the *end* of year 1, and so forth.

TIME VALUE OF MONEY

When the money consequences of an alternative occur in a short period of time—say, less than one year—we might simply add up the various sums of money and obtain the net result. But we cannot treat money this way over longer periods of time. This is because money today does not have the same value as money at some future time.

Consider this question: Which would you prefer, $100 today or the assurance of receiving $100 a year from now? Clearly, you would prefer the $100 today. If you had the money today, rather than a year from now, you could use it for the year. And if you had no use for it, you could lend it to someone who would pay interest for the privilege of using your money for the year.

EQUIVALENCE

In the preceding section we saw that money at different points in time (for example, \$100 today or \$100 one year hence) may be equal in the sense that it is \$100, but \$100 a year hence is *not* an acceptable substitute for \$100 today. When we have acceptable substitutes, we say they are *equivalent* to each other. Thus at 8% interest, \$108 a year hence is equivalent to \$100 today.

Example 6.2

At a 10% per year (compound) interest rate, \$500 now is *equivalent* to how much three years hence?

Solution

A value of \$500 now will increase by 10% in each of the three years.

$$\begin{aligned} \text{Now} &= \$500.00 \\ \text{End of 1st year} &= 500 + 10\%(500) = \$550.00 \\ \text{End of 2nd year} &= 550 + 10\%(550) = \$605.00 \\ \text{End of 3rd year} &= 605 + 10\%(605) = \$665.50 \end{aligned}$$

Thus \$500 now is *equivalent* to \$665.50 at the end of three years. Note that interest is charged each year on the original \$500 plus the unpaid interest.

Equivalence is an essential factor in engineering economics. Suppose we wish to select the better of two alternatives. First, we must compute their cash flows. For example:

	Alternative	
Year	***A***	***B***
0	−\$2000	−\$2800
1	+800	+1100
2	+800	+1100
3	+800	+1100

The larger investment in alternative *B* results in larger subsequent benefits, but we have no direct way of knowing whether it is better than alternative *A*. So we do not know which to select. To make a decision, we must resolve the alternatives into *equivalent* sums so that they may be compared accurately.

COMPOUND INTEREST

To facilitate equivalence computations, a series of compound interest factors will be derived here, and their use will be illustrated in examples.

Symbols and Functional Notation

i = effective interest rate per interest period. In equations, the interest rate is stated as a decimal (i.e., 8% interest is 0.08).

n = number of interest periods. Usually, the interest period is one year, in which case n would be number of years.

P = a present sum of money.

F = a future sum of money. The future sum F is an amount n interest periods from the present that is equivalent to P at interest rate i.

A = an end-of-period cash receipt or disbursement (annuity) in a uniform series continuing for n periods. The entire series is equivalent to P or F at interest rate i.

G = uniform period-by-period increase in cash flows; the uniform gradient.

r = nominal annual interest rate.

From Table 6.1 we can see that the functional notation scheme is based on writing (to find/given, i, n). Thus, if we wished to find the future sum F, given a uniform series of receipts A, the proper compound interest factor to use would be $(F/A, i, n)$.

Table 6.1 Periodic compounding: Functional notation and formulas

Factor	Given	To Find	Functional Notation	Formula
Single payment				
Compound amount factor	P	F	$(F/P, i\%, n)$	$F = P(1+i)^n$
Present worth factor	F	P	$(P/F, i\%, n)$	$P = F(1+i)^{-n}$
Uniform payment series				
Sinking fund factor	F	A	$(A/F, i\%, n)$	$A = F\left[\frac{i}{(1+i)^n - 1}\right]$
Capital recovery factor	P	A	$(A/P, i\%, n)$	$A = P\left[\frac{i(1+i)^n}{(1+i)^n - 1}\right]$
Compound amount factor	A	F	$(F/A, i\%, n)$	$F = A\left[\frac{(1+i)^n - 1}{i}\right]$
Present worth factor	A	P	$(P/A, i\%, n)$	$P = A\left[\frac{(1+i)^n - 1}{i(1+i)^n}\right]$
Uniform gradient				
Gradient present worth	G	P	$(P/G, i\%, n)$	$P = G\left[\frac{(1+i)^n - 1}{i^2(1+i)^n} - \frac{n}{i(1+i)^n}\right]$
Gradient future worth	G	F	$(F/G, i\%, n)$	$F = G\left[\frac{(1+i)^n - 1}{i^2} - \frac{n}{1}\right]$
Gradient uniform series	G	A	$(A/G, i\%, n)$	$A = G\left[\frac{1}{i} - \frac{n}{(1+i)^n - 1}\right]$

Single-Payment Formulas

Suppose a present sum of money P is invested for one year at interest rate i. At the end of the year, the initial investment P is received together with interest equal to Pi, or a total amount $P + Pi$. Factoring P, the sum at the end of one year is $P(1 + i)$. If the investment is allowed to remain for subsequent years, the progression is as follows:

Amount at Beginning of the Period	+	Interest for the Period	=	Amount at End of the Period
1st year, P	+	Pi	=	$P(1 + i)$
2nd year, $P(1 + i)$	+	$Pi(1 + i)$	=	$P(1 + i)^2$
3rd year, $P(1 + i)^2$	+	$Pi(1 + i)^2$	=	$P(1 + i)^3$
nth year, $P(1 + i)^{n-1}$	+	$Pi(1 + i)^{n-1}$	=	$P(1 + i)^n$

The present sum P increases in n periods to $P(1 + i)^n$. This gives a relation between a present sum P and its equivalent future sum F:

$$\text{Future sum} = (\text{present sum})(1 + i)^n$$
$$F = P(1 + i)^n$$

This is the **single-payment compound amount formula**. In functional notation it is written

$$F = P(F/P, i, n)$$

The relationship may be rewritten as

$$\text{Present sum} = (\text{Future sum})\ (1 + i)^{-n}$$
$$P = F(1 + i)^{-n}$$

This is the **single-payment present worth formula**. It is written

$$P = F(P/F, i, n)$$

Example 6.3

At a 10% per year interest rate, \$500 now is *equivalent* to how much three years hence?

Solution

This problem was solved in Example 6.2. Now it can be solved using a single-payment formula. $P = \$500$, $n = 3$ years, $i = 10\%$, and F = unknown:

$$F = P(1 + i)^n = 500(1 + 0.10)^3 = \$665.50$$

This problem also may be solved using a compound interest table:

$$F = P(F/P, i, n) = 500(F/P, 10\%, 3)$$

From the 10% compound interest table, read $(F/P, 10\%, 3) = 1.331$.

$$F = 500(F/P, 10\%, 3) = 500(1.331) = \$665.50$$

Example 6.4

To raise money for a new business, a small startup company asks you to lend it some money. The entrepreneur offers to pay you \$3000 at the end of four years. How much should you give the company now if you want to realize 12% interest per year?

Solution

P = unknown, F = \$3000, n = 4 years, and i = 12%:

$$P = F(1 + i)^{-n} = 3000(1 + 0.12)^{-4} = \$1906.55$$

Alternative computation using a compound interest table:

$$P = F(P/F, i, n) = 3000(P/F, 12\%, 4) = 3000(0.6355) = \$1906.50$$

Note that the solution based on the compound interest table is slightly different from the exact solution using a hand-held calculator. In engineering economics, the compound interest tables are considered to be sufficiently accurate.

Uniform Payment Series Formulas

Consider the situation shown in Figure 6.2. Using the single-payment compound amount factor, we can write an equation for F in terms of A:

$$F = A + A(1 + i) + A(1 + i)^2 \qquad \textbf{(i)}$$

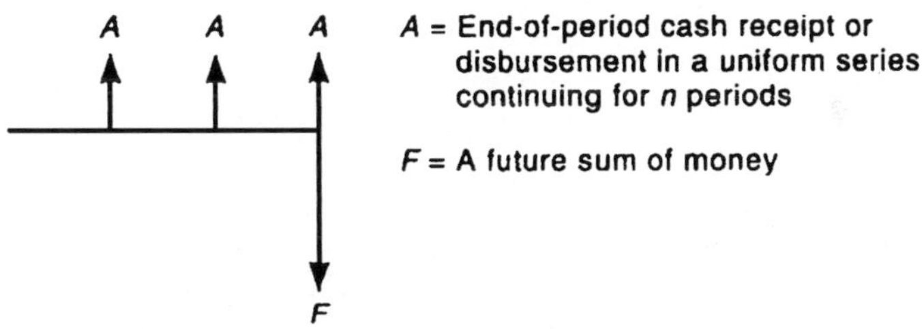

Figure 6.2 Cash flow diagram—uniform payment series

In this situation, with n = 3, Equation (i) may be written in a more general form:

$$F = A + A(1 + i) + A(1 + i)^{n-1} \qquad \textbf{(ii)}$$

Multiply Eq. (ii) by $(1 + i)$ $\quad (1 + i)F = A(1 + i) + A(1 + i)^{n-1} + A(1 + i)^n \qquad$ **(iii)**

Subtract Eq. (ii) yields: $\quad iF = -A + A(1 + i)^n$

This produces the **uniform series compound amount formula**:

$$F = A\left[\frac{(1+i)^n - 1}{i}\right].$$

Solving this equation for A produces the **uniform series sinking fund formula**:

$$A = F\left[\frac{i}{(1+i)^n - 1}\right].$$

Since $F = P(1 + i)^n$, we can substitute this expression for F in the equation and obtain the **uniform series capital recovery formula**:

$$A = P\left[\frac{i(1+i)^n}{(1+i)^n - 1}\right].$$

Solving the equation for P produces the **uniform series present worth formula**:

$$P = A\left[\frac{(1+i)^n - 1}{i(1+i)^n}\right].$$

In functional notation, the uniform series factors are:

Compound amount (F/A, i, n)

Sinking fund (A/F, i, n)

Capital recovery (A/P, i, n)

Present worth (P/A, i, n)

Example 6.5

If $100 is deposited at the end of each year in a savings account that pays 6% interest per year, how much will be in the account at the end of five years?

Solution

A = $100, F = unknown, n = 5 years, and i = 6%:

$$F = A(F/A, i, n) = 100(F/A, 6\%, 5) = 100(5.637) = \$563.70$$

Example 6.6

A fund established to produce a desired amount at the end of a given period, by means of a series of payments throughout the period, is called a **sinking fund**. A sinking fund is to be established to accumulate money to replace a $10,000 machine. If the machine is to be replaced at the end of 12 years, how much should be deposited in the sinking fund each year? Assume the fund earns 10% annual interest.

Solution

$$\text{Annual sinking fund deposit } A = 10{,}000(A/F, 10\%, 12)$$
$$= 10{,}000(0.0468) = \$468$$

Example 6.7

An individual is considering the purchase of a used automobile. The total price is $6200. With $1240 as a down payment, and the balance paid in 48 equal monthly payments with interest at 1% per month, compute the monthly payment. The payments are due at the end of each month.

Solution

The amount to be repaid by the 48 monthly payments is the cost of the automobile *minus* the $1240 down payment.

P = $4960, A = unknown, n = 48 monthly payments, and i = 1% per month:

$$A = P(A/P, 1\%, 48) = 4960(0.0263) = \$130.45$$

Example 6.8

A couple sell their home. In addition to cash, they take a mortgage on the house. The mortgage will be paid off by monthly payments of \$450 for 50 months. The couple decides to sell the mortgage to a local bank. The bank will buy the mortgage, but it requires a 1% per month interest rate on its investment. How much will the bank pay for the mortgage?

Solution

$A = \$450$, $n = 50$ months, $i = 1\%$ per month, and P = unknown:

$$P = A(P/A, i, n) = 450(P/A, 1\%, 50) = 450(39.196) = \$17{,}638.20$$

Uniform Gradient

At times, one will encounter a situation where the cash flow series is not a constant amount A; instead, it is an increasing series. The cash flow shown in Figure 6.3 may be resolved into two components (Figure 6.4). We can compute the value of P^* as equal to P' plus P, and we already have the equation for P':$P' = A(P/A, i, n)$.

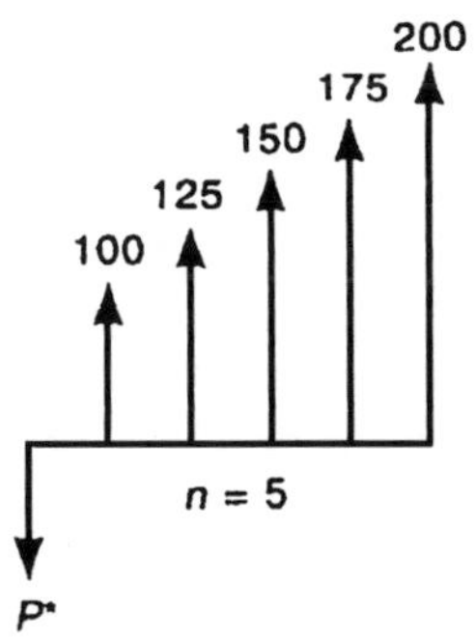

Figure 6.3 Cash flow diagram—uniform gradient

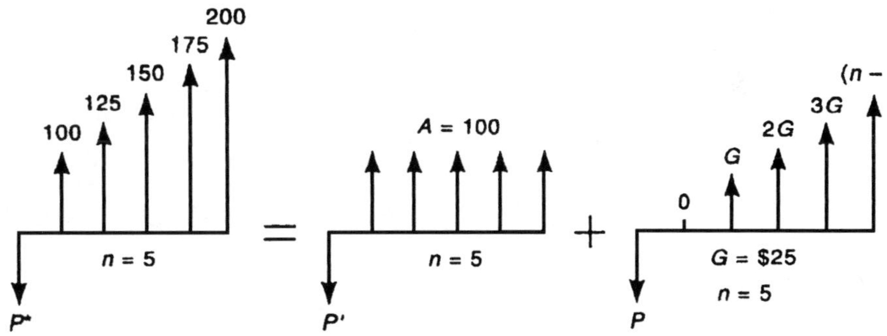

Figure 6.4 Uniform gradient diagram resolved

The value for P in the right-hand diagram is

$$P = G\left[\frac{(1+i)^n - 1}{i^2(1+i)^n} - \frac{n}{i(1+i)^n}\right]$$

This is the **uniform gradient present worth formula**. In functional notation, the relationship is $P = G(P/G, i, n)$.

Example 6.9

The maintenance on a machine is expected to be \$155 at the end of the first year, and it is expected to increase \$35 each year for the following seven years (Exhibit 1). What sum of money should be set aside now to pay the maintenance for the eight-year period? Assume 6% interest.

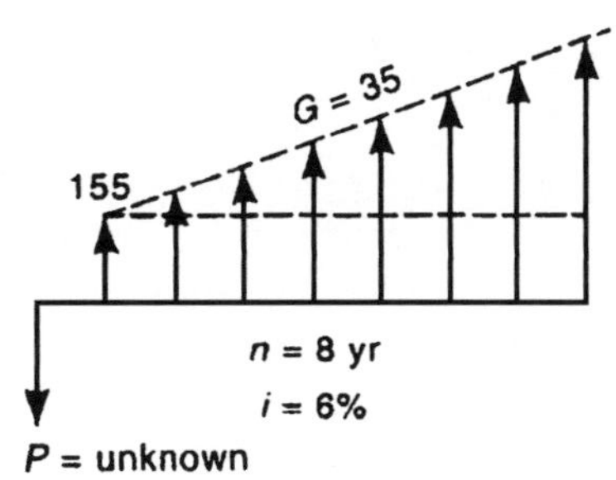

Exhibit 1

Solution

$$P = 155(P/A, 6\%, 8) + 35(P/G, 6\%, 8)$$
$$= 155(6.210) + 35(19.841) = \$1656.99$$

In the gradient series, if—instead of the present sum, P—an equivalent uniform series A is desired, the problem might appear as shown in Figure 6.5. The relationship between A' and G in the right-hand diagram is

$$A' = G\left[\frac{1}{i} - \frac{n}{(1+i)^n - 1}\right]$$

In functional notation, the uniform gradient (to) uniform series factor is: $A' = G(A/G, i, n)$.

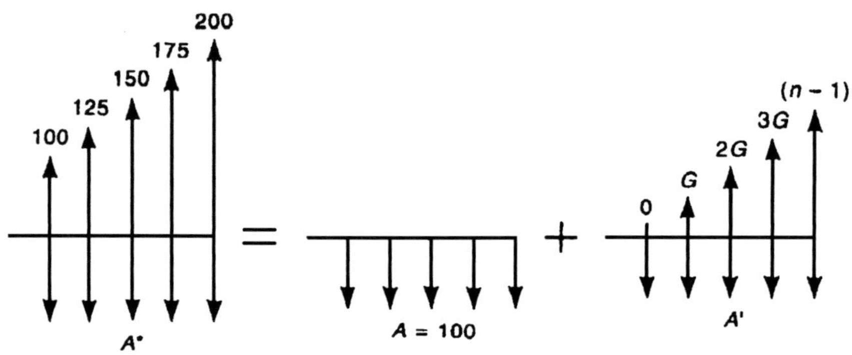

Figure 6.5 Uniform series, uniform gradient cash flow diagram

The **uniform gradient uniform series factor** may be read from the compound interest tables directly, or computed as

$$(A/G, i, n) = \frac{1 - n(A/F, i, n)}{i}$$

Note carefully the diagrams for the uniform gradient factors. The first term in the uniform gradient is zero and the last term is $(n - 1)G$. But we use n in the equations and function notation. The derivations (not shown here) were done on this basis, and the uniform gradient compound interest tables are computed this way.

Example 6.10

For the situation in Example 6.9, we wish now to know the uniform annual maintenance cost. Compute an equivalent A for the maintenance costs.

Solution

Refer to Exhibit 2. The equivalent uniform annual maintenance cost is

$$A = 155 + 35(A/G, 6\%, 8) = 155 + 35(3.195) = \$266.83$$

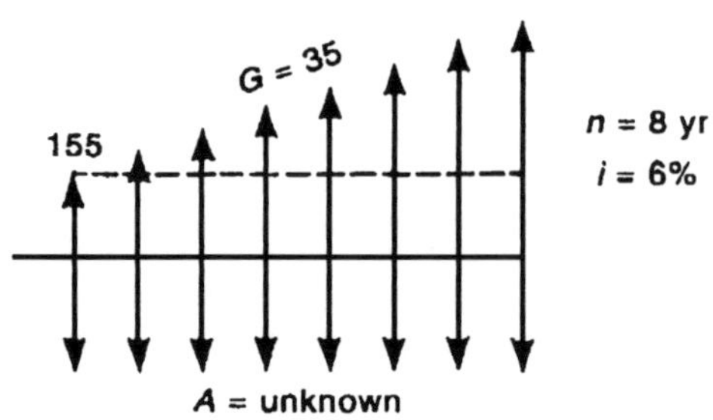

Exhibit 2

Standard compound interest tables give values for eight interest factors: two single payments, four uniform payment series, and two uniform gradients. The tables do *not* give the uniform gradient future worth factor, $(F/G, i, n)$. If it is needed, it may be computed from two tabulated factors:

$$(F/G, i, n) = (P/G, i, n)(F/P, i, n)$$

For example, if $i = 10\%$ and $n = 12$ years, then $(F/G, 10\%, 12) = (P/G, 10\%, 12)\ (F/P, 10\%, 12) = (29.901)(3.138) = 93.83$.

A second method of computing the uniform gradient future worth factor is

$$(F/G,i,n) = \frac{(F/G,i,n)-n}{i}$$

Using this equation for $i = 10\%$ and $n = 12$ years, $(F/G, 10\%, 12) = [(F/A, 10\%, 12) - 12]/0.10 = (21.384 - 12)/0.10 = 93.84$.

Table 6.2 Continuous compounding: Functional notation and formulas

Factor	Given	To Find	Functional Notation	Formula
Single payment				
Compound amount factor	P	F	$(F/P, r\%, n)$	$F = P\,[e^{rn}]$
Present worth factor	F	P	$(P/F, r\%, n)$	$P = F\,[e^{-rn}]$
Uniform payment series				
Sinking fund factor	F	A	$(A/F, r\%, n)$	$A = F\left[\frac{e^{r}-1}{e^{rn}-1}\right]$
Capital recovery factor	P	A	$(A/P, r\%, n)$	$A = P\left[\frac{e^{r}-1}{1-e^{-rn}}\right]$
Compound amount factor	A	F	$(F/A, r\%, n)$	$F = A\left[\frac{e^{rn}-1}{e^{r}-1}\right]$
Present worth factor	A	P	$(P/A, r\%, n)$	$P = A\left[\frac{1-e^{-rn}}{e^{r}-1}\right]$

r = nominal annual interest rate, n = number of years.

Example 6.11

Five hundred dollars is deposited each year into a savings bank account that pays 5% nominal interest, compounded continuously. How much will be in the account at the end of five years?

Solution

$A = \$500$, $r = 0.05$, $n = 5$ years.

$$F = A(F/A, r\%, n) = A\left[\frac{e^{rn}-1}{e^{r}-1}\right] = 500\left[\frac{e^{0.05(5)}-1}{e^{0.05}-1}\right] = \$2769.84$$

NOMINAL AND EFFECTIVE INTEREST

Nominal interest is the annual interest rate without considering the effect of any compounding. **Effective interest** is the annual interest rate taking into account the effect of any compounding during the year.

Non-annual Compounding

Frequently an interest rate is described as an annual rate, even though the interest period may be something other than one year. A bank may pay 1% interest on the amount in a savings account every three months. The *nominal* interest rate in this situation is $4 \times 1\% = 4\%$. But if you deposited \$1000 in such an account, would you have $104\%(1000) = \$1040$ in the account at the end of one year? The answer is no, you would have more. The amount in the account would increase as follows.

Amount in Account

Beginning of year:	1000.00
End of three months:	1000.00 + 1%(1000.00) = 1010.00
End of six months:	1010.00 + 1%(1010.00) = 1020.10
End of nine months:	1020.10 + 1%(1020.10) = 1030.30
End of one year:	1030.30 + 1%(1030.30) = 1040.60

At the end of one year, the interest of \$40.60, divided by the original \$1000, gives a rate of 4.06 percent. This is the *effective* interest rate.

$$\text{Effective interest rate per year:} \quad i_{eff} = (1 + r/m)^m - 1$$

where r = nominal annual interest rate
m = number of compound periods per year
r/m = effective interest rate per period

Example 6.12

A bank charges 1.5% interest per month on the unpaid balance for purchases made on its credit card. What nominal interest rate is it charging? What is the effective interest rate?

Solution

The nominal interest rate is simply the annual interest ignoring compounding, or $12(1.5\%) = 18\%$.

$$\text{Effective interest rate} = (1 + 0.015)^{12} - 1 = 0.1956 = 19.56\%$$

SOLVING ENGINEERING ECONOMICS PROBLEMS

The techniques presented so far illustrate how to convert single amounts of money, and uniform or gradient series of money, into some equivalent sum at another point in time. These compound interest computations are an essential part of engineering economics problems.

The typical situation is that we have a number of alternatives; the question is, which alternative should we select? The customary method of solution is to express each alternative in some common form and then choose the best, taking both the monetary and intangible factors into account. In most computations an interest rate must be used. It is often called the **minimum attractive rate of return (MARR)**, to indicate that this is the smallest interest rate, or rate of return, at which one is willing to invest money.

Criteria

Engineering economics problems inevitably fall into one of three categories:

1. *Fixed input.* The amount of money or other input resources is fixed. *Example*: A project engineer has a budget of $450,000 to overhaul a plant.
2. *Fixed output.* There is a fixed task or other output to be accomplished. *Example*: A mechanical contractor has been awarded a fixed-price contract to air-condition a building.
3. *Neither input nor output fixed.* This is the general situation, where neither the amount of money (or other inputs) nor the amount of benefits (or other outputs) is fixed. *Example*: A consulting engineering firm has more work available than it can handle. It is considering paying the staff to work evenings to increase the amount of design work it can perform.

There are five major methods of comparing alternatives: present worth, future worth, annual cost, rate of return, and benefit-cost analysis. These are presented in the sections that follow.

Present Worth

Present worth analysis converts all of the money consequences of an alternative into an equivalent present sum. The criteria are:

Category	Present Worth Criterion
Fixed input	Maximize the present worth of benefits or other outputs
Fixed output	Minimize the present worth of costs or other inputs
Neither input nor output fixed	Maximize present worth of benefits minus present worth of costs, or maximize net present worth

Appropriate Problems

Present worth analysis is most frequently used to determine the present value of future money receipts and disbursements. We might want to know, for example, the present worth of an income-producing property, such as an oil well. This should provide an estimate of the price at which the property could be bought or sold.

An important restriction in the use of present worth calculation is that there must be a common analysis period for comparing alternatives. It would be incorrect, for example, to compare the present worth (PW) of cost of pump A, expected to last 6 years, with the PW of cost of pump B, expected to last 12 years (Figure 6.6). In situations like this, the solution is either to use some other analysis technique (generally, the annual cost method is suitable in these situations) or to restructure the problem so that there is a common analysis period.

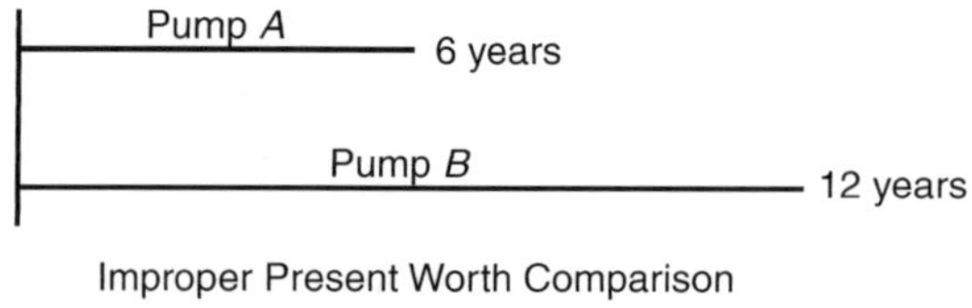

Figure 6.6 Improper present worth comparison

In this example, a customary assumption would be that a pump is needed for 12 years and that pump A will be replaced by an identical pump A at the end of 6 years. This gives a 12-year common analysis period (Figure 6.7). This approach is easy to use when the different lives of the alternatives have a practical least-common-multiple life. When this is not true (for example, the life of J equals 7 years and the life of K equals 11 years), some assumptions must be made to select a suitable common analysis period, or the present worth method should not be used.

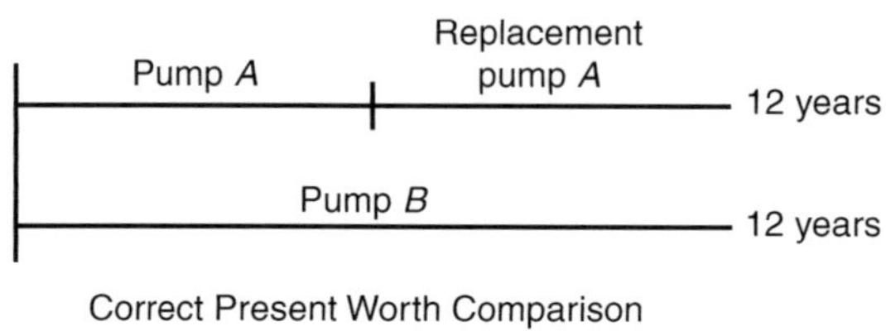

Figure 6.7 Proper present worth comparison

Example **6.13**

Machine X has an initial cost of \$10,000, an annual maintenance cost of \$500 per year, and no salvage value at the end of its 4-year useful life. Machine Y costs \$20,000, and the first year there is no maintenance cost. Maintenance is \$100 the second year, and it increases \$100 per year thereafter. The machine has an anticipated \$5000 salvage value at the end of its 12-year useful life. If the minimum attractive rate of return (MARR) is 8%, which machine should be selected?

Solution

The analysis period is not stated in the problem. Therefore, we select the least common multiple of the lives, or 12 years, as the analysis period.

Present worth of cost of 12 years of machine X:

$$\begin{aligned} \text{PW} &= 10{,}000 + 10{,}000(P/F, 8\%, 4) + 10{,}000(P/F, 8\%, 8) + 500(P/A, 8\%, 12) \\ &= 10{,}000 + 10{,}000(0.7350) + 10{,}000(0.5403) + 500(7.536) = \$26{,}521 \end{aligned}$$

Present worth of cost of 12 years of machine Y:

$$\begin{aligned} \text{PW} &= 20{,}000 + 100(P/G, 8\%, 12) - 5000(P/F, 8\%, 12) \\ &= 20{,}000 + 100(34.634) - 5000(0.3971) = \$21{,}478 \end{aligned}$$

Choose machine Y, with its smaller PW of cost.

Example 6.14

Two alternatives have the following cash flows:

	Alternative	
Year	***A***	***B***
0	–\$2000	–\$2800
1	+800	+1100
2	+800	+1100
3	+800	+1100

At a 4% interest rate, which alternative should be selected?

Solution

The net present worth of each alternative is computed:

$$\text{Net present worth (NPW)} = \text{PW of benefit} - \text{PW of cost}$$
$$\text{NPW}_A = 800(P/A, 4\%, 3) - 2000 = 800(2.775) - 2000 = \$220.00$$
$$\text{NPW}_B = 1100(P/A, 4\%, 3) - 2800 = 1100(2.775) - 2800 = \$252.50$$

To maximize NPW, choose alternative *B*.

Infinite Life and Capitalized Cost

In the special situation where the analysis period is infinite ($n = \infty$), an analysis of the present worth of cost is called **capitalized cost**. There are a few public projects where the analysis period is infinity. Other examples are permanent endowments and cemetery perpetual care.

When n equals infinity, a present sum P will accrue interest of Pi for every future interest period. For the principal sum P to continue undiminished (an essential requirement for n equal to infinity), the end-of-period sum A that can be disbursed is Pi (Figure 6.8).

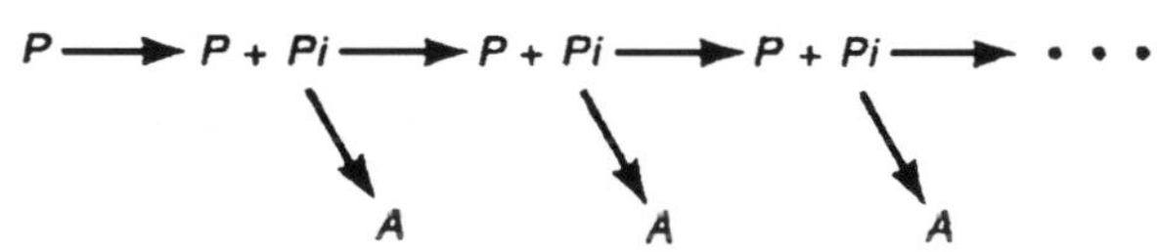

Figure 6.8 Infinite life, capitalized cost diagram

When $n = \infty$, the fundamental relationship is

$$A = Pi$$

Some form of this equation is used whenever there is a problem involving an infinite analysis period.

Example 6.15

In his will, a man wishes to establish a perpetual trust to provide for the maintenance of a small local park. If the annual maintenance is \$7500 per year and the trust account can earn 5% interest, how much money must be set aside in the trust?

Solution

When $n = \infty$, $A = Pi$ or $P = A/i$. The capitalized cost is $P = A/i = \$7500/0.05 = \$150{,}000$.

Future Worth or Value

In present worth analysis, the comparison is made in terms of the equivalent present costs and benefits. But the analysis need not be made in terms of the present—it can be made in terms of a past, present, or future time. Although the numerical calculations may look different, the decision is unaffected by the selected point in time. Often we do want to know what the future situation will be if we take some particular course of action now. An analysis based on some future point in time is called **future worth analysis**.

Category	Future Worth Criterion
Fixed input	Maximize the future worth of benefits or other outputs
Fixed output	Minimize the future worth of costs or other inputs
Neither input nor output fixed	Maximize future worth of benefits minus future worth of costs, or maximize net future worth

Example 6.16

Two alternatives have the following cash flows:

	Alternative	
Year	***A***	***B***
0	–$2000	–$2800
1	+800	+1100
2	+800	+1100
3	+800	+1100

At a 4% interest rate, which alternative should be selected?

Solution

In Example 6.14, this problem was solved by present worth analysis at year 0. Here it will be solved by future worth analysis at the end of year 3.

Net future worth (NFW) = FW of benefits – FW of cost

$$\begin{aligned}\text{NFW}_A &= 800(F/A, 4\%, 3) - 2000(F/P, 4\%, 3)\\ &= 800(3.122) - 2000(1.125) = +\$247.60\end{aligned}$$

$$\begin{aligned}\text{NFW}_B &= 1100(F/A, 4\%, 3) - 2800(F/P, 4\%, 3)\\ &= 1100(3.122) - 2800(1.125) = +\$284.20\end{aligned}$$

To maximize NFW, choose alternative *B*.

Annual Cost

The annual cost method is more accurately described as the method of equivalent uniform annual cost (EUAC). Where the computation is of benefits, it is called the method of equivalent uniform annual benefits (EUAB).

Criteria

For each of the three possible categories of problems, there is an annual cost criterion for economic efficiency.

Category	Annual Cost Criterion
Fixed input	Maximize the equivalent uniform annual benefits (EUAB)
Fixed output	Minimize the equivalent uniform annual cost (EUAC)
Neither input nor output fixed	Maximize EUAB – EUAC

Application of Annual Cost Analysis

In the section on present worth, we pointed out that the present worth method requires a common analysis period for all alternatives. This restriction does not apply in all annual cost calculations, but it is important to understand the circumstances that justify comparing alternatives with different service lives.

Frequently, an analysis is done to provide for a more-or-less continuing requirement. For example, one might need to pump water from a well on a continuing basis. Regardless of whether each of two pumps has a useful service life of 6 years or 12 years, we would select the alternative whose annual cost is a minimum. And this still would be the case if the pumps' useful lives were the more troublesome 7 and 11 years. Thus, if we can assume a continuing need for an item, an annual cost comparison among alternatives of differing service lives is valid. The underlying assumption in these situations is that the shorter-lived alternative can be replaced with an identical item with identical costs, when it has reached the end of its useful life. This means that the EUAC of the initial alternative is equal to the EUAC for the continuing series of replacements.

On the other hand, if there is a specific requirement to pump water for ten years, then each pump must be evaluated to see what costs will be incurred during the analysis period and what salvage value, if any, may be recovered at the end of the analysis period. The annual cost comparison needs to consider the actual circumstances of the situation.

Examination problems are often readily solved using the annual cost method. And the underlying *continuing requirement* is usually present, so an annual cost comparison of unequal-lived alternatives is an appropriate method of analysis.

Example 6.17

Consider the following alternatives:

	A	*B*
First cost	\$5000	\$10,000
Annual maintenance	500	200
End-of-useful-life salvage value	600	1000
Useful life	5 years	15 years

Based on an 8% interest rate, which alternative should be selected?

Solution

Assuming both alternatives perform the same task and there is a continuing requirement, the goal is to minimize EUAC.

Alternative A:

$$\begin{aligned}\text{EUAC} &= 5000(A/P, 8\%, 5) + 500 - 600(A/F, 8\%, 5)\\ &= 5000(0.2505) + 500 - 600(0.1705) = \$1650\end{aligned}$$

Alternative B:

$$\begin{aligned}\text{EUAC} &= 10{,}000(A/P, 8\%, 15) + 200 - 1000(A/F, 8\%, 15)\\ &= 10{,}000(0.1168) + 200 - 1000(0.0368) = \$1331\end{aligned}$$

To minimize EUAC, select alternative B.

Rate of Return Analysis

A typical situation is a cash flow representing the costs and benefits. The rate of return may be defined as the interest rate where PW of cost = PW of benefits, EUAC = EUAB, or PW of cost – PW of benefits = 0.

Example 6.18

Compute the rate of return for the investment represented by the following cash flow table.

Year:	0	1	2	3	4	5
Cash flow:	–\$595	+250	+200	+150	+100	+50

Solution

This declining uniform gradient series may be separated into two cash flows (Exhibit 3) for which compound interest factors are available.

Note that the gradient series factors are based on an *increasing* gradient. Here the declining cash flow is solved by subtracting an increasing uniform gradient, as indicated in the figure.

PW of cost – PW of benefits = 0

$$595 - [250(P/A, i, 5)] - 50(P/G, i, 5) = 0$$

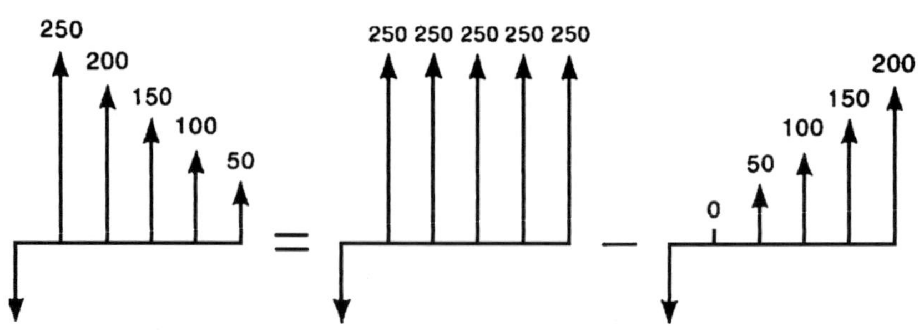

Exhibit 3

Try $i = 10\%$:

$$595 - [250(3.791) - 50(6.862)] = -9.65$$

Try $i = 12\%$:

$$595 - [250(3.605) - 50(6.397)] = +13.60$$

The rate of return is between 10 and 12%. It may be computed more accurately by linear interpolation:

$$\text{Rate of return} = 10\% + (2\%)\left(\frac{9.65 - 0}{13.60 + 9.65}\right) = 10.83\%$$

Two Alternatives

Compute the incremental rate of return on the cash flow representing the difference between the two alternatives. Since we want to look at increments of *investment*, the cash flow for the difference between the alternatives is computed by taking the higher initial-cost alternative minus the lower initial-cost alternative. If the incremental rate of return is greater than or equal to the predetermined minimum attractive rate of return (MARR), choose the higher-cost alternative; otherwise, choose the lower-cost alternative.

Example **6.19**

Two alternatives have the following cash flows:

	Alternative	
Year	***A***	***B***
0	–$2000	–$2800
1	+800	+1100
2	+800	+1100
3	+800	+1100

If 4% is considered the minimum attractive rate of return (MARR), which alternative should be selected?

Solution

These two alternatives were previously examined in Examples 6.14 and 6.16 by present worth and future worth analysis. This time, the alternatives will be resolved using a rate-of-return analysis.

Note that the problem statement specifies a 4% MARR, whereas Examples 6.14 and 6.16 referred to a 4% interest rate. These are really two different ways of saying the same thing: the minimum acceptable time value of money is 4%.

First, tabulate the cash flow that represents the increment of investment between the alternatives. This is done by taking the higher initial-cost alternative minus the lower initial-cost alternative:

	Alternative		Difference between alternatives
Year	***A***	***B***	***B* – *A***
0	–$2000	–$2800	–$800
1	+800	+1100	+300
2	+800	+1100	+300
3	+800	+1100	+300

Then compute the rate of return on the increment of investment represented by the difference between the alternatives:

$$\text{PW of cost} = \text{PW of benefits}$$
$$800 = 300(P/A, i, 3)$$
$$(P/A, i, 3) = 800/300 = 2.67$$
$$i = 6.1\%$$

Since the incremental rate of return exceeds the 4% MARR, the increment of investment is desirable. Choose the higher-cost alternative *B*.

Before leaving this example, one should note something that relates to the rates of return on alternative *A* and on alternative *B*. These rates of return, if calculated, are:

	Rate of Return
Alternative *A*	9.7%
Alternative *B*	8.7%

The correct answer to this problem has been shown to be alternative *B*, even though alternative *A* has a higher rate of return. The higher-cost alternative may be thought of as the lower-cost alternative plus the increment of investment between them. Viewed this way, the higher-cost alternative *B* is equal to the desirable lower-cost alternative *A* plus the difference between the alternatives.

The important conclusion is that computing the rate of return for each alternative does *not* provide the basis for choosing between alternatives. Instead, incremental analysis is required.

Example 6.20

Consider the following:

	Alternative	
Year	***A***	***B***
0	–$200.0	–$131.0
1	+77.6	+48.1
2	+77.6	+48.1
3	+77.6	+48.1

If the MARR is 10%, which alternative should be selected?

Solution

To examine the increment of investment between the alternatives, we will examine the higher initial-cost alternative minus the lower initial-cost alternative, or *A* – *B*.

	Alternative		Increment
Year	***A***	***B***	***A* – *B***
0	–$200.0	–$131.0	–$69.0
1	+77.6	+48.1	+29.5
2	+77.6	+48.1	+29.5
3	+77.6	+48.1	+29.5

Solve for the incremental rate of return:

$$\text{PW of cost} = \text{PW of benefits}$$
$$69.0 = 29.5(P/A, i, 3)$$
$$(P/A, i, 3) = 69.0/29.5 = 2.339$$

From compound interest tables, the incremental rate of return is between 12 and 18%. This is a desirable increment of investment; hence we select the higher-initial-cost alternative *A*.

Three or More Alternatives

When there are three or more mutually exclusive alternatives, proceed with the same logic presented for two alternatives. The components of incremental analysis are listed below.

Step 1. Compute the rate of return for each alternative. Reject any alternative where the rate of return is less than the desired MARR. (This step is not essential, but helps to immediately identify unacceptable alternatives.)

Step 2. Rank the remaining alternatives in order of increasing initial cost.

Step 3. Examine the increment of investment between the two lowest-cost alternatives as described for the two-alternative problem. Select the better of the two alternatives and reject the other one.

Step 4. Take the preferred alternative from step 3. Consider the next higher initial-cost alternative, and proceed with another two-alternative comparison.

Step 5. Continue until all alternatives have been examined and the best of the multiple alternatives has been identified.

Example 6.21

Consider the following:

	Alternative	
Year	***A***	***B***
0	–$200.0	–$131.0
1	+77.6	+48.1
2	+77.6	+48.1
3	+77.6	+48.1

If the MARR is 10%, which alternative, if any, should be selected?

Solution

One should carefully note that this is a *three-alternative* problem, where the alternatives are *A*, *B*, and *Do nothing*. In this solution we will skip step 1. Reorganize the problem by placing the alternatives in order of increasing initial cost:

	Alternative		
Year	***Do Nothing***	***B***	***A***
0	0	– $131.0	–$200.0
1	0	+ 48.1	+77.6
2	0	+ 48.1	+77.6
3	0	+ 48.1	+77.6

Examine the *B – Do nothing* increment of investment:

Year	*B – Do Nothing*
0	–\$131.0 – 0 = – \$131.0
1	+ 48.1 – 0 = + 48.1
2	+ 48.1 – 0 = + 48.1
3	+ 48.1 – 0 = + 48.1

Solve for the incremental rate of return:

$$\begin{aligned} \text{PW of cost} &= \text{PW of benefits} \\ 131.0 &= 48.1(P/A, i, 3) \\ (P/A, i, 3) &= 131.0/48.1 = 2.723 \end{aligned}$$

From compound interest tables, the incremental rate of return is about 5%. Since the incremental rate of return is less than 10%, the *B – Do nothing* increment is not desirable. Reject alternative *B*.

Year	*A – Do Nothing*
0	–\$200.0 – 0 = –\$200.0
1	+77.6 – 0 = +77.6
2	+77.6 – 0 = +77.6
3	+77.6 – 0 = +77.6

Next, consider the increment of investment between the two remaining alternatives. Solve for the incremental rate of return:

$$\begin{aligned} \text{PW of cost} &= \text{PW of benefits} \\ 200.0 &= 77.6(P/A, i, 3) \\ (P/A, i, 3) &= 200.0/77.6 = 2.577 \end{aligned}$$

The incremental rate of return is 8%, less than the desired 10%. Reject the increment and select the remaining alternative: *Do nothing*.

If you have not already done so, you should go back to Example 6.20 and see how the slightly changed wording of the problem has radically altered it. Example 6.20 required a choice between two undesirable alternatives. This example adds the *Do nothing* alternative, which is superior to *A* and *B*.

BENEFIT-COST ANALYSIS

Generally, in public works and governmental economic analyses, the dominant method of analysis is the **benefit-cost ratio (B/C)**. It is simply the ratio of benefits divided by costs, taking into account the time value of money.

$$\text{B/C} = \frac{\text{PW of benefits}}{\text{PW of cost}} = \frac{\text{Equivalent uniform annual benefits}}{\text{Equivalent uniform annual cost}}$$

For a given interest rate, a B/C ratio ≥1 reflects an acceptable project. The B/C analysis method is parallel to rate-of-return analysis. The same kind of incremental analysis is required.

Example **6.22**

Solve Example 6.20 by benefit-cost analysis.

Solution

	Alternative		Increment
Year	***A***	***B***	***A – B***
0	–\$200.0	–\$131.0	–\$69.0
1	+77.6	+48.1	+29.5
2	+77.6	+48.1	+29.5
3	+77.6	+48.1	+29.5

The benefit-cost ratio for the $A - B$ increment is

$$\text{B/C} = \frac{\text{PW of benefits}}{\text{PW of cost}} = \frac{29.5(P/A, 10\%, 3)}{69.0} = \frac{73.37}{69.0} = 1.06$$

Since the B/C ratio exceeds 1, the increment of investment is desirable. Select the higher-cost alternative A.

Breakeven Analysis

In business, *breakeven* is defined as the point where income just covers costs. In engineering economics, the breakeven point is defined as the point where two alternatives are equivalent.

Example **6.23**

A city is considering a new \$50,000 snowplow. The new machine will operate at a savings of \$600 per day compared with the present equipment. Assume that the MARR is 12%, and the machine's life is ten years with zero resale value at that time. How many days per year must the machine be used to justify the investment?

Solution

This breakeven problem may be readily solved by annual cost computations. We will set the equivalent uniform annual cost (EUAC) of the snowplow equal to its annual benefit and solve for the required annual utilization. Let X = breakeven point = days of operation per year.

$$\text{EUAC} = \text{EUAB}$$
$$50{,}000(A/P, 12\%, 10) = 600X$$
$$X = 50{,}000(0.1770)/600 = 14.8 \text{ days/year}$$

BONDS

A **bond** is a form of debt represented by a certificate. The bond will specify the amount of the debt, the interest rate of the bond, how often the interest is paid, and when the debt will be repaid.

Bond Value

Bond value is the present worth of all the future interest payments plus the future repayment of the debt, computed at some selected interest rate.

Example 6.24

A \$5000 bond is being offered for sale. It has a stated interest rate of 7%, paid annually. At the end of eight years the \$5000 debt will be repaid along with the last interest payment. It you want an 8% rate of return on this investment (bond yield), how much would you be willing to pay for the bond (bond value)?

Solution

The bond pays 7% × \$5000 = \$350 at the end of every year and will repay the \$5000 debt at the end of eight years.

$$\begin{aligned}\text{Bond value} &= \text{PW of all future benefits}\\ &= 350(\text{P/A},8\%,8) + 5000(\text{P/F},8\%,8)\\ &= 350(5.747) + 5000(0.5403) = \$4712.95\end{aligned}$$

Bond Yield

Bond yield is the interest rate at which the benefits of owning the bond are equivalent to the cost of the bond.

Example 6.25

If the bond in Example 6.24 can actually be purchased for \$4200, what is the bond yield?

Solution

Set the cost of the bond equal to the PW of the bond benefits and solve for the unknown interest rate. The resulting i^* is the bond yield.

$$\$4200 = 350(P/A,i,8) + 5000(P/F,i,8)$$

The equation must be solved by trial and error. Try $i = 10\%$.

$$\$4200 \stackrel{?}{=} 350(5.335) + 5000(0.467) = 4202.25$$

We see that i^* is very close to 10%. No further computations are required. The bond yield is very close to 10%.

PAYBACK PERIOD

Payback period is the period of time required for the profit or other benefits of an investment to equal the cost of the investment.

Example 6.26

A project has the following costs and benefits.

Year	Costs	Benefits
0	$1400	
1	500	$200
2	300	100
3–10		400/year

What is the payback period?

Solution

The total cost is $2200. At the end of year 6 the total benefits will be 200 + 100 + 400 + 400 + 400 + 400 = $1900. And at the end of year 7 benefits will be 1900 + 400 = $2300. The payback period is where benefits equal cost. Since a cash flow table is normally based on the end-of-year convention (See Example 6.1), the payback period is at the end of year 7 and the answer is 7 years.

If, on the other hand, the problem had been stated in words something like ". . .and the benefits from the third year on are $400/year," you would assume the benefits occur uniformly throughout the year. In this situation, the correct answer would be 6.75 years.

VALUATION AND DEPRECIATION

Depreciation of capital equipment is an important component of many after-tax economic analyses. For this reason, one must understand the fundamentals of depreciation accounting.

Notation

BV = book value
C = cost of the property (basis)
D_j = depreciation in year j
S_n = salvage value in year n

Depreciation is the systematic allocation of the cost of a capital asset over its useful life. **Book value** is the original cost of an asset minus the accumulated depreciation of the asset.

$$\text{Book value (BV)} = C - \Sigma(D_j)$$

In computing a schedule of depreciation charges four items are considered:

1. Cost of the property, C (called the *basis* in tax law)
2. Type of property. Property is classified either as **tangible** (such as machinery) or **intangible** (such as a franchise or a copyright) and as either **real property** (real estate) or **personal property** (everything not real property)
3. Depreciable life in years, n
4. Salvage value of the property at the end of its depreciable (usable) life, S_n

Straight-Line Depreciation

Depreciation charge in any year is given by

$$D_j = \frac{C - S_n}{n}$$

An alternate computation of the depreciation charge in year j is

$$D_j = \frac{C - \text{Depreciation taken to beginning of year } j - S_n}{\text{Remaining useful life at beginning of year } j}$$

Double-Declining-Balance Depreciation

DDB depreciation in any year, $D_j = \frac{2}{n}$ (C – depreciation in years prior to j).

For 150% declining-balance depreciation, replace the 2 in the equation with 1.5.

Modified Accelerated Cost Recovery System Depreciation

The modified accelerated cost recovery system (MACRS) depreciation method generally applies to property placed in service after 1986. To compute the MACRS depreciation for an item one must know the following:

1. Cost (basis) of the item
2. Property class. All tangible property is classified in one of six classes (3, 5, 7, 10, 15, and 20 years), based on the life over which it is depreciated (see Table 6.3). Residential real estate and nonresidential real estate are in two separate real property classes of 27.5 years and 39 years, respectively.
3. Depreciation computation
   - The 3-, 5-, 7-, and 10-year property classes use double-declining-balance depreciation with conversion to straight-line depreciation in the year that increases the deduction.
   - The 15- and 20-year property classes use 150% declining-balance depreciation with conversion to straight-line depreciation in the year that increases the deduction.
   - In MACRS the salvage value is assumed to be zero.

Half-Year Convention

Except for real property, a half-year convention is used. Under this convention all property is considered to be placed in service in the middle of the tax year, and a half year of depreciation is allowed in the first year. For each of the remaining years, one is allowed a full year of depreciation. If the property is disposed of prior to the end of the recovery period (property class life), a half year of depreciation is allowed in that year. If the property is held for the entire recovery period, a half year of depreciation is allowed for the year following the end of the recovery period (see Table 6.3).

Table 6.3 Modified ACRS (MACRS) depreciation for personal property—half-year convention

		Applicable Percentage for the Class of Property		
Recovery Year	**3-year Recovery**	**5-year Recovery**	**7-year Recovery**	**10-year Recovery**
1	33.33	20.00	14.29	10.00
2	44.45	32.00	24.49	18.00
3	14.81†	19.20	17.49	14.40
4	7.41	11.52†	12.49	11.52
5		11.52	8.93†	9.22
6		5.76	8.92	7.37
7			8.93	6.55†
8			4.46	6.55
9				6.56
10				6.55
11				3.28

† Use straight-line depreciation for the year marked and all subsequent years.

Example 6.27

A \$5000 computer has an anticipated \$500 salvage value at the end of its five-year depreciable life. Compute the depreciation schedule by MACRS depreciation. Do the MACRS computation by hand, and then compare the results with the values from Table 6.3.

Solution

The depreciation method is double declining balance with conversion to straight line for the computer's five-year property class and the half-year convention is used. Salvage value S_n is assumed to be zero for MACRS. Using the equation for DDB depreciation in any year:

Year

$$1\left(\frac{1}{2}\text{year}\right) \quad D_1 = \frac{1}{2}\times\frac{2}{5}(5000 - 0) = \$1000$$

$$2 \quad D_2 = \frac{2}{5}(5000 - 1000) = 1600$$

$$3 \quad D_3 = \frac{2}{5}(5000 - 2600) = 960$$

$$4 \quad D_4 = \frac{2}{5}(5000 - 3560) = 576$$

$$5 \quad D_5 = \frac{2}{5}(5000 - 4136) = 346$$

$$6\left(\frac{1}{2}\text{year}\right) \quad D_6 = \frac{1}{2}\times\frac{2}{5}(5000 - 4482) = 104$$

$$\overline{\$4586}$$

The computation must now be modified to convert to straight-line depreciation at the point where the straight-line depreciation will be larger. Using the alternate straight-line computation,

$$D_5 = \frac{5000 - 4136 - 0}{1.5 \text{ years remaining}} = \$576$$

This is more than the \$346 computed using DDB; hence, switch to the straight-line method for year 5 and beyond.

$$D_6\left(\frac{1}{2}\text{year}\right) = \frac{1}{2}(576) = \$288$$

Answers:

Year	Depreciation (MACRS)
1	\$1000
2	1600
3	960
4	576
5	576
6	288
	\$5000

The computed MACRS depreciation is identical with that obtained from Table 6.3.

INFLATION

Inflation is characterized by rising prices for goods and services, while deflation produces a decrease in prices. An inflationary trend makes future dollars have less purchasing power than present dollars. This benefits long-term borrowers of money because they may repay a loan of present dollars in the future with dollars of reduced buying power. The help to borrowers is at the expense of lenders. Deflation has the opposite effect. Money borrowed at a point in time followed by a deflationary period subjects the borrower to loan repayment with dollars of greater purchasing power than those he borrowed. This is to the lenders' advantage at the expense of borrowers.

Price changes occur in a variety of ways. One method of stating a price change is a uniform rate of price change per year.

f = general inflation rate per interest period
i = effective interest rate per interest period

The following example problem will illustrate the computations.

Example 6.28

A mortgage will be repaid in three equal payments of \$5000 at the end of years 1, 2, and 3. If the annual inflation rate, f, is 8% during this period, and the investor wishes a 12% annual interest rate (i), what is the maximum amount he would be willing to pay for the mortgage?

Solution

The computation is a two-step process. First, the three future payments must be converted into dollars with the same purchasing power as today's (year 0) dollars.

Year	Actual Cash Flow		Multiplied by		Cash Flow Adjusted to Today's (Year 0) Dollars
0	—	×	—	=	—
1	+5000	×	$(1 + 0.08)^{-1}$	=	+ 4630
2	+5000	×	$(1 + 0.08)^{-2}$	=	+ 4286
3	+5000	×	$(1 + 0.08)^{-3}$	=	+ 3969

The general form of the adjusting multiplier is

$$(1 + f)^{-n} \text{ or } (P/F, f, n)$$

Now that the problem has been converted to dollars of the same purchasing power (today's dollars in this example), we can proceed to compute the present worth of the future payments at the desired 12% interest rate.

Year	Actual Cash Flow		Multiplied by		Present Worth
0	—	×	—	=	—
1	+4630	×	$(1 + 0.12)^{-1}$	=	+4134
2	+4286	×	$(1 + 0.12)^{-2}$	=	+3417
3	+3969	×	$(1 + 0.12)^{-3}$	=	+2825
					$10,376

The investor would pay $10,376.

Alternate Solution

Instead of doing the inflation and interest rate computations separately, one can compute a combined equivalent interest rate per interest period, d.

$$d = (1 + f)(1 + i) - 1 = i + f + (i \times f)$$

For this cash flow, $d = 0.12 + 0.08 + 0.12(0.08) = 0.2096$. Since we do not have 20.96% interest tables, the problem must be calculated using present-worth equations.

$$\begin{aligned} PW &= 5000(1 + 0.2096)^{-1} + 5000(1 + 0.2096)^{-2} + 5000(1 + 0.2096)^{-3} \\ &= 4134 + 3417 + 2825 = \$10{,}376 \end{aligned}$$

Example 6.29

One economist has predicted that there will be a 7% per year inflation of prices during the next ten years. If this proves to be correct, an item that presently sells for $10 would sell for what price ten years hence?

Solution

$f = 7\%$, $P = \$10$

$F = ?$, $n = 10$ years

Here, the computation is to find the future worth F, rather than the present worth, P.

$$F = P(1 + f)10 = 10(1 + 0.07)10 = \$19.67$$

Effect of Inflation on a Rate of Return

The effect of inflation on the computed rate of return for an investment depends on how future benefits respond to the inflation. If benefits produce constant dollars, which are not increased by inflation, the effect of inflation is to reduce the before-tax rate of return on the investment. If, on the other hand, the dollar benefits increase to keep up with the inflation, the before-tax rate of return will not be adversely affected by the inflation.

This is not true when an after-tax analysis is made. Even if the future benefits increase to match the inflation rate, the allowable depreciation schedule does not increase. The result will be increased taxable income and income tax payments. This reduces the available after-tax benefits and, therefore, the after-tax rate of return.

Example 6.30

A man bought a 5% tax-free municipal bond. It cost \$1000 and will pay \$50 interest each year for 20 years. The bond will mature at the end of 20 years and return the original \$1000. If there is 2% annual inflation during this period, what rate of return will the investor receive after considering the effect of inflation?

Solution

$$d = 0.05,\ i = \text{unknown},\ f = 0.02$$

Combined effective interest rate/interest period,

$$d = i + f + (i \times f)$$
$$0.05 = i + 0.02 + 0.02i$$
$$1.02i = 0.03,\ i = 0.294 = 2.94\%$$

REFERENCE

Newnan, D. G., et al. *Engineering Economic Analysis*, 6th ed. San Jose, CA: Engineering Press, 2000.

PROBLEMS

6.1 A retirement fund earns 8% interest, compounded quarterly. If $400 is deposited every three months for 25 years, the amount in the fund at the end of 25 years is nearest to:

a. $50,000 c. $100,000
b. $75,000 d. $125,000

6.2 The repair costs for some handheld equipment are estimated to be $120 the first year, increasing by $30 per year in subsequent years. The amount a person needs to deposit into a bank account paying 4% interest to provide for the repair costs for the next five years is nearest to:

a. $500 c. $700
b. $600 d. $800

6.3 One thousand dollars is borrowed for one year at an interest rate of 1% per month. If this same sum of money were borrowed for the same period at an interest rate of 12% per year, the saving in interest charges would be closest to:

a. $0 c. $5
b. $3 d. $7

6.4 How much should a person invest in a fund that will pay 9%, compounded continuously, if he wishes to have $10,000 in the fund at the end of ten years?

a. $4000 c. $6000
b. $5000 d. $7000

6.5 A store charges 1.5% interest per month on credit purchases. This is equivalent to a nominal annual interest rate of:

a. 1.5% c. 18.0%
b. 15.0% d. 19.6%

6.6 A small company borrowed $10,000 to expand its business. The entire principal of $10,000 will be repaid in two years, but quarterly interest of $330 must be paid every three months. The nominal annual interest rate the company is paying is closest to:

a. 3.3% c. 6.6%
b. 5.0% d. 13.2%

6.7 A store's policy is to charge 3% interest every two months on the unpaid balance in charge accounts. The effective interest rate is closest to:

a. 6% c. 15%
b. 12% d. 19%

6.8 The effective interest rate on a loan is 19.56%. If there are 12 compounding periods per year, the nominal interest rate is closest to:

a. 1.5% c. 9.0%
b. 4.5% d. 18.0%

6.9 A deposit of \$300 was made one year ago into an account paying monthly interest. If the account now has \$320.52, the effective annual interest rate is closest to:

a. 7%
b. 10%
c. 12%
d. 15%

6.10 If the effective interest rate per year is 12%, based on monthly compounding, the nominal interest rate per year is closest to:

a. 8.5%
b. 9.3%
c. 10.0%
d. 11.4%

6.11 If 10% nominal annual interest is compounded daily, the effective annual interest rate is nearest to:

a. 10.00%
b. 10.38%
c. 10.50%
d. 10.75%

6.12 An individual wishes to deposit a certain quantity of money now so that he will have \$500 at the end of five years. With interest at 4% per year, compounded semiannually, the amount of the deposit is nearest to:

a. \$340
b. \$400
c. \$410
d. \$416

6.13 A steam boiler is purchased on the basis of guaranteed performance. A test indicates that the operating cost will be \$300 more per year than the manufacturer guaranteed. If the expected life of the boiler is 20 years, and the time value of money is 8%, the amount the purchaser should deduct from the purchase price to compensate for the extra operating cost is nearest to:

a. \$2950
b. \$3320
c. \$4100
d. \$5520

6.14 A consulting engineer bought a fax machine with one year's free maintenance. In the second year the maintenance cost is estimated at \$20. In subsequent years the maintenance cost will increase \$20 per year (that is, third year maintenance will be \$40, fourth year maintenance will be \$60, and so forth). The amount that must be set aside now at 6% interest to pay the maintenance costs on the fax machine for the first six years of ownership is nearest to:

a. \$101
b. \$164
c. \$229
d. \$284

6.15 An investor is considering buying a 20-year corporate bond. The bond has a face value of \$1000 and pays 6% interest per year in two semiannual payments. Thus the purchaser of the bond will receive \$30 every six months, and in addition he will receive \$1000 at the end of 20 years, along with the last \$30 interest payment. If the investor believes he should receive 8% annual interest, compounded semiannually, the amount he is willing to pay for the bond (bond value) is closest to:

a. \$500
b. \$600
c. \$700
d. \$800

6.16 Annual maintenance costs for a particular section of highway pavement are \$2000. The placement of a new surface would reduce the annual maintenance cost to \$500 per year for the first five years and to \$1000 per year for the next five years. The annual maintenance after ten years would again be \$2000. If maintenance costs are the only saving, the maximum investment that can be justified for the new surface, with interest at 4%, is closest to:

a. \$5500 c. \$10,000
b. \$7170 d. \$10,340

6.17 A project has an initial cost of \$10,000, uniform annual benefits of \$2400, and a salvage value of \$3000 at the end of its ten-year useful life. At 12% interest the net present worth (NPW) of the project is closest to:

a. \$2500 c. \$4500
b. \$3500 d. \$5500

6.18 A person borrows \$5000 at an interest rate of 18%, compounded monthly. Monthly payments of \$167.10 are agreed upon. The length of the loan is closest to:

a. 12 months c. 24 months
b. 20 months d. 40 months

6.19 A machine costing \$2000 to buy and \$300 per year to operate will save labor expenses of \$650 per year for eight years. The machine will be purchased if its salvage value at the end of eight years is sufficiently large to make the investment economically attractive. If an interest rate of 10% is used, the minimum salvage value must be closest to:

a. \$100 c. \$300
b. \$200 d. \$400

6.20 The amount of money deposited 50 years ago at 8% interest that would now provide a perpetual payment of \$10,000 per year is nearest to:

a. \$3000 c. \$50,000
b. \$8000 d. \$70,000

6.21 An industrial firm must pay a local jurisdiction the cost to expand its sewage treatment plant. In addition, the firm must pay \$12,000 annually toward the plant operating costs. The industrial firm will pay sufficient money into a fund that earns 5% per year to pay its share of the plant operating costs forever. The amount to be paid to the fund is nearest to:

a. \$15,000 c. \$60,000
b. \$30,000 d. \$240,000

6.22 At an interest rate of 2% per month, money will double in value in how many months?

a. 20 months c. 24 months
b. 22 months d. 35 months

6.23 A woman deposited \$10,000 into an account at her credit union. The money was left on deposit for 80 months. During the first 50 months the woman earned 12% interest, compounded monthly. The credit union then changed its interest policy so that the woman earned 8% interest compounded quarterly during the next 30 months. The amount of money in the account at the end of 80 months is nearest to:

a. \$10,000
b. \$12,500
c. \$15,000
d. \$20,000

6.24 An engineer deposited \$200 quarterly in her savings account for three years at 6% interest, compounded quarterly. Then for five years she made no deposits or withdrawals. The amount in the account after eight years is closest to:

a. \$1200
b. \$1800
c. \$2400
d. \$3600

6.25 A sum of money, Q, will be received six years from now. At 6% annual interest the present worth of Q is \$60. At this same interest rate the value of Q ten years from now is closest to:

a. \$60
b. \$77
c. \$90
d. \$107

6.26 If \$200 is deposited in a savings account at the beginning of each year for 15 years and the account earns interest at 6%, compounded annually, the value of the account at the end of 15 years will be most nearly:

a. \$4500
b. \$4700
c. \$4900
d. \$5100

6.27 The maintenance expense on a piece of machinery is estimated as follows:

Year	1	2	3	4
Maintenance	\$150	\$300	\$450	\$600

If interest is 8%, the equivalent uniform annual maintenance cost is closest to:

a. \$250
b. \$300
c. \$350
d. \$400

6.28 A payment of \$12,000 six years from now is equivalent, at 10% interest, to an annual payment for eight years starting at the end of this year. The annual payment is closest to:

a. \$1000
b. \$1200
c. \$1400
d. \$1600

6.29 A manufacturer purchased \$15,000 worth of equipment with a useful life of six years and a \$2000 salvage value at the end of the six years. Assuming a 12% interest rate, the equivalent uniform annual cost (EUAC) is nearest to:

a. \$1500
b. \$2500
c. \$3500
d. \$4500

6.30 Consider a machine as follows:

Initial cost: \$80,000
End-of-useful-life salvage value: \$20,000
Annual operating cost: \$18,000
Useful life: 20 years

Based on 10% interest, the equivalent uniform annual cost for the machine is closest to:

a. \$21,000 c. \$25,000
b. \$23,000 d. \$27,000

6.31 Consider a machine as follows:

Initial cost: \$80,000
Annual operating cost: \$18,000
Useful life: 20 years

What must be the salvage value of the machine at the end of 20 years for the machine to have an equivalent uniform annual cost of \$27,000? Assume a 10% interest rate. The salvage value S_{20} is closest to:

a. \$10,000 c. \$30,000
b. \$20,000 d. \$40,000

6.32 Twenty-five thousand dollars is deposited in a savings account that pays 5% interest, compounded semiannually. Equal annual withdrawals are to be made from the account beginning one year from now and continuing forever. The maximum amount of the equal annual withdrawals is closest to:

a. \$625 c. \$1250
b. \$1000 d. \$1265

6.33 An investor is considering the investment of \$10,000 in a piece of land. The property taxes are \$100 per year. The lowest selling price the investor must receive if she wishes to earn a 10% interest rate after keeping the land for ten years is:

a. \$20,000 c. \$23,000
b. \$21,000 d. \$27,000

6.34 The rate of return for a \$10,000 investment that will yield \$1000 per year for 20 years is closest to:

a. 1% c. 8%
b. 4% d. 12%

6.35 An engineer invested \$10,000 in a company. In return he received \$600 per year for six years and his \$10,000 investment back at the end of the six years. His rate of return on the investment was closest to:

a. 6% c. 12%
b. 10% d. 15%

6.36 An engineer made ten annual end-of-year purchases of $1000 of common stock. At the end of the tenth year, just after the last purchase, the engineer sold all the stock for $12,000. The rate of return received on the investment is closest to:

a. 2% c. 8%
b. 4% d. 10%

6.37 A company is considering buying a new piece of machinery.

Initial cost: $80,000
End-of-useful-life salvage value: $20,000
Annual operating cost: $18,000
Useful life: 20 years

The machine will produce an annual saving in material of $25,700. What is the before-tax rate of return if the machine is installed? The rate of return is closest to:

a. 6% c. 10%
b. 8% d. 15%

6.38 Consider the following situation: Invest $100 now and receive two payments of $102.15—one at the end of year 3 and one at the end of year 6. The rate of return is nearest to:

a. 6% c. 10%
b. 8% d. 18%

6.39 Two mutually exclusive alternatives are being considered:

Year	A	B
0	–$2500	–$6000
1	+746	+1664
2	+746	+1664
3	+746	+1664
4	+746	+1664
5	+746	+1664

The rate of return on the difference between the alternatives is closest to:

a. 6% c. 10%
b. 8% d. 12%

6.40 A project will cost $50,000. The benefits at the end of the first year are estimated to be $10,000, increasing $1000 per year in subsequent years. Assuming a 12% interest rate, no salvage value, and an eight-year analysis period, the benefit-cost ratio is closest to:

a. 0.78 c. 1.28
b. 1.00 d. 1.45

6.41 Two alternatives are being considered:

	A	B
Initial cost	$500	$800
Uniform annual benefit	$140	$200
Useful life, years	8	8

The benefit-cost ratio of the difference between the alternatives, based on a 12% interest rate, is closest to:

a. 0.60 c. 1.00
b. 0.80 d. 1.20

6.42 An engineer will invest in a mining project if the benefit-cost ratio is greater than 1.00, based on an 18% interest rate. The project cost is $57,000. The net annual return is estimated at $14,000 for each of the next eight years. At the end of eight years the mining project will be worthless. The benefit-cost ratio is closest to:

a. 0.60 c. 1.00
b. 0.80 d. 1.20

6.43 A city has retained your firm to do a benefit-cost analysis of the following project:

Project cost: $60,000,000
Gross income: $20,000,000 per year
Operating costs: $5,500,000 per year
Salvage value after ten years: None

The project life is ten years. Use 8% interest in the analysis. The computed benefit-cost ratio is closest to:

a. 0.80 c. 1.20
b. 1.00 d. 1.60

6.44 A piece of property is purchased for $10,000 and yields a $1000 yearly profit. If the property is sold after five years, the minimum price to break even, with interest at 6%, is closest to:

a. $5000 c. $7700
b. $6500 d. $8300

6.45 Given two machines:

	A	B
Initial cost	$55,000	$75,000
Total annual costs	$16,200	$12,450

With interest at 10% per year, at what service life do these two machines have the same equivalent uniform annual cost? The service life is closest to:

a. four years c. six years
b. five years d. eight years

6.46 A machine part that is operating in a corrosive atmosphere is made of low-carbon steel. It costs $350 installed, and lasts six years. If the part is treated for corrosion resistance it will cost $700 installed. How long must the treated part last to be as economical as the untreated part, if money is worth 6%?

a. 8 years
b. 11 years
c. 15 years
d. 17 years

6.47 A firm has determined that the two best paints for its machinery are Tuff-Coat at $45 per gallon and Quick at $22 per gallon. The Quick paint is expected to prevent rust for five years. Both paints take $40 of labor per gallon to apply, and both cover the same area. If a 12% interest rate is used, how long must the Tuff-Coat paint prevent rust to justify its use?

a. Five years
b. Six years
c. Seven years
d. Eight years

6.48 Two alternatives are being considered:

	A	B
Cost	$1000	$2000
Useful life in years	10	10
End-of-useful-life salvage value	$100	$400

The net annual benefit of alternative A is $150. If interest is 8%, what must be the net annual benefit of alternative B for the two alternatives to be equally desirable?

a. $150
b. $200
c. $225
d. $275

6.49 A $5000 municipal bond is offered for sale. It will provide 8% annual interest by paying $200 to the bondholder every six months. At the end of ten years, the $5000 will be paid to the bondholder along with the final $200 interest payment. If you consider 12% nominal annual interest, compounded semiannually, an appropriate bond yield, the amount you would be willing to pay for the bond is closest to:

a. $2750
b. $3850
c. $5000
d. $7400

6.50 A municipal bond is being offered for sale for $10,000. It is a zero-coupon bond, that is, the bond pays no interest during its 15-year life. At the end of 15 years the owner of the bond will receive a single payment of $26,639. The bond yield is closest to:

a. 4%
b. 5%
c. 6%
d. 7%

6.51 A firm is considering purchasing $8000 of small hand tools for use on a production line. It is estimated that the tools will reduce the amount of required overtime work by $2000 the first year, with this amount increasing by $1000 per year thereafter. The payback period for the hand tools is closest to:

a. 2.00 years
b. 2.50 years
c. 2.75 years
d. 3.00 years

6.52 Special tools for the manufacture of finished plastic products cost $15,000 and have an estimated $1000 salvage value at the end of an estimated three-year useful life and recovery period. The third-year straight-line depreciation is closest to:

a. $3000
b. $3500
c. $4000
d. $4500

6.53 Refer to the facts of Problem 6.52. The first-year MACRS depreciation is closest to:

a. $3000
b. $3500
c. $4000
d. $5000

6.54 An engineer is considering the purchase of an annuity that will pay $1000 per year for ten years. The engineer feels he should obtain a 5% rate of return on the annuity after considering the effect of an estimated 6% inflation per year. The amount he would be willing to pay to purchase the annuity is closest to:

a. $1500
b. $3000
c. $4500
d. $6000

6.55 An automobile costs $20,000 today. You can earn 12% tax-free on an *auto purchase account*. If you expect the cost of the auto to increase by 10% per year, the amount you would need to deposit in the account to provide for the purchase of the auto five years from now is closest to:

a. $12,000
b. $14,000
c. $16,000
d. $18,000

6.56 An engineer purchases a building lot for $40,000 cash and plans to sell it after five years. If he wants an 18% before-tax rate of return, after taking the 6% annual inflation rate into account, the selling price must be nearest to:

a. $55,000
b. $65,000
c. $75,000
d. $125,000

6.57 A piece of equipment with a list price of $450 can actually be purchased for either $400 cash or $50 immediately plus four additional annual payments of $115.25. All values are in dollars of current purchasing power. If the typical customer considered a 5% interest rate appropriate, the inflation rate at which the two purchase alternatives are equivalent is nearest to:

a. 5%
b. 6%
c. 8%
d. 10%

SOLUTIONS

6.1 d.

$$F = A(F/A,i,n) = 400(F/A,2\%,100)$$
$$= 400(312.23) = \$124{,}890$$

6.2 d.

$$P = A(P/A,i,n) + G(P/G,i,n)$$
$$= 120(P/A,4\%,5) + 30(P/G,4\%,5)$$
$$= 120(4.452) + 30(8.555) = \$791$$

6.3 d.

At i = 1%/month: $F = 1000(1 + 0.01)^{12} = \$1126.83$
At i = 12%/year: $F = 1000(1 + 0.12)^{1} = 1120.00$
Saving in interesting charges = 1126.83 – 1120.00 = \$6.83

6.4 a.

$$P = Fe^{-rn} = 10{,}000e^{-0.09(10)} = 4066$$

6.5 c. The nominal interest rate is the annual interest rate ignoring the effect of any compounding. Nominal interest rate = 1.5% × 12 = 18%.

6.6 d. The interest paid per year = 330 × 4 = 1320. The nominal annual interest rate = 1320/10,000 = 0.132 = 13.2%.

6.7 d.

$$i_e = (1 + r/m)^m - 1 = (1 + 0.03)6 - 1 = 0.194 = 19.4\%$$

6.8 d.

$$i_e = (1 + r/m)^m - 1$$
$$r/m = (1 + i^e)^{1/m} - 1 = (1 + 0.1956)1/12 - 1 = 0.015$$
$$r = 0.015(m) = 0.015 \times 12 = 0.18 = 18\%$$

6.9 a.

$$i_e = 20.52/300 = 0.0684 = 6.84\%$$

6.10 d.

$$i^e = (1 + r/m)^m - 1$$
$$0.12 = (1 + r/12)12 - 1$$
$$(1.12)1/12 = (1 + r/12)$$
$$1.00949 = (1 + r/12)$$
$$r = 0.00949 \times 12 = 0.1138 = 11.38\%$$

6.11 c.

$$i^e = (1 + r/m)^m - 1 = (1 + 0.10/365)^{365} - 1 = 0.1052 = 10.52\%$$

6.12 c.

$$P = F(P/F,i,n) = 500(P/F,2\%,10) = 500(0.8203) = \$410$$

6.13 a.

$$P = 300(P/A,8\%,20) = 300(9.818) = \$2945$$

6.14 c. Using single payment present worth factors:

$$P = 20(P/F,6\%,2) + 40(P/F,6\%,3) + 60(P/F,6\%,4) + 80(P/F,6\%,5) + 100(P/F,6\%,6) = \$229$$

Alternate solution using the gradient present worth factor:

$$P = 20(P/G,6\%,6) = 20(11.459) = \$229$$

6.15 d.

$$\text{PW} = 30\,(P/A,4\%,40) + 1000(P/F,4\%,40) = 30(19.793) + 1000(0.2083) = \$802$$

6.16 d. Benefits are \$1500 per year for the first five years and \$1000 per year for the subsequent five years.

As Exhibit 6.16 indicates, the benefits may be considered as \$1000 per year for ten years, plus an additional \$500 benefit in each of the first five years.

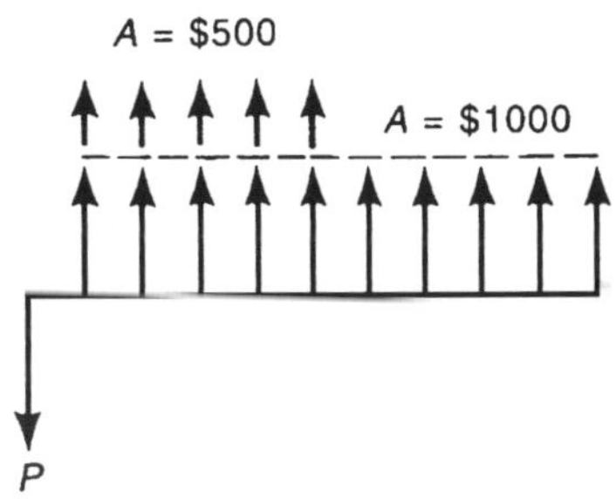

Exhibit 6.16

$$\begin{aligned}\text{maximum investment} &= \text{present worth of benefits}\\ &= 1000(P/A,4\%,10) + 500(P/A,4\%,5)\\ &= 1000(8.111) + 500(4.452) = \$10{,}337\end{aligned}$$

6.17 c.

$$\begin{aligned}\text{NPW} &= \text{PW of benefits} - \text{PW of cost}\\ &= 2400(P/A,12\%,10) + 3000(P/F,12\%,10) - 10{,}000 = \$4526\\ &= 2400(5.65) + 3000(.32) - 10{,}000 = \$4526\end{aligned}$$

6.18 d.

$$\begin{aligned}\text{PW of benefits} &= \text{PW of cost}\\ 5000 &= 167.10(P/A,1.5\%,n)\\ (P/A,1.5\%,n) &= 5000/167.10 = 29.92\end{aligned}$$

From the $1\frac{1}{2}\%$ interest table, $n = 40$.

6.19 c.

$$\begin{aligned} NPW &= \text{PW of benefits} - \text{PW of cost} = 0 \\ &= (650 - 300)(P/A,10\%,8) + S_8\,(P/F,10\%,8) - 2000 = 0 \\ &= 350(5.335) + S_8(0.4665) - 2000 = 0 \\ S_8 &= 132.75/0.4665 = \$285 \end{aligned}$$

6.20 a. The amount of money needed now to begin the perpetual payments is $P' = A/i = 10{,}000/0.08 = 125{,}000$. From this we can compute the amount of money, P, that would need to have been deposited 50 years ago:

$$P = 125{,}000(P/F,8\%,50) = 125{,}000(0.0213) = \$2663$$

6.21 d.

$$P = A/i = 12{,}000/0.5 = \$240{,}000$$

6.22 d.

$$\begin{aligned} 2 &= 1(F/P,i,n) \\ (F/P,2\%,n) &= 2 \end{aligned}$$

From the 2% interest table, n = about 35 months.

6.23 d. At the end of 50 months

$$F = 10{,}000(F/P,1\%,50) = 10{,}000(1.645) = \$16{,}450$$

At the end of 80 months

$$F = 16{,}450(F/P,2\%,10) = 16{,}450(1.219) = \$20{,}053$$

6.24 d.

$$\begin{aligned} FW &= 200(F/A,1.5\%,12)(F/P,\ 1.5\%,20) \\ &= 200(13.041)(1.347) = \$3513 \end{aligned}$$

6.25 d. The present amount $P = 60$ is equivalent to Q six years hence at 6% interest. The future sum F may be calculated by either of two methods:

$$F = Q(F/P,6\%,4) \text{ and } Q = 60(F/P,6\%,6)$$

or

$$F = P(F/P,6\%,10)$$

Since P is known, the second equation may be solved directly.

$$F = P(F/P,6\%,10) = 60(1.791) = \$107$$

6.26 c.

$$F' = A(F/A,i,n) = 200(F/A,6\%,15) = 200(23.276) = \$4655.20$$
$$F = F'\,(F/P,i,n) = 4655.20(F/P,6\%,1) = 4655.20(1.06) = \$4935$$

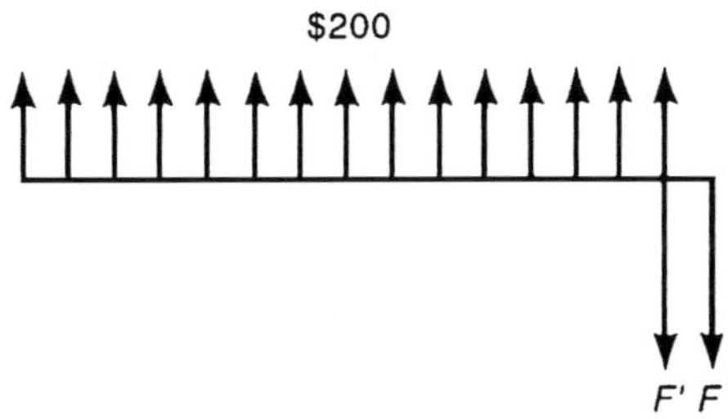

Exhibit 6.26

6.27 c.

$$\text{EUAC} = 150 + 150(A/G,8\%,4) = 150 + 150(1.404) = \$361$$

6.28 b.

$$\text{Annual payment} = 12{,}000(P/F,10\%,6)(A/P,10\%,8)$$
$$= 12{,}000(0.5645)(0.1874) = \$1269$$

6.29 c.

$$\text{EUAC} = 15{,}000(A/P,12\%,6) - 2000(A/F,12\%,6)$$
$$= 15{,}000(0.2432) - 2000(0.1232) = \$3402$$

6.30 d.

$$\text{EUAC} = 80{,}000(A/P,10\%,20) - 20{,}000(A/F,10\%,20)$$
$$+ \text{ annual operating cost}$$
$$= 80{,}000(0.1175) - 20{,}000(0.0175) + 18{,}000$$
$$= 9400 - 350 + 18{,}000 = \$27{,}050$$

6.31 b.

$$\text{EUAC} = \text{EUAB}$$
$$27{,}000 = 80{,}000(A/P,10\%,20) + 18{,}000 - S_{20}(A/F,10\%,20)$$
$$= 80{,}000(0.1175) + 18{,}000 - S_{20}(0.0175)$$
$$S_{20} = (27{,}400 - 27{,}000)/0.0175 = \$22{,}857$$

6.32 d. The general equation for an infinite life, $P = A/i$, must be used to solve the problem.

$$ie = (1 + 0.025)2 - 1 = 0.050625$$

The maximum annual withdrawal will be $A = Pi = 25{,}000(0.050625)$ = \$1266.

6.33 d.

$$\text{Minimum sale price} = 10{,}000(F/P,10\%,10) + 100(F/A,10\%,10)$$
$$= 10{,}000(2.594) + 100(15.937) = \$27{,}530$$

6.34 c.

$$\text{NPW} = 1000(P/A,i,20) - 10{,}000 = 0$$
$$(P/A,i,20) = 10{,}000/1000 = 10$$

From interest tables: $6\% < i < 8\%$.

6.35 a. The rate of return was 600/10,000 = 0.06 = 6%.

6.36 b.

$$F = A(F/A,i,n)$$
$$12{,}000 = 1000(F/A,i,10)$$
$$(F/A,i,10) = 12{,}000/1000 = 12$$

In the 4% interest table: $(F/A,4\%,10) = 12.006$, so $i = 4\%$.

6.37 b.

$$\text{PW of cost} = \text{PW of benefits}$$
$$80{,}000 = (25{,}700 - 18{,}000)(P/A,i,20) + 20{,}000(P/F,i,20)$$

Try $i = 8\%$.

$$80{,}000 \stackrel{?}{=} 7700(9.818) + 20{,}000(0.2145) = 79{,}889$$

Therefore, the rate of return is very close to 8%.

6.38 d.

$$\text{PW of cost} = \text{PW of benefits}$$
$$100 = 102.15(P/F,i,3) + 102.15(P/F,i,6)$$

Solve by trial and error. Try $i = 12\%$.

$$100 \stackrel{?}{=} 102.15(0.7118) + 102.15(0.5066) = 124.46$$

The PW of benefits exceeds the PW of cost. This indicates that the interest rate i is too low. Try $i = 18\%$.

$$100 \stackrel{?}{=} 102.15(0.6086) + 102.15(0.3704) = 100.00$$

Therefore, the rate of return is 18%.

6.39 c. The difference between the alternatives:

$$\text{Incremental cost} = 6000 - 2500 = \$3500$$

$$\text{Incremental annual benefit} = 1664 - 746 = \$918$$
$$\text{PW of cost} = \text{PW of benefits}$$
$$3500 = 918(P/A,i,5)$$
$$(P/A,i,5) = 3500/918 = 3.81$$

From the interest tables, i is very close to 10%.

6.40 c.

$$\text{B/C} = \frac{\text{PW of benefits}}{\text{PW of cost}} = \frac{10{,}000\,(\text{P/A}, 12\%, 8) + 1000\,(\text{P/G}, 12\%, 8)}{50{,}000}$$

$$= \frac{10{,}000\,(4.968) + 1000\,(14.471)}{50{,}000} = 1.28$$

6.41 c.

$$B/C = \frac{\text{PW of benefits}}{\text{PW of cost}} = \frac{60(P/A,12\%,8)}{300} = \frac{60(4.968)}{300} = 0.99$$

Alternate solution:

$$B/C = \frac{\text{EUAB}}{\text{EUAC}} = \frac{60}{300(A/P,12\%,8)} = \frac{60}{300(0.2013)} = 0.99$$

6.42 a.

$$B/C = \frac{\text{PW of benefits}}{\text{PW of cost}} = \frac{14,000(P/A,18\%,8)}{57,000} = \frac{14,000(4.078)}{57,000} = 1.00$$

6.43 d.

$$B/C = \frac{\text{EUAB}}{\text{EUAC}} = \frac{20,000,000 - 5,500,000}{60,000,000(A/P,8\%,10)} = 1.62$$

6.44 c.

$$F = 10,000(F/P,6\%,5) - 1000(F/A,6\%,5)$$
$$= 10,000(1.338) - 1000(5.637) = \$774$$

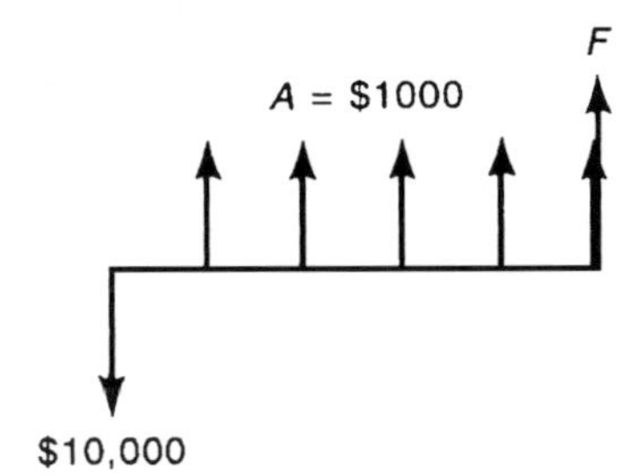

Exhibit 6.44

6.45 d.

$$\text{PW of cost}_A = \text{PW of cost}_B$$
$$55,000 + 16,200(P/A,10\%,n) = 75,000 + 12,450(P/A,10\%,n)$$
$$(P/A,10\%,n) = (75,000 - 55,000)/(16,200 - 12,450)$$
$$= 5.33$$

From the 10% interest tables, $n = 8$ years.

6.46 c.

$$\text{EUAC}_{\text{untreated}} = \text{EUAC}_{\text{treated}}$$
$$350(A/P,6\%,6) = 700(A/P,6\%,n)$$
$$350(0.2034) = 700(A/P,6\%,n)$$
$$(A/P,6\%,n) = 71.19/700 = 0.1017$$

From the 6% interest table, $n = 15+$ years.

6.47 d.

$$\text{EUAC}_{\text{T-C}} = \text{EUAC}_{\text{Quick}}$$
$$(45 + 40)(A/P,12\%,n) = (22 + 40)(A/P,12\%,5)$$
$$(A/P,12\%,n) = 17.20/85 = 0.202$$

From the 12% interest table, $n = 8$.

6.48 d. At breakeven,

$$\text{NPWA} = \text{NPWB}$$
$$150(P/A,8\%,10) + 100(P/F,8\%,10) - 1000 = \text{NAB}(P/A,8\%,10) + 400(P/F,8\%,10) - 2000$$
$$52.82 = 6.71(\text{NAB}) - 1814.72$$

Net annual benefit (NAB) = (1814.72 + 52.82)/6.71 = \$278

6.49 b. The number of six-month compounding periods in this problem is 20. So $n = 20$ and $12\%/2 = 6\%$ is the interest rate for the six-month interest period.

$$\begin{aligned}\text{Bond value} &= \text{PW of all future benefits}\\ &= 200(P/A,6\%,20) + 5000(P/F,6\%,20)\\ &= 200(11.470) + 5000(0.3118) = \$3853\end{aligned}$$

6.50 d. We know $P = 10{,}000$, $F = 26{,}639$, $n = 15$, and i = bond yield. Using the equation for the single payment compound amount:

$$F = P(1 + i)^n$$
$$26{,}639 = 10{,}000(1 + i)^{15}$$
$$2.66391/15 = (1 + i)$$
$$1.0675 = 1 + i$$
$$i = 0.0675 = 6.75\%$$

6.51 c. The annual benefits are \$2000, \$3000, \$4000, \$5000, and so on. The payback period is the time when \$8000 of benefits are received. This will occur in 2.75 years.

6.52 d.

$$D_3 = (C - S)/n = (15{,}000 - 1000)/3 = \$4666$$

6.53 d. From the modified ACRS table (Table 6.3) read for the first recovery year and three-year recovery the MACRS depreciation is 33.33% × 15,000 = \$5000.

6.54 d.

$$d = i + f + (i \times f) = 0.05 + 0.06 + 0.05(0.06) = 0.113 = 11.3\%$$

$$P = A(P/A,11.3\%,10) = 1000\left[\frac{(1+0.113)^{10} - 1}{0.113(1+0.113)^{10}}\right]$$

$$= 1000\left[\frac{1.9171}{0.3296}\right] = \$5816$$

6.55 d.

Cost of auto five years hence (F) = $P(1 + \text{inflation rate})^n$

= 20,000(1 + 0.10)5 = 32,210

Amount to deposit now to have \$32,210 available five years hence:

$P = F(P/F,i,n) = 32{,}210\ (P/F,12\%,5) = 32{,}210(0.5674) = \$18{,}276$

6.56 d.

Selling price (F) = 40,000(F/P,18%,5)(F/P,6%,5)
= 40,000(2.288)(1.338) = \$122,500

6.57 b.

PW of cash purchase = PW of installment purchase
400 = 50 + 115.25(P/A,d,4)
(P/A,d,4) = 350/11.25 = 3.037

From the interest tables, d = 12%.

$d = i + f + i\,(f)$

$0.12 = 0.05 + f + 0.05\text{f}$

$f = 0.07/1.05 = 0.0667 = 6.67\%$

CHAPTER 7

Electricity and Magnetism

Lincoln D. Jones and Thyagarajan Srinivasan

OUTLINE

This chapter assumes the reader completed an introductory electrical engineering or circuits course but needs to review points of theory and application. For instance, the solution for a single voltage source in a series circuit might easily come to mind, but if one changed the circuit by inserting another source (multiple source network) that may be acting as a load, the method of solution may not be readily apparent without a short review.

This chapter presents the portions of theory that are relevant to the FE/EIT exam. An attempt is made to reduce the more detailed theory to the level for the expected problem and to leave out the more abstract formal development of a solution. For example, rather than setting up a three-phase ac problem for a complete solution—including possibly an unbalanced network—this review will assume the more likely case of all three legs of the circuit being part of a balanced network and proceed with a much simpler method of solution.

The objective of this presentation will be to jog the memory so that the reader can feel comfortable with the simpler type of *quick solution* problems expected on the examination. It should be noted that multiple-choice answers for any such problems are significantly different, which allows one to quickly locate the correct answer if the simplest possible solution can be found. As an example, in a series ac circuits problem one can frequently plot a phasor solution with ruler and protractor that is accurate enough to pick the correct answer. One needs to be careful to not solve for more than what is asked.

ELECTRICAL QUANTITIES AND DEFINITIONS

Electrostatics

The subject of electric fields is sometimes omitted in a formal course in electrical engineering (the knowledge is frequently assumed from a prerequisite physics course). Thus, one needs to review electric fields and flux densities attributable to electrical charges. These fields, forces, and flux densities require three-dimensional vector notation. However, most problems will reduce in complexity to two dimensions; this reduction will permit quick graphical solutions.

Whereas the forces exerted between charges depend on whether the charges are in motion, one may start with stationary electric charges that produce electric fields. These fields may be defined in terms of the forces they produce on one another.

The smallest amount of charge that can exist is the charge of one electron, which is 1.602×10^{-19} coulombs (C). One coulomb of charge is thus equivalent to 6.24×10^{18} electrons. This is the amount of charge that is necessary to develop a force of one newton in an electric field of one volt per meter. An electric field, **E** (boldface for vector quantity), is the amount of force (F) that would be exerted on a *positive charge* (assuming it is concentrated at a point) *of one coulomb* if it were placed in that field:

$$\mathbf{F} = Q\mathbf{E}$$

The electric field is not thought of as a point but rather as being distributed throughout a small region. This is a vector quantity; the direction of force on the one coulomb would be toward the point source of the field if that point were a negative charge. The units of measurement are newtons per unit of charge; alternately, it may be given in terms of volts/meter since

$$\frac{\text{force}}{\text{charge}} = \frac{\text{force} \times \text{distance}}{\text{charge} \times \text{distance}} = \frac{\text{energy}}{\text{charge} \times \text{distance}} = \frac{\text{voltage}}{\text{distance}}$$

For point charges, the force is directly proportional to the product of the two charges and inversely proportional to the square of the distance between them (similar to the laws of gravity).

$$\mathbf{F}_2 = \frac{Q_1 Q_2}{4\pi\varepsilon r^2}\mathbf{a}_{r12}, \quad \text{where} \tag{7.1a}$$

$\mathbf{F}_2$ = the force on charge 2 due to charge 1

Q_i = the ith point charge

r = the distance between charges 1 and 2

$\mathbf{a}_{r12}$ = a unit vector directed from 1 to 2

ε = the permittivity (or dielectric constant) of the medium

The constant of proportionality depends on the medium between the two charges. This constant is $1/(4\pi\varepsilon)$ or approximately 9×10^9 for free space. If the medium is not free space, the 9×10^9 is merely divided by the relative permittivity to give the simplified *in line* form as

$$F = (9 \times 10^9/\varepsilon_{rel})Q_1Q_2/r^2 \text{ newtons} \tag{7.1b}$$

$$\varepsilon = \varepsilon_{rel}\varepsilon_0 \text{ F/m} \tag{7.2}$$

and ε_0 = permittivity of free space = 8.85×10^{-12} farads/meter. Permittivity of air is approximately the same as that of free space or a vacuum (that is, $\varepsilon_{rel} = 1.0006$), but if the medium happens to be caster oil, ε_{rel} is almost 5.

A convenient way of thinking of these fields is to imagine flux lines radiating either away from or toward a point source, as shown in Figure 7.1. If one imagines a positive point source charge at the center of a sphere and arrows pointed away from the center, these arrows would be the flux lines (the bigger the charge, the more arrows).

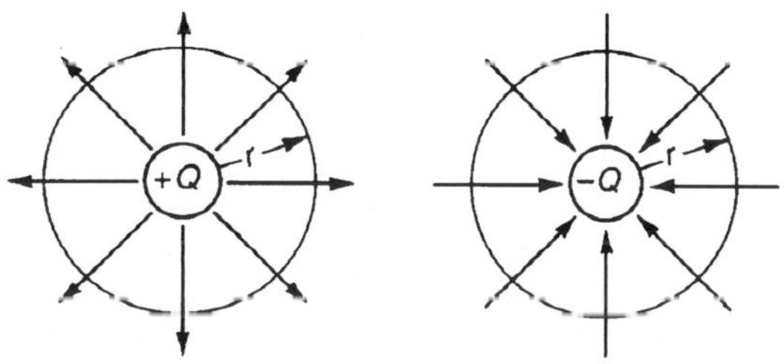

Figure 7.1 Electric flux lines around a charge

The flux density on the surface of the sphere (whose center is located at the point charge) would then be the number of arrows through a unit area on the sphere.

$$D = Q/A \tag{7.3a}$$

$$Q = \mathbf{D} \bullet \mathbf{A} \tag{7.3b}$$

where A is the area of the sphere ($4\pi r^2$) and Q is the quantity of charge in coulombs (assuming the area is normal to the flux). Equation (7.3b) is presented for those familiar with the dot and cross product vector notation (this notation guarantees the portion of the surface being considered is normal to the flux). If one were to divide the area of the sphere into very small areas, dA's, the sum of areas times the amount of flux through each area would be the amount of charge at the center of the sphere. More formally, as the size of each area approaches zero, and as one integrates over the entire area of the sphere, the total charge enclosed is Q; this is known as Gauss's Law, Equations (7.4a) and (7.4b). But, by using the dot-product notation, one is not limited to a spherical shape with the charge at the center; the formal law is then given as Equation (7.4b) or (7.4c):

$$Q = \sum D\ dA \text{ (for entire surface)} \tag{7.4a}$$

$$Q = \oint_s \mathrm{D} \bullet dA \tag{7.4b}$$

$$Q/\varepsilon = \oint_s \mathrm{E} \bullet dA \tag{7.4c}$$

The field strength, **E**, at the sphere's surface is then proportional to D,

$$\mathbf{E} = (D/\varepsilon)\mathbf{a}_r \tag{7.5}$$

where $\mathbf{a}_r$ is a unit radial vector direction. If there are a number of charges throughout a region, it is usually easier to use the flux density concept in solving problems. In this chapter, unit vectors will be denoted as **a** so that e can be reserved for other purposes.

Example 7.1

Assume three point charges, A, B, and C, as shown in Exhibit 1. Points A and B are 2 meters apart; Point C is on a perpendicular bisector between A and B and is 1 meter lower. Point A has 4×10^{-6} coulombs of negative charge, Point B has 10×10^{-6} coulombs of positive charge, and Point C has no charge yet. Determine the force and the field on B (in this case, due only to the charge on A) and the flux density at Point C.

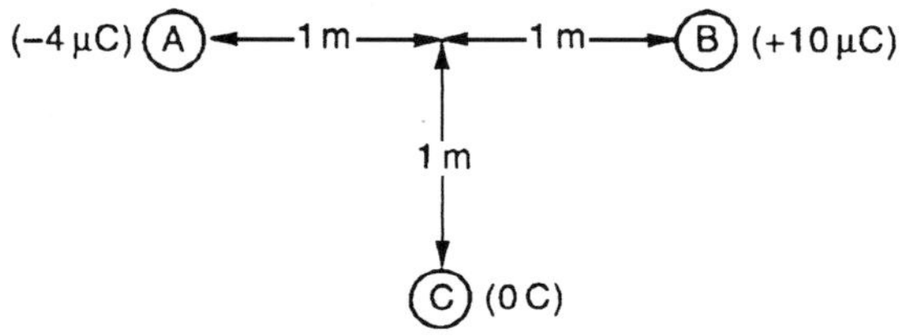

Exhibit 1 Point charge locations

Solution

The force on B due to A is directly proportional to the charges on A and B and inversely proportional to the square of the distance of separation. Using Equation (7.1),

$$\mathbf{F} = (9 \times 10^9)(-4 \times 10^{-6})(10 \times 10^{-6})/(2^2) = -9 \times 10^{-2} \text{ newtons}$$

with the direction of the force toward each other.

$$\mathbf{E} \text{ (at B)} = \mathbf{F}/Q_B = 9 \times 10^{-2}/10^{-5} = 9 \times 10^3 \text{ V/m}$$

(If the medium happened to be a special oil with relative permittivity 5.0, both **F** and **E** would only be one-fifth as large.) A more orderly solution (especially if several charges are involved—or none at C in this case) is to use the flux density relationship to find the individual Ds, then convert to Es. All that is now necessary to find the net **E** is to use vector summation of the flux densities and divide by ε. To find the flux density at C attributable to charge A, imagine a sphere passing through C with its center at A; repeat for B.

$$D_{CA} \text{ (at C attributable to A)} = Q_A/(4\pi r^2)$$

$$\mathbf{E}_{CA} = (D/\varepsilon)\mathbf{a}_{rCA} = (Q_A/r^2)(1/4\pi\varepsilon)\mathbf{a}_{rCA} \text{ V/m}$$

Recall that $1/(4\pi\varepsilon)$ for free space is 9×10^9, then

$$\mathbf{E}_{CA} = \left[(4\times10^{-6})/\sqrt{2^2}\right](9\times10^9)\text{a},_{CA} = 1.8\times10^4\,\mathbf{a},_{rCA} \text{ V/m}$$

$$\mathbf{E}_{CB} = \left[(10\times10^{-6})/\sqrt{2^2}\right](9\times10^9)\text{a},_{CA} = 4.5\times10^4\,\mathbf{a},_{rCA} \text{ V/m}$$

To find the E_C net one may use the more formal procedure of finding the rectangular components of each of the field vectors since only two dimensions are involved. Then, summing the horizontal and vertical components, the net field vector is the square root of the sum of the squares. If one assumes the reference vector, $\mathbf{a}_r$, has the horizontal component a and the vertical component jb (the j implies the 90° or vertical axis), then

$$\mathbf{E}_{CA} = |\mathbf{E}_{CA}|(\cos\theta + j\sin\theta) = 1.8 \times 10^4 (\cos 135° + j \sin 135°)$$
$$= (-1.27 + j1.27) \times 10^4 \text{ V/m}$$

$$\mathbf{E}_{CB} = 4.5 \times 10^4 (\cos 225° + j \sin 225°) = (-3.18 - j3.18) \times 10^4 \text{ V/m}$$

$$\mathbf{E}_{Cnet} = (-4.45 - j1.91) \times 10^4 = 4.84 \times 10^4 \angle 203.2° \text{ V/m}$$

where the angle θ is the angle between the vector **E** and a horizontal reference line. However, solving for the vector field at C may be done faster by using a graphical solution, assuming reasonable accuracy. The points A, B, and C are set up (to their own scale); then, working with Point C, one places the vectors (using any suitable scale) that represent the two other fields. The reference direction and magnitude is then pictured as shown in Exhibit 2.

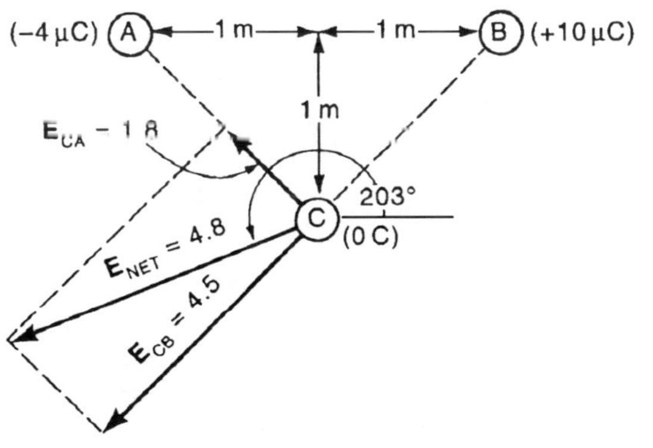

Exhibit 2 Sample problem vector fields (×104) at C

Now, assume the charge at C is $+5 \times 10^{-6}$ coulombs; find the net vector field at Point B. To find this field at Point B, merely remove the charge at B.

$$\mathbf{E}_{BA} = (Q_A/r^2)(9 \times 10^9)\mathbf{a}_{rBA} = 0.90\times10^4\,\mathbf{a}_{rBA} \text{ V/m}$$

$$\mathbf{E}_{BC} = (Q_C/r^2)(9 \times 10^9)\mathbf{a}_{rBC} = 2.25\times10^4\,\mathbf{a}_{rBC} \text{ V/m}$$

Again, solve graphically for the solution (Exhibit 3).

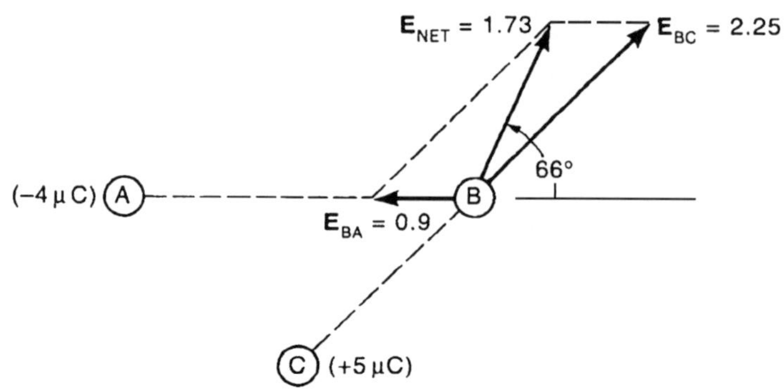

Exhibit 3 The electric field at B (×104)

The actual solution, using vector notation, is found to be **E** (at B) = 1.73×10^4 $\angle 66.5°$ volts/meter.

A more complex problem involving the *x*, *y*, and *z* axes is no more complicated except for the bookkeeping difficulty of solving a three-dimensional problem.

Before leaving the subject of static electric fields, the subject of energy storage in a capacitor (a pair of parallel plates separated by a distance, *d*) will be introduced. Capacitance is the ratio of the total stored charge uniformly distributed on the plates to the voltage difference between the plates. The separation distance, *d*, between plates will be considered to be small compared to the plate area (so that fringing may be neglected), then the field and flux lines will be perpendicular to the plates. The capacitance is

$$C = q_c(t)/v_c(t) \text{ farads (coulombs/volt)} \quad \textbf{(7.6a)}$$

and for non-time-varying quantities

$$C = Q/(Ed) = AD/(Ed) = A(E\varepsilon)/(Ed) = A\varepsilon/d \quad \textbf{(7.6b)}$$

Permittivity may now be defined in a slightly different manner, where ε equals the charge induced on one square meter of the capacitor plates by an electric field intensity of one volt/meter. Thus, for a free space separation of the plates, 8.85×10^{-12} coulombs is induced on one square meter of a conducting plate by an electric field having an intensity of one volt per meter.

Magnetic Quantities and Definitions

Magnetic effects are related to the motion of charges, or currents. From the previous section on electrostatics, a force of one newton is produced by a charge of one coulomb in an electric field intensity of one volt per meter. Electric current, on the other hand, can be thought of as moving charges. **Current** is the time rate of change of the electric charge passing through a surface area. This definition is expressed as

$$i = dq/dt \quad \textbf{(7.7)}$$

The unit of current is the **ampere**, A, which equals one coulomb per second.

From the concept of moving charges, one can begin to understand magnetic fields. For permanent magnets (due to *static* magnetic fields) one recalls from physics that for some materials the molecular structure has the electron orbits of the atoms aligned. These tiny moving charges of electrons produce tiny currents.

This alignment results in magnetic fields; actually ferromagnetic materials can be thought of as a large number of magnetic domains, with the domains being mostly aligned. However, when these (magnetic) domains are in disarray or randomly aligned, the material is unmagnetized.

Oersted, in 1819, observed that a magnetic flux existed about a wire carrying an electric current. (Flux lines can almost be visualized by observing the pattern of sprinkled iron filings on a piece of paper held over one pole of a magnet.) A few years after Oersted observed the effect of magnetic flux, Ampere found that wire coils carrying a current acted in the same manner as magnets. Simply stated, a coil of several turns of wire produced a stronger magnetic flux than only one turn of wire for the same current. And, if there were a ferromagnetic material to carry (or to provide a path for) the magnetic flux, the flux strength would be much greater.

Consider a toroidal ferromagnetic ring wrapped with a coil of several turns of wire, with the coil connected to a variable current source. Assume the current is zero and the material is not magnetized to begin with, then with an increase of current, an increase in flux results. One tends to think that the relationship would be linear; but the relationship is actually rather complex for a ferromagnetic material. As the current is further increased, the flux tends to level off. The leveling off is caused by most of the magnetic domains in the material aligning themselves in the same direction; consequently, increasing the current beyond the *knee of the bend* does not significantly increase the flux. If the current is decreased, most (but not all) of the magnetic domains return to random directions for zero current. The *going up* path is not necessarily the same as the *going down* path; these paths are known as the hysteresis curve for a particular material. When the current reaches zero on the *coming down* side, the material is left partially magnetized; this is called the residual magnetism. Again, assuming the material is unmagnetized to begin with, the flux is caused by the current through the number of turns of wire. This product of the current and the number of turns of wire, or amp-turns ($A \bullet t$), is frequently called the magneto-motive force (mmf) and indicated with a script F. The mmf is the source that causes the magnetic flux (ϕ). A plot of these two quantities is given in Figure 7.2(b); this plot is referred to as the hysteresis curve.

If the width of the hysteresis curve were small (that is, the residual magnetic flux is small), only the first quadrant of the hysteresis curve is needed and can be drawn as a single line; then, only one curve per kind of magnetic material is needed.

Before a numerical problem can be solved, one needs to quantify and further define several terms. Rather than using flux, ϕ (units of webers), it is more appropriate to use flux density, B (units of tesla or webers/unit area). Flux density is the amount of flux passing through a unit area (normal to the flux). Also, rather than using the straight magneto-motive force designation, a more useful one is mmf/length, called the magnetic field intensity, H. These standardized quantities will allow a plot of B vs. H; this plot is always given for solid magnetic materials (that is, before any air gap might be cut into them).

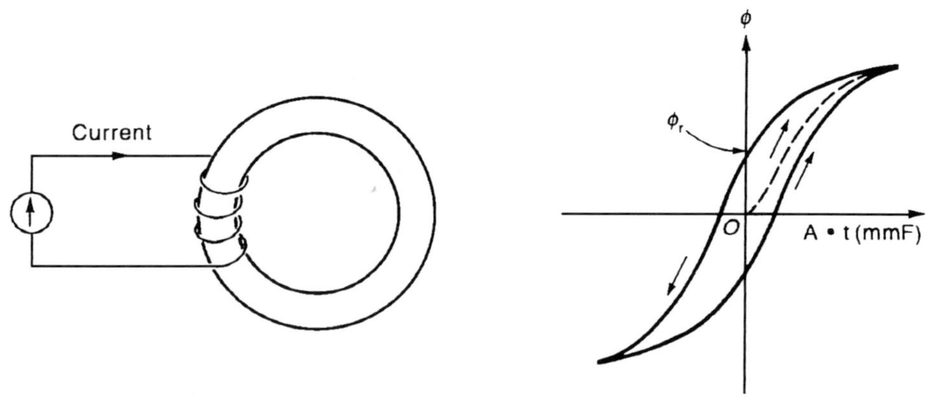

Figure 7.2 Magnetic characteristics

One needs to define further the direction of magnetic flux and fields; this is difficult to do without the concept of magnetic poles. To show this effect, consider an air gap cut into the ferromagnetic material. However, analogous to electrical charges being positive or negative, one can think of magnetic poles (they always occur in pairs) as north and south (recall, like poles repel and unlike ones attract). As an example, consider a freely suspended magnet (say, a magnetized needle in a compass) in the earth's magnetic field; one end of the magnet points toward the geographical north. By common usage, this end of the magnet is referred to as the *north-seeking pole*. For the toroidal ring with the air gap cut in it, if the positive direction of the current enters the top end of the coil, the direction of the flux in the ring will be clockwise, and the top of the air gap is the north pole, whereas the bottom is the south pole.

In the previous example, the magnetic flux in the air gap is concentrated and may be very high. The lines of magnetic flux produced in the material and the air gap are the same and continuous when there is a current in the wire. Without the ferromagnetic material in the path, the relationship between the current (causing the magnetic flux) and the resulting magnetic field would be linear but very weak. The linear constant of proportionality is called the permeability of free space, μ_0 (as in the air gap). This constant is

$$\mu_0 = 4\pi \times 10^{-7} \text{ F/m} \tag{7.8a}$$

For ferromagnetic materials, the relative permeability is nonlinear; for approximate calculations, it is sometimes linearized in the region of the curve before magnetic saturation is reached. The slope of the curve in the saturation region approaches that of free space or air, thus

$$\mu_r \mu_0 \rightarrow \mu_0 \tag{7.8b}$$

The relationship between B and H may then be expressed as the slope of a B vs. H curve:

$$\mu = \mu_r \mu_0 = B/H \tag{7.8c}$$

when B and H are normal to each other.

As stated previously, flux lines are continuous (that is, the lines of flux in the ring and air gap are the same), whereas the flux density in the ring may or may not be the same as that in the air gap—frequently they are considered the same by neglecting the fringing effect in the gap. Also, the cause of the flux (that is, the amp-turns) is thought of as being distributed along the whole ring, thus it is appropriate to use field intensity, H (amp-turns/meter or mmf/m). The length for this example is merely the mean circumference of the ring (the width of the air gap usually being negligible). Kirchhoff's voltage law in electrical series circuits states that the net voltage drop in a loop equals zero; one may use the same analogy for series magnetic circuits. That is, the net mmf in a magnetic series loop equals zero. Stated another way, the mmf source must equal the sum of the mmf drops (or losses) in the series circuit. Thus, summing these mmf drops in the iron and the air gap is all that is needed to find the mmf for the source.

Although the previous example involved a coil of wire, a more fundamental problem is one involving a long straight wire.

For a wire on the z-axis carrying a current,

$$\boldsymbol{H} = \frac{\boldsymbol{B}}{\mu} = \frac{\boldsymbol{I}\mathbf{a}_\theta}{2\pi r} \tag{7.9a}$$

where

$\boldsymbol{H}$ = the magnetic field strength (amps/meter)
$\boldsymbol{B}$ = the magnetic flux density (tesla)
$\mathbf{a}_\theta$ = the unit vector in positive θ direction in cylindrical coordinates
$\boldsymbol{I}$ = the current
μ = the permeability of the medium (for air, $\mu = \mu_0 = 4\pi \times 10^{-7}$ H/m)

The flux density at some radial distance, r, from the center line of the wire is given in Equation (7.9b).

$$B = (\mu i)(2\pi r) \tag{7.9b}$$

For the direction of the current shown in Figure 7.3, the flux density direction is shown with the head of an arrow as $(\odot)$ and the tail of an arrow as $(\otimes)$; this notation will be the same for current.

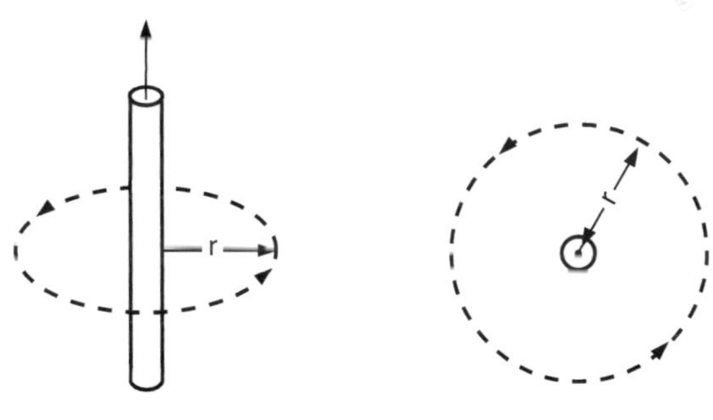

(a) Side view (b) Top view

Figure 7.3 The flux density around a wire

The direction of the flux may be remembered by using the right-hand rule: place the thumb in the direction of the current; the partially closed fingers will point in the direction of the magnetic flux. Now, if the long wire were formed into a circular coil of radius r, one could use calculus to consider a differential length of the wire and integrate around the closed loop to find the flux density within the loop. If there were several turns, N, for the loop, the equation for the flux density would be

$$B = (\mu Ni)/r \tag{7.10}$$

The right-hand rule may also be used to find the direction that a current would flow if a voltage is induced (or generated) in a wire by the relative motion of a magnetic field and a wire (see Figure 7.4). Consider a straight wire being moved within a magnetic field (assume the wire and motion is normal to the magnetic field). A voltage induced in the wire will be proportional to the strength of the field, the length (l) of wire within the field, and the velocity, v, of the wire.

$$e \text{ (induced voltage)} = Blv \tag{7.11a}$$

For a coil of N turns enclosing flux ϕ

$$e = -N\, d\phi/dt \tag{7.11b}$$

where ϕ = the flux (webers) enclosed by the N conductor turns and

$$\phi = \int_A B \cdot dA$$

The direction of the flux is from the north pole face to the south pole face. It will help to think of the wire within the field as being a voltage generator (the induced voltage); then if there is a closed path (perhaps through an external resistive load) so that a current could flow, the polarity of the *generator* must be plus where the current (thumb direction) leaves the magnetic field and minus where it enters the field.

In this section, vector notation involving three-dimensional space has been kept to a minimum; the FE examination will probably not include this added complication for magnetic fields. Should such a problem occur, one could still attempt to tailor the problem using vector components. As an example, assume one is asked to find an induced voltage for the motion of a wire in a magnetic field as depicted in Figure 7.4 using Equation (7.11a). However, assume also that the wire velocity direction is not normal to the magnetic field but at an angle θ from the horizontal. The solution and the equation would be the same, except that one must use the vertical component of the velocity rather than the magnitude of the entire velocity.

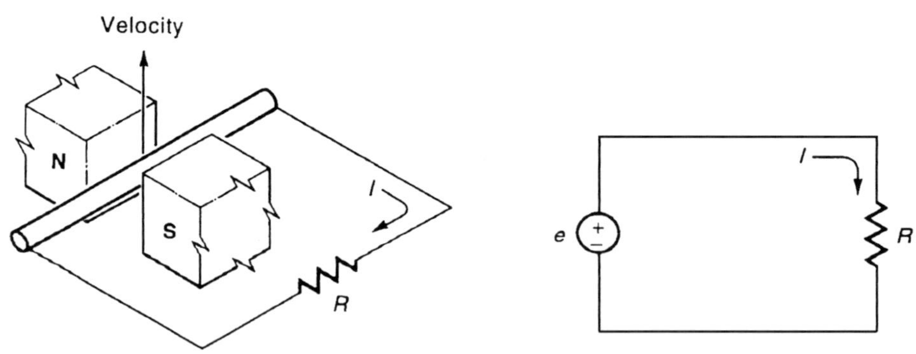

(a) Motion of a wire (b) Equivalent circuit

Figure 7.4 Induced voltage

Power, Energy, and Measurements

Before the discussion of power, energy, and measurements is undertaken, three items need to be reviewed. The first is notation, the second is the calculation of resistance, and the third is a more complete definition of voltage.

Notation

In this section, the notation for time-dependent quantities will normally be given by lowercase, italic letters (for example, for voltage, current, and such, $v(t)$, $i(t)$, where the (t) may or may not be included). A constant value, such as a battery voltage, will be given in uppercase, italic letters (for example, V and I); the effective or rms (root mean square) values and other quantities that have a nonvarying value (such as the average power) will also be expressed as an uppercase letter. Where confusion is possible, a subscript is normally used. Further discussion of notation will be presented where appropriate.

Resistance

Most reviews assume that one is familiar with simple wire resistance. Although the actual resistance of a wire conductor is usually considered to be negligible in circuit analysis, it is easily calculated. The parameters are shown in Figure 7.5.

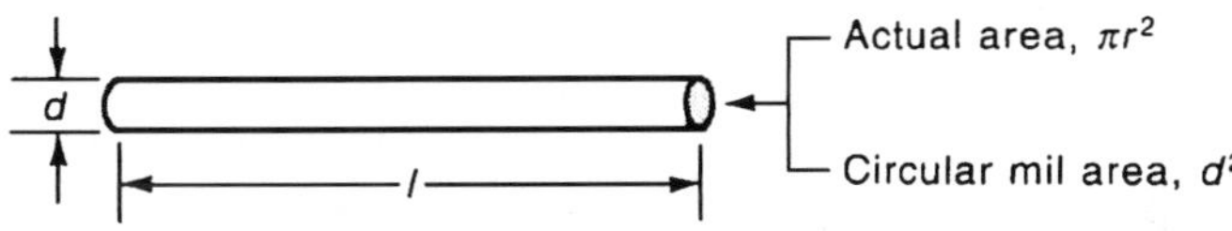

Figure 7.5 Wire resistance parameters

The resistance is proportional to its length and inversely proportional to its cross-sectional area; the constant of proportionality is the resistivity, ρ:

$$R = \frac{\rho L}{A} \tag{7.12}$$

The resistivity of a particular conductor is normally given for a standard temperature of 20°C. However, the units depend on the length and area. For the MKS system of units, the value of ρ for copper is $1.7 \times 10^{-8}\ \Omega \bullet \text{m}$ (for aluminum, it is almost twice this value).

Example 7.2

As an example, for a 5-meter length of 12-gauge (approximate diameter of 2 mm) copper wire at room temperature (near 20°C) find the resistance, R.

Solution

$$R = (1.7 \times 10^{-8})(5/[\pi(0.001)^2]) = 0.027\ \Omega$$

On the other hand, for another temperature, the resistivity is modified by the temperature coefficient, α, to be

$$\rho = \rho_0[1 + \alpha(\mathrm{T} - \mathrm{T}_0)] \tag{7.13a}$$

or

$$R = R_0[1 + \alpha(\mathrm{T} - \mathrm{T}_0)] \tag{7.13b}$$

where $(\mathrm{T} - \mathrm{T}_0)$ is the change of temperature from 20°C.

Generally, these two equations will solve most resistive type problems. However, here is one note of caution: unfortunately, the inch, foot, and circular mil units for area are still in use. These units can be confusing since the area of the wire may be given in circular mils; this area omits π in the true area, πr^2, computation and, instead, includes it in the resistivity constant. In this case, the resistivity for copper wire is 10.4 Ω circ-mils/foot. Actually, the computation is made easier; assume a length of 5 feet for a wire diameter of 0.03 inches (or 30 mils):

$$R = (10.4)(5)/(30)^2 = 0.0578\ \Omega$$

The key to this type of problem is to check carefully by dimensional analysis to determine what units are being used. Wire tables, listing these parameters, should be furnished with the EIT examination when they are needed.

Voltage

Whereas a simple definition of **voltage** is the potential difference that will cause a current of one ampere to flow through a resistance of one ohm, a more formal definition is *a charge of one coulomb receives or delivers an energy of one joule moving through a voltage of one volt.* The instantaneous voltage is defined by Equation (7.14).

$$v = dw/dq \tag{7.14}$$

In other words, if a unit quantity of electricity (coulomb) gives up energy equal to one joule as it proceeds from one point to another, the difference in potential is one volt. The joule is sometimes called the coulomb-volt. Likewise, the energy acquired by an electron when it is raised through a difference of potential of one volt is called an electron-volt. The definition of voltage leads directly into the subject of power and energy.

Power and Energy

Power (instantaneous, i.e., lowercase italic notation) is the rate of change of energy, dw/dt, or

$$p(t) = dw/dt = (dw/dq)(dq/dt) = v(t)(i) \tag{7.15}$$

The total energy is then given in Equation (7.16). (The function of time, t, is implied and will be dropped for simplicity.)

$$w = \int_0^T p\, dt = \int_0^T vi\, dt \tag{7.16}$$

Energy Storage

Energy may be stored in some electrical circuit elements. For electrical circuits, there are three basic elements: resistance, inductance, and capacitance. For the resistor, electrical energy cannot be stored since it is turned into heat; however, both the inductor and capacitor are capable of energy storage (see Figure 7.6).

The inductor, L (measured in units of henries, H), usually a coil of wire that can produce a magnetic field, stores energy in the magnetic field; the capacitor stores energy within its electrical field (i.e., within its dielectric medium). The circuit symbols and voltage-current relationships are given in Figure 7.7. Equations (7.17) and (7.18) give the energy stored as

$$v_L = L(di_L/dt) \tag{7.17a}$$

$$i_C = C(dv_C/dt) \tag{7.17b}$$

$$w_L = (1/2)Li_L^2 \tag{7.18a}$$

$$w_C = (1/2)Cv_C^2 \tag{7.18b}$$

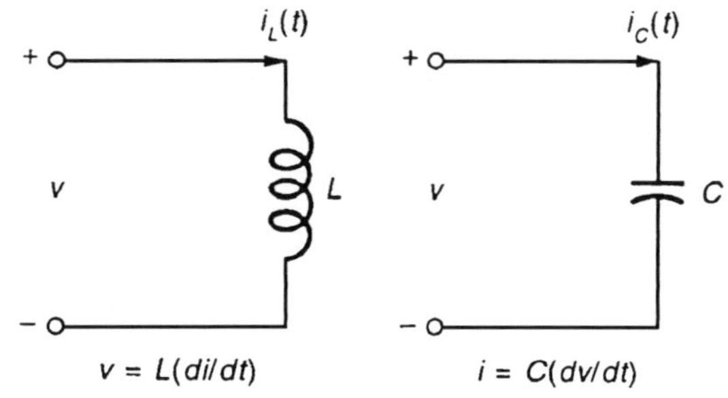

Figure 7.6 Energy storage elements

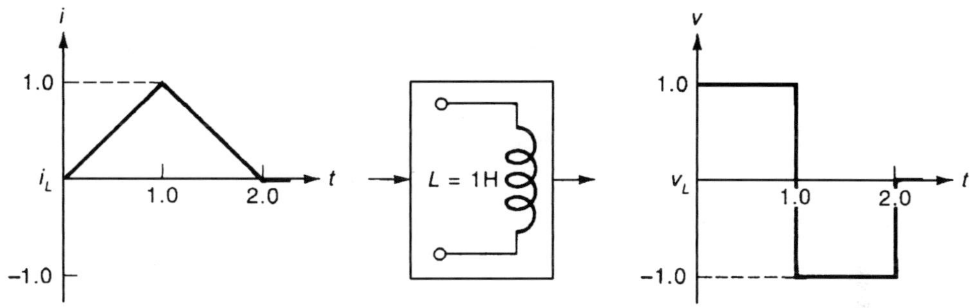

Figure 7.7 Inductor current-voltage relationship

The voltage across an inductor is a function of the rate of change of current. If there is no rate of change (or direct current), then the voltage must be zero, and the inductor acts like a short circuit or zero resistance. Actually, there is always some wire resistance, but usually this is negligible. For current to flow in a capacitor, there must be a rate of change of voltage. If there is no rate of change, there is no current, and the capacitor acts like an open circuit (this will be discussed in more detail later). As an example, assume the current (from a variable current source) through an inductor is as given in Figure 7.7(a); then the voltage across the inductor must be as shown in Figure 7.7(b).

As shown in Figure 7.8, the power *taken* (absorbed or stored) by the inductor is a function of time until $t = 1$ second and is positive; for the current with a negative slope, power is *given up* (or returned to the circuit source) from $t = 1$ second until $t = 2$ seconds. On the other hand, the energy stored (also a function of time) follows the square law given directly in Equation (7.18a). The resulting curves are found by splitting Equation (7.16):

$$w = \int_0^2 p\,dt = \int_0^1 vi\,dt + \int_1^2 vi\,dt$$

Average Voltage and Current

All DC quantities, being constant values, are represented by uppercase letters. However, for changing quantities one must be careful with notation. The formal definition involves the time period of interest, T. The average voltage (or current) is given by

$$V_{\text{avg}} = (1/T)\int_0^T v\,dt \tag{7.19}$$

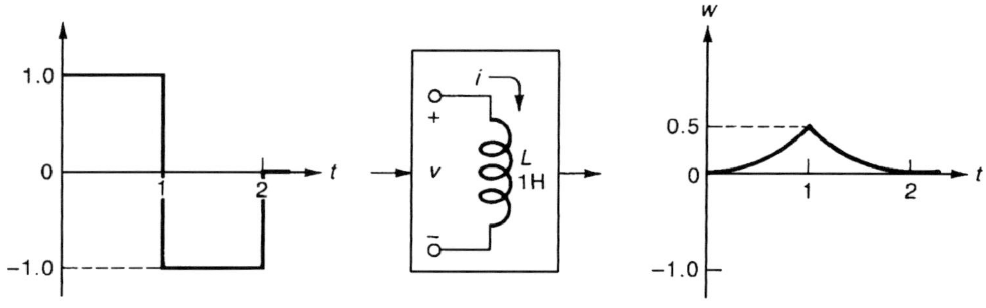

(a) Power (b) Energy

Figure 7.8 Power and energy plot for the current/voltage curves of Figure 7.7

or the area under the voltage curve divided by the period of time. Of course, the same holds for current.

If the positive and negative areas balance over a specific period of time, then the net area is zero. It should then be obvious that the average voltage from a sine wave generator over a full period, T (or 2π), is zero:

$$V_{avg} = (1/T)\int_0^T V_{max} \sin \omega t \, dt = 0$$

Over a *half period*, $(1/2)T$, the average voltage is

$$V_{avg} = (2/T)\int_0^{T/2} V_{max} \sin \omega t \, dt = 2V_{max}/\pi$$

(*Caution*: a half wave over a full period has an average value of V_{max}/π.)

Average Power

Since power is defined as the product of voltage and current, then for simple DC voltage or current sources, the average power is simply the product of the current and voltage. This is written as uppercase, italic P (without subscript) as $P = VI$. (Generally, use this equation only for DC quantities, as ac or other types of wave forms may have a phase shift. This will be discussed later.)

Effective Values

For most wave forms, other than straight DC, one must define an effective or rms value of current and voltage since an average value of current or voltage could be zero for a sinusoidal signal. Another way of stating this is the following: "What equivalent value of current will cause the same heating power in a resistor as would a DC current?" The effective (or rms) value of current or voltage is defined as

$$I_{eff} = I_{rms} = I = \sqrt{(1/T)\int_0^T i^2 \, dt} \tag{7.20}$$

If $i = I_{max} \sin \omega t$ or $i = I_{max} \cos \omega t$, then

$$I = 0.707 I_{max} = I_{max}/\sqrt{2} \tag{7.21a}$$

$$V = 0.707 V_{max} = V_{max}/\sqrt{2} \tag{7.21b}$$

Although Equation (7.21) and Figure 7.9 are for sinusoids, the effective or rms values may be found for any periodic wave form.

For sinusoids, the average power may involve phase angles (more on this later); however, one is safe in finding the power developed as heat in a resistor by using

$$P_{avg} = P = I^2 R = \left(I_m / \sqrt{2}\right)^2 \; R \text{ watts} \tag{7.22}$$

For a short example, assume a known sinusoidal current of i = 14.1 cos ωt flows through a resistor of 25 ohms. The effective, or rms, current is easily computed from Equation (7.20) or (7.21) and the power by Equation (7.22):

$$P = \left(14.1/\sqrt{2}\right)^2 (25) = 2500 \text{ watts}$$

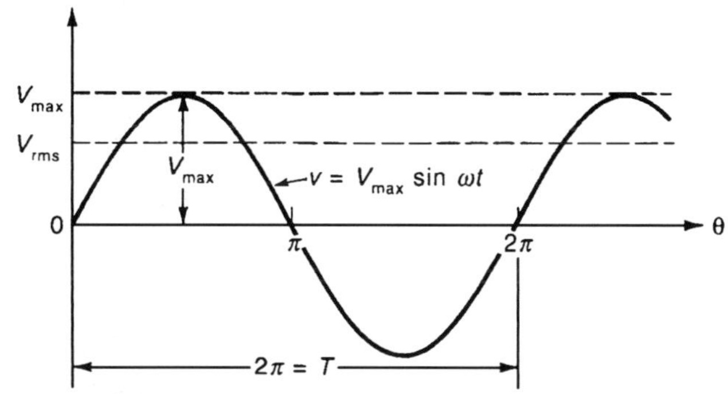

Figure 7.9 Effective or rms value for a sinusoid

If the sinusoid is replaced with another kind of periodic wave form, Equation (7.20) would, of course, still be used. However, sometimes a table of *standard wave forms* with a list of effective or rms values is available.

DC CIRCUITS AND RESISTANCE

Ohm's and Kirchhoff's Laws for a Single Source Network

The solutions of all circuits problems whether they are dc, ac, single source, or multiple source, involve the use of these laws. The first is Kirchhoff's voltage law (KVL):

For any closed loop in a circuit, the voltage algebraically sums to zero. $\Sigma v = 0$	or	The voltage rises equal the voltage drops in a closed loop. $\Sigma V_{rises} = \Sigma V_{drops}$

Another way of stating this relationship is to say the sum of the voltage rises is equal to the sum of the voltage drops in any loop. Both of these statements are true of both dc and instantaneous values. However, one needs to be careful when dealing with ac analysis with possibly different phase angles; this relationship will be discussed later.

The second is Kirchhoff's current law (KCL); this law involves any junction, or node, in an electrical circuit:

For any node in an electrical circuit, the net current algebraically sums to zero. $\Sigma i = 0$	or	The total current entering a junction equals the total current leaving. $\Sigma I_{in} = \Sigma I_{out}$

Restated, the sum of the currents entering a junction must equal the sum of the currents leaving the junction. Again, this is true of both dc and instantaneous values; however, for ac circuit analysis care is required in its application.

Ohm's law for resistance ($R = v/i$) is well known and is the same for dc, instantaneous quantitities, and ac values. For a series circuit (where the current is the same in every element) the equivalent resistance merely becomes

$$R_T = R_1 + R_2 + \cdots + R_n = \sum R \tag{7.23a}$$

For parallel circuits (where the voltage across every element is the same) the equivalent resistance, R_T, is

$$1/R_T = 1/R_1 + 1/R_2 + \cdots + 1/R_n = \sum G \tag{7.23b}$$

The conductance G is simply the inverse of the resistance (whose units are siemens or mhos, $1/\Omega$). From Equation (7.23b), one can quickly determine that the equivalent resistance for two resistors (or impedances) in parallel is $R_T = R_1R_2/(R_1 + R_2)$.

Example 7.3

A simple example follows; for this type of circuit problem, it is usually quicker to convert the parallel resistors to an equivalent value. The simpler series circuit is easily solved: by finding the voltage drop across any element, the voltage across the equivalent resistance is found. Assume the question is to find the power dissipated in the 6-Ω resistor in Exhibit 4.

Solution

The equivalent parallel resistance is found to be 3/2 Ω. This value yields a total series resistance of 4 Ω as shown in Exhibit 4b. The series current is easily found to be 3 A; the voltage drop across the parallel resistance is $V_{eq} = I(R_T) = (3)(3/2) = 9/2$ volts. The current through the 6-Ω resistor is $I = V_{eq}/R_6 = (9/2)/6 = 3/4$ A. The power dissipated in the 6-Ω resistor is $P_{6\Omega} = I^2R = (3/4)^2 6 = 27/8 = 3.375$ watts.

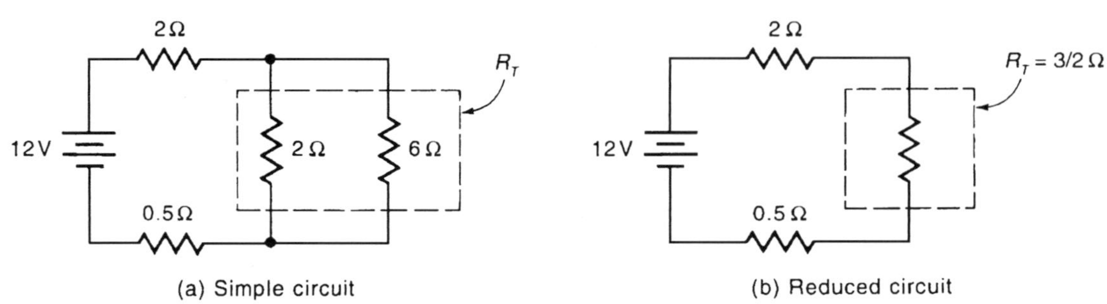

Exhibit 4 A simple voltage source circuit

Here, the power from the source equals the sum of the power dissipated in each of the resistors. It is instructive to compute the sum of the power lost; this is

$$P_{source} = V_{source} I = (12)(3) = 36 \text{ watts}$$

$$\sum P_{lost} = I^2 \left(\sum R_{ser}\right) = 9(2 + 3/2 + 1/2) = 36 \text{ watts}$$

For this problem, a somewhat more formal method is to use the voltage dividing equation after R_T is found (all elements are now in series). The voltage across R_T may be calculated to be $V_T = (R_T / \Sigma R)V_{bat}$; then, the current through R_6 is determined as before.

Example 7.4

Consider another example with a current source: here, the series portion of the circuit should be reduced to one resistor; with all of the resistors now in parallel, the voltage across this group is easily found. Now the question is to find the amount of power dissipated in the 2-ohm resistor in Exhibit 5.

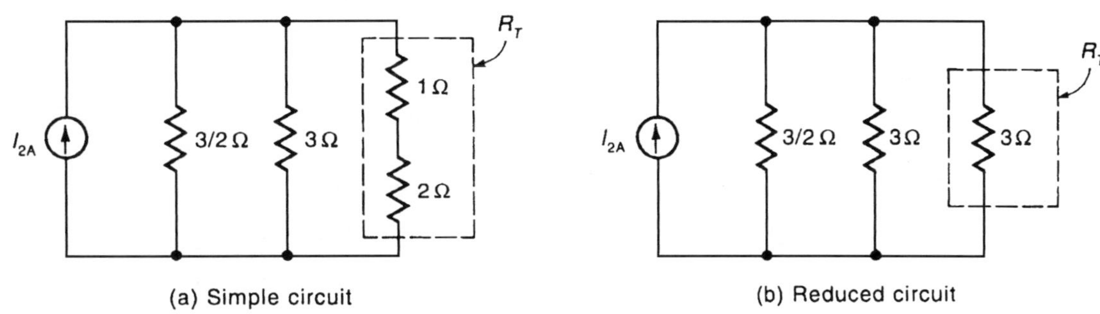

Exhibit 5 A simple single current source circuit

Solution

After finding the one series resistor, R_s, that is equivalent to the two resistances in series, all resistances are now in parallel. Then using Equation (7.23b), or merely taking two resistors at a time, one easily computes the total equivalent one resistance to be 3/4 Ω. The voltage across the parallel circuit is $RI = (3/4)(2) = 3/2$ volts. The current through the series equivalent branch is $I_{se} = V_{par} / R_{se} = (3/2)/3 = 1/2$ amperes; the power dissipated in the 2-Ω resistor is $P = I^2R = (1/2)^2 2 = 0.5$ watts.

Again, a quick check is to find if the power taken from the source ($P_s = VI = (3/2)2 = 3$ watts) and the power dissipated in the resistors are equal:

$$\sum P_{lost} = V^2(1/R_{3/2} + 1/R_3 + 1/R_{se}) = (3/2)^2(2/3 + 1/3 + 1/3) = 3 \text{ watts}$$

For this problem, as before, there is a somewhat more formal method of finding the current through R_{se} directly; $I_{se} = (G_{se}/\Sigma G)I_{tot}$. Knowing the branch current, the power is easily found.

Multiple Source Networks and Theorems

Although the term *multiple source circuits* implies multiple power output, this implication may be incorrect as the voltage or current source may actually absorb power. If the current is flowing into a voltage *source*—from plus to minus—the *source* is actually a *load* taking power (that is, a battery being charged). Therefore, it is very important to determine the direction of current (or voltage polarity as the case may be). The procedure is to assume current directions and carefully label these on a circuit diagram; then let the mathematics determine the actual direction. One way to do this is to assume a current through each element (of course series elements will have the same current) of a circuit. This is in contrast to assuming loop or mesh currents; however, both methods amount to the same thing. A short example (see Figure 7.10) will show the method. (To keep the circuit diagrams uniform and to later allow for time varying voltage sources, a circle is used to indicate the voltage source rather than showing a battery.)

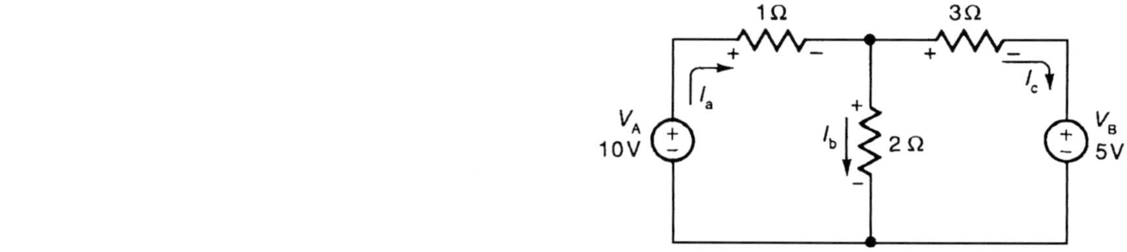

Figure 7.10 A two-loop network

After labeling the assumed current directions, place a plus sign where the current enters the resistor and a negative sign where it leaves. Then, by Kirchhoff's voltage law, sum the voltages around each loop:

$$V_A - R_1 I_a - R_2 I_b = 0 \quad \text{(i)}$$
$$V_B + R_3 I_c - R_2 I_b = 0 \quad \text{(ii)}$$

Since three variables are present, another equation is needed; Kirchhoff's current law will produce the third equation (actually J – 1 node junction equations are needed; here J = 2). Summing currents at one junction (say, the upper middle node) gives $I_a = I_b + I_c$.

Rewriting the equations for loops one and two (here, replacing all I_a), yields

$$V_A = R_1(I_b + I_c) + R_2 I_b, \qquad 10 = 3I_b + I_c \quad \text{(i)}$$
$$V_B = R_2 I_b - R_3 I_c, \qquad 5 = 2I_b - 3I_c \quad \text{(ii)}$$

Solving for the currents, the results are

$$I_a = 3.64 \text{ A}, \quad I_b = 3.18 \text{ A}, \quad I_c = 0.455 \text{ A}$$

Since all of the currents are positive, the assumed directions of the currents are correct; therefore, the voltage *source* B is really a load (perhaps a battery being charged). As a check, the sum of the voltages around the outside loop should equal zero (starting at the upper left-hand corner): Does 1(3.64) + 3(0.455) + 5 – 10 = 0? Or, 3.64 + 1.365 + 5 – 10 = 0.005 (acceptable).

Another way of analyzing the circuit is to use the *mesh* technique of analysis. Notice in the previous two-loop problem, the current through the middle resistance that was identified as I_b is nothing more than $I_a - I_c$. If we now consider going around the entire left loop as I_a and around the entire right loop as I_c, the currents mesh at the middle resistance. Because voltages are summed around each loop, the net voltage across the middle resistor (when going around the left loop) is $RI_a - RI_c$; the loop currents are in opposite directions. The same kind of equation is written for the right-hand loop. Thus one current variable (I_b) is eliminated. There are still other ways of analyzing the circuit, as will be pointed out later.

Example **7.5**

Another example that involves mixed sources of current and voltages is shown in Exhibit 6. Find the voltage across the current source for this circuit configuration.

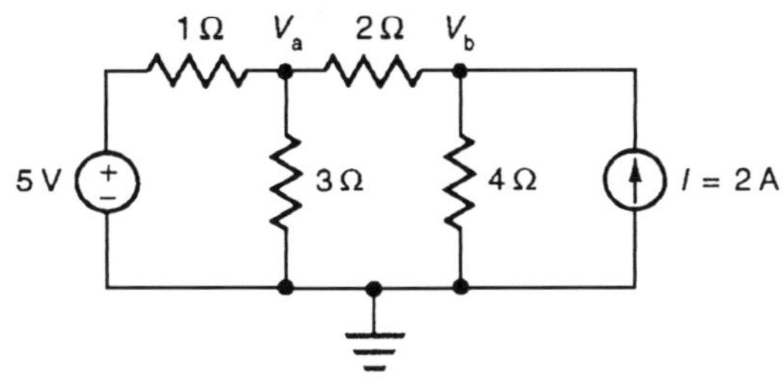

Exhibit 6 Mixed source circuit

Solution

Here, the node-voltage method of analysis will be used. This method finds voltages with respect to some common reference point called a ground node (see Exhibit 6). By wisely selecting this common point to include as many junctions as possible, the number of simultaneous equations may be reduced. The node voltage will be considered as a voltage rise from the common reference node (that is, current directions may then be assumed away from the node while writing the equations for that particular node).

$$\text{At } V_a, \sum I \qquad 0 = (V_a - 5)/1 + (V_a - V_b)/2 + (V_a - 0)/3$$

$$\text{At } V_b, \sum I \qquad 0 = (V_b - V_a)/2 + (V_b - 0)/4 + (-2)$$

Simplifying (multiplying through by 6 and 4, respectively) and collecting terms,

$$30 = 11V_a - 3V_b$$
$$8 = -2V_a + 3V_b$$

Solving these equations for the node voltages gives $V_a = 4.22$ V and $V_b = 5.48$ V. From these voltages, any desired currents or powers may be found. To check the solution, find the sum of the power for the current and voltage sources and com pare this result with the sum of the power dissipated by the resistors ($P_{\text{sources}} = 14.9$ watts $= \Sigma P_{\text{resistors}}$). Either of the apparent sources could actually be a load by taking power from the other source. If the voltage source is a battery and it is found that the actual current flows through the battery from plus to minus, then the battery is being charged and acts as a load.

Network Reduction, Thevenin and Norton Equivalents

Most circuits may be greatly reduced in complexity and thus yield a simpler solution. This is especially so if a variable load is involved and requires a number of repeated calculations. Consider the circuit shown in Figure 7.11; many calculations would be involved if one were asked to compute the power dissipated in R_L for several different values of R_L. If one could reduce the network (all of that circuitry to the left of × – ×) to one simple voltage source in series with one resistance [see Figure 7.11(b)], then calculating the current to R_L would be simple and easy.

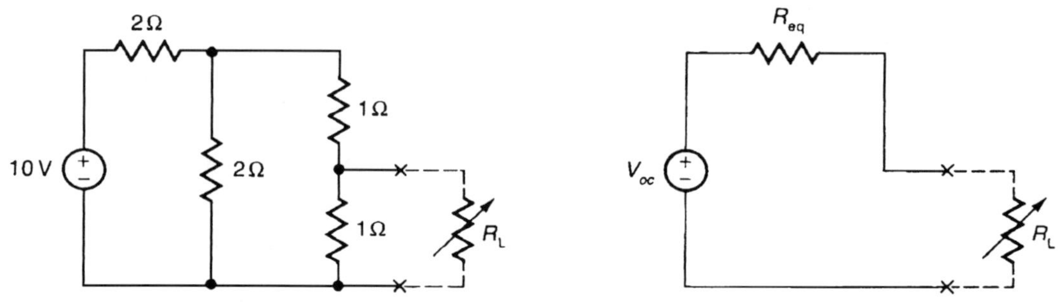

(a) Original circuit (b) Reduced circuit

Figure 7.11 Circuit for Thevenin's reduction

Reducing the network is done with Thevenin's theorem. One can find the Thevenin voltage by merely opening the network at the point of interest (that is, temporarily removing R_L) and measuring the open circuit voltage at × – ×; the open circuit voltage is the Thevenin voltage. Next, place a short circuit (an ammeter, assumed to have zero resistance) across this section and measure the short circuit current. The Thevenin resistance is given as

$$R_{eq} = \frac{V_{OC}}{I_{SC}}$$

To find a Thevenin equivalent circuit on an examination, one cannot measure the various voltages or currents but must calculate them. Calculating the open circuit voltage does not present a problem, but more needs to be said about calculating the resistance.

There are at least two ways to calculate the Thevenin resistance. One is to calculate the short circuit current and proceed as before. The other method is to replace all voltage sources with their internal resistance (usually zero) and any current source with its internal resistance (usually infinity) and then calculate the resistance one would see looking back into the circuit (to the left of × – × in Figure 7.11). However, caution is needed if any dependent sources are present. For the values given in Figure 7.11, the calculated open circuit voltage is 5/3 volts, and the short circuit current is 5/2 amperes; thus the Thevenin equivalent circuit is as given in Figure 7.12(a).

The current source that is equivalent to the Thevenin circuit is Norton's equivalent circuit. Norton's circuit is the current source equivalency for Thevenin's voltage source [see Figure 7.12(b)]. The value of the resistance is the same except that it is usually stated in terms of conductance ($1/R$) and the units are mho (old) or siemens (new). The current source may be found directly from Thevenin's circuit by finding the short circuit current directly from Thevenin's circuit. If one is asked for a Norton equivalent circuit, one can first find Thevenin's circuit then convert to Norton's circuit.

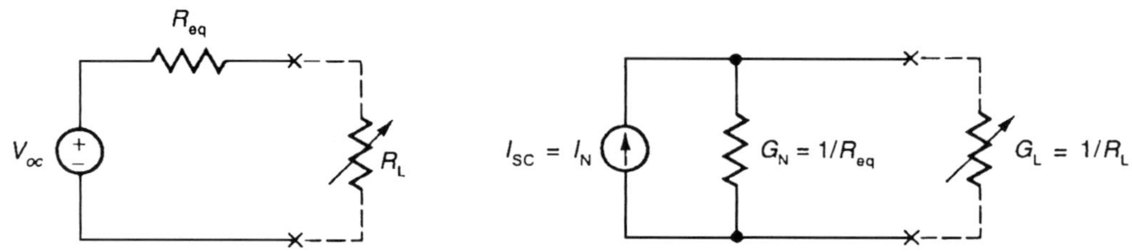

(a) Thevenin's (b) Norton's

Figure 7.12 Equivalent circuits

Assume one wishes to determine the maximum power that could be dissipated in a variable load resistor that is nested in a multiple source circuit. First, it is necessary to know the size of the resistor. The **maximum power transfer theorem** (for dc circuits) states that maximum power is extracted from a circuit when the circuit is converted to Thevenin's equivalent circuit and the load resistance is equal to Thevenin's resistance. A consequence of this theorem is that half of the power is dissipated in the load and the other half in Thevenin's equivalent resistance. Consider the circuit of Figure 7.13; all that is necessary is to isolate the load resistor from being *buried* in the circuit. This is done by rearranging the circuit as shown in Figure 7.13(b). One way is to move the load resistor to the right side of the circuit and then find Thevenin's equivalent circuit for the portion on the left.

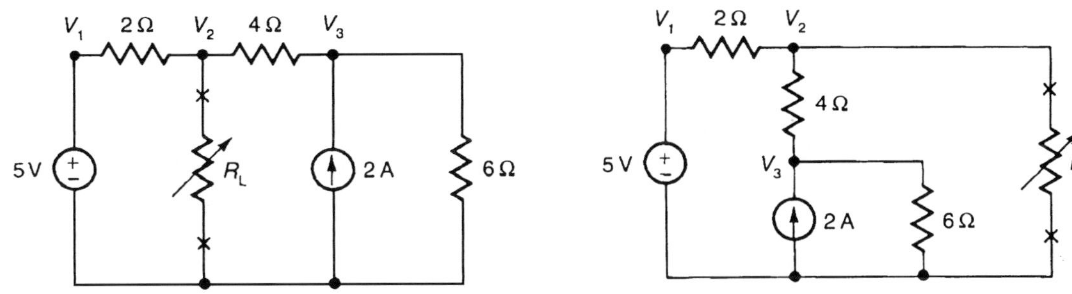

(a) Original circuit (b) Rearranged circuit

Figure 7.13 A maximum power transfer problem

For the rearranged circuit of Figure 7.13, the open circuit voltage (that is, with R_L temporarily disconnected) will be found by the node method of analysis:

$$\text{At } V_2, \sum I = (V_2 - V_1)/2 + (V_2 - V_3)/4 = 0$$

$$\text{At } V_3, \sum I = (V_3 - V_2)/4 - 2 + V_3/6 = 0$$

Solving (with $V_1 = 5$ volts) for $V_2 = V_T$ yields 6.116 volts. Thevenin's resistance can now be found by using the simpler method of merely replacing all independent voltage sources with zero and all independent current sources with an open circuit; then, looking back into the circuit × – ×, the equivalent resistance can now be found. This resistance yields R_{oc} = 1.667 Ω. From the maximum power relationship, $R_L = R_{oc}$. The power dissipated in R_L = 5.61 watts; see Figure 7.14.

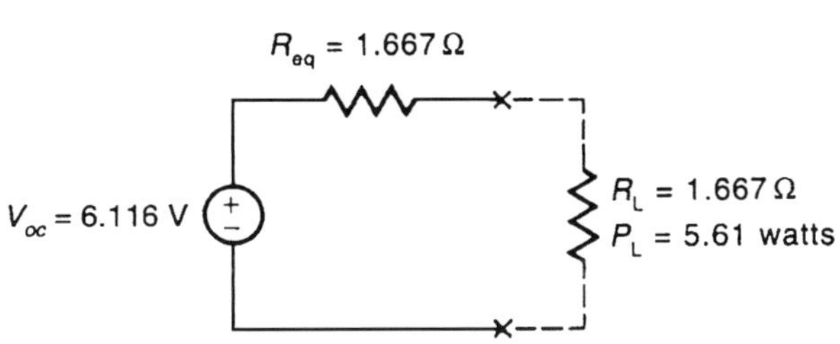

Figure 7.14 Final load resistance using Thevenin's circuit

Transient Response for a Single Energy Storage Element

If an energy storage element, such as a capacitor or inductor, is present in a circuit with one or more resistors, the problem solution will be a function of time. The voltage across a capacitor or current through an inductor cannot change instantaneously but takes time (unless infinite pulses are involved). It is assumed that the circuit is connected to either a voltage or current source or has some initial values.

Capacitive Circuits

If a capacitor (assume initially uncharged) is being charged from a voltage source through a resistor as shown in Figure 7.15(a), then it takes time for it to reach the charging voltage. The voltage vs. time solution is plotted in Figure 7.15(b). Here it is obvious that the amount of time for the capacitor voltage to reach its final value is infinite. On the other hand, if the capacitor already has an initial charge (or voltage) and is being discharged through a resistor, it again takes time for it to discharge [refer to Figure 7.16(b)]; this amount of time is also infinite.

Rather than use infinity as the changing time, it is more convenient to define a time constant, τ, to obtain a practical result. The time constant is some value of t that makes the exponent, x, of e^x, equal to -1. As an example, if $x = -t/RC$, then

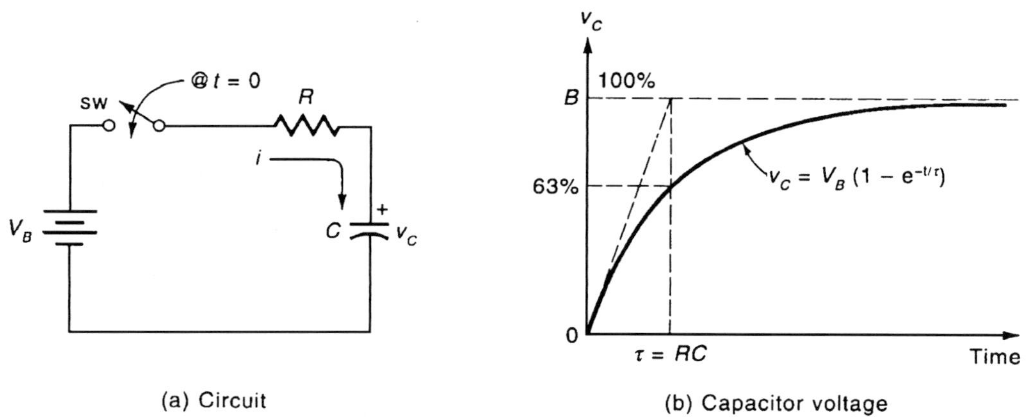

Figure 7.15 The charging of a capacitor, τ = RC

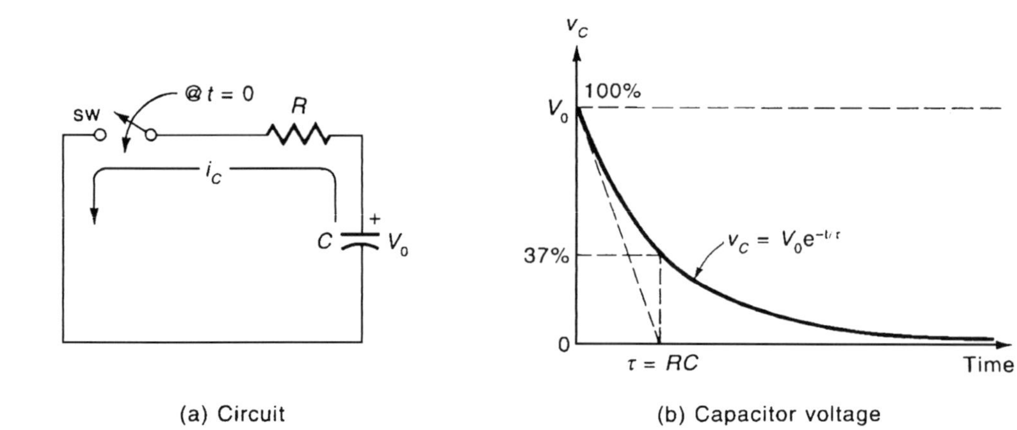

Figure 7.16 The discharging of a capacitor, τ = RC

the time constant is equal to the value of RC. This value of e^{-1} is 0.3679, or approximately 37%; it is so often used in electrical transient analysis that it probably should be committed to memory. In Figure 7.15, V_B is the final value of the voltage, and at $t = \tau$ the parenthetical quantity is approximately 63% so the variable is within 37% of its final value. Also note that in Figure 7.16, the voltage decreases from the initial value, V_0, by 63% (or is within 37% of the final value) at $t = \tau$. If the initial slope of the curves in both Figures 7.15 and 7.16 were extrapolated forward, the intersection of the asymptotes with the horizontal axis would occur at $t = \tau$. To be realistic, *within* 37% is not good enough for engineering purposes, and a more practical approach is to consider the voltage at *almost* its final value in five time constants (well within 1%).

The equations in Figure 7.15 and Figure 7.16 are the solutions of the integral-differential equations that describe the circuit(s). Kirchhoff's voltage law states that sums of voltages around any closed loop must be equal to zero at *any* instant of time. In Figure 7.15, the switch closes at $t = 0$; the voltage around the loop (for $t > 0$) is $V_B = Ri + v_c$, where v_c is q/C. Since current is defined as dq/dt, v_c is the integral of the present expression. If the initial voltage (or charge) on the capacitor was zero at the instant before the switch was closed, then $V_B = Ri + q/C = Ri + (I/C) \int_0^t i\,dt$.

The solution to this equation is

$$v_c(t) = v_c(0)e^{-t/RC} + V(I - e^{-t/RC}) \tag{7.24a}$$

and Equation (7.24a) frequently reduces to Equation (7.24b):

$$v_C = v_B(1 - e^{-t/\tau}) \tag{7.24b}$$

where $\tau = RC$. If t goes to infinity, then $v_C \rightarrow V_B$. If the initial value was not zero, as in Figure 7.16, it must also be considered. The differential equation used to obtain the solution for Figure 7.16(b) may be found by the node voltage method; use the lower junction of RC as the reference node, and the currents to the top voltage node are given by $i_C + i_R = 0 = C(dv/dt) + v/R$. Knowing the initial voltage, $V_C(0^-)$, the voltage across the capacitor just before the switch was closed, and that $V_C(0^-) = V_C(0) = V_C(0^+)$, the differential equation is easily solved. The solution, where $\tau = RC$, is $v_C = V_C(0)e^{-t/\tau}$.

A somewhat more comprehensive example involving a multiple time constant circuit is presented in Figure 7.17. The switch is assumed to be open for a long time (so long that any previous voltage across the capacitor has been reduced to zero because of R_3). The switch is then moved to the middle position for five seconds ($0 < t < 5$); it will then be switched to the lower position and remain there; this lower position will have a time, t', that starts at the time of switching to this new position. It is desired to know the current through the capacitor at $t = 9.8$ s ($t' = 4.8$ s). Although this problem may seem more comprehensive than most, the reader should follow the details of the solution to note simplification techniques. First, it should be obvious that the problem may be broken into two parts. The first part spans the first 5 seconds, and the second runs from $t = 5$ (or $t' > 0$) seconds. When the switch is in the middle position, one needs to make a Thevenin equivalent circuit (refer to Figure 7.18) to the left of $\times - \times$ and then solve for the capacitor voltage at $t = 5$s as done on the previous problem. This voltage at $t = 5$ s is easily computed from the capacitor charge equation as $v_C(t = 5) = 7.5(1 - e^{-5/3}) = 6.083$ volts.

One could almost obtain this value from a sketch, as at one time constant (3 seconds) the voltage would be 63% of 7.5 volts (4.72 volts), and at two time constants (6 seconds) the voltage would be 87% of 7.5 volts (6.49 volts). Thus at 5/3 time constant, a plot would yield about 6.1 volts. Of course the final voltage of Figure 7.18 is the initial voltage when the switch is moved from the middle to the bottom position in Figure 7.19. The voltage at $t' = 4.8$ ($t = 9.8$) seconds is 37% of the initial value, $0.37 \times 6.08 = 2.25$ volts. The current at this instant of time is $v_C/R_{Eq} = 1.87\ \mu A$.

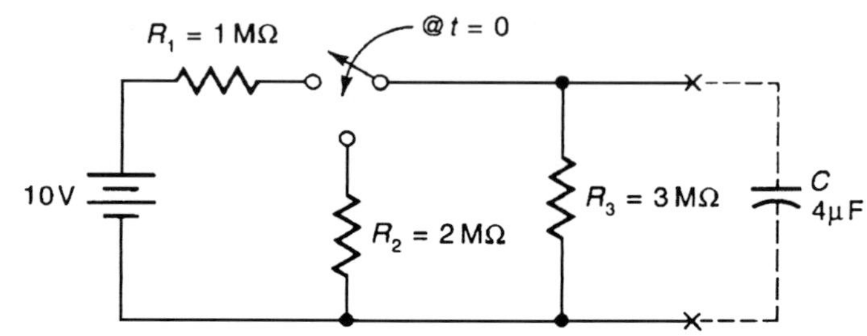

Figure 7.17 Multiple time constant circuit

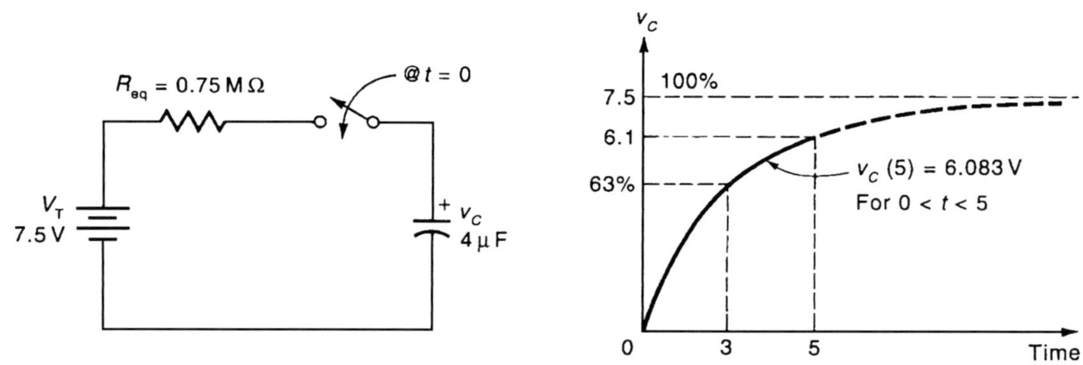

(a) Thevenin's equivalent (b) Capacitor voltage

Figure 7.18 The charging of a capacitor during $0 < t < 5$ seconds, for $\tau = 3$

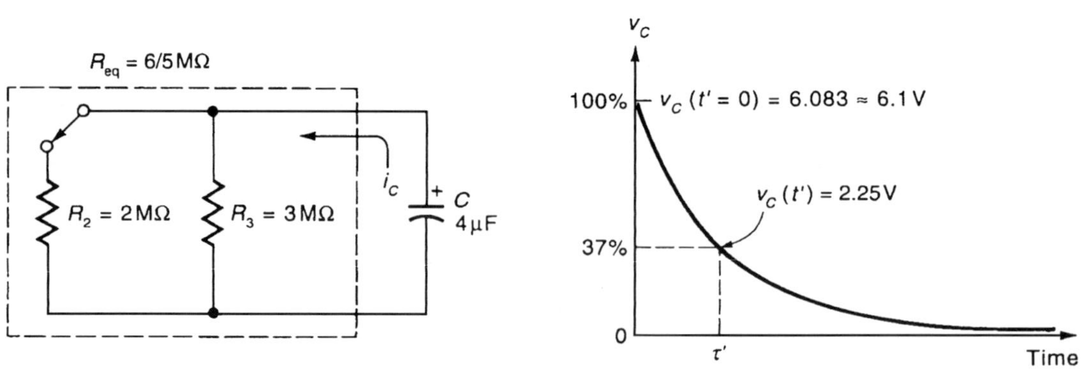

(a) Equivalent circuit (b) Capacitor voltage

Figure 7.19 The discharging of a capacitor for $t > 5$ (or $t' > 0$) seconds, $\tau' = 4.8$

Inductive Circuits

When an inductor (instead of a capacitor) is in a circuit, the current through the inductor takes time to change (like a voltage across a capacitor) and the circuit may be treated much like the previous voltage-capacitor relationship,

$$i(t) = i(0)e^{-Rt/L} + V/R(1 - e^{-Rt/L}) \qquad \textbf{(7.25)}$$

An example *RL* circuit problem is presented in Figure 7.20. Here, the time constant, τ, is *L*/*R*. Assume zero initial current in the inductor. A voltage will be induced across an inductor equal to $L(di/dt)$ because of the rate of change of magnetic flux through a coil or around a wire. The polarity of this voltage depends on whether the flux is increasing or decreasing. Unlike a capacitor, where the capacitor acts as an open circuit (that is, no current flow) after the *voltage* stabilizes, the inductor acts like a short circuit after the *current* stabilizes. Thus the final value of the current in Figure 7.20 is the applied voltage divided by the series resistance.

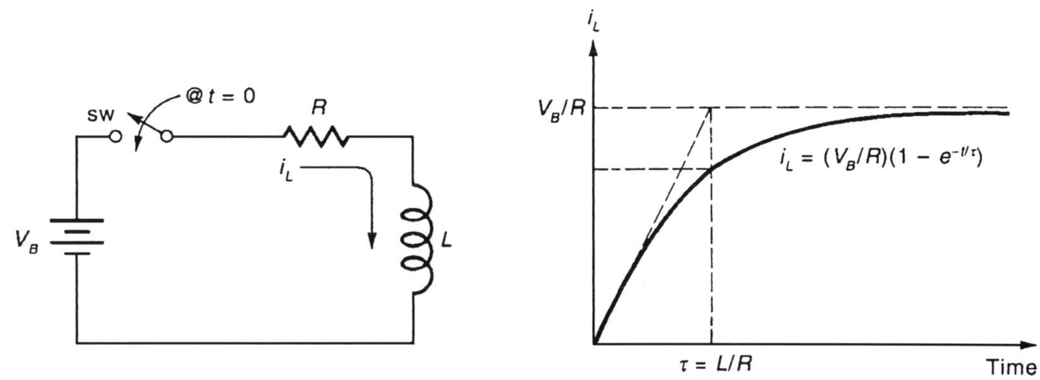

(a) Circuit (b) Inductor current

Figure 7.20 An RL circuit, τ = L/R

Example 7.6

An inductor has an initial current caused by the switch in Exhibit 7 being connected to a voltage source for a long period of time and then suddenly switched to the open position at $t = 0$. Find the current at $t = 0.1$ seconds and the voltage across the inductor at the same time.

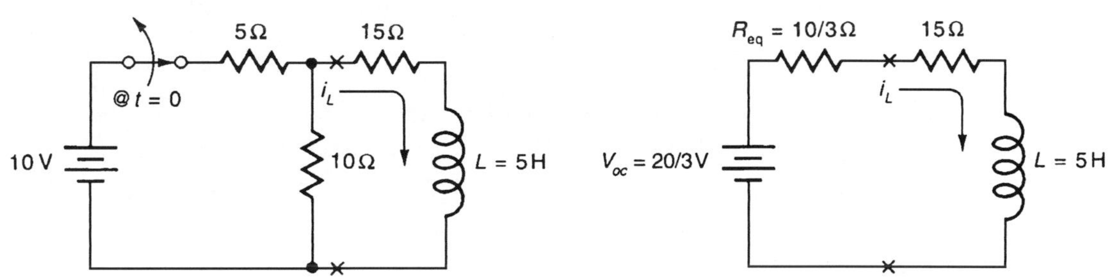

(a) Circuit (b) Thevenin's equivalent

Exhibit 7 Inductor problem before the switch opens

Solution

Before the switch is open, the easy way to solve the problem is to make a Thevenin equivalent circuit of that portion of the circuit to the left of × – ×. While the switch is connected to the voltage source, the polarity of the voltage across the inductor is positive at the upper terminal of the inductor (while the voltage source is providing an increasing current in the inductor). Just before the switch opens the current in the inductor is $I(0^-)$. Because current cannot change instantaneously, $I_0(0^+)$ is the same current immediately after the switch is opened. However, the voltage polarity switches immediately as the energy stored in the inductor causes the current to continue to flow in the same direction [that is, $L(di/dt)$ acts like a voltage source that is strong enough to keep the current flowing]. The current plot is given in Exhibit 8. The voltage is applied for a long period of time, so any rates of change, $L(di/dt)$, are completed; the inductor acts like a short circuit current and is easily found to be 0.3636. At $t(0)$, this current is the initial current $I(0) = I_0$ for the equation in Exhibit 8(c). The current at $t = 0.1$ second is found directly from the exponential equation where the time constant, $\tau = L/R$, has a resistance equal to the sum of the two resistors in series, $R = 25\ \Omega$, so $\tau = (5/25) = 0.2$ second. This current is $i = I_0 e_{-t/\tau} = 0.3636e^{-(0.1/0.2)} = 0.220$ A.

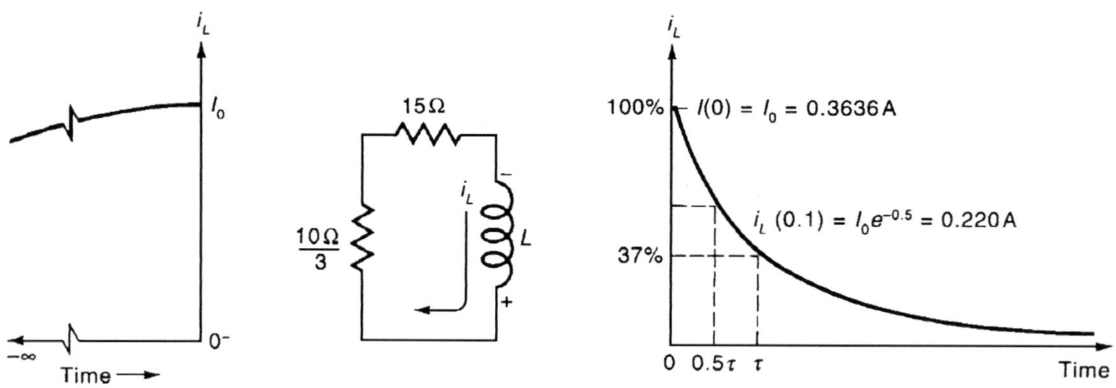

(a) Before $t = 0$ (b) Circuit, $t > 0$ (c) After $t = 0$

Exhibit 8 Current relationship in an inductor

The voltage across $L(di/dt)$ must be the same as the voltage drop across the resistors, since the loop is closed. Thus at $t = 0.2$ second, the voltage is $iR = 5.50$ volts; the polarity of this voltage is with the plus sign at the bottom of the inductance.

For an *RLC* circuit that involves two energy storage elements, *L* and *C*, the solution is complex. The governing equation is a second order differential equation whose solution may involve complex conjugate parameters.

AC CIRCUITS

In this review, the emphasis is on easily visualized graphical solutions rather than formal mathematical techniques. The review is based on assumptions that should be understood before the material is presented. The assumptions are the following:

- All sources are sinusoidal and are of the same frequency in any circuit unless specifically noted. For certain kinds of circuits, such as filters and resonant circuits, frequency changes will be made and will be considered in steady state and of the same frequency.

- Voltages and/or currents are considered at steady state. Any transients that may have resulted from the sudden closing of a switch are not considered. This assumption greatly simplifies the analysis; differential equation solutions will be bypassed and are replaced with phase-shifting operators. Although switch closing may be part of a circuit, it is assumed that one waits a short period of time (actually five time constants or more) for any transient or nonsinusoidal effects to die out before analyzing the circuit.
- Current or voltage sources are single phase unless otherwise noted (such as in three-phase circuit analysis).
- Circuits are linear. If a sinusoidal signal is applied to a circuit, then a current or voltage measured anyplace in the circuit will also be sinusoidal (with no harmonics). The only difference between an input and an output sinusoid will be a possible change in magnitude and phase.

An oversimplified pictorial (see Figure 7.21) will help one visualize a sinusoidal voltage source along with some of the above assumptions. The oversimplification is that of showing only one loop of wire in a magnetic field producing a pure sine wave. The frequency of the sine wave is directly proportional to shaft speed. And, the time for one revolution for a single pair of poles (called the period, T) is the inverse of the frequency. The wave is periodic and continuous and may be represented with a phasor.

Consider how one could draw a sine wave rather accurately by placing the tip of a pencil at the center of an X–Y plot (start with it laying horizontally to the right of the origin) and then allowing it to rotate counterclockwise through 360°. If one were to view the eraser from afar to the left and then to project horizontally the tip of the eraser onto the vertical axis of a time plot, the projection would trace out a pure sine wave (see Figure 7.22); the pencil could be thought of as a phasor, a two-dimensional vector. This phasor (length) represents the maximum value (height) of the sine wave (it is usually more convenient to substitute the rms value rather than the maximum value for the length). This maximum value will, at first, be used to demonstrate how the rotating phasor is visualized. When using these phasors, one's real concern is a time *snapshot* of the relative positions of the phasors. Also, rather than plotting the voltage vs. θ as in Figures 7.21 and 7.22, it is more convenient to plot it against ωt or $2\pi ft$. The angle the phasor passes through, with respect to some reference position or another phasor, represents a phase angle.

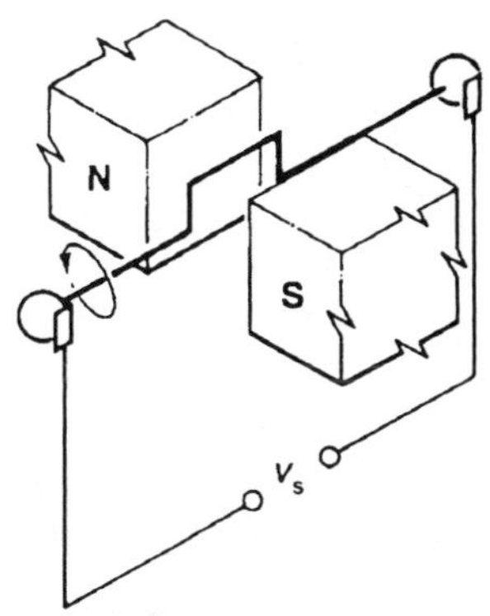

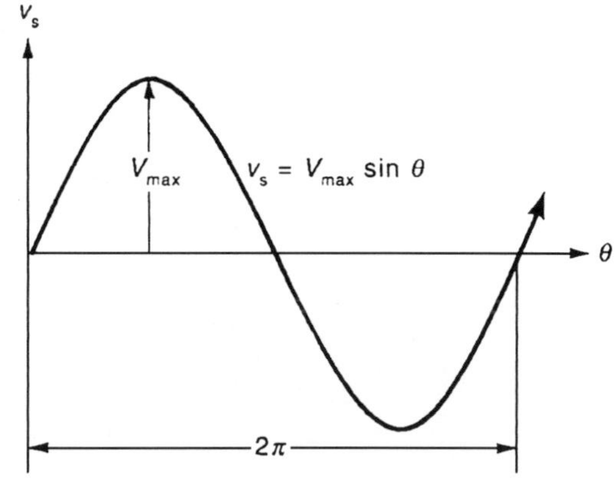

Figure 7.21 Simplified ac voltage source

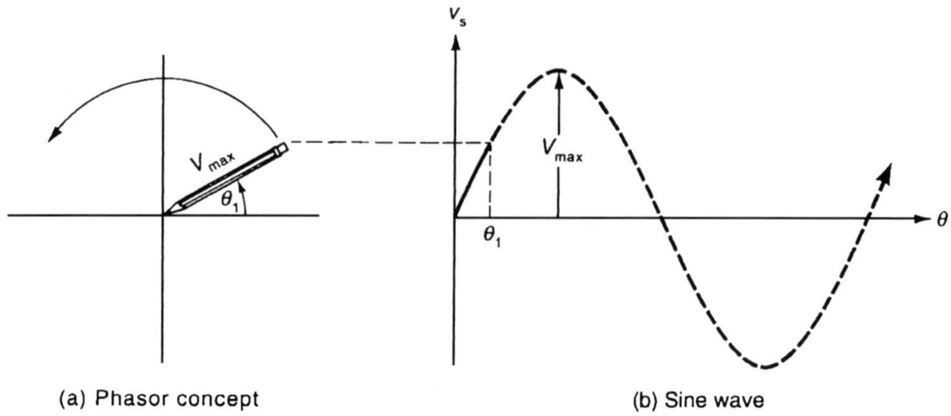

Figure 7.22 Graphical production of a sine wave

As an example of the simplicity of ac circuit analysis when using this phasor representation, consider summing two ac voltage sources graphically as in Figure 7.23(b). Here, one would sum the two voltages and continually draw the sum of the two waves; the result is

$$v_{total} = V_{max1} \sin \omega t + V_{max2} \sin(\omega t + \alpha) = V_{total} \sin(\omega t - \theta)$$

This graphic summing produces the desired total voltage and also displays the phase relationship with respect to either of the original sine waves. However, in Figure 7.23(c), the parallelogram formed from the two phasors yields the resulting phasor (the length is $v_{max\ total}$) much easier than summing the two sine waves. Furthermore, it gives the phase angle directly. The result is the same, in terms of relative magnitude and phase angle, for any instant of time. A formal mathematical description is

$$v_{total} = Im[V_{max\ total} e^{j(\omega t - \theta)}] \tag{7.26}$$

The notation *Im* implies the imaginary part of the expression that follows. A much simpler descriptive notation that most engineers use is to describe the representative phasor by a magnitude and its phase angle as $\mathbf{V}_{max\ total} \angle{-\theta}$. In electrical engineering, a positive rotation of the phasor is considered to be counterclockwise. Thus, in Figure 7.23, the parallelogram resultant phasor may be listed as being θ degrees behind (or lagging) the phasor V_1 or as ϕ degrees ahead (or leading) V_2.

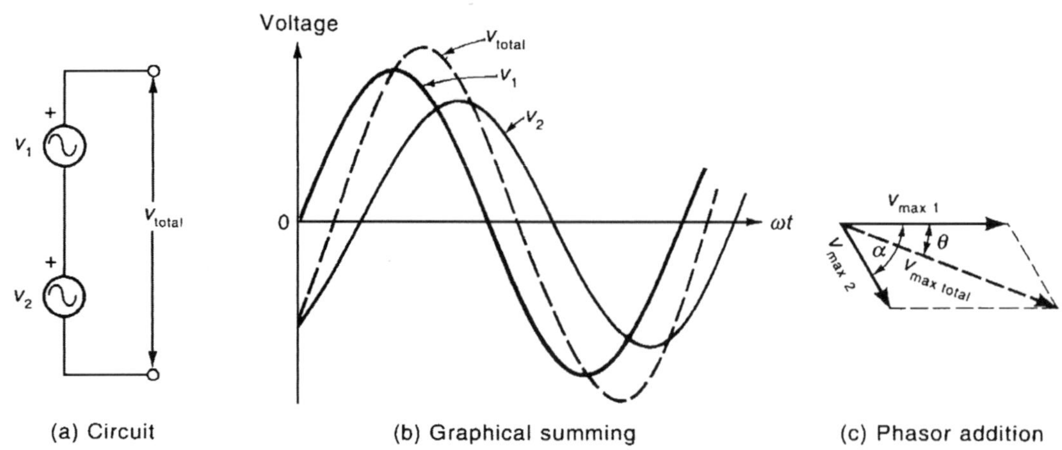

Figure 7.23 The addition of two voltage sources

Phasor Manipulation

Although adding two phasors is easily accomplished with the parallelogram, adding more than two might better be done by breaking each phasor into its x and y components and then summing the components. Multiplying phasors is done by multiplying the magnitudes and summing the angles, and division is done by dividing the magnitudes and subtracting the angles. As an example, two phasors **A** = $10\angle 0°$ and **B** = $5\angle 90°$ sum to **A+B** = $11.2\angle 26.6°$. And **AB** = $50\angle 90°$; or **A/B** = $2\angle -90°$. The concept of rotation is important. Assigning phasor **B** to be **B** $10\angle 180°$, then **B** = –**A**; one could say that **B** is the same as **A** if **A** were rotated through 180°. Here, multiplying phasor **A** by –1 is considered as operating on **A** to rotate it through 180°. One could then say that multiplying a phasor by the square root of –1 is also an operator that rotates the phasor through 90°. This imaginary value, the square root of –1, is referred to as j. The operator j then rotates the phasor by 90°. As an example, the only difference between a sine wave and a cosine wave (for the same amplitude) is that a sine wave is shifted through 90° to become a cosine wave,

$$\cos\theta = \sin(\theta + 90°) = j\sin\theta \tag{7.27}$$

Because of the effect of Equation (7.27), it will be more convenient to use the real (x-axis) and imaginary j (y-axis) notation when dealing with phasors.

For most ac voltages or currents, one is usually interested in rms values rather than peak or maximum quantities. It was previously shown that the rms [or in Equations (7.20) and (7.21) *what a meter would read*] value of a sine wave is its maximum value divided by the square root of two (or $V_{rms} = 0.707V_{max}$). Thus, one usually begins with rms magnitudes when using phasors. This makes voltage, current, and (especially) power calculations much simpler; the notation is even easier (the rms subscripts are dropped). In ac circuits there are really only three important devices: resistors (energy dissipating element), capacitors (energy storage element), and inductors (also an energy storage element).

> Since all phasors are relative, plot phasor lengths as rms values rather than peak values.

Resistors

As in dc circuits, this resistive element dissipates power equal to the current squared times the resistance. This is not the only equation for power, but *it is the safest equation to use in ac circuits* unless one is very careful with notation and is experienced in ac circuit theory. Also, the voltage across a resistor is always in phase with the current through the resistor. The resistor stores no energy but dissipates the power in terms of heat.

> Current and voltage are in phase for a resistor: $\mathbf{V}_R = R\mathbf{I}$.

By this notation, voltages and currents in bold indicate that they may be complex values.

CAPACITORS

The capacitor stores energy for half a cycle and gives it up on the other half, the net (real) energy is zero. However, it is sometimes convenient to refer to the product of voltage and current as imaginary power (using the symbol Q and calling it reactive power). The result of this energy interchange is that the current through the capacitor has a 90° phase difference for the two phasors; the current is ahead of the voltage (or leads the voltage). Whereas the resistance, R, impedes the flow of current, the impending quantity for a capacitor is the reactance, called X_c. Not only does X_c impede the current, but it also causes a phase shift of –90° so that the voltage across the capacitor is lagging the current by 90°. These descriptive words may be replaced by the symbol $-j$.

> The voltage drop across a capacitor is 90° behind the current.
>
> $$\mathbf{V}_C = -jX_C\mathbf{I}, \qquad \text{where } X_C = \frac{1}{2\pi fC}$$

The instantaneous voltage across a capacitor for a given current is

$$v = 1/C \int i \, dt \tag{7.28a}$$

and if the given current is $i = I_{max} \sin \omega t$ then

$$v = 1/C \int I_{max} \sin \omega t \, dt \tag{7.28b}$$

and the solution is

$$v = [-1/(\omega C)]\, I_{max} \cos \omega t = [1/(\omega C)] I_{max} (-j) \sin \omega t \tag{7.28c}$$

Using phasor notation, one can show that Equation (7.28c) is equivalent to

$$\mathbf{V}_C = -jX_C\mathbf{I}, \quad \text{where } X_C = 1/(\omega C) \tag{7.29}$$

It is obvious that a numerical value of X_C is a function of frequency. In fact, if the frequency is infinite, X_C is zero as in a short circuit; and if the frequency is zero, X_C is infinity as in an open circuit. This relationship between current and voltage is shown graphically in Figure 7.24.

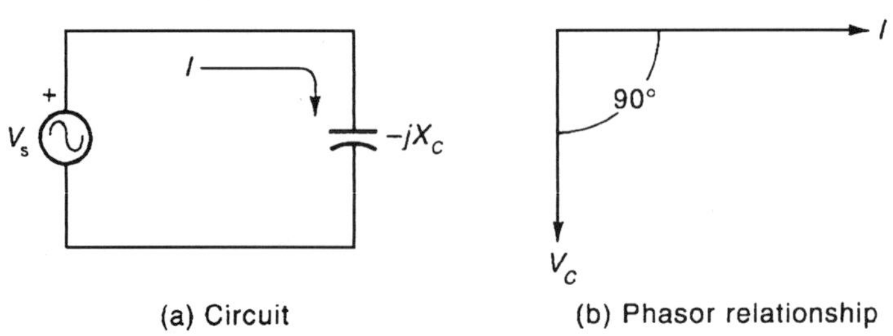

Figure 7.24 Voltage-current relationship for a capacitor. (The voltage and current are not to the same scale.)

Example 7.7

Consider a series *R–C* circuit whose measured current is 5 amperes (rms). Determine the voltage source required to produce this current. Assume that the voltage source is known to have a frequency of 60 Hz, with $R = 10$ ohms and $C = (0.1/377)$ farads. The factor $2\pi f$ for 60 Hz is 377.

Solution

The effect of the capacitor is to create a 90° phase shift between the current and voltage while the resistor produces no phase shift. The instantaneous Kirchhoff voltage equation is

$$v_{\text{total}} = Ri + v_C = Ri + (1/C)\int i\,dt \qquad \textbf{(7.30a)}$$

Rather than solving the above equation, phasors will be used here in the form

$$\mathbf{V}_s = \mathbf{V}_R + \mathbf{V}_C = R\mathbf{I} - jX_C\mathbf{I} \qquad \textbf{(7.30b)}$$

Now R and jX_C should be viewed as operators (they operate on the current to produce a voltage—and the j is to produce a 90 degree phase shift). Then X_C is calculated from $1/(2\pi fC)$ to be 10 ohms.

$$\mathbf{V}_s = 10\mathbf{I} - j10\mathbf{I} \qquad \textbf{(7.30c)}$$

Remember, the $-j10$ operator multiplies the current phasor by ten and shifts it through –90°. With the two voltage phasors, one only needs the parallelogram construction to get the voltage source phasor along with its angular relationship to the current, as shown in Exhibit 9.

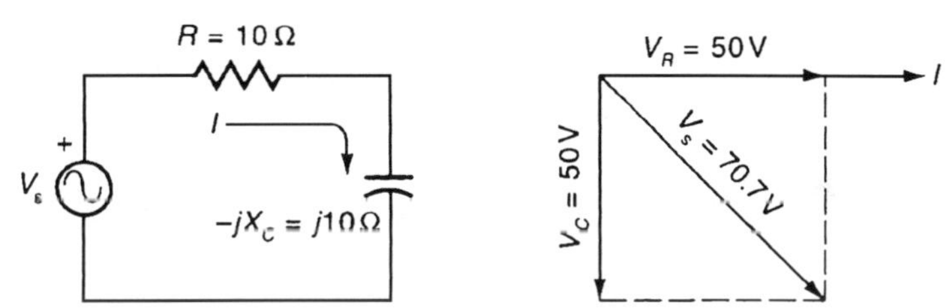

(a) *R–C* circuit (b) Phasor relationship

Exhibit 9 Voltage relationship for an *R–C* circuit

The voltage source is 70.7 volts with the voltage phasor lagging 45° behind the current. Or if the voltage source is considered to be the reference, one could say that the current leads the voltage by 45°. Observe that all of the (real) power from the source is dissipated by heat loss in the resistor (no real power is dissipated in the capacitor):

$$P_R = I^2R = 5^2(10) = 250 \text{ watts}$$

The power from the source, P_s, is the product of voltage and current times the power factor, which is defined as the cosine of the angle between the voltage and current. Thus

$$P_s = VI\cos 45° = 70.7 \times 5 \times 0.707 = 250 \text{ watts}$$

The reactive power, Q, is calculated as follows:

$$P_{\text{reactive}} = Q = VI \sin\theta = 70.7 \times 5 \times 0.707 = 250 \text{ VARs}$$

(*Note*: P_s and Q match only because θ is 45°.)

Inductors

Like the capacitor, the inductor stores energy for half a cycle and returns it to the circuit for the next half cycle. In circuit analysis, the inductor behaves in a similar manner as a capacitor except the sign is reversed; here the term is $+jX_L$ where X_L is directly proportional to the frequency and is equal to $\omega_L = 2\pi fL$. For the inductor, the voltage leads the current by 90°, or the current lags the voltage.

> A voltage drop across an inductor leads the current by 90°.
>
> $$\mathbf{V}_L = jX_L\mathbf{I}, \quad \text{where } X_L = 2\pi fL$$

Also, at zero frequency, the reactance is zero; and for an infinite frequency, the reactance is infinite. The phase relationship is shown in Figure 7.25.

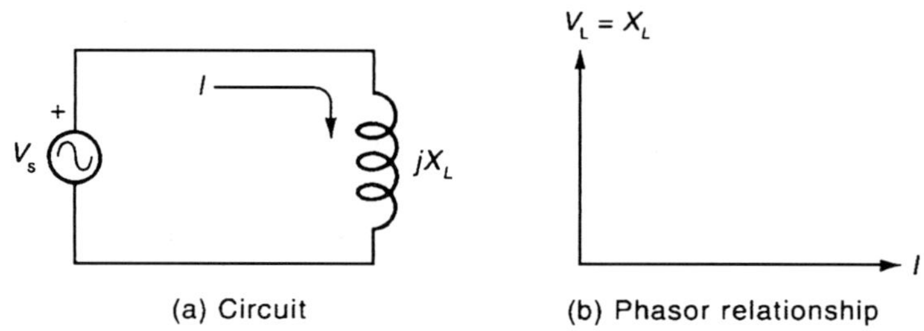

Figure 7.25 Voltage-current relationship for an inductor

Example 7.8

Consider the circuit of Exhibit 10, where the current is, as before, 5 amperes, R is 10 Ω, and L = 10/377 henries.

Solution

Here X_L is ohms (Exhibit 10). The answers are unchanged from Example 7.7 except for the phase relationships.

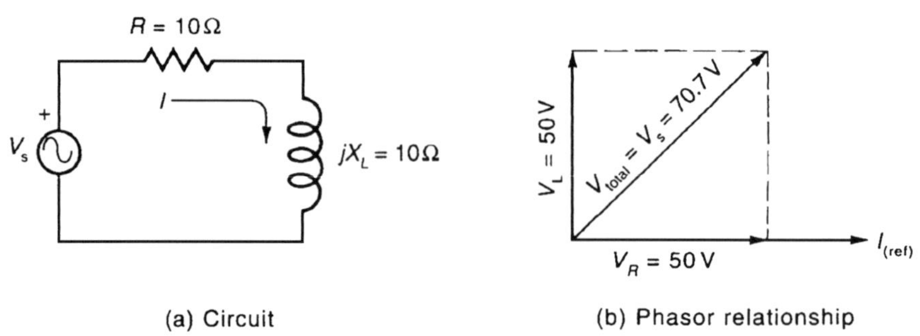

Exhibit 10 An *R-L* circuit voltage-current relationship

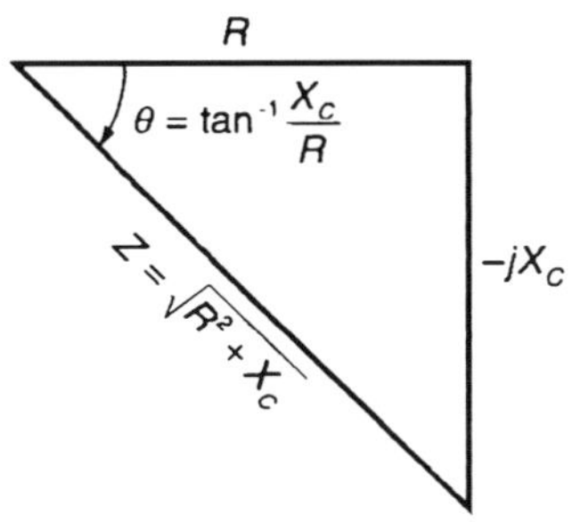

Figure 7.26 The impedance relationship

Impedance Relationship

In general, ac voltages and currents are related by a quantity called impedance (Z). It is usually complex, and it is sometimes referred to as Ohm's law for ac circuits. This association is easily explained by reference to either of the two previous example problems. For instance, in Exhibit 9 and Equation (7.30), the current was assumed to be known. If the current were unknown and the source voltage was given as 70.7 volts, then $\mathbf{V}_s = 70.7\angle 0°$ volts can be the known reference phasor. From Equation (7.30b), the unknown current is easily factored out of the expression to give $\mathbf{V}_s = (R - jX_C)\mathbf{I}$.

The parenthetical quantity—or the operators—for this problem, as in Equation (7.30c), is the complex impedance. It has a right-triangle relationship as in Figure 7.26 and is given by $\mathbf{Z} = R - jX_C = 10 - j10 = 10\sqrt{2}\angle{-45°}$.

The $10 - j10$ is known as the rectangular form, and the $10\sqrt{2}\angle - 45°$ is known as the polar form. The rectangular form is usually used when adding or subtracting is involved, and the polar form is used when one is multiplying or dividing. To finish the problem, the current is found as $\mathbf{I} = \mathbf{V}_s/\mathbf{Z} = (70.7\angle 0°)/(10\sqrt{2}\angle - 45°) = \angle + 45°$ A.

Example 7.9

Another series circuit problem will expand on the complex impedance concept; see Exhibit 11. Here, all of the resistances and reactances and the source voltage are given for a particular frequency. It is desired to find the power dissipated in the 20-ohm load resistance and also to find the total power from the source voltage. Since the circuit is a series one, the current is common for all elements; therefore the current should be found. The voltage source is 100 volts.

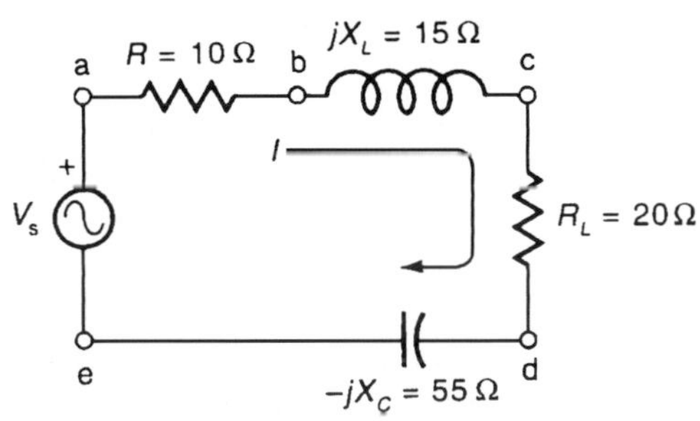

Exhibit 11 A series circuit with known parameters

Solution

Find the total impedance to find the current (see Exhibit 12): $\mathbf{Z} = 10 + j15 + 20 - j55 = (10 + 20) + j(15 - 55) = 30 - j40 = 50 \angle -53.1°\Omega$.

From this complex impedance, current and power can be found directly:

$$\mathbf{I} = \mathbf{V}/\mathbf{Z} = (100\angle 0°)/(50\angle{-53.1°}) = 2 \angle +53.1° \text{ amperes}$$

$$P_{RL} = I^2R_L = 2^2(20) = 80 \text{ watts}$$

$$P_s = V_sI \cos\theta = 100 \times 2 \cos(-53.1°) = 120 \text{ watts}$$

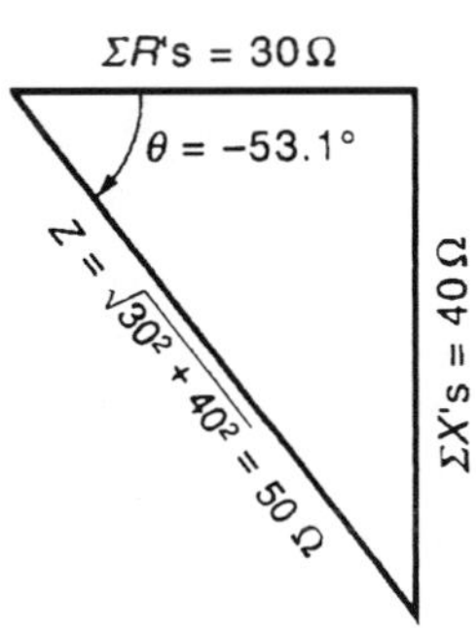

Exhibit 12 Impedance triangle

Check: $P_s = \Sigma I^2R = 2^2(10 + 20) = 120$ watts.

Example 7.10

Suppose, in Example 7.9, one is asked to find the magnitude and phase of the voltage across two points—say the voltage from point b to point d.

Solution

The phasor plot readily yields this answer. Since the current is common to all elements, the current should be made the reference phasor. Again, the currents and voltages are not necessarily plotted to the same scale. Rather than having all phasors emanating from the origin, it is convenient to plot the voltages in a cumulative fashion, from head to tail (see Exhibit 13).

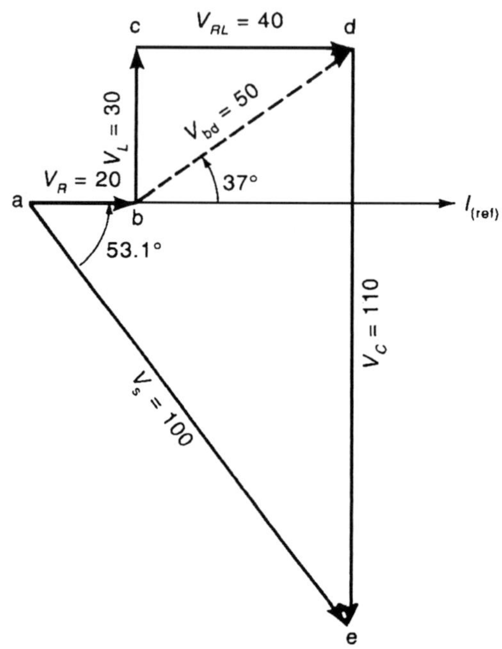

Exhibit 13 Plot of voltages vs. current

The voltage, V_{bd}, may be read directly from the plot to be 50 volts at an angle of 37° ahead of the current. An interesting aspect of this problem is brought out if the frequency of the source is doubled. Doubling the frequency causes no change in the resistive elements but dramatic changes in the reactive ones. For the inductor, X_L is $15 \times 2 = 30$ ohms and for the capacitor $X_C = 0.5 \times 55 = 27.5$ ohms. The impedance then becomes $\mathbf{Z} = 10 + j30 + 20 - j27.5 = 30 + j2.5 = 30.1\angle 4.76°$ ohms. The current is $\mathbf{I} = (100\angle 0°)/30.1\angle 4.76° = 3.32\angle -4.76°$ A, and the power dissipated in the 20-ohm resistor is

$$P_{RL} = I^2 R_L = 3.32^2 \times 20 = 220 \text{ watts}$$

At this point, several practical observations are in order. For approximate answers, if either the real or imaginary part is more than ten times the other, the hypotenuse is almost equal to the longest leg, and the angle is within a few degrees of being zero. Using this approximation for impedance, the approximate current is 3.3 amperes and P_{RL} is 222 watts. Another observation is that by changing the frequency, the power output has gone from only 80 watts to well over 200. There is a specific frequency to get maximum current and power; this is called the resonant frequency.

Resonant Frequency

For series circuits, the resonant frequency occurs when all of the reactive components cancel so that $X_L = X_C$, $\omega L = 1/\omega C$. For this particular frequency, the subscript notation may be given as *res*, *0*, or *n*.

Series Resonant Frequency

$$\omega_0 = 2\pi f_0 = \frac{1}{\sqrt{LC}}$$

The series resonant frequency concept leads directly to the maximum power transfer theorem, which should now be obvious. Maximum power occurs when the reactive components in an ac series circuit cancel.

Parallel ac circuits are no more difficult than series ones but may require slightly more bookkeeping; also, since many reciprocals are involved, additional definitions may be used to identify these reciprocals. As an example, the reciprocal of impedance, Z, is admittance, Y; the units of Y are siemens (old notation mhos) abbreviated S. The admittance may be further expanded to have a real and imaginary part. The real part is called the conductance, G, and the imaginary part is called the susceptance, B; both parts have the units of siemens. However, for simple problems, these definitions are optional; consider the example shown in Figure 7.27.

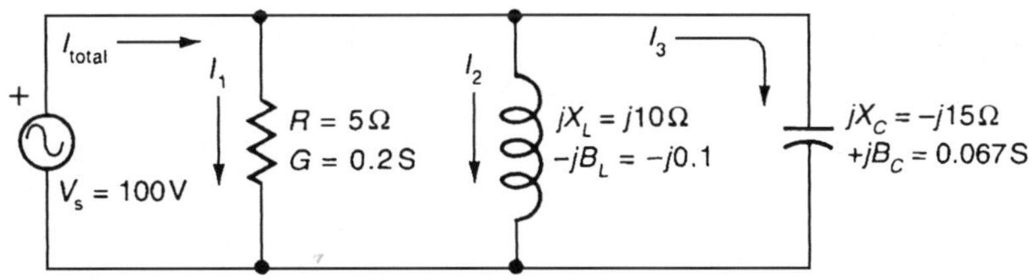

Figure 7.27 A parallel circuit

Example **7.11**

Solution

The equivalent impedance for this problem may be found by either dividing the source voltage by the total current (see Exhibit 14) or by taking the reciprocal of the total admittance.

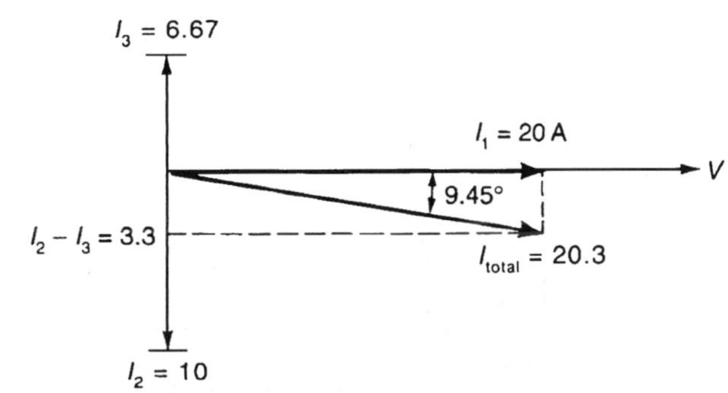

Exhibit 14 The phasor relationship for Figure 7.27

$$\mathbf{I}_{total} = \mathbf{I}_1 + \mathbf{I}_2 + \mathbf{I}_3$$

where

$$\mathbf{I}_1 = \mathbf{V}/R = (100\angle 0°)/5 = 20\angle 0° \text{ A}$$
$$\mathbf{I}_2 = \mathbf{V}/(jX_L) = (100\angle 0°)/(10\angle 90°) = 10\angle -90° \text{ A}$$
$$\mathbf{I}_3 = \mathbf{V}/(-jX_C) = (100\angle 0°)/(15\angle -90°) = 6.67\angle 90° \text{ A}$$
$$\mathbf{I}_{total} = 20 - j10 + j6.67 = 20 - j3.33 = 20.28\angle -9.46°$$

$$\mathbf{Z} = \mathbf{V}/\mathbf{I} = (100\angle 0°)/(20.28\angle -9.46°) = 4.93\angle 9.46° \text{ ohms}$$

However, using admittances, Y may be found directly:

$$G = 0.20 \text{ S}, \quad -jB_L = -j0.10 \text{ S}, \quad jB_C = j0.066 \text{ S}$$
$$\mathbf{Y} = G - jX_L + jX_C = 0.20 - j0.10 + j0.066 = 0.20 - j0.03 = 0.203\angle -9.45° \text{ S}$$
$$\mathbf{Z} = 1/\mathbf{Y} = 1/(0.203\angle -9.45°) = 4.93\angle 9.45° \text{ ohms}$$

Example 7.12

A final example of a combined series and parallel circuit is given in Exhibit 15. For this kind of problem, the reader is urged to make two separate plots to find the individual branch currents, then add the two current phasors on a third plot to find the total current. Find the power taken from the source.

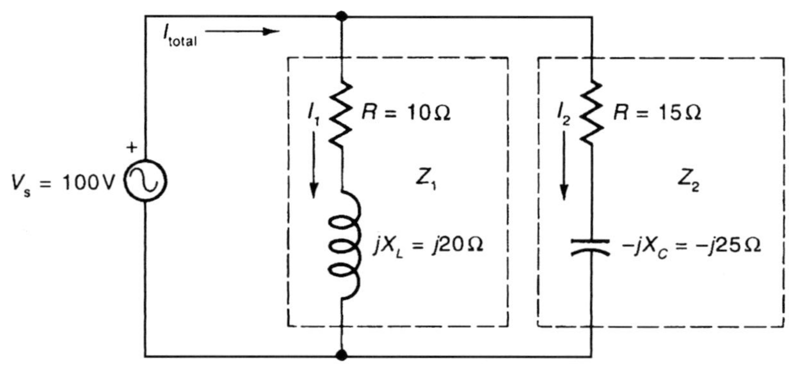

Exhibit 15 Combined series and parallel circuit

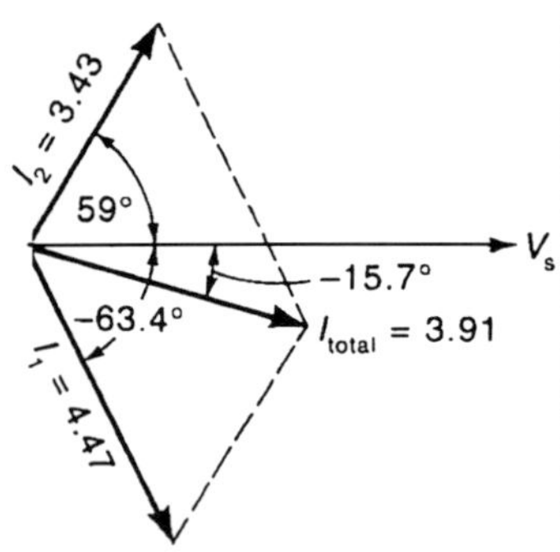

Exhibit 16 Phasor diagram of currents

Solution

First, find the currents:

$$\mathbf{I}_1 = \mathbf{V}/\mathbf{Z}_1 = (100\angle 0°)/(10 + j20) = 4.47\angle -63.4° = 2.00 - j4.00$$
$$\mathbf{I}_2 = \mathbf{V}/\mathbf{Z}_2 = (100\angle 0°)/(15 - j25) = 3.43\angle 59.0° = 1.76 + j2.94$$
$$\mathbf{I}_{total} = \mathbf{I}_1 + \mathbf{I}_2 = (2.00 + 1.76) + j(-4.00 + 2.94) = 3.76 - j1.06 = 3.91\angle -15.7°$$

See Exhibit 16 for the phasor diagram. The power from the source is now found to be

$$P_s = \mathbf{V}\mathbf{I}_{total} \cos\theta = 100 \times 3.91 \times \cos(-15.7) = 376 \text{ watts}$$

(*Check*: $P_{total} = (\mathbf{I}_1)^2 R_1 + (\mathbf{I}_2)^2 R_2 = 200 + 176 = 376$ watts.)

Although calculations are straightforward, the reader is urged to sketch the phasor diagram. Again, for a multiple-choice problem, the graphical solution may be sufficiently accurate for one to select the correct answer.

Quality Factor

For parallel resonance as for series resonance circuits, Z still equals R at resonant frequency. For the series circuit the inductive and capacitive reactances cancel; and for parallel circuits, the inductive and capacitive susceptances (in siemens) cancel. In describing the behavior of these resonant circuits another descriptive quantity is frequently used—especially so for parallel circuits—which is called the *quality factor*, Q, of the circuit. For a series circuit Q is a measure of the energy stored in an inductor compared to the energy dissipated in the resistance; this ratio may also be given as

$$Q_{series} = \omega_o L/R = 1/(\omega_o CR) \quad \textbf{(7.31)}$$

where Q is dimensionless and may be quite high for a sharp resonant peak as a narrow bandwidth circuit. On the other hand, for a parallel circuit, Q may be defined (see Figure 7.27) as

$$Q_{parallel} = 1/Q_{series} = \omega_o RC = R/(\omega_o L) \quad \textbf{(7.32)}$$

and the bandwidth, BW, may easily be found as

$$BW = \omega_o/Q_{parallel} \text{ (rad/s)} \quad \textbf{(7.33)}$$

Transfer Function

The transfer function is the ratio of desired response of a system to the input (or excitation) when all the initial conditions are zeros.

Example 7.13

Find the transfer function $T_{(s)} = \dfrac{V_2(s)}{V_1(s)}$ of the circuit shown in Exhibit 17.

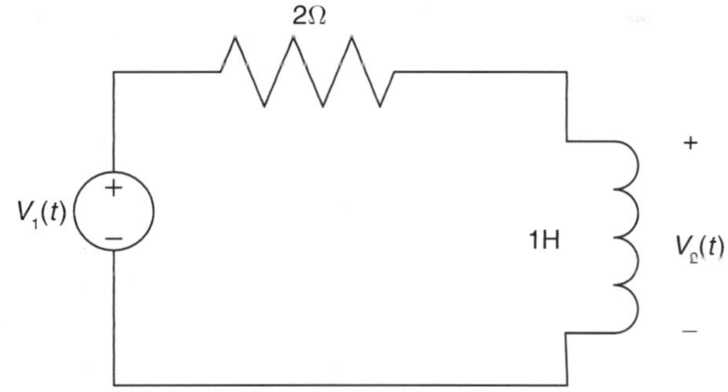

Exhibit 17

Solution

The s-domain circuit is drawn, as shown in Exhibit 18, by replacing the inductor with s •, the source by $V_1(s)$, and maintaining the resistor as such. Using voltage division, $T(s) = \dfrac{V_2(s)}{V_1(s)} = \dfrac{s}{s+2}$.

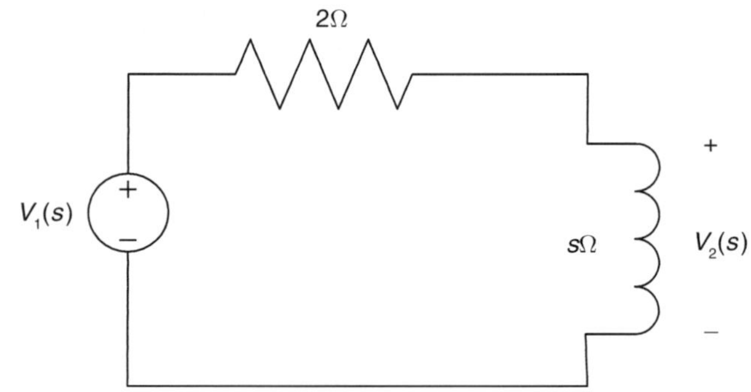

Exhibit 18

For a transfer function H(s), a plot of |H(jω)| versus ω and a plot of the phase angle of H(jω) versus ω are called the frequency response plots. For convenience, a plot of 20 log |H(jω)| dB (decibel) versus log ω, and a plot of the phase angle of H(jω) versus log ω are usually drawn. These are called Bode plots.

ELECTRIC MACHINES

AC Machines

The speed of rotation of a synchronous motor and that of a rotating magnetic field, in either a synchronous motor or an induction motor, are decided by the equation

$$n_s = \frac{120f}{p} \tag{7.34}$$

where f = line frequency in Hz, p = the number of poles, and the speed is in rpm. For three-phase induction motors, the actual speed of rotation will always be less than the speed of the rotating field. This difference is expressed by a factor called slip, as given by the equation

$$slip = \frac{n_s - n}{n_s} \tag{7.35}$$

DC Machines

For direct current machines, the voltage induced in the armature, E_a, is directly proportional to the product of speed, n, and the magnetic flux, ϕ, generated by the field. The magnetic flux is directly proportional to the field current, I_f. The relation between E_a and I_f is determined experimentally.

Example 7.14

Exhibit 19 shows the equivalent circuit of a separately excited DC generator where R_a is the armature resistance, R_L is the load resistance, and E_a is the induced EMF of the armature. It is determined experimentally that the induced EMF is 100 V for a field current of 1 A on no load at 1000 rpm. Find the load voltage, V_L, at 800 rpm if the field current is 1 A, R_a is 1 Ω, and R_L is 10 Ω.

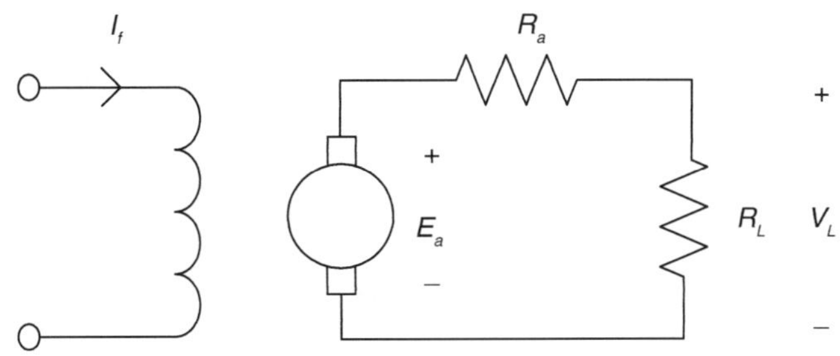

Exhibit 19

Solution

Since the induced EMF $E_a = K_a n\phi$, and flux $\phi = K_f I_f$, where K_a and K_f are constants, at any given I_f, E_a is proportional to speed n. As the field current is fixed, E_a is directly proportional to speed. Then, at 800 rpm, $E_a = 100\dfrac{800}{1000} = 80$ V.

Using KVL, $80 = I_a(1 + 10)$; $I_a = 7.27$ A; load voltage $V_L = 10I_a = 72.7$ V.

ELECTRONIC CIRCUITS

A thorough review of electronics is not undertaken here. But two topics—the solid state diode and the operational amplifier—have been selected as deserving attention. The complexity of certain aspects of both solid state theory and integrated circuit theory is well beyond the scope of this review, so only the idealized devices will be dealt with.

Diodes

The solid state theory of diode operation depends on the particular kind of diode being considered. For example, for the junction diode, knowledge of solid state theory and the behavior of majority and minority carriers in the presence of an electric field is desirable. But when an idealized diode is treated, the full theory is not essential. For a p–n junction diode, a very simplistic explanation of the operation at the junction between a *p* and an *n* type semiconductor is that in each type of material there are many free charges available; these are holes (p, positive) and electrons (n, negative), respectively. If the diode is biased in favor of forcing positive charges near the boundary (the positive voltage being connected to the p material and the negative terminal to the n material), a rapid recombining of the charges takes place [see Figure 7.28(a)]; this is the direction of easy current flow. On the other hand, if the diode is reverse biased [Figure 7.28(b)], the free charges are attracted toward their bias polarities, leaving a dearth of charges at the junction for very little recombination and almost no current flow. When the bias is in

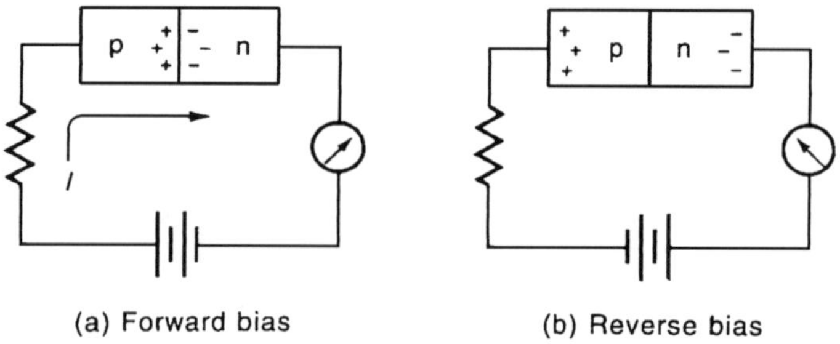

Figure 7.28 Charges inside a junction diode

the forward direction, the voltage necessary to cause the charges to recombine is fairly small but still could be significant (see Figure 7.29). The voltage is near half of one volt (approximately 0.4 V for germanium and near 0.7 for silicon) and is nonlinear. In the reverse direction, the current is essentially zero until breakdown voltage is reached; this breakdown is referred to as Zener voltage, which ranges from a few volts to several hundred volts. When designing circuits that use these diodes, care must be taken to work well within the reverse peak breakdown voltage. If one ignores the nonlinearity by assuming that the half volt is negligible for the forward direction, then the symbol shown in Figure 7.29(b) represents the ideal diode (the arrowhead side is the anode, and the other side is the cathode). These diodes, of course, have many applications; in this review, the applications will be limited to the ideal case in pure rectifying circuits.

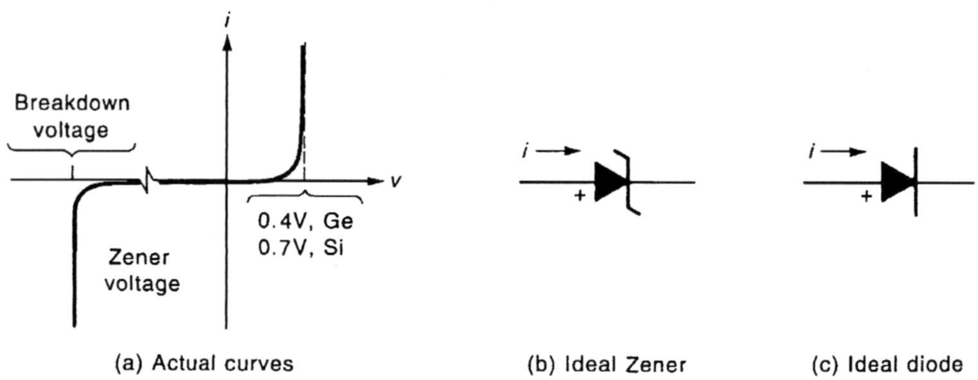

Figure 7.29 Junction diode voltage–current characteristics

Rectifying Circuits

Most diode rectifying circuits are used to convert ac voltages to *pulsed dc* voltages (see Figure 7.30). The diode is used to convert an ac voltage to a half-wave sinusoid so that the meter measurement found is the average current. Rectification is not limited to sinusoid; the following table (see Table 7.1) shows a sample of various kinds of signals and kinds of rectification, either full- or half-wave, relationships.

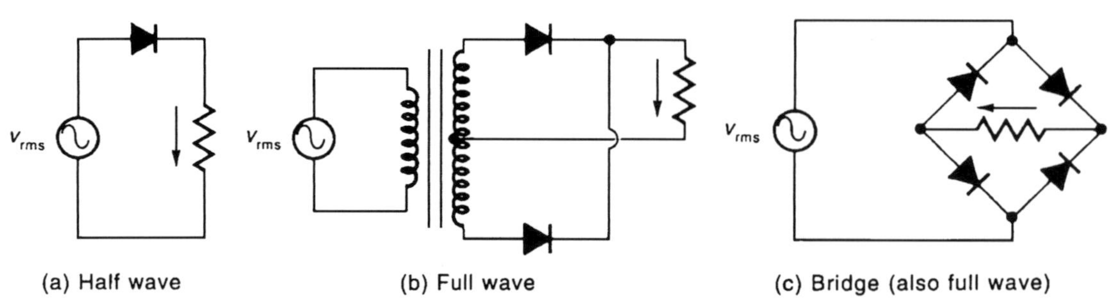

Figure 7.30 Rectifier circuits

Table 7.1 Rectified values for three signals

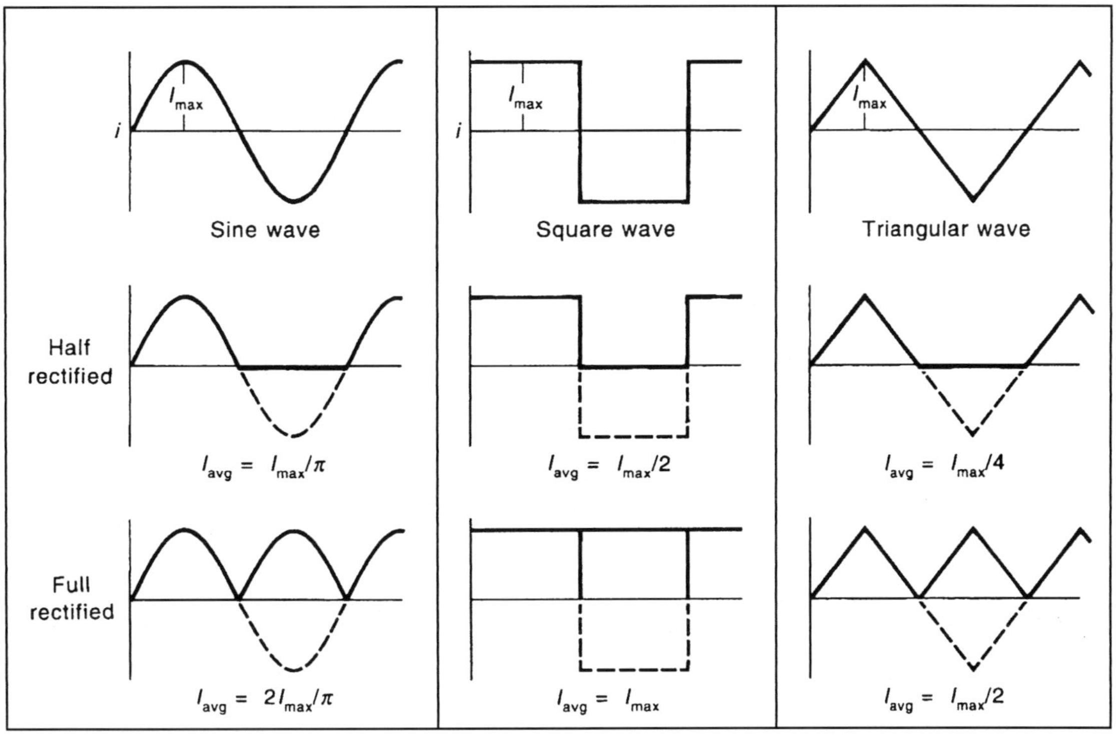

Example 7.15

A triangular wave form with a peak value of 50 volts is rectified by a half-wave rectifier and goes to a load resistor of 100 ohms (see Exhibit 20); the ammeter in the circuit is a dc type that measures average current. What is this current?

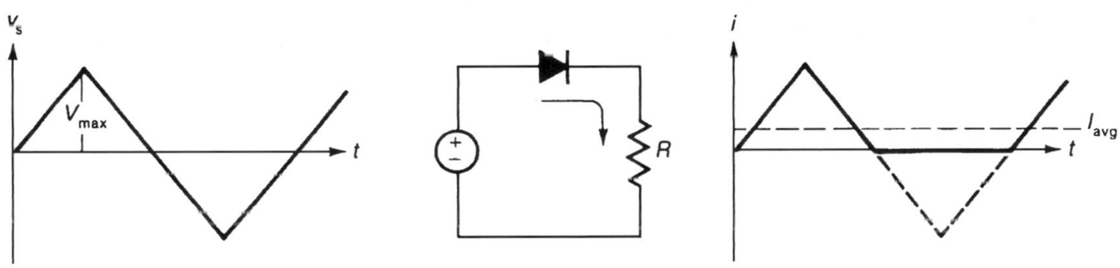

Exhibit 20 Half-wave rectifying circuit

Solution

The current is v/R and is found from the equation for average values as

$$I_{avg} = 1/R\left(1/T\int_0^T v\,dt\right) = 1/R\left[1/T\int_0^{T/2} v\,dt + 1/T\int_{T/2}^T 0\,dt\right] \qquad \textbf{(7.36)}$$

The integral is the area of the triangle over a half period divided by a full period, T.

$$I_{avg} = [(V_{max}/R)(0.5)(0.5T)]/T = (50/100)(1/4) = 0.125 \text{ A}$$

Or, the current can be found by reading directly from Table 7.1; the average reading is one-quarter of the peak (or maximum) value of the current.

Operational Amplifiers

The operational amplifier is a high gain differential amplifier circuit (see the following list) that has been highly developed over the years. Since the cost has gone from a few hundred dollars (old vacuum tube era) to a few cents for highly developed integrated circuits, the applications for this device cover almost all areas of engineering instrumentation. This discussion focuses on an ideal operational amplifier that is treated externally as a black box. This operational amplifier (referred to as an *op-amp*) differs from a normal amplifier. For example, a home hi-fi audio amplifier's frequency response is considered good if it amplifies voltages over a frequency range from approximately 20 Hz to 20,000 Hz. An op-amp has much higher amplification. It also has several other significant characteristics:

- The amplification is usually of the order 100,000 or more. It is based on the input voltage being the difference between two very small voltages, $v_d = v_p - v_n = v_1 - v_2$. These small voltages are designated as positive and negative; however, the actual polarity of the applied voltages could be either. This implies that the output voltage is zero if the two small input voltages are equal.
- The amplification is flat over a frequency range from zero Hz (that is, *down to dc*) to some very high frequency (high compared to the highest frequency being amplified). Before the low-drift transistor, great care and clever circuit design were required to obtain a *dc amplification*.
- The currents to the actual input pins (both positive and negative) are very small and may usually be neglected; this means that the input resistance (or impedance) is very high. Also, the output resistance (or impedance) is very low; again, "low" means relatively small compared with the external circuitry values.
- The device is linear over a known range, which means that the superposition theorem applies over a range of voltages (for example, if the input voltage were doubled, the output voltage would double).

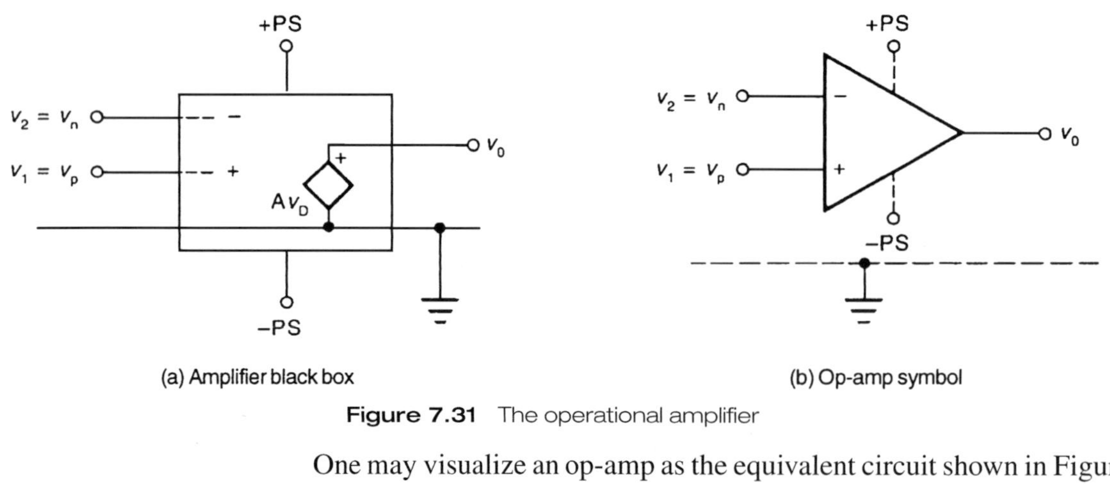

Figure 7.31 The operational amplifier

One may visualize an op-amp as the equivalent circuit shown in Figure 7.31(a). In this figure, the positive and negative power supply (+PS, –PS) connections are shown, but, as in Figure 7.31(b), they may not be shown. Since input voltages may be positive or negative, the power supply voltages are both relative to some common point, usually referred to as ground (signal ground). The power supply voltages must be larger than the largest expected output voltage within the limits of the op-amp. These constraints are normally assumed and are not usually stated or even shown on a diagram [see Figure 7.31(b)]; the triangular symbol may have the vertical side shown as slightly rounded in some diagrams.

The op-amp is used in circuits designed for use in either the inverting or the noninverting mode. Consider the inverting mode circuit (see Figure 7.32); here, since the two voltage input pins are assumed to go to an open circuit, the input currents for both the plus and minus pins are zero, and the voltage difference between these two pins is essentially zero. Since the gain is so high and output voltages are in the several-volts range, the input voltages are in the microvolt range. This makes the circuit analysis especially easy because if one knows one input voltage, the other is essentially the same (actually only a few microvolts different). So, if one input happens to be grounded [as in Figure 7.32(a)], the other input is almost zero. The equations for the output voltage given in Figure 7.32(a) are easily obtained by realizing that the current into the op-amp itself is zero. By summing currents at Node 1, i_a must equal $-i_f$ and the voltage at the junction is near zero volts. Since the node voltage is essentially zero, i_1 is v_a/R_1 and i_f is v_0/R_f. Thus, the circuit amplifies (by a factor of $-R_f/R_1$) as it inverts and has a relatively low current input at v_a (to the external circuit if R_1 is high) and whose output acts almost like an ideal voltage source. For more than one input, one simply sums the currents at Node 1 to yield the equation in Figure 7.32(b).

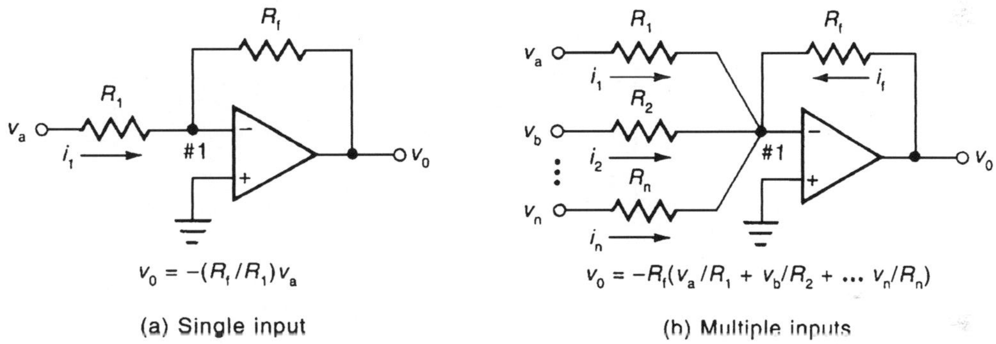

Figure 7.32 The op-amp in the inverting mode

As an example problem [for Figure 7.32(b)], four different transducers produce a possible maximum ac voltage of 0.1, 1.0, 5.0, and 10 volts, respectively. Each transducer is to be recorded on a one channel recorder whose desirable input signal level is one volt but whose input impedance will not allow a direct connection to the transducers. Since only one input at a time may be recorded, it is necessary only to determine each resistance ratio for a summing op-amp circuit. Summing is not required, but multiplying by a constant is. The ratios are easily found; in changing the level of the first voltage from 0.1 to 1, the ratio is ten. The common feedback resistor is arbitrarily chosen to be 100k ohms (the typical range of values runs from a few kilo-ohms to several megohms), then R_1 must be 10k ohms. Calculate the numerical values for the other resistor ratios shown in the circuit of Figure 7.33. When using the op-amp as a summer (all inputs connected at the same time), the instantaneous algebraic sum of the input voltages multiplied by their resistor ratio values will be summed together. They should not exceed the specified linear output voltage.

The other standard circuit configuration for the op-amp is the noninverting mode. For this, the negative input pin is normally fed from the output [perhaps through a resistive network as in Figure 7.34(b)], and the plus pin terminal is connected as an input. For the direct-feedback design [see Figure 7.34(a)], the output voltage—being also the minus terminal input—must be within a few microvolts of the plus terminal. Thus, this circuit has an output that is essentially the same as the input, and this circuit has a gain of unity with no change in polarity. The circuit is usually referred to as a buffer or voltage follower; the advantages are that it takes almost no power from an input source (that is, it does not "load" the input), and the output is almost as though it were an ideal voltage source. For the circuit shown in Figure 7.34(b), the minus input voltage must follow a certain percentage of the output and is considered a noninverting amplifier whose gain is v_0/v_{in} = Gain $= 1 + R_2/R_1$. As an example, if a nonverting op-amp with a gain of three is desired, the resistor ratio must be two. If R_2 is 10 kΩ, then R_1 must be 5 kΩ.

The key to solving these operational amplifier circuits is that the current input to the op-amp itself is considered to be zero. And, if either the plus or minus input voltage is known (or is a ratio of some other voltage), the other input voltage is considered to be the same. The application of Kirchhoff's laws to the rest of the external circuit will usually yield the correct answer.

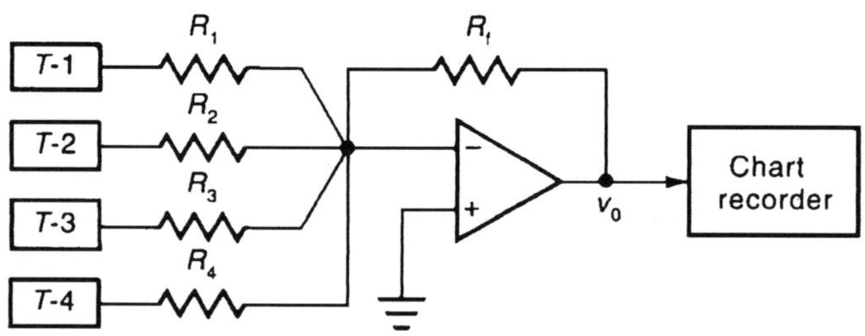

Figure 7.33 An op-amp circuit for matching voltage levels of various transducers. (*Caution:* only one switch to be closed at any one time.)

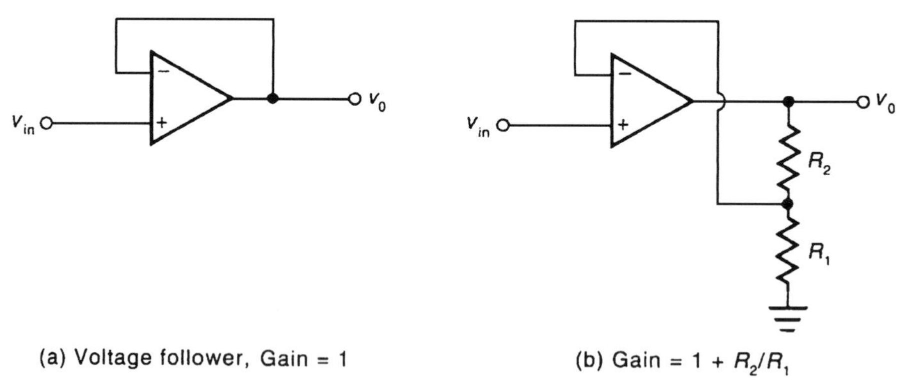

Figure 7.34 Noninverting op-amp circuits

TRANSFORMERS

The transformer is principally used in ac circuits to convert voltages from one level to another through the medium of magnetic fields. There are many other uses for the transformer, but they are specialized (such as pulse transformers). For ac circuits (implying sinusoidal wave forms), the ideal transformer is considered as lossless with 100% efficiency; actually, the efficiency of a typical transformer is greater than 90%. For these ideal devices, the product of the input volt-amps equals the product of output volt-amps (for larger power transformers, $kVA_{in} = kVA_{out}$). The product is the apparent power (VA) rather than real power (watts). The nameplate rating of the transformer is important. The manufacturer gives the normal operating conditions; these nameplate ratings include the frequency, the voltage and the voltage ratio, and the kVA rating. The voltage ratios are the same as the turns ratio, *a*, and the current ratios are inversely related to the turns ratio. Whether a voltage is stepped up or down depends on which side one considers as primary and as secondary. For this discussion, assume the left side is primary (1) and the right side is secondary (2).

Example 7.16

For example, consider a transformer with a nameplate rating of 5 k*VA*, 60 Hz, and 880:220 V. The primary side might come from an 880-volt source, and the secondary would be at 220 volts.

Solution

The voltage/current rating is always given in rms values, and the turns ratio, of course, is 4:1. The current (or load) on the secondary side could be as high as $I_2 = VA/V_2 = 5000/220 = 22.7$ amperes, whereas the primary side would be $I_1 = 5000/880 = 5.68$ amperes, or 1/4 of 22.7 amperes. Note (from Exhibit 21) that a resistive load on the secondary side is $R_L = V/I = 220/22.7 = 9.69$ ohms, whereas it would be 880/5.68 = 155 ohms on the primary side. This equivalent resistance on the left side is a^2 times R_L:

$$R'_L = a^2 R_L = (4)^2 \times 9.69 = 155\ \Omega$$

Here, $kVA_{in} = kVA_{out}$ and $kW_{in} = kW_{out}$ because the load is a pure resistive one (the power factor of the load is unity). If the load were $5 + j5$ (or $Z_L = \sqrt{2} \times 5\angle 45°$), the current would be

$$I_2 = V_2/Z_L = 220/\left(\sqrt{2} \times 5\angle 45°\right) = 31.1\angle -45°\ \text{A}$$

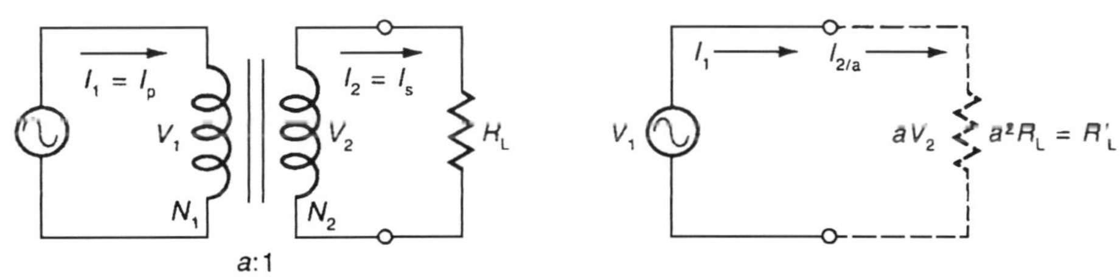

(a) Circuit with load (b) Circuit with equivalent load

Exhibit 21 A typical two-winding, loaded transformer

The current on the left side is $I_1 = I_2/a = 7.78$ amperes. These exceed the nameplate values, and a larger transformer would have to be selected. The equivalent impedance, Z'_L, of the load if reflected to the left side is

$$Z'_L = a^2 Z_L = (4)^2 5 + j(4)^2 5 = 80 + j80\ \Omega$$

Caution is needed here because the power-in/power-out relationship is misleading, $P = VI\cos\theta = 220\,(31.1)\,(0.707) = 4837$ W, or less than 5 kW; the transformer rating is 40 kVA, not necessarily 40 kW.

This chapter has reviewed selected electrical fundamentals. The depth of coverage has been limited. The reader who has more study time should select a text written about electrical engineering for all engineers. Of the many books available, several stand out. Any edition of the following should be available at a library or bookstore:

- Carlson and Gisser, *Electrical Engineering Concepts and Applications*, Addison-Wesley.
- Clement and Johnson, *Electrical Engineering Science*, McGraw-Hill.
- Smith, *Circuits, Devices and Systems*, Wiley.

SELECTED SYMBOLS AND ABBREVIATIONS

Symbol or Abbreviation	Definition
A	area
A	amperes
A • t	amp-turns
ac	alternating current
B	magnetic flux density; susceptance
C	coulomb
C	capacitance
dc	direct current
E	electric field
e	induced voltage
ε	permittivity
F	farad
F	force
G	conductance
H	henry
H	magnetic field intensity
I_c	core loss
I_e	exciting current
I_f	field current
I_i	current
I_m	magnetizing current
K	dielectric constant
KCL	Kirchhoff's Current Law
KVL	Kirchhoff's Voltage Law
L	inductance
l	length
ma	milliamp
mv	millivolts
mmf	magneto-motive force
N	number of turns
N_{ag}	newton air gap
n	speed
n_s	synchronous speed
P_i, p	power

(Continued)

(Continued)

Symbol or Abbreviation	Definition
ϕ	magnetic flux
Q	point change, reactive power
R	resistance
R_{eq}	Thevenin's resistance
rms	root mean square
S	siemens
s	slip
T	tesla
T	period
T_d	developed torque
θ	angle
V, v	voltage
v	velocity
V_i, V_{oc}	terminal voltage, Thevenin's voltage
W	work
w	total energy
Y	admittance
Z, z	impedance
N_P	number of pole pairs

PROBLEMS

7.1 For a parallel plate capacitor separated by an air gap of 1 cm and with an applied dc voltage across the plates of 500 volts, determine the force on an electron mass of 18.2×10^{-31} kg inserted in the space. The mass of an electron is 9.1×10^{-31} kg.

a. 3.2×10^{-14} N
b. 1.6×10^{-14} N
c. 9.1×10^{-31} N
d. 1.6×10^{-19} N

7.2 Assume a point charge of 0.3×10^{-3} C at an origin. What is the magnitude of the electric field intensity at a point located 2 meters in the x-direction, 3 meters in the y-direction, and 4 meters in the z-direction from the origin?

a. 500 kV/m
b. 5 kV/m
c. 93 kV/m
d. 9.3 MV/m

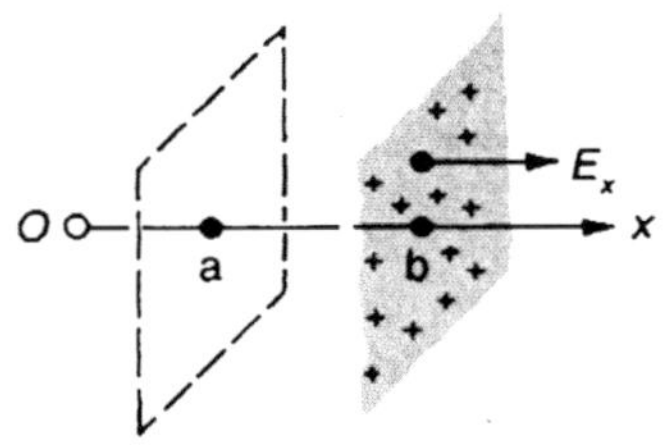

Exhibit 7.3 Charge placement

7.3 An infinite sheet of charge with a positive charge density, σ, has an electric field of:

$$\mathbf{E} = [\sigma/(2\varepsilon_0)]_{ax} \text{ for } x > b$$

If a second sheet of charge with a charge density of $-\sigma$ is then placed (see Exhibit 7.3), what is the electric field for $a < x < b$?

a. $(\sigma/\varepsilon_0)_{ax}$
b. 0
c. $(-\sigma/\varepsilon_0)_{ax}$
d. $(\sigma/2\varepsilon_0)_{ax}$

7.4 Two equal charges of 10 μC are located one meter apart on a horizontal line, and another charge of 5 μC is placed one meter below the first charge (forming a right triangle). What is the magnitude of the force on the 5 μC charge?

a. 0.09×10^{6} N
b. 12.6×10^{4} N
c. 6.39×10^{4} N
d. 63×10^{-2} N

7.5 For a coil of 100 turns wound around a toroidal core of iron with a relative permeability of 1000, find the current needed to produce a magnetic flux density of 0.5 tesla in the core. The dimensions of the core are given in Exhibit 7.5.

a. 390 A
b. 39 A
c. 1.2 A
d. 12.2 A

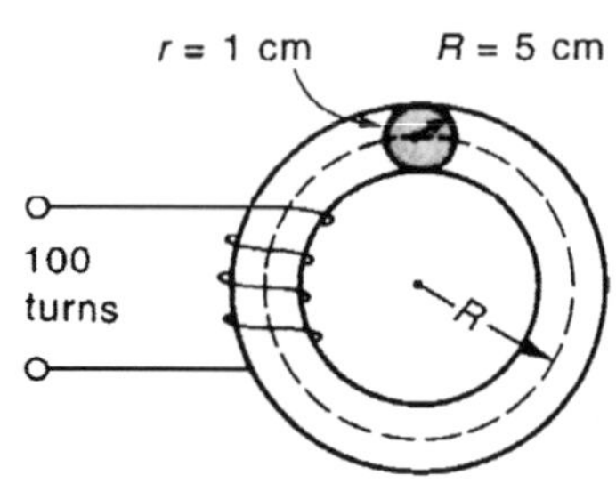

Exhibit 7.5

7.6 Two long straight wires, bundled together, have a magnetic flux density around them. One wire carries a current of 5 amperes, and the other carries a current of 1 ampere in the opposite direction. Determine the magnitude of the flux density at a point 0.2 meters away (i.e., normal to the wires).

a. $2\pi \times 10^{-6}$ T   c. 4×10^{-6} T
b. $4\pi \times 10^{-6}$ T   d. $16\pi \times 10^{-6}$ T

7.7 The root mean square of $i(t)$ for Exhibit 7.7 is:

a. 1.6 A
b. 3.0 A
c. 2.0 A
d. 0.0 A

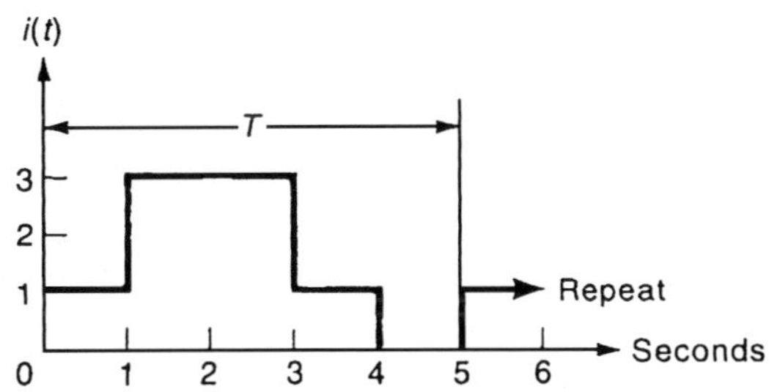

Exhibit 7.7 A current wave

7.8 A current phases through a 0.2-henry inductor. The current increases in a linear fashion from a value of zero at $t = 0$ to 20 A at $t = 20$ seconds. What is the amount of energy stored in the inductor at 10 seconds?

a. 0 J   c. 40 J
b. 10 J   d. 1 J

7.9 A sine wave of 10 volts (rms) is applied to a 10-ohm resistor through a half-wave rectifier. What is the average value of the current (i.e., what would a dc meter read)?

a. 0.32 A   c. 1.0 A
b. 0.45 A   d. 0.9 A

7.10 Two resistors of 2 ohms each are connected in parallel, another resistor of 1 ohm is connected in series with the parallel combination, and the resistive combination is connected to a 2-volt source. How much power is dissipated in either one of the parallel resistors?

a. 0.25 W   c. 1.0 W
b. 0.5 W   d. 1.5 W

7.11 A current source of 2 amperes is connected to four resistors, all in parallel. The resistors have values of 1, 2, 3, and 4 ohms, respectively. How much power is dissipated in the 2-ohm resistor?

a. 8.0 W   c. 2.0 W
b. 4.0 W   d. 0.46 W

7.12 Three resistors of 2 ohms each are connected in a "T" arrangement, and each side of the T is connected to its own battery voltage source. The battery on the left side is 1 volt (+ on top) and the battery on the right side is 2 volts (– on top). What is the power dissipated in the resistor in the middle leg?

a. 0 W
b. 4.5 W
c. 0.056 W
d. 0.89 W

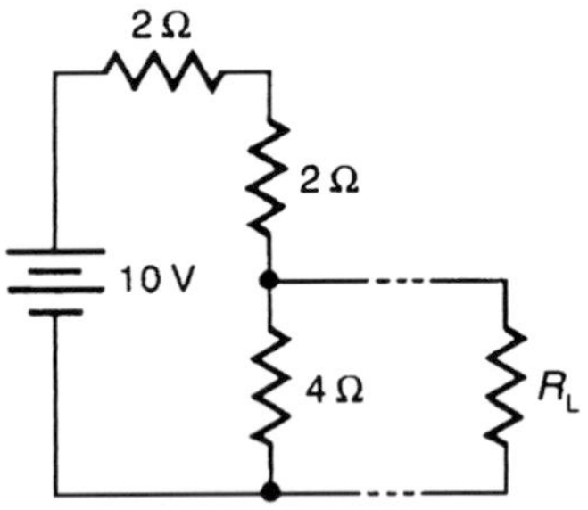

Exhibit 7.13 Original circuit

7.13 For the circuit in Exhibit 7.13, the load resistor, R_L, might "see" a Thevenin equivalent circuit in its place. What are the values of the equivalent circuit?

a. $V_{oc} = 5$ V, $R_{eq} = 2\ \Omega$
b. $V_{oc} = 5$ V, $R_{eq} = 1\ \Omega$
c. $V_{oc} = 4$ V, $R_{eq} = 2\ \Omega$
d. $V_{oc} = 1$ V, $R_{eq} = 2\ \Omega$

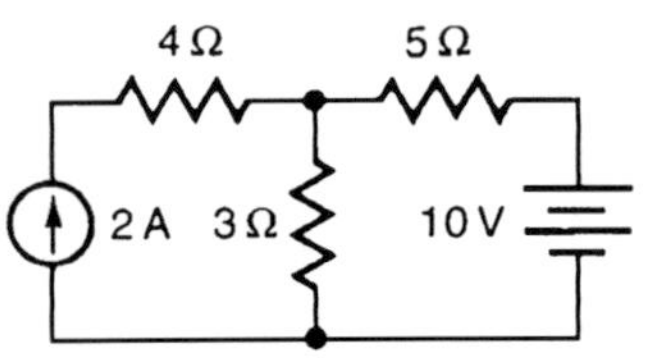

Exhibit 7.14 Circuit diagram

7.14 For the circuit shown in Exhibit 7.14, determine the magnitude of the voltage across the current source.

a. 16 V
b. 30 V
c. 20 V
d. 23 V

7.15 A charged 100-pF capacitor has initial voltage across it of 10 volts; a 100 kΩ resistor is suddenly connected across the capacitor (at $t = 0$) to discharge it. What is the magnitude of the instantaneous current at 20 microseconds?

a. 13.5 μA
b. 36.8 μA
c. 13.5 pA
d. 36.8 pA

7.16 Three 2-ohm resistors are arranged in "T" configuration. Connected to the left side is a 10-volt battery and switch; to the right is a 1-farad capacitor. The switch closes at $t = 0$. Find the time for the capacitor to reach 63% of its final voltage.

a. 0.2 s
b. 2.0 s
c. 0.3 s
d. 3.0 s

7.17 Three 2-ohm resistors are arranged in a "T" configuration. Connected to the left side is a 10-volt battery; to the right side is connected a switch and a 1-henry inductor. Assume the switch closes at $t = 0$. Find the amount of time for the inductor to reach 63% of its final current.

a. 0.2 s
b. 2.0 s
c. 0.3 s
d. 3.0 s

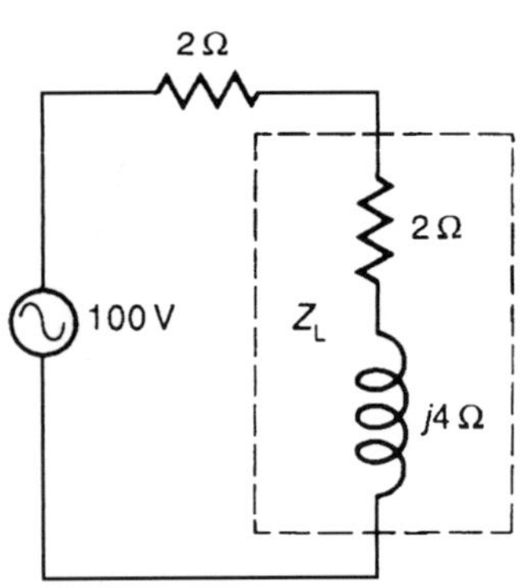

Exhibit 7.18 An ac circuit

7.18 For the circuit in the Exhibit 7.18, determine the power dissipated in the impedance, Z_L.

a. 1250 W
b. 625 W
c. 312 W
d. 1.7 kW

7.19 A resistance of 10 Ω, an inductor of 10/377 henries, and a capacitor of 20/377 farads are all connected in series to a 60-hertz, 100-volt source. Determine the magnitude of the current.

a. 3.3 A
b. 0.38 A
c. 10 A
d. $10\sqrt{2}$ A

7.20 If in Problem 7.19, all parameters were the same except that the frequency of the voltage source were doubled to 120 Hz, what would be the magnitude of the current?

a. 3.3 A
b. 0.38 A
c. 10 A
d. 5.5 A

7.21 A 10-ohm resistor and a capacitor with a capacitive reactance of $-j10\ \Omega$ are connected in parallel. The parallel combination is connected in series through an inductive reactance of $+j10\ \Omega$ to a 100-volt ac source. Determine the magnitude of the current from the source.

a. 6.7 A
b. 7.1 A
c. 10 A
d. $10\sqrt{2}$ A

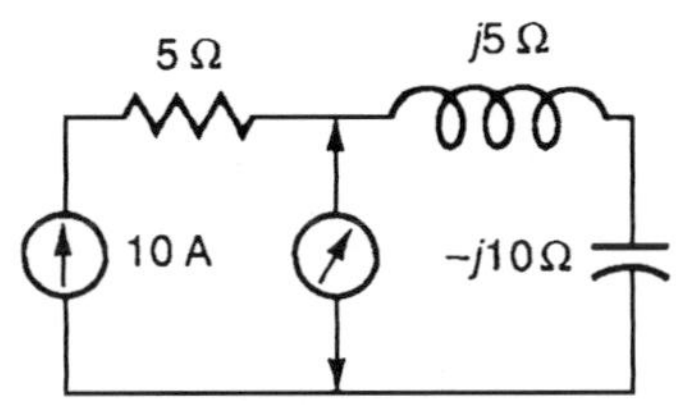

Exhibit 7.22 An ac circuit

7.22 An ac current source of 10 amperes is connected to a series circuit as shown in Exhibit 7.22. What would a voltmeter measure if the meter were connected across the inductor and capacitor combination?

a. 200 V
b. 150 V
c. 70 V
d. 50 V

7.23 For an *RLC* series circuit with $R = 10\ \Omega$, $L = 0.1$ henry, and $C = 0.1$ farad, all connected to an ac voltage source of 100 volts whose frequency could be varied, determine the frequency if maximum power is to be dissipated in *R*.

a. 10 Hz
b. 0 Hz
c. 1.6 Hz
d. 16 Hz

7.24 Find the transfer function $T(s) = \dfrac{V_2(s)}{V_1(s)}$ of the circuit shown in Exhibit 7.24.

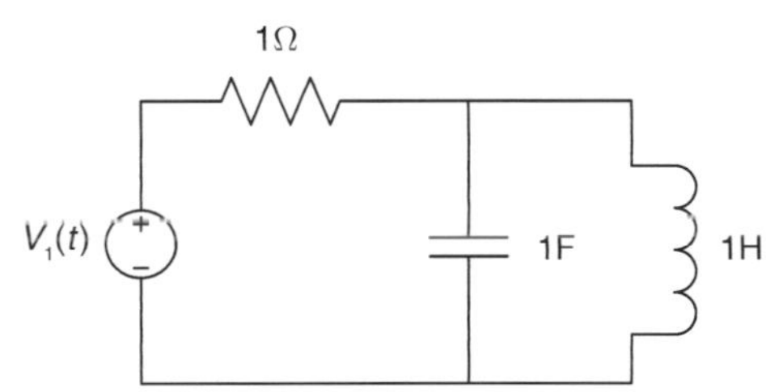

Exhibit 7.24

a. $\dfrac{s}{s^2 + 1}$

b. $\dfrac{s}{s^2 + s + 1}$

c. $\dfrac{1}{s^2 + s + 1}$

d. $\dfrac{1}{s^2 + 1}$

7.25 Find the transfer function $T(s) = \dfrac{V_0(s)}{V_{in}(s)}$ of the circuit shown in Exhibit 7.25. Assume that the op-amp is ideal.

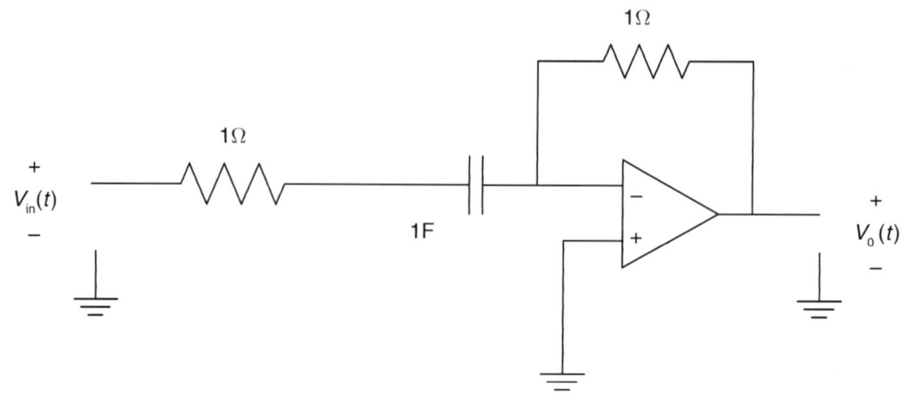

Exhibit 7.25

a. $-\dfrac{s}{s^2+1}$

b. $-\dfrac{1}{s^2+s+1}$

c. $-\dfrac{s}{s+1}$

d. $-\dfrac{1}{s+1}$

7.26 A low frequency ac voltage of 5 volts (rms) is applied to the input of the op-amp circuit in Exhibit 7.26. Determine the magnitude of the output voltage (rms).

a. 5 V
b. 2.5 V
c. 10 V
d. 20 V

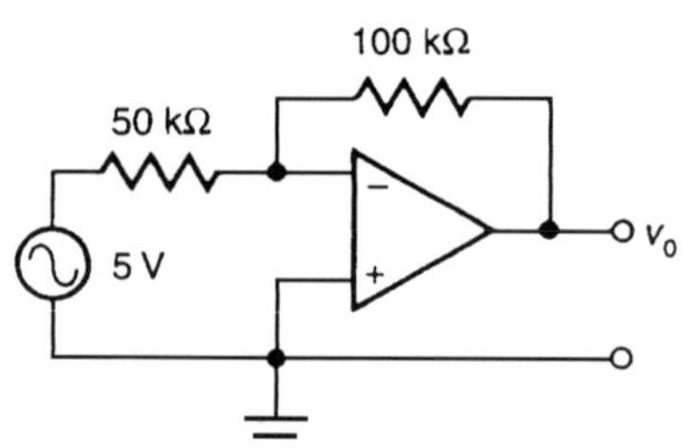

Exhibit 7.26 An op-amp circuit

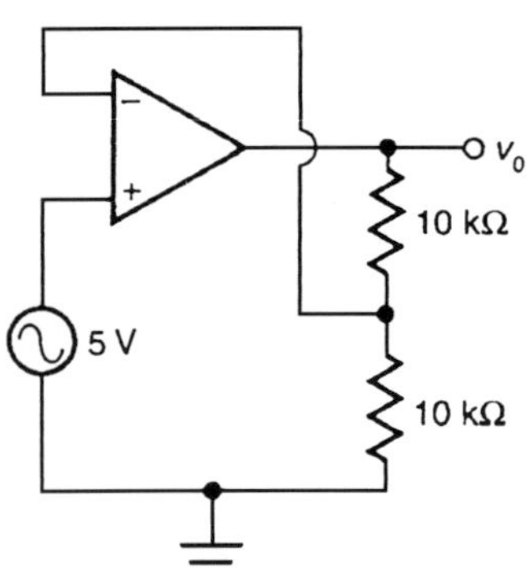

Exhibit 7.27 An op-amp circuit

7.27 A low frequency ac voltage of 5 volts (rms) is applied to the input of the op-amp circuit in Exhibit 7.27. Determine the magnitude of the output voltage (rms).

a. 5 V
b. 2.5 V
c. 10 V
d. 20 V

7.28 The inputs for the op-amp circuits are given in Exhibit 7.28. What is the output voltage from the last op-amp?

a. +1 V c. +3 V
b. –1 V d. –3 V

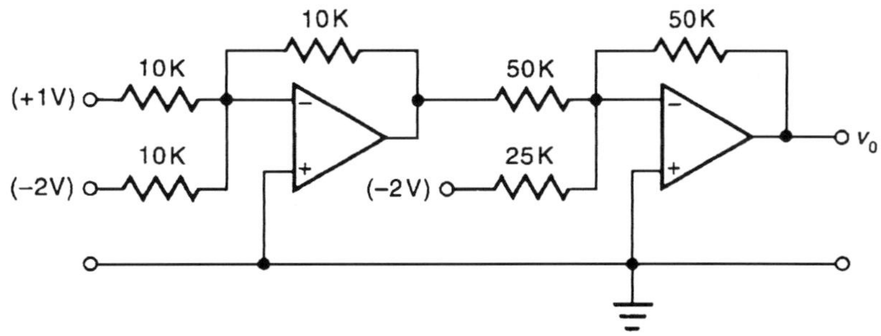

Exhibit 7.28 Summing circuits

7.29 A square wave of 1 V (or 2 V peak-to-peak) is applied to a bridge rectifier circuit where a 1-ohm resistor is connected to the output. What is the average current through the resistor?

a. 0.5 A c. πA
b. π/2 A d. 1.0 A

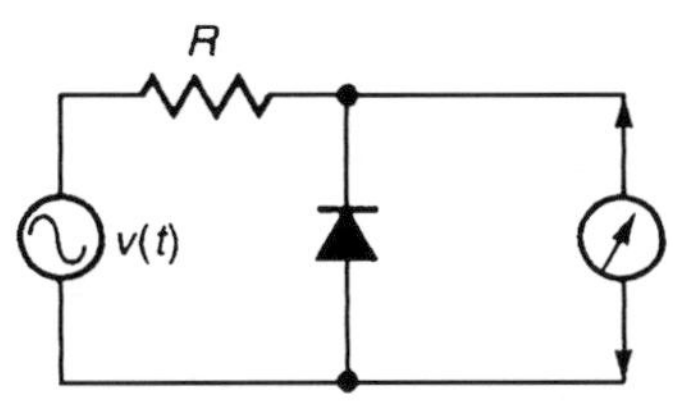

Exhibit 7.30 Clipping circuit

7.30 A sine wave voltage $v(t) = 14 \sin 2\pi ft$ is applied to the clipping circuit of Exhibit 7.30. A dc voltmeter (average reading instrument) is connected to the output. What does the meter read?

a. 2 V c. 7.5 V
b. 4.5 V d. 9 V

7.31 For an ideal, two-winding transformer whose nameplate reads 5 kVA, 400:200 V, and 60 Hz, determine the magnitude of a load that could be connected to the low voltage side for rated conditions. Assume the load impedance is $\mathbf{Z}_L = Z\angle 45°$, where Z is:

a. 4 Ω c. 16 Ω
b. 8 Ω d. 25 Ω

SOLUTIONS

7.1 b. The "mass" of 2 electrons has a charge $Q = 3.2 \times 10^{-19}$ C, thus the electric field is $\mathbf{E} = 500$ V/0.01 m $= 50 \times 10^3$ V/m. The force is then $\mathbf{F} = Q\mathbf{E} = (3.2 \times 10^{-19}) \times (50 \times 10^3) = 1.6 \times 10^{-14}$ N.

7.2 c. The magnitude of the length of the resultant vector, R, in the x, y, z plane is

$$R = \sqrt{2^2 + 3^2 + 4^2} = \sqrt{29}$$

The magnitude of the electric field, $\mathbf{E}$, is $Q/(4\pi\varepsilon R^2) = (0.3\times10^{-3})/(4\pi \times 8.85 \times 10^{-12} \times 29) = 93{,}000$ V/m.

7.3 c. On the b plane (for that plane alone), $E^+ = (-\sigma/2e)_{ax}$, and for the negatively charged plane at the a plane (again for that plane alone, but acting to the right of a), is the same as before, therefore:

$$\mathbf{E} = \mathbf{E}^+ + \mathbf{E}^- = (-\sigma/\varepsilon_0)_{ax} \text{ V/m}$$

7.4 d. From a sketch of the vectors in Exhibit 7.4, the value of $\mathbf{E}_{net}$ is 12.6×10^4 V/m. $\mathbf{F}$ is then found to be

$$\mathbf{E} = \frac{10\times10^{-6}}{4\pi(8.85\times10^{-12})\,r^2}$$

$$\mathbf{F} = Q_C\mathbf{E}_{net} = 5 \times 10^{-6}\,(12.6 \times 10^4) = 63 \times 10^{-2} \text{ N}$$

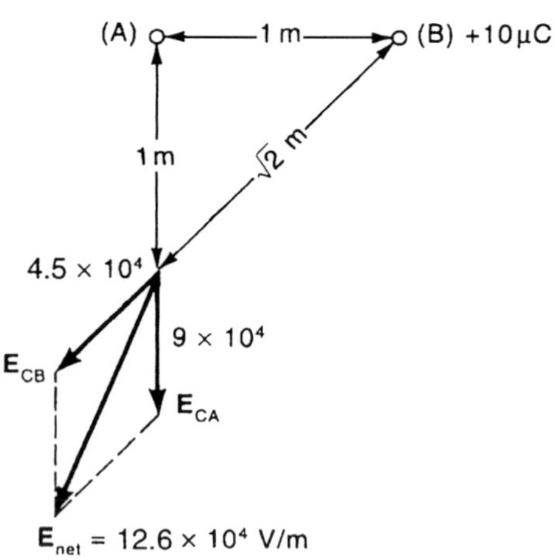

Exhibit 7.4 Vector field

7.5 c. The cross-sectional area of the iron core is $\pi r^2 = \pi \times 10^{-4}$ m². The path length is $l = 2\pi R = 2\pi \bullet 5 \times 10^{-2} = 0.1\pi$ m. The permeability is $\mu = \mu_0\mu_r = 4\pi \bullet 10^{-7} \times 10^3 = 4\pi \bullet 10^{-4}$.

Hence, $H = B/\mu = 0.5/(4\pi \bullet 10^{-4}) = 390$ A • t/m, and since H = mmf/length, mmf = $H \times$ length = 390 (0.1π) = 122.5 A • t. Thus, I = mmf/turns = 122.5/100 = 1.225 A.

7.6 c. Assume the two wires are bundled close together (with respect to the 0.2-meter position); then the net current to the right is $I_{net} = 5 - 1 = 4$ A. The flux density is given by $B = (\mu I)/(2\pi r) = (4\pi \times 10^{-7})(4)/(2\pi\ 0.2) = 4 \times 10^{-6}$ tesla.

7.7 c.

$$I_{rms} = \sqrt{(1/T)\int_0^T i^2\,dt} = \sqrt{(1/5)[(1)^2(1) + (3)^2(2) + (1)^2(1) + 0]} = 2\text{ A}$$

7.8 b. The current at 10 seconds is 10 A. And from the equation for energy storage in an inductor, it is found that $W = (1/2)\,Li^2 = (1/2)(0.2)(10)^2 = 10$ J.

7.9 b. From the figure and knowing the effective value for a rectified sine wave, one may compute V_{avg} and convert to I_{avg} from Ohm's law:

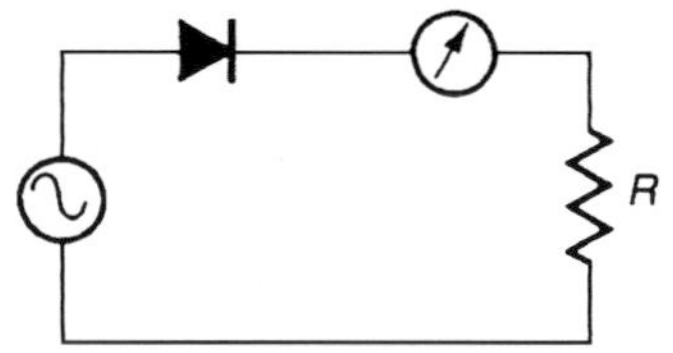

Exhibit 7.9

$$V_{max} = \sqrt{2}V_{rms} = \sqrt{2}(10\text{ V}),\quad V_{avg} = V_{max}/\pi = \sqrt{2}(10/\pi) = 4.49$$

$$I_{avg} = V/R = 4.5/10 = 0.45\text{ A}$$

7.10 b. In the Exhibit 7.10, the resistance of the parallel combination is 1 Ω; then the series combination is 2 Ω. The current I_1 is $V/R = 2/2 = 1$ A, and the current for the parallel branch is

$$I_2 = \left[G_2/\left(\sum G\right)\right]I_1 = (1/2)/[(1/2)(1/2)]1$$

$$= 0.5\text{ A},\ P_2 = I_2^2 R = (0.5)^2\ 2 = 0.5\text{ watts}$$

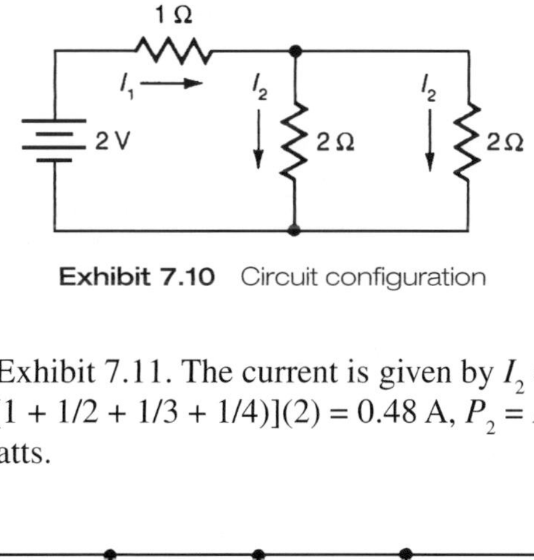

Exhibit 7.10 Circuit configuration

7.11 d. Refer to Exhibit 7.11. The current is given by $I_2 = [G_2/\Sigma G)]I_{total} = [(1/2)/(1 + 1/2 + 1/3 + 1/4)](2) = 0.48$ A, $P_2 = I^2R = (0.48)^2(2) = 0.46$ watts.

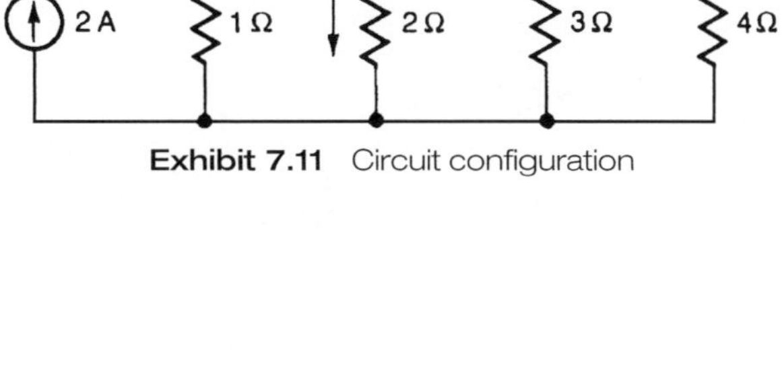

Exhibit 7.11 Circuit configuration

7.12 **c.** From the Exhibit 7.12, set up mesh currents, I_1 and I_2, then sum voltages around each loop:

Loop 1: $1 = 2I_1 + 2(I_1 - I_2)$;
Loop 2: $2 = 2(I_2 - I_1) + 2I_2$

By Cramer's Rule,

$$I_1 = \frac{\begin{vmatrix} 1 & -2 \\ 2 & 4 \end{vmatrix}}{\begin{vmatrix} 4 & -2 \\ -2 & 4 \end{vmatrix}} = 2/3\ A \qquad I_2 = \frac{\begin{vmatrix} 4 & 1 \\ -2 & 2 \end{vmatrix}}{\begin{vmatrix} 4 & -2 \\ -2 & 4 \end{vmatrix}} = 5/6\ A$$

$I_{mid\text{-}leg} = I_1 - I_2 = -1/6$ A; $\quad P_{mid\text{-}leg} = I^2R = (-1/6)^2 2 = 0.056$ W

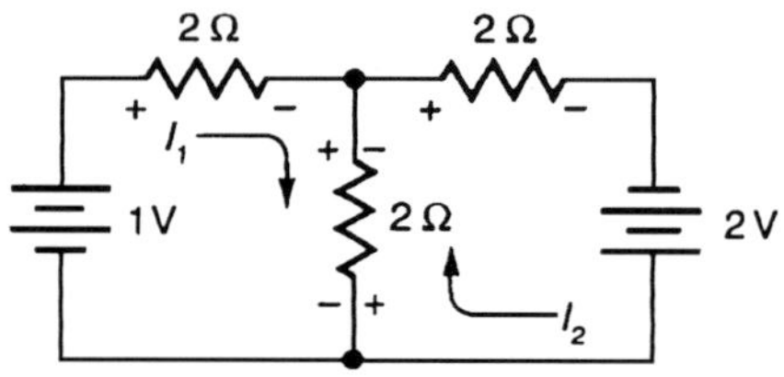

Exhibit 7.12 Circuit configuration

7.13 **a.** To find the Thevenin voltage, find the voltage across the 4-ohm resistor with R_L removed, which yields 5 volts. Thevenin resistance may be found by either of two methods. One method is to look back into the circuit with all active sources replaced with their own impedances and to calculate this output impedance; the other method is to short out the terminals of interest. For this second method, the impedance is

$$R_{eq} = V_{oc}/I_{sc} = 5/(10/4) = 2\ \Omega$$

7.14 **a.** First find the node voltage, V_1, at the intersection of the three resistors by summing the currents (start by assuming all currents flow away from the node). Assume that I_1 is to the left, I_3 to the right, and I_2 down.

$$\sum I = 0: (-2) + V_1/3 + (V_1 - 10)/5 = 0,\ 8V_1 = 60,\ V_1 = 7.5\text{ V};$$
$$V_A = V_1 - I_1R_4 = 7.5 - (-2)4 = 15.5\text{ V}$$

7.15 a. Refer to Exhibit 7.15. The time constant, τ, for an R–C circuit is RC. The voltage across the capacitor will discharge at a rate given by

$$v_C = V_C e^{-t/\tau} = 10\ e^{-20/10} = 1.35\ \text{V}, \quad i = v/R = 1.35/10^5 = 13.5 \times 10^{-6}\ \text{A}$$

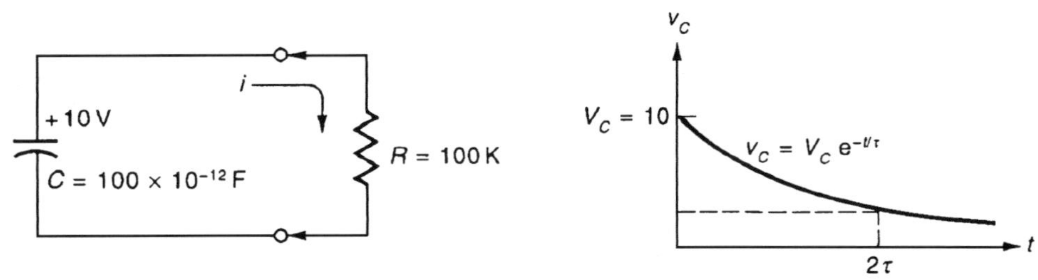

(a) R–C circuit (b) Voltage discharge curve

Exhibit 7.15 R–C circuit

7.16 d. First construct Thevenin's circuit so that all elements will be in series; this makes the R of the RC circuit especially easy to determine. Over one time constant, the voltage will build up to 63% of its final value.

$$V_{oc} = 5\ \text{V}, \quad R_{eq} = 5\ \Omega, \quad \tau = R_{ser}\ C = 3\ (1) = 3\ \text{s}$$

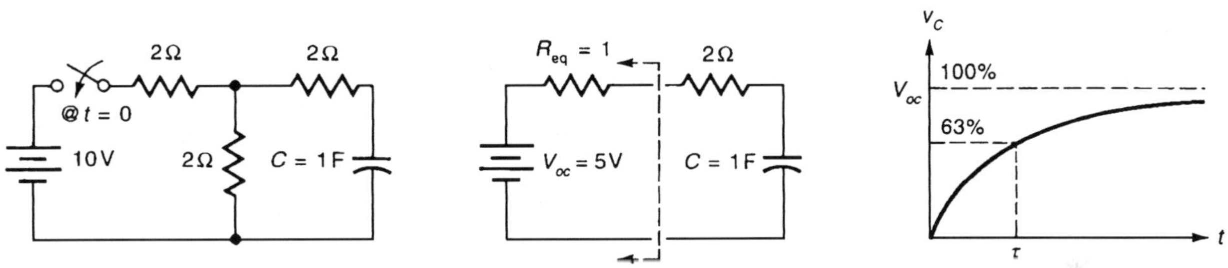

(a) Original circuit (b) Thevenin's circuit (c) Charging curve

Exhibit 7.16 Simplifying an R–C circuit

7.17 c. Make a Thevenin's equivalent circuit for finding the series resistance. The time constant for an LR circuit is $\tau = L/R = 1/3$ s. For a charging circuit, in one time constant, the current builds up to 63% of its final value.

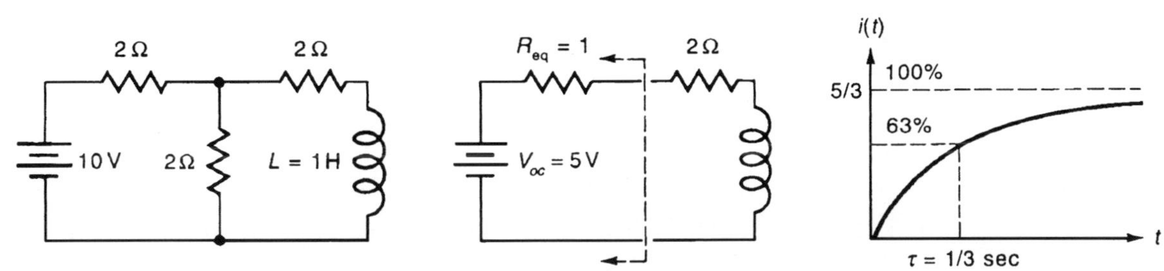

(a) Original circuit (b) Thevenin's circuit (c) Charging curve

Exhibit 7.17 Simplifying an L-R circuit

7.18 b. For a series circuit, $\mathbf{Z}_{\text{total}} = R_2 + \mathbf{Z}_L$, then

$$\mathbf{I} = \mathbf{V}/\mathbf{Z} = 100\angle 0°/(4 + j4) = 100\angle 0°/\left[\sqrt{2}(4\angle 45°)\right] = 17.7\ \angle -45°$$

$$P_2 = I^2R = (17.7)^2(2) = 625 \text{ watts}$$

7.19 c. First it is necessary to compute the reactances (recall that $2\pi f = 377$ for 60 Hz): $X_L = 2\pi fL = (377)(10/377) = 10\ \Omega$, and $X_C = 1/(2\pi fC)$ $= 10\ \Omega$. Thus

$$\mathbf{Z} = 10 + j10 - j10 = 10\angle 0°$$
$$\mathbf{I} = \mathbf{V}/\mathbf{Z} = 100\angle 0°/10\angle 0° = 10 \text{ A}$$

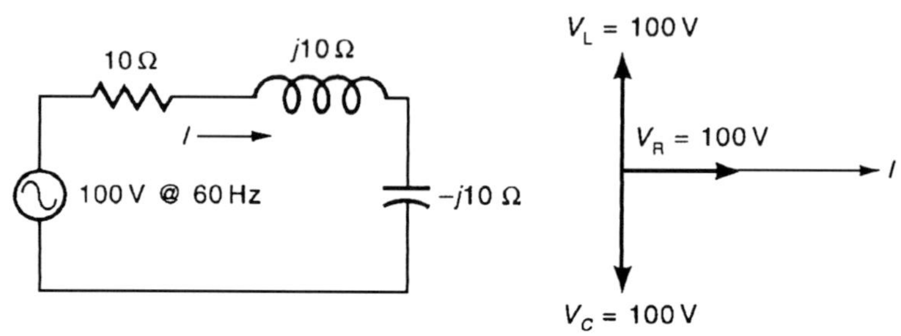

(a) Circuit diagram (b) Phasor diagram

Exhibit 7.19 An *R–L–C* circui

7.20 d. If the frequency is doubled, the reactances will change to $X_L = 2\ (10)$ $= 20\ \Omega$, $X_C = 0.5\ (10) = 5\ \Omega$; hence, $Z = 10 + j20 - j5 = 18.2\angle 56.3°$. $|I| = |V/Z|$ (answer only requires magnitude) $= 100/18.2 = 5.5$ A.

7.21 d. First find the series equivalent to the parallel portion of the circuit:

$$Z_P = (10)(-j10)/(10 - j10) = \left(10/\sqrt{2}\right)\angle -45° = 5\ j5\ \Omega;$$

$$Z_{\text{total}} = j10 + 5 - j5 = 5\sqrt{2}\ \angle 45°\ I = V/Z = 100/\left(5\sqrt{2}\right) = 10\sqrt{2}\ \text{A}$$

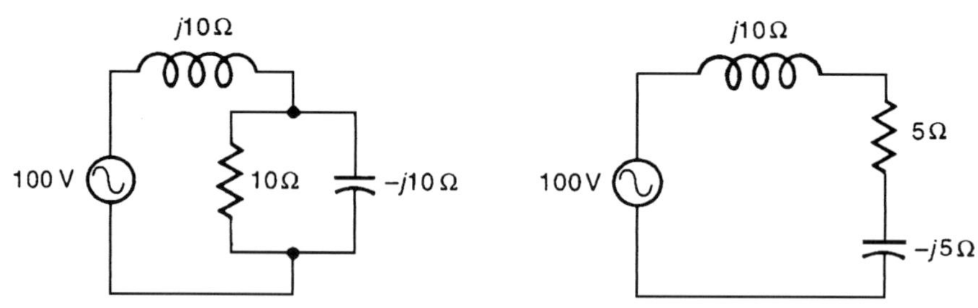

(a) Original circuit (b) Modified circuit

Exhibit 7.21 Reduced circuit

7.22 d. The impedance for the *L–C* portion of the circuit is $\mathbf{Z}_{L\text{-}C} = j5 - j10$ $= -j5 = 5\angle -90°$; $V_{L\text{-}C} = IZ_{L\text{-}C} = (10)(5) = 50$ V.

7.23 c. Maximum current occurs if X_L and X_C just cancel and Z is left with only resistance: $X_L = 2\pi fL$, $X_C = 1/(2\pi fC)$; $f_{res} = 1/(2\pi\sqrt{LC}) = 1.6$ Hz.

7.24 b. The frequency-domain circuit is shown in Exhibit 7.24a.

Exhibit 7.24a

Combining the impedances in parallel, $Z_{eq} = \dfrac{s\left(\dfrac{1}{s}\right)}{s+\dfrac{1}{s}} = \dfrac{s}{s^2+1}$

Using voltage division rule, $T(s) = \dfrac{Z_{eq}}{1+Z_{eq}} = \dfrac{s}{s^2+s+1}$

7.25 c. As the op-amp is ideal, $T(s) = \dfrac{V_0(s)}{V_{in}(s)} = -\dfrac{Z_2}{Z_1} = -\dfrac{1}{1+1/s} = -\dfrac{s}{s+1}$

7.26 c. The op-amp circuit produces a gain of $-(R_f/R_1)$ for an inverting amplifier; and here, only the magnitude is requested so the sign is not important.

$G = (100\text{k}/50\text{k}) = 2$; $V_0 = 2V_{in}$, $V_0 = 2\ (5) = 10$ V

7.27 c. This op-amp circuit produces a gain of $1 + (R_2/R_1)$ for a noninverting amplifier.

$G = 1 + (10\text{k}/10\text{k}) = 2$; $V_0 = 2\ (5) = 10$ V

7.28 c. The output from the first summing op-amp is $v_0 = -[(R_f/R_1)v_1 + (R_f/R_2)v_2] = -[(10\text{k}/10\text{k})1 + (10\text{k}/10\text{k})(-2)] = +1$ V. The output from the second summing op-amp is $v_0 = -[(R_f/R_3)v_3 + (R_f/R_4)v_4] = -[(50\text{k}/50\text{k})1 + (50\text{k}/25\text{k})(-2)] = +3$ V.

7.29 d. Refer to Exhibit 7.29. The bridge rectifier is a full-wave type, and, for a square wave, the result of taking the average is the maximum of the input itself. $I = V/R = 1/1 = 1$ A (average)

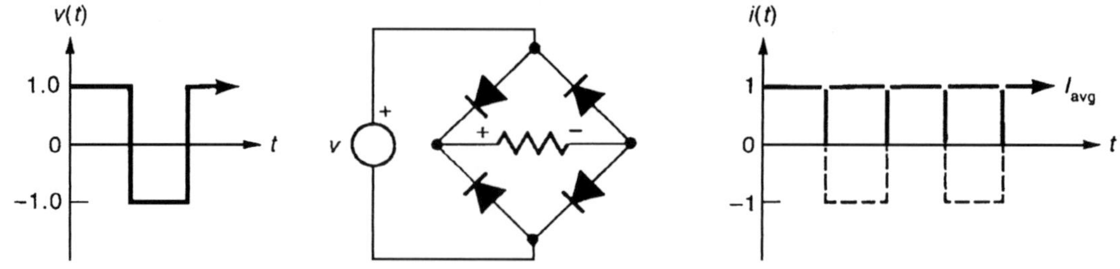

(a) Circuit (b) Full wave rectification

Exhibit 7.29 A full-wave bridge rectifier circuit

7.30 b. Since the output is unloaded (that is, the meter resistance is assumed to be >> R), the output is "shorted out" by the ideal diode for the negative half cycle; thus the output voltage appears as an average value. For a half-wave rectifier, the voltage is $V_{avg} = V_{max}/\pi = 14/\pi = 4.5$ V.

7.31 b. Assume the load side is V_2, then under full load conditions (rated current at rated voltage), the current is $I_2 = 5000$ VA/200 V = 25 A; $Z_L = 200$ V/25 A = 8 Ω regardless of phase.

CHAPTER 8

Statics

Charles E. Smith

OUTLINE

Statics is concerned with the forces of interaction between bodies or within bodies of mechanical systems that have no significant accelerations. Typical engineering problems require the analyst to predict forces induced at certain points by known forces applied at other points.

INTRODUCTORY CONCEPTS IN MECHANICS

Newton's Laws of Motion

Every element of a mechanical system must satisfy **Newton's second law of motion**, which states that the resultant force f acting on the element is related to the acceleration a of the element by

$$f = ma$$

where m represents the mass of the element. This entire chapter deals with the special case in which $a = 0$. **Newton's third law** requires that the force exerted on a body A by a body B is of equal magnitude and opposite direction to the force exerted on body B by body A. A careful, unambiguous account of this law is essential for successful analysis of forces; rules for ensuring that such an analysis is done properly will be reviewed in the Equilibrium section.

Newton's Law of Gravitation

Every pair of material elements is attracted toward one another by a pair of *gravitational* forces, the magnitude of which is given by

$$f_g = \frac{\gamma m_1 m_2}{r^2} \qquad \textbf{(8.1)}$$

where γ is the **universal gravitational constant** (about 6.7×10^{-11} N•m²/kg²), m_1 and m_2 are the masses of the elements, and r is the distance between them. Because very large masses are necessary to make these forces significant, f_g can often be neglected. A notable exception is the force exerted by the earth (which has a mass of about 6×10^{24} kg) on objects near its surface. In this case, the gravitational force has a magnitude given by

$$f_g = mg \qquad \textbf{(8.2)}$$

where m is the mass of the attracted object and g (which is related to the earth's mass and radius) has a value that varies between 9.78 and 9.83 N/kg with geographic location. In technically correct terminology, the word *weight* refers to the force of gravity by the earth. However, *weight* has other, closely related meanings that can be a source of serious confusion to the analyst of mechanical systems.

Dimensions and Units of Measurement

Every quantity in mechanics can be expressed in terms of three fundamental quantities. In the SI system of units, these are *mass*, *length*, and *time*, and units are the kilogram (kg), the meter (m), and the second (s), respectively. In the SI system, Newton's second law provides a definition of a fourth unit in terms of the three fundamental ones. The **newton** (N) is defined as the force required to accelerate a 1-kilogram body at the rate of one meter per second per second (m/s^2). This can be expressed symbolically as N = kg • m/s^2. See Table 8.1.

Considerable confusion results from the introduction and widespread use (especially outside the United States) of the **kilogram-force (kgf)**, defined to be 9.80665 N.

Errors stemming from force values given in this noncoherent unit can be avoided by converting to newtons before doing further calculation. For example,

Table 8.1 Units common to mechanics

Quantity	SI Unit	Coherent U.S. Engineering Unit	Common Noncoherent Unit
Mass	kilogram (kg)	slug = lbf • s^2/ft ≈ 14.594 kg	pound-mass (lbm) $\approx \frac{1}{32.174}$ slug ≈ 0.4536 kg
Length	meter (m)	foot (ft) = 0.3048 m	inch (in.) = 25.4 mm
Time	second (s)	second (s)	minute (min) = 60 s
Force	newton (N) N = kg • m/s^2	pound-force (lbf) ≈ 4.448 N	kilogram-force (kgf) or kilopound (kp) = 9.80665 N

an estimate of the acceleration imparted by a force of 217 kgf to a 100-kg body would proceed as follows:

$$f = (217\ \text{kgf})\ (9.80665\ \text{N/kgf}) = 2128\ \text{N}$$

$$a = \frac{f}{m} = \frac{2128\ \text{N}}{100\ \text{kg}} = 21.28 \left(\text{N/kg} = \frac{\text{kg} \bullet \text{m/s}^2}{\text{kg}} = \frac{\text{m}}{\text{s}^2} \right)$$

Errors in unit conversion, and many errors of analysis, will be revealed by the practice of appending unit symbols to *every* number and algebraically reducing combinations of symbols resulting from multiplication and division.

VECTOR GEOMETRY AND ALGEBRA

Handling many of the problems that arise in mechanics can be greatly simplified by means of the operations of vector analysis. Their use will be successful if attention is given to the geometric meaning of each operation.

A **vector** represents a physical quantity that can be characterized by a magnitude and a direction in space, which will be taken to be three-dimensional here. We use an arrow to depict each vector, with the length of the arrow proportional to the magnitude represented, and the orientation representing the direction of the vector. Boldface letters are used in text and equations to represent vectors. The **magnitude** of vector **a** is written as |**a**| and sometimes as a.

Addition and Subtraction

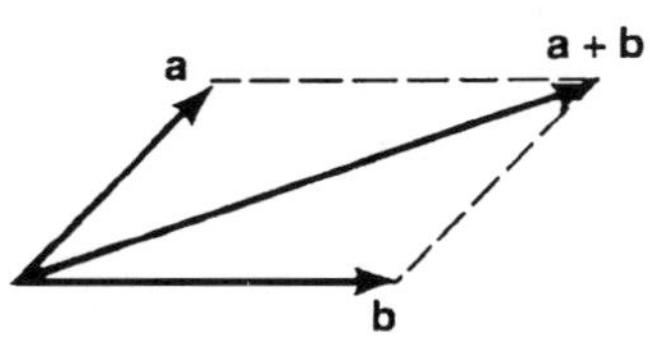

Figure 8.1

The **sum of two vectors** is a vector determined according to the so-called parallelogram law, as illustrated in Figure 8.1. The sum is the vector represented by the diagonal of the parallelogram formed by the two vectors placed with their tails coincident. The commutative law,

$$\mathbf{a} + \mathbf{b} = \mathbf{b} + \mathbf{a}$$

and the associative law,

$$\mathbf{a} + (\mathbf{b} + \mathbf{c}) = (\mathbf{a} + \mathbf{b}) + \mathbf{c}$$

both follow from this definition.

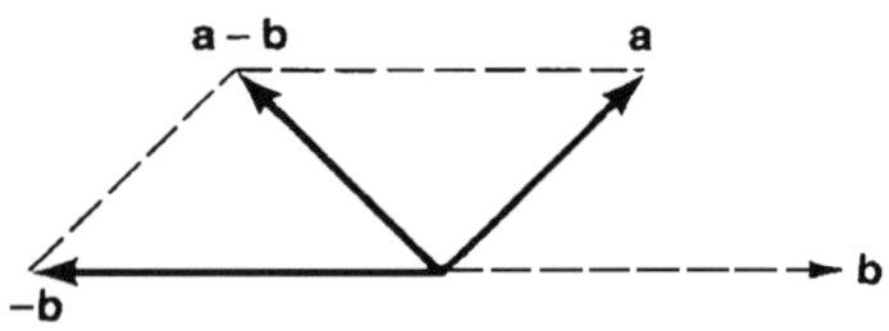

Figure 8.2

The negative of a vector **b** is defined to be of the same magnitude as, but of the opposite direction to, **b**, and it is written as –**b**. Subtraction of two vectors is then defined by

$$\mathbf{a} - \mathbf{b} = \mathbf{a} + (-\mathbf{b})$$

as indicated in Figure 8.2.

Multiplication by a Scalar

The product of a scalar, p, and a vector, **a**, is the vector written as $p\mathbf{a}$, and it is defined to have a magnitude of $p|\mathbf{a}|$, and direction the same as, or the opposite of, **a,** depending on whether p is positive or negative. The following laws can be readily verified from these definitions:

$$p(q\mathbf{a}) = (pq)\mathbf{a}$$
$$(p+q)\mathbf{a} = p\mathbf{a} + q\mathbf{a}$$
$$p(\mathbf{a}+\mathbf{b}) = p\mathbf{a} + p\mathbf{b}$$

Addition of two or more vectors is sometimes called the **composition** of the vectors. A reversal of this process is called the **resolution** of a vector, that is, determining a set of vectors (usually in prescribed directions) the sum of which will be the given vector.

Example 8.1

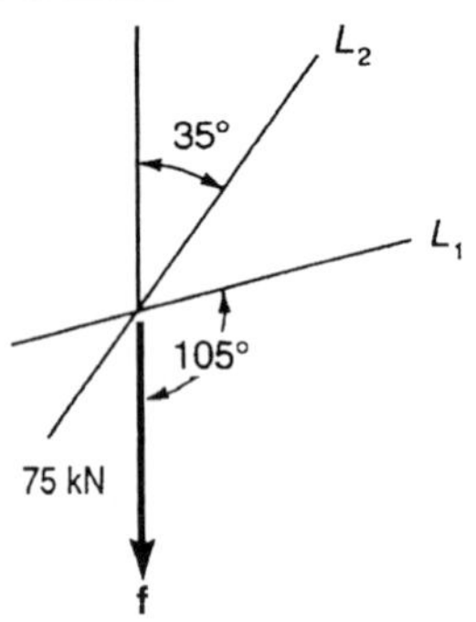

Exhibit 1

Resolve the vector **f** of magnitude 75 kN into two vectors (components) in the directions of lines L_1 and L_2 shown in Exhibit 1.

Solution

The given vector is to be the diagonal of the parallelogram having the desired components as sides. The parallelogram can be completed by drawing lines through the head of the vector, parallel to the given lines. The magnitudes of the two components can then be determined by applying the trigonometric law of sines:

$$|\mathbf{f}_1| = \frac{\sin 35^\circ}{\sin 40^\circ}(75\ \text{kN}) = 66.9\ \text{kN}$$

$$|\mathbf{f}_2| = \frac{\sin 105^\circ}{\sin 40^\circ}(75\ \text{kN}) = 112.7\ \text{kN}$$

Dot Product

The **dot product** of two vectors **a** and **b** is a scalar (sometimes called the scalar product or inner product) that is equal to the product of the magnitudes of the vectors and the cosine of the angle θ between the vectors. It is written as

$$\mathbf{a} \bullet \mathbf{b} = |\mathbf{a}||\mathbf{b}| \cos\theta \quad \textbf{(8.3)}$$

A special case is the dot product of a vector with itself,

$$\mathbf{a} \bullet \mathbf{a} = |\mathbf{a}||\mathbf{a}| \cos\theta$$

which provides a way of expressing the magnitude of a vector:

$$|\mathbf{a}| = \sqrt{\mathbf{a} \bullet \mathbf{a}} \quad \textbf{(8.4)}$$

The commutative law,

$$\mathbf{a} \bullet \mathbf{b} = \mathbf{b} \bullet \mathbf{a}$$

and the distributive law,

$$\mathbf{a} \bullet (\mathbf{b} + \mathbf{c}) = \mathbf{a} \bullet \mathbf{b} + \mathbf{a} \bullet \mathbf{c}$$

can both be verified from the above definitions.

Unit Vectors and Projections

An extremely useful tool is the **unit vector**. It has a magnitude of 1 and is designated here by the symbol **e**. Unit vectors can be introduced (defined) by giving the direction in terms of the geometry of the application, or defined in terms of a specified vector by multiplying the vector by the reciprocal of its magnitude. For example, the unit vector in the direction of **a** is given by

$$\mathbf{e}_a = \frac{1}{|\mathbf{a}|}\mathbf{a}$$

Rearranged, this relationship expresses the vector **a** in terms of its magnitude and a unit vector that gives its direction:

$$\mathbf{a} = |\mathbf{a}|\mathbf{e}_a \quad \textbf{(8.5)}$$

The **projection** of vector **a** onto a line L is the vector from the projection of the tail of **a** onto L to the projection of the head of **a** onto L, as indicated in Figure 8.3.

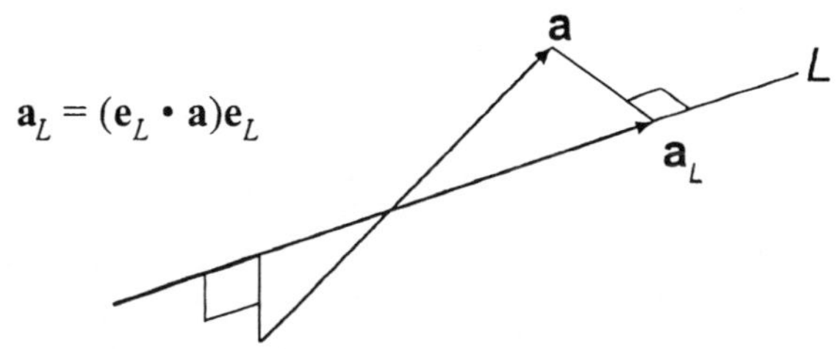

Figure 8.3

The magnitude of this projection will be the product of $|\mathbf{a}|$ and the cosine of the angle between $\mathbf{a}$ and L. With unit vector $\mathbf{e}_L$, defined to be parallel to L, the projection of $\mathbf{a}$ onto L can be expressed as

$$\mathbf{a}_L = (\mathbf{e}_L \bullet \mathbf{a})\mathbf{e}_L$$

That is, vector projection onto a line in a selected direction can be evaluated by dot-multiplying the vector by a unit vector in that direction. As indicated in the next section, this approach provides the most direct means of obtaining equivalent scalar relationships from a vector relationship.

Vector and Scalar Equations

Many physical laws, such as Newton's laws of motion, are best expressed by vector equations. In general, a vector equation can provide up to three independent scalar equations. In Example 8.1, the directions of the lines L_1 and L_2 might be determined by the orientations of two members of a structural truss, and the direction of $\mathbf{f}$ by a gravitational force. Then the relationship $\mathbf{f}_1 + \mathbf{f}_2 = \mathbf{f}$ might be a requirement of equilibrium and geometry, from which the magnitudes of $\mathbf{f}_1$ and $\mathbf{f}_2$ are to be determined in terms of the magnitude of $\mathbf{f}$. With the unit vectors $\mathbf{e}_1$ and $\mathbf{e}_2$ introduced as shown, the equation

$$f_1\mathbf{e}_1 + f_2\mathbf{e}_2 = \mathbf{f}$$

brings the unknowns f_1 and f_2 into evidence and makes available two scalar equations for their determination. A corresponding scalar equation can be obtained by dot-multiplying each member of the vector equation by a selected vector; for example, with the unit vectors already introduced, we can write the two equations

$$f_1\mathbf{e}_1 \bullet \mathbf{e}_1 + f_2\mathbf{e}_1 \bullet \mathbf{e}_2 = \mathbf{e}_1 \bullet \mathbf{f}$$
$$f_1\mathbf{e}_2 \bullet \mathbf{e}_1 + f_2\mathbf{e}_2 \bullet \mathbf{e}_2 = \mathbf{e}_2 \bullet \mathbf{f}$$

reduce them to

$$f_1 + (\cos 40°)f_2 = (75\text{ kN})\ \cos 105°$$
$$(\cos 40°)f_1 + f_2 = (75\text{ kN})\ \cos 145°$$

and solve these for

$$f_1 = 66.9\text{ kN} \qquad f_2 = -112.7\text{ kN}$$

The vector equation could also be dot-multiplied by the unit vectors $\mathbf{e}_a$ and $\mathbf{e}_b$ defined to be perpendicular to f_1 and f_2, as shown in Figure 8.4. This gives

$$f_1\mathbf{e}_a \bullet \mathbf{e}_1 + f_2\mathbf{e}_a \bullet \mathbf{e}_2 = \mathbf{e}_a \bullet \mathbf{f}$$
$$f_1\mathbf{e}_b \bullet \mathbf{e}_1 + f_2\mathbf{e}_b \bullet \mathbf{e}_2 = \mathbf{e}_b \bullet \mathbf{f}$$

which reduce to

$$(\cos 130°)f_2 = (75\text{ kN})\cos 15°$$
$$(\cos 50°)f_1 = (75\text{ kN})\cos 55°$$

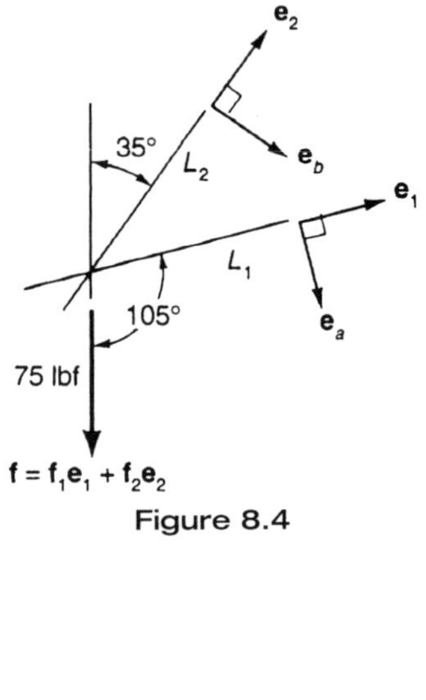

Figure 8.4

The unit vectors $\mathbf{e}_a$ and $\mathbf{e}_b$ are better choices for the purpose of evaluating f_1 and f_2 because each is perpendicular to one of the vectors with unknown magnitudes, so that we can eliminate an unknown from each equation.

The Cross Product

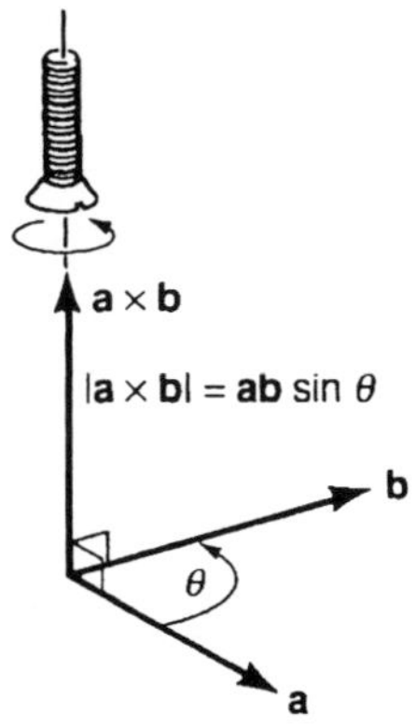

Figure 8.5

The **cross product** of two vectors **a** and **b** (sometimes called the vector product) is defined to be a vector that is perpendicular to the plane of **a** and **b**, with magnitude equal to the product of the magnitudes of **a** and **b** and the sine of the angle θ between **a** and **b** and with direction determined by the **right-hand rule**. The right-hand rule states that the vector's direction coincides with that of the advancement of a right-hand screw, with the axis of the screw being oriented perpendicular to **a** and **b**, and turned in the direction **a**-toward-**b**, as shown in Figure 8.5. Note that the cross product is *not* commutative; instead

$$\mathbf{b} \times \mathbf{a} = -\mathbf{a} \times \mathbf{b}$$

However, the definitions of cross-multiplication and addition can be used to show that the distributive law

$$\mathbf{a} \times (\mathbf{b} + \mathbf{c}) = \mathbf{a} \times \mathbf{b} + \mathbf{a} \times \mathbf{c}$$

is valid, and that the cross product is associative with respect to multiplication by scalars:

$$(p\mathbf{a}) \times (q\mathbf{b}) = (pq)(\mathbf{a} \times \mathbf{b})$$

Example 8.2

It is desired to resolve a given vector **a** into two components, one parallel to a second given vector **b** and one perpendicular to **b**. The results are to be expressed in terms of **a** and **b**, using vector operations defined in the preceding pages.

Solution

The component parallel to **b** will be the projection of **a** onto the line parallel to **b**, which is expressible in terms of a unit vector in the direction of **b**:

$$\mathbf{a}_{\parallel} = (\mathbf{e}_b \bullet \mathbf{a})\mathbf{e}_b = \frac{(\mathbf{b} \bullet \mathbf{a})\mathbf{b}}{\mathbf{b} \bullet \mathbf{b}}$$

The component perpendicular to **b** can be determined from the fact that **a** is to be the sum of this and the component just calculated:

$$\mathbf{a}_{\perp} = \mathbf{a} - \mathbf{a}_{\parallel}$$

Alternatively, the perpendicular component can be calculated by

$$\mathbf{a}_{\perp} = \frac{(\mathbf{b} \times \mathbf{a}) \times \mathbf{b}}{\mathbf{b} \bullet \mathbf{b}}$$

Verifying this last relationship from the definition of the cross product provides useful practice in relating the operation to geometry. The vector **a** can now be expressed in terms of the components parallel and perpendicular to **b**:

$$\mathbf{a} = \frac{(\mathbf{b} \bullet \mathbf{a})\mathbf{b}}{\mathbf{b} \bullet \mathbf{b}} + \frac{(\mathbf{b} \times \mathbf{a}) \times \mathbf{b}}{\mathbf{b} \bullet \mathbf{b}}$$

Rectangular Cartesian Components

A special way of resolving vectors consists of forming three mutually perpendicular components. The directions are chosen (usually with consideration of the geometry of the problem at hand) and three mutually perpendicular unit vectors $\mathbf{e}_x$, $\mathbf{e}_y$, and $\mathbf{e}_z$ are defined to be parallel to these directions. The rectangular Cartesian components of a vector **a** are then the projections of **a** onto lines in the selected directions, expressed as

$$\mathbf{a} = a_x\mathbf{e}_x + a_y\mathbf{e}_y + a_z\mathbf{e}_z \qquad \textbf{(8.6)}$$

in which $a_x = \mathbf{e}_x \bullet \mathbf{a}$, $a_y = \mathbf{e}_y \bullet \mathbf{a}$, and $a_z = \mathbf{e}_z \bullet \mathbf{a}$. These relationships, together with the associative and distributive laws mentioned previously, lead to the following formulas:

$$\mathbf{a} \bullet \mathbf{b} = a_x b_x + a_y b_y + a_z b_z \qquad \textbf{(8.7)}$$

$$|\mathbf{a}| = \sqrt{a_x^2 + a_y^2 + a_z^2} \qquad \textbf{(8.8)}$$

$$\begin{aligned} \mathbf{a} \times \mathbf{b} = {} & (a_y b_z - a_z b_y)\mathbf{e}_x \\ & + (a_z b_x - a_x b_z)\mathbf{e}_y \\ & + (a_x b_y - a_y b_x)\mathbf{e}_z \end{aligned} \qquad \textbf{(8.9)}$$

Each of these equations depends on the fact that the unit vectors are mutually perpendicular; in addition, the expression for the cross product is valid only if the unit vectors form a *right-handed* set; that is,

$$\mathbf{e}_x = \mathbf{e}_y \times \mathbf{e}_z, \quad \mathbf{e}_y = \mathbf{e}_z \times \mathbf{e}_x, \quad \text{and} \quad \mathbf{e}_z = \mathbf{e}_x \times \mathbf{e}_y$$

Unit vectors defining the three dimensions in Cartesian coordinates are also often represented as **i**, **j**, and **k**.

Example 8.3

Determine the lengths of the guylines $O'P$ and $O'Q$, shown in Exhibit 2, and the angle between them.

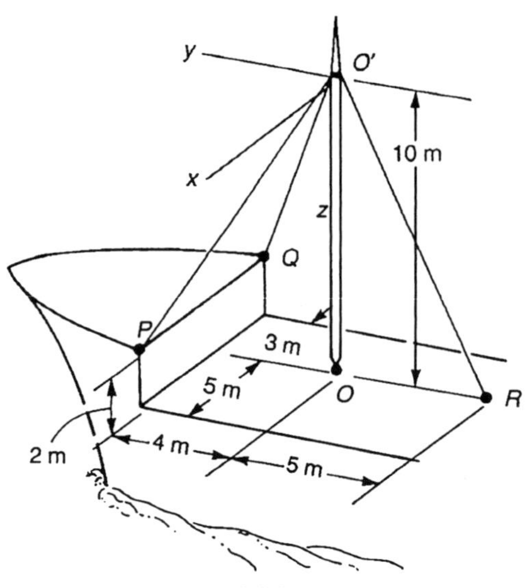

Exhibit 2

Solution

Let position vectors **a** and **b** extend from point O' to points P and Q, respectively, and resolve these into components along the x, y, and z axes shown:

$$\mathbf{a} = (5\text{ m})\mathbf{e}_x + (4\text{ m})\mathbf{e}_y + (8\text{ m})\mathbf{e}_z$$
$$\mathbf{b} = (-3\text{ m})\mathbf{e}_x + (4\text{ m})\mathbf{e}_y + (8\text{ m})\mathbf{e}_z$$

The required lengths can then be determined from Equation (8.8):

$$|\mathbf{a}| = \sqrt{(5\text{ m})^2 + (4\text{ m})^2 + (8\text{ m})^2} = 10.25\text{ m}$$

$$|\mathbf{b}| = \sqrt{(-3\text{ m})^2 + (4\text{ m})^2 + (8\text{ m})^2} = 9.43\text{ m}$$

The required angle is a factor in the definition of $\mathbf{a} \bullet \mathbf{b}$, and in fact it will be the only unknown in this equation after $\mathbf{a} \bullet \mathbf{b}$ is evaluated. From Equation (8.7), we have

$$\mathbf{a} \bullet \mathbf{b} = (5\text{ m})(-3\text{ m}) + (4\text{ m})(4\text{ m}) + (8\text{ m})(8\text{ m}) = 65\text{ m}^2$$

Then Equation (8.3) gives

$$\cos\theta = \frac{\mathbf{a} \bullet \mathbf{b}}{|\mathbf{a}||\mathbf{b}|} = \frac{65\text{ m}^2}{(10.25\text{ m})(9.43\text{ m})} = 0.672$$

from which

$$\theta = 47.7°$$

FORCE SYSTEMS

A body may have several forces acting on it simultaneously. To account for these forces in an organized way, some general properties of a set of forces, or a force system, will prove useful.

Types of Forces

Normally, forces are distributed over some region of the body they act upon; however, some simplification can often be gained without significant loss of accuracy by considering that a force is concentrated at a single point called the **point of application** of the force. However, in some circumstances, it may be necessary to account for the way the forces are distributed over a region of the body. Thus, we make a distinction between **concentrated forces** and **distributed forces**.

A second distinction that is often important is between **surface forces**, or actions that take place where surfaces contact, and **body forces**, which are distributed throughout a body, as in the case of gravity.

Point of Application and Line of Action

In addition to the vector value of a force (that is, magnitude and direction), the point at which a force acts on the body is important to the way the body responds. For this reason, analysis usually requires not only a **force** vector to specify the magnitude and direction of the force, but also a **position** vector to specify the location of the *point of application* of the force. The **line of action** of a force is the line parallel to the force vector and through the point of application. This line is important to the understanding of *moments* of forces.

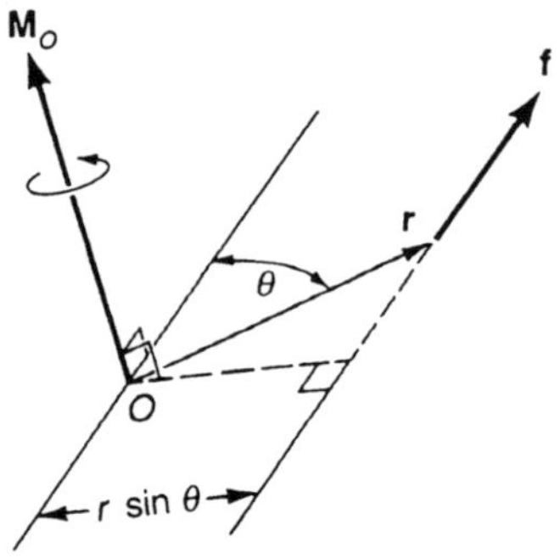

Figure 8.6

Moments of Forces

The **moment** about a point O of a force **f** is the vector defined as

$$\mathbf{M}_O = \mathbf{r} \times \mathbf{f} \qquad \textbf{(8.10)}$$

where **r** is a position vector from O to any point on the line of action of **f**. Reference to Figure 8.6 and the definition of the cross product reveal that the magnitude of the moment is

$$|\mathbf{M}_O| = df \qquad \textbf{(8.11)}$$

where f is the magnitude of **f** and $d = r \sin \theta$ is the perpendicular distance from O to the line of action of **f**.

Example 8.4

Suppose the guyline $O'P$ in Exhibit 3 has a tension of 800 N. What is the moment about O of the force from this cable acting on the mast?

Solution

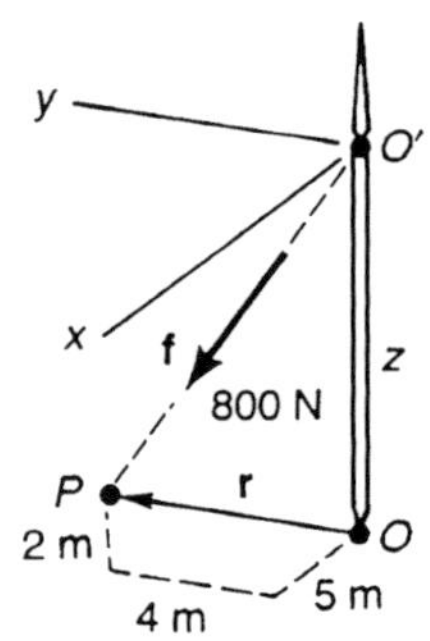

Exhibit 3

Because point P is on the line of action of this force, a position vector, **r**, from O to P can be used for evaluation of the moment about O. Referring to the axes shown in Exhibit 3, the x-y-z resolution of this vector may be written as

$$\mathbf{r} = (5\text{ m})\mathbf{e}_x + (4\text{ m})\mathbf{e}_y + (-2\text{ m})\mathbf{e}_z$$

The resolution of the force can be determined by multiplying its magnitude by the unit vector in the direction of $O'P$, components of which can be obtained by dividing the vector **a** from the preceding example by its magnitude:

$$\mathbf{f} = 800\text{N}\left(\frac{5}{10.25}\mathbf{e}_x + \frac{4}{10.25}\mathbf{e}_y + \frac{8}{10.25}\mathbf{e}_z\right)$$

$$= (390.4\mathbf{e}_x + 312.3\mathbf{e}_y + 624.6\mathbf{e}_z)\text{N}$$

The moment can now be evaluated with reference to Equation (8.9):

$$\begin{aligned}\mathbf{M}_O &= [(4\text{ m})(624.6\text{ N}) - (-2\text{ m})(312.3\text{ N})]\mathbf{e}_x \\ &\quad + [(-2\text{ m})(390.4\text{ N}) - (5\text{ m})(624.6\text{ N})]\mathbf{e}_y \\ &\quad + [(5\text{ m})(312.3\text{ N}) - (4\text{ m})(390.4\text{ N})]\mathbf{e}_z \\ &= (3123\mathbf{e}_x - 3904\mathbf{e}_y)\text{ N}\bullet\text{m}\end{aligned}$$

Observe that point O' is also on the line of action of **f**, so that a position vector $\mathbf{r} = (-10\text{ m})\mathbf{e}_z$ could have been used instead of the one above, and with less arithmetic.

The moment about an axis Oi is defined as the projection onto the axis of the moment about some point on the axis. To express this, we define the positive sense along the axis with the unit vector $\mathbf{e}_i$ and write

$$\mathbf{M}_{Oi} = (\mathbf{e}_i \bullet \mathbf{M}_O)\,\mathbf{e}_i \qquad \textbf{(8.12)}$$

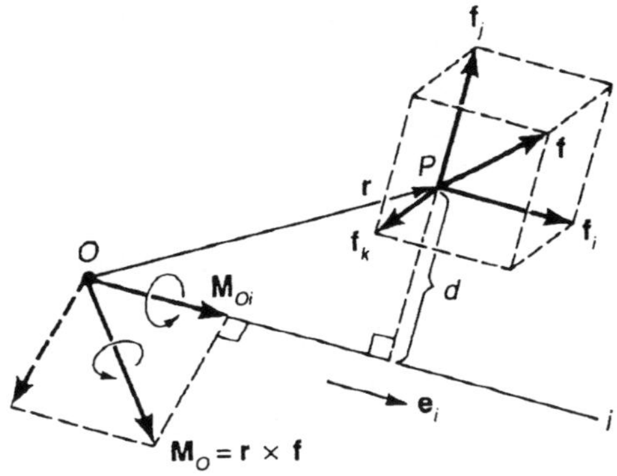

Figure 8.7

Although the moment about the axis can be computed according to this definition, an alternative form is often easier to use and provides a different interpretation. Substitution of Equation (8.10) into Equation (8.12) and use of the vector identity $\mathbf{a} \bullet (\mathbf{b} \times \mathbf{c}) = (\mathbf{a} \times \mathbf{b}) \bullet \mathbf{c}$ leads to $\mathbf{M}_{Oi} = [(\mathbf{e}_i \times \mathbf{r}) \bullet \mathbf{f}\,]\,\mathbf{e}_i$.

Now if $\mathbf{f}$ is resolved into a component $\mathbf{f}_i$ parallel to the axis Oi, a component $\mathbf{f}_j$ perpendicular to O_i and in the plane of $\mathbf{r}$ and O_i, and a component $\mathbf{f}_k$ perpendicular to this plane, several important facts become apparent. Referring to Figure 8.7, note that because both $\mathbf{f}_i$ and $\mathbf{f}_j$ are perpendicular to $\mathbf{e}_i \times \mathbf{r}$, neither of these components contributes to $\mathbf{M}_{Oi'}$. Also, because $\mathbf{e}_i \times \mathbf{r}$ has the magnitude $d = |\mathbf{r}| \sin \angle \frac{\mathbf{r}}{\mathbf{i}}$, we can express the magnitude of the moment component as

$$|\mathbf{M}_{Oi}| = df_k$$

where d is the perpendicular distance from point P to the axis Oi.

The *sense* of $\mathbf{M}_{Oi}$ (that is, whether it is directed in the positive or negative i-direction) is readily determined from the direction of $\mathbf{f}_k$ and the right-hand rule. Alternatively, the sense may be determined by the sign of the factor $\mathbf{e}_i \bullet \mathbf{M}_O$ in Equation (8.12). Note that the same value of d would be obtained regardless of where the point O is on the i-axis, and recall that the position vector $\mathbf{r}$ in the definition $\mathbf{M}_O = \mathbf{r} \times \mathbf{f}$ can be from O to any point on the line of action of $\mathbf{f}$. This means that Equation (8.12) will yield the value $\mathbf{M}_{Oi}$ with $\mathbf{r}$ as a position vector from *any* point on the axis O_i to *any* point on the line of action of $\mathbf{f}$.

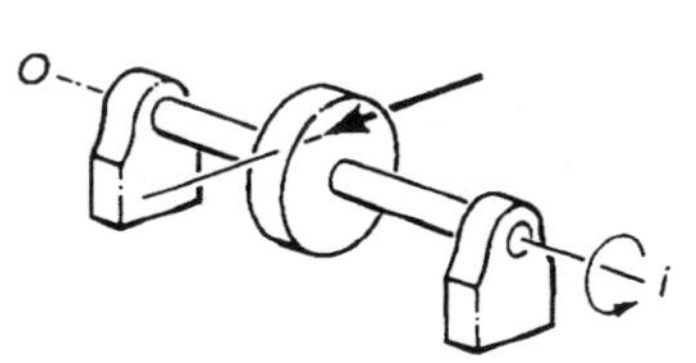

Figure 8.8

The moment about an axis is a measure of the tendency of the force(s) to cause rotation about the axis. For example, if a rotor is mounted in bearings and subjected to a set of forces, the moment of these forces about the axis of the bearings is found to be directly related to the rate of change of rotational speed. Neither forces parallel to the axis nor forces with lines of action passing through the axis will affect the rotation.

Resultant Forces and Moments

If there are several forces $\mathbf{f}_1, \mathbf{f}_2, \ldots, \mathbf{f}_n$, each with its own line of action, the **resultant force** is defined as

$$\mathbf{f} = \sum_{i=1}^{n} \mathbf{f}_i$$

and the **resultant moment** about a point O is defined as

$$\mathbf{M}_O = \sum_{i=1}^{n} \mathbf{r}_i \times \mathbf{f}_i$$

where $\mathbf{r}_i$ is a position vector from O to any point on the line of action of $\mathbf{f}_i$.

Couples

A special set of forces, called a **couple**, has zero resultant force but a nonzero resultant moment. An example is a pair of forces of equal magnitude, opposite directions, and separate lines of action.

Moments about Different Points

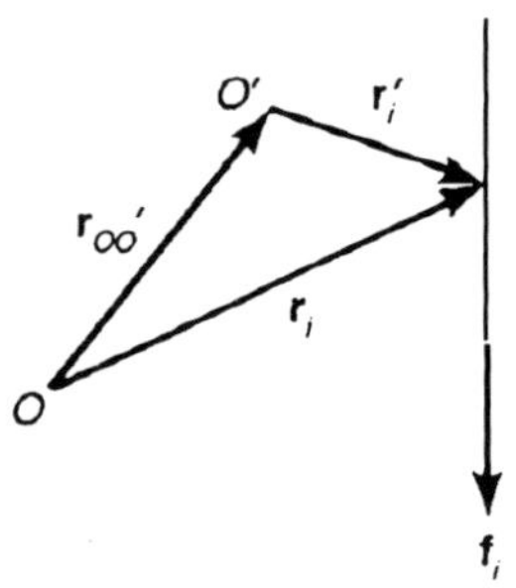

Figure 8.9

The resultant moment of a set of forces about two different points, O and O', are related as follows. With $\mathbf{r}_{oo'}$ designating the position vector from O to O', the position vectors from these two points to a point on the line of action of the force $\mathbf{f}_i$ are shown in Figure 8.9 and are related by

$$\mathbf{r}_i = \mathbf{r}_{i'} + \mathbf{r}_{oo'}$$

The moment about O can then be expressed as

$$\begin{aligned} \mathbf{M}_O &= \mathbf{r}_1 \times \mathbf{f}_1 + \mathbf{r}_2 \times \mathbf{f}_2 + \ldots + \mathbf{r}_n \times \mathbf{f}_n \\ &= (\mathbf{r}'_1 + \mathbf{r}_{OO'}) \times \mathbf{f}_1 + (\mathbf{r}'_2 + \mathbf{r}_{OO'}) \times \mathbf{f}_2 + \ldots + (\mathbf{r}_n + \mathbf{r}_{OO'}) \times \mathbf{f}_n \\ &= \mathbf{r}'_1 \times \mathbf{f}_1 + \mathbf{r}'_2 \times \mathbf{f}_2 + \ldots + \mathbf{r}'_n \times \mathbf{f}_n + \mathbf{r}_{OO'} \times (\mathbf{f}_1 + \mathbf{f}_1 + \ldots + \mathbf{f}_n) \\ &= \mathbf{M}_{O'} + \mathbf{r}_{OO'} \times \mathbf{f} \end{aligned} \qquad \textbf{(8.13)}$$

That is, the moment about O is equal to that about O' plus the moment that a force equal to the resultant of the given forces would have about O if the line of action of this force passed through O'. For the special case in which $\mathbf{f} = 0$, Equation (8.13) shows that the moment of a couple about every point is the same.

Equivalent Force Systems

Two sets of forces are said to be **equivalent** if each has the same resultant force and the same resultant moment about some point. If these conditions are met, then Equation (8.13) can be used to show that the sets will have the same resultant moment about *any* point.

Example 8.5

A 7-kN force acts on the end of the beam, its line of action passing along the center of the web of the channel section. If a bracket were attached to the end of the beam, allowing this force to be applied 80 mm to the left of the center of the web (see Exhibit 4), what horizontal forces applied along the flanges would need to be added so that, together with the displaced 7-kN force, they would form a set equivalent to the 7-kN force acting along the web?

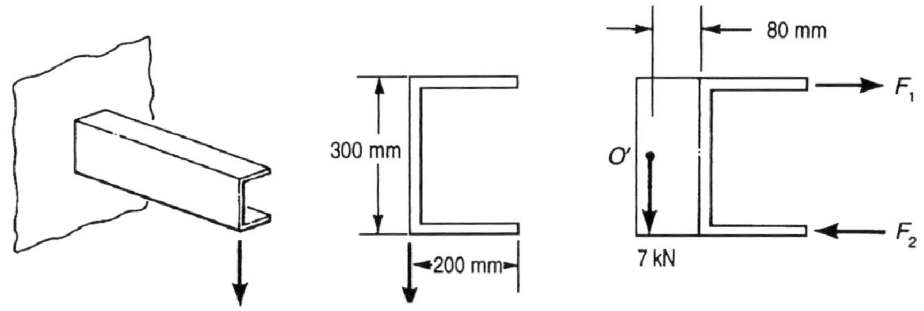

Exhibit 4

Solution

Since the resultants of the two sets must be the same (7-kN downward), the resultant of F_1 and F_2 must be zero, which in turn means that F_1 and F_2 must have equal magnitudes and opposite directions. To determine the magnitude, the resultant moment of each set can be equated; a convenient point about which to evaluate these moments is located where the lines of action of F_2 and the force through O' intersect. The moment about this point of the original force is (0.08 m)(7 kN) = 0.56 kN• m acting clockwise. The resultant moment of the equivalent set about this point is (0.3 m)F_1, also in the clockwise direction. Equivalence requires that (0.3 m)F_1 = 0.56 kN• m from which F_1 = 1.867 kN. Thus, the equivalent set consists of the vertical, 7-kN force through O', together with a couple that has a clockwise-acting moment of 0.56 kN• m.

EQUILIBRIUM

Newton's Laws require for every body or system of bodies that

$$\sum_i \mathbf{f}_i = \sum_j m_j \mathbf{a}_j$$

and

$$\sum_i \mathbf{r}_i \times \mathbf{f}_i = \sum_j \mathbf{r}_j \times m_j \mathbf{a}_j$$

in which $\mathbf{f}_i$ is one of the forces acting on the system (from an *external* source); m_j and $\mathbf{a}_j$ are the mass and acceleration, respectively, of the *j*th material element; $\mathbf{r}_i$ and $\mathbf{r}_j$ are position vectors from *any* selected point to, respectively, a point on the line of action of $\mathbf{f}_i$ and the *j*th material element; and the sums are to include *all* the external forces and material elements. A system will be in static equilibrium whenever the accelerations are all zero. In these cases the laws require that the resultant of all forces from external sources be zero, and that the resultant moment of these forces about any point be zero.

Free-Body Diagrams

In spite of the simplicity of equilibrium relationship, they can easily be misapplied. Experience has repeatedly shown that nearly all such errors stem from lack of attention to the appropriate **free-body diagram**. The free-body diagram must show clearly what body, bodies, or parts thereof are being considered as the system,

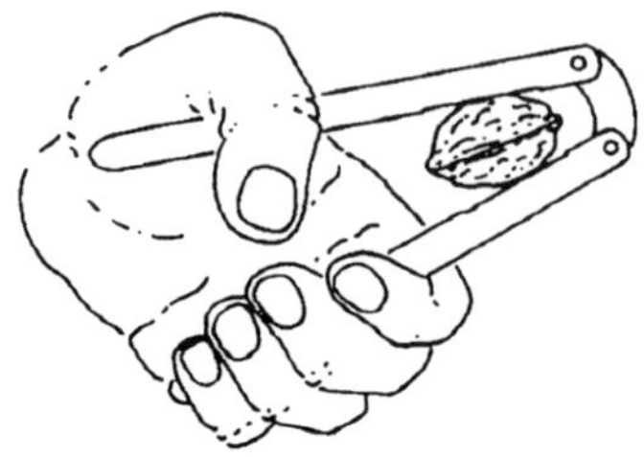

Figure 8.10

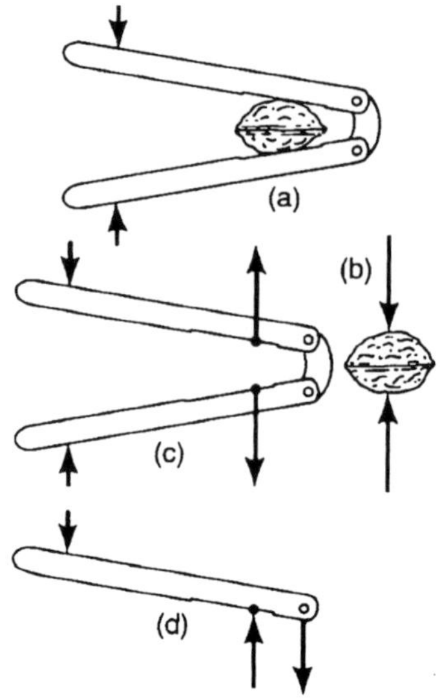

Figure 8.11

and all of the forces acting *on* the system from sources *outside* the system. For illustration, consider the device shown in Figure 8.10. Several different free-bodies are possibly useful and will be constructed.

First, consider the system consisting of the nutcracker together with the walnut. This system is shown isolated from all other objects, with arrows depicting the forces that come from objects *external* to the system. Assuming forces of gravity to be negligible, the only external body that exerts forces on this system is the hand. These forces are shown in Figure 8.11(a). The interaction between the nut and the cracker is *not* shown on this free body, since it is internal to this system.

To expose the force tending to break the nut, we might consider free-bodies of the nut and of the nutcracker, shown in Figures 8.11(b) and (c). In Figure 8.11(c) the hand and the walnut are external to the nutcracker. Therefore, arrows depicting the forces from the nut as well as from the hand are included in this free body.

Another possibly useful free body is that of the upper handle. Objects external to this are the hand, the walnut, and the connecting pin. This free body appears in Figure 8.11(d).

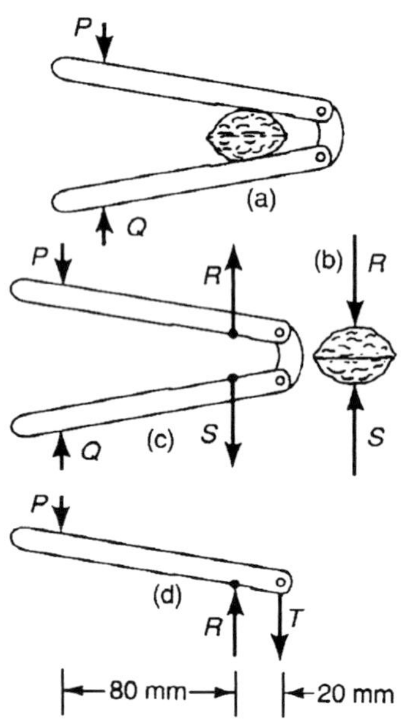

Figure 8.12

Here is a summary of the procedure: First, a sketch must be made clearly showing the system to be considered for equilibrium. The system boundary is normally chosen so that it passes through a point where a force interaction of particular interest occurs. Next, *all* forces acting on the system, from bodies *external* to the system, must be properly represented. Force interactions between bodies *within* the system are *not* considered.

To show clearly the physical significance of quantities in equilibrium equations, symbols must accompany the arrows representing the forces that appear in these equations. In Figure 8.12, the letters P, Q, R, S, and T have been chosen to indicate the magnitudes of several forces. Each letter represents a **scalar** multiplier of a unit vector in the direction of the arrow; since this scalar can take on a positive or negative value, a force in the same or opposite direction indicated by the arrow can be represented. If an analysis leads to the values, say, $P = 80$ N and $R = 400$ N, the forces acting on the upper handle would be 80 N *downward* on the right-hand end and 400 N *upward* where the walnut makes contact at R. If different circumstances led to, say, $P = -15$ N and $R = -75$, these values would imply that the forces acting on the upper handle are 15 N *upward* on the right-hand end and 75 N *downward* where the walnut makes contact. (A little adhesive between the walnut and the nutcracker would make this possible.)

Observe that the forces on the walnut have the same labels as their counterparts on the nutcracker, and the arrows have opposite directions. We have in this way implied satisfaction of Newton's third law without further fuss. Other relatively simple aspects of force analysis can be treated as the free-bodies are constructed; for example, unless the walnut is to accelerate, it is evident from a glance at its free body that $R = S$. Writing this equation could be circumvented by simply labeling both arrows with the same letter.

Equations of Equilibrium

With free-body diagrams properly drawn and labeled, equations of equilibrium can be written for any of the chosen bodies. For example, the vertical force equilibrium of the free body of Figure 8.12(a) requires that $Q - P = 0$. Each of the other two equations of force equilibrium, involving components in the horizontal direction and components perpendicular to the plane of the sketch, is the trivial equation $0 = 0$. Moment equilibrium shows that the lines of action of the two forces must coincide, a fact that has already been incorporated into the diagram. Similar analyses of the free-bodies of Figure 8.12(b) and (c) lead to

$$S - R = 0$$

and

$$Q - P + R - S = 0$$

Now suppose that the reason for this analysis is to obtain an estimate of how hard one must squeeze in order to crack a nut, given the cracking requires 245 N applied to the nut. Then $R = 245$ N, and the three equations above contain the three unknowns, P, Q, and S. Unfortunately, attempts to solve these for P or Q will fail, because the three equations are not independent, because the last equation can be deduced from the first two by addition. Equilibrium of still another body must be considered in order to obtain an independent equation. The free body of the handle in Figure 8.12(d) can provide two more equations: the vertical component for force equilibrium,

$$T + P = 245 \text{ N} \qquad \textbf{(a)}$$

and an equation of moment equilibrium. Summing moments about the point of contact with the walnut leads to

$$-(20 \text{ mm})T + (80 \text{ mm})P = 0 \qquad \textbf{(b)}$$

To solve for P, we can multiply Equation (a) by 20 mm and add Equation (b) to the result, yielding

$$(100 \text{ mm})P = (20 \text{ mm})\,(245 \text{ N}) \qquad \textbf{(c)}$$

This gives $P = 49$ N.

A more direct analysis stems from considering moments about a different point on the handle, resulting in Equation (c) as the first equation written.

The following examples provide further illustration of the use of the basic laws of static equilibrium.

Example 8.6

Neglecting gravity forces in Exhibit 5(a), except those on the 300-kg load, determine the forces in the cable and in the boom.

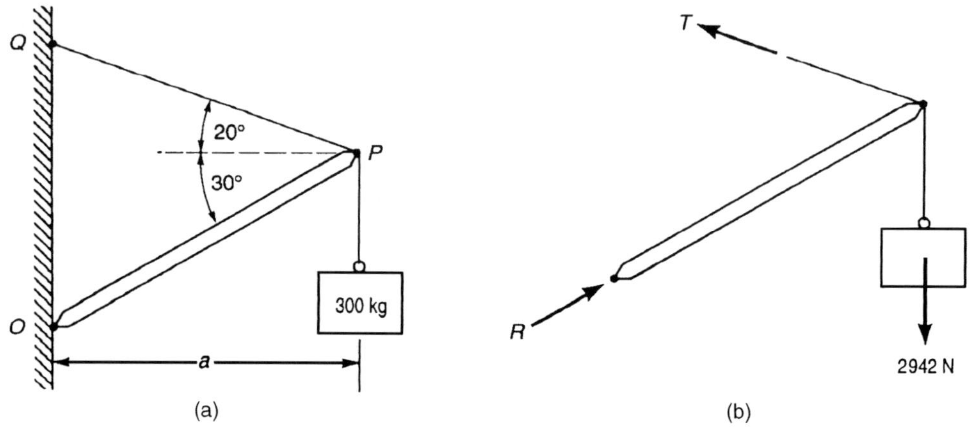

Exhibit 5

Solution

First, a free-body diagram is drawn (Exhibit 5(b)), with the system boundary passing through the support point O and through the cable segment PQ. Vanishing of moment about point P requires that the reaction at O must be directed along the boom. The force **T** can be separated into its x and y components. These two force components generate additive moments, $\mathbf{r} \times \mathbf{f}$, both of which are positive due to the right-hand rule. No moment is added for the force **R** because it is applied at the pivot point (the position vector at which **R** is applied is of length 0). The equation of moments about O can be expressed as

$$(T \sin 20° - 2942 \text{ N})a + (T \cos 20°)(a \tan 30°) = 0$$

from which

$$T = \frac{2942 \text{ kN}}{\sin 20° + \cos 20° \tan 30°} = 3.33 \text{ kN}$$

Equilibrium of horizontal forces,

$$R \cos 30° - T \cos 20° = 0$$

leads to the magnitude of the reaction at O:

$$R = \frac{(3.33 \text{ N}) \cos 20°}{\cos 30°} = 3.61 \text{ kN}$$

Example 8.7

Neglecting gravity forces except those on the 2-Mg load shown in Exhibit 6(a), determine the tension in cable AB, which is holding up the crane boom.

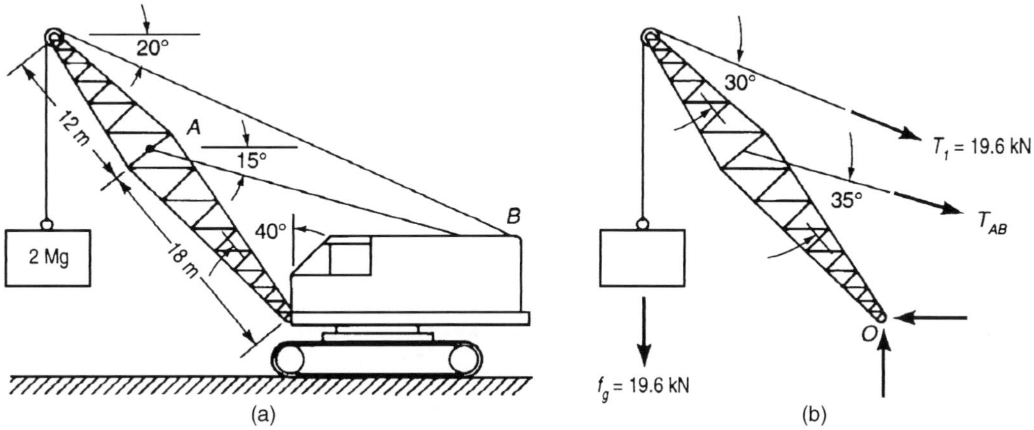

Exhibit 6

Solution

The boundary of the free body passes through the two upper cable segments and the support point, labeled O in Exhibit 6(b). By considering moment equilibrium of a system consisting of the pulley and a portion of the cable, including the section that is in contact with the pulley (not shown), we find that $T_1 = f_g = 19.6$ kN.

To avoid introducing the unknown reaction at O into the analysis, consider moments of forces about this point. With the radius of the pulley denoted as r, the equation of moments about O is

$$f_g[(12\text{ m} + 18\text{ m})\sin 40° + r] - T_1[(12\text{ m} + 18\text{ m})\sin 30° + r] - T_{AB}\sin 35°(18\text{ m}) = 0$$

With the value of $T_1 = f_g$ substituted, this is readily solved for the tension in the supporting cable:

$$T_{AB} = \frac{(19.6\text{ kN})(30\text{ m})(\sin 40° - \sin 30°)}{(18\text{ m})\sin 35°} = 8.13\text{ kN}$$

Example 8.8

Gravity forces on the structural members are negligible compared with P and Q in Exhibit 7(a). Evaluate all the forces acting on each of the three members in the A-frame.

Solution

Free-bodies of the entire frame and of each individual member are shown in Exhibit 7(b)–(e). The roller support at D means that no horizontal force can be transmitted from the ground at that point. From the free body in view (b) we can consider moments about point E,

$$(3a)Q + (3a\tan 30°)P - (6a\tan 30°)R_D = 0$$

horizontal forces,

$$R_{Ex} - Q = 0$$

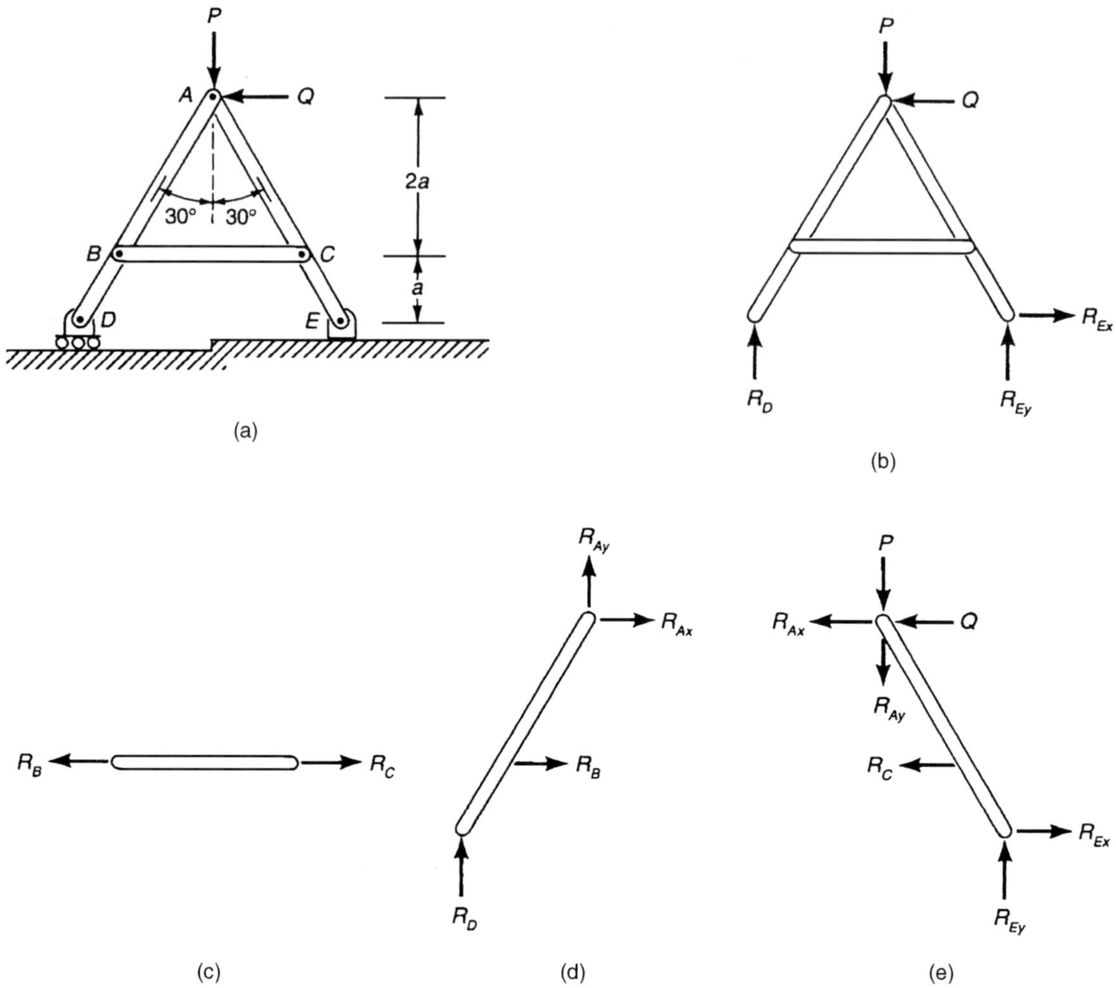

Exhibit 7

and moments about point D:

$$(6a \tan 30°)R_{Ey} + (3a)Q - (3a \tan 30°)P = 0$$

Note that $6a \tan 30°$ equals the distance from point O to point E.

These equations can then be solved for the support reactions:

$$R_D = \frac{1}{2}(P + \cot 30° Q)$$

$$R_{Ex} = Q$$

$$R_{Ey} = \frac{1}{2}(P - \cot 30° Q)$$

As a check, it might be a good idea to consider vertical forces. Moment equilibrium of the free body in view (c) implies that the lines of action of R_B and R_C are along the bar. Its horizontal equilibrium gives us

$$R_C - R_B = 0$$

Now, turning to the free body in view (d), we can write equations of moment equilibrium about points A as

$$(2a)R_B - (3a \tan 30°)R_D = 0$$

and horizontal and vertical force equilibrium as

$$R_{Ax} + R_B = 0$$
$$R_{Ay} + R_D = 0$$

With values of R_D, R_{Ax}, and R_{Ay} above, these equations give the following values of the remaining unknown reactions:

$$R_B = R_C = \frac{3}{4}(\tan 30° P + Q)$$

$$R_{Ax} = -\frac{3}{4}(\tan 30° P + Q)$$

$$R_{Ay} = -\frac{1}{2}(P + \cot 30°\ Q)$$

Example 8.9

The cables OA, OB, and OC support a suspended block shown in Exhibit 8. Determine the tension in each cable in terms of the gravitational force f_g.

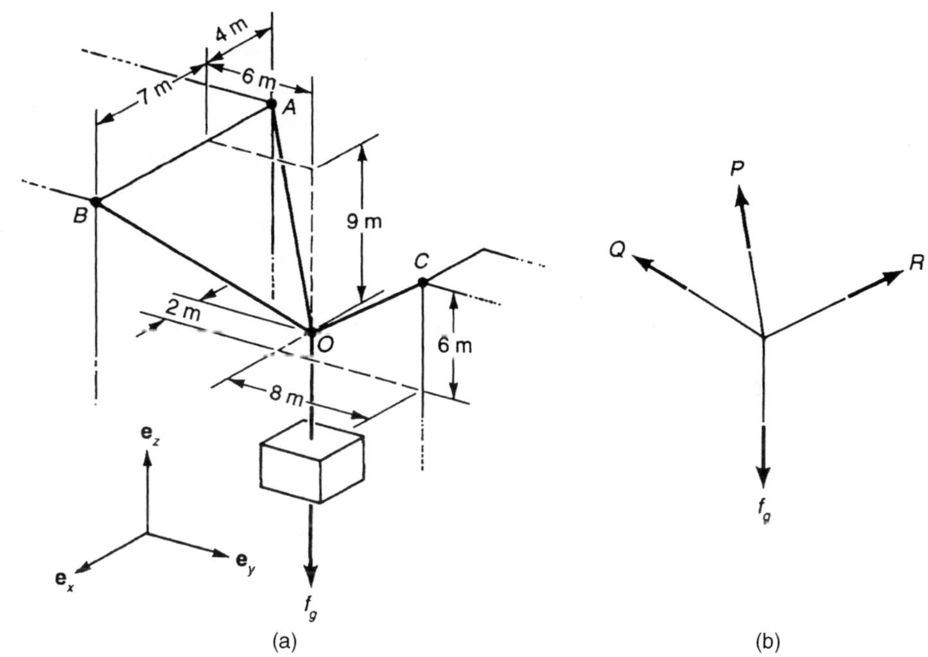

Exhibit 8

Solution

In Exhibit 8(b), the desired tensions P, Q, and R are shown on the free body of the portion of the structure in the neighborhood of point O. To resolve the forces in the directions of $\mathbf{e}_x$, $\mathbf{e}_y$, and $\mathbf{e}_z$, we need to find the direction cosines between OA, OB, and OC and these directions. This is done, as in Example 8.3, by dividing the projection of the cable onto the axis by the length of the cable; for instance, the direction cosine between OA and $\mathbf{e}_x$ is

$$\cos\angle_{OA}^{e_x} = \frac{-4m}{\sqrt{(4m)^2 + (6m)^2 + (9m)^2}} = -0.347$$

The results are summarized in the following equations:

$$\begin{aligned}\mathbf{e}_{OA} &= -0.347\mathbf{e}_x - 0.520\mathbf{e}_y + 0.780\mathbf{e}_z\\ \mathbf{e}_{OB} &= 0.543\mathbf{e}_x - 0.466\mathbf{e}_y + 0.699\mathbf{e}_z\\ \mathbf{e}_{OC} &= 0.196\mathbf{e}_x + 0.784\mathbf{e}_y + 0.588\mathbf{e}_z\end{aligned}$$

Force equilibrium requires that

$$P\mathbf{e}_{OA} + Q\mathbf{e}_{OB} + \mathrm{R}\mathbf{e}_{OC} - f_g\mathbf{e}_z = 0$$

Dot-multiplying this equation by $\mathbf{e}_x$, $\mathbf{e}_y$, and $\mathbf{e}_z$ yields the following relations:

$$\begin{aligned}-0.347P + 0.543Q + 0.196R &= 0\\ -0.520P - 0.466Q + 0.784R &= 0\\ 0.780P + 0.699Q + 0.588R &= f_g\end{aligned}$$

which can be solved for the desired forces:

$$\begin{aligned}P &= 0.660f_g\\ Q &= 0.217f_g\\ R &= 0.567f_g\end{aligned}$$

TRUSSES

A **truss** is a structure that is built with interconnected axial force members. Each such member is a straight rod that can transmit force along its axis. This limitation is the result of interconnections that are all of the ball-and-socket type; that is, they constrain the end points of the connected members against relative position change but allow the members complete freedom to rotate about the connection point. Also, external forces are applied only at these joints.

The symbol T_{IJ} will be used here to denote the tensile force in the member that connects joints I and J. This means that if T_{IJ} takes on a positive value, the member is in tension, and if T_{IJ} takes on a negative value, the member is in compression.

Equations from Joints

One approach to determining the forces in individual members within a truss is to isolate, as a free body, the portion of the truss in the neighborhood of each joint and write equations of force equilibrium for each. This procedure is illustrated for the truss shown in Figure 8.13.

First, a free body is drawn for the joint G (Figure 8.14), and the corresponding force equilibrium equation is written:

$$T_{GE}\mathbf{e}_{GE} + T_{GF}\mathbf{e}_{GF} - (50\text{ kN})\mathbf{e}_y = 0$$

To evaluate T_{GE}, the equation may be dot-multiplied by $\mathbf{e}_a$, which is defined to be perpendicular to the other unknown force, as shown in Figure 8.14:

$$\mathbf{e}_a \bullet \mathbf{e}_{GE}T_{GE} + \mathbf{e}_a \bullet \mathbf{e}_{GF}T_{GF} - \mathbf{e}_a \bullet \mathbf{e}_y(50\text{ kN}) = 0$$

$$\left[\left(\frac{1}{\sqrt{17}}\right)\left(-\frac{4}{\sqrt{5}}\right) + \left(\frac{4}{\sqrt{17}}\right)\left(\frac{3}{5}\right)\right]T_{GE} - \frac{4}{\sqrt{17}}(50\text{ kN}) = 0$$

$$T_{GE} = 125\text{ kN}$$

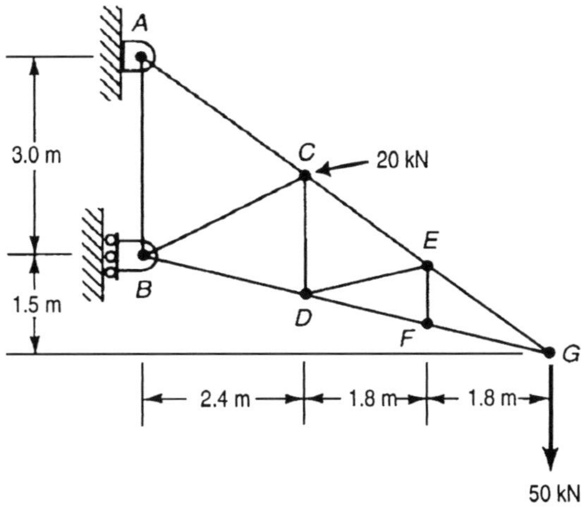

Figure 8.13

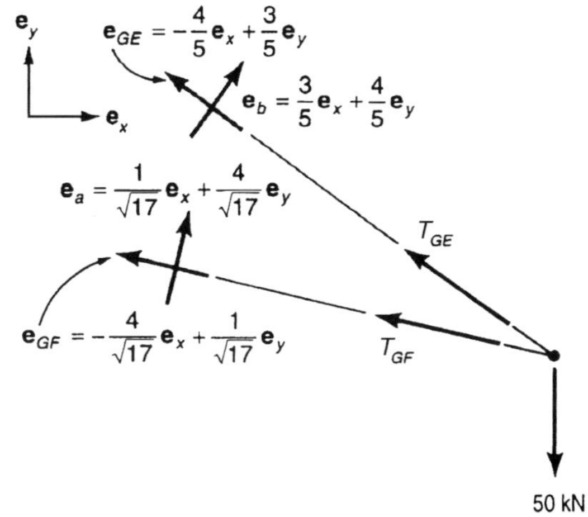

Figure 8.14

Dot-multiplication by $\mathbf{e}_b$ will similarly yield the value of T_{GF}:

$$\mathbf{e}_b \bullet \mathbf{e}_{GE} T_{GE} + \mathbf{e}_b \bullet \mathbf{e}_{GF} T_{GF} - \mathbf{e}_b \bullet \mathbf{e}_y (50 \text{ kN}) = 0$$

$$\left[\left(\frac{3}{5}\right)\left(-\frac{4}{\sqrt{17}}\right) + \left(\frac{4}{5}\right)\left(\frac{1}{\sqrt{17}}\right)\right] T_{GF} - \frac{4}{5}(50 \text{ kN}) = 0$$

$$T_{GF} = -103.1 \text{ kN}$$

Next, consider a free-body diagram of the neighborhood of joint F (Figure 8.15). Considering projections of forces perpendicular to the line DFG, it becomes evident without writing equations that $T_{FE} = 0$. In view of this, and considering forces parallel to DFG, it becomes evident that $T_{FD} = T_{FG} = -103.1$ kN.

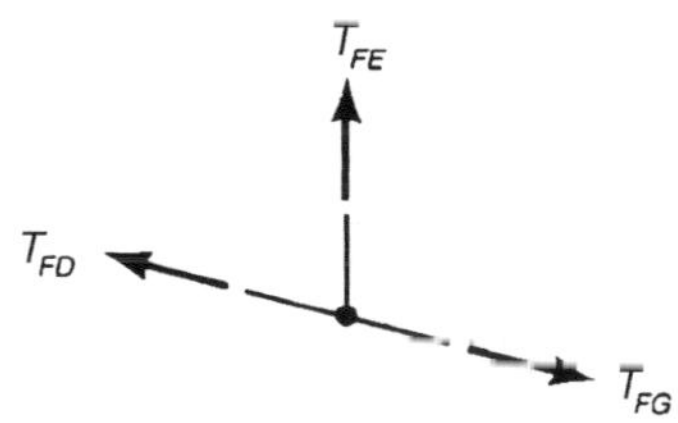

Figure 8.15

We can next proceed to joint E, where T_{ED} and T_{EC} are now the only unknowns. Once these forces are evaluated, the equilibrium equations for joint D contain only two unknowns. Proceeding in this manner, we can evaluate the remainder of the internal forces and the reactions at the supports.

By proper selection of the order in which joints of the truss are considered, it is usually possible to work through a planar truss in the manner indicated. If the values of all the forces are not required, however, use of the equations from joints may be much less efficient than the approach explained next.

Equations from Sections

The forces external to *any* portion of a system in equilibrium have zero resultants of force and moment. This is the basis for the procedure illustrated now. Figure 8.16 shows free-body diagrams of three different portions of the truss in

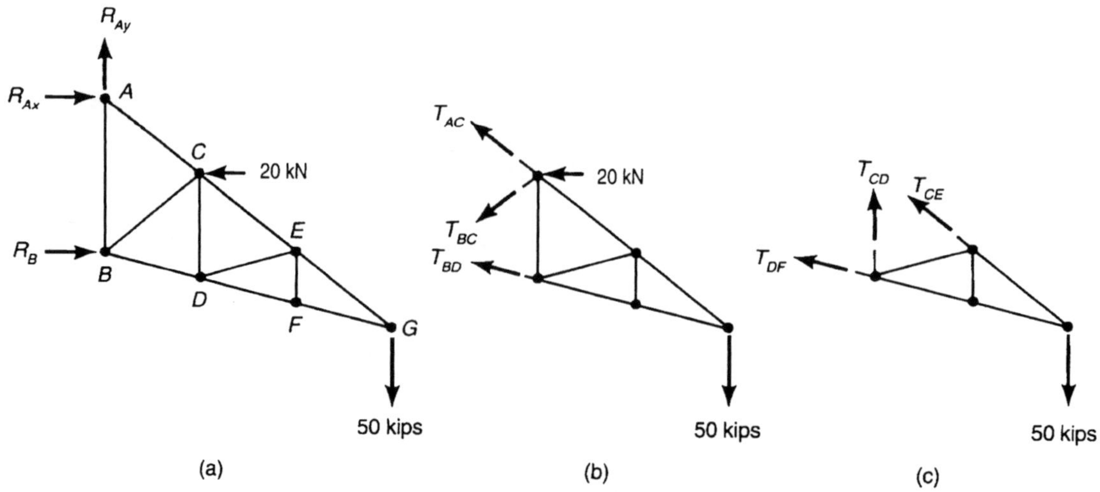

Figure 8.16

Figure 8.13. Referring to the free body of the entire structure, Figure 8.16(a), we can write the resultant moment about A as

$$(3.0 \text{ m})R_B - (1.8 \text{ m})(20 \text{ kN}) - (6.0 \text{ m})\ (50 \text{ kN}) = 0$$

from which

$$R_B = 112 \text{ kN}$$

Then summation of horizontal force components gives

$$R_{Ax} = -92 \text{ kN}$$

and summation of vertical force components gives

$$R_{Ay} = 50 \text{ kN}$$

Next, for the free body shown in Figure 8.16(b), we can sum moments about point B,

$$(3.0 \text{ m})\frac{4}{5}T_{AC} + (1.2 \text{ m})(20 \text{ kN}) - (6.0 \text{ m})(50 \text{ kN}) = 0$$

to obtain

$$T_{AC} = 115 \text{ kN}$$

Then we can sum moments about point G,

$$(3.6 \text{ m})\left(\frac{1}{\sqrt{5}}T_{BC}\right) + (2.7 \text{ m})\left(\frac{2}{\sqrt{5}}T_{BC} + 20 \text{ kN}\right) = 0$$

to obtain

$$T_{BC} = 13.42 \text{ kN}$$

and sum moments about point C,

$$-(1.8 \text{ m})\frac{4}{\sqrt{17}}T_{BD} - (3.6 \text{ m})(50 \text{ kN}) = 0$$

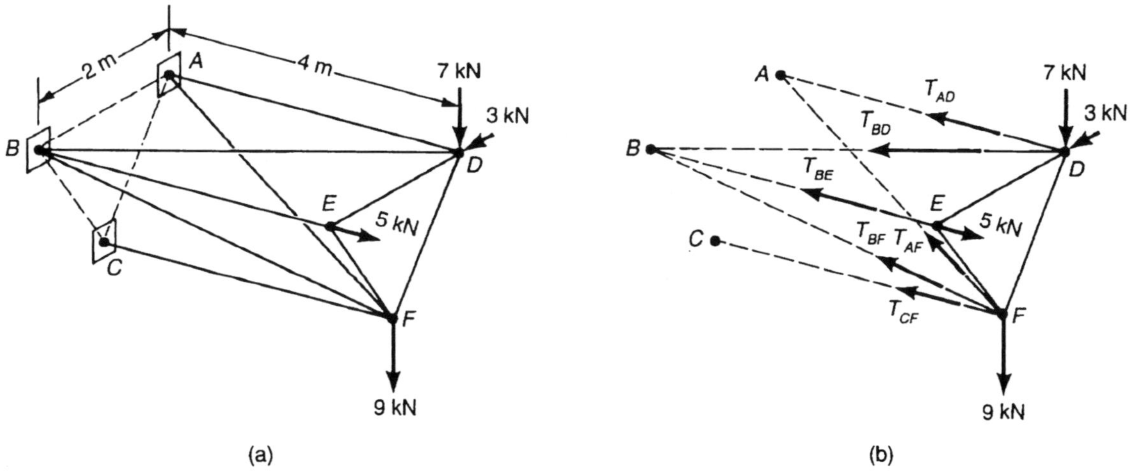

Figure 8.17

to obtain

$$T_{BD} = -103.1 \text{ kN}$$

Now, considering the free body in Figure 8.16(c), we can obtain T_{CD} by summing moments about point G:

$$T_{CD} = 0$$

By continuing in this fashion, we can evaluate the remaining forces in a fairly efficient manner.

The strategy in this method is to isolate a portion of the structure, with the boundary passing through the member in which a force is to be evaluated, and, with the free body completed, to find a direction for force reckoning or an axis for moment reckoning that will yield an equation with as few unknown forces as possible.

For further illustration, consider the truss shown in Figure 8.17(a). The vertical triangles ABC and DEF are both equilateral, and the rectangle $ABED$ is in a horizontal plane. The force in each of the nine members is to be evaluated.

Consider first the free body of the portion of the truss shown in Figure 8.17(b). To analyze this section, we select various axes, about which moments involve only one unknown force. The moment about each axis is evaluated most readily by resolving each force as indicated in Figure 8.17, keeping in mind that the moment of a force about an axis is zero if the line of action intersects the axis or is parallel to the axis:

$$M_{DF} = \left(\sqrt{3} \text{ m}\right)(T_{BE} - 5 \text{ kN}) = 0$$

$$T_{BE} = 5 \text{ kN}$$

$$M_{CF} = \left(\sqrt{3} \text{ m}\right)\left(\frac{2}{\sqrt{20}} T_{BD} + 3 \text{ kN}\right) - (1 \text{ m})(7 \text{ kN}) = 0$$

$$T_{BD} = \sqrt{5}\left(\frac{7}{\sqrt{3}} - 3\right) \text{kN} = 2.33 \text{ kN}$$

$$M_{FE} = \left(\sqrt{3}\text{ m}\right)\left(T_{AD} + \frac{4}{\sqrt{20}}T_{BD}\right) = 0$$

$$T_{AD} = -2\left(\frac{7}{\sqrt{3}} - 3\right)\text{kN} = 2.08\text{ kN}$$

$$M_{BE} = \left(\sqrt{3}\text{ m}\right)\left(\frac{2}{\sqrt{20}}T_{AF}\right) - (1\text{ m})(9\text{ kN}) - (2\text{ m})(7\text{ kN}) = 0$$

$$T_{AF} = 23\sqrt{\frac{5}{3}}\text{ kN} = 29.7\text{ kN}$$

$$M_{DA} = \left(\sqrt{3}\text{ m}\right)\left(\frac{2}{\sqrt{20}}T_{BF}\right) - (1\text{ m})(9\text{ kN}) = 0$$

$$T_{BF} = 9\sqrt{\frac{5}{3}}\text{ kN} = 11.62\text{ kN}$$

$$M_{BA} = \left(\sqrt{3}\text{ m}\right)T_{CF} + (4\text{ m})(7\text{ kN}) + (4\text{ m})(9\text{ kN}) = 0$$

$$T_{CF} = -\frac{64}{\sqrt{3}}\text{ kN} = -37\text{ kN}$$

To determine the forces in the remaining three members, it is a straightforward matter to isolate joints D and E and sum forces:

$$T_{DF} = -\frac{14}{\sqrt{3}}\text{ kN} = -8.08\text{ kN}$$

$$T_{DE} = T_{EF} = 0$$

As with planar trusses, three-dimensional trusses can be analyzed by writing equations of force equilibrium for each joint, or by considering a larger portion of the truss. The procedure for writing equilibrium equations for a joint is illustrated in the previous section and typically leads to a set of simultaneous equations for the unknown forces. Often, as in the preceding example, the task of solving simultaneous equations can be avoided by considering an entire section of the truss and finding axes about which only one unknown force has a nonzero moment.

COUPLE-SUPPORTING MEMBERS

The loads that a rigid bar can carry are not limited to axial forces. If lateral forces are applied, equilibrium requires that forces across a section have a nonzero moment about the center of the section, as demonstrated in Figure 8.18. This figure shows a sketch and free-body diagrams: one of the entire and others of two separated portions of the beam. Equilibrium of the portion on the left indicates at once that the forces across the plane of separation must form a lateral force and a couple. The detailed distribution of interaction forces may be fairly complicated and cannot be deduced from equilibrium alone; however, it is often possible to evaluate the resultant force and the moment of the equivalent couple from statics.

Twisting and Bending Moments

In general, the moment of the couple at a section can have any direction, depending on how the external loads are applied. For example, the bracket in Figure 8.19 has moment components at section A in each of three directions aligned with the axis

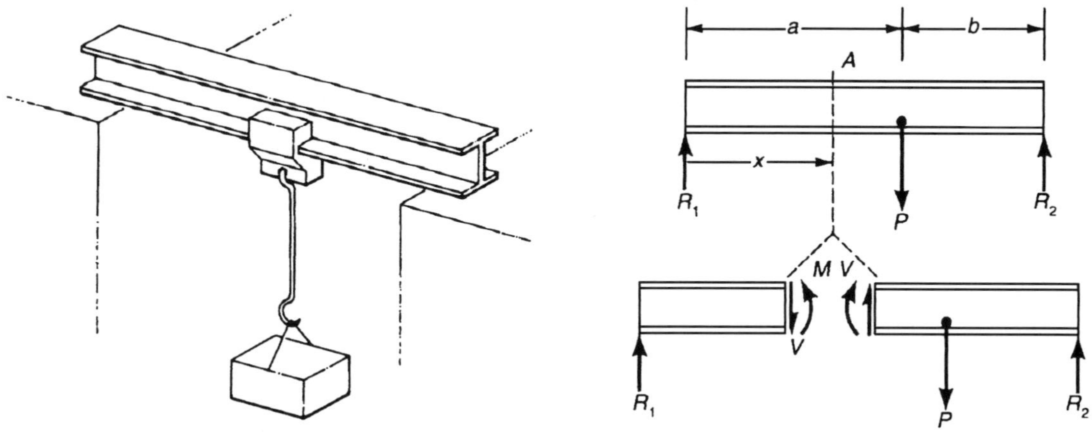

Figure 8.18

of the bracket. The moment vector is usually resolved into a component parallel to the axis of the bar as well as one or two components perpendicular to the axis. (This facilitates the analysis of strength and deformation of the bar.) The component of moment parallel to the axis of the bar is called the **twisting moment**, and the components perpendicular to the axis are called **bending moments**, after the types of deformation they produce, as shown in Figure 8.20.

The resultant force acting at a section of a bar is similarly resolved. The component parallel to the axis of the rod is called the **axial force**, and the components perpendicular to the axis are called **shearing forces**.

Evaluation of these force and moment components is accomplished by using the same basic ideas already examined: A free body of a portion of the member on either side of the section of interest is isolated and properly labeled, and equations of equilibrium are written and solved.

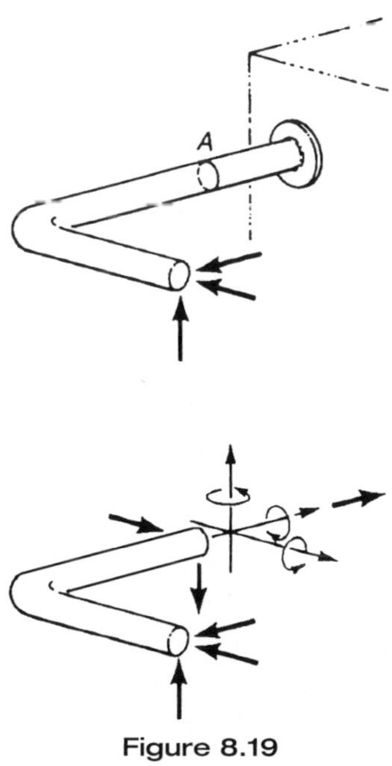

Figure 8.19

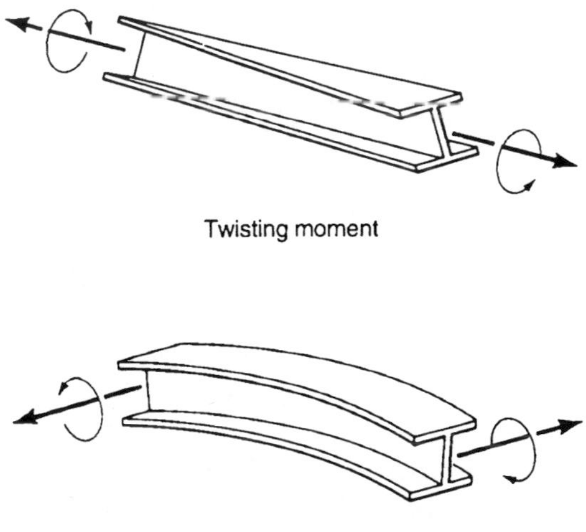

Figure 8.20

For example, equilibrium of the portion of the beam to the left of section A in Figure 8.18 yields

$$V = R_1 \quad \text{and} \quad M = R_1 x$$

and equilibrium of the entire beam gives values of the support reactions in terms of their applied load P as

$$R_1 = \frac{bP}{a+b} \qquad R_2 = \frac{aP}{a+b}$$

Thus, in terms of the applied load, the shear and bending moment are

$$V = \frac{bP}{a+b} \qquad x < a$$

$$M = \frac{bPx}{a+b} \qquad x < a$$

The qualification $x < a$ is necessary because the analysis was done for sections to the left of the applied load. A similar analysis for sections on the other side of the load results in

$$V = \frac{-aP}{a+b} \qquad x > a$$

$$M = \frac{a(a+b-x)P}{a+b} \qquad x > a$$

As another example, consider the force and moment components at the support O for the automobile torsion bar in Figure 8.21. Force equilibrium of the free body requires that

$$R + (6.8\text{ kN})\left(-\frac{8}{17}\mathbf{e}_x + \frac{15}{17}\mathbf{e}_z\right) + (2.5\text{ kN})\mathbf{e}_y = 0$$

or

$$R = (3.2\mathbf{e}_x - 2.5\mathbf{e}_y - 6.0\mathbf{e}_z)\text{ kN}$$

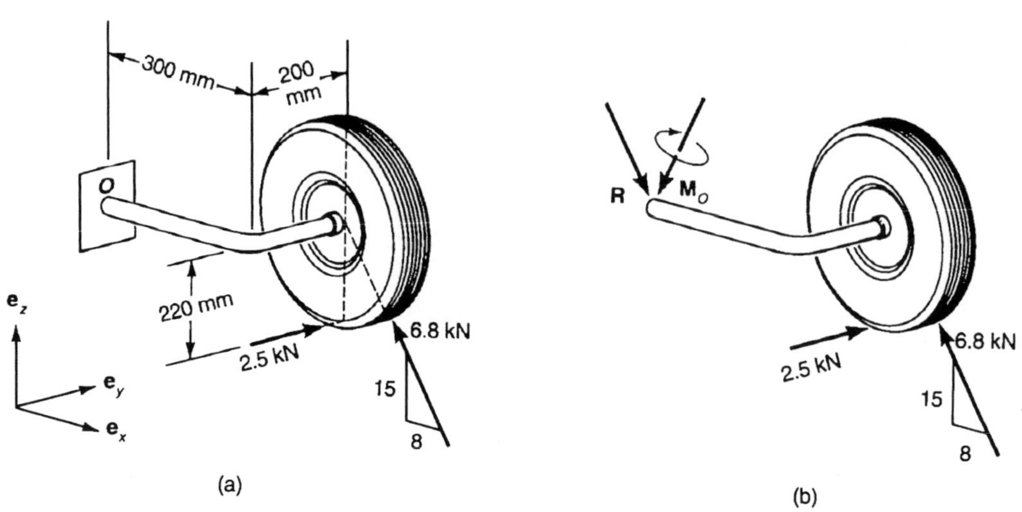

Figure 8.21

Therefore, there is a compressive axial force of 3.2 kN and a resultant shearing force of $\sqrt{(2.5)^2+(6.0)^2}$ kN $= 6.5$ kN at the section O. Next, moment equilibrium requires that

$$\mathbf{M}_O + [(0.3\text{ m})\mathbf{e}_x + (0.2\text{ m})\mathbf{e}_y] \times [(6.0\text{ kN})\mathbf{e}_z - (3.2\text{ kN})\mathbf{e}_x] \\ + [(0.3\text{ m})\mathbf{e}_x + (0.2\text{ m})\mathbf{e}_y - (0.22\text{ m})\mathbf{e}_z] \times (2.5\text{ kN})\mathbf{e}_y = 0$$

or

$$\mathbf{M}_O = (-1.75\mathbf{e}_x + 1.80\mathbf{e}_y - 1.39\mathbf{e}_z)\text{ kN}\bullet\text{m}$$

This indicates the presence of a twisting moment

$$\mathbf{M}_t = 1.75\text{ kN}\bullet\text{m}$$

and a bending moment resultant

$$M_b = \sqrt{(1.80)^2 + (1.39)^2}\text{ kN}\bullet\text{m} \\ = 2.27\text{ kN}\bullet\text{m}$$

SYSTEMS WITH FRICTION

Friction forces act tangentially to the surfaces on which two objects make contact. The ratio of the tangential force to the normal force at the contact surfaces is called the **coefficient of friction**. In general, this ratio depends on several variables, such as the surface materials, the surface finishes, the presence of any surface films, the velocity of sliding, and the temperature. The ratio of tangential force necessary to initiate sliding from a state of rest to the normal force is called the **coefficient of static friction**, whereas the force ratio as sliding continues is called the **coefficient of sliding friction**. Typically, the coefficient of static friction is somewhat greater than the coefficient of sliding friction. Furthermore, a decrease in sliding friction force with an increase in the speed of sliding has been observed for some materials, although this variation is usually small enough that it can be neglected.

In spite of the complexity of the mechanism of friction (Figure 8.22), the approximation known as *Coulomb friction* has been found to lead to predictions

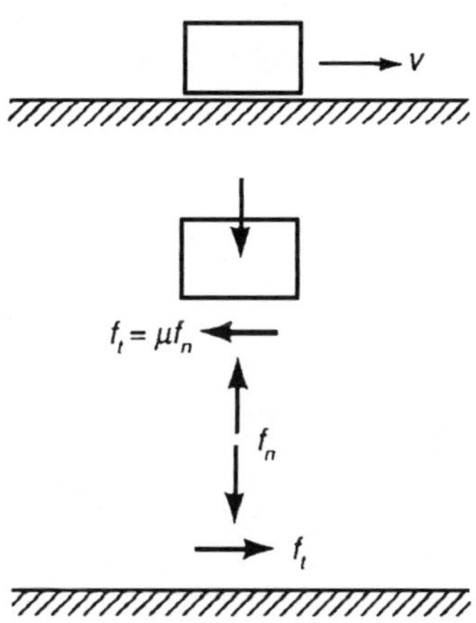

Figure 8.22

of acceptable accuracy for many dry surfaces. The so-called **Coulomb's Law of Friction** states that whenever sliding takes place, the tangential component of force between the surfaces is proportional to the normal component and acts in a direction to oppose the motion; that is,

$$\mathbf{f}_t = \mu_1 f_n \mathbf{e}_v \qquad v \neq 0 \tag{8.14a}$$

in which $\mathbf{e}_v$ is a unit vector in the direction on the relative velocity, v, of the objects on which $\mathbf{f}_t$ acts, and the coefficient of sliding friction μ_1 depends on the surfaces in contact but not on the magnitude of the normal force or on the velocity. When the surfaces are not sliding, the friction force can have any direction required for equilibrium, but its magnitude is limited by

$$f_t \leq \mu_0 f_n \qquad v = 0 \tag{8.14b}$$

in which μ_0 is the coefficient of static friction. When applied forces induce a friction force that reaches this limit, motion is incipient, meaning that any change tending to increase the friction force will cause acceleration.

Example 8.10

What is the magnitude P of the force required to move the 40-kg block up the 15° incline in Exhibit 9? Also, in the absence of P, will the block remain stationary on the incline? The coefficients of static and sliding friction are both 0.3.

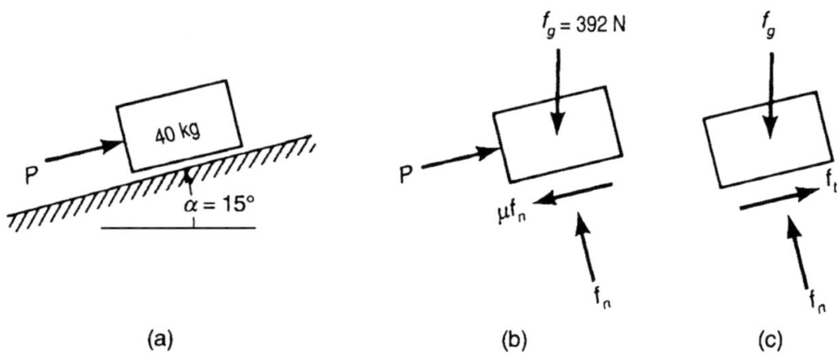

Exhibit 9

Solution

Referring to the free-body diagram (Exhibit 9(b)), we can write equations of equilibrium in the direction normal to the incline,

$$f_n - f_g \cos\ \alpha = 0$$

and in the direction along the incline,

$$P - f_g \sin\ \alpha - \mu f_n = 0$$

Elimination of f_n from these two equations results in

$$\begin{aligned} P &= f_g(\sin\alpha + \mu\ \cos\alpha) \\ &= (392\ \text{N})(\sin 15° + 0.30\ \cos 15°) \\ &= 215\ \text{N} \end{aligned}$$

The free-body diagram in Exhibit 9(c) depicts the situation in the absence of the force *P*. Observe the reversal of the direction of the friction force. Now, if the block is to remain at rest,

$$f_n = f_g \cos\alpha$$
$$f_t = f_g \sin\alpha$$

But the friction force is limited by

$$f_t \leq \mu f_n$$

Substitution of the equilibrium equations into this inequality gives

$$f_g \sin\alpha \leq \mu f_g \cos\alpha$$

or

$$\tan\alpha \leq \mu$$

Because $\tan\alpha = 0.27$, which is less than $\mu = 0.3$, the block will remain at rest.

Example 8.11

In the absence of *P*, the angle of the incline in the previous example is slowly increased until the block begins to slide downward. If the coefficient of static friction is $\mu_0 = 0.47$ and the coefficient of sliding friction is $\mu_1 = 0.44$, what will be the acceleration of the block after it breaks loose?

Solution

Prior to breakaway of the block,

$$f_t \leq \mu_0 f_n$$

Or, with the equilibrium relationships from the free-body diagram (Exhibit 9(c)),

$$f_g \sin\alpha \leq \mu_0 f_g \cos\alpha$$

The equality occurs when the critical angle, α_c, is reached:

$$\tan\alpha_c = \mu_0 = 0.47$$

from which

$$\alpha_c = 25.2°$$

After the block breaks loose,

$$f_t = \mu_1 f_n$$

Now, the component of acceleration perpendicular to the inclined plane remains zero, so that

$$f_n - f_g \cos\alpha_c = 0$$

But, in the direction parallel to the plane, the static equilibrium relationship must be replaced with Newton's second law,

$$f_g \sin\alpha_c - f_t = ma$$

in which a is the downward tangential acceleration. Combining these relationships leads to

$$a = \frac{f_g}{m}(\sin\alpha_c - \mu_1 \cos\alpha_c)$$

But, since $f_g = mg$,

$$a = g\sin\alpha_c\left(1 - \frac{\mu}{\tan\alpha_c}\right) = g\sin\alpha_c\left(1 - \frac{\mu_1}{\mu_2}\right) = (9.81\ \text{m/s}^2)\sin 25.2°\left(1 - \frac{0.44}{0.47}\right)$$

$$= 0.27\ \text{m/s}^2$$

DISTRIBUTED FORCES

When forces are distributed throughout some region (as in the case of gravity) or over a surface (as in the case of pressure on the wall of a water tank), the fundamental ideas illustrated previously apply. A feature not yet illustrated, however, is the computational detail of summation, which takes the form of integration.

Example 8.12

The beam in Exhibit 10 supports a load that varies in intensity along the length as indicated. The intensity (force per unit length of beam) has the values w_A and w_B at the two ends and varies linearly between these points.

In terms of w_A, w_B, and L, what is the resultant of the downward forces, and what are the magnitudes R_A and R_B of the reactions at the supports?

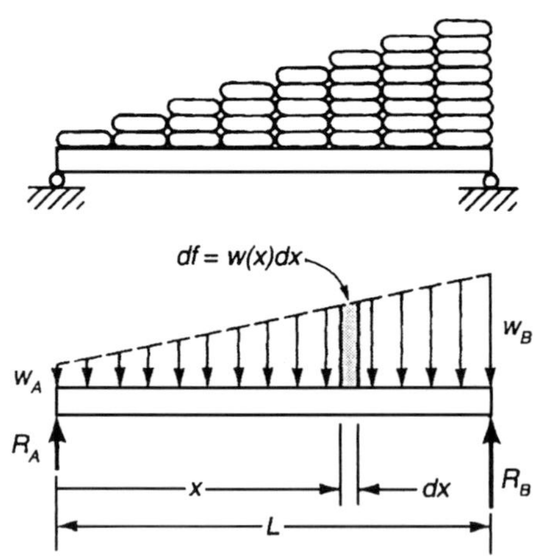

Exhibit 10

Solution

First, we write an expression for the load intensity as a function x, the distance along the span measured from the left-hand support:

$$w(x) = w_A + (w_B - w_A)\frac{x}{L}$$

Next, consider the force in the shaded portion of the load diagram, acting between the points given by x and $x + dx$. The force in this region will be the product of the intensity (force per unit length) and the length dx:

$$df = w(x)\,dx$$

Summing all such forces gives their resultant:

$$\begin{aligned} f &= \int w(x)\,dx \\ &= \int_0^L \left[w_A + (w_B - w_A)\frac{x}{L} \right] dx \\ &= \frac{1}{2}(w_A + w_B)L \end{aligned}$$

To evaluate R_B, we equate to zero the sum of moments about A of all the forces:

$$\begin{aligned} M_A &= R_B L - \int x\,[w(x)]dx \\ &= R_B L - \int_0^L x\left[w_A + (w_B - w_A)\frac{x}{L} \right] dx \\ &= R_B L - \frac{1}{6}(w_A + 2w_B)L^2 = 0 \end{aligned}$$

This yields

$$R_B = \frac{1}{6}(w_A + 2w_B)L$$

To evaluate R_A, we can use the fact that the sum of all vertical forces must be zero:

$$\begin{aligned} R_A &= f - R_B \\ &= \frac{1}{2}(w_A + w_B)L - \frac{1}{6}(w_A + 2w_B)L \\ &= \frac{1}{6}(2w_A + w_B) \end{aligned}$$

Example 8.13

The uniform, slender, semicircular arch in Exhibit 11 is acted on by gravity and the reactions from the supports.

The free-body diagram shows the desired bending moment, M_b, as the reaction from the other half of the arch. Because of horizontal equilibrium, no axial force exists at this section. The vertical shearing force is also zero at this section because Newton's third law would require a shearing force in the opposite direction on the other half of the arch, and this pair of forces would be inconsistent with the symmetry of the system.

Because the arch is slender, the forces of gravity may be treated as distributed along a circular *line*. Let the cross-sectional area be denoted by A and the density (mass per unit volume) by ρ. Then the volume of the shaded element of the arch will be equal to $Aa\,d\theta$, and the magnitude of the force of gravity acting on it will be

$$df_g = \rho(Aa\,d\theta)g$$

The resultant of the gravitational forces then has the magnitude

$$\begin{aligned} f_g &= \int_0^{\pi/2} \rho Aga\,d\theta \\ &= \frac{1}{2}\pi a\rho Ag \end{aligned}$$

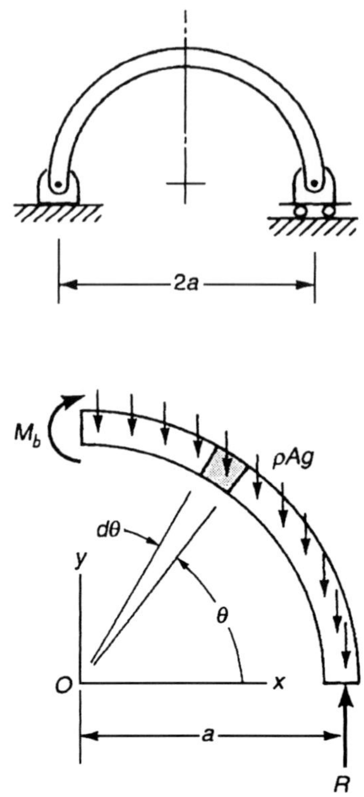

Exhibit 11

and vertical force equilibrium gives the support reaction as

$$R = \frac{1}{2}\pi a \rho A g$$

Now, the resultant moment of the forces of gravity about the point O will be in the clockwise direction and of magnitude

$$\begin{aligned} M_{Og} &= \int a \cos\theta \, df_g \\ &= \int_0^{\pi/2} a \cos\theta \, \rho A g a \, d\theta \\ &= a^2 \rho A g \end{aligned}$$

Finally, moment equilibrium about point O requires that

$$aR - M_{Og} - M_b = 0$$

which, together with the above results, gives

$$M_b = \left(\frac{\pi}{2} - 1\right) a^2 \rho A g$$

In an arch with a large radius, this bending moment could well cause failure of the structure.

Single Force Equivalents

The construction and evaluation of integrals such as those in Examples 8.12 and 8.13 can be circumvented if knowledge of an equivalent discrete force is available. This information, for a variety of special cases, is available in tabulated formulas, these formulas having been determined by integration. Proper use of such formulas requires an understanding of the following concepts.

Center of Mass and Center of Gravity

Consider the forces of gravity distributed throughout an arbitrary body, as shown in Figure 8.23. The force acting on the element with mass *dm* will be

$$d\mathbf{f}_g = dm\, g\mathbf{e}_g$$

and the resultant force will be given by

$$\begin{aligned} \mathbf{f}_g &= \int d\mathbf{f}_g \\ &= \int dm\, g\mathbf{e}_g \\ &= \left(\int dm\right) g\mathbf{e}_g = mg\mathbf{e}_g \end{aligned}$$

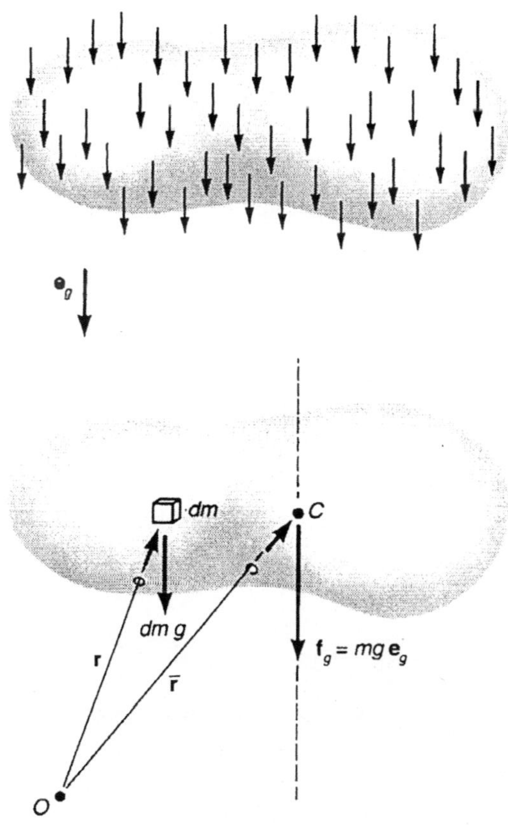

Figure 8.23

where m is the total mass of the body. The resultant moment about a point O is given by

$$\begin{aligned} M_{og} &= \int \mathbf{r} \times d\mathbf{f}_g \\ &= \int \mathbf{r} \times \left(g\mathbf{e}_g \; dm\right) \\ &= \left(\int \mathbf{r} \; dm\right) \times g\mathbf{e}_g \end{aligned}$$

where $\mathbf{r}$ is the position vector locating the mass element. Now, to compose the forces into a single, equivalent resultant, the line of action of the equivalent force $\mathbf{f}_g$ must pass through a point located by the position vector $\mathbf{r}_f$, satisfying the moment equivalence:

$$\mathbf{r}_f \times \mathbf{f}_g = \mathbf{M}_{Og}$$

Thus

$$\mathbf{r}_f \times \left(mg\mathbf{e}_g\right) = \left(\int \mathbf{r} \; dm\right) \times g\mathbf{e}_g$$

$$\mathbf{r}_f \times \mathbf{e}_g = \overline{\mathbf{r}} \times \mathbf{e}_g$$

where

$$\overline{\mathbf{r}} = \frac{1}{m}\int \mathbf{r} \; dm \tag{8.15}$$

The vector $\overline{\mathbf{r}}$ locates an important point C, called the **center of mass** of the body. By the choice of $\mathbf{r}_f = \overline{\mathbf{r}}$, moment equivalence will be satisfied for any $\mathbf{e}_g$, which implies that the line of action of the equivalent force passes through C regardless of how the body is oriented.

The center of mass C is of fundamental importance in the study of dynamics. In statics its significance stems from the fact that the resultant moment about C, of the forces of uniform gravity, is zero. That is, the body could be statically balanced by supporting it at this point only. For this reason it is also called the center of gravity.

Example **8.14**

Find the location of the center for gravity for the portion of the arch isolated as a free body in Example 8.13.

Solution

Using the center of the circle as a reference point, we can locate the shaded mass element with the position vector

$$\mathbf{r} = a \cos\,\theta\, \mathbf{e}_x + a \sin\theta\, \mathbf{e}_y$$

Then, from the definition for the center of mass, Equation (8.15),

$$\begin{aligned}\bar{r} &= \frac{1}{m}\int \left(a\cos\theta\,\mathbf{e}_x + a\sin\theta\,\mathbf{e}_y\right)\left(\rho A a\,d\theta\right) \\ &= \frac{\rho A a^2}{m}\int_0^{\pi/2} \left(\cos\theta\,\mathbf{e}_x + \sin\theta\,\mathbf{e}_y\right) d\theta \\ &= \frac{2a}{\pi}\left(\mathbf{e}_x + \mathbf{e}_y\right)\end{aligned}$$

Centroids

The mass, dm, of the element used in the preceding integrals can be expressed in terms of the density, ρ, and the corresponding element of volume, dV, as $dm = \rho\,dV$. Then the position vector locating the center of mass can be written as

$$\bar{\mathbf{r}} = \frac{\int \mathbf{r}\rho\,dV}{\int \rho\,dV}$$

Now, if the density is uniform throughout the body, ρ can be brought outside the integrals with the result

$$\bar{\mathbf{r}} = \frac{1}{V}\int \mathbf{r}\,dV$$

The location of the point C here depends entirely on geometry, because all contributions having to do with material have been canceled. The point C, located according to this equation, is called the **centroid of the volume** V. Similarly, the **centroid of a surface area** A is defined as the point located by the position vector

$$\bar{\mathbf{r}} = \frac{1}{A}\int \mathbf{r}\,dA$$

and the **centroid of a line segment** of length L is defined as the point located by the position vector

$$\bar{\mathbf{r}} = \frac{1}{L}\int \mathbf{r}\,dL$$

The calculation in Example 8.14 was for a uniform mass per unit length of the arch. With this density canceled out, the center of mass of the arch segment is also the centroid of a quarter-segment of a circular line.

The integrals $\int \mathbf{r}\,dV, \int \mathbf{r}\,dA$, and $\int \mathbf{r}\,dL$ are called the first moments of the volume, area, and line, respectively, about the reference point O. In carrying out the calculations, it is often convenient to work with one rectangular Cartesian component of the position vector at a time, that is, to evaluate separately the component equivalents obtained by dot-multiplying the vector definition by a unit vector in each direction:

$$\bar{x} = \mathbf{e}_x \bullet \bar{\mathbf{r}} = \frac{1}{V}\int x\, dv$$

$$\bar{y} = \mathbf{e}_y \bullet \bar{\mathbf{r}} = \frac{1}{V}\int y\, dv$$

$$\bar{z} = \mathbf{e}_z \bullet \bar{\mathbf{r}} = \frac{1}{V}\int z\, dv$$

Example 8.15

Determine the location of the centroid of the shaded triangle of Exhibit 12 in terms of the dimensions a, b, and c.

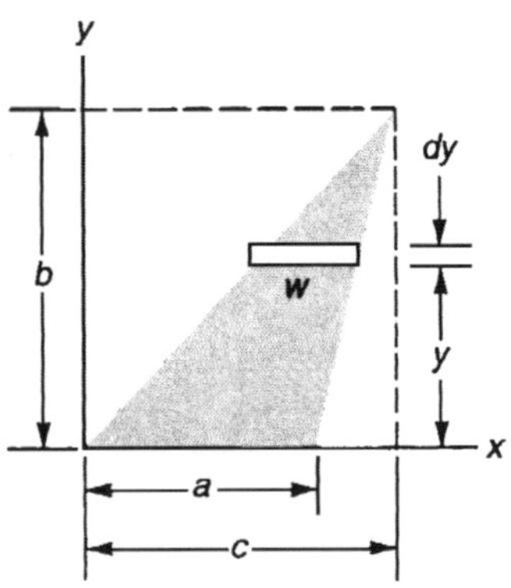

Exhibit 12

Solution

The area and first moment of area can be computed by evaluating the contribution to these quantities from the unshaded element in Exhibit 12 and summing these contributions by integration with respect to y. The width, w, of the element can be expressed in terms of y after observing that the entire triangle is similar to the triangle that lies above the unshaded element:

$$\frac{w}{b-y} = \frac{a}{b}$$

or

$$w = a\left(1 - \frac{y}{b}\right)$$

Thus the area of the element is

$$dA = w\, dy = a\left(1 - \frac{y}{b}\right)dy$$

and the area of the triangle is

$$A = \int_0^b a\left(1 - \frac{y}{b}\right)dy = \frac{1}{2}ab$$

This result is very well known. The center of the element is located on the straight line that connects the apex of the triangle with the midpoint of the base, that is, on the line that has the equation

$$x = \frac{a}{2} + \left(c - \frac{a}{2}\right)\frac{y}{b}$$

Therefore, the x-component of the first moment of area of the triangle is

$$\int x\, dA = \int_0^b \left[\frac{a}{2} + \left(c - \frac{a}{2}\right)\frac{y}{b}\right] a\left(1 - \frac{y}{b}\right)dy = \frac{1}{6}ab(a+c)$$

With this and the previously discussed value of area, the x-coordinate of the centroid is determined as

$$\bar{x} = \frac{1}{A}\int x\, dA = \frac{ab(a+c)/6}{ab/2} = \frac{1}{3}(a+c)$$

A similar calculation yields the y-component of the first moment of area as

$$\int y\, dA = \int_0^b ya\left(1-\frac{y}{b}\right)dy = \frac{ab^2}{6}$$

from which we find the y-coordinate of the centroid to be

$$\bar{y} = \frac{1}{A}\int y\, dA = \frac{ab^2/6}{ab/2} = \frac{1}{3}b$$

A composite volume, area, or line may be built up from several parts, for which the location of each centroid is known. In this case, a summation having the same form as the integral definition can readily be shown as valid. In the case of an area, A, that is a composite formed from n areas A_1, A_2, …, A_n and having centroids located by $\bar{\mathbf{r}}_1, \bar{\mathbf{r}}_2, \ldots, \bar{\mathbf{r}}_n$, the centroid of the composite is located at

$$\bar{\mathbf{r}} = \frac{1}{A}\sum_{i=1}^{n} A_i \bar{\mathbf{r}}_i$$

where

$$A = \sum_{i=1}^{n} A_i$$

Example 8.16

Locate the centroid of the plane area shown shaded in Exhibit 13.

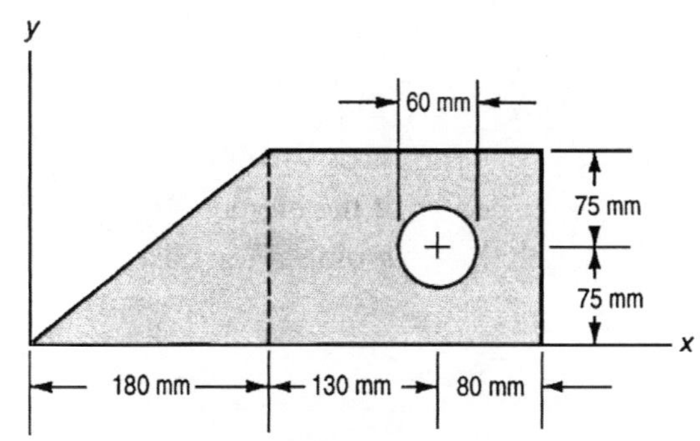

Exhibit 13

Solution

The areas and coordinates of centroids of individual parts are as follows:

	A, m^2	$\bar{x}$ **, mm**	$\bar{y}$ **, mm**
Triangle	0.0135	120	50
Rectangle	0.0315	285	75
Circle	–0.00283	310	75

The coordinates locating the centroid are then

$$\bar{x} = \frac{(120)(0.0135) + (285)(0.0315) + (310)(-0.00283)}{0.0135 + 0.0315 - 0.00283} = 231 \text{ mm}$$

$$\bar{x} = \frac{(50)(0.0135) + (75)(0.0315) + (75)(-0.00283)}{0.0135 + 0.0315 - 0.00283} = 67 \text{ mm}$$

Second Moments of Area

The geometric properties defined in the previous two sections are associated with uniformly distributed forces. Forces that vary *linearly* over a plane area lead to additional geometric properties that are useful in evaluating moment resultants. One example is fluid pressure acting on submerged, plane surfaces; another is the bending stress induced in beams. The latter is covered in the chapter on mechanics of materials. Consider the flat surfaces shown in Figure 8.24 with a force intensity that varies linearly with the distance from the line *Op*. With the force per unit area denoted as σ_z and the distance from *Op* as *q*, the linear variation is expressed as

$$\sigma_z = kq$$

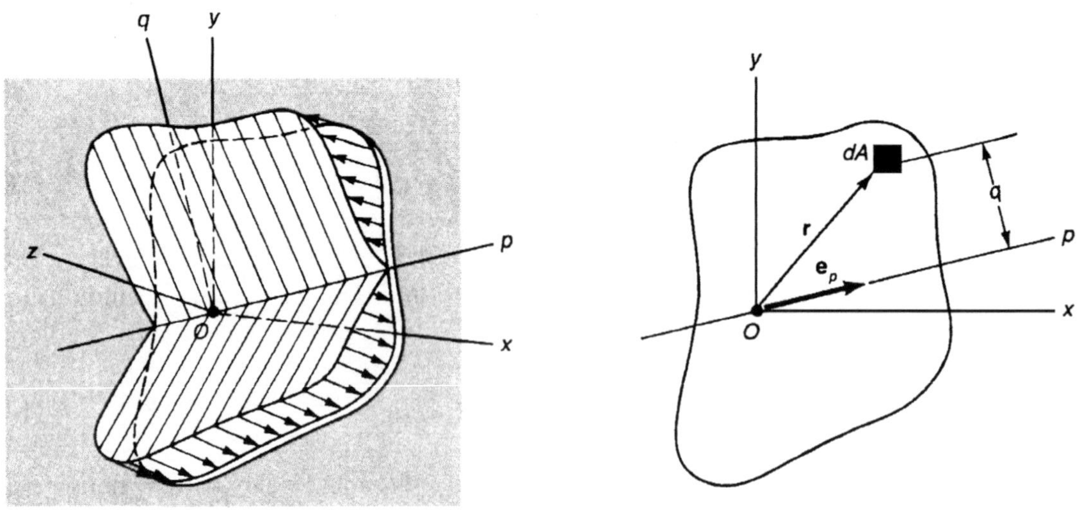

(a) (b)

Figure 8.24

where k is a proportionality constant. (In the analysis of beam bending, this constant is the product of Young's modulus and the curvature of the deformed axis of the beam.) Observe from Figure 8.24(b) that the distance from axis Op to the area element dA is given by

$$q = r \sin \angle^{\mathbf{r}} \mathbf{e}_p = (\mathbf{e}_p \times \mathbf{r}) \bullet \mathbf{e}_z$$

so that the force acting on the area element can be expressed as

$$d\mathbf{f} = \sigma_z \; dA \; \mathbf{e}_z = k(\mathbf{e}_p \times \mathbf{r}) dA$$

The resultant moment about O is then

$$\begin{aligned} \mathbf{M}_O &= \int_A \mathbf{r} \times d\mathbf{f} \\ &= k \int_A \mathbf{r} \times (\mathbf{e}_p \times \mathbf{r}) \; dA \\ &= k \int_A \left[(\mathbf{r} \bullet \mathbf{r}) \mathbf{e}_p - (\mathbf{e}_p \bullet \mathbf{r}) \mathbf{r} \right] dA \end{aligned}$$

In terms of the rectangular Cartesian components in the directions of x and y,

$$\mathbf{r} = x\mathbf{e}_x + y\mathbf{e}_y$$
$$\mathbf{e}_p = p_x \mathbf{e}_x + p_y \mathbf{e}_y$$

The moment can be expressed as

$$\mathbf{M}_O = k \int_A [(x^2 + y^2)\mathbf{e}_p - (xp_x + yp_y)\mathbf{r}] dA$$

Components of the moment are

$$\mathbf{M}_{Ox} = \mathbf{e}_x \bullet \mathbf{M}_O = k \int [(x^2 + y^2) p_x - (xp_x + yp_y) x] \, dA$$

and

$$\begin{aligned} M_{Oy} &= k \left[\left(-\int_A xy \; dA \right) p_x + \left(\int_A x^2 \; dA \right) p_y \right] \\ &= k \left[\left(\int_A y^2 \; dA \right) p_x + \left(-\int_A xy \; dA \right) p_y \right] \end{aligned}$$

The integrals that appear in these expressions are called **second moments of area** and are fundamental to the relation between the moment and the orientation of the zero-force line Op. The integrals

$$I_{xx} = \int_A y^2 dA \quad \text{and} \quad I_{yy} = \int_A x^2 dA \qquad \textbf{(8.16)}$$

are often called **moments of inertia** of the area about the x and y axes, respectively. (This terminology stems from analogous integrals related to mass distribution in the study of the kinetics of rigid bodies.) The integral

$$I_{xy} = -\int_A xy\, dA \tag{8.17}$$

is called the product of inertia of the area with respect to the x and y axes. (Many define it as the negative of this definition; then the corresponding terms in the next equation require opposite signs.) The quantities I_{xx}, I_{yy}, and I_{xy} are collectively called *second moments of area.*

In terms of the notation just introduced above, the relationship that gives the moment resultant in terms of the placement of the zero-force line Op is

$$\begin{Bmatrix} M_{Ox} \\ M_{Oy} \end{Bmatrix} = k \begin{bmatrix} I_{xx} & I_{xy} \\ I_{yx} & I_{yy} \end{bmatrix} \begin{Bmatrix} p_x \\ p_y \end{Bmatrix} \tag{8.18}$$

This is fundamental to the analyses of bending moment and deformation of elastic beams that are usually studied in mechanics of materials. Here, we examine only the evaluation of the moments and products of inertia.

The values of I_{xx}, I_{yy}, and I_{xy} depend on the placement of the origin O and the orientation of the x and y axes with respect to the area concerned. For special orientations of the x and y axes, called *principal directions*, the value of I_{xy} is zero. This situation will occur if the orientation is such that the area is symmetric with respect to either the x or y axis. (When no such symmetry exists, the determination of the principal directions requires computations that will not be addressed here.)

Parallel Axis Formulas

Often a value of one of the second moments of area is known for a given placement of the origin coordinates, but another value is needed for a different origin. Substituting the coordinate change indicated in Figure 8.25 into Equations (8.16) and (8.17) leads to the following *parallel axis formulas*, which make this evaluation much easier than carrying out an integration:

$$\begin{aligned} I_{xx}^O &= I_{xx}^C + A\bar{y}^2 \\ I_{yy}^O &= I_{yy}^C + A\bar{x}^2 \\ I_{xy}^O &= I_{xy}^C + A\overline{xy}^2 \end{aligned} \tag{8.19}$$

Here, the superscript C indicates the value with the origin at the centroid, and the superscript O indicates an arbitrary origin. A is the area, and (x, y) are the coordinates of C with respect to O.

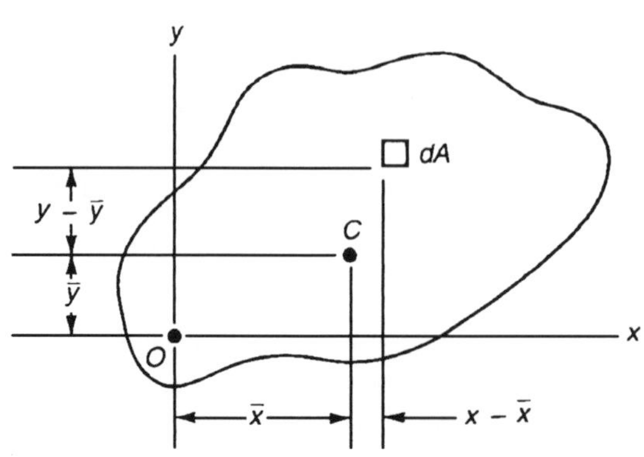

Figure 8.25

Example 8.17

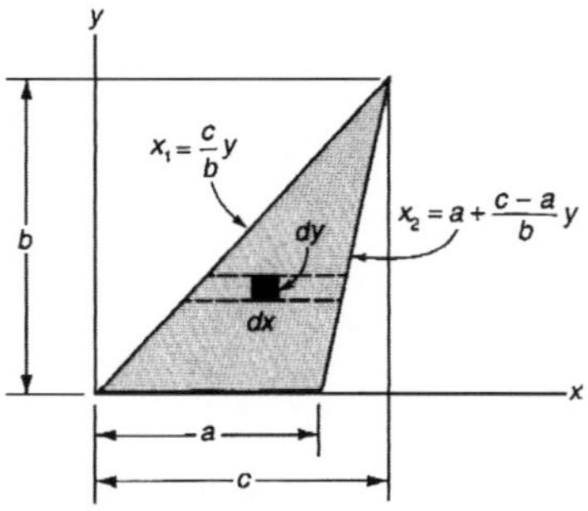

Exhibit 14

In terms of the dimensions shown in Exhibit 14, evaluate the second moments of area of the triangle with respect to its centroid.

Solution

The area of the shaded element is $dx\,dy$. The line that bounds the triangle on the left is

$$x_1 = \frac{c}{b} y$$

and the line that bounds it on the right is

$$x_2 = a + \frac{c-a}{b} y$$

These two equations will form the limits of the first integration (with respect to x) in each case. The moments and product of inertia, with respect to the axes with origin at 0, have the values

$$I_{xx} = \int_0^b \int_{x_1}^{x_2} y^2\, dx\, dy$$

$$= \int_0^b [y^2 x]_{cy/b}^{a+(c-a)y/b}\, dy$$

$$= \int_0^b y^2 \left(a - \frac{ay}{b}\right) dy = \frac{ab^3}{12}$$

$$I_{xy} = -\int_0^b \int_{x_1}^{x_2} xy\, dx\, dy$$

$$= -\int_0^b \left[\frac{x^2}{2}\right]_{cy/b}^{a+(c-a)y/b} y\, dy$$

$$= -\frac{a}{2}\int_0^b \left[a + 2(c-a)\frac{y}{b} - (2c-a)\frac{y^2}{b^2}\right] y\, dy$$

$$= -\frac{a}{2}\left[\frac{ab^2}{2} + \frac{2}{3}(c-a)b^2 - \frac{1}{4}(2c-a)b^2\right]$$

$$= -\frac{ab^2}{24}(a+2c)$$

$$I_{yy} = \int_0^b \int_{x_1}^{x_2} x^2 \, dx \, dy$$

$$= \frac{1}{3} \int_0^b [x^3]_{cy/b}^{a+(c-a)y/b} \, dy$$

$$= \frac{ab}{3} \int_0^b \left[a^2 \left(1 - \frac{y}{b}\right)^3 + 3ac\left(1 - \frac{y}{b}\right)^2 \frac{y}{b} + 3c^2 \left(1 - \frac{y}{b}\right) \frac{y^2}{b^2} \right] \frac{dy}{b}$$

$$= \frac{ab}{12}(a^2 + ac + c^2)$$

As in Example 8.15, the coordinates of the centroid are $x = (a + c)/3$ and $y = b/3$. Use of the parallel axis formulas (8.19) then gives

$$I_{xx}^C = I_{xx}^O - A\bar{y}^2$$

$$= \frac{ab^3}{12} - \frac{ab}{2}\left(\frac{b}{3}\right)^2 = \frac{ab^3}{36}$$

$$I_{yy}^C = I_{yy}^O - A\bar{x}^2$$

$$= \frac{ab}{12}\left(a^2 + ac + c^2\right) - \frac{ab}{c}\left(\frac{a+c}{3}\right)^2 = \frac{ab}{36}\left(a^2 - ac + c^2\right)$$

$$I_{xy}^C = I_{xy}^O - A\bar{x}\bar{y}$$

$$= \frac{ab^2}{24}\left(a + 2c\right) + \frac{ab}{c}\left(\frac{a+c}{3}\right)\left(\frac{b}{3}\right) = \frac{ab^2}{72}\left(a - 2c\right)$$

PROBLEMS

8.1 A 70-kg astronaut is "floating" inside a spaceship that is in a circular orbit at an altitude of 207 km above the earth, where the gravitational field intensity is 9.2 N/kg. What is the magnitude of the force of gravity on the astronaut?

a. Zero c. 70 kgf
b. 70 N d. 644 N

8.2 In the SI system of units, a pressure of 14.7 lbf/in.^{2} is:

a. 101 kPa c. 6.67 kg/in.^{2}
b. 101 Pa d. 4.77×10^4 kg/m^2

8.3 In the SI system of units, a fuel economy of 29 mi/gal is:

a. 12.33 km/L
b. 68.2 km/L
c. 46.7 km/L
d. 46.7 km/gal

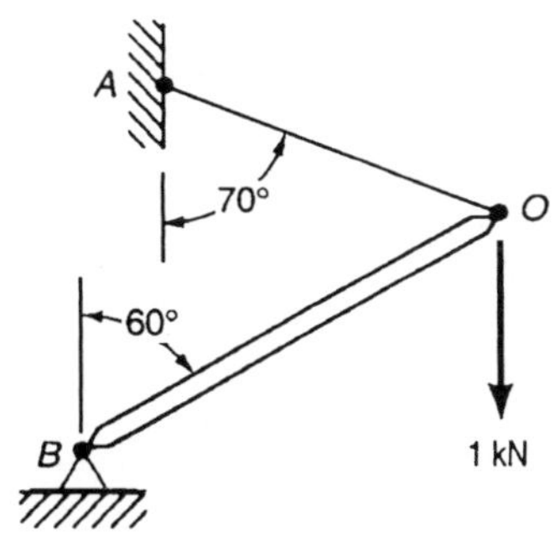

Exhibit 8.4

8.4 The components parallel to *OA* and *OB*, of the vertical 1-kN force in Exhibit 8.4 have magnitudes of:

a. 0.34 and 0.50 kN c. 1.13 and 1.23 kN
b. 0.34 and 0.66 kN d. 0.5 and 0.5 kN

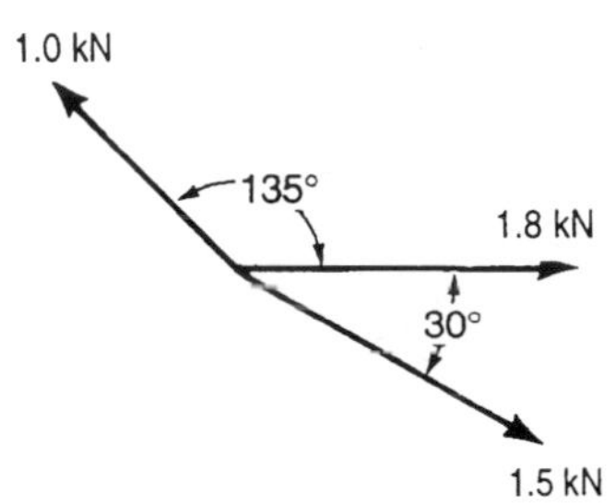

Exhibit 8.5

8.5 The resultant of the three forces in Exhibit 8.5 has a magnitude of:

a. 4.3 kN c. 2.3 kN
b. 3.0 kN d. 2.39 kN

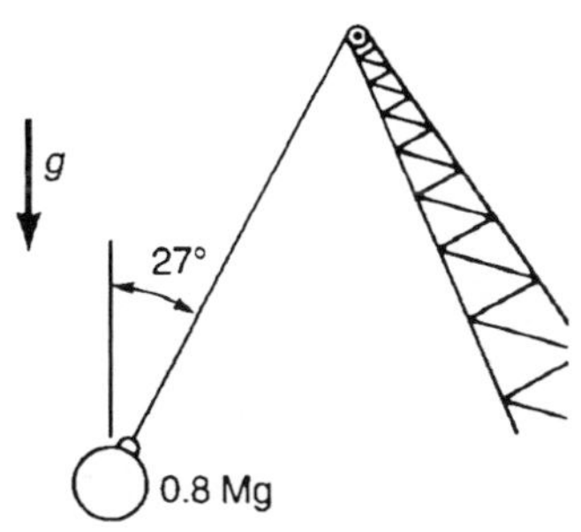

Exhibit 8.6

8.6 At a certain instant, the tension in the cable on which the destruction ball is suspended is 9.3 kN (see Exhibit 8.6). At this instant, the acceleration of the ball is:

a. 10.36 m/s^2 c. 1.21 m/s^2
b. 5.30 m/s^2 d. 9.81 m/s^2

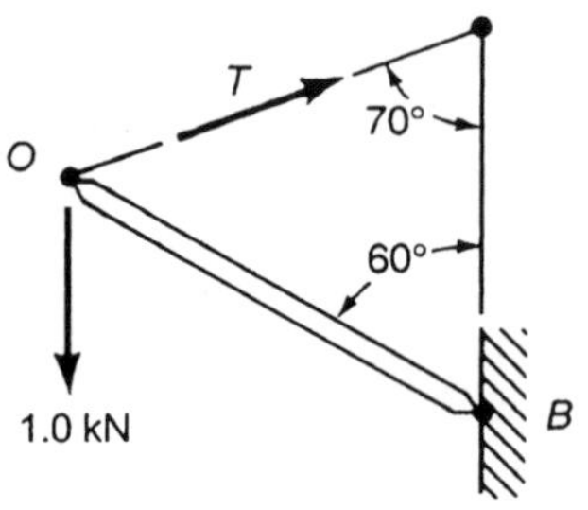

Exhibit 8.7

8.7 If the resultant of the 1-kN vertical force and the tensile force T from the cable is to be in the direction of the boom OB (see Exhibit 8.7), what must be the magnitude of T?

a. 2.9 kN c. 1.1 kN
b. 1.0 kN d. 0.94 kN

8.8 Evaluate the magnitude of the resultant force on the doorknob in Exhibit 8.8. The three components are mutually perpendicular.

a. 13 N c. 17 N
b. 19 N d. 5 N

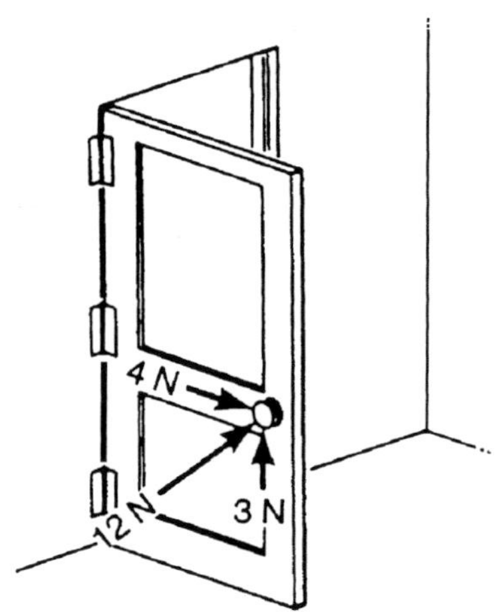

Exhibit 8.8

8.9 To raise the load, the hydraulic cylinder exerts a force of 50 kN in the direction of its axis, AB. For the position shown in Exhibit 8.9, the components of this force parallel and perpendicular to OB are:

a. 43.8 and 24.2 kN c. 23.1 and 26.9 kN
b. 41 and 28.7 kN d. 25 and 25 kN

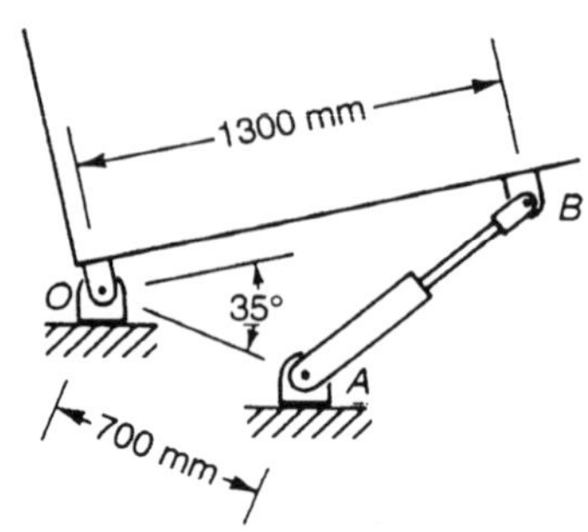

Exhibit 8.9

8.10 Refer to Exhibit 8.10. The moment about *O* of the 250-N force has a magnitude of:

a. 63.0 N•m c. 79.5 N•m
b. 77.0 N•m d. 23 N•m

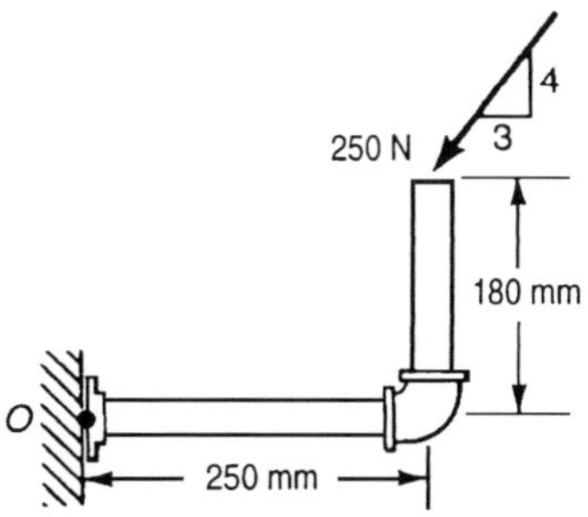

Exhibit 8.10

8.11 The connecting rod in Exhibit 8.11 exerts a force of 4.5 kN on the crank. The moment about *O* of this force has a magnitude of:

a. 408 N•m
b. 450 N•m
c. 318 N•m
d. 190 N•m

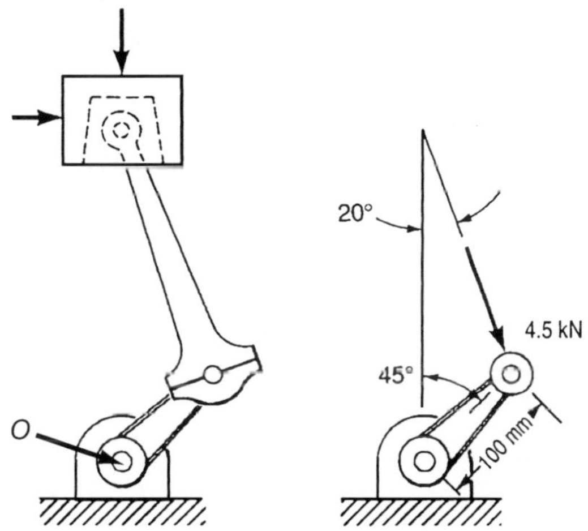

Exhibit 8.11

8.12 The tension in the line *AB* of Exhibit 8.12 is 3.50 kN. What must be the tension in the line *BC* if the moment about *O* of the force that the cable *BC* exerts on the spreader bar *OB* is equal and opposite to that of the force that the cable *AB* exerts on the spreader bar?

a. 3.50 kN c. 6.58 kN
b. 2.86 kN d. 4.29 kN

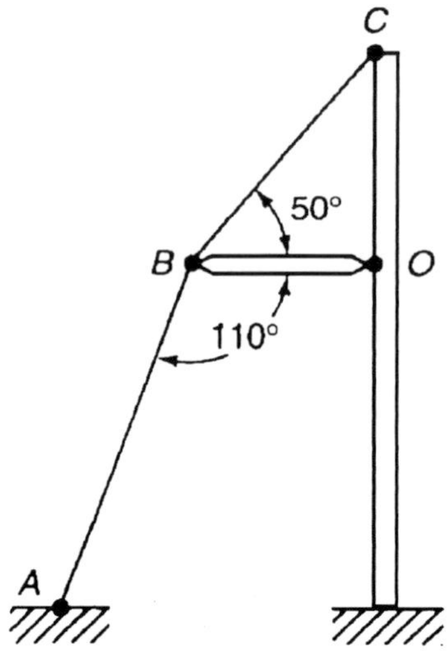

Exhibit 8.12

8.13 The plumber in Exhibit 8.13 exerts a vertical downward force of 1 kN on the wrench handle. The moment about *C* of this force has a magnitude of:

a. 500 N•m c. 900 N•m
b. 750 N•m d. 1250 N•m

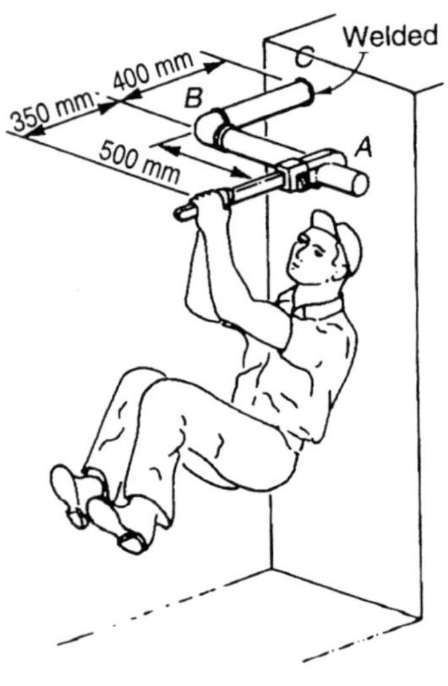

Exhibit 8.13

8.14 The moment about the axis *CB* of the previous problem has a magnitude of:

a. 500 N•m c. 900 N•m
b. 750 N•m d. 1250 N•m

8.15 The brake is set on the wheel in Exhibit 8.15, and it will not slip until the moment about the center of the wheel of forces acting on the lug wrench reaches 150 N• m. Will the brake slip?

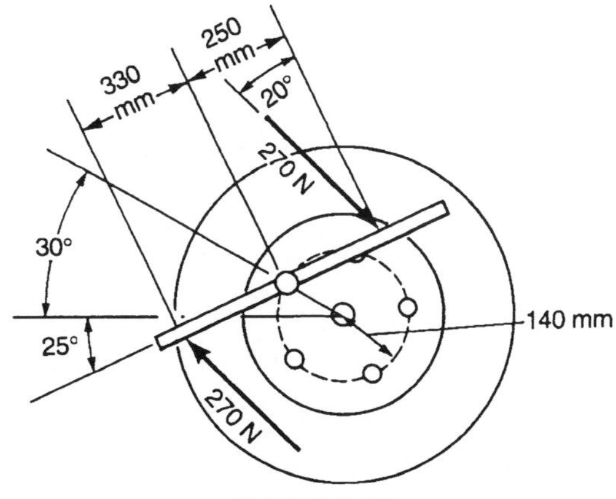

Exhibit 8.15

a. 147.2 N• m; No
b. 156.6 N• m; Yes
c. 1335.6 N• m; No
d. 313.2 N• m; Yes

8.16 The tension in the vertical line *AC* is 2 kN and that in the line *BC* is 6 kN (see Exhibit 8.16). The magnitude of the resultant force exerted by the two lines at *C* is:

a. 8.0 kN
b. 6.8 kN
c. 4.0 kN
d. 6.3 kN

Exhibit 8.16

8.17 The moment of **f** about the axis *AB* in Exhibit 8.17 has the magnitude:

a. (144/65 m)f
b. (12/5 m)f
c. (29/13 m)f
d. (12 m)f

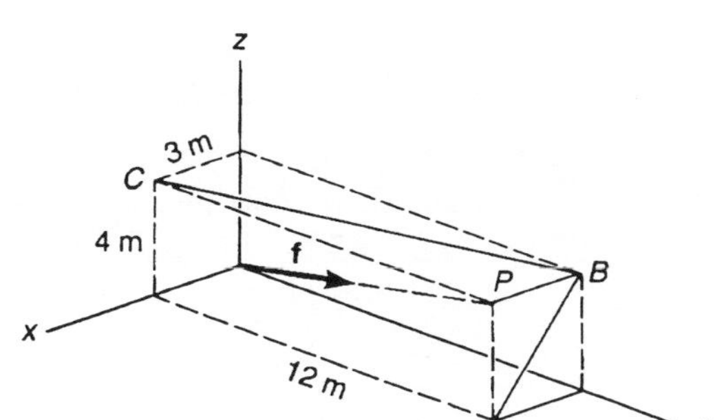

Exhibit 8.17

8.18 The surfaces are smooth where the drum makes contact (see Exhibit 8.18). The reaction at the contact point on the right is:

a. 1.77 kN c. 2.29 kN
b. 3.06 kN d. 2.70 kN

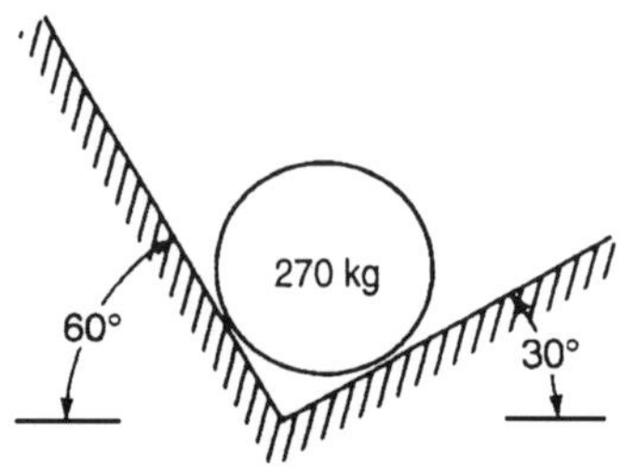

Exhibit 8.18

8.19 Two wheels, each of radius a but of different mass, are connected by the rod of length R in Exhibit 8.19. The assembly is free to roll in the circular trough. The angle θ for equilibrium is given by:

a. $\tan\theta = \dfrac{f_{g_1} - f_{g_2}}{f_{g_1} + f_{g_2}} \tan 30^\circ$

b. $\sin\theta = \dfrac{f_{g_1} - f_{g_2}}{f_{g_1} + f_{g_2}} \sin 30^\circ$

c. $\tan\theta = \dfrac{f_{g_1} - f_{g_2}}{f_{g_1} + f_{g_2}} \sin 30^\circ$

d. $\sin\theta = \dfrac{f_{g_1} - f_{g_2}}{f_{g_1} + f_{g_2}} \tan 30^\circ$

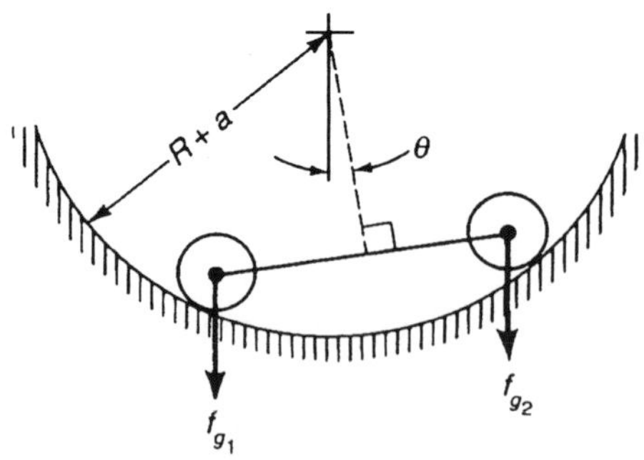

Exhibit 8.19

Exhibit 8.20

8.20 Determine the force with which the 80-kg man in Exhibit 8.20 must pull on the rope to support himself. The force is closest to:

a. 785 N c. 471 N
b. 628 N d. 157 N

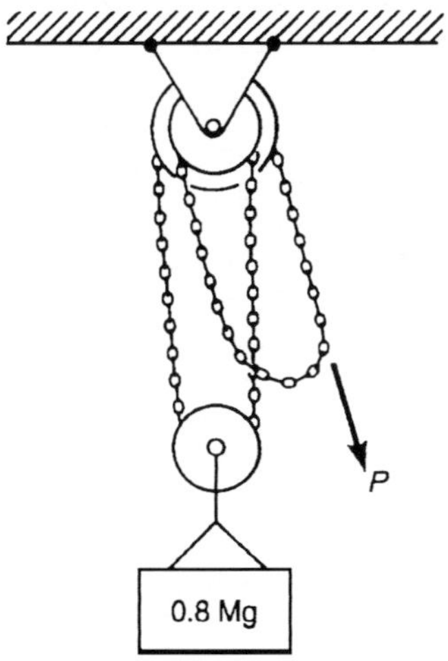

Exhibit 8.21

8.21 Each of the tracks in the upper pulley unit is recessed to fit the chain, so as to prevent slipping (see Exhibit 8.21). The smaller track has a radius equal to 0.9 times that of the larger track. Evaluate the force *P* necessary to lift the block by means of the differential chain hoist. The force is nearest to:

a. 0.392 kN c. 3.92 kN
b. 0.784 kN d. 1.96 kN

8.22 The weight of the linkage in Exhibit 8.22 is negligible compared with f_g. What is the value of *P* necessary to maintain equilibrium?

a. $0.333f_g$ c. $0.577f_g$
b. $0.500f_g$ d. $0.144f_g$

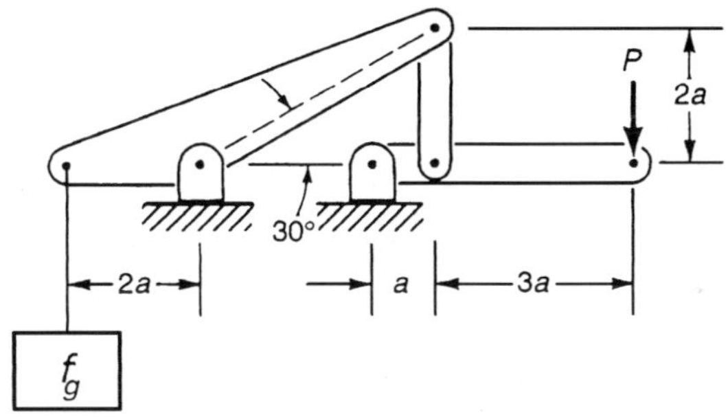

Exhibit 8.22

8.23 Neglecting the mass of the structure of Exhibit 8.23, the tension in the bar *AB*, induced by the 320-kg lifeboat, is nearest to:

a. 3.14 kN c. 1.57 kN
b. 0.32 kN d. 12.55 kN

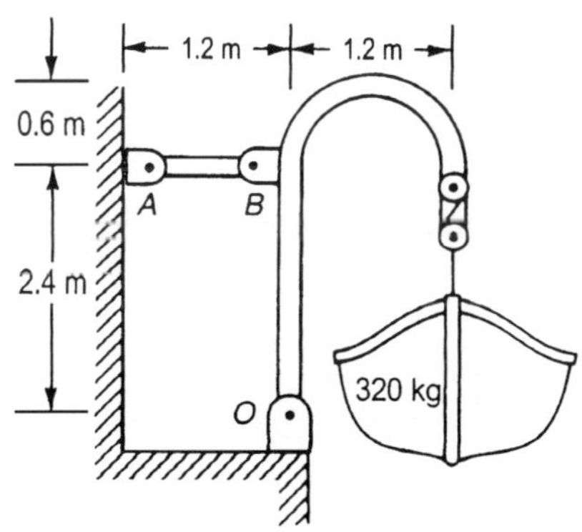

Exhibit 8.23

8.24 Until a clamp is tightened, the drill press table is free to slide along the column (see Exhibit 8.24). Estimate the coefficient of friction required so that the collar will be self-locking against the column under the action of the thrust from the drill. Neglect gravity.

a. 0.21 c. 0.63
b. 0.42 d. 0.84

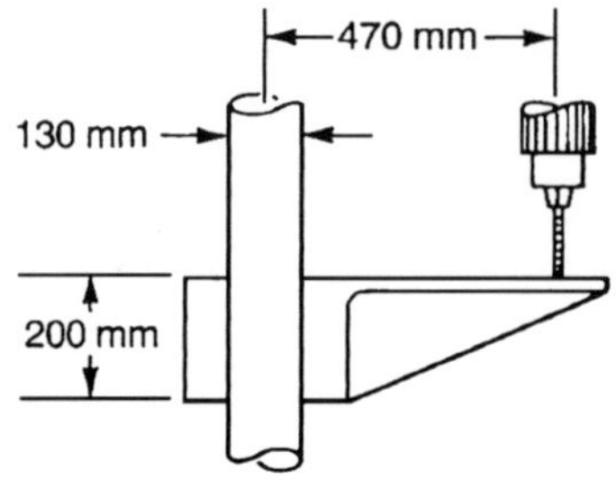

Exhibit 8.24

8.25 The combined reaction at the two rear wheels of the car in Exhibit 8.25 has the magnitude:

a. 5.0 kN c. 7.0 kN
b. 4.0 kN d. 6.3 kN

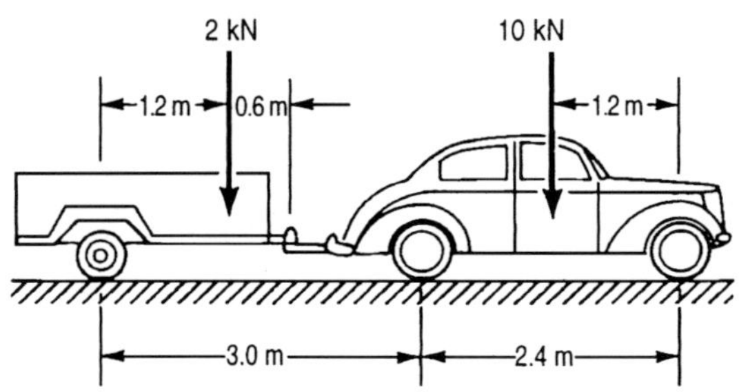

Exhibit 8.25

8.26 In a trailer "load-leveler" hitch, the angle bar slips into the cylindrical socket at *A*, forming a thrust bearing where the bar bottoms (see Exhibit 8.26(a)). The end, *B*, is then attached by a short chain to the towing vehicle, as in Exhibit 8.26(b). The pretension in the chain is 1.7 kN. The reaction at *C* (both trailer wheels) has magnitude:

a. 8.34 kN c. 6.20 kN
b. 8.05 kN d. 5.25 kN

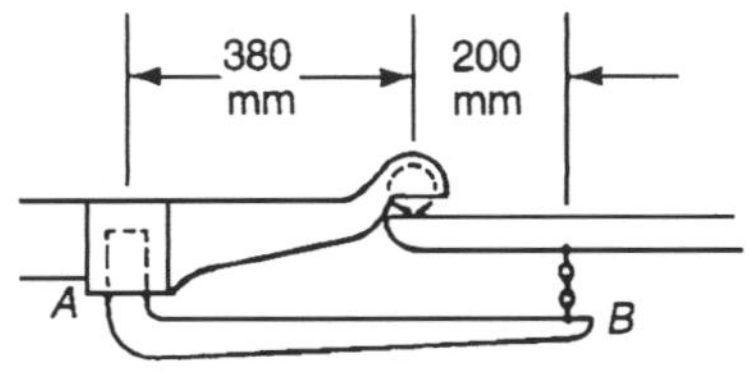

Exhibit 8.26a

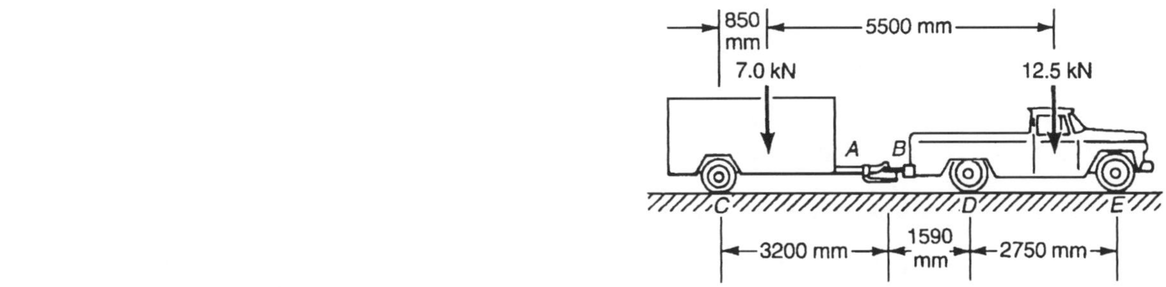

Exhibit 8.26b

8.27 The reaction at the rear wheels, *D*, of the vehicle of Problem 8.26 has magnitude:

a. 8.34 kN c. 6.20 kN
b. 8.05 kN d. 5.25 kN

8.28 Evaluate the cutting force at *C* in terms of the force *P* on the handles of the compound snips (Exhibit 8.28).

a. 2*P* c. 6*P*
b. 4*P* d. 8*P*

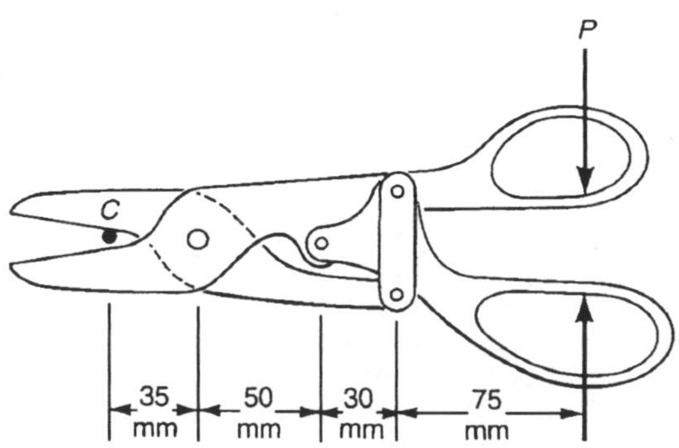

Exhibit 8.28

8.29 The magnitude of the force on the nut at *C* exerted by the jaws of the self-locking pliers (Exhibit 8.29) is:

a. 240 N c. 1600 N
b. 480 N d. 2020 N

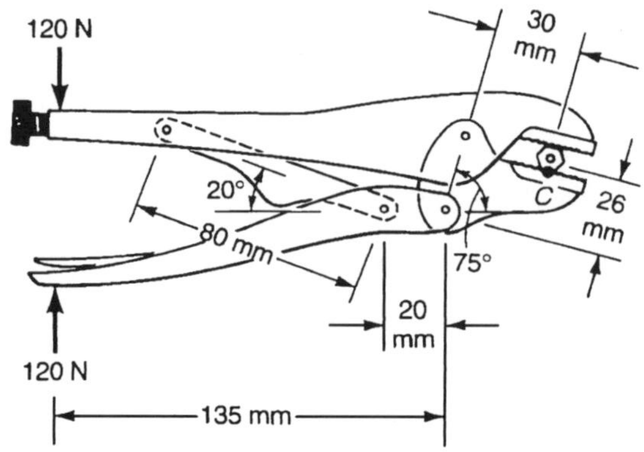

Exhibit 8.29

8.30 The moment of the couple transmitted through the shaft at section *A* (Exhibit 8.30) has magnitude:

a. 9.60 N•m c. zero
b. 15.0 N•m d. 30 N•m

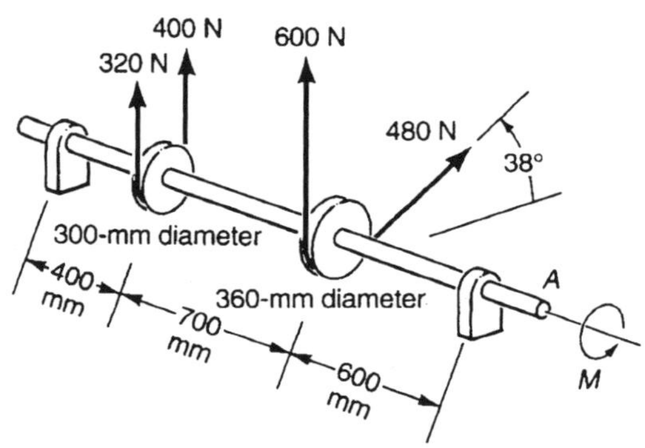

Exhibit 8.30

8.31 The reaction at the bearing near *A* in Problem 8.30 has magnitude:

a. 245 N c. 788 N
b. 745 N d. 1296 N

8.32 The slider A has a mass of 4 kg and is constrained to slide without friction along the fixed vertical rod (Exhibit 8.32). The mass of the wire AB is negligible. The slider B is constrained to slide along the horizontal rod without friction. What must be the magnitude F of the force applied to the slider B to maintain equilibrium?

a. 39.2 N c. 98 N
b. 19.6 N d. 49 N

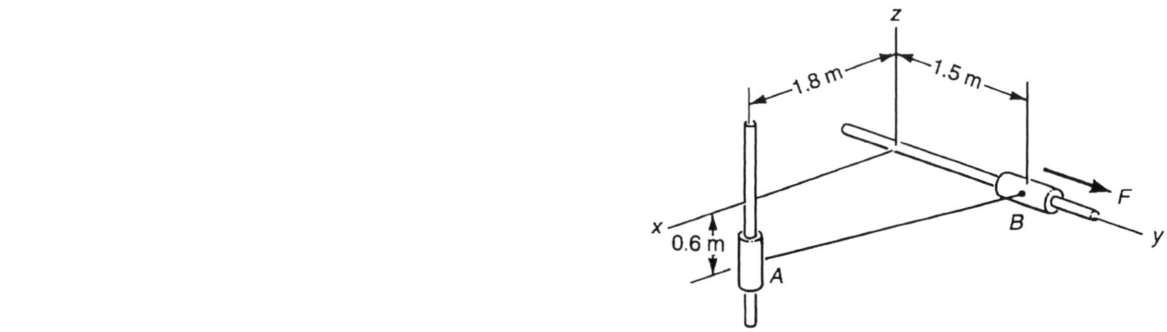

Exhibit 8.32

8.33 The turnbuckle in the guyline AE of Exhibit 8.33 is to be tightened such that the vertical component of the reaction at O is 800 N. What must be the tension in AE?

a. 424 N c. 636 N
b. 212 N d. 267 N

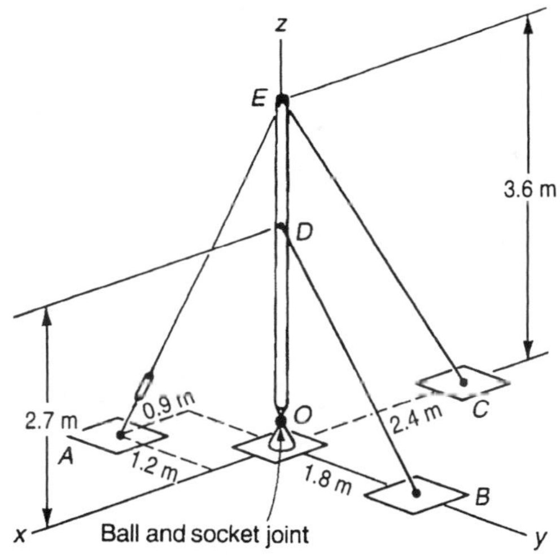

Exhibit 8.33

8.34 Determine the tensions in the cables *AB* and *CD* of Exhibit 8.34. They are closest to:

a. 2.27 kN, 1.45 kN
b. 0.75 kN, 0.75 kN
c. 0.45 kN, 1.05 kN
d. 1.87 kN, 1.20 kN

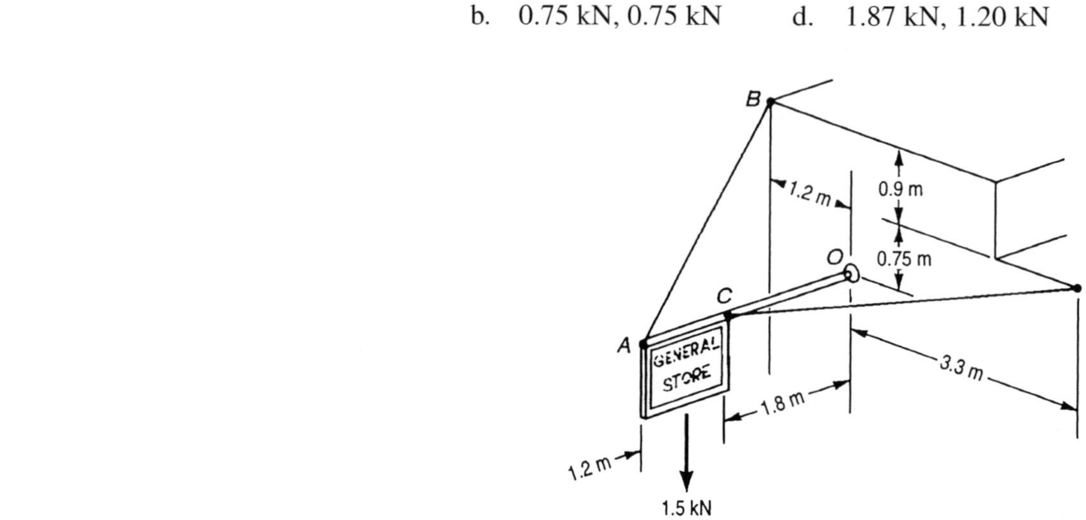

Exhibit 8.34

8.35 The 40-kg rectangular plate in Exhibit 8.35 is held by hinges along its edge *OA* and by the wire *BD*. The gravity force f_g acts through the geometric center of the plate. What is the tension of the wire?

a. 392 N
b. 196 N
c. 36.5 N
d. 358 N

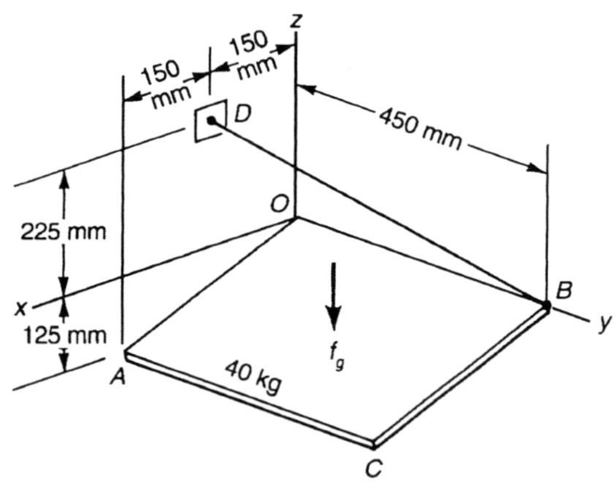

Exhibit 8.35

8.36 The force in the member *BC* of Exhibit 8.36 is:

a. 12 kN, tension
b. 17 kN, tension
c. 17 kN, compression
d. 12 kN, compression

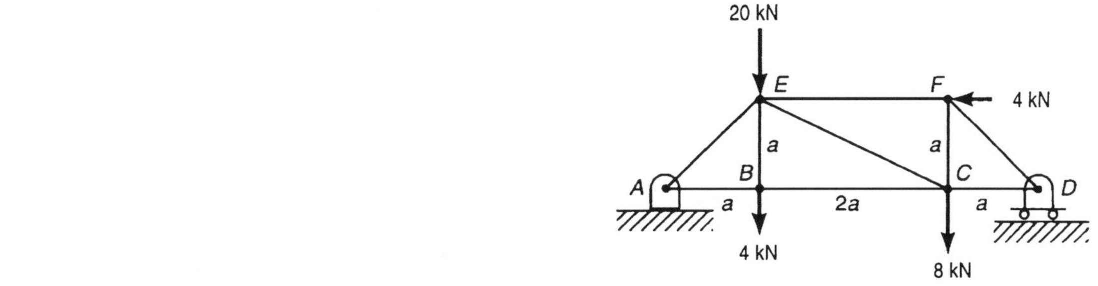

Exhibit 8.36

8.37 The force in the member *ML* of Exhibit 8.37 is:

a. 131 kN, tension
b. 131 kN, compression
c. 100 kN, tension
d. 106 kN, tension

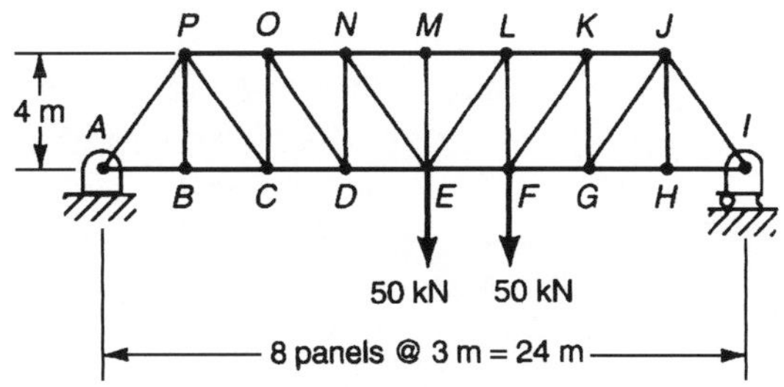

Exhibit 8.37

8.38 The force in the member *CF* of Exhibit 8.38 is:

a. 60.6 kN, tension
b. 60.6 kN, compression
c. 43.3 kN, tension
d. 43.3 kN, compression

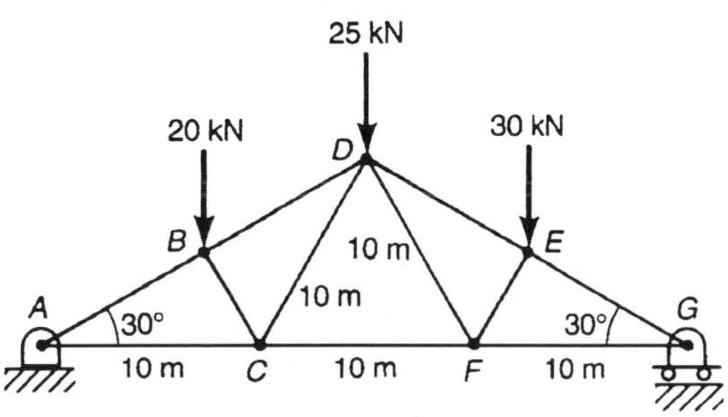

Exhibit 8.38

8.39 Members *AD*, *BE*, and *CF* are perpendicular to the plane *ABC* (see Exhibit 8.39). The force in the member *AD* is:

a. 6 kN, compression
b. 6 kN, tension
c. 8 kN, tension
d. 8 kN, compression

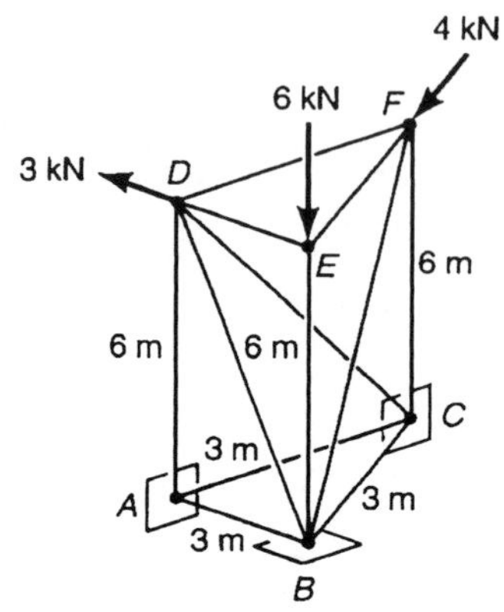

Exhibit 8.39

8.40 The force in member *BE*, shown in Exhibit 8.40, is:

a. 0.87*P*, compression
b. *P*, compression
c. 0.87*P*, tension
d. *P*, tension

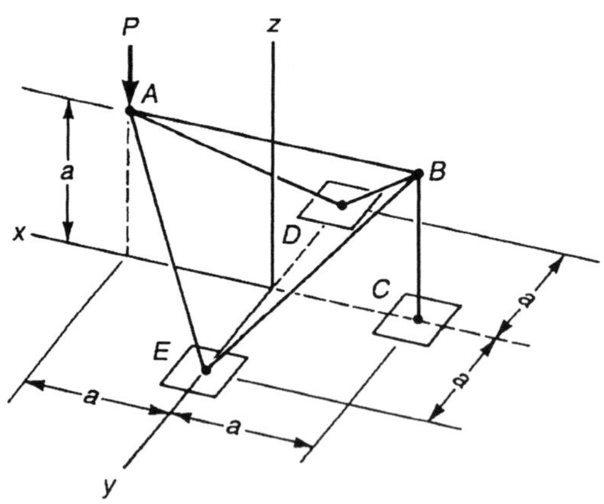

Exhibit 8.40

8.41 The contents of the crate in Exhibit 8.41 are such that the center of gravity is at the geometric center. In the absence of P, the crate will:

a. remain stationary
b. slide downward
c. tip over
d. slide upward

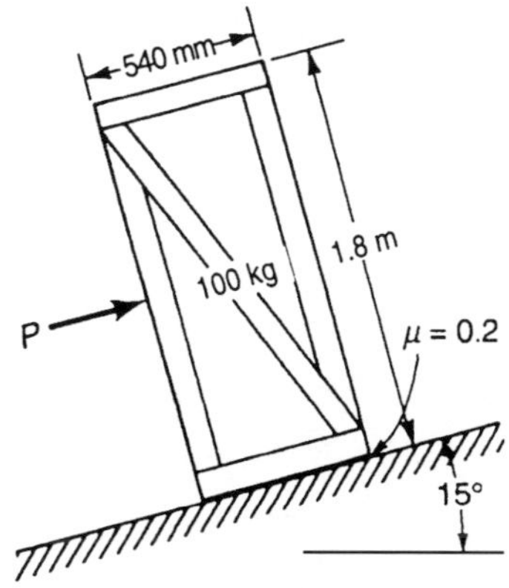

Exhibit 8.41

8.42 In Exhibit 8.41, the greatest distance from the incline to the line of action of P, such that the force can slide the crate up the incline without it, is:

a. 1.8 m
b. 0.9 m
c. 1.35 m
d. 1.09 m

8.43 The contents of the crate in Exhibit 8.43 are such that the center of gravity coincides with the geometric center. The greatest value of the angle that will allow the force to slide the crate without tipping it is given by:

a. $\tan^{-1}\left(\frac{h}{b}\right)$
b. $\tan^{-1}\left(\frac{b}{h}\right)$
c. $\tan^{-1}\left(\frac{1}{\mu}-\frac{2h}{b}\right)$
d. $\tan^{-1}\left(\frac{1}{\mu}+\frac{2h}{b}\right)$

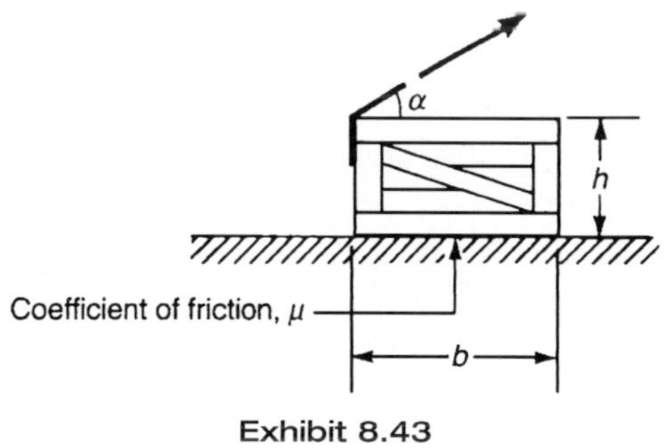

Exhibit 8.43

8.44 If the coefficient of friction between all surfaces in Exhibit 8.44 is 0.27, what will be the minimum value of P necessary to initiate motion?

a. 103 N c. 40.5 N
b. 81.0 N d. 51.5 N

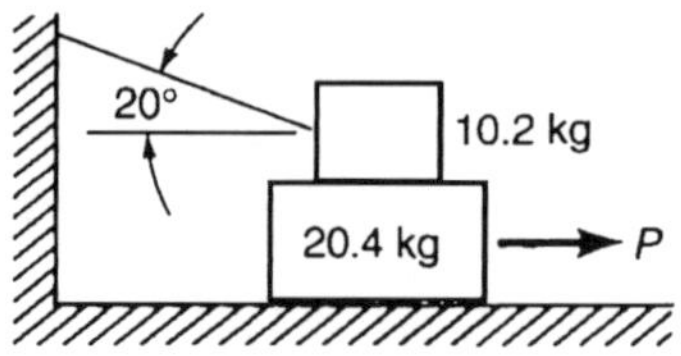

Exhibit 8.44

8.45 If the coefficient of friction between all surfaces in Exhibit 8.45 is 0.27, what will be the minimum value of P necessary to initiate motion?

a. 205 N c. 94.5 N
b. 103 N d. 9.63 N

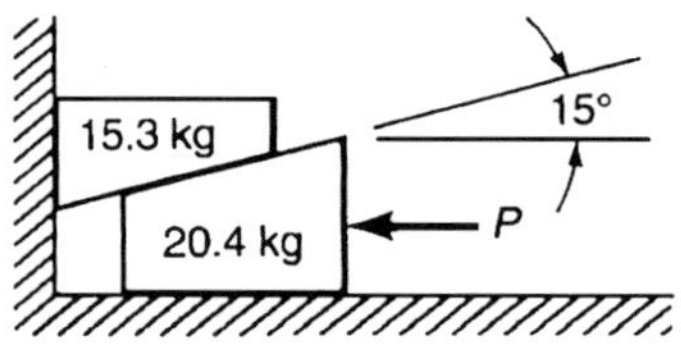

Exhibit 8.45

8.46 The forklift in Exhibit 8.46 is being used to roll the 2-Mg drum up the 40-degree incline while the height of the forks remains constant. The coefficient of friction is 0.45 between the vertical rails and the drum, and 0.30 between the incline and the drum. What horizontal thrust must the vehicle apply to the drum to move it?

a. 85 kN c. 78.2 kN
b. 62.7 kN d. 103 kN

Exhibit 8.46

8.47 Determine the minimum coefficient of friction between the block and the weightless bar in Exhibit 8.47 necessary to prevent collapse.

a. 0.15 c. 0.42
b. 0.27 d. 0.72

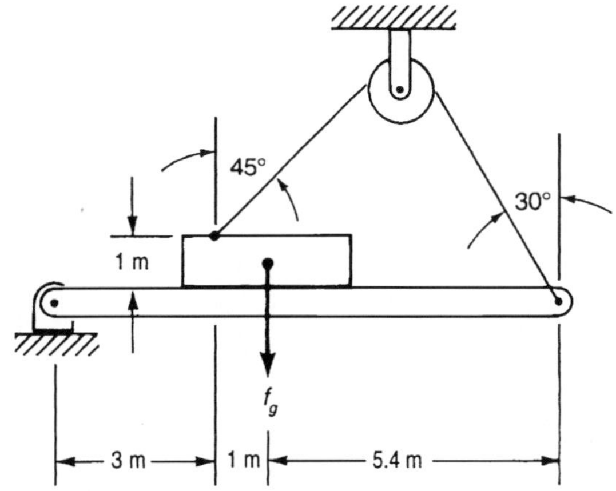

Exhibit 8.47

8.48 The small rollers in Exhibit 8.48 are intended to prevent clockwise rotation of the large drum. The coefficient of friction between the rollers and drum, and between the rollers and walls, is μ. Determine the minimum distance d such that the friction will effect a self-locking mechanism against clockwise rotation. Gravity is negligible.

a. $d > (a+b)\mu$ c. $d > \dfrac{2a+(1-\mu^2)b}{1+\mu^2}$

b. $d > \dfrac{a+b}{\mu}$ d. $d > \dfrac{(a+b)\mu}{1+\mu^2}$

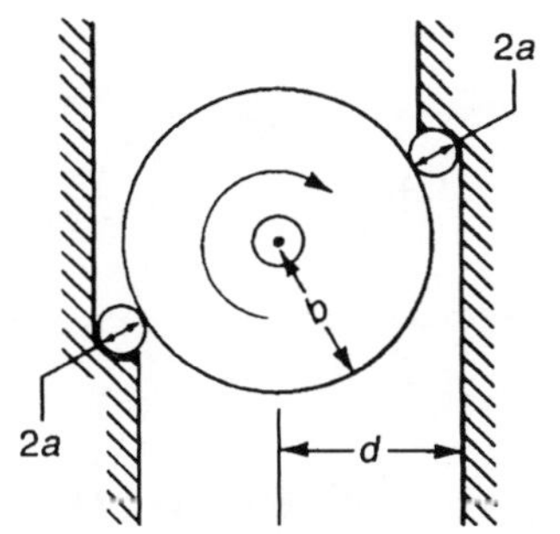

Exhibit 8.48

SOLUTIONS

8.1 d. $f_g = (9.2 \text{ N/kg})(70 \text{ kg}) = 644 \text{ N}$

8.2 a. $P = (14.7 \text{ lbf/in.}^2)(39.37 \text{ in./m})^2(4.448 \text{ N/lbf}) = 101{,}000 \text{ N/m}^2$
$= 101 \text{ kPa}$

8.3 a. Fuel economy = (29 mi/gal)

$$\frac{(5280 \text{ ft/mi})(0.3048 \text{ m/ft})}{(1000 \text{ m/km})(3.7854 \text{ L/gal})} = 12.33 \text{ km/L}$$

8.4 c.

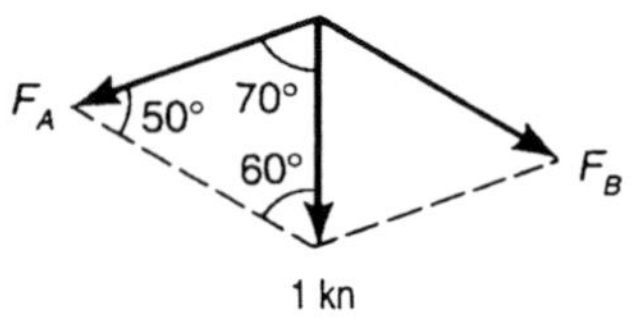

Exhibit 8.4a

$$\frac{F_A}{\sin 60°} = \frac{F_B}{\sin 70°} = \frac{1 \text{ kN}}{\sin 50°}$$

$$F_A = \frac{\sin 60°}{\sin 50°}(1 \text{ kN}) = 1.13 \text{ kN}$$

$$F_B = \frac{\sin 70°}{\sin 50°}(1 \text{ kN}) = 1.23 \text{ kN}$$

8.5 d.

$$F_x = 1.8 \text{ kN} + (1.5 \text{ kN}) \cos 30° + (1.0 \text{ kN}) \cos 135° = 2.39 \text{ kN}$$
$$F_y = (1.0 \text{ kN}) \sin 135° - (1.5 \text{ kN}) \sin 30° = -0.043 \text{ kN}$$
$$F = \sqrt{F_x^2 + F_y^2} = 2.39 \text{ kN}$$

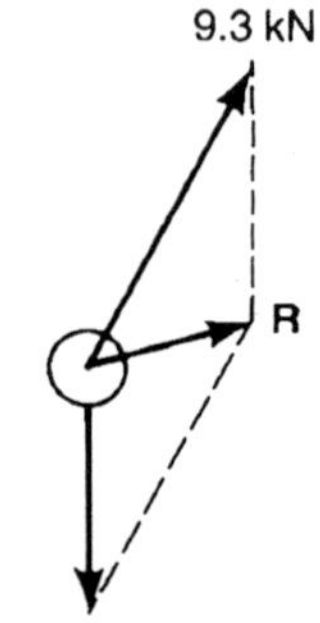

Exhibit 8.6a

8.6 b.
The force of gravity on the ball is $f_g = (800 \text{ kg})(9.81 \text{ N/kg}) = 7.85 \text{ kN}$.
The net force, $\mathbf{R} = x\,\mathbf{e}_x + y\,\mathbf{e}_y = (9.3 \text{ kN} \sin 27°)\,\mathbf{e}_x + (9.3 \text{ kN} \cos 27° - 7.85 \text{ kN})\,\mathbf{e}_y$

$$|\mathbf{R}| = R = \sqrt{x^2 + y^2}$$

$$R = \sqrt{(9.3 \text{ kN})^2 \sin^2 27° + (9.3 \text{ kN})^2 \cos^2 27° - 2(9.3 \text{ kN})(\cos 27°)(7.85 \text{ kN}) + (7.85 \text{ kN})^2}$$

$$R = 4.24 \text{ kN}$$

$$a = \frac{4.24 \text{ kN}}{0.8 \text{ Mg}} = 5.30 \text{ m/s}^2$$

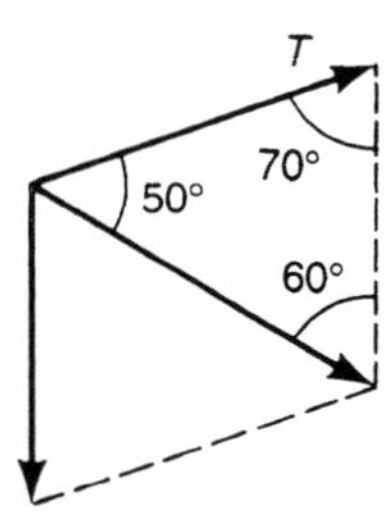

Exhibit 8.7a

8.7 c.

$$\frac{T}{\sin 60°} = \frac{1 \text{ kN}}{\sin 50°}$$

$$T = 1.13 \text{ kN}$$

8.8 a.

$$F = \sqrt{(4)^2 + (12)^2 + (3)^2} = 13 \text{ N}$$

8.9 a.

First use geometry to solve the triangle.

$$\overline{AB} = \sqrt{(0.7)^2 + (1.3)^2 - 2(0.7)(1.3)\cos 35^\circ} \text{ m} = 0.83 \text{ m}$$

$$\frac{\sin_B \angle_A^O}{0.7 \text{ m}} = \frac{\sin 35^\circ}{0.83 \text{ m}};\ {}_B\angle_A^O = 28.92^\circ$$

Then, compute the forces.

$$(50 \text{ kN})\cos\ 28.9^\circ = 43.8 \text{ kN}$$

$$(50 \text{ kN})\sin\ 28.9^\circ = 24.2 \text{ kN}$$

8.10 d.

Note that the direction of the 250-N force, **f**, is described by a 3-4-5 triangle. Separate **f** into its x and y components.

$$\mathbf{f} = (-250 \text{ N } (3/5))\ \mathbf{e}_x + (-250 \text{ N } (4/5))\ \mathbf{e}_y$$
$$M_0 = \mathbf{r} \times \mathbf{f}$$
$$\mathbf{r} = 0.25 \text{ m } \mathbf{e}_x + 0.18 \text{ m } \mathbf{e}_y$$
$$M_0 = (0.18 \text{ m})(150 \text{ N}) - (0.25 \text{ m})(200 \text{ N}) = -23.0 \text{ N}\bullet\text{m}$$

8.11 a.

$$M_0 = (0.1 \text{ m}) \cos 25^\circ\ (4.5 \text{ kN}) = 408 \text{ N}\bullet\text{m}$$

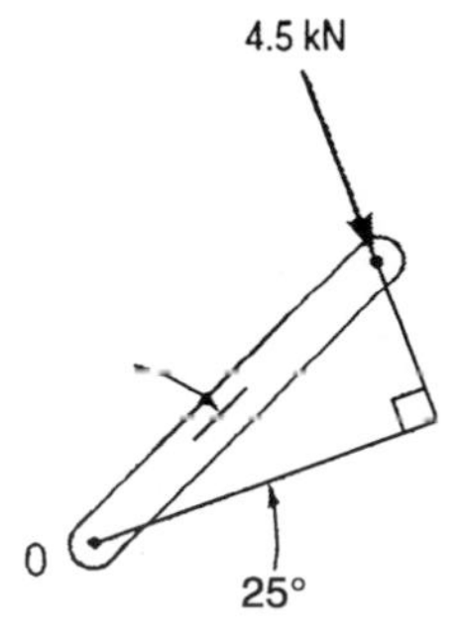

Exhibit 8.11a

8.12 d.

$$lT_{BC} \sin 50^\circ = l(3.5 \text{ kN}) \sin 110^\circ$$
$$T_{BC} = 4.29 \text{ kN}$$

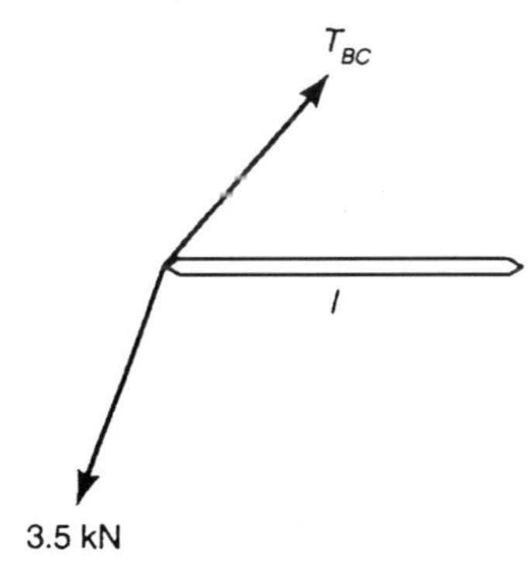

Exhibit 8.12a

8.13 c.

$$r = \sqrt{(0.75 \text{ m})^2 + (0.5 \text{ m})^2} = 0.9014 \text{ m}$$
$$M_C = (0.9014 \text{ m})(1 \text{ kN}) = 901.4 \text{ N}\bullet\text{m}$$

8.14 a.

$$M_{CB} = (0.5 \text{ m})(1 \text{ kN}) = 500 \text{ N}\bullet\text{m}$$

8.15 a.

$$M = (250 \text{ mm} + 330 \text{ mm}) \cos 20° \,(270 \text{ N}) = 147.2 \text{ N} \bullet \text{m}$$

8.16 b.

$$\overline{CB} = \sqrt{(2.4 \text{ m})^2 + (3.6 \text{ m})^2 + (1.2 \text{ m})^2} = 4.49 \text{ m}$$

$$R_x = \frac{2.4}{4.49}(6 \text{ kN}) = 3.207 \text{ kN}$$

$$R_y = \frac{3.6}{4.49}(6 \text{ kN}) = 4.811 \text{ kN}$$

$$R_z = \frac{1.2}{4.49}(6 \text{ kN}) + 2 \text{ kN} = 3.604 \text{ kN}$$

$$R = \sqrt{R_x^2 + R_y^2 + R_z^2} = 6.81 \text{ kN}$$

8.17 a.

$$M_{AB} = \mathbf{e}_{AB} \bullet (\mathbf{r}_{AP} \times \mathbf{f})$$

$$\mathbf{e}_{AB} = \frac{-3\mathbf{e}_x + 4\mathbf{e}_z}{\sqrt{(-3)^2 + 4^2}}$$

$$\mathbf{r}_{AP} = (4 \text{ m})\mathbf{e}_z$$

$$\mathbf{f} = \left(\frac{3\mathbf{e}_x + 12\mathbf{e}_y + 4\mathbf{e}_z}{\sqrt{3^2 + 12^2 + 4^2}}\right) f$$

$$M_{AB} = \left(-\frac{3}{5}\mathbf{e}_x + \frac{4}{5}\mathbf{e}_z\right) \bullet \left[(4 \text{ m})\mathbf{e}_z \times \left(\frac{3}{13}\mathbf{e}_x + \frac{12}{13}\mathbf{e}_y + \frac{4}{13}\mathbf{e}_z\right) f\right]$$

$$= \begin{vmatrix} -\frac{3}{5} & 0 & \frac{4}{5} \\ 0 & 0 & 4 \text{ m} \\ \frac{3}{13} & \frac{12}{13} & \frac{4}{13} \end{vmatrix} f = \left(\frac{144}{65} \text{ m}\right) f$$

8.18 c.

$$R_R - (2648 \text{ N}) \cos 30° = 0$$
$$R_R = 2.29 \text{ kN}$$

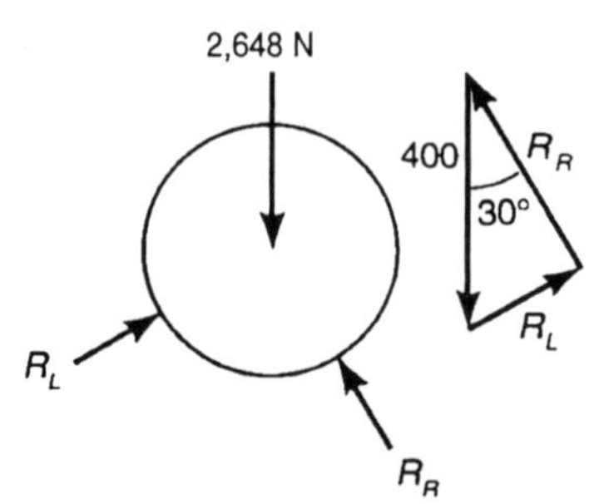

Exhibit 8.18a

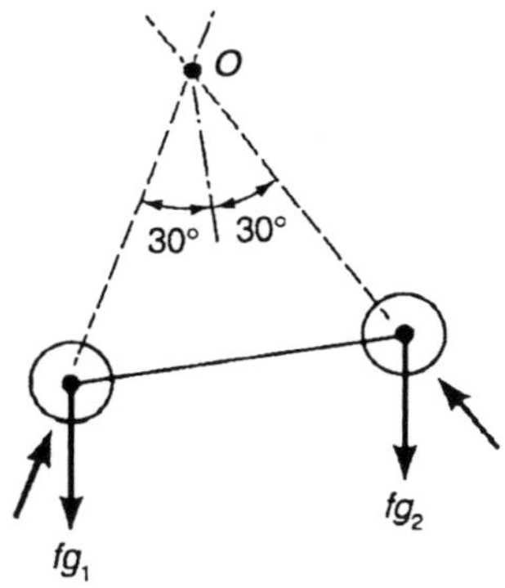

Exhibit 8.19a

8.19 a.

$$M_0 = R \sin(30° - \theta) f_{g_1} - R \sin(30° + \theta) f_{g_2} = 0$$

$$f_{g_1} (\sin 30° \cos\theta - \cos 30° \sin\theta) - f_{g_2} (\sin 30° \cos\theta + \cos\theta + \cos 30° \sin\theta) = 0$$

Dividing by cos 30° cos θ,

$$f_{g_1}(\tan 30° - \tan\theta) = f_{g_2}(\tan 30° + \tan\theta)$$

$$\tan\theta = \frac{f_{g_1} - f_{g_2}}{f_{g_1} + f_{g_2}} \tan 30°$$

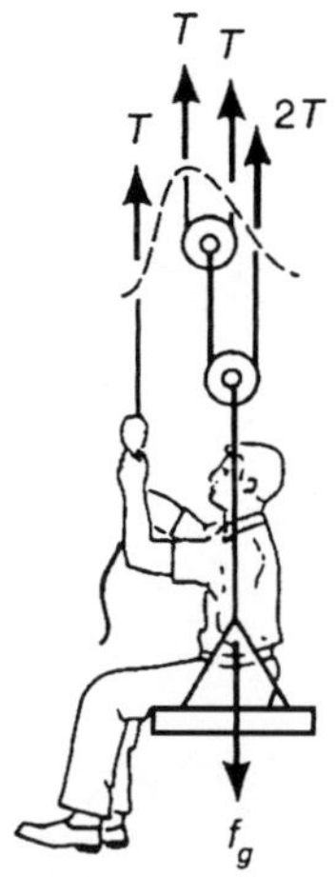

Exhibit 8.20a

8.20 d.

$$T + T + T + 2T - f_g = 0$$

$$T = \frac{1}{5} f_g$$

$$= \frac{1}{5}(80 \text{ kg})(9.8 \text{ N/kg}) = 157 \text{ N}$$

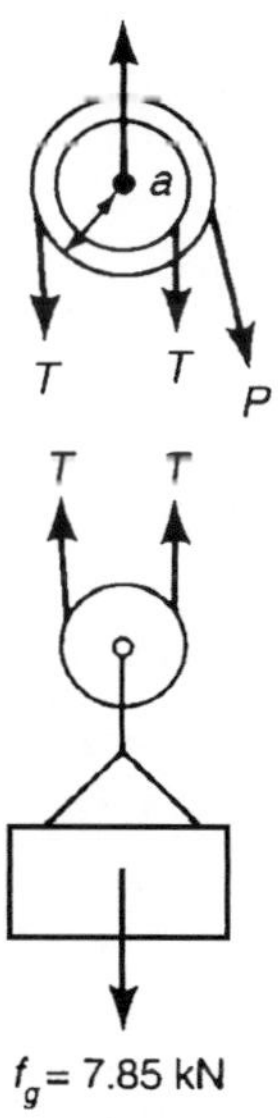

Exhibit 8.21a

8.21 a. The force of gravity on the block is

$$f_g = 0.8 \text{ Mg } (9.81 \text{ m/s}^2) = 7.85 \text{ kN}$$

$$aT - 0.9aT - aP = 0$$

$$P = 0.1T$$

$$2T = f_g$$

$$P = 0.1\left(\frac{1}{2} f_g\right) = 0.05 f_g = 0.392 \text{ kN}$$

8.22 d.

$$\sum M_A = aQ - 4aP = 0 \quad Q = 4P$$

$$\sum M_B = 2af_g - 2a \cot 30° \quad Q = 0$$

$$f_g - \sqrt{3}\, Q$$

$$P = \frac{Q}{4} = \frac{f_g}{4\sqrt{3}} = 0.144 f_g$$

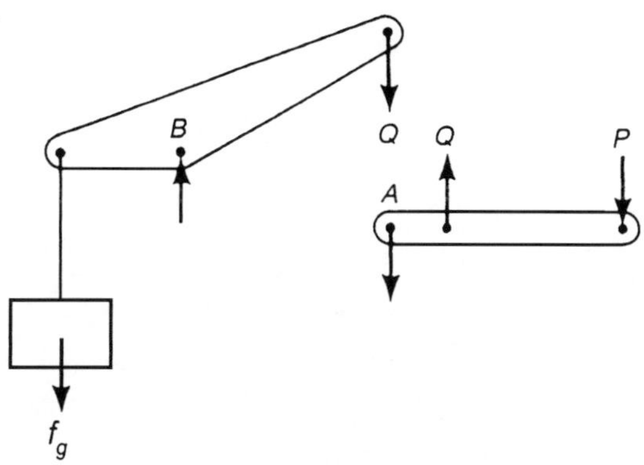

Exhibit 8.22a

8.23 c.

$$\sum M_O = (2.4 \text{ m})T_{AB} - (1.2 \text{ m})f_g = 0$$

$$T_{AB} = \frac{1.2}{2.4}(320 \text{ kg})(9.8 \text{ N/kg}) = 1.57 \text{ kN}$$

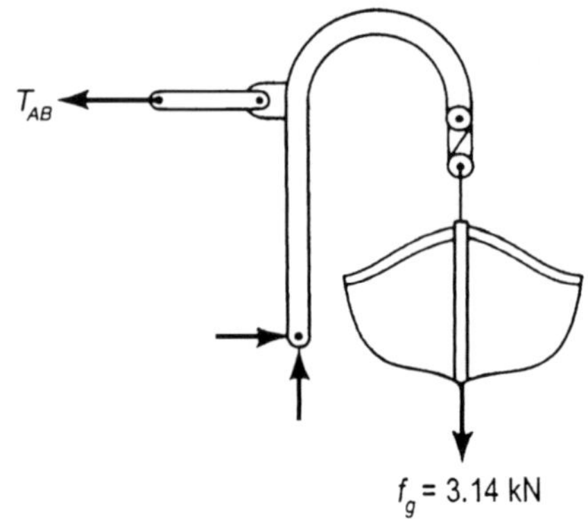

Exhibit 8.23a

8.24 a. Neglecting the force of gravity on the table, $N(200 \text{ mm}) + \mu N(130 \text{ mm}) = P(535 \text{ mm})$ by summing moments about point A. Thus,

$$2\mu N \geq P = \left(\tfrac{200}{535} + \tfrac{130\mu}{535}\right)N$$

$$\mu \geq \frac{\frac{200}{535}}{2 - \frac{130}{535}} = \frac{10}{47} = 0.213$$

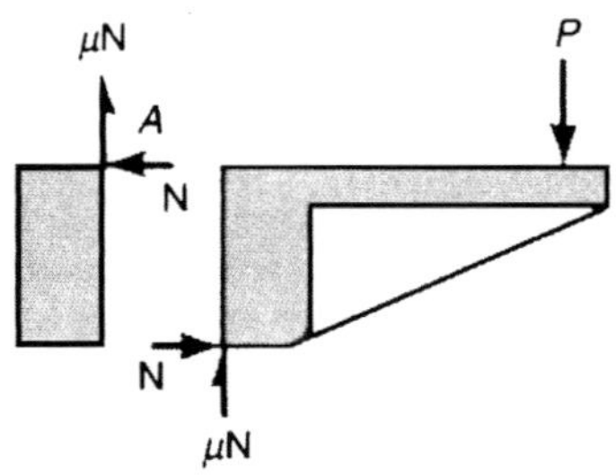

Exhibit 8.24a

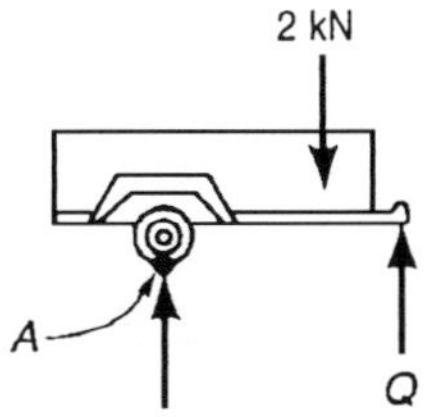

Exhibit 8.25a

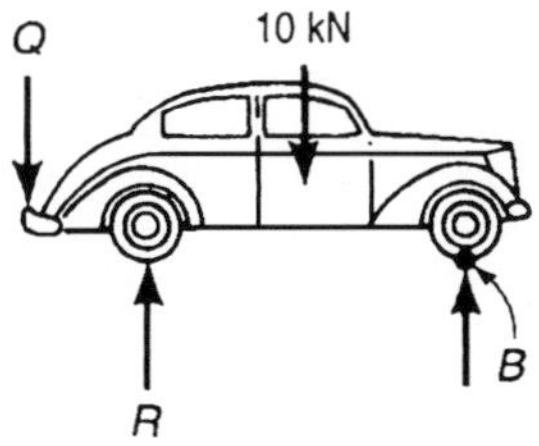

Exhibit 8.25b

8.25 c.

$$\sum M_A = (1.8\text{ m})Q - (1.2\text{ m})(2\text{ kN}) = 0$$

$$Q = \frac{1.2}{1.8}(2\text{ kN}) = 1.33\text{ kN}$$

$$\sum M_B = (3.6\text{ m})Q - (2.4\text{ m})R + (1.2\text{ m})(10\text{ kN}) = 0$$

$$R = \frac{(3.6\text{ m})(1.33\text{ kN}) + (1.2\text{ m})(10\text{ kN})}{2.4\text{ m}} = 7.0\text{ kN}$$

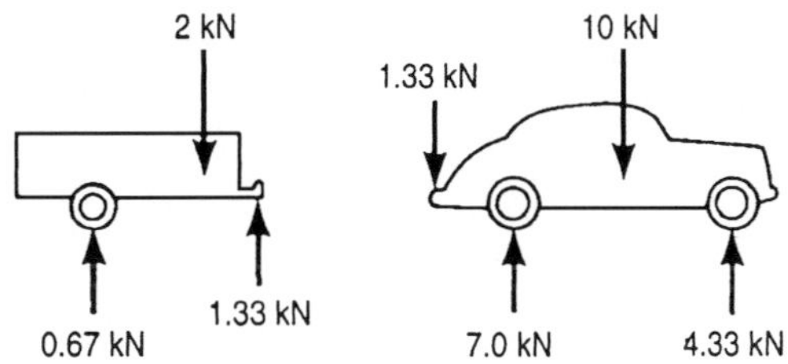

Exhibit 8.25c

8.26 d.

$$\sum M_A = (0.58\text{ m})(1.7\text{ kN}) - M = 0$$

$$M = 0.986\text{ kN}\bullet\text{m}$$

$$\sum F_U = 1.7\text{ kN} - P = 0$$

$$P = 1.7\text{ kN}$$

$$\sum M_C = R(3200\text{ mm}) + 986\text{ N}\bullet\text{m} + (1.7\text{ kN})(2820\text{ mm}) - (7.0\text{ kN})(850\text{ mm}) = 0$$

$$R = 0.053\text{ kN}$$

$$R_C = 7.0\text{ kN} - 1.7\text{ kN} - \text{R} = 5.25\text{ kN}$$

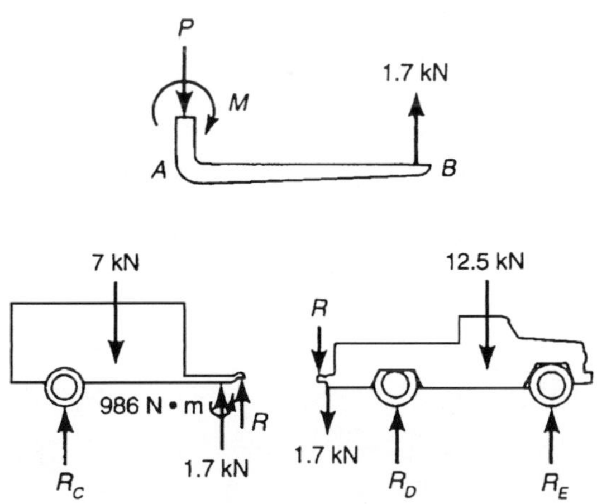

Exhibit 8.26c

8.27 c.

$$\sum M_E = (12.5\text{ kN})(1190\text{ mm}) + (1.7\text{ kN})(4140\text{ mm}) + R(4340\text{ mm}) - R_D(2750\text{ mm}) = 0$$

$$R_D = 8.05\text{ kN}$$

8.28 d.

$$\sum M_A = (105\text{ mm})P - (30\text{ mm})Q = 0$$

$$Q = 3.5\text{P}$$

$$\sum M_B = (80\text{ mm})Q - (35\text{ mm})R = 0$$

$$R = \frac{80}{35}(3.5P) = 8.0P$$

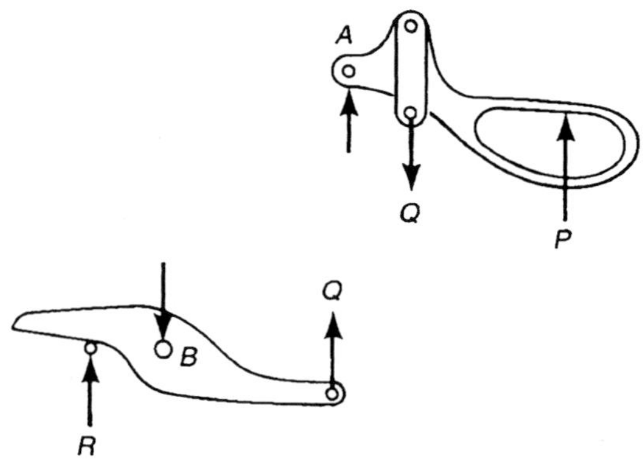

Exhibit 8.28a

8.29 d.

$$h = 115 \tan 20° = 135 \tan(20° - \phi)$$

$$\phi = 20° - \tan^{-1}\left(\frac{115}{135}\tan 20°\right) = 2.77°$$

$$\overline{OA} = \frac{115 \text{ mm}}{\cos 20°} = 122.4 \text{ mm}$$

$$\sum M_A = \overline{OA} \sin \phi \, R - (115 \text{ mm})(120 \text{ N}) = 0$$

$$R = \frac{115(120)}{122.4 \ \sin 2.77°} = 2330 \text{ N}$$

$$\sum M_B = (26 \text{ mm})R \sin(85° + \phi) - (30 \text{ mm})Q = 0$$

$$Q = \frac{26}{30}(2330 \text{ N})\sin 87.77° = 2020 \text{ N}$$

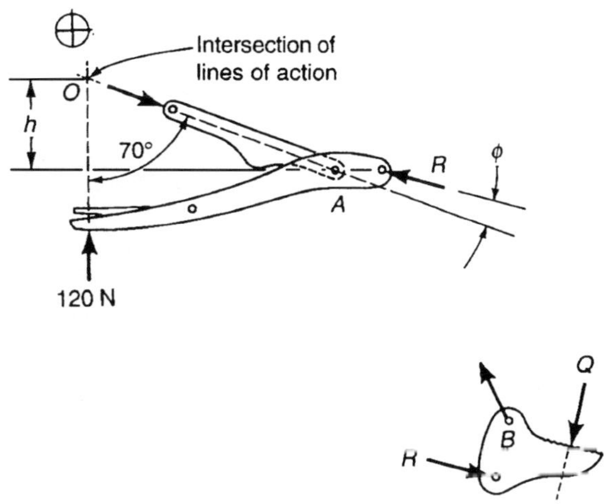

Exhibit 8.29a

8.30 a.

$$M = (600 \text{ N})(0.18 \text{ m}) - (480 \text{ N})(0.18 \text{ m}) + (320 \text{ N})(0.15 \text{ m}) - (400 \text{ N})(0.15 \text{ m}) = 9.60 \text{ N}\bullet\text{m}$$

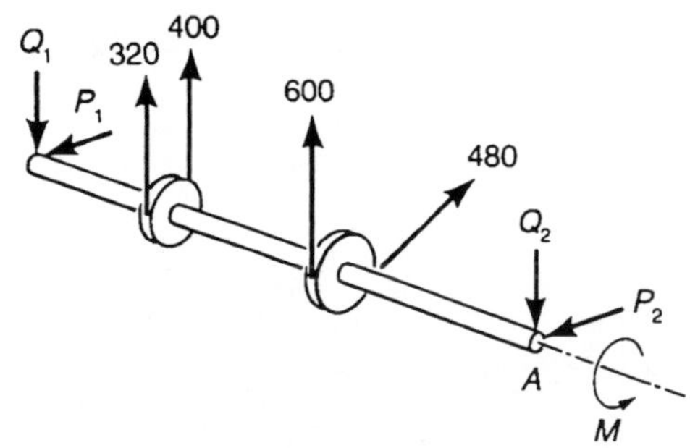

Exhibit 8.30a

8.31 c.

$$P_2(1.7\text{ m}) = (480\text{ N})(\cos 38°)(1.1\text{ m}),\ P_2 = 245\text{ N}$$

$$Q_2(1.7\text{ m}) = (720\text{ N})(0.4\text{ m}) + (600\text{ N} + 480\text{ N}\sin 38°)(1.1\text{ m}),\ Q_2 = 749\text{ N}$$

$$R_2 = \sqrt{P_2^2 + Q_2^2} = 788\text{ N}$$

8.32 c.

$$\mathbf{T} = T\mathbf{e}_{AB} = T\,\frac{-1.8\mathbf{e}_x + 1.5\mathbf{e}_y + 0.6\mathbf{e}_z}{\sqrt{5.85}}$$

$$\text{Vertical forces at } A:\quad \frac{0.6}{\sqrt{5.85}}T = mg$$

$$y\text{-forces at B:}\quad F\,\frac{1.5}{\sqrt{5.85}}T$$

$$\text{Combine:}\quad F = \frac{1.5}{0.6}mg = 98\text{ N}$$

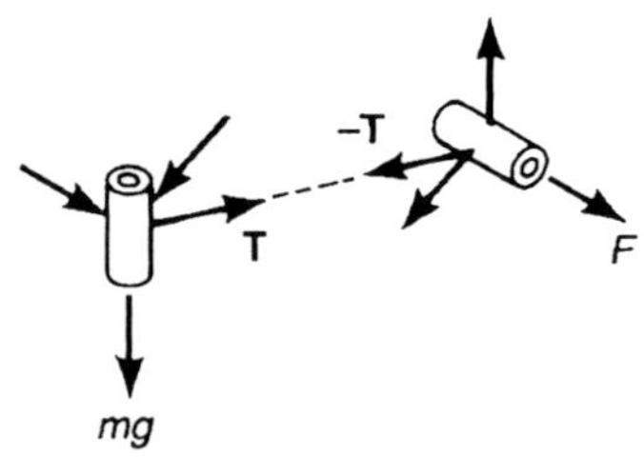

Exhibit 8.32a

8.33 a.

$$d = \frac{4}{5}(1.8\text{ m}) = 1.44\text{ m}$$

$$\mathbf{T}_{AE} = T_{AE}\left(\frac{3}{13}\mathbf{e}_x - \frac{4}{13}\mathbf{e}_y - \frac{12}{13}\mathbf{e}_z\right)$$

$$\sum M_{BC} = -(2.94\text{ m})\left(\frac{12}{13}T_{AE}\right) + (1.44\text{ m})R_z = 0$$

$$T_{AE} = \frac{26}{49}R_z = 424\text{ N}$$

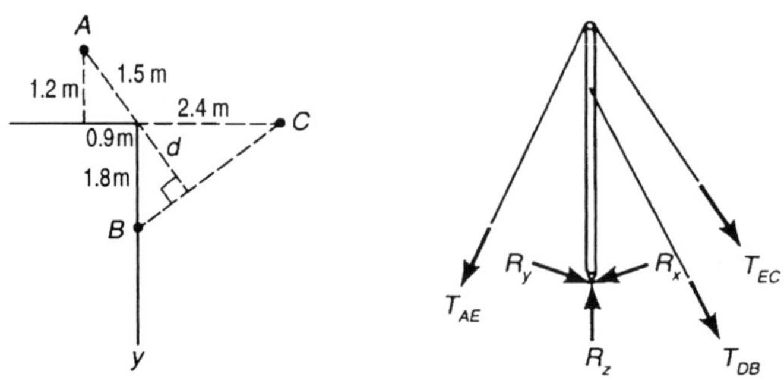

Exhibit 8.33a

8.34 a.

$$\mathbf{e}_{OD} = \frac{3.3\mathbf{e}_y + 0.75\mathbf{e}_z}{\sqrt{11.4525}} \qquad \mathbf{T}_{CD} = \boldsymbol{T}_{CD}\frac{-1.8\mathbf{e}_x + 3.3\mathbf{e}_y + 0.75\mathbf{e}_z}{\sqrt{14.6925}}$$

$$\mathbf{e}_{OB} = \frac{-1.2\mathbf{e}_y + 1.65\mathbf{e}_z}{\sqrt{4.1625}} \qquad \mathbf{T}_{AB} = \boldsymbol{T}_{AB}\frac{-3\mathbf{e}_x + 1.2\mathbf{e}_y + 1.65\mathbf{e}_z}{\sqrt{13.1625}}$$

$$M_{OD} = \mathbf{e}_{OD} \bullet \left[(3\text{ m})\mathbf{e}_x \times \mathbf{T}_{AB}\right] + \mathbf{e}_{OD} \bullet \left[(2.4\text{ m})\mathbf{e}_x \times (-1.5\text{ kN }\mathbf{e}_z)\right]$$

$$= \ (-1.5484\text{ m})T_{AB} + 3.5128\text{ kN}\bullet\text{m} \ = \ 0$$

$$T_{AB} = \frac{3.5128\text{ kN}\bullet\text{m}}{1.5484\text{ m}} = 2.27\text{ kN}$$

$$M_{OB} = \mathbf{e}_{OB} \bullet \left[(1.8\text{ m})\mathbf{e}_x \times \mathbf{T}_{CD}\right] + \mathbf{e}_{OB} \bullet \left[(2.4\text{ m})\mathbf{e}_x \times (-1.5\text{ kN }\mathbf{e}_z)\right]$$

$$= \ (-1.4604\text{ m})\boldsymbol{T}_{CD} + 2.1174\text{ kN}\bullet\text{m} \ = \ 0$$

$$\mathbf{T}_{CD} = \frac{2.1174\text{ kN}\bullet\text{m}}{1.4604\text{ m}} = 1.45\text{ kN}$$

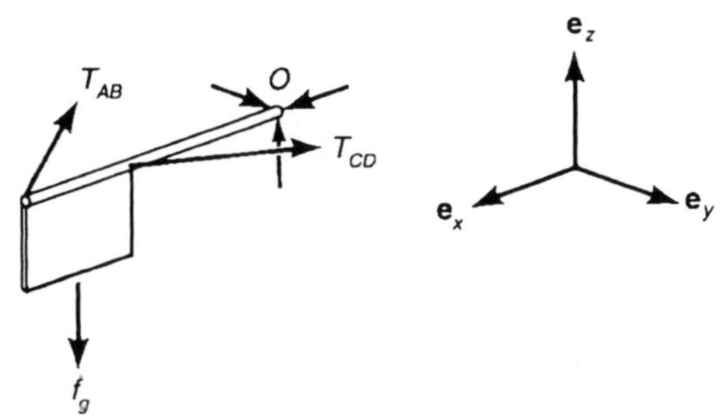

Exhibit 8.34a

8.35 d.

$$M_{OA} = \mathbf{e}_{OA} \bullet (0.45 \text{ m } \mathbf{e}_y \times \mathbf{T}) - (0.225 \text{ m})\frac{12}{13} f_g$$

$$= \frac{0.45 \text{ m}}{13}\left(\frac{46T}{7} - 6f_g\right) = 0$$

$$T = \frac{21}{23} f_g = \frac{21}{23} (40 \text{ kg})(9.8 \text{ N/kg}) = 358 \text{ N}$$

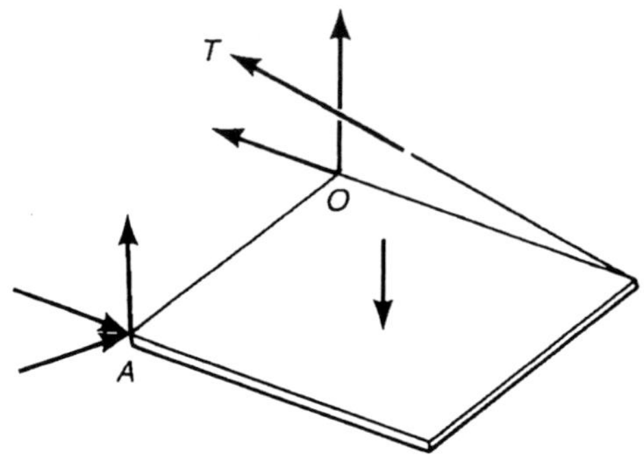

Exhibit 8.35a

8.36 b.

$$\sum M_A = a(4 \text{ kN}) + 4aR_D - a(24 \text{ kN}) - 3a(8 \text{ kN}) = 0$$

$$R_D = 11.0 \text{ kN}$$

$$\sum M_E = 3a(11 \text{ kN}) - 2a(8 \text{ kN}) - aT_{BC} = 0$$

$$T_{BC} = 17 \text{ kN (tension)}$$

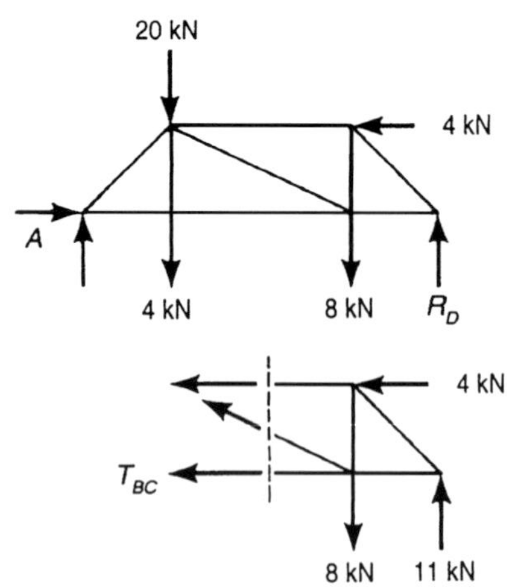

Exhibit 8.36a

8.37 b.

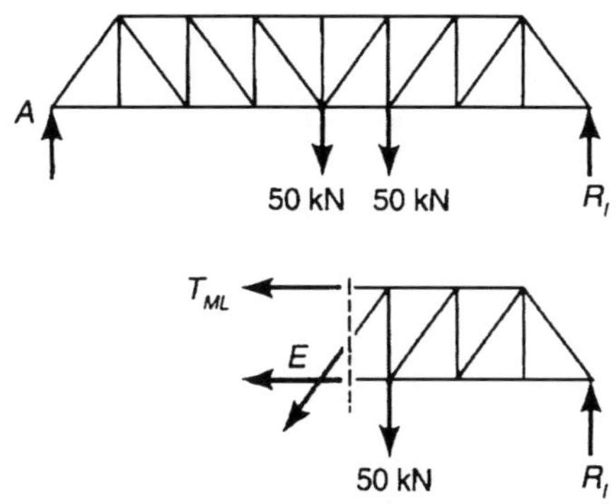

Exhibit 8.37a

$$\sum M_A = (24\text{ m})R_I - (12\text{ m})(50\text{ kN}) - (15\text{ m})(50\text{ kN}) = 0$$

$$R_I = 56.25\text{ kN}$$

$$\sum M_E = (4\text{ m})T_{ML} + (12\text{ m})(56.25\text{ kN}) - (3\text{ m})(50\text{ kN}) = 0$$

$$T_{ML} = -131.25\text{ kN}$$

8.38 c.

$$\sum M_A = (30\text{ m})R_G - (22.5\text{ m})(30\text{ kN}) - (15\text{ m})(25\text{ kN}) - (7.5\text{ m})(20\text{ kN}) = 0$$

$$R_G = 40\text{ kN}$$

$$\sum M_D = (15\text{ m})(40\text{ kN}) - (7.5\text{ m})(30\text{ kN}) - (5\sqrt{3}\text{ m})T_{CF} = 0$$

$$T_{CF} = 43.3\text{ kN}$$

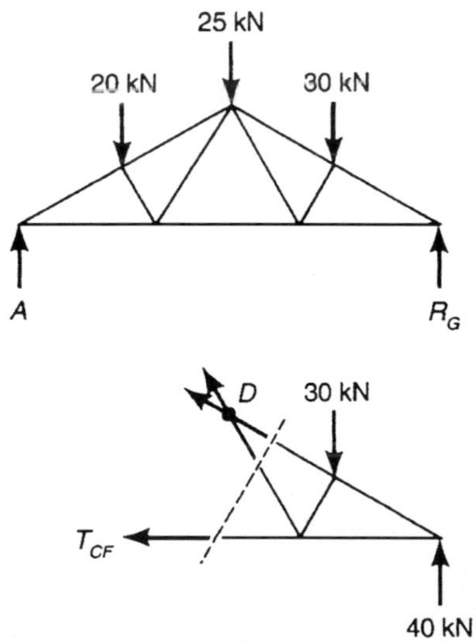

Exhibit 8.38a

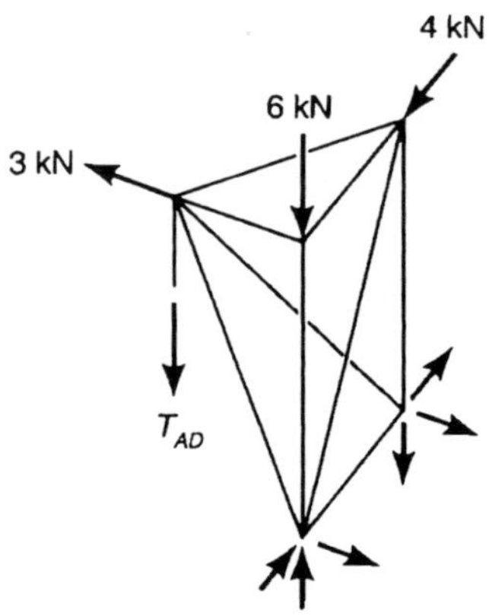

Exhibit 8.39a

8.39 a.

$$\sum M_{CB} = (6\text{ m})\frac{\sqrt{3}}{2}(3\text{ kN}) + \left(\frac{3\sqrt{3}}{2}\text{ m}\right)T_{AD} = 0$$

$$T_{AD} = -6\text{ kN (compression)}$$

8.40 a.

$$\sum M_{DE} = aP - aT_{AB} = 0; \qquad T_{AB} = P$$

$$\sum M_{DC} = \mathbf{e}_{CD} \bullet [a\mathbf{e}_z \times (P\mathbf{e}_x + \mathbf{T}_{BE})] = 0$$

$$\mathbf{T}_{BE} = T_{BE}\frac{\mathbf{e}_x + \mathbf{e}_y - \mathbf{e}_z}{\sqrt{3}}$$

$$(\mathbf{e}_{CD} \bullet \mathbf{e}_y)P = -\mathbf{e}_{CD} \bullet \left(\mathbf{e}_z \times T_{BE}\frac{\mathbf{e}_x + \mathbf{e}_y - \mathbf{e}_z}{\sqrt{3}}\right)$$

$$\frac{\mathbf{e}_x - \mathbf{e}_y}{\sqrt{2}} \bullet \mathbf{e}_y P = \frac{\mathbf{e}_x - \mathbf{e}_y}{\sqrt{2}} \bullet \frac{\mathbf{e}_x - \mathbf{e}_y}{\sqrt{3}} T_{BE}$$

$$T_{BE} = -\frac{\sqrt{3}}{2}P, \text{ or } -0.87P \text{ (compression)}$$

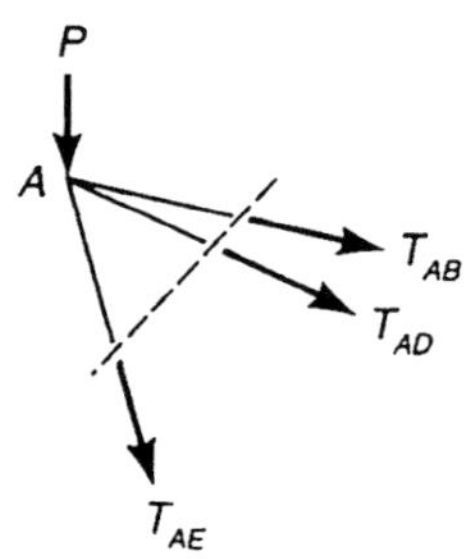

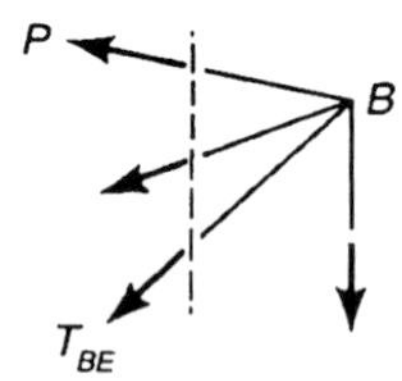

Exhibit 8.40a

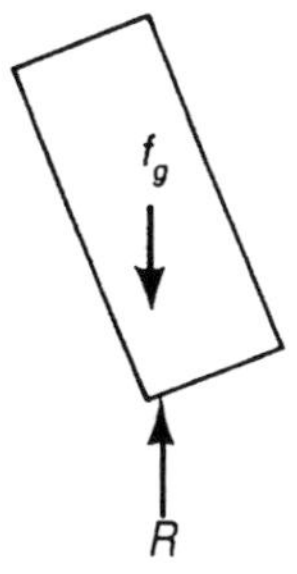

Exhibit 8.41a

8.41 b. The line of action of f_g (and hence of R) passes through the bottom at a distance of 28.8 mm from the lower corner. Hence the crate will not tip over: tan 15° = 0.27 > μ, so the crate will slide.

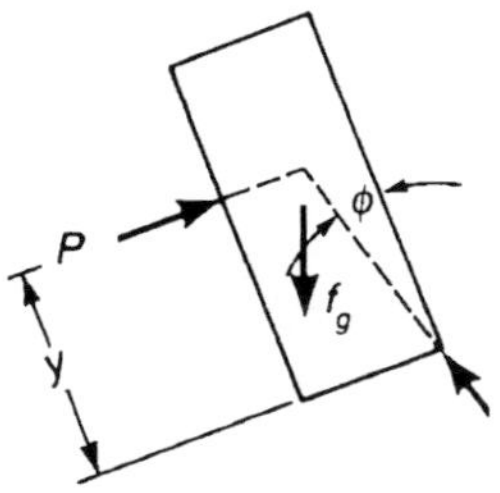

Exhibit 8.42

8.42 d. Vanishing resultant moment requires that the lines of action of the three forces intersect, so that

$$\tan\phi = \mu = \frac{0.27\text{ m} - (y - 0.9\text{ m})\tan 15^\circ}{y}$$

$$y = \frac{0.27\text{ m} + (0.9\text{ m})\tan 15^\circ}{\mu + \tan 15^\circ} = 1.09\text{ m}$$

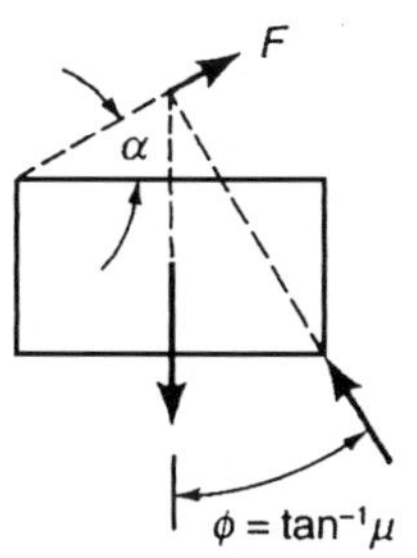

Exhibit 8.43a

8.43 c. Vanishing resultant moment requires that the lines of action of the three forces intersect, so that

$$\frac{b}{2}\tan\alpha = \frac{b}{2}\cot\phi - h = \frac{b}{2\mu} - h$$

$$\alpha = \tan^{-1}\left(\frac{1}{\mu} - \frac{2h}{b}\right)$$

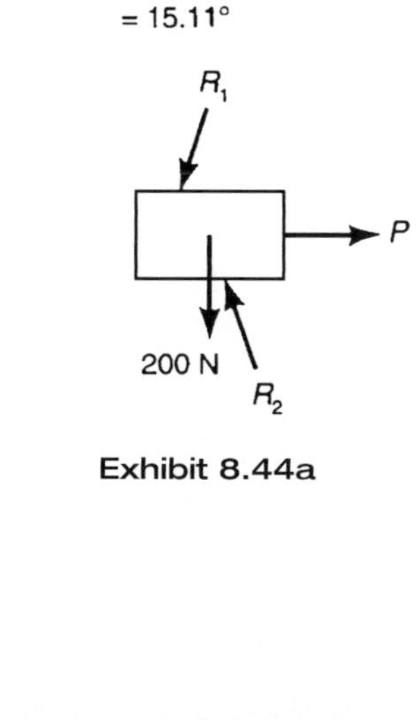

Exhibit 8.44a

8.44 a. Sum forces that are parallel to T on the upper block:

$R_1 \cos(20° - 15.11°) = (100\text{ N})\cos 20°$; $R_1 = 94.3$ N

Sum forces that are perpendicular to R_2 on the lower block:

$P \cos 15.11° - (200\text{ N}) \sin 15.11° - (94.3\text{ N})\sin 30.22° = 0$

$$P = 0.27(200\text{ N}) + \frac{\sin 45.22^\circ}{\cos 15.11^\circ}(206\text{ N}) = 103\text{ N}$$

8.45 a.

$$\phi = \tan^{-1}(0.27) = 15.11°$$

Sum forces that are perpendicular to R_0:

$$R_1 \cos(15° + 2\phi) = (150 \text{ N}) \cos \phi; \; R_1 = 206 \text{ N}$$

Sum forces that are perpendicular to R_2:

$$P \cos \phi - (200 \text{ N}) \sin \phi - R_1 \sin(15° + 2\phi) = 0$$

$$P = 0.27(200 \text{ N}) + \frac{\sin 45.22°}{\cos 15.11°}(206 \text{ N}) = 103 \text{ N s}$$

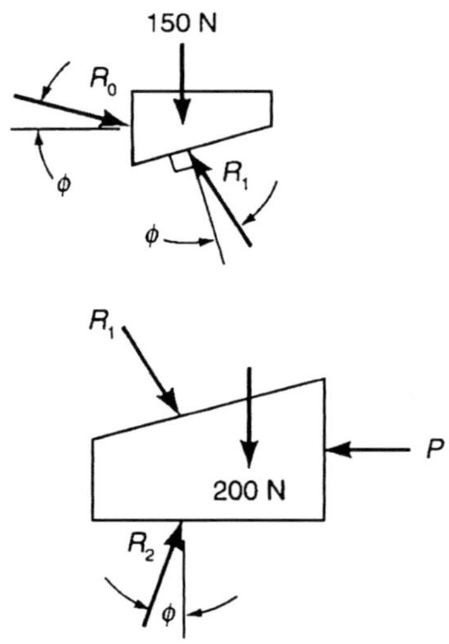

Exhibit 8.45a

8.46 b. Vanishing resultant moment requires that the lines of action of the three forces intersect, so that

$$a \tan \phi_1 = a \cos \alpha - \frac{a \sin \alpha}{\tan(\alpha + \phi_2)}$$

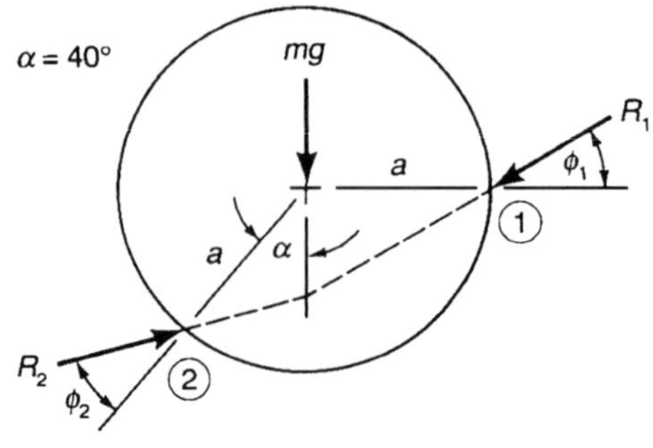

Exhibit 8.46a

If slip occurs at ②, $\phi_2 = \tan^{-1} 0.3 = 16.70°$.

$$\tan \phi_1 = \cos 40° - \frac{\sin 40°}{\tan 56.70°} = 0.344$$

Since this is less than $\mu_1 = 0.45$, slip *does* occur at ②.

$$\sum M_2 = a \cos\alpha \, R_1 \cos \phi_1 - (a + a \sin\alpha) R_1 \sin\phi_1 - a \sin\alpha \, \text{mg} = 0$$

$$R_1 \cos\phi_1 = \frac{\text{mg} \sin\alpha}{\cos\alpha - (1 + \sin\alpha)\tan\phi_1} = 3.20 \text{ mg} = 62.7 \text{ kN}$$

8.47 **c.** $\sum F_x = 0$ on the block leads to

$$T \sin 45° - \mu N = 0$$
$$-T\sqrt{2} + Ne = 0$$

Combine:

$$e = 2\mu$$

$$(T \cos 30°)(9.4) - \text{N}(4 + e) = 0$$

$$\frac{T}{N} = \frac{(4+e)}{(9.4 \cos 30°)} = \frac{\mu}{\sin 45°}$$

$$\mu = \frac{2}{\frac{9.4 \cos 30°}{2 \sin 45°} - 1} = 0.42$$

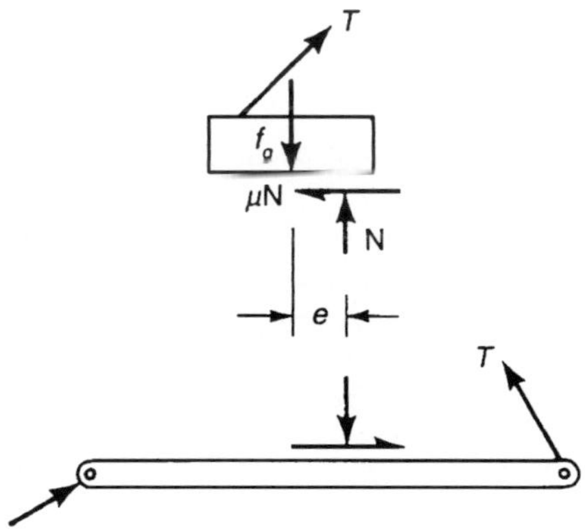

Exhibit 8.47a

8.48 c.

$$d = a + (b + a)\cos 2\phi$$

$$\cos 2\phi = \frac{1 - \tan^2 \phi}{1 + \tan^2 \phi}$$

To prevent slip, $\tan \phi < \mu$, so that

$$d > a + (b + a)\frac{1 - \mu^2}{1 + \mu^2} = \frac{2a + (1 - \mu^2)b}{1 + \mu^2}$$

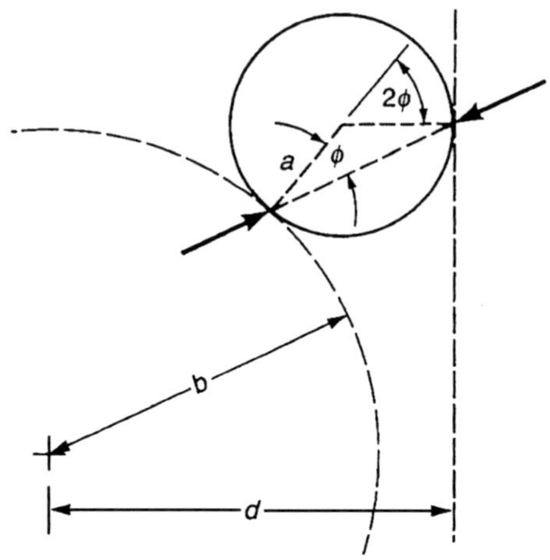

Exhibit 8.48a

CHAPTER 9

Dynamics, Kinematics, and Vibrations

Charles E. Smith, Fidelis O. Eke, and Jerry H. Hamelink

OUTLINE

MASS MOMENTS OF INERTIA

The mass moment of inertia of a body is a property that determines how much resistance to angular acceleration it has. This mass moment is called the second moment about an axis. The equation is:

$$I = \int r^2 \, dm$$

We can determine the mass moment of a cylinder as shown in Figure 9.1.

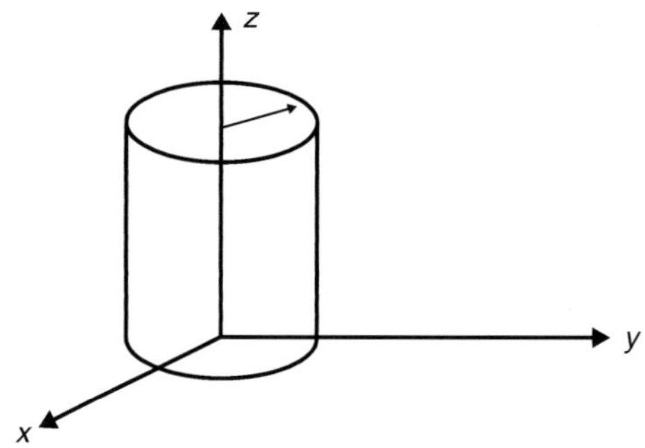

Figure 9.1

Consider an elemental moment of inertia of the cylinder:

$$d\,I_z = \int r^2 dm = \rho 2\pi h r^3 dr$$

And for uniform density of the cylinder, the equation becomes:

$$I_z = \rho 2\pi h \int_0^R r^3 dr = 2\rho\pi h \frac{R^4}{4} = \frac{\rho\pi R^4 h}{2}$$

Then, to find the mass of the circular cylinder:

$$m = \int_m dm = 2\pi\rho h \int^R r dr = \rho\pi h R^2$$

Therefore:

$$I_Z = \frac{1}{2} m R^2$$

Example 9.1

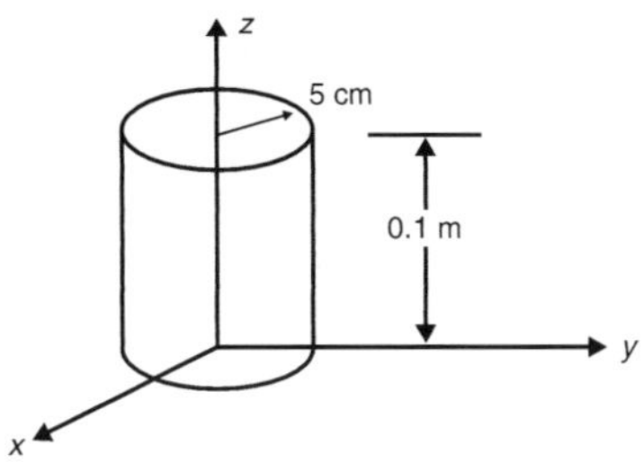

Exhibit 1

Consider a cylinder made with an alloy having a density of 400 kg/m^3. The dimensions of the body are shown in Exhibit 1. Determine the mass moment of inertia of the cylinder.

Solution

Determine the mass of the cylinder.

$$\text{mass} = \rho \pi h R^2 = \left(400 \frac{\text{kg}}{\text{m}^3}\right)(\pi)(0.1\text{ m})(0.05\text{ m})^2 = 0.314\text{ kg}$$

Determine the mass moment of inertia about the Z axis.

$$I_z = \frac{1}{2}\text{ mass} \times R^2 = \frac{1}{2}(0.314\text{ kg})(0.05\text{ m})^2 = 3.9 \times 10^{-4}\text{ kg–m}^2$$

Example 9.2

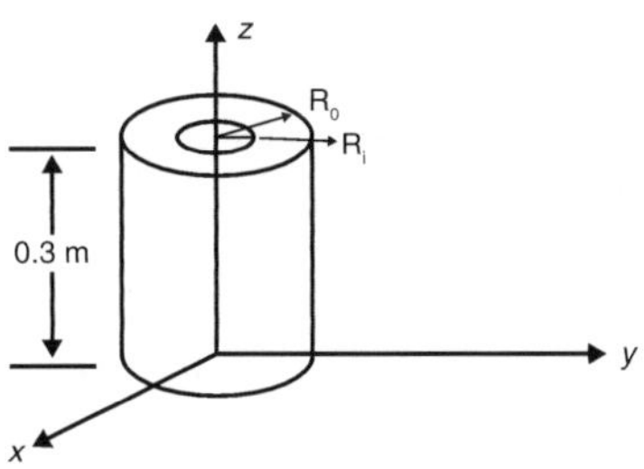

Exhibit 2

Consider the homogeneous hollow right circular cylinder, as shown in Exhibit 2, with the following dimensions:

$\rho = 7000$ kg/m^3

Inner radius = 5 cm

Outer radius = 8 cm

$h = 0.3$ m

Determine the mass moment of inertia with respect to the geometric axis.

Solution

Determine the mass of the cylinder = mass of outer cylinder – mass of inner cylinder.

$$\text{mass} = \pi \rho h(R_0^2 - R_i^2) = \left(7000 \frac{\text{kg}}{\text{m}^3}\right)(\pi)(0.3\text{ m})(0.08^2 - 0.05^2)$$

$$\text{mass} = 25.73\text{ kg}$$

Determine the mass moment of inertia with respect to the geometric axis.

$$I_z = \frac{1}{2}\text{mass}\left(R_0^2 - R_1^2\right) = \frac{25.73}{2}\text{kg}\left(0.08^2 - 0.05^2\right)\text{m}^2 = 0.05017\text{ kg–m}^2$$

The analysis of a mechanical system having elements under acceleration must consider these accelerations along with the related forces. In such analysis, the force side of Newton's second law, $f = ma$, and the third law of action and reaction are dealt with in exactly the same manner as in statics. But it is the relationships among positions, velocities, and accelerations that complete the discipline of dynamic analysis. The following two sections review these relationships, and the remainder of the chapter deals with their incorporation into Newton's laws of motion.

KINEMATICS OF A PARTICLE

Consider a point P that moves along a smooth path as indicated in Figure 9.2. The position of the point may be specified by the vector $\mathbf{r}(t)$, defined to extend from an arbitrarily selected, fixed point O to the moving point P. The **velocity v** of the point is defined to be the derivative with respect to t of $\mathbf{r}(t)$, written as

$$\mathbf{v} = \frac{d\mathbf{r}}{dt} \qquad \textbf{(9.1)}$$

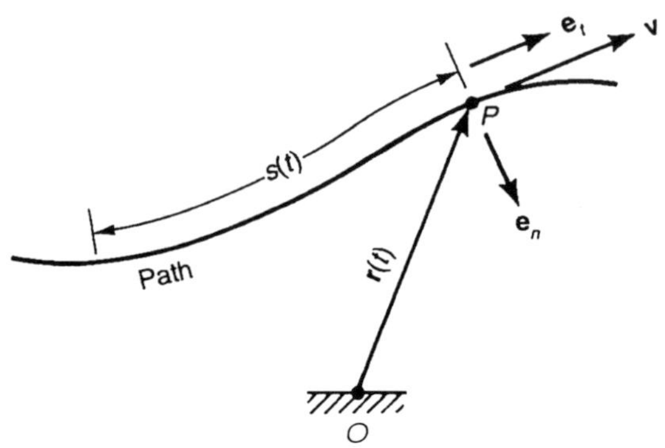

Figure 9.2

Although this definition is sometimes used for evaluation [that is, by differentiating a specific expression for $\mathbf{r}(t)$], it will often be more direct to use other relationships. It follows from the above definition that the velocity vector is tangent to the path of the particle; thus, upon introduction of a unit vector $\mathbf{e}_t$, defined to be tangent to the path, the velocity can also be expressed as

$$\mathbf{v} = v\mathbf{e}_t \qquad \textbf{(9.2)}$$

The position of P can also be specified in terms of the distance $s(t)$ traveled along the path from an arbitrarily selected reference point. Then an incremental change in position may be approximated as $\Delta\mathbf{r} \approx \Delta s\mathbf{e}_t$, in which the accuracy increases as the increments Δt and $\Delta\mathbf{r}$ approach zero. This leads to still another way of expressing the velocity as

$$\mathbf{v} = \frac{ds}{dt}\mathbf{e}_t \qquad \textbf{(9.3)}$$

The scalar

$$v = \frac{ds}{dt} = \dot{s} \qquad \textbf{(9.4)}$$

can be either positive or negative, depending on whether the motion is in the same or the opposite direction as that selected in the definition of $\mathbf{e}_t$.

The **acceleration** of the point is defined as the derivative of the velocity with respect to time:

$$\mathbf{a} = \frac{d\mathbf{v}}{dt} \qquad \textbf{(9.5)}$$

A useful relationship follows from application to Equation (9.2) of the rules for differentiating products and functions of functions:

$$\frac{d\mathbf{v}}{dt} = \dot{v}\mathbf{e}_t + \mathbf{v}\frac{ds}{dt}\frac{d\mathbf{e}_t}{ds}$$

As the direction of $\mathbf{e}_t$ varies, the square of its magnitude, $|\mathbf{e}_t|^2 = \mathbf{e}_t \bullet \mathbf{e}_t$, remains fixed and equal to 1, so that

$$\frac{d}{ds}|\mathbf{e}_t|^2 = \frac{d}{ds}(\mathbf{e}_t \bullet \mathbf{e}_t) = 2\mathbf{e}_t \bullet \frac{d\mathbf{e}_t}{ds} = 0$$

This shows that $d\mathbf{e}_t/ds$ is either zero or perpendicular to $\mathbf{e}_t$. With another unit vector $\mathbf{e}_n$ defined to be in the direction of $d\mathbf{e}_t/ds$, this vector may be expressed as

$$\frac{d\mathbf{e}_t}{ds} = \kappa\mathbf{e}_n$$

The scalar κ is called the local **curvature** of the path; its reciprocal, $\rho = 1/\kappa$, is called the local **radius of curvature** of the path. In the special case in which the path is straight, the curvature and hence $d\mathbf{e}_t/ds$ are zero. These lead to the following expression for the **acceleration** of the point:

$$\mathbf{a} = \dot{v}\mathbf{e}_t + \frac{v^2}{\rho}\mathbf{e}_n \tag{9.6}$$

The two terms express the **tangential** and **normal** (or **centripetal**) components of acceleration.

If a driver of a car with sufficient capability "steps on the gas," a positive value of $\dot{v}$ is induced, whereas if he "steps on the brake," a negative value is induced. If the path of the car is straight (zero curvature or "infinite" radius of curvature), the entire acceleration is $\dot{v}\mathbf{e}_t$. If the car is rounding a curve, there is an additional component of acceleration directed laterally, toward the center of curvature of the path. These components are indicated in Figure 9.3, a view of the plane of $\mathbf{e}_t$ and $d\mathbf{e}_t/ds$.

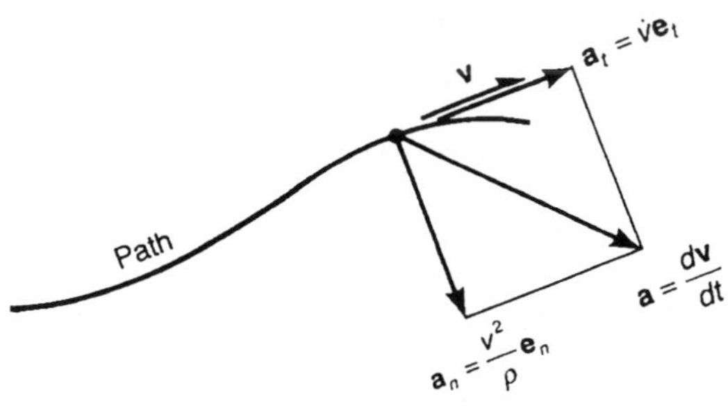

Figure 9.3

Example 9.3

At a certain instant, the velocity and acceleration of a point have the rectangular Cartesian components given by

$$\mathbf{v} = (3.5\mathbf{e}_x - 7.2\mathbf{e}_y + 9.6\mathbf{e}_z)\ \text{m/s}$$
$$\mathbf{a} = (-20\mathbf{e}_x + 20\mathbf{e}_y + 10\mathbf{e}_z)\ \text{m/s}^2$$

At this instant, what are the rate of change of speed dv/dt and the local radius of curvature of the path?

Solution

The rectangular Cartesian components of the unit tangent vector can be determined by dividing the velocity vector by its magnitude:

$$\mathbf{e}_t = \frac{\mathbf{v}}{|\mathbf{v}|} = \frac{3.5\mathbf{e}_x - 7.2\mathbf{e}_y + 9.6\mathbf{e}_z}{\sqrt{(3.5)^2 + (-7.2)^2 + (9.6)^2}} = 0.280\mathbf{e}_x - 0.576\mathbf{e}_y + 0.768\mathbf{e}_z$$

The rate of change of speed can then be determined as the projection of the acceleration vector onto the tangent to the path:

$$\begin{aligned}\dot{v} &= \mathbf{e}_t \bullet \mathbf{a} = [(0.280)(-20) + (-0.576)(20) + (0.768)(10)]\ \text{m/s}^2 \\ &= -9.44\ \text{m/s}^2\end{aligned}$$

The negative sign indicates the projection is opposite to $\mathbf{e}_t$ (which was defined by the above equation to be in the same direction as the velocity). This means that the speed is *decreasing* at 9.44 m/s. One sees from Figure 9.3 that the normal component of acceleration has magnitude

$$a_n = \sqrt{|\mathbf{a}|^2 - \dot{v}^2} = \sqrt{(-20)^2 + (20)^2 + (10)^2 - (-9.44)^2\,\text{m/s}^2} = 28.5\ \text{m/s}^2$$

which, from Equation (9.6), is related to the speed and radius of curvature by $a_n = v^2/\rho$. Rearrangement of this equation gives the radius of curvature as

$$\rho = \frac{v^2}{a_n} = \frac{[(3.5)^2 + (-7.2)^2 + (9.6)^2]\ \text{m}^2/\text{s}^2}{28.5\ \text{m/s}^2} = 5.48\ \text{m}$$

Relating Distance, Velocity, and the Tangential Component of Acceleration

The basic relationships among tangential acceleration a_t, velocity $v\mathbf{e}_t$, and distance s are

$$\frac{dv}{dt} = a_t \quad \text{or} \quad v = v_0 + \int a_t\, dt \tag{9.7}$$

$$\frac{ds}{dt} = v \quad \text{or} \quad s = s_0 + \int v\, dt \tag{9.8}$$

in which v_0 and s_0 are constants of integration. An alternative relationship comes from writing $dv/dt = (ds/dt)(dv/ds) = v\, dv/ds$:

$$v\frac{dv}{ds} = a_t \quad \text{or} \quad v^2 = v_0^2 + 2\int a_t\, ds \tag{9.9}$$

Equations (9.7) and (9.8) are useful in dealing with the *time* histories of acceleration, velocity, and distance, whereas Equation (9.9) is helpful in dealing with the manner in which velocity and acceleration vary with distance.

Example 9.4

The variation of tangential acceleration with time is given in Exhibit 3. If a point with an initial velocity of 24 m/s is subjected to this acceleration, what will be its velocity at $t = 6$ s, 10 s, and 15 s, and what will be the values of s at $t = 4$ s, 7.6 s, and 15 s?

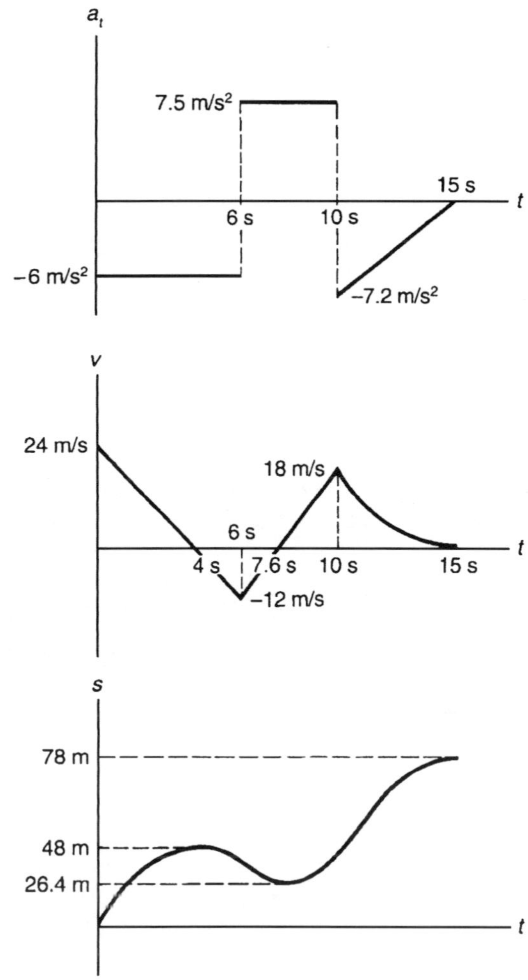

Exhibit 3

Solution

Equation (9.7) has the following graphical interpretations: At each point, the slope of the v-t curve is equal to the ordinate on the a_t-t curve. During any interval, the change in the value of v is equal to the area under the a_t-t curve for the same interval. With these rules and the given initial value of v, the variation of v with t can be plotted, and values of v can be calculated for each point.

The reader should use these rules to verify all details of the v-t curve shown. Equation (9.8) indicates that identical rules for slopes, ordinates, and areas relate the curve of distance s to that of velocity v, so the same procedure can be used to construct the s-t curve from the v-t curve. Again, the reader should verify all details of this curve.

Example 9.5

The tangential acceleration of the pendulum bob shown in Exhibit 4 varies with position according to $a = -g \sin(s/l)$, in which g is the local acceleration of gravity. If a speed v_0 is imparted at the vertical position (where $s = 0$), what will be the maximum value of s reached?

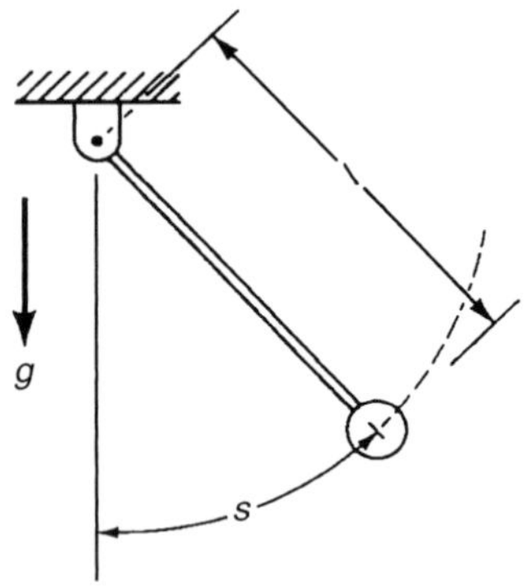

Exhibit 4

Solution

Because the relationship between tangential acceleration and *position* is given, Equation (9.9) will prove useful. The integrated form leads to

$$v^2 = v_0^2 + 2\int_0^s g \sin\left(\frac{s}{l}\right) ds$$
$$= v_0^2 + 2gl\left(1 - \cos\frac{s}{l}\right)$$

which gives the velocity v in the terms of any position s. Since $v = \dot{s}$, the maximum s will occur when $v = 0$, and the corresponding s is easily isolated from the above equation after setting $v = 0$:

$$s_{\max} = l \cos^{-1}\left(1 - \frac{v_0^2}{2gl}\right)$$

Observe that if $v_0^2 > 2gl$, no real value of $s_{\max}$ exists, because v never reaches zero in that case.

Constant Tangential Acceleration

When the tangential acceleration is constant, Equations (9.7) through (9.9) reduce to

$$v = v_0 + a_t t \tag{9.10}$$

$$s = s_0 + v_0 t + \frac{1}{2} a_t t^2 \tag{9.11}$$

$$v^2 = v_0^2 + 2a_t s \tag{9.12}$$

Rectilinear Motion

In the special case in which the path is a straight line, the unit tangent vector $\mathbf{e}_t$ is constant, and the curvature $1/\rho$ is zero throughout. The acceleration is then given by $(dv/dt)\mathbf{e}_t$, and the subscript on the symbol a_t may be dropped without ambiguity.

Example 9.6

A particle is launched vertically upward with an initial speed of 10 m/s and subsequently moves with constant downward acceleration of magnitude 9.8 m/s^2. What is the maximum height reached by the particle? How long does it take to return to the original launch position? And how fast is it traveling at its return to the launch position?

Solution

In this case the path will be straight and the acceleration is constant. With $\mathbf{e}_t$ defined as upward, the constant scalars appearing in Equations (9.10) through (9.12) have the values $v_0 = 10$ m/s and $a_t = a = -9.8$ m/s^2, so that these equations become

$$v = 10 \text{ m/s} - (9.8 \text{ m/s}^2)t \quad \textbf{(i)}$$

$$s = (10 \text{ m/s})t - \frac{1}{2}(9.8 \text{ m/s}^2)t^2 \quad \textbf{(ii)}$$

$$v^2 = (10 \text{ m/s})^2 - 2(9.8 \text{ m/s}^2)s \quad \textbf{(iii)}$$

The maximum height reached can be obtained by setting $v = 0$ in (iii), which gives

$$s_{\max} = \frac{(10 \text{ m/s})^2}{2(9.8 \text{ m/s}^2)} = 5.1 \text{ m}$$

The time required to reach this height can be obtained by setting $v = 0$ in (i), which gives

$$t_1 = \frac{10 \text{ m/s}}{9.8 \text{ m/s}^2} = 1.02 \text{ s}$$

Finally, setting $s = 0$ in (iii) yields the two values of v that specify the velocity at the launch position:

$$v = \pm 10 \text{ m/s}$$

The positive value gives the upward initial velocity, and the negative value gives the equal-magnitude, downward velocity of the particle when it returns to the launch position.

Rectangular Cartesian Coordinates

Multidimensional motion can be analyzed in terms of components associated with a set of fixed unit vectors $\mathbf{e}_x$, $\mathbf{e}_y$, and $\mathbf{e}_z$, which are defined to be mutually perpendicular. For some aspects of analysis, it is also important that they form a "right-handed" set, or $\mathbf{e}_z = \mathbf{e}_x \times \mathbf{e}_y$, $\mathbf{e}_x = \mathbf{e}_y \times \mathbf{e}_z$, and $\mathbf{e}_y = \mathbf{e}_z \times \mathbf{e}_x$. In terms of these unit vectors, the position, velocity, and acceleration can be expressed as

$$\begin{aligned} \mathbf{r} &= x\mathbf{e}_x + y\mathbf{e}_y + z\mathbf{e}_z \\ \mathbf{V} &= v_x\mathbf{e}_x + v_y\mathbf{e}_y + v_z\mathbf{e}_z \\ \mathbf{A} &= a_x\mathbf{e}_x + a_y\mathbf{e}_y + a_z\mathbf{e}_z \end{aligned}$$

with

$$\begin{aligned} v_x &= \dot{x} \\ a_x &= \dot{v}_x = \ddot{x}, \quad \text{etc.} \end{aligned}$$

Example 9.7

A wheel rolls without slipping along a straight surface with the orientation of the wheel given in terms of the angle $\theta(t)$. See Exhibit 5. Express the velocity and acceleration of the point P on the rim of the wheel in terms of this angle, its derivatives, and the radius b of the wheel.

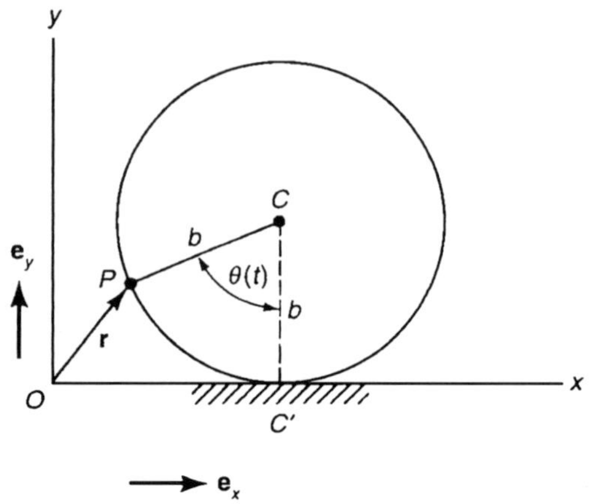

Exhibit 5

Solution

The origin for the x-y coordinates of P is the location of P when $\theta = 0$. Because the wheel rolls without slipping, the distance OC' is equal to the length of the circular arc PC'. The x-coordinate of P is then $OC' - b\sin\theta = b\theta - b\sin\theta$. The y-coordinate of P is that of C (i.e., b) minus $b\cos\theta$. In terms of these coordinates, the position vector from O to P may be expressed as

$$\mathbf{r} = b(\theta - \sin\theta)\mathbf{e}_x + b(1 - \cos\theta)\mathbf{e}_y$$

The velocity is then determined by differentiation of this expression:

$$\mathbf{v} = b\dot{\theta}[(1 - \cos\theta)\mathbf{e}_x + \sin\theta\,\mathbf{e}_y]$$

The acceleration is determined by another differentiation:

$$\mathbf{a} = b\ddot{\theta}[(1 - \cos\theta)\mathbf{e}_x + \sin\theta\,\mathbf{e}_y] + b\dot{\theta}^2(\sin\theta\,\mathbf{e}_x + \cos\theta\,\mathbf{e}_y)$$

These expressions may be simplified somewhat by rewriting them in terms of the unit vectors $\mathbf{e}_r$ and $\mathbf{e}_\theta$ as defined in Exhibit 6. These unit vectors are given in terms of the original horizontal and vertical unit vectors by

$$\mathbf{e}_r = -\sin\theta\,\mathbf{e}_x - \cos\theta\,\mathbf{e}_y$$
$$\mathbf{e}_\theta = -\cos\theta\,\mathbf{e}_x + \sin\theta\,\mathbf{e}_y$$

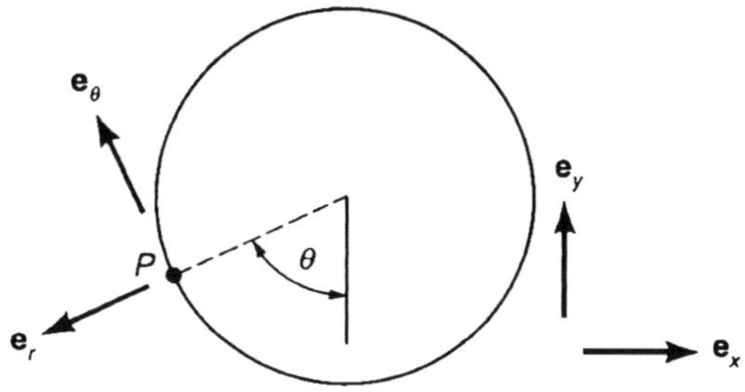

Exhibit 6

The above expressions for velocity and acceleration can now be written as

$$\mathbf{v} = b\dot{\theta}(\mathbf{e}_x + \mathbf{e}_\theta)$$
$$\mathbf{a} = b\ddot{\theta}(\mathbf{e}_x + \mathbf{e}_\theta)b\dot{\theta}^2\mathbf{e}_r$$

Further simplification is possible upon examination of the sum $\mathbf{e}_x + \mathbf{e}_\theta$, shown in Exhibit 7. The magnitude of this sum is 2 sin (θ/2), and its direction is perpendicular

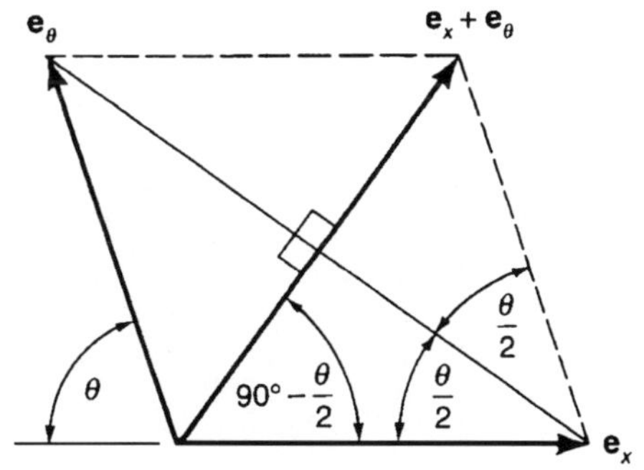

Exhibit 7

to the line connecting points P and C'. The velocity can thus be expressed as $\mathbf{v} = (2b \sin \theta/2\, \dot{\theta})\mathbf{e}_t$ in which the unit tangent vector is perpendicular to the line PC'. The acceleration can be simplified correspondingly to

$$\mathbf{a} = 2b\sin\frac{\theta}{2}\ddot{\theta}\mathbf{e}_t - b\dot{\theta}^2\mathbf{e}_r$$

Several steps were taken to reach the results in Example 9.7. The position vector was expressed in terms of the geometric constraints on the rolling of the wheel, differentiation led to expressions for the velocity and acceleration, and the introduction of auxiliary unit vectors and several trigonometric relationships simplified several expressions.

As mentioned earlier, direct use of the definitions expressed by Equations (9.1) and (9.5) may not be the easiest means of evaluating velocities and accelerations. Indeed, we will now review some kinematic relationships for rigid bodies that will make much shorter work of this example.

Circular Cylindrical Coordinates

Figure 9.4 shows a coordinate system that is useful for a number of problems in particle kinematics. The x and y coordinates of the rectangular Cartesian system are replaced with the distance r and the angle ϕ, while the definition of the z-coordinate remains unchanged. Two of the unit vectors associated with the rectangular Cartesian system are also replaced with $\mathbf{e}_r = \cos\phi\,\mathbf{e}_x + \sin\phi\,\mathbf{e}_y$ and $\mathbf{e}_\phi = -\sin\phi\,\mathbf{e}_x + \cos\phi\,\mathbf{e}_y$.

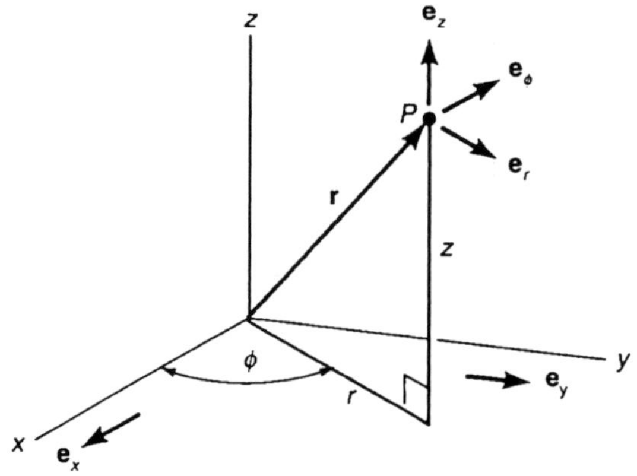

Figure 9.4

Since the angle ϕ varies, these two unit vectors also vary; their derivatives may be obtained by differentiating the above expressions:

$$\frac{d\mathbf{e}_r}{dt} = (-\mathbf{e}_x \sin\phi + \mathbf{e}_y \cos\phi)\frac{d\phi}{dt} = \dot{\phi}\mathbf{e}_\phi$$
$$\frac{d\mathbf{e}_\phi}{dt} = (-\mathbf{e}_x \cos\phi - \mathbf{e}_y \sin\phi)\frac{d\phi}{dt} = -\dot{\phi}\mathbf{e}_r$$

These are used along with the expression $\mathbf{r} = r\,\mathbf{e}_r + z\,\mathbf{e}_z$ for position to obtain expressions for velocity and acceleration:

$$\begin{aligned}\mathbf{v} = \dot{\mathbf{r}} &= \dot{r}\mathbf{e}_r + r\dot{\mathbf{e}}_r + \dot{z}\mathbf{e}_z \\ &= \dot{r}\mathbf{e}_r + r\dot{\phi}\mathbf{e}_\phi + \dot{z}\mathbf{e}_z\end{aligned} \tag{9.13}$$

$$\begin{aligned}\mathbf{a} = \dot{\mathbf{v}} &= \ddot{r}\mathbf{e}_r + \dot{r}\dot{\mathbf{e}}_r + (\dot{r}\dot{\phi} + r\ddot{\phi})\mathbf{e}_\phi + r\dot{\phi}\dot{\mathbf{e}}_\phi + \ddot{z}\mathbf{e}_z \\ &= (\ddot{r} - r\dot{\phi}^2)\mathbf{e}_r + (r\ddot{\phi} + 2\dot{r}\dot{\phi})\mathbf{e}_\phi + \ddot{z}\mathbf{e}_z\end{aligned} \tag{9.14}$$

Example 9.8

In Exhibit 8, the slider moves along the rod as it rotates about the fixed point O. At a particular instant, the slider is 200 mm from O, moving outward at 3 m/s relative to the rod; this relative speed is increasing at 130 m/s². At the same instant, the rod is rotating at a constant rate of 191 rpm. Evaluate the velocity and acceleration of the slider, and determine the rate of change of speed of the slider.

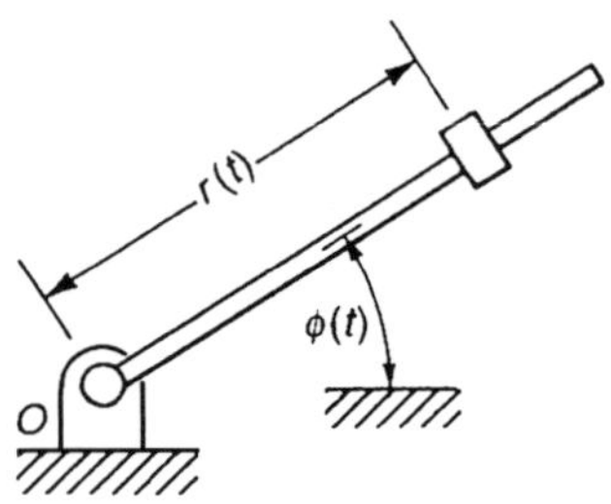

Exhibit 8

Solution

The angular speed of the rod is

$$\dot{\phi} = (191\text{ rpm})\,\frac{2\pi\text{ rad/rev}}{60\text{ s/min}} = 20.00\text{ rad/s}$$

and its angular acceleration $\ddot{\phi}$ is zero. Other values to be substituted into Equations (9.13) and (9.14) are r = 0.2 m, $\dot{r} = 3$ m/s, and $\ddot{r} = 130$ m/s².

Substitution into Equations (9.13) and (9.14) leads directly to the following radial and transverse components of velocity and acceleration:

$$\mathbf{v} = (3\ \mathbf{e}_r + 4\ \mathbf{e}_\phi)\ \text{m/s}$$
$$\mathbf{a} = (50\ \mathbf{e}_r + 120\ \mathbf{e}_\phi)\ \text{m/s}^2$$

Now the radial and transverse components of the unit vector tangent to the path can be obtained by dividing the velocity vector by its magnitude:

$$\mathbf{e}_t = \frac{3\mathbf{e}_r + 4\mathbf{e}_\phi}{\sqrt{(3)^2 + (4)^2}} = 0.6\mathbf{e}_r + 0.8\mathbf{e}_\phi$$

The rate of change of speed is the projection of the acceleration vector onto the tangent to the path, which can be obtained by dot-multiplying the acceleration with the unit tangent vector:

$$\begin{aligned}\dot{v} = a_t &= \mathbf{e}_t \bullet \mathbf{a} \\ &= (0.6)(50\ \text{m/s}^2) + (0.8)(120\ \text{m/s}^2) \\ &= 126\ \text{m/s}^2\end{aligned}$$

Circular Path

When the path is circular, r is constant, and Equations (9.13) and (9.14) reduce to

$$\mathbf{v} = r\dot{\phi}\mathbf{e}_\phi$$
$$\mathbf{a} = -r\dot{\phi}^2\mathbf{e}_r + r\ddot{\phi}\mathbf{e}_\phi$$

Comparing these with Equations (9.2) and (9.6) (with $\rho = r$),

$$\mathbf{v} = v\mathbf{e}_t$$

$$\mathbf{a} = \dot{v}\mathbf{e}_t + \frac{v^2}{r}\mathbf{e}_n$$

we see that, for circular path motion, $\mathbf{e}_t = \mathbf{e}_\phi$, $\mathbf{e}_n = -\mathbf{e}_r$, and

$$v = r\dot{\phi} \qquad \textbf{(9.15)}$$

$$a_n = r\dot{\phi}^2 \qquad \textbf{(9.16)}$$

Example 9.9

A satellite is to be placed in a circular orbit over the equator at such an altitude that it makes one revolution around the earth per sidereal day (23.9345 hours). The gravitational acceleration is $(3.99 \times 10^{14}\ \text{m}^3/\text{s}^2)/r^2$, where r is the distance from the center of the earth. What is the altitude at which the satellite must be placed to achieve this period of orbit?

Solution

The angular speed of the line from the center of the earth to the satellite is

$$\dot{\phi} = \frac{2\pi \text{ rad}}{(23.9345 \text{ h})(3600 \text{ s/h})} = 7.292 \times 10^{-5} \text{ rad/s}$$

The acceleration has no tangential component, but the radial component in terms of the orbit radius and the angular speed will be $a_n = r(7.292 \times 10^{-5}\text{s}^{-1})^2$. This acceleration is imparted by the earth's gravitational attraction, so that $r(7.292 \times 10^{-5}\text{s}^{-1})^2 = (3.99 \times 10^{14}\ \text{m}^3/\text{s}^2)/r^2$. This equation is readily solved for r, resulting in

$$r = \sqrt[3]{\frac{3.99 \times 10^{14}\,\text{m}^3/\text{s}^2}{(7.292 \times 10^{-5}\text{s}^{-1})^2}} = 42.2 \times 10^6 \text{ m}$$

The altitude will then be the difference between this value and the size of earth's radius, which is about 6.4×10^6 m: altitude = 35.8×10^6 m.

RIGID BODY KINEMATICS

The analysis of numerous mechanical systems rests on the assumption that the bodies making up the system are *rigid*. If the forces involved and the materials and geometry of the bodies are such that there is little deformation, the resulting predictions can be expected to be quite accurate.

The Constraint of Rigidity

If a body is **rigid**, the distance between each pair of points remains constant as the body moves. This constraint may be expressed in terms of a position vector $\mathbf{r}_{PQ}$ from a point P of the body to a point Q of the body, as indicated in Figure 9.5. If the magnitude of $\mathbf{r}_{PQ}$ is constant, then

$$\frac{d}{dt}|\mathbf{r}_{PQ}|^2 = \frac{d}{dt}(\mathbf{r}_{PQ} \bullet \mathbf{r}_{PQ}) = 2\mathbf{r}_{PQ} \bullet \frac{d\mathbf{r}_{PQ}}{dt} = 0 \qquad \textbf{(i)}$$

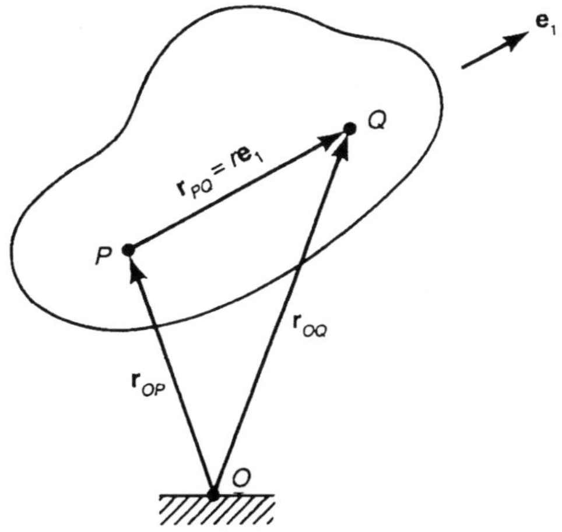

Figure 9.5

which indicates that $\dot{\mathbf{r}}_{PQ}$ is perpendicular to $\mathbf{r}_{PQ}$. Now, with a selected, fixed point designated as O, and vectors $\mathbf{r}_{OP}$ and $\mathbf{r}_{OQ}$ defined as indicated in Figure 9.5, differentiation of the vector relationship $\mathbf{r}_{PQ} = \mathbf{r}_{OQ} - \mathbf{r}_{OP}$ leads to the relationship

$$\frac{d\mathbf{r}_{PQ}}{dt} = \mathbf{v}_Q - \mathbf{v}_P \qquad \textbf{(ii)}$$

in which $\mathbf{v}_P$ and $\mathbf{v}_Q$ designate the velocities of P and Q, respectively. Finally, if we define $\mathbf{e}_1$ to be the unit vector in the direction of $\mathbf{r}_{PQ}$, so that

$$\mathbf{r}_{PQ} = r\mathbf{e}_1 \qquad \textbf{(iii)}$$

then substitution of (ii) and (iii) into (i) leads to

$$2r\mathbf{e}_1 \bullet (\mathbf{v}_Q - \mathbf{v}_P) = 0$$

or

$$\mathbf{e}_1 \bullet \mathbf{v}_Q = \mathbf{e}_1 \bullet \mathbf{v}_P \qquad \textbf{(9.17)}$$

This shows that *the projections of the velocities of any two points of a rigid body onto the line connecting the two points must be equal.* This is intuitively plausible; otherwise the distance between the points would be changing. This frequently provides the most direct way of evaluating the velocities of various points within a mechanism.

Example **9.10**

As the crank OQ in Exhibit 9 rotates clockwise at 200 rad/s, the piston P moves vertically. What will be the velocity of the piston at the instant when the angle θ is 50 degrees?

Solution

Since point Q must follow a circular path, its speed may be determined from Equation (9.15): $v_Q = (0.075 \text{ m})(200 \text{ s}^{-1}) = 15$ m/s, with the direction of $\mathbf{v}_Q$ as indicated in the figure. Because the cylinder wall constrains the piston, its velocity is vertical. The connecting rod PQ is rigid, so the velocities of the points P and Q must

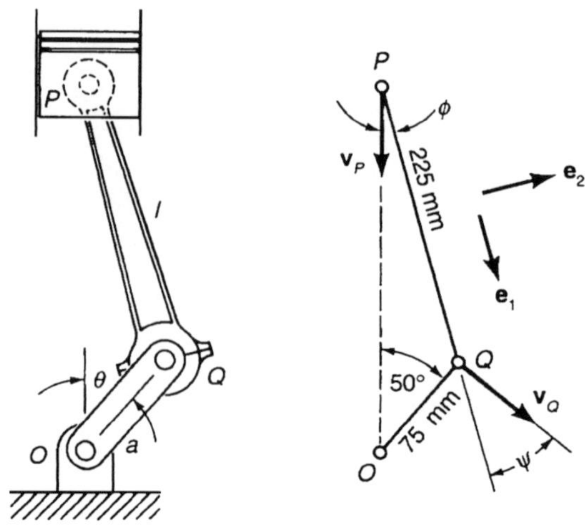

Exhibit 9

satisfy $v_P \cos \phi = v_Q \cos \psi$. The trigonometric rule of sines, applied to the triangle *OPQ*, gives

$$\sin\phi = \frac{a}{l}\sin\theta = \frac{75}{225}\sin 50^\circ$$

which yields $\phi = 14.8°$. The other required angle is then $\psi = 90° - \theta - \phi = 25.2°$. Once these angles are determined, the constraint equation yields the speed of the piston:

$$v_P = \frac{\cos\psi}{\cos\phi} v_Q = 14.04 \text{ m/s}$$

The Angular Velocity Vector

If a rigid body is in *plane motion*, that is, if the velocities of all points of the body lie in a fixed plane, then its orientation may be specified by the angle θ between two fixed lines, one of which passes through the body, as indicated in Figure 9.6. The rate of change of this angle is central to the analysis of the velocities of various points of the body.

To determine this relationship, consider Figure 9.7, which shows a position vector from the point *P* to point *Q*, both fixed in the moving body. Two configurations are shown, one at time *t* and another after an arbitrary change during a time increment Δt. $\mathbf{e}_1$ is defined to be the unit vector in the direction of $\mathbf{r}_{PQ}(t)$, and $\mathbf{e}_2$ is

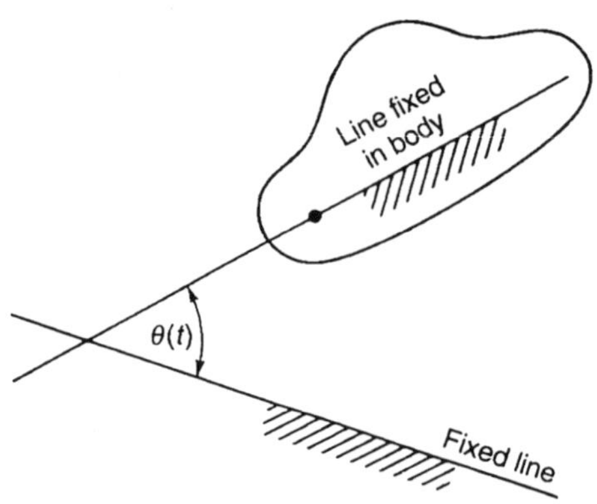

Figure 9.6

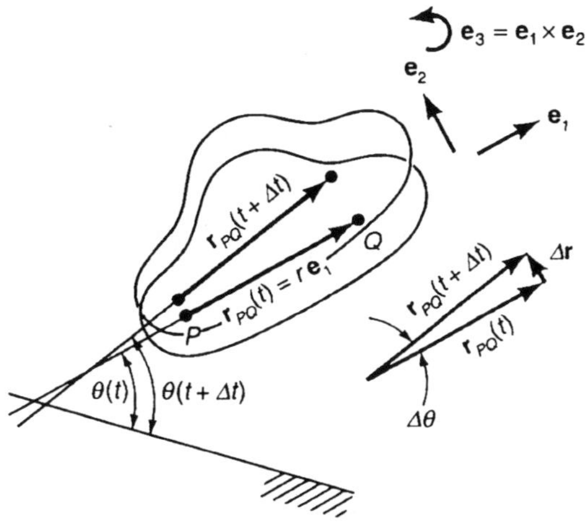

Figure 9.7

defined to be the unit vector of P and Q, 90° counterclockwise from $\mathbf{e}_1$. Both $\mathbf{e}_1$ and $\mathbf{e}_2$ are further assumed to lie in the plane of motion; this assumption is convenient but not limiting. The vector diagram in Figure 9.7 shows the change in $\mathbf{r}_{PQ}$ to be given by the approximation $\Delta\mathbf{r} \approx r\Delta\mathbf{e}_2$. Dividing both sides by the time increment Δt and letting this increment approach zero leads to the relation

$$\frac{d\mathbf{r}_{PQ}}{dt} = r\dot{\theta}\mathbf{e}_2$$

The scalar $\dot{\theta}$ will be denoted also by ω. Note from the definition of θ that a positive value of ω indicates a counterclockwise rotation, whereas a negative value of ω indicates a clockwise rotation. Another useful form of this relation may be written in terms of the **angular velocity vector**,

$$\omega = \omega\,\mathbf{e}_3$$

where $\mathbf{e}_3$ is defined to be $\mathbf{e}_1 \times \mathbf{e}_2$, oriented perpendicular to the plane of Figure 9.7. With this definition,

$$\frac{d\mathbf{r}_{PQ}}{dt} = \boldsymbol{\omega} \times \mathbf{r}_{PQ}$$

Note that this relation is valid for *any* two points in the body. (A pair of points different from those shown in Figure 9.7 might give rise to a different angle, but as the body moves, *changes* in this angle would equal *changes* in θ.)

It can be shown that for the most general motion of a rigid body (not restricted to planar motion) there also exists a unique angular velocity vector for which the same relation holds. However, in nonplanar motion, the angular velocity ω is not straightforwardly related to the rate of change of an angle, and its calculation requires a more extensive analysis than in the case of planar motion.

When $\dot{\mathbf{r}}_{PQ}$ is replaced with $\mathbf{v}_Q - \mathbf{v}_P$ according to Equation (ii) of the previous section, the important velocity relationship

$$\mathbf{v}_Q = \mathbf{v}_P + \omega \times \mathbf{r}_{PQ} \qquad \textbf{(9.20)}$$

is obtained, which, for planar motion, becomes

$$\mathbf{v}_Q = \mathbf{v}_P + r\omega\mathbf{e}_2 \qquad \textbf{(9.21)}$$

In the special case in which $\omega = 0$, this indicates that all points have the same velocity, a motion called **translation**. In the special case in which $\mathbf{v}_P = \mathbf{0}$, the motion is simply rotation about a fixed axis through P. Thus, in the general case, the two terms on the right of Equation (9.20) can be seen to express a superposition of a translation and a rotation about P. But since P can be selected *arbitrarily*, there are as many combinations of a translation and a corresponding "center of rotation" as the analyst wishes to consider!

In all of the these cases, the angular velocity is a property of the *body's* motion, and Equation (9.21) relates the velocities of *any* two points of a body experiencing planar motion. Dot-multiplication of each member of Equation (9.21) with $\mathbf{e}_2$ leads to the following means of evaluating the angular velocity of a plane motion in terms of the velocities of two points:

$$\omega = \frac{\mathbf{e}_2 \bullet \mathbf{v}_Q - \mathbf{e}_2 \bullet \mathbf{v}_P}{r} \quad \textbf{(9.22)}$$

That is, ω will be the difference between the magnitudes of the projections of the velocities of P and Q onto the perpendicular to the line connecting P and Q, divided by the distance between P and Q.

Example **9.11**

What will be the angular velocity of the connecting rod in Example 9.10, at the instant when the angle θ is 50 degrees?

Solution

Referring to Exhibit 9 for the definition of $\mathbf{e}_2$, we see that

$$\omega = \frac{v_Q \sin\psi + v_Q \sin\phi}{l}$$

$$= \frac{(15\ \text{m/s})\sin 25.2° + (14.04\ \text{m/s})\sin 14.8°}{0.225\ \text{m}} = 44.3\ \text{rad/s}$$

The positive value indicates that the rotation is counterclockwise at this instant.

Instantaneous Center of Zero Velocity

For planar motion with $\omega \neq 0$, there always exists a point C' of the body (or an imagined extension of the body) that has zero velocity. If point P of Equation (9.21) is selected to be this special point, the equation reduces to $\mathbf{v}_Q = \mathbf{v}_{C'} + \omega \times \mathbf{r}_{C'Q} = r\omega\mathbf{e}_2$ where r is now the distance from C' to Q and $\mathbf{e}_2$ is perpendicular to the line connecting C' and Q. This latter property can be used to locate C' if the directions of the velocities of two points of the body are known.

Example **9.12**

What is the location of the instantaneous center C' of the connecting rod in Examples 9.10 and 9.11? Use this to verify the previously determined values of the angular velocity of the connecting rod and the velocity of point P.

Solution

The velocity of any point of the connecting rod must be perpendicular to the line from C' to that point. Hence C' must lie at the point of intersection of the horizontal line through P and the line through Q perpendicular to $\mathbf{v}_Q$ (i.e., on the line through O and Q), as shown in Exhibit 10. The pertinent distances can be found as follows:

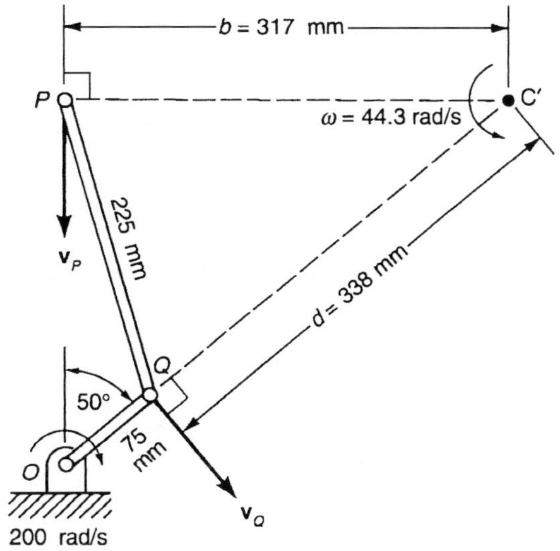

Exhibit 10

$$OP = (75\text{ mm})\cos 50^\circ + (225\text{ mm})\cos 14.8^\circ = 266\text{ mm}$$
$$PC' = OP\tan 50^\circ = 317\text{ mm}$$
$$QC' = OP\sec 50^\circ - 75\text{ mm} = 338\text{ mm}$$

The angular velocity of the connecting rod is then

$$\omega = \frac{v_Q}{QC'} = \frac{15\text{ m/s}}{0.339\text{ m}} = 44.3\text{ rad/s}$$

and the velocity of P is then

$$v_P = PC'\,\omega = (0.317\text{ m})(44.\,3\text{ s}^{-1}) = 14.04\text{ m/s}$$

in agreement with values the previously obtained.

Example 9.13

Using the properties of the instantaneous center, determine the velocity of the point P on the rim of the rolling wheel in Example 9.7.

Solution

Since the wheel rolls without slipping, the point of the wheel in contact with the flat surface has zero velocity and is therefore its instantaneous center. The angular speed of the wheel is $\dot{\theta}$, and the distance from C' to P is readily determined from Exhibit 11:

$$r = 2b\sin\frac{\theta}{2}$$

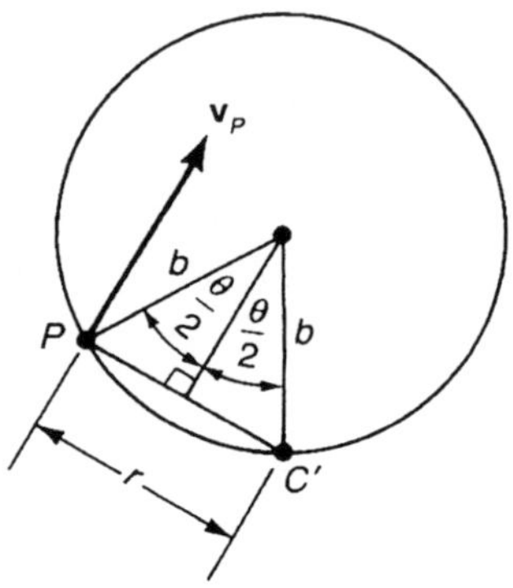

Exhibit 11

The velocity of point P then has the magnitude

$$v_P = r\omega = 2b\sin\frac{\theta}{2}\dot{\theta}$$

and the direction shown in Exhibit 11. This direction should be evident by inspection once it is realized that a positive $\dot{\theta}$ corresponds to clockwise rotation. The reader may find it instructive to recall the conventions for the choice of $\mathbf{e}_2$ and positive ω used in the derivation leading to Equation (9.21) and verify the agreement. Note the simplicity of this analysis as compared with the one expressing the position of P in a rectangular Cartesian coordinate system.

Accelerations in Rigid Bodies

Formally differentiating Equation (9.20) and substituting for $\dot{\mathbf{r}}_{PQ}$ using Equation (9.19) leads to

$$\mathbf{a}_Q = \mathbf{a}_P + (\alpha \times \mathbf{r}_{PQ}) + \omega \times (\omega \times \mathbf{r}_{PQ}) \qquad \textbf{(9.23)}$$

in which the vector $\alpha = d\omega/dt$ is called the **angular acceleration** of the body. For planar motion, $\alpha = \alpha\mathbf{e}_3 = \dot{\omega}\mathbf{e}_3$ and $\omega \times (\omega \times \mathbf{r}) = -\omega^2\mathbf{r}$ so that

$$\mathbf{a}_Q = \mathbf{a}_P + r\alpha\mathbf{e}_2 - r\omega^2\mathbf{e}_1 \qquad \textbf{(9.24)}$$

where $\mathbf{e}_1$ and $\mathbf{e}_2$ are defined as indicated in Figure 9.7.

Equivalent relationships, analogous to Equation (9.17) and Equation (9.22) for velocity, can be obtained by dot-multiplying this equation by $\mathbf{e}_1$ and by $\mathbf{e}_2$:

$$\mathbf{e}_1 \bullet \mathbf{a}_Q = \mathbf{e}_1 \bullet \mathbf{a}_P - r\omega^2 \qquad \textbf{(9.25)}$$

$$\alpha = \frac{\mathbf{e}_2 \bullet \mathbf{a}_Q - \mathbf{e}_2 \bullet \mathbf{a}_P}{r} \qquad \textbf{(9.26)}$$

Example **9.14**

If the speed of the crank in Examples 9.8 through 9.10 is constant, what are the acceleration $\mathbf{a}_P$ of the piston and the angular acceleration α of the connecting rod at the instant when the angle θ is 50 degrees (Exhibit 12)?

Solution

When the crank speed is constant, the acceleration of Q is entirely centripetal, of magnitude

$$a_Q = r\omega^2$$

$$a_Q = (0.075 \text{ m})(200 \text{ s}^{-1})^2 = 3000 \text{ m/s}^2$$

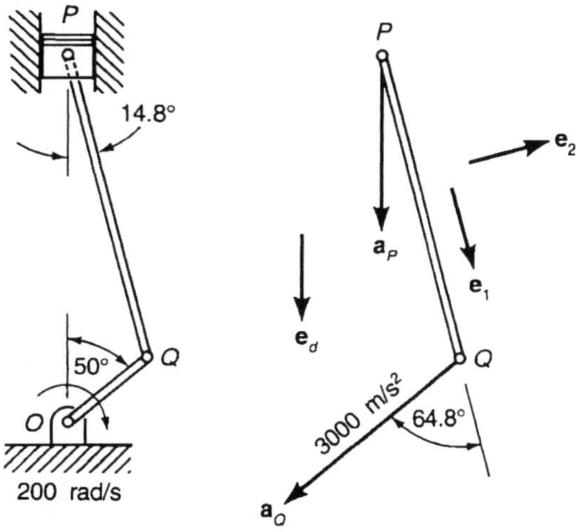

Exhibit 12

and directed toward the center of curvature O of the path of Q. The acceleration of P is vertically upward or downward. To determine the direction, we define a downward unit vector $\mathbf{e}_d$ and let $\mathbf{a}_p = a_p\mathbf{e}_d$ (see Exhibit 12). A positive value of a_P then indicates a downward acceleration and a negative value an upward acceleration. These expressions for a_Q and $\mathbf{a}_P$ are substituted into Equation (9.25), along with the previously determined angular velocity of the rod, giving

$$(3000\ \text{m/s}^2)\cos 64.8° = a_p \cos 14.8° - (0.225\ \text{m})(44.3\ \text{s}^{-1})^2$$

which yields

$$a_P = 1779\ \text{m/s}^2$$

The angular acceleration α of the rod can then be determined from Equation (9.26):

$$\alpha = \frac{(3000\ \text{m/s}^2)\cos 154.8° \quad (1779\ \text{m/s}^2)\cos 104.8°}{0.225\ \text{m}} = -10{,}050\ \text{rad/s}^2$$

The negative value indicates that the angular acceleration is clockwise; that is, the 44.3-rad/s counterclockwise angular velocity is rapidly decreasing at this instant.

NEWTON'S LAWS OF MOTION

Every element of a mechanical system must satisfy Newton's second law of motion; that is, the resultant force $\mathbf{f}$ acting on the element is related to the acceleration $\mathbf{a}$ of the element by

$$\mathbf{f} = m\mathbf{a} \tag{9.27}$$

in which m represents the mass of the element. Newton's third law requires that the force exerted on a body A by a body B is of equal magnitude and opposite direction to the force exerted on body B by body A. These laws and their logical consequences provide the basis for relating motions to the forces that cause them.

Applications to a Particle

A **particle** is an idealization of a material element in which its spatial extent is disregarded, so that the motion of all of its parts is completely characterized by the path of a geometric *point*. When the accelerations of various parts of a system differ significantly, the system is considered to be composed of a number of particles and analyzed as described in the next section.

Example 9.15

An 1800-kg aircraft in a loop maneuver follows a circular path of radius 3 km in a vertical plane. At a particular instant, its velocity is 210 m/s directed 25 degrees above the horizontal as shown in Exhibit 13. If the engine thrust is 16 kN greater than the aerodynamic drag force, what is the rate of change of the aircraft's speed, the magnitude of the aircraft's acceleration, and the aerodynamic lift force?

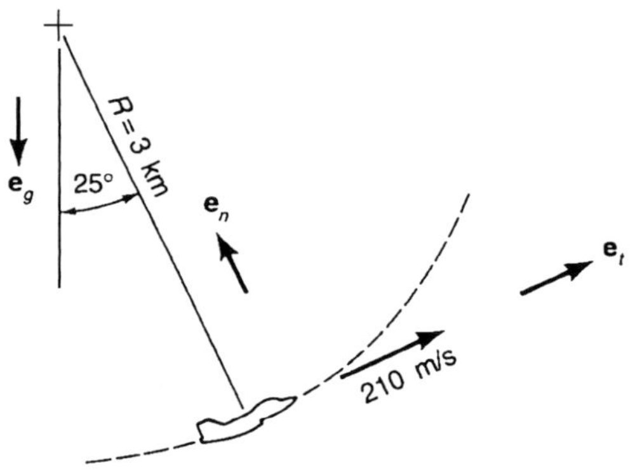

Exhibit 13

Solution

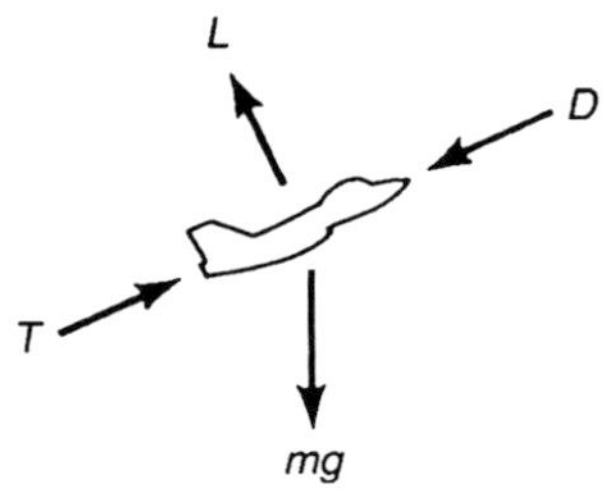

Exhibit 14

Since the dimensions of the aircraft are small compared with the radius of the path, all of its material elements can be considered to have essentially the same motion, so treating the aircraft as a particle as described above is reasonable.

The forces acting on the aircraft are shown on the free-body diagram, Exhibit 14. The thrust **T**, the drag **D**, and the lift **L** all result from aerodynamic pressure from the surrounding air and engine gas. The lift is defined to be the component of the total force that is perpendicular to the flight path, and arises primarily from the wings. The force of gravity, mg, is the only other force arising from a source external to the free body. The left-hand side of Equation (9.27) is the resultant of these forces, whereas the right-hand side is obtained from Equation (9.6). Thus, Newton's second law is written in this case as

$$(T-D)\mathbf{e}_t + L\mathbf{e}_n + mg\mathbf{e}_g = m\left(\dot{v}\mathbf{e}_t + \frac{v^2}{R}\mathbf{e}_n\right)$$

Two independent equations arise from this two-dimensional vector equation. Dot multiplication with $\mathbf{e}_t$ yields $(T - D) + mg\mathbf{e}_t \bullet \mathbf{e}_g = m\dot{v}$ because $\mathbf{e}_t$ and $\mathbf{e}_n$ are always perpendicular vectors and thus their dot product is 0. This equation, rearranged, leads to the rate of change of speed:

$$\dot{v} = \frac{T-D}{m} - g\sin 25° = \frac{16{,}000\ \text{N}}{1800\ \text{kg}} - (9.81\ \text{m/s}^2)\sin 25° = 4.74\ \text{m/s}^2$$

Dot multiplication with $\mathbf{e}_n$ yields

$$L + mg\mathbf{e}_n \bullet \mathbf{e}_g = \frac{mv^2}{R}$$

which then allows us to determine the magnitude of lift force,

$$L = m\left(g\cos 25^\circ + \frac{v^2}{R}\right) = (1800\text{ kg})\left[(9.81\text{ m/s}^2)\cos 25^\circ + \frac{(210\text{ m/s})^2}{3000\text{ m}}\right] = 42.5\text{ kN}$$

The magnitude of the acceleration is then determined by combining the tangential and normal components found above:

$$|\mathbf{a}| = \sqrt{(4.74\text{ m/s}^2)^2 + \left[\frac{(210\text{ m/s})^2}{3000\text{ m}}\right]^2} = 15.45\text{ m/s}^2$$

Example 9.16

Two blocks are interconnected by an inextensible, massless line through the pulley arrangement shown in Exhibit 15. The inertia and friction of the pulleys are negligible. The coefficient of friction between the block of mass m_1 and the horizontal surface is μ. What is the acceleration of the block of mass m_2 as it moves downward?

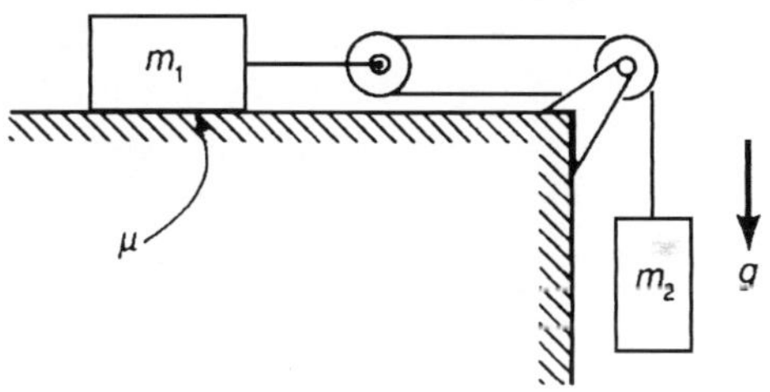

Exhibit 15

Solution

Since the motion of each block is a translation, the acceleration of each element in a block is the same; hence their spatial extensions may be ignored and each block may be treated as a particle.

Moment equilibrium of each pulley requires that the tension be the same in each part of the longer line around the pulleys. Denoting this tension by T, the two free-body diagrams in Exhibit 16 are used to write expressions for Newton's second law. Since the acceleration of the block on the left has no vertical component, $R_1 - m_1g = 0$. Denoting its rightward acceleration by a_1, consideration of the horizontal forces leads to $2T - \mu R_1 = m_1a_1$. Denoting the downward acceleration of the other block by a_2, application of Newton's second law to the other free body yields $m_2g - T = m_2a_2$. To relate the two accelerations, consider the pulley connected to the horizontal block. The instantaneous center of this pulley is at its rim and directly below its center. Thus, the speed of the pulley's upper rim is twice that of its center, so that the speed of the vertically moving block is twice that of the horizontally moving block. Since this is true at all times, we have $a_2 = 2a_1$.

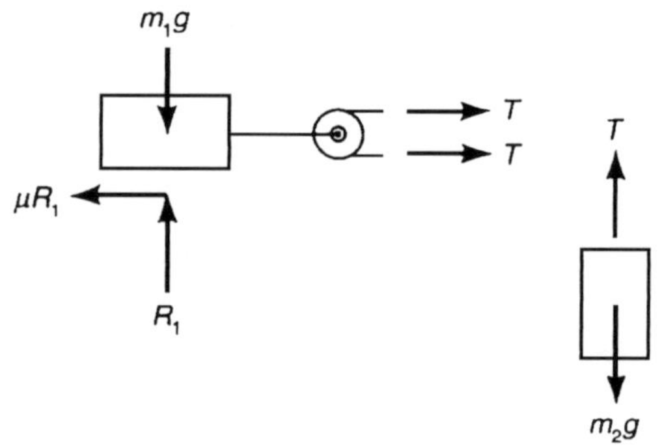

Exhibit 16

Eliminating R, T, and a_1 from these four equations now leads to

$$a_2 = \frac{4m_2 - 2\mu m_1}{4m_2 + m_1} g$$

Observe that if $m_2 > \mu m_1/2$, the acceleration is downward, whereas if $m_2 < \mu m_1/2$, the acceleration is upward, implying that the downward velocity will reach zero, after which time the friction force will no longer equal μR_1.

Systems of Particles

A mechanical **system** is any collection of material elements of fixed identity whose motion we may wish to consider. Such a system is treated as a collection of particles in which the individual particles must obey Newton's laws of motion.

It proves to be very useful to separate the forces acting on a system into those arising from sources outside the system and arising from the interaction between members of the system, as shown in Figure 9.8. That is, the resultant force on the ith particle is written as

$$\mathbf{f}_i = \mathbf{f}_{ie} + \sum_j \mathbf{f}_{i/j}$$

in which $\mathbf{f}_{ie}$ represents the resultant of all forces on the ith particle arising from sources external to the system, and $\mathbf{f}_{i/j}$ represents the force exerted on ith particle by the jth particle. With this notation, Newton's third law may be expressed as $\mathbf{f}_{j/i} = -\mathbf{f}_{i/j}$.

Now, each particle moves according to Newton's second law:

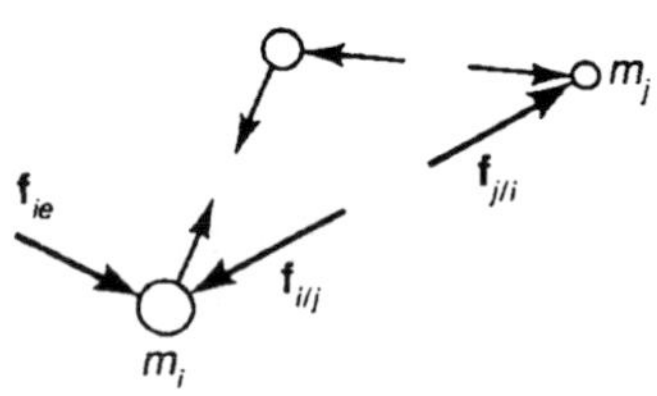

Figure 9.8

$$\mathbf{f}_{ie} + \sum_j \mathbf{f}_{i/j}$$

in which m_i denotes the mass of the ith particle and $\mathbf{a}_i$ its acceleration. There are as many such equations as there are particles in the system; if all such equations are added, the result is

$$\sum_i \mathbf{f}_{ie} + \sum_i \sum_j \mathbf{f}_{i/j} = \sum_i m_i \mathbf{a}_i$$

In view of Newton's third law, the internal forces can be grouped as pairs of oppositely directed forces of equal magnitude, and so their sum vanishes, leaving

$$\sum_i \mathbf{f}_{ie} = \sum_i m_i \mathbf{a}_i \qquad \textbf{(9.28a)}$$

That is, the resultant of all *external* forces is equal to the sum of the products of the individual masses and their corresponding accelerations.

Linear Momentum and Center of Mass

The right-hand member of Equation (9.28a) can be expressed alternatively in terms of the **linear momentum** of the system, which is defined as

$$\mathbf{p} = \sum_i m_i \mathbf{v}_i \tag{9.29}$$

in which $\mathbf{v}_i$ denotes the velocity of the *i*th particle. Differentiation of this equation results in

$$\frac{d\mathbf{p}}{dt} = \sum_i m_i \mathbf{a}_i$$

which is the same expression appearing in Equation (9.28a). Hence an alternative to Equation (9.28a) is

$$\sum_i \mathbf{f}_{ie} = \frac{d\mathbf{p}}{dt} \tag{9.28b}$$

which states that the sum of the external forces is equal to the time rate of change of the linear momentum of the system.

The **center of mass** of the system is a point C located, relative to an arbitrarily selected reference point O, by the position vector $\mathbf{r}_C$, which satisfies the defining equation

$$m\mathbf{r}_C = \sum_i m_i \mathbf{r}_i \tag{9.30}$$

in which m denotes the total mass of the system and $\mathbf{r}_i$ is a position vector from O to the *i*th particle. Differentiation of this equation leads to

$$m\mathbf{v}_C = \sum_i m_i \mathbf{v}_i \tag{9.31}$$

which shows that the linear momentum is the product of the total mass and the velocity of the center of mass. Another differentiation yields

$$m\mathbf{a}_C = \sum_i m_i \mathbf{a}_i$$

which provides still another way of expressing Equation (9.28a):

$$\sum_i \mathbf{f}_{ie} = m\mathbf{a}_C \tag{9.28c}$$

This is sometimes called the **principle of motion of the mass center**. It indicates that the mass center responds to the resultant of external forces exactly as would a single particle having a mass equal to the total mass of the system.

Example 9.17

A motor inside the case shown in Exhibit 17 drives the eccentric rotor at a constant angular speed ω. The distance from the rotor bearing to its center of mass is e, the mass of the rotor is m_r, and the mass of the nonrotating housing is $m - m_r$. (That is, the total mass of the rotor and housing together is m.) The housing is free to translate horizontally, constrained by the rollers, and under the influence of a spring of stiffness k and a dashpot that transmits a force to the housing of magnitude c times the speed of the housing in the direction opposite to that of the velocity of the housing. Write the differential equation that governs the extension $x(t)$ of the spring from its relaxed position.

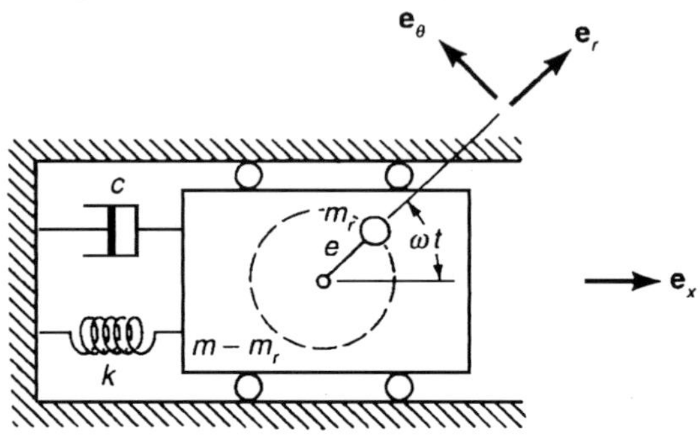

Exhibit 17

Solution

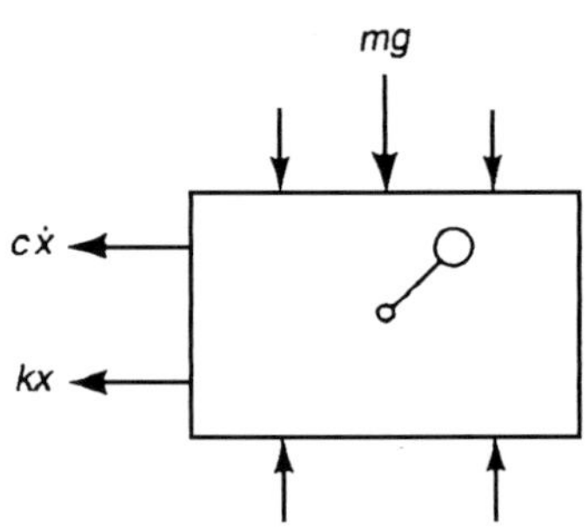

Exhibit 18

Consider the system consisting of the housing and rotor together. The free-body diagram (Exhibit 18) shows forces acting on this system from sources *external* to it. It does *not* include the torque necessary to maintain constant rotor speed nor the reaction at the bearing, these being internal, action-reaction pairs.

Note that when x is positive (the spring extended), the force exerted by the spring on the housing acts to the left, and when x is negative (the spring compressed), this force acts to the right. Both situations are depicted properly by the label kx on the arrow; that is, this indicates that the force equals $-kx\mathbf{e}_x$ in all cases. The same consideration applies to the arrow and label representing the force from the dashpot. The sum of all external forces then is

$$\sum_i \mathbf{f}_{ie} = -(kx + c\dot{x})\mathbf{e}_x + f_y\mathbf{e}_y$$

The acceleration of the housing is simply $\ddot{x}\mathbf{e}_x$, while that of the mass center of the rotor can be most readily determined by using Equation (9.24), letting P be the center of the bearing and noting that $\alpha = 0$. This gives the acceleration of the mass center of the rotor as

$$\mathbf{a} = \ddot{x}\mathbf{e}_x - e\omega^2\mathbf{e}_r$$

Substitution into Equation (9.28a) results in

$$-(kx + c\dot{x})\mathbf{e}_x + f_y\mathbf{e}_y = (m - m_r)\ddot{x}\mathbf{e}_x + m_r(\ddot{x}\mathbf{e}_x - e\omega^2\mathbf{e}_r)$$

The forces f_y are neither known nor of interest for our purpose; they may be eliminated from the equation by dot-multiplying each member with $\mathbf{e}_x$, which leads to

$$-(kx + c\dot{x}) = m\ddot{x} - m_r e\omega^2 \cos \omega t$$

A "standard" form of this equation is obtained by placing the dependent variable and its derivatives on one side and the known function of time on the other:

$$m\ddot{x} + c\dot{x} + kx = m_r e\omega^2 \cos \omega t$$

The same result can be obtained using either Equation (9.28b) or (9.28c).

Impulse and Momentum

The integral with respect to time of the resultant of external forces is called the **impulse** of this resultant force:

$$\mathbf{g} = \int_{t_1}^{t_2} \sum_i \mathbf{f}_{ie} \, dt$$

Integration of both members of Equation (9.28b) results in the following integrated form:

$$\mathbf{g} = \mathbf{p}(t_2) - \mathbf{p}(t_1) \qquad \textbf{(9.32)}$$

This states that impulse of the resultant external force is equal to the change in momentum of the system. Since it is a vector equation, we may obtain up to three independent relationships from it, one for each of three dimensions.

A special case occasionally arises, in which one or more components of the impulse are absent. The corresponding components of momentum then remain constant, and are said to be **conserved**.

Example 9.18

A rocket is simulated by a vehicle that is accelerated by the action of the passenger throwing rocks in the rearward direction as the vehicle moves along the roadway as shown in Exhibit 19. At a certain time, the mass of the vehicle, passenger, and supply of rocks is m, and all are moving at speed v. The passenger then launches a rock of mass m_1 with a rearward velocity of magnitude v_e *relative to the vehicle*. What is the increase Δv in the speed of the vehicle resulting from this action?

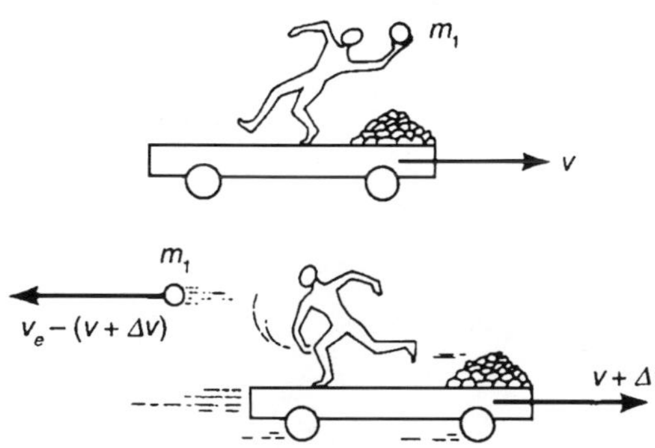

Exhibit 19

Solution

Assuming there is negligible friction at the wheels, the system consisting of passenger, vehicle, and rocks has no external forces acting on it in the direction of travel. The horizontal component of momentum is therefore conserved; that is, it is the same after the rock is launched as it was prior to the launching. After this is written in detail as

$$p_{\text{final}} - p_{\text{rock}} = p_{\text{initial}}$$

$$(m - m_1)(v + \Delta v) - m_1[v_e - (v + \Delta v)] = mv$$

the equation can then be solved for the increase in vehicle speed:

$$\Delta v = \frac{m_1}{m} v_e$$

Moments of Force and Momentum

Equations (9.28) and (9.29) are valid regardless of the lines of action of the forces $\mathbf{f}_{ie}$. For example, the acceleration of the center of mass and the change in momentum of the system shown in Figure 9.9 will be the same for each of the different points of application of the force. However, there are characteristics of the motions induced by these forces that *do* depend on the lines of action, some of which are revealed by considering *moments* of forces.

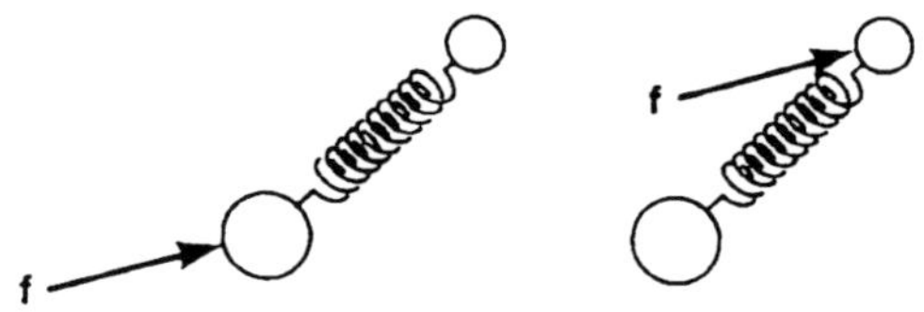

Figure 9.9

Consider again the external and internal forces acting on two typical particles of a system (Figure 9.10). Let the position vectors $\mathbf{r}_i$ and $\mathbf{r}_j$ locate the ith and jth particles, respectively, with respect to a selected point O. If the equation expressing Newton's second law for the ith particle is cross-multiplied by $\mathbf{r}_i$ and all such equations are added, the result is

$$\sum_i \mathbf{r}_i \times \mathbf{f}_{ie} + \sum_i \sum_j \mathbf{f}_{i/j} = \sum_i \mathbf{r}_i \times m_i \mathbf{a}_i$$

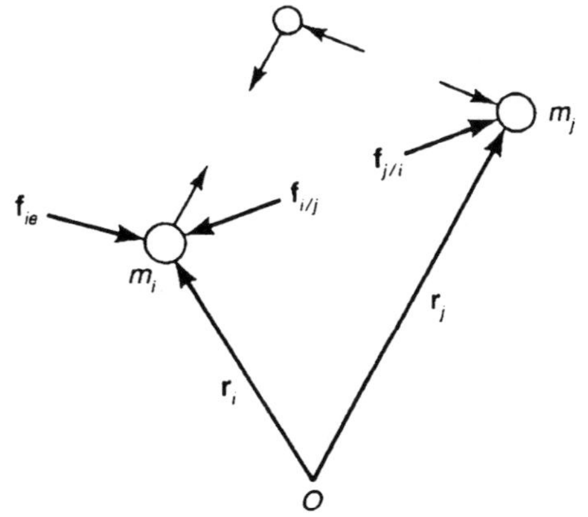

Figure 9.10

Now, if the forces $\mathbf{f}_{i/j}$ and $\mathbf{f}_{j/i} = -\mathbf{f}_{i/j}$ have a common line of action, then

$$\mathbf{r}_i \times \mathbf{f}_{i/j} + \mathbf{r}_j \times \mathbf{f}_{j/i} = 0$$

That is, the **moments** of the members of each action-reaction pair cancel one another, leaving the **moment equation** for the system:

$$\mathbf{M}_O = \sum_i \mathbf{r}_i \times m_i \mathbf{a}_i \tag{9.33a}$$

in which the moment of the external forces is evaluated as in the previous chapter:

$$\mathbf{M}_O = \sum_i \mathbf{r}_i \times \mathbf{f}_{ie}$$

Example **9.19**

The pendulum in Exhibit 20 consists of a stiff rod of negligible mass with two masses attached, and it swings in the vertical plane about the frictionless hinge at O under the influence of gravity. What will be the angular acceleration of the pendulum in terms of angular displacement θ and the other parameters indicated in the sketch?

Exhibit 20

Solution

The free-body diagram shows the forces external to the system consisting of the rod together with the two particles. Since the reaction at the support is unknown and is of no interest for our purpose, a good strategy would be to consider moments about this point. Referring to the free-body diagram, we evaluate the resultant moment as usual:

$$\mathbf{M}_O = (r_i \mathbf{e}_r) \times (m_1 g \mathbf{e}_g) + (r_2 \mathbf{e}_r) \times (m_2 g \mathbf{e}_g) = -(m_1 r_1 + m_2 r_2) g \sin\theta \mathbf{e}_3$$

Since each particle follows a circular path with center at O, their accelerations may be expressed as $\mathbf{a}_i = r_i \dot{\theta}^2 \mathbf{e}_r + r_i \ddot{\theta} \mathbf{e}_\theta$. The right-hand member of the moment law, Equation (9.33a), is then evaluated in this case as

$$\begin{aligned}\sum \mathbf{r}_i \times m_i \mathbf{a}_i &= (r_1 \mathbf{e}_r) \times m_1 r_1 (-\dot{\theta}^2 \mathbf{e}_r + \ddot{\theta} \mathbf{e}_\theta) + (r_2 \mathbf{e}_r) \times m_2 r_2 (-\dot{\theta}^2 \mathbf{e}_r + \ddot{\theta} \mathbf{e}_\theta) \\ &= (m_1 r_1^2 + m_2 r_2^2) \ddot{\theta} \mathbf{e}_3\end{aligned}$$

Substitution into the moment law, Equation (9.33a), results in

$$-(m_1 r_1 + m_2 r_2) g \sin\theta \mathbf{e}_3 = (m_1 r_1^2 + m_2 r_2^2) \ddot{\theta} \mathbf{e}_3$$

or

$$\ddot{\theta} = -\frac{m_1 r_1 + m_2 r_2}{m_1 r_1^2 + m_2 r_2^2} g \sin\theta$$

The **moment of momentum** or **angular momentum about point** O is defined as

$$\mathbf{H}_O = \sum_i \mathbf{r}_i \times m_i \mathbf{v}_i \tag{9.34}$$

Now, if O is fixed in the inertial frame, then $\mathbf{v}_i = \mathbf{r}_i$, and it follows that

$$\frac{d\mathbf{H}_O}{dt} = \sum_i (\dot{\mathbf{r}}_i \times m_i \mathbf{v}_i + \mathbf{r}_i \times m_i \dot{\mathbf{v}}_i) = \sum_i \mathbf{r}_i \times m_i \mathbf{a}_i$$

This provides an alternative way of writing the moment law, as

$$\mathbf{M}_O = \frac{d\mathbf{H}_O}{dt} \tag{9.33b}$$

In the preceding example, the angular momentum about O is

$$\mathbf{H}_O = (r_1 \mathbf{e}_r) \times (m_1 r_1 \dot{\theta} \mathbf{e}_\theta) + (r_2 \mathbf{e}_r) \times (m_2 r_2 \dot{\theta} \mathbf{e}_\theta) = \left(m_1 r_1^2 + m_2 r_2^2\right) \dot{\theta} \mathbf{e}_3$$

Differentiating this expression and substituting for the right-hand member of Equation (9.33b) leads to the result achieved using Equation (9.33a).

As with forces and linear momentum, there are situations in which the moment of external forces vanishes. Then, Equation (9.33b) implies that the angular momentum about O remains constant, or is *conserved*.

Example 9.20

Suppose the pendulum of Example 9.19 is suspended at rest when it is struck by a small projectile, which becomes imbedded in the lower ball (Exhibit 21). What angular velocity ω is imparted to the pendulum?

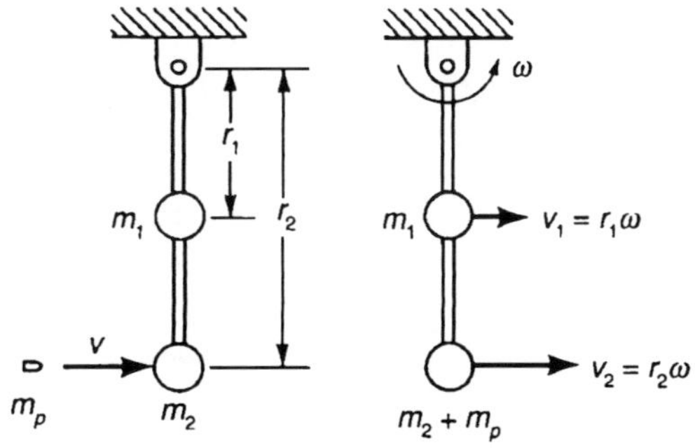

Exhibit 21

Solution

During the collision, which can induce a large reaction at the support as well as between the projectile and ball, the moment of forces external to the system—the pendulum and the projectile—will be zero. Hence, the angular momentum of this system prior to impact will equal that immediately after impact:

$$r_2 m_p v \mathbf{e}_3 = r_1 m_1 v_1 \mathbf{e}_3 + r_2 (m_2 + m_p) v_2 \mathbf{e}_3 = [m_1 r_1^2 + (m_2 + m_p) r_2^2] \omega \mathbf{e}_3$$

Hence,

$$\omega = \frac{r_2 m_p v}{m_1 r_1^2 + (m_2 + m_p) r_2^2}$$

WORK AND KINETIC ENERGY

The integration in Equation (9.9) has extensive implications that will be examined in this section.

A Single Particle

If the form of the tangential acceleration indicated in Equation (9.9) is merged with Equation (9.6), the result can be used to express Newton's second law as

$$\mathbf{f} = m\left(v\frac{dv}{ds}\mathbf{e}_t + \frac{v^2}{\rho}\mathbf{e}_n \right) \tag{i}$$

Now, if each member is dot-multiplied by an increment of change of position, $d\mathbf{r} = ds\mathbf{e}_t$, and the resulting scalars are integrated, they become

$$\int_{\mathbf{r}_1}^{\mathbf{r}_2} \mathbf{f} \bullet d\mathbf{r} = \int_{s_1}^{s_2} f_t \, ds = W_{1-2}$$

and

$$m\int_{v_1}^{v_2} v\,dv = \tfrac{1}{2}mv_2^{\,2} - \tfrac{1}{2}mv_1^{\,2}$$

The integral W_{1-2} is called the **work** done on the particle by the force **f** as the particle moves from position 1 to position 2. The scalar $T = \frac{1}{2}mv^2$ is called the **kinetic energy** of the particle. Since (i) holds throughout any interval, a consequence of Newton's second law is the **work-kinetic energy relationship**:

$$W_{1-2} = T_2 - T_1 \tag{9.35}$$

Work is considered positive when the kinetic energy of the system has been increased. When enough information is available to permit evaluation of the work integral, this provides a useful way of predicting the change in the speed of the particle.

Example 9.21

A 3.5-Mg airplane is to be launched from the deck of an aircraft carrier with the aid of a steam-powered catapult. The force that the catapult exerts on the aircraft varies with the distance s along the deck as shown in Exhibit 22. If other forces are negligible, what value of the constant f_0 is necessary for the catapult to accelerate the aircraft from rest to a speed of 160 km/h at the end of the 30-m travel?

$$f(s) = \frac{f_0}{1 + \dfrac{s}{30\text{m}}}$$

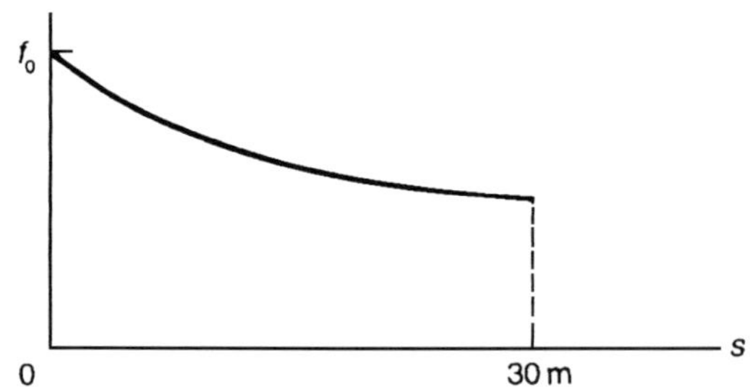

Exhibit 22

Solution

Letting d stand for the 30-m travel, the work done on the aircraft will be

$$W = \int_0^d \frac{f_0\,ds}{1+\frac{s}{d}} = (f_0 d)\ln\left(1+\frac{s}{d}\right)\Bigg|_0^d = (f_0 d)\ln 2 = (20.8\text{ m})f_0$$

This will equal the change in kinetic energy, which is initially zero.

$$\Delta T = (0.5)mv^2 - 0$$

$$\Delta T = \frac{1}{2}(3500\text{ kg})\left[\left(160\,\frac{\text{km}}{\text{h}}\right)\frac{1000\text{ m/km}}{3600\text{ s/h}}\right]^2 = 3.46\times10^6\text{ J}$$

The work-kinetic energy relationship

$$W = \Delta T$$

$$(20.8 \text{ m}) f_0 = 3.46 \times 10^6 \text{ N}\bullet\text{m}$$

implies that the constant f_0 must have the value

$$f_0 = 166 \text{ kN}$$

Work of a Constant Force

A commonly encountered force of constant magnitude and direction is that of gravity near the earth's surface. When a constant force acts on a particle as it moves, the work done by the force can be evaluated as indicated in Figure 9.11. The increment of work as the particle undergoes an increment $d\mathbf{r}$ of displacement can be expressed as

$$dW = \mathbf{f}_0 \bullet d\mathbf{r} = |\mathbf{f}_0||d\mathbf{r}| \cos \measuredangle_{\mathbf{f}_0}^{d\mathbf{r}} = f_0 dq$$

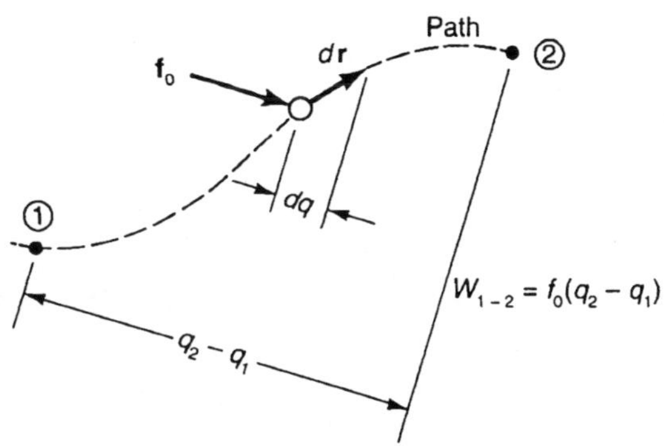

Figure 9.11

in which dq is the component of the displacement increment that is parallel to the force. Since the force is constant,

$$W_{1-2} = f_0 \int_1^2 dq = f_0(q_2 - q_1) \qquad \textbf{(9.36)}$$

Thus, any movement of the particle that is perpendicular to the direction of the force has no effect on the work. In other words, the work is the same as would have been done if the particle had moved rectilinearly through a distance of $(q_2 - q_1)$ in a direction parallel to the force.

Example **9.22**

How fast must the toy race car be traveling at the bottom of the hill to be able to coast to the top of the hill (see Exhibit 23)?

Solution

As the car moves up the hill, the work done by the force of gravity will be $W_g = -mgh$. The work done by friction forces may be negligible if the wheels are well made. If this is the case, the work done by all forces is approximately that due to

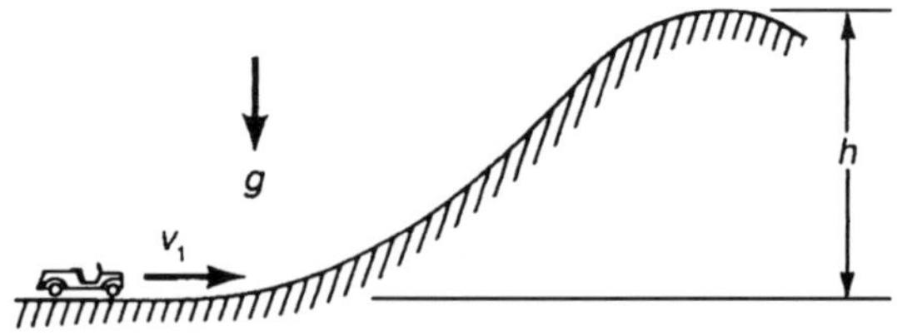

Exhibit 23

gravity. The speed of the car as it nears the hilltop can approach zero, so the work-kinetic energy equation may be written as

$$-mgh = T_2 - T_1 = 0 - \frac{1}{2}mv_1^2$$

which implies a minimum required speed of

$$v_1 = \sqrt{2gh}$$

With friction, the required speed will be somewhat greater.

Distance-Dependent Central Force

A force that remains directed toward or away from a fixed point is called a **central force**. Examples of forces for which the magnitude depends only on the distance from the particle to a fixed point are the force of gravitational attraction and the force from an elastic, tension-compression member with one end pinned to a fixed support. Figure 9.12 shows a particle P moving with such a central force acting on it; the dependence on distance is expressed by the function $f(r)$, with the convention that a positive value of f indicates an attractive force and a negative value of f a repulsive force.

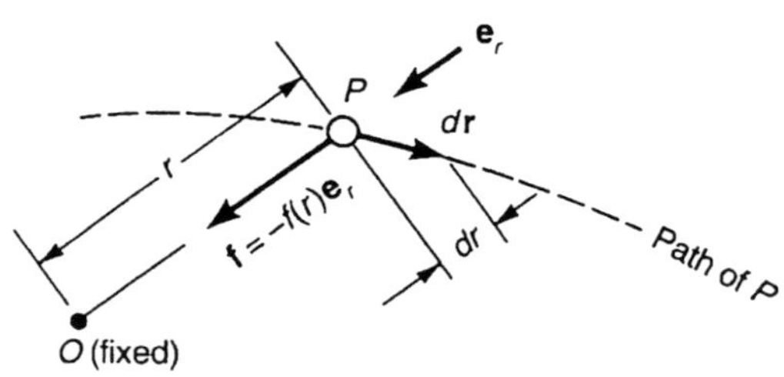

Figure 9.12

Now, the increment of work may be written as

$$dW = \mathbf{f} \bullet d\mathbf{r} = -f(r)\mathbf{e}_r \bullet d\mathbf{r}$$

Referring to the figure, we see that $\mathbf{e}_r \bullet d\mathbf{r} = |d\mathbf{r}| \cos \measuredangle_{\mathbf{er}}^{d\mathbf{r}}$ is equal to the change dr in radial distance r. Thus,

$$W_{1-2} = -\int_{r_1}^{r_2} f(r)\,dr \tag{9.37}$$

Similar to the case of the constant force, the work done by a central force through an arbitrary motion is the same as would be done for a rectilinear motion, but in the radial direction.

Example 9.23

The elastic spring in Exhibit 24 has a linear force-displacement characteristic; that is, it exerts a force equal to the stiffness k times the amount it is stretched from its relaxed length l_0. As the particle moves from position 1 to position 2, what is the work done by the spring force on the particle?

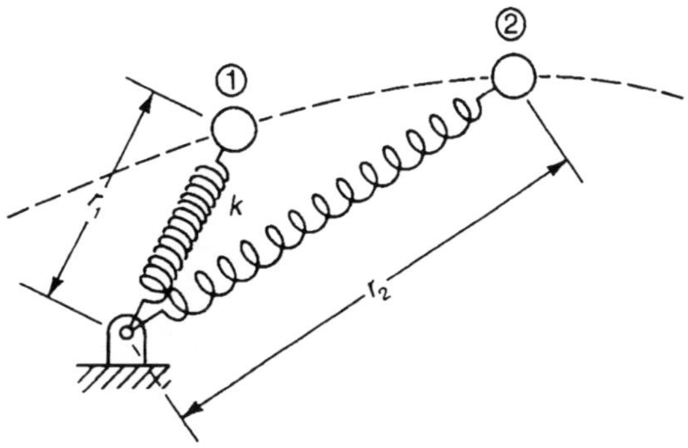

Exhibit 24

Solution

This is a case of a central distance-dependent force with $f(r) = -k(r - l_0)$. Equation (9.37) then becomes

$$W_{1-2} = -k\int_{r_1}^{r_2} (r - l_0)\,dr$$

A more convenient form results if we introduce the amount of spring extension as $\delta = (r - l_0)$. The integral then becomes

$$W_{1-2} = -k\int_{\delta_1}^{\delta_2} \delta\,d\delta = -\frac{k}{2}\left(\delta_2^2 - \delta_1^2\right)$$

Example 9.24

A 0.6-kg puck slides on a horizontal surface without friction under the influence of the tension in a light cord that passes through a small hole at O (see Exhibit 25). A spring under the surface imparts a tension in the cord that is proportional to the distance from the hole to the puck; its stiffness is $k = 30$ N/m. At a certain instant, the puck is 200 mm from the hole and moving at 2 m/s in the direction indicated in the top view. If the spring is in its relaxed position when the puck is at the hole, what is the maximum distance from the hole reached by the puck?

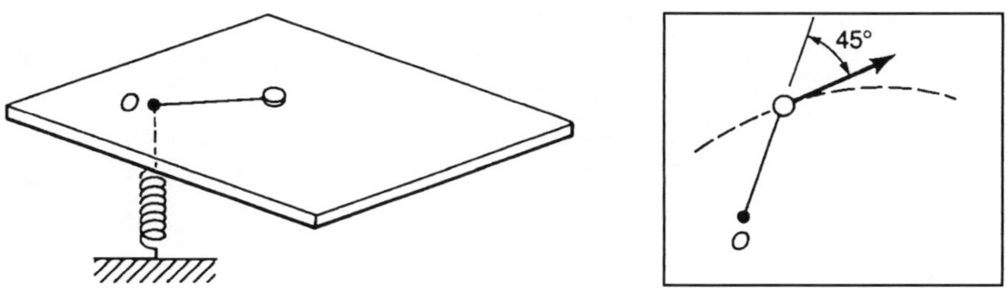

Exhibit 25

Solution

With the initial and maximum distances denoted by r_1 and r_2, respectively, the work done on the puck by the force from the cord from the initial position to that of maximum distance will be

$$W_{1-2} = -\frac{k}{2}\left(r_2^2 - r_1^2\right)$$

Since this is the only force that does work, this value must equal the change in kinetic energy:

$$-\frac{k}{2}\left(r_2^2 - r_1^2\right) = \frac{m}{2}\left(v_2^2 - v_1^2\right)$$

The moment about O of the force from the cord is zero, so the angular momentum about the hole is conserved. Because there is no radial component of velocity at the maximum distance, the angular momentum there is simply $r_2 m v_2$. Hence,

$$r_1\ m v_1\ \sin\ 45° = r_2 m v_2$$

These two equations contain the unknowns r_2 and v_2. Isolating v_2 from the latter, substituting this expression into the energy equation, and rearranging leads to the equation

$$\left(\frac{r_2}{r_1}\right)^4 - \left(1 + \frac{m v_1^2}{k r_1^2}\right)\left(\frac{r_2}{r_1}\right)^2 + \frac{m v_1^2}{k r_1^2}\sin^2 45° = 0$$

When the given values are substituted, the quadratic formula yields

$$\left(\frac{r_2}{r_1}\right) = 1.618$$

as the largest root so that the maximum distance reached is r_2 = 1.618 (200 mm) = 324 mm.

Example **9.25**

A torpedo expulsion device operates by means of gas expanding within a tube that holds the torpedo. When test-fired with the tube firmly anchored, a 550-kg torpedo leaves the tube at 20 m/s. In operation, a 30-Mg submarine is traveling at 5 m/s when it expels a 550-kg torpedo in the forward direction. What are the speeds of the submarine and torpedo immediately after expulsion?

Solution

Considering the two-body system consisting of the submarine and torpedo, let us assume that the external forces remain in balance during expulsion. Then $W_e = 0$. Assuming also that the gas pressure depends only on the position of the torpedo relative to the submarine, the work W_{12} done by the internal forces will be the same during actual operation as during the test-firing. The work-kinetic energy relation ship for the test-firing yields $W_{12} = ½ (550 \text{ kg})(20 \text{ m/s})^2 = 110 \text{ kJ}$, and for the operating condition the relationship is

$$110 \text{ kJ} = \frac{1}{2}(30,000 \text{ kg})\left[v_1^2 - (5 \text{ m/s})^2\right] + \tfrac{1}{2}(550 \text{ kg})\left[v_2^2 - (5 \text{ m/s})^2\right]$$

Also, if the external forces are in balance, momentum will be conserved:

$$(30,550 \text{ kg})(5 \text{ m/s}) = (30,000 \text{ kg})v_1 + (550 \text{ kg})v_2$$

These two relationships give the desired speeds as

$$v_1 = 4.6 \text{ m/s}, \ v_2 = 24.8 \text{ m/s}$$

The 19.8-m/s boost in speed given the torpedo is slightly less than when it is fired from the firmly anchored tube. However, the speed of the torpedo relative to the submarine is

$$v_2 - v_1 = 20.2 \text{ m/s}$$

or slightly higher than in the fixed-tube test.

Two special cases of the work done by internal forces are of interest. The simpler is that in which the particles are constrained so that the distances between all pairs remain fixed, that is, the case of a rigid body. In this case, all the $dr_{i/j}$ are zero and the work-kinetic energy equation, Equation (9.38), reduces to $W_e = \Delta T$.

Another special case occurs when the force T_{ij} depends only on the distance $r_{i/j}$. That is, the force is not a function of relative velocity or previous history of deformation. Then the work integral $-\int T_{ij} dr_{i/j}$ is a function only of the distance between the particles and does not depend on the manner in which the particles move to reach a particular configuration. This would be the case, for example, with elastic spring interconnections or gravitational interactions. In this case, we can define the potential functions as

$$V_{ij}(r_{i/j}) = \int_{(r_{i/j})_0}^{r_{i/j}} T_{ij}(\rho_{i/j}) d\rho_{i/j}$$

and if their sum is denoted by

$$V = \sum_{i-j} V_{ij}$$

the work-energy integral becomes

$$W_e = \Delta T + \Delta V$$

That is, when the internal forces are all conservative, the work done by the external forces is equal to the change in total mechanical energy within the system.

KINETICS OF RIGID BODIES

If a system of particles is structurally constrained so that the distance between every pair of particles remains constant as the system moves, it forms a rigid body. Thus, the laws of kinetics in the previous section are applicable, along with the kinematics relationships reviewed earlier. Of the kinematics relationships, Equation (9.28c) will be useful in the form given, whereas the moment equation, Equation (9.33a), must be specialized to relate accelerations to angular velocity and angular acceleration.

In the general (three-dimensional) case, both the moment equation and kinematics relationships become considerably more complicated than they are for planar motion, and will be outside the scope of this review.

Moment Relationships for Planar Motion

Figure 9.13 shows the plane of the motion of a rigid body, with a point P (to be selected by the analyst for moment reference) along with an element of mass dm, which in the following is analogous to the mass m_i in the earlier analysis of a set of particles. The summation appearing in Equation (9.33a) will be written as an integral in this case, because the body is viewed as having continuously distributed mass. The element of mass is located relative to P by the position vector $\mathbf{r} = r\mathbf{e}_1 + z\mathbf{e}_3$. Its acceleration is related to that of point P through Equation (9.24), and the moment equation, Equation (9.33a), may be written as

$$\begin{aligned}\mathbf{M}_P &= \int_m (r\mathbf{e}_1 + z\mathbf{e}_3)\times(\mathbf{a}_P + r\alpha\mathbf{e}_2 - r\omega^2\mathbf{e}_1)\,dm \\ &= \left(\int_m \mathbf{r}\,dm\right)\times\mathbf{a}_P + \left(\int_m r^2\,dm\right)\alpha\mathbf{e}_3 - \alpha\int_m zr\mathbf{e}_1\,dm - \omega^2\int_m zr\mathbf{e}_2\,dm\end{aligned}$$

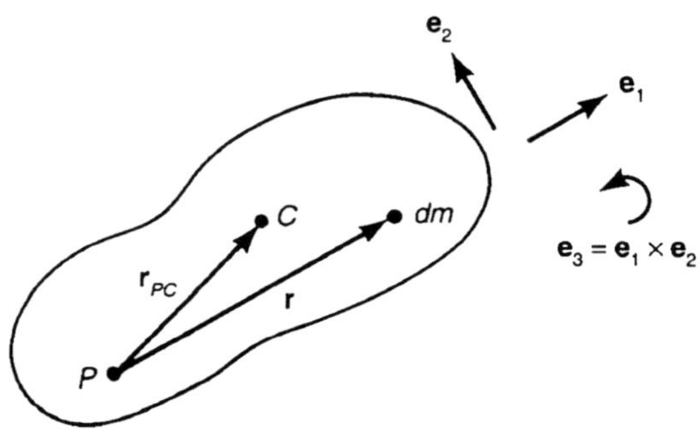

Figure 9.13

If the body's mass is distributed symmetrically with respect to the plane of motion through P, the last two integrals in the last line will vanish; without such symmetry, these terms indicate the possibility of components of moment in the plane of motion. Thus, even for *plane* motion, the distribution of mass may imply forces *perpendicular* to the plane of motion. These will not be pursued in detail here, but the reader must be aware of this possibility. The other two integrals will be of concern. The first is exactly the expression one would write to determine the location of the center of mass from point P:

$$\int_m \mathbf{r}\,dm = m\mathbf{r}_{PC}$$

The second integral is called the **moment of inertia** of the body about the axis through P and perpendicular to the plane of motion:

$$\int_m r^2\,dm = I_P$$

The moment about this axis is thus related to accelerations through

$$M_{P3} = \mathbf{e}_3 \bullet \mathbf{M}_P = I_P\alpha + \mathbf{e}_3 \bullet (m\mathbf{r}_{PC} \times \mathbf{a}_P) \quad \textbf{(9.40a)}$$

By using Equation (9.24) to relate the acceleration of P to that of the mass center C, it is possible to express this moment law in the alternative form

$$M_{P3} = I_C\alpha + \mathbf{e}_3 \bullet (m\mathbf{r}_{PC} \times \mathbf{a}_C) \quad \textbf{(9.40b)}$$

where I_C is the moment of inertia of the body about an axis through C perpendicular to the plane of motion. The two moments of inertia are related through the **parallel axis formula**

$$I_P = I_C + md^2$$

in which d is the distance between P and C.

Two special cases warrant attention. If P is chosen to be the mass center C, then $\mathbf{r}_{PC} = 0$, and the relationship is

$$M_{C3} = I_C\alpha$$

If the body is hinged about a fixed support and P is selected to be on the axis of the hinge, then

$\mathbf{a}_P = 0$, and the relationship is

$$M_{P3} = I_P\alpha$$

These last two relationships indicate the moment of inertia of the body is the property that provides resistance to changes in the angular velocity, much as mass provides resistance to changes in the velocity of a particle. For bodies of simple geometry, the integrals have been evaluated in terms of mass and the geometry, and results can be found in tabulated summaries. More complicated bodies can require tedious work to estimate the moment of inertia, or there are experiments based on the implications of Equation (9.40) that can be used to determine it. It is common to specify the moment of inertia by giving the mass of the body and its **radius of gyration**, k_P, defined by $I_P = mk_P^2$.

Example 9.26

A 23-kg rotor has a 127 mm radius of gyration about its axis of rotation. What average torque about its fixed axis of rotation is required to bring the rotor from rest to a speed of 200 rpm in 6 seconds?

Solution

The moment of inertia of the rotor is

$$I = (23 \text{ kg})(0.127 \text{ m})^2 = 0.371 \text{ kg} \bullet \text{m}^2$$

and the average angular acceleration is

$$\alpha = \frac{(200 \text{ rpm})}{6 \text{ s}} \frac{(2\pi \text{ rad/r})}{(60 \text{ s/min})} = 3.49 \text{ rad/s}^2$$

For fixed-axis rotation,

$$M = I\alpha = (0.371 \text{ kg} \bullet \text{m}^2)(3.49 \text{ s}^{-2}) = 1.29 \text{ N} \bullet \text{m}$$

Example 9.27

The car with rear-wheel drive in Exhibit 26 has sufficient power to cause the drive wheels to slip as it accelerates. The coefficient of friction between the drive wheels and roadway is μ. What is the acceleration in terms of g, μ, and the dimensions shown?

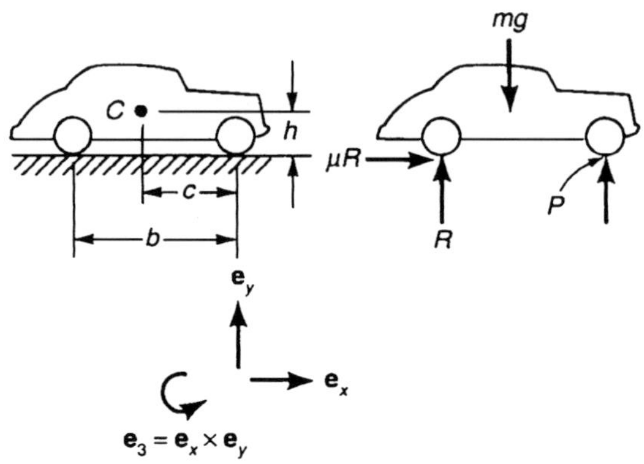

Exhibit 26

Solution

Assuming the car does not rotate, every point will have the same acceleration, $\mathbf{a} = a_x\mathbf{e}_x$. The free-body diagram shows the forces external to the car, with the label on the horizontal force at the drive wheels accounting for the fact that the friction limit has been reached there. Equation (9.28c) implies that $\mu R = ma_x$. Since the reaction R in this equation is unknown, another relationship must be introduced. Of several that could be written (for example, forces in another direction, moments about a selected point), it would be best if no additional unknowns are introduced. Thus, to avoid bringing the unknown reaction at the front wheels into the analysis, consider moments about point P, which are related by the moment law, Equation (9.40b):

$$-bR + c\,mg = I(0) + \mathbf{e}_3 \bullet [m(-c\mathbf{e}_x + h\mathbf{e}_y) \times a_x\mathbf{e}_x] = -mha_x$$

Eliminating R between this equation and the friction equation leads to

$$a_x = \frac{\mu c}{b - \mu h} g$$

Example 9.28

The uniform slender rod in Exhibit 27 slides along the wall and floor under the effects of gravity. If friction is negligible, what is the angular acceleration in terms of g, l, and the angle θ?

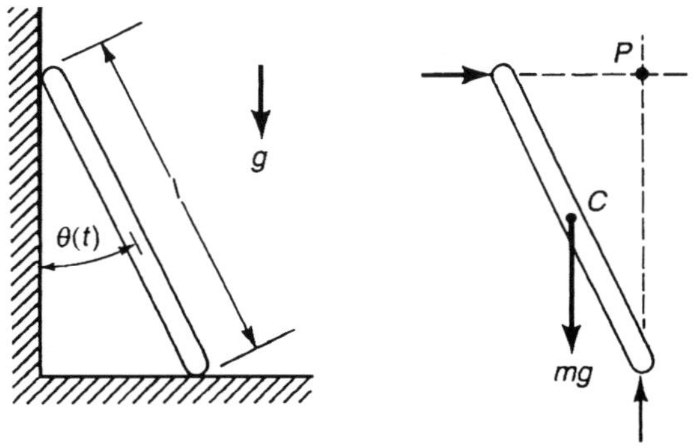

Exhibit 27

Solution

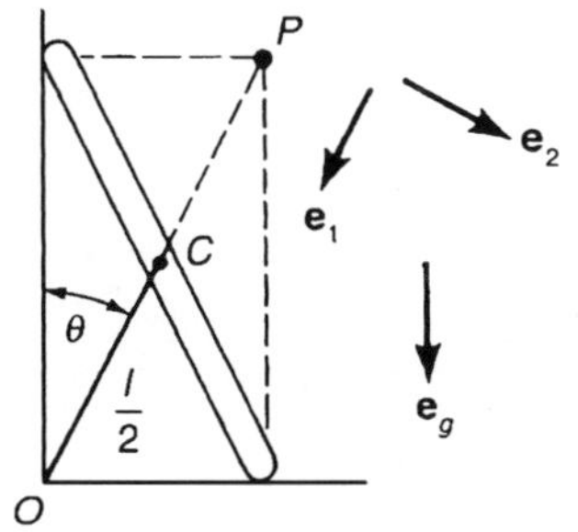

Exhibit 28

As indicated on the free-body diagram, the reactions from the wall and floor are horizontal and vertical since there is no friction. Because neither of these is known, a good strategy would be to avoid dealing with them; to this end, consider moments about point P, which will be related by Equation (9.40b). To evaluate the acceleration of the mass center C, observe that it follows a circular path with radius $l/2$ and center at O (see Exhibit 28). Thus, the acceleration of C is

$$\mathbf{a}_C = \frac{l}{2}(\dot{\theta}^2 \mathbf{e}_1 + \ddot{\theta}\mathbf{e}_2)$$

Referring to the free-body diagram, we can write the moment of forces about P as

$$\mathbf{M}_P = \frac{l}{2}\mathbf{e}_1 \times mg\mathbf{e}_g = \frac{1}{2}mgl\sin\theta\mathbf{e}_3$$

Any of a number of references gives the moment of inertia of a slender, uniform rod about an axis through its center as

$$I_C = \frac{1}{12}ml^2$$

Substituting the above into the moment law, Equation (9.40b), yields

$$\frac{1}{2}mgl\sin\theta = \frac{1}{12}ml^2\ddot{\theta} + \mathbf{e}_3 \bullet \left[m\frac{l}{2}\mathbf{e}_1 \times \frac{l}{2}(\dot{\theta}^2\mathbf{e}_1 + \ddot{\theta}\mathbf{e}_2)\right] = \frac{ml^2}{3}\ddot{\theta}$$

which leads to the desired angular acceleration:

$$\ddot{\theta} = \frac{3g}{2l}\sin\theta$$

Example 9.29

The uniform, slender beam of length l is suspended by the two wires in the configuration shown in Exhibit 29 when the wire on the left is cut. Immediately after the wire is severed (that is, while the velocities of all points are still zero), what is the tension in the remaining wire?

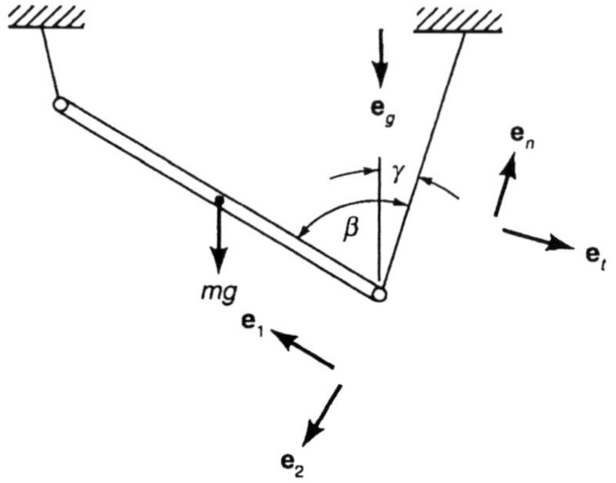

Exhibit 29

Solution

The free-body diagram (Exhibit 30) shows the desired force and the only other force acting on the beam, that of gravity. Summing moments about the center of mass gives a relationship between the desired tension and the angular acceleration of the bar:

$$\frac{l}{2}T\sin\beta = \frac{ml^2}{12}\alpha \qquad \textbf{(i)}$$

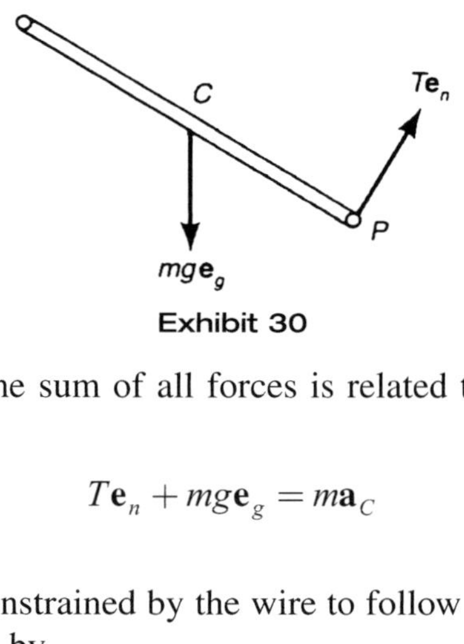

Exhibit 30

We also know that the sum of all forces is related to the acceleration of the mass center by

$$T\mathbf{e}_n + mg\mathbf{e}_g = m\mathbf{a}_C \qquad \textbf{(ii)}$$

Since the end P is constrained by the wire to follow a circular path, its acceleration may be expressed by

$$\mathbf{a}_P = a_t\mathbf{e}_t + \frac{v_P^2}{R}\mathbf{e}_n \qquad \textbf{(iii)}$$

in which a_t is another unknown quantity. Finally, this acceleration is related to that of the center of mass by

$$\mathbf{a}_C = \mathbf{a}_P + \frac{l}{2}\alpha\mathbf{e}_2 - \frac{l}{2}\omega^2\mathbf{e}_1 \qquad \textbf{(iv)}$$

Since velocities are still zero, the centripetal terms v_P^2/R and $\frac{l}{2}\omega^2$ are both zero. With this simplification, Equations (ii), (iii), and (iv) readily combine to give

$$T\mathbf{e}_n + mg\mathbf{e}_g = m\left(a_t\mathbf{e}_t + \frac{l}{2}\alpha\mathbf{e}_2\right)$$

To avoid dealing with the unknown a_t, we may dot-multiply each term in this equation by $\boldsymbol{e}_n$ with the result

$$T + mg\mathbf{e}_n \bullet \mathbf{e}_g = \frac{1}{2}ml\alpha\mathbf{e}_n \bullet \mathbf{e}_2$$

Referring to the specified geometry, the dot products are evaluated as

$$\mathbf{e}_n \bullet \mathbf{e}_g = \cos\ (180° - \gamma) = -\cos\ \gamma$$

$$\mathbf{e}_n \bullet \mathbf{e}_2 = \cos\ (90° + \beta) = -\sin\ \beta$$

and the equation can be written as

$$T + \frac{1}{2}ml\alpha \sin\beta = mg\cos\gamma \qquad \textbf{(v)}$$

Now α is readily eliminated by substituting the expression for α obtained from Equation (i) into Equation (v), leading to the desired value of the tension:

$$T = \frac{mg\ \cos\gamma}{1 + 3\sin^2\beta}$$

Work and Kinetic Energy

If a rigid body has a number of forces $f_1, f_2, \ldots, f_n$ applied at points $P_1, P_2, \ldots, P_n$, the time rate at which these forces do work on the body (that is, the power transmitted to the body) can be evaluated as

$$\frac{dW}{dt} = \mathbf{f}_1 \bullet \frac{d\mathbf{r}_1}{dt} + \mathbf{f}_2 \bullet \frac{d\mathbf{r}_2}{dt} + \cdots + \mathbf{f}_n \bullet \frac{d\mathbf{r}_n}{dt} = \sum_i \mathbf{f}_i \bullet \mathbf{v}_i$$

But $\mathbf{v}_i$, the velocity of point P_i, can be related to the velocity of a selected point P of the body:

$$\mathbf{v}_i = \mathbf{v}_P + \omega \times \mathbf{r}_{Pi}$$

so that the power can also be expressed as

$$\frac{dW}{dt} = \sum_i \mathbf{f}_i \bullet (\mathbf{v}_P + \omega \times \mathbf{r}_{Pi}) = \left(\sum_i \mathbf{f}_i\right) \bullet \mathbf{v}_P + \left(\sum_i \mathbf{r}_{Pi} \times \mathbf{f}_i\right) \bullet \omega$$

$$= \mathbf{f} \bullet \mathbf{v}_P + \mathbf{M}_P \bullet \omega \qquad \textbf{(9.41)}$$

in which **f** is the resultant of all of the forces. *P* may be selected as any point of the body, and M_p is the resultant moment about *P*. For example, as a rotor turns about a fixed axis, there may be forces from the support bearings in addition to an accelerating torque about the axis of rotation. If the point *P* is selected to be on the axis of rotation, $\mathbf{v}_p$ will be zero, and the power transmitted to the rotor (which will, of course, induce a change in its kinetic energy) is simply the dot-product of the torque and the angular velocity. Negative or positive values are possible, depending on the angle between $\mathbf{M}_p$ and ω (that is, whether the moment component is in the same or the opposite direction as the rotation).

The kinetic energy of a rigid body is the sum of the kinetic energies of its individual elements, whose velocities can be related to the velocity of a selected point *P* and the angular velocity ω. Referring to Figure 9.13, we write this for plane motion as

$$T=\frac{1}{2}\int_m |\mathbf{v}|^2\, dm=\frac{1}{2}\int_m (\mathbf{v}_P+\omega\times\mathbf{r})\bullet(\mathbf{v}_P+\omega\times\mathbf{r})dm$$
$$=\frac{1}{2}\left(\int_m dm\right)v_P^2+\mathbf{v}_P\bullet\left[\omega\times\left(\int_m \mathbf{r}dm\right)\right]+\frac{1}{2}\int_m |\omega\times\mathbf{r}|^2 dm$$

The integral in the first term is simply the mass *m* of the body. The integral in the second term is related to the mass and position of the center of mass by

$$\int_m \mathbf{r}dm=m\mathbf{r}_{PC}$$

For plane motion, $\omega\times\mathbf{r}=\omega r\mathbf{e}_2$, so the last term becomes

$$\int_m |\omega\times\mathbf{r}|^2 dm=\omega^2\int_m r^2 dm=I_P\omega^2$$

The expression for kinetic energy for plane motion of a rigid body is then

$$T=\frac{1}{2}mv_P^2+m\mathbf{v}_P\bullet(\omega\times\mathbf{r}_{PC})+\frac{1}{2}I_P\omega^2 \qquad \textbf{(9.42)}$$

Example 9.30

The wheel in Exhibit 31 is released from rest and rolls down the hill with sufficient friction to prevent slipping. Its mass is *m*, its radius is *r*, and its central radius of gyration is *k*. After the center of the wheel has dropped a vertical distance *h*, what is the speed of the center of mass of the wheel?

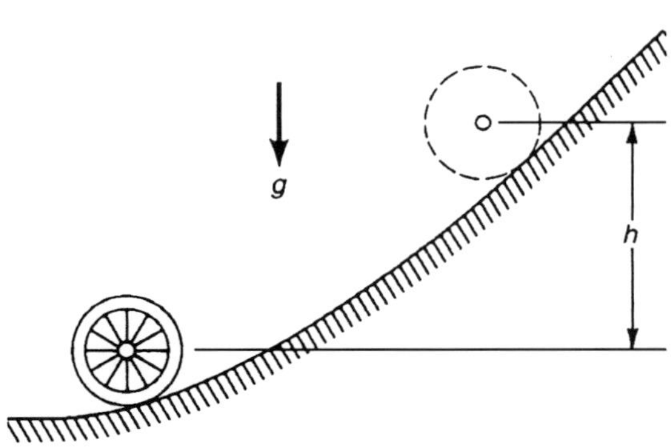

Exhibit 31

Solution

Since the velocity of the contact point is zero, the work of the force there is zero. The work of the force of gravity is then the total work done on the wheel and is simply $W = mgh$. Since the contact point is the instantaneous center, the speed of the center of the wheel is readily related to the angular velocity, $\omega = v_C/r$. The kinetic energy can be written from Equation (9.2), with P selected as any point on the wheel. If we choose P to be the center of mass,

$$T = \frac{1}{2}mv_C^2 + 0 + \frac{1}{2}mk^2\omega^2 = \frac{1}{2}\left(1 + \frac{k^2}{r^2}\right)mv_C^2$$

If, instead, we choose P to be the instantaneous center, the kinetic energy is

$$T = \frac{1}{2}m(0)^2 + 0 + \frac{1}{2}(mk^2 + mr^2)\omega^2 = \frac{1}{2}\left(1 + \frac{k^2}{r^2}\right)mv_C^2$$

Since the work must equal the change in kinetic energy,

$$mgh = \frac{1}{2}\left(1 + \frac{k^2}{r^2}\right)mv_C^2$$

and the speed of the center will be

$$v_C = \sqrt{\frac{2gh}{1 + \frac{k^2}{r^2}}}$$

SELECTED SYMBOLS AND ABBREVIATIONS

Symbol or Abbreviation	Description
$\mathbf{a}$	acceleration
$\mathbf{a}_t$	tangential component of acceleration
$\mathbf{e}_t$	unit vector tangent to path
$\mathbf{e}_n$	unit vector in principal normal direction
$\mathbf{e}_i$	unit vector in direction indicated by the specific value of *i*
$\mathbf{f}$	resultant force
g	gravitational field intensity
$\mathbf{g}$	impulse resultant force
$\mathbf{H}_o$	angular momentum about *O*
I_P	moment of inertia about *P*
κ	local curvature of path
k_P	radius of gyration about *P*
M_i	mass of *i*th particle
M	total mass
$\mathbf{M}_P$	moment of forces about *P*
N	coefficient of kinetic fraction
ρ	radius of curvature of path
$\mathbf{r}$	position vector
s	distance along path
t	time
$\boldsymbol{T}$	kinetic energy
$\mathbf{v}$	velocity
$\boldsymbol{W}$	work
α	angular acceleration
μ	coefficient of sliding friction
ω	angular velocity

PROBLEMS

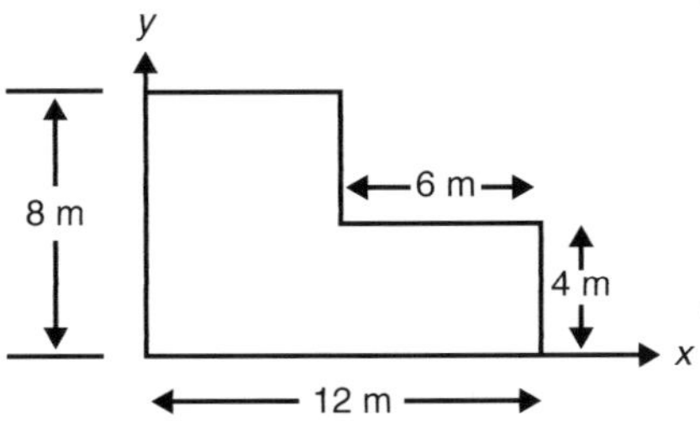

Exhibit 9.1

9.1 The uniform density flat plate shown in Exhibit 9.1 has a mass of 480 kg. The density of the body is most nearly:

a. 10 kg/m^2 c. 6.7 kg/m^2
b. 4.6 kg/m^2 d. 11.8 kg/m^2

9.2 Assuming the density of the plate in Exhibit 9.1 is 9 kg/m^2, the mass moment of inertia of the body is most nearly:

a. 10,400 kg-m^2 c. 9,600 kg-m^2
b. 18,300 kg-m^2 d. 7,400 kg-m^2

9.3 A particle is thrown vertically upward from the edge *A* of the ditch shown in Exhibit 9.3. If the initial velocity is 4 m/s, and the particle is known to hit the bottom, *B*, of the ditch exactly 6 seconds after it was released at *A*, determine the depth of this ditch. Neglect air resistance.

a. 24.0 m c. 200 m
b. 152.6 m d. 176.6 m

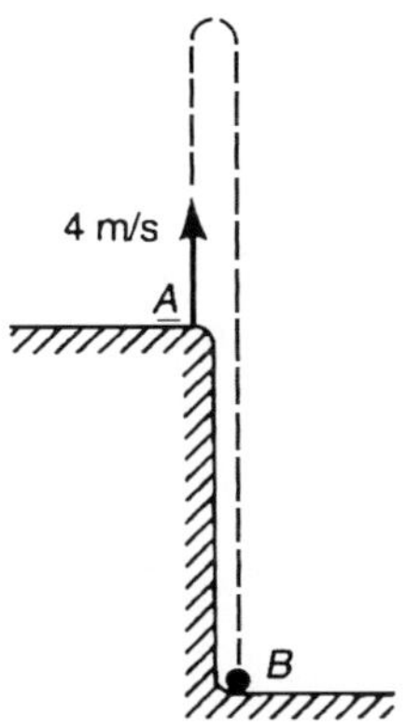

Exhibit 9.3

9.4 The slider *P* in Exhibit 9.4 is driven by a complex mechanism in such a way that (i) it remains on a straight path throughout; (ii) at the instant *t* = 0, the slider is located at A, (iii) at any general instant of time, the velocity of *P* is given by $v = (3t2 - t + 2)$ m/s. Determine the distance of *P* from point *O* when t = 2 s.

a. 26 m c. 6 m
b. 10 m d. 12 m

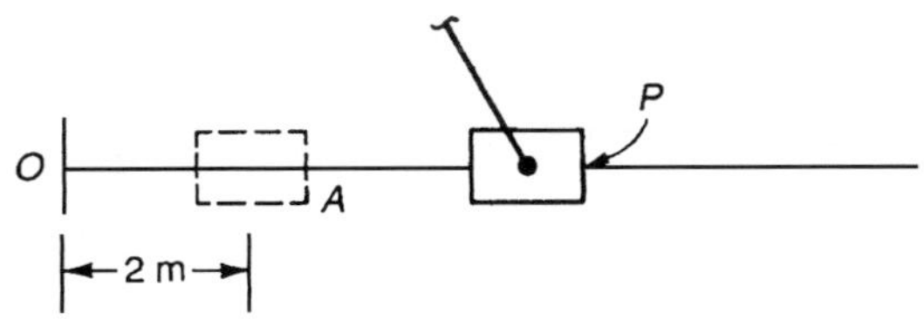

Exhibit 9.4

9.5 A particle in rectilinear motion starts from rest and maintains the acceleration profile shown in Exhibit 9.5. The displacement of the particle in the first 8 seconds is:

a. 4 m c. 24 m
b. 28 m d. 20 m

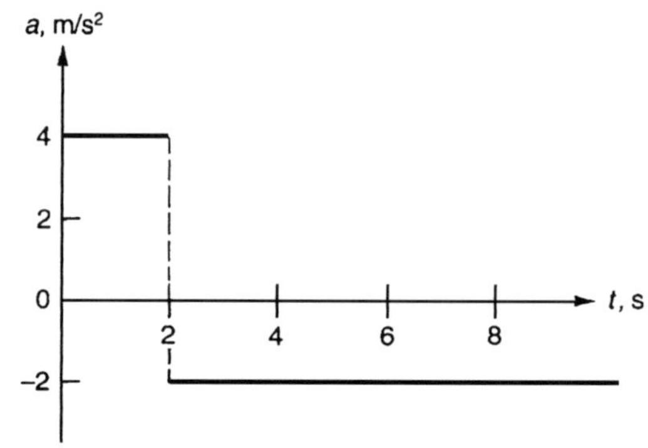

Exhibit 9.5

9.6 A ball is thrown by a player (Exhibit 9.6) from a position 2 m above the ground surface with a velocity of 40 m/s inclined at 60° to the horizontal. Determine the maximum height, H, the ball will attain.

a. 63.2 m c. 30.6 m
b. 61.2 m d. 31 m

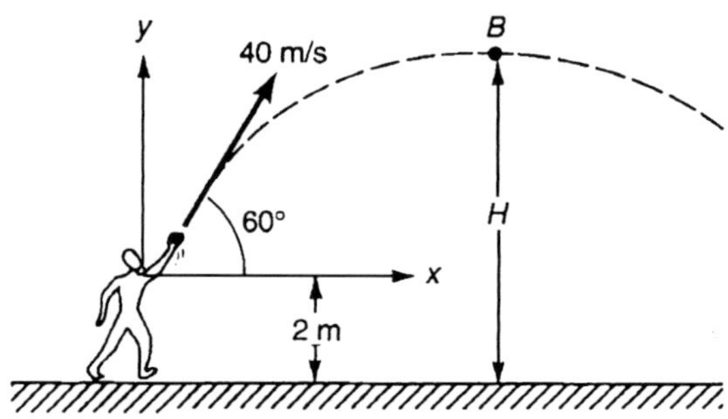

Exhibit 9.6

9.7 A golf ball (Exhibit 9.7) is struck horizontally from point A of an elevated fairway. Determine the initial speed that must be imparted to the ball if the ball is to strike the base of the flag stick on the green 140 meters away. Neglect air friction.

a. 34.3 m/s c. 90 m/s
b. 103 m/s d. 19.2 m/s

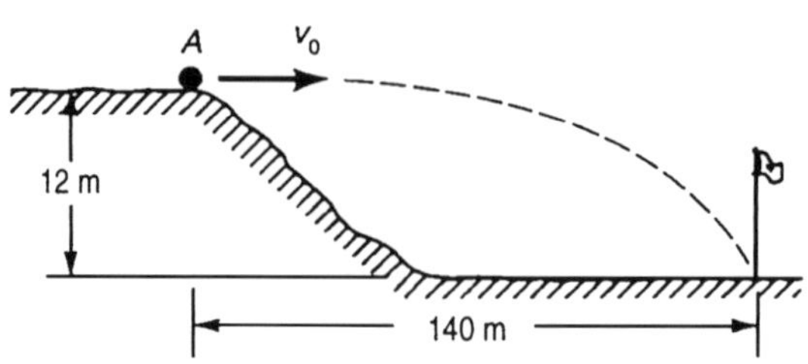

Exhibit 9.7

9.8 In Exhibit 9.8, the rod R rotates about a fixed axis at O. A small collar B is forced down the rod (toward O) at a constant speed of 3 m/s relative to the rod. If the value of θ at any given instant is $\theta = (t^2 + t - 2)$ rad, find the magnitude of the acceleration of B at time $t = 1$ second, when B is known the be 1 meter away from O.

a. 8.0 m/s^2 c. 18.4 m/s^2
b. 20.2 m/s^2 d. 3.0 m/s^2

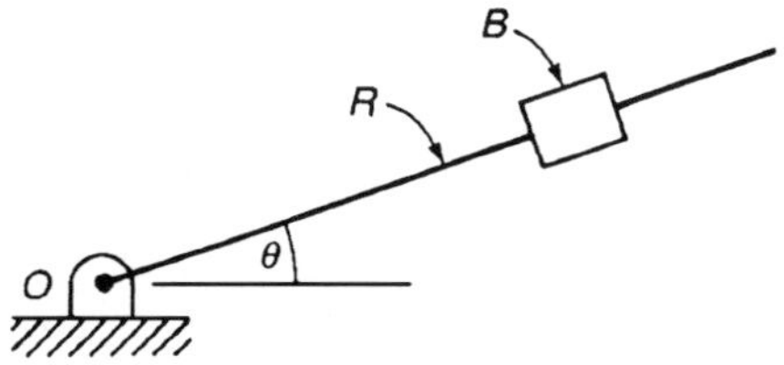

Exhibit 9.8

9.9 A rocket (Exhibit 9.9) is fired vertically upward from a launching pad at B, and its flight is tracked by radar from point A. Find the magnitude of the velocity of the rocket when $\theta = 45°$ if $\dot{\theta} = 0.1$ rad/s.

a. 36 m/s   c. 90 m/s
b. 180 m/s   d. 360 m/s

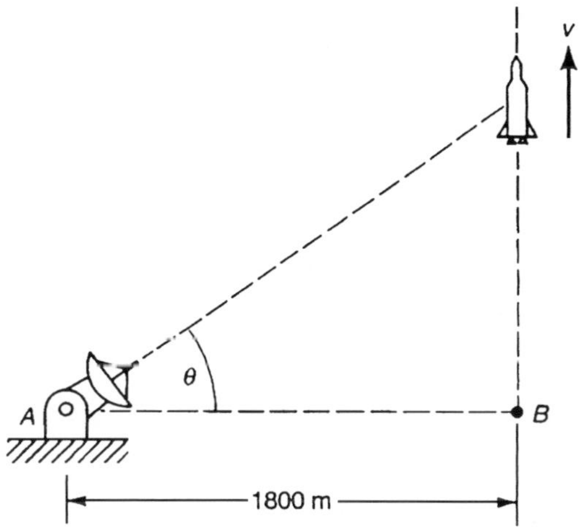

Exhibit 9.9

9.10 A particle is given an initial velocity of 50 m/s at an angle of 30° with the horizontal as shown in Exhibit 9.10. What is the radius of curvature of its path at the highest point, C?

a. 19.5 m   c. 221 m
b. 255 m   d. 191 m

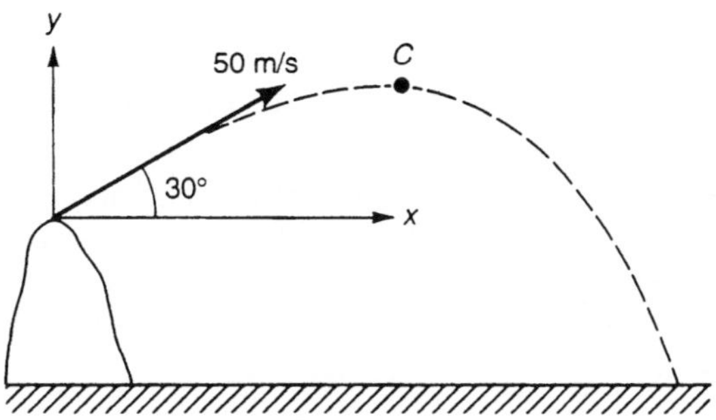

Exhibit 9.10

9.11 An automobile moves along a curved path that can be approximated by a circular arc of radius 110 meters. The driver keeps his foot on the accelerator pedal in such a way that the speed increases at the constant rate of 3 m/s². What is the total acceleration of the vehicle at the instant when its speed is 20 m/s?

a. 22.0 m/s² c. 3.0 m/s²
b. 3.6 m/s² d. 4.7 m/s²

9.12 A pilot testing an airplane at 800 kph wishes to subject the aircraft to a normal acceleration of 5 *gs* in order to fulfill the requirements of an on-board experiment. Find the radius of the circular path that would allow the pilot to do this.

a. 502 m c. 1007 m
b. 3308 m d. 1453 m

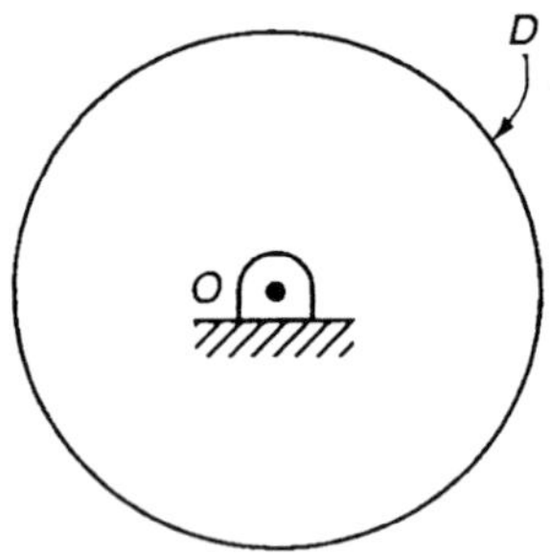

Exhibit 9.13

9.13 At the instant $t = 0$, the disk D in Exhibit 9.13 is spinning about a fixed axis through O at an angular speed of 300 rpm. Bearing friction and other effects are known to slow the disk at a rate that is k times its instantaneous angular speed, where k is a constant with the value $k = 1.2\ \text{s}^{-1}$. Determine when (from $t = 0$) the disk's spin rate is cut in half.

a. 6.5 s c. 0.8 s
b. 13.1 s d. 0.6 s

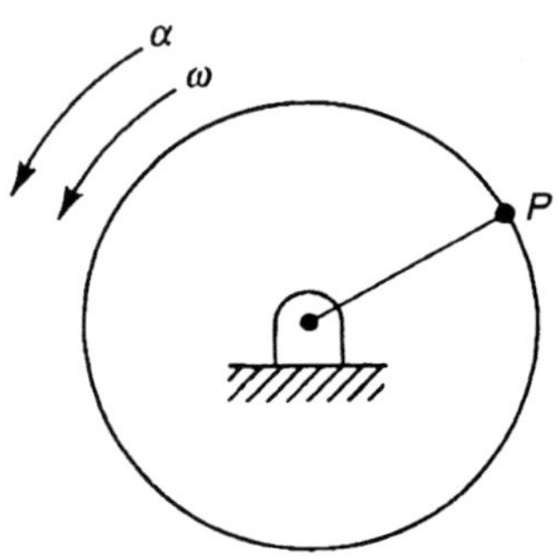

Exhibit 9.14

9.14 In Exhibit 9.14, a flywheel 2 m in radius is brought uniformly from rest up to an angular speed of 300 rpm in 30 s. Find the speed of a point P on the periphery 5 seconds after the wheel started from rest.

a. 10.5 m/s c. 62.8 m/s
b. 5.2 m/s d. 100.0 m/s

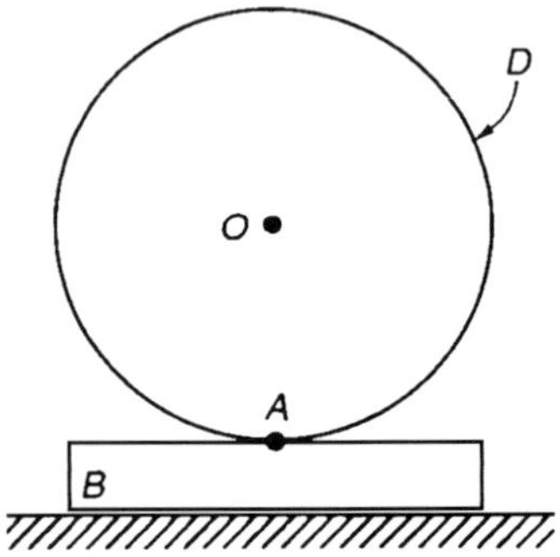

Exhibit 9.15

9.15 The block B (Exhibit 9.15) slides along a straight path on a horizontal floor with a constant velocity of 2 m/s to the right. At the same time, the disk, D, of 3-m diameter rolls without slip on the block. If the velocity of the center, O, of the disk is directed to the left and remains constant at 1 m/s, determine the angular velocity of the disk.

a. 0.3 rad/s counterclockwise
b. 2.0 rad/s counterclockwise
c. 0.7 rad/s counterclockwise
d. 0.7 rad/s clockwise

9.16 In Exhibit 9.16, the disk, D, rolls without slipping on a horizontal floor with a constant clockwise angular velocity of 3 rad/s. The rod, R, is hinged to D at A, and the end, B, of the rod touches the floor at all times. Determine the angular velocity of R when the line OA joining the center of the disk to the hinge at A is horizontal as shown.

a. 0.6 rad/s counterclockwise
b. 0.6 rad/s clockwise
c. 3.0 rad/s counterclockwise
d. 3.0 rad/s clockwise

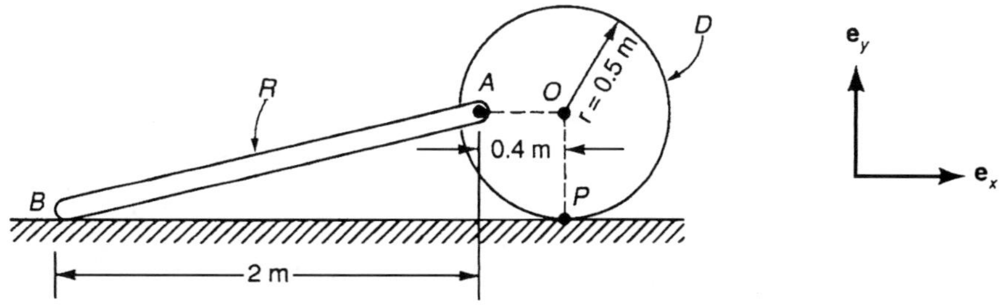

Exhibit 9.16

9.17 The fire truck in Exhibit 9.17 is moving forward along a straight path at the constant speed of 50 km/hr. At the same time, its 2-meter ladder OA is being raised so that the angle θ is given as a function of time by $\theta = (0.5t^2 - t)$ rad, where t is in seconds. The magnitude of the acceleration of the tip of the ladder when $t = 2$ seconds is:

a. 0
b. 4.0 m/s^2
c. 2.0 m/s^2
d. 2.8 m/s^2

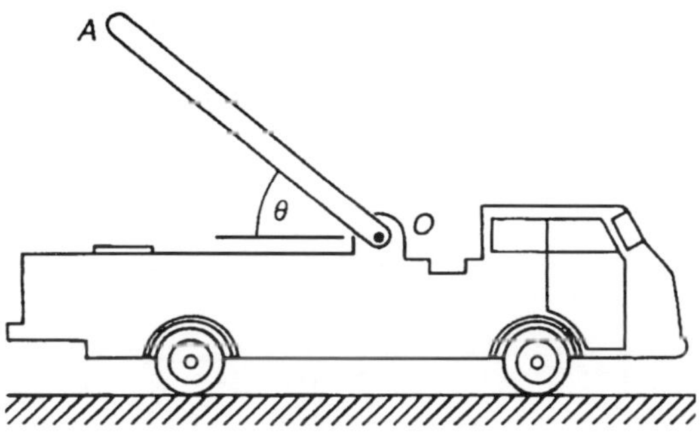

Exhibit 9.17

9.18 In Exhibit 9.18, the block B is constrained to move along a horizontal rectilinear path with a constant acceleration of 2 m/s^2 to the right. The slender rod, R, of length 2 m is pinned to B at O and can swing freely in the vertical plane. At the instant when $\theta = 0°$ (rod is vertical), the angular velocity of the rod is zero but its angular acceleration is 2.5 m/s^2 clockwise. Find the acceleration of the midpoint G of the rod at this instant ($\theta = 0°$).

a. 3.0 m/s^2 ←
b. 0.5 m/s^2 →
c. 2.5 m/s^2 ←
d. 2.5 m/s^2 →

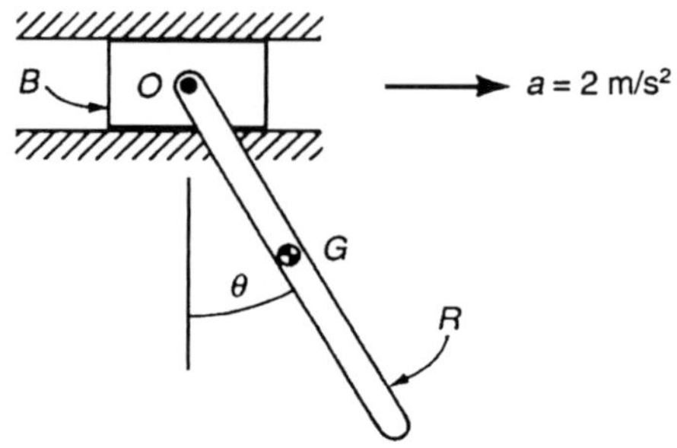

Exhibit 9.18

9.19 The block, B, in Exhibit 9.19, contains a square-cut circular groove. A particle, P, moves in this groove in the clockwise direction and maintains a constant speed of 6 m/s relative to the block. At the same time, the block slides to the right on a straight path at the constant speed of 10 m/s. Find the magnitude of the absolute velocity of P at the instant when $\theta = 30°$.

a. 8.7 m/s
b. 16 m/s
c. 4 m/s
d. 14 m/s

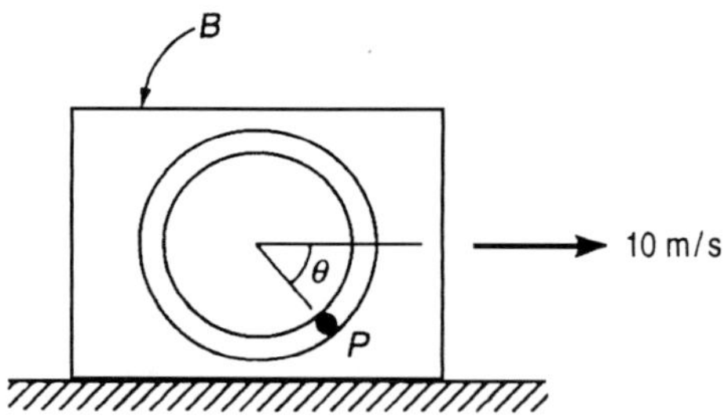

Exhibit 9.19

9.20 In Exhibit 9.20, a pin moves with a constant speed of 2 m/s along a slot in a disk that is rotating with a constant clockwise angular velocity of 5 rad/s. Calculate the absolute acceleration of this pin when it reaches the position C (directly above O). The unit vectors $\mathbf{e}_x$ and $\mathbf{e}_y$ are fixed to the disk.

a. $17.5\mathbf{e}_y$ m/s^2 c. $-2.5\mathbf{e}_y$ m/s^2
b. $-17.5\mathbf{e}_y$ m/s^2 d. $-22.5\mathbf{e}_y$ m/s^2

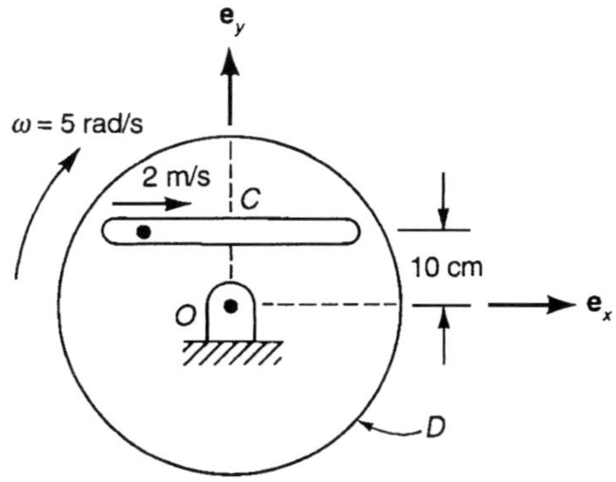

Exhibit 9.20

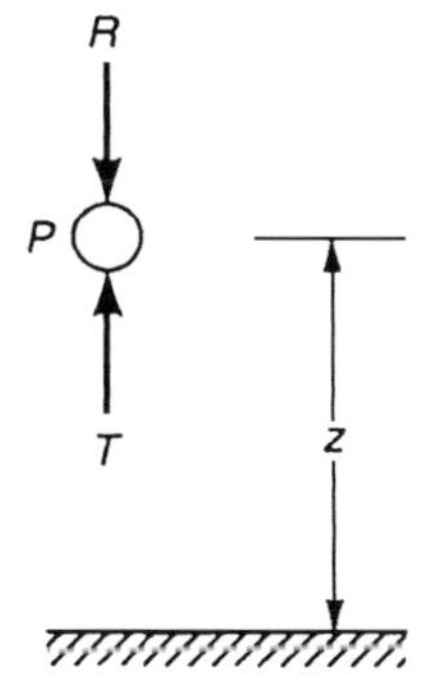

Exhibit 9.21

9.21 In Exhibit 9.21, a particle P of mass 5 kg is launched vertically upward from the ground with an initial velocity of 10 m/s. A constant upward thrust $T = 100$ newtons is applied continuously to P, and a downward resistive force $R = 2z$ newtons also acts on the particle, where z is the height of the particle above the ground. Determine the maximum height attained by P.

a. 6.0 m c. 15.8 m
b. 45.5 m d. 55.5 m

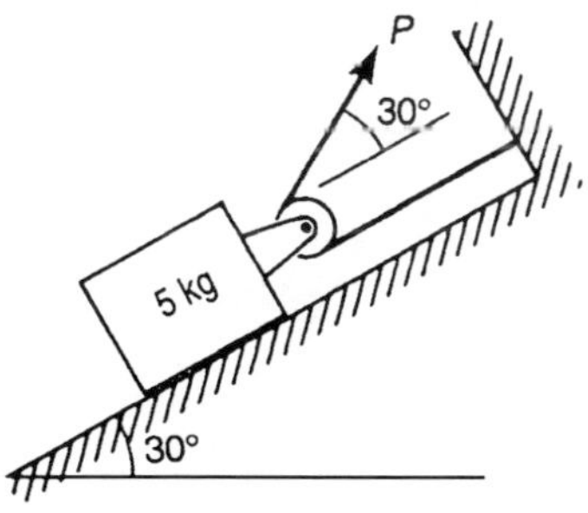

Exhibit 9.22

9.22 Determine the force P required to give the block shown in Exhibit 9.22 an acceleration of 2 m/s^2 up the incline. The coefficient of kinetic friction between the block and the incline is 0.2.

a. 39.2 N c. 44.6 N
b. 21.9 N d. 49.8 N

9.23 In Exhibit 9.23 the rod R rotates in the vertical plane about a fixed axis through the point O with a constant counterclockwise angular velocity of 5 rad/s. A collar B of mass 2 kg slides down the rod (toward O) so that the distance between B and O decreases at the constant rate of 1 m/s. At the instant when $\theta = 30°$ and $r = 400$ mm, determine the magnitude of the applied force P. The coefficient of kinetic friction between B and R is 0.1.

a. 9.9 N
b. 11.9 N
c. 10.5 N
d. 0.3 N

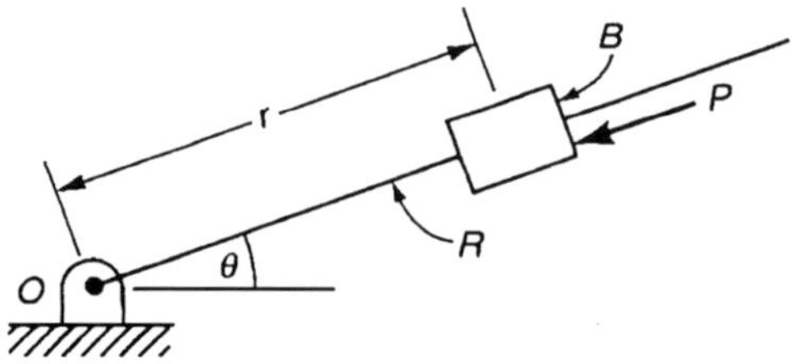

Exhibit 9.23

9.24 The 3-kg collar in Exhibit 9.24 slides down the smooth circular rod. In the position shown, its velocity is 1.5 m/s. Find the normal force (contact force) the rod exerts on the collar.

a. 12.2 N↖
b. 24.4 N↘
c. 19.2 N↖
d. 12.2 N↘

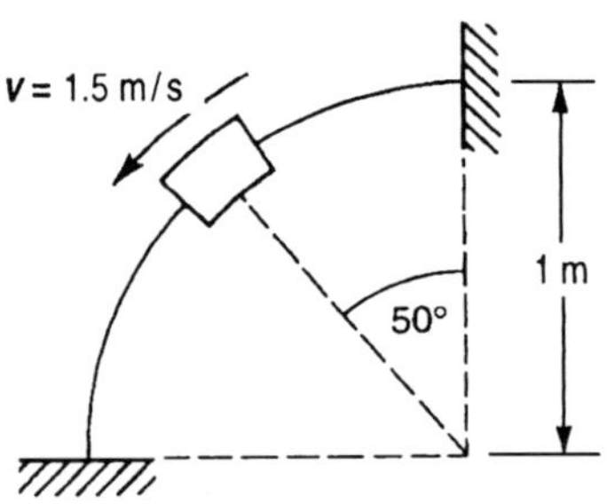

Exhibit 9.24

9.25 Forklift vehicles, Exhibit 9.25, tend to roll over if they are driven too fast while turning. For a vehicle of mass m with a mass center that describes a circle of radius R, find the relationship between the forward speed u and the vehicle dimensions and path radius at the onset of tipping.

a. $u = (Rgb/H)^{0.5}$
b. $u = (RgH/b)^{0.5}$
c. $u = (gh)^{0.5}$
d. $u = (bg)^{0.5}$

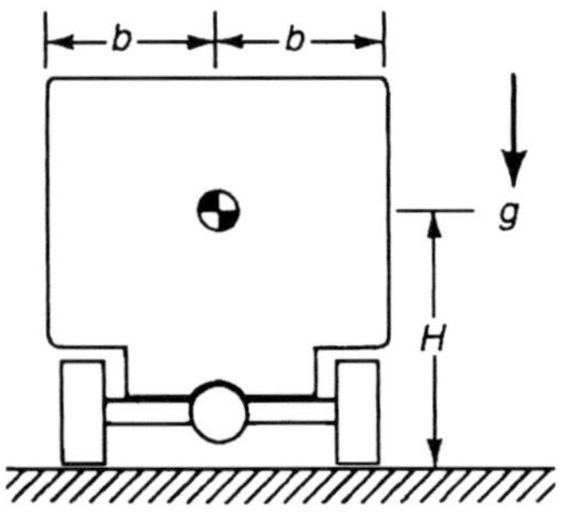

Exhibit 9.25

9.26 A toy rocket of mass 1 kg is placed on a horizontal surface, and the engine is ignited (Exhibit 9.26). The engine delivers a force equal to $(0.25 + 0.5t)$ N, where t is time in seconds, and the coefficient of friction between the rocket and the surface is 0.01. Determine the velocity of the rocket 7 seconds after ignition.

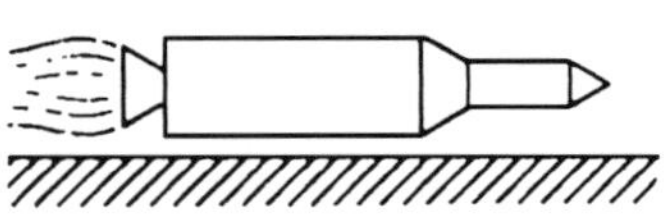

Exhibit 9.26

a. 14.0 m/s
b. 3.7 m/s
c. 13.3 m/s
d. 26.3 m/s

9.27 A 2000-kg pickup truck is traveling backward down a 10° incline at 80 km/hr when the driver notices through his rearview mirror an object on the roadway. He applies the brakes, and this results in a constant braking (retarding) force of 4000 N. How long does it take the truck to stop?

a. 11.1 s c. 2.3 s
b. 74.9 s d. 13.0 s

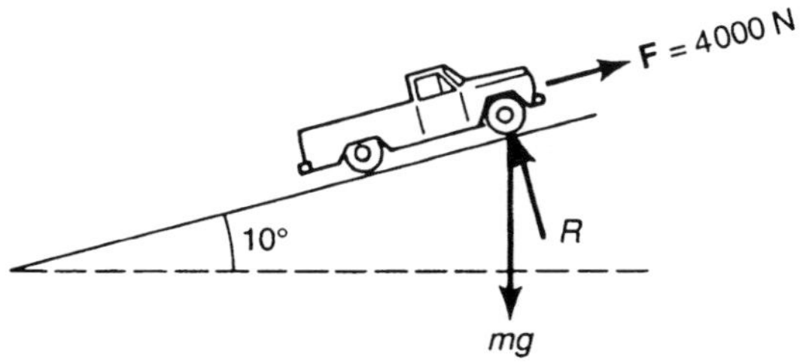

Exhibit 9.27

9.28 In Exhibit 9.28, a particle C of mass 2 kg is sliding down a smooth incline with a velocity of 3 m/s when a horizontal force $P = 15$ N is applied to it. What is the distance traveled by C between the instant when P is first applied and the instant when the velocity of C becomes zero?

a. 1.7 m c. 2.8 m
b. 5.6 m d. 0.9 m

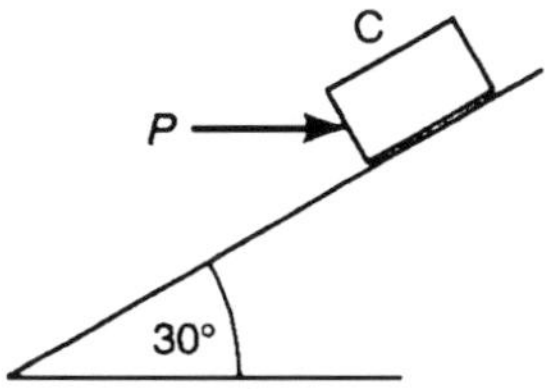

Exhibit 9.28

9.29 A particle moves in a vertical plane along the path ABC shown in Exhibit 9.29. The portion AB of the path is a quarter-circle of radius r and is smooth. The portion BC is horizontal and has a coefficient of friction μ. If the particle has mass m and is released from rest at A, determine the horizontal distance H that the particle will travel along BC before coming to rest.

a. μr c. r/μ
b. $2r$ d. $2r/\mu$

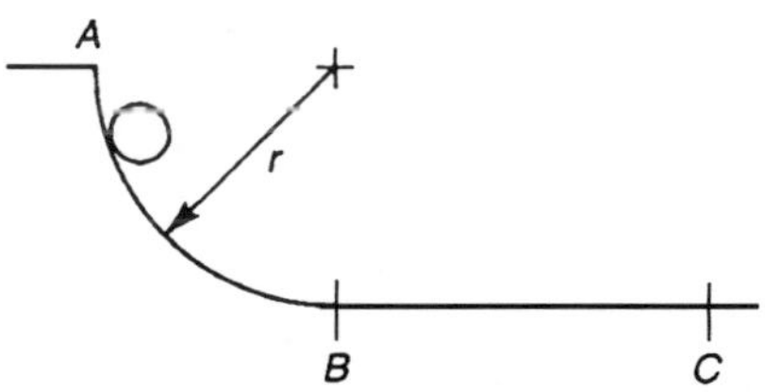

Exhibit 9.29

9.30 In Exhibit 9.30, a block of mass 2 kg is pressed against a linear spring of constant $k = 200$ N/m through a distance Δ on a horizontal surface. When the block is released at A, it travels along the straight horizontal path ADB and traverses point B with a velocity of 1 m/s. If the coefficient of kinetic friction between the block and the floor is 0.2, find Δ.

a. 0.22 m c. 0.26 m
b. 0.12 m d. 0.08 m

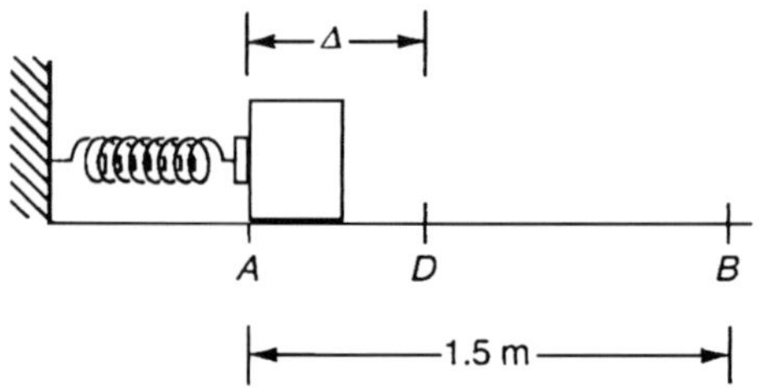

Exhibit 9.30

9.31 In Exhibit 9.31, a 6-kg block is released from rest on a smooth inclined plane as shown. If the spring constant $k = 1000$ N/m, determine how far the spring is compressed. Assume the acceleration of gravity, $g = 10$ m/s^2.

a. 0.40 m c. 0.83 m
b. 0.45 m d. 3.96 m

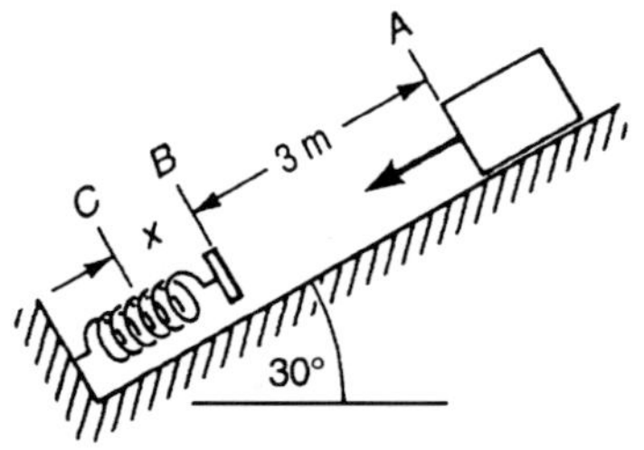

Exhibit 9.31

9.32 A train of joyride cars full of children in an amusement park is pulled by an engine along a straight-level track. It then begins to climb up a 5° slope. At a point B, 50 m up the grade when the velocity is 32 km/h, the last car uncouples without the driver noticing (Exhibit 9.32). If the total mass of the car with its passengers is 500 kg and the track resistance is 2% of the total vehicle weight, calculate the total distance up the grade where the car stops at point C.

a. 260 m c. 48.7 m
b. 37.6 m d. 87.6 m

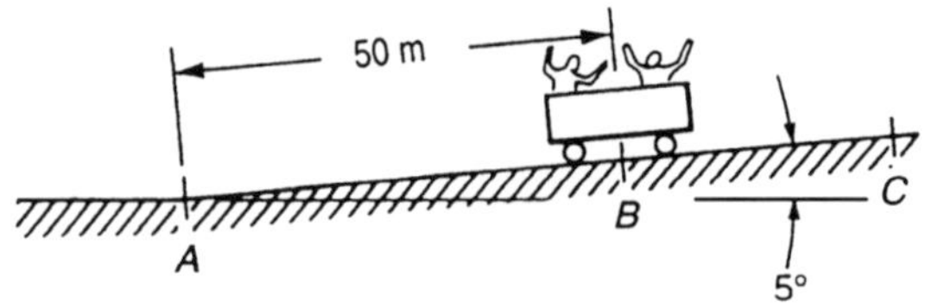

Exhibit 9.32

9.33 Two identical rods, each of mass 4 kg and length 3 m, are rigidly connected as shown in Exhibit 9.33. Determine the moment of inertia of the rigid assembly about an axis through the point A and perpendicular to the plane of the paper.

a. 19 kg-m^2 c. 18 kg-m^2
b. 23 kg-m^2 d. 15 kg-m^2

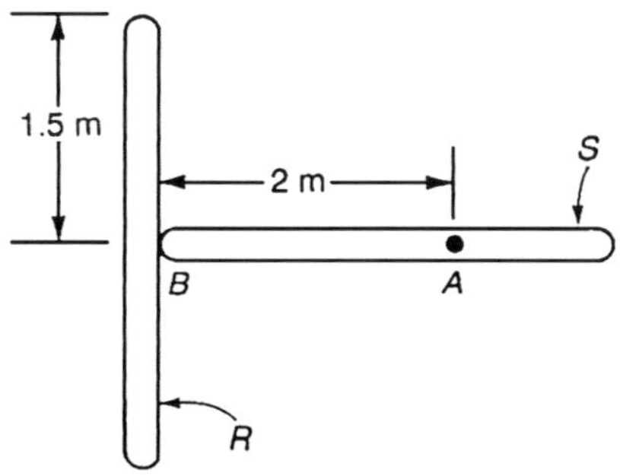

Exhibit 9.33

9.34 A torque motor, represented by the box in Exhibit 9.34, is to drive a thin steel disk of radius 2 m and mass 1.5 kg around its shaft axis. Ignoring the bearing friction about the shaft and the shaft mass, find the angular speed of the disk after applying a constant motor torque of 5 N-m for 5 seconds. The initial angular velocity of the shaft is 1 rad/s.

a. 8.3 rad/s c. 5.2 rad/s
b. 7.3 rad/s d. 9.3 rad/s

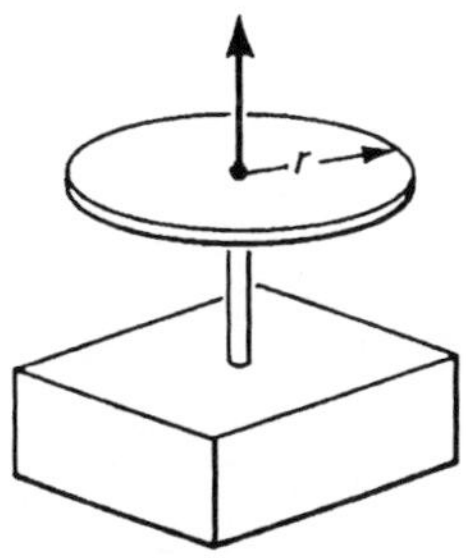

Exhibit 9.34

9.35 In Exhibit 9.35, the uniform slender rod R is hinged to a block B that can slide horizontally. Determine the horizontal acceleration α that must be given to B in order to keep the angle θ constant at 10°, balancing the rod in a tilted position.

a. 1.73 m/s^2 c. 9.81 m/s^2
b. 0 d. 9.66 m/s^2

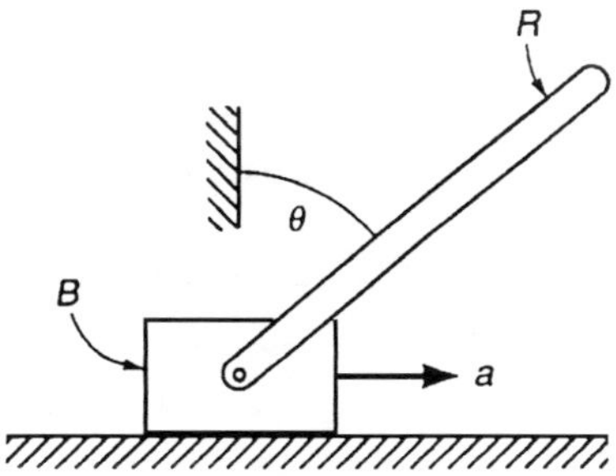

Exhibit 9.35

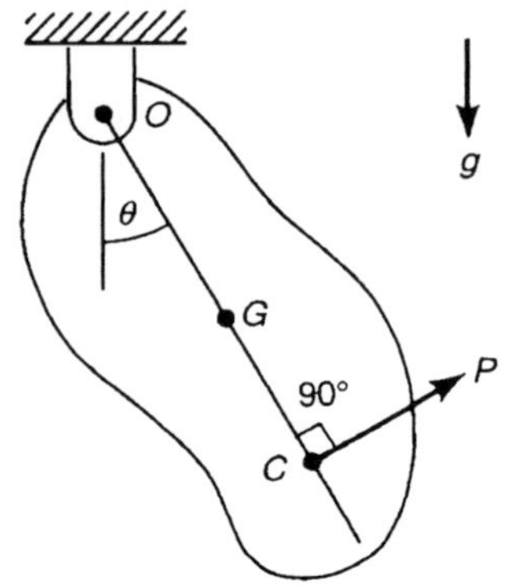

Exhibit 9.36

9.36 In Exhibit 9.36, a force P of constant magnitude is applied to the physical pendulum at point C and remains perpendicular to OC at all times. The pendulum moves in the vertical plane and has mass 3 kg; its mass center is located at G, and the distances are $OG = 1.5$ m, $OC = 2$ m. Also, $P = 10$ N and the radius of gyration of the pendulum about an axis through C and perpendicular to the plane of motion is 0.8 m. Determine the angular acceleration of the pendulum when $\theta = 30°$.

a. 5.31 rad/s² counterclockwise
b. 5.31 rad/s² clockwise
c. 0.26 rad/s² counterclockwise
d. 0.26 rad/s² clockwise

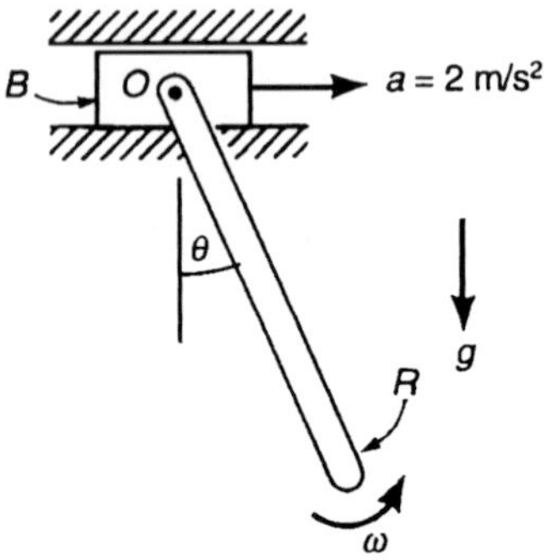

Exhibit 9.37

9.37 In Exhibit 9.37, the block B moves along a straight horizontal path with a constant acceleration of 2 m/s² to the right. The uniform slender rod R of mass 1 kg and length 2 m is connected to B through a frictionless hinge and swings freely about O as B moves. Determine the horizontal component of the reaction force at O on the rod when $\theta = 30°$ and $\omega = 2$ rad/s counterclockwise.

a. 4.31 N → c. 6.31 N →
b. 4.31 N ← d. 6.31 N ←

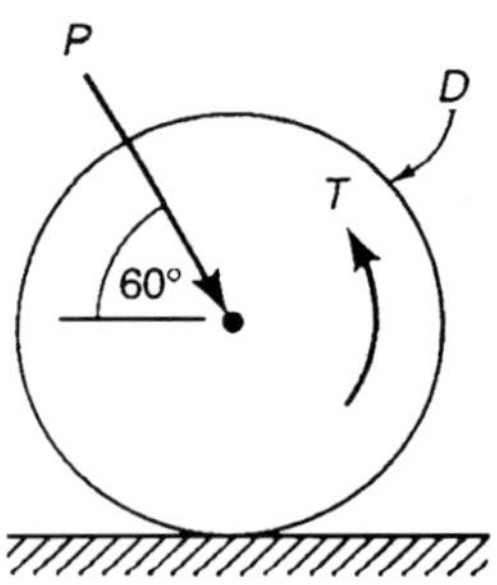

Exhibit 9.38

9.38 In Exhibit 9.38, a homogeneous cylinder rolls without slipping on a horizontal floor under the influence of a force $P = 6$ N and a torque $T = 0.5$ N-m. The cylinder has radius 1 m and mass 2 kg. If the cylinder started from rest, what is its angular velocity after 10 seconds?

a. 8.3 rad/s c. 1.7 rad/s
b. 6.8 rad/s d. 0.68 rad/s

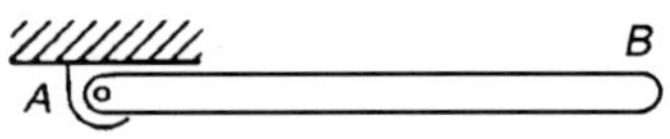

Exhibit 9.39

9.39 A slender rod of length 2 m and mass 3 kg is released from rest in the horizontal position (Exhibit 9.39) and swings freely (no hinge friction). Find the angular velocity of the rod when it passes a vertical position.

a. 4.43 rad/s c. 7.68 rad/s
b. 3.84 rad/s d. 5.43 rad/s

9.40 The uniform slender bar of mass 2 kg and length 3 m is released from rest in the near-vertical position as shown in Exhibit 9.40, where the torsional spring is undeformed. The rod is to rotate clockwise about *O* and come gently to rest in the horizontal position. Determine the stiffness *k* of the torsional spring that would make this possible. The hinge is smooth.

a. 47.8 N-m/rad
b. 37.5 N-m/rad
c. 0.7 N-m/rad
d. 23.8 N-m/rad

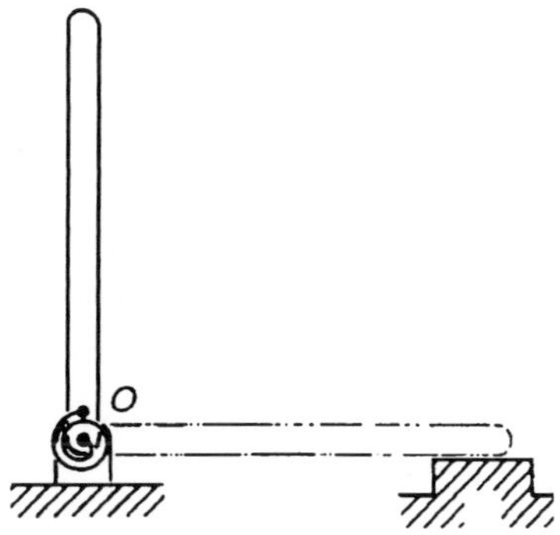

Exhibit 9.40

9.41 A solid homogeneous cylinder is released from rest in the position shown in Exhibit 9.41 and rolls without slip on a horizontal floor. The cylinder has a mass of 12 kg. The spring constant is 2 N/m, and the unstretched length of the spring is 3 m. What is the angular velocity of the cylinder when its center is directly below the point *O*?

a. 1.33 rad/s
b. 1.63 rad/s
c. 1.78 rad/s
d. 2.31 rad/s

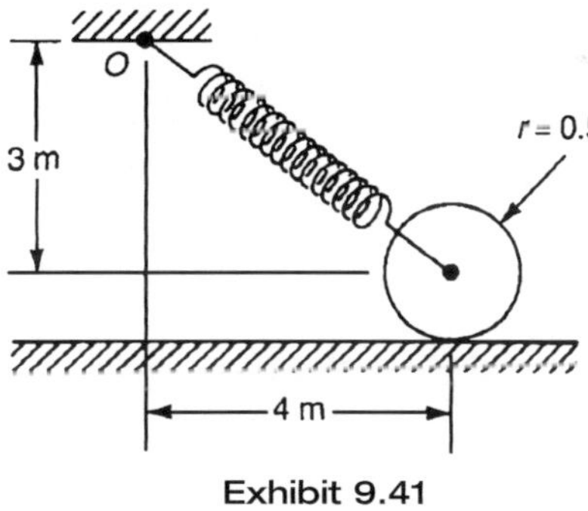

Exhibit 9.41

SOLUTIONS

9.1 c. The density of the plate is:

$$\rho = \frac{\text{mass}}{\text{area}} = \frac{480 \text{ kg}}{\left[(8\times 6)+(4\times 6)\right]\text{m}^2} = \frac{480 \text{ kg}}{72 \text{ m}^2} = 6.67 \text{ kg/m}^2$$

9.2 a. Assuming that the density of the plate is 9 kg/m^2, the mass moment of inertia of the flat plate may be found by:

$$I_x = \frac{1}{3}mh^2 = \frac{1}{3}(\rho A h^2) = \frac{1}{3}\rho(6)(8)^3 + \frac{1}{3}\rho(6)(4)^3$$

$$= \frac{1}{3}(9 \text{ kg/m}^2)(6 \text{ m})(8 \text{ m})^3 + \frac{1}{3}(9 \text{ kg/m}^2)(6 \text{ m})(4 \text{ m})^3$$

$$= 9216 + 1152 = 10{,}368 \text{ kg} - \text{m}^2$$

9.3 b. Let the depth of the ditch be h, and set up a vertical s-axis, positive upwards with the origin at A (Exhibit 9.3a). Then, for motion between A and B, $s_0 = 0$, $s = -h$, $v_0 = 4$ m/s, $a = -9.81$ m/s^2, and $t = 6$ s. Substituting these values in the relationship $s = s_0 + v_0 t + (at^2)/2$ yields $-h = 0 + 4(6) - 9.81(6)^2/2$ or $h = 152.6$ m.

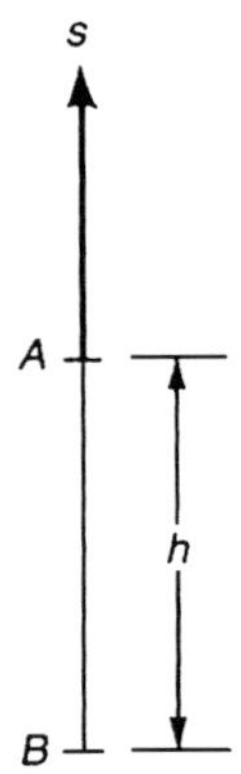

Exhibit 9.3a

9.4 d. Set up a horizontal s-axis with origin at O, and positive to the right (Exhibit 9.4a). Then, $sO = 2$ m, and $v = ds/dt$, so

$$\int_{s_0}^{s} ds = \int_0^t v\,dt = \int_0^t (3t^2 - t + 2)\,dt$$

or $s = s_0 + t^3 - t^2/2 + 2t$. Substituting values into the equation yields $s(2\text{s}) = 12$ m.

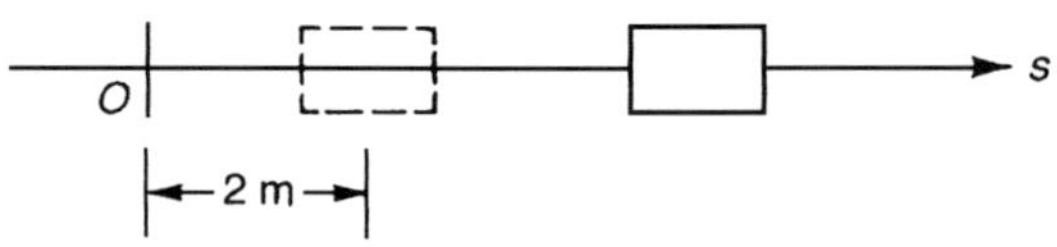

Exhibit 9.4a

9.5 d. The velocity-time curve for this particle is shown below the acceleration curve in Exhibit 9.5a. (Velocity at any given instant t_1 equals the area under the acceleration curve from time 0 to the time t_1, plus the initial velocity). Any area above the $a = 0$ line is counted as positive, and any area below is counted as negative.

The displacement D is the total area under the velocity curve between $t = 0$ and $t = 8$ s. The area of a triangle is $0.5(b)(h)$. Hence, $D = 0.5(6)(8) - 0.5(2)(4) = 20$ m.

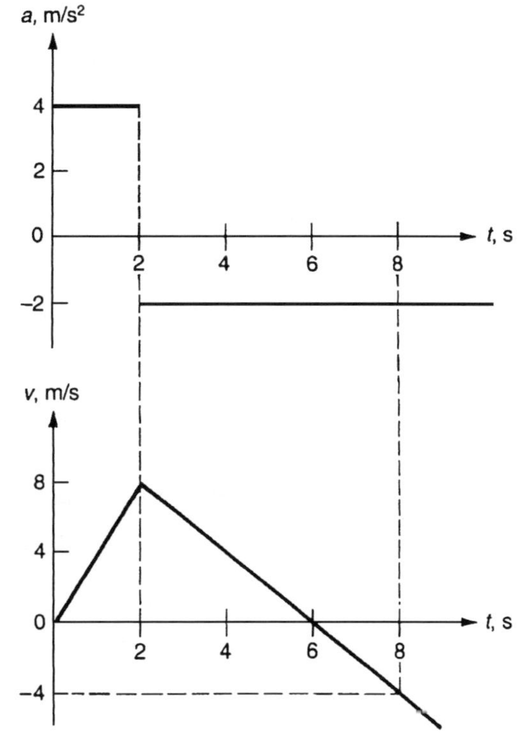

Exhibit 9.5a

9.6 a. When the ball reaches its highest position, B, the vertical component of the velocity of the ball is zero. Applying Equation (9.12) vertically,

$$v_y^2 = v_{yo}^2 + 2a_y(y - y_0)$$
$$0 = (40\ \sin\ 60°)^2 - 2(9.81)(y - y_0)$$

Thus, $y - y_0 = 61.2$ m and $H = 63.2$ m.

9.7 **c.** Refer to Exhibit 9.7a.

Horizontal Motion:

$$x = x_0 + v_{x0}t + (a_x t^2)/2$$
$$140 = 0 + v_0 t + 0$$
$$t = 1.56 \text{ s}$$

Vertical Motion:

$$y = y_0 + y_{y0}t + (a_y t^2)/2$$
$$-12 = 0 + 0 - 0.5(9.81)t^2$$
$$v_o = 140/\text{t} = 140/1.56 = 89.7 \text{ m/s}$$

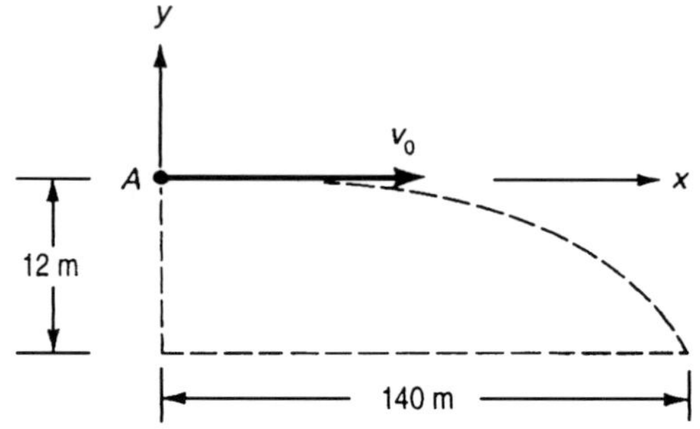

Exhibit 9.7a

9.8 **c.** In Exhibit 9.8a, we have

$$r = 1 \text{ m}; \qquad \dot{r} = -3 \text{ m/s}; \qquad \ddot{r} = 0$$

Since $\theta = (t^2 + t - 2)$ rad, differentiation gives

$$\dot{\theta} = (2t + 1) \text{ rad/s; at } t = 1 \text{ s, } \dot{\theta} = 2(1) + 1 = 3 \text{ rad/s}$$

Also, $\ddot{\theta} = 2 \text{ rad/s}^2$.

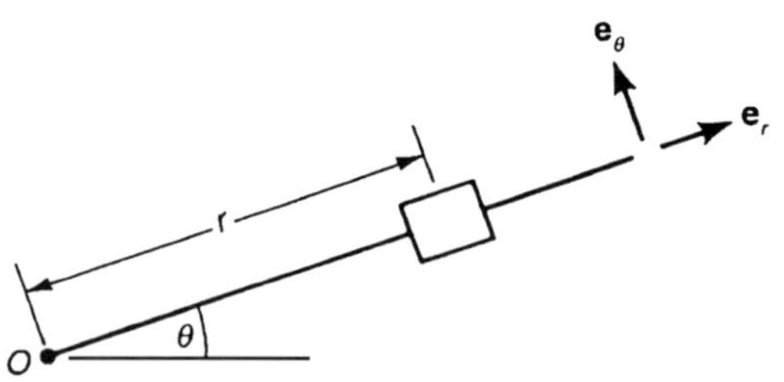

Exhibit 9.8a

Acceleration of the collar is therefore given by

$$\begin{aligned}\mathbf{a} &= (\ddot{r} - r\dot{\theta}^2)\mathbf{e}_r + (r\ddot{\theta} + 2\dot{r}\dot{\theta})\mathbf{e}_\theta \\ &= [0 - 1(3)^2]\mathbf{e}_r + [1(2) + 2(-3)(3)]\mathbf{e}_\theta \ \mathbf{m/s^2} \\ &= [-9\mathbf{e}_r - 16\mathbf{e}_\theta]\ \mathbf{m/s^2}\end{aligned}$$

Hence, $|a| = [(-9)^2 + (-16)^2]^{0.5} = 18.4 \text{ m/s}^2$.

9.9 d. Refer to Exhibit 9.9a. The velocity is

$$v = \left(v_r^2 + v_\theta^2\right)^{0.5} \quad \text{where } v_r = \dot{r} \text{ and } v_\theta = r\dot{\theta}$$

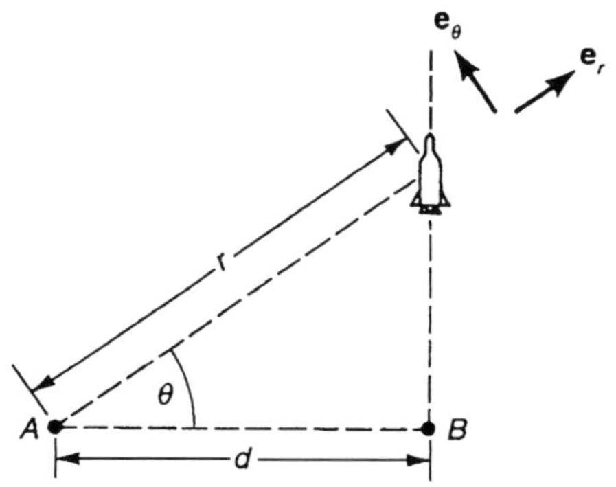

Exhibit 9.9a

Now, $r = d/\cos\theta$ so

$$v_r = \dot{r} = \frac{d\dot{\theta}\sin\theta}{\cos^2\theta} = \frac{d\dot{\theta}\tan\theta}{\cos\theta}$$

and

$$v_\theta r\dot{\theta} = \frac{d\dot{\theta}}{\cos\theta}$$

Thus

$$v^2 = v_r^2 + v_\theta^2 = \frac{d^2\dot{\theta}^2(\tan^2\theta + 1)}{\cos^2\theta} = \frac{d^2\dot{\theta}^2}{\cos^4\theta}; \quad \ddot{r} = 0$$

and

$$v = \frac{d\dot{\theta}}{\cos^2\theta} = \frac{1800(0.1)(2)}{1} \text{ m/s} = 360 \text{ m/s}$$

9.10 d. At the highest point, the vertical component of velocity is zero, and the acceleration is normal to the path. Thus, $v = v_x = v_{x0} + a_x t = 50$ cos 30° + 0. Also, the normal component of the acceleration is $a_n = -a_y = 9.81$ m/s². But $a_n = v^2/\rho$. Hence $\rho = v^2/a_n$ = (50 cos 30°)²/9.81 = 191 m.

9.11 d. The tangential acceleration is given as $a_t = 3$ m/s²; the normal acceleration is $a_n = v^2/\rho$ = [(20)²/110] m/s² = 3.6 m/s². Hence, the total acceleration is a = [(3)² + (3.6)²]$^{0.5}$ m/s² = 4.7 m/s².

9.12 c. The normal acceleration is given by $a_n = v^2/\rho$. So, $\rho = v^2/a_n$. Substituting values (converted to consistent units), we obtain

$$\rho = \frac{[800(1000)]^2}{[60(60)]^2(5)(9.81)} = 1007 \text{ m}$$

9.13 d. At any given instant, the angular acceleration of the disk is $\alpha = -k\omega = d\omega/dt$. So,

$$\int_{\omega_0}^{\omega} \frac{d\omega}{\omega} = -k\int_0^t dt$$

$$\ln \omega|_{\omega_0}^{\omega} = \ln(\omega/\omega_0) = -kt \text{ and } t = -(1/k)\ln(\omega/\omega_0) = -(1/1.2)\ln(0.5) = 0.6\,s$$

9.14 a. Initially, $\omega_0 = 0$. At $t = 30$ s, $\omega = 300$ rpm $= [300\,(2\pi)/60]$ rad/s $= 10\pi$ rad/s. For uniformly accelerated rotational motion,

$$\begin{aligned} \omega &= \omega_0 + \alpha t \\ \alpha &= (\omega - \omega_0)/t = (10\pi - 0)/30 \text{ rad/s}^2 = \pi/3 \text{ rad/s}^2 \\ \omega(5) &= \omega_0 + \alpha t = 0 + (\pi/3)(5) \text{ rad/s} = 5\pi/3 \text{ rad/s} \\ v_P(5) &= \omega(5)r = (5\pi/3)\,(2) \text{ m/s} = 10.5 \text{ m/s} \end{aligned}$$

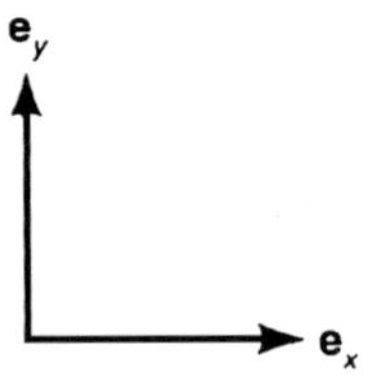

Exhibit 9.15a

9.15 b. Adopt the coordinate system of Exhibit 9.15a. Rolling of the disk without slip on B implies that the velocity of the point A, viewed as a point on D, equals the velocity of A viewed as a point on B. Hence, $\mathbf{v}_A = 2\mathbf{e}_x$ m/s. Since A and O are points of the same rigid body D, $\mathbf{v}_O = \mathbf{v}_A + \omega \times \mathbf{r}_{AO}$, or $-1\mathbf{e}_x = 2\mathbf{e}_x + \omega\mathbf{e}_z \times (1.5)\mathbf{e}_y = (2 - 1.5\omega)\mathbf{e}_x$. Hence, $-1 = 2 - 1.5\omega$ or $\omega = 2$ rad/s. The positive sign indicates a counterclockwise rotation.

9.16 a. In the current configuration, $\mathbf{v}_P = 0$. P and A are points on D, so

$$\begin{aligned} \mathbf{v}_A &= \mathbf{v}_P + \omega_D \times \mathbf{r}_{PA} = 0 - 3\mathbf{e}_z \times (0.5\mathbf{e}_y - 0.4\mathbf{e}_x) \\ &= (1.5\mathbf{e}_x + 1.2\mathbf{e}_y) \text{ m/s} \end{aligned}$$

Similarly, because B and A are points on R, $\mathbf{v}_B = \mathbf{v}_A + \omega_R \times \mathbf{r}_{AB}$. Therefore,

$$\begin{aligned} v_B\mathbf{e}_x &= 1.5\mathbf{e}_x + 1.2\mathbf{e}_y + \omega_R\mathbf{e}_z \times (-2\mathbf{e}_x - 0.5\mathbf{e}_y) \\ &= (1.5 + 0.5\omega_R)\mathbf{e}_x + (1.2 - 2\omega_R)\mathbf{e}_y \end{aligned}$$

Equating the coefficients of $\mathbf{e}_y$ yields $0 = 1.2 - 2\omega_R$, or $\omega_R = 0.6$ rad/s. The positive sign indicates that ω_R is in the positive $\mathbf{e}_z$ direction, so the rotation is counterclockwise.

9.17 d. Referring to the coordinate system in Exhibit 9.17a, the accelerations are

$$\begin{aligned} \mathbf{a}_A &= \mathbf{a}_O + \alpha \times \mathbf{r}_{OA} + \omega \times (\omega \times \mathbf{r}_{OA}) \\ \mathbf{a}_O &= 0 \text{ (constant velocity)} \end{aligned}$$

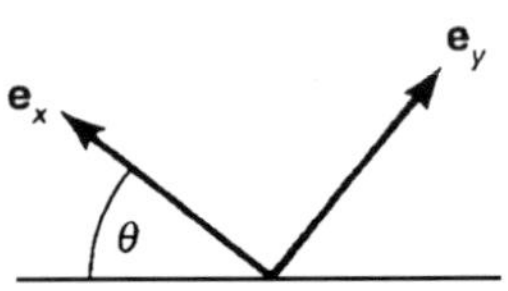

Exhibit 9.17a

Since $\theta = (0.5t^2 - t)$ rad, $\dot{\theta} = (t - 1)$rad/s and $\ddot{\theta} = 1$ rad/s^2.

So, at $t = 2$ s, $\omega = \dot{\theta}\,\mathbf{e}_z = 1\mathbf{e}_z$ rad/s. Since $\boldsymbol{a} = \ddot{\theta}\,\mathbf{e}_z = 1\mathbf{e}_z$ rad/s^2,

$$\mathbf{a}_A = 0 + \mathbf{e}_z \times (2\mathbf{e}_x) + \mathbf{e}_z \times [\mathbf{e}_z \times (2\mathbf{e}_x)] = (2\mathbf{e}_y - 2\mathbf{e}_x) \text{ m/s}^2$$

Hence, $|\mathbf{a}_A| = (2^2 + 2^2)^{0.5}$ m/s^2 $= 2.8$ m/s^2.

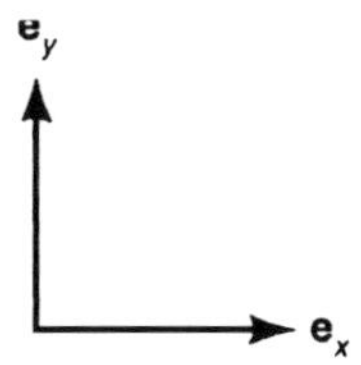

Exhibit 9.18a

9.18 **b.** With the coordinate system in Exhibit 9.18a, we can write

$$\mathbf{a}_O = 2\mathbf{e}_x\ \text{m/s}^2;\ \ \omega = 0;\ \ \alpha = -2.5\mathbf{e}_z\ \text{rad/s}^2$$

Now, $\mathbf{a}_G = \mathbf{a}_O + \mathbf{a}\times\mathbf{r}_{OG} + \omega\times(\omega\times\mathbf{r}_{OG})$ where $\mathbf{r}_{OG} = -1\mathbf{e}_y$ m. Hence

$$\mathbf{a}_G = [2\mathbf{e}_x - 2.5\mathbf{e}_z\times(-1)\mathbf{e}_y + 0]\ \text{m/s}^2 = -0.5\mathbf{e}_x\ \text{m/s}^2$$

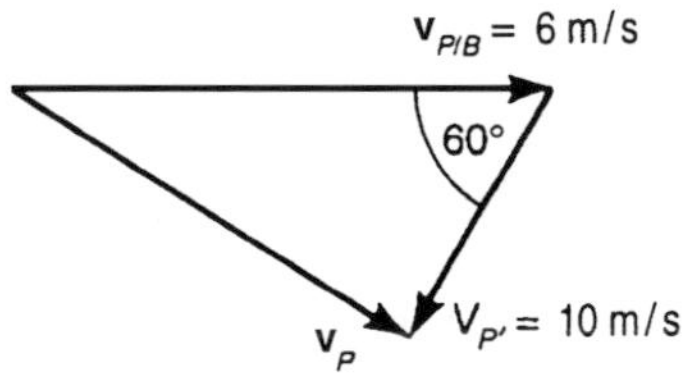

Exhibit 9.19a

9.19 **a.** We know that $\mathbf{v}_P = \mathbf{v}_{P/B} + \mathbf{v}_{P'}$ where $\mathbf{v}_{P/B}$ is the velocity of P relative to B and $\mathbf{v}_{P'}$ is the velocity of the point P' of the block that coincides with P at the instant under consideration (coincident point velocity). Here, $|\mathbf{v}_{P/B}| = 6$ m/s and $|\mathbf{v}_{P'}| = 10$ m/s (velocity of block). Because v_P is the vector sum of $\mathbf{v}_{P/B}$ and $\mathbf{v}_{P'}$, as shown in Exhibit 9.19a, we can use the law of cosines,

$$(\mathbf{v}_P)^2 = 10^2 + 6^2 - 2(10)(6)\cos 60° = 76$$
$$\mathbf{v}_P = 8.7\ \text{m/s}$$

9.20 **d.**

$$\mathbf{a}_P = \mathbf{a}_{P/D} + \mathbf{a}_{P¢}\rho' + \mathbf{a}_C$$

Here,

$\mathbf{a}_{P/D}$ = relative acceleration = 0

$\mathbf{a}_{P'}$ = acceleration of the point of D that is coincident with P at the instant under consideration:

$$\begin{aligned}\mathbf{a}_{P'} &= \mathbf{a}_O + \alpha\times\mathbf{r}_{OC} + \omega\times(\omega\times\mathbf{r}_{OC})\\ &= 0 + 0 + (-5\mathbf{e}_z)\times[(\ 5\mathbf{e}_z)\times(0.1\mathbf{e}_y)]\\ &= -2.5\mathbf{e}_y\ \text{m/s}^2\end{aligned}$$

$\mathbf{a}_C$ = Coriolis acceleration = $2\omega\times\mathbf{v}_{P/D}$

$\mathbf{a}_C = 2(-5\mathbf{e}_z)\times 2\mathbf{e}_x = -20\mathbf{e}_y$ m/s^2

Finally,

$$\mathbf{a}_P = [-2.5\mathbf{e}_y - 20\mathbf{e}_y]\ \text{m/s}^2 = -22.5\mathbf{e}_y\ \text{m/s}^2$$

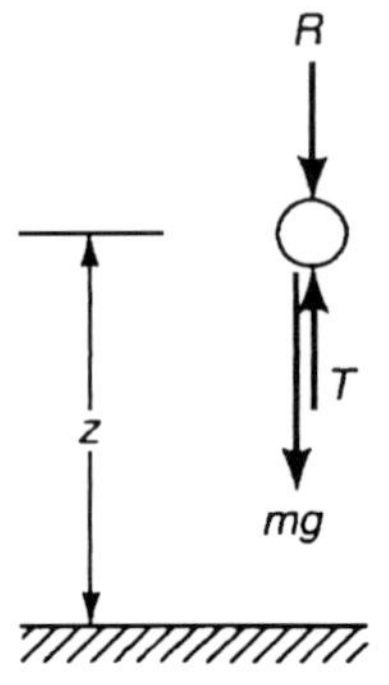

Exhibit 9.21a

9.21 d. The free-body diagram is shown in Exhibit 9.21a. Apply Newton's second law:

$$\sum F_z = ma_z$$
$$T - R - mg = ma$$

$$a = [(T-R)/m] - g = [(100-2z)N/5\text{ kg}] - 9.81\text{ m/s}^2$$

so

$$a = 10.2 - 0.4z = \frac{v\,dv}{dz}$$

and

$$\int_0^H (10.2 - 0.4z)\,dz = \int_{10}^0 v\,dv$$

where H is the highest height attained. Note also that $v = 0$ at this height. Integration yields,

$$10.2H - 0.2H^2 = [v^2/2]_{10}^0 = -50$$

or

$$0.2H^2 - 10.2H - 50 = 0$$

Solving this quadratic (and discarding the negative value) gives the maximum height,

$$H = 55.5\text{ m}$$

9.22 b. Refer to Exhibit 9.22a. Apply Newton's second law in the x and y directions.

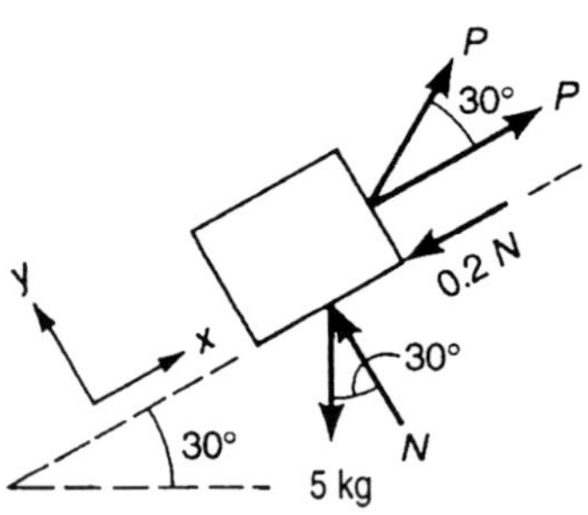

Exhibit 9.22a

$$\sum F = ma_x \qquad \textbf{(i)}$$

$$P + P\cos 30° - 0.2N - 5(9.81)\sin 30° = 5(2)$$

$$\sum F_y = 0$$
$$N + P\sin 30° - 5(9.81)\cos 30° = 0 \qquad \textbf{(ii)}$$

Solving Equations (i) and (ii) simultaneously yields $P = 21.9$ N.

9.23 **a.** In Exhibit 9.23a, apply Newton's second law in the radial and transverse directions.

$$\sum F_\theta = ma_\theta$$

$$N - mg\cos\theta = m(r\ddot{\theta} + 2\dot{r}\dot{\theta}) \quad \textbf{(i)}$$

$$\sum F_r = ma_r$$

$$\mu N - P - mg\sin\theta = m(\ddot{r} - r\dot{\theta}^2) \quad \textbf{(ii)}$$

Substitute values into Equations (i) and (ii):

$$N - 2(9.81)\ \cos 30° = 2[0 + 2(-1)(5)] \quad \textbf{(iii)}$$

$$0.1N - P - 2(9.81)\sin 30° = 2[0 - 0.4(5)^2] \quad \textbf{(iv)}$$

From Equation (iii), $N = -3.0$ newtons. Substituting this value into Equation (iv) gives $P = 9.9\text{N}$.

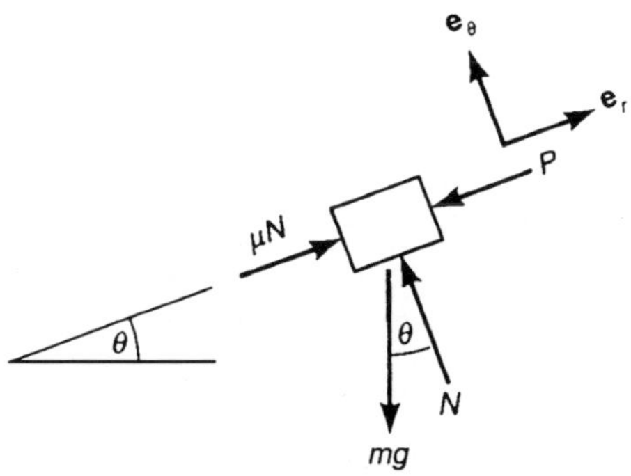

Exhibit 9.23a

9.24 **a.** Apply Newton's second law to the diagrams in Exhibit 9.24:

$$\sum F_n = ma_n$$

$$mg\cos\theta - N = mv^2/\rho$$

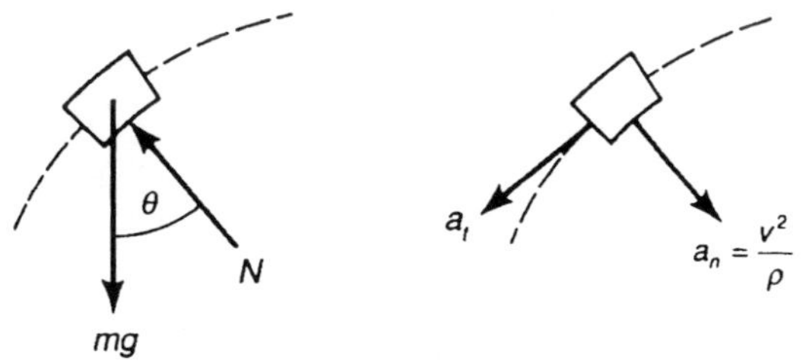

Exhibit 9.24a

or

$$\begin{aligned} N &= m[g\cos\theta - v^2/\rho] \\ &= (3)[9.81\cos 50° - 1.5^2/1] \\ &= 12.2\ \text{N} \end{aligned}$$

The positive sign indicates that N is directed as shown in the free-body diagram, Exhibit 9.24a.

9.25 a. At the onset of tipping, the free-body diagram and the inertia force diagram are as shown in Exhibit 9.25a. Take moments about point A:

$$mgb = m(u^2/R)H$$

Hence,

$$u = (Rgb/H)^{0.5}$$

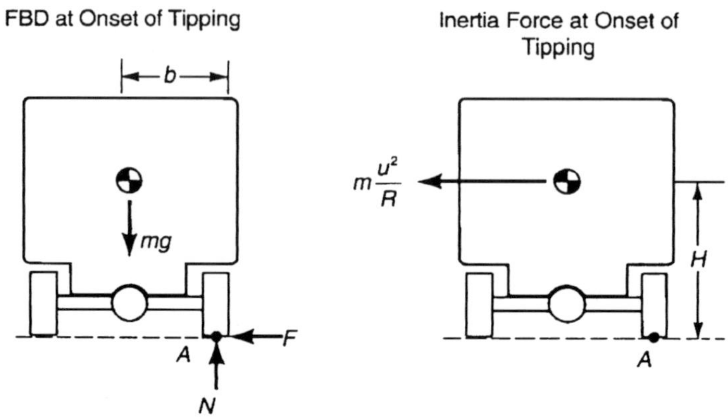

Exhibit 9.25a

9.26 c. In Exhibit 9.26a, the sum of the forces in the vertical direction yields

$$N = mg$$

Apply the impulse-momentum principle in the horizontal direction [Equation (9.32)]:

$$\int_{t_1}^{t_2} F_{\text{horiz}}\,dt = mv_2 - mv_1$$

$$F_{\text{horiz}} = (0.25 + 0.5t) - 0.1N$$

Thus

$$\int_0^7 \left[0.25 + 0.5t - (0.01)(9.81)\right]dt = (1)v_2 - 0$$

or

$$0.25t + 0.25t^2 - 0.098t\Big|_0^7 = v_2 = 13.3 \text{ m/s}$$

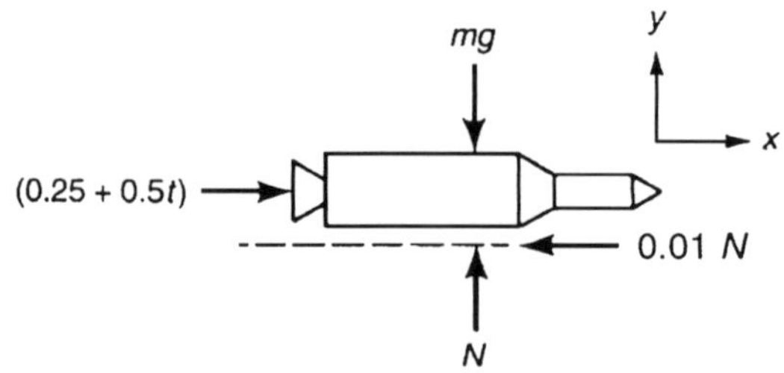

Exhibit 9.26a

9.27 **b.** Apply the impulse-momentum principle between the instant t_1 when the brakes are applied and the instant t_2 when the truck comes to a stop:

$$\int_{t_1}^{t_2} \mathbf{F}dt = m\mathbf{v}_2 - m\mathbf{v}_1$$

In the direction tangent to the road surface,

$$\int_0^t (mg \sin 10° - F)dt = 0 - mv_1$$

or

$$[2000(9.81)\sin 10° - 4000]t = -2000\frac{80(1000)}{60(60)}$$

so that $t = 74.9$ s.

9.28 **c.** Refer to Exhibit 9.28a. The subscript 1 is used for the instant when the force P is used first applied, and the subscript 2 is used for the instant when the block comes to rest. Apply the work-energy principle between 1 and 2:

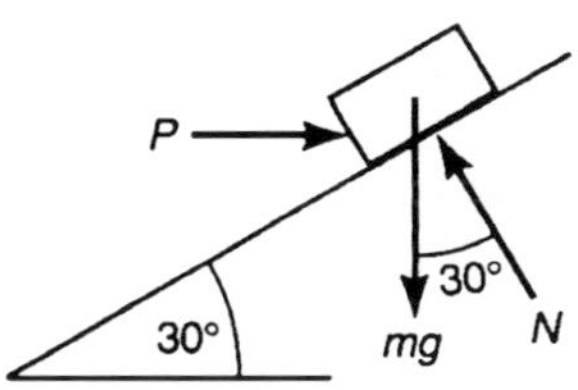

Exhibit 9.28a

$$W_{1-2} = T_2 - T_1$$
$$(mg \sin 30° - P\cos 30°)\Delta x = 0 - \tfrac{1}{2}mv_1^2$$

which yields

$$\Delta x = \frac{\frac{1}{2}mv_1^2}{P\cos 30° - mg\sin 30°} = \frac{0.5(2)3^2}{15\cos 30° - 2(9.81)\sin 30°} = 2.83 \text{ m}$$

9.29 **c.** Consult Exhibit 9.29a. Apply the work-energy principle between A and B, and then between B and C.

$A \rightarrow B$: $W_{A-B} = T_B - T_A$

$$mgr = \frac{1}{2}mv_B^2 - 0 \quad \textbf{(i)}$$

$B \rightarrow C$: $W_{B-C} = T_C - T_B$

$$N = mg \quad \textbf{(ii)}$$

$$-\mu NH = 0 - \frac{1}{2}mv_B^2 \quad \textbf{(iii)}$$

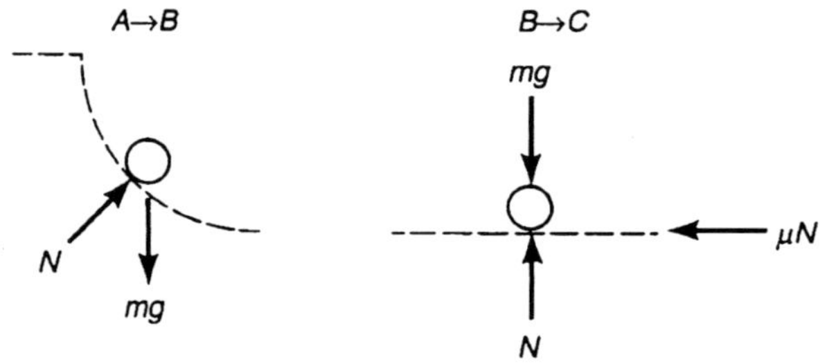

Exhibit 9.29a

Substituting Equations (i) and (ii) into Equation (iii),

$$-\mu mgH = -mgr, \quad \text{or} \quad H = r/\mu$$

9.30 c. Let D be the position at which the spring has its natural (unstretched) length. Apply the work-energy principle (Exhibit 9.30a) from A to D:

$$W_{A-D} = T_D - T_A$$

$$\frac{1}{2}k\Delta^2 - \mu N\Delta = \frac{1}{2}mv_D^2 - 0$$

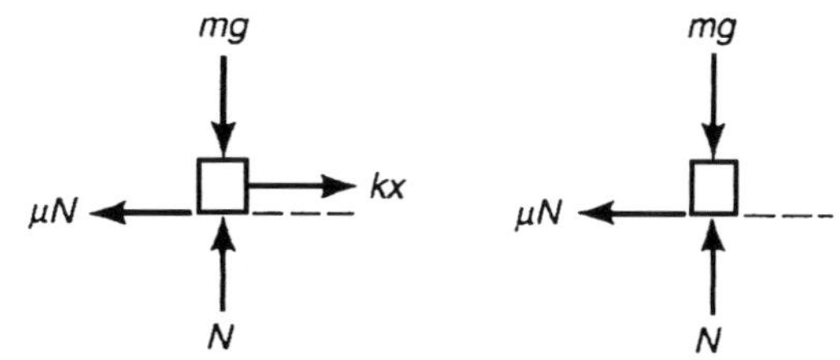

Exhibit 9.30a

Since $N = mg$, we have

$$\frac{1}{2}mv_D^2 = \frac{1}{2}k\Delta^2 - \mu mg\Delta \qquad \textbf{(i)}$$

Now apply the work-energy principle from D to B:

$$W_{D-B} = T_B - T_D$$

$$-\mu N(1.5 - \Delta) = \frac{1}{2}mv_B^2 - \frac{1}{2}mv_D^2$$

Again, since $N = mg$, and $\frac{1}{2}mv_D^2$ is given by Equation (i), we have

$$-\mu mg(1.5 - \Delta) = \frac{1}{2}mv_B^2 - \frac{1}{2}k\Delta^2 + \mu mg\Delta$$

or

$$\Delta = \sqrt{\frac{2\left(1.5\mu mg + 0.5mv_B^2\right)}{k}} = 0.26 \text{ m}$$

9.31 b. Apply the work-energy principle between A and B, and then between B and C.

$$W_{A-B} = T_B - T_A$$

With the forces shown in Exhibit 9.31a,

$$mg \sin 30°(3) = \frac{1}{2}mv_B^2 - 0$$

$$W_{B-C} = T_C - T_B$$

$$-\frac{1}{2}kx^2 + mg \sin 30°(x) = 0 - \frac{1}{2}mv_B^2 = -mg \sin 30°(3)$$

Substituting values, we obtain the quadratic equation $500x^2 - 30x - 90 = 0$, which can be solved to yield $x = 0.45$ m.

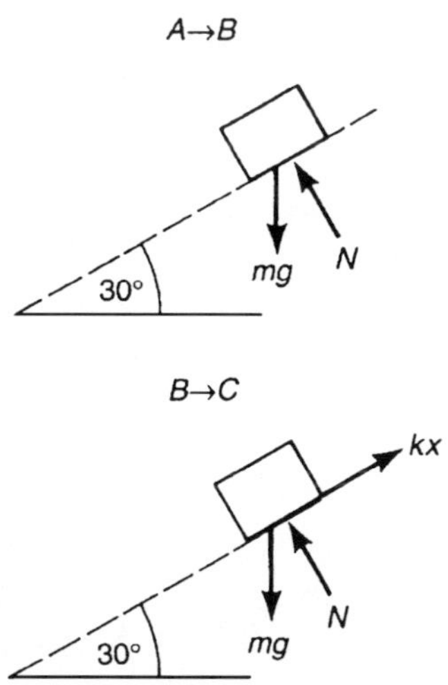

Exhibit 9.31a

9.32 d. Using the diagram in Exhibit 9.32a, apply the work-energy principle between B and C:

$$W_{B-C} = T_c - T_B$$

$$-mg \sin 5°(x) - 0.02\, mgx = 0 - \frac{1}{2} m v_B^2$$

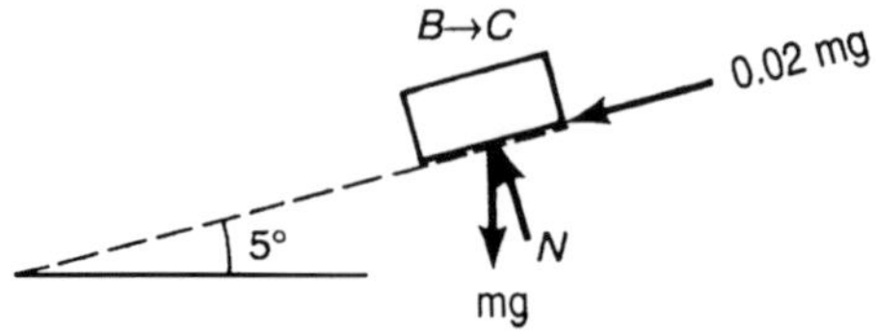

Exhibit 9.32a

or

$$x = \frac{\frac{1}{2} m v_B^2}{mg \sin 5° + 0.02\, mg} = \frac{0.5(500)\left[\frac{(32 \times 1000)}{60 \times 60}\right]^2}{500(9.81)\sin 5° + 0.02(500)9.81} = 37.6 \text{ m}$$

Total distance up the grade is (50 + 37.6) m = 87.6 m.

9.33 b. Consult Exhibit 9.33a. The moment of inertia of each rod about its mass center is

$$I_{R/B} = I_{S/O} = \frac{1}{12} ml^2 = \frac{1}{12}(4)(3)^2 = 3 \text{ kg-m}^2$$

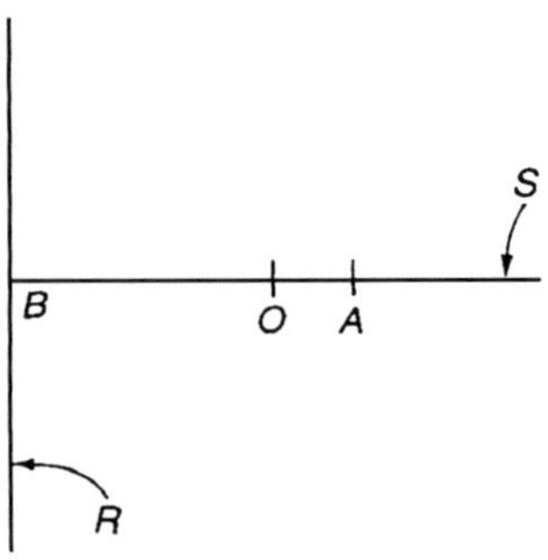

Exhibit 9.33a

Here, O is the mass center of S, and B is the mass center of R. Apply the parallel axes theorem:

$$I_{R/A} = I_{R/B} + m(2)^2 = 3 + 4(2)^2 = 19 \text{ kg-m}^2$$

$$I_{S/A} = I_{S/O} + m(2 - 1.5)^2 = 3 + 4(0.5)^2 = 4 \text{ kg-m}^2$$

And, for the assemblage,

$$I_A = I_{R/A} + I_{S/A} = (19 + 4) \text{kg-m}^2 = 23 \text{ kg-m}^2$$

9.34 d. Since $M = I\alpha = (1/2)mr^2\alpha$, $\alpha = M/(0.5mr^2)$. Substituting values, we have $\alpha = 1.67$ rad/s^2. Because this angular acceleration is constant, the final angular velocity is given by

$$\omega = \omega_0 + \alpha t = 1 + 1.67(5) = 9.3 \text{ rad/s}$$

9.35 a. When the desired configuration is achieved, the rod is in translation. The free-body diagram and the inertia force diagram for the rod are shown in Exhibit 9.35a. Taking moments about point O,

$$mg(l/2)\sin\theta = ma(l/2)\cos\theta$$

where l is the length of the rod. Thus,

$$a = g\tan\theta = 9.81\tan 10° = 1.73\ \text{m/s}^2$$

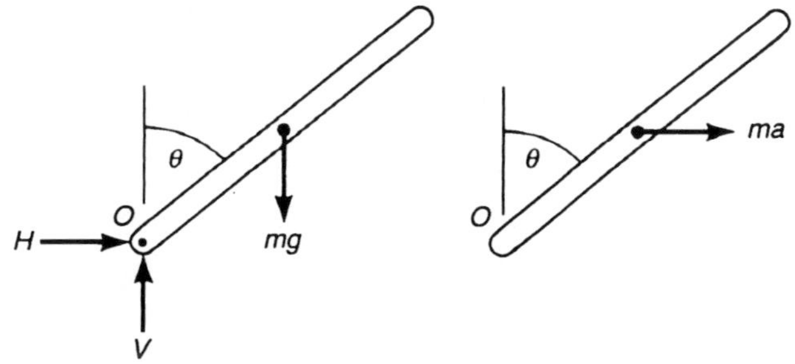

Exhibit 9.35a

9.36 d. From the diagrams in Exhibit 9.36a, and taking moments about O,

$$PH - mgl\ \sin\theta = I_G\alpha + ml^2\alpha$$

so that

$$\alpha = (PH - mgl\ \sin\theta)/(I_G + ml^2)$$

Now,

$$I_C = I_G + m(H-l)^2$$

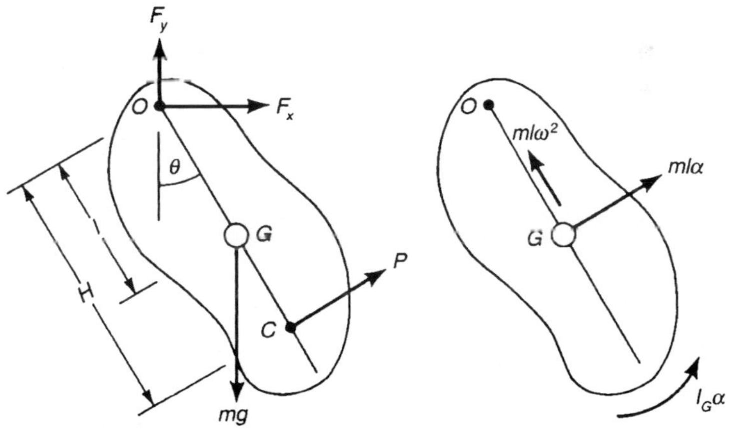

Exhibit 9.36a

from the parallel axis theorem. Thus,

$$I_G = I_C - m(H-l)^2 = mk^2 - m(H-l)^2$$

and

$$\alpha = \frac{PH - mgl\ \sin\theta}{m[k^2 - (H-l)^2] + ml^2}$$

Substitute values to get

$$\alpha = \frac{10(2) - 3(9.81)(1.5)(0.5)}{3[(0.8)^2 - (0.5)^2 + (1.5)^2]} = 0.26\ \text{rad/s}^2$$

9.37 **b.** C is the center of mass of the rod in Exhibit 9.37a. Summing moments about O gives

$$-mg(l/2)\sin\theta = (1/12)ml^2\alpha + m(l/2)^2\alpha + ma(l/2)\cos\theta \quad \textbf{(i)}$$

Summing forces along the horizontal, gives

$$H = ma + m(l/2)\,\alpha\cos\theta - m\omega^2(l/2)\sin\theta \quad \textbf{(ii)}$$

From Equation (i),

$$\alpha = -\frac{3}{2}\frac{(g\sin\theta\cos\theta)}{l}$$

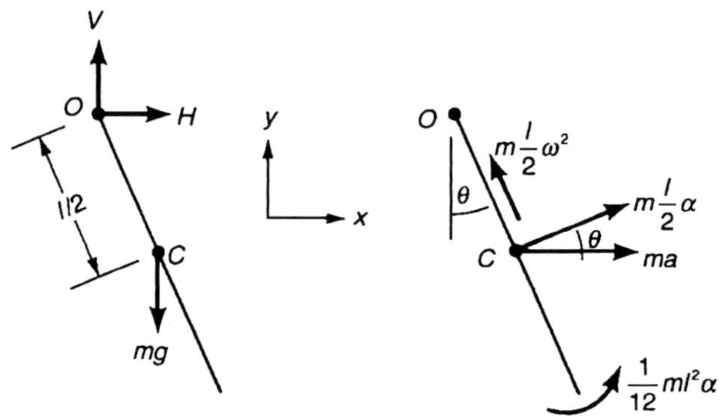

Exhibit 9.37a

Substituting values,

$$\alpha = -4.98 \text{ rad/s}^2 \quad \textbf{(iii)}$$

Substituting Equation (iii) and the given values into Equation (ii) yields

$$H = (1)(2) + (1)(1)(-4.98)\cos 30° - (1)(2)^2(1)\sin 30° = -4.31 \text{ N}$$

9.38 a.

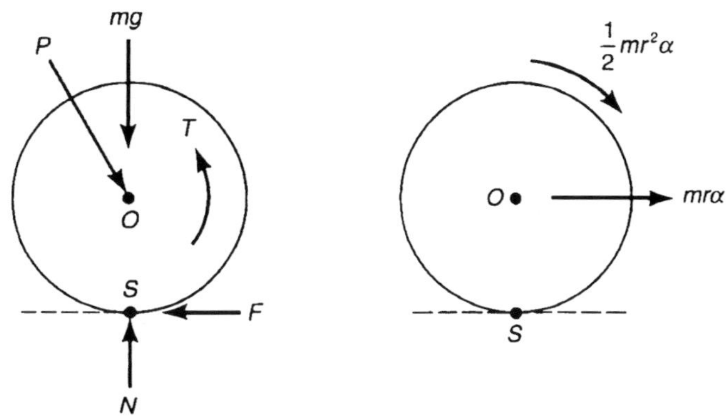

Exhibit 9.38a

Using the free-body diagram shown in Exhibit 9.38a, take moments about the contact point S:

$$rP\cos 60° - T = mr^2\alpha + (1/2)\,mr^2\alpha = (3/2)\,mr^2\alpha$$

or

$$\alpha = (rP\cos 60° - T)/(1.5mr^2) = \text{constant}$$

Substituting values, we find

$$\alpha = 0.83\ \text{rad/s}^2$$

With a constant angular acceleration, the angular velocity is

$$\omega = \omega_0 + \alpha t = 0 + 0.83(10) = 8.33\ \text{rad/s}$$

9.39 **b.** Exhibit 9.39a shows the forces acting on the rod as it swings from position 1 (horizontal) to position 2 (vertical). The work-energy principle gives

$$W_{1\to 2} = T_2 - T_1$$

That is,

$$mg\frac{l}{2} = \frac{1}{2}I_A\omega_2^2 - 0 = \frac{1}{2}\times\frac{1}{3}ml^2\omega_2^2$$

and

$$\omega^2 = \sqrt{\frac{3g}{l}} = \sqrt{\frac{(3)(9.81)}{2}} = 3.84 \text{ rad/s}$$

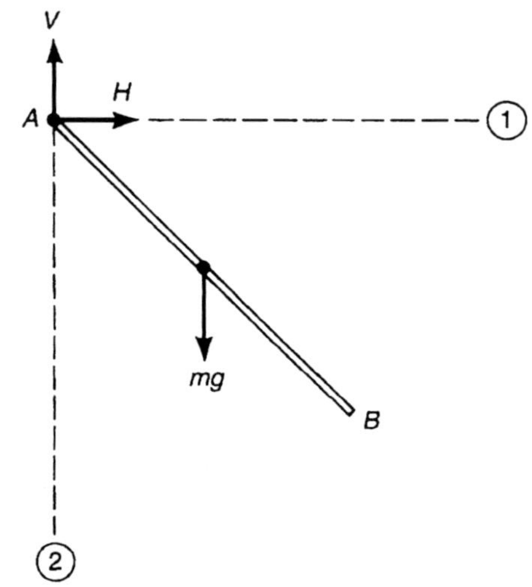

Exhibit 9.39a

9.40 d. Exhibit 9.40a shows the forces and torque acting on the rod as it rotates from position 1 (vertical) to position 2 (horizontal). The work-energy relation gives

$$W_{1-2} = T_2 - T_1$$

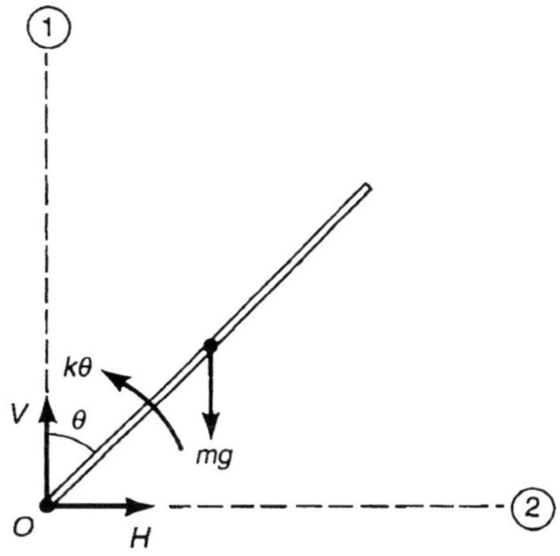

Exhibit 9.40a

That is,

$$mg\frac{l}{2} + \frac{l}{2}k(\theta_1^2 - \theta_2^2) = \frac{1}{2}I_0\omega_2^2 - \frac{1}{2}I_0\omega_1^2$$

Now, $\theta_1 = 0,\ \theta_2 = \pi/2\text{k}$, and $\omega_2 = \omega_2 = 0$. Thus,

$$k = \frac{mgl}{\theta_2^2 - \theta_2^2} = \frac{2(9.8)(3)}{(\pi/2)^2} = 23.8 \text{ N-m/rad}$$

9.41 a. The forces acting on the cylinder during this motion are shown in Exhibit 9.41a. Applying the work-energy principle,

$$W_{1-2} = T_2 - T_1$$

F and *R* do no work because their point of application has zero velocity (rolling without slip); *mg* does no work because its point of application moves perpendicular to the force. Work done by the spring force is

$$W_{sp}\frac{1}{2}k\left(\Delta_1^2 - \Delta_2^2\right) = W_{1-2}$$

where

$$\Delta_1 = [(3^2 + 4^2)^{0.5} - 3]\,\text{m} = 2\text{ m}, \quad \text{and} \quad \Delta_2 = 0$$

$$T_1 = 0, \text{ and } T_2 = \frac{1}{2}m(v_P)^2 + \frac{1}{2}I_P\omega^2 = \frac{1}{2}m(\omega r)^2 + \frac{1}{2}\times\frac{1}{2}mr^2\omega^2 = \frac{3}{4}mr^2\omega^2$$

Substituting into the work-energy principle yields

$$W_{1-2} = \frac{1}{2}k\Delta_t^2 = \frac{3}{4}mr^2\omega^2$$

and

$$\omega = \left(\frac{2}{3}\frac{k}{m}\right)^{0.5}\frac{\Delta_1}{r} = \left(\frac{2(2)}{3(12)}\right)^{0.5} \times \frac{2}{0.5} = 1.33 \text{ rad/s}$$

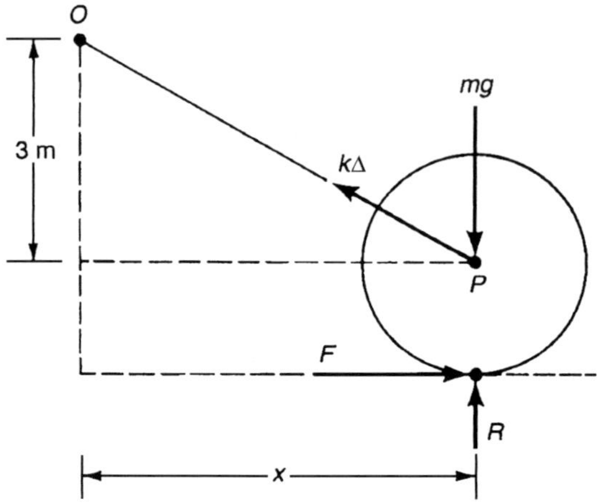

Exhibit 9.41a

CHAPTER 10

Mechanics of Materials

James R. Hutchinson and Jerry H. Hamelink

OUTLINE

The subject of stress analysis is usually studied in courses entitled "Mechanics of Materials" or "Strength of Materials" and is applied in problems of machine design to determine whether the device will withstand the loads to which it may be subjected. Stress analysis is actually an expansion of the more general subject of mechanics and includes the study of the effects of the elastic properties and strengths of the structural materials being considered.

Problems of stress analysis include both static and dynamic loading conditions. They may also include environmental conditions such as temperature, corrosion, radiation, and other factors that can influence material properties in the short or long term.

The simplest type of a stress problem is a pair of static forces acting in opposite directions along the longitudinal axis at the ends of a straight bar of constant cross section. In this case the stress is equal to the force divided by the cross-sectional area of the bar.

Example **10.1**

A mass of 100 kg is suspended vertically on the end of a 5.00-mm-diameter wire. What is the stress in the wire?

Solution

A 100-kg mass exerts a gravitational force of $100g = 980.66$ N. The cross-sectional area of a 5.00-mm-diameter wire is equal to 19.64×10^{-6} m^2 the stress in the wire is

$$\sigma = 980.66/(19.64 \times 10^{-6}) = 49.93 \text{ MPa}$$

Stress in a member is always accompanied by strain in the ratio

$$\varepsilon = \sigma/E$$

where

ε = strain, m/m
σ = stress, Pa
E = Modulus of Elasticity, Pa

Example 10.2

In Example 10.1, if the supporting wire is 10.00-m long, how much will it stretch (elongate) when the mass is attached to it?

Solution

The modulus of elasticity of steel is given in the *FE Handbook* as 2.1×10^{11} Pa

$$\varepsilon = (49.93 \times 10^6)/(2.1 \times 10^{11}) = 2.378 \times 10^{-4} \text{ m/m}$$

The total strain (elongation) is then

$$\Delta L = \varepsilon \times L = (10)(2.378 \times 10^{-4}) = 2.378 \times 10^{-3} \text{ m, or } 2.378 \text{ mm}$$

THERMAL STRESS

Most metals expand when their temperature is increased. If a member is constrained from elongating, it will be subject to strain. Since stress and strain are proportional, a strain will result in a stress, or will be evidence of a stress. The stress in a constrained member is calculated as if the member is allowed to expand and then is physically compressed the amount of the expansion.

$$\sigma = E \times \varepsilon$$

The thermal expansion, or compression, of a member is determined by multiplying the thermal coefficient of expansion by the temperature differential. The thermal coefficient of expansion for steel is $\alpha = 11.7 \times 10^{-6}$ m/m·°C. In most cases it will simplify the calculations if the strain due to stress and the strain due to temperature are calculated separately and are then added algebraically to obtain the net, or final, result. This can best be illustrated by means of examples.

Example 10.3

A 1-m-long copper bar having a circular cross section of 2.5 cm in diameter is arranged as shown in Exhibit 1, with a 0.025 mm gap between its end and a rigid wall at room temperature. If the temperature increase 35 °C, find the stress in the bar if its thermal coefficient of expansion is 16.7×10^{-6} m/m·°C and its modulus of elasticity is 1.2×10^{11} Pa.

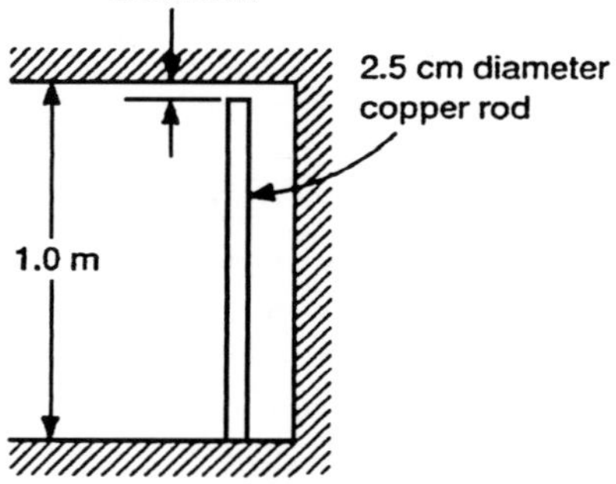

Exhibit 1

Solution

The unrestrained increase in length due to the increase in temperature would have been

$$\Delta L = (1.0)(35)(16.7 \times 10^{-6}\ \text{m/m}\cdot{}^\circ\text{C}) = 0.000585\ \text{m, or } 0.585\ \text{mm}$$

The bar can expand 0.025 mm and then will be restrained. This is the same (stresswise) as if it had been allowed to increase 0.585 mm and was then compressed.

$$0.585 - 0.025 = 0.560\ \text{mm}$$

The resulting stress would be

$$S = E \times \varepsilon = (1.2 \times 10^{11}\ \text{Pa})(0.560/1000) = 67.2\ \text{MPa}$$

There are a few assumptions that would have to be made, such as

1. The walls must be perfectly rigid, i.e., undergo neither expansion nor deformation.
2. The bar (column) must remain straight.
3. The yield point of the copper must not be exceeded.

Example **10.4**

An exhaust manifold on a diesel engine is held rigidly at both ends. Under extended operation at high power output the manifold may become cherry red in color, indicating that it has reached a very high temperature, say 705 °C. If the manifold were bolted in place at a temperature of 25 °C, what would be the apparent stress in the manifold at the higher temperature? Assume the manifold is held rigidly and the rest of the engine assembly does not change its dimensions.

Solution

The easiest way to look at a problem like this is to assume that the heated member is free to expand to the higher temperature and then is physically forced to its former dimension. The theoretical longitudinal strain of the steel manifold would equal be

$$\begin{aligned}\varepsilon &= \Delta L/L = \Delta T \times \alpha \\ &= (705\,^\circ\text{C} - 25\,^\circ\text{C})(11.7 \times 10^{-6}\ \text{m/m}\cdot{}^\circ\text{C}) = 0.00796\ \text{m/m}\end{aligned}$$

It would then be compressed an equal amount when the temperature of the manifold cooled to the ambient temperature. This gives an apparent stress of

$$\sigma = \varepsilon \times E = (0.00796)(2.1 \times 10^{11}) = 1.67 \text{ GPa}$$

This is well above the yield point of stainless steel, so the manifold would yield in compression as it was heated. Since the length of the manifold between the supports would now be shorter, it would yield in tension when the manifold cooled. After a number of such cycles the manifold would fail. This has been termed "low-cycle fatigue."

HOOP STRESS

If the wall thickness of a cylindrical vessel is smaller than one-twentieth of the radius it is considered to be a thin-walled vessel. The hoop stress can then be calculated with the equation

$$\sigma = PD/2t$$

where

P = internal pressure, Pa
D = internal diameter, meters
t = wall thickness, meters

Similarly, if the pressure vessel were a closed vessel, the longitudinal force acting on the cylindrical shell would be equal to the internal area times the pressure, and the longitudinal stress would be equal to

$$PD^2(\pi/4)/\pi Dt = PD/4t, \text{ or just half of the hoop stress}$$

Example 10.5

If a 750-mm-diameter cylinder with a 5.00-mm-thick wall is pressurized to 1.5 MPa, what is the hoop stress?

Solution

$$PD/2t = \sigma = (1.5 \times 10^6)(0.750)/2(0.005) = 112.5 \text{ MPa}$$

Similarly the longitudinal stress is equal to one half of the hoop stress, or

$$\sigma = 112.5/2 = 56.25 \text{ MPa.}$$

These constitute the two principal stresses in the wall of the vessel. The maximum shear stress, from the diagram of Mohr's circle in the *FE Handbook*, is equal to one half of the algebraic difference between the two principal stresses.

$$\tau_{max} = [(112.5 - 56.25) \times 10^6]/2 = 28.125 \text{ MPa}$$

MOHR'S CIRCLE

This brings up the topic of Mohr's circle and combined stresses. Mohr's circle was developed as a simplified method for determining the maximum stress resulting from a number of differently applied loads or combined stresses. We shall restrict ourselves here to a brief discussion of one method of determining the principal stresses and the maximum shearing stress for a condition of combined loading, a method that is based on Mohr's circle.

Example 10.6

Derive the equation for the maximum shearing stress in a member that is subjected to combined tension, torsion, and circumferential tension, specifically, a pipe under pressure with torsion and axial tension applied, using the diagram in Exhibit 2 (Mohr's circle). The equation is to be derived for the maximum shearing stress, from the information in the diagram, in terms of $S_{s,max}$ = maximum shearing stress, S_s = stress due to torsion, S_x = stress due to pure tension, and S_y = stress due to circumferential tension. In addition, sketch a stress diagram for a unit area of the external pipe surface.

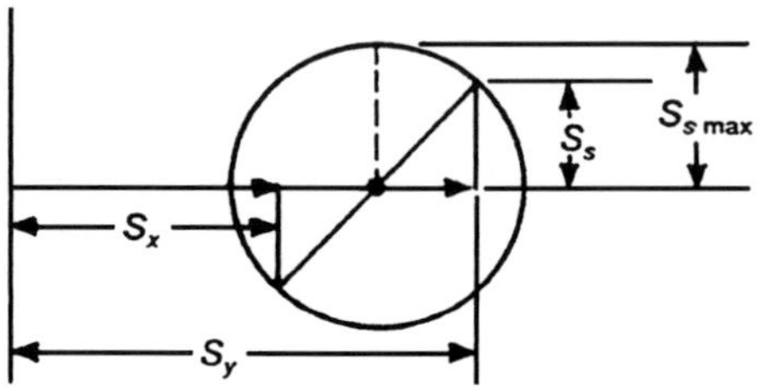

Exhibit 2

Solution

This example illustrates a relatively simple application of Mohr's circle. The subject can be examined in a more general sense to obtain maximum benefit from the example in describing the application of the principles of Mohr's circle.

First, determine the stresses acting at a section on the outer surface of the pipe (see Exhibit 3).

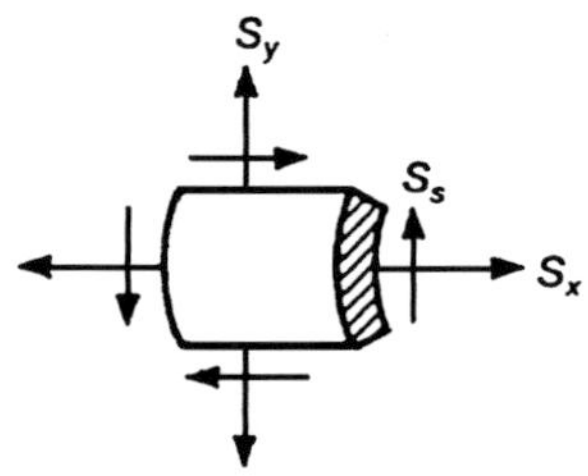

Exhibit 3

$S_x = F/A$ (axial load divided by the cross-sectional area)
$S_y = Pd/2t$ (assuming a thin-walled tube)
$S_s = TC/J$

Having determined these stresses, we can construct Mohr's circle for the given case. The general procedure is shown in Exhibit 4. First, lay out the coordinate axes. From the origin, point 0, lay off a distance equal to $(S_x + S_y)/2$ along the abscissa.

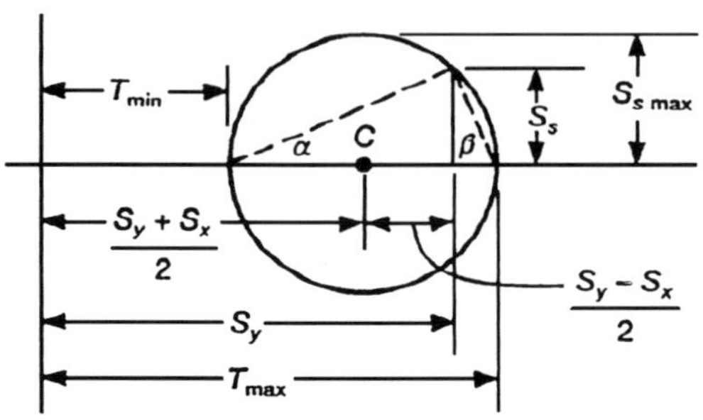

Exhibit 4

This gives the location of C, the center of the circle. Note here that $S_x + S_y$ is the algebraic sum of the two perpendicular stresses; if one of these stresses had been compressive, we would have had the arithmetic difference instead of the arithmetic sum. At a distance from the origin equal to S_y draw a vertical line, and at a distance equal to S_s above the abscissa mark a point on this line. This point lies on the circumference of the circle. Now, draw the circle.

The maximum shearing stress is equal to the radius of the circle. From the figure, the trigonometric relationship is

$$S_{s,\max} = \sqrt{\left[S_s^2 + \{(S_y - S_x)/2\}^2\right]}$$

The principal stresses can also be quickly obtained and are shown in Exhibit 5, which shows a section of the tube surface with the applied stresses. The minimum and maximum stresses, $T_{\min}$ and $T_{\max}$ are shown, as well as the directions of the stresses by the angles α and β.

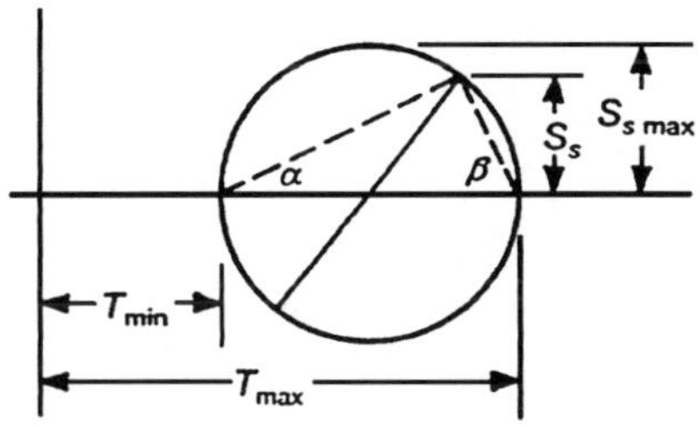

Exhibit 5

The lines drawn from the ends of the horizontal diameter to the construction point on the circumference give these angles, and the principal stresses act in directions that are perpendicular to these lines. That is, $T_{\min}$ acts at an angle of β, and $T_{\max}$ acts at an angle of α, to the direction of S_y. The maximum shearing stress acts on a plane at an angle of 45° to these two principal planes.

A stress diagram for a unit area of the external tube surface, Exhibit 6, shows the directions of the different stresses and would ordinarily be drawn before Mohr's circle was constructed.

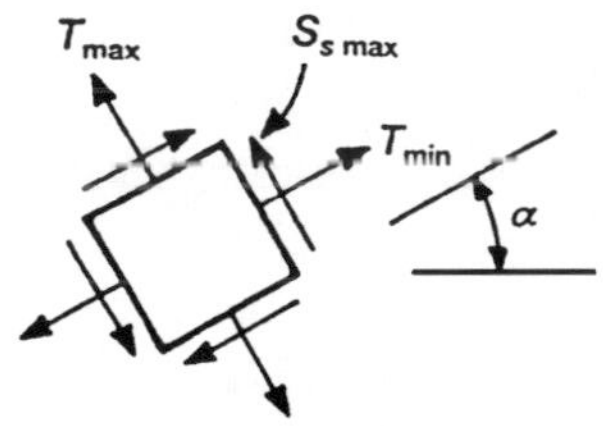

Exhibit 6

To sum up, the equations for determining the maximum stresses can be listed as follows

$$\text{Principal stress (1)} = (S_x + S_y)/2 + \sqrt{\left[S_x^2 + \{(S_y - S_x)/2\}^2\right]}$$

$$\text{Principal stress (2)} = (S_x + S_y)/2 - \sqrt{\left[S_s^2 + \{(S_y - S_x)/2\}^2\right]}$$

$$\text{Maximum sheer stress} = \sqrt{\left[S_s^2 + \{(S_y - S_x)/2\}^2\right]}$$

The equation for the principal stress can be used to check the maximum stress obtained in Example 10.5.

From the example, $S_y = 56.25$ MPa and $S_x = 112.5$ MPa. Since there is no applied shear stress, the maximum principal stress would be

$$S_{s,\max} = (S_x + S_y)/2 + \sqrt{\{(S_y - S_x)/2\}^2} = 84.375 + 28.125 = 112.5 \text{ MPa}$$

SHRINK FIT

Parts are often assembled by heating the outer member, and thus expanding it, while simultaneously cooling the inner member and shrinking it, if necessary. This puts the outer member in tension and the inner member in compression.

Example 10.7

A 1.00-m cast-iron wheel has a 1.00-mm-thick steel ring shrunk onto it as a tire. The interference fit is 0.1000 mm. What is the stress in the ring? What is the contact pressure?

Solution

The diametral strain is 0.100/1000 = 0.000100 m/m. The circumferential strain would be the same, though the circumferential elongation would be equal to π times the diametral elongation.

$$\text{Stress, } \sigma = E \times \varepsilon = (2.1 \times 10^{11})(1 \times 10^{-4}) = 21.0 \text{ MPa stress in the ring}$$

$$\text{Hoop stress } \sigma = PD/2t, \text{ so}$$

$$P = 2t\sigma/D = 2(0.001)(21.0 \times 10^{6})/1.0 = 42{,}000 \text{ Pa}$$

The contact pressure would thus be equal to 42.0 kPa.

It is assumed that the cast iron wheel does not shrink since the ring is very thin, by comparison, and would exert minimal force on the wheel.

Example 10.8

A steel liner is assembled in an aluminum pump housing by heating the housing and cooling the liner in dry ice. At 20 °C, before assembly, the outside diameter of the liner is 8.910 cm, and the inside diameter of the housing is 8.890 cm. After assembly and inspection, several units were rejected because of poor liners. It is desired to salvage the housing by heating the complete unit to a temperature that would cause a difference (clearance) in diameter of 0.050 mm between the liner and the housing and permit free removal of the liner. Determine the temperature at which this may be possible. The thermal coefficient of linear expansion for steel, α_S, is equal to 11.7×10^{-6} m/m · °C, and that for aluminum, α_A, is equal to 28.8×10^{-6} m/m · °C.

Solution

At 20 °C the interference is 8.910 – 8.890 = 0.020 cm. A clearance of 0.050 mm is desired, so the expansion of the aluminum housing must be 0.025 cm more than the expansion of the steel liner

$$\Delta D = 0.020 \text{ cm (interference)} + 0.005 \text{ cm (clearance)} = 0.025 \text{ cm}$$

The unit expansion of the aluminum must be 0.025/8.890 = 0.00281 m/m greater than that of the steel. The differential expansion is equal to

$$a_A - \alpha_S = 0.0000288 - 0.0000117 = 17.1 \times 10^{-6}\ \text{m/m}\cdot{}^\circ\text{C}$$

The required temperature increase would then be

$$\Delta T = (0.00281\ \text{m/m})/[17.1 \times 10^{-6}\ \text{m/m}\cdot{}^\circ\text{C}] = 164\,^\circ\text{C}$$

Therefore, the assembly would have to be heated to a temperature of 184 °C. To check

$$164(11.7 \times 10^{-6}\ \text{m/m}\cdot{}^\circ\text{C}) = 0.001919\ \text{m/m, so the liner would expand to } 8.910 + (8.910)(0.001919) = 8.927\ \text{cm}$$

The housing would have a thermal strain of

$$164(28.8 \times 10^{-6}\ \text{m/m}\cdot{}^\circ\text{C}) = 0.004723\ \text{m/m, so it would expand to } 8.890 + (8.890 \times 0.004723) = 8.932\text{, giving a clearance of } 8.932 - 8.927 = 0.005\ \text{cm, which checks.}$$

TORSION

Pure torsional loading (couple only, no bending) produces a shearing stress in a shaft with the magnitude of the stress, in any cross section, being proportional to the distance from the center of the shaft (Figure 10.1). This follows from the fact that the deformation at any point is equal to $\rho d\theta$, and, since stress is proportional to strain, the stress increases from zero at the center to a maximum at the outside. Let S_s be the maximum shearing stress at the outer fiber. Then S_s/r will be the stress one meter from the center and $(\rho/r) \times S_s$ will be the stress at a distance ρ from the center. Force is equal to stress times area, so the force resisting the applied torque, due to the area dA, is equal to $(\rho/r)S_s\,dA$; the resisting moment about the axis of the bar due to this force is equal to $(\rho^2/r)S_s\,dA$, and $(S_s/r) \times \int \rho^2\,dA$ is the total moment about the axis due to all of the internal shearing forces. The quantity $\int \rho^2\,dA$ is, we recall from Mechanics (see *FE Handbook*), the moment of inertia of the cross-sectional area about the axis through its center or the *polar moment of inertia of the area*, which is ordinarily represented by the symbol J. This gives us the relationship that torque, T, is equal to $(S_s/r) \times J$, and, therefore, the maximum stress in the shaft due to the applied torque is $S_s = rT/J$. The stress at any other distance ρ from the center of the shaft is equal to

$$(\rho/r) \times (rT/J) = \rho T/J$$

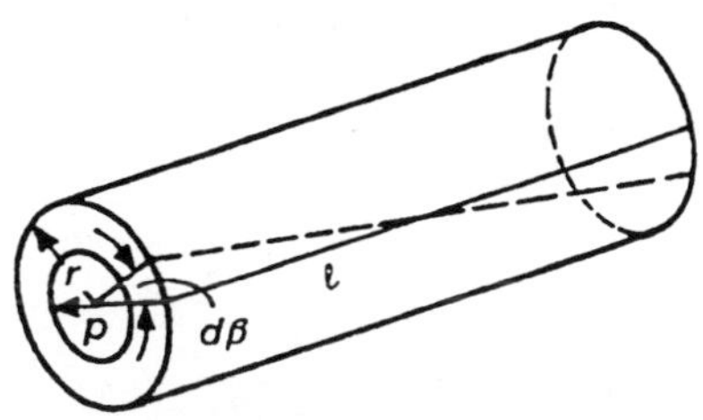

Figure 10.1

Within the proportional limit, $\varepsilon_s = S_s/E_s$, which gives the relationship $S_s = E_s\varepsilon_s = (E_s r\theta)/l$, since, from Figure 10.2, $\varepsilon_s = \theta/L$. This also gives us $T = (E_s J\theta)/L$.

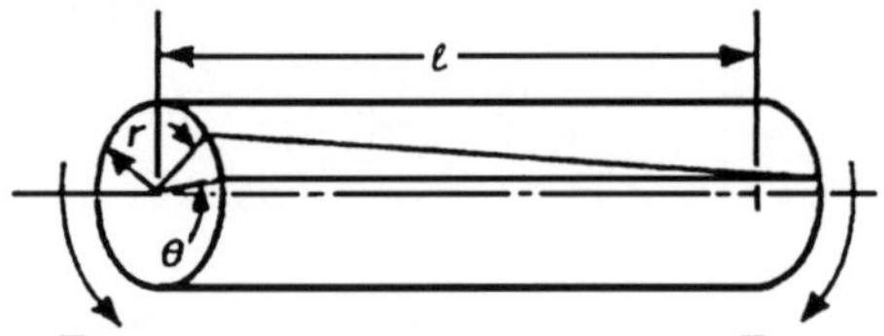

Figure 10.2

Example 10.9

A solid round shaft 9.0 cm in diameter transmits 75 kW of power at 200 rpm. What is the maximum torsional stress in the shaft? The appropriate equations are given in the *FE Handbook*.

Solution

The maximum stress will occur in the outer fibers. The power transmitted by a shaft is equal to $2\pi NTq$. In Exhibit 7, torque is equal to tension, T, times the radius, r, which gives Newton-meters or joules, and joules/sec is equal to watts. So N, in this equation must be equal to revolutions per second, or 3.333 rev/s. Then we have that the torque is

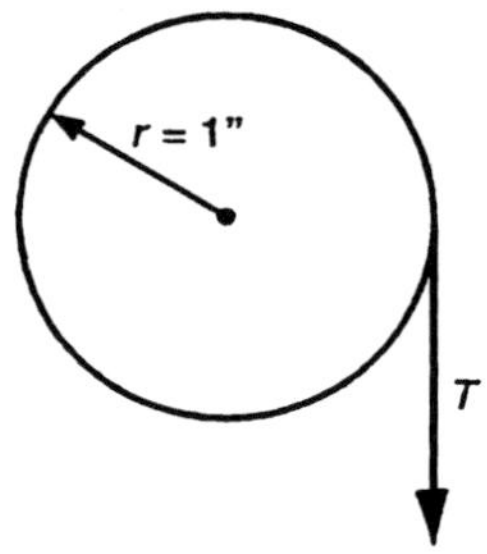

Exhibit 7

$$Tq = 75{,}000/2\pi(3.333) = 3581.3 \text{ J}$$

The polar moment of inertia of a circular section is

$$J = (\pi r^4)/2 = \pi(0.045)^4/2 = 0.00000644 \text{ m}^4$$

The maximum shear stress is thus

$$S_{s,\max} = (0.045)(3581.3)/0.00000644 = 25.02 \text{ MPa}$$

Example 10.10

For a hollow shaft of the same outer diameter, but with an inner diameter equal to half of the outer diameter, or 4.5 cm, what would be the increase in stress?

Solution

Again, the maximum stress will occur in the outermost fibers and will be equal to rT/J where r is equal to the distance from the neutral axis to the outermost fibers, T is equal to the torque, and J is equal to the polar moment of inertia

$$J = \pi/32 \times (OD^4 - ID^4) = (0.0982)[(65.61 - 4.10) \times 10^{-6}\ \text{J}] = 6.040 \times 10^{-6}\ \text{m}^4$$

which is equal to 6.040/6.44 = 0.938, or 93.8%, of that of a solid shaft. The maximum shear stress is equal to

$$S_{s,\max} = OD/2 \times T/J = (0.045)(3581.3)/(6.040 \times 10^{-6}) = 26.68\ \text{MPa}$$

which gives an increase of 1.66 MPa, or an increase of 6.6%.

Example 10.11

Referring to Exhibit 8, what would be the torque M_B such that the maximum unit shear stress would be the same in both parts. Disregard stress concentrations.

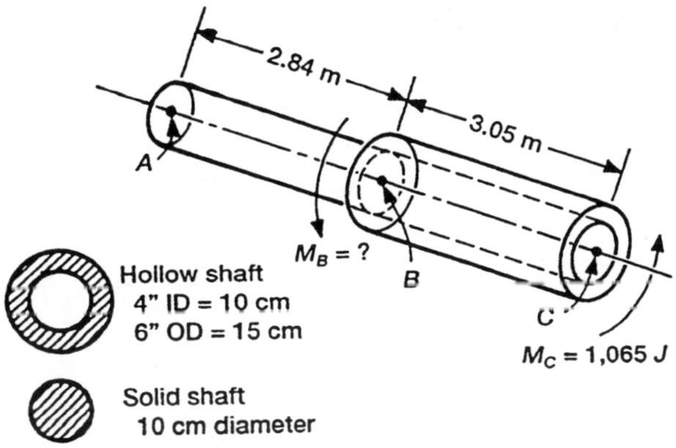

Exhibit 8

Solution

The maximum stress due to torsional loading is equal to $r \times T/J$, where J is equal to the polar moment of inertia. For the hollow shaft

$$J = \pi/32 \times (OD^4 - ID^4) = (0.09817)[(506.25 - 100.0) \times 10^{-6}] = 39.882 \times 10^{-6}\ \text{m}^4$$

The maximum shear stress in the hollow shaft is then

$$S_{s,\max} = (0.075)(1065)/(39.882 \times 10^{-6}\ \text{m}^4) = 2.00\ \text{MPa}$$

For the solid shaft

$$J = (0.09817)(0.10)^4 = 9.817 \times 10^{-6}\ \text{m}^4$$

For the shear stress in the solid shaft to be equal to 2.00 MPa, the applied torque would need to be

$$T = (2.00 \times 10^6)(9.817 \times 10^{-6}\ \text{m}^4)/0.05 = 392.7\ \text{J}$$

The required torque MB would then be

$$M_a = M_C - T = 1065 - 392.7 = 672.3\ \text{J}$$

Example 10.12

What would be the angle of twist in the overall length of the shaft in Example 10.11 if the shaft were made of steel, using the torques as defined and calculated above?

Solution

The angle of twist is given in the *FE Handbook* as

$$\phi = T \times L/(G \times J)$$

where
- ϕ is in radians
- T is in joules
- L is in meters
- G is the shear modulus, which is equal to 8.3×10^{10} Pa (see *FE Handbook*)
- J is the polar moment of inertia, as above

From A to B

$$\phi = (392.3)(2.84)/(8.3 \times 10^{10})(9.817 \times 10^{-6}) = 0.00137 \text{ radians}$$

From B to C

$$\phi = (1065)(3.05)/(8.3 \times 10^{10})(39.882 \times 10^{-6}) = 0.000981 \text{ radians}$$

The total twist would be equal to 0.00235 radians, or $0.00235(180/\pi) =$ 0.135 degrees.

Example 10.13

A motor running at 1750 rpm transmits 10 kW of power through a 25.0-mm-diameter shaft that is 2.0 m long. What is the angle of twist in the shaft?

Solution

$$\text{Watts} = \text{J/s} = \text{N} \times \text{m/s} = 2\pi NT, \text{ where } N \text{ is rev/s} = 1750/60 = 29.167$$
$$T = \text{Watts}/2\pi N = 10{,}000/(2\pi)(29.167) = 54.57 \text{ N} \times \text{m}$$
$$J = \pi(0.025)^4/32 = 3.835 \times 10^{-8} \text{ m}^4$$
$G = 8.3 \times 10^{10}$ Pa, from the *FE Handbook*
$$\phi = (54.57)(2.00)/(3.835)(8.3)(100) = 0.0343 \text{ radians}$$
$$\text{Angle of twist} = (0.0343)(180/\pi) = 1.965°$$

Example 10.14

What would the shaft diameter in Example 10.13 have to be to reduce the angle of twist to 1.00°?

Solution

The polar moment of inertia would have to be increased in the ratio of 1.9647:1.000 so that the new shaft diameter would be

$$D = (25.0)(1.9647)^{0.25} = (25.0)(1.1839) = 29.60 \text{ mm}$$

Example 10.15

Given the bi-metal shaft shown in Exhibit 9, if a torque of 2.00 kJ is applied at the point of juncture, and both ends are restrained at their ends, what would be the reaction where the steel section is attached to the wall? What would be the stress in the aluminum section?

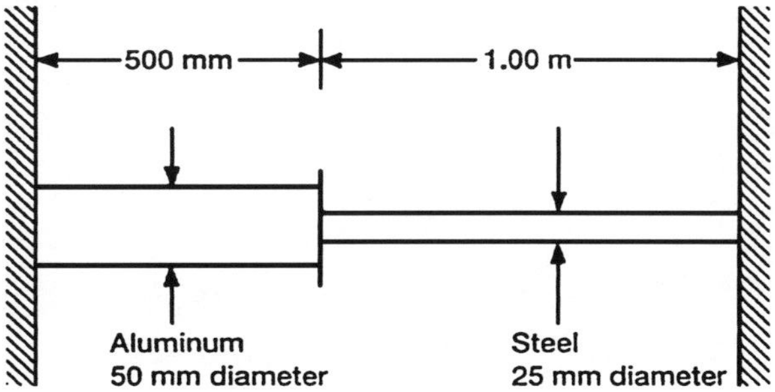

Exhibit 9

Solution

From the *FE Handbook* the total angle of twist of a shaft subjected to a torque

$$\phi = TL/GJ$$

where

T = torque, joules
L = length of shaft, meters
G = shear modulus, Pa
J = polar moment of inertia, m^4

The total angle of twist would be the same for both the steel and the aluminum sections.

For the aluminum section

$$\begin{aligned} J &= \pi D^4/32 = \pi(0.050)^4/32 \\ &= 6.136 \times 10^{-7}\ m^4 \\ L &= 0.500\ m \\ G &= 2.8 \times 10^{10}\ Pa \\ \phi &= T_{Al}(2.910 \times 10^{-5}) \end{aligned}$$

For the steel section

$$\begin{aligned} J &= 3.835 \times 10^{-4}\ m^4 \\ L &= 1.000\ m \\ G &= 8.3 \times 10^{10}\ Pa \\ \phi &= T_{Steel}(31.42 \times 10^{-5}) \end{aligned}$$

The total angles of twist are equal so,

$$T_{Steel}(31.42) = 2.910T_{Al} \text{ and}$$
$$T_{Steel} = 0.0926T_{Al}$$

The sum of the torques in the steel and the aluminum sections is equal to the total applied torque.

$$1.0926T_{Al} = 2000 \text{ J}$$
$$T_{Al} = 1831 \text{ J}$$
$$T_{Steel} = 2000 - 1831 = 169 \text{ joules}$$

The stress due to applied torque, $\sigma = Tc/J$.
For the aluminum section

$$\sigma = (1831)(0.025)/(6.136 \times 10^{-7}) = 74.6 \text{ MJ}$$

BEAMS

The general relationship for determining the tensile or compressive stress in a beam due to bending is

$$S = (Mc)/I$$

where
M = moment, J
I = moment of inertia of the cross section
c = distance from the neutral axis to the outermost fiber

The stress at any other point than the outer fiber is proportional to its distance from the center and would be equal to

$$S = (y/c)(Mc/I) = (My)/I$$

The values of c and I can be determined from the geometry of the beam. The moment, M, depends upon the loading of the beam. As can be deduced, the maximum stress will occur at the location of the greatest moment (for a beam of constant cross section).

The transverse shear stress at any point is equal to the magnitude of the shearing force divided by the cross-sectional area at that point. It is seldom possible to determine by a glance just where the maximum shear and the maximum moment will occur in a loaded beam, so it is usually desirable to construct both the shear diagram and the moment diagram for the beam under consideration; then the points of maximum shear and maximum moment will be readily apparent.

SHEAR DIAGRAM

The shear at any section of a beam is equal to the algebraic sum of all the external forces on *either* side of the section; it is considered positive if the segment of the beam on the left of the cross section tends to move up with respect to the segment on the right. The algebraic sum of all of the forces on both sides of the section will, of course, equal zero, since the system is in static equilibrium. It is usually simplest to add all the shearing forces to the left of the section being considered; this sum is equal to the shearing force acting at that point. If this process is repeated for different points over the length of the beam, a complete shear diagram may be constructed.

MOMENT DIAGRAM

The moment diagram may be constructed in a similar manner; one must remember that the bending moment at any section of a beam is the algebraic sum of the moments of all the external forces on either side of the section. When calculating the bending moment, use the segment (or side) for which the arithmetic will be the simplest. A positive moment is one that tends to cause the beam to be concave on the upper side (top fibers in compression). An easy method for arriving at the correct sign in calculating bending moment is to give plus signs to the moments of upward forces and minus signs to the moments of downward forces. This method will give the correct sign to the algebraic sum of the moments, whether the forces to the right of the section or to the left of the section are used.

Examples of shear and moment diagrams for a few of the more common types of loading are shown in Figure 10.3.

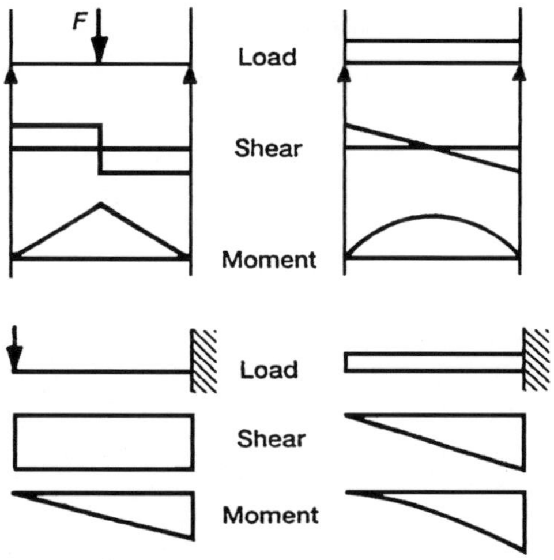

Figure 10.3

For combined loading the shear and moment diagrams may be constructed by adding (algebraically) the individual shear and moment diagrams due to the individual loadings. This will give the total values for shear and moment for all parts of the beams. Each load will produce the same effect as if it had acted alone; it is unaffected by the other loads. Another method is to calculate the moments at different points on the beam and draw a smooth curve. Care must be taken, however, to make certain that the points of maximum total shear and maximum total moment are included. These points of maximum magnitude will be more easily determined by sketching the component shear and moment diagrams and determining the total shear and moment for all points of maximum magnitude (both positive and negative) on the component diagrams.

There are a few relationships that are of value in constructing and in checking shear and moment diagrams. One of these is that the derivative of the moment, M, with respect to distance, is equal to the shear V, or $V = (dM)/(dx)$. This means that the slope of the moment curve at any point is equal to the magnitude of the shear at that point. The relationship may also be written as

$$\int dM = \int V(x)\,dx, \qquad \text{from point 1 to point 2, or}$$
$$M_2 - M_1 = \int V(x)\,dx, \qquad \text{from point 1 to point 2,}$$

which means that the difference between the values of the moments at points 1 and 2 is equal to the area under the shear diagram between points 1 and 2. Similarly, we have

$$W = (dV)/(dx)$$

where W is the load at any point

$$V_2 - V_1 = \int W(x)\, dx$$

($W(x)$ means W as a function of x.)

Example 10.16

A wooden beam is made up of three timbers: two 10 by 15 cm and one 10 by 20 cm, as shown in Exhibit 10. If this section is used as a uniformly loaded beam in a simple span 10 meters long, what maximum total load could it sustain, assuming a maximum bending stress of 8.3 MPa?

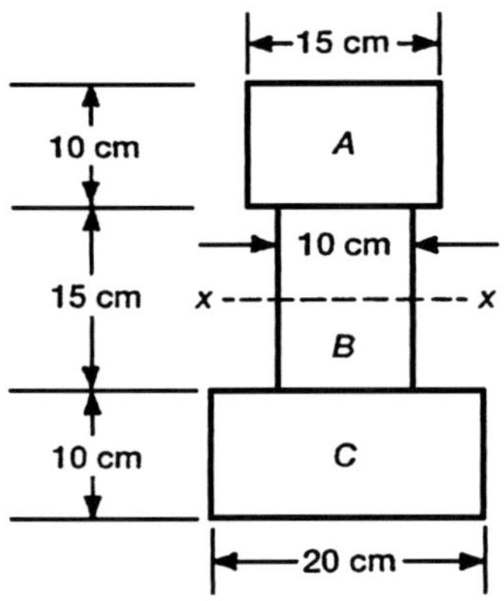

Exhibit 10

Solution

First, determine the location of the neutral axis. This can be obtained by taking moments about the lower edge.

$$\begin{aligned} y &= \text{(moments of individual areas)/(total area)} \\ &= [(10)(20)(5) + (10)(15)(17.5) + (10)(15)(30)]/(200 + 150 + 150) \\ &= 16.25 \text{ cm from the bottom edge} \end{aligned}$$

The moment of inertia, I, for a rectangular area is equal to $bh^3/12$, and the parallel axis theorem states that the moment of inertia of a section, I, about an axis other than that which passes through the centroid of the section is equal to the I about the centroidal axis plus the area of the section times the square of the distance between the axis through the centroid and the new axis. This is explained in the *FE Handbook*.

$$I_A = (0.15)(0.10)^3/12 = 12.5 \times 10^{-6} \text{ m}^4$$

Similarly,

$$I_B = (0.10)(0.15)^3/12 = 28.125 \times 10^{-6} \text{ m}^4\text{, and}$$
$$I_C = (0.20)(0.10)^3/12 = 16.667\ 10^{-6} \text{ m}^4$$

The moment of inertia about the beam centroidal axis is equal to

$$I = [(12.5 \times 10^{-6}) + (0.015)(0.1375^2)] + [(28.125 \times 10^{-6}) + (0.015)(0.0125^2)] + [(16.667 \times 10^{-6}) + (0.020)(0.125^2)]$$

$$I = [(12.5 + 284.59) + (25.125 + 2.34) + (16.667 + 312.4)] \times 10^{-6} = 656.6 \times 10^{-6} \text{ m}^4$$

$$\text{Stress} = Mc/I$$

where $M = (wL^2)/8$ (see *FE Handbook*)

$$c = 35 - 16.25 = 18.75 \text{ cm}, \quad \text{or} \quad 0.1875 \text{ m}$$

The total load, as determined by compression in the top portion of the beam, is equal to

$$P = wL = (SI/c) \times 8/L = (8.3 \times 10^6 \text{ Pa})(656.6 \times 10^{-6})/(0.1875)(8/10) = 23{,}252 \text{ N} = 2371 \text{ kg}$$

If the load is determined by tension in the bottom fibers, the distance c would be equal to 0.165 meters, and the permissible load would be

$$P = [(8.3)(6565.6)/0.1625](8/10) = 26{,}830 \text{ N}, \quad \text{or} \quad 2736 \text{ kg}$$

Example 10.17

Construct the shear and moment diagrams for the beam shown in Exhibit 11, and show the positions and magnitude of the maximum points. What is the maximum moment in the beam?

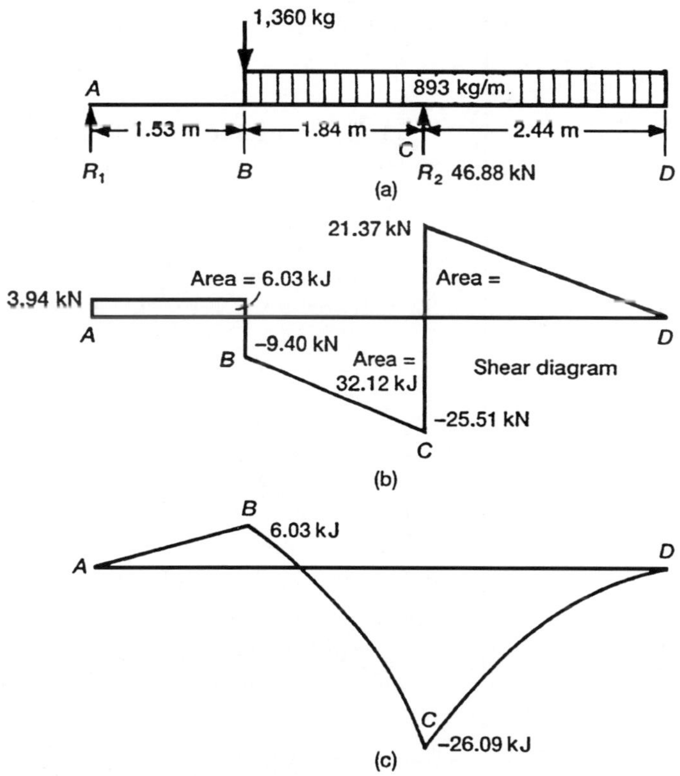

Exhibit 11

Solution

To make certain that the location and magnitude of the maximum moment are determined accurately, it is best to construct the shear and moment diagrams. The diagrams will be constructed just below the beam diagram so that locations of the important points will be shown in their proper positions. First it is necessary to determine the reactions at R_1 and R_2. The loads are given in kg, these must be converted to newtons. The uniform loading of 893 kg/m equals 893g or 8758 N/m and the concentrated mass loading of 1360 kg exerts a force of 13.34 kN.

Taking moments about R_1 gives

$$R_2 = [(1.53 \times 13{,}340) + (1.53 + 2.14) \times (8{,}758 \times 4.28)]/3.37 = 46.88 \text{ kN}$$

The total load on the beam is equal to

$$13{,}340 + 8{,}758 \times 4.28 = 50.82 \text{ k}$$

so the reaction at the left end is equal to

$$R_1 = 50.82 - 46.88 = 3.94 \text{ kN}$$

The force at point A is equal to 3.94 kN up, so this provides a positive shear force up which remains constant to point B. At point B there is an applied mass of 1360 kg which produces a force of 13,338 N down, so the sharing force becomes 3940 – 13,338 equals –9398 N, or a negative shearing force of 9,398 N. The downward, or negative, shear force increases uniformly by $893g = 8758$ N/m for 1.84 meters to point C. At point C the magnitude of the shearing force is equal to

$$V = -9{,}398 - 8{,}758 \times 1.84 = -25{,}513 \text{ N}$$

At point C there is an upward force of 46.88 kN, so the magnitude of the shear force becomes

$$V = -25.51 + 46.88 = 21.37 \text{ kN}$$

This shear force will reduce linearly by the applied load of 8.76 kN/m for 2.44 meters to point D, where it becomes zero, which checks.

From the previously given relationship

$$M_2 - M_1 = \int V(x)\,dx \text{ from Points 1 to 2 (see } \textit{FE Handbook}\text{)}$$

This means that the difference between the values of the moments at Points 1 and 2 is equal to the area under the shear diagram between Points 1 and 2.

Calculating the different areas gives

From A to B, A = 3.90 kN × 1.53 m = 6.03 kJ
From B to C, A = 1.84 × (9.40 + 25.51)/2 = 32.12 kJ
From C to D, A = 2.44 × 21.37/2 = 26.07 kJ

The moment at point B *is* equal to 6.03 kJ and the moment at point C is equal to 6.03 – 32.12 = –26.09 kJ. The moment at point D, the end of the beam, is equal to 26.09 – 26.07 = 0.02, which checks within the computational accuracy. The maximum moment is equal to –26.09 kJ and occurs at the location of the support, R_2.

Example 10.18

What would be the maximum stress in a cantilever beam holding a mass of 500 kg at its end if it is 3.00 m long and is made of a 200 mm wide by 300 mm high wooden beam?

$$\text{Stress, } \sigma = Mc/I$$

The maximum stress would occur where the moment was a maximum.

$$M_{max} = 500 \times 9.8066 \times 3.00 = 14{,}710 \text{ J}$$

The moment of inertia of the beam,

$$I = bh^3/12 = 0.200 \times 0.300^3/12 = 4.500 \times 10^{-4} \text{ m}^4$$

The distance from the neutral axis to the outermost fiber,

$$c = 0.150 \text{ m}$$

The maximum stress is thus equal to

$$\sigma = 4.903 \text{ MPa}$$

As noted in the *FE Handbook* the moment of inertia of an area about an axis that is parallel to, but does not pass through its centroid equals its moment of inertia about its centroidal axis plus its area times the square of the distance between the two axes.

Example 10.19

What is the moment of inertia about its centroidal axis of a beam composed of two 200 mm by 300 mm beams connected in the form of a tee, see Exhibit 12.

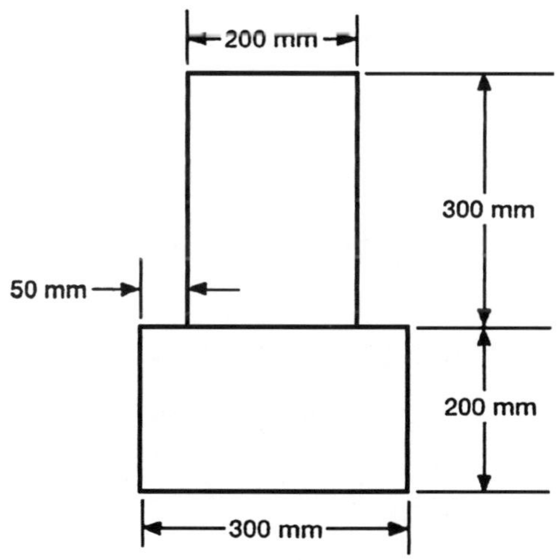

Exhibit 12

Solution

First it is necessary to determine the location of the centroidal axis. Take moments of the areas about a horizontal axis through the top edge of the composite beam and divide by the total cross-sectional area.

$$c = (0.150 \times 0.060 + 0.400 \times 0.060)/(0.060 + 0.060) = 0.275 \text{ m}$$

The centroid of the composite beam is 0.275 m down from the top.

The moment of inertia of the vertical member about its horizontal centroidal axis, parallel to the axis through the centroid of the composite beam is

$$I = bh^3/12 = 450 \times 10^{-6}\ \text{m}^4$$

and about the new axis,

$$I = 0.060 \times (0.275 - 0.150)^2 + 450 \times 10^{-6} = 1387.5 \times 10^{-6}\ \text{m}^4$$

The moment of inertia of the cross member about its centroidal axis,

$$I = 0.300 \times 0.200^3/12 = 200 \times 10^{-6}\ \text{m}^4$$

and about the new axis,

$$I = (0.125^2 \times 0.060) + 200 \times 10^{-6} = 1137.5\ 10^{-6}\ \text{m}^4$$

The total moment of inertia of the composite beam about its centroidal axis,

$$I = (1387.5 \times 10^{-6}\ \text{m}^4) + (1137.5 \times 10^{-6}\ \text{m}^4) = 2525 \times 10^{-6}\ \text{m}^4$$

Example 10.20

What would be the moment of inertia of an aluminum box-beam shown in Exhibit 13? What would be the maximum deflection of such a beam if it were uniformly loaded cantilever beam 3.00 m long, with a distributed loading of 100 kg/m?

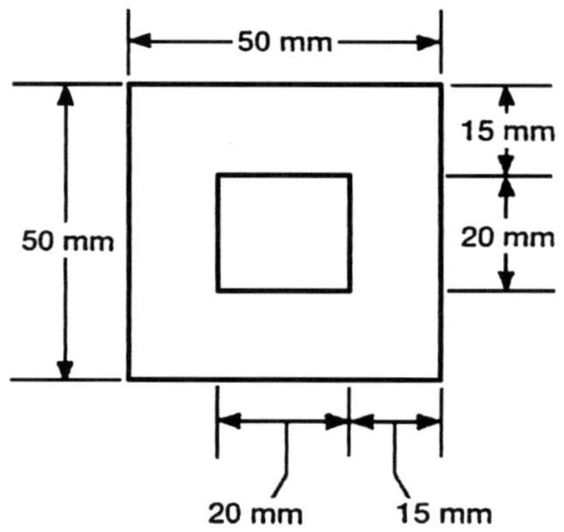

Exhibit 13

Solution

$I = bh^3/12$, which reduces to $b^4/12$ for a square cross section. For this case, the moment of inertia of the section would equal the I of the 50-mm square minus the I of the 20-mm core section.

$$I = (0.050^4 - 0.020^4)/12 = 5.075 \times 10^{-7}\ \text{m}^4$$

The loading would be equal to $100 \times 9.8066 = 980.66$ N/m.

The formula for the maximum deflection of a uniformly loaded cantilever beam as given in the *FE Handbook* is

$$\delta_{max} = w_0 \times L^4/(8EI)$$
$$\delta_{max} = (\text{N/m}) \times L^4/8EI = 980.66 \times 81/(8 \times 5.075 \times 10^{-7} \times 6.9 \times 10^{10})$$

where

$w_0 = 980.66$ N/m
$\delta = 0.284$ m, or 28.4 cm

What would be the maximum stress in the beam?

$$\delta = Mc/I = [L \times L/2]c/I$$
$$M = 980.66\ L^2/2$$
$$\delta = (4413)(0.025)/(5.075 \times 10^{-7}) = 217.4 \text{ Pa}$$

Example **10.21**

What concentrated load at the end of the cantilever beam in Example 10.20 would produce the same deflection?

Solution

From the *FE Handbook* the deflection of the end of an end-loaded cantilever beam equals

$$\delta = PL^3/(3EI)$$

and for a uniformly loaded beam

$$\delta = L^4/8EI, \text{ so}$$
$$P = (3EI/L^3)(L^4/8EI) = (3/8)L = (0.375)(980.66)(3.0) = 1103 \text{ N}$$

or a concentrated load of 113 kg as opposed to a uniformly distributed load of 300 kg.

SHEAR STRESSES IN BEAMS

A loaded beam will strain in some fashion. When it does, it flexes and the longitudinal fibers are placed in tension or compression every place except at the neutral axis. The effect of flexure on a simple beam is shown in Figure 10.4. If the beam were made of two parallel pieces, originally of the same length which were placed in contact, but not fixed to one another, the ends of the top member would overlap the ends of the bottom member as shown in Figure 10.5. The bottom-most fibers of the top member would strain in tension and elongate. The uppermost fibers of the bottom member would strain in compression and reduce in length. If the two members were joined together in such a manner that the differential strain were prevented, there would be a definite longitudinal shearing stress at the junction, the neutral axis, of the beam shown in Figure 10.5.

The longitudinal shear stress is related to the transverse (vertical in this case) shear stress by the relationship

$$S_s = VQ/(I \times t) \text{ Pa} \quad \text{or,}$$
$$\text{Shear flow} = VQ/I \text{ N/m}$$

where

V = vertical shear load at the point being considered.
$Q = \int fy\, dA = yA$

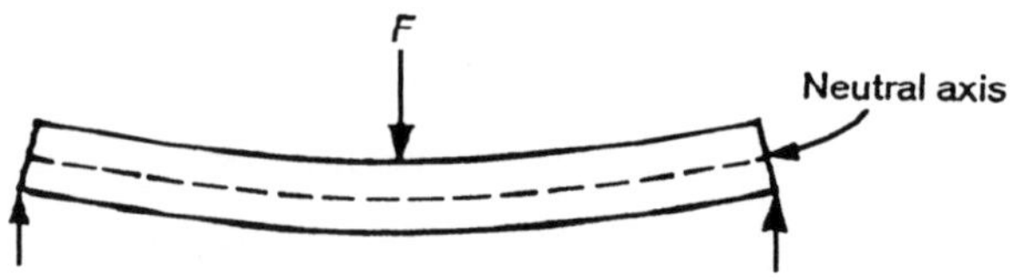

Figure 10.4

Figure 10.5

where

y = the distance from the neutral axis of the beam to the neutral axis of area A

A = the area of the section above the plane at which the shear stress acts

I = transverse moment of inertia

t = length (width) of section being considered.

Example 10.22

Take, for example, a rectangular cross-section beam as shown in Exhibit 14. What is the maximum longitudinal shear in the beam?

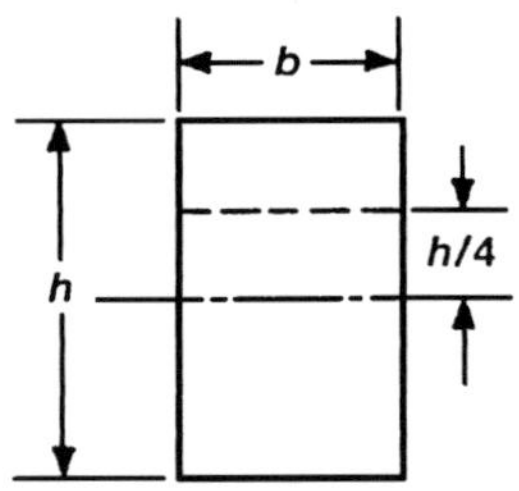

Exhibit 14

Solution

The longitudinal shear stress will be the greatest at the neutral axis. Using the above relationship for longitudinal shear,

$$Q = yA = (h/4) \times (bh/2) = bh^2/8$$
$$I = bh^3/12$$
$$t = b$$
$$S_s = VQ/It = V \times (bh^2/8)/[(bh^3/12) \times b] = 3V/2bh$$

The transverse shear stress equals V/bh, so the longitudinal shear stress at the neutral axis is 50 percent greater than the transverse shear stress for a rectangular beam.

For a circular beam, see Exhibit 15.

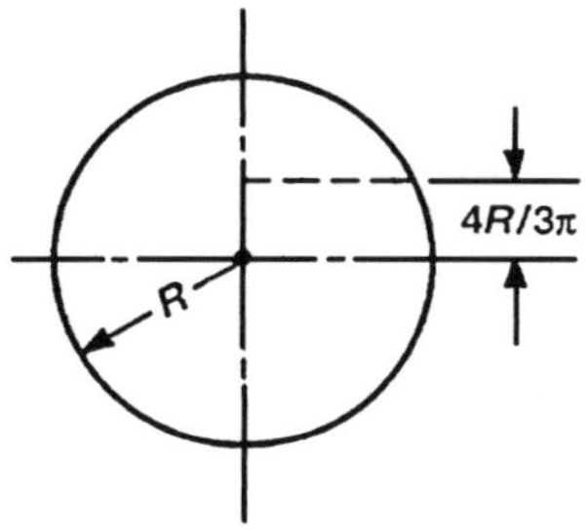

Exhibit 15

$$Q = yA = (4\pi/3\pi) \times (\pi r^2/2) = 2r^3/3$$
$$I = \pi r^4/4$$
$$t = 2r$$
$$S_s = VQ/It = V \times (2r^3/3)/[(\pi r^4/4) \times 2r] = 4V/(3\pi r^2)$$

So, the longitudinal shear at the neutral axis of a beam with a circular cross section is one-third greater than the transverse shear.

Example **10.23**

Two wooden planks 5 cm by 15 cm by 2.0 meters long are layered as shown in Figure 10.5. They support a mass 28 kg at a point midway between the end supports. It is desired to stiffen the beam since the assembly sags too much in the center, so the two planks are to be fastened together with nails. If one nail is capable of withstanding a shearing force of 650 newtons, what should be the spacing of the nails?

Solution

First determine the transverse shear stress. Since it is a simply supported beam the reactions at each end will hold half the mass, or the force at each end will equal $14g = 137.4$ N. The transverse shear stress of the assembled beam will equal $137.4/(0.10)(0.15) = 9153$ Pa. The shear stress at the neutral axis of the beam will thus equal $(1.50)(9153) = 13{,}730$ Pa. The assembled beam is 0.15 meters wide, so the longitudinal shear force will equal $(0.15)(13{,}730) = 2060$ N per meter of length. This means that $2060/650 = 3.17$ nails will be required for every meter of beam length, or one nail every 31.5 cm.

COMPOSITE BEAMS

The beams so far considered have all been of one homogeneous material. There are many beams, however, that are made of two materials, principally reinforced-concrete beams and steel-reinforced wood beams. Such beams are called "composite beams"; because of the difference in the moduli of elasticity of the two materials, they require a slightly different type of analysis than that used for homogeneous beams. One type of composite beam is shown in Figure 10.6. It is made up of a 15- by 20-cm timber, 3.0 meters in length, with a 15-cm-wide strip of steel 3.2 mm thick attached to the bottom edge. To withstand the applied load, the beam will deflect, the upper part of the wooden portion will compress, and the lower part of the wood and the steel will strain in tension. The stress in any part of the beam will equal the unit deformation in that particular portion times the modulus of elasticity of the material of which that portion is made. It is assumed that the steel will strain the same amount as the assumed wood fibers on the bottom of the beam. Then the steel will be stressed in the ratio E_s/E_w times the stress in the assumed bottom-most wood fibers. Since $E_s = 2.1 \times 10^{11}$ Pa and $E_w = 7.0 \times 10^9$, the stress in the steel will be thirty times the stress in the equivalent bottom wood fibers. This means that the tensile load in the steel will be considerably greater than that in the wood, and the neutral axis will be below the geometric center of the beam.

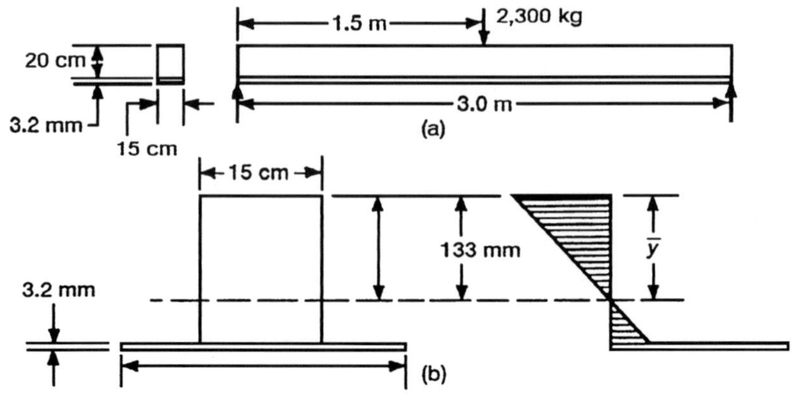

Figure 10.6

Another way of looking at this problem is that since, Force = stress × area and Stress = $E \times \varepsilon$, the steel strip could, theoretically, be replaced by a piece of wood with an area equal to $E_s/E_w \times$ area of steel, or thirty times the area (width) of the steel, but with the same thickness. The equivalent wooden beam is shown in the figure, and the internal-force diagram showing the distribution of $S \times A$ over the cross section of the beam is also shown.

$$E_s/E_w = n$$

Example 10.24

What is the stress in the steel band shown in Figure 10.6?

Solution

To determine the stress in the, wood and in the steel, we need to know the distance to the neutral axis y. This can be determined by taking moments about a horizontal axis through the upper edge of the beam; one must remember that a slice of the equivalent beam section would balance on a fulcrum placed at the neutral axis (for an all-wood beam).

$$(15)(20)(10) + (0.32)(450)(20 + 0.16) = [300 + (450)(0.32)]y$$
$$y = 5903.04/444.0 = 13.295 \text{ cm}$$

$S = Mc/I$ where I is the moment of inertia of the equivalent bema about the neutral axis. Utilizing the parallel-axis theorem,

$$I = (0.15)(0.20)^3/12 + 0.300(0.13295 - 0.100)^2 + (4.50)(0.0032)(0.2016 - 0.1330)^2$$
$$I = 0.000100 + 0.00003257 + 0.00006777 = 0.00020034 \text{ m}^4$$

$$\text{Maximum moment, } M = (2300g/2)(3/2) = 16.917 \text{ kJ}$$
$$\text{For steel strip, } c = 0.2030 - 0.1330 = 0.0702 \text{ m}$$

The tensile stress in the bottom fibers of the equivalent wooden beam would equal

$$S = (16{,}917)(0.0702)/0.000200 = 5.938 \text{ MPa}$$

and the stress in the steel would thus be equal to

$$S = 30 \times 5.938 = 178 \text{ MPa}$$
$$n = 210/7 = 30$$

The commonest type of composite beam is the reinforced-concrete beam, which can also be handled by the method described. One difference, however, is that the concrete is assumed to withstand no tensile stress and the portion of the concrete on the tension side of the neutral axis is disregarded in the stress calculations. A reinforced-concrete beam is shown in Figure 10.7.

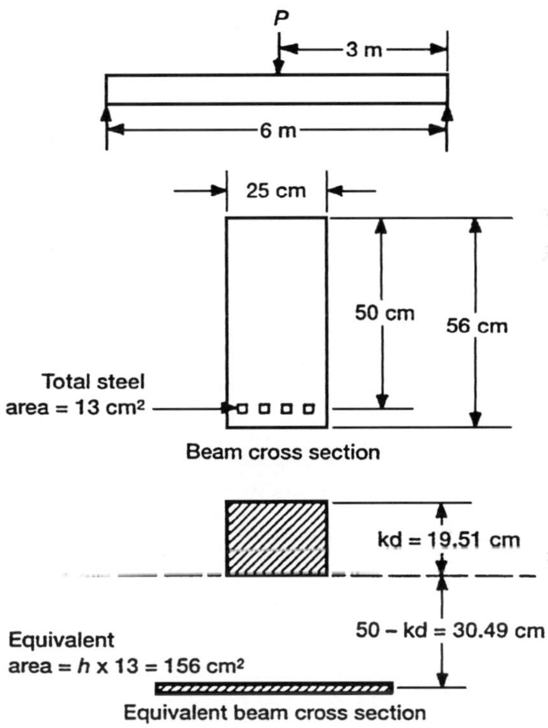

Figure 10.7

Example **10.25**

Determine the allowable concentrated load, P, which the reinforced concrete beam show in Figure 10.7 will hold if $n = 12$ and the maximum allowable stresses are 6.90 MPa in compression for the concrete and 138 MPa in tension for the steel.

$$n = E_{steel}/E_{concrete}$$

Solution

The equivalent beam cross section is drawn; as the concrete will take only compressive loading, the effective concrete area will extend only to the neutral axis. The location of the neutral axis is found by taking moments about the neutral axis. The concrete section is shown shaded in the figure. It is assumed that the portion of the concrete that will be stressed in compression will occupy the top "kd" distance of the beam with an area of $kd \times 25$ cm^2. The moment of this area about the neutral axis must equal the moment of the "tensile" concrete ($n \times$ the area of the steel reinforcing bars) about the neutral axis.

$$25kd(kd/2) = 156(50 - kd)$$

where 156 equals the area of the steel reinforcement times n, or

$$12 \times 13 = 156$$
$$(kd)^2 + 12.48\ kd - 624 = 0$$

Using the general solution for a quadratic equation in the *FE Handbook* gives

$$kd = \left[-12.48 \pm \sqrt{(156 + 2{,}496)}\right]/2$$
$$kd = (-12.48 + 51.50)/2 = 19.51 \text{ cm}$$
$$I = (25 \times 19.51^3)/12 + (25 \times 19.51) \times (19.51/2)^2 + 156 \times (50 - 19.51)^2$$
$$= 206{,}909 \text{ cm}^4 = 2.07 \times 10^{-3} \text{ m}^4$$

Note that the moment of inertia of the steel portion about its own neutral axis is disregarded since it is so small compared with the other components.

$$S = MC/I$$
$$M_{\text{concrete}} = 6.90 \times 10^6 \times .00207/0.1951 = 73{,}209 \text{ Nm}$$
$$M = (P/2) \times 3 \text{ so,}$$
$$P = 2M/3$$

P, as limited by compressive stress in the concrete, is equal to 48,810 N or 4977 kg.

$$M_{\text{steel}} = 138 \times 10^6 \times 0.00207/(n \times 0.3049) = 78{,}075 \text{ Nm}$$
$$P = 52{,}050 \text{ N, or } 5307 \text{ kg}$$

Note the use of the n in the denominator for determining the allowable moment for the steel.

The concentrated load, P, is limited by the stress in the concrete and is equal to 4977 kg.

RADIUS OF GYRATION

The moment of inertia is defined as $\int r^2 dm$ or $\int r^2 dA$ and is discussed in Chapter 2. If all the mass, or all the area, were concentrated at a distance "r" from the point about which the moment of inertia was to be calculated, then "I" would become just r^2m or r^2A and "r" would equal the radius of gyration. The radius of gyration is usually indicated by "k" rather than "r".

$$I = k^2m, \text{ or}$$
$$I = k^2A$$

For a circular cross section the radius of gyration equals $D/4$.

$$I = \pi D^4/64 = AD^2/16$$

and, since $I = k^2A$ where k = radius of gyration, $k = D/4$.

It should be noted that, in general, the mass of a body cannot be considered as acting at its center of mass for the purpose of calculating the moment of inertia. This is discussed in Chapter 2.

The moment of inertia about an axis in the plane of an area is called the "plane moment of inertia"; the moment of inertia of an area about an axis perpendicular to the area is termed the "polar moment of inertia" which is discussed in Chapter 2.

COLUMNS

A column with a high slenderness ratio, *L*/*r*, length divided by the radius of gyration of the column, is prone to buckling, whereas a short column can be treated by normal methods. If the *L*/*r* ratio is greater than 30 it is safer to treat the column as a "long column" and use Euler's Formula to determine whether or not the loading is critical.

Example 10.26

A 50-mm diameter, 3.00 m long, steel rod is pinned at both ends and is unstressed at 20 °C. What is the highest temperature to which the bar may be heated before it will buckle? The thermal coefficient of expansion of steel is $11.7 \times 10^{-6}/°C$.

Solution

The radius of gyration equals 0.050/4 = 0.0125 and the *L*/*R* ratio equals 3.00/0.0125 = 240 so it is a "long column" and should be treated by Euler's formula, which, from the *FE Handbook* is

$$P_{cr} = \pi^2 EI/(kL)^2, \text{ and}$$
$$P_{cr}/A = \pi^2 E/[k(L/r)]^2$$

The value for the constant *k* from the table in the *FE Handbook* is equal to 1.00.

$$L/r = 240.0$$
$$P_{cr}/A = 9.870 \times 2.1 \times 10^{11}/(1 \times 240.0)^2 = 35.98 \text{ MPa}$$

The compressive stress induced in the steel rod due to an increase in temperature will equal

$$S = \Delta T(11.7 \times 10^{-6})(2.1 \times 10^{11}) \text{ Pa, so}$$
$$\Delta T = 35.98/2.457 = 14.64 \text{ °C, and}$$
$$T = 34.64 \text{ °C}$$

RIVETED JOINTS

Riveted joints can be generally divided into two general classes—joints in which the line of action of the applied force passes through the centroid of the rivet pattern and those in which the force exerts a moment on the rivet pattern. In a simple continuous joint such as is shown in Figure 10.8 the load is assumed to be distributed equally among the rivets making up the joint. In the sample shown one-quarter of the load would be assumed to be held by each rivet. The small moment force due to the displacement of the lapping plates is ignored. However, if the load is offset from a line of action passing through the centroid of the rivet pattern as shown in Figure 10.9, the force due to the effect of the moment must also be included.

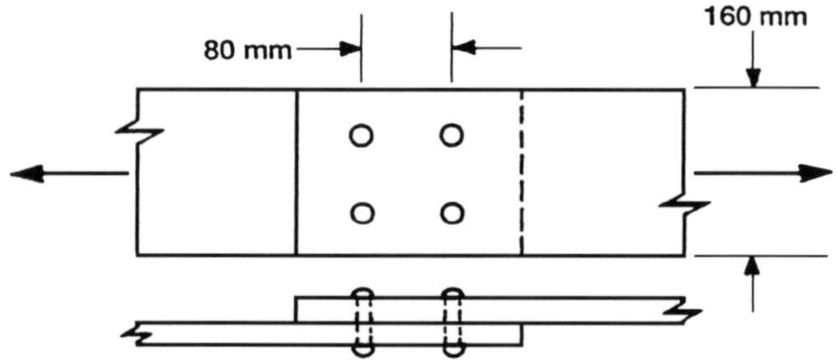

Figure 10.8

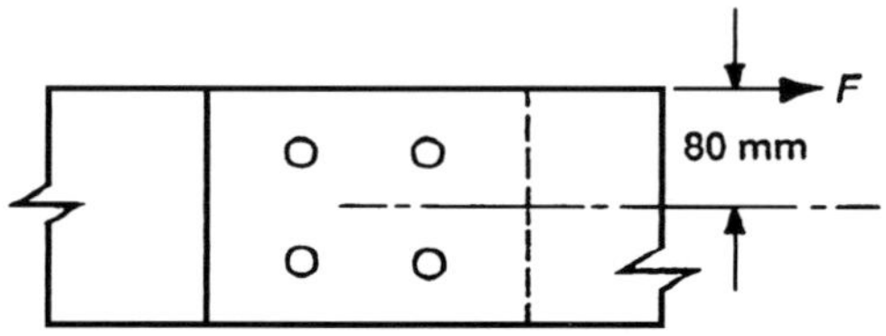

Figure 10.9

Example 10.27

The rivets in Figure 10.9 are 12.0 mm in diameter and form a square pattern with sides of 80 mm located centrally in the 160 mm wide plates. The 80 mm is measured between the centers of the rivets at the corners of the square. The load is shown applied at the edge of the 160 mm wide plate. This load will apply a moment load to the rivet pattern equal to $F \times 0.080$ which will be resisted equally by each of the rivets. Each rivet is $0.040 \times \sqrt{2} = 0.05657$ m from the centroid of the rivet pattern. The resisting force required to counteract the moment would equal

$$0.080 \times F/(4)(0.05657) = 0.3535\ F \text{ each rivet}$$

this force would add geometrically to the direct force of $F/4$ applied to each rivet (see Exhibit 16). This would increase the force on two of the rivets and reduce the force on two others. The maximum force would equal (Exhibit 17)

$$([(0.250 + 0.707(0.3535)]^2 + [(0.707)(0.3535)]^2)^{1/2} = 0.559\ F$$

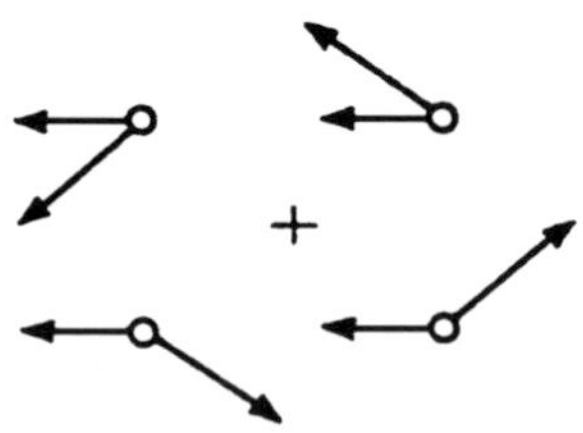

Exhibit 16

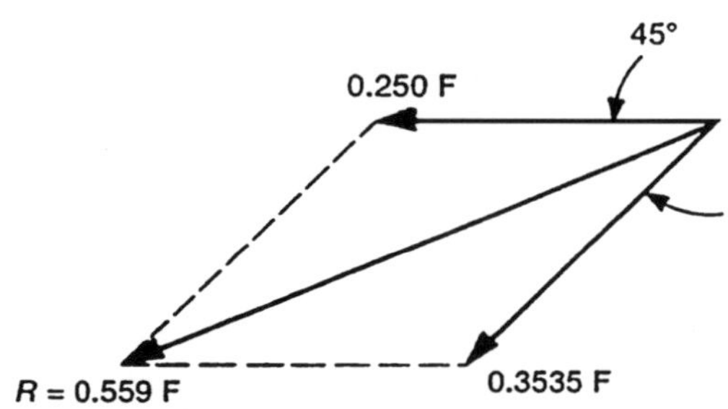

Exhibit 17

The eccentricity of the load application would more than double the maximum stress in the highest stressed rivet in this particular case.

WELDED JOINTS

The size of a weld is generally specified by the size of the fillet. For the joint shown in Figure 10.10 the fillet is given as 6.0 mm. The shear area, however, is the thickness of the throat of the fillet or, for this case

$$A = 6 \times 0.707 \times \text{length}$$

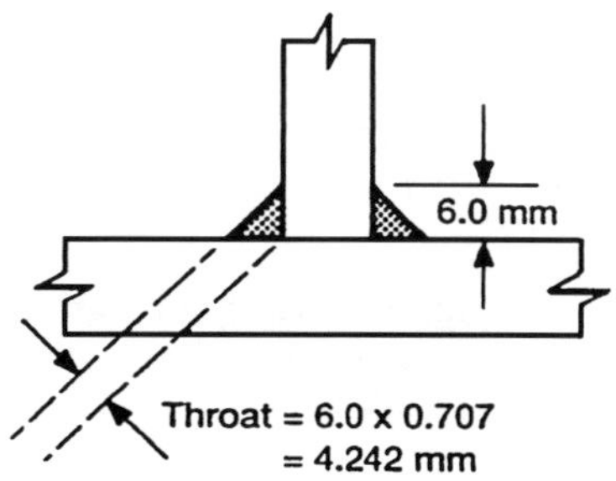

Figure 10.10

Example 10.28

For the joint shown in Figure 10.10, what is the maximum force which could be applied to the vertical member per centimeter of width if the allowable shear stress of the weld material is 100 MPa?

Solution

For each centimeter length of the joint there would be

$$2 \times 0.006 \times 0.707 \times 0.010 = 84.84 \times 10^{-6}\ \text{m}^2 \text{ of weld area}$$
$$F = 84.84 \times 100.00 = 8{,}480\ \text{N}$$

MECHANICS OF MATERIALS

Mechanics of materials deals with the determination of the internal forces (stresses) and the deformation of solids such as metals, wood, concrete, plastics, and composites. In mechanics of materials there are three main considerations in the solution of problems:

1. Static equilibrium
2. Force-deformation relations
3. Compatibility

Equilibrium refers to the equilibrium of forces. For the purposes of this chapter it is assumed that the system is in static equilibrium (i.e., not moving). The laws of statics must hold for the body and all parts of the body. *Force-deformation relations* refer to the relation of the applied forces to the deformation of the body. If certain forces are applied, then certain deformations will result. *Compatibility* refers to the compatibility of deformation. Upon loading, the parts of a body or structure must not come apart. These three principles will be emphasized throughout.

AXIALLY LOADED MEMBERS

If a force P is applied to a member as shown in Figure 10.11(a), then a short distance away from the point of application the force becomes uniformly distributed over the area as shown in Figure 10.11(b). The force per unit area is called the axial, or normal, stress and is given the symbol σ. Thus,

$$\sigma = \frac{P}{A}$$

The original length between two points A and B is L as shown in Figure 10.11(c). Upon application of the load P, the length L grows by an amount ΔL. The final length is $L + \Delta L$ as shown in Figure 10.11(d). A quantity measuring the intensity of deformation and being independent of the original length L is the strain ε, defined as

$$\varepsilon = \frac{\Delta L}{L} = \frac{\delta}{L}$$

where ΔL is denoted as δ.

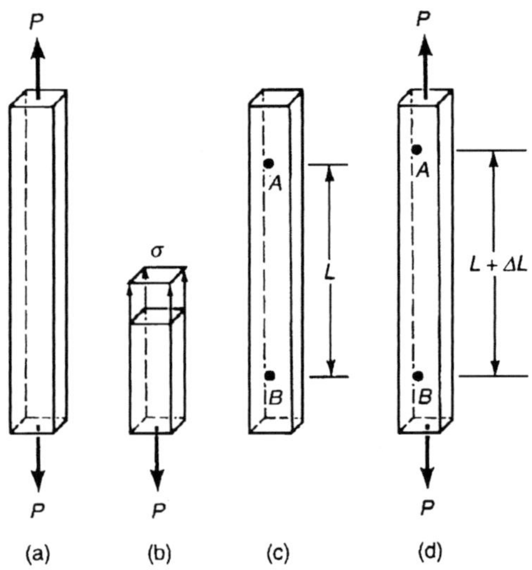

Figure 10.11 Axial member under force P

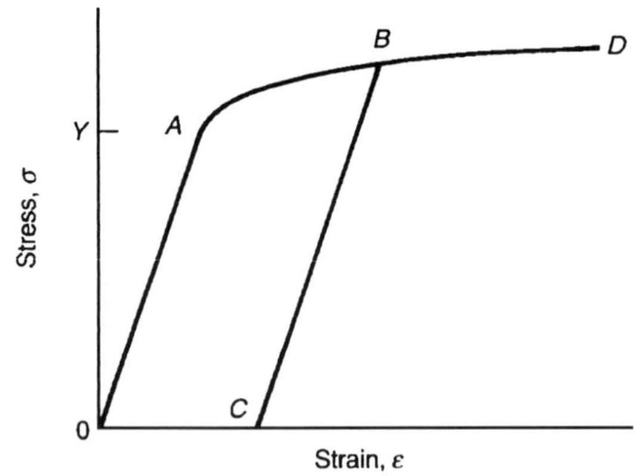

Figure 10.12 Stress-strain curve for a typical material

The relationship between stress and strain is determined experimentally. A typical plot of stress versus strain is shown in Figure 10.12. On initial loading, the plot is a straight line until the material reaches yield at a stress of Y. It is noted that in this example the yield strength point A is also the proportional limit (i.e., the maximum stress at which the stress-strain curve remains linear). If the stress remains less than yield, then subsequent loading and reloading continues along that same straight line. If the material is allowed to go beyond yield, then during an increase in the load the curve goes from A to D. If unloading occurs at some point B, for example, then the material unloads along the line BC, which has approximately the same slope as the original straight line from 0 to A. Reloading would occur along the line CB and then proceed along the line BD. It can be seen that if the material is allowed to go into the plastic region (A to D) it will have a permanent strain offset on unloading.

Modulus of Elasticity

The region of greatest concern is that below the yield point. The slope of the line between 0 and A is called the modulus of elasticity and is given the symbol E, so

$$\sigma = E\varepsilon$$

This is Hooke's law for axial loading; a more general form will be considered in a later section. The modulus of elasticity is a function of the material alone and not a function of the shape or size of the axial member.

The relation of the applied force in a member to its axial deformation can be found by inserting the definitions of the axial stress and the axial strain into Hooke's law, which gives

$$\frac{P}{A} = E\frac{\delta}{L}$$

or

$$\delta = \frac{PL}{AE}$$

In the examples that follow, wherever it is appropriate, the three steps of (1) static equilibrium, (2) force-deformation, and (3) compatibility will be explicitly stated.

Example 10.29

The steel rod shown in Exhibit 18 is fixed to a wall at its left end. It has two applied forces. The 3 kN force is applied at the Point B and the 1 kN force is applied at the Point C. The area of the rod between A and B is $A_{AB} = 1000\ \text{mm}^2$, and the area of the rod between B and C is $A_{BC} = 500\ \text{mm}^2$. Take $E = 210$ GPa. Find (a) the stress in each section of the rod and (b) the horizontal displacement at the points B and C.

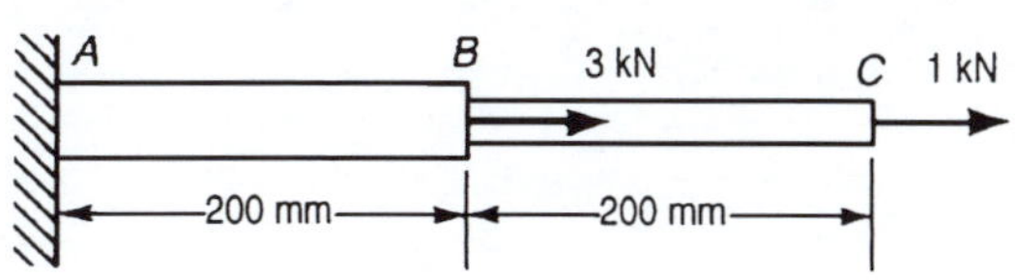

Exhibit 18

Solution—Static Equilibrium

Draw free-body diagrams for each section of the rod (Exhibit 19). From a summation of forces on the member BC, $F_{BC} = 1$ kN. Summing forces in the horizontal direction on the center free-body diagram, $F_{BA} = 3 + 1 = 4$ kN. Summing forces on the left free-body diagram gives $F_{AB} = F_{BA} = 4$ kN. The stresses then are

$$\sigma_{AB} = 4 \text{ kN}/1000 \text{ mm}^2 = 4 \text{ MPa}$$
$$\sigma_{BC} = 1 \text{ kN}/500 \text{ mm}^2 = 2 \text{ MPa}$$

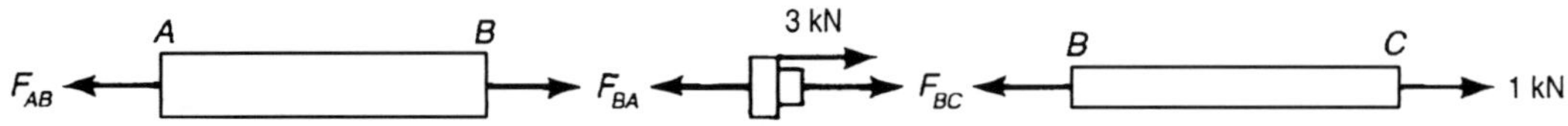

Exhibit 19

Solution—Force-Deformation

$$\delta_{AB} = \left(\frac{PL}{AE}\right)_{AB} = \frac{(4 \text{ kN})(200 \text{ mm})}{(1000 \text{ mm}^2)(210 \text{ GPa})} = 0.00381 \text{ mm}$$

$$\delta_{BC} = \left(\frac{PL}{AE}\right)_{BC} = \frac{(1 \text{ kN})(200 \text{ mm})}{(500 \text{ mm}^2)(210 \text{ GPa})} = 0.001905 \text{ mm}$$

Solution—Compatibility

Draw the body before loading and after loading (Exhibit 20).

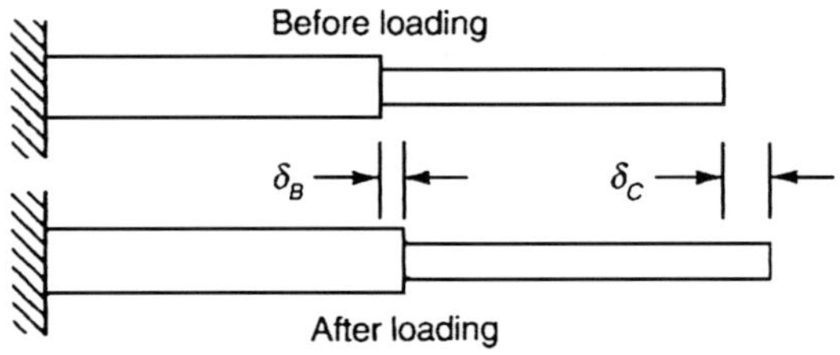

Exhibit 20

It is then obvious that

$$\delta_B = \delta_{AB} = 0.00381 \text{ mm}$$

$$\delta_C = \delta_{AB} + \delta_{BC} = 0.00381 + 0.001905 = 0.00571 \text{ mm}$$

In this first example the problem was statically determinate, and the three steps of static equilibrium, force-deformation, and compatibility were independent steps. The steps are not independent when the problem is statically indeterminate, as the next example will show.

Example 10.30

Consider the same steel rod as in Example 10.29 except that now the right end is fixed to a wall as well as the left (Exhibit 21). It is assumed that the rod is built into the walls before the load is applied. Find (a) the stress in each section of the rod, and (b) the horizontal displacement at the point B.

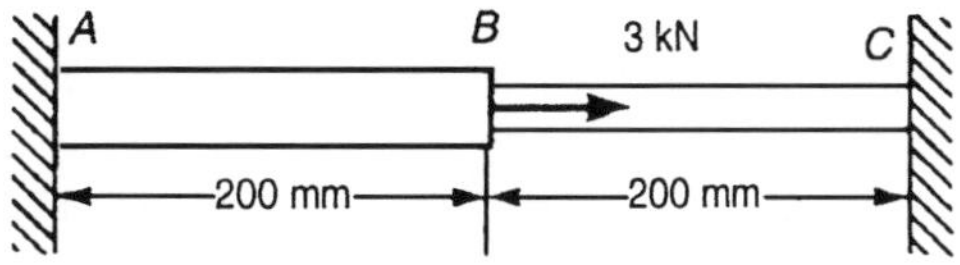

Exhibit 21

Solution—Equilibrium

Draw free-body diagrams for each section of the rod (Exhibit 22). Summing forces in the horizontal direction on the center free-body diagram:

$$-F_{AB} + F_{BC} + 3 = 0$$

It can be seen that the forces cannot be determined by statics alone so that the other steps must be completed before the stresses in the rods can be determined.

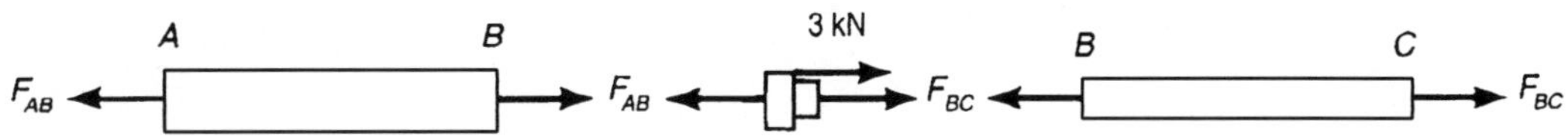

Exhibit 22

Solution—Force-Deformation

$$\delta_{AB} = \left(\frac{PL}{AE}\right)_{AB} = \frac{F_{AB}L}{A_{AB}E}$$

$$\delta_{BC} = \left(\frac{PL}{AE}\right)_{BC} = \frac{F_{BC}L}{A_{BC}}$$

The static equilibrium, force-deformation, and compatibility equations can now be solved as follows (see Exhibit 23). The force-deformation relations are put into the compatibility equations (recall: $A_{AB} = 2A_{BC}$):

$$\frac{F_{AB}L}{2A_{BC}E} = -\frac{F_{BC}L}{A_{BC}E}$$

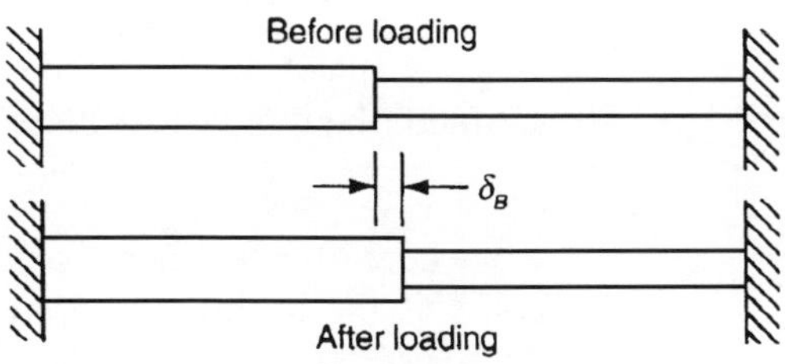

Exhibit 23

Then, $F_{AB} = -2F_{BC}$. Insert this relationship into the equilibrium equation:

$$-F_{AB} + F_{BC} + 3 = 0 = 2F_{BC} + F_{BC} + 3; \quad F_{BC} = -1 \text{ kN and } F_{AB} = 2 \text{ kN}$$

The stresses are

$$\sigma_{AB} = 2 \text{ kN/1000 mm}^2 = 2 \text{ MPa (tension)}$$
$$\sigma_{BC} = -1 \text{ kN/500 mm}^2 = -2 \text{ MPa (compression)}$$

The displacement at B is

$$\delta_A = \delta_{AB} = F_{AB}L/(AE) = (2 \text{ kN})(200 \text{ mm})/[(1000 \text{ mm}^2)(210 \text{ GPa})] = 0.001905 \text{ mm}$$

Poisson's Ratio

The axial member shown in Figure 10.11 also has a strain in the lateral direction. If the rod is in tension, then stretching takes place in the axial or longitudinal direction while contraction takes place in the lateral direction. The ratio of the magnitude of the lateral strain to the magnitude of the longitudinal strain is called Poisson's ratio ν.

$$\nu = -\frac{\text{Lateral strain}}{\text{Longitudinal strain}}$$

Poisson's ratio is a dimensionless material property that never exceeds 0.5. Typical values for steel, aluminum, and copper are 0.30, 0.33, and 0.34, respectively.

Example **10.31**

A circular aluminum rod 10 mm in diameter is loaded with an axial force of 2 kN. What is the decrease in diameter of the rod? Take $E = 70$ GN/m^2 and $\nu = 0.33$.

Solution

The stress is $\sigma = P/A = 2 \text{ kN}/(\pi 5^2 \text{ mm}^2) = 0.0255 \text{ GN/m}^2 = 25.5 \text{ MN/m}^2$.

The longitudinal strain is $\varepsilon_{\text{lon}} = \sigma/\text{E} = (25.5 \text{ MN/m}^2)/(70 \text{ GN/m}) = 0.000364$.

The lateral strain is $\varepsilon_{\text{lat}} = -\nu\ \varepsilon_{\text{lon}} = -0.33(0.000364) = -0.000120$.

The decrease in diameter is then $-\text{D}\ \varepsilon_{\text{lat}} = -(10 \text{ mm})(-0.000120) = 0.00120 \text{ mm}$.

Thermal Deformations

When a material is heated, thermal strains are created. If the material is unrestrained, the thermal strain is

$$\varepsilon_t = \alpha(t - t_0)$$

where α is the linear coefficient of thermal expansion, t is the final temperature, and t_0 is the initial temperature. Since strain is dimensionless, the units of α are °F^{-1} or °C^{-1} (sometimes the units are given as in/in/°F or m/m/°C, which amounts to the same thing). The total strain ε_T is equal to the strain from the applied loads plus the thermal strain. For problems where the load is purely axial, this becomes

$$\varepsilon_T = \frac{\sigma}{E} + \alpha(t - t_0)$$

The deformation δ is found by multiplying the strain by the length L:

$$\delta = \frac{PL}{AE} + \alpha L(t - t_0)$$

Example 10.32

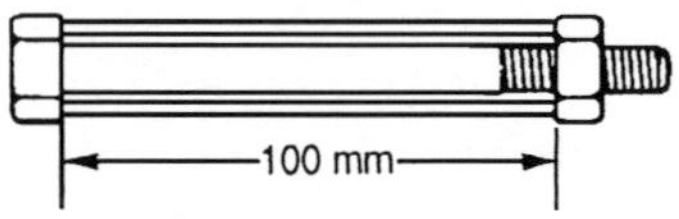

Exhibit 24

A steel bolt is put through an aluminum tube as shown in Exhibit 24. The nut is threaded on the bolt until it just makes contact with the tube (i.e., no torque). The temperature of the entire assembly is then raised by 60°C. Because the coefficient of thermal expansion of aluminum is greater than that of steel, the bolt will be put in tension and the tube in compression. Find the force in the bolt and the tube. For the steel bolt, take $E = 210$ GPa, $\alpha = 12 \times 10^{-6}\ °C^{-1}$ and $A = 32\ mm^2$. For the aluminum tube, take $E = 69$ GPa, $\alpha = 23 \times 10^{-6}\ °C^{-1}$ and $A = 64\ mm^2$.

Solution—Equilibrium

Draw free-body diagrams (Exhibit 25). From equilibrium of the bolt it can be seen that $P_B - P_T = 0$.

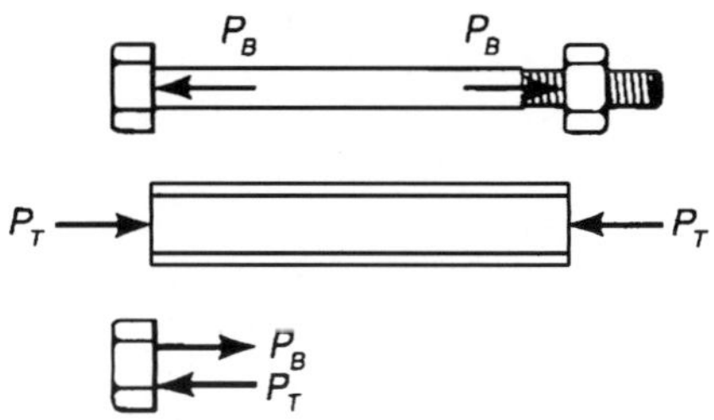

Exhibit 25

Solution—Force-Deformation

Note that both members have the same length and the same force magnitude, P.

$$\delta_B = \frac{PL}{A_B E_B} + \alpha_B L(t - t_0)$$

$$\delta_T = -\frac{PL}{A_T E_T} + \alpha_T L(t - t_0)$$

The minus sign in the second expression occurs because the tube is in compression.

Solution—Compatibility

The tube and bolt must both expand the same amount; therefore,

$$\delta_B = \delta_T$$

$$\delta_B = \delta_T = \frac{P \times (100 \text{ mm})}{32 \text{ mm}^2 \times 210 \text{ GPa}} + 12 \times 10^{-6} \frac{1}{°\text{C}} \times 100 \text{ mm} \times 60 \text{ °C}$$

$$= -\frac{P \times (100 \text{ mm})}{64 \text{ mm}^2 \times 69 \text{ GPa}} + 23 \times 10^{-6} \frac{1}{°\text{C}} \times 100 \text{ mm} \times 60 \text{ °C}$$

Solving for P gives $P = 1.759$ kN.

Variable Load

In certain cases the load in the member will not be constant but will be a continuous function of the length. These cases occur when there is a distributed load on the member. Such distributed loads most commonly occur when the member is subjected to gravitation, acceleration, or magnetic fields. In such cases, $\delta = \frac{PL}{AE}$ holds only over an infinitesimally small length $L = dx$. $\delta = \frac{PL}{AE}$ then becomes

$$d\delta = \frac{P(x)}{AE} dx$$

or equivalently

$$\delta = \int_0^L \frac{P(x)}{AE} dx$$

Example **10.33**

An aluminum rod is hanging from one end. The rod is 1 m long and has a square cross-section 20 mm by 20 mm. Find the total extension of the rod resulting from its own weight. Take $E = 70$ GPa and the unit weight $\gamma = 27$ kN/m³.

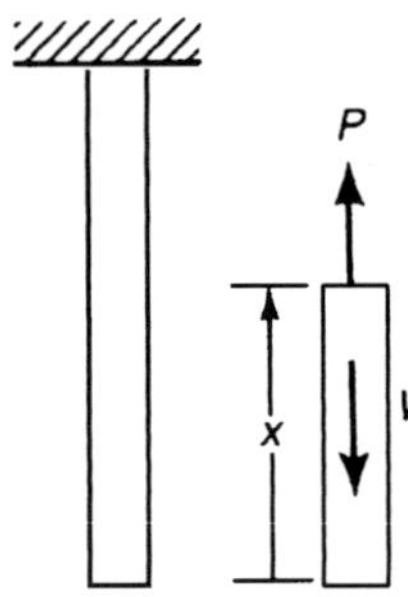

Exhibit 26

Solution—Equilibrium

Draw a free-body diagram (Exhibit 26). The weight of the section shown in Exhibit 26 is

$$W = \gamma V = \gamma Ax = P$$

which clearly yields P as a function of x, and gives

$$\delta = \int_0^L \frac{\gamma Ax}{AE} dx = \frac{\gamma}{E} \int_0^L x dx = \frac{\gamma L^2}{2E} \frac{\left(27 \frac{\text{kN}}{\text{m}^3}\right)(\text{lm})^2}{2\left(70 \frac{\text{GN}}{\text{m}^2}\right)} = 0.1929 \ \mu\text{m}$$

THIN-WALLED CYLINDER

Consider the thin-walled circular cylinder subjected to a uniform internal pressure q as shown in Figure 10.13. A section of length a is cut out of the vessel in (a). The cut-out portion is shown in (b). The pressure q can be considered as acting across

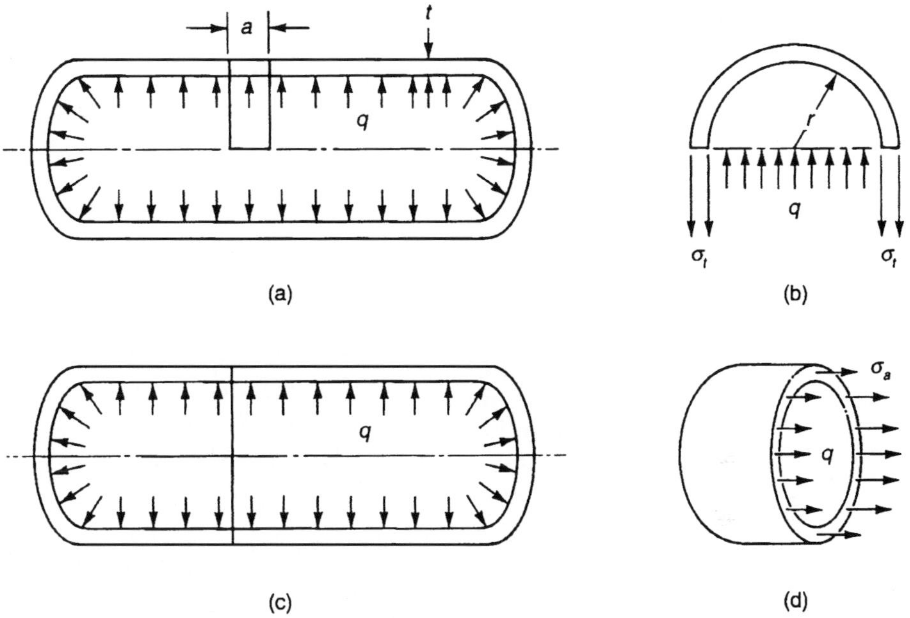

Figure 10.13

the diameter as shown. The tangential stress σ_t is assumed constant through the thickness. Summing forces in the vertical direction gives

$$qDa - 2\sigma_t\, ta = 0$$

$$\sigma_t = \frac{qD}{2t}$$

where D is the inner diameter of the cylinder and t is the wall thickness. The axial stress σ_a is also assumed to be uniform over the wall thickness. The axial stress can be found by making a cut through the cylinder as shown in (c). Consider the horizontal equilibrium for the free-body diagram shown in (d). The pressure q acts over the area πr^2 and the stress σ_a acts over the area πDt, which gives

$$\sigma_a \pi Dt = q\pi\left(\frac{D}{2}\right)^2$$

so

$$\sigma_a = \frac{qD}{4t}$$

Example 10.34

Consider a cylindrical pressure vessel with a wall thickness of 25 mm, an internal pressure of 1.4 MPa, and an outer diameter of 1.2 m. Find the axial and tangential stresses.

Solution

$$q = 1.4 \text{ MPa}; \; D = 1200 - 50 = 1150 \text{ mm}; \; t = 25 \text{ mm}$$

$$\sigma_t = \frac{qD}{2t} = \frac{1.4 \text{ MPa} \times 1150 \text{ mm}}{2 \times 25 \text{ mm}} = 32.2 \text{ MPa}$$

$$\sigma_a = \frac{qD}{4t} = \frac{1.4 \text{ MPa} \times 1150 \text{ mm}}{4 \times 25 \text{ mm}} = 16.1 \text{ MPa}$$

GENERAL STATE OF STRESS

Stress is defined as force per unit area acting on a certain area. Consider a body that is cut so that its area has an outward normal in the x direction as shown in Figure 10.14. The force ΔF that is acting over the area ΔA_x can be split into its components ΔF_x, ΔF_y, and ΔF_z. The stress components acting on this face are then defined as

$$\sigma_x = \lim_{\Delta A_x \to 0} \frac{\Delta F_x}{\Delta A_x}$$

$$\tau_{xy} = \lim_{\Delta A_x \to 0} \frac{\Delta F_x}{\Delta A_x}$$

$$\tau_{xz} = \lim_{\Delta A_x \to 0} \frac{\Delta F_z}{\Delta A_x}$$

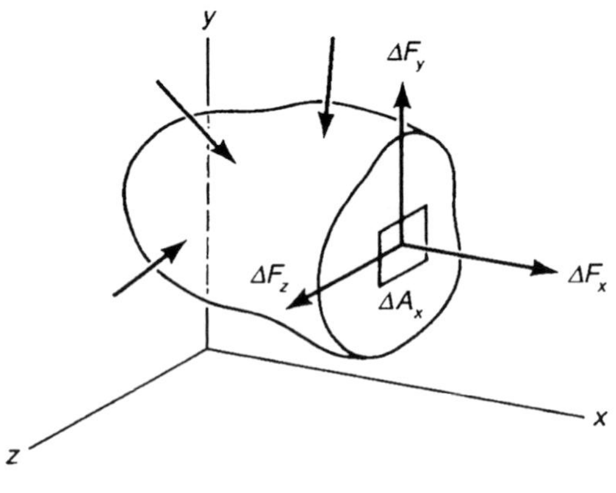

Figure 10.14 Stress on a face

The stress component σ_x is the normal stress. It acts normal to the x face in the x direction. The stress component τ_{xy} is a shear stress. It acts parallel to the x face in the y direction. The stress component τ_{xz} is also a shear stress and acts parallel to the x face in the z direction. For shear stress, the first subscript indicates the *face* on which it acts, and the second subscript indicates the *direction* in which it acts. For normal stress, the single subscript indicates both face and direction. In the general state of stress, there are normal and shear stresses on all faces of an element as shown in Figure 10.15.

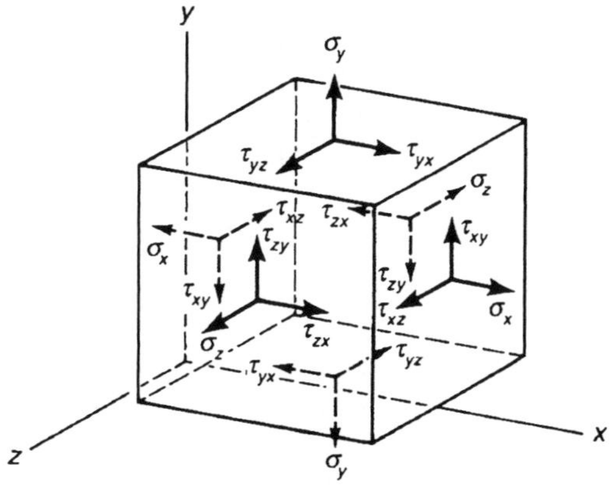

Figure 10.15 Stress at a point (shown in positive directions)

From equilibrium of moments around axes parallel to x, y, and z and passing through the center of the element in Figure 10.15, it can be shown that the following relations hold:

$$\tau_{xy} = \tau_{yx}; \qquad \tau_{yz} = \tau_{zy}; \qquad \tau_{zx} = \tau_{xz}$$

Thus, at any point in a body the state of stress is given by six components: $\sigma_x, \sigma_y, \sigma_z, \tau_{xy}, \tau_{yz}$, and τ_{zx}. The usual sign convention is to take the components shown in Figure 10.15 as positive. One way of saying this is that normal stresses are positive in tension. Shear stresses are positive on a positive face in the positive direction. A **positive face** is defined as a face with a positive outward normal.

PLANE STRESS

In elementary mechanics of materials, we usually deal with a state of plane stress in which only the stresses in the x-y plane are nonzero. The stress components σ_z, τ_{xz}, and τ_{yz} are taken as zero.

Mohr's Circle—Stress

In plane stress, the three components σ_x, σ_y, τ_{xy} define the state of stress at a point, but the components on any other face have different values. To find the components on an arbitrary face, consider equilibrium of the wedges shown in Figure 10.16.

Summation of forces in the x' and y' directions for the wedge shown in Figure 10.16(a) gives

$$\sum F_{x'} = 0 = \sigma_{x'}\Delta A - \sigma_x \Delta A(\cos\theta)^2 - \sigma_y \Delta A(\sin\theta)^2 - 2\tau_{xy}\Delta A \sin\theta \cos\theta$$

$$\sum F_{y'} = 0 = \tau_{x'y'}\Delta A + \left(\sigma_x - \sigma_y\right)\Delta A \sin\theta\cos\theta - \tau_{xy}\Delta A\left[\left(\cos\theta\right)^2 - \left(\sin\theta\right)^2\right]$$

Canceling ΔA from each of these expressions and using the double angle relations gives

$$\sigma_{x'} = \frac{\sigma_x + \sigma_y}{2} + \frac{\sigma_x - \sigma_y}{2}\cos 2\theta + \tau_{xy}\sin 2\theta$$

$$\tau_{x'y'} = -\frac{\sigma_x - \sigma_y}{2}\sin 2\theta + \tau_{xy}\cos 2\theta$$

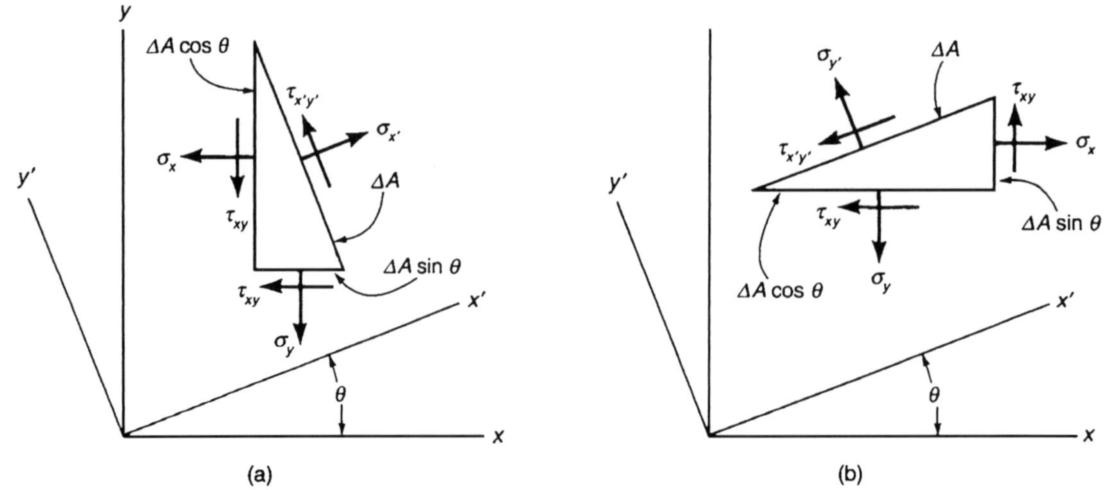

Figure 10.16 Stress on an arbitrary face

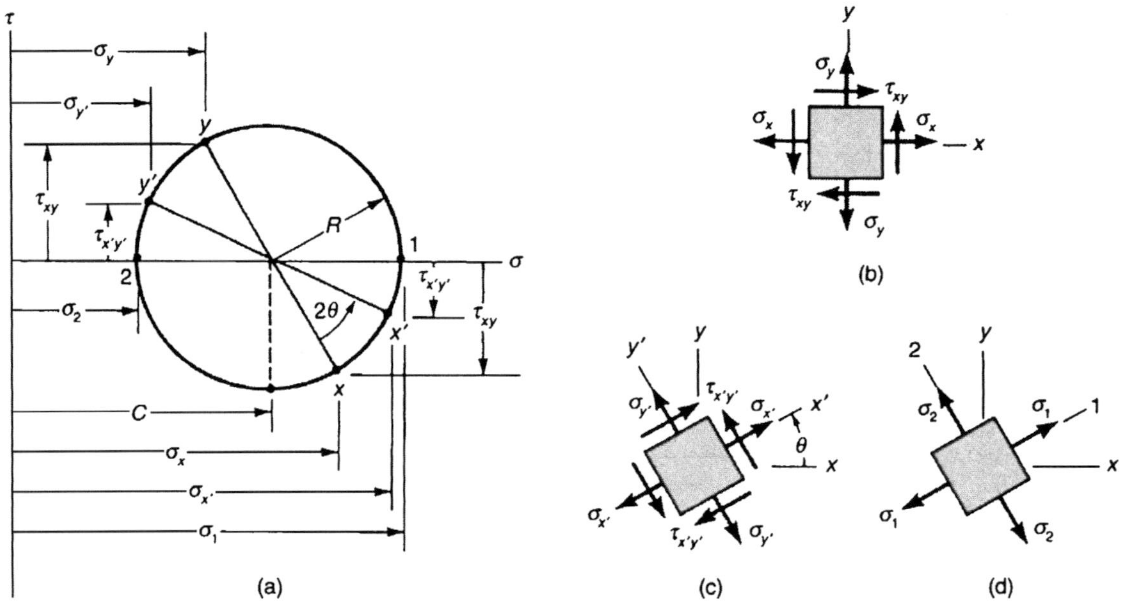

Figure 10.17 Mohr's circle for the stress at a point

Similarly, summation of forces in the y' direction for the wedge shown in Figure 10.16(b) gives

$$\sigma_{y'} = \frac{\sigma_x + \sigma_y}{2} - \frac{\sigma_x - \sigma_y}{2}\cos 2\theta - \tau_{xy}\sin 2\theta$$

The three previous equations are the parametric equations of Mohr's circle; Figure 10.17(a) shows the general Mohr's circle; Figure 10.17(b) shows the stress on the element in an x-y orientation; Figure 10.17(c) shows the stress in the same element in an x'-y' orientation; and Figure 10.17(d) shows the stress on the element in the 1-2 orientation. Notice that there is always an orientation (for example, a 1-2 orientation) for which the shear stress is zero. The normal stresses σ_1 and σ_2 on these 1-2 faces are the principal stresses, and the 1 and 2 axes are the principal axes of stress. In three-dimensional problems the same is true. There are always three mutually perpendicular faces on which there is no shear stress. Hence, there are always three principal stresses.

To draw Mohr's circle knowing σ_x, σ_y, and τ_{xy}:

1. Draw vertical lines corresponding to σ_x and σ_y as shown in Figure 10.18(a) according to the signs of σ_x and σ_y (to the right of the origin if positive and to the left if negative).
2. Put a point on the σ_x vertical line a distance τ_{xy} below the horizontal axis if τ_{xy} is positive (above if τ_{xy} is negative) as in Figure 10.8(a). Name this point x.
3. Put a point on the σ_y vertical line a distance τ_{xy} in the opposite direction as on the σ_x vertical line also as shown in Figure 10.18(a). Name this point y.
4. Connect the two points x and y, and draw the circle with diameter xy as shown in Figure 10.18(b).

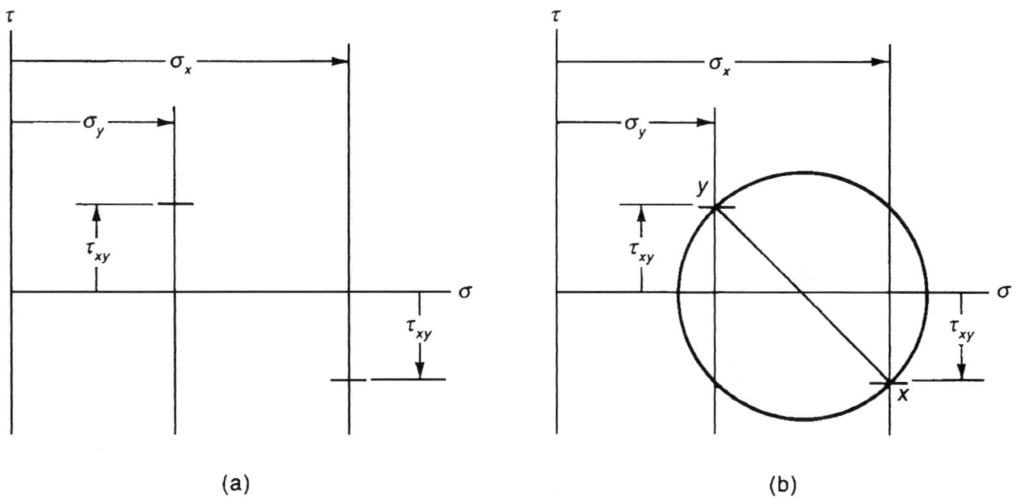

Figure 10.18 Constructing Mohr's circle

Upon constructing Mohr's circle you can now rotate the xy diameter through an angle of 2θ to a new position $x'y'$, which can determine the stress on any face at that point in the body as shown in Figure 10.17. Note that rotations of 2θ on Mohr's circle correspond to θ in the physical plane; also note that the direction of rotation is the same as in the physical plane (that is, if you go clockwise on Mohr's circle, the rotation is also clockwise in the physical plane). The construction can also be used to find the principal stresses and the orientation of the principal axes.

Problems involving stress transformations can be solved with the parametric equations, from construction of Mohr's circle, or from some combination. As an example of a combination, it can be seen that the center of Mohr's circle can be represented as

$$C = \frac{\sigma_x + \sigma_y}{2}$$

The radius of the circle is

$$R = \sqrt{\left(\frac{\sigma_x - \sigma_y}{2}\right)^2 + \tau_{xy}^{\,2}}$$

The principal stresses then are

$$\sigma_1 = C + R; \qquad \sigma_2 = C - R$$

Example 10.35

Given $\sigma_x = -3$ MPa; $\sigma_y = 5$ MPa; $\tau_{xy} = 3$ MPa. Find the principal stresses and their orientation.

Solution

Mohr's circle is constructed as shown in Exhibit 27. The angle 2θ was chosen as the angle between the y axis and the 1 axis clockwise from y to 1 as shown in the circle. The angle θ in the physical plane is between the y axis and the 1 axis also clockwise from y to 1. The values of σ_1, σ_2, and 2θ can all be scaled from the circle. The values can also be calculated as follows:

$$R = \sqrt{\left(\frac{\sigma_x - \sigma_y}{2}\right)^2 + \tau_{xy}^{\ 2}} = \sqrt{\left(\frac{-3-5}{2}\right)^2 + 3^2} = 5 \text{ MPa}$$

$$C = \frac{\sigma_x + \sigma_y}{2} = \frac{-3+5}{2} = 1 \text{ MPa}$$

$$\sigma_1 = C + R = 6 \text{ MPa}$$

$$\sigma_2 = C - R = -4 \text{ MPa}$$

$$2\theta = \tan^{-1}(3/4) = 36.87°; \qquad \theta = 18.43°$$

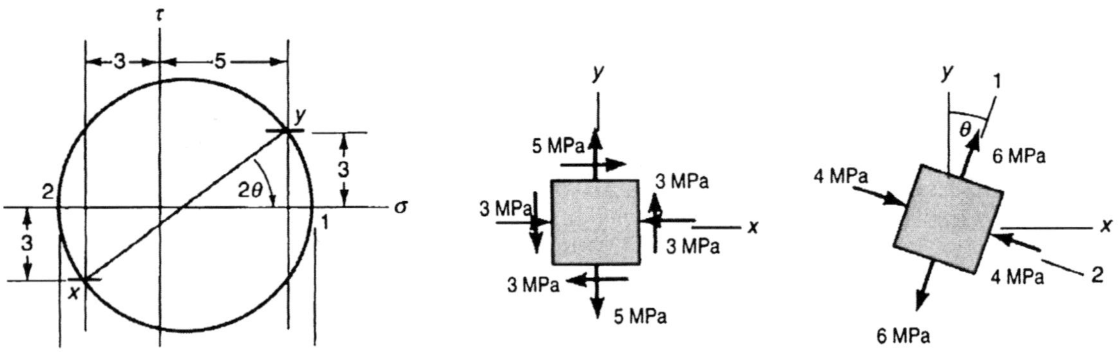

Exhibit 27

STRAIN

Axial strain was previously defined as

$$\varepsilon = \frac{\Delta L}{L}$$

In the general case, there are three components of axial strain, ε_x, ε_y and ε_z. Shear strain is defined as the decrease in angle of two originally perpendicular line segments passing through the point at which strain is defined. In Figure 10.19, AB is vertical and BC is horizontal. They represent line segments that are drawn before loading. After loading, points A, B, and C move to A', B', and C', respectively. The angle between $A'B'$ and the vertical is α, and the angle between B' and C' and the horizontal is β. The original right angle has been decreased by $\alpha + \beta$, and the shear strain is

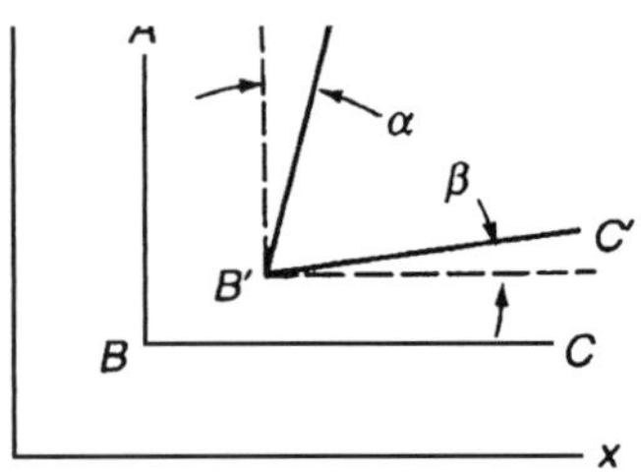

Figure 10.19 Definition of shear strain

$$\gamma_{xy} = \alpha + \beta$$

In the general case, there are three components of shear strain, γ_{xy}, γ_{yz}, and γ_{zx}.

Plane Strain

In two dimensions, strain undergoes a similar rotation transformation as stress. The transformation equations are

$$\varepsilon_{x'} = \frac{\varepsilon_x + \varepsilon_y}{2} + \frac{\varepsilon_x - \varepsilon_y}{2}\cos 2\theta + \frac{\gamma_{xy}}{2}\sin 2\theta$$

$$\frac{\gamma_{x'y'}}{2} = -\frac{\varepsilon_x - \varepsilon_y}{2}\sin 2\theta + \frac{\gamma_{xy}}{2}\cos 2\theta$$

$$\varepsilon_{y'} = \frac{\varepsilon_x + \varepsilon_y}{2} - \frac{\varepsilon_x - \varepsilon_y}{2}\cos 2\theta - \frac{\gamma_{xy}}{2}\sin 2\theta$$

These equations are the same as Mohr's parametric equations for stress, except that the σ_x has been replaced with $\varepsilon_{x'}$, σ_y with $\varepsilon_{y'}$, and τ_{xy} with $\gamma_{xy}/2$. Therefore, Mohr's circle for strain is treated the same way as that for stress, except for the factor of two on the shear strain.

Example 10.36

Given that $\varepsilon_x = 600\ \mu$; $\varepsilon_y = -200\ \mu$; $\gamma_{xy} = +800\ \mu$, find the principal strains and their orientation. The symbol μ signifies 10^{-6}.

Solution

From the Mohr's circle shown in Exhibit 28, it is seen that $2\theta = 45°$; so, $\theta = 22.5°$ counterclockwise from x to 1. The principal strains are $\varepsilon_1 = 766\ \mu$ and $\varepsilon_2 = -366\ \mu$.

The principal strains can also be found by computation in the same way as principal stresses,

$$R = \sqrt{\left(\frac{\varepsilon_x - \varepsilon_y}{2}\right)^2 + \left(\frac{\gamma_{xy}}{2}\right)^2} = \sqrt{\left(\frac{600 + 200}{2}\right)^2 + \left(\frac{-800}{2}\right)^2} = 565.7\ \mu$$

$$C = \frac{\varepsilon_x + \varepsilon_y}{2} = \frac{600 - 200}{2} = 200\ \mu$$

$$\varepsilon_1 = C + R = 766\ \mu$$

$$\varepsilon_2 = C - R = -366\ \mu$$

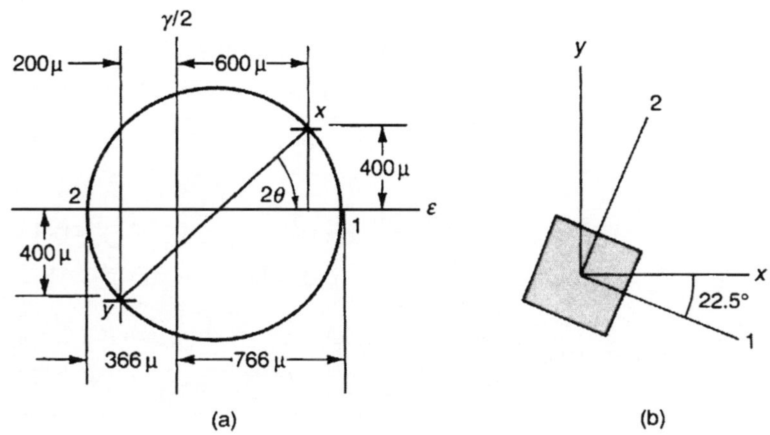

Exhibit 28

HOOKE'S LAW

The relationship between stress and strain is expressed by Hooke's law. For a linear elastic isotropic material it is

$$\varepsilon_x = \frac{1}{E}(\sigma_x - v\sigma_y - v\sigma_z)$$

$$\varepsilon_y = \frac{1}{E}(\sigma_y - v\sigma_z - v\sigma_x)$$

$$\varepsilon_z = \frac{1}{E}(\sigma_z - v\sigma_x - v\sigma_y)$$

$$\gamma_{xy} = \frac{1}{G}\tau_{xy}$$

$$\gamma_{xy} = \frac{1}{G}\tau_{xy}$$

$$\gamma_{zx} = \frac{1}{G}\tau_{zx}$$

Further, there is a relationship between E, G, and v, which is

$$G = \frac{E}{2(1+v)}$$

Thus, for an isotropic material there are only two independent elastic constants. An **isotropic material** is one that has the same material properties in all directions. Notable exceptions to isotropy are wood- and fiber-reinforced composites.

Example 10.37

A steel plate in a state of plane stress is known to have the following strains: $\varepsilon_x = 650\ \mu$, $\varepsilon_y = 250\ \mu$, and $\gamma_{xy} = 400\ \mu$. If $E = 210$ GPa and $v = 0.3$, what are the stress components, and what is the strain ε_z?

Solution

In a state of plane stress, the stresses $\sigma_z = 0$, $\tau_{xz} = 0$, and $\tau_{yz} = 0$. From Hooke's law,

$$\varepsilon_x = \frac{1}{E}(\sigma_x - \nu\sigma_y - 0)$$

$$\varepsilon_y = \frac{1}{E}(\sigma_y - \nu\sigma_x - 0$$

Using these relations to solve for stress gives

$$\sigma_x = \frac{E}{1-\nu^2}(\varepsilon_x + \nu\varepsilon_y) = \frac{210\text{ GPa}}{1-0.3^2}[650\ \mu + 0.3(250\ \mu)] = 167.3\text{ Mpa}$$

$$\sigma_y = \frac{E}{1-\nu^2}(\varepsilon_y + \nu\varepsilon_x) = \frac{210\text{ GPa}}{1-0.3^2}[250\ \mu + 0.3(650\ \mu)] = 102.7\text{ Mpa}$$

From Hooke's law, the strain γ_{xy} is

$$\gamma_{xy} = \frac{\tau_{xy}}{G};\ G = \frac{E}{2(1+v)};\ \tau_{xy} = \frac{E\gamma_{xy}}{2(1+v)} = \frac{(210\text{ GPa})\ (400\ \mu)}{2(1+0.3)} = 32.3\text{ Mpa}$$

The strain in the z direction is

$$\varepsilon_z = \frac{1}{E}(0 - \nu\sigma_x - \nu\sigma_y) = \frac{-\nu}{E}(\sigma_x + \sigma_y)$$
$$= \frac{-0.3}{210\ \text{GPa}}(167.3\ \text{MPa} + 102.7\ \text{MPa}) = -386\ \mu$$

ELASTIC AND PLASTIC DEFORMATION

Figure 10.20 shows the conventional stress-strain diagram for a ductile material. The elastic region is the area in which the material will stretch when loaded and relax back to its original size and shape when unloaded. Hooke's law for axial loading ($\sigma = E\varepsilon$) refers to the relationship between stress and strain in the elastic region of the stress-strain diagram.

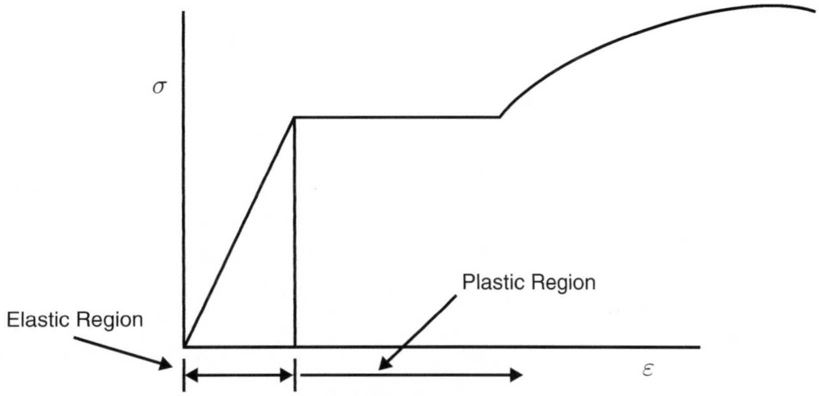

Figure 10.20

The constant of proportionality of Hooke's Law is obtained from the linear portion of the stress-strain diagram. The majority of our engineering work is in the area of elastic deformation, especially in the area of repetitive loads.

Example 10.38

A steel bar made up of two segments, as shown in Exhibit 29, is loaded with a 100 kN tension load. Determine the elongation of the bar assuming that it is stretched only in the elastic deformation region.

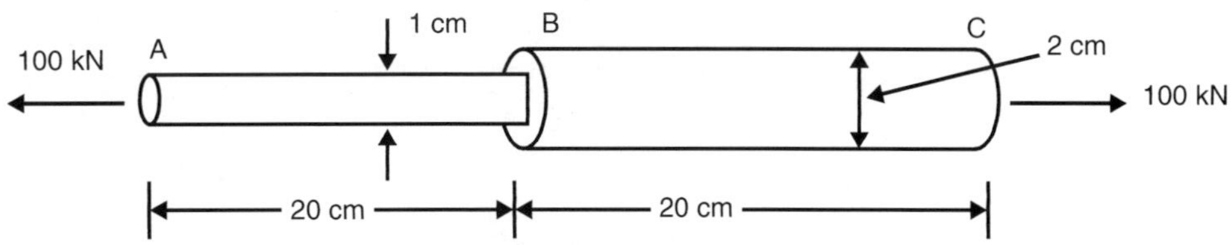

Exhibit 29

Solution

$$\delta_{AB} = \frac{PL}{AE} = \frac{(100\ \text{kN})(20\ \text{cm})}{\frac{\pi d^2}{4}\text{cm}^2\,(220\ \text{GPa})}$$

$$\delta_{AB} = \frac{(100\times10^3\ \text{N})(0.20\ \text{m})}{\frac{\pi}{4}0.01^2\,\text{m}^2\left(220\times10^9\right)\frac{\text{N}}{\text{m}^2}} = 0.00116\ \text{m} = 0.116\ \text{cm}$$

$$\delta_{BC} = \frac{PL}{AE} = \frac{(100\times10^3\ \text{N})(0.20\ \text{m})}{\frac{\pi}{4}0.02^2\,\text{m}^2\left(220\times10^9\right)\frac{\text{N}}{\text{m}^2}} = 0.000289\ \text{m} = 0.0289\ \text{cm}$$

The total elastic deformation is

$$\delta_{AB} + \delta_{BC} = 0.116 + 0.0289 = 0.1449\ \text{cm}$$

The plastic region of the conventional stress-strain diagram in Figure 10.20 consists of a large region beyond the knee of the elastic region. That is, the plastic region extends beyond the proportional limit. Once the specimen is loaded beyond the proportional limit, the majority of the elongation cannot be relaxed.

Figure 10.21 shows a stress-strain diagram for an alloy. The upper curve is the approximate curve obtained by running a loading test on the material. The stress is given as the vertical line running from 50 MPa to 350 MPa, and the approximate strain ranges from 0 to 0.40 mm/mm. The diagram also has an inner curve, which is a magnification of the extremely left portion of the first curve. The strain scale of the magnified portion runs from 0 to 0.0040 mm/mm. This means the scale is magnified by 100.

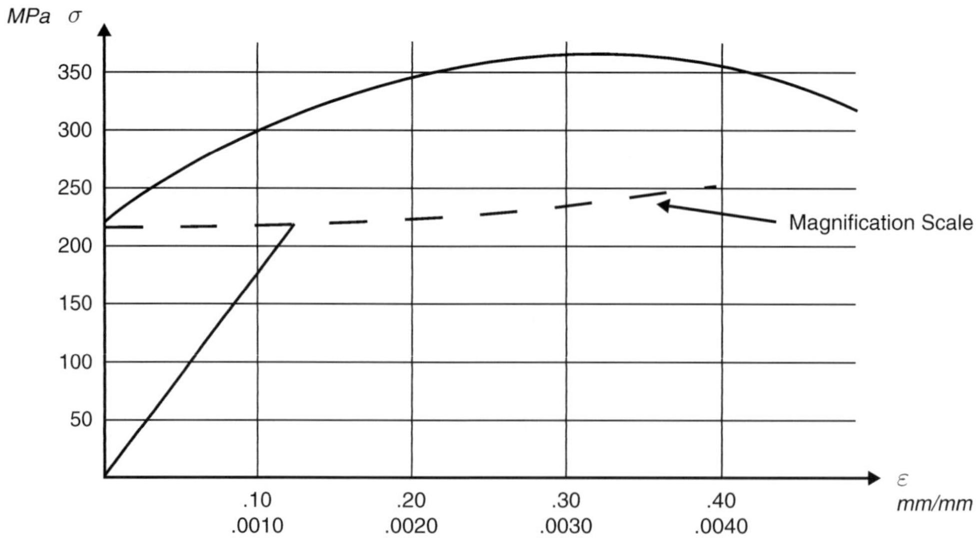

Figure 10.21

Example 10.39

Using the stress-strain diagram in Figure 10.21, determine (i) modulus of elasticity and (ii) the proportional limit.

Solution

(i) Modulus of elasticity: Using the straight-line portion of the magnified scale:

$$E = \frac{\sigma}{\varepsilon} = \frac{180 \text{ MPa}}{0.0010 \text{ mm/mm}} = 180 \text{ GPa}$$

(ii) Proportional limit: This point may be selected on the magnification scale where the knee occurs:

$$\sigma_{PL} = 220 \text{ MPa}$$

Example 10.40

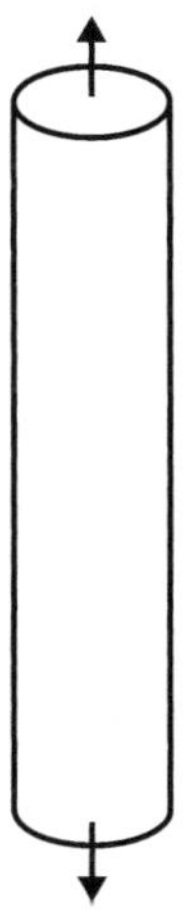

Exhibit 30

Consider that a bar made of the alloy in Example 10.39 is loaded to 300 MPa in tension, as shown in Exhibit 30. Determine (i) the modulus of resilience; (ii) the elastic recovery of the bar; and (iii) the permanent set.

Solution

(i) For modulus of resilience:

$$\mu_r = \frac{1}{2}\sigma_{PL}\varepsilon_{PL} = \frac{1}{2}(220 \text{ MPa})(0.0012 \text{ mm/mm}) = 0.132 \text{ MJ/m}^3$$

(ii) For the elastic recovery of the bar:

$$\text{Elastic recovery} = \frac{300 \times 10^6}{E} = \frac{300 \times 10^6}{180 \times 10^9} = 0.001667 \text{mm/mm}$$

(ii) Permanent set = 0.085 – 0.001667 = 0.083 mm/mm

STATICALLY INDETERMINATE STRUCTURES

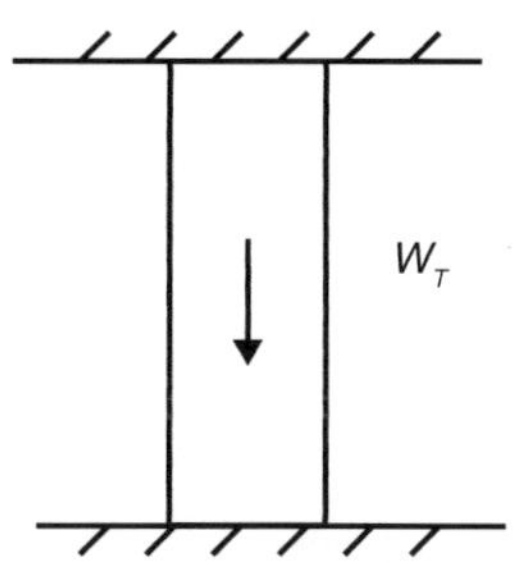

Figure 10.22

A statically indeterminate structure is one for which there are not enough equilibrium equations to determine the reactions. For example, consider the 100 kg post anchored between the ceiling and floor in Figure 10.22. To determine the reactions F_A and F_B, draw the free-body diagram shown in Figure 10.23.

Using the equation $\Sigma Fy = 0$,

$$F_B + F_A = W_t$$

There are not enough equilibrium equations available to solve this problem, so additional equations must be written. These additional equations involve the geometry of the deformation in the members of the structure.

For this axially loaded structure, we note that the structure does not change in length upon loading it; that is,

$$\delta_{AB} = 0$$

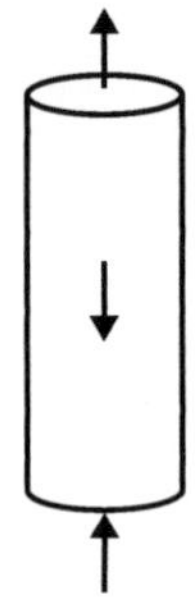

Figure 10.23

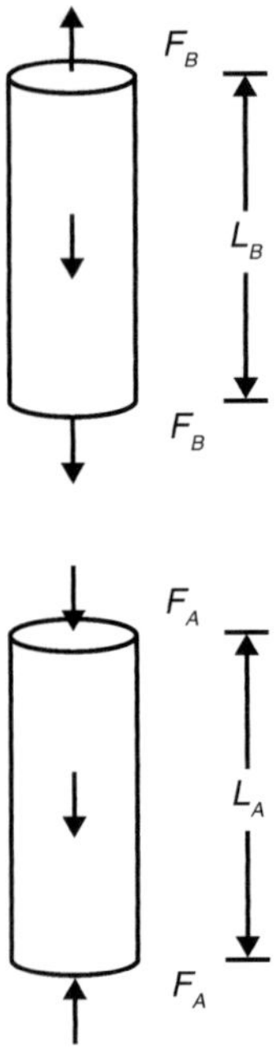

Figure 10.24

Let us cut the axially loaded beam into two parts (see Figure 10.24) and apply the loads.

Using the elongation equation:

$$\delta_B = \frac{PL}{AE} = \frac{F_B L_B}{AE} = 0$$

$$\delta_A = \frac{PL}{AE} = \frac{F_A L_A}{AE} = 0$$

$$F_A + F_B = P = W_t$$

We can then determine:

$$F_A = P\left(\frac{L_B}{L}\right) \qquad F_B = P\left(\frac{L_A}{L}\right)$$

The post is cut into two pieces:

$$L_B = 1.5 \text{ m} \qquad L_A = 2.0 \text{ m}$$

Now determine F_A and F_B:

$$F_A = \frac{PL_B}{L} = \frac{(100 \text{ kg})\left(9.81 \frac{\text{N}}{\text{kg}}\right)(1.5 \text{ m})}{3.5 \text{ m}} = 420 \text{ N}$$

$$F_B = \frac{PL_A}{L} = \frac{(100)(9.81)(2.0)}{3.5} = 560 \text{ N}$$

Example 10.41

A column constructed from concrete and one steel rod has an applied load of 400 kN. Assume that 80% of the load is carried by the concrete and 20% by the steel. See Exhibit 31. Determine the diameter of the rod.

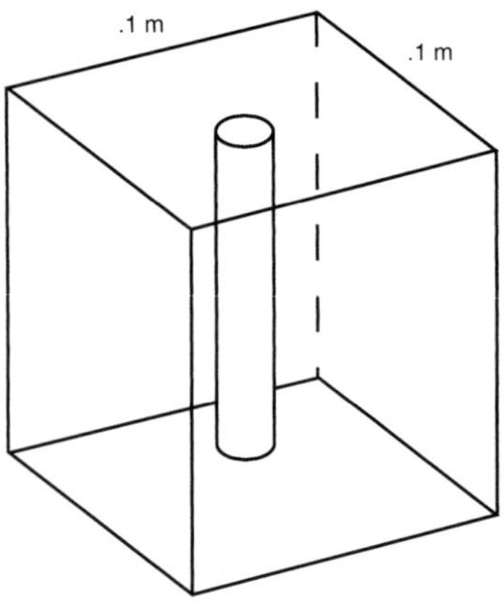

Exhibit 31

Solution

$$E_{\text{concrete}} = 25 \text{ GPa} \qquad E_{\text{steel}} = 180 \text{ GPa}$$

Knowing that $\delta_{\text{concrete}} = \delta_{\text{steel}}$

$$P_{\text{steel}} = 80 \text{ kN} \qquad P_{\text{concrete}} = 320 \text{ kN}$$

$$\left(\frac{P_{\text{concrete}} L}{AE}\right)_{\text{concrete}} = \left(\frac{P_{\text{steel}} L}{AE}\right)_{\text{steel}}$$

$$\frac{320 \times 10^3}{\left[.1^2 - \frac{\pi}{4} d^2\right] 25 \times 10^9} = \frac{80 \times 10^3}{\left[\frac{\pi}{4} d^2\right] 180 \times 10^9}$$

$$565.5\, d^2 = .25 \qquad d = 20.67 \text{ m}$$

TORSION

Torsion refers to the twisting of long members. Torsion can occur with members of any cross-sectional shape, but the most common is the circular shaft. Another fairly common shaft configuration, which has a simple solution, is the hollow, thin-walled shaft.

Circular Shafts

Figure 10.25(a) shows a circular shaft before loading; the r-θ-z cylindrical coordinate system is also shown. In addition to the outline of the shaft, two longitudinal lines, two circumferential lines, and two diametral lines are shown scribed on the shaft. These lines are drawn to show the deformed shape loading. Figure 10.25(b) shows the shaft after loading with a torque T. The **double arrow notation** on T indicates a moment about the z-axis in a right-handed direction. The effect of the torsion is that each cross-section remains plane and simply rotates with respect to other cross-sections. The angle ϕ is the twist of the shaft at any position z. The rotation $\phi(z)$ is in the θ direction.

The distance b shown in Figure 10.25(b) can be expressed as $b = \phi r$ or as $b = \gamma z$. The shear strain for this special case can be expressed as

$$\gamma_{\phi z} = r\frac{\phi}{z}$$

For the general case where ϕ is not a linear function of z the shear strain can be expressed as

$$\gamma_{\phi z} = r\frac{d\phi}{dz}$$

where $d\phi/dz$ is the twist per unit length or the rate of twist.

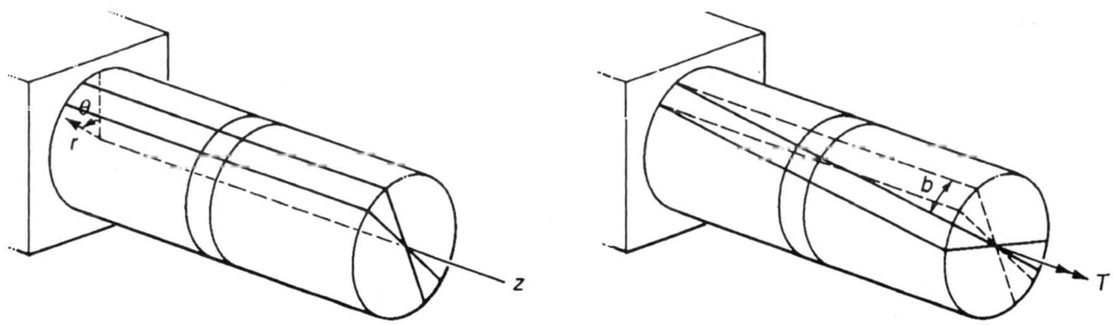

(a) Before loading

(b) After loading

Figure 10.25 Torsion in a circular shaft

The application of Hooke's law gives

$$\tau_{\phi z} = G\gamma_{\phi z} = Gr\frac{d\phi}{dz}$$

The torque at the distance z along the shaft is found by summing the contributions of the shear stress at each point in the cross-section by means of an integration:

$$T = \int_A \tau_{\phi z} r\, dA = G\frac{d\phi}{dz}\int_A r^2 dA = GJ\frac{d\phi}{dz}$$

where $J = \int_A r^2 dA$ is the polar moment of inertia of the cross-sectional area of the shaft about the axis of the shaft. For a solid shaft with an outer radius of r_o the polar moment of inertia is

$$J = \frac{\pi r_o^4}{2}$$

For a hollow circular shaft with outer radius r_o and inner radius r_i, the polar moment of inertia is

$$J = \frac{\pi}{2}\left(r_o^4 - r_i^4\right)$$

Note that the J that appears is the polar moment of inertia only for the special case of circular shafts (either solid or hollow). For any other cross-section shape, this is valid only if J is redefined as a torsional constant *not equal* to the polar moment of inertia. The two equations can be combined to give

$$\tau_{\phi z} = \frac{Tr}{J}$$

The maximum shear stress occurs at the outer radius of the shaft and at the location along the shaft where the torque is maximum.

$$\tau_{\phi z\,\max} = \frac{T_{\max} r_o}{J}$$

The angle of twist of the shaft is

$$\phi = \int_0^L \frac{T}{GJ} dz$$

For a uniform circular shaft with a constant torque along its length, this equation becomes

$$\phi = \frac{TL}{GJ}$$

Example 10.42

The hollow circular steel shaft shown in Exhibit 32 has an inner diameter of 25 mm, an outer diameter of 50 mm, and a length of 600 mm. It is fixed at the left end and subjected to a torque of 1400 N • m as shown in Exhibit 32. Find the maximum shear stress in the shaft and the angle of twist at the right end. Take $G = 84$ GPa.

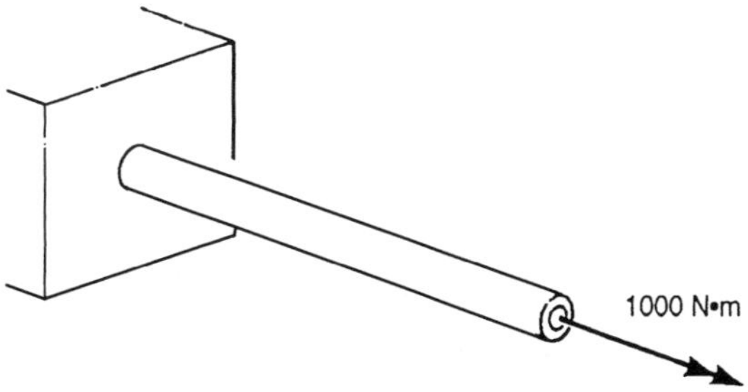

Exhibit 32

Solution

$$J = \frac{\pi}{2}\left(r_o^4 - r_i^4\right) = \frac{\pi}{2}\left[(25\text{ mm})^4 - (12.5\text{ mm})^4\right] = 575\times 10^3\text{ mm}^4$$

$$\tau_{\theta z\,\max} = \frac{T_{\max} r_o}{J} = \frac{(1400\text{N} \bullet \text{m})(25\text{ mm})}{575\times 10^3\text{mm}^4} = 60.8\text{ MPa}$$

$$\phi = \frac{TL}{GJ} = \frac{(1400\text{N} \bullet \text{m})(600\text{ mm})}{(84\text{ GPa})(575\times 10^3\text{mm}^4)} = 0.01738\text{ rad}$$

Hollow, Thin-Walled Shafts

In hollow, thin-walled shafts, the assumption is made that the shear stress τ_{sz} is constant throughout the wall thickness t. The shear flow q is defined as the product of τ_{sz} and t. From a summation of forces in the z direction, it can be shown that q is constant—even with varying thickness. The torque is found by summing the contributions of the shear flow. Figure 10.26 shows the cross-section of the thin-walled tube of nonconstant thickness. The z coordinate is perpendicular to the plane of the paper. The shear flow q is taken in a counter-clockwise sense. The torque produced by q over the element ds is

$$dT = q\,r\,ds$$

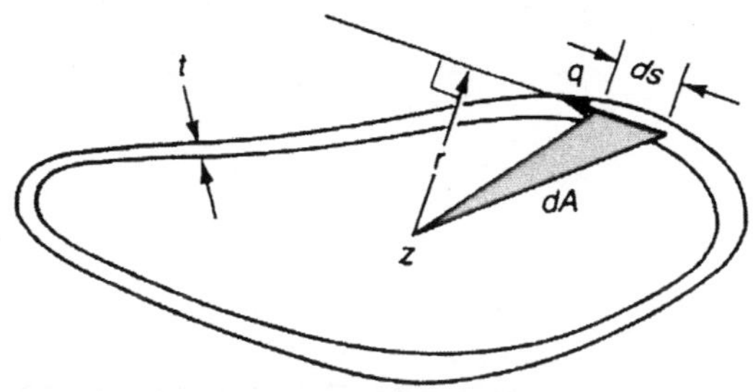

Figure 10.26 Cross-section of thin-walled tube

The total torque is, therefore,

$$T = \oint qr\,ds = q\oint r\,ds$$

When ds is small, the area dA is the area of the triangle of base ds and height r, which can be approximated by

$$dA = \tfrac{1}{2}(\text{base})(\text{height}) = \frac{r\,ds}{2}$$

so that

$$\oint r\,ds = 2A_m$$

where A_m is the area enclosed by the wall (including the hole). It is best to use the centerline of the wall to define the boundary of the area; hence A_m is the mean area. The expression for the torque is

$$T = 2A_m q$$

and from the definition of q the shear stress can be expressed as

$$\tau_{sz} = \frac{T}{2A_m t}$$

Example 10.43

A torque of 10 kN • m is applied to a thin-walled rectangular steel shaft whose cross-section is shown in Exhibit 33. The shaft has wall thicknesses of 5 mm and 10 mm. Find the maximum shear stress in the shaft.

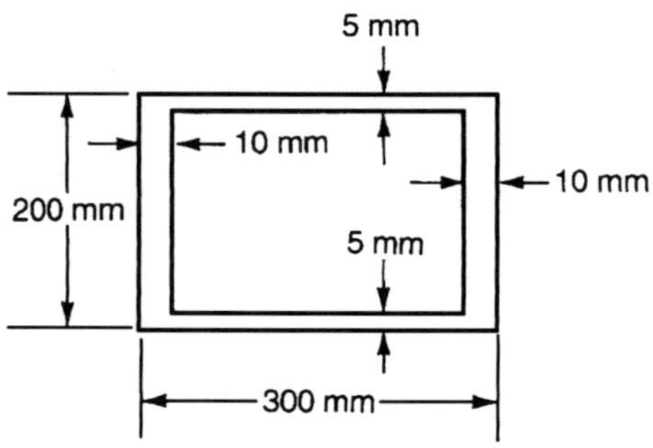

Exhibit 33

Solution

$$A_m = (200 - 5)(300 - 10) = 56{,}550 \text{ mm}^2$$

The maximum shear stress will occur in the thinnest section, so $t = 5$ mm.

$$\tau_{sz} = \frac{T}{2A_m t} = \frac{10 \text{ kN} \bullet \text{m}}{2(56{,}550 \text{ mm}^2)(5 \text{ mm})} = 17.68\frac{\text{MN}}{\text{m}^2}$$

BEAMS

Shear and Moment Diagrams

Shear and moment diagrams are plots of the shear forces and bending moments, respectively, along the length of a beam. The purpose of these plots is to clearly show maximums of the shear force and bending moment, which are important in the design of beams. The most common sign convention for the shear force and bending moment in beams is shown in Figure 10.27. One method of determining the shear and moment diagrams is by the following steps:

1. Determine the reactions from equilibrium of the entire beam.
2. Cut the beam at an arbitrary point.

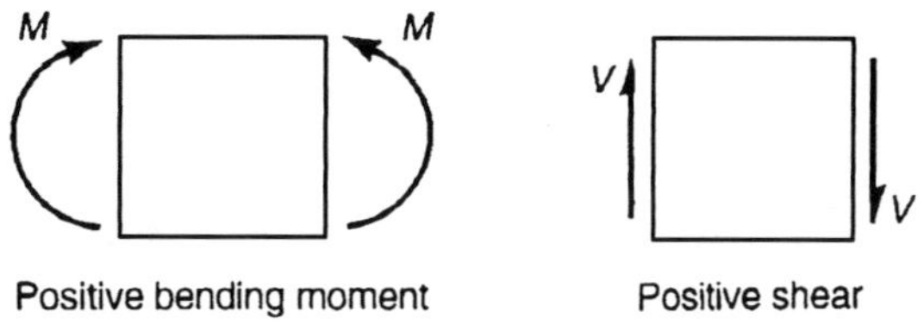

Figure 10.27 Sign convention for bending moment and shear

3. Show the unknown shear and moment on the cut using the positive sign convention shown in Figure 10.27.
4. Sum forces in the vertical direction to determine the unknown shear.
5. Sum moments about the cut to determine the unknown moment.

Example 10.44

For the beam shown in Exhibit 34, plot the shear and moment diagram.

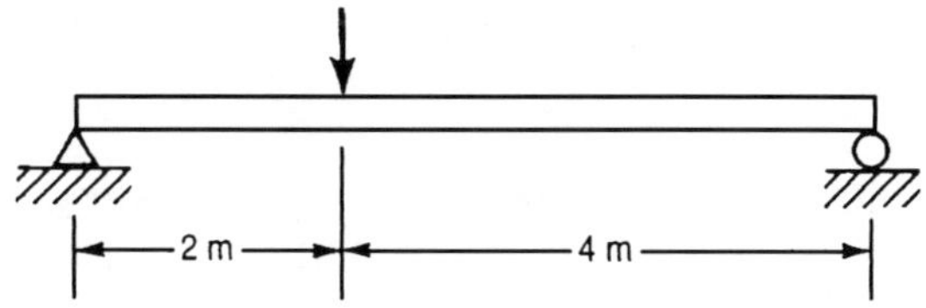

Exhibit 34

Solution

First, solve for the unknown reactions using the free-body diagram of the beam shown in Exhibit 35(a). To find the reactions, sum moments about the left end,

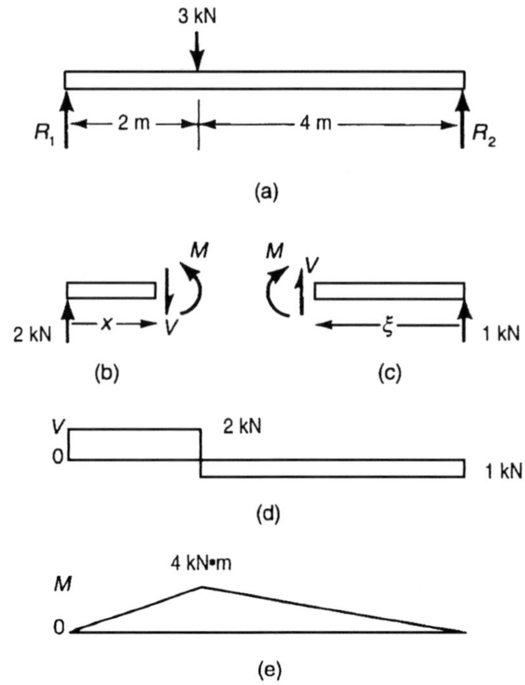

Exhibit 35

which gives

$$6R_2 - (3)(2) = 0 \quad \text{or} \quad R_2 = 6/6 = 1 \text{ kN}$$

Sum forces in the vertical direction to get

$$R_1 + R_2 = 3 = R_1 + 1 \quad \text{or} \quad R_1 = 2 \text{ kN}$$

Cut the beam between the left end and the load as shown in Exhibit 35(b). Show the unknown moment and shear on the cut using the positive sign convention shown in Figure 10.27. Sum the vertical forces to get

$$V = 2 \text{ kN (independent of } x)$$

Sum moments about the cut to get

$$M = R_1x = 2x$$

Repeat the procedure by making a cut between the right end of the beam and the 3-kN load, as shown in Exhibit 35(c). Again, sum vertical forces and sum moments about the cut to get

$$V = 1 \text{ kN (independent of } \xi\text{), and } M = 1\xi$$

The plots of these expressions for shear and moment give the shear and moment diagrams shown in Exhibit 35(d) and 35(e).

It should be noted that the shear diagram in this example has a jump at the point of the load and that the jump is equal to the load. This is always the case. Similarly, a moment diagram will have a jump equal to an applied concentrated moment. In this example, there was no concentrated moment applied, so the moment was everywhere continuous.

Another useful way of determining the shear and moment diagram is by using differential relationships. These relationships are found by considering an element of length Δx of the beam with a distributed applied load q per unit length. The forces on that element are shown in Figure 10.28. Summation of forces in the y direction gives

$$q\Delta x + V - V - \frac{dV}{dx}\Delta x = 0$$

which gives

$$\frac{dV}{dx} = q$$

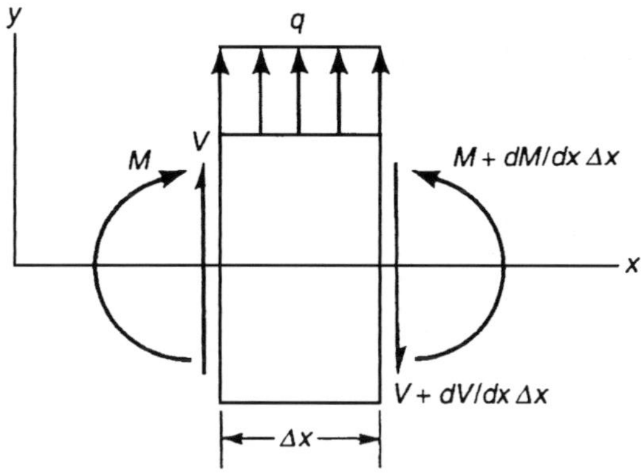

Figure 10.28

Summing moments and neglecting higher order terms gives

$$-\mathrm{M} + \mathrm{M} + \frac{dM}{dx}\Delta x - V\Delta x = 0$$

which gives

$$\frac{dM}{dx} = V$$

Integral forms of these relationships are expressed as

$$V_2 - V_1 = \int_{x_1}^{x_2} q\, dx$$

$$M_2 - M_1 = \int_{x_1}^{x_2} V\, dx$$

Example 10.45

The simply supported uniform beam shown in Exhibit 36 carries a uniform load of w_0. Plot the shear and moment diagrams for this beam.

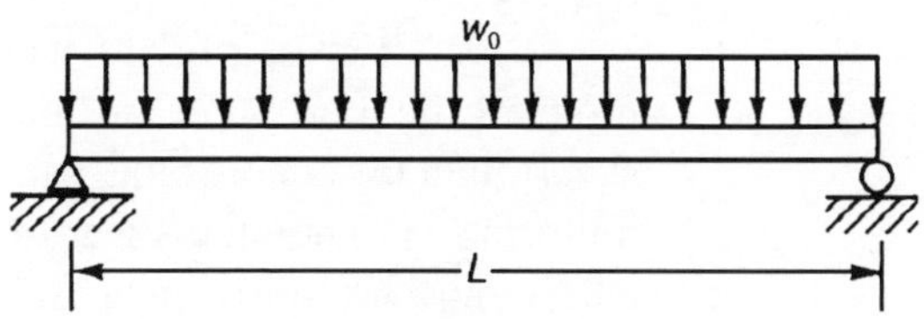

Exhibit 36

Solution

As before, the reactions can be found first from the free-body diagram of the beam shown in Exhibit 37(a). It can be seen that, from symmetry, $R_1 = R_2$. Summing vertical forces then gives

$$R = R_1 = R_2 = \frac{w_0 L}{2}$$

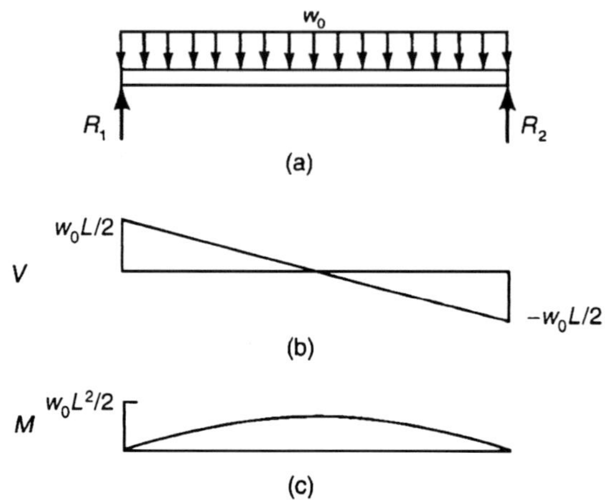

Exhibit 37

The load $q = -w_0$, so it reads

$$V(x) - V_0 = \int_0^x q dx = qx$$

where V_0 is the shear force at $x = 0$.

$$V = V_0 - \int_0^x w_0 \, dx = \frac{w_0 L}{2} - w_0 x$$

Noting that the moment at $x = 0$ is zero, gives

$$M(x) = M_0 + \int_0^x \left(\frac{w_0 L}{2} - w_0 x \right) dx = 0 + \frac{w_0 L x}{2} - \frac{w_0 x^2}{2} = \frac{w_0 x}{2}(L - x)$$

It can be seen that the shear diagram is a straight line, and the moment varies parabolically with x. Shear and moment diagrams are shown in Exhibit 37(b) and Exhibit 37(c). It can be seen that the maximum bending moment occurs at the center of the beam where the shear stress is zero. The maximum bending moment always has a relative maximum at the place where the shear is zero because the shear is the derivative of the moment, and relative maxima occur when the derivative is zero.

Often, it is helpful to use a combination of methods to find the shear and moment diagrams. For instance, if there is no load between two points, then the shear diagram is constant, and the moment diagram is a straight line. If there is a uniform load, then the shear diagram is a straight line, and the moment diagram is parabolic. The following example illustrates this method.

Example 10.46

Draw the shear and moment diagrams for the beam shown in Exhibit 38(a).

Solution

Draw the free-body diagram of the beam as shown in Exhibit 38(b). From a summation of the moments about the right end,

$$10\,R_1 = (4)(7) + (3)(2) = 34; \quad \text{so } R_1 = 3.4\,\text{kN}$$

From a summation of forces in the vertical direction,

$$R_2 = 7 - 3.4 = 3.6\,\text{kN}$$

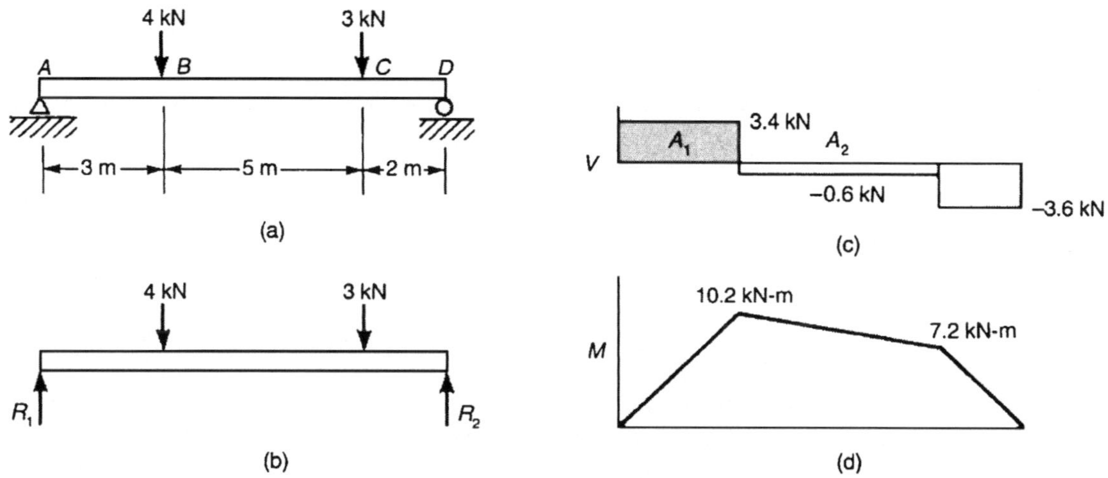

Exhibit 38

The shear in the left portion is 3.4 kN, the shear in the right portion is −3.6 kN, and the shear in the center portion is 3.4 − 4 = 0.6 kN. This is sufficient information to draw the shear diagram shown in Exhibit 38(c). The moment at A is zero, so the moment at B is the shaded area A_1 and the moment at C is $A_1 - A_2$.

$$M_B = A_1 = (3.4\,\text{kN})(3\,\text{m}) = 10.2\,\text{kN} \bullet \text{m}$$

$$M_C = A_1 - A_2 = (3.4\text{kN})(3\text{m}) - (0.6\text{kN})(5\text{m}) = 7.2\text{kN} \bullet \text{m}$$

The moments at A and D are zero, and the moment diagram consists of straight lines between the points A, B, C, and D. There is, therefore, enough information to plot the moment diagram shown in Exhibit 38(d).

Stresses in Beams

The basic assumption in elementary beam theory is that the beam cross-section remains plane and perpendicular to the neutral axis as shown in Figure 10.29 when the beam is loaded. This assumption is strictly true only for the case of pure bending (constant bending moment and no shear) but gives good results even when shear is present. Figure 10.29 shows a beam element before and after loading. It can be seen that there is a line of length ds that does not change length due to deformation. This line is called the neutral axis. The distance y is measured from this neutral axis. The strain in the x direction is $\Delta L/L$. The change in length $\Delta L = -y d\phi$ and the length is ds, so

$$\varepsilon_x = -y\frac{d\phi}{ds} = -\frac{y}{\rho} = -\kappa y$$

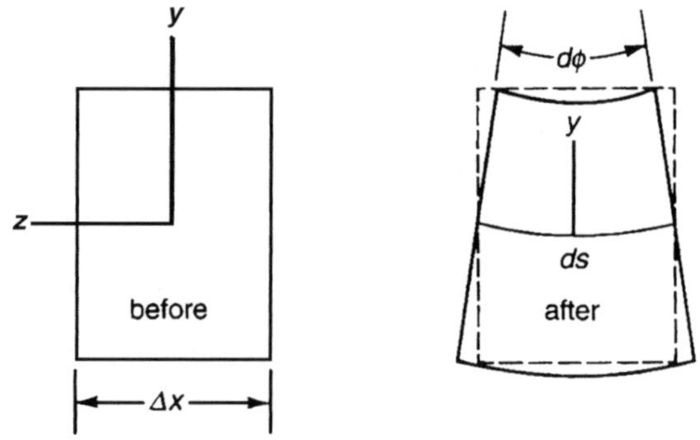

Figure 10.29

where ρ is the radius of curvature of the beam and κ is the curvature of the beam. Assuming that σ_y and σ_z are zero, Hooke's law yields

$$\sigma_x = -E\kappa y$$

The axial force and bending moment can be found by summing the effects of the normal stress σ_x,

$$P = \int_A \sigma_x \, dA = -E\kappa \int_A y \, dA$$

$$M = -\int_A y\sigma_x \, dA = E\kappa \int_A y^2 dA = EI\kappa$$

where I is the moment of inertia of the beam cross-section. If the axial force is zero (as is the usual case) then the integral of $y\,dA$ is zero. That means that y is measured from the centroidal axis of the cross-section. Since y is also measured from the neutral axis, the neutral axis coincides with the centroidal axis. The bending stress σ_x can be expressed as

$$\sigma_x = -\frac{My}{I}$$

The maximum bending stress occurs where the magnitude of the bending moment is a maximum and at the maximum distance from the neutral axis. For symmetrical beam sections the value of $y_{max} = \pm C$, where C is the distance to the extreme fiber, so the maximum stress is

$$\sigma_x = \pm\frac{MC}{I} = \pm\frac{M}{S}$$

where S is the section modulus ($S = I/C$).

Example 10.47

A 100 mm × 150 mm wooden cantilever beam is 2 m long. It is loaded at its tip with a 4-kN load. Find the maximum bending stress in the beam shown in Exhibit 39. The maximum bending moment occurs at the wall and is $M_{max} = 8$ kN • m.

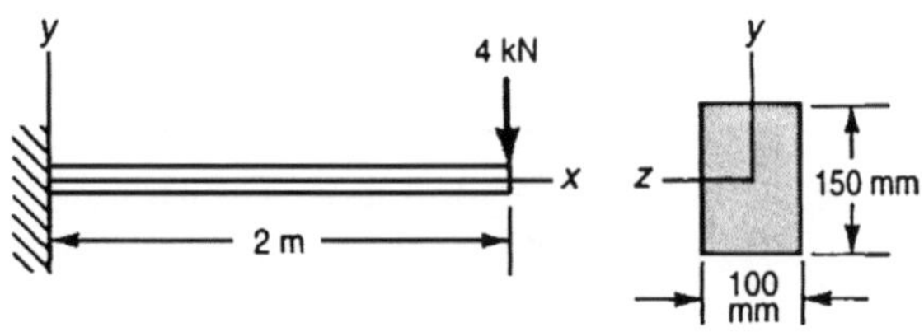

Exhibit 39

Solution

$$I = \frac{bh^3}{12} = \frac{100(150)^3}{12} = 28.1 \times 10^6 \text{ mm}^4$$

$$\sigma_{x\max} = \frac{|M|_{\max} c}{I} = \frac{(8 \text{ kN} \bullet \text{m})(75 \text{ mm})}{28.1 \times 10^6 \text{ mm}^4} = 21.3 \text{ MPa}$$

Shear Stress

To find the shear stress, consider the element of length Δx shown in Figure 10.30(a). A cut is made in the beam at $y = y_1$. At that point the beam has a thickness b. The shaded cross-sectional area above that cut is called A_1. The bending stresses acting on that element are shown in Figure 10.30(b). In general, the stresses are slightly larger at the right side than at the left side so that a force per unit length q is needed for equilibrium. Summation of forces in the x direction for the free-body diagram shown in Figure 10.30(b) gives

$$-F = q\Delta x = \int_{A_1} \sigma \, dA - \int_{A_1} \left(\sigma + \frac{d\sigma}{dx} \Delta x \right) dA = -\int_{A_1} \frac{d\sigma}{dx} \Delta x \, dA$$

From the expression for the bending stress ($\sigma = -My/I$) it follows that

$$\frac{d\sigma}{dx} = -\left(\frac{dM}{dx} \right) \frac{y}{I} = -V \frac{y}{I}$$

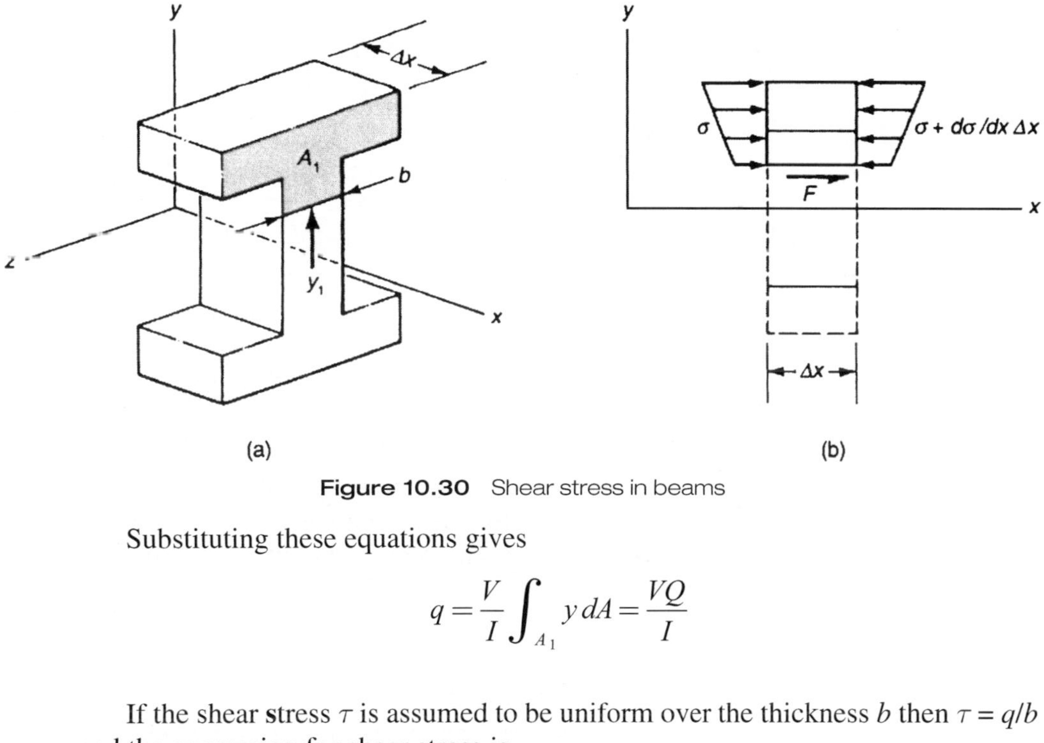

Figure 10.30 Shear stress in beams

Substituting these equations gives

$$q = \frac{V}{I} \int_{A_1} y \, dA = \frac{VQ}{I}$$

If the shear stress τ is assumed to be uniform over the thickness b then $\tau = q/b$ and the expression for shear stress is

$$\tau = \frac{VQ}{Ib}$$

where V is the shear in the beam, Q is the moment of area above (or below) the point in the beam at which the shear stress is sought ($y = y_1$), I is the moment of inertia of the entire beam cross-section, and b is the thickness of the beam cross-section at the point where the shear stress is sought ($y = y_1$). The definition of Q is

$$Q = \int_{A_1} y\,dA = A_1 y$$

Example 10.48

The cross-section of the beam shown in Exhibit 40 has an applied shear of 10 kN. Find (a) the shear stress at a point 20 mm below the top of the beam and (b) the maximum shear stress from the shear force.

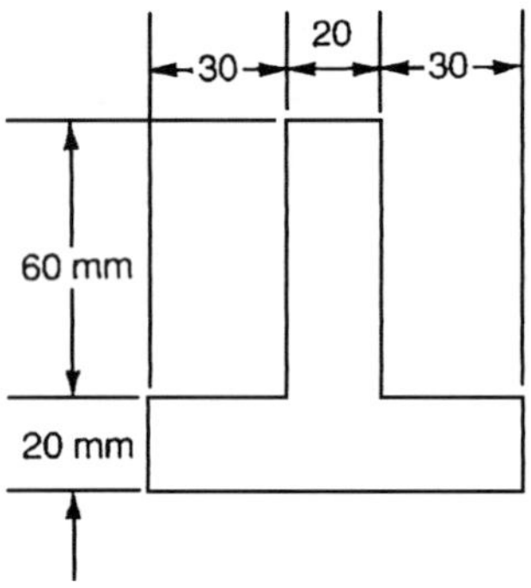

Exhibit 40

Solution

The section is divided into two parts by the dashed line shown in Exhibit 41(a). The centroids of each of the two sections are also shown in Exhibit 41(a). The centroid of the entire cross-section is found as follows:

$$\bar{y} = \frac{\sum_{n=1}^{N} \bar{y}_n A_n}{\sum_{n=1}^{N} A_n} = \frac{(60)(20)(30+20)+(80)(20)(10)}{(60)(20)+(80)(20)} = 27.14 \text{ mm (from bottom)}$$

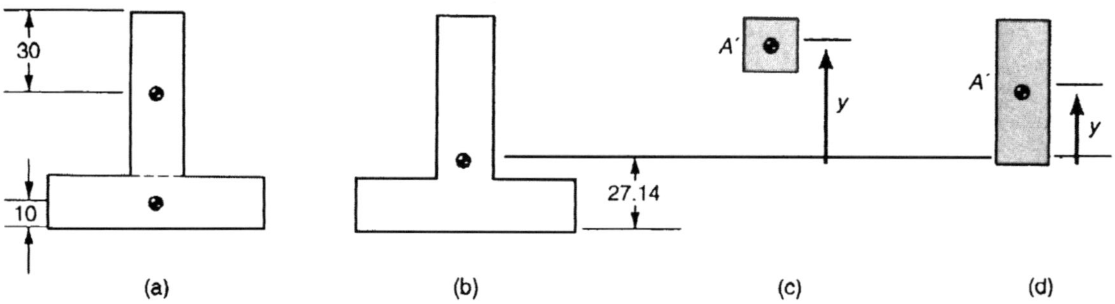

Exhibit 41

Exhibit 41(b) shows the location of the centroid.

The moment of inertia of the cross-section is found by summing the moments of inertia of the two sections taken about the centroid of the entire section. The moment of inertia of each part is found about its own centroid; then the parallel axis theorem is used to transfer it to the centroid of the entire section.

$$\begin{aligned} I &= \sum_{n=1}^{N} I_n + A_n \bar{y}_n \\ &= \frac{(20)(60)^3}{12} + (20)(60)(50-27.14)^2 + \frac{(80)(20)^3}{12} - (20)(80)(27.14-10)^2 \\ &= 1.510 \times 10^6 \text{ mm}^4 \end{aligned}$$

For the point 20 mm below the top of the beam, the area A' and the distance y are shown in Exhibit 41(c). The distance y is from the neutral axis to the centroid of A'. The value of Q is then

$$Q = \int_{A'} y\,dA = A'y = (20)(20)(70 - 27.14) = 17{,}140\ \text{mm}^3$$

$$\tau = \frac{VQ}{Ib} = \frac{(10\text{kN})(17{,}140\ \text{mm}^3)}{(1.510 \times 10^6\ \text{mm}^4)(20\ \text{mm})} = 0.00568\frac{\text{kN}}{\text{mm}^2} = 5.68\ \text{MPa}$$

The maximum Q will be at the centroid of the cross-section. Since the thickness is the same everywhere, the maximum shear stress will appear at the centroid. The maximum moment of area Q_{max} is

$$Q = \int_{A'} y dA = A'\,y_1 = (20)(80 - 27.14)\frac{(80 + 27.14)}{2} = 56{,}600\ \text{mm}^3$$

$$\tau = \frac{VQ}{Ib} = \frac{(10\text{kN})(56{,}600\ \text{mm}^3)}{(1.510 \times 10^6\ \text{mm}^4)(20\ \text{mm})} = 0.01875\frac{\text{kN}}{\text{mm}^2} = 18.75\ \text{MPa}$$

Deflection of Beams

The beam deflection in the y direction will be denoted as y, while most modern texts use v for the deflection in the y direction. The *Fundamentals of Engineering Supplied-Reference Handbook* uses the older notation. The main assumption in the deflection of beams is that the slope of the beam is small. The slope of the beam is dy/dx. Since the slope is small, the slope is equal to the angle of rotation in radians.

$$\frac{dy}{dx} = \text{rotation in radians}$$

Because the slope is small it also follows that

$$\kappa = \frac{1}{\rho} \approx \frac{d^2y}{dx^2}$$

From Equation (10.62) this gives

$$\frac{d^2y}{dx^2} = \frac{M}{EI}$$

This equation, together with two boundary conditions, can be used to find the beam deflection. Integrating twice with respect to x gives

$$\frac{dy}{dx} = \int \frac{M}{EI}dx + C_1$$

$$y = \iint \frac{M}{EI}dx + C_1x + C_2$$

where the constants C_1 and C_2 are determined from the two boundary conditions. Appropriate boundary conditions are on the displacement y or on the slope dy/dx. In the common problems of uniform beams, the beam stiffness EI is a constant and can be removed from beneath the integral sign.

Example 10.49

The uniform cantilever beam shown in Exhibit 42 has a constant, uniform, downward load w_0 along its length. Find the deflection and slope of this beam.

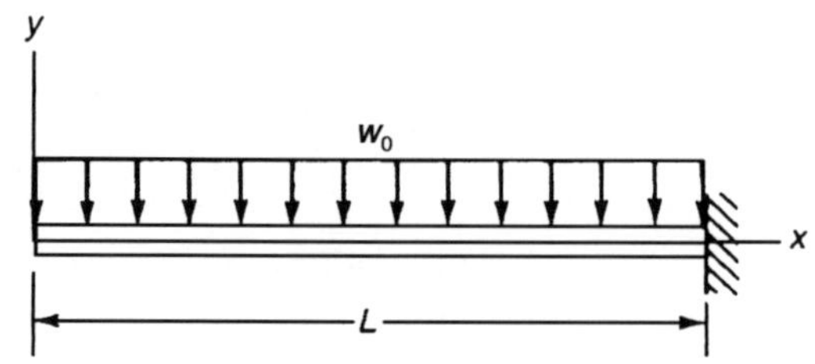

Exhibit 42

Solution

The moment is found by drawing the free-body diagram shown in Exhibit 43. The uniform load is replaced with the statically equivalent load w_0x at the position $x/2$. Moments are then summed about the cut giving

$$M = -w_0\frac{x^2}{2}$$

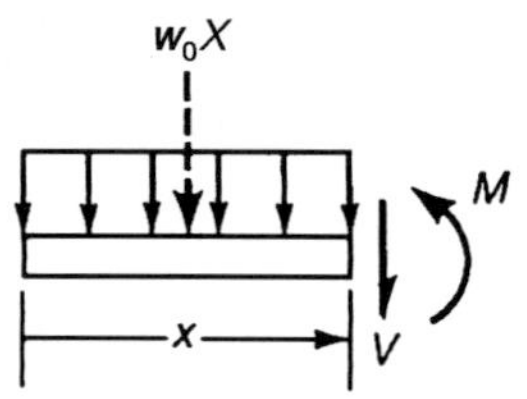

Exhibit 43

Integrating twice with respect to x,

$$\frac{dy}{dx} = \int \frac{M}{EI}\,dx + C_1 = \frac{1}{EI}\int\left(-w_0\frac{x^2}{2}\right)dx + C_1 = -\frac{1}{6}\frac{w_0x^3}{EI} + C_1$$

$$y = \int\left(-\frac{1}{6}\frac{w_0x^3}{EI}\right)dx + C_1x + C_2 = -\frac{1}{24}\frac{w_0x^4}{EI} + C_1x + C_2$$

At $x = L$ the displacement and slope must be zero so that

$$y(L) = 0 = -\frac{1}{24}\frac{w_0L^4}{EI} + C_1L + C_2$$

$$\frac{dy}{dx}(L) = 0 = -\frac{1}{6}\frac{w_0L^3}{EI} + C_1$$

Therefore,

$$C_1 = \frac{1}{6}\frac{w_0L^3}{EI}; \qquad C_2 = -\frac{1}{8}\frac{w_0L^4}{EI}$$

Inserting C_1 and C_2 into the previous expressions gives

$$y = -\frac{w_0}{24EI}(x^4 - 4xL^3 + 3L^4)$$

$$\frac{dy}{dx} = \frac{w_0}{6EI}(L^3 - x^3)$$

Fourth-Order Beam Equation

The second-order beam can be combined with the differential relationships between the shear, moment, and distributed load. Differentiate the equation with respect to x,

$$\frac{d}{dx}\left(EI\frac{d^2y}{dx^2}\right) = \frac{dM}{dx} = V$$

Differentiate again with respect to x.

$$\frac{d^2}{dx^2}\left(EI\frac{d^2y}{dx^2}\right) = \frac{dV}{dx} = q$$

For a uniform beam with a constant Young's Modulus (that is, constant EI) the fourth-order beam equation becomes

$$EI\frac{d^4y}{dx^4} = q$$

This equation can be integrated four times with respect to x. Four boundary conditions are required to solve for the four constants of integration. The boundary

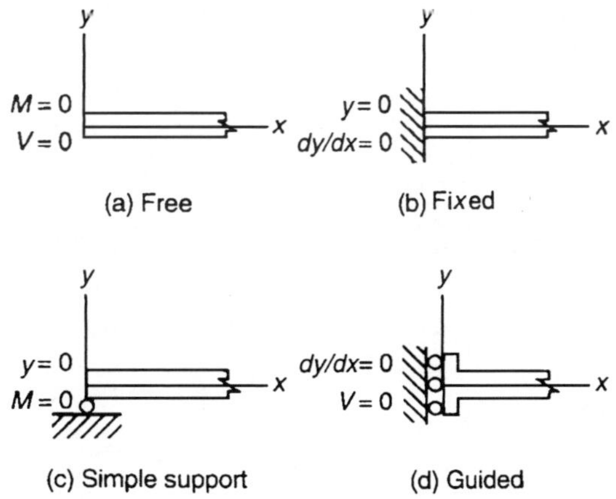

Figure 10.31 Boundary conditions for beams

conditions are on the displacement, slope, moment, and/or shear. Figure 10.31 shows the appropriate boundary conditions on the end of a beam, even with a distributed loading. If there is a concentrated force or moment applied at the end of a beam, that force or moment enters the boundary condition. For instance, an upward load of P at the left end for the free or guided beam would give $V(0) = P$ instead of $V(0) = 0$.

Example 10.50

Consider the uniformly loaded uniform beam shown in Exhibit 44. The beam is clamped at both ends. The uniform load w_0 is acting downward. Find an expression for the displacement as a function of x.

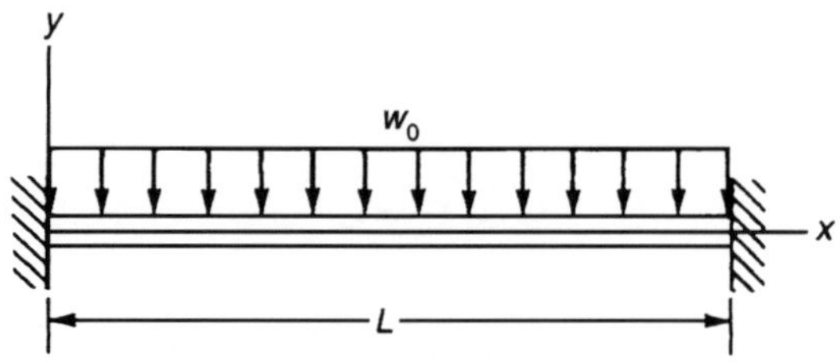

Exhibit 44

Solution

The differential equation is

$$EI\frac{d^4y}{dx^4} = q = -w_0\,(\text{constant})$$

Integrate four times with respect to x.

$$\begin{aligned}
V &= EI\frac{d^3y}{dx^3} = -w_0x + C_1 \\
M &= EI\frac{d^2y}{dx^2} = -w_0\frac{x^2}{2} + C_1x + C_2 \\
EI\frac{dy}{dx} &= -w_0\frac{x^3}{6} + C_1\frac{x^2}{2} + C_2x + C_3 \\
EIy &= -w_0\frac{x^4}{24} + C_1\frac{x^3}{6} + C_2\frac{x^2}{2} + C_3x + C_4
\end{aligned}$$

The four constants of integration can be found from four boundary conditions. The boundary conditions are

$$y(0) = 0;\quad \frac{dy}{dx}(0) = 0;\quad y(L) = 0;\quad \frac{dy}{dx}(L) = 0$$

These lead to the following:

$$\begin{aligned}
EIy(0) &= 0 = C_4 \\
EI\frac{dy}{dx}(0) &= 0 = C_3 \\
EIy(L) &= 0 = -w_0\frac{L^4}{24} + C_1\frac{L^3}{6} + C_2\frac{L^2}{2} \\
EI\frac{dy}{dx}(L) &= 0 = -w_0\frac{L^3}{6} + C_1\frac{L^2}{2} + C_2L
\end{aligned}$$

Solving the last two equations for C_1 and C_2 gives

$$C_1 = \frac{1}{2}w_0L;\qquad C_2 = -\frac{1}{12}w_0L^2$$

Inserting these values into the equation for y gives

$$y = -\frac{w_0 x^2}{EI}\left(\frac{1}{24}x^2 - \frac{1}{12}xL + \frac{1}{24}L^2\right)$$

Some solutions for uniform beams with various loads and boundary conditions are shown in Table 10.1.

Table 10.1 Deflection and slope formulas for beams

	Beam	Deflection, v	Slope, v'
1.	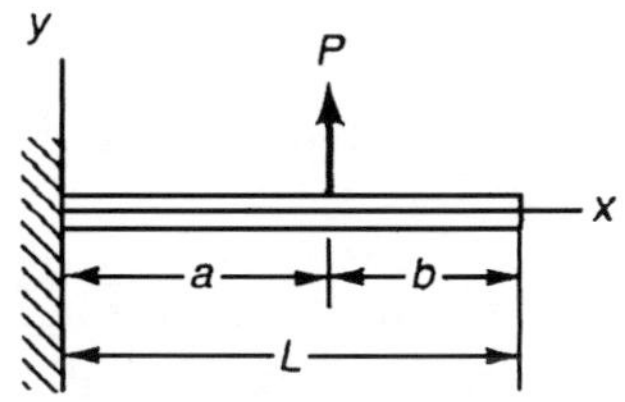	For $0 \le x \le a$ $y = \frac{Px^2}{6EI}(3a - x)$ For $a \le x \le L$ $y = \frac{Pa^2}{6EI}(3x - a)$	For $0 \le x \le a$ $\frac{dy}{dx} = \frac{px}{2EI}(2a - x)$ For $a \le x \le L$ $\frac{dy}{dx} = \frac{Pa^2}{2EI}a$
2.	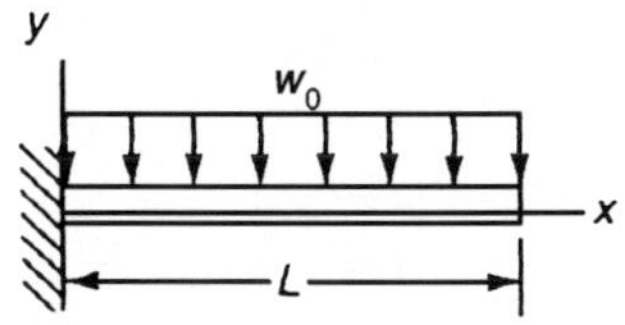	$y = -\frac{w_0 x^2}{24\,EI}(x^2 - 4Lx + 6L^2)$	$\frac{dy}{dx} = -\frac{w_0 x}{6EI}(x^2 - 12Lx + 12L^2)$
3.	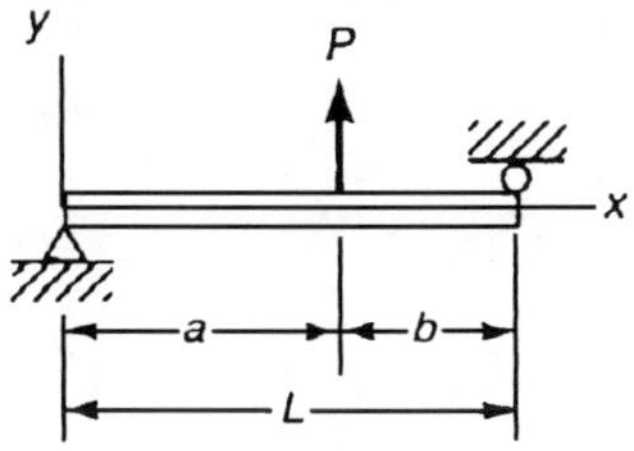	For $0 \le x \le a$ $y = \frac{Pbx}{6LEI}(L^2 - b^2 - x^2)$ For $a \le x \le L$ $y = \frac{Pa(L - x)}{6LEI}(2Lx - a^2 - x^2)$	For $0 \le x \le a$ $\frac{dy}{dx} = \frac{Pb}{6LEI}(L^2 - b^2 - 3x^2)$ For $a \le x \le L$ $\frac{dy}{dx} = \frac{Pa}{6LEI}(2L^2 + a^2 - 6Lx + 3x^2)$
4	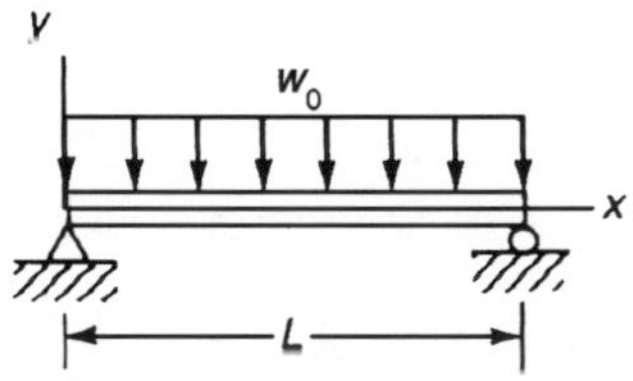	$y = -\frac{w_0 x}{24EI}(L^3 - 2Lx^2 + x^3)$	$\frac{dy}{dx} = -\frac{w_0}{24EI}(L^3 - 6Lx^2 + 4x^3)$
5.	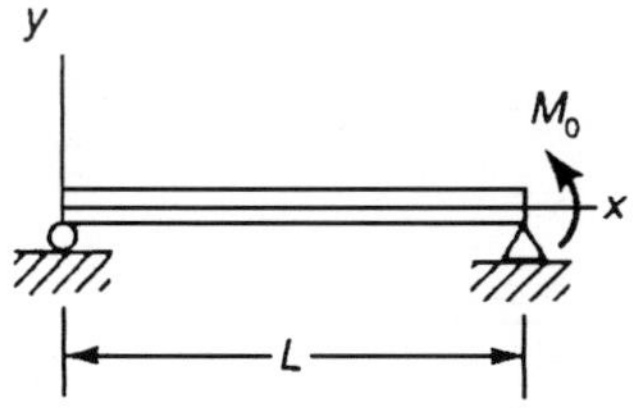	$y = -\frac{M_0 x}{6EIL}(L^2 - x^2)$	$\frac{dy}{dx} = -\frac{M_0}{6EIL}(L^2 - 3x^2)$

Superposition

In addition to the use of second-order and fourth-order differential equations, a very powerful technique for determining deflections is the use of superposition. Because all of the governing differential equations are linear, solutions can be directly superposed. Use can be made of tables of known solutions, such as those in Table 10.1, to form solutions to many other problems. Some examples of superposition follow.

Example 10.51

Find the maximum displacement for the simply supported uniform beam loaded by two equal loads placed at equal distances from the ends as shown in Exhibit 45.

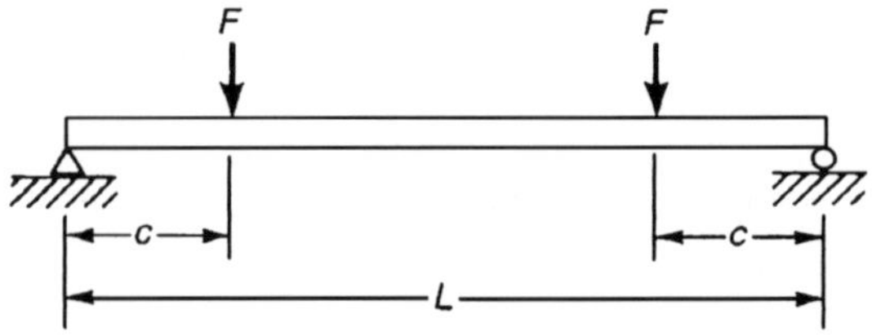

Exhibit 45

Solution

The solution can be found by superposition of the two problems shown in Exhibit 46. From the symmetry of this problem, it can be seen that the maximum deflection will be at the center of the span. The solution for the beam shown in Exhibit 46(a) is found as case 3 in Table 10.1. In Exhibit 46(a) the center of the span is to the left of the load F so that the formula from the table for $0 \le x \le a$ is chosen. In the formula, $x = L/2$, $c = b$, and $P = -F$ so that

$$y_a\left(\frac{L}{2}\right) = \frac{Pbx}{6LEI}\left(L^2 - b^2 - x^2\right) = -\frac{Fc\left(\frac{L}{2}\right)}{6LEI}\left[L^2 - c^2 - \left(\frac{L}{2}\right)^2\right] = -\frac{Fc}{48EI}\left(3L^2 - 4c^2\right)$$

The central deflection of the beam in Exhibit 46(b) will be the same, so the maximum downward deflection, Δ, will be

$$\delta = 2y_a\left(\frac{L}{2}\right) = -\frac{Fc}{24EI}(3L^2 - 4c^2)$$

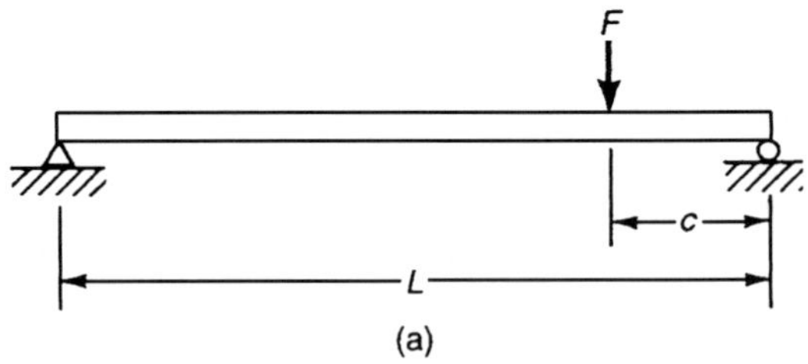

(a)

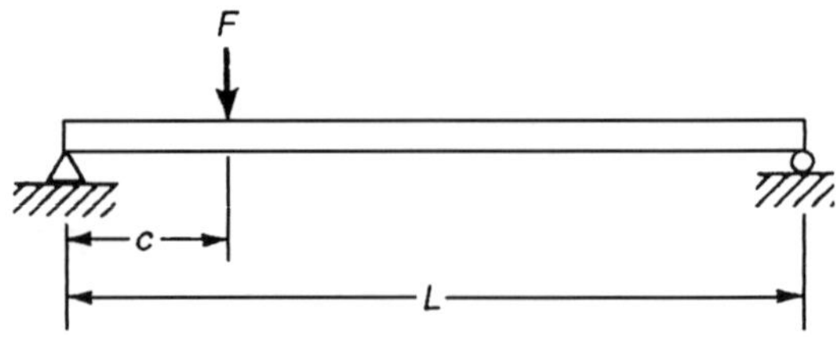

(b)

Exhibit 46

Example 10.52

Find an expression for the deflection of the uniformly loaded, supported, cantilever beam shown in Exhibit 47.

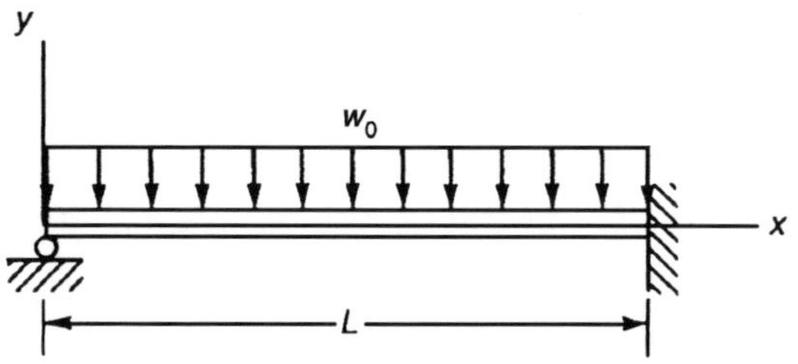

Exhibit 47

Solution

Superpose cases 4 and 5 as shown in Exhibit 48 so that the moment M_0 is of the right magnitude and direction to suppress the rotation at the right end. The rotation

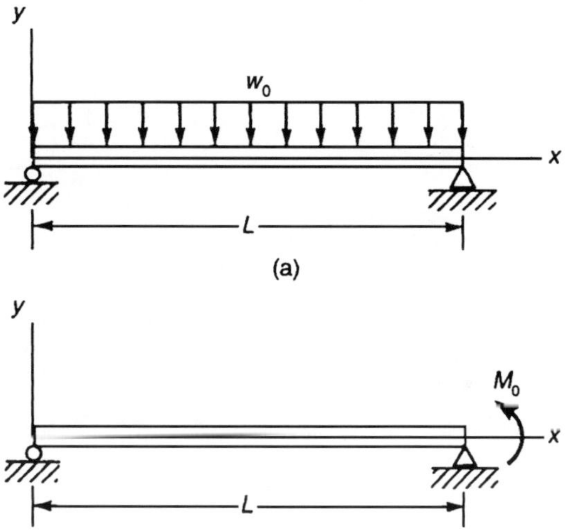

Exhibit 48

for each case from Table 10.1 is

$$\left(\frac{dy}{dx}\right)_4\bigg|_{x=L} = -\frac{w_0}{24EI}(L^3 - 6L^3 + 4L^3) = \frac{w_0 L^3}{24EI}$$

$$\left(\frac{dy}{dx}\right)_5\bigg|_{x=L} = -\frac{M_0}{6EIL}(L^2 - 3L^2) = \frac{M_0 L}{3EI}$$

Setting the rotation at the end equal to zero gives

$$\left(\frac{dy}{dx}\right)_4\bigg|_{x=L} + \left(\frac{dy}{dx}\right)_5\bigg|_{x=L} = 0 = \frac{w_0 L^3}{24EI} + \frac{M_0 L}{3EI}$$

$$M_0 = -\frac{w_0 L^2}{8}$$

Substituting this expression into the formulas in the table and adding gives

$$y = -\frac{w_0 x}{24EI}(L^3 - 2Lx^2 + x^3) + \frac{w_0 L^2}{8}\frac{x}{6EI}(L^2 - x^2) = -\frac{w_0 x}{48EI}(L^3 - 3Lx^2 + 2x^3)$$

COMBINED STRESS

In many cases, members can be loaded in a combination of bending, torsion, and axial loading. In these cases, the solution of each portion is exactly as before; the effects of each are simply added. This concept is best illustrated by an example.

Example 10.53

In Exhibit 49, there is a thin-walled, aluminum tube AB, which is attached to a wall at A. The tube has a rectangular cross-section member BC attached to it. A vertical load is placed on the member BC as shown. The aluminum tube has an outer diameter of 50 mm and a wall thickness of 3.25 mm. Take $P = 900$ N, $a = 450$ mm, and $b = 400$ mm. Find the state of stress at the top of the tube at the point D. Draw Mohr's circle for this point, and find the three principal stresses.

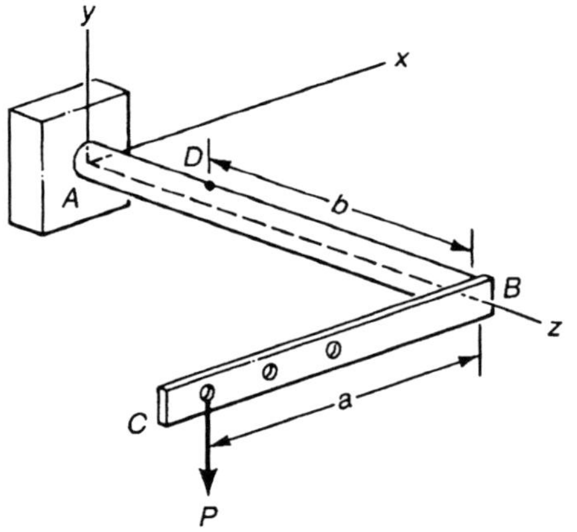

Exhibit 49

Solution

Cut the tube at the Point D. Draw the free-body diagram as in Exhibit 50(a). From that free-body diagram, a summation of moments at the cut about the z-axis gives

$$T = Pa = (900\text{ N})(450\text{ mm}) = 405\text{ N} \bullet \text{m}$$

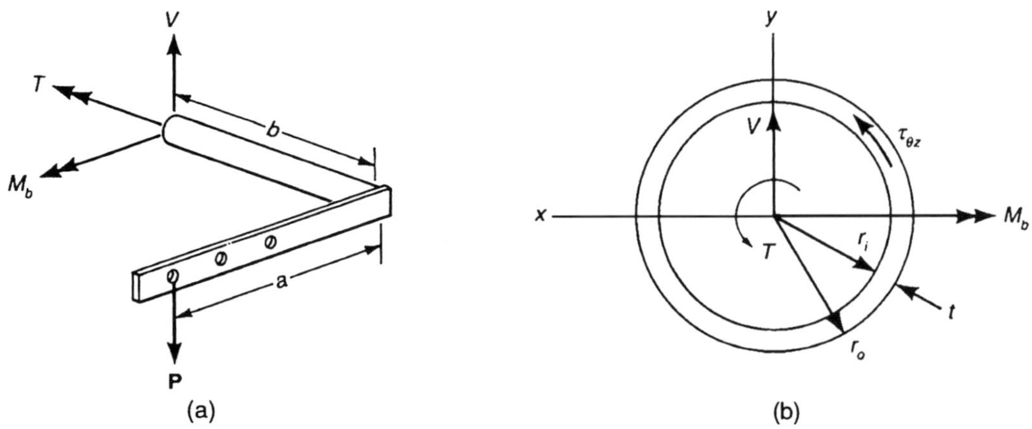

Exhibit 50

A summation of moments at the cut about an axis parallel with the x-axis gives

$$M_b = Pb = (900\text{ N})(400\text{ mm}) = 360\text{ N} \bullet \text{m}$$

A summation of vertical forces gives

$$V = P$$

Exhibit 50(b) shows the force and moments acting on the cross-section. The bending and shearing stresses caused by these loads are

$$\sigma_z = \frac{M_b y}{I_{xx}} \qquad (\text{from } M_b)$$

$$\tau_{\theta z} = \frac{Tr}{I_z} \qquad (\text{from } T)$$

$$\tau_{zy} = \frac{VQ}{I_{xx} b} \qquad (\text{from } V)$$

The shearing stress attributed to V will be zero at the top of the beam and can be neglected. The moments of inertia are

$$I_{xx} = \frac{\pi\left(r_o^4 - r_i^4\right)}{4} = \frac{\pi(25^4 - 21.75^4)}{4} = 131\times10^3\text{ mm}^4$$

$$I_z = \frac{\pi\left(r_o^4 - r_i^4\right)}{2} = 2I_{xx} = 262\times10^3\text{ mm}^4$$

At the top of the tube r = 25 mm and y = 25 mm, so the stresses are

$$\sigma_z = \frac{M_b y}{I_{xx}} = \frac{(360\text{ N}\bullet\text{m})(25\text{mm})}{131\times10^3\text{ mm}^4} = 68.7\text{ MPa}$$

$$\tau_{\theta z} = \frac{Tr}{I_z} = \frac{(405\text{ N}\bullet\text{m})(25\text{mm})}{262\times10^3\text{ mm}^4} = 38.6\text{ MPa}$$

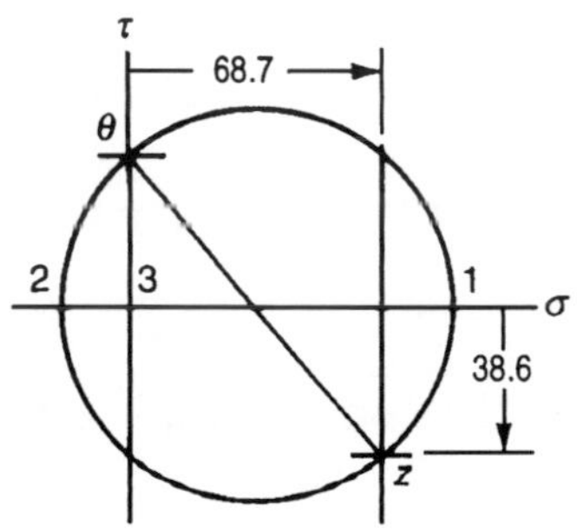

Exhibit 51

The Mohr's circle plot for this is shown in Exhibit 51

$$R = \sqrt{\left(\frac{\sigma_z - \sigma_\theta}{2}\right)^2 + \tau_{\theta z}^2} = \sqrt{\left(\frac{68.7 - 0}{2}\right)^2 + 38.6^2} = 51.7\text{MPa}$$

$$C = \frac{\sigma_z \sigma_\theta}{2} = \frac{68.7}{2} = 34.4\text{MPa}$$

$$\sigma_1 = C + R = 86.1\text{MPa}$$

$$\sigma_2 = C - R = -17.3\text{MPa}$$

Because this is a state of plane stress, the third principal stress is

$$\sigma_3 = 0$$

COLUMNS

Buckling can occur in slender columns when they carry a high axial load. Figure 10.32(a) shows a simply supported slender member with an axial load. The beam is shown in the horizontal position rather than in the vertical position for convenience. It is assumed that the member will deflect from its normally straight configuration as shown. The free-body diagram of the beam is shown in Figure 10.32(b). Figure 10.32(c) shows the free-body diagram of a section of the beam. Summation of moments on the beam section in Figure 10.32(c) yields

$$M + Py = 0$$

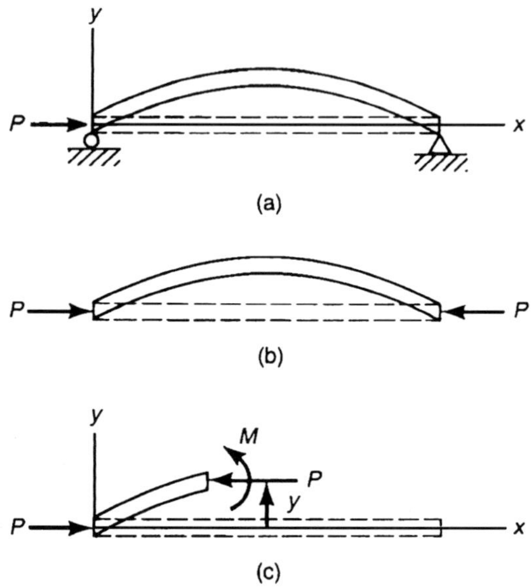

Figure 10.32 Buckling of simply supported column

Since M is equal to EI times the curvature, the equation for this beam can be expressed as

$$\frac{d^2y}{dx^2} + \lambda y = 0$$

where

$$\lambda^2 = \frac{P}{EI}$$

The solution satisfying the boundary conditions that the displacement is zero at either end is

$$v = \sin(\lambda x), \text{ where } \lambda = n\pi/L \; n = 1, 2, 3\ldots$$

The lowest value for the load P is the buckling load, so $n = 1$ and the critical buckling load, or Euler buckling load, is

$$P_{cr} = \frac{\pi^2 EI}{L^2}$$

For other than simply supported boundary conditions, the shape of the deflected curve will always be some portion of a sine curve. The simplest shape consistent with the boundary conditions will be the deflected shape. Figure 10.33 shows a sine curve and the beam lengths that can be selected from the sine curve. The critical buckling load can be redefined as

$$P_{cr} = \frac{\pi^2 EI}{L_e^2} = \frac{\pi^2 EI}{(kL)^2} = \frac{\pi^2 E}{(k\,L/r)^2}$$

where the radius of gyration r is defined as $\sqrt{I/A}$. The ratio L/r is called the slenderness ratio.

From Figure 10.33, it can be seen that the values for L_e and k are as follows:

For simple supports: $L = L_e$; $L_e = L$; $k = 1$

For a cantilever: $L = 0.5L_e$; $L_e = 2L$; $k = 2$

For both ends clamped: $L = 2L_e$; $L_e = 0.5L$; $k = 0.5$

For supported-clamped: $L = 1.43L_e$; $L_e = 0.7L$; $k = 0.7$

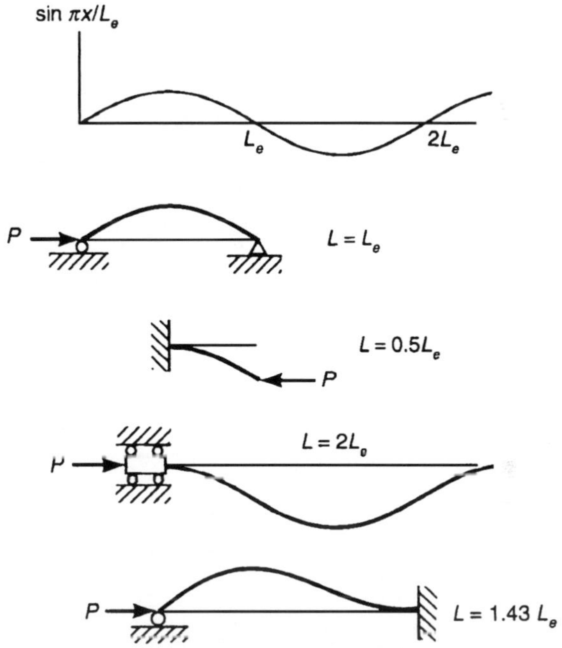

Figure 10.33 Buckling of columns with various boundary conditions

In dealing with buckling problems, keep in mind that the member must be slender before buckling is the mode of failure. If the beam is not slender, it will fail by yielding or crushing before buckling can take place.

Example 10.54

A steel pipe is to be used to support a weight of 130 kN as shown in Exhibit 52. The pipe has the following specifications: $OD = 100$ mm, $ID = 90$ mm, $A = 1500$ mm^2, and $I = 1.7 \times 10^6$ mm^4. Take $E = 210$ GPa and the yield stress $Y = 250$ MPa. Find the maximum length of the pipe.

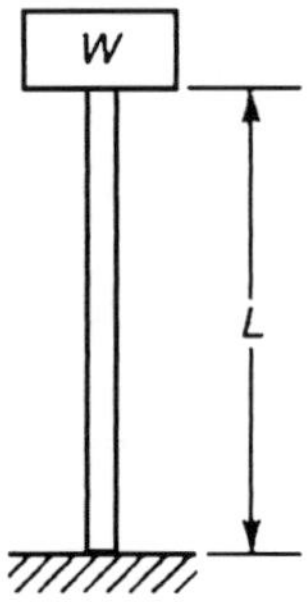

Exhibit 52

Solution

First, check to make sure that the pipe won't yield under the applied weight. The stress is

$$\sigma = \frac{P}{A} = \frac{130\,\text{kN}}{1500\,\text{mm}^2} = 86.7\,\text{MPa} < Y$$

This stress is well below the yield, so buckling will be the governing mode of failure. This is a cantilever column, so the constant k is 2. The critical load is

$$P_{\text{cr}} = \frac{\pi^2 EI}{(2L)^2}$$

Solving for L gives

$$L = \pi\sqrt{\frac{EI}{4P}} = \pi\sqrt{\frac{(210\,\text{GPa})(1.7\times10^6\ \text{mm}^4)}{4(130\,\text{kN})}} = 2.60\,\text{m}$$

The maximum length is 2.6 m.

SELECTED SYMBOLS AND ABBREVIATIONS

Symbol or Abbreviation	Description
σ	stress
ε	strain
v	Poisson's ratio
kip	kilopound
E	modulus of elasticity
δ	deformation
W	weight
P	load
P, p	pressure
I	moment of inertia
τ	shear stress
T	torque
A	area
M	moment
V	shear
L	length
F	force

PROBLEMS

10.1 A 45,340-kg mass is supported by a 102-mm diameter steel cylinder surrounded by a copper tube of equal length with an outside diameter of 203 mm, which fits snugly around the steel cylinder. The assembly is shown in Exhibit 10.1. The weight is distributed uniformly over the surfaces of the supporting members. What is the stress in the copper tube?

$$E(\text{steel}) = 2.1 \times 10^{11}\ \text{Pa}$$
$$E(\text{copper}) = 1.12 \times 10^{11}\ \text{Pa}$$

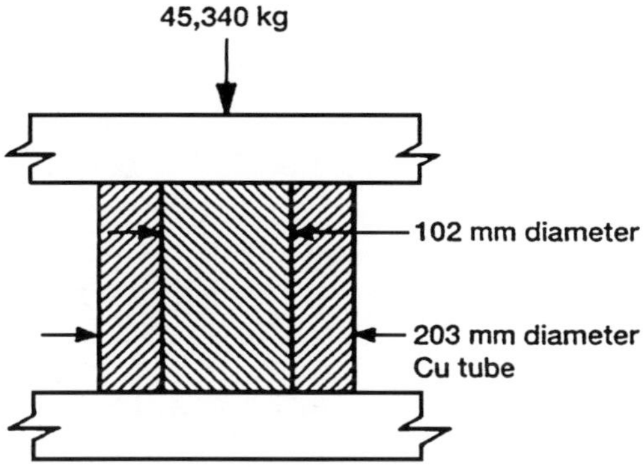

Exhibit 10.1

a. 10.58 MPa
b. 10.92 MPa
c. 11.26 MPa
d. 11.67 MPa

10.2 A 5.0-cm-diameter steel rod 3.0 meters long is pinned at both ends, and is unstressed at 15 °C. See Exhibit 10.2. Which of the following most nearly equals the highest temperature to which the rod may be heated before it will buckle?

$$\alpha = 11.7 \times 10^{-6}\ \text{m/m} \cdot {}^\circ\text{C}$$

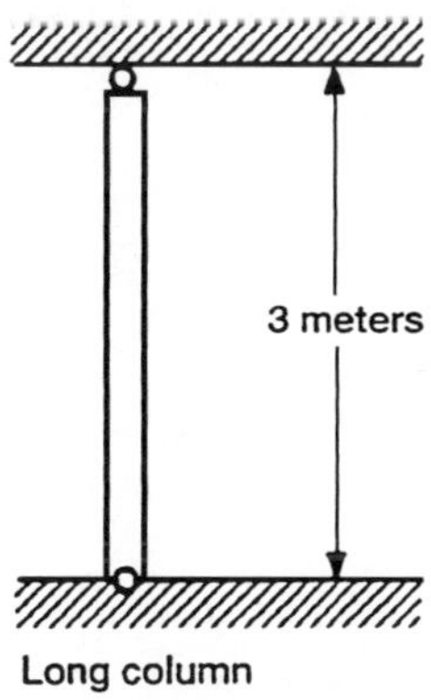

Exhibit 10.2

a. 29.6 C
b. 30.5 C
c. 31.6 C
d. 45.6 C

10.3 Three bars of different materials, as shown in Exhibit 10.3, are to be compressed equally by a plate that is to remain horizontal. Which of the following most nearly is equal to the distance x?

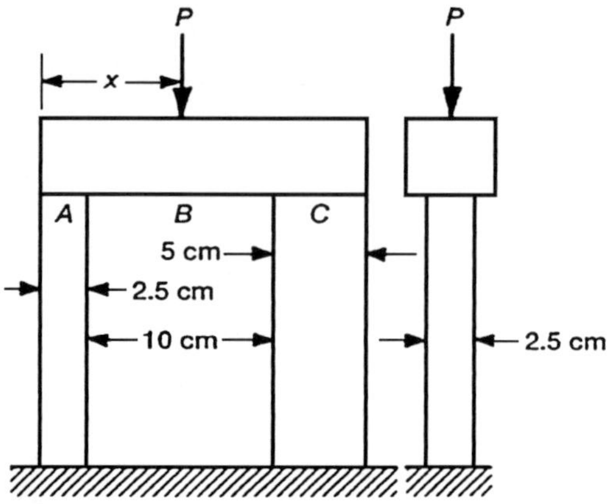

Exhibit 10.3

Bar	E(GPa)	Allowable Stress (MPa)
A	210	138
B	70	69
C	140	104

a. 6.0 cm
b. 6.5 cm
c. 7.3 cm
d. 7.9 cm

10.4 A 10-cm-diameter, horizontal, solid steel shaft is 2.50 meters long and is rigidly held at both ends. A torque of 16,000 N · m is applied to a pulley that is keyed to the shaft 60 cm from the left end. Which of the following most nearly equals the maximum torsional stress in the shaft?

a. 58 MPa
b. 62 MPa
c. 66 MPa
d. 69 MPa

10.5 The shaft shown in Exhibit 10.5 is made of one material and is held rigidly at the ends. Which of the following most nearly equals the maximum end torque reaction?

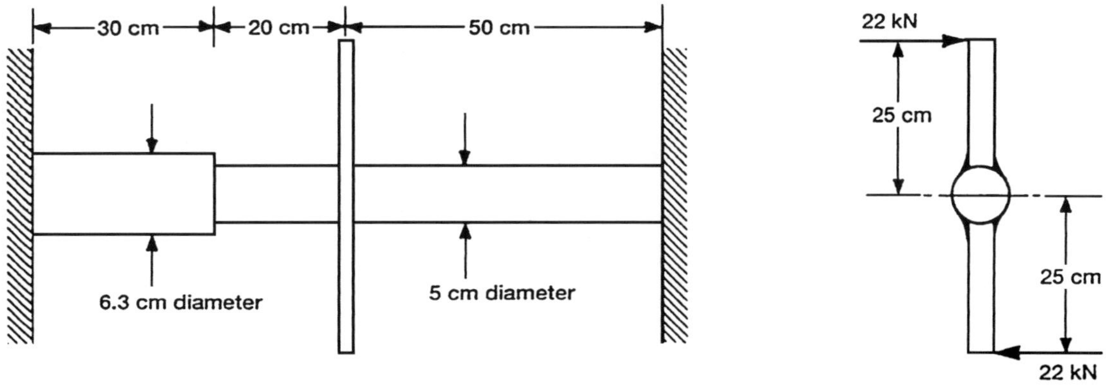

Exhibit 10.5

a. 4.3 kJ
b. 4.9 kJ
c. 6.3 kJ
d. 6.7 kJ

10.6 A massless, bimetallic, cylindrical bar is twisted by a torque of 1900 N · m at the change of section. Which of the following most nearly is equal to the maximum end reaction? (See Exhibit 10.6.)

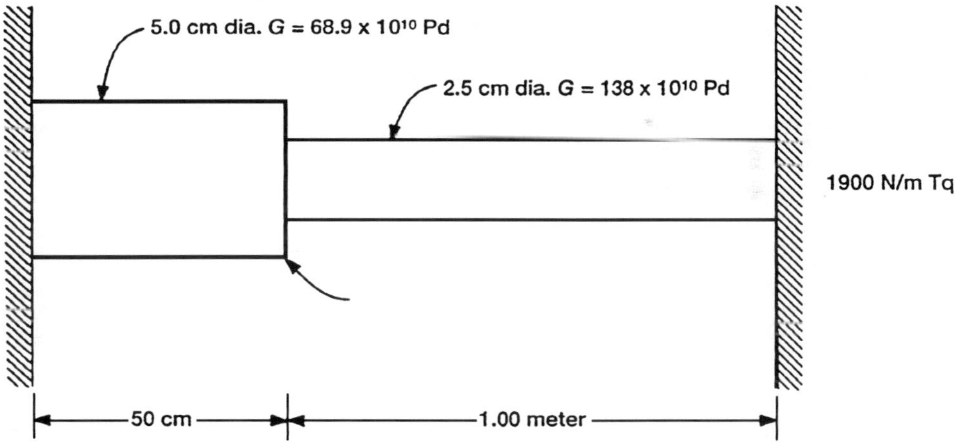

Exhibit 10.6

a. 0.9 kJ
b. 1.3 kJ
c. 1.8 kJ
d. 2.2 kJ

10.7 A mass of 450 kg was originally held up by two 2.5-cm-diameter steel tie rods as shown in Exhibit 10.7a. When it was desired to increase the mass to 900 kg a proposal was made to add two 2.5-cm-diameter cables as shown. Assume no stress before the weight was applied, and no slack in the system. Which of the following most nearly is equal to the amount of the added load that would be held by the cables? For steel cable $E =$ 84 GPa

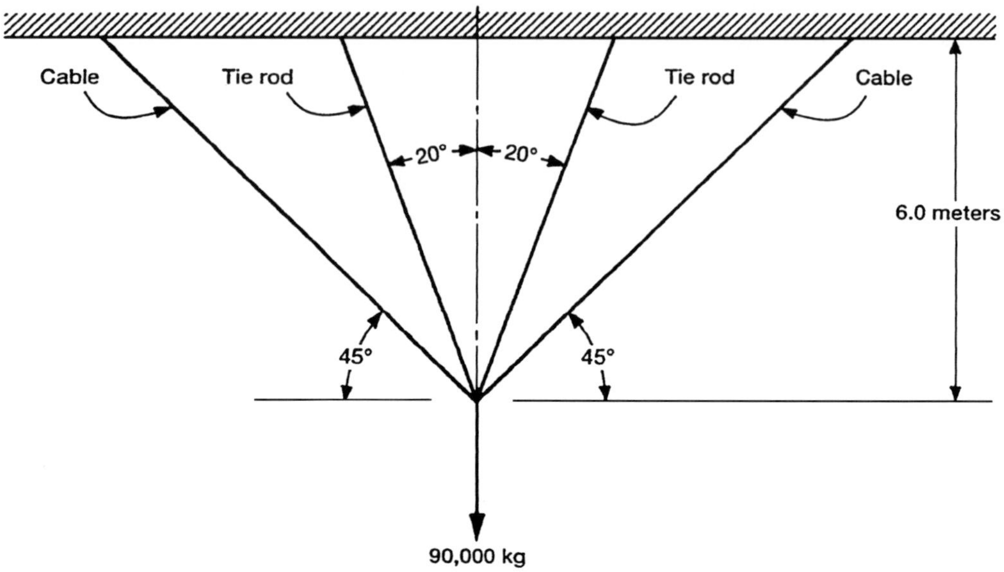

Exhibit 10.7a

a. 29%
b. 21%
c. 18%
d. 15%

10.8 A 25.0-mm-thick steel plate 0.25 m wide is loaded as shown in Exhibit 10.8. What is the maximum stress in the plate?

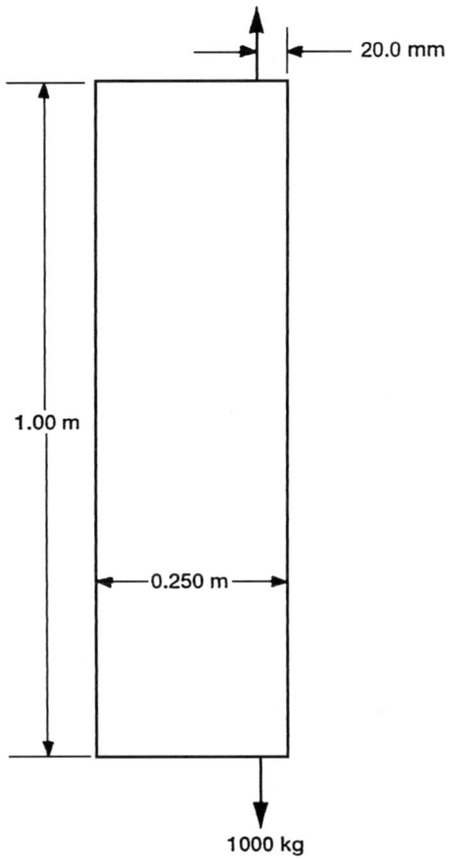

Exhibit 10.8

a. 1.60 Mpa
b. 2.91 MPa
c. 4.22 MPa
d. 5.55 MPa

10.9 A closed-end tube with a 152.4-mm *OD* and a 2.54-mm wall thickness contains a fluid at a pressure of 3.448 MPa. It is subjected to a torsional load of 678 N · m. What is the maximum principal stress in the tube?
a. 93 MPa
b. 98 Mpa
c. 101 MPa
d. 105 MPa

10.10 Mohr's circle is used to determine
a. Shear stress
b. Poisson's ratio
c. Residual stress
d. Combined stress

10.11 What would be the maximum stress in a cantilever beam holding a mass of 500 kg at its end if it is 3.00 meters long and is made of a 200-mm wide by 300-mm high wooden beam?

a. 6.8 MPa
b. 6.1 Mpa
c. 5.5 MPa
d. 4.9 MPa

10.12 A water tank is filled to a depth of ten meters. It is 35 meters in diameter and is made of steel 8 mm thick. What is the stress in the tank wall at the bottom?

a. 214 MPa
b. 206 Mpa
c. 199 MPa
d. 192 MPa

10.13 What is the magnitude of the force acting on member *CD* in Exhibit 10.13?

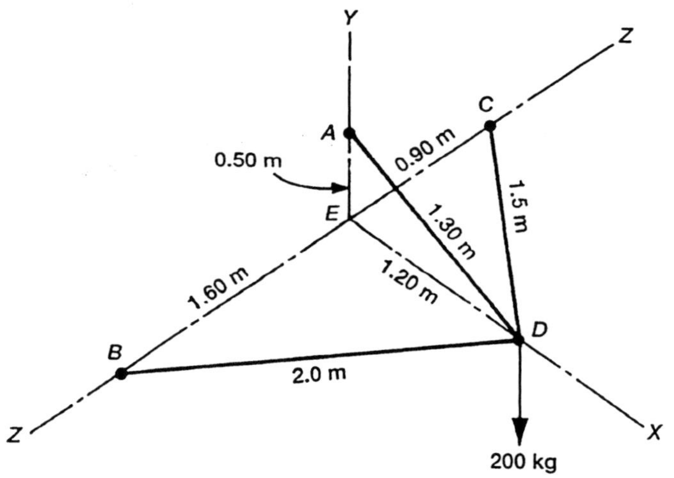

Exhibit 10.13

a. 2.9 kN
b. 4 kN
c. 3.8 kN
d. 4.3 kN

10.14 Given the beam shown in Exhibit 10.14, loaded as shown, which of the following most nearly is equal to the maximum moment, measured in N·m?

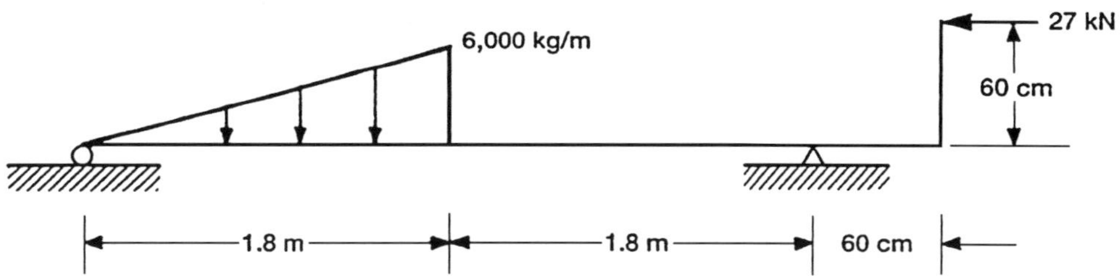

Exhibit 10.14

a. 41 kJ
b. 44 kJ
c. 48 kJ
d. 52 kJ

10.15 A cantilever beam supports a uniformly varying load as shown in Exhibit 10.15. Which of the following most nearly is equal to the maximum bending stress?

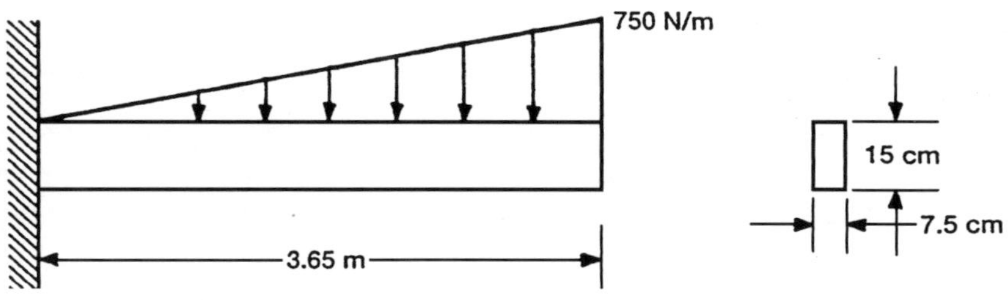

Exhibit 10.15

a. 8.7 MPa
b. 9.2 MPa
c. 10.6 MPa
d. 11.8 MPa

10.16 Which of the following most nearly is equal to the maximum moment for the beam that is loaded as shown in Exhibit 10.16?

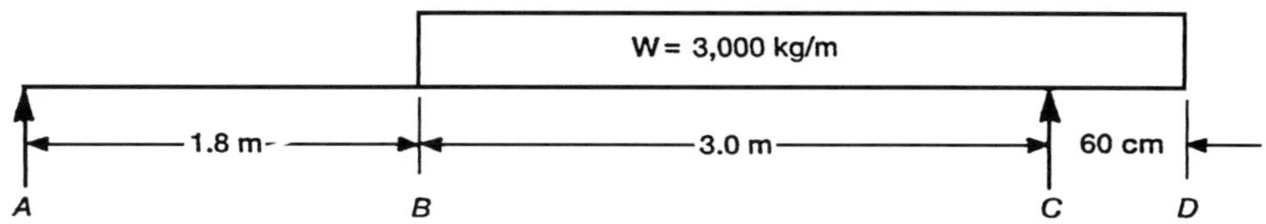

Exhibit 10.16

a. 56.5 kJ
b. 59.7 kJ
c. 62.1 kJ
d. 64.3 kJ

10.17 For the beam in Exhibit 10.17, which is loaded as shown, which of the following most nearly is equal to the maximum moment in the beam?

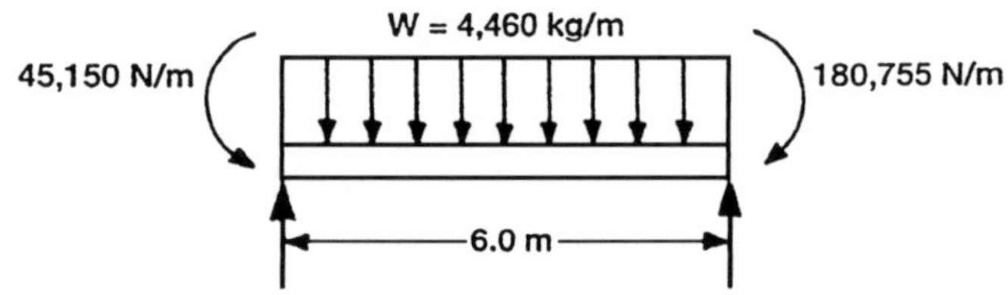

Exhibit 10.17

a. 180.8 kJ
b. 153.7 kJ
c. 132.6 kJ
d. 89.8 kJ

10.18 A section of a rectangular water tank is as shown in Exhibit 10.18. The beams are in pairs and are spaced 1.0 meter apart along the length of the tank. Assume that the section shown is near the middle portion of the tank and that no end (of the tank) effects act on the section shown. Which of the following most nearly equals the maximum moment in the beam shown in the figure?

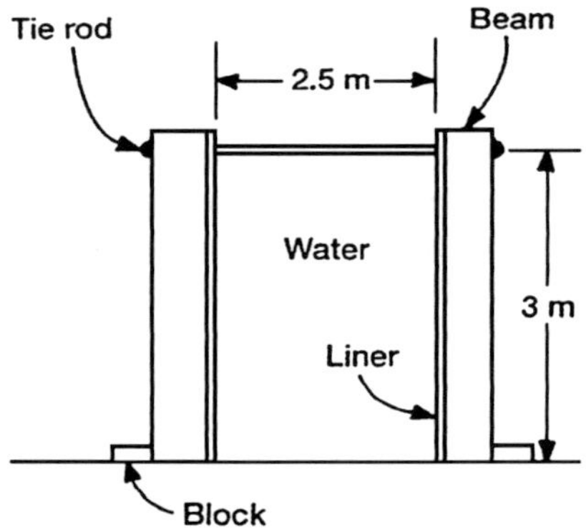

Exhibit 10.18

a. 29.4 kJ
b. 22.7 kJ
c. 16.3 kJ
d. 13.1 kJ

10.19 Which of the following most nearly is equal to the maximum moment that occurs in the beam that is loaded as shown in Exhibit 10.19?

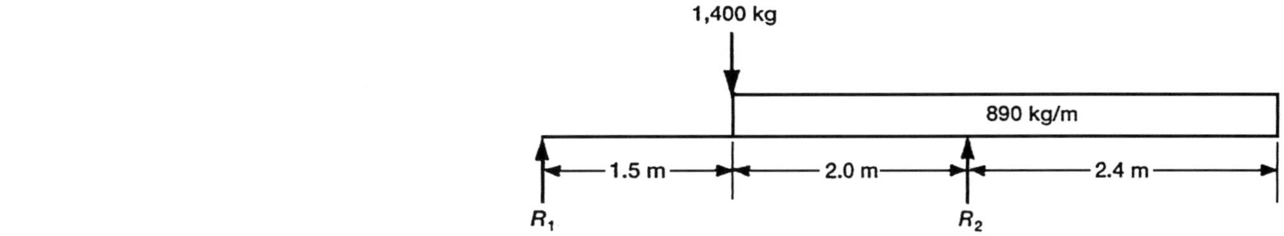

Exhibit 10.19

a. 8.5 kJ
b. 12.7 kJ
c. 18.3 kJ
d. 25.1 kJ

10.20 The composite beam shown in Exhibit 10.20 is subjected to a maximum moment of 17,000 N · m. Which of the following most nearly is equal to the maximum tensile stress in the steel? $E_{wood} = 8.4$ GPa.

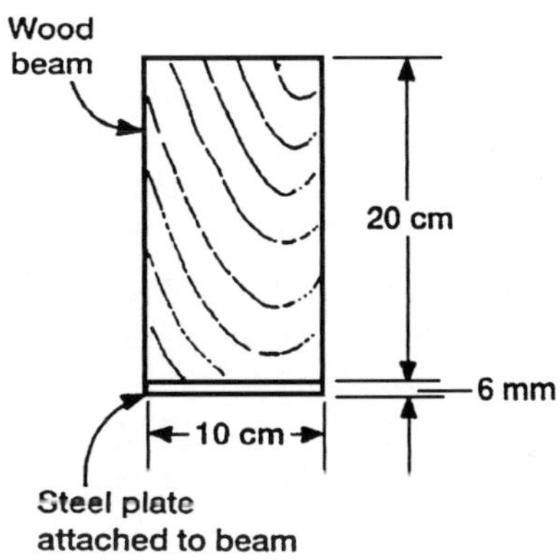

Exhibit 10.20

a. 167 MPa
b. 123 MPa
c. 92 MPa
d. 7 MPa

10.21 Which of the following most nearly is equal to the deflection of point A in Exhibit 10.21 if the parts are all of steel?

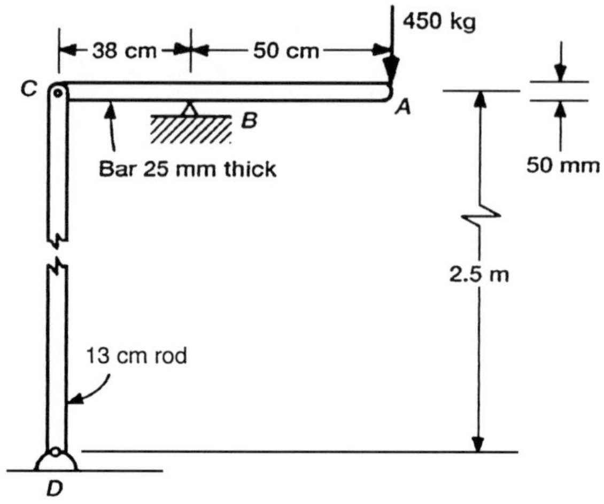

Exhibit 10.21

a. 4.1 mm
b. 5.2 mm
c. 6.6 mm
d. 7.3 mm

10.22 A short steel column with a 25-cm *ID*, and a 30-cm *OD* is filled with concrete. If the maximum allowable stress for the steel is 276 MPa and that for the concrete is 24 MPa, which of the following most nearly is equal to the maximum axial load that the composite column can safely hold? $E_{concrete} =$ 19.6 GPa.

a. 684 Mg
b. 652 Mg
c. 621 Mg
d. 598 Mg

10.23 Which of the following most nearly is equal to the maximum mass P that could be supported by the steel beam shown in Exhibit 10.23 if the compressive stress is limited to 13.8 MPa in the plastic support block?

Assume $E_{cable} = 84$ GPa and $E_{plastic} = 10.5$ GPa.

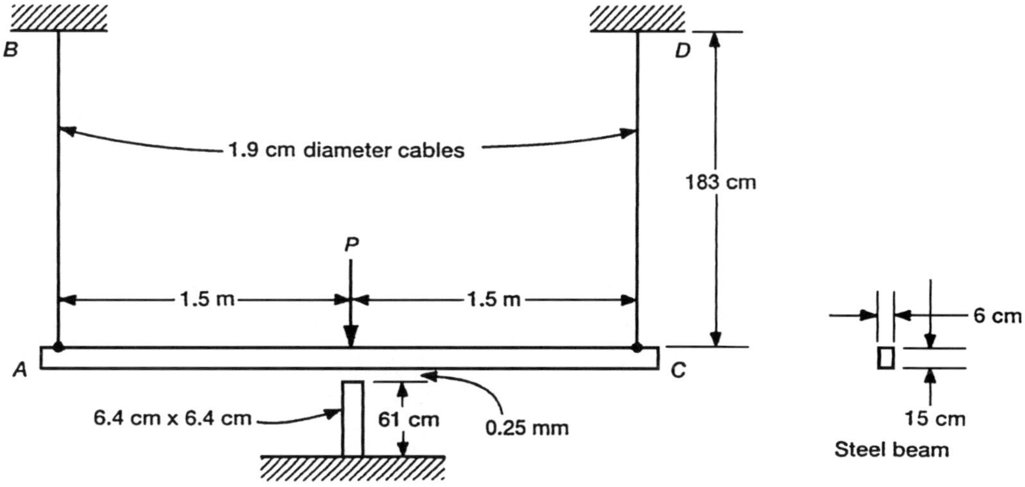

Exhibit 10.23

a. 4.9 Kg
b. 5.5 Kg
c. 5.9 Kg
d. 6.3 Kg

10.24 A beam is simply supported at the two ends and is uniformly loaded along its length with a load of 225 kg/m. If the beam is 4.5 meters long and is made of a 20-cm by 20-cm wooden timber, which of the following most nearly is equal to the deflection at a point 122 cm from one end? $E_{wood} =$ 5.0 GPa.

a. 5.9 mm
b. 6.4 mm
c. 6.9 mm
d. 7.2 mm

10.25 A reinforced concrete beam, 30 cm wide by 76 cm deep and 6 meters long, is simply supported at the ends. The beam is reinforced with four bars, each with a cross-sectional area of 3.9 cm^2 and placed 8 cm from the bottom of the beam. The beam supports its own mass plus an applied load of 3000 kg per linear meter. Which of the following most nearly is equal to the stress in the steel rebar?

$$n = 10 \text{ and}$$
$$\rho_{concrete} = 2370 \text{ kg/m}^3.$$

a. 141 MPa
b. 149 MPa
c. 151 MPa
d. 158 MPa

10.26 The stepped circular aluminum shaft in Exhibit 10.26 has two different diameters: 20 mm and 30 mm. Loads of 20 kN and 12 kN are applied at the end of the shaft and at the step. The maximum stress is most nearly:

a. 23.4 MPa c. 28.3 MPa
b. 26.2 MPa d. 30.1 MPa

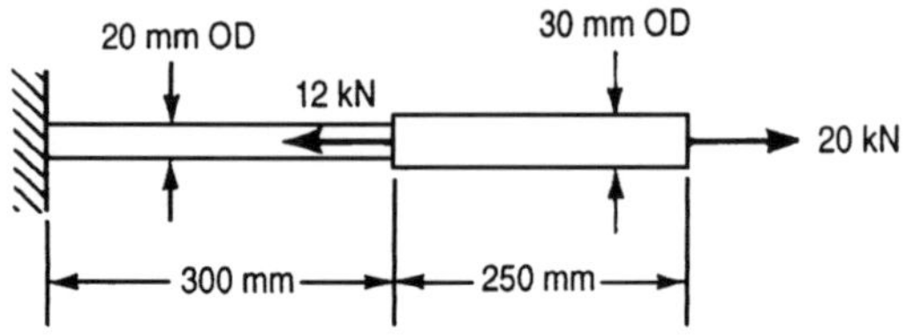

Exhibit 10.26

10.27 For the same shaft as in Problem 10.26 take $E = 69$ GPa. The end deflection is most nearly:

a. 0.18 mm c. 0.35 mm
b. 0.21 mm d. 0.72 mm

10.28 The shaft in Exhibit 10.28 is the same aluminum stepped shaft considered in problems 10.26 and 10.27, except now the right-hand end is also built into a wall. Assume that the member was built in before the load was applied. The maximum stress is most nearly:

a. 12.2 MPa c. 13.1 MPa
b. 12.7 MPa d. 15.2 MPa

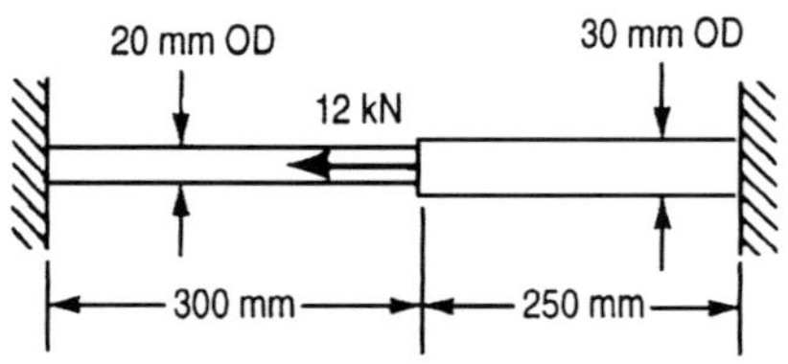

Exhibit 10.28

10.29 For the same shaft as in Problem 10.28 the deflection of the step is most nearly:

a. 0.038 mm c. 0.064 mm
b. 0.042 mm d. 0.086 mm

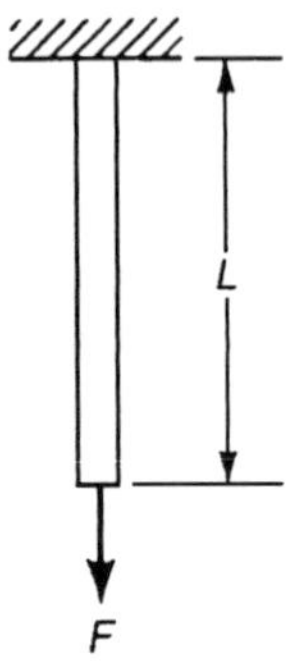

Exhibit 10.30

10.30 The uniform rod shown in Exhibit 10.30 has a force F at its end which is equal to the total weight of the rod. The rod has a unit weight γ. The total deflection of the rod is most nearly:

a. $1.00\ \gamma\ L^2/E$ c. $1.50\ \gamma\ L^2/E$
b. $1.25\ \gamma\ L^2/E$ d. $1.75\ \gamma\ L^2/E$

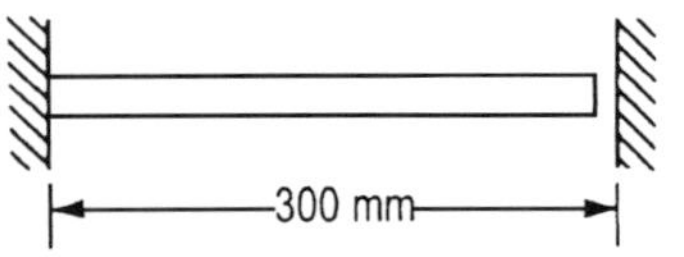

Exhibit 10.31

10.31 At room temperature, 22°C, a 300-mm stainless steel rod (Exhibit 10.31) has a gap of 0.15 mm between its end and a rigid wall. The modulus of elasticity $E = 210$ GPa. The coefficient of thermal expansion $\alpha = 17 \times 10^{-6}$/°C. The area of the rod is 650 mm². When the temperature is raised to 100 °C, the stress in the rod is most nearly:

a. 175 MPa (tension) c. −17.5 MPa (compression)
b. 0 MPa d. −175 MPa (compression)

10.32 A steel cylindrical pressure vessel is subjected to a pressure of 21 MPa. Its outer diameter is 4.6 m, and its wall thickness is 200 mm. The maximum principal stress in this vessel is most nearly:

a. 183 MPa c. 362 MPa
b. 221 MPa d. 432 MPa

10.33 A pressure vessel shown in Exhibit 10.33 is known to have an internal pressure of 1.4 MPa. The outer diameter of the vessel is 300 mm. The vessel is made of steel; $v = 0.3$ and $E = 210$ GPa. A strain gage in the circumferential direction on the vessel indicates that, under the given pressure, the strain is 200×10^{-6}. The wall thickness of the pressure vessel is most nearly:

a. 3.2 mm c. 6.4 mm
b. 4.3 mm d. 7.8 mm

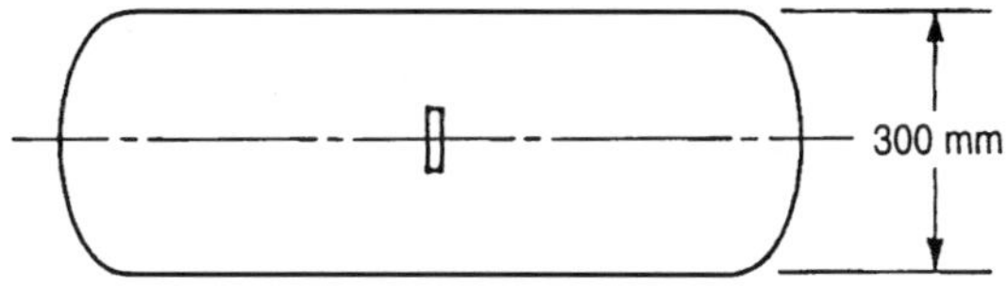

Exhibit 10.33

10.34 An aluminum pressure vessel has an internal pressure of 0.7 MPa. The vessel has an outer diameter of 200 mm and a wall thickness of 3 mm. Poisson's ratio is 0.33 and the modulus of elasticity is 69 GPa for this material. A strain gage is attached to the outside of the vessel at 45° to the longitudinal axis as shown in Exhibit 10.34. The strain on the gage would read most nearly:

a. 40×10^{-6} c. 80×10^{-6}
b. 60×10^{-6} d. 160×10^{-6}

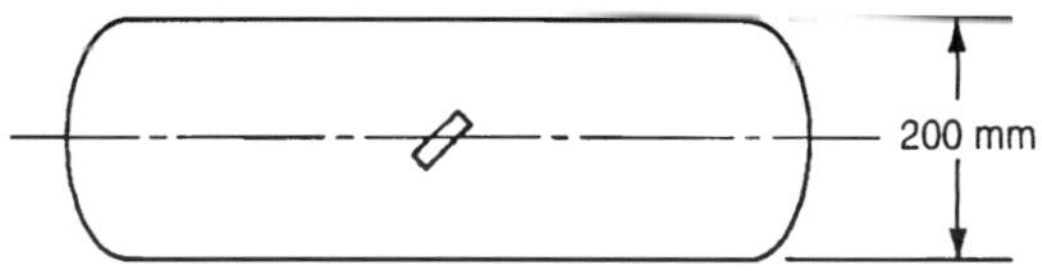

Exhibit 10.34

10.35 If $\sigma_x = -3$ MPa, $\sigma_y = 5$ MPa, and $\tau_{xy} = -3$ MPa, the maximum principal stress is most nearly:

a. 4 MPa c. 6 MPa
b. 5 MPa d. 7 MPa

10.36 Given that $\sigma_x = 5$ MPa, $\sigma_y = -1$ MPa, and the maximum principal stress is 7 MPa, the shear stress τ_{xy} is most nearly:

a. 1 MPa c. 3 MPa
b. 2 MPa d. 4 MPa

10.37 Given $\varepsilon_x = 800\ \mu$, $\varepsilon_y = 200\ \mu$, and $\gamma_{xy} = 400\ \mu$, the maximum principal strain is most nearly:

a. 840 μ c. 900 μ
b. 860 μ d. 960 μ

10.38 A steel plate in a state of plane stress has the same strains as in Problem 10.37: $\varepsilon_x = 800\ \mu$, $\varepsilon_y = 200\ \mu$, and $\gamma_{xy} = 400\ \mu$. Poisson's ratio $v = 0.3$ and the modulus of elasticity $E = 210$ GPa. The maximum principal stress in the plane is most nearly:

a. 109 MPa c. 173 MPa
b. 132 MPa d. 208 MPa

10.39 A stepped steel shaft shown in Exhibit 10.39 has torques of 10 kN • m applied at the end and at the step. The maximum shear stress in the shaft is most nearly:

a. 760 MPa c. 870 MPa
b. 810 MPa d. 930 MPa

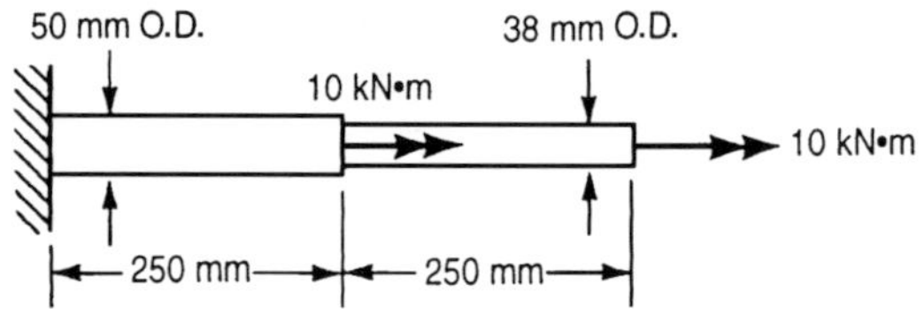

Exhibit 10.39

10.40 The shear modulus for steel is 83 MPa. For the same shaft as in Problem 10.39, the rotation at the end of the shaft is most nearly:

a. 0.014° c. 1.4°
b. 0.14° d. 14°

10.41 The same stepped shaft as in problems 10.39 and 10.40 is now built into a wall at its right end before the load is applied (Exhibit 10.41). The maximum stress in the shaft is most nearly:

a. 130 MPa c. 230 MPa
b. 200 MPa d. 300 MPa

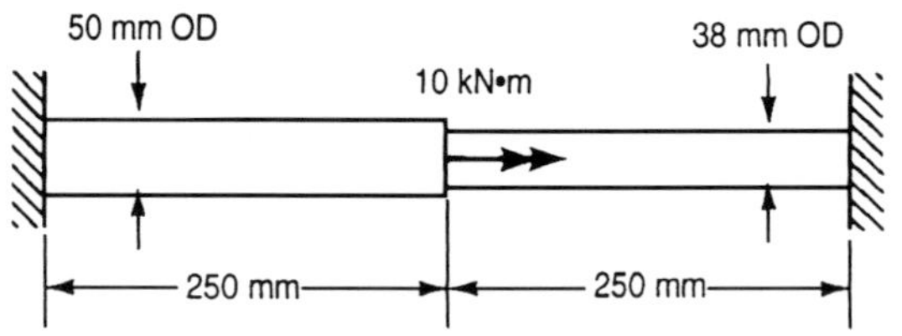

Exhibit 10.41

10.42 For the same shaft as in Problem 10.41 the rotation of the step is most nearly:

a. 0.2° c. 1.8°
b. 1.1° d. 2.1°

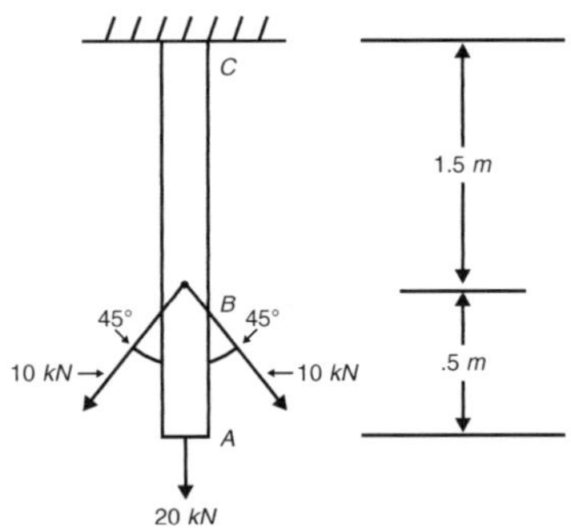

Exhibit 10.43

Use the following information for problems 10.43 and 10.44.

A steel rod with modulus of elasticity of 180 GPa is subjected to the loads as shown in Exhibit 10.43. The cross-sectional area of the rod is 1 cm².

10.43 The displacement of point *B* is most nearly:

a. 4.60 mm c. 2.85 mm
b. 7.30 mm d. 1.08 mm

10.44 The displacement of point *A* is most nearly:

a. 2.85 mm c. 5.61 mm
b. 8.20 mm d. 3.85 mm

Use the following information for problems 10.45 through 10.46.

The assembly shown in Exhibit 10.45 consists of three steel rods with $E = 180$ GPa and a rigid bar BCD.

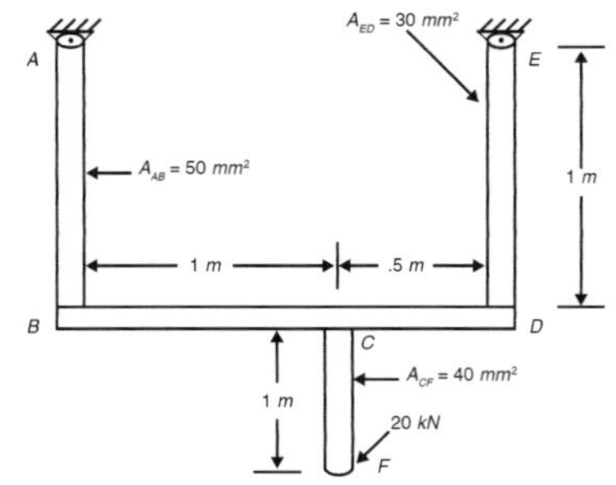

Exhibit 10.45

10.45 The axial force at D in kN is most nearly:
a. 25 ↓ c. 20 ↑
b. 13.3 ↓ d. 16.5 ↑

10.46 The axial force at B in kN is most nearly:
a. 6.7 ↓ c. 6.7 ↑
b. 8.5 ↑ d. 9.3 ↓

Use the free-body diagram in Exhibit 10.47 for problems 10.47 through 10.49.

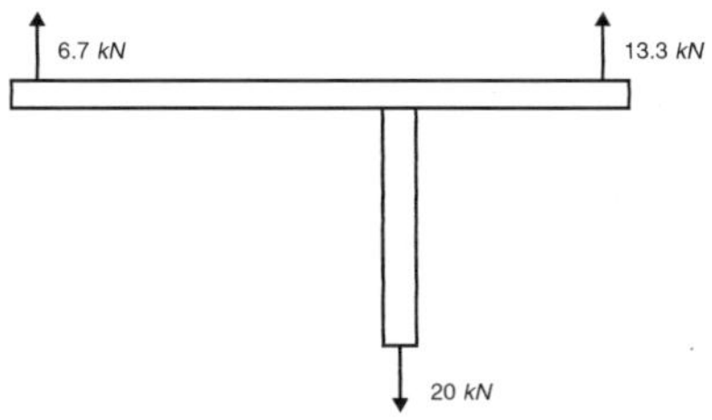

Exhibit 10.47

10.47 The displacement of point D is most nearly:
a. 4.6 mm c. 6.2 mm
b. 2.5 mm d. 1.8 mm

10.48 The displacement of point B is most nearly:
a. 0.74 mm c. 0.55 mm
b. 0.40 mm d. 0.38 mm

10.49 The displacement of point F is most nearly:
a. 2.8 mm c. 6.2 mm
b. 3.5 mm d. 4.7 mm

Use Exhibit 10.50 and the following information for the solution of problems 10.50 through 10.53.

Exhibit 10.50 shows a stress-strain diagram for a steel alloy rod, which is loaded in tension and is 1 cm in diameter. The inner curve shown is the magnified curve in which the strain scale is larger by a factor of 100 for the same stress scale. The outer curve is the typical approximate stress-strain curve.

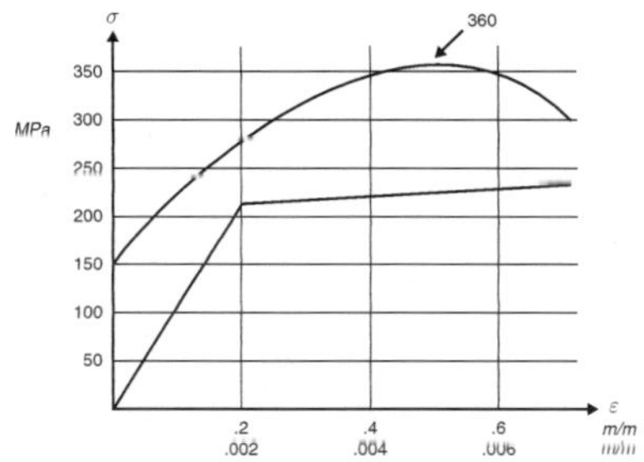

Exhibit 10.50

10.50 The modulus of elasticity of the rod in GPa is most nearly:
a. 120 c. 113
b. 430 d. 265

10.51 The yield load is most nearly:
a. 18 kN c. 25 kN
b. 12 kN d. 30 kN

10.52 The ultimate load is most nearly:
a. 15 kN c. 32kN
b. 21.6 kN d. 28.3 kN

10.53 The modulus of resilience is most nearly:
a. 1.2 MJ/m³ c. 2.4 MJ/m³
b. 0.23 MJ/m³ d. 6.8 MJ/m³

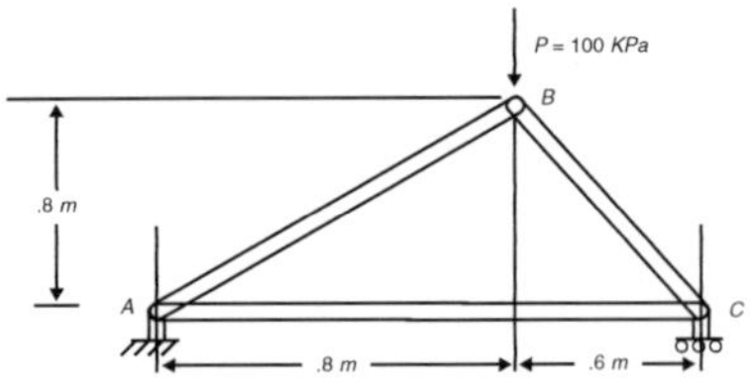

Exhibit 10.54

Use Exhibit 10.54 and the following information for problems 10.54 through 10.56.

A truss is constructed of three members of a certain alloy that has a modulus of elasticity of 150 GPa. Each of the members has a cross-sectional area of 2 cm^2.

10.54 The load in bar *BC* in kN is most nearly:
a. 55 c. 71
b. 80 d. 93

10.55 The stress in bar *AB* in MPa is most nearly:
a. 300 c. 250
b. 350 d. 270

10.56 The elongation of bar *AC* in mm is most nearly:
a. 2.0 c. 2.8
b. 3.5 d. 1.5

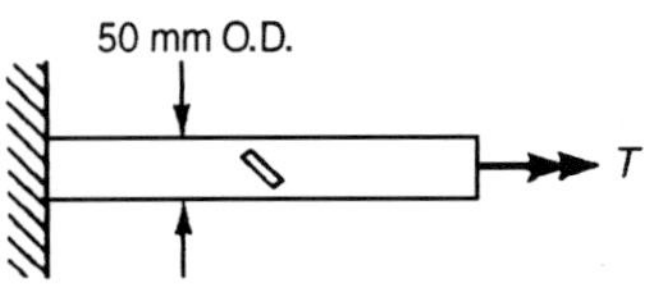

Exhibit 10.57

10.57 A strain gage shown in Exhibit 10.57 is placed on a circular steel shaft that is being twisted with a torque *T*. The gage is inclined 45° to the axis. If the strain reads $\varepsilon_{45} = 245\ \mu$, the torque is most nearly:
a. 1000 N • m c. 1570 N • m
b. 1230 N • m d. 2635 N • m

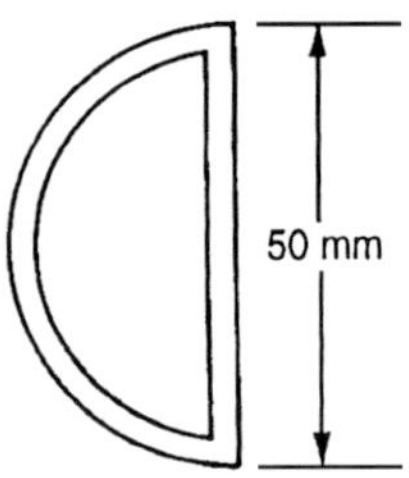

Exhibit 10.58

10.58 A shaft whose cross section is in the shape of a semicircle is shown in Exhibit 10.58 and has a constant wall thickness of 3 mm. The shaft carries a torque of 300 N • m. Neglecting any stress concentrations at the corners, the maximum shear stress in the shaft is most nearly:
a. 32 MPa c. 59 MPa
b. 48 MPa d. 66 MPa

10.59 The maximum magnitude of shear in the beam shown in Exhibit 10.59 is most nearly:
a. 40 kN c. 60 kN
b. 50 kN d. 75 kN

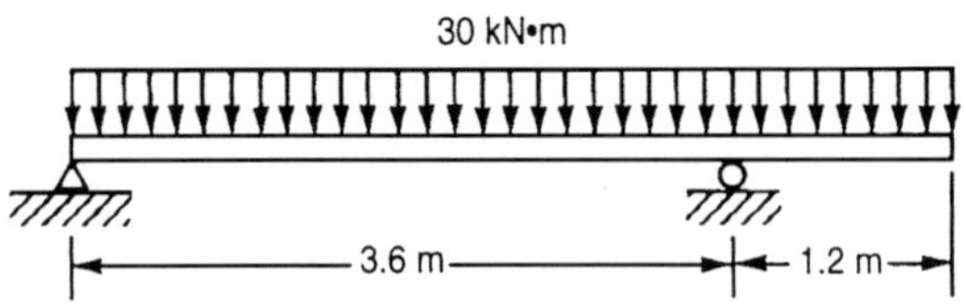

Exhibit 10.59

10.60 For the same beam as in Problem 10.59, the magnitude of the largest bending moment is most nearly:
a. 21.0 kN • m c. 38.4 kN • m
b. 26.3 kN • m d. 42.1 kN • m

10.61 The shear diagram shown in Exhibit 10.61 is for a beam that has zero moments at either end. The maximum concentrated force on the beam is most nearly:

a. 60 kN upward c. 0
b. 30 kN upward d. 30 kN downward

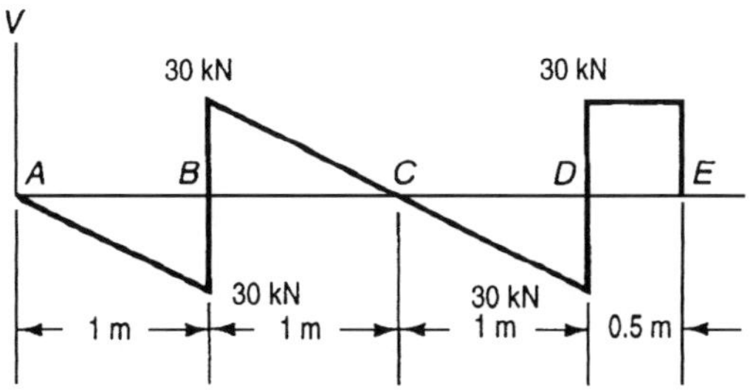

Exhibit 10.61

10.62 For the same beam as in Problem 10.61 the largest magnitude of the bending moment is most nearly:

a. 0 c. 12 kN • m
b. 8 kN • m d. 15 kN • m

10.63 The 4-m long, simply supported beam shown in Exhibit 10.63 has a section modulus $Z = 1408 \times 10^3$ mm^3. The allowable stress in the beam is not to exceed 100 MPa. The maximum load, w (including its own weight), that the beam can carry is most nearly:

a. 50 kN • m c. 60 kN • m
b. 40 kN • m d. 70 kN • m

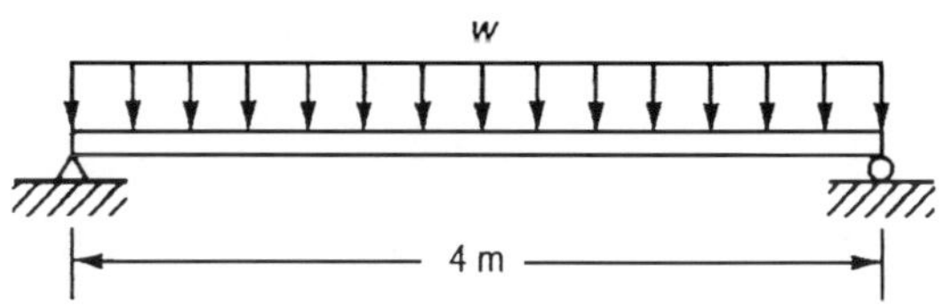

Exhibit 10.63

10.64 The standard wide flange beam shown in Exhibit 10.64 has a moment of inertia about the z-axis of $I = 365 \times 10^6$ mm^4. The maximum bending stress is most nearly:

a. 4.5 MPa c. 6.5 MPa
b. 5.0 MPa d. 8 MPa

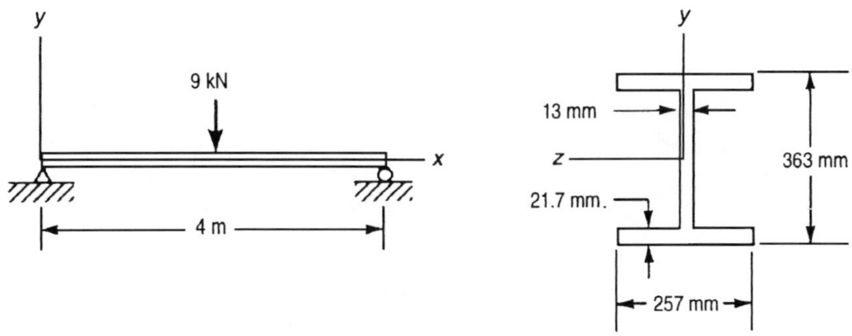

Exhibit 10.64

10.65 For the same beam as in Problem 10.64, the maximum shear stress τ_{xy} in the web is most nearly:

a. 1 MPa
b. 1.5 MPa
c. 2.0 MPa
d. 2.5 MPa

10.66 The deflection at the end of the beam shown in Exhibit 10.66 is most nearly:

a. 0.330 FL^3/EI (downward)
b. 0.380 FL^3/EI (downward)
c. 0.410 FL^3/EI (downward)
d. 0.440 FL^3/EI (downward)

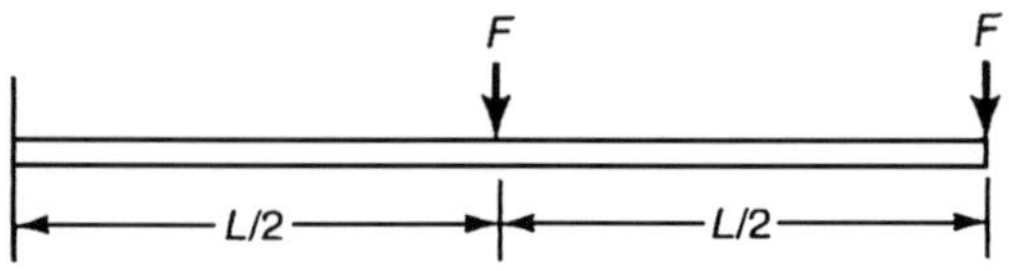

Exhibit 10.66

10.67 A uniformly loaded beam (Exhibit 10.67) has a concentrated load wL at its center that has the same magnitude as the total distributed load w. The maximum deflection of this beam is most nearly:

a. 0.029 wL^4/EI (downward)
b. 0.034 wL^4/EI (downward)
c. 0.043 wL^4/EI (downward)
d. 0.056 wL^4/EI (downward)

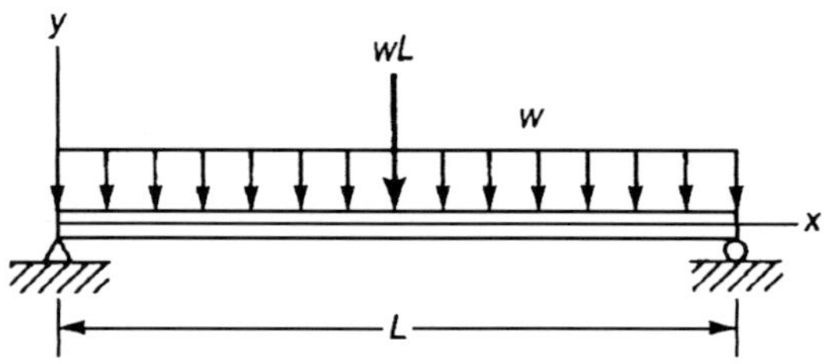

Exhibit 10.67

10.68 The reaction at the center support of the uniformly loaded beam shown in Exhibit 10.68 is most nearly:

a. 0.525 wL
b. 0.550 wL
c. 0.575 wL
d. 0.625 wL

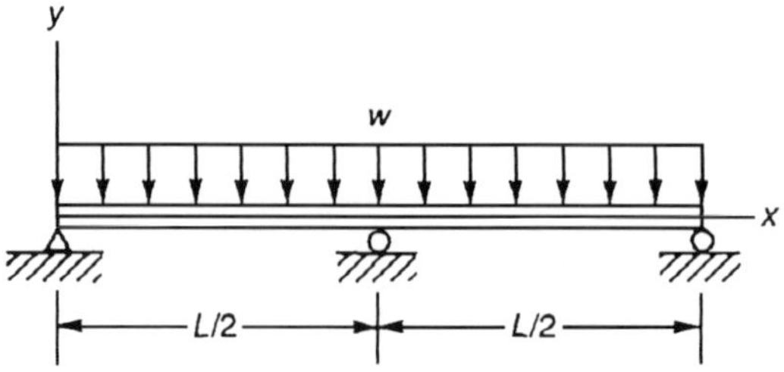

Exhibit 10.68

10.69 A solid circular rod has a diameter of 25 mm (Exhibit 10.69). It is fixed into a wall at *A* and bent 90° at *B*. The maximum bending stress in the section *BC* is most nearly:

a. 21.7 MPa c. 32.6 MPa
b. 29.3 MPa d. 45.7 MPa

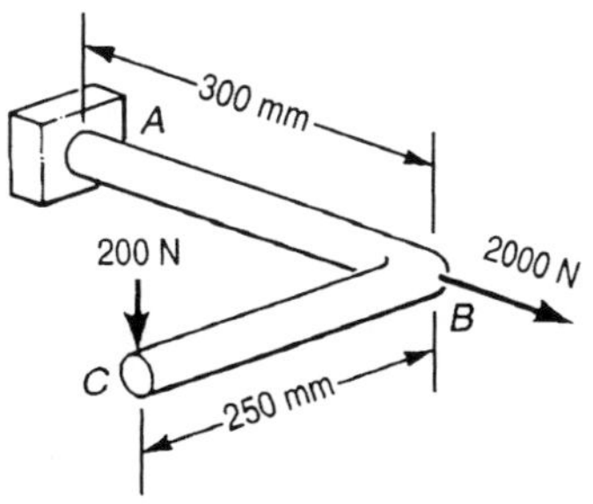

Exhibit 10.69

10.70 For the same member as in Problem 10.69, the maximum bending stress in the section AB is most nearly:

a. 21 MPa c. 31 MPa
b. 25 MPa d. 39 MPa

10.71 For the same member as in Problem 10.69, the maximum shear stress due to torsion in the section *AB* is most nearly:

a. 15.2 MPa c. 17.4 MPa
b. 16.3 MPa d. 18.5 MPa

10.72 For the same member as in Problem 10.69, the maximum stress due to the axial force in the section *AB* is most nearly:

a. 4 MPa c. 6 MPa
b. 5 MPa d. 8 MPa

10.73 For the same member as in Problem 10.69, the maximum principal stress in the section *AB* is most nearly:

a. 17 MPa c. 39 MPa
b. 27 MPa d. 44 MPa

10.74 A truss is supported so that it can't move out of the plane (Exhibit 10.74). All members are steel and have a square cross section 25 mm by 25 mm. The modulus of elasticity for steel is 210 GPa. The maximum load *P* that can be supported without any buckling is most nearly:

a. 14 kN
b. 25 kN
c. 34 kN
d. 51 kN

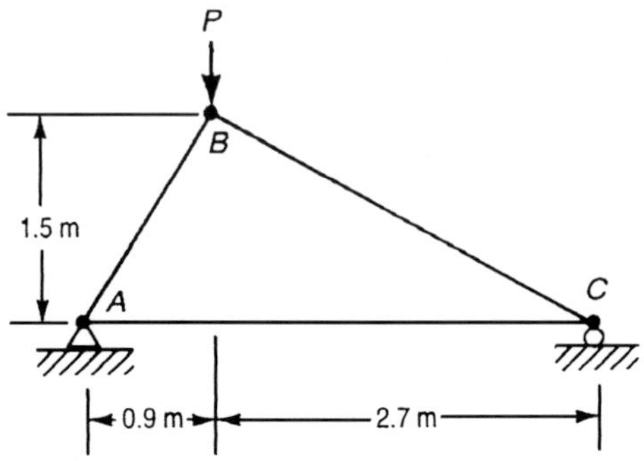

Exhibit 10.74

10.75 A beam is pinned at both ends (Exhibit 10.75). In the x-y plane it can rotate about the pins, but in the x-z plane the pins constrain the end rotation. In order to have buckling equally likely in each plane, the ratio b/a is most nearly:

a. 0.5
b. 1.0
c. 1.5
d. 2.0

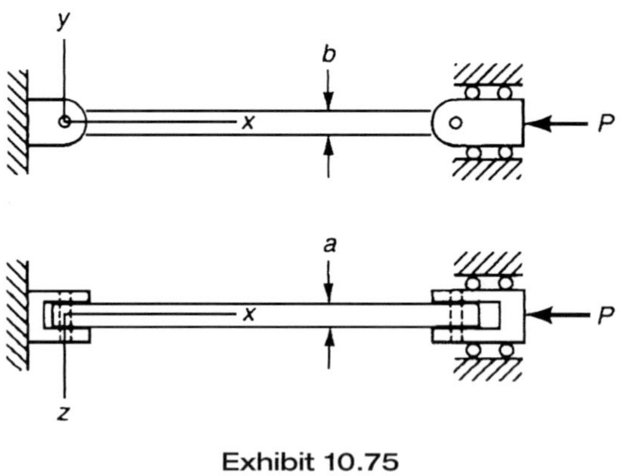

Exhibit 10.75

SOLUTIONS

10.1 c. The area of the steel core cylinder is equal to $(0.102)^2 \times \pi/4 =$ 0.008171 m^2

The outer area of the copper sleeve equals is 0.032366 m^2, so the area of the copper put into compression is equalto 0.02419 m^2.

The load to be held is equal to 45,340 kg = 444,650 N. Then,

$$\varepsilon\,[(0.008171)(2.1) + (0.02419)(1.12 \times 10^{11})] = 444{,}650 \text{ N, and}$$
$$\varepsilon = 4.4465/(0.044251 \times 10^6) = 100.5 \times 10^{-6} \text{ m/m}$$

The stress in the copper tube is

$$\sigma = (100.5 \times 10^{-6})(1.12 \times 10^{11}) = 11.26 \text{ MPa}$$

10.2 a. Euler's formula for long columns is, from the *FE Handbook*

$$P_{cr} = \pi^2 EI/(kl)^2$$

For this case, from the tables given in the *FE Handbook*

$$k = 1, \quad \text{and}$$
$$I = \pi r^4/4$$

for a solid circular bar, or, as given in the *FE Handbook* in the section on columns

$$I = r^2 A$$

where r = radius of gyration, which is equal to the radius of a circular bar divided by 2, which again gives

$$I = \pi r^4/4$$

where r is the radius of the solid round bar.

$$I = 0.30680 \times 10^{-6} \text{ m}^4$$

The critical load is then

$$P_{cr} = (9.870)(2.1)(0.03068 \times 10^6)/(3^2) = 70.205 \text{ kN}$$

which indicates a stress of

$$\sigma = 70{,}205/0.001963 = 35.764 \text{ MPa}$$

which corresponds to a strain of

$$\varepsilon = 35{,}764{,}000/(2.1 \times 10^{11}) = 170.304 \times 10^{-6} \text{ m/m}$$

The increase in temperature to produce this amount of strain is

$$\Delta T = \varepsilon/\alpha = (170.304 \times 10^{-6})/(11.7 \times 10^{-6}) = 14.55\,°\text{C}$$
$$T = 15 + 14.6 = 29.6\,°\text{C}$$

10.3 d. The strains in the three blocks will be equal, so the forces exerted by each block are as follows

$$P_A = \varepsilon(210 \times 10^9)(625 \times 10^{-6}) = (131.25 \times 10^6)\varepsilon$$
$$P_B = \varepsilon(70 \times 10^9)(2500 \times 10^{-6}) = (175.00 \times 10^6)\varepsilon$$
$$P_C = \varepsilon(105 \times 10^9)(1250 \times 10^{-6}) = (131.25 \times 10^6)\varepsilon$$

The total force is then

$$P_{total} = (437.5 \times 10^6)\varepsilon$$

Take moments about the left end

$$M = P_A(0.0125) + P_B(0.075) + P_C(0.15) = P_{total}x$$
$$x = [(1.641 + 13.125 + 19.6875) \times 10^6]/(437.5 \times 10^6)$$
$$= 0.0788 \text{ m}, \quad \text{or} \quad 7.88 \text{ cm}$$

10.4 b. The angle of twist in the 60-cm length will be the same as the angle of twist in the 190-cm length.

$$\phi_1 = \phi_2, \quad \text{and}$$
$$\phi = TL/ZGJ \text{ (See } \textit{FE Handbook}\text{)}$$
$$J = \pi a^4/2 = 9.817 \times 10^{-6} \text{ m}^4$$
$$G = 8.3 \times 10^{10} \text{ Pa, as given in the } \textit{FE Handbook}$$
$$T_1 + T_2 = 16{,}000 \text{ J}$$
$$T_1(0.60) = T_2(1.90), \text{ so}$$
$$T_1 = 3.167T_2$$
$$T_1 + T_2 = 4.167\,T_2$$
$$T_2 = 16{,}000/4.167 = 3{,}840 \text{ J}, \quad \text{and}$$
$$T_1 = 12{,}160 \text{ J}$$

Torsional stress is equal to Tc/J, so the maximum stress occurs in the 60-cm length of the shaft and is

$$\sigma = (12{,}160)(0.05)/(9.817 \times 10^{-6}) = 61.933 \text{ MPa}$$

10.5 d.

$$\text{Torque} = 2(22{,}000)(0.25) = 11{,}000 \text{ N} \times \text{m}$$

The angle of twist of the section to the left of the applied torque will be equal to the angle of twist of the section to the right of the applied torque, or $\phi_L = \phi_R$.

$$J = \pi r^4/2 \text{ for a solid round bar.}$$

For a 6.3-cm-diameter bar

$$J = \pi(0.0315)^4/2 = 1.546 \times 10^{-6} \text{ m}^4$$

For the 5-cm-diameter section

$$J = 0.6136 \times 10^{-6} \text{ m}^4$$

From the *FE Handbook*

$$\phi = TL/GJ, \text{ and}$$
$$G = 83 \text{ GPa}$$
$$T_L + T_R = \text{applied torque}$$
$$T_L(0.30)/(1.546 \times 10^{-6})\text{G} + T_L(0.20)/(0.6136 \times 10^{-6})G$$
$$= T_R(0.50)/(0.6136 \times 10^{-6})G$$
$$T_L(0.5200)/83{,}000 = T_R(0.8149)/83{,}000$$
$$T_L = 1.567T_R$$
$$2.567T_R = 11.00 \text{ kJ}$$
$$T_R = 4{,}285 \text{ J, and}$$
$$T_L = 6{,}715 \text{ J}$$

10.6 c. The angles of twist for the left and right sections will be equal, or $\phi_L = \phi_R$.

$\phi = TL/JG$, as given in the *FE Handbook.*

$$J = \pi r^4/2$$

For the 5.0-cm-diameter section

$$J = 0.6136 \times 10^{-6} \text{ m}^4$$

and for the 2.5-cm diameter section

$$J = 0.0383 \times 10^{-6} \text{ m}^4$$
$$T_L + T_R = \text{applied torque} = 1900 \text{ J}$$
$$T_L(0.500)/(68.9 \times 10^{10})(0.6136 \times 10^{-6})$$
$$= T_R(1.00)/(138 \times 10^{10})(0.0383 \times 10^{6})$$
$$T_L = (0.1892/0.01184)T_R = 15.98T_R$$
$$16.98T_R = 1900 \text{ J}$$
$$T_R = 112 \text{ J}$$
$$T_L = 1790 \text{ J}$$

10.7 a. First, sketch a figure like that shown in Exhibit 10.7b. When the 90,000-kg load is applied the distance H will increase, and the tie rod L and the cable M will strain. Distances D and E will remain constant. The stresses in L and M will be proportional to the strains in these two members.

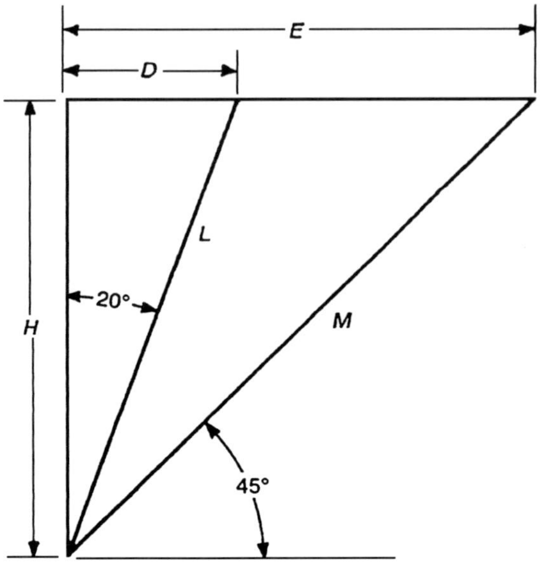

Exhibit 10.7b

The relative strains can be calculated as follows

$$L^2 = D^2 + H^2$$
$$M^2 = E^2 + H^2$$

Differentiating both equations gives

$$2LdL = 2HdH, \quad \text{and}$$
$$2MdM = 2Hdh$$

Remembering that D and E remain constant,

$$\delta_L = dL/L = (H/L^2)dH, \quad \text{and}$$
$$\delta_M = dM/M = (H/M^2)dH$$

then

$$dH = (L^2/H)(dL/L) = (M^2/H)(dM/M)$$
$$\delta_L = (M^2/L^2)\delta_M$$
$$M = H/(\cos 45°) = 1.4142H$$
$$L = H/(\cos 20°) = 1.0642H$$
$$\delta_L = 1.766\delta_M$$

The system is symmetrical about the centerline, so L and M, together, will support half of the load.

$$F_L \cos 20° + F_M \cos 45° = 45{,}000g$$
$$F_L = \delta_L E_L A_L$$

where $A_L = 490.9 \times 10^{-6}$ m^2

Similarly,

$$F_M = \delta_M E_M A_M$$

where A_M is also equal to 490.9×10^{-6} m^2.

$$1.766\delta_M(210 \times 10^9)(490.9 \times 10^{-6})\cos 20°$$
$$+ \delta_M(84 \times 10^9)(490.9 \times 10^{-6})\cos 45° = 441{,}300 \text{ N}$$
$$(171{,}076 + 29{,}150)\delta_M(1000) = 441{,}300$$
$$\delta_M = 0.002204 \text{ m/m}$$
$$\delta_L = 1.766\delta_M = 0.003892 \text{ m/m}$$

$$F_M = \delta_L E_M A_M = 0.002204(84 \times 10^9)(490.9 \times 10^{-6})$$
$$= 90{,}880 \text{ N force in each cable}$$

The vertical component is equal to 90,880 cos 45° = 64,250 N.

$$F_L = \delta_M E_L A_L = 0.003892(210 \times 10^9)(490.9 \times 10^{-6})$$
$$= 401{,}220 \text{ N force in each rod}$$

The vertical component is equal to 401,220 cos 20° = 377,020 N

The cables support 64,250/441,300 = 0.1456, or the cables support 14.6% of the total load and, thus, 14.6% of the added load.

10.8 d. The stress will consist of two components: direct stress and moment stress, since the plate is loaded eccentrically. The direct stress is

$$\sigma = (1000)(9.807)/(0.250)(0.025) = 1.569 \text{ MPa}$$

The load is located 0.125 – 0.020 = 0.105 meters away from the centroid of the plate, so the load will produce a moment equal to 0.105(9807) = 1030 N × m in the plate. The equation for stress due to an applied moment is $\sigma = Mc/I$ where $I = bh^3/12$. For this case

$$I = (0.025)(0.250)^3/12 = 32.55 \times 10^{-6} \text{ m}^4$$

The maximum bending stress,

$$\sigma = Mc/I = (1030)(0.125)/(32.55 \times 10^{-6}) = 3.955 \text{ MPa}$$

This would add to the direct stress, giving a maximum stress in the outer fibers of 5.55 MPa.

10.9 c. The maximum principal stress will be calculated using the method of Mohr's circle, as given in the *FE Reference Handbook* supplied by the NCEES. Assume a thin-walled tube, then $\sigma = PD/2t$.

The tube ID = 152.4 – 5.08 = 147.32 mm or 0.14732 m

Hoop stress, $\sigma_y = (3.448 \times 10^6)(0.14732)/0.00508 = 99.992$ MPa

$$\text{Longitudinal stress, } \sigma_x = \text{PA}_{\text{ID}}/\text{Area of tube wall}$$
$$A_{\text{ID}} = (\pi/4)(0.14732)^2 = 0.017046 \text{ m}^2$$
$$\text{Area of tube wall} = (\pi/4)(OD^2 - ID^2) = 0.001195 \text{ m}^2$$
$$\sigma_x = (3.448 \times 10^6)(0.017046)/0.001195 = 49.184 \text{ MPa}$$
$$\text{Shear stress } S_s = Tc/J$$
$$\text{where } J = (\pi/32)(OD^4 - ID^4) = 6.7152 \times 10^{-6} \text{ m}^4$$
$$S_s = (678)(0.0762)/(6.7152 \times 10^{-6}) = 7.6935 \text{ MPa}$$

From the figure of Mohr's circle in the *FE Handbook*

$$\tau_{max} = \sqrt{\left[S_s^2 + \{(\sigma_x - \sigma_y)/2\}^2\right]} = \sqrt{(7.6935^2 + 25.404^2)} = 26.54 \text{ MPa}$$

$$\sigma_1 = (\sigma_x + \sigma_y)/2 + \tau_{max} = 74.588 + 26.45 = 101 \text{ MPa}$$

10.10 d. Mohr's circle is used to determine the stress resulting from the stresses on two other surfaces that are perpendicular to each other.

10.11 d. Stress, $\sigma = Mc/I$

where

$$M = (500)(9.8066)(3.00) = 14{,}710\ \text{N}\cdot\text{m}, \quad \text{or} \quad \text{J}$$
$$I = bh^3/12 = (0.200)(0.300)^3/12 = 4.500 \times 10^{-4}\ \text{m}^4$$
$$c = 0.150\ \text{in}$$

$$\sigma = (14{,}710)(0.150)/(4.500 \times 10^{-4}) = 4.903\ \text{MPa}$$

10.12 a.

$$\text{Hoop stress} = PD/2t$$

$$\text{Pressure} = \rho \times \text{g} \times \text{depth} = (1000)(9.8066)(10)$$
$$= 98.07\ \text{kPa at the bottom of the tank}$$
$$\text{Stress} = (98{,}070)(35.0)/0.16 = 214.5\ \text{MPa}$$

10.13 c. The force exerted by a 200-kg mass = (200)(9.8066) = 1961 N

$$\text{Force in member } AD = (13/5)(1961) = 5099\ \text{N}$$
$$ED = X \text{ component of force in } AD = (1.2/1.3)(5099) = 4707\ \text{N}$$

The Z components of forces in member CD and BD in the X-Z plane are equal—

$$(0.9/1.5)F_{CD} = (1.6/2.0)F_{BD,} \text{ so}$$
$$F_{BD} = 0.75(F_{CD})$$

The sum of the X components of F_{CD} and F_{BD} is equal to the X component of the force in F_{AD}

$$(1.2/1.5)F_{CD} + (1.2/2.0)F_{BD} = (1.2/1.3)F_{AD}$$
$$0.800\ F_{CD} + (0.600)(0.750)F_{CD} = 4707\ \text{N}$$
$$F_{CD} = 4707/1.250 = 3765.6\ \text{N}$$

10.14 a. Determine the reaction at the support points. Take moments about the right support

$$M = (0.600)(27{,}000) + (½)(6000)g(1.80)(1.8 + 0.600) = 3.600R_L$$
$$R_L = 39{,}408\ \text{N}$$

The area of the triangular load distribution at the left end of the beam is equal to (½)(6000)g(1.8) m = 52,958 N, and its center of force is 1.8 + 1.8/3 = 2.4 m from the right support.

From $\sum F_y = 0$

$$R_R = 52{,}958 - 39{,}408 = 13{,}550\ \text{N}$$

The maximum moment occurs at the point where the shear is zero.

The shear, starting at the left support, would equal 39,408 minus the applied load. The total applied load to the left of any point on the beam would be equal to $(X/2)X(6000)g/1.8) = 16{,}345X^2$ where X is the distance from the left end of the beam, since the magnitude of the load in the first 1.8 meters of the beam to the right of the left support is equal to $X(6000)g/1.8$ for the first 1.8 meters of the beam. The total load acting down on the beam to the left of a point X is equal to the area of the load triangle from the left end of the beam to that point. The point at which the applied load is equal to 39,408 N, the magnitude of the left support, is $39{,}408 = 16{,}345X^2$, giving $X = 1.553$ meters. At this point the magnitude of the shear is equal to zero, so the moment is a maximum. Summing the moments to the left of the point gives

$$M = (-39{,}408)(1.553/3) + (39{,}408)(1.553) = 40.800 \text{ N} \times \text{m}, \text{ or } 40{,}800 \text{ J}$$

10.15 d. The maximum bending stress is equal to Mc/I where M is the maximum moment, c is equal to the distance from the neutral axis to the outermost fibers of the beam, and I is equal to the moment of inertia of the beam for a beam of constant cross-section. For this case

$$I = bh^3/12 = (0.075)(0.150)^3/12 = 21.094 \times 10^{-6} \text{ m}^4$$
$$c = 0.075$$
$$M = (3.65/2)(750)(\tfrac{2}{3})(3.65) = 3331 \text{ N} \times \text{m}$$
$$\text{Stress} = (3331)(0.075)/(21.094 \times 10^{-6}) = 11.84 \text{ MPa}$$

10.16 b. First, determine the reactions at the two support points, A and C. Take moments about the left support

$$M = 4.8R_C = (3.6)(3.6)(3000)g$$
$$R_c = 79{,}440 \text{ N}$$

From $\sum y = 0$

$$R_A = (3.6)(3000)g - 79{,}440 = 26{,}520 \text{ N}$$

Determine the point of zero shear. Starting from the left end of the applied load

$$26{,}520 = 3000gX$$
$$X = 0.901 \text{ meters}$$

So the point of zero shear occurs at a distance of 1.8 + 0.9016 = 2.701 m from the left support. The moment is a maximum at the point where the shear is zero. Summing the moments to the left of the point of zero shear gives

$$M = -(0.901)(3000)g(0.901/2) + (26{,}520)(2.701)$$
$$= 59{,}689 \text{ N} \times \text{m}, \quad \text{or} \quad 59{,}689 \text{ J}$$

10.17 d. First, determine the reactions at the two supports. Calculate the effects of the end moments. Treating the left end moment alone

$$R_R = 45{,}150/6.0 = -7527 \text{ N, giving}$$
$$R_L = 7527 \text{ N from } \sum y = 0$$

Similarly, treating the right end moment alone

$$R_L = -180{,}755/6.0 = -30{,}126 \text{ N, and}$$
$$R_R = 30{,}126$$

Treating the applied load alone

$$R_L = (6.0)(4460)g/2 = 131{,}218 \text{ N, and}$$
$$R_R = 131{,}218 \text{ N}$$

Summing the three partial loads gives

$$R_L = 131{,}218 + 7527 - 30{,}126 = 108{,}619 \text{ N}$$

Similarly

$$R_R = 131{,}218 + 30{,}126 - 7527 = 153{,}817 \text{ N}$$

To find the point of zero shear, sum the applied load from the left support

$$108{,}619 - 4460gX = 0$$

Which gives $X = 2.483$ meters. Sum the moments to the left of this point

$$M = (108{,}619)(2.483) - (4460)g(2.483)(2.483/2)$$
$$= 134{,}869 \text{ N} \times \text{m due to the load.}$$

Subtracting the moment applied at the left support gives

$$M = 134{,}869 - 45{,}150 = 89{,}719 \text{ N} \times \text{m}$$

Check by summing the moments about the right end

$$M = (3.517)(153{,}817) - (4.460)\text{g}(3.517)(3.517/2)$$
$$= 270{,}462 \text{ N} \times \text{m due to the load.}$$

Subtracting the moment applied at the right support gives

$$M = 270{,}460 - 180{,}755 = 89{,}707 \text{ N} \times \text{m},$$

which is the maximum moment at the mid-portion of the beam; however, the applied moment at the right support of 180,755 N × m is greater.

10.18 **c.** The loading on the vertical support beam would increase linearly with depth, from zero at the top. The pressure of the water increases from zero at the top to $(3.0)(1000)g = 29{,}421$ Pa at the bottom of the tank. Each vertical beam would withstand the force exerted by a 1.0-m length of the tank wall. The beam would thus be loaded with a varying load, ranging from zero at the top to 29,421 N/m at the lower end due to the pressure of the water. The beam is hinged at the top and bottom, and is thus simply supported at the ends. Calculate the end reactions.

The total load withstood by the beam due to the triangular load pattern is equal to one-half the length times the height = (½)(3.0)(29,421) = 44,132 N. The centroid of a triangle is one-third the distance from the base to the apex, so the reaction at the top of the beam, taking moments about the bottom end, would be

$$R_T = (44{,}132)(3)(1/3)/3 = 14{,}711 \text{ N}$$

Similarly, taking moments about the top support, gives

$$R_B = (44{,}132)(3)(2/3)/3 = 29{,}422 \text{ N}$$

Taking the top support as a reference point, the loading on the beam at a depth D would be

$$P_D = (44{,}132/3)D$$

The point of zero shear can be determined

$$29{,}422 - (44{,}132/3)D(\tfrac{1}{2})D = 0$$
$$D^2 = 29{,}422/7{,}355$$
$$D = 2.00 \text{ m}$$

Sum the moments acting on the beam about a point 2.00 m below the surface of the water.

Loading on the beam at a depth of 2.00 m is

$$w = (2.00)(1000)g(1.00) = 19{,}614 \text{ N/m}$$

This will decrease linearly to zero at the surface, giving a triangular load. Total force over the 2.00 top meters of depth is

$$P = (19{,}614)(2.00)(\tfrac{1}{2}) = 19{,}614 \text{ N.}$$

The moment arm to the center of force triangle = 2.00/3 = 0.667 m. Summing the moments acting on the beam at 2.00 meters depth yields

$$M = (2.00)(14{,}711) - (0.667)(19{,}614) = 29{,}422 - 13{,}083 = 16{,}340$$

10.19 **d.** Determine reactions at the two support points. Sum moments about R_1

$$M = R_2(3.5) - (1400)(1.50)g - (4.40)(890)g(3.70) = 0$$
$$R_2 = 162{,}690/3.50 = 46{,}483 \text{ N}$$
$$R_1 = [1400 + (4.40)(890)]g - 46{,}483 = 52{,}134 - 46{,}483 = 5651 \text{ N}$$

Find the point of zero shear. The shear load to the right of R_1 would be constant at 5651 *N* to the point of the applied load of 1400 kg and would then change abruptly to 5651 – 1400g = –8079 N, passing through zero. The moment at this point would be equal to (1.50)(5651) = 8477 N × m.

There would also be an abrupt change in the shear loading at support R_2, where the shear loading would again pass through zero. At this point the moment would be

$$M = (2.40)(890)g(1.20) = 25{,}137 \text{ N} \times \text{m},$$

which is larger than the moment at the other point of zero shear, so this moment would control, 25,137 N × m = 25,137 J.

10.20 **a.** First, locate the neutral axis and calculate n

$$n = 210/8.4 = 25$$

Take moments about the upper edge.

$$y = [(10)(20)(10) + (20.3)(0.60)(10)(25)]/[(200 + (0.60)(250)]$$
$$= 5045/350 = 14.414 \text{ cm} = 0.14414 \text{ m}$$
$$I = bh^3/12 + A_w s_1^2 + (\text{Equivalent Area})s_2^2$$
$$\text{Equivalent Area} = (10.0)(0.60)(25) = 150 \text{ cm}^2$$

Using the parallel axis theorem (see *FE Handbook*) and ignoring the moment of inertia of the wood-equivalent area because it is so small

$$I = (10)(20)^3/12 + [200(14.414 - 10)^2] + [150(20.3 - 14.414)^2]$$
$$= 6667 + 3897 + 5197 = 15{,}761 \text{ cm}^4$$
$$= 1.5761 \times 10^{-4} \text{ m}^4$$
$$\text{Stress} = Mc/I$$

where c = distance from neutral axis to the outermost fibers.

$c = 0.1441$ m to the top fibers in the wood, or
$c = 20.6 - 14.414 = 6.186$ cm $= 0.06186$ m to the outermost steel fibers

If the strain were in the wood the stress would be

$$S = (17{,}000)(0.06186)/(1.5761 \times 10^{-4}) = 6.672 \text{ MPa}$$

However, the strain is in steel, so the stress is

Stress = $n \times$ wood stress = (25)(6.672) = 166.8 MPa = stress in steel.

10.21 c. The beam will bend, and the rod will stretch. The force acting on the rod is

$$F_R(38) = 450g(50)$$
$$F_R = 450g(50/38) = 5807 \text{ N}$$
$$\text{Strain in the rod} = L \times S/E$$
$$\delta_L = (5807)/(132.7)/(210 \text{ GPa}) = (250)(43.760 \text{ MPa}/210{,}000 \text{ MPa})$$
$$= 0.0521 \text{ cm}$$

Treat the beam as two cantilever beams with the ends at the fulcrum.

For a cantilever beam with load concentrated at the free end, the deflection is

$$\delta = PL^3/(3EI) \text{ (see } \textit{FE Handbook}\text{)}$$
$$I = bh^3/12 = (0.025)(0.050)^3/12 = 0.2604 \times 10^{-6} \text{ m}^4$$

Deflection at the left end is

$$\delta_L = (5807)(0.38)^3(3)(210 \text{ GPa})(0.2604 \times 10^{-6})$$
$$= (318.64/164{,}000) = 1.943 \text{ mm}$$

Deflection at the right end is

$$\delta_R = 450g(0.50)^3/164{,}000 = 3.364 \text{ mm}$$

Total equivalent deflection at the left end is

$$\delta_{L,\text{equiv.}} = 0.521 + 1.943 = 2.464 \text{ mm}$$

This would give a deflection of the right end in inverse proportion to the lengths of the two sections of the beam as for similar triangles

$$\delta_{R,\text{equiv.}} (50/38)(2.464) = 3.242 \text{ mm}$$

The right end of beam would deflect 3.242 mm due to bending of the left end and the extension of the rod.

Deflection at Point A is

$$\delta_A = 3.242 + 3.364 = 6.606 \text{ mm}$$

10.22 a.

Area of steel = $(30^2 - 25^2)(\pi/4) = 216 \text{ cm}^2 = 0.0216 \text{ m}^2$
Area of concrete = 0.0491 m^2

Allowable stress in the steel is equal to 275 MPa, and strain = 275/210,000 = 0.00131 m/m.

Allowable stress in the concrete = 24 MPa, and strain = 24/19,600 = 0.00122 m/m.

Therefore, the stress in the concrete controls.

$$\text{Load held by steel} = (0.00122)(210\text{ GPa})(0.0216) = 5{,}533{,}920\text{ N}$$
$$\text{Load held by concrete} = (0.00122)(19.6\text{ GPa})(0.0491) = 1{,}174{,}080\text{ N}$$

The total permissible load is equal to 6,708,000 N, or 684,000 kg.

10.23 d. First, calculate the permissible deflection and load ability of the plastic block

$$A = (6.4)(6.4) = 40.96\text{ cm}^2 = 0.004096\text{ m}^2$$
$$P = (0.004096)(13{,}800 \times 10^6) = 56{,}525\text{ N}$$
$$\delta = (0.61)(13.8 \times 10^6)/(10{,}500 \times 10^6) = 8.017 \times 10^{-4}\text{ m}$$

The beam and cables can thus deflect 0.00025 + 0.0008017 = 0.001052 m. See the *FE Handbook* for the equation for beam deflection at the center with an applied load, P

$$\delta = PL^3/(48EI)$$
$$I = bh^3/12 = (0.06)(0.15)^3/12 = 16.875 \times 10^{-6}\text{ m}^4$$

Beam deflection

$$\delta_{beam} = P(27)/(48)(210{,}000)(16.875)$$
$$= P(0.1587 \times 10^{-6})\text{ m where } P \text{ is in N.}$$
$$\delta_{cable} = LP/AE$$
$$A = 2(0.019)^2(\pi/4) = 568.2 \times 10^{-6}\text{ m}^2 \text{ for the two cables}$$
$$\delta_{cable} = 1.83P/(568.2)(84{,}000) = P(0.03834 \times 10^{-6})\text{ m}$$

The beam and cables constitute two springs in series, where

$$I/k - 1/k_1 + 1/k_2 = (0.1587 \mid 0.0383) \times 10^{-6} = 0.1970 \times 10^{-6}\text{ m/N}$$

The load to deflect the beam-cable assembly 0.01052 m is

$$P = 0.001052/(0.1970 \times 10^{-6}) = 5340\text{ N}$$

Add to this the load withstood by the plastic block to get a total load of 5340 + 56,525 = 61,865 N, or 6308 kg

10.24 b. The beam deflection formula given in the *FE Handbook* is

$$\delta = (w_0x/24EI)/(L^3 - 2Lx^2 + x^3)$$

where

w_0 is the load per unit length = 225g = 2207 N/m
x = 1.22 meters
L = 4.5 m
$I = (0.20)(0.20)^3/12 = 133.3 \times 10^{-6}\text{ m}^4$
$\delta = [(2{,}207)(1.22)/(24)(10{,}500)(133.3)](91.125 - 13.396 + 1.816)$
$\delta = (80.155 \times 10^{-6})(79.545) = 0.00638\text{ m, or 6.38 mm}$

10.25 d. Concrete is assumed to take no load in tension. The equivalent "tensile concrete" area (area of steel rebar times n) is equal to 4(3.90)(10) = 165 cm² or 0.0165 m² with the center of the section taken as 68 cm, or 0.68 m, from the top of the beam. Determine the location of the neutral axis.

Taking moments about the top of the beam gives

$$y = [(0.30y)(y/2) + (0.68)(0.0165)]/(0.30y + 0.0165)$$

which reduced to

$$y^2 + 0.115y - 0.0747 = 0$$

Solving for y, using the general solution for a quadratic equation in the *FE Handbook*, gives

$$y = 0.2180 \text{ m}$$
$$I = (bh^3/12) + A_1S_1^{\,2} + A_2S_2^{\,2}$$

where S is equal to the distance from the CG of the area to the neutral axis.

$$I = [(0.30)(0.218)^3/12] + (0.30)(0.218)(0.218/2)^2 + 0.0165(0.680 - 0.218)^2 = 0.004558 \text{ m}^4$$

The maximum moment is

$$M_{max} = wL^2/8$$
$$w = (0.30)(0.76)(2370 + 3000)g = 34{,}720 \text{ N/m}$$
$$M_{max} = (34{,}720)(6)^2/8 = 156{,}240 \text{ J}$$

Stress in equivalent "tensile concrete" = (156,240)(0.462)/0.004558 = 15.837 MPa

Tensile stress in rebar = 10(15.837) = 158.37 MPa, where $n = 10$

10.26 c. Draw free-body diagrams. Equilibrium of the center free-body diagram gives

$$F_1 = 20 - 12 = 8 \text{ kN}$$

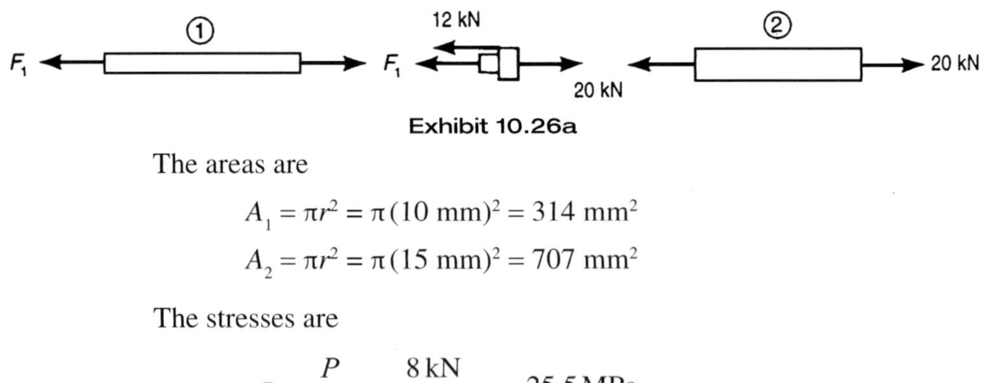

Exhibit 10.26a

The areas are

$$A_1 = \pi r^2 = \pi(10 \text{ mm})^2 = 314 \text{ mm}^2$$
$$A_2 = \pi r^2 = \pi(15 \text{ mm})^2 = 707 \text{ mm}^2$$

The stresses are

$$\sigma_1 = \frac{P}{A} = \frac{8 \text{ kN}}{314 \text{ mm}^2} = 25.5 \text{ MPa}$$
$$\sigma_2 = \frac{P}{A} = \frac{20 \text{ kN}}{707 \text{ mm}^2} = 28.3 \text{ MPa}$$

10.27 b. The force-deformation equations give

$$\delta_1 = \frac{P_1L_1}{A_1E_1} = \frac{(8 \text{ kN})(300 \text{ mm})}{(314 \text{ mm}^2)(69 \text{ GPa})} = 0.1107 \text{ mm}$$
$$\delta_2 = \frac{P_2L_2}{A_2E_2} = \frac{(20 \text{ kN})(250 \text{ mm})}{(707 \text{ mm}^2)(69 \text{ GPa})} = 0.1025 \text{ mm}$$

Compatibility of deformation gives

$$\delta_{\text{end}} = \delta_1 + \delta_2 = 0.1107 + 0.1025 = 0.213\,\text{mm}$$

10.28 **c.** Draw the free-body diagrams. From the center free-body diagram, summation of forces yields

$$F_2 = 12\text{ kN} + F_1$$

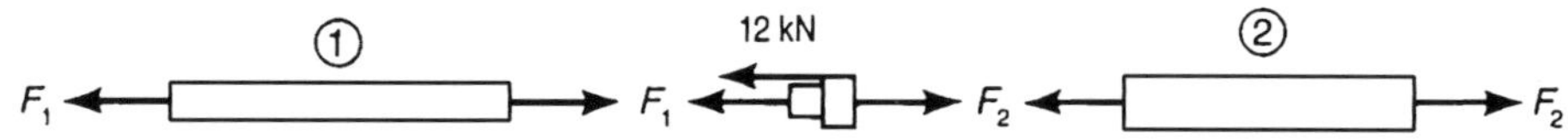

Exhibit 10.28a

Force-deformation relations are

$$\delta_1 = \frac{P_1 L_1}{A_1 E_1} = \frac{(F_1\text{ kN})(300\,\text{mm})}{(314\,\text{mm}^2)(69\,\text{GPa})} = 0.01384\,F_1$$

$$\delta_2 = \frac{P_2 L_2}{A_2 E_2} = \frac{(F_2\text{ kN})(200\,\text{mm})}{(707\,\text{mm}^2)(69\,\text{GPa})} = 0.00410\,F_2$$

Compatibility gives

$$\delta_{\text{end}} = 0 = \delta_1 + \delta_2 = 0.01384\,F_1 + 0.00410\,F_2$$

Substitution of the equilibrium relation, $F_2 = 12 + F_1$, into the above equation gives

$$0 = 0.01384\,F_1 + 0.00410\,(12 + F_1)$$
$$F_1 = -2.74\text{ kN},\ F_2 = 9.26\text{ kN}$$

The stresses are

$$\sigma_1 = \frac{P}{A} = \frac{(-2.74\,\text{kN})}{(314\,\text{mm}^2)} = -8.73\,\text{MPa};\qquad \sigma_2 = \frac{P}{A} = \frac{(9.26\,\text{kN})}{(707\,\text{mm}^2)} = 13.10\,\text{MPa}$$

10.29 **a.** The same three-step process as in Problem 10.28 must be carried out. Since this process has already been completed, the results can be used. The deflection can be expressed as

$$\delta = \delta_1 = -\delta_2 = 0.01384\,F_1 = 0.01384\,(-2.74) = -0.0380\text{ mm}$$

10.30 **c.** Draw a free-body diagram. Summation of forces in the vertical direction gives

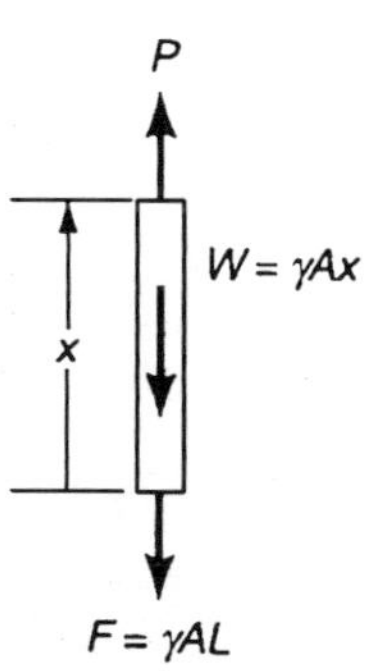

Exhibit 10.30a

$$P = \gamma AL + \gamma Ax$$

$$\delta = \int_0^L \frac{P}{AE}\,dx = \int_0^L \frac{\gamma A(L+x)}{AE}\,dx = \frac{\gamma}{E}\left(L^2 + \frac{L^2}{2}\right) = \frac{3\gamma L^2}{2E}$$

10.31 d. A force will develop in the rod if it attempts to grow more than 0.15 mm. Assuming that it does grow that amount, the displacement is

$$\delta = \frac{PL}{AE} + \alpha L(t - t_o) = 0.15\,\text{mm}$$

$$= \frac{P(300\,\text{mm})}{(650\,\text{mm})(210\,\text{GPa})} + \left(17\times10^{-6}\frac{1}{°\text{C}}\right)(300\,\text{mm})(100\,°\text{C} - 22\,°\text{C})$$

$$0.15\text{ mm} = 0.00220\,P + 0.3978$$

$$P = -112.7\text{ kN}$$

$$\sigma = \frac{P}{A} = \frac{-112.7\,\text{kN}}{650\,\text{mm}^2} = -173.5\,\text{MPa}$$

10.32 b. In a cylindrical pressure vessel the three principal stresses are

$$\sigma_t = qD/2t; \quad \sigma_a = qD/4t; \quad \sigma_r \approx 0$$

The maximum is σ_t, which gives

$$\sigma_t = \frac{qD}{2t} = \frac{(21\,\text{MPa})(4600\,\text{mm} - 400\,\text{mm})}{2(200\,\text{mm})} = 221\,\text{MPa}$$

10.33 b. The stresses in the pressure vessel are, as in the last problem,

$$\sigma_t = \frac{qD}{2t}; \qquad \sigma_a = \frac{qD}{4t}; \qquad \sigma_r \approx 0$$

The wall thickness is usually thin enough so that it can be assumed that $D_i \approx D_o$

The tangential strain can be found from Hooke's law:

$$\varepsilon_t = \frac{1}{E}(\sigma_t - \nu\sigma_\theta - \nu\sigma_r) = \frac{1}{E}\left(\frac{qD}{2t} - \nu\frac{qD}{4t} - \nu 0\right) = 0.425\frac{qD}{Et}$$

The thickness is then

$$t = 0.425\frac{pD}{E\varepsilon_t} = 0.425\frac{(1.4\,\text{MPa})(300\,\text{mm})}{(210\,\text{GPa})(200\times10^{-6})} = 4.25\,\text{mm}$$

10.34 d. Draw Mohr's circle for the stress. At 45° in the physical plane (90° on Mohr's circle) the two normal stresses are $3qD/8t$. Hooke's law gives

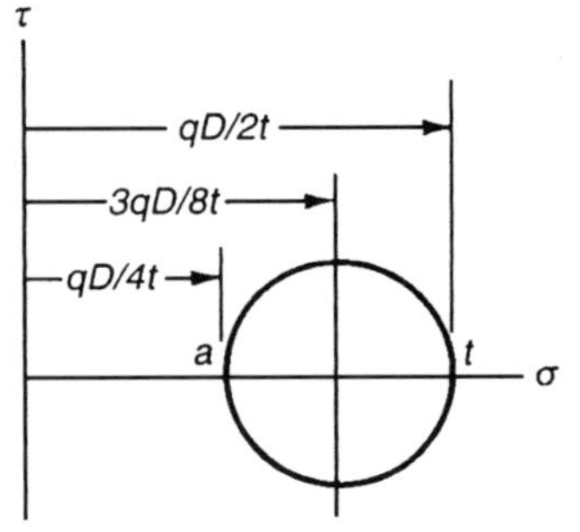

Exhibit 10.34

$$\varepsilon_{45} = \frac{1}{E}(\sigma_{45} - \nu\sigma_{-45} - \nu\sigma_r) = \frac{1}{E}\left(\frac{3qD}{8t} - \nu\frac{3qD}{8t} - \nu 0\right)$$

$$= \frac{(1-\nu)}{E}\left(\frac{3qD}{8t}\right)$$

$$\varepsilon_{45} = \frac{(1-0.33)}{(69\,\text{GPa})}\frac{3(0.7\,\text{MPa})(200\,\text{mm} - 6\,\text{mm})}{8(3\,\text{mm})} = 164.8\times10^{-6}$$

10.35 c. Draw Mohr's circle. The maximum principal stress is 6 MPa. As an alternative,

$$R=\sqrt{\left(\frac{\sigma_x-\sigma_y}{2}\right)^2+\tau_{xy}^2}=\sqrt{\left(\frac{-3-5}{2}\right)^2+(-3)^2}=5$$

$$C=\frac{\sigma_x+\sigma_y}{2}=\frac{-3+5}{2}=1$$

$$\sigma_1=R+C=6\text{ MPa}$$

Exhibit 10.35a

10.36 d. Draw Mohr's circle. The center of the circle is

$$C=\frac{\sigma_x+\sigma_y}{2}=\frac{5-1}{2}=2$$

The radius is then $R=7-2=5$. The shear stress can be found from the Mohr's circle or from the expression

$$R=\sqrt{\left(\frac{\sigma_x-\sigma_y}{2}\right)^2+\tau_{xy}^2};\qquad \tau_{xy}^2=R^2-\left(\frac{\sigma_x\quad\sigma_y}{2}\right)^2$$

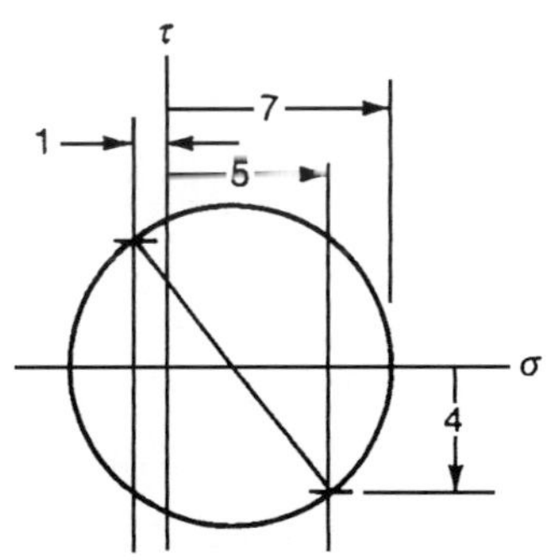

Exhibit 10.36a

In either case, the shear stress $\tau_{xy}=4$ MPa.

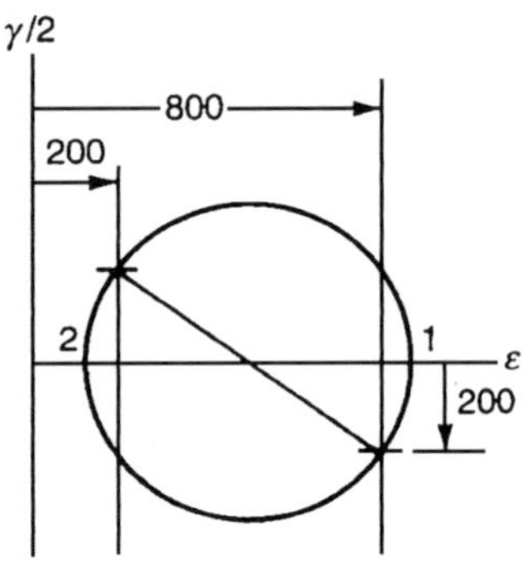

Exhibit 10.12a

10.37 b. Draw Mohr's circle. ε_1 can be scaled from the circle or computed as follows:

$$R=\sqrt{\left(\frac{\varepsilon_x-\varepsilon_y}{2}\right)^2+\left(\frac{\gamma_{xy}}{2}\right)^2}=\sqrt{\left(\frac{800-200}{2}\right)^2+\left(\frac{400}{2}\right)^2}=361\,\mu$$

$$C=\frac{\varepsilon_x+\varepsilon_y}{2}=\frac{800+200}{2}=500\,\mu$$

$$\varepsilon_1=C+R=500\,\mu+361\,\mu=861\,\mu$$

$$\varepsilon_2=C-R=500\mu-361\mu=139\mu$$

ε_1 is the maximum.

10.38 d. Problems of this type can be done by using Hooke's law first and then Mohr's circle or by using Mohr's circle first and then applying Hooke's law. Since Mohr's circle was already drawn for this problem in the previous solution, the second approach will be followed. The principal strains were found to be the following: $\varepsilon_1 = 861$ μ; $\varepsilon_2 = 139$ μ. Hooke's law in plane stress is

$$\varepsilon_1=\frac{1}{E}(\sigma_1-v\sigma_2);\qquad \varepsilon_2=\frac{1}{E}(\sigma_2-v\sigma_1)$$

Inverting these relationships gives

$$\sigma_1=\frac{E}{1-v^2}(\varepsilon_1+v\varepsilon_2);\qquad \sigma_2=\frac{E}{1-v^2}(\varepsilon_2+v\varepsilon_1)$$

The maximum principal stress is σ_1, which is

$$\sigma_1=\frac{E}{1-v^2}(\varepsilon_1+v\varepsilon_2)=\frac{210\,\text{GPa}}{1-0.3^2}[861\times10^{-6}+0.3(139\times10^{-6})]=208\text{ MPa}$$

10.39 d. Draw free-body diagrams. The torque in shaft 1 is $T_1 = 10 + 10 = 20$ kN • m. The torque in shaft 2 is $T_2 = 10$ kN • m.

$$\tau_1=\frac{T_1r_1}{J_1}=\frac{(20\text{ kN}\bullet\text{m})(25\text{ mm})}{0.5\pi(25\text{ mm})^4}=815\text{ MPa}$$

$$\tau_2=\frac{T_2r_2}{J_2}=\frac{(10\text{ kN}\bullet\text{m})(19\text{ mm})}{0.5\pi(19\text{ mm})^4}=928\text{ MPa}$$

The largest stress is 928 MPa.

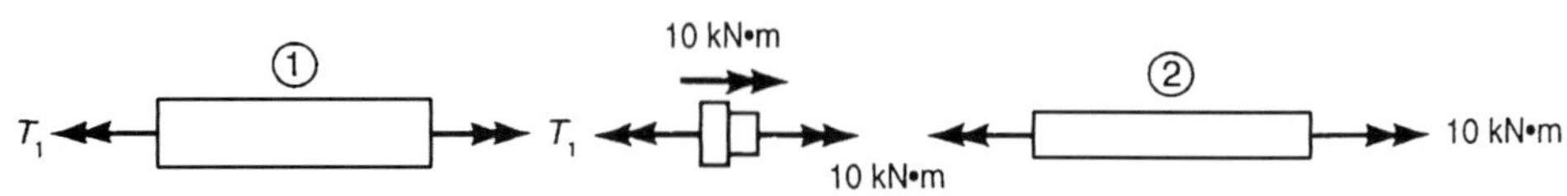

Exhibit 10.39a

10.40 d. From the force-deformation relations,

$$\phi_1 = \frac{T_1 L_1}{GJ_1} = \frac{(20\,\text{kN} \bullet \text{m})(250\,\text{mm})}{(83\,\text{GPa})0.5\pi(25\,\text{mm})^4} = 0.982\,\text{rad} = 5.63°$$

$$\phi_2 = \frac{T_2 L_2}{GJ_2} = \frac{(10\,\text{kN} \bullet \text{m})(250\,\text{mm})}{(83\,\text{GPa})0.5\pi(19\,\text{mm})^4} = 0.1471\,\text{rad} = 8.43°$$

From compatibility,

$$\phi = \phi_1 + \phi_2 = 5.63° + 8.43° = 14.06°$$

10.41 d. Draw the free-body diagrams. Equilibrium of the center free body gives

$$T_1 = T_2 + 10$$

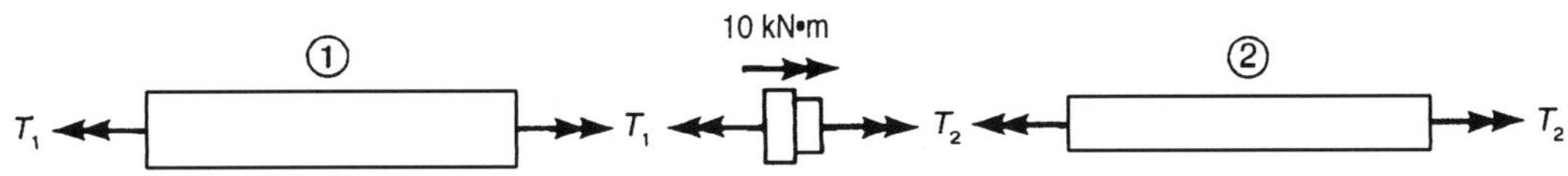

Exhibit 10.41a

The force-deformation relations are

$$\phi_1 = \frac{T_1 L_1}{GJ_1} = \frac{(10 + T_2)(250\,\text{mm})}{(83\,\text{GPa})0.5\pi(25\,\text{mm})^4} = 49.1 \times 10^{-3} + 4.91 \times 10^{-3} T_2$$

$$\phi_2 = \frac{T_2 r_2}{J_2} = \frac{T_2(250\,\text{mm})}{(83\,\text{GPa})\ 0.5\pi(19\,\text{mm})^4} = 14.7 \times 10^{-3}\ T_2$$

Compatibility requires that

$$\phi_1 + \phi_2 = 0 = 49.1 \times 10^{-3} + (4.91 \times 10^{-3} + 14.7 \times 10^{-3})\,T_2$$

Solving for the torques gives

$$T_2 = \frac{5.305}{2.207} = -2.50\,\text{kN} \bullet \text{m}$$

$$T_1 = T_2 + 10 = -2.50 + 10 = 7.50\,\text{kN} \bullet \text{m}$$

The stresses then are

$$\tau_1 = \frac{T_1 r_1}{J_1} = \frac{(7.50\,\text{kN} \bullet \text{m})(25\,\text{mm})}{0.5\pi(25\,\text{mm})^4} = 306\,\text{MPa}$$

$$\tau_2 = \frac{T_2 r_2}{J_2} = \frac{(-2.5\,\text{kN} \bullet \text{m})(19\,\text{mm})}{0.5\pi(19\,\text{mm})^4} = -232\,\text{MPa}$$

10.42 **d.** The same three-step process as in Problem 10.41 must be carried out. Since this process has already been completed, the results can be used. The rotation can be expressed as

$$\phi = \phi_1 = -\phi_2 = \frac{T_1 L_1}{GJ_1} = \frac{(7.50\text{kN} \bullet \text{m})(250\,\text{mm})}{(83\,\text{GPa})0.5\pi(25\,\text{mm})^4} = 0.0368\ \text{rad} = 2.11°$$

10.43 **c.** To determine displacement of B, calculate the load at B (Exhibit 10.43a).

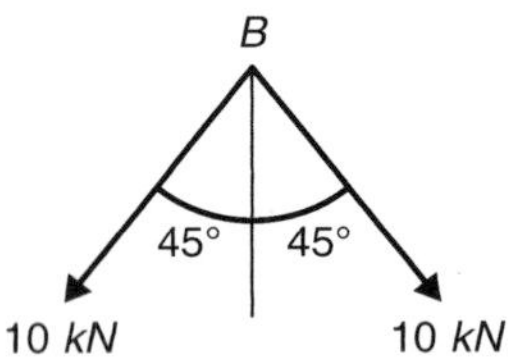

Exhibit 10.43a

(Load at B) = 2 (10 sin 45) = 14.14 kN

Therefore, the load at C = 14.14 + 20 = 34.14 kN

$$\delta_{CB} = \frac{PL_{CB}}{AE} = \frac{\left(34.14 \times 10^3\ \text{N}\right)(1.5\ \text{m})}{\left(\frac{1}{100}\right)^2 \text{m}^2 \left(180 \times 10^9\right)\ \text{N/m}^2} = 0.002845\ \text{m} = 2.85\ \text{mm}$$

10.44 **d.** Determine displacement at A:

δ_A = 2.845 + 1 = 3.85 mm

$$\delta_A = \delta_B + \delta_{AB} = 2.845\ \text{mm} + \frac{\left(20 \times 10^3\ \text{N}\right)(0.5\ \text{m})}{\left(\frac{1}{100}\right)^2 \text{m}^2 \left(180 \times 10^9\right) \text{N/m}^2}$$

10.45 **b.** Using the free-body diagram of $BCDF$ (Exhibit 10.45a):

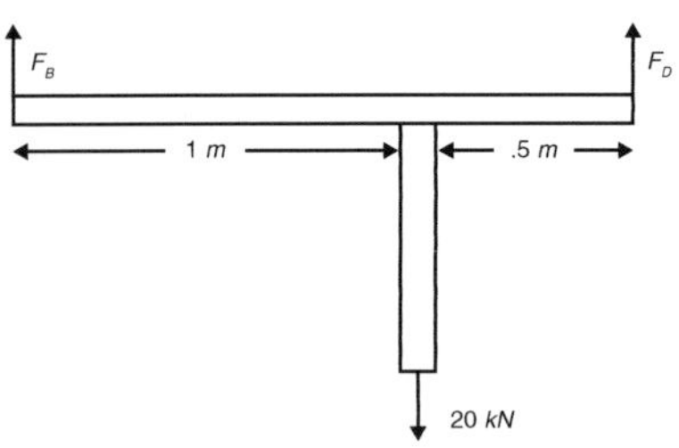

Exhibit 10.45a

$M_B = 0 = 20$ kN (1 m) – F_D (1.5 m)

$F_D = 13.3$ kN ↑

$\Sigma F_y = 0 = F_D + F_B - F_F$

10.46 **c.** To determine the load at D, we must write an equilibrium equation of forces:

13.3 ↑ + F_B – 20 = 0 F_B = 6.7 kN ↑

10.47 **b.** The displacement of point D is found:

$$\delta_D = \frac{PL}{AE} = \frac{\left(13.3 \times 10^3\ \text{kN}\right)(1\ \text{m})}{\left(30 \times 10^{-6}\right)\text{m}^2 \left(180 \times 10^9\right)\text{N/m}^2}$$

$$\delta_D = 0.00246\ \text{m} = 2.46\ \text{mm}$$

10.48 **a.** The displacement of point B is found:

$$\delta_B = \frac{PL}{AE} = \frac{\left(6.7\times10^3\ \text{kN}\right)(1\ \text{m})}{\left(50\times10^{-6}\,\text{m}^2\right)\left(180\times10^9\ \text{N/m}^2\right)}$$

$$\delta_B = 0.000744\ \text{m} = 0.744\ \text{mm}$$

10.49 **d.** To determine the displacement of F, draw a displacement triangle from B to D, as shown in Exhibit 10.49.

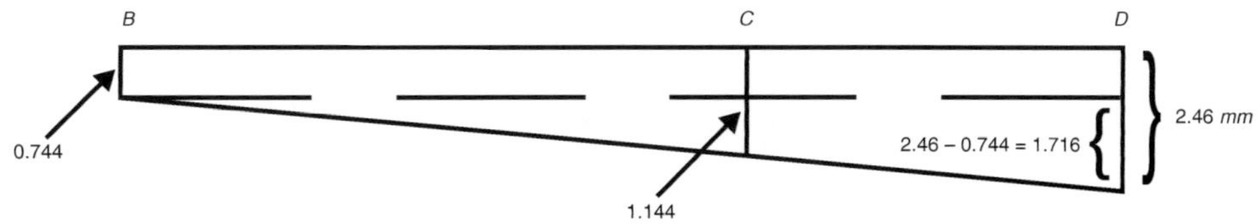

Exhibit 10.49

From the triangle, we see that point C is displaced:

$$\delta_C = 0.744\ \text{mm} + 1.144\ \text{mm} = 1.888\ \text{mm}$$

$$\delta_{F/C} = \frac{PL}{AE} = \frac{\left(20\times10^3\ \text{N}\right)(1\ \text{m})}{\left(40\times10^{-6}\right)\ \text{m}^2\ \left(180\times10^9\right)\text{N/m}^2} = 0.00277\ \text{m}$$

$$\delta_F = 1.888 + 2.77 = 4.65\ \text{mm}$$

10.50 **c.** Find the modulus of elasticity by using the values from the magnified stress-strain portion of the curve.

$$E = \frac{\sigma}{\varepsilon} = \frac{225\ \text{MPa}}{0.002\ \text{mm}} = 112.5\ \text{GPa}$$

10.51 **a.** The yield load can be obtained by taking the yield stress off the curve and multiplying by the area of the bar.

$$P_y = \sigma_y A$$

$$P_y = \left(225\times10^6\ \text{N/m}^2\right)(\pi/4)(0.01)^2\ \text{m}^2 = 17.7\ \text{kN}$$

10.52 **d.** Obtain the ultimate load by taking the ultimate stress off the curve and multiplying by the area of the bar.

$$\text{P}_{ult} = \sigma_{ult} A = (360\times10^6\,\text{Pa})(\pi/4)(0.01)^2 = 28.3\ \text{kN}$$

10.53 b. The modulus of resilience may also be obtained using values from the stress-strain curve.

$$\mu_r = \frac{1}{2}\frac{\sigma_{PL}^2}{E} = \left(\frac{1}{2}\right)\frac{\left(225\times10^6\right)^2}{112.5\times10^9} = 0.225 \text{ MJ/m}^3$$

10.54 c. Sketch a free-body diagram (Exhibit 10.54a) and use the equilibrium equations to determine the load on *BC*.

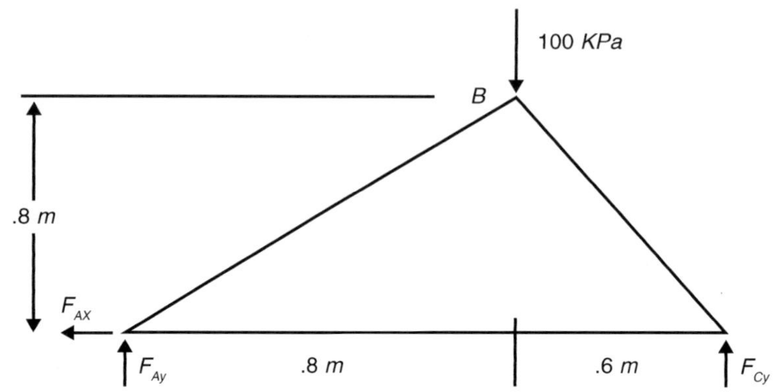

Exhibit 10.54a

$\Sigma F_X = 0 = -F_{AX} = 0; \qquad F_{AX} = 0$

$M_A = 0 = 100 \text{ kPa } (0.8 \text{ m}) - F_{Cy}\ (1.4 \text{ m})$

Therefore:

$F_{Cy} = 57 \text{ kN}; \qquad F_{Ay} = 43 \text{ kN}$

Making a free-body diagram of bar *BC* (Exhibit 10.54b), calculate the axial load of *BC*.

$4/5\ F_{BC} = F_{Cy} = 57 \text{ kN, so } F_{BC} = 71.25 \text{ kN}$

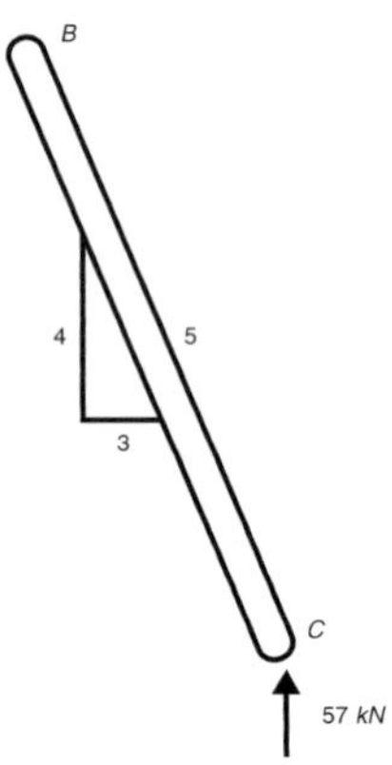

Exhibit 10.54b

10.55 a. Taking a free-body diagram of *AB* (Exhibit 10.55), we solve for the stress in *AB*

$$F_{AB} = \frac{F_{AY}}{\sin\ 45} = \frac{43}{.707} = 60.8 \text{ kN}$$

$$\sigma_{AB} = \frac{F_{AB}}{A} = \frac{60.8 \text{ kN}}{2 \text{ cm}^2} = \frac{60.8\times10^3 \text{ N}}{0.0002 \text{ m}^2} = 304 \text{ MPa}$$

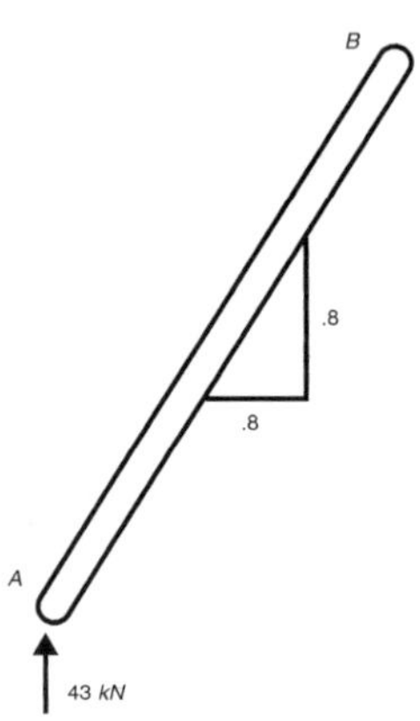

Exhibit 10.55

10.56 **a.** To determine the elongation of bar AC, use equation:

$$\delta_{AC} = \frac{F_{AC}L_{AC}}{AE}$$

Knowing that $F_{AC} = 43$ kN,

$$\delta_{AC} = \frac{\left(43\times10^{3}\ \text{N}\right)\left(1.4\ \text{m}\right)\left(1000\ \text{mm/m}\right)}{\left(0.0002\ \text{m}^{2}\right)\left(150\times10^{9}\ \text{N/m}^{2}\right)} = 2.0\ \text{mm}$$

10.57 **a.** For a torsion problem, the shear strain is

$$\gamma_{\phi z} = \frac{\tau_{\phi z}}{G} = \frac{Tr}{GJ}$$

Other shear strains in the $r - \phi$ orientation are zero. Mohr's circle for this state of strain is shown in Exhibit 10.57(a). From Mohr's circle,

$$\varepsilon_{45} = \frac{\gamma_{\phi z}}{2} = \frac{\tau_{\phi z}}{2G} = \frac{Tr}{2GJ}$$

$$T = \frac{2GJ\varepsilon_{45}}{r} = \frac{2(83\,\text{GPa})[0.5\pi(25\,\text{mm})^{4}](245\times10^{-6})}{25\,\text{mm}} = 998\,\text{N}\bullet\text{m}$$

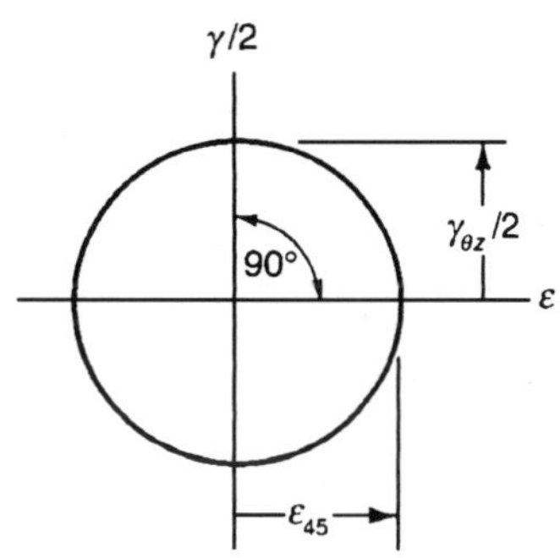

Exhibit 10.57a

10.58 **d.** In thin-walled shafts, the shear stress is

$$\tau_{sz} = \frac{T}{2At}$$

The area A is the cross-sectional area of the shaft including the hole, so

$$A = \frac{\pi r^{2}}{2} = \frac{\pi(25\,\text{mm} - 3\,\text{mm})^{2}}{2} = 760\ \text{mm}^{2}$$

$$\tau_{sz} = \frac{T}{2At} = \frac{300\,\text{N}\bullet\text{m}}{2(760\,\text{mm}^{2})(3\,\text{mm})} = 65.7\ \text{MPa}$$

10.59 **c.** Draw the free-body diagram of the beam, replacing the distributed load with its statically equivalent loads. Summation of moments about the left end gives

$$0 = -3.6\,R_2 + (108)(1.8) + (36)(4.2)$$
$$R_2 = 96 \text{ kN}$$

Summation of forces in the vertical direction gives

$$R_1 = 144 - 96 = 48 \text{ kN}$$

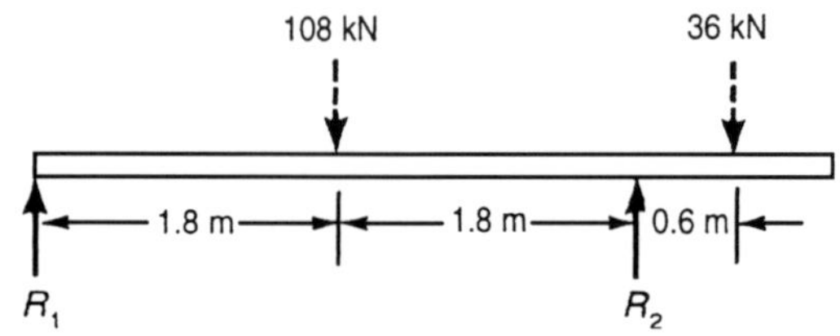

Exhibit 10.59a

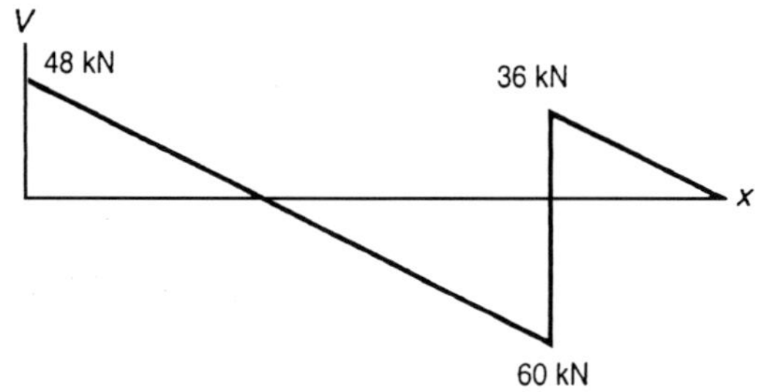

Exhibit 10.59b

This is enough information to plot the shear diagram (Exhibit 10.59b). The largest magnitude of shear is 60 kN.

10.60 **c.** The maximum bending moment occurs where the shear is zero. From the shear diagram, the distance to the zero from the left end can be found by similar triangles.

$$\frac{48}{x} = \frac{108}{3.6}; \qquad x = \frac{(48)(3.6)}{108} = 1.6\,\text{m}$$

The areas of the shear diagrams are the changes in moment (Exhibit 10.60a).

$$A_1 = \frac{(48\,\text{kN})(1.6\,\text{m})}{2} = 38.4\,\text{kN} \bullet \text{m}$$

$$A_2 = \frac{(36\,\text{kN})(1.2\,\text{m})}{2} = 21.6\,\text{kN} \bullet \text{m}$$

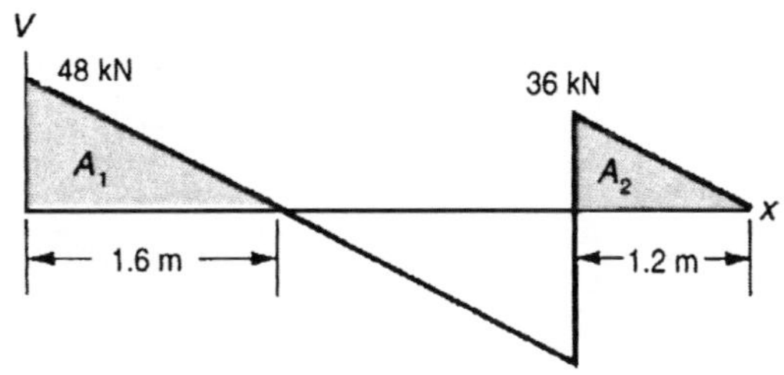

Exhibit 10.60a

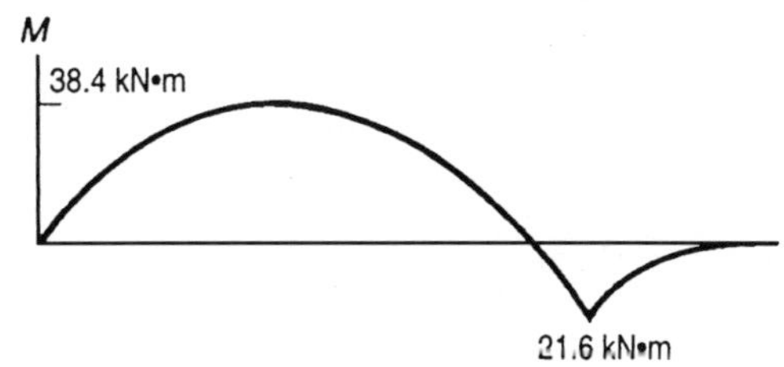

Exhibit 10.60b

The moment diagram is shown in Exhibit 10.60b. The maximum bending moment is 38.4 kN • m.

10.61 **a.** There is a jump at *B* and *D* of 60 kN upward and a downward jump of 30kN at *E*. These jumps correspond to concentrated forces.

10.62 **d.** The areas of the shear diagrams (Exhibit 10.62) are the changes in moment. Since the moments are zero on either end,

$$M_B = A_1 = \frac{(30\,\text{kN})(1\,\text{m})}{2} = 15\ \text{kN} \bullet \text{m}$$

$$M_C = A_1 + A_2 = \frac{(30\,\text{kN})(1\,\text{m})}{2} + \frac{(30\,\text{kN})(1\,\text{m})}{2} = 0$$

$$M_D = A_3 = (30\,\text{kN})(-0.5\,\text{m}) = -15\ \text{kN} \bullet\ \text{m}$$

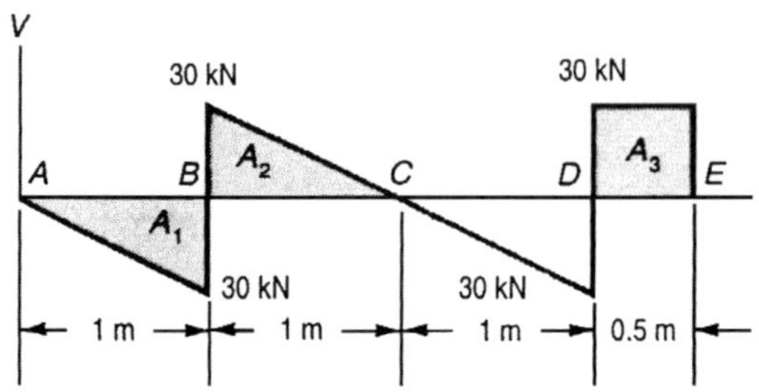

Exhibit 10.62

The largest magnitude of the bending moment is therefore 15 kN • m.

10.63 **d.** It is obvious that each support will carry half of the load, so the reactions are $wL/2$. The shear diagram is shown in Exhibit 10.63a. The maximum bending moment is

$$M = A_1 = \frac{1}{2}\left(\frac{wL}{2}\right)\left(\frac{L}{2}\right) = \frac{wL^2}{8}$$

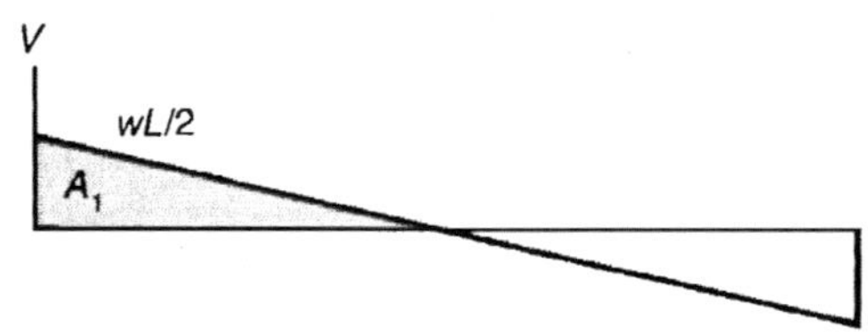

Exhibit 10.63a

The maximum bending stress is

$$\sigma_{\max} = \frac{M}{Z} = \frac{wL^2}{8Z} = 100\,\text{MPa}$$

$$w = \frac{(100\,\text{MPa})\,8Z}{L^2} = \frac{(100\,\text{MPa})(8)(1408 \times 10^3\ \text{mm}^3)}{(4\,\text{m})^2} = 70.4\ \text{kN/m}$$

10.64 **a.** Draw the free-body diagram and the shear and moment diagrams as shown in Exhibit 10.64a. The maximum bending stress is

$$\sigma_{max} = \frac{M_{max}c}{I} = \frac{(9\text{kN} \bullet \text{m})\left(\frac{363\text{ mm}}{2}\right)}{\left(365 \times 10^6\text{ mm}^4\right)} = 4.48\text{ MPa}$$

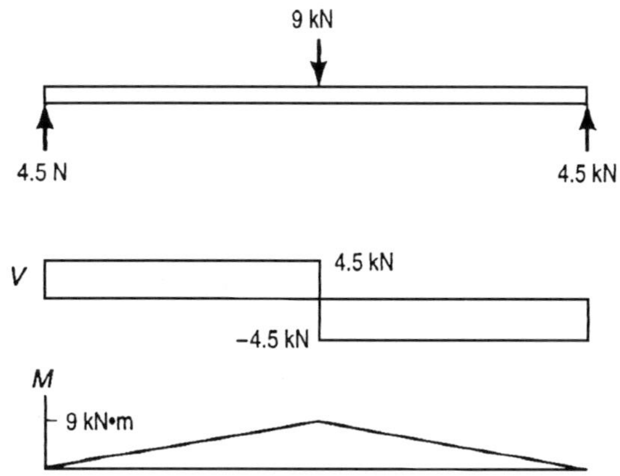

Exhibit 10.64a

10.65 **a.** From the previous problem, the maximum shear in the beam is 4.5 kN. The maximum shearing stress will take place at the centroid (Exhibit 10.65), so a cut must be made there in order to calculate Q. The moment of the area Q is, therefore,

$$Q = A_1\overline{y}_1 + A_2\overline{y}_2$$

$$Q = (257\text{ mm})(21.7\text{ mm})\left(\frac{363\text{ mm}}{2} - \frac{21.7\text{ mm}}{2}\right) + \ldots \mid \left(\frac{363\text{ mm}}{2} - 21.7\text{ mm}\right)$$

$$\times(13\text{ mm})\left(\frac{\frac{363\text{ mm}}{2} - 21.7\text{ mm}}{2}\right)$$

$$Q = 1.117 \times 10^6\text{ mm}^3$$

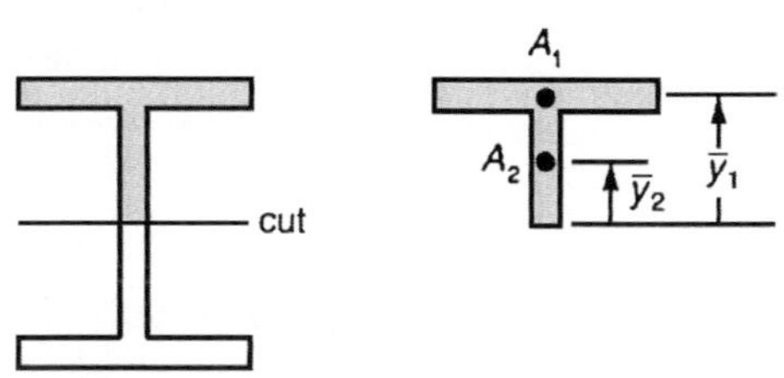

Exhibit 10.65

The maximum shear stress is then

$$\tau = \frac{VQ}{Ib} = \frac{(4.5\text{ kN})(1.117 \times 10^6\text{ mm}^3)}{(365 \times 10^6\text{ mm}^4)(13\text{ mm})} = 1.060\text{ MPa}$$

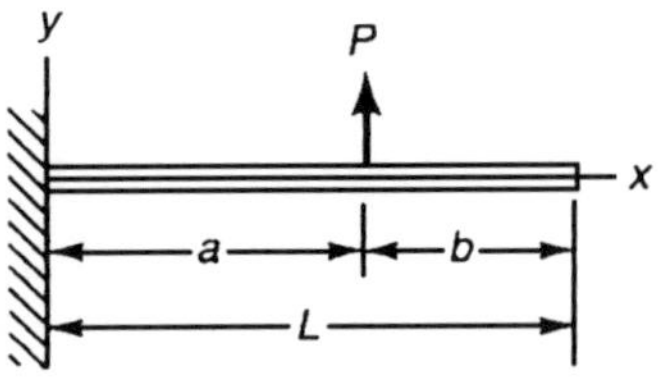

Exhibit 10.66a

10.66 d. From Table 10.1, Beam Type 1 (Exhibit 10.66a), for a $\leq x \leq L$

$$y = \frac{Pa^2}{6EI}(3x - \mathrm{a})$$

For the load at the half-way point, $a = L/2$, $x = L$, and $P = -F$. For the load at the end, $a = L$, $x = L$, and $P = -F$. Therefore,

$$y = \frac{-F\left(\frac{L}{2}\right)^2}{6EI}\left[3L - \left(\frac{L}{2}\right)\right] + \frac{-FL^2}{6EI}(3L - L) = -0.4375\frac{FL^3}{EI}$$

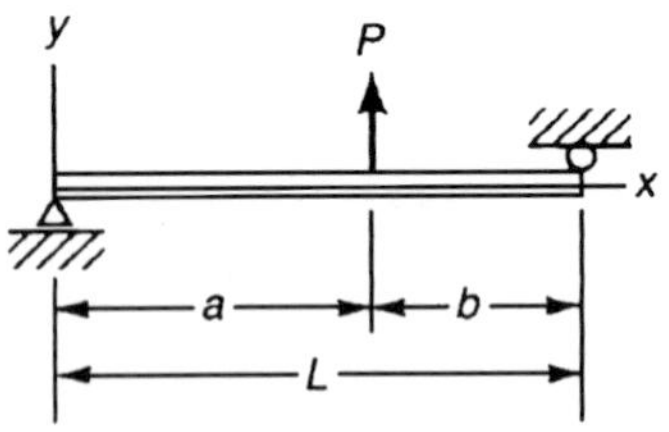

Exhibit 10.67a

10.67 b. The maximum deflection for this beam will take place at the center of the beam. This problem can be solved with the superposition of the following cases from Table 10.1 (Exhibits 10.67a and b).

For $0 \leq x \leq a$,

$$y = \frac{Pbx}{6LEI}(L^2 - b^2 - x^2)$$

For this problem, $P = -wL$, $a = b = L/2$, and $x = L/2$.

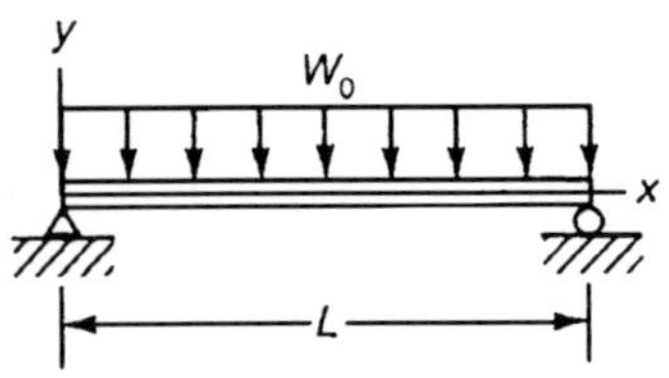

Exhibit 10.67b

$$y = -\frac{wx}{24EI}(L^3 - 2Lx^2 + x^3)$$

For this problem, $x = L/2$. The total deflection is, therefore,

$$y = \frac{Pbx}{6LEI}\left(L^2 - b^2 - x^2\right) - \frac{wx}{24EI}\left(L^3 - 2Lx^2 + x^3\right)$$

$$y = \frac{(-wL)\left(\frac{L}{2}\right)\left(\frac{L}{2}\right)}{6LEI}\left[L^2 - \left(\frac{L}{2}\right)^2 - \left(\frac{L}{2}\right)^2\right] - \frac{w\left(\frac{L}{2}\right)}{24EI}\left[L^3 - 2L\left(\frac{L}{2}\right)^2 + \left(\frac{L}{2}\right)^3\right]$$

$$v = -0.0339\frac{wL^4}{EI} = -\frac{13wL^4}{384EI}$$

10.68 d. This problem can be solved from superposition of the same two cases as used in Problem 10.67. For the concentrated load solution, $b = L/2$ and P is left as an unknown. In both, $x = L/2$. The center support means the beam does not deflect in the center. Therefore,

$$y = 0\frac{P\left(\frac{L}{2}\right)\left(\frac{L}{2}\right)}{6LEI}\left[L^2 - \left(\frac{L}{2}\right)^2 - \left(\frac{L}{2}\right)^2\right] - \frac{w\left(\frac{L}{2}\right)}{24EI}\left[L^3 - 2L\left(\frac{L}{2}\right)^2 + \left(\frac{L}{2}\right)^3\right]$$

$$y = 0 = \frac{PL^3}{48\ EI} = -\frac{5wL^4}{384EI}$$

$$P = \frac{5}{8}wL$$

10.69 c. Draw the free-body diagram (Exhibit 10.69a). From a summation of moments about the cut at B, the maximum bending moment in BC is the moment $M = 200\text{ N} \times 250\text{ mm}$ or $50\text{ kN} \bullet \text{m}$. The maximum bending stress is

$$\sigma = \frac{Mc}{I} = \frac{(50\,\text{kN} \bullet \text{mm})(12.5\,\text{mm})}{0.25\,\pi(12.5\,\text{mm})^4} = 32.6\text{ MPa}$$

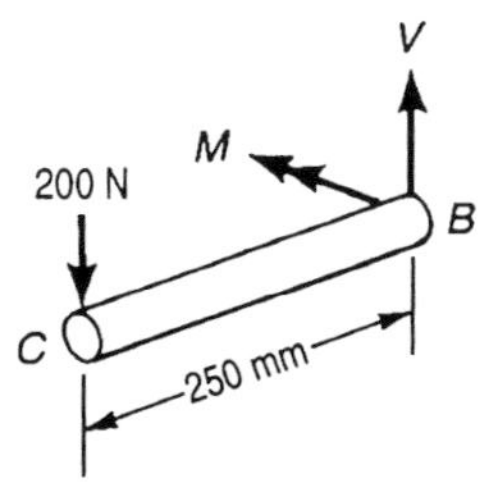

Exhibit 10.69a

10.70 d. Draw the free-body diagram (Exhibit 10.70). The maximum stresses in section AB will occur at A. Summation of forces in the vertical direction gives $V_A = 200$ N. Summation of forces along the direction of the rod AB gives $P = 2000$ N. Summation of moments along the rod AB gives $T_A = 200\text{ N} \times 250\text{ mm}$ or $50\text{ kN} \bullet \text{mm}$. Summation of moments at the cut perpendicular to the rod AB gives $M_A = 200\text{ N} \times 300\text{ mm} = 600\text{ kN} \bullet \text{mm}$. The maximum bending stress is

$$\sigma = \frac{Mc}{I} = \frac{(60\,\text{kN} \bullet \text{mm})(12.5\,\text{mm})}{0.25\,\pi(12.5\,\text{mm})^4} = 39.1\,\text{MPa}$$

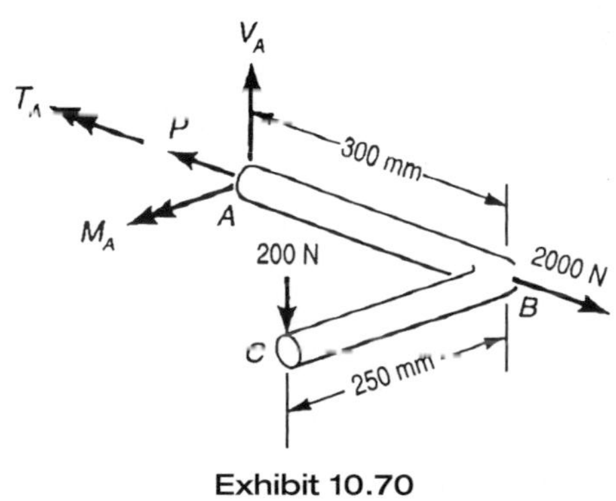

Exhibit 10.70

10.71 b. From the free-body diagram in Problem 10.70, the maximum torque is $50\text{ kN} \bullet \text{mm}$. The maximum shear stress is, therefore,

$$\tau_{\max} = \frac{T_{\max} r_0}{J} = \frac{(50\,\text{kN} \bullet \text{mm})(12.5\,\text{mm})}{0.5\,\pi(12.5\,\text{mm})^4} = 16.30\text{ MPa}$$

10.72 a. From the free-body diagram in Problem 10.70, the maximum axial force is 2000 N. The maximum stress due to this force is, therefore,

$$\sigma = \frac{P}{A} = \frac{2000\,\text{N}}{\pi(12.5\,\text{mm})^2} = 4.07\,\text{MPa}$$

10.73 d. The stresses were found in the previous three problems. There is an axial stress due to both bending and axial loads. This stress is

$$\sigma = 39.1\text{ MPa} + 4.07\text{ MPa} = 43.2\text{ MPa}$$

The shear stress is 16.3 MPa. These are the only nonzero stresses. The maximum principal stress can be calculated as follows:

$$R = \sqrt{\left(\frac{\sigma_x - \sigma_y}{2}\right)^2 + \tau_{xy}^2} = \sqrt{\left(\frac{43.2 - 0}{2}\right)^2 + 16.3^2} = 22.0\text{ MPa}$$

$$C = \frac{\sigma_x + \sigma_y}{2} = \frac{43.2 + 0}{2} = 21.6\text{ MPa}$$

$$\sigma_1 = C + R = 22.0 + 21.6 = 43.6\text{ MPa}$$

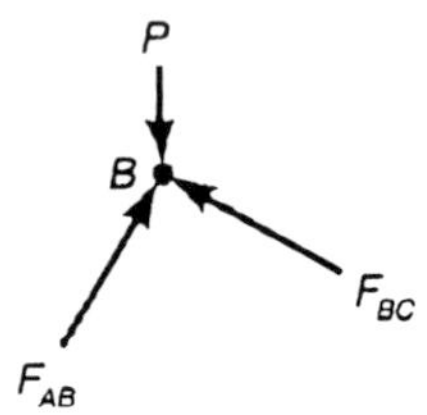

Exhibit 10.74a

10.74 a. Draw the free-body diagram of the joint B (Exhibit 10.74a). Summation of forces in the vertical direction gives

$$P = F_{AB}\frac{1.5}{\sqrt{1.5^2 + 0.9^2}} + F_{BC}\frac{1.5}{\sqrt{2.7^2 + 1.5^2}}$$

Summation of forces in the horizontal direction gives

$$F_{AB}\frac{0.9}{\sqrt{1.5^2 + 0.9^2}} = F_{BC}\frac{2.7}{\sqrt{2.7^2 + 1.5^2}}$$

Solving for F_{AB} and F_{BC} gives

$$F_{AB} = 0.875\,P; \quad F_{BC} = 0.515\,P$$

The member AC is in tension and does not need to be considered. The moment of inertia for both members is

$$I = \frac{bh^3}{12} = \frac{(25\text{ mm})(25\text{ mm})^3}{12} = 32{,}600\text{ mm}^4$$

The critical buckling load for member AB is

$$P_{cr} = F_{AB} = \frac{\pi^2 EI}{L^2} = \frac{\pi^2(210\text{ GPa})(32{,}600^4)}{(1.5\text{ m})^2 + (0.9\text{ m})^2} = 22.0\text{ kN}$$

The load P for buckling to occur in AB is

$$P = \frac{22\text{ kN}}{0.875} = 25.2\text{ kN}$$

The critical buckling load for member BC is

$$P_{cr} = F_{BC} = \frac{\pi^2 EI}{L^2} = \frac{\pi^2(210\text{ GPa})(32{,}600\text{ mm}^4)}{[(2.7\text{ m})^2 + (1.5\text{ m})^2]} = 7.07\text{ kN}$$

The load P for buckling to occur in BC is

$$P = \frac{7.07\text{ kN}}{0.515} = 13.7\text{ kN}$$

10.75 **d.** To buckle in the x-y plane the critical buckling load is

$$P_{cr} = \frac{\pi^2 EI}{L^2} \frac{\pi^2 E\left(\dfrac{ab^3}{12}\right)}{L^2}$$

To buckle in the x-z plane the critical buckling load is

$$P_{cr} = \frac{4\pi^2 EI}{L^2} \frac{4\pi^2 E\left(\dfrac{ba^3}{12}\right)}{L^2}$$

Equating these two representations of P_{cr} gives

$$ab^3 = 4ba^3; \quad b^2 = 4a^2; \quad \frac{b}{a} = 2$$

CHAPTER 11

Material Properties and Processing

Brian Flinn

OUTLINE

All engineering products are made of materials. Thus, engineers become directly involved with materials, whether they be design engineers, production engineers, or applications engineers. Their familiarity with a wide spectrum of materials becomes particularly important as they advance through management and into administration, where they must oversee the activities of additional engineers on their technical staffs.

The way that an engineering product performs in service is a consequence of the combination of the components of the product. Thus, a cellular phone must have the diodes, resistors, capacitors, and other components that function together to meet its design requirements. Likewise, a competitively produced car must possess a carefully designed engine with its numerous parts, as well as safety features and operating characteristics that meet customer approval. Materials are pertinent to each and every design consideration.

Just as it is to be expected that the internal circuitry of a four-function hand calculator will differ from the internal circuitry of its multifunctional scientific counterpart, the internal structure of a steel gear differs from that of the sheet steel to be used in an automotive fender. Their roles, and therefore their properties, are designed to be different.

The variations in the internal structures of materials that lead to property differences include variations in atomic coordination and electronic energies, differences in internal geometries (microstructures), and the incorporation of larger structures, sometimes called macrostructures. Each of these is considered in the following sections, along with procedures for obtaining desired structures and properties.

ATOMIC ORDER IN ENGINEERING MATERIALS

Atoms, Ions, and Electrons

There is an order within atoms. Each atom has an integer number of protons. That number is called the **atomic number**. The natural elements possess, progressively, 1 to 92 of these protons, which carry a positive charge. A neutral atom has a number of electrons equal to the number of protons. Each electron is negative with a charge of 1.6×10^{-17} coulombs. Electrons are only allowed in given orbitals (called shells) that correspond to specific allowed energy levels.

With the exception of the principal isotope of hydrogen, each atom possesses neutrons. While these are charge-neutral, they add to the mass of each atom. The protons and neutrons reside in the nucleus of the atom. Figure 11.1 shows the Bohr model of a sodium atom. The lighter elements contain approximately equal numbers of protons and neutrons; however, in heavier elements the number of neutrons exceeds the number of protons. Furthermore, the number of neutrons per atom is not fixed. Thus, we encounter several **isotopes** for most atoms.

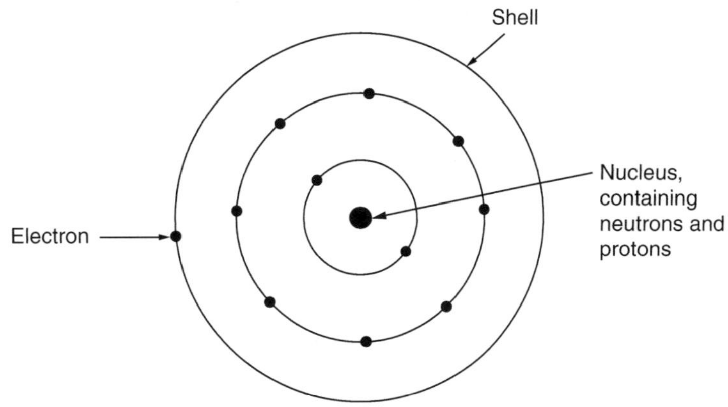

Figure 11.1 Bohr model of a sodium atom

Table 11.1 Data for selected atoms

Element	Protons Electrons Atomic No.	Neutrons (in natural isotopes)	Atomic Mass Unit	Grams per Avogadro's Number*
Hydrogen	1	0 or 1	1.008	1.008
Carbon	6	6 or 7	12.011	12.011
Oxygen	8	8, 9, or 10	15.995	15.995
Chlorine	17	18 or 20	35.453	35.453
Iron	26	28, 30, 31, or 32	55.847	55.847
Gold	79	118	196.97	196.97
Uranium	92	142, 143, or 146	238.03	238.03

*Avogadro's number = 6.022×10^{23}

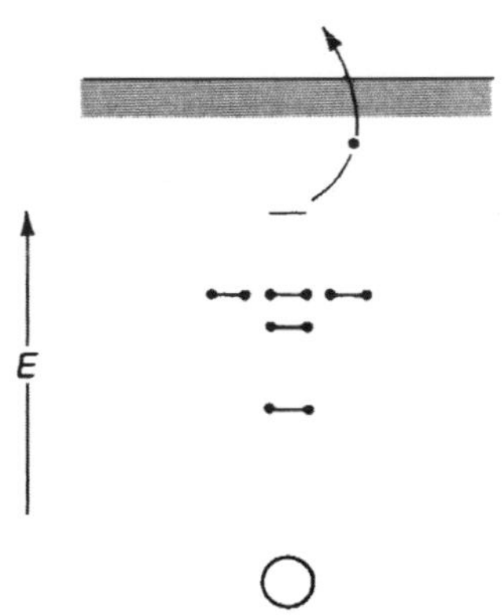

Figure 11.2 Ionization energy (schematic for sodium). Electrons reside at specific energy **states** (levels). Energy must be supplied to remove an electron from an atom, producing a positive ion. Two electrons (of opposite magnetic spins) may reside in each state.

Since the mass of an electron is appreciably less than 1% of that of protons and neutrons, the mass of an atom is directly related to the combined number of the latter two. By definition, an **atomic mass unit** (amu) is 1/12 of the mass of a carbon isotope that has six protons and six neutrons, C^{12}. The **atomic mass** of an element is equal to the number of these atomic mass units. (Selected values are listed in Table 11.1.) Thus while there are integer numbers of neutrons, protons, and electrons in each atom, the mass generally is not an integer, because more than one isotope is typically present.

A limited number of electrons may be accepted or released by an atom, thus introducing a charge on the atom (due the difference in the number of protons and electrons). A charged atom is an **ion**. Negative ions that have accepted extra electrons are called **anions**. Positive ions that have released electrons are called **cations**. Because they are charged, ions respond to electric fields. These fields may involve macroscopic dimensions (in electroplating baths); or they may involve interatomic distances (in molecules). Unlike charges attract; like charges repel.

Energy is required to remove an electron from a neutral atom. Figure 11.2 shows this schematically for a sodium atom. Conversely, fixed quantities of energy are released when electrons are captured by a positive ion. These energy levels (**states**) associated with an atom are fixed. Furthermore each state may accept only two electrons, and these must have opposite magnetic characteristics. Electronic, magnetic, and optical properties of materials must be interpreted accordingly.

Molecules

Atoms can join to one another; this is called **bonding**. Strong attractive forces can develop between atoms by three mechanisms; (1) coulombic attraction between oppositely charged ions, forming ionic bonds; (2) sharing of electrons to fill outer shells, creating covalent bonds; and (3) formation of ion cores surrounded by valence electrons that have been excited above the Fermi level and have become free electrons, forming metallic bonds. More detailed examples of these three primary types of bonds follow.

Ionic bonds form between metallic and nonmetallic atoms. The metallic atoms release their valence electrons to become cations (which are positively charged) and the nonmetallic atoms accept them to become anions. Ionic bonds are nondirectional. For example, a sodium ion that has lost an electron (Na^+) associates with as many negatively charged chlorine ions (Cl^-) as space will allow. And each Cl^- ion will become *coordinated* with as many Na^+ ions as necessary to balance the charge. The resulting structure will continue to grow in three dimensions until all available ions are positioned. Energy is released with each added ion.

Covalent bonds form between atoms that share valence electrons in order to fill their outer shells. In the simplest case, two hydrogen atoms release energy as they combine to produce a hydrogen molecule:

$$2H \rightarrow H2 \quad \text{or} \quad 2H \rightarrow H—H \tag{11.1}$$

The bond between the two involves a pair of shared electrons. This mechanism is common among many atomic pairs. In this case only one pair of atoms is involved. The covalent bonds of molecules are stereospecific; that is, they are between specific atoms and are therefore directional bonds.

In polymers a string or network of thousands or millions of atoms is bonded together. Examples include polyethylene, which has the structure shown in Figure 11.3(a), in which there is a backbone of carbon atoms that are covalently bonded; that is, they share pairs of valence electrons. Since each carbon atom has four valence electrons, it can form four covalent bonds, thus adding two hydrogens at the side of the chain in addition to the two bonds along the chain. Polyvinyl chloride, Figure 11.3(b), is related but has one of the four hydrogen atoms replaced by a chlorine atom.

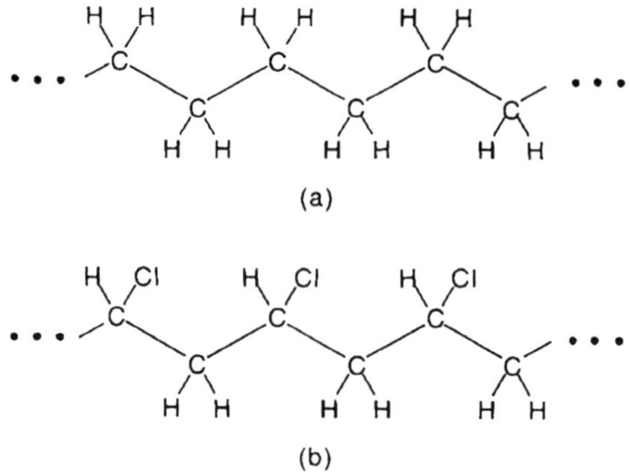

Figure 11.3 Covalent bonds: (a) polyethylene; (b) polyvinyl chloride

Metallic bonds form between the elements on the left side of the periodic table known as metals. Metallic bonds can form between atoms of the same element, to form a pure metal such as gold (Au), or between different metal atoms to form an alloy, such as brass, a mixture of copper (Cu) and zinc (Zn). The basis for the metallic bond is the formation of ion cores created when the metal's valence electrons are no longer associated with a specific atom. These electrons (called free electrons) move freely around surrounding metal ion cores, shielding them from each other. These freely moving electrons are often referred to as a *sea of electrons*. Metallic bonds are nondirectional.

The type and strength of bonding between atoms and molecules determines many properties of materials. As a general rule (and with other things being equal), the stronger the atomic bonding, the higher the melting temperature, hardness, and elastic modulus of materials. Ionically bonded materials are usually electrical and thermal insulators, while materials with metallic bonds have high electrical and thermal conductivities.

Crystallinity

The repetition of atomic coordinations in three dimensions produces a periodic structure that is called a **crystal**. The basic building block of a crystal is called a unit cell. Figure 11.4 illustrates the unit cell structure of iron. Each atom in this metal has eight nearest neighbors, which are symmetrically coordinated to give a **body-centered cubic** (BCC) crystal. About 30% of the metals have this structure. The atoms of aluminum and copper, among another 30% of the metals, become coordinated with 12 nearest neighbors with the result that **face-centered cubic** (FCC) crystal lattices are formed, as shown in Figure 11.5. A third group of metals form **hexagonal crystals** with **close packing** (HCP), as shown in Figure 11.6. As in the FCC crystals, each atom is coordinated with 12 nearest neighbors in HCP crystals.

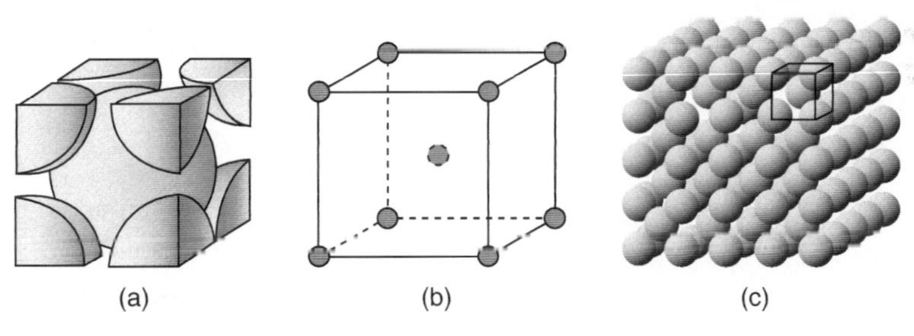

Figure 11.4 Body-centered cubic crystal structure: (a) a hard-sphere unit cell representation; (b) a reduced-sphere unit cell; (c) an aggregate of many atoms

It is possible to have a very high degree of perfection in crystals. For example, the repetition dimension (**lattice constant**) of the FCC lattice of pure copper is constant to the fifth significant figure (and to the sixth if the thermal expansion is factored in). This high degree of ordering provides a quantitative base for anticipating properties. Included are density calculations, certain thermal properties, and some of the effects of alloying.

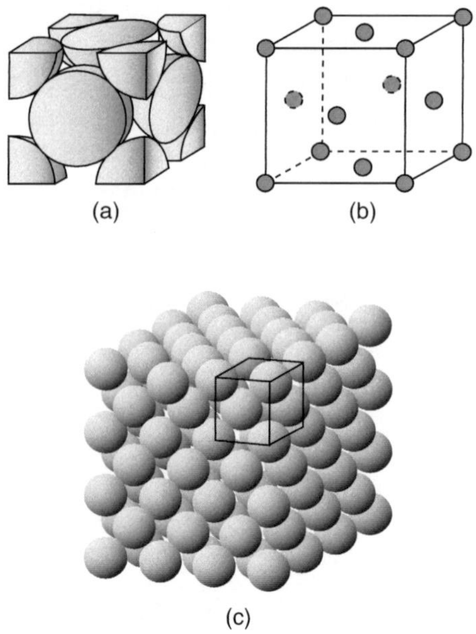

Figure 11.5 Face-centered cubic crystal structure: (a) a hard-sphere unit cell representation; (b) a reduced-sphere unit cell; (c) an aggregate of many atoms

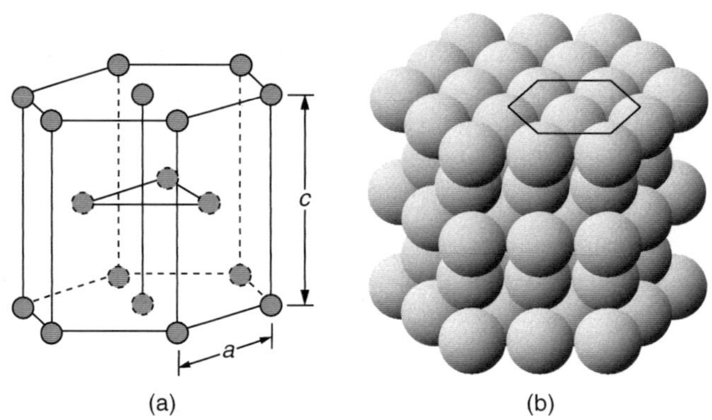

Figure 11.6 Hexagonal close-packed crystal structure: (a) a reduced-sphere unit cell (*a* and *c* represent the short and long edge lengths, respectively); (b) an aggregate of many atoms

Directions and Planes

Many properties are **anisotropic**; that is, they differ with direction and orientation. We can identify crystal directions by selecting any zero location (the origin) and determining the *x*, *y*, and *z* coordinates for any point along the direction ray. A corner of the unit cell is often used as the origin. The unit length is the edge length of the unit cell. The direction of easy magnetization in iron is parallel to one of the crystal axes. This is labeled the [100] direction, because that direction passes from the origin through a point that is one unit along the *x*-axis and zero units along the other two axes. Figure 11.7(a) shows a ray in the [120] direction and a ray in the $\left[1\bar{1}0\right]$ direction, where the overbar indicates a negative direction. Parallel directions carry the same label. We use square brackets, [ ], for closures for direction rays.

In cubic crystals, each of the four directions that are diagonal through the cube are identical (because the three axes are identical). We label these four directions as a *family* with pointed arrows for closures, <111>.

Figure 11.7(b) identifies the three shaded planes as (001), (210), and (111). Here the labeling procedure (known as indexing) is somewhat more complicated than for directions (but leads to simplified mathematics for complex calculations). As an example we will draw the (210) plane. We first invert the three indices, 1/**2**,

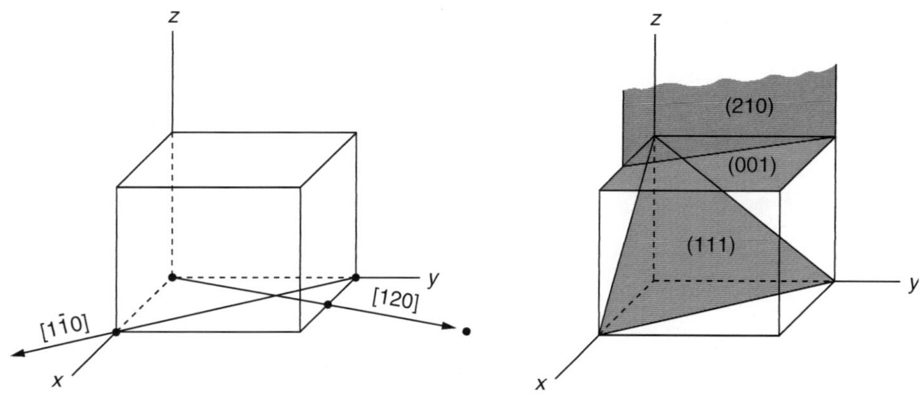

(a) (b)

Figure 11.7 Crystal notation. (a) Directions, [120] and [1$\bar{1}$0]. The *x*, *y*, and *z* coordinates are used for crystal directions. Square brackets are used as closures. Negative coordinates are indicated with an overbar. (b) Planes, (001), (210), and (111). The reciprocals of the axial intercepts are used for crystal planes. Parentheses are used as closures. Any point may be selected as an arbitrary origin.

1/**1**, 1/**0**. These are the intercept dimensions of the plane across the three axes; specifically, 0.5 on the *x*-axis, 1.0 on the *y*-axis, and infinity along the *z*-axis. An adjacent parallel plane with intercepts of 1, 2, and ∞ carries the same (210) index. We use parentheses, ( ), as closures for the indices of individual planes, and braces, { }, as closures for a family of comparable planes. To index an unknown plane, the procedure is reversed: (1) Choose an origin that the plane does not pass through. (2) Determine the intercepts of the plane on the *x*, *y*, and *z* axes. (3) Take reciprocals of these intercepts. (4) Clear fractions and enclose in parentheses ( ).

Characteristics of Ordered Solids

There are several useful properties of unit cells that can be determined through geometric relations and 3-D visualization. These include:

- Number of atoms per unit cell
- Number of nearest-neighbor atoms (coordination number)
- Lattice parameter (spacing of atoms)
- Distance of nearest approach of atoms
- Atomic packing factor (the volume of atoms per unit volume of the solid)
- Density

Some of these relationships are given in Table 11.2 and in the NCEES *Fundamentals of Engineering Supplied-Reference Handbook* and therefore do not need to be memorized, but it is useful to see how these are determined. The following examples will illustrate these relationships.

Table 11.2 Characteristics of selected crystal structures

Unit Cell	Number of Atoms per Unit Cell	Coordination Number	Lattice Parameter	Packing Factor
BCC	2	8	$a = 4R/\sqrt{3}$	0.68
FCC	4	12	$a = 2R/\sqrt{2}$	0.72
HCP		12		0.72

Figure 11.4 showed a unit cell of α iron with an atom at its center. Inasmuch as each of the eight corner atoms is shared by the eight adjacent unit cells, we can note that there are (1 + 8/8 = 2) atoms per unit cell. From Table 11.1, each iron atom has a mass of 55.85 g per 0.6022×10^{24} atoms. Since iron forms a body-centered cubic crystal, its unit cell has a mass of 1.855×10^{-26} g. X-ray diffraction techniques give a lattice constant value of 2.866×10^{-18} m. As a result, the mass per unit volume, that is, the **density**, may be calculated to be nearly 7.88 g/cm^3. Careful density measurements give a value of slightly more than 7.87 g/cm^3 at ambient temperatures. The close agreement for the simple property of density implies that the concept of the crystal structure is valid.

The relationship between the size of the atoms (atomic radius, r) and lattice parameter (a) for BCC crystals is demonstrated in Figure 11.8. Note that the atoms touch along the body diagonal. By inspection this gives a known length of the body diagonal of $4r$.

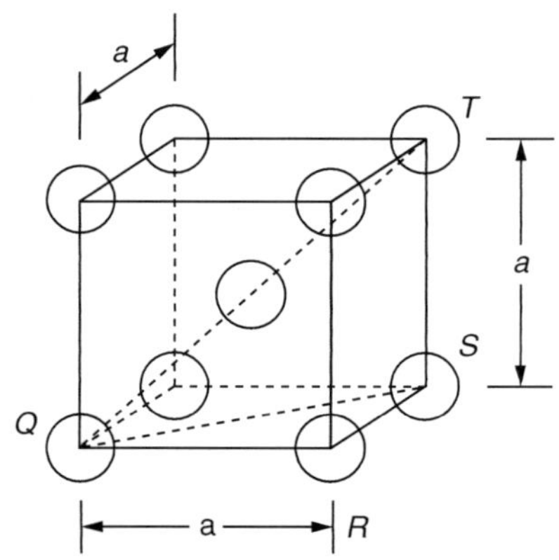

Figure 11.8 BCC unit cell

Using the triangle QRS

$$\| \overline{QS}^2 \| = a^2 + a^2 = 2a^2$$

and for triangle QST

$$\| \overline{QT}^2 \| = \| \overline{TS}^2 \| + \| \overline{QS}^2 \|$$

But $\overline{QT} = 4r$, r being the atomic radius. Also, $\overline{TS} = a$. Therefore,

$$(4r)^2 = a^2 + 2a^2$$

or

$$a = \frac{4r}{\sqrt{3}}$$

The relationship between r and a for FCC crystals can be similarly determined by noting that the atoms touch along the face diagonal as shown in Figure 11.9.

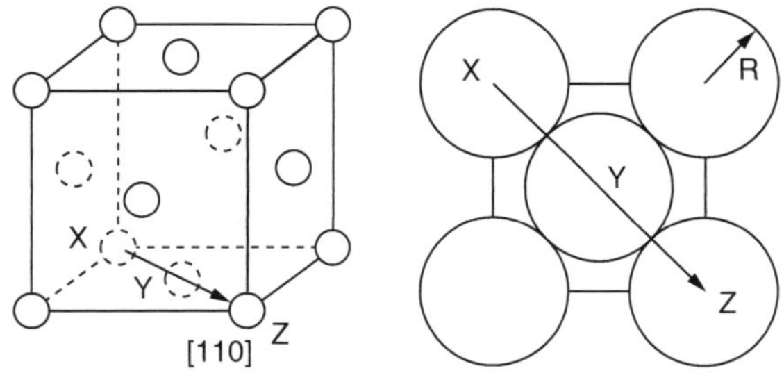

(a) (b)

Figure 11.9 (a) Reduced-sphere FCC unit cell with the [110] direction indicated. (b) The bottom face-plane of the FCC unit cell in (a) on which is shown the atomic spacing in the [110] direction, through atoms labeled X, Y, and Z.

The plastic deformation of these solids is also related to crystal structure. To illustrate, in Figure 11.10 the {111} plane of an FCC structure is shown. The {111} planes of aluminum and other FCC metals are the most densely packed planes; each atom of those planes is surrounded by six other atoms in the same plane. There is no arrangement where more atoms could have been included. This is not true for other planes within the FCC crystal. Since there is a fixed number of atoms per unit volume, it is apparent that the interplanar spacings between parallel (111) planes through the centers of atoms must be greater than between planes of other orientations. It is thus not surprising that sliding (**slip**) occurs there at lower stresses than on other planes.

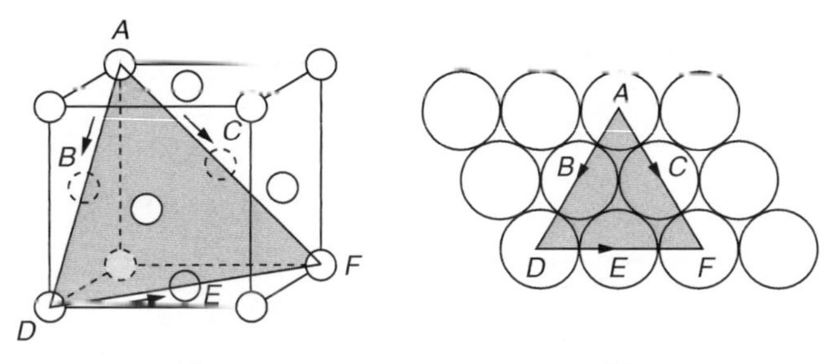

(a) (b)

Figure 11.10 Figure 11.10 (a) A {111}<110> slip system shown within an FCC unit cell. (b) The (111) plane from (a) and three <110> slip directions (as indicated by arrows) within that plane comprise possible slip systems.

Not only is the required shear stress for slip low on the {111} planes, the <110> directions on those planes require less stress for slip than other directions. This is because the step distance between like crystal positions is the shortest, specifically, $2r$ where r is the atomic radius. We speak of the <110>{111} **slip system** for FCC metals. In a BCC metal, <111> {110} is the prominent slip system. Note especially that the {111} planes and the <110> directions are operative in FCC metals, whereas it is the {110} planes and the <111> directions in BCC metals. Hexagonal metals such as Mg, Zn, and Ti do not deform as readily as BCC and FCC metals because there are fewer combinations of directions and planes for slip by shear stresses.

Example 11.1

Carbon (12.011 amu) contains C^{12} and C^{13} isotopes with masses of 12.00000 amu and 13.00335 amu, respectively. What are the percentages of each?

Solution

$$12.011 = x\,(12.00000) + (1 - x)(13.00335)$$
$$1.00335\,x = 0.99335$$
$$x = 98.9\%$$

Carbon is 98.9% C^{12} and 1.1% C^{13}.

Example 11.2

Aluminum has a face-centered cubic unit cell, that is, an atom at each corner of the unit cell and an atom at the center of each face (see Figure 11.5). The Al–Al distance (= 2*r*) is 0.2863 nm. Calculate the density of aluminum. (The mass of an aluminum atom is 26.98 amu.)

Solution

$$\text{Volume} = [2(0.2863 \times 10^{-9}\text{ m})/\sqrt{2}]^3 = 6.638 \times 10^{-29}\text{ m}^3$$
$$\text{Mass} = (8/8 + 6/2\text{ atoms})(26.98\text{ g}/6.022 \times 10^{23}\text{ atoms}) = 1.792 \times 10^{-22}\text{ g}$$
$$\text{Density} = (1.792 \times 10^{-22}\text{ g})/(6.638 \times 10^{-29}\text{ m}^3) = 2.700 \times 10^{6}\text{ g/m}^3 = 2.700\text{ g/cm}^3$$
$$\text{Actual density} = 2.699\text{ g/cm}^3$$

Example 11.3

What is the repeat distance along a <211> direction of a copper crystal that is face-centered cubic and has a unit cell dimension (lattice constant) of 0.3615 nm?

Solution

Select the center of any atom as the origin. Make a sketch of a cubic unit cell with that origin arbitrarily set at the lower left rear corner, as shown in Exhibit 1. One of the <211> directions, with coefficients of 2, 1, and 1, exits the first unit cell through the center of its front face, where another atom is centered (with no other intervening atoms).

$$d^2 = a^2 + \left(\frac{a}{2}\right)^2 + \left(\frac{a}{2}\right)^2 = \frac{6}{4}\,(0.3615\text{ nm})^2;\ d = 0.4427\text{ nm}$$

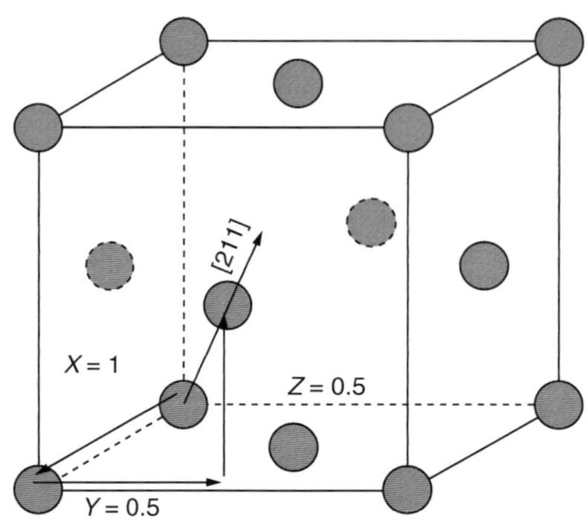

Exhibit 1 FCC unit cell

Example **11.4**

Assuming spherical atoms, calculate the packing factor of a BCC metal.

Solution

The packing factor is the volume of atoms per unit volume of the solid. Based on Figure 11.4, there are (8/8 + 1) atoms/unit cell.

$$\text{Volume of 2 atoms in the BCC unit cell} = 2 \times (4\pi/3)r^3 = 8.38r^3$$

Since the cube diagonal is $4r$,

$$\text{Volume of unit cell} = a^3 = \left(\frac{4r}{\sqrt{3}}\right)^3 = 12.32r^3$$

$$\text{Packing factor} = \frac{8.38r^3}{12.32r^3} = 68\%$$

Example **11.5**

How many atoms are there per mm^2 on one of the {110} planes of copper (FCC)?

Solution

From Example 11.3, $a_{Cu} = 0.3615$ nm. A {110} plane lies diagonally through the unit cell and is parallel to one of the axes. There are (4/4 + 2/2 atoms) in an area measuring a by $a\sqrt{2}$. The number of atoms is 2 atoms/[(0.3615 × 10^{-6} mm)^2 $\sqrt{2}$] = 10.8 × 10^{12}/mm^2.

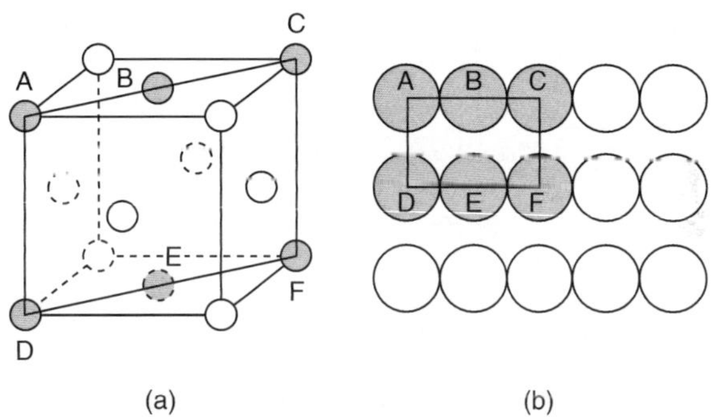

Exhibit 2 (a) Reduced-sphere FCC unit cell with (110) plane. (b) Atomic packing of an FCC (110) plane. Corresponding atom positions from (a) are indicated.

ATOMIC DISORDER IN SOLIDS

In the previous section, we paid attention to the orderly combinations that can exist in engineering materials. A variety of properties and behaviors are closely related to that ordering. Examples that were cited included density, slip systems, and molecular melting.

However, no solid has perfect order. There are always imperfections present, and these may be highly significant. A few missing potassium atoms in a compound such as KBr do not detectably affect the density; however, their absence introduces color. Likewise, the absence of a partial plane of atoms in a metal significantly modifies the shear stress required by a slip system for plastic deformation. Also, a rubber is vulcanized by the joining (**crosslinking**) of adjacent molecules with only a minor compositional change (sulfur addition). As a final example, the thermal conductivity is doubled in diamond if the 1% of naturally present C^{13} has been removed.

Crystal defects can be characterized as point, line (one-dimensional), plane (two-dimensional), or volume (three-dimensional) defects.

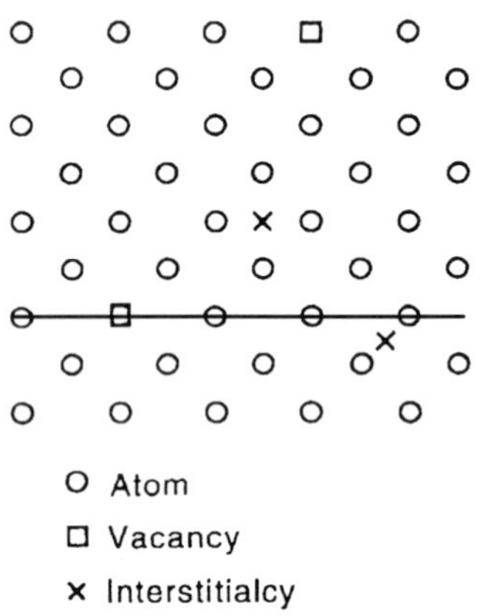

Figure 11.11 Point imperfections. These defects can originate from imperfect crystallization, or through a relocation of energized atoms.

Point Imperfections

Imperfections may be atomically local in nature. Missing atoms (**vacancies**), extra atoms (**interstitials**), **displaced** atoms, and impurity atoms are called **point imperfections** (Figure 11.11). Their existence facilitates the transport of atoms (**diffusion**), thus becoming important in materials processing. In service, the presence of point imperfections scatters internal waves and thus reduces energy transport. These include elastic waves for thermal conductivity, light waves for optical transparency, and electron waves for electrical conduction.

Lineal Imperfections

The principal defect of this type is a **dislocation**. It is most readily visualized (Figure 11.12) as a partial displacement along a slip system or an extra half-plane of atoms. Dislocations facilitate plastic deformation by slip; however, increased numbers of dislocations lead to dislocation tangles or *traffic snarls* and therefore to interference of slip. Thus, ductility decreases and strength increases. Dislocations may develop during initial crystal growth as well as from plastic deformation.

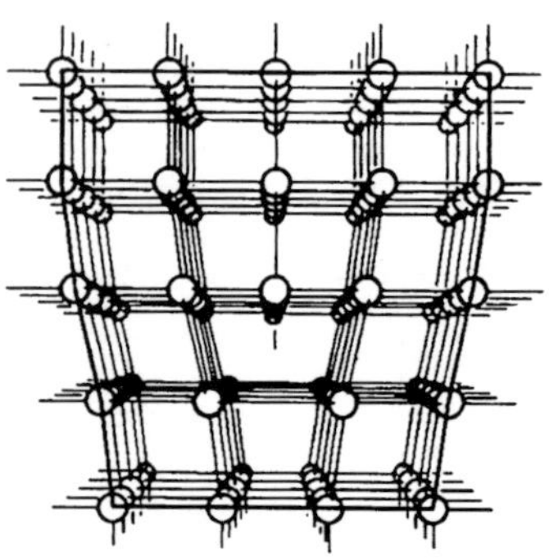

Figure 11.12 Dislocation, ⊥ (schematic). These lineal imperfections facilitate slip within crystals. Excessive numbers of defects, however, lead to their entanglement and a resistance to deformation. The resulting increase in strength is called **work hardening**.

Boundaries

No liquid or solid is infinite; each has a **surface**. The resulting two-dimensional boundary has a different structure and bonding than that encountered in the underlying material. Since atoms are absent from one side of the boundary, the atoms at these surfaces possess additional energy, and subsurface distortions are introduced.

Grain Boundaries

Boundaries also occur where two growing crystals meet. These are called grain boundaries. There are atoms on each side of the boundary; however, any misorientation between the two crystal grains leads to local inefficiencies in atomic packing. As a result some atom-to-atom distances are compressed; others are stretched. Both distortions increase the energy of the atoms along the grain boundaries. This **grain-boundary energy** introduces reactive sites for structural modification during processing and in service. The imperfect grain boundary is also an *avenue* for atomic diffusion within solid materials (Figure 11.13). Further, the mismatch at a grain boundary blocks slip that might otherwise occur, particularly at ambient temperatures where atom-by-atom mobility is limited.

Solutions

Both sugar and salt dissolve in water, each producing a **solution** (commonly called a syrup and a brine, respectively). There are many familiar solutions. Lower melting temperatures, increased conductivity, and altered viscosities commonly result.

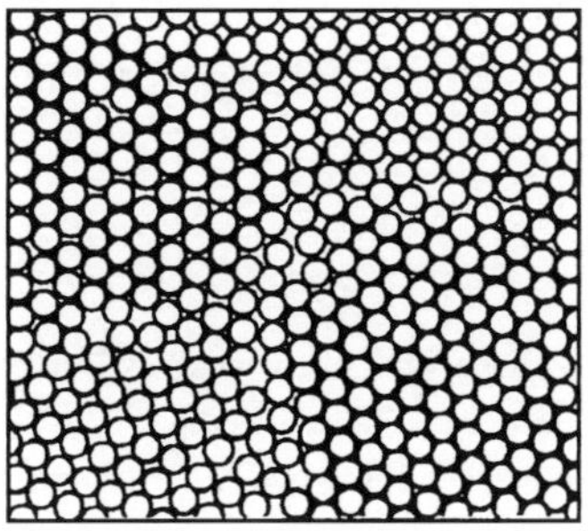

Figure 11.13 Grain boundaries (schematic). Most materials contain a multitude of grains, each of which is a separate crystal. The boundary between grains is a zone of mismatch. Atoms along grain boundaries possess added +energy because they are not as efficiently coordinated with their neighbors.

Solid Solutions

Impurities, both unwanted and intentional, may also dissolve into a solid. A crystal cannot be perfect when foreign atoms are present. Common brass (70Cu–30Zn) is a familiar example. Zinc atoms simply substitute for copper atoms to produce a **substitutional solid solution**. It has the face-centered cubic crystal structure of pure copper; however, with approximately one-third of the copper atoms replaced by zinc atoms, we can anticipate certain changes. First, the size of the unit cell is increased because zinc atoms are approximately 4% larger than copper atoms. (Fifteen percent mismatch is the practical limit for extensive substitutional solid solution.) Second, charge transport is greatly reduced because the electrons are scattered as they travel toward the positive electrode through locally varying electrical fields. Thus, the electrical and thermal **conductivities** are decreased. Also, atoms of a different size immobilize dislocation movements and in turn produce **solution hardening**.

There are some important situations in which small atoms can be positioned among larger atoms. The most widely encountered example is the solution of carbon in face-centered cubic iron at elevated temperatures. The result is an **interstitial solid solution**. Iron changes to body-centered cubic at ambient temperatures. The BCC iron cannot accommodate many carbon atoms in its interstices. This loss of interstitial solubility plays a major role in the various heat treatments of steel.

Amorphous Solids

Materials lose their crystallinity when they melt. The long-range order of the crystals is not maintained. As a liquid, the material is amorphous (literally, without form). Those materials that are closely packed—for example, metals—will expand on melting [Figure 11.14(a)]. Being thermally agitated, the atoms do not maintain close coordination with neighboring atoms (or molecules). A few materials, such as ice, which has stereospecific bonds, lose volume on melting and become more dense [Figure 11.14(b)].

In certain cases, crystallization may be avoided during cooling. An amorphous solid can result. Two common examples are window glass and the candy part of peanut brittle. In neither case is there time for the relatively complex crystalline structures to form. Very rapid cooling rates are required to avoid crystallization when the crystalline structures are less complex. For example, metals must be quenched a thousand degrees in milliseconds to avoid crystallization. The amorphous materials that result are considered to be **vitreous**, or glass-like.

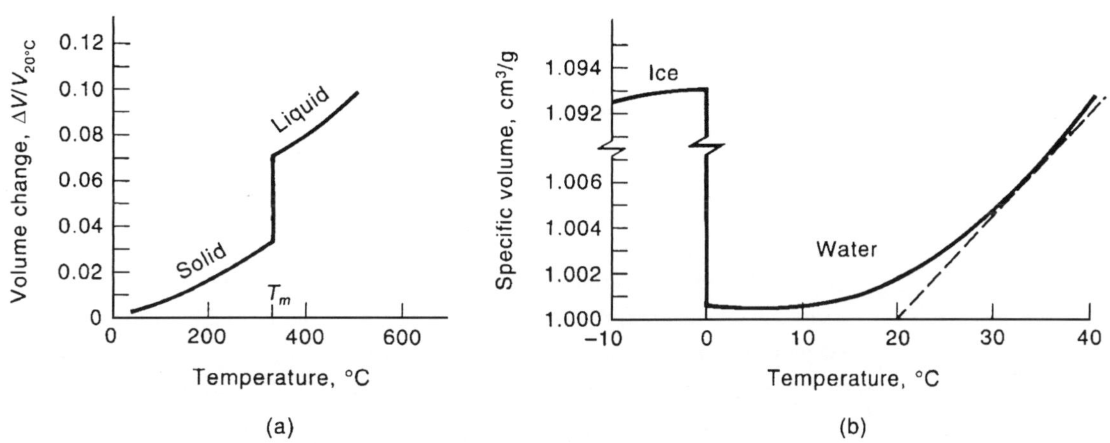

Figure 11.14 Volume changes during heating. (a) Lead (FCC); (b) H2O. Melting destroys the efficient packing of metallic atoms within solids, so most metals expand when melted. The crystalline structure of ice, silicon, and a number of related materials with stereospecific bonds have low, inefficient atomic packing within solids. Therefore, they lose volume when melted.

Example 11.6

Sterling silver contains 92.5% silver and 7.5% copper by weight. What percentage of the atoms on a (111) plane are silver? Copper?

Solution

The alloy is a random solid solution; therefore the percentage of atoms on the (111) plane or any other plane will be the same as throughout the alloy. Change weight percent to atom percent. The atomic masses are 107.87 and 63.54 amu, respectively.

Basis: 1000 amu = 925 amu Ag + 75 amu Cu

Ag: 925 amu/(107.87 amu/Ag) = 8.58 Ag atoms

Cu: 75 amu/(63.54 amu/Cu) = 1.18 Cu atoms

Total: = 9.76 atoms

Ag atoms = 8.58/9.76 = 87.9%; Cu atoms = 12.1%

MICROSTRUCTURES OF SOLID MATERIALS

The atomic coordination within solids is on the nanometer scale. The resulting structures involve either crystalline solids, or amorphous solids such as the glasses. As discussed, certain properties arise from these atom-to-atom relationships. Other properties arise from longer-range structures, generally with micrometer to millimeter dimensions, called **microstructures**.

Atomic Movements in Materials

Our initial examination of crystals implied that an atom becomes permanently coordinated with adjacent atoms and remains fixed in position. This is not entirely true. In the first place, there is thermal vibration of the atoms within the crystal. Thus, while the lattice constant and the mean interatomic distances are fixed to several significant figures, the instantaneous interatomic distances vary. The amplitude of vibration increases with temperature. At the melting temperature, the crystal is literally *shaken apart*. As the melting point is approached, a measurable fraction of atoms jump out of their crystalline positions. They may return, or they may move to other sites, producing the vacancies and interstitials discussed in the previous section.

Diffusion

Within a single-component material, such as pure copper, there is equal probability that like numbers of copper atoms will jump in each of the coordinate directions. Thus, there is no net change.

Imagine, however, one location, x_1, in nickel containing 2000 atoms of copper for every mm^3 of nickel, whereas 1 mm to the right at x_2 there are 1000 atoms of copper for every mm^3. Although all copper atoms have the same probability for jumping in either direction, there is a net movement of copper atoms to the right simply because there are unequal numbers of copper atoms in the two locations. There is a copper **concentration gradient**, $\Delta C/\Delta x$. In this case

$$\Delta C/\Delta x = (C_2 - C_1 \text{ Cu/mm}^3)/(x^2 - x^1 \text{ mm}) = -(1000 \text{ Cu/mm}^3)/\text{mm} \quad \textbf{(11.2)}$$

The rate of diffusion, called the **flux**, J, is proportional to the concentration gradient

$$J = -D \frac{dC}{dx} \tag{11.3}$$

where D is the *diffusivity*, also called the **diffusion coefficient**. (Its units are m^2s^{-1} since the units for flux and concentration gradient are $m^{-2}s^{-1}$ and m^{-4}, respectively.)

The diffusion coefficient can be calculated by

$$D = D_o e^{-Q/(RT)}$$

where D_o (with units of m^2s^{-1}) is the proportionality constant, Q (with units of $J(mole)^{-1}$) is the activation energy, R is the gas constant, and T is absolute temperature.

Among the various factors that affect the diffusivity are (1) the size of the diffusing atom, (2) the crystal structure of the matrix, (3) bond strength, and (4) temperature. Comparisons are made in Table 11.3. The diffusivity of the C in Fe_{FCC} is higher than for the Fe in Fe_{FCC} because the diffusing carbon atom is smaller than the iron atom. The diffusivity of the C in Fe_{BCC} is higher than for the C in Fe_{FCC} because the latter contains more iron atoms per m^3. The FCC packing factor is higher than the BCC form. The diffusivity for Fe in Fe_{FCC} is lower than for Cu in Cu_{FCC} because the iron atoms are more strongly bonded than the copper atoms. (The melting temperatures provide evidence of this.) In each case, higher diffusion coefficients accompany higher temperatures.

The process engineer obtains many of the properties required of a material by heat-treating procedures. Diffusion plays the predominant role in achieving the required microstructures.

Table 11.3 Selected diffusion coefficients

Diffusing Atom	Host Structure	Diffusion Coefficient, D, m^2/s 500°C (930°F)	1000°C (1830°F)
Fe	FCC Fe	2×10^{-23}	2×10^{-16}
C	FCC Fe	5×10^{-15}	3×10^{-11}
C	BCC Fe	1×10^{-12}	2×10^{-9}
Cu	FCC Cu	1×10^{-18}	2×10^{-13}

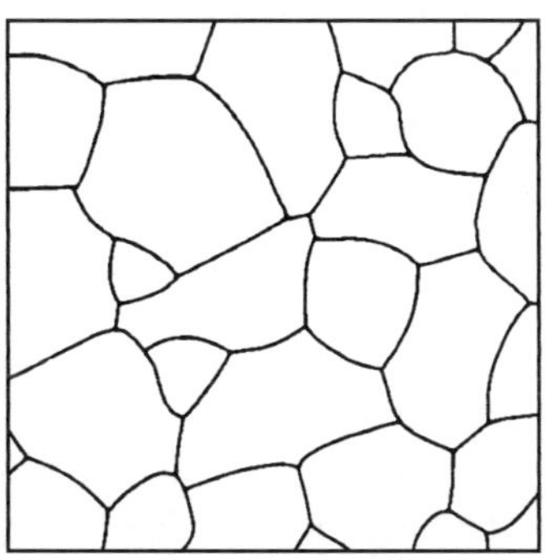

Figure 11.15 Grains (schematic). Each grain is a separate crystal. There is a surface of mismatch between grains because adjacent grains have unlike orientations. The **grain boundary area** is inversely related to grain size.

Structures of Single-Phase Solids

Many materials possess only one structure. Examples include copper wire, transparent polystyrene cups, and Al_2O_3 substrates for electronic circuits. The wire contains only face-centered-cubic crystals. The polystyrene is an amorphous solid with minimal crystallinity. The substrates have numerous crystals, all with the same crystalline structure. We speak of single phases because none of these materials contains a second structure.

Grains

Each of the individual crystals in a copper wire is called a **grain**. Recall from the previous section that adjacent crystals may be misoriented with respect to each other, and that there is a boundary between them. This is shown schematically in Figure 11.15.

The **grain size** is an important structural parameter, because the grain boundary area varies inversely with the grain size. Diffusion is faster along grain boundaries because there is less-perfect packing of the atoms and, consequently, a more open structure. At ambient temperatures, grain boundaries interfere with plastic deformation, thus increasing the strength. At elevated temperatures the grain boundaries contribute to creep and therefore are to be minimized. Grain boundaries also serve as locations that initiate structural changes within a solid.

Grain growth may occur in a single-phase material. The driving force is the fact that the atoms at the grain boundary possess extra energy. Grain growth reduces this excess energy by minimizing the boundary area. Higher temperatures increase the rate of grain growth because the atoms migrate faster. However, since the growth rate is inversely related to grain size, we see a decrease in the rate of growth with time.

The texture of a single-phase solid can also depend on **grain shape** and **orientation**. Even a noncrystalline material may possess a structure. For example, the molecular chains within a nylon fiber have been aligned during processing to provide greater tensile strength.

Phase Diagrams

Many materials possess more than one phase. A simple and obvious example is a cup containing both ice and water. While ice and water have the same composition, the structures of the two phases are different. There is a **phase boundary** between the two phases. A less obvious, but equally important material is the steel used as a bridge beam. The steel in the beam contains two phases: nearly pure iron (body-centered cubic), and an iron carbide, Fe_3C.

In these two examples, we have a mixture of two phases. Solutions are phases with more than one component. In the previous section, we encountered brass, a solid solution with copper and zinc as its components.

Although the steel beam just cited contains a mixture of two phases at ambient temperatures, it contained only one phase when it was red-hot during the shaping process. At that temperature, the iron was face-centered-cubic and was thereby able to dissolve all of the carbon into its interstices. No Fe_3C remained. The phases within a material can be displayed on a **phase diagram** as a function of temperature and composition.

Phase diagrams are useful to the engineer because they indicate the temperature and composition requirements for attaining the required internal structures and accompanying service properties. The phase diagram shows us (1) *what* phases to expect, (2) the *composition* of each phase that is present, and (3) the *quantity* of each phase within a mixture of phases.

What Phases?

Sterling silver contains 92.5Ag–7.5Cu (weight percent). Using the Ag–Cu phase diagram of Figure 11.16, we observe that this composition is liquid above 910°C; below 740°C, it contains a mixture of the α and β solid structures. The former is a solid solution of silver plus a limited amount of copper; the latter is a solid solution of copper plus a limited amount of silver. Between 740 and 810°C, only one phase is present, α. In that temperature range, all of the copper can be dissolved in the α solid solution. From 810 to 910°C, the alloy changes from no liquid to all liquid.

Phase Compositions?

Pure silver has a face-centered-cubic structure. Copper forms the same crystalline structure. Not surprisingly, copper atoms can be substituted for silver. But the solid solution in α is limited, because the silver atom is 13% larger in diameter than is copper. However, the solubility increases with temperature to 8.8 weight-percent copper at 780°C. This is shown by the boundary of the silver-rich shaded area of Figure 11.16. Within that shaded area, we have a single phase called α.

Conversely, the FCC structure of copper can dissolve silver atoms in solid solution. At room temperature the solubility is very small. Again, as the temperature is raised the solubility limit increases to 8 weight-percent silver (92 wt. % Cu) at 780°C. This copper-rich phase is called β.

Silver and copper are mutually soluble in a liquid solution. Above 1100°C there is no limit to the solubility. As the temperature is reduced below the melting point of copper (1084°C), the copper solubility limit decreases from 100% to 28%. Excess copper produces the β phase. Likewise, below the melting point of silver (962°C), the silver solubility limit decreases from 100% to 72%. Excess silver produces the α phase. The two solubility curves cross at approximately 72Ag–28Cu and 780°C. We call this low-melting liquid **eutectic**.

How Much of Each Phase in a Mixture?

At 800°C, a 72Ag–28Cu alloy is entirely liquid. At the same temperature, but with added copper, the solubility limit is reached at 32% Cu. Beyond that limit, still at 800°C, any additional copper precipitates as β. Halfway across the two-phase field of (L + β), there will be equal quantities of the two phases. Additional copper in the alloy increases the amount of β. Within this two-phase region the liquid remains saturated with 32% Cu, and the solid β is saturated with 8% Ag (thus, 92% Cu).

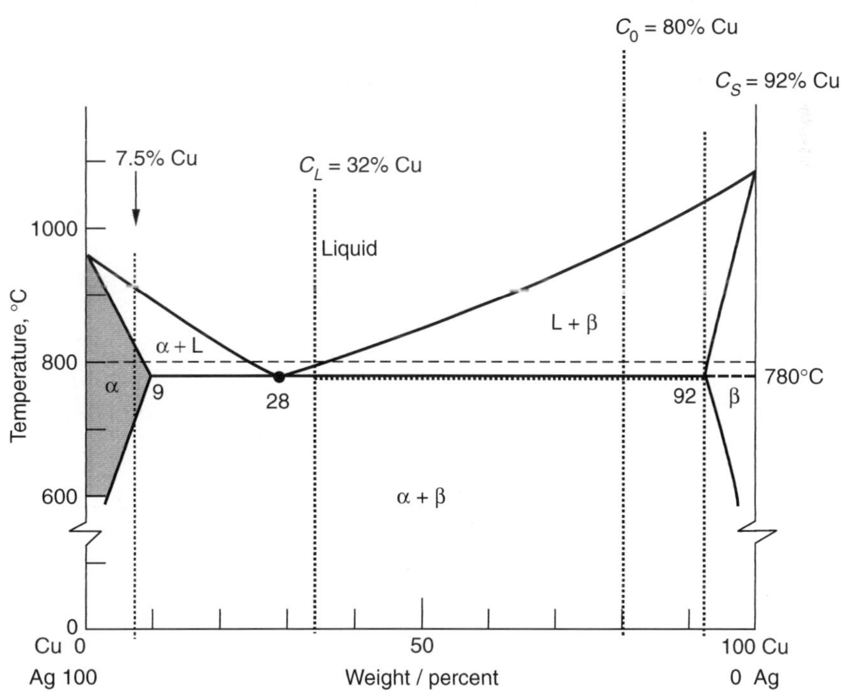

Figure 11.16 Silver-copper phase diagram. Sterling silver contains 7.5% copper; therefore, it has a single phase, α, in the 740–810°C range.

The *composition* of each phase is dictated by the solubility limits. (In a one-phase field, the solubility limits are not factors; the alloy composition and the phase composition are identical.) The *amount* of each phase in a two-phase field is determined by interpolation between the solubility limits using what is called the lever law. For example, at 800°C, an alloy consisting of 80% Cu and 20% Ag would consist of a liquid phase containing 32% Cu and a solid phase of 92% Cu. The weight fraction of liquid phase, W_L, and solid phase, W_s, is determined by

In addition $W_L + W_S = 1$.

$$W_L = \frac{C_s - C_O}{C_S - C_L} \qquad \text{and} \qquad W_S = \frac{C_O - C_L}{C_S - C_L}$$

Reaction Rates in Solids

A phase diagram is normally an equilibrium diagram; that is, all reactions have been completed. The time required to reach equilibrium generally increases with decreasing temperature. Thus, a material may not always possess the expected phases with predicted amounts or compositions. Even so, the phase diagram is valuable. For example, sterling silver (92.5Ag–7.5Cu) contains only one phase when equilibrated at 775°C (Figure 11.16). Slow cooling to room temperature precipitates β as a minor second phase, as would be expected when plenty of time is available. Rapid cooling, however, traps the copper atoms within the a solid solution. This situation is used to advantage, because the solid solution is stronger than the ($\alpha + \beta$) combination. Also, a single-phase alloy corrodes less readily. This explains why the *impure* sterling silver is commonly preferred over pure silver.

The selection of compositions and processing treatments is generally based on a knowledge of equilibrium diagrams plus a knowledge of how equilibrium is circumvented.

Microstructures of Multiphase Solids

The microstructure of a single-phase, crystalline solid includes the *size*, *shape*, and *orientation* of the grains. Variations in these properties are also found in multiphase solids. In addition, the microstructure of a multiphase solid may also vary in the *amount* of each phase and the *distribution* of the phases. In an equilibrated microstructure, the amounts of the phases may be predicted directly from the phase diagrams using the lever law. From Figure 11.17, a 1080 steel (primarily iron, with 0.80 percent carbon) will have twice as much Fe_3C (W_{Fe3C}) at room temperature as a 1040 steel (with 0.40 percent carbon). The Fe_3C is a hard phase. Therefore, with all other factors equal, we expect a 1080 steel to be harder than a 1040 steel, and it is.

$$W_{Fe_3C} = \frac{C_0 - C_\alpha}{C_{Fe_3C} - C_\alpha}$$

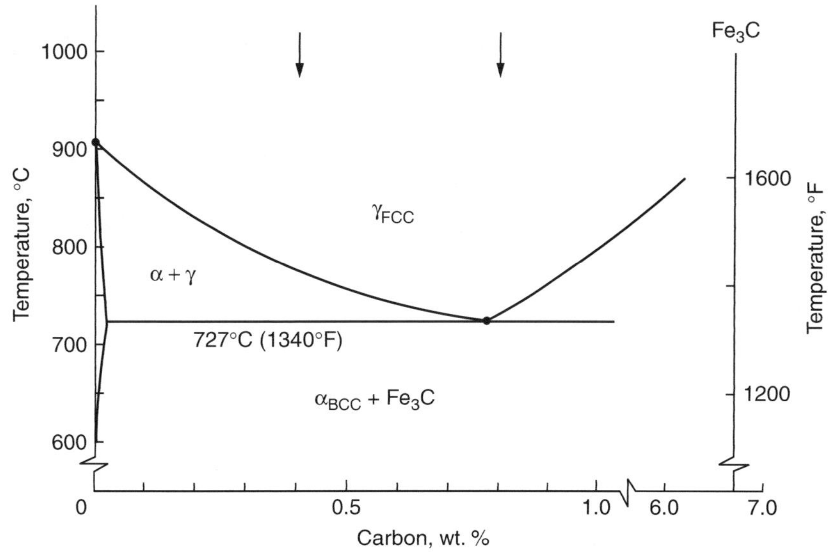

Figure 11.17 Portion of the Fe–Fe_3C phase diagram. Most steels are heat treated by initially forming austenite, γ, which is Fe_{FCC}. It changes to a mixture of ferrite, α, and carbide at lower temperatures.

The distribution of phases within microstructures is more difficult to quantify, so only descriptive examples will be cited. Similar to sterling silver, aluminum will dissolve several percent of copper in solid solution at 550°C. If it is cooled slowly, the copper precipitates as a minor, hard, brittle compound ($CuAl_2$) along grain boundaries of the aluminum. The alloy is weak and brittle and has little practical use. If the same alloy is cooled rapidly from 550°C, trapping the copper atoms within the aluminum grains as shown in Figure 11.18(a), the quenched solid solution is stronger and more ductile and has commercial uses. If the quenched alloy is reheated to 100°C, the $CuAl_2$ precipitates, as expected from the phase diagram shown in Figure 11.18(b) and (c). In this case, however, the precipitate is very finely dispersed within the grains of aluminum. The alloy retains its toughness, because the brittle $CuAl_2$ does not form a network for fracture paths. In addition, the strength is increased because the submicroscopic hard particles interfere with the deformation of the aluminum along slip planes. Alloys of this type are used in airplane construction because they are light, strong, and tough. Observe that all of the examples in this paragraph are for the same alloy. The properties have been varied through microstructural control.

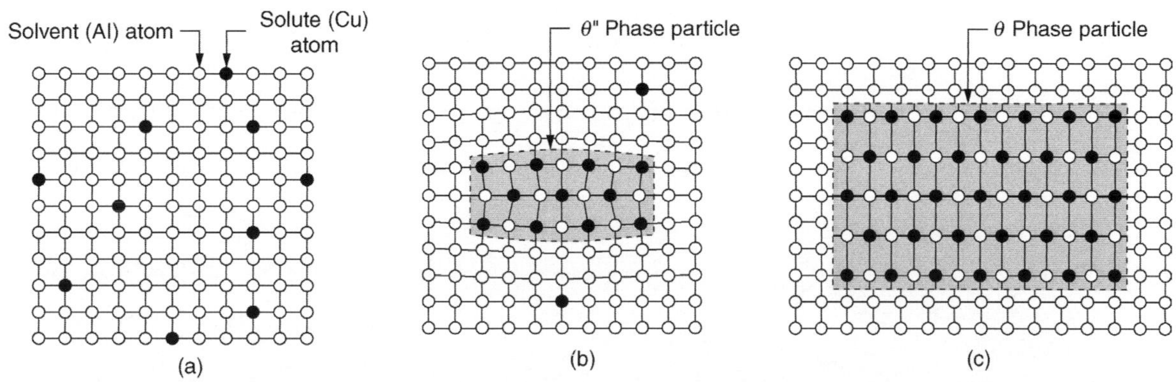

Figure 11.18 Schematic depiction of several stages in the formation of the equilibrium precipitate (*q*) phase. (a) A supersaturated a solid solution; (b) a transition, θ'', precipitate phase; (c) the equilibrium θ' phase, within the a-matrix phase. Actual phase particle sizes are much larger than shown here.

Two distinct microstructures may be produced in a majority of steels. In one, called **spheroidite**, the hard Fe_3C is present as rounded particles in a matrix of ductile ferrite (α). In the other, called **pearlite**, Fe_3C and ferrite form fine alternating layers, or lamellae. Spheroidite is softer but tougher; pearlite is harder and less ductile (Figure 11.19). The mechanical properties are controlled by the spacing of the Fe_3C, because the hard Fe_3C phase stops dislocation motion. The closer the spacing between Fe_3C particles, the greater the strength or hardness, but the lower the ductility.

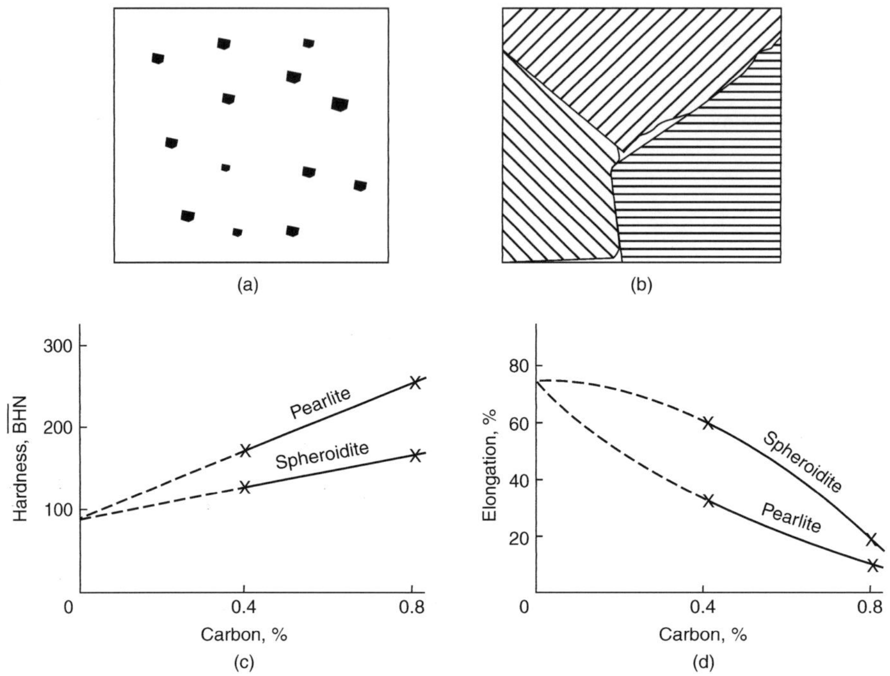

Figure 11.19 Phase distributions (schematic). The two samples are of the same steel, both containing ferrite (white) and carbide (black) but heat treated differently. Spheroidite is shown in (a), and pearlite in (b). The hardness of the two structures is shown in (c), and ductility is shown in (d).

Example **11.7**

At 780°C, an Ag–Cu liquid of eutectic composition (Figure 11.16) solidifies to solid α and solid β. (i) What are the compositions of the two solid phases just below the eutectic temperature? (ii) How much of each of these two phases exists per 100 grams of alloy?

Solution

(i) From Figure 11.16, the α phase is 91Ag–9Cu, and the β phase is 8Ag–92Cu.

(ii) Interpolation along the 780°C tie-line between the solubility limits yields

$$\beta\text{: (using Cu) } [(28 - 9)/(92 - 9)](100\text{ g}) = 23\text{ g}$$

$$\alpha\text{: (using Cu) } [(92 - 28)/(92 - 9)](100\text{ g}) = 77\text{ g}$$

or

$$\alpha\text{: (using Ag) } [(72 - 8)/(92 - 9)](100\text{ g}) = 77\text{ g}$$

Example 11.8

The phases and microstructures of an SAE 1040 steel may be related to the Fe-Fe_3C phase diagram (Figure 11.17). (i) Assume equilibrium. What are the phases and their weight percents at 728°C? At 726°C? (ii) Pearlite has the eutectoid composition of 0.77% C. What percent of the steel will *not* be contained in the pearlite?

Solution

(i) At 728°:

$$\alpha\ (0.02\%\ \text{C}): (0.77 - 0.40)/(0.77 - 0.02) = 49\%\ \alpha$$

$$\gamma\ (0.77\%\ \text{C}): (0.40 - 0.02)/(0.77 - 0.02) = 51\%\ \gamma$$

$$\text{Fe}_3\text{C contains } 12/[12 + 3(55.85)] = 6.7\%\ \text{C}$$

At 726°:

$$\alpha\ (0.02\%\ \text{C}): (6.7 - 0.40)/(6.7 - 0.02) = 94\%\ \alpha$$

$$\text{Fe}_3\text{C}: (0.40 - 0.02)/(6.7 - 0.02) = 6\%\ \text{Fe}_3\text{C}$$

(ii) All of the pearlite, α + Fe_3C, comes from the γ. This FCC phase changes to pearlite at 727°C. The α that was present at 728°C remains unchanged. Therefore, the answer is $P = 51\%$; and unchanged $\alpha = 49\%$.

MATERIALS PROCESSING

For many engineers, materials are first encountered in terms of handbook data or stockroom inventories. However, materials always have a prior history. They must be obtained from natural sources, then subjected to compositional modifications and complex shaping processes. Finally, specific microstructures are developed to achieve the properties that are necessary for extended service.

Extraction and Compositional Adjustments

Few materials are used in their natural form. Wood is cut and reshaped into lumber, chipboard, or plywood; most metals must be extracted from their ores; rubber latex is useless unless it is vulcanized.

Extraction from Ores

Most ores are oxides or sulfides that require chemical reduction. Commonly, oxide ores are chemically reduced with carbon- or CO-containing gas. For example,

$$\text{Fe}_2\text{O}_3 + 3\ \text{CO} \rightarrow 2\ \text{Fe} + 3\ \text{CO}_2 \qquad \textbf{(11.4)}$$

Elevated temperatures are used to speed up the reactions and more completely reduce the metal. If the metal is melted, it is more readily separated from the accompanying gangue materials. The reduced product is a carbon-saturated metallic iron. As such, it has only limited applications. Normally, further processing is required.

Refining

Dissolved impurities must be removed. Even if the above ore were of the highest quality iron oxide, it would be necessary to refine it, because the metallic product is saturated with carbon. In practice, small but undesirable quantities of silicon and other species are also reduced and dissolved in the iron. They are removed by closely controlled **reoxidation** at chemically appropriate temperatures (followed by **deoxidation**). The product is a **steel**. Alloying additions are made as specified to create different types of steel.

Chemicals from which a variety of plastics are produced are refined from petroleum. The principal step of petroleum refinement involves selective distillation of liquid petroleum. Lightweight fractions are removed first. Controlled temperatures and pressures distill the molecular fractions that serve as precursors for polymers. Residual fractions are directed to other products.

Polymerization makes macromolecules out of the smaller molecules that are the product of distillation. **Addition polymerization** involving a C=C→C—C reaction is encountered in the polymerization of ethylene, $H_2C{=}CH_2$; vinyl chloride, $H_2C{=}CHCl$; and styrene, $H_2C{=}CH(C_6H_5)$.

$$C{=}C \rightarrow C{-}C{-} \quad \textbf{(11.5)}$$

Shaping Processes

The earliest cultural ages of human activities produced artifacts that had been shaped from stone, bronze, or iron. In modern technology, we speak of casting, deformation, cutting, and joining in addition to more specialized shaping procedures.

Casting

The concept of casting is straightforward. A liquid is solidified within a mold of the required shape. For metal casting, attention must be given to volume changes; in most cases there is shrinkage. In order to avoid porosity, provision must be made for feeding molten metal from a **riser**. **Segregation** may occur at the solidifying front because of compositional differences in the $(\alpha + L)$ range. (See the earlier discussion of phase diagrams and the discussion of annealing processes in a following subsection.)

A number of ceramic products are made by **slip casting**. The slip is a slurry of fine powders suspended in a fluid, usually water. The mold is typically of gypsum plaster with a porosity that absorbs water from the adjacent suspension. When the shell forming inside the mold is sufficiently thick, the remaining slip is drained. Subsequent processing steps are **drying** and **firing**. The latter high-temperature step bonds the powder into a coherent product.

The casting process is also used in forming polymeric products. Here the solidification is accomplished by polymerization. There is a chemical reaction between the small precursor molecules of the liquid to produce macromolecules and a resulting solid.

Deformation Processes

Deformation processes include forging, rolling, extrusion, and drawing, plus a number of variants. In each case, a force is applied, and a dimensional change results. **Forging** involves shape change by impact. **Rolling** may be used for sheet products as well as for products with constant cross sections, for example, structural beams. **Extrusion** is accomplished through open or closed dies. The former requires that the product be of uniform cross section, such as plastic pipe or siderails for aluminum ladders. Closed dies are molds into which the material is forced. These forming processes can be done at ambient temperature (cold working) or elevated temperatures ($T > 0.4T_m$ in Kelvin; hot working). T_m is the melting temperature. The forming temperature has a strong effect on mechanical properties. Products formed at ambient temperature have high dislocation densities and hence greater strength and hardness but lower ductility than hot-worked products.

In general, ceramics do not lend themselves to the above deformation processes because they lack ductility. Major exceptions are the glasses, which deform not by crystalline slip but by viscous flow.

Cutting

Chiseling and sawing are cutting processes that predate history. Current technology includes **machining** in which a cutting tool and the product move with respect to each other. Depth and rates of cut are adjustable to meet requirements. **Grinding** is a variant of machining that is used for surface removal.

Joining

The process of **welding** produces a joint along which the abutting material has been melted and filled with matching metal. **Soldering** and **brazing** processes use fillers that have a lower melting point than the adjoining materials, which remain solid. There are glass solders as well as metallic ones. Adhesives have long been used for joining wood and plastic components, and many have now been developed to join metals and ceramics.

Annealing Processes

Annealing processes involve reheating a material sufficiently that internal adjustments may be made between atoms or between molecules. The temperature of annealing varies with (1) the material, (2) the amount of time available, and (3) the structural changes that are desired.

Homogenization

The dynamics of processing will produce segregation. For example, when an 80Cu–20Ni alloy starts to solidify at 1200°C, the first solid contains 30% Ni. When solidification is completed, the final liquid has only 12% Ni. Uniformity can be obtained if the alloy is reheated to a temperature at which the atoms can relocate by diffusion. There is a time-temperature relationship (log t vs. $1/T$). In this case an increase in temperature from 500 to 550°C reduces the necessary annealing time by a factor of eight.

Recrystallization

Networks of dislocations are introduced when most metals are plastically deformed at ambient temperatures (cold working). The result is a work hardening and loss of ductility. Whereas the resulting increase in strength is often desired, the property changes resulting from dislocations make further deformation processing more difficult. Annealing will remove the dislocations and restore the initial workable characteristics by forming new, strain-free crystals.

A one-hour heat treatment is a common shop practice because it allows for temperature equalization as well as scheduling requirements. For that time frame, it is necessary to heat a metal to approximately 40 percent of its melting temperature (on the absolute scale). Thus, copper that melts at 1085°C (1358 K) may be expected to recrystallize in the hour at 270°C (545 K). The recrystallization time is shorter at higher temperatures and longer at lower temperatures.

Grain Growth

Extended annealing, beyond that required for recrystallization, produces grain growth and therefore coarser grains. Normally, this is to be avoided. However, grain growth has merit in certain applications because grain boundaries hinder magnetic domain boundary movements, reducing creep. So, coarse grains are preferred in the sheet steel used in transformers, for example. At high temperatures, grain boundaries permit creep to occur under applied stresses, producing changes in dimensions.

Residual Stresses

Expansions and contractions occur within materials during heat-treating operations. These are isotropic for many materials; or they may vary with crystal orientation. Also, differential expansions exist between the two or more phases in a multiphase material. The latter, plus the presence of thermal gradients, introduce internal stresses, which can lead to delayed fracture if not removed.

It is generally desirable to eliminate these residual stresses by an annealing process called **stress relief**. The required temperature is less than that for recrystallization because atomic diffusion is generally not necessary; rather, adjustments are made through the local movement of dislocations. Stress relief is performed on metals before the final machining or grinding operation. Annealing is always performed on glass products, because any residual surface tension easily activates cracks in this nonductile material.

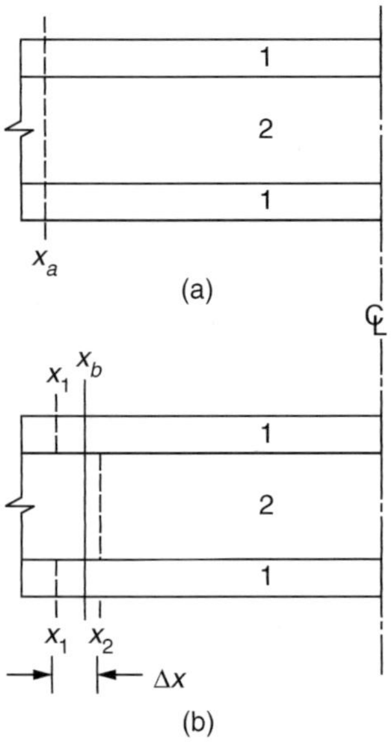

Figure 11.20 Induced stresses (sandwich glass). (a) The previously bonded composite containing glasses 1 and 2 is annealed. There are no internal stresses at x_a. (b) If the three layers were separate, the outer layers, which have a lower thermal coefficient, would contract less during cooling than the center glass (x_1 vs. x_2). Since the layers are bonded together, the restricted contraction of the center (x_a to x_b rather than x_a to x_2) induces compression within the surface layers (x_1 to x_b).

Induced Stresses

In apparent contradiction to the last statement, residual compressive stresses may be prescribed for certain glasses, since glass like most nonductile materials is strong in **compression** but weak in **tension**. As an example, a familiar dinnerware product is made from a *sandwich* glass sheet in which the *bread* layers have a lower thermal expansion than does the *meat* layer. The processing involves heating the dinnerware to relieve all stresses (annealing). As the dinnerware pieces are cooled, the center layer tries to contract more than the surface layers, placing the surfaces under compression (Figure 11.20). Any tension encountered in service must overcome the residual compression before a crack can propagate.

Time-Dependent Processes

We have seen in a previous section that sterling silver is solution treated to dissolve all of the copper within the silver-rich α phase. The single phase is preserved by rapidly cooling the alloy. This avoids the precipitation of the copper-rich β phase, as required for equilibrium. The cooling rate need not be drastic for sterling silver because it takes a minute or more to nucleate and grow the precipitate, β.

Even more time is available for the production of a **silicate glass**—a supercooled liquid. The necessary bond breaking and rearrangements are very slow. The available processing times can approach an hour or more. However, in order to produce a **metallic glass**, the cooling rate must approach 1000°C per *millisecond*. Otherwise, individual metal atoms rapidly order themselves into one of the crystalline patterns described earlier.

Martensitic Reactions

Most steel processing treatments initially heat the steel to provide a single-phase microstructure of γ or **austenite** (Fe_{FCC}). This is face-centered-cubic and dissolves all of the carbon that is present. Normal cooling produces **pearlite**, a lamellar microstructure containing layers of α or ferrite (Fe_{BCC}) and **carbide** (Fe_3C), as shown previously on the Fe–Fe_3C phase diagram (Figure 11.17).

$$Fe_{FCC} \xrightarrow{\text{cooling}} Fe_{BCC} + Fe_3C \tag{11.6}$$

If **quenched**, a different structure forms:

$$Fe_{FCC} \xrightarrow{\text{quenching}} \text{Martensite} \tag{11.7}$$

Martensite is a transition phase. It offers an interesting possibility for many applications because it is much harder than pearlite. Unfortunately, martensite is also very brittle, and its usefulness in steel is severely limited. Martensite will exist almost indefinitely at ambient temperatures, but reheating the steel provides an opportunity for the completion of the reaction of Equation (11.6).

$$\text{Martensite} \xrightarrow{\text{tempering}} Fe_3C \tag{11.8}$$

The reheating process is called **tempering**. The product, which has a microstructure of finely dispersed carbide particles in a ferrite matrix, is both hard and tough. It is widely used and is called **tempered martensite**.

Hardenability

Tempered martensite is the preferred microstructure of many high-strength steels. Processing requires a sufficiently rapid quench to obtain the intermediate martensite, Equation (11.7), followed by tempering, Equation (11.8). This means severe quenching for products of Fe–C steels that are larger than needles or razor blades. Even then, martensite forms only at the quenched surface. The subsurface metal transforms directly to the ferrite and carbide, Equation (11.6).

The reaction rate of Equation (11.6) can be decreased by the presence of various alloying elements in the steel. Thus, gears and similar products commonly contain fractional percentages of nickel, chromium, molybdenum, or other metals. Quenching severity can be reduced, and larger components can be hardened throughout. These alloying elements delay the formation of carbide because, not only must small carbon atoms relocate, it is also necessary for the larger metal atoms to choose between residence in the ferrite or in the carbide.

Surface Modification

Products may be treated so that their surfaces are modified and therefore possess different properties than the original material. Chrome plating is a familiar example.

Carburizing

Strength and hardness of a steel increase with carbon content. Concurrently, ductility and toughness decrease. With these variables, the engineer must consider trade-offs when specifying steels for mechanical applications. An alternate possibility is to alter the surface zone. A common example is the choice of a low-carbon, tough steel that has had carbon diffused into the subsurface (<1 mm) to produce hard carbide particles. Wear resistance is developed without decreasing bulk toughness.

Nitriding

Results similar to carburizing are possible for a steel containing small amounts of aluminum. Nitrogen can be diffused through the surface to form a subscale containing particles of aluminum nitride. Since AlN has structure and properties that are related to diamond, wear resistance is increased for the steel.

Shot Peening

Superficial deformation occurs when a ductile material is impacted by sand or by hardened shot. A process employing shot peening places the surface zone in local compression and therefore lowers the probability for fracture initiation during tensile loading.

Example **11.9**

Assume a single spherical shrinkage cavity forms inside a 2-kg lead casting as it is solidified at 327°C. What is the initial diameter of the cavity after solidification? (The greatest density of molten lead is 10.6 g/cm^3.)

Solution

Refer to Figure 11.14(a). Based on a unit volume at 20°C, lead shrinks from 1.07 to 1.035 during solidification.

The volume of molten lead is (2000 g)/(10.6 g/cm^3) = 188.7 cm^3.

Volume of solid lead at 327°C: (188.7 cm^3)(1.035/1.07) = 182.5 cm^3

$$\text{Shrinkage: } 188.7\ \text{cm}^3 - 182.5\ \text{cm}^3 = \pi d^3/6;\ d = 2.28\ \text{cm}$$

Example **11.10**

How much energy is involved in polymerizing one gram of ethylene (C_2H_4) into polyethylene, Equation (11.5). The double carbon bond possesses 162 kcal/mole, and the single bond has 88 kcal/mole.

Solution

As shown in Equation (11.5), one double carbon bond changes to two single bonds/mer:

$$(1\ \text{g})/(24 + 4\ \text{g/mole}) = 0.0357\ \text{moles}$$

The energy required to break 0.0357 moles of double bonds is

$$(1)(0.0357)(162\ \text{kcal}) = 5.79\ \text{kcal}$$

The energy released in joining twice as many single bonds is

$$(2)(0.0357)(-88\ \text{kcal}) = -6.29\ \text{kcal}$$

The net energy change is –6.29 + 5.79 kcal = 500 cal released/g.

Example 11.11

There are 36 equiaxed grains per mm² observed at a magnification of 100 in a selected area of copper. The copper is heated to double the average grain diameter. (i) How many grains exist per mm²? (ii) What will be the percentage (increase, decrease) in grain boundary area?

Solution

(i) Doubling a lineal dimension decreases the number of grains by a factor of four, so there are 9 grains per mm².

(ii) Surface area is a function of the lineal dimension squared:

$$a_1/a_2 \propto d_1^2/d_2^2$$

Thus, $a_1/a_2 = 0.25$, or a 75% decrease in grain boundary area.

MATERIALS IN SERVICE

Products of engineering are made to be used. Conditions that are encountered in service most often vary tremendously from those present in the stockroom: static and dynamic loads, elevated and subambient temperatures, solar and nuclear radiation exposure, and many other reactions with the surrounding environment. All of these situations can lead to deterioration and even to failure. The design engineer should be able to anticipate the conditions of failure.

Mechanical Failure

Under ambient conditions, excessive loads can lead to bending or to cracking, the principal modes of mechanical failure. The former depends upon geometry and the stress level. **Stress** is defined as load divided by cross-sectional area, $\sigma = F/A$. Cracking (and succeeding fracture) includes those two considerations, plus the loading rate.

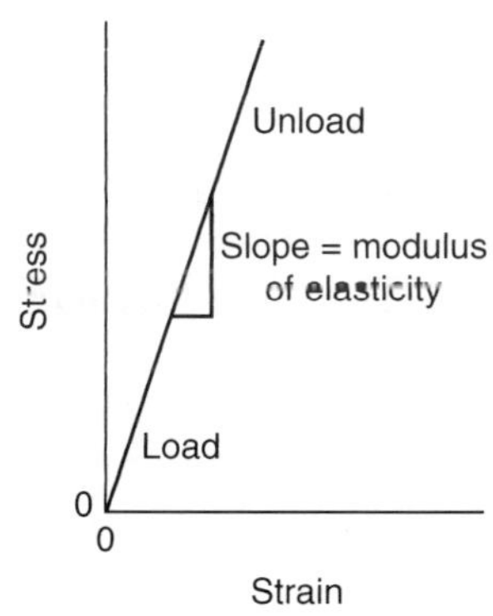

Figure 11.21 Schematic stress-strain diagram showing linear elastic deformation for loading and unloading cycles

Yield Strength

When solids are stressed, strain occurs. Strain is the change in length divided by the original length, $\varepsilon = \Delta L/L_0$. The ratio of stress to strain is called the **elastic modulus**, $E = \sigma/\varepsilon$, and is initially constant in most solids as shown in Figure 11.21. The interatomic spacings are altered as the load is increased. Initial slip starts at a threshold level called the **yield strength**, σ_y, shown in Figure 11.22. Higher stresses will produce a permanent distortion, which will be called failure if the product was designed to maintain its initial shape. The toughness of a material is related to the energy or work required for fracture, a product of both strength and ductility. In a tensile test, the energy is the area under the stress-strain curve. Ductile fracture involves significant plastic deformation and hence absorbs much more energy than brittle fracture, which has little or no plastic deformation. A complete stress-strain curve is shown in Figure 11.23.

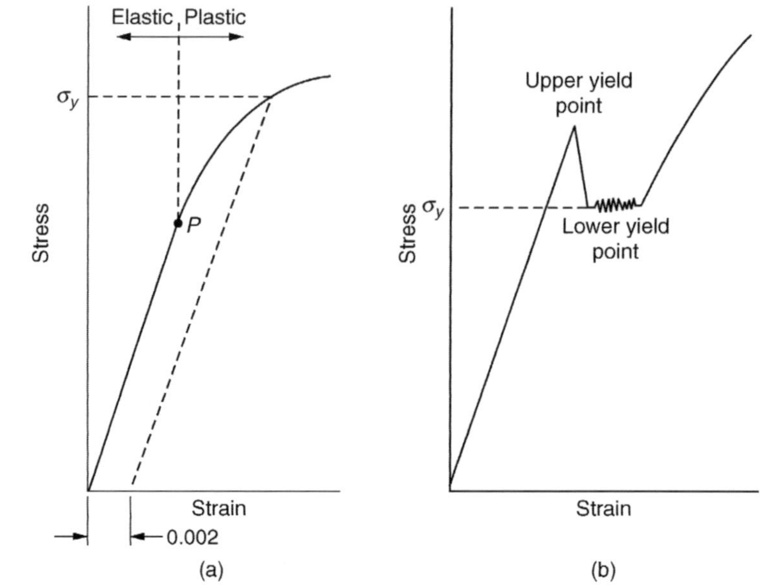

Figure 11.22 (a) Typical stress-strain behavior for a metal, showing elastic and plastic deformations, the proportional limit *P*, and the yield strength σ_y, as determined using the 0.002 strain offset method; (b) representative stress-strain behavior found for some steels, demonstrating the yield point phenomenon

Fracture

Breakage always starts at a location of **stress concentration**. This may be at a *flaw* of microscopic size, such as an abrasion scratch produced while cleaning eyeglasses, or it may be of larger dimensions, such as a hatchway on a ship.

With a crack, the **stress intensity factor**, K_I, is a function of the applied stress, σ, and of the square root of the crack length a:

$$K_I = y\sigma\sqrt{\pi a} \tag{11.9}$$

The proportionality constant, y, relates to cross-sectional dimensions and is generally near unity.

Fracture toughness is the **critical** stress intensity factor, K_{Ic}, to propagate fracture and is a property of the material. This corresponds to the yield strength, σ_y, being the critical stress to initiate slip. However, stronger materials generally have lower fracture toughness and *vice versa*.

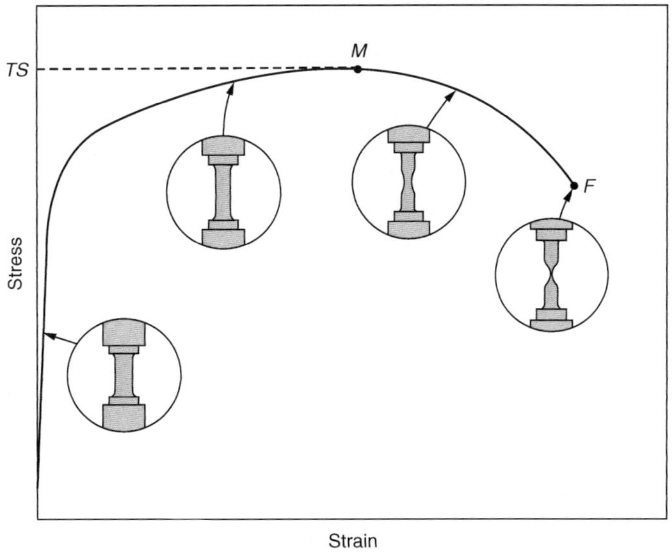

Figure 11.23 Typical engineering stress-strain behavior to fracture, point *F*. The tensile strength *TS* is indicated at point *M*. The circular insets represent the geometry of the deformed specimen at various points along the curve.

To illustrate the relationship between strength and toughness, consider a steel that has a yield strength, σ_y, of 1200 MPa, and a critical stress intensity factor, K_{Ic}, of 90 MPa • $m^{1/2}$.

In the presence of a 2-mm crack, a stress of 1135 MPa would be required to propagate a fracture, according to Equation (11.9). This is below the yield stress. If the value of K_{Ic} had been 100 MPa • $m^{1/2}$, fracture would not occur; rather, the metal would deform at 1200 MPa. A 2.2-mm crack would be required to initiate fracture without yielding.

Fatigue

Cyclic loading reduces permissible design stresses, as illustrated in Figure 11.24. Minute structural changes that introduce cracks occur during each stress cycle. The crack extends as the cycles accumulate, leading to eventual fracture. When this delayed behavior was first observed, it was assumed that the material got tired; hence the term *fatigue*. Steels and certain other materials possess an *endurance limit*, below which unlimited cycling can be tolerated.

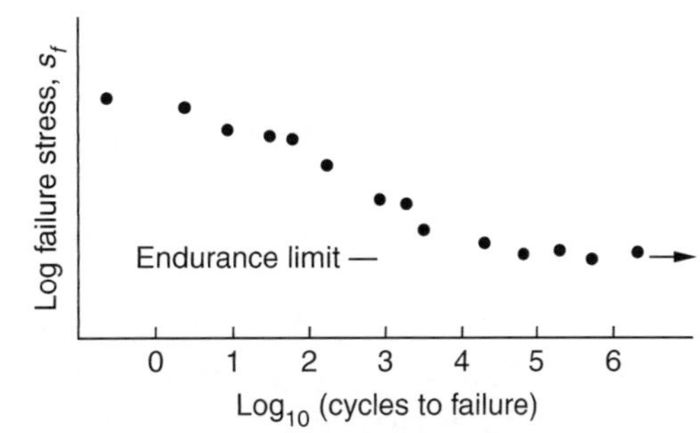

Figure 11.24 Cyclic fatigue. The stress for failure, sf, decreases as the number of loading cycles is increased. Most steels exhibit an endurance limit, a stress level for unlimited cycling. (A static tensile test is only one-fourth of a cycle.)

Example 11.12

(i) What is the maximum static force that can be supported without permanent deformation by a 2-mm-diameter wire that has a yield strength of 1225 MPa?

(ii) The elastic deformation at this threshold stress is 0.015 m/m. What is its elastic modulus?

Solution

(i) $\sigma_y = F/A$

$F = \pi(2 \text{ mm}/2)^2(1225 \text{ MPa}) = 3800 \text{ N}$

(ii) $E = \sigma/\varepsilon = 1225 \text{ MPa}/0.015 \text{ m/m} = 82{,}000 \text{ MPa}$

Example 11.13

The value of K_{Ic} for steel is 186 MPa•m$^{1/2}$. What is the maximum tolerable crack length, a, when the steel carries a nominal stress of 800 MPa? (Assume 1.1 as the proportionality constant.)

Solution

$$186 \text{ MPa}\bullet\text{m}^{1/2} = 800 \text{ MPa } (1.1)(\pi a)^{1/2}$$

$$a = 0.014 \text{ m} = 14 \text{ mm}$$

Thermal Failure

Melting is the most obvious change in a material at elevated temperatures. Overheating, short of melting, can also introduce microstructural changes. For example, the tempered martensite of a tool steel is processed so that it has a very fine dispersion of carbide particles in a tough ferrite matrix. It is both hard and tough and serves well for machining purposes. However, an excessive cutting speed raises the temperature of the cutting edge, causing the carbide particles to grow and softening the steel; if heating continues, failure eventually occurs by melting at the cutting edge. *High-speed* tools incorporate alloy additions, such as vanadium and chromium, that form carbides that are more stable than iron carbide. Thus, they can tolerate the higher temperatures that accompany faster cutting speeds.

Creep

As the name implies, **creep** describes a slow (<0.001%/hr) dimensional change within a material. It becomes important in long-term service of months or years. Slow viscous flow is commonly encountered in plastic materials. Refractories (temperature-resistant ceramics) will slowly slump when small amounts of liquid accumulate.

In metals, creep occurs when atoms become sufficiently mobile to migrate from compressive regions of the microstructure into tensile regions. Grain boundary areas are heavily involved. For this reason, coarser grained metals are advantageous for high-temperature applications. Three stages of creep are identified in Figure 11.25. Following the initial elastic strain, Stage 1 of creep is fairly rapid as stress variations are equalized. In Stage 2, the creep rate, dL/dt, is essentially constant. Design considerations are focused on this stage. Stage 3, which accelerates when the cross-sectional area starts to be reduced, leads to eventual rupture.

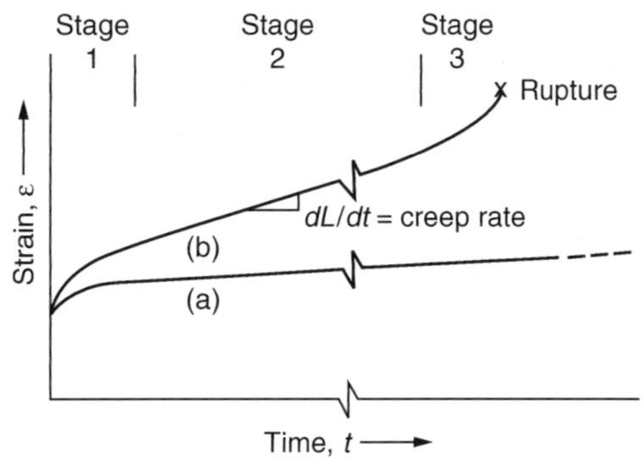

Figure 11.25 Creep. (*a*) Low stresses and/or low temperatures; (*b*) high stresses, high temperatures. The initial strain is elastic, followed by rapid strain adjustments (Stage 1). Design calculations are commonly based on the steady-state strain rate (Stage 2). Strain accelerates (Stage 3) when the area reduction becomes significant. Rupture occurs at *x*.

Spalling

Spalling is thermal cracking. It is the result of stress caused by differential volume changes during processing or service. As discussed earlier, stresses can be introduced into a material (1) by thermal gradients, (2) by anisotropic volume changes, or (3) by differences in expansion coefficients in multiphase materials. Cyclic heating and cooling lead to **thermal fatigue** when the differential stresses produce localized cracking.

The spalling resistance index (SRI) of a material is increased by higher thermal conductivities, k, and greater strengths, S; it is reduced with greater values of the thermal expansion coefficients, α, and higher elastic moduli. In functional form,

$$\mathrm{SRI} = f(kS/\alpha E) \tag{11.10}$$

Low-Temperature Embrittlement

Many materials display an abrupt drop in ductility and toughness as the temperature is lowered. In glass and other amorphous materials, this change is at the temperature below which atoms or molecules cannot relocate in response to the applied stresses. This is the **glass-transition temperature**, T_g. Metals are crystalline and do not have a glass transition. However, steels and a number of other metals have a **ductility-transition temperature** below which fracture is nonductile. The impact energy required for fracture can drop by an order of magnitude at this transition temperature. Thus it becomes a very significant consideration in design for structural applications. The **ductile-to-brittle transition temperature** (DBTT) is often measured using the Charpy Impact test, with samples soaked at different temperatures immediately prior to testing. Representative curves and the influence of carbon content on the DBTT in steel are shown in Figure 11.26.

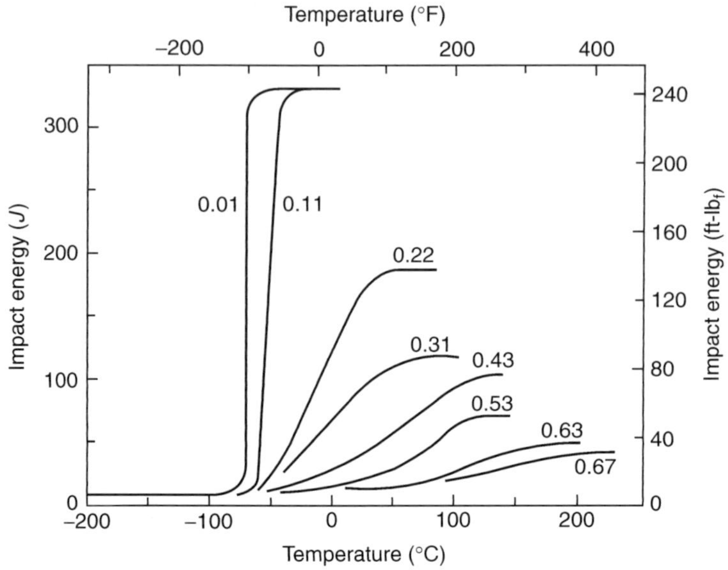

Figure 11.26 Influence of carbon content on the Charpy V-notch energy-versus-temperature behavior for steel

Radiation Damage

Unlike heat, which energizes all of the atoms and molecules within a structure, radiation introduces energy at *pinpoints* called **thermal spikes**. Individual bonds are broken, specific atoms are dislodged, molecules are ruptured, and electrons are energized. Each of these actions disorders the structure and alters the properties of the material. As expected, slip and deformation are resisted. Therefore, while strength and hardness increase, ductility and toughness decrease. Electrical and thermal conductivities drop within metals, because there is more scattering of the electrons as they move along the voltage or thermal gradient. These property changes, among others, are considered to be damaging changes, especially when they contradict carefully considered design requirements.

Damage Recovery

Partial correction of radiation damage is possible by annealing. Reheating a material allows internal readjustments, since atoms that have been displaced are able to relocate into a more ordered structure. It is similar to recrystallization after work hardening, where dislocations involving lineal imperfections composed of many atoms must be removed, except that radiation damage involves *pinpoint* imperfections, which means that it can be removed at somewhat lower temperatures than those required for recrystallization.

Chemical Alteration

Oxidation

Materials can be damaged by reacting chemically with their environments. All metals except gold oxidize in ambient air. Admittedly, some oxidize very slowly at ordinary temperatures. Others, such as aluminum, form a protective oxide surface that inhibits further oxidation. However, all metals—including gold—will oxidize significantly at elevated temperatures or in chemical environments that consume the protective oxidation.

Several actions are required for oxide scale to accumulate on a metal surface. Using iron as an example, the iron atom must be ionized to Fe^{2+} before it or the electrons move through the scale to produce oxygen ions, O^{2-}, at the outer surface. There, FeO or Fe_3O_4 accumulates. As the scale thickens, the oxidation rate decreases. However, exceptions exist. For example, the volume of MgO is less than the volume of the original magnesium metal. Therefore, the scale cracks and admits oxygen directly to the underlying metal. Also, an Al_2O_3 scale is insulating so the ionization steps are precluded.

Moisture

Moisture can produce chemical **hydration**. As examples, MgO reacts with water to produce $Mg(OH)_2$, and Fe_2O_3 can be hydrated to form $Fe(OH)_3$. Water can be absorbed into materials. Small H_2O molecules are able to diffuse among certain large polymeric molecules. Consequently, polymers such as the aramids, which we normally consider to be very strong, are weakened—a fact that the design engineer must consider in specifications.

Corrosion

Metallic corrosion is familiar to every reader who owns a car, since rust—$Fe(OH)_3$—is the most obvious product of corrosion. Oxidation produces positive ions and electrons:

$$M \rightarrow M^{n+} + ne^- \quad \textbf{(11.11)}$$

The reaction stops unless the electrons are removed. Oxygen accompanied by water (Figure 11.27) is a common consumer of the electrons:

$$O_2 + 2\,H_2O + 4e^- \rightarrow 4\,(OH)^- \quad \textbf{(11.12)}$$

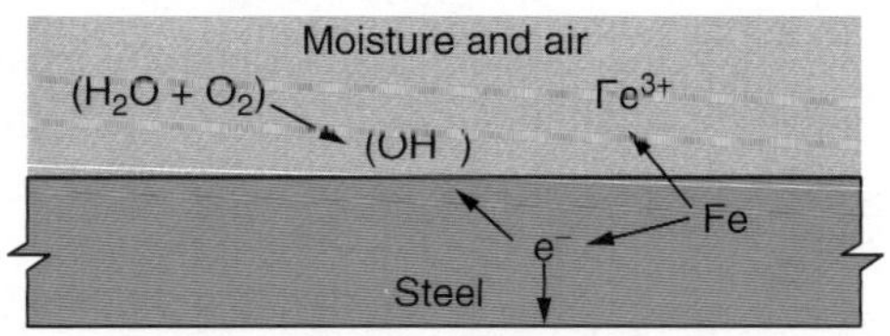

Figure 11.27 Rust formation. Electrons are removed from iron atoms and react with water and oxygen to produce $Fe(OH)_3$—rust.

Alternatively, if ions of a metal with a low oxidation potential are present, they can be reduced, consuming electrons from the preceding corrosion reaction, Equation (11.11). Copper is cited as a common example:

$$Cu^{2+} + 2e^- \rightarrow Cu^0 \quad \textbf{(11.13)}$$

Electroplating uses this reaction advantageously to deposit metals from a solution by the addition of electrons. The relative reactivity of metals with respect to standard electrodes is represented by the electromotive force (emf) series given in Table 11.4. However, in real environments such as sea water the galvanic series, as shown in Table 11.5, is more commonly used to determine the likelihood of corrosion.

Table 11.4 The standard emf series

	Electrode Reaction	Standard Electrode Potential V^0(V)
↑	$Au^{3+} + 3^{e-} \rightarrow Au$	+1.420
	$O_2 + 4H^- + 4e^- \rightarrow 2H_2O$	+1.229
	$Pl^{2}+ + 2e^- \rightarrow Pl$	+1.200
	$Ag^+ + e^- \rightarrow Ag$	+0.800
Increasingly inert (cathodic)	$Fe^{3+} + e^- \rightarrow Fe^{2+}$	+0.771
	$O_2 + 2H_2O + 4e^- \rightarrow 4(OH^-)$	+0.401
	$Cu^{2+} + 2e^- \rightarrow Cu$	+0.340
	$2H^+ + 2e^- \rightarrow H_2$	0.000
	$Pb^{2+} + 2e^- \rightarrow Pb$	–0.126
	$Sn^{2+} + 2e^- \rightarrow Sn$	–0.136
	$Ni^{2}+ + 2e^- \rightarrow Ni$	–0.250
	$Co^{2+} + 2e^- \rightarrow Co$	–0.277
	$Cd^{2+} + 2e^- \rightarrow Cd$	–0.403
	$Fe^{2+} + 2e^- \rightarrow Fe$	–0.440
Increasingly inert (cathodic)	$Cr^{3+} + 3e^- \rightarrow Cr$	–0.744
	$Zn^{2+} + 2e^- \rightarrow Zn$	–0.763
	$Al^{2+} + 3e^- \rightarrow Al$	–1.662
	$Mg^{2+} + 2e^- \rightarrow Mg$	–2.363
	$Na^+ + e^- \rightarrow Na$	–2.714
↓	$K^+ + e^- \rightarrow K$	–2.924

Corrosion Control

Corrosion is minimized by a variety of means. Some involve the avoidance of one or more of the above reactions. Feed water for steam-power boilers is deaerated; surfaces are painted to limit the access of water and air; junctions between unlike metals are electrically insulated. Other control procedures induce a reverse reaction: sacrificial anodes such as magnesium are attached to the side of a ship; iron sheet is galvanized with zinc. Each corrodes preferentially to the underlying steel and forces the iron to assume the role of Equation (11.13). Corrosion may also be restricted by an impressed voltage. Natural gas lines utilize this procedure by connecting a negative dc voltage to the pipe.

COMPOSITES

Composites are not new. Straw in brick and steel reinforcing rods in concrete have been used for a long time. But there is current interest in the development of new composites by designing materials appropriately. It is possible to benefit from the positive features of each of the contributors in order to optimize the properties of the composite.

The internal structures of composites may be viewed as enlarged poly-component micro-structures, which were previously discussed. Attention is given to size, shape, amounts, and distribution of the contributing materials. However, there is commonly a significant difference in processing composites. Typically, the internal structure of a composite is a function of mechanical processing steps—mixing, emplacement, surface deposition, and so on—rather than thermal processing. These processing differences suggest a different approach in examining property-structure relationships.

Table 11.5 The galvanic series

↑	Platinum
	Gold
	Graphite
	Titanium
	Silver
	316 Stainless steel (passive)
	304 Stainless steel (passive)
	Inconel (80Ni–13Cr–7Fe) (passive)
	Nickel (passive)
	Monel (70Ni–30Cu)
Increasingly inert (cathodic)	Copper–nickel alloys
	Bronzes (Cu–Sn alloys)
	Copper
	Brasses (Cu–Zn alloys)
	Inconel (active)
	Nickel (active)
	Tin
	Lead
Increasingly inert (cathodic)	316 Stainless steel (active)
	304 Stainless steel (active)
	Cast iron
	Iron and steel
	Aluminum alloys
	Cadmium
↓	Commercially pure aluminum
	Zinc
	Magnesium and magnesium alloys

Reinforced Materials

In familiar composites, such as reinforced concrete, the steel carries the tensile load. Also, there must be bonding between the reinforcement and the matrix. In a *glass* fishing rod, the glass fibers are all oriented longitudinally within the polymer matrix. If it is fractured, the break exhibits a splintered appearance with a noticeable amount of fiber pull-out.

Steel reinforcing bars (rebar) are used to reinforce concrete to meet strength and cost factors. Positioning is dictated by stress calculations. Rebar surfaces are commonly merloned (ridged circumferentially) for better anchorage.

Glass is widely used in fiber-reinforced plastics (FRP). The positioning of the fibers varies with the product. Continuous fibers are used in such structures as fuel storage tanks or rocket casings. However, chopped fibers are required when FRP are processed within molds. A matrix-to-glass bond must be achieved through chemical coatings at the interfaces.

Silicon carbide (SiC) and alumina (Al_2O_3) fibers are used increasingly in high-temperature composites with either metallic or ceramic matrices.

Inert fillers such as silica flour and wood flour serve multiple purposes in many polymers. They add strength, rigidity, and hardness to the product. In addition, they generally are less expensive than the matrix polymer.

Reinforcing bars or fibers are expected to carry the bulk of the tensile load. To be effective, the reinforcement must have a higher elastic modulus, E_r, than does the matrix because the reinforcement and the matrix undergo the same tensile strain when loaded ($\varepsilon_r = \varepsilon_m$). Therefore, $(E/\sigma)_r = (E/\sigma)_m$; and

$$\sigma_r / \sigma_m = E_r / E_m \tag{11.14}$$

Mixture Rules

We commonly study and analyze properties of composites in terms of the properties of the contributing materials. Mixture rules can then be formulated in which the properties are a function of the amounts and geometric distributions of each of the contributors. The simplest mixture rules are based on volume fraction, *f*. For example, the density of the mixture, ρ_m, is the volume-weighted average:

$$\rho_m = f_1\rho_1 + f_2\rho_2 + \ldots \tag{11.15a}$$

or simply

$$\rho_m = \sum f_i \rho_i \tag{11.15b}$$

Likewise, for *heat capacity*, *c*:

$$c_m = \sum f_i c_i \tag{11.16}$$

Directional Properties

Many composites are anisotropic. Laminates of two or more layers are bidirectional, as is a matted FRP that uses chopped fibers. The previously cited *glass* fishing rod has a structure with uniaxial anisotropy.

When considering conductivity, elastic moduli, strength, or other properties that are directional, attention must be given to the anisotropy of the composite. Consider the electrical resistance across a laminate that contains alternate layers of two materials with different resistances, R_1 and R_2. In a direction perpendicular to the laminate, the plies are in series. Thus the relationship is

$$R_\perp = L_1 R_1 + L_2 R_2 + \cdots \tag{11.17}$$

where L_i is the thickness of each ply. For unit dimensions, volume fractions, f_i, and the resistivities, ρ_i, are applicable:

$$\rho_\perp = f_1\rho_1 + f_2\rho_2 + \cdots \tag{11.18}$$

In contrast, for directions parallel to the laminate, resistivities follow the physics analog for parallel circuits:

$$\frac{1}{\rho_\parallel} = \frac{f_1}{\rho_1} + \frac{f_2}{\rho_2} + \cdots \tag{11.19}$$

The mixture rules for conductivities, either thermal, *k*, or electrical, *σ*, are inverted to those for resistivities:

$$\frac{1}{k_\perp} = \frac{f_1}{k_1} + \frac{f_2}{k_2} + \cdots \tag{11.20}$$

and

$$k_\parallel = f_1 k_1 + f_2 k_2 + \cdots \tag{11.21}$$

More elaborate mixture rules must be used when the structure of the composite is geometrically more complex, such as the use of particulate fillers or chopped fibers. In general, however, the property of the composite falls between the calculated values using the parallel and series versions of the preceding equations.

Preparation

Composites receive their name from the fact that two or more distinct starting materials are combined into a unified material. The application of a protective coating to metal or wood is one form of composite processing. The process is not as simple as it first appears because priming treatments are commonly required after the surface has been initially cleared of contaminants and moisture; otherwise, peeling and other forms of deterioration may develop. Electroplating, Equation (11.13), commonly requires several intermediate surface preparations so that the final plated layer meets life requirements.

Wood paneling combines wood and polymeric materials into large sheets. **Plywood** not only has dimensional merit, it transforms the longitudinal anisotropy into a more desirable two-dimensional material. Related products include chipboard and similar composites. Composite panels include those with veneer surfaces where appearance and technical properties are valued.

The concept of uniformly mixing particulate **fillers** into a composite is simple. The resulting product is isotropic.

Several considerations are required for the use of **fibers** in a composite. Must the fiber be continuous? What are the directions of loading? What is the shear strength between the fiber and the matrix? Chopped fibers provide more reinforcement than do particles, and at the same time permit molding operations. The **aspect ratio** (L/D) must be relatively low for die molding. Higher ratios, and therefore more reinforcement, are used in sheet molding. (Sheet molded products are used where strength is not critical in the third dimension.) **Continuous fibers** maximize the mechanical properties of FRP composites. Their uses, however, are generally limited to products that permit parallel layments.

Example **11.14**

A rod contains 40 volume percent longitudinally aligned glass fibers within a plastic matrix. The glass has an elastic modulus of 70,000 MPa; the plastic, 3500 MPa. What fraction of a 700-N tensile load is carried by the glass?

Solution

Based on Equation (11.14):

$$(F_{gl}/0.4A)/(F_p/0.6A) = (70{,}000 \text{ MPa} /3500 \text{ MPa}) = 20$$

$$F_{gl} = 20(0.4/0.6)\, F_p = 13.3\, F_p$$

$$F_{gl}/(F_{gl} + F_p) = (13.3\, F_p)/(13.3\, F_p + F_p) = 93\%$$

Example **11.15**

An electric highline cable contains one cold-drawn steel wire and six annealed aluminum wires, all with a 2-mm diameter. (The steel provides the required strength; the aluminum, the conductivity.) Using the following data, calculate (i) the resistivity and (ii) the elastic modulus of the composite wire.

$$\text{Steel: } \rho = 17 \times 10^{-6}\ \Omega \bullet \text{cm},\ E = 205{,}000 \text{ MPa}$$

$$\text{Aluminum: } \rho = 3 \times 10^{-6}\ \Omega \bullet \text{cm},\ E = 70{,}000 \text{ MPa}$$

Solution

(i) From Equation (11.19):

$$1/\rho_{\|} = (1/7)/(17 \times 10^{-6}\ \Omega \bullet \text{cm}) + (6/7)/(3 \times 10^{-6}\ \Omega \bullet \text{cm})$$
$$= 0.294 \times 10^{6}\ \Omega^{-1} \bullet \text{cm}^{-1}$$

$$\rho_{\|} = 3.4 \times 10^{-6}\ \Omega \bullet \text{cm}$$

(ii) We must write a mixture rule for the elastic modulus of a composite in parallel. Let A be the area of one wire.

Since $\varepsilon_c = \varepsilon_{Al} = \varepsilon_{St}$, $[(F/A)/E]_C = [(F/A)/E]_{Al} = [(F/A)/E]_{St}$

Also, $F_C = F_{Al} + F_{St} = F_C(f_{Al}A_C/A_C)(E_{St}/E_C) + F_C(f_{St}A_C/A_C)(E_{St}/E_C)$

Canceling,

$$E_C = f_{Al}E_{Al} + f_{St}E_{St} = (6/7)(70{,}000\ \text{MPa}) + (1/7)(205{,}000\ \text{MPa}) = 89{,}000\ \text{MPa}$$

The apparent modulus will be lower because there will also be cable extension by the straightening of the cable wire.

STEEL

Steel is a heat-treatable alloy, primarily an alloy of iron and carbon. The lower limit for a mild steel is about 0.05%, though the lower limit might be defined as that value below which iron carbide cannot be thrown out of solid solution, which is below 0.01%. The upper theoretical limit is 1.7%, which corresponds to the point in the iron-carbon equilibrium diagram beyond which iron carbide cannot be wholly held in solution at any temperature. The lowest carbon alloys are only slightly, and usually only with difficulty, affected by heat treatment, and are termed *irons* rather than *steels*.

Other elements are almost always included in ordinary steel, some by design, and some because it is too difficult or expensive to remove them. Excess amounts of sulphur and phosphorous, for example, are quit deleterious. Too much sulphur results in a condition termed hot-shortness, meaning that the steel becomes brittle at high temperatures. Too much phosphorous, on the other hand, causes cold-shortness. Steel is not a uniform substance like pure gold, but is made up of components—ferrite and cementite. Ferrite is BCC iron in the crystalline form that exists at room temperature in slowly cooled carbon steel. Cementite is a compound of carbon and iron, Fe_3C. When a carbon steel is cooled at a slow rate from a red heat, cementite and ferrite form a laminar composition, which resembles mother of pearl and is termed pearlite. Pearlite contains 0.85% C which is the eutectoid composition of steel. A hypoeutectoid steel is one containing less than 0.85% C and when cooled will contain free ferrite. A slowly cooled hypereutectoid steel will contain pearlite and excess cementite.

AUSTENITE

When carbon steel is heated above its transformation temperature the, pearlite changes to austenite. Austenite has a face centered cubic lattice and thus contains more atoms per cell, though the packing factor is greater for the FCC cell than for the BCC cell. Austenite is nonmagnetic and is more dense than ferrite. Austenite may exist at room temperature only when its transformation has been fully suppressed. Manganese, nickel, and chromium are used to suppress the transformation. Stainless steels, for example, which contain more than 6% nickel and more than 24% of nickel and chromium combined, are austenitic at room temperature, and are nonmagnetic.

MARTENSITE

When austenite is cooled rapidly to a temperature of 200 to 300 °C, it forms a very hard structure called martensite. This is a skewed tetragonal lattice that forms a needlelike structure. Martensite is less dense than pearlite, so there is a slight increase in volume when a steel is fully hardened. If the cooling is too rapid, the metal does not have time to absorb the increase in volume, and may rupture. Even if the cooling rate is lower than the rate that causes rupture, the increase in volume may still result in severe internal stresses, which must be relieved if the part is to be safely loaded. Such parts are heated to a temperature below the lower transformation temperature and "stress relieved."

Elements are added to steel to increase its toughness, hardness, corrosion resistance, or other property. As an example, plain carbon steel with a BCC structure becomes very brittle at low temperatures. At −0 °C, a wrought-iron pipe becomes as brittle as glass and will shatter if stuck by a hammer. But austenitic steels do not exhibit the low-temperature embrittlement phenomenon.

Example 11.16

A design for an army tank for use in Antarctica is designed with cleats of high-carbon steel. The ambient temperature is expected to drop as low as −0 °C. What should be recommended by a design review?

Solution

Since the tank treads will be in contact with the ground, could reach as low a temperature as the forecast minus 80 °C, and will also be subject to continuous shock loading there is a high probability that the treads, being made of a BCC steel, would fracture. They should be replaced with an abrasion-resistant nonbrittle metal.

NONFERROUS METALS AND ALLOYS

This classification includes all metals in which iron is not present in large quantities. It includes such metals as copper, aluminum, magnesium, and zinc. Also included are more exotic metals such as silver, gold, tungsten, and others that, though used in relatively small amounts, are very important for specialized purposes. Nonferrous metals are generally used for parts requiring special fabrication and where the ease of fabrication outweighs the higher material costs. Other factors such as density, stiffness, corrosion resistance, electric properties, and color can also affect selection.

MODULUS OF ELASTICITY

The modulus of elasticity is a measure of the elastic deformation of a metal when it is stressed in tension or compression within its elastic limit. It is measured in Pa and equals pascals divided by m/m elongation.

Example 11.17

A 100 kg mass is supported by a 100 m long steel wire 2.5 mm diameter. How much would the wire stretch (strain) elastically?

Solution

The stress in the wire would equal

$$S = 100 \times 9.8066/(4.909 \times 10^{-6}) = 199.8 \text{ MPa}$$
$$\text{Strain, } \varepsilon = 199{,}800{,}000/(2.1 \times 10^{11}) = 95.14 \times 10^{-5} \text{ m/m}$$

For 100 m length the elongation would equal 95 mm

The stiffness of a part can sometimes be very important, and stiffness is a function of the modulus of elasticity, E. For example, the modulus of elasticity for steel, E_s, equals 2.1×10^{11} Pa, whereas E_a, the modulus of elasticity for aluminum is only 6.9×10^{10}, or 67% less.

Example 11.18

A steel cantilever beam holds 150 kg on the end. It is three meters long and ten cm high by five cm wide. To save mass it is proposed to replace the steel beam with an aluminum one of the same length and same proportions. To give the same deflection, (a) what would be the deflection of the end of the beam, (b) what would be the dimensions of the aluminum beam, and (c) what would be the percentage mass reduction? The density of steel is 7.9 g/cm^3 and that of aluminum is 2.7 g/cm^3.

Solution

a). The deflection of the beam would equal $PL^3/(3EI)$

For the steel beam $I = b \times h^3/12 = 4.167 \times 10^{-6}$ m^4

$$\Delta h = 150 \times 9.8066 \times 3.00^3/(3 \times 2.1 \times 10^{11} \times 4.167 \times 10^{-6})$$
$$\Delta h = 0.01513 \text{ m or } 1.51 \text{ cm}$$

b). For the aluminum beam to have the same stiffness as the steel beam, the product of $E \times I$ for the aluminum beam must be the same as for the steel beam. So $I_{al} = 21/6.9 \times I_s$

$$I_{al} = 12.682 \times 10^{-6} = b \times (2b)^3/12$$

Which gives $b = (19.023 \times 10^{-6})^{1/4} = 0.06604$

c). The dimensions of the aluminum for the same stiffness would be 6.604 cm × 13.208 cm

The ratio of the masses would equal

$$6.604 \times 13.208 \times 2.70/(5.0 \times 10.0 \times 7.87) = 0.5985$$

There would be a 40 percent reduction in the mass of the beam.

CORROSION

The ability of a metal, or design, to resist corrosion is a very important factor in selecting a particular metal for a specific application. Corrosion causes millions of dollars of structural damage to metal components, and millions more due to surface damage or tarnishing to decorative items. The proper selection of materials or protective systems is very important to a mechanical design.

Corrosion may be defined as the destructive chemical or electrochemical reaction of a material and its environment. The most common type of corrosion is rust. Rust is a reddish-brown or orange coating on iron or steel due to oxidation of the surface when exposed to air or moisture. It is made up principally of hydrated ferric oxide. Rust is formed by the galvanic action of the iron and small particles of other metals contained in it in contact with acidic solutions formed by the dissolution of carbon dioxide in rainwater and dew. Moisture is always present in the atmosphere, so the potential for electrochemical corrosion always exists.

ELECTROCHEMICAL CORROSION

If two dissimilar metals in contact are joined by an electrolyte, they will form a galvanic corrosion cell and one will be dissolved or corroded. An electrolyte is a substance capable of carrying ions. It is typically an aqueous solution and can be formed from condensed moisture, dew, or rain and contaminated by small amounts of dirt, salts, acids, or alkalis from the atmosphere. The anode and cathode of a corrosion cell can even be parts of the same component with different potentials. The anodic material, the one with the highest potential in reference to hydrogen, as shown in the table of "Standard Oxidation Potentials for Corrosion Reactions" in the *FE Handbook*, will be corroded. For example if aluminum rivets should be used to join steel sheets, the rivets would be corroded and could fail if not protected. Conversely, if two aluminum plates should be bolted together with iron bolts, pitting due to corrosion would occur in the plates next to the bolts.

It is noted from the table that zinc has a higher oxidation potential than iron, and this explains why steel items are sometimes zinc coated—the zinc will corrode in preference to the steel and leave the steel uncorroded. The fact that the cathode is resistant to corrosion is made use of in protecting ship hulls by attaching sacrificial anodes of metals such as magnesium to the hull. Cathodic resistance to corrosion is also made use of in protecting buried pipelines by connecting them to the negative side of a DC current supply.

NONMETALLIC MATERIALS

There are many nonmetallic materials of importance to engineers, but most of them lie within the province of civil, chemical, and electrical engineers. However, mechanical engineers will find it of value to be aware of some of the properties of nonmetallic materials that can be of benefit in different mechanical designs. Many plastics can prove quite useful to the mechanical engineering designs. Plastics have the advantage of being more formable than any metallic material. In addition, plastics can have low density and can have special properties such as specially desired heat or electrical insulation, corrosion resistance, low friction coefficient, resistance to deterioration by moisture, good color range, or other property not obtainable in the required amount with a metal. A plastic is defined in the broadest sense as any non-metallic material that can be molded to shape. Plastics can be molded, cast, or extruded. They can also be used for coatings and films. Plastics are generally lumped into two general classifications—thermoplastic and thermosetting. Thermoplastics are those which can begin to soften at a temperature as low as 60 °C and then can be molded without any change in chemical structure. Thermosetting materials undergo a chemical change when molded and cannot be re-softened by heating. Thermosetting plastics usually require considerably higher mold temperatures than thermoplastics, and the finished part can usually withstand much higher temperatures without deforming.

SELECTED SYMBOLS AND ABBREVIATIONS

Symbol or Abbreviation	Definition
A	area
a	unit cell length
a	crack length
amu	atomic mass unit
D	diffusion coefficient
D_o	proportionality constant
E	elastic modulus
ε	engineering strain
F	force
f	volume fraction
K_I	stress intensity factor
P	density
Q	activation energy
R	gas constant
r	atomic radius
σ	stress
σ_y	yield strength
SRI	Spalling Resistance Index
Tg	glass-transition temperature
Tm	melting temperature
V^0	standard electrode potential
W_L	weight fraction of liquid phase
W_S	weight fraction of solid phase
y	proportionality constant

REFERENCE

Van Vlack, L. *Elements of Materials Science and Engineering*, 6th ed. Addison-Wesley, 1989.

PROBLEMS

11.1 For a neutral atom:

a. the atomic mass equals the mass of the neutrons plus the mass of the protons
b. the atomic number equals the atomic mass
c. the number of protons equals the atomic number
d. the number of electrons equals the number of neutrons

11.2 All isotopes of a given element have:

a. the same number of protons
b. the same number of neutrons
c. equal numbers of protons and neutrons
d. the same number of atomic mass units

11.3 Which of the following statements is *FALSE*?

a. An anion has more electrons than protons.
b. Energy is released when water is solidified to ice.
c. Energy is required to remove an electron from a neutral atom.
d. Energy is released when a H_2 molecule is separated into two hydrogen atoms.

11.4 Select the correct statement.

a. Crystals possess long-range order.
b. Within a crystal, like ions attract and unlike ions repel.
c. A body-centered cubic metallic crystal (for example, iron) has nine atoms per unit cell.
d. A face-centered cubic metallic crystal (for example, copper) has 14 atoms per unit cell.

11.5 In a cubic crystal, a is the edge of a unit cell. The shortest repeat distance in the [111] direction of a body-centered cubic crystal is:

a. $a\sqrt{2}$
b. $2a$
c. $a\sqrt{3}/2$
d. $a\sqrt{3}/4$

11.6 All but which of the following data are required to calculate the density of aluminum in g/m^3?

a. Avogadro's number, which is 6.0×10^{23}
b. atomic number of Al, which is 13
c. crystal structure of Al, which is face-centered cubic
d. atomic mass of Al, which is 27 amu

11.7 The atomic packing factor of gold, an FCC metal, is:

a. $(4\pi r^3/3)/(4r/\sqrt{3})^3$
b. $4(4\pi r^3/3)/(4r/\sqrt{2})^3$
c. $4(4\pi r^3/3)/(r/\sqrt{2})^3$
d. $4(2r/\sqrt{2})^3/(4\pi r^3/3)$

11.8 Ethylene is C_2H_4. To meet bonding requirements, how many bonds are present?

a. 6 single
b. 4 single and 2 double
c. 1 double and 4 single
d. 12 single

11.9 Gold is FCC and has a density of 19.3 g/cm^3. Its atomic mass is 197 amu. Its atomic radius, *r*, may be calculated using which of the following?

a. $19.3\ \text{g/cm}^3 = (197)(6.02 \times 10^{23})/[(4r/)\sqrt{2^3}]$
b. $19.3\ \text{g/cm}^3 = 2\ (197/6.02 \times 10^{23})/[(4r/)\sqrt{2^3}]$
c. $19.3\ \text{g/cm}^3 = 4\ (197/6.02 \times 10^{23})/[(4r/)\sqrt{2^3}]$
d. $19.3\ \text{g/cm}^3 = 6\ (197)(6.02 \times 10^{23})/[(4r/)\sqrt{2^3}]$

11.10 Each of the following groups of plastics is thermoplastic *EXCEPT*:

a. polyvinyl chloride (PVC) and a polyvinyl acetate
b. phenolics, melamine, and epoxy
c. polyethylene, polypropylene, and polystyrene
d. acrylic (Lucite) and polyamide (nylon)

11.11 Styrene resembles vinyl chloride, C_2H_3C1, except that the chlorine is replaced by a benzene ring. The mass of each mer is:

a. 8(12) + 9(1) amu
b. 26 + 78 amu
c. 27 + 6(12) + 6(1) amu
d. 2(12) + 3(1) + 77 amu

11.12 The $<1\bar{1}0>$ family of directions in a cubic crystal include all but which of the following? (An overbar is a negative coefficient.)

a. [110]
b. $[0\bar{1}1]$
c. [101]
d. $[1\bar{1}1]$

11.13 The {112} family of planes in a cubic crystal includes all but which of the following directions?

a. (212)
b. (211)
c. $(1\bar{1}2)$
d. (121)

11.14 Crystal imperfections include all but which of the following?

a. Dislocations
b. Displaced atoms
c. Interstitials
d. Dispersions

11.15 A dislocation may be described as a:

a. displaced atom
b. shift in the lattice constant
c. slip plane
d. lineal imperfection

11.16 A grain within a microstructure is:
- a. a particle the size of a grain of sand
- b. the nucleus of solidification
- c. a particle the size of a grain of rice
- d. an individual crystal

11.17 Which of the following does *NOT* apply to a typical brass?
- a. An alloy of copper and zinc
- b. A single-phase alloy
- c. An interstitial alloy of copper and zinc
- d. A substitutional solid solution

11.18 Sterling silver, as normally sold:
- a. is pure silver
- b. is a supersaturated solid solution of 7.5% copper in silver
- c. is 24-carat silver
- d. has higher conductivity than pure silver

11.19 Atomic diffusion in solids matches all but which of the following generalities?
- a. Diffusion is faster in FCC metals than in BCC metals.
- b. Smaller atoms diffuse faster than do larger atoms.
- c. Diffusion is faster at elevated temperatures.
- d. Diffusion flux is proportional to the concentration gradient.

11.20 The proportionality constant for a particular gas diffusing through copper at 1000°C is 0.022 cm^2/s and the activation energy is 97 kJ/mole. Find the diffusion coefficient.
- a. 1.6×10^{-6} cm^2/s
- b. 1.9×10^{-6} cm^2/s
- c. 2.3×10^{-6} cm^2/s
- d. 4.4×10^{-6} cm^2/s

11.21 Grain growth involves all but which of the following?
- a. Reduced growth rates with increased time
- b. An increase in grain boundary area per unit volume
- c. Atom movements across grain boundaries
- d. A decrease in the number of grains per unit volume

11.22 Imperfections within metallic crystal structures may be any but which of the following?
- a. Lattice vacancies and extra interstitial atoms
- b. Displacements of atoms to interstitial site (Frenkel defects)
- c. Lineal defects or slippage dislocations caused by shear
- d. Ion pairs missing in ionic crystals (Shottky imperfections)

11.23 All but which of the following statements about solid solutions are correct?

a. In metallic solid solutions, larger solute atoms occupy the interstitial space among solvent atoms in the lattice sites.
b. Solid solutions may result from the substitution of one atomic species for another, provided radii and electronic structures are compatible.
c. Defect structures exist in solid solutions of ionic compounds when there are differences in the oxidation state of the solute and solvent ions, because vacancies are required to maintain an overall charge balance.
d. Order-to-disorder transitions that occur at increased temperatures in solid solutions result from thermal agitation that dislodges atoms from their preferred neighbors.

11.24 In ferrous oxide, $Fe_{1-x}O$, 2% of the cation sites are vacant. What is the Fe^{3+}/Fe^{2+} ratio?

a. 2/98
b. 0.04/0.94
c. 0.04/0.96
d. 0.06/0.94

11.25 A solid solution of MgO and FeO contains 25 atomic percent Mg^{2+} and 25 atomic percent Fe^{2+}. What is the weight fraction of MgO? (Mg: 24; Fe: 56; and O: 16 amu)

a. 40/(40 + 72)
b. 24/(24 + 56), or 4/80
c. 25/(25 + 25)
d. (25 + 25)/(50 + 50)

11.26 The boundary between two metal grains provides all but which of the following?

a. An impediment to dislocation movements
b. A basis for an increase in the elastic modulus
c. A site for the nucleation of a new phase
d. Interference to slip

11.27 If 5% copper is added to silver:

a. the hardness is decreased
b. the strength is decreased
c. the thermal conductivity is decreased
d. the electrical resistivity is decreased

11.28 All but which of the following statements about diffusion and grain growth are correct?

a. Atoms can diffuse both within grains and across grain (crystal) boundaries.
b. The activation energy for diffusion through solids is inversely proportional to the atomic packing factor of the lattice.
c. Grain growth results from local diffusion and minimizes total grain boundary area. Large grains grow at the expense of small ones, and grain boundaries move toward their centers of curvature.
d. Net diffusion requires an activation energy and is irreversible. Its rate increases exponentially with temperature. It follows the diffusion equation, in which flux equals the product of diffusivity and the concentration gradient.

11.29 Refer to the accompanying Mg-Zn phase diagram, Exhibit 11.29. Select an alloy of composition C (71Mg–29Zn) and raise it to 575°C so that only liquid is present. Change the composition to 60Mg–40Zn by adding zinc. When this new liquid is cooled, what will be the first solid to separate?

a. A solid intermetallic compound
b. A mixture of solid intermetallic compound and solid eutectic C (71Mg–29Zn)
c. A solid eutectic C (71Mg–29Zn)
d. A solid solution containing less than 1% intermetallic compound dissolved in Mg

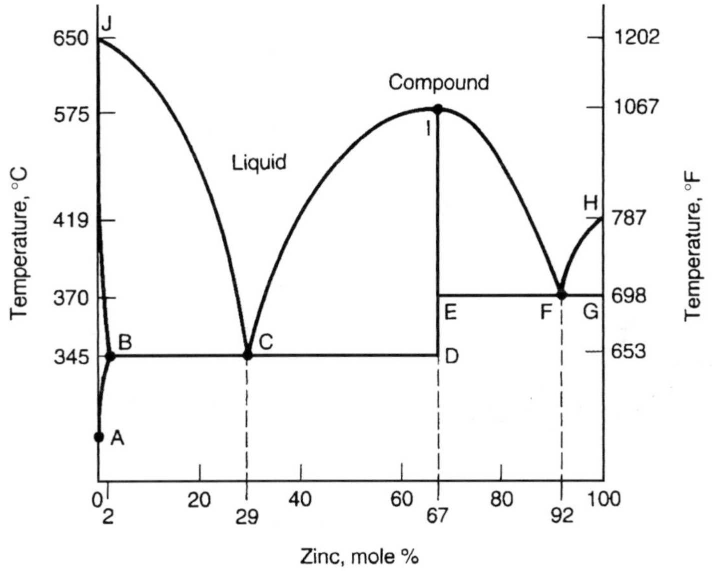

Exhibit 11.29 Magnesium-zinc phase diagram

11.30 Refer to the Mg-Zn phase diagram of Exhibit 11.29. Which of the following compounds is present?

a. Mg_3Zn_2
b. Mg_2Zn_3
c. MgZn
d. $MgZn_2$

11.31 Refer to Exhibit 11.31, a schematic sketch of the Fe-Fe_3C phase diagram. All but which of the following statements are *TRUE*?

a. A eutectoid reaction occurs at location C, 727°C (1340°F).
b. The eutectic composition is 99.2 weight percent Fe and 0.8 weight percent C.
c. A peritectic reaction occurs at K, 1500°C (2732°F).
d. A eutectic reaction occurs at G, 1130°C (2202°F).

11.32 Refer to Exhibit 11.31, the Fe-Fe_3C phase diagram. Pearlite contains ferrite (α) and carbide (Fe_3C). The weight fraction of carbide in pearlite is:

a. 0.8%
b. BC/CD
c. CD/BD
d. BC/BD

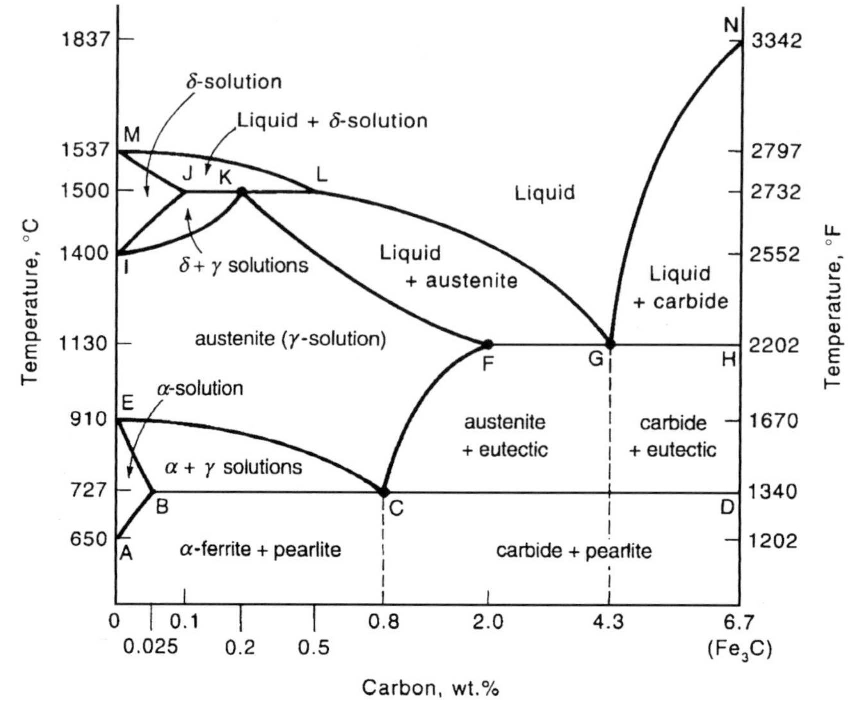

Exhibit 11.31 Iron-iron carbide phase diagram (schematic)

11.33 Consider the Ag-Cu phase diagram (Figure 11.16). Silver-copper alloys can contain *approximately* half liquid and half solid at all but which of the following situations?

a. 40Ag–60Cu at 781°C
b. 81Ag–19Cu at 781°C
c. 20Ag–80Cu at 910°C
d. 50Ag–50Cu at 779°C

11.34 Refer to the Ag-Cu phase diagram of Figure 11.16. Which of the following statements is *FALSE*?

a. The solubility limit of copper in the liquid at 900°C is approximately 62%.
b. The solubility limit of silver in b at 700°C is approximately 5%.
c. The solubility limit of silver in b at 900°C is approximately 7%.
d. The solubility limit of copper in b at 800°C is approximately 92%.

11.35 Other factors being equal, diffusion flux is facilitated by all but which of the following?

a. Smaller grain sizes
b. Smaller solute (diffusing) atoms
c. Lower concentration gradients
d. Lower-melting solvent (host structure)

11.36 Consider copper. All but which of the following statements are applicable for grain growth?

a. Atoms jump the boundary from large grains to small grains.
b. Grain size varies inversely with boundary area.
c. Grain growth occurs because the boundary atoms possess higher energy than interior atoms.
d. Grain growth occurs because larger grains have less boundary area.

11.37 A phase diagram can provide answers for all but which of the following questions?

a. What are the direction of the planes at a given temperature?
b. What phases are present at a given temperature?
c. What are the phase compositions at a given temperature?
d. How much of each phase is present at a given temperature?

11.38 In the Ag-Cu system (Figure 11.16), three equilibrated phases may be present at:

a. 930°C c. 830°C
b. 880°C d. 780°C

11.39 During heating, a 72Ag–28Cu alloy (Figure 11.16) may have any but which of the following equilibrium relationships?

a. ~25% β at 600°C, where β contains ~2% Ag
b. less than 50% β (95Cu–5Ag) at 700°C
c. ~74% α(93Ag–7Cu) at 750°C
d. α(26% Cu) and β(32% Cu) at 800°C

11.40 During cooling, a (20Ag–80Cu) alloy (Figure 11.16) has all but which of the following equilibrium situations?

a. The first solid forms at 980°C
b. The first solid contains 5% Ag
c. The second solid appears at 780°C
d. The last liquid contains 32% Cu

11.41 Add copper to 100 g of silver at 781°C (Figure 11.16). Assume equilibrium. Which statement is *TRUE*?

a. Liquid first appears with the addition of exactly 9 g of copper.
b. The last α disappears when approximately 39 g of copper has been added.
c. The solubility limit of copper in solid silver (α) is 28% copper.
d. Solid β first appears with the addition of 92 g of copper.

11.42 All but which of the following statements about strain hardening is correct?

a. Strain hardening is produced by cold working.
b. Strain hardening is relieved during annealing above the recrystallization temperature.
c. With more strain hardening, more time-temperature exposure is required for relief.
d. Strain hardening is relieved during recrystallization. Recrystallization produces less strained and more ordered structures.

11.43 Which process is used for the high-temperature shaping of many materials?

a. Reduction c. Polymerization
b. Recrystallization d. Extrusion

11.44 All but which of the following processes strengthens metals?

a. Precipitation processes that produce submicroscopic particles during a low-temperature heat treatment
b. Increasing the carbon content of low-carbon steels
c. Annealing above the recrystallization temperature
d. Mechanical deformation below the recrystallization temperature (cold working)

11.45 All but which of the following statements about the austentite-martensite-bainite transformations are correct?

a. Pearlite is a stable lamellar mixture consisting of BCC ferrite (α) plus carbide (Fe_3C). It forms through eutectoid decomposition during slow cooling of austenite. Most alloying elements in steel retard this transformation.
b. Martensite has a body-centered structure of iron that is tetragonal and is supersaturated with carbon. It forms by shear during the rapid quenching of austenite (FCC iron).
c. Tempering of martensite is accomplished by reheating martensite to precipitate fine particles of carbide within a ferrite matrix, thus producing a tough, strong structure.
d. Bainite and tempered martensite have distinctly different microstructures.

11.46 Steel can be strengthened by all but which of the following practices?

a. Annealing
b. Quenching and tempering
c. Age or precipitation hardening
d. Work hardening

11.47 Residual stresses can produce any but which of the following?

a. Warpage
b. Distortion in machined metal parts
c. Cracking of glass
d. Reduced melting temperatures

11.48 The reaction ($\gamma \rightarrow \alpha + Fe_3C$) is most rapid at:

a. the eutectoid temperature
b. 10°C above the eutectoid temperature
c. the eutectic temperature
d. 100°C below the eutectoid temperature

11.49 Grain growth, which reduces boundary area, may be expected to:

a. decrease the thermal conductivity of ceramics
b. increase the hardness of a solid
c. decrease the creep rate of a metal
d. increase the recrystallization rate

11.50 Rapid cooling can produce which one of the following in a material such as sterling silver?
a. Homogenization
b. Phase separation
c. Grain boundary contraction
d. Supersaturation

11.51 Martensite, which may be obtained in steel, is a:
a. supersaturated solid solution of carbon in iron
b. supercooled iron carbide, Fe_3C
c. undercooled FCC structure of austenite
d. superconductor with zero resistivity at low temperature

11.52 Alloying elements produce all but which one of the following effects in steels?
a. They alter the number of atoms in a unit cell of austenite.
b. They increase the depth of hardening in quenched steel.
c. They increase the hardness of ferrite in pearlite.
d. They retard the decomposition of austenite.

11.53 Hardenability may be defined as:
a. resistance to indentation
b. the hardness attained for a specified cooling rate
c. another measure of strength
d. rate of increased hardness

11.54 The linear portion of the stress-strain diagram of steel is known as the:
a. irreversible range
b. scant modulus
c. modulus of elasticity
d. elastic range

11.55 The ultimate (tensile) strength of a material is calculated from:
a. the applied force divided by the true area at fracture
b. the applied force times the true area at fracture
c. the tensile force at the initiation of slip
d. the applied force and the original area

11.56 All but which one of the following statements about slip are correct?
a. Slip occurs most readily along crystal planes that are least densely populated.
b. Slip, or shear along crystal planes, results in an irreversible plastic deformation or permanent set.
c. Ease of slippage is directly related to the number of low-energy slip planes within the lattice structure.
d. Slip is impeded by solution hardening, with odd-sized solute atoms serving as anchor points around which slippage does not occur.

11.57 When a metal is cold worked more severely, all but which one of the following generally occur?

a. The recrystallization temperature decreases.
b. The tensile strength increases.
c. Grains become equiaxed.
d. Slip and/or twinning occur.

11.58 All but which of the following statements about the rusting of iron are correct?

a. Contact with water or oxygen is required for rusting to occur.
b. Halides aggravate rusting, a process that involves electrochemical oxidation-reduction reactions.
c. Contact with a more electropositive metal restricts rusting.
d. Corrosion occurs in oxygen-rich areas.

11.59 All but which of the following statements about mechanical and thermal failure is *TRUE*?

a. Creep is time-dependent, plastic deformation that accelerates at increased temperatures. Stress rupture is the failure following creep.
b. Ductile fracture is characterized by significant amounts of energy absorption and plastic deformation (evidenced by elongation and reduction in cross-sectional area).
c. Fatigue failure from cyclic stresses is frequency-dependent.
d. Brittle fracture occurs with little plastic deformation and relatively low energy absorption.

11.60 The stress intensity factor is calculated from:

a. yield stress and crack depth
b. applied stress and crack depth
c. tensile stress and strain rate
d. crack depth and strain rate

11.61 Service failure from applied loads can occur in all but which of the following cases?

a. Cyclic loading, tension to compression
b. Glide normal to the slip plane
c. Cyclic loading, low tension to higher tension
d. Stage 2 creep

11.62 Brittle failure becomes more common when:

a. the endurance limit is increased
b. the glass-transition temperature is decreased
c. the critical stress intensity factor is increased
d. the ductility-transition temperature is increased

11.63 Where applicable, all but which of the following procedures may reduce corrosion?

a. Avoidance of bimetallic contacts
b. Sacrificial anodes
c. Aeration of feed water
d. Impressed voltages

11.64 A fiber-reinforced rod contains 50 volume percent glass fibers ($E = 70$ GPa, $\sigma_y = 700$ MPa) and 50 volume percent plastic ($E = 7$ GPa). The glass carries what part of a 5000-N tensile load?

a. [(7000 MPa)(0.5)]/[70,000 MPa)(0.5)] = 0.1; $F_{gl} = (0.1)(5000\text{ N}) = 500\text{ N}$

b. [(70,000)(0.5) + 7000(0.5)] = 5000/x; $x = 0.0002$

c. $[(F_{gl}/0.5A)/(F_p / 0.5A)] = (70/7) = 10$; $F_{gl}/(F_p + F_{gl}) = 10F_p/(10F_p + F_p) = 0.91$

d. (70)/[70 + 2(7)] = 0.83

11.65 A design engineer should know what a metal's electromotive potential is because it tells

a. if the metal can be heat treated

b. what the atomic structure is

c. how easy the metal can be formed

d. its resistance to corrosion

11.66 An extensometer is attached to a 15.0 cm length of a test piece of unknown metal in a laboratory. The test section is one cm square. It was then subjected to a tensile force of 20.0 kN and the elongation of the test section was measured. It was found to have increased in length by 0.008475 mm. What was the modulus of elasticity of the metal?

a. 21×10^{10} Pa

b. 6.9×10^{10} Pa

c. 35.4×10^{10} Pa

d. 14.8×10^{10} Pa

11.67 Plastically deforming steel at room temperature

a. increases the ultimate strength.

b. increases the corrosion resistance.

c. makes it nonmagnetic.

d. increases the yield strength.

SOLUTIONS

11.1 **c.** Each step through the periodic table introduces an additional proton and electron to a neutral atom.

11.2 **a.** The number of protons are fixed for an individual element. If the number of protons (and electrons) were varied, the chemical properties would be affected.

11.3 **d.** To separate H_2 into hydrogen atoms, the H-to-H bond would have to be broken, thus requiring energy.

11.4 **a.** Unlike ions attract. FCC metals possess four atoms per unit cell; BCC metals have two.

11.5 **c.** The [111] direction passes diagonally through the unit cell. The distance is $a\sqrt{3}$, which equals two repeat distances.

11.6 **b.** The mass is determined from 27 amu per 6.0×10^{23} atoms. Each cell of four atoms has a volume of $(4r/\sqrt{2})^3$.

11.7 **b.** Assuming spherical atoms, there are four atoms of radius r per unit cell. The cube edge is $4r/\sqrt{2}$.

11.8 **c.** There is a double bond between the two carbons. Each hydrogen is held with a single bond.

11.9 **c.** Density is mass/volume. The mass per FCC unit cell is 4 Au $\times$ (197 g/6.02×10^{23} Au). The volume per FCC unit cell of a metal is (face diagonal/$\sqrt{2}$)^3 or $(4r/\sqrt{2})^3$.

11.10 **b.** Thermoplastic materials are polymerized but soften for molding at elevated temperatures. The polymeric molecules are linear. Thus they include the ethylene-type compounds that are bifunctional (two reaction sites per mer).

Thermosetting materials develop three-dimensional structures that become rigid during processing. For example, phenol is trifunctional and thus forms a network structure. Reheating does not soften them.

11.11 **d.** Benzene is C_6H_6; however, in styrene, one hydrogen is absent at the connection to the C_2H_3–base.

11.12 **d.** Since a cubic crystal has interchangeable x-, y-, and z-axes, the <110> family includes all directions with permutations of 1, 1, and 0 (either + or –). (This is not necessarily true for noncubic crystals.)

11.13 **a.** Since a cubic crystal has interchangeable x-, y-, and z-axes, the {112} family includes all planes with index permutations of 1, 1, and 2 (either + or –). (This is not necessarily true for noncubic crystals.)

11.14 **d.** (b) and (c) involve individual atoms (point imperfections). (a) is a lineal imperfection. Boundaries result from a two-dimensional mismatch of crystal structures.

11.15 **d.** There are two types of dislocations: (1) an edge dislocation may be described as an edge of a missing half-plane of atoms; (2) a screw dislocation is the core of a helix.

11.16 **d.** Unless special efforts are made to grow single crystals, many crystals are nucleated and grow until they encounter neighboring crystals. Each grain is individually oriented.

11.17 **c.** Zinc is sufficiently near copper in size and electrical behavior to proxy for copper in the crystal structure. It is too big for the interstices.

11.18 **b.** The 7.5% copper replaces silver atoms. If it is cooled rapidly, the copper is retained in solid solution. The copper atoms interfere with electron movements within the silver.

11.19 **a.** FCC metals have a higher packing factor than do BCC metals; therefore, with other factors equal, diffusion is reduced.

11.20 **c.**

$$D = D_o e^{-Q/(RT)} = \left(0.022\ \text{cm}^2/\text{s}\right) e^{\frac{-97000\ \text{J/mol}}{8.314\ \text{J/(mol-K)}(1000+273\ \text{K})}} = 2.3 \times 10^{-6}\ \text{cm}^2/\text{s}$$

11.21 **b.** As the grains grow, their volume increases by the third power. Their surface area increases by the square.

11.22 **d.** Metallic crystals are not ionic and do not have discrete ions.

11.23 **a.** The interstitial sites are smaller than the atoms in metals.

11.24 **b.** To balance the charge, each missing Fe^{2+} ion must be compensated by two Fe^{3+} ions. Therefore, out of 100 cation sites, two are vacant, four are Fe^{3+}, and thus 94 are Fe^{2+}.

11.25 **a.** Using a computational basis of four atoms, (1 Mg^{2+} + 1 O^{2-}) + (1 Fe^{2+} + 1 O^{2-}) = (24 + 16) + (56 + 16) = (40 + 72).

11.26 **b.** The elastic strains between atoms along the boundary follow the same relationships as the strains among atoms within the grains.

11.27 **c.** Solid solution increases strength (solution hardening). It also decreases conductivity (and increases resistivity). Sterling silver is 92.5Ag–7.5Cu.

11.28 **b.** When atoms are moved from one site to another, bonds are broken and reconstituted. During transition, an activation energy is required to distort the lattice. Small solute atoms, low-melting-point solvents, and lower atomic packing factors in a lattice all require a lower activation energy. Hence activation energy for diffusion is *directly* proportional to the packing factor.

11.29 **a.** On cooling, curve CI is encountered at approximately 420°C (790°F). That curve is the solubility limit of Zn in that liquid. Zinc in excess of the solubility limit separates as the intermetallic compound, $MgZn_2$, which is plotted as the vertical line EI.

11.30 **d.** A ratio of 67 Zn atoms to 33 Mg atoms is 2-to-1; therefore, $MgZn_2$.

11.31 **b.** The eutectic composition is that of a low-melting liquid saturated with two solids. The 0.8 weight percent composition is a solid, not a liquid, at 727°C.

11.32 **d.** The (ferrite + carbide) area extends across the lower part of the phase diagram from nil carbon to 6.7 carbon. At the left side there is no Fe_3C; at the right side there is only carbide (Fe_3C contains 6.7% carbon). The amount of carbide between the two extremes may be determined by linear interpolation.

11.33 **d.** The eutectic temperature is 780°C (1445°F). That is the lowest temperature at which liquid can exist of equilibrium. At 779°C there are approximately equal amounts of α and β, the two solid solutions.

11.34 **d.** The curves of a phase diagram are solubility limits for the phases within the single-phase regions. Since copper is the solvent for the β structure, β has no upper limit of copper solubility (other than 100%).

11.35 **c.** The diffusion flux is proportional to the concentration gradient. (The other choices cited reduce the activation energy needed for diffusion.)

11.36 **a.** Boundary atoms possess higher energy. Therefore, the boundary is reduced and the grain size is increased at temperatures where the atoms can move. The net movement of the atoms is to the larger grains (with less boundary area).

11.37 **a.** Phase diagrams cannot be used to predict crystalline properties.

11.38 **d.** Above the eutectic temperature, (α + L) can be present concurrently, as can (L + β), but not (α + β). Below the eutectic temperature, (α + β) may be present, but no liquid. As the eutectic temperature is passed during cooling or heating, all three phases may coexist.

11.39 **d.** At 800°C, a 72Ag–28Cu alloy is fully liquid, which therefore has 72–28 composition.

11.40 **d.** During equilibrium cooling, the final liquid for this alloy does not disappear until the eutectic temperature is reached at 780ºC. At that temperature, the liquid composition is 72Ag–28Cu.

11.41 **b.** All α disappears on the right side of the (α + L) field, where the composition is 72Ag–28Cu, or 100g Ag to 39 g Cu.

11.42 **c.** As the temperature is increased, the atoms gain additional energy and can relocate, eliminating the strain energy that accompanies dislocations. *Less time* is required at higher temperature. *Less time* is also required for a highly cold-worked material because there is additional stored energy present.

11.43 **d.** Reduction and polymerization involve chemical reactions. Tempering and recrystallization involve reheating but no shape change.

11.44 **c.** Strength and hardness are increased at the expense of ductility and toughness (opposite of brittleness). The increase is facilitated by microstructures that interfere with dislocation movements. These include a high density of dislocations from plastic deformation, and the presence of many fine, hard particles. Annealing removes dislocations and permits the agglomeration of particles into fewer large particles.

11.45 **d.** The production of tempered martensite is indicated in (c) above. Bainite is formed by isothermally decomposing austenite directly to a microstructure of fine carbide particles within a ferrite matrix. Although the processing differs, the resulting microstructure and properties are nearly identical.

11.46 **a.** Annealing removes the hardness that was introduced by cold work. Quenched and tempered steels are harder with higher carbon contents, because more hard carbide particles are present. Alloying elements perform several hardening functions: They solution-harden the ferrite matrix; they slow down grain growth; and they delay the formation of pearlite, thus permitting more martensite with slower cooling rates (in turn, more tempered martensite may be realized farther below the quenched surface).

11.47 **d.** Stresses will relax below the melting temperature. Tensile stresses facilitate the cracking of brittle materials; compression limits cracking.

11.48 **d.** The reaction occurs only below the eutectoid temperature.

11.49 **c.** While grain boundaries interfere with slip at low temperatures, they facilitate creep at elevated temperatures.

11.50 **d.** The processing step of rapid cooling, such as quenching, retains the structures that existed at higher temperatures, even though a solubility limit is exceeded.

11.51 **a.** Martensite is a transition phase between austenite and ferrite, which retains carbon interstitially. Given an opportunity, the carbon separates as Fe_3C.

11.52 **a.** Alloying elements can dissolve substitutionally in austenite, which remains FCC.

11.53 **b.** Hardenability may be described as the ability, or the ease, by which martensitic hardness is obtained at various cooling rates (as located on an end-quenched, or Jominy, test).

11.54 **d.** The ratio of stress-to-strain is defined as the elastic modulus.

11.55 **d.** $\sigma_u = F/A_0$

11.56 **a.** Slip occurs most readily in directions that have the shortest steps, and along planes that are farthest apart. The latter are automatically the planes that are most densely populated.

11.57 **c.** Cold working—such as rolling, forging, drawing, or extrusion—deforms the material at temperatures below the recrystallization temperature. Strain hardening occurs, increasing both the yield and ultimate strength. Internal strains and minute cracks are introduced as slip or twinning occurs. Ductility, elongation, and the recrystallization temperature are decreased. A preferred grain orientation is introduced in the direction of elongation, and the grains are flattened.

11.58 **d.** Since oxygen is required for rust formation, oxygen-depleted areas become the anode and are corroded. This may lead to pitting, particularly if rust or other corrosion products are accumulated locally to prevent the access of oxygen. Iron and other metals may be protected from corrosion by the presence of a more electropositive metal such as magnesium or zinc. This is the reason for coating steel with zinc to produce galvanized sheet.

11.59 **c.** Although fatigue strength is not sensitive to temperature or loading rates, it is very sensitive to surface imperfections from which cracks originate and propagate.

11.60 **b.** The stress intensity factor is proportional to the applied stress and the square root of the crack depth.

11.61 **b.** Glide occurs parallel to the slip plane.

11.62 **d.** Brittleness exists below T_{DT}.

11.63 **c.** Corrosion commonly occurs in the combined presence of oxygen and water. Protection may be obtained by making a cathode out of the critical part or by avoiding air.

11.64 c. With equal strains, $(\sigma_{gl}/\sigma_p) = E_{gl}/E_p = 10$. Likewise with equal areas, $F_{gl} = 10F_p$, and $F_{gl}/(F_{gl} + F_p) = 10/(10 + 1)$.

11.65 d. The electromotive potential of a metal indicates how susceptible it will be to corrosion in an application.

11.66 c. The modulus of elasticity is measured in pascals, or newtons per square meter. The test showed a strain of

$\varepsilon = 0.008475/150 = 0.000565$ mm/mm for a stress of
$S = 20{,}00/0.0001 = 200$ MPa $\quad S = \varepsilon \times E$ so
$E = 20 \times 10^7/(5.65 \times 10^{-4}) = 3.54 \times 10^{11}$ Pa

11.67 d. Cold working, or plastically deforming steel at room temperature increases the yield strength and reduces the ductility.

CHAPTER 12

Fluid Mechanics

Gary R. Crossman

OUTLINE

Fluid mechanics is the study of fluids at rest or in motion. The topic is generally divided into two categories: *liquids* and *gases*. Liquids are considered to be incompressible, and gases are compressible. The treatment of incompressible fluids and compressible fluids each has its own group of equations. However, there are times when a gas may be treated as incompressible (or at least uncompressed). For example, the flow of air through a heating duct is one such case. This chapter will concentrate on incompressible fluids.

FLUID PROPERTIES

Thermodynamic properties are important in incompressible fluid mechanics. Those of particular importance are density, specific gravity, specific weight, viscosity, and pressure. Temperature is also important but is primarily used in finding other properties such as density and viscosity in tables or graphs.

Density

The **density**, ρ, is the mass per unit volume and is the reciprocal of the specific volume, a property used in thermodynamics:

$$\rho = \frac{m}{V} = \frac{1\,\text{kg}}{v\,\text{m}^3}$$

Specific Gravity

The **specific gravity,** SG, is defined by the following equation:

$$\text{SG} = \frac{\rho\left(\dfrac{\text{kg}}{\text{m}^3}\right)}{\dfrac{1000\ \text{kg}}{\text{m}^3}}$$

where 1000 kg/m3 is the density of water at 4°C.

In many cases the specific gravity of a liquid is known or found from tables and must be converted to density using this equation.

Specific Weight

The **specific weight**, γ, of a fluid is its weight per unit volume and is related to the density as follows:

$$\gamma = \frac{W}{V} = \rho\left(\frac{g}{g_c}\right)\frac{\text{N}}{\text{m}^3}$$

where g = local acceleration of gravity, $\frac{\text{m}}{\text{s}^2}$, and g_c = gravitational constant:

$$g_c = \frac{\text{kg}\bullet\text{m}}{\text{N}\bullet\text{s}^2}$$

The density of water at 4°C is 1000 kg/m³. Its specific weight at sea level ($g = 9.81$ m/s²) is calculated as follows:

$$\gamma = \rho\frac{g}{g_c} = 1000\frac{\text{kg}}{\text{m}^3} = \frac{9.81\frac{\text{m}}{\text{s}^2}}{\frac{\text{kg}\bullet\text{m}}{\text{N}\bullet\text{s}^2}} = 9810\frac{\text{N}}{\text{m}^3}$$

The density, specific gravity, and specific weight of a liquid are generally considered to be constant, with little variation, over a wide temperature range.

Viscosity

The **viscosity** of a fluid is a measure of its resistance to flow; the higher the viscosity the more resistance to flow. Water has a relatively low viscosity, and heavy fuel oils have a high viscosity. The **dynamic (absolute) viscosity**, μ, of a fluid is defined as the ratio of shearing stress to the rate of shearing strain. In equation form:

$$\mu = \frac{\tau}{\frac{dV}{dy}} \frac{\text{N}\bullet\text{s}}{\text{m}^2}\left(\frac{\text{kg}}{\text{m}\bullet\text{s}}\right)$$

where τ = shearing stress (force per unit area), N/m², and dV/dy = rate of shearing strain, $1/s$.

Fluids may be classified as Newtonian or non-Newtonian. Newtonian fluids are those in which dV/dy in the above equation can be considered to be constant for a given temperature. Thus the shearing stress, τ (horizontal force divided by the surface area), of a plate on a thin layer, δ, of a Newtonian fluid, as shown in Figure 12.1, may be found from

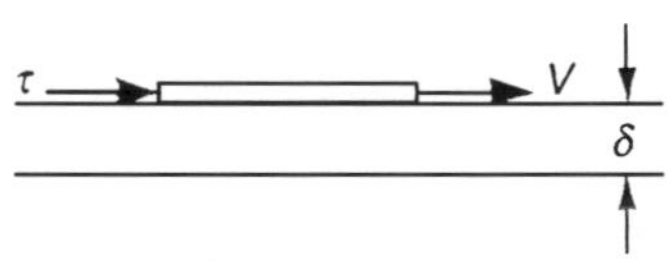

Figure 12.1

$$\tau = \mu\frac{dV}{dy} = \frac{\mu V}{\delta}$$

where V = velocity, m/s, and δ = thickness, m.

Most common fluids (liquids), such as water, oil, gasoline, and alcohol, are classified as Newtonian fluids.

The **kinematic viscosity** is defined by

$$v = \frac{\mu}{\rho}\frac{\text{m}^2}{\text{s}}$$

Both the dynamic and kinematic viscosities are highly dependent on temperature. The viscosity of most liquids decreases significantly (orders of magnitude) with increases in temperature, while the viscosity of gases increases mildly with increases in temperature. The viscosity of any gas is less than the viscosity of any liquid. Viscosities are generally found in tables and graphs.

The definition of viscosity assists in the development of the engineering definition of a fluid as follows:

> A fluid is a substance that will deform readily and continuously when subjected to a shear force, no matter how small the force.

Pressure

Pressure, p, is the force per unit area of a fluid on its surroundings or vice versa. Pressure may be specified using two different datums. Absolute pressure, P_{abs}, is measured from absolute zero or a complete vacuum (void). At absolute zero there are no molecules and a negative absolute pressure does not exist. Absolute pressures are needed for ideal gas relations and in compressible fluid mechanics. Gage pressure, p_{gage}, on the other hand, uses local atmospheric pressure as its datum. Gage pressures may be positive (above atmospheric pressure) or negative (below atmospheric pressure). Negative gage pressure is also called vacuum. A complete vacuum occurs at a negative gage pressure that is equivalent to the atmospheric pressure or at absolute zero.

The relationship between absolute pressure and gage pressure is as follows:

$$P_{abs} = p_{gage} + p_{atm} \frac{\text{N}}{\text{m}^2} (\text{Pa})$$

Actually, the pressure is usually expressed in kN/m^2 or kPa but should be converted to these units for use in most equations.

Example 12.1

A pressure gage measures 50 kPa vacuum in a system. What is the absolute pressure if the atmospheric pressure is 101 kPa?

Solution

Change vacuum to a negative gage pressure:

$$p_{abs} = -50 \text{ kPa} + 101 \text{ kPa} = 51 \text{ kPa} \bullet 1000 = 51{,}000 \text{ Pa}$$

Most pressure-measuring devices measure gage pressure. For incompressible fluid dynamics, gage pressure may be used in most equations. This capability simplifies equations significantly when one or more pressures in the system are atmospheric or $p_{gage} = 0$.

Surface Tension

Surface tension is another property of liquids. It is the force that holds a water droplet or mercury globule together, since the cohesive forces of the liquid are more than the adhesive forces of the surrounding air. The surface tension (or surface tension coefficient), σ, of liquids in air is available in tables and can be used to calculate the internal pressure, p, in a droplet from

$$p = \frac{4\sigma}{d}$$

where σ = surface tension of the liquid, kN/m, and d = droplet diameter, m. Values of surface tension for various liquids are found in tables as a function of the surrounding medium (air, etc.) and the temperature.

Surface tension is also the property that causes a liquid to rise (or fall) in a capillary tube. The amount of rise (or fall) depends on the liquid and the capillary tube material. When *adhesive* forces dominate, the liquid will rise—as with water. When cohesive forces dominate, it will fall—as with mercury. The capillary rise, h, can be calculated from the following equation:

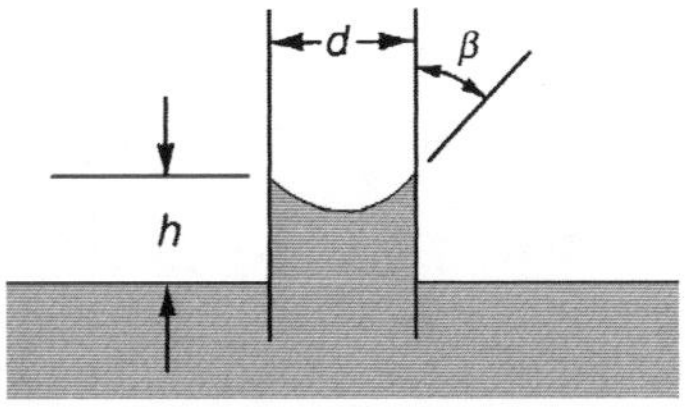

Figure 12.2

$$h = \frac{4\sigma \cos\beta}{\gamma d}$$

where β = angle made by the liquid with the tube wall, and d = diameter of capillary tube, as shown in Figure 12.2.

The angle, β, varies with different liquid/tube material combinations and is found in tables. β is within the range 0 to 180°. When $\beta > 90°$, h will be negative.

FLUID STATICS

Pressure-Height Relationship

For a static liquid, the pressure increases with depth (decreases with height) according to the following relationship

$$p_2 - p_1 = -\gamma(Z_2 - Z_1) = \gamma h$$

where h = depth from Point 1 to Point 2.

If p_1 is at the surface of a liquid that is open to the atmosphere, then the gage pressure at Point 2 is found from

$$p_2 = p = \gamma h$$

Example 12.2

Calculate the gage pressure at a depth of 100 meters in seawater, for which $\gamma = 10.1\ \text{kN/m}^3$.

Solution

$$p = \gamma h = \left(\frac{10.1\ \text{kN}}{\text{m}^3}\right)\left(100\ \text{m}\right) = 1010\frac{\text{kN}}{\text{m}^2} = 1010\ \text{kPa}$$

Manometers

A manometer is a device used to measure moderate pressure differences using the pressure-height relationship. The simplest manometer is the U-tube shown in Figure 12.3. The pressure difference between System 1 and System 2 is found from

$$p_1 - p_2 = \gamma_m h_m + \gamma_2 h_2 - \gamma_1 h_1$$

where γ_m, γ_1, and γ_2 = specific weight of manometer fluid, fluid in System 1, and fluid in System 2, respectively, and h_m, h_1, h_2 = depths as shown. If the fluids in both systems are gases and the manometer fluid is any liquid, then $\gamma_m >> \gamma_1$ or γ_2 and the equation simplifies to

$$p_1 - p_2 = \gamma_m h_m = \gamma h$$

If System 2 were the atmosphere ($p_2 = 0_{gage}$) then

$$p_1 = \gamma h = \text{gage pressure in System 1}$$

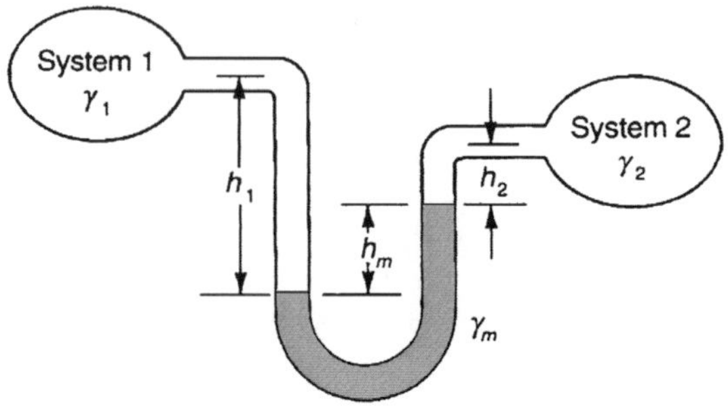

Figure 12.3

If System 1 were the atmosphere ($p_1 = 0$) then

$$p_2 = -\gamma h = \text{gage pressure in System 2}$$

The gage pressure in System 2 would be negative or a vacuum. Manometers are commonly used to measure system pressures between –101.3 kPa and +101.3 kPa. In many cases, particularly where the gage pressure is negative, the pressure may be given in millimeters of a fluid, and the equation above used to convert it to standard units.

Example 12.3

A system gage pressure is given as 500 millimeters of mercury vacuum (mm Hg vac). What is the gage pressure in kPa? The specific gravity of mercury is 13.6.

Solution

The pressure is $p = p_2 = -\gamma h$ since vacuum is a negative gage pressure:

$$\gamma_m = (13.6)\left(9.81\frac{\text{kN}}{\text{m}^3}\right) = 133.4\frac{\text{kN}}{\text{m}^3}$$

$$p = -\gamma h = -133.4\frac{\text{kN}}{\text{m}^3} \times 0.5\,\text{m} = -66.7\ \text{kPa}$$

The conversion factor from millimeters of mercury to N/m^2 (pascals) is 133.4.

A barometer is a special type of mercury manometer. In this case one leg of the U-tube is very wide. If we can adjust the scale on the narrow leg so that zero is at the level of the large leg, then the narrow leg will read the atmospheric pressure impinging on the wide leg corrected by the vapor pressure of the mercury. A barometer is shown in Figure 12.4.

There are several other types of manometers. A compound manometer consists of more than one U-tube in series between one system and another. The equation for $p_1 - p_2$ may be developed by starting at System 2 and adding γh's going downward and subtracting γh's going upward until System 1 is reached as follows:

$$p_2 + \sum \gamma h \text{ (downward)} - \sum \gamma h \text{ (upward)} = p_1$$

An inclined manometer is used to measure small pressure differentials. The measurement along the manometer must be multiplied by the sine of the angle of incline. An inclined manometer is generally "single leg," similar to the barometer previously described, and is shown in Figure 12.5. The pressure difference is found from $p_1 - p_2 = \gamma_m L \sin \alpha$, where L = length along manometer leg and α = angle of inclination.

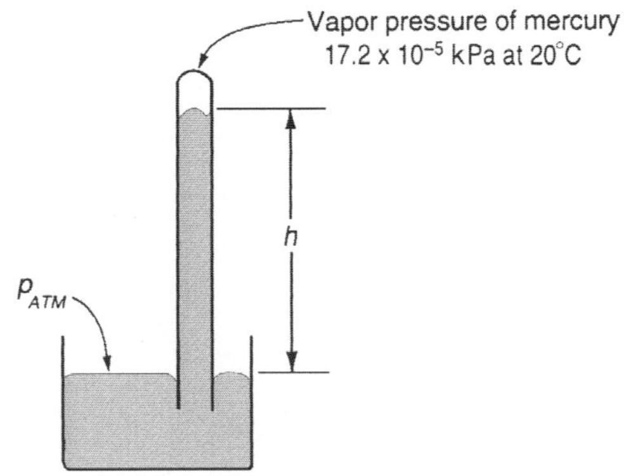

Figure 12.4

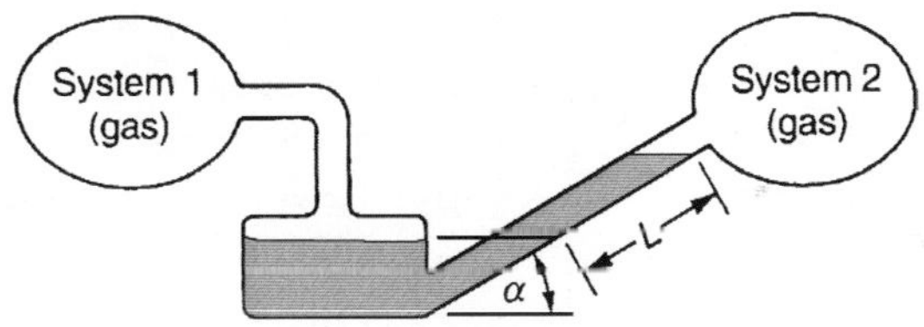

Figure 12.5

Forces on Flat Submerged Surfaces

A flat surface of arbitrary shape below a liquid surface is shown in Figure 12.6. Thew resultant force, F, on one side of the flat surface acts perpendicular to the surface. Its magnitude and location may be calculated from the following equations:

$$F = (p_0 + \gamma h_c)A$$

and

$$h_p = h_c + \frac{I_c \sin^2 \alpha}{\left(\frac{p_0}{\gamma} + h_c\right)A}$$

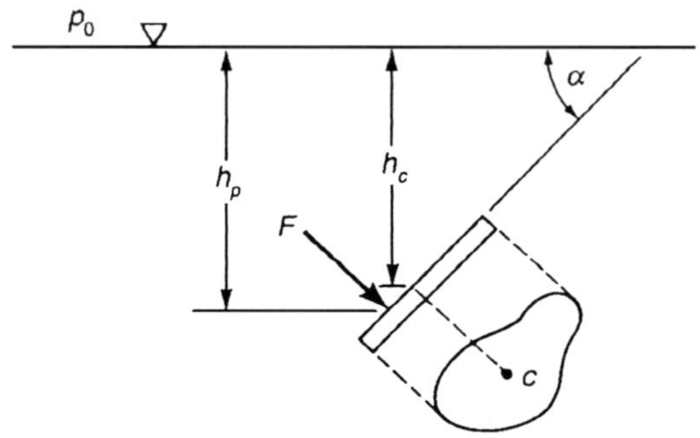

Figure 12.6

where

F = resultant force on the flat surface, N
p_0 = gage pressure on the surface, Pa
γ = specific weight of the fluid, N/m^3
h_c = vertical distance from fluid surface to the centroid of the flat surface area, m
A = area of flat surface, m^2
h_p = vertical distance from fluid surface to the center of pressure of the flat surface (where the equivalent, concentrated force acts), m
I_c = moment of inertia of the flat surface about a horizontal axis through its centroid, m^4
α = angle that the inclined flat surface makes with the horizontal surface

The values of h_c, the location of the centroid axis from the base $\bar{y}$, and the moment of inertia about the centroid I_c for common geometric shapes, such as rectangles, triangles, and circles, may be determined from existing tables. Typical values are presented in Table 12.1.

For the common case when p_0 is atmospheric pressure ($p_0 = 0$), the equations simplify to

$$F = \gamma h_c A$$

and

$$h_p = h_c + \frac{I_c \sin^2 \alpha}{h_c A}$$

From the above equations it is apparent that the center of pressure is always below the centroid except when the surface is horizontal ($\alpha = 0$). In that case the center of pressure is at the centroid. The deeper the flat surface is located below the fluid surface, the closer the center of pressure is to the centroid.

The pressure profile on the flat surface is generally trapezoidal (triangular, if the flat surface pierces the surface of a fluid exposed to atmospheric pressure). The slope of the pressure profile is equivalent to the specific weight of the fluid. Examples are shown in Figure 12.7.

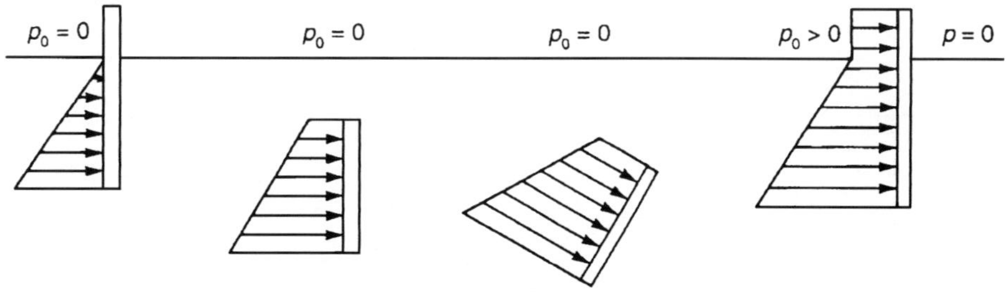

Figure 12.7

Table 12.1 Areas, centroids, and moments of inertia for selected areas

Section	Area of Section, A	Distance to Centroidal Axis, $\bar{y}$	Moment of Inertia about Centroidal Axis, I_c
Rectangle 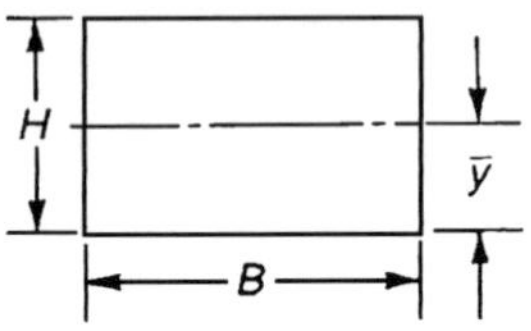	BH	$H/2$	$BH^3/12$
Triangle 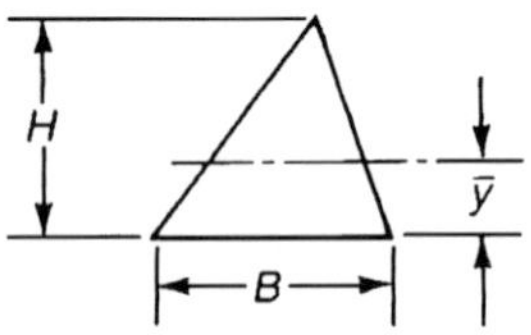	$BH/2$	$H/3$	$BH^3/36$
Circle 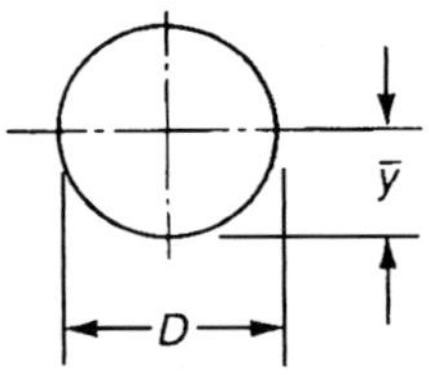	$\pi D^2/4$	$D/2$	$\pi D^4/64$
Ring 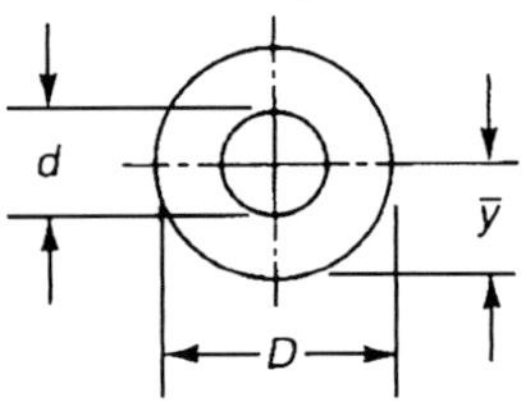	$\dfrac{\pi(D^2 - d^2)}{4}$	$D/2$	$\dfrac{\pi(D^4 - d^4)}{64}$
Semicircle 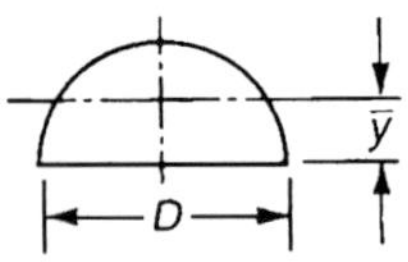	$\pi D^2/8$	$0.212D$	$(6.86 \times 10^{-3})\, D^4$
Quadrant 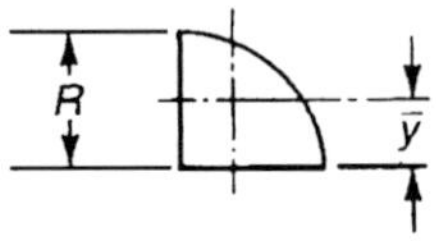	$\pi D^2/16$ $\pi R^2/4$	$0.212D$ $0.424R$	$(3.43 \times 10^{-3})D^4$ $(5.49 \times 10^{-2})R^4$
Trapezoid 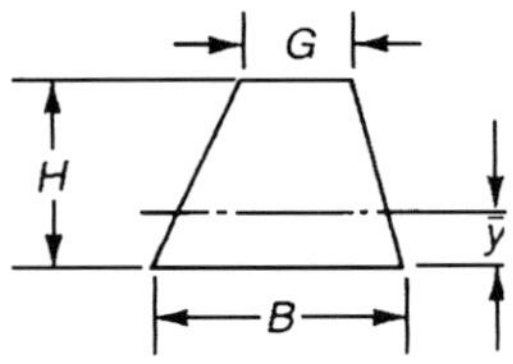	$\dfrac{H(G+B)}{2}$	$\dfrac{H(2G+B)}{3(G+B)}$	$\dfrac{H^3(G^2+4GB+B^2)}{36(G+B)}$

Example 12.4

A vertical side of a saltwater tank contains a round viewing window 60 cm in diameter with its center 5 meters below the liquid surface. If the specific weight of the saltwater is 10 kN/m^3, find the force of the water on the window and where it acts.

Solution

Assume atmospheric pressure on the liquid surface, $p_0 = 0$.

$$d = 60\,\text{cm} = 0.6\,\text{m}, \quad A = \frac{\pi(0.6)^2}{4} = 0.283\,\text{m}^2$$

$$F = \gamma h_c A = 10\frac{\text{kN}}{\text{m}^3} \times 5\,\text{m} \times 0.283\,\text{m}^2 = 14.14\,\text{kN}$$

$$I_c = \frac{\pi d^4}{64} = \frac{\pi(0.6\,\text{m})^4}{64} = 0.00636\,\text{m}^4$$

$$\alpha = 90^\circ, \quad \sin\alpha = 1$$

$$h_p = h_c + \frac{I_c \sin^2\alpha}{h_c A} = 5\,\text{m} + \frac{0.00636\,\text{m}^4(1)^2}{5\,\text{m}\bullet 0.283\,\text{m}^2} = 5.0045\,\text{m}$$

Example 12.5

In many cases problems involving fluid forces on flat surfaces are combined with a statics problem. The fluid force is just another force to be added into the statics equation.

In the previous example, suppose the circular window were hinged at the top with some sort of clamp at the bottom (Exhibit 1). What force, P, would be required of the clamp to keep the window closed?

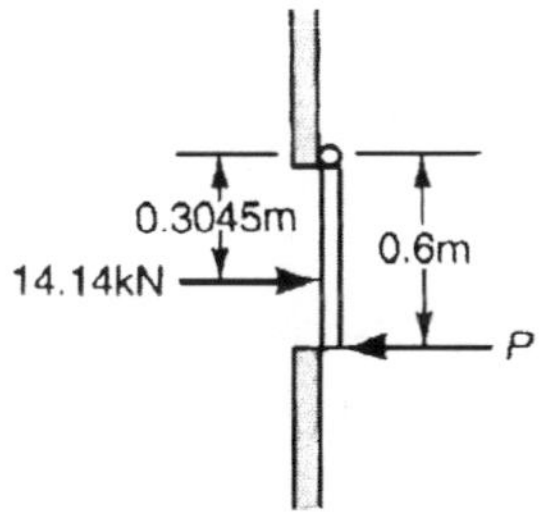

Exhibit 1

Solution

From Example 12.4 calculations, the force of the water is 14.14 kN located 5.0045 m below the fluid surface. The hinge is located 5 m – 0.6 m/2 = 4.7 m below the water surface. Thus the force location is 5.0045 m – 4.7 m = 0.3045 m below the hinge.

Summing moments about hinge,

$$\sum M_H = 0.6\,\text{m}\bullet P - 0.3045\,\text{m}\bullet 14.14\,\text{kN} = 0$$

$$P = \frac{0.3045\,\text{m}\bullet 14.14\,\text{kN}}{0.6\,\text{m}} = 7.18\,\text{kN}$$

Buoyancy

In addition to the force of gravity, or weight, all objects submerged in a fluid are acted on by a buoyant force, F_B. The buoyant force acts upward and is equal to the weight of the fluid displaced by the object. This is known as Archimedes' principle. The upward buoyant force also acts through the center of gravity (or centroid) of the displaced volume, known as the center of buoyancy, B. Thus

$$F_B = \gamma_f V_D$$

where F_B = buoyant force, N; γ_f = specific weight of the fluid, N/m^3; and V_D = volume displaced by the object, m^3.

For a freely floating object (no external forces) the weight of the object (acting downward) is equal to the buoyant force on the object (acting upward) or

$$W = F_B = \gamma_f V_D$$

This equation is useful in determining what part of an object will float below the surface of a liquid. For objects partially submerged in a liquid and a gas, the buoyant force of the gas is usually neglected. However, the buoyant force on a totally submerged body in a gas is very important in the study of balloons, dirigibles, and so on.

Example 12.6

A wooden cube that is 15 centimeters on each side with a specific weight of 6300 N/m^3 is floating in fresh water (γ = 9,810 N/m^3) (Exhibit 2). What is the depth of the cube below the surface?

Solution

$$W = F_B = \gamma_f V_D$$

There are actually two buoyant forces on the cube, that of the water on the volume below the surface and that of the air on the volume above the surface. Neglecting the buoyant force of the air and rearranging the equation:

$$V_D = \frac{W}{\gamma_f} = \frac{\gamma_C V_C}{\gamma_f} = \frac{(6300\ \text{N/m}^3)(0.15\ \text{m})^3}{9810\ \text{N/m}^3} = 0.00217\ \text{m}^3$$

$$V_D = (0.15)^2 \bullet d = 0.00217\ \text{m}^3$$

$$d = \frac{0.00217\ \text{m}^3}{.0225\ \text{m}^2} = .0964\ \text{m} = 9.64\ \text{cm}$$

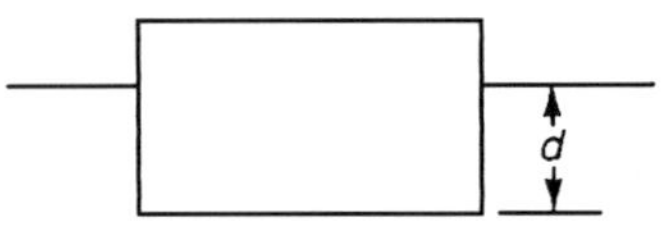

Exhibit 2

Neutral buoyancy exists when the buoyant force equals the weight when an object is completely submerged. The object will remain at whatever location it is placed below the fluid surface.

In the case of an object floating at the interface of two liquids, the total buoyant force is equal to the sum of the buoyant forces on the object created by each fluid on that part that is immersed. When external forces also act on a submerged or partially submerged object, they must be included in the force balance on the object. The force balance equation then becomes

$$W + \sum F_{\text{ext}}(\text{down}) = F_B + \sum F_{\text{ext}}(\text{up})$$

If weight is added to an object internally, or possibly on top of a partially submerged object, it will only affect the weight of the object. But if the weight is added externally, beneath the surface of the fluid, its buoyant force as well as its weight must be considered.

Example 12.7

If, in Example 12.6, a concrete weight (anchor) is added to the bottom of the cube externally, what anchor volume, V_A, would be required to make the cube float neutrally (below the surface). The specific weight of the concrete, γ_c, is 24 kN/m³.

Solution

Let the subscript C denote the properties of the cube and subscript A denote those of the anchor. Summing forces vertically,

$$W_C + W_A = F_{BC} + F_{BA}$$

$$\gamma_C V_C + \gamma_A V_A = \gamma_f V_D + \gamma_f V_A$$

Solving for V_A,

$$V_A = \frac{\gamma_f V_D - \gamma_C V_C}{\gamma_A - \gamma_f}$$

But for neutral buoyancy, the displaced volume, V_D, is equal to the total volume of the cube, V_C, and

$$V_A = \frac{(\gamma_f - \gamma_C) \bullet V_C}{\gamma_A - \gamma_f} = \frac{(9810 - 6300)\frac{\text{N}}{\text{m}^3} \bullet (0.15 \text{ m})^3}{(24{,}000 - 9810)\frac{\text{N}}{\text{m}^3}} = 8.34 \times 10^{-4} \text{ m}^3$$

THE FLOW OF INCOMPRESSIBLE FLUIDS

The Continuity Equation

Most problems in fluid mechanics involve steady flow, meaning that the amount of mass in a system does not change with time. This is generally written as

$$\dot{m}_1 = \dot{m}_2 = \dot{m} = \text{mass rate}$$

where the subscript 1 denotes the entrance and the subscript 2 denotes the exit of the system. The mass rate may be written in terms of fluid properties:

$$\dot{m} = \rho A V$$

where ρ = fluid density, A = cross sectional area of flow, and V = average velocity of the fluid. Thus,

$$\rho_1 A_1 V_1 = \rho_2 A_2 V_2$$

and, since $\rho_1 = \rho_2 = \rho$ for an incompressible fluid, then

$$A_1 V_1 = A_2 V_2 = Q = \text{volume flow rate}$$

This equation is useful in determining one velocity when another velocity in the system is known.

Example 12.8

Water is flowing in a 5 centimeter diameter pipe at a velocity of 5 m/s (Exhibit 3). The pipe expands to a 10-centimeter diameter pipe. Find the velocity in the 10-centimeter diameter pipe and the flow rate in liters (L) per minute.

Solution

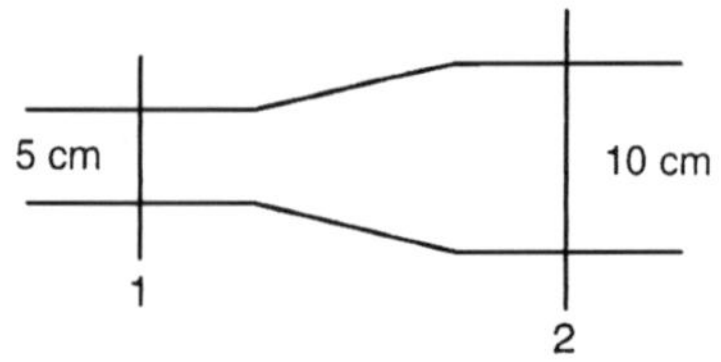

Exhibit 3

$$A_1V_1 = A_2V_2$$

$$V_2 = \frac{A_1}{A_2}V_1 = \frac{\pi d_1^2/4}{\pi d_2^2/4}V_1 = \left(\frac{d_1}{d_2}\right)^2 V_1 = \left(\frac{5\text{ cm}}{10\text{ cm}}\right)^2\left(5\frac{\text{m}}{\text{s}}\right) = 1.25\text{ m/s}$$

$$Q = A_1V_1 = \frac{\pi d_1^2}{4}V_1 = \frac{\pi(.05\text{ m})^2}{4}\bullet 5\frac{\text{m}}{\text{s}} = 0.00982\frac{\text{m}^3}{\text{s}}$$

$$Q = 0.00982\frac{\text{m}^3}{\text{s}}\frac{1000\text{ L}}{\text{m}^3}\bullet\frac{60\text{ s}}{\text{min}} = 589\text{ L/min}$$

In most cases the flow area may be calculated using the diameter. In some cases the nominal pipe size is known, such as a 4-inch Schedule 40 pipe. The exact inside dimensions of Schedule 40 and other pipes as well as dimensions for steel and copper tubing can be found in existing tables.

Reynolds Number

The **Reynolds number**, Re, is a dimensionless flow parameter that helps describe the nature of flow. It is sometimes defined as the ratio of dynamic forces to viscous forces. In terms of other fluid properties, it is defined as

$$\text{Re} = \frac{\rho Vd}{\mu} = \frac{Vd}{\nu}$$

where V = fluid velocity, and d = characteristic length (diameter for pipes).

If the Reynolds number is below 2300, flow is laminar and occurs in layers with no mixing of adjacent fluid. Re = 2300 is known as the critical Reynolds number. Above the critical Reynolds number mixing begins to occur, and the flow becomes turbulent. As the Reynolds number increases, the flow becomes more turbulent.

For pipe flow with a circular cross section the Reynolds number may also be calculated from

$$\text{Re} = \frac{4\rho Q}{\pi d\mu} = \frac{4Q}{\pi d\nu}$$

The Reynolds number is a significant indicator of the influence of friction on the flow that occurs in pipes and other conduits as well as through flow meters. It is also important in the application of dynamic similarity to modeling and many other areas of fluid mechanics.

Example 12.9

For the pipe in Example 12.8, calculate the Reynolds number in the 5-centimeter diameter section of pipe. The kinematic viscosity of the water is 1.12×10^{-6} m²/s.

Solution

$$\text{Re} = \frac{Vd}{v} = \frac{5\frac{\text{m}}{\text{s}} \bullet (0.05\ \text{m})}{1.12 \times 10^{-6}\ \frac{\text{m}^2}{\text{s}}} = 2.2 \times 10^5$$

The flow is well into the turbulent regime.

The Energy Equation

The energy equation in fluid mechanics is similar to that used in thermodynamics. Instead of each energy term having the traditional units such as kJ/kg, energy is expressed in meters of head. For instance, kinetic energy is called velocity head. The general energy equation between two points in a system for incompressible steady flow (mass and energy in the system or at a point do not vary with time) is given by the following expression:

$$\frac{p_1}{\gamma} + \frac{V_1^2}{2g} + Z_1 + h_A - h_R = \frac{p_2}{\gamma} + \frac{V_2^2}{2g} + Z_2 + h_f$$

where

$\frac{p_1}{\gamma}, \frac{p_2}{\gamma}$ = pressure heads at Points 1 and 2

$\frac{V_1^2}{2g}, \frac{V_2^2}{2g}$ = velocity heads at Points 1 and 2

Z_1, Z_2 = potential or elevation heads at Points 1 and 2

h_A, h_R = the head added (pump) or removed (turbine) mechanically

h_f = head loss from friction in the pipe and fittings between Points 1 and 2

The energy equation in fluid mechanics assumes no heat transfer or changes in temperature. This equation, including its reduced forms, will solve most energy-related problems in fluid mechanics when used in conjunction with the continuity equation.

Bernoulli's Equation

Whereas Bernoulli's equation is usually derived from momentum principles using vector calculus, it can also be produced from the energy equation by introducing two additional restrictions to those of incompressible, steady flow and no heat transfer. If we restrict the energy equation to systems with no mechanical energy addition or removal (no pump or turbine) and with no (or negligible) friction losses, Bernoulli's equation is produced,

$$\frac{p_1}{\gamma} + \frac{V_1^2}{2g} + Z_1 = \frac{p_2}{\gamma} + \frac{V_2^2}{2g} + Z_2$$

Bernoulli's equation can be used to solve a variety of problems.

Example 12.10

Referring again to Example 12.8, calculate the pressure just after the expansion to the 10-centimeter diameter pipe if the pressure in the 5-centimeter pipe is 300 kPa. Friction is negligible. The specific weight of water is 9.81 kN/m³.

Solution

For the horizontal orientation $Z_1 = Z_2$ and Bernoulli's equation reduces to

$$\frac{p_1}{\gamma} + \frac{V_1^2}{2g} = \frac{p_2}{\gamma} + \frac{V_2^2}{2g}$$

$$p_2 = p_1 + \gamma\left(\frac{V_1^2 - V_2^2}{2g}\right)$$

$$p_2 = 300 \text{ kPa} + 9.81\frac{\text{kN}}{\text{m}^3}\left[\frac{(5^2 - 1.25^2)\frac{\text{m}^2}{\text{s}^2}}{2 \bullet 9.81\frac{\text{m}}{\text{s}^2}}\right] = 311.7 \text{ kPa}$$

Other Forms of the Energy Equation

An important rearrangement of the energy equation is to solve for the head added by a pump or removed by a turbine. For the head added by a pump:

$$h_A = \frac{p_2 - p_1}{\gamma} + \frac{V_2^2 - V_1^2}{2g} + Z_2 - Z_1 + h_f$$

Example 12.11

A pump is being used to deliver 130 L/min of hot water from a tank through 15 meters of 2.5-cm diameter, smooth pipe, exiting through a 1.0 cm diameter nozzle 3 meters above the level of the tank as shown in Exhibit 4. The head loss from friction of the pipe is 8.33 m. The specific weight of the hot water is 9.53 kN/m³. Calculate the head delivered to the water by the pump.

Solution

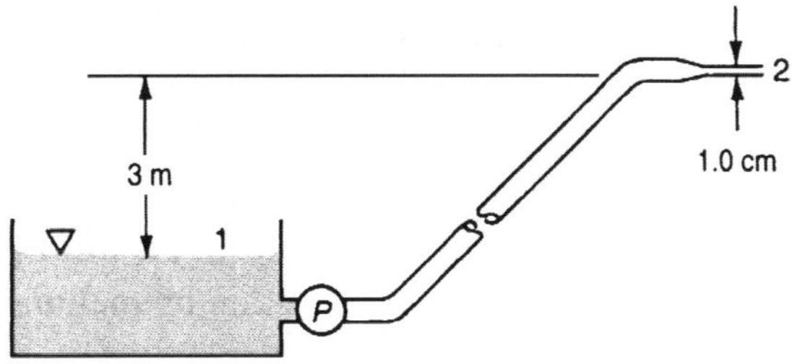

Exhibit 4

Select Points 1 and 2 as shown.

$$h_A = \frac{p_2 - p_1}{\gamma} + \frac{V_2^2 - V_1^2}{2g} + Z_2 - Z_1 + h_f$$

$$p_2 = p_1 = 0, \quad V_1 = 0, \quad Z_2 = 3\text{ m}, \quad Z_1 = 0, \quad h_f = 8.33\text{ m}$$

$$V_2 = \frac{Q}{A_2} = \frac{130\dfrac{\text{L}}{\text{min}} \bullet \dfrac{1\text{ m}^3}{1000\text{ L}} \bullet \dfrac{\text{min}}{60\text{ s}}}{\dfrac{\pi(.01)^2}{4}\text{m}^2} = 27.6\frac{\text{m}}{\text{s}}$$

$$h_A = \frac{\left(27.6\dfrac{\text{m}}{\text{s}}\right)^2}{2 \bullet 9.81\dfrac{\text{m}}{\text{s}^2}} + 3\text{ m} + 8.33\text{ m} = 50.2\text{ m}$$

In this problem the head of the pump serves three purposes: to increase the velocity of the water, raise its level, and overcome friction.

Pump and Turbine Power and Efficiency

The power delivered by a pump to a fluid or removed by a turbine from the fluid is given by the following:

$$P - \gamma Q h_A - \gamma Q h_R$$

The term γQ is the weight rate of flow. In the SI system, the units of power will usually be kN-m/s or kilowatts.

In selecting a pump or turbine, its efficiency is important. The efficiency may be calculated from

$$\eta_P = P / \dot{W} \bullet 100$$

$$\eta_T = \dot{W} / P \bullet 100$$

where η_p, η_T = pump and turbine efficiency, respectively, %, P = fluid power, and $\dot{W}$ = mechanical (or shaft) power actually delivered to the pump or by the turbine.

Example 12.12

From Example 12.11, calculate the power delivered to the water by the pump. If the efficiency of the pump is 60%, calculate the mechanical power delivered to the pump ($\dot{W}$).

Solution

$$P = \gamma Q h_A = 9.53 \frac{\text{kN}}{\text{m}^3} \bullet 130 \frac{\text{L}}{\text{min}} \bullet \frac{1\,\text{m}^3}{1000\,\text{L}} \bullet \frac{\text{min}}{60\,\text{s}} \bullet 50.2\ \text{m} = 1.04\ \text{kW}$$

$$\eta = \frac{P}{\dot{W}} \bullet 100 \quad \text{or} \quad \dot{W} = \frac{P \bullet 100}{\eta} = \frac{1.04\ \text{kW} \bullet 100}{60} = 1.73\ \text{kW}$$

Head Loss from Friction in Pipes

Most of the terms in the energy equation will be known or calculated from the energy equation in conjunction with the continuity equation. Even the head loss from friction may be calculated if all other parameters are known. For example, if one wished to know the friction loss in a particular horizontal length of pipe, or in a fitting with equal entrance and exit areas, it could be calculated from the energy equation. Thus, $Z_1 = Z_2$, and since $A_1 = A_2$, then $V_1 = V_2$ and the energy equation becomes

$$\frac{p_1}{\gamma} = \frac{p_2}{\gamma} + h_f$$

If the pressure drop ($p_1 - p_2$) were known or measured, then

$$h_f = \frac{p_1 - p_2}{\gamma}$$

For most applications of the energy equation, the head loss from friction must be known and substituted into the energy equation to solve for an unknown pressure or height, pump or turbine head, or flow rate. For pipe flow, the head loss may be calculated from the Darcy equation:

$$h_f = f \frac{L}{d} \frac{V^2}{2g}$$

where h_f is the head loss from friction in a pipe of length L and diameter d, and f is the friction factor that is a function of the Reynolds number, Re, and the pipe relative roughness, ε / d.

The friction factor, f, can be found from the Moody diagram where f is plotted as a function of the Reynolds number and appears as a family of curves for different values of relative roughness, ε / d. *Relative roughness* is the roughness factor of the pipe, ε, divided by the pipe diameter, d. Typical roughness factors are shown in Table 12.2. Glass and plastic have the smallest roughness factors and are shown by the "smooth" curve on the Moody diagram. The Moody diagram is presented in Figure 12.8.

For Reynolds numbers below 2300 (laminar flow) the friction factor is independent of the relative roughness and may be calculated from the following relationship:

$$f = \frac{64}{\text{Re}}$$

Table 12.2 Moody pipe roughness

Material	Roughness, ε (m)
Glass, plastic	Smooth
Copper, brass, lead (tubing)	1.5×10^{-6}
Cast iron—uncoated	2.4×10^{-4}
Cast iron—asphalt coated	1.2×10^{-4}
Commercial steel or welded steel	4.6×10^{-5}
Wrought iron	4.6×10^{-5}
Riveted steel	1.8×10^{-3}
Concrete	1.2×10^{-3}

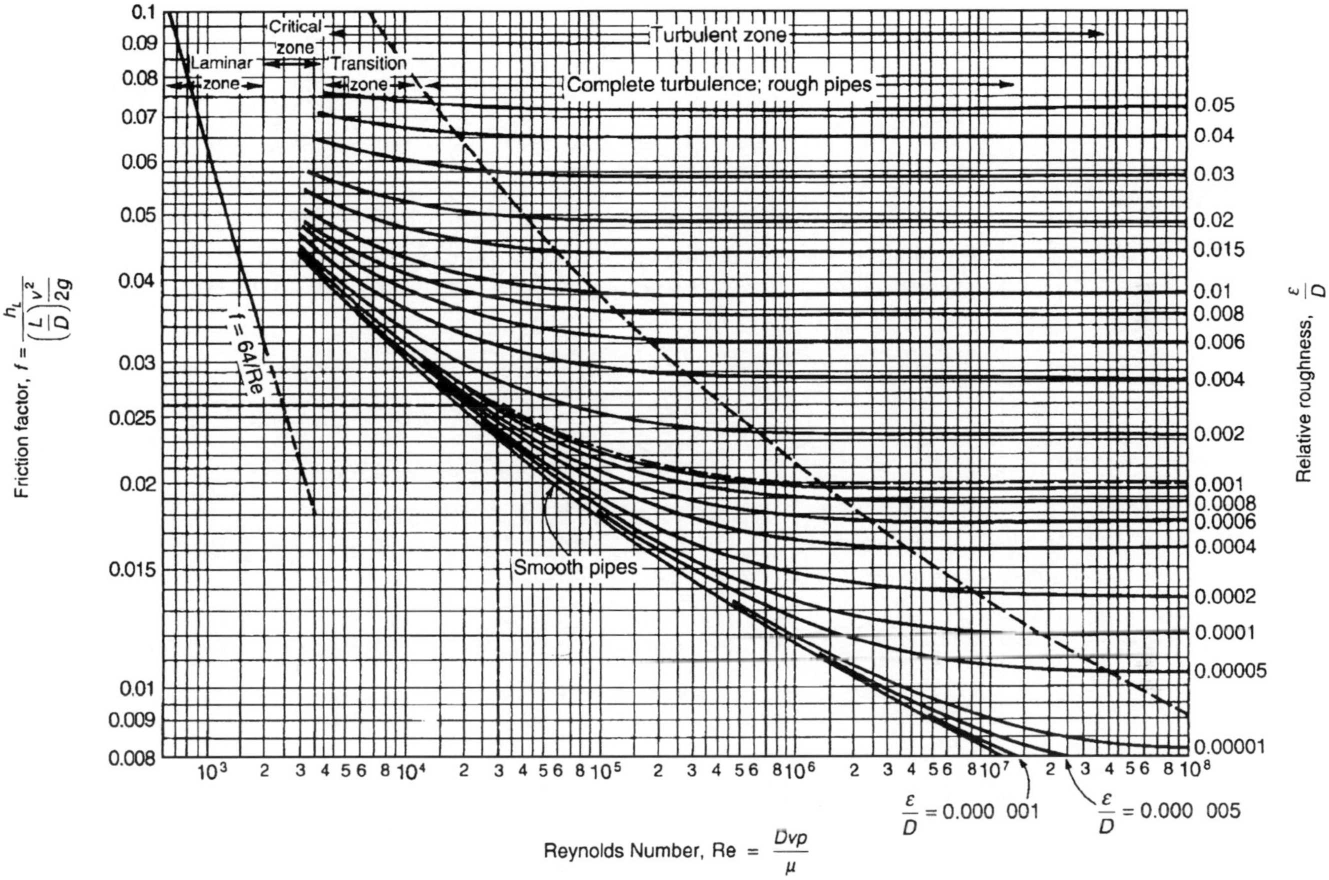

Figure 12.8 Moody diagram used with permission from *Transactions of the ASME*, 1944, vol. 66 by L. F. Moody

Substituting this value into the Darcy equation allows direct calculation of the head loss for laminar flow:

$$h_f = \frac{32\mu LV}{\gamma d^2}$$

This relation is known as the Hagen-Poiseuille equation. It may also be written in terms of flow rate, Q, where $V = Q/A = 4Q/\pi d^2$. Then

$$h_f = \frac{128\mu LQ}{\pi\gamma d^4}$$

Thus, the head loss from friction for laminar flow can be seen to be proportional to flow rate and inversely proportional to the diameter raised to the fourth power. If the diameter is doubled, and the flow remains laminar, the head loss will decrease by a factor of 16.

If the head loss from friction for turbulent flow is rewritten in terms of flow rate, Q, substitution of $4Q/\pi d^2$ for V yields

$$h_f = \frac{8fLQ^2}{\pi^2 g d^5}$$

Thus, the head loss from friction for turbulent flow is approximately proportional to the square of the flow rate and approximately inversely proportional to the diameter raised to the fifth power. If the diameter is doubled, the head loss will decrease by a factor of approximately 32. The proportionality is approximate because the friction factor, f, may change slightly with changes in flow rate and diameter.

Example 12.13

From Example 12.11 using the energy equation, the head loss from friction for 15 meters of 2.5-cm-diameter smooth pipe was given as 8.33 m. Show how this value was calculated. The dynamic viscosity, v, is 3.56×10^{-7} m²/s.

Solution

$$h_f = f\frac{L}{d}\frac{V^2}{2g}, \quad L = 15 \text{ m}, \ d = 2.5 \text{ cm} = .025 \text{ m}$$

$$V = \frac{Q}{A} = \frac{130\dfrac{\text{L}}{\text{min}} \bullet \dfrac{1 \text{ m}^3}{1000 \text{ L}} \bullet \dfrac{\text{min}}{60 \text{ s}}}{\dfrac{\pi(.025)^2}{4}\text{m}^2} = 4.41 \text{ m/s}$$

$$\text{Re} = \frac{Vd}{v}\frac{4.41\dfrac{m}{s} \bullet .025 \text{ m}}{3.56 \times 10^{-7}\dfrac{\text{m}^2}{\text{s}}} + 3.1 \times 10^5$$

Now enter the Moody diagram for this Reynolds number, and reflect off the ε/d = smooth line to read $f = 0.014$.

Hence,

$$h_f = 0.014 \bullet \frac{15 \text{ m}}{.025 \text{ m}} \bullet \frac{\left(4.41\dfrac{\text{m}}{\text{s}}\right)^2}{2 \bullet 9.81\dfrac{\text{m}}{\text{s}^2}} = 8.33 \text{ m}$$

Minor Losses

Flow losses from friction in pipe fittings, contractions, and enlargements are collectively known as minor losses. In problems where the pipe length is large, the minor losses in the system may be neglected. Minor losses are generally denoted by one of three methods:

(1) An equivalent length of pipe, L_e, is chosen, and the amount of the loss is calculated from

$$h_f = f\frac{L_e}{d}\frac{V^2}{2g}$$

(2) An equivalent length of pipe in diameters, $(L/d)_e$, is chosen, and the amount of loss is calculated from

$$h_f = f\left(\frac{L}{d}\right)_e\frac{V^2}{2g}$$

(3) A loss coefficient, C (or K), is chosen, and the amount of loss is calculated from

$$h_f = C\frac{V^2}{2g}$$

In more recent years the use of loss coefficients (C values) has become predominant and will be used in Fundamentals of Engineering exams. The C values may be a function of flow rate, fitting geometry, and/or diameter ratio (as in the case of contractions or enlargements) but in many cases may be considered constant over a wide range of conditions.

The equivalent lengths and C values for fittings and contractions are tabulated in catalogues, handbooks, and manuals. Representative values are listed in Table 12.3 and are illustrated in Figure 12.9.

The total friction loss in a piping system is the sum of the pipe losses and all minor losses.

Table 12.3 Resistance in valves and fittings expressed as equivalent length in pipe diameters $(L/d)_e$, and loss coefficient, C

Type	Equivalent Length in Pipe Diameters, $(L/d)_e$	Loss Coefficient, C
Globe valve—fully open	340	6.80
Angle valve—fully open	145	2.90
Gate valve—fully open	13	0.26
Check valve—swing type	135	2.70
Check valve—ball type	150	3.00
Butterfly valve—fully open	40	0.80
90° standard elbow	30	0.60
90° long-radius elbow	20	0.40
90° street elbow	50	1.00
45° standard elbow	16	0.32
45° street elbow	26	0.52
Close return bend	50	1.00
Standard tee with flow-through run	20	0.40
Standard tee with flow-through branch	60	1.20

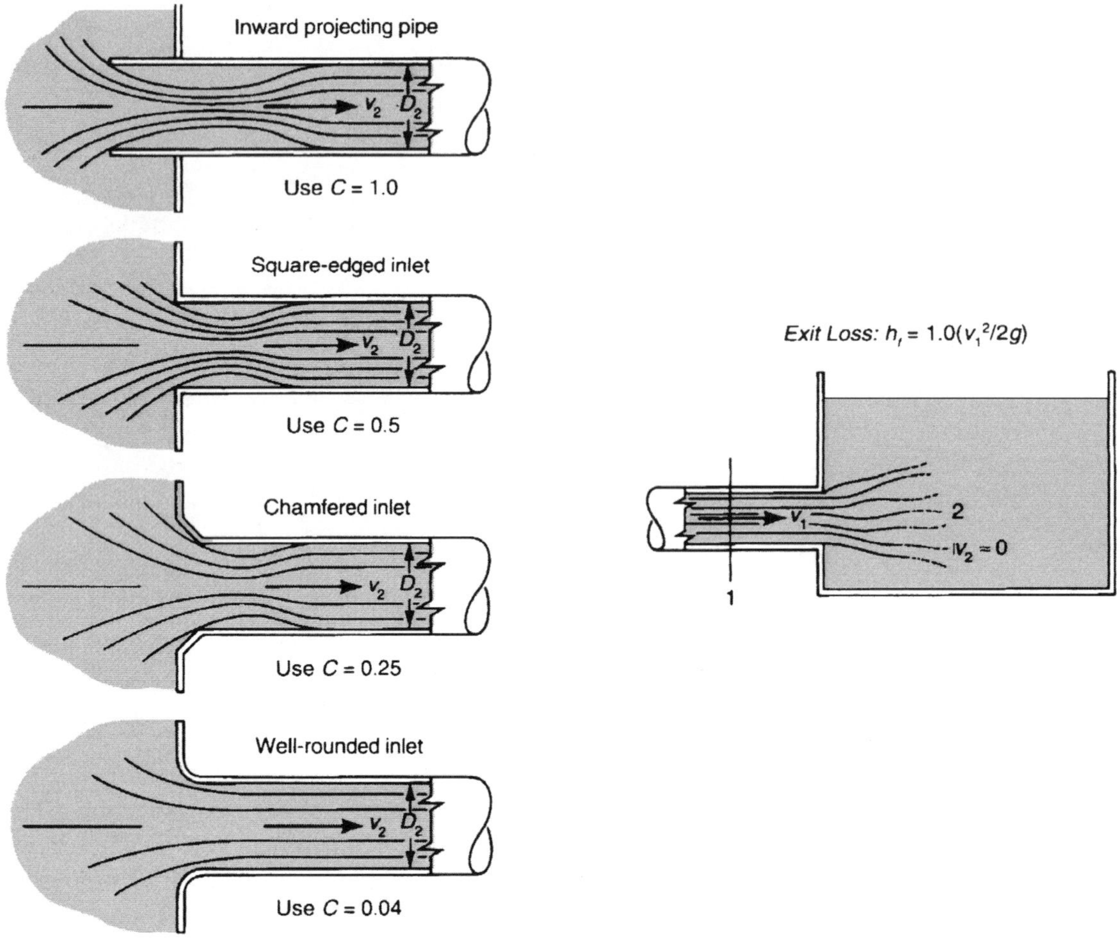

Figure 12.9 Entrance loss coefficients

Example **12.14**

For Example 12.13, suppose there were three elbows ($C = 0.6$), two gate valves ($C = 0.26$), one globe valve ($C = 6.8$), and a square-edged entrance from the tank ($C = 0.5$). Calculate the head loss created by the minor losses for the existing flow conditions. Recalculate the head delivered by the pump considering both pipe and local losses.

Solution

Minor losses:

$$h_{fm} = \underset{\text{elbows}}{3C\frac{V^2}{2g}} + \underset{\text{gate vlvs.}}{2C\frac{V^2}{2g}} + \underset{\text{globe vlv.}}{C\frac{V^2}{2g}} + \underset{\text{entrance}}{C\frac{V^2}{2g}}$$

$$h_{fm} = [3(0.6) + 2(0.26) + 6.8 + 0.5]\frac{V^2}{2g}$$

$$h_{fm} = (9.62)\frac{\left(4.41\dfrac{\text{m}}{\text{s}}\right)^2}{2 \bullet 9.81\dfrac{\text{m}}{\text{s}^2}} = 9.54\ \text{m}$$

Total losses:

$$h_f = 8.33\text{m(pipe)} + 9.54\text{m(minor)} = 17.87\text{m}$$

$$h_A = \frac{\left(27.6\frac{\text{m}}{\text{s}}\right)^2}{2 \bullet 9.81\frac{\text{m}}{\text{s}^2}} = 3\text{m} + 17.87\text{m} = 59.7\text{m}$$

Problems to determine the pump or turbine head, a pressure, or an elevation are the simplest to solve. Since the flow rate and pipe diameter are known, the Reynolds number and relative roughness can be calculated directly; the friction factor can then be found from the Moody diagram, and the head loss is calculated for substitution into the energy equation. Problems where the flow rate is being sought—and other parameters including the pipe diameter are known—require a single iteration process (for turbulent flow). Since the Reynolds number cannot be initially calculated, an initial friction factor must be assumed—then corrected—to determine the flow rate. Problems where the pipe diameter is being sought, but the flow rate and other parameters are known, require an iteration process. After simplification of the energy equation, different pipe diameters are assumed and friction factors are determined from the Moody diagram, then both are substituted into the energy equation until the equation is satisfied.

Flow in Noncircular Conduits

The same fundamental equations for Reynolds number, relative roughness, and head loss from friction may be used for noncircular conduits. In place of the diameter, an equivalent diameter (or characteristic length) is used. The equivalent diameter is defined by

$$d_e = 4R_H = 4\frac{A}{WP}$$

where d_e = equivalent diameter, R_H = hydraulic radius = $\frac{A}{WP}$, A = cross-sectional area, and WP = wetted perimeter.

Example 12.15

Calculate the equivalent diameter of a rectangular conduit 0.6 meters wide and 0.3 meters high.

Solution

$$A = 0.6\,\text{m} \bullet 0.3\,\text{m} = 0.18\,\text{m}^2$$
$$WP = 2(0.6\,\text{m} + 0.3\,\text{m}) = 1.8\,\text{m}$$
$$d_e = 4\frac{A}{WP} = 4 \bullet \frac{0.18\,\text{m}^2}{1.8\,\text{m}} = 0.4\,\text{m}$$

Parallel Pipe Flow

The text to this point has addressed only flow in series piping systems; that is, all flow was considered to go through each pipe and fitting. If the flow divides

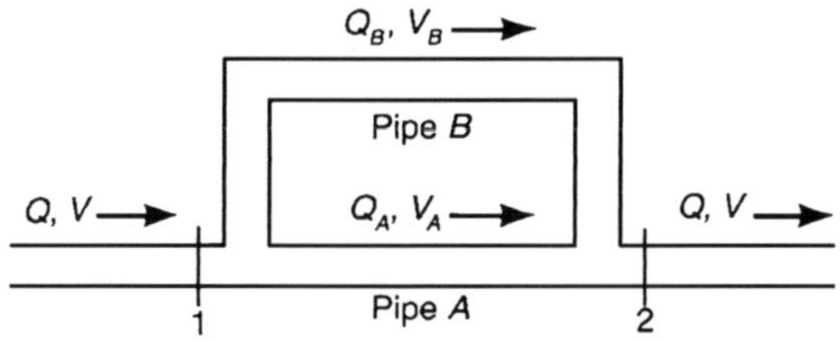

Figure 12.10

into two parallel branches and returns again to a single pipe as shown in Figure 12.10, the flow will divide so that the head loss in each branch is the same, or

$$h_{f1-2} = h_{fA} = h_{fB}$$

In the case when minor losses can be neglected, this relation becomes

$$h_{f1-2} = f_A \frac{L_A}{d_A} \frac{V_A^2}{2g} = f_B \frac{L_B}{d_B} \frac{V_B^2}{2g}$$

In addition, the continuity equation requires that

$$Q = QA + QB$$

Generally the flow rate, Q, or velocity, V, is known. Thus we have two equations and two unknowns, V_A and V_B. Values of f_A and f_B can be estimated and the two equations are then solved simultaneously for V_A and V_B. Using V_A and V_B to calculate the Reynolds numbers, corrected values of f_A and f_B can be found from the Moody diagram, and V_A and V_B are then recalculated. Then the head loss can be found from the above equation. For a flow that divides into three parallel branches, the same analysis can be used, where

$$h_{f1-2} = h_{fA} = h_{fB} = h_{fC}$$
$$Q = Q_A + Q_B + Q_C$$

In this case, we have three equations, and three unknowns.

FORCES ATTRIBUTABLE TO CHANGE IN MOMENTUM

The force created by the change in momentum of a fluid undergoing steady flow is given by the impulse momentum equation

$$F = \Delta(\dot{m}V) = \rho Q V_2 - \rho Q V_1$$

where F = resultant force on the fluid stream, ρQ = mass rate of flow, V_1, V_2 = inlet and exit velocities, respectively, $\rho Q V_1$ = momentum per second at the inlet, and $\rho Q V_2$ = momentum per second at the exit.

Forces on Bends

The magnitude and direction of the resultant force in flow through a bend will depend on the change in velocity magnitude and/or the change in direction of the flow. If the flow occurs in a two-dimensional bend, the equation can be rewritten in scalar form. Using Figure 12.11,

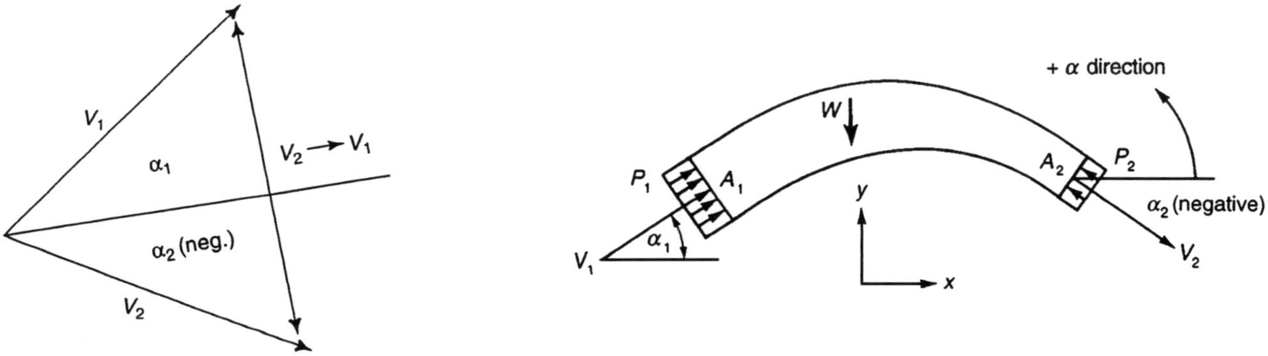

Figure 12.11

$$F_x = p_2 A_2 \cos\alpha_2 - p_1 A_1 \cos\alpha_1 + \rho Q(V_2 \cos\alpha_2 - V_1 \cos\alpha_1)$$

$$F_y - W = -p_2 A_2 \sin\alpha_2 - p_1 A_1 \sin\alpha_1 + \rho Q(V_2 \sin\alpha_2 - V_1 \sin\alpha_1)$$

$$F = \sqrt{F_x^2 + F_y^2}$$

where

F_x, F_y = resultant force component on the fluid by the bend in the x and y directions, respectively
p_1, p_2 = inlet and exit pressure on the fluid in the bend
A_1, A_2 = inlet and exit cross-sectional areas, respectively
α_1, α_2 = angles that the direction of flow makes with the positive x-direction at the entrance and exit, respectively
F = magnitude of resultant force
W = weight of the fluid in the bend

The equation may be simplified somewhat if the x-direction is chosen as the direction of the entering flow. The $\alpha_1 = 0$ and α_2 would be measured relative to that x-axis and

$$F_x = p_2 A_2 \cos\alpha_2 - p_1 A_1 + \rho Q(V_2 \cos\alpha_2 - V_1)$$

$$F_y - W = p_2 A_2 \sin\alpha_2 + \rho Q V_2 \sin\alpha_2$$

It is noted that the second equation, as written, requires that the y-axis remain parallel with gravity, since the coefficient of the weight W is unity. In many cases the weight of the fluid in the bend may not be significant. In addition to the impulse-momentum equations, it may be necessary simultaneously to utilize the energy and continuity equations in solving such problems.

Example 12.16

The 45° reducing bend discharges 0.008 m³/s of water into the atmosphere, as shown in Exhibit 5. The entrance diameter of the bend is 50 mm, and the exit diameter is 30 mm. Neglect the small elevation change, the weight of the fluid in the bend, and friction. Calculate the magnitude and force of the water on the bend. The density of the water is 1000 kg/m³, and its specific weight is 9810 N/m³.

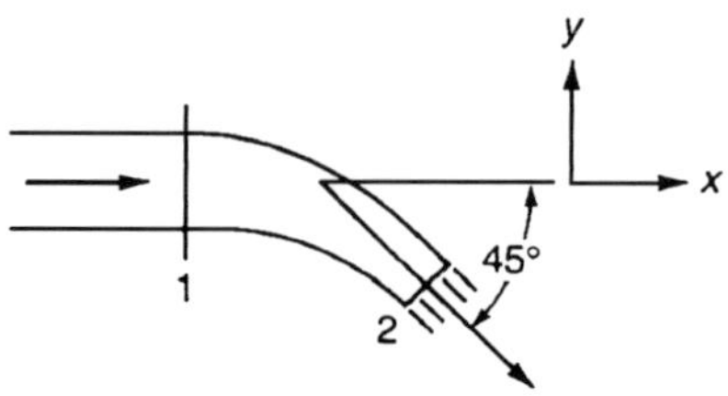

Exhibit 5

Solution

Select sections 1 and 2 as shown. Then $\alpha_1 = 0$, $\cos\alpha_1 = 1$, $\sin\alpha_1 = 0$, $\alpha_2 = -45°$, $\cos\alpha_2 = 0.707$, $\sin\alpha_2 = -0.707$, $p_2 = 0$, and the force-momentum equation becomes

$$F_x = -p_1A_1 + \rho Q[V_2(0.707) - V_1]$$
$$F_y = \rho Q[V_2(-0.707)]$$

$$\rho Q = 1000\frac{\text{kg}}{\text{m}^3} \bullet 0.008\frac{\text{m}^3}{\text{s}} = 8.0\frac{\text{kg}}{\text{s}}$$

$$A_1 = \frac{\pi(0.05\,\text{m})^2}{4} = 0.00196\,\text{m}^2, \quad A_2 = \frac{\pi(0.03\,\text{m})^2}{4} = 0.00071\ \text{m}^2$$

$$V_1 = \frac{Q}{A_1} = \frac{0.008\ \text{m}^3/\text{s}}{0.00196\ \text{m}^2} = 4.08\,\text{m/s}, \quad V_2 = \frac{0.008\,\text{m}^3/\text{s}}{0.00071\,\text{m}^2} = 11.3\ \text{m/s}$$

The pressure p_1 may be found from the energy equation:

$$\frac{p_1}{\gamma} + \frac{V_1^2}{2g} + Z_1 = \frac{p_2}{\gamma} + \frac{V_2^2}{2g} + Z_2$$

$$p_1 = \gamma\left(\frac{V_2^2 - V_1^2}{2g}\right) = 9810\frac{\text{N}}{\text{m}^3} \bullet \frac{\left(11.3^2 - 4.08^2\right)\frac{\text{m}^2}{\text{s}^2}}{2 \bullet 9.81\frac{\text{m}}{\text{s}^2}} = 55{,}500\frac{\text{N}}{\text{m}^2}$$

$$F_x = -55{,}500\frac{\text{N}}{\text{m}^2} \bullet 0.00196\text{m}^2 + 8.0\frac{\text{kg}}{\text{s}}\left(11.3\frac{\text{m}}{\text{s}} \bullet 0.707 - 4.08\frac{\text{m}}{\text{s}}\right)$$

$$F_x = -108.8\text{N} + 31.3\text{N} = -77.5\text{N}$$

$$F_y = 8.0\frac{\text{kg}}{\text{s}} \bullet 11.3\frac{\text{m}}{\text{s}}(-0.707) = -63.9\text{N}$$

$$F = \sqrt{(-77.5)^2 + (-63.9)^2} = 100.4\text{N}$$

$$\text{at } \theta = -90.0° - \arctan\frac{-77.5}{-63.9} = -140.5°$$

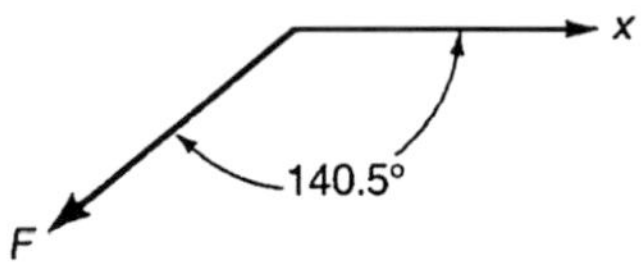

Exhibit 6

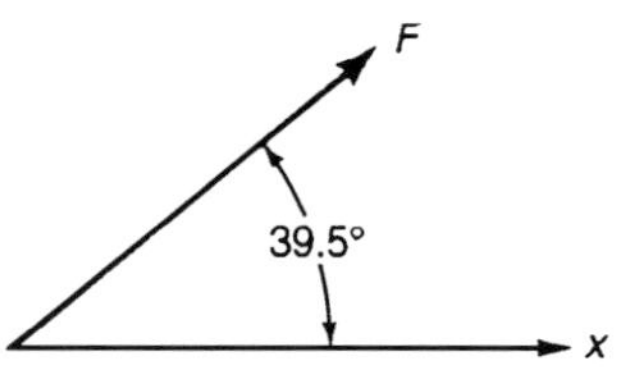

Exhibit 7

This is the force of the bend on the fluid (Exhibit 6). The force of the fluid on the bend is equal and opposite (Exhibit 7).

Jet Engine and Rocket Thrust

The impulse-momentum relationship may be used to calculate the thrust created by a jet engine. Since the inlet and exit pressures are zero, the thrust may be calculated from a combination of equations for both the air and fuel:

$$F = \rho_a Q_a (V_2 - V_1) + \rho_f Q_f V_2$$

where V_1 = entering velocity relative to the engine (for engines in flight, this is the velocity of the aircraft), V_2 = exit velocity relative to the engine, ρ_a, ρ_f = density of the air and fuel, respectively, Q_a, Q_f = flow rate of the air and fuel, respectively, and F = thrust.

This equation assumes that the fuel enters the engine perpendicular to the thrust direction and leaves with the exhaust flow.

In the case of rocket propulsion, since the fuel and oxidizer initially are at rest, the equation for thrust becomes

$$F = \rho_m Q_m V_2$$

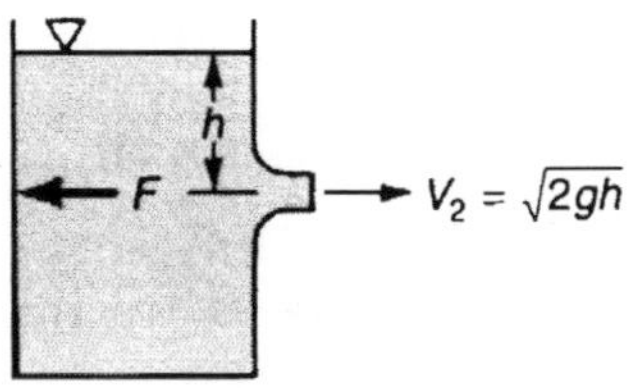

Figure 12.12

where ρ_m = density of the fuel-oxidizer mixture, and Q_m = flow rate of the fuel-oxidizer mixture.

A particular example of jet propulsion is shown Figure 12.12. The thrust is given by

$$F = \rho Q V_2 = \rho (A_2 V_2) V_2 = \rho A_2 V_2^2$$

but by Bernoulli's equation $V_2 = \sqrt{2gh}$. Substituting,

$$F = \rho A_2 (2gh) = 2\gamma A_2 h$$

where γ = specific weight of the liquid, A_2 = area of nozzle exit, h = height of fluid surface above the nozzle, and F = thrust or propulsion force.

Forces on Stationary Vanes

The impulse-momentum equation can also be used to determine forces on stationary vanes. Consider the stationary vane shown in Figure 12.13.

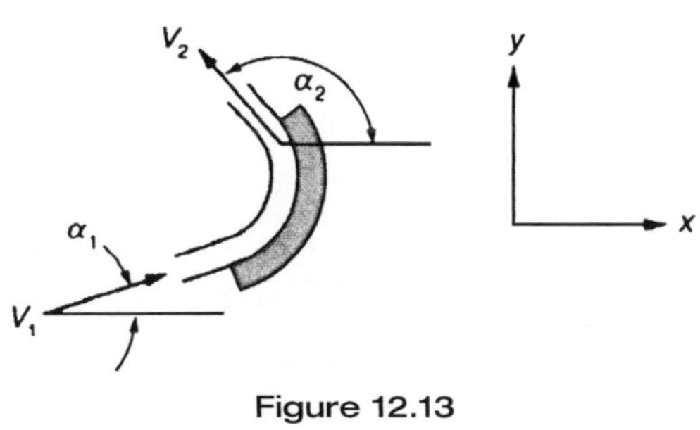

Figure 12.13

Since the fluid is open to the atmosphere, $p_1 = p_2 = 0$. If friction is neglected, it can be shown from Bernoulli's equation that $V_1 = V_2 = V$. The impulse-momentum equation reduces to

$$F_x = \rho Q(V_1 \cos \alpha_1 - V_2 \cos \alpha_2) = \rho QV(\cos \alpha_1 - \cos \alpha_2)$$
$$F_y = \rho QV(\sin \alpha_1 - \sin \alpha_2)$$

where F_x, F_y = *force of the fluid on the vane* in the x and y directions, respectively; α_1, α_2 = angle between the positive x-direction and the entrance and exit velocities, respectively; ρQ = mass flow rate; and V = fluid velocity.

If we reorient the vane so that the entrance velocity is in the x-direction, as shown in Figure 12.14, the equation can be rewritten as

$$F_x = \rho QV(1 - \cos \alpha_2)$$
$$F_y = -\rho QV \sin \alpha_2$$

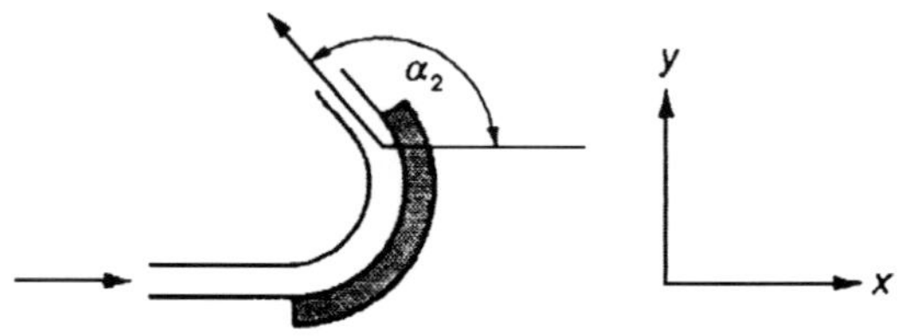

Figure 12.14

Example 12.17

Water impinges upon a stationary vane with a velocity of 15 m/s through a cross-sectional area of 6 square centimeters. The vane is oriented so that the fluid enters the vane cavity in the x-direction and leaves at an angle of 60° (Exhibit 8). The density of the water is 1000 kg/m³. Calculate the force on the vane.

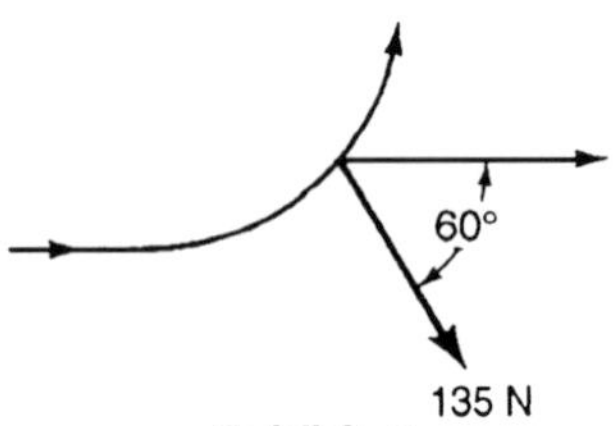

Exhibit 8

Solution

$$Q = AV = 6\,\text{cm}^2 \bullet \left(\frac{1\,\text{m}}{100\,\text{cm}}\right)^2 \bullet 15\,\text{m/s} = 0.009\frac{\text{m}^3}{\text{s}}$$

$$F_y = \rho QV(1 - \cos \alpha_2) = \frac{1000\,\text{kg}}{\text{m}^3} \bullet 0.009\frac{\text{m}^3}{\text{s}} \bullet 15\frac{\text{m}}{\text{s}} \bullet (1 - \cos 60°)$$

$$F_x = 67.5\frac{\text{kg} - \text{m}}{\text{s}^2} = 67.5\,\text{N}$$

$$F_y = -\rho QV \sin \alpha_2 = \frac{1000\,\text{kg}}{\text{m}^3} \bullet 0.009\frac{\text{m}^3}{\text{s}} \bullet 15\frac{\text{m}}{\text{s}} \bullet \sin 60°$$

$$F_y = -116.9\ \text{N}$$

$$F = \sqrt{(67.5)^2 + (-116.9)^2} = 135.0\,\text{N}$$

$$\text{at } \theta = \arctan\frac{-116.9}{67.5} = -60°$$

Forces on Moving Vanes

If the fluid enters the vane cavity in the x-direction and the vane is also moving in the x-direction with a velocity, v, as shown in Figure 12.15(a), then the force of the fluid on the moving vane in the direction of motion may be calculated from

$$F_x = \rho Q'(V - v)(1 - \cos\alpha_2)$$
$$F_y = -\rho Q'(V - v)\sin\alpha_2$$
$$Q' = A(V - v)$$

where v = velocity of the vane, and $V - v$ = fluid velocity relative to the vane.

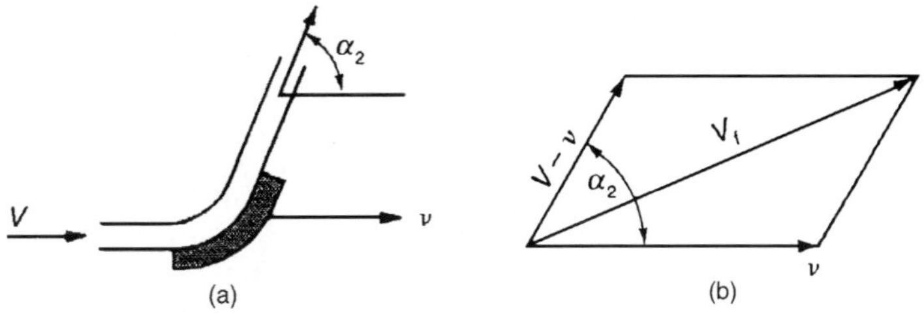

Figure 12.15

The final direction and magnitude of the jet is shown by V_f in the vector diagram [Figure 12.15(b)].

An impulse turbine contains a series of moving vanes as described above, one immediately replacing another as the turbine rotor rotates. See Figure 12.16.

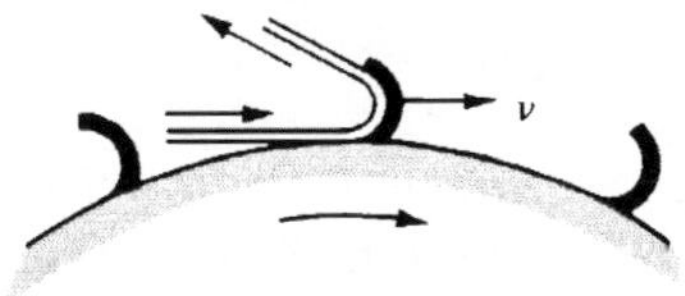

Figure 12.16

The x-direction force of the fluid on the series of vanes is given by

$$F = \rho Q(V - v)(1 - \cos\alpha_2)$$

The power delivered to the turbine, P, is given by

$$P = Fv = \rho Qv(V - v)(1 - \cos\alpha_2)$$

The maximum power for a given discharge angle, α_2, occurs when $V = 2v$ and is given by

$$p_{\max} = \frac{\rho QV^2}{4}(1 - \cos\alpha_2)$$

The discharge angle that produces the maximum power possible is $\alpha_2 = 180°$. This maximum power is calculated from

$$P_{\max} = \frac{\rho QV^2}{2} = g_c\gamma\, Q\frac{V^2}{2g}$$

VELOCITY AND FLOW MEASURING DEVICES

Pitot Tubes

For liquids flowing at relatively low pressures the mean static pressure may be measured using a piezometric tube indicated in Figure 12.17 by h_1. The stagnation pressure (i.e., the pressure at which the velocity is zero) is indicated by h_2.

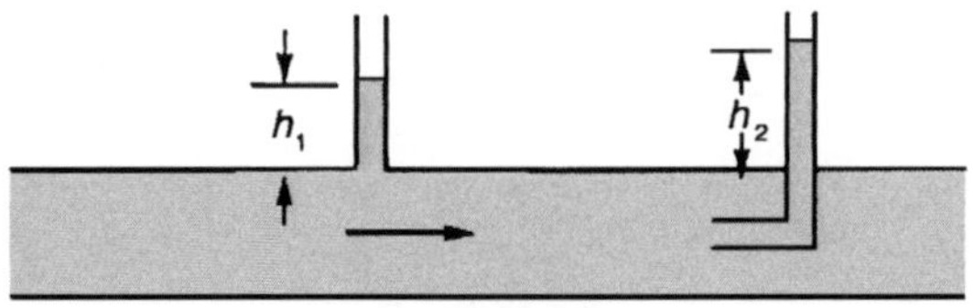

Figure 12.17

The relationship of each measurement to pressure, specific weight, and velocity is shown in the following two equations:

$$h_1 = \frac{p}{\gamma}, \quad h_2 = \frac{p_s}{\gamma} = \frac{p}{\gamma} + \frac{V^2}{2g}$$

Combining these relations, the velocity in the duct is

$$V = \sqrt{2g\left(\frac{p_s - p}{\gamma}\right)} = \sqrt{2g(h_2 - h_1)}$$

where V = velocity, γ = specific weight, p = static pressure, and p_s = stagnation pressure.

The combination of the two tubes as a single device is known as a pitot tube. If a manometer is connected between the static and stagnation pressure taps, velocities at moderate pressures may be calculated using the following equation:

$$V = \sqrt{2gh_m\left(\frac{\gamma_m}{\gamma} - 1\right)}$$

where h_m = height indicated by the manometer, and γ_m = specific weight of manometer fluid.

Example 12.18

A pitot tube is used to measure the mean velocity in a pipe where water is flowing. A manometer containing mercury is connected to the pitot tube and indicates a height of 150 mm. The specific weights of the water and mercury are 9810 N/m^3 and 133,400 N/m^3, respectively. Calculate the velocity of the water.

Solution

$$V = \sqrt{2gh_m\left(\frac{\gamma_m}{\gamma} - 1\right)}$$

$$V = \sqrt{2 \bullet 9.81\frac{\text{m}}{\text{s}^2} \bullet 0.15\text{m} \bullet \left(\frac{133{,}400\frac{\text{N}}{\text{m}^3}}{9810\frac{\text{N}}{\text{m}^3}} - 1\right)} = 6.09\frac{\text{m}}{\text{s}}$$

The pitot tube equation may also be used for compressible fluids with Mach numbers less than or equal to 0.3.

Flow Meters

There are three commonly used meters that measure flow rate in fluid systems: venturi meters, flow nozzles, and orifice meters. All three operate on the same basic principle, their equations being developed by combining the Bernoulli and continuity equations. The three meters are shown in Figure 12.18.

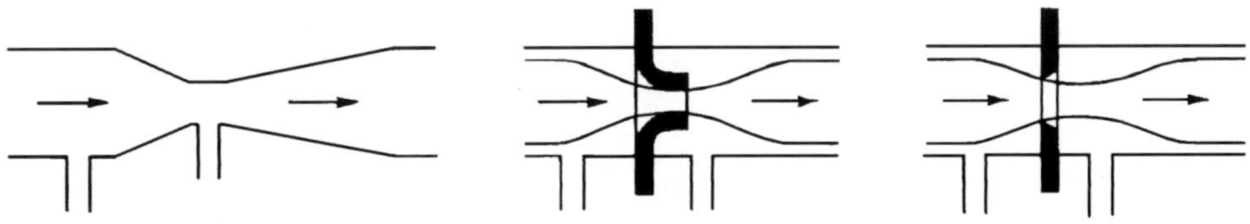

Figure 12.18

The equation for flow rate is given by

$$Q = \frac{c_v c_c}{\sqrt{1 - c_c^2 \left(\frac{A_2}{A_1}\right)^2}} \bullet A_2 \bullet \sqrt{2g\left(\frac{p_1 - p_2}{\gamma} + Z_1 - Z_2\right)}$$

or if a manometer is used between the pressure taps:

$$Q = \frac{c_v c_c}{\sqrt{1 - c_c^2 \left(\frac{A_2}{A_1}\right)^2}} \bullet A_2 \bullet \sqrt{2gh_m\left(\frac{\gamma_m}{\gamma} - 1\right)}$$

where

Q = flow rate

$p_1 - p_2$ = pressure difference between a point before the entrance to the meter, and the point of narrowest flow cross section in the meter

$Z_1 - Z_2$ = height difference between a point before the entrance to the meter and the point of narrowest flow cross section in the meter

A_1 = area of entrance

A_2 = area of narrowest flow cross section, except in the orifice, where it is the orifice area

h_m = height indicated by manometer

γ_m = specific weight of manometer fluid

γ = specific weight of fluid

c_v = coefficient of velocity

c_c = coefficient of contraction

For the orifice meter (and sometimes the flow nozzle) the coefficient terms in the equation are combined as follows:

$$c = \frac{c_v c_c}{\sqrt{1 - c_c^2 \left(\frac{A_2}{A_1}\right)^2}}$$

and the flow rate equation is then written as

$$Q = cA_2\sqrt{2g\left(\frac{p_1 - p_2}{\gamma} + Z_1 - Z_2\right)} = cA_2\sqrt{2gh_m\left(\frac{\gamma_m}{\gamma} - 1\right)}$$

where c = orifice (or flow nozzle) coefficient.

The following values of c_c, c_v, and c are used for the various flow meters:

Venturi:	$c_c = 1,\ 0.95 < c_v < 0.99;$	c_v (nominal) = 0.984
Flow nozzle:	$c_c = 1,\ 0.95 < c_v < 0.99;$	c_v (nominal) = 0.98
Orifice:	$c_c = 0.62,\ c_v = 0.98;$	c (nominal) = 0.61

Actual values of c for orifice meters (and flow nozzles) vary with the diameter ratio, $d_0{:}d_1$, and the Reynolds number and are found on existing graphs. Curves for the values of c_v for venturi meters and flow nozzles are also available.

Flow from a Tank

The flow from a tank through various types of exit configurations can be calculated by using the energy equation and experimentally determined configuration coefficients. Consider a tank as shown in Figure 12.19. For frictionless flow, the flow-rate can be calculated using the energy and continuity equations, which reduce to

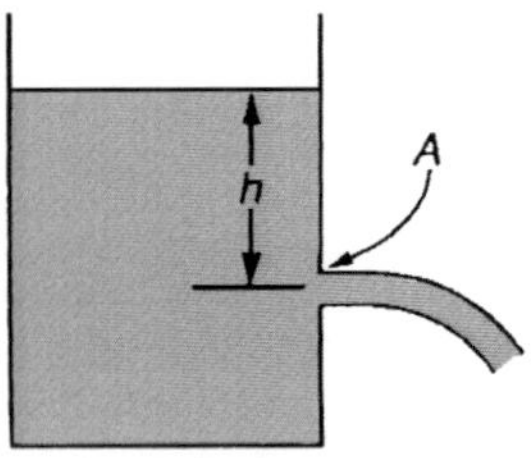

Figure 12.19

$$Q = AV = A\sqrt{2gh}$$

Considering friction, the flow rate may be calculated from

$$Q = cA\sqrt{2gh}, \quad c = c_v c_c$$

where Q = flow rate from the tank, h = height of water level above the exit, and c = coefficient of discharge for the exit.

If the friction in the exit is neglected, then $c = 1$.

For a sharp-edged orifice, $c_v = 0.98$, $c_c = 0.62$, and $c = 0.61$.

For a rounded exit, $c_v = 0.98$, $c_c = 1.00$, and $c = 0.98$.

For a short tube exiting from the tank, $c_v = 0.80$, $c_c = 1.00$, and $c = 0.80$.

For a re-entrant pipe, $c_v = 0.98$, $c_c = 0.52$, and $c = 0.51$.

For the special case when the flow from the tank discharges beneath the surface of the same fluid outside the tank, the flow rate from the tank is given by

$$Q = cA\sqrt{2gh(h_1 - h_2)}$$

where h_1 = height of fluid above exit in tank, and h_2 = height of fluid above exit outside tank.

SIMILARITY AND DIMENSIONLESS NUMBERS

The Reynolds number is a dimensionless number defined as the ratio of inertial forces to viscous forces. To test a model of some prototype, such as an air foil or length of pipe, the Reynolds number of the model must be equal to the Reynolds number of the prototype, or

$$(\mathrm{Re})_m = (\mathrm{Re})_p$$

$$\left(\frac{\rho V l}{\mu}\right)_m = \left(\frac{\rho V l}{\mu}\right)_p$$

where l = characteristic length.

A model is generally smaller geometrically than its prototype. Thus, if the characteristic length, l, of a model is to be one-tenth that of the prototype then one or more of the other terms in the Reynolds number must be adjusted to retain the Reynolds number the same for the model. For instance, the velocity, V, could be increased by a factor of ten using the same fluid or the fluid could be changed (liquid to gas) such that μ/ρ decreases by a factor of 10 with the same velocity, or a combination of the two. This condition is known as **dynamic similarity** of the prototype and model.

In fact, there are several independent force ratios that should be maintained in developing a model, depending on what forces are predominant in the situation. These force ratios involve pressure, inertia, viscosity, gravity, elasticity, and surface tension. The force ratios required for dynamic similarity are defined as follows, where the subscript m denotes model and p denotes prototype:

Inertia Force/Pressure Force Ratio

$$\left(\frac{\rho V^2}{p}\right)_m = \left(\frac{\rho V^2}{p}\right)_p$$

Inertia Force/Viscous Force Ratio

$$\left(\frac{\rho V l}{\mu}\right)_m = \left(\frac{\rho V l}{\mu}\right)_p = \text{Reynolds number, Re}$$

Inertia Force/Gravity Force Ratio

$$\left(\frac{V^2}{lg}\right)_m = \left(\frac{V^2}{lg}\right)_p = \text{Froude number, F}$$

Inertia Force/Elastic Force Ratio

$$\left(\frac{\rho V^2}{E}\right)_m = \left(\frac{\rho V^2}{E}\right)_p = \text{Cauchy number, } C_a$$

(E = modulus of elasticity of fluid)

Inertia Force/Surface Tension Force

$$\left(\frac{\rho l V^2}{\sigma}\right)_m = \left(\frac{\rho l V^2}{\sigma}\right)_p = \text{Weber number, } W_e$$

In many applications, one or more of the force ratios may be neglected because the forces are negligible.

INCOMPRESSIBLE FLOW OF GASES

The relationships thus far developed are primarily for the flow of incompressible fluids (liquids). Many are also applicable to the flow of compressible fluids (gases), for example, the continuity equation, equation for the Reynolds number, and so forth. The energy equation for incompressible flow also may be used under the following conditions:

1. The change in pressure in the pipe length is less than 10 percent of the inlet pressure. The density and specific weight at inlet conditions (pressure and temperature) should be used.
2. The change in pressure in the pipe length is between 10 and 40 percent of the inlet pressure. The density and specific weight at the average of the inlet and outlet conditions should be used. In some problems the outlet (or inlet) pressure is sought. In this case, the inlet (or outlet) conditions are used to find initial values of density and specific weight, and the approximate outlet (or inlet) pressure is then calculated. An iterative process ensues.

It may be necessary to utilize the perfect gas law from thermodynamics to calculate various properties, particularly the density of the gas given the pressure and temperature. The perfect gas law may be written in the following form

$$\rho = \frac{P}{RT}$$

where $R = \overline{\text{R}}/\text{MW}$ = gas constant
MW = gas molecular weight
$\overline{\text{R}}$ = universal gas constant

It should also be noted that the speed of sound, c, in a perfect gas is given by

$$c = \sqrt{kRT}$$

where k = ratio of specific heats, c_p/c_v
c_p = specific heat at constant pressure
c_v = specific heat at constant volume

It is apparent from the equation above that the speed of sound (acoustic velocity) in a gas depends only on its temperature.

The mach number Ma is the ratio of the actual fluid velocity to the speed of sound:

$$\text{Ma} = V/c$$

The accuracy of utilizing incompressible fluid flow equations for the flow of gases decreases with increasing velocities and their use is not recommended for mach numbers greater than 0.2.

HYDROSTATICS

Fluid mechanics involves the study of liquids, vapors, and gasses (highly superheated vapors). Vapors and gasses usually are assumed to exert constant pressure throughout the contained volume. This of course does not hold true for our atmosphere because, as we know, the barometric pressure decreases as we go higher into the atmosphere, hence, pressurization of the inside of aircraft. Hydrostatics, however, deals mainly with the action of liquids. This is when the total pressure at any given point in the liquid is equal to the pressure at the surface plus the pressure exerted by the head of the liquid.

Manometry

Manometers are often used to measure pressures and differential pressures. They use a column of liquid as the indication of pressure.

$$\text{Pressure due to liquid column} = \rho g h$$

where

$\rho =$ Density kg/m^3
$g =$ Acceleration due to gravity 9.81 m/s^2
$h =$ Vertical height of column of liquid m

Therefore pressure is

$$P = [(\rho \text{ kg/m}^3)(g \text{ m/s}^2)(h \text{ m})][\text{m}^2/\text{m}^2] = \text{N/m}^2 = \text{Pa}$$

Example 12.19

A U-tube manometer is used to measure the pressure in an air line. The 15 cm of liquid in the U-tube manometer is mercury, which has a specific gravity of 13.6 and is open to an atmosphere that is 1.2 bar. Determine the absolute pressure in the line. Refer to Exhibit 9.

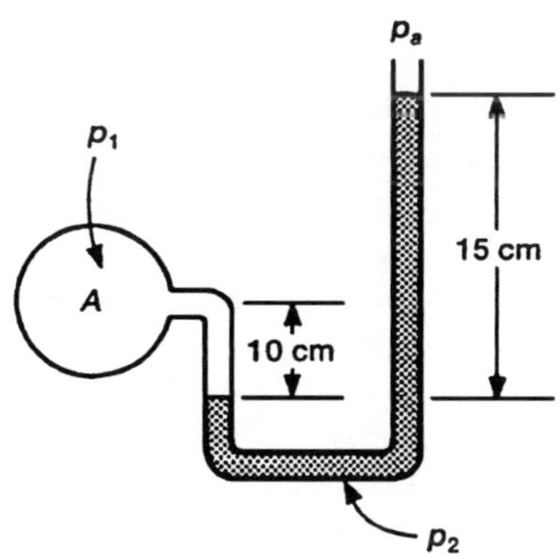

Exhibit 9

Solution

The generic equation for determining pressure may be written as:

$$P_{\text{absolute}} = P_{\text{gage}} + P_{\text{atmosphere}}$$

Keep in mind that the absolute pressure is the total pressure, and is equal to the pressure indicated by the gage added to the pressure of the atmosphere surrounding the gage. The generic equation may now be written as

$$P_{abs} = \rho gh + 1.2 \text{ bar} = 13.6(1000 \text{ kg/m}^3)(9.81 \text{ m/s}^2)(0.15 \text{ m}) + 1.2 \text{ bar}$$
$$P_{abs} = (20{,}012 \text{ kg/m}\cdot\text{s}^2)(\text{m}^2/\text{m}^2) + 1.2 \text{ bar}$$
$$P_{abs} = 20{,}012 \text{ N/m}^2 + 1.2 \times 10^5 = 1.4 \times 10^5 \text{ N/m}^2 = 1.4 \times 10^5 \text{ Pa} = 1.4 \text{ bar}$$

(Note that the density of mercury is equal to the specific gravity of mercury times the density of water where the density of water is taken as 1000 kg/m³.)

Example **12.20**

A company uses a number of multi-pass heat exchangers that use water as the cooling fluid. It is necessary to determine the power needs for the pumping of the water through the heat exchangers. A differential manometer is used to determine the pressure differential between the incoming and outgoing water. The manometer system is shown in Exhibit 10 with the distances shown. The second liquid used is mercury, with a specific gravity of 13.6. Determine the pressure drop in the heat exchanger in Pa.

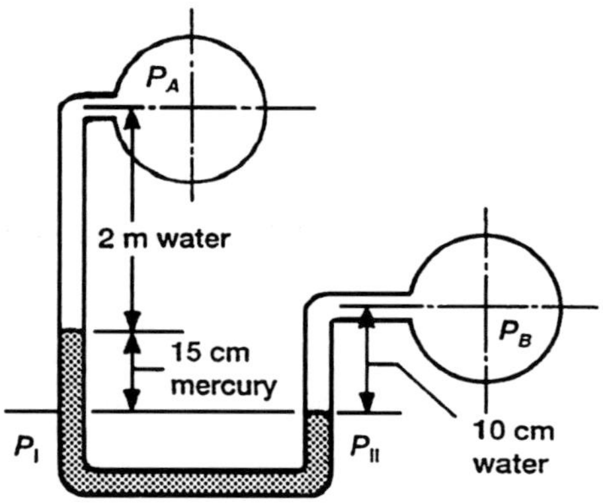

Exhibit 10

Solution

This heat exchanger is a little more complex than the former one. First, draw a horizontal line at the lowest interface between the two fluids. Then you may eliminate everything below this line. Next, equate the pressures at the slice of the interface section.

$$P_I = (\rho gh)_{hg} + (\rho gh)_{water} + P_A$$
$$P_{II} = (\rho gh)_{water} + P_B$$

Combining the two equations

$$(13.6)(1000 \text{ kg/m}^3)(9.81 \text{ m/s}^2)(0.15 \text{ m}) + (1000 \text{ kg/m}^3)(9.81 \text{ m/s}^2)(2 \text{ m}) + P_A$$
$$= (1000 \text{ kg/m}^3)(9.81 \text{ m/s}^2)(0.10 \text{ m}) + P_B$$

Solving for the differential pressure

$$P_A - P_B = 981 - 19{,}620 - 20{,}012 = -38{,}651 \text{ Pa}$$

This indicates that the pressure drop is 38.6 kPa.

Forces on Submerged Areas

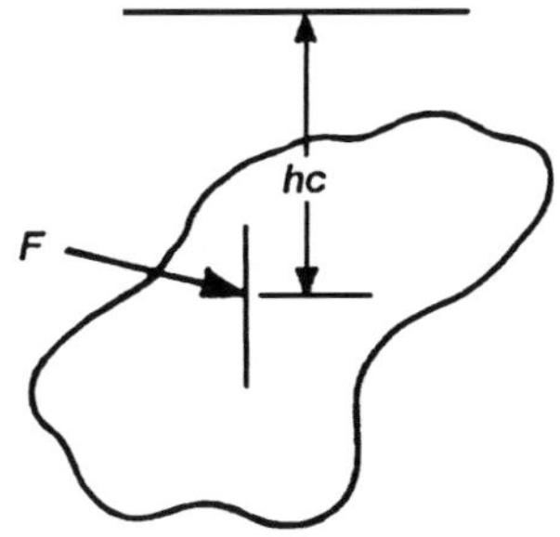

Figure 12.19

Pressure is a non-directional force and acts perpendicularly on a surface (refer to Figure 12.19). It is also noted that this force varies directly with the depth or the distance down from the surface. The vertical distance from the surface of the liquid to the centroid of the body is labeled h_c. The force acting on the body is equal to

$$F = \gamma h_c A$$

where

F = Total force acting on the body
γ = Specific weight of the liquid
h_c = Distance from the surface of the liquid to the centroid of the body
A = Area of the body

Example 12.21

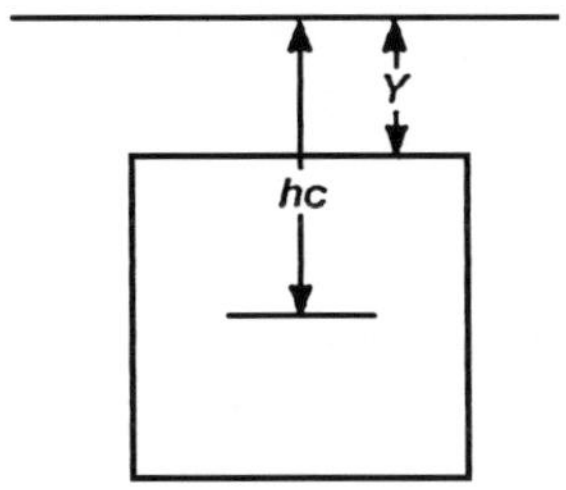

Exhibit 11

A 6-m × 6-m plate is submerged in water as shown in Exhibit 11 with the top edge of the plate

Determine the force, in kN, acting on the plate for the cases indicated

a). At the surface

b). 5 meters down

c). 100 meters down

Solution

a). $F = \gamma h_c A = (1000 \text{ kg/m}^3)(9.81 \text{ m/s}^2)(3 \text{ m})(6 \text{ m} \times 6\text{m})$
$= 1{,}059{,}480 \text{ kg} \cdot \text{m/s}^2$ or, force in kN = 1,059 kN

b). $F = (1000 \text{ kg/m}^3)(9.81 \text{ m/s}^2)(5 + 3)\text{m } (6 \text{ m} \times 6 \text{ m}) = 2{,}825 \text{ kN}$

c). $F = (1000 \text{ kg/m}^3)(9.81 \text{ m/s}^2)(100 + 3)\text{m } (6 \text{ m} \times 6 \text{ m}) = 36{,}375 \text{ kN}$

The center of pressure of a submerged body is that point where there would be no moment due to the pressure forces on the body. The equation for the calculation to the center of pressure is

$$h_p = h_c + (I_c / h_c A)$$

where

h_p = Distance to *center of pressure* of the body
h_c = Distance to *center of gravity* of body
I_c = *moment of inertia* of body
A = Area of the body

Determine the distance to the center of pressure for each of cases above:

Calculating the moment of inertia and the area

$I_c = bh^3/12 = [(6)(6)^3]/12 = 108 \text{ m}^4$ $\quad A = bh = (6)(6) = 36 \text{ m}^2$

Therefore

a). $h_p = 3 + (108)/(3)(36) = 4$ m

b). $h_p = (3 + 5) + 1 = 9$ m

c). $h_p = (3 + 100) + 1 = 104$ m

Buoyancy

Buoyancy is the ability of a fluid to support a body in the fluid. A body has the buoyant force acting on its center of gravity equal to the displaced fluid. The buoyant force is equal to the mass of the fluid times the gravitational attraction.

Example 12.22

A 20-m × 30-m flat-bottom car ferry that is used to transport automobiles across the Mississippi has a mass of 35,000 kg and can carry sixty 5,000-kg automobiles. How far will the ferry sink into the water?

Solution

The buoyant force is equal to the displaced liquid

$$F = mg = \text{Displaced volume} \times \text{Specific weight}$$
$$F = [35{,}000 + 60(5000)]\text{kg } (9.81 \text{ m/s}^2) = 3.286 \times 10^6 \text{ kg} \cdot \text{m/s}^2$$

Equating to displacement volume times specific weight

$$F = 3.286 \times 10^6 \text{ N} = (20 \text{ m})(30 \text{ m})(\text{Vertical displacement } Y \text{ m}) \times (1000 \text{ kg/m}^3)(9.81 \times \text{m/s}^2)$$
$$Y = 0.56 \text{ m}$$

Example 12.23

A 1.5-m-diameter and 4-m-tall cylinder with a mass of 2000 kg is closed at one end (Exhibit 12). It is placed in water such that the open end is down into the water. How far above the water will the tank float? Assume that the air and water are at the same temperature of 20 °C.

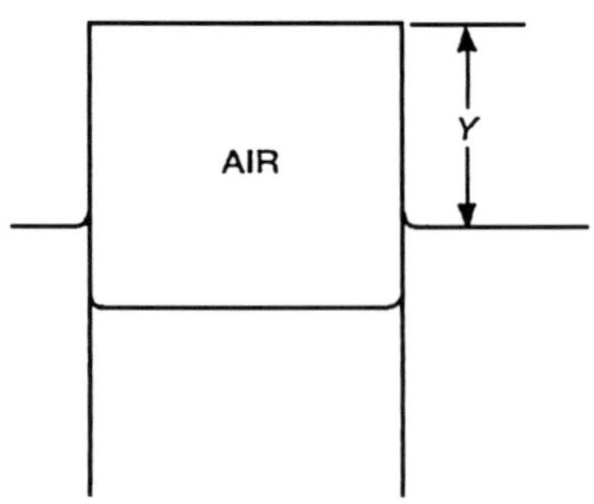

Exhibit 12

Solution

Since the tank is open downward the tank contains both air and water. Assuming, first, that the tank itself has insignificant displacement of water, the buoyant force is equal to the volume of water displaced by the entrapped air in the tank. The buoyant force has to be equal to the weight of the container. We can assume that the air in the tank has negligible weight.

$F =$ (mass of tank)(gravity) $=$ (Volume of displaced water)(weight of water)

$F = (2000 \text{ kg})(9.81 \text{ m/s}^2) = (\pi D^2 m^2/4)(\text{displacement m})(1000 \text{ kg/m}^3)(9.81 \text{ m/s}^2)$

$19{,}620 \text{ N} = \pi(1.5 \text{ m})^2/4(Y) \text{ m } (9{,}810 \text{ kg/m}^2\cdot\text{s}^2)$

$Y = 1.13$ m air in the tank will displace enough water for tank to float.

The pressure in the tank times the area of the opening of the tank is equal to the buoyant force, or

$$F = 19{,}620 \text{ N} = P_g = P_g\,\pi D^2/4 = P_g\,(3.14)(1.5)^2\text{m}^2/4$$
$$P_g = 11{,}103 \text{ Pa}$$
$$P_{tot} = P_g + P_{atm} = 11{,}103 + 101{,}300 = 112{,}403 \text{ Pa}$$

The volume of air in the tank may be obtained by using Charles' and Boyle's laws:

$$[P_1V_1]/T_1 = [P_2V_2]/T_2$$

and, for $T_1 = T_2$

$$P_1V_1 = P_2V_2$$

And solving for V_2

$$V_2 = V_1P_1/P_2 = V_1(101{,}300)/112{,}403 = 0.901V_1$$

This means that the column of air for the constant-area drum will be equal to 0.901 × 4 m, or equal to 3.60 m. The height of the tank sticking out of the water shall then be equal to 3.60 m minus the length of air needed to displace the equivalent amount of water

$$h_{\text{out of water}} = 3.60 \text{ m} - 1.13 \text{ m} = 2.47 \text{ m}$$

HYDRODYNAMICS

Hydrodynamics is a study of liquids in motion. Since this study is about liquids, which are incompressible, two general conditions must be satisfied. The first condition that must be satisfied is the continuity equation. This simply means that mass flow in must equal to the mass flow out.

For incompressible, i.e., constant, density

$$A_1V_1 = A_2V_2$$

The second condition is that the summation of energy entering and leaving a system must be equal. This summation of energy is a simplification of the *first law of thermodynamics* and is known as the *Bernoulli equation*. This equation may be written as

$$P_1/\rho g + V_1^2/2g + Z_1 + h_A = P_2/\rho g + V_2^2/2g + Z_2 + h_L$$

where

P	= Pressure	Pa
ρ	= Density	kg/m^3
g	= Acceleration due to gravity	m/s^2
V	= Velocity of the fluid	m/s
Z	= Elevation	m
h_A	= Energy added to fluid	m
h_L	= Energy lost from the fluid	m
A	= Crossectional area of pipe	m^2

Example 12.24

A water tank is used to supply the water for a village. The tank level is maintained at an elevation of 520 meters above sea level and is allowed to flow 200 m through a 4-cm pipe to a 1-cm-diameter nozzle at an elevation of 450 meters above sea level. If there is no friction, what is the velocity of flow at the exit of the nozzle?

Solution

Write the continuity equation relating the flow in the pipe to the flow in the nozzle

$$(\rho AV)_2 = (\rho AV)_3$$
$$\pi D_2^2 V_2 = \pi D_3^2 V_3, \text{ therefore:}$$
$$V_3 = [D_2/D_3]^2 V_2$$
$$V_3 = (4/1)^2 V_2 = 16V_2$$

Write the Bernoulli equation for the pipe, noting that at the water tank end there is no velocity, and at both ends there is no pressure. Take the datum to be 450 meters above sea level. Write the equation, assuming that the flow is out of the full pipe, then make the continuity equation substitution to solve for the velocity

$$P_1/\rho g + V_1^2/2g + Z_1 = P_2/\rho g + V_2^2/2g + Z_2$$
$$70 \text{ m} = V_2^{\ 2}/9.81 \text{ m/s}^2 \quad \text{Or} \quad V_2 = 26.2 \text{ m/s}$$

Therefore, the velocity out of the nozzle is

$$V_3 = (16)(26.2) = 419.2 \text{ m/s}$$

Example 12.25

Gasoline with a specific gravity of 0.80 is flowing through a 0.050-m-diameter pipe that has a venturi throat of 0.025 m (see Exhibit 13). The differential pressure between the throat and the pipe as shown in the Figure is 38 cm. The c_v of the venturi meter is 0.97, and the meter is on a 30° positive angle.

Determine

a). Difference in pressure between pipe and venturi throat, in meters of gasoline

b). Velocity at the throat ideal

c). Flow rate of gasoline, in l/min

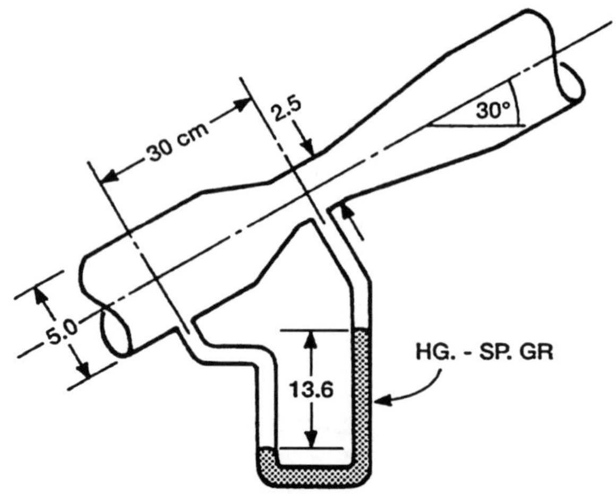

Exhibit 13

Solution

Write Bernoulli's equation for the venturi meter

$$P_1/\rho g + V_1^2/2g + Z_1 = P_2/\rho g + V_2^2/2g + Z_1 + \text{head loss due to friction}$$

Since we are only looking at a very short device there is not going to be any head loss due to friction. We do have to consider the venturi velocity coefficient, however.

Density of gasoline = Specific gravity of gasoline times density of water
= 800 kg/m^3
Density of mercury = Specific gravity of mercury times density of water
= 13,600 kg/m^3

a). The pressure difference is therefore due to the difference in the columns

$$[P_1 - P_2]/(800\ \text{kg/m}^3)(9.81\ \text{m/s}^2) = [(0.38\ \text{m})(13{,}600 - 800)\text{kg/m}^3] \times [9.81\ \text{m/s}^2]/(800\ \text{kg/m}^3)(9.81\ \text{m/s}^2) = 6.08 \text{ meters of gasoline}$$

b). The difference due to elevation change

$$Z_1 - Z_2 = -0.30\ \text{m} \times \sin 30° = -0.15 \text{ meters of gasoline}$$

Substituting these values into the Bernoulli equation

I)

$$6.08\ \text{m} - 0.15\ \text{m} = \left[V_2^2 - V_1^2\right]/(2)(9.81\ \text{m/s}^2)$$

$$\left(V_2^2 - V_1^2\right) = 116.3\ \text{m}^2/\text{s}^2$$

Using the continuity equation to determine the relation between the velocity in the pipe and the velocity in the throat of the venturi

II)

$$(\rho AV)_1 = (\rho AV)_2, \text{ with } \rho_1 = \rho_2$$

$$\left(\pi D_1^2/4\right)(V_1) = \left(\pi D_2^2/4\right)(V_2)$$

$$V_1 = V_2(0.025/0.050)^2 = V_2/4$$

Substituting equation II into equation I and solving for V_2

$$V_2^2 - (V_2/4)^2 = 116.3\ \text{m}^2/\text{s}^2$$

$$V_2 = 11.14\ \text{m/s}$$

c). To find the flow rate of gasoline we must multiply the c_v times the velocity in the pipe to get the pipe velocity.

$$V_1 = (V_2\text{m/s})/4 = 11.14/4 = 2.785\ \text{m/s ideal}$$

$$V_{1\ \text{actual}} = 0.97(2.785) = 2.70\ \text{m/s}$$

$$\text{Volume flow rate} = VA = (2.70\ \text{m/s})(\pi)(D_1)^2/4\ \text{m}^2 = (2.70)(\pi)(0.050)^2/4\ \text{m}^3/\text{s}$$

$$= 0.00530\ \text{m}^3/\text{s}(100\ \text{cm/m})^3(60\ \text{s/min})/(1000\ \text{cm}^3/1) = 318\ \text{l/min}$$

Dimensional Analysis

Dimensional analysis is a method by which one may predict physical parameters that affect fluid flow phenomena. By Newton's law we may operate in two different, yet related, systems. These systems are the MLT system and the FLT system.

$$F = ma = \mathrm{ML/T^2}$$

where

F is force with F as the primary dimension
m is mass with M as the primary dimension
a is acceleration with $\mathrm{L/T^2}$ as the primary dimensions

Example 12.26

Determine the generic equation for the drag on a body, considering the following factors that may influence drag

$$F_D = f(D, V, \rho, \mu)$$

where

F_D	= Drag force	F
D	= Diameter of pipe	L
V	= Velocity of the fluid	L/T
ρ	= Density of the fluid	$\mathrm{M/L^3}$
μ	= Viscosity of the fluid	M/LT

Solution

After choosing the dimensions from the FLT or MLT systems, assign the dimensions and set the exponents.

$$\mathrm{ML/T^2} = F_D = \mathrm{C}\mathrm{L}^a(\mathrm{L/T})^b(\mathrm{M/L^3})^c(\mathrm{M/LT})^d$$

Set up the table of exponents for

M: $1 = c + d$	$c = 1 - d$
L: $1 = a + b - 3c - d$	$a = 1 - (2 - d) + 3(1 - d) + d = 2 - d$
T: $-2 = -b - d$	$b = 2 - d$

$$F_D = CD^{2-d}\,V^{2-d}\,r^{1-d}\,\mu^d$$

Grouping in terms of like exponents

$$F_D = C(DV)^2[1/(\rho DV)/\mu](\rho)$$

Rearranging the terms, noting that $(\rho DV)/\mu = \mathrm{Re}$

$$\text{Drag force } F_D = f(\mathrm{Re})D^2V^2\rho$$

Momentum Force

When a fluid flows through a pipe that rounds a bend, forces are exerted by the fluid upon whatever resists the bend from movement. The forces are the summation of momentum in the various directions. The generic equation is

$$mV_1 + \int F dt = mV_2$$

Example 12.27

A jet of water that is 5 cm in diameter and is moving to the right at 30 m/s hits a stationary flat plate held normal to the flow. What force is required to hold the plate stationary?

Solution

A summation of momentum equation in the *x* direction should be used. Note that when the flow hits the stationary plate all *x*-direction velocity is gone because the water fans out into the *y*- and *z*-directions.

Writing the momentum equation in the *x* direction

$$mV_{x1} + F_x dt = mV_{x2}$$
$$[1000\ \text{kg/m}^3][(\pi D^2)/4]\ \text{m}^2[30\ \text{m/s}]^2 + F_x = 0$$
$$F_x = -1767\ \text{N} \quad \text{or} \quad 1.77\ \text{kN to the left}$$

Example 12.28

Water flows in a 6-cm-diameter stream at 50 m/s to the right. The stream of water hits a stationary, curved plate, and the stream is deflected 60° upward and to the right. Determine the *x* and *y* forces required to hold the plate stationary.

Solution

Again, the momentum equation is to be used; however, this time it is to be written in the *x* direction and in the *y* direction.

$$mV_{x1} + F_x dt = mV_{x2}$$
$$(1000\ \text{kg/m}^3)[(\pi D^2)/4]\ \text{m}^2(50\ \text{m/s})^2 + F_x = (1000)[(\pi D^2)/4][(50\ \text{m/s})(\cos 60°)]^2$$
$$F_x = -7.069 + 1.767 = -5.30\ \text{kN to the left.}$$
$$mV_{y1} + F_y dt - mV_{y2}$$
$$0 + F_y = (1000)[(\pi D^2)/4](50\ \text{m/s}\ \sin 60°)^2$$
$$F_y = 5.301\ \text{kN} \quad \text{Upward}$$

FLOW THROUGH PIPES

If we go back to the Bernoulli equation that was introduced and defined in the section on *hydrodynamics*, we notice that h_L is the energy lost by the fluid between the two points in question, 1 and 2. This energy loss is due to fittings and friction in the line, such as elbows, meters, valves, and normal flow. The losses are calculated using the Darcy–Weisbach equation, which is shown below.

$$h_L = f(L/D)(V^2/2g)$$

where

h_L	= Head loss	meters
f	= Friction factor	no units
L	= Length of the pipe	meters
D	= Diameter of the pipe	meters
V	= Velocity of flow in pipe	m/s
g	= Acceleration due to gravity	m/s^2

Note that the value D is given as the diameter of the pipe. If the pipe is really a duct then the value of D is a characteristic dimension and, for flow in open channels, amounts to four times the *hydraulic radius*. The *hydraulic radius* is equal to the *cross-sectional area of flow* divided by the *wetted perimeter*.

The term f is called the friction factor and is dependent upon the *Reynolds number* and the roughness of the pipe. If the Reynolds number is less than 2,000 the flow is called *laminar flow*, and the friction factor $f = 64/\text{Re}$. When the Reynolds number is above 10,000 the flow is considered to be *turbulent*, and the value of f is taken from the Moody diagram. Between the two numbers the flow is considered to be in transition.

Example **12.29**

How much power, in kW, does it require to pump 250 m³/hr of oil with a specific gravity of 0.85 and a viscosity of 0.011 kg/m·s through 5 km of 30 cm diameter commercial steel pipe. The pipe rises some 150 m and has 5 elbows and two gate valves with coefficients of 0.85 and 1.25, respectively.

Solution

Starting with the Bernoulli equation

$$P_1/\rho g + V_1^2/2g + Z_1 + h_A = P_2/\rho g + V_2^2/2g + Z_2 + h_L$$

Then, noting that since this is a pump power problem, everything on the left hand side of the equation may be simplified down to the energy added (h_A) to the oil. Assuming that the flow is open to the atmosphere, or that the flow just makes it to the end of the pipe, we can also drop P_2.

$$h_A = V_2^2/2g + Z_2 + h_L$$

Solving for the terms in the Bernoulli equation

$$V_2^2/2\text{g} = \{[(250\ \text{m}^3/\text{hr})/(3600\ \text{s/hr})]/[(\pi D^2/4)\ \text{m}^2]\}^2/[2(9.81\ \text{m/s}^2)]$$

$$V_2^2/2\text{g} = \{(0.0694\ \text{m}^3/\text{s})/[\pi(0.30)^2/4]\ \text{m}^2\}^2/[19.62\ \text{m/s}^2] = 0.050\ \text{m}$$

where the velocity is equal to 0.982 m/s

$Z_2 = 150$ m

h_L = Head loss due to elbows, gate valves, and the friction flow in the pipe.

$$h_{L(\text{elbows})} = C_{\text{elbows}}\,(\text{Number of elbows})(V^2/2g) = (0.85)(5)(0.50) = 2.125\ \text{m}$$

$$h_{L(\text{Gate valves})} = C_{\text{valves}}\,(\text{Number of valves})(V^2/2g) = (1.25)(2)(0.50) = 1.25\ \text{m}$$

$h_{L(\text{Due to friction})} = f\,(L/D)(V^2/2g)$ Where f for commercial steel pipe has to be determined after the Reynolds number has been calculated.

$$\text{Re} = \rho DV/\mu = (1000\ \text{kg/m}^3)(0.85)(0.30\ \text{m})(0.982\ \text{m/s})/(0.011\ \text{kg/m}\cdot\text{s}) = 22{,}764$$

The Reynolds number is greater than 2000, and, as a consequence, we have turbulent flow. We then must go to the Moody diagram in the *FE Supplied-Reference Handbook* and select the value of f for commercial steel pipe. The handbook lists the roughness as $e = 0.046$ mm. Calculating the relative roughness as e/D we get

$$e/D = (0.046\ \text{mm})/(300\ \text{mm}) = 0.0001533$$

The Moody diagram lists the friction factor for an e/D of 0.0001533 and Re = 22,764 to be 0.0265. Now we can calculate the friction head loss in the pipe.

$$h_L = [(0.0265)(5000\ \text{m})/(0.30\ \text{m})](0.050\ \text{m}) = 22.08\ \text{m}$$

The values may then be put in the simplified Bernoulli equation

$$h_A = 0.050\ \text{m} + 150\ \text{m} + 2.125\ \text{m} + 1.25\ \text{m} + 22.08 = 175.5\ \text{m oil}$$

Therefore, the work added to the oil is

$$\text{Power} = P(\text{AV}) = (\rho g h)(250\ \text{m}^3/\text{hr})/(3600\ \text{s/hr}) = (\rho g h)(0.0694\ \text{m}^3/\text{s})$$
$$\text{Power} = (1000\ \text{kg/m}^3)(0.85)(9.81\ \text{m/s}^2)(175.5\ \text{m}) \times (0.0694\ \text{m}^3/\text{s})[(\text{N})/(\text{kg}\cdot\text{m/s}^2)](\text{kJ}/10^3\ \text{N}\cdot\text{m})$$
$$\text{Power} = 101.56\ \text{kW}$$

OPEN-CHANNEL FLOW

Open-channel flow is the type of flow that has a free surface in contact with atmospheric pressure. The flow is caused by a sloping of the channel. Accurate solutions are very difficult to obtain because of the varied conditions under which the data is obtained.

Example **12.30**

What flow can be expected in a 3-m-wide rectangular, cement-lined channel with a slope of 5 m in a distance of 15,000 m. The water is to flow 0.5 m deep.

Solution

Take the Manning equation from the *Handbook* to solve for the velocity. Then take the velocity and multiply by the area of flow to obtain the flow rate.

$$V - (1/n)R^{2/3}S^{1/2}$$

where

V	= velocity	m/s
n	= roughness coefficient	0.015
R	= Hydraulic radius	m
S	= Slope of channel	m/m

$$V = (1/0.015)[(3 \times 0.5\ \text{m}^2)/(0.5 + 3 + 0.5\ \text{m})]^{2/3}(5\ \text{m}/15{,}000\ \text{m})^{1/2} = 0.633\ \text{m/s}$$
$$\text{Flow rate} = AV = (3 \times 0.5\ \text{m}^2)(0.633\ \text{m/s}) = 0.95\ \text{m}^3/\text{s}$$

FANS

A fan is a device that moves gasses or vapors from one location to another. Since fans are usually low-velocity devices, the gas, which usually is air, can be considered to be incompressible for the majority of engineering calculations. The general characteristics of fan operation are the following:

a). Volumetric fan output varies directly with the fan speed of rotation for a given fan.

b). The pressure, or head, of the fan varies directly with the square of the speed of the fan.

c). The power required to run a given fan varies directly with the cube of the speed.

d). For a given installation and constant fan speed, the pressure output and the operating power required will be proportional to the density of the gas.

e). For a constant mass flow rate of gas, the fan speed, the volumetric output, and the pressure vary inversely with the density of the gas. In addition, the power required varies inversely with the square of the density of the gas.

f). At a constant pressure, the speed, volumetric output, and power vary inversely with the square root of the density of the gas.

Example 12.31

A coal-fired boiler requires 1700 m^3/min of air at a pressure of 15 cm of water for the combustion process. The fan has a mechanical efficiency of 58% at the given conditions. How large of a motor, in kW, is required to power this fan?

Solution

Generally, we may conclude that fans are low-pressure devices and, as a consequence, that the air is incompressible. The power output, we learned back in thermodynamics, is equal to pressure times the volume rate of flow.

$$\text{Power} = \text{Pressure} \times \text{Volume rate of flow}$$

where

$$\text{Pressure} = \rho g h = (1000\ \text{kg/m}^3)(9.81\ \text{m/s}^2)(0.15\ \text{m}) = 1471\ \text{N/m}^2$$

Therefore

$$\text{Power} = (1471\ \text{N/m}^2)(1700\ \text{m}^3/\text{min})/(60\ \text{s/min}) = 41{,}678\ \text{N}\cdot\text{m/s}$$
$$\text{Power} = 41.7\ \text{kW}$$

and

$$\text{Motor kW} = (\text{Air power})/\eta = 41.7/0.58 = 71.9\ \text{kW motor needed.}$$

Example 12.32

An existing ventilating fan delivers 400 m^3/min of air against a back pressure of 5 cm water when the fan is operating at 450 rpm. The plant wishes to increase the air flow to 500 m^3/min.

Determine

a). Air power produced by the fan

b). New speed to obtain the increased air flow

c). The new back pressure

d). The new air power at the increased flow rate

Solution

a). Knowing that power is equal to pressure times volume rate of flow we get

$$\text{Power} = (\rho gh)(\text{Volume flow rate}) = (1000 \text{ kg/m}^3)(9.81 \text{ m/s}^3)(0.05 \text{ m})(400 \text{ m}^3\text{/min})$$
$$\text{Power} = 196{,}200 \text{ kg}\cdot\text{m}^2\text{/s}^2\cdot\text{min} = 196{,}200 \text{ N}\cdot\text{m/min}$$
$$\text{Power} = (196{,}200 \text{ N}\cdot\text{m/min})/60 \text{ s/min} = 3270 \text{ N}\cdot\text{m/s or J/s}$$

Therefore

$$\text{Power} = 3.27 \text{ kW}$$

b). Knowing that the volumetric air flow rate is directly proportional to the fan speed we get

$$(\text{Speed})_{\text{new}}/(\text{Speed})_{\text{old}} = (\text{Flow rate})_{\text{new}}/(\text{Flow rate})_{\text{old}}$$
$$(\text{Speed})_{\text{new}} = [(500 \text{ m}^3\text{/min})/(400 \text{ m}^3\text{/min})](450 \text{ rpm})$$
$$(\text{Speed})_{\text{new}} = 563 \text{ rpm}$$

c). Knowing that the back pressure is directly proportional to the square of the speed of the fan

$$(\text{Pressure})_{\text{new}}/(\text{Pressure})_{\text{old}} = [(\text{Speed})_{\text{new}}/(\text{Speed})_{\text{old}}]^2$$
$$(\text{Pressure})_{\text{new}} = [(563 \text{ rpm})/(450 \text{ rpm})]^2(5 \text{ cm}) = 7.83 \text{ cm}$$

d). Using the equation for power, Power = Pressure × Volume flow rate, at the new conditions

$$\text{Power} = (1000 \text{ kg/m}^3)(9.81 \text{ m/s}^2)(0.0783 \text{ m})(500 \text{ m}^3\text{/min})$$
$$= 384{,}062 \text{ N}\cdot\text{m/min} = 6.4 \text{ kW}$$

A second method is to use the power law, which is

$$\text{Power}_2 = \text{Power}_1(V_1/V_2)^3 = 3.27(563/450)^3 = 6.4 \text{ kW}$$

which is the same value as calculated by the first method.

Example 12.33

A blower delivers 350 m³/min of air at standard atmosphere and 20 °C when rotating at 500 rpm, with a back pressure of 5 cm of water gage pressure. If the air temperature rises to 90 °C and the speed of the fan remains constant, determine:

a). New back pressure in cm of water

b). Ratio of the new power required to the original power required

Solution

We should examine these changes in light of the fan laws. The volumetric flow rate remains constant because the speed of rotation is unchanged. We find, however, that the pressure head is proportional to the density of the gas and that the operating power is also proportional to the density of the gas.

a). Using the perfect gas law, written in terms of density

$$\rho = P/RT$$

Letting the atmospheric pressure remain the same along with the gas constant

$$\rho_1/\rho_2 = [P/RT]_1/[P/RT]_2 = T_2/T_1 = (90 + 273)/(20 + 273) = 1.24$$

The new density is

$$\rho_2 = \rho_1/1.24$$

Therefore, using the fan laws

$$P_2 = P_1/(\rho_2/\rho_1) = 5\ \text{cm}(\rho_1/1.24)(1/\rho_1) = 4.03\ \text{cm H}_2\text{O}$$

b). Using the density ratio calculated above

$$[\text{Power}_2/\text{Power}_1] = \rho_2/\rho_1 = 1/1.24 = 0.806$$

This means that the power requirement of the 90 °C air is only 80.6% of the required power needed if the air were at 20 °C.

Example 12.34

Using the same fan installation as given in Example 12.33, it is desired to maintain the same mass flow rate of air at 90 °C as the mass flow rate at 20 °C. Determine

a). New fan speed

b). New volumetric flow rate

c). New back pressure

d). New power required

Solution

Going back to the fan laws it is noted that, for a constant mass flow rate the fan speed, the fan volumetric flow rate and the back pressure vary inversely with the density of the gas. In addition, the power required varies inversely with the square of the density of the gas.

a). New fan speed = (Old fan speed)(ρ_1/ρ_2) = (500 rpm)(1.24) = 620 rpm

b).

$$\begin{aligned}\text{New volumetric flow rate} &= (\text{Old volumetric flow rate})(\rho_1/\rho_2)\\ &= (350\ \text{m}^3/\text{min})(1.24) = 434\ \text{m}^3/\text{min}\end{aligned}$$

c).

$$\begin{aligned}(\text{New back pressure}) &= (\text{Old back pressure})(\rho_1/\rho_2)\\ &= (5\ \text{cm})(1.24) = 6.2\ \text{cm}\end{aligned}$$

d).

$$\begin{aligned}\text{New power required} &= \text{Old power required } (\rho_1/\rho_2)^2\\ &= [(\rho g h_1)(\text{Volumetric flow rate})](1.24)^2\\ &= (1000\ \text{kg/m}^3)(9.81\ \text{m/s}^2)(0.05\ \text{m})(350\ \text{m}^3/\text{min})(1.24)^2\\ &= 263{,}967\ \text{N}\cdot\text{m/min} = 4.4\ \text{kW}\end{aligned}$$

PUMPS

A pump is a device that is used to increase the pressure of a liquid. Pumps fall into different categories, such as axial-flow pumps, centrifugal-flow pumps, piston pumps, and other types of positive-displacement pumps. A typical centrifugal-flow pump has the liquid entering at the central portion of the pump, or hub, and then the liquid is accelerated outward by means of the rotating impeller. Some of the kinetic energy imparted to the liquid results in a pressure rise of the liquid. The efficiency of a typical pump is equal to

$$\eta_p = \text{(Fluid power out of the pump)/(Power input to the pump)}$$

Example 12.35

A 12-kW pump running at 1500 rpm has an inlet diameter of 25 cm, and the discharge line is 15 cm in diameter. The output from the pump is 3100 l/min of 20 °C water at 110 kPa. The pressure at the suction side is a negative 12 cm of mercury and the centerline of the pump discharge is 1 meter above the centerline of the intake pipe. Determine the efficiency of the pump.

Solution

Write the Bernoulli equation in order to find the energy added to the water. The efficiency may then be obtained using the above mentioned efficiency equation.

$$(P/\rho g)_1 + Z_1 + V_1^2/2g + E = (P/\rho g)_2 + Z_2 + V_2^2/2g + F$$

where, in addition to the normal terms for the Bernoulli equation covered in the *Fluid Mechanics* chapter, E = Energy added to the system F = Friction loss in the system, which is equal to zero in this case

Solving for each term of the Bernoulli equation

$$\begin{aligned}(P/\rho g)_1 &= [(-12 \text{ cm Hg})(13.6 \text{ cm } H_2O/\text{cm Hg})(1000 \text{ kg/m}^2) \\ &\quad \times (9.81 \text{ m/s}^2)]/[(1000 \text{ kg/m}^2)(9.81 \text{ m/s}^2)] \\ &= -1.632 \text{ m } H_2O\end{aligned}$$

$$\begin{aligned}(P/\rho g)_2 &= [(110{,}000 \text{ Pa})(1 \text{ N/m}_2/\text{Pa})(1 \text{ kg}\cdot\text{m/s}^2)/1 \text{ N}]/[(1000 \text{ kg/m}^2)(9.81 \text{ m/s}^2)] \\ &= 11.2 \text{ m } H_2O \\ Z_2 - Z_1 &- 1.0 \text{ m } H_2O\end{aligned}$$

To determine the velocity head we must use both the continuity equation and the kinetic energy of the fluid.

$$A_1V_1 = A_2V_2 = 3100 \text{ l/min} = 3.1 \text{ m}^3\text{/min} = 0.05167 \text{ m}^3\text{/s}$$
$$A_1V_1 = (\pi D_1^2/4)V_1 - [\pi(0.25 \text{ m})^2]/4V_1 = 0.05167 \text{ m}^3\text{/s}, \quad \text{or} \quad V_1 = 1.053 \text{ m/s}$$

Therefore, the velocity head entering is

$$V_1^2/2g = (1.053 \text{ m/s})^2/[2(9.81 \text{ m/s}^2)] = 0.0565 \text{ m } H_2O$$

Now, examining the fluid leaving the pump

$$A_1V_1 = A_2V_2$$

Therefore

$$\begin{aligned}V_2 &= (A_1/A_2)V_1 \\ &= [(\pi D_1^2/4)/(\pi D_2^2/4)]V_1 = (0.25/0.15)^2(1.053) = 2.925 \text{ m/s}\end{aligned}$$

Therefore, the velocity head leaving is

$$V_2^2/2g = (2.925)^2/[2(9.81)] = 0.436 \text{ m } H_2O$$

Now that all of the terms have been calculated we may put them together into the Bernoulli equation to determine the amount of energy added

$$-1.632 + Z_1 + 0.0565 + E = 11.2 + (Z_1 + 1.0) + 0.436$$

Therefore

$$E = 14.21 \text{ m } H_2O \text{ is added to the system}$$

Now we need to calculate the power of the pump if the system were 100% efficient.

$$\begin{aligned}\text{Power} &= \text{Pressure} \times \text{Volume flow rate} \\ &= (14.21 \text{ m})(1000 \text{ kg/m}^3)(9.81 \text{ m/s}^2)(3.1/60)\text{m}^3/\text{s} \\ &= 7202 \text{ kg}\cdot\text{m}^2/\text{s}^3 = 7202 \text{ N}\cdot\text{m/s} = 7202 \text{ J/s}, \quad \text{or} \quad 7.202 \text{ kW}\end{aligned}$$

Therefore, the efficiency is

$$\eta = (7.202 \text{ kW})/(12 \text{ kW}) = 60\%$$

PUMP AFFINITY LAWS

The pump affinity laws give approximate results regarding the operation of centrifugal pumps. For pumps of constant speed

a). Capacity varies directly with the impeller diameter.

b). Head varies directly with the square of the impeller diameter.

c). Horsepower varies directly with the cube of the impeller diameter.

In addition, the laws for the operation of fans also hold true. Specifically:

a). The capacity of the flow varies directly with the speed of the impeller.

b). The pressure head varies directly with the square of the speed.

c). The power required varies directly with the cube of the speed.

Example 12.36

Using the basic information given in Example 12.35, replace the 1500-rpm motor with a 1760-rpm motor and determine

a). New flow rate

b). New output pressure

c). New power requirement

Assume that the efficiency of the unit remains constant.

Solution

We may use the affinity laws for the solution of the problem.

a). Capacity varies directly with impeller rpm.

$$\begin{aligned}\text{Capacity}_2 &= \text{Capacity}_1[(\text{rpm}_2)/(\text{rpm}_1)] = 3.1 \text{ m}^3/\text{min } (1760/1500) \\ &= 3.63 \text{ m}^3/\text{min}\end{aligned}$$

b). Pressure varies directly with the square of the impeller rpm.

$$P_2 = P_1[(\text{rpm}_2)/(\text{rpm}_1)]^2 = (110\ \text{kPa})[1760/1500]^2 = 151.4\ \text{kPa}$$

c). Power required varies directly with cube of the impeller rpm.

$$\text{Power}_2 = \text{Power}_1[(\text{rpm}_2)/(\text{rpm}_1)]^3 = 12\ \text{kW}(1760/1500)^3 = 19.38\ \text{kW}$$

CAVITATION

When the local pressure of a liquid falls below its vapor pressure the liquid will vaporize, and, by vaporization, it will form a bubble of vapor that will push the liquid away from that location. This bubble formed is called a cavity or a void. When the outer, cooler liquid absorbs the heat from the cavity the vapor in the cavity returns to a liquid form. In general, it will occupy about 1/1000 of the space, and the collapse is essentially instantaneous and can actually be quite violent. This cavitation can cause extensive damage in impellers and even in other parts on the pump. In addition, the collapse may be quite noisy and cause a sever reduction in the efficiency of the pump. Consequently, care should be made to keep the suction side of the pump at a pressure higher than the vapor pressure of the liquid.

Example 12.37

A pump is used to draw 20 °C water out of a mine shaft. The velocity in the intake line to the pump is an average of 3 m/s. At a particular point in the intake line the pressure is 10.0 kPa. How far above this point would cavitation in the line start?

Solution

We may determine the vapor pressure of the water from the saturated water tables in the *Handbook*. At 20 °C the saturation pressure, i.e., vapor pressure, is equal to 2.239 kPa absolute. In the same table, at the same conditions, the specific volume ($v = 1/\rho$) is equal to 0.001002 m^3/kg. Therefore, density, ρ, is equal to 998 kg/m^3.

The total pressure in the line at the reference point is equal to the gage pressure plus the atmospheric pressure. Therefore, assuming that the atmospheric pressure is standard, the total pressure is

$$P_{\text{total}} = P_{\text{gage}} + P_{\text{atmospheric}} = 10\ \text{kPa} + 101.3\ \text{kPa} = 111.3\ \text{kPa}$$

In order for the water to flash into vapor, the total pressure would have to be equal to the vapor pressure, or 2.239 kPa. This means we would have to have a drop in pressure created by a negative head of water equivalent to 111.3 kPa − 2.239 kPa = 109.06 kPa. Using the manometer equation

$$P = \rho g h$$

or

$$\begin{aligned} h &= P/\rho g \\ &= [(111.3\ \text{kPa})(1000\ \text{N/m}^2)/\text{kPa}]/[(998\ \text{kg/m}^3)(9.81\ \text{m/s}^2)(\text{N/kg}\cdot\text{m/s}^2)] \\ &= 11.37\ \text{m} \end{aligned}$$

Therefore, the pump intake cannot be located any higher than this particular point in order to prevent cavitation in the line.

NPSH (NET POSITIVE SUCTION HEAD)

Cavitation is also able to occur in the impeller of a centrifugal pump due to the rotating impeller. To prevent this situation from occurring, a pump manufacturer will test the pump to determine how much pressure is needed at the intake to prevent cavitation during pump operation. This pressure is called *Net Positive Suction Head (NPSH)*. The equation defining the NPSH is equal to

$$\text{NPSH} = P_i/\rho g + V_i^2/2g - P_v/\rho g$$

where

P_i = Inlet pressure to the pump (absolute pressure/total pressure)
ρ = Density at inlet conditions
V_i = Velocity at the inlet to the pump
P_v = Vapor pressure of the liquid being pumped

Example 12.38

Liquid propane is contained in a storage tank at a pressure of 1.4 MPa, which is the equilibrium vapor pressure of the propane at the pumping conditions. The propane has a specific gravity of 58. If the level of the liquid propane is 4.00 meters above the centerline of the pump, and if the frictional losses between the supply and the pump is 2.00 meters, determine the NPSH when the velocity in the suction line is 1.5 m/s.

Solution

Starting with the generic NPSH equation given in the *Handbook* and calculating each of the terms we get

$$\begin{aligned}\text{NPSH} &= P_i/\rho g + V_i^2/2g - P_v/\rho g \\ &= [1.4\text{ MPa} + 4.00\text{ m} - 2.00\text{ m}] + (1.5\text{ m/s})^2/2g - 1.4\text{ MPa} \\ &= 2.00\text{ m} + (1.5\text{ m/s})^2/2(9.81\text{ m/s}^2) = 2.11\text{ m} \qquad \text{available}\end{aligned}$$

Example 12.39

A pump discharges 7500 l/min of a certain brine solution (s.g. of 1.20) to an evaporating pond. The intake line is 30 cm in diameter and is at the same level as the pump discharge line, which is 20 cm in diameter. The pressure at the inlet of the pump is –150 mm of mercury. The pressure gage connected to the pipe discharge reads 140 kPa, and its center is 1.50 m above the center of the discharge flange. The vapor pressure of the brine solution is 35 mm of Hg.

a). If the pump efficiency is 82%, how much power is required?

b). If vapor pressure of the brine is 35.3 mm Hg, what is the NPSH?

c). If the flow rate is increased to 10,000 l/min, what is the new NPSH?

Solution

a). Starting with the Bernoulli equation to determine the energy added to the brine we get

$$P_1/\rho g + V_1^2/2g + Z_1 + E = P_2/\rho g + V_2^2/\rho g + Z_2 + F$$

Calculating the values for each of the above terms (noting that F = 0)

$$\begin{aligned} P_1/\rho g &= -[(150/760)(101.3\ \text{kPa})(10^3\ \text{N/m}^2)/\text{kPa}]/\\ &\quad [(1.20)(1000\ \text{kg/m}^3)(9.81\ \text{m/s}^2)\text{N/kg}\cdot\text{m/s}^2]\\ &= -1.698\ \text{m}\\ V_1^2/2g &= [(AV)/A]^2/2g\\ &= [(7.5\ \text{m}^3/\text{min})/(\pi D_1^2/4\ \text{m}^2)(60\ \text{s/min})]^2/2(9.81\ \text{m/s}^2)\\ &= [(0.125\text{m}^3/\text{s})/\pi(0.30)^2/4\ \text{m}^2]^2/(19.81\ \text{m/s}^2)\\ &= (1.77\ \text{m/s})^2/19.81\ \text{m/s}^2 = 0.158\ \text{m}\\ Z_2 - Z_1 &= 0 \end{aligned}$$

The pressure head developed by the pump will be a function of the gage pressure at its location plus the effect of the location of the pressure gage.

$$\begin{aligned} P_2/\rho g &= P_{\text{gage}} + \text{Elevation of the gage}\\ &= [(140{,}000\ \text{Pa})(\text{N/m}^2)/(\text{Pa})]/[(1000\ \text{kg/m}^3)\\ &\quad (1.2)(9.81\ \text{m/s}^2)\text{N}/(\text{kg}\cdot\text{m/s}^2)] + 1.5\ \text{m} = 13.39\ \text{m} \end{aligned}$$

It is safe to assume that the density of the brine is constant across the pump; therefore, the velocity is merely a function of the area.

$$A_1V_1 = A_2V_2$$

or,

$$\begin{aligned} V_2 &= V_1(A_1/A_2)\\ &= (1.77\ \text{m/s})(D_1/D_2)^2\\ V_2 &= (1.77\ \text{m/s})(30/20)^2 = 3.98\ \text{m/s} \end{aligned}$$

Calculating the pressure head on the pump outlet side

$$V_2^2/2g = (3.98\ \text{m/s})^2/[2(9.81\ \text{m/s}^2)] = 0.807\ \text{m}$$

Therefore

$$\begin{aligned} E &= V_2^2/2g - V_1^2/2g + P_2/\rho g - P_1/\rho g\\ E &= 0.807 - 0.158 + 13.39 - (-1.698) = 15.737\ \text{m brine} \end{aligned}$$

The power required to drive the pump may be obtained using the equation in the Fluid Mechanics section of the *Handbook*

$$\begin{aligned} W &= Q\gamma h/\eta\\ &= \{[(7.5\ \text{m}^3/\text{min})/60\ \text{s/min}](1000\ \text{kg/m}^3)(9.81\ \text{m/s}^2)(1.2)(15.737\ \text{m})\}/0.82\\ &= (28{,}240\ \text{kg}\cdot\text{m}^2/\text{s}^2)(\text{N/kg}\cdot\text{m/s}^2)(\text{J/N}\cdot\text{m}) = 28.24\ \text{kW} \end{aligned}$$

b).

$$\text{NPSH} = P_{\text{absolute}}/\gamma + V_1^2/2g - h_v$$

Assuming that the atmospheric pressure is standard and equal to 1.013 Bar

$$\text{NPSH} = [P_{\text{pump inlet}} + P_{\text{barometric}}]/\gamma + V_1^2/2g - P_v/\gamma$$

Solving for each of the terms

$$P_v = (35/760)(1.013 \text{ bar}) = 0.04665 \text{ bar}$$
$$\gamma_{brine} = \rho g = [(1000 \text{ kg/m}^3)(1.2)](9.81 \text{ m/s}^2) = 11{,}772 \text{ kg/m}^2 \cdot \text{s}^2$$
$$h_v = [(0.04665 \text{ bar})(10^5 \text{N/m}^2)/\text{bar}][(\text{kg} \cdot \text{m/s}^2)/\text{N}]/11{,}772 \text{ kg/m}^2 \cdot \text{s}^2 = 0.396 \text{ m}$$

$h_{\text{pump inlet}}$ has been previously calculated and is equal to -1.698 m

$$h_{\text{barometric}} = [(1.013 \text{ bar})(10^5 \text{ N/m}^2)/(\text{bar})][(\text{kg} \times \text{m/s}^2)/(\text{N})]/(11{,}772 \text{ kg/m}^2/\text{s}^2)$$
$$h_{\text{barometric}} = 8.61 \text{ m}$$

$h_{\text{velocity head}}$ previously calculated as 0.159 m

$$\text{NPSH} = 8.61 - 1.698 + 0.159 - 0.396 = 6.675 \text{ m}$$

c). The flow rate is increased to 10 m^3/min; determine the new NPSH.

Knowing that the vapor pressure remains constant and that the velocity head is increased as the ratio of the square of the flow rates

$$h_{\text{velocity head}} = (10/7.5)^2(0.159) = 0.283 \text{ m}$$
$$h_v = 0.396 \text{ m}$$

According to the affinity laws regarding pump operation, the head varies directly with the square of the flow rate. Therefore

$$h_{\text{pump inlet}} = (10/7.5)^2\,(-1.698) = -3.02 \text{ m}$$
$$\text{NPSH}_{\text{new}} = 8.61 - 3.02 + 0.283 - 0.396 = 5.477 \text{ m}$$

COMPRESSORS

Oftentimes compressors are positive-displacement gas/vapor "pumps" using reciprocating pistons. The capacity of the compressor is usually stated as the quantity of gas handled at intake conditions.

Example 12.40

Assuming isothermal compression with a compressor having no clearance, determine the work required to compress 150 m^3/hr of *standard air* to 700 kPa gage in a single-stage compressor?

Solution

One may take the isothermal work equation from the *Handbook*

$$W = mRT \ln(P_2/P_1)$$

Using the perfect gas relation, $PV = mRT$ and $P_1 = 101{,}300$ Pa as standard pressure, work may be calculated as

$$W = P_1V_1\ln(P_2/P_1)$$
$$= [(101{,}300 \text{ Pa})(\text{N/m}^2)/\text{Pa}][(150 \text{ m}^3/\text{hr})/(3600 \text{ s/hr})]\ln(101.3/801.3)$$
$$= -8729.3 \text{ N} \cdot \text{m/s} = 8.73 \text{ kW of work done on the gas.}$$

SELECTED SYMBOLS AND ABBREVIATIONS

Symbol or Abbreviation	Description
α	angle with horizontal
β	angle with vertical
C	loss coefficient
C_a	Cauchy number
c	discharge coefficient, speed of sound
c_c	contraction coefficient
c_p	gas specific heat at constant pressure
c_v	velocity coefficient, gas specific heat at constant volume
δ	small thickness
d	diameter
d_e	equivalent diameter
ε	roughness factor
F_B	buoyant force
f	Darcy friction factor
F	Froude number
γ	specific weight
g_c	gravitational constant
h	depth of fluid
h_A	pump head
h_f	head loss from friction
h_R	turbine head
I	moment of inertia
k	ratio of specific heats
l	characteristic length
$\dot{m}$	mass flow rate
Ma	mach number
MW	molecular weight
μ	dynamic viscosity
η	efficiency
v	kinematic viscosity
P	power to or from fluid
p	pressure
Q	volume rate of flow
ρ	density
R	gas constant
$\overline{R}$	universal gas constant
Re	Reynolds number
R_H	hydraulic radius
σ	surface tension

(Continued)

(Continued)

Symbol or Abbreviation	Description
s	seconds
SG	specific gravity
τ	shear stress
T	absolute temperature, °R, °K
V	fluid velocity, volume
V_D	displaced volume
v	vane velocity, specific volume
$\dot{W}$	mechanical (shaft) power
W_c	Weber number
WP	Wetted perimeter
Z	elevation

PROBLEMS

12.1 Kinematic viscosity may be expressed in units of:

a. m^2/s c. kg • s/m
b. s^2/m d. kg/s

12.2 The absolute viscosity of a fluid varies with pressure and temperature and is defined as a function of:

a. density and angular deformation rate
b. density, shear stress, and angular deformation rate
c. density and shear stress
d. shear stress and angular deformation rate

12.3 An open chamber rests on the ocean floor in 50 m of sea water (SG = 1.03). The air pressure in kilopascals that must be maintained inside to exclude water is nearest to:

a. 318 c. 505
b. 431 d. 661

12.4 What is the static gage pressure in pascals in the air chamber of the container in Exhibit 12.4? The specific weight of the water is 9810 N/m³.

a. −14,700 Pa c. 0
b. −4500 Pa d. +4500 kPa

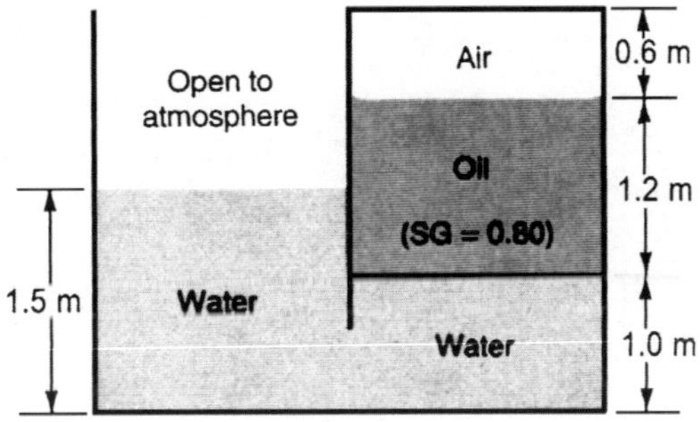

Exhibit 12.4

12.5 The pressure in kilopascals at a depth of 100 meters in fresh water is nearest to:

a. 268 kPa c. 981 kPa
b. 650 kPa d. 1620 kPa

12.6 What head, in meters of air, at ambient conditions of 100 kPa and 20 °C, is equivalent to 15 kPa?

a. 49 c. 257
b. 131 d. 1282

12.7 With a normal barometric pressure at sea level, the atmospheric pressure at an elevation of 1200 meters is nearest to:

a. 87.3 kPa c. 115.3 kPa
b. 83 kPa d. 101.3 kPa

12.8 The funnel in Exhibit 12.8 is full of water. The volume of the upper part is $0.165m^3$ and of the lower part is $0.057m^3$. The force tending to push the plug out is:

a. 1.00 kN
b. 1.47 kN
c. 1.63 kN
d. 2.00 kN

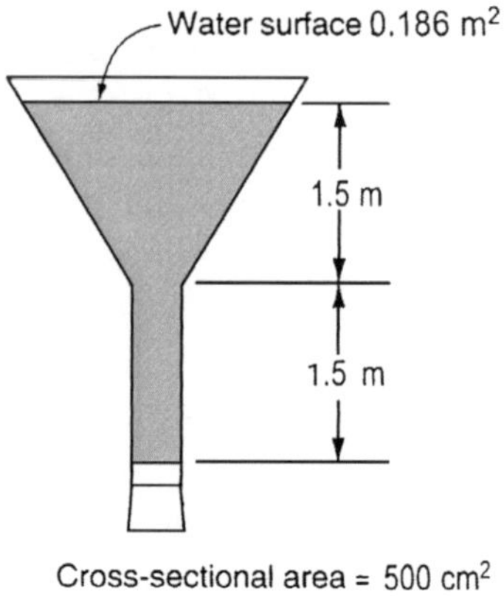

Exhibit 12.8

12.9 An open-topped cylindrical water tank has a horizontal circular base 3 meters in diameter. When it is filled to a height of 2.5 meters, the force in Newtons exerted on its base is nearest to:

a. 17,340
b. 34,680
c. 100,000
d. 170,000

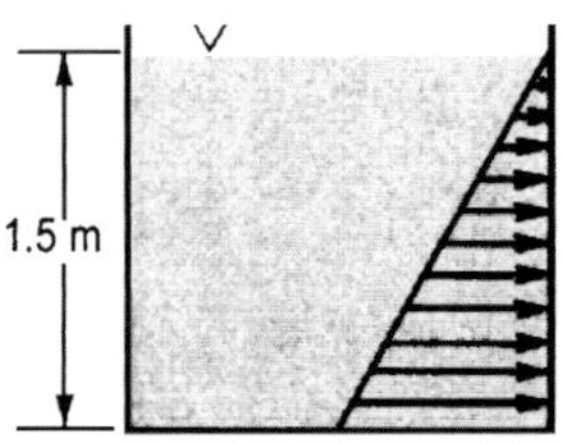

Exhibit 12.10

12.10 A cubical tank with 1.5 meter sides is filled with water (see Exhibit 12.10). The force, in kilonewtons, developed on one of the vertical sides is nearest to:

a. 4.1
b. 8.3
c. 16.5
d. 33.0

12.11 A conical reducing section (see Exhibit 12.11) connects an existing 10-centimeter-diameter pipeline with a new 5-centimeter-diameter line. At 700 kPa under no-flow conditions, what tensile force in kilonewtons is exerted on the reducing section?

a. 5.50
b. 2.07
c. 1.37
d. 4.13

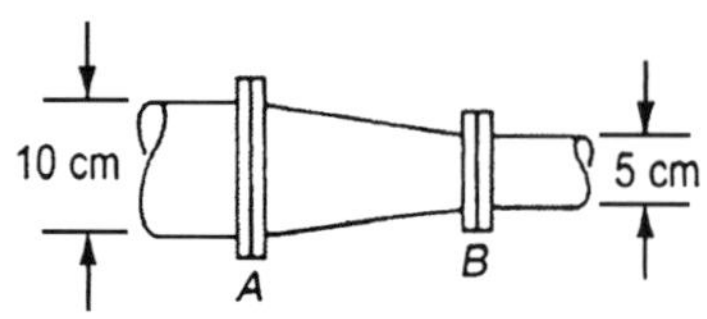

Exhibit 12.11

12.12 A circular access (see Exhibit 12.12) port 60 cm in diameter seals an environmental test chamber that is pressurized to 100 kPa above the external pressure. What force in newtons does the port exert upon its retaining structure?

a. 7100
b. 9500
c. 14,100
d. 28,300

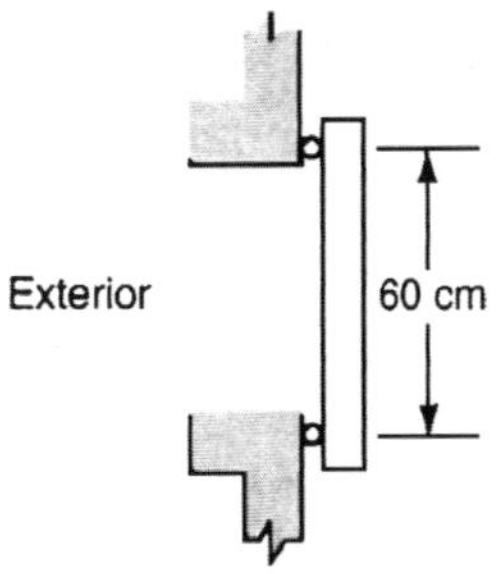

Exhibit 12.12

12.13 A gas bubble rising from the ocean floor is 2.5 centimeters in diameter at a depth of 15 meters. Given that the specific gravity of seawater is 1.03, the buoyant force in newtons being exerted on the bubble at this instant is nearest to:

a. 0.0413
b. 0.0826
c. 0.164
d. 0.328

12.14 The ice in an iceberg has a specific gravity of 0.922. When floating in seawater (SG = 1.03), the percentage of its exposed volume is nearest to:

a. 5.6
b. 7.4
c. 8.9
d. 10.5

12.15 A cylinder of cork is floating upright in a container partially filled with water. A vacuum is applied to the container that partially removes the air within the vessel. The cork will:

a. rise somewhat in the water
b. sink somewhat in the water
c. remain stationary
d. turn over on its side

12.16 A floating cylinder 8 cm in diameter and weighing 9.32 newtons is placed in a cylindrical container that is 20 cm in diameter and partially full of water. The increase in the depth of water when the float is placed in it is:

a. 10 cm
b. 5 cm
c. 3 cm
d. 2 cm

12.17 A block of wood floats in water (see Exhibit 12.17) with 15 centimeters projecting above the water surface. If the same block were placed in alcohol of specific gravity 0.82, the block would project 10 centimeters above the surface of the alcohol. The specific gravity of the wood block is:

a. 0.67
b. 3.00
c. 0.55
d. 0.60

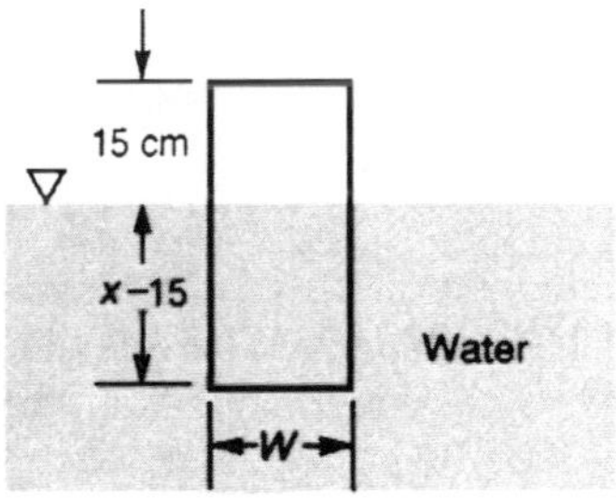

Exhibit 12.17

12.18 The average velocity in a full pipe of incompressible fluid at Section 1 in Exhibit 12.18 is 3 m/s. After passing through a conical section that reduces the stream's cross-sectional area at Section 2 to one-fourth of its previous value, the velocity at Section 2, in m/s, is:

a. 1.0 c. 3
b. 1.5 d. 12

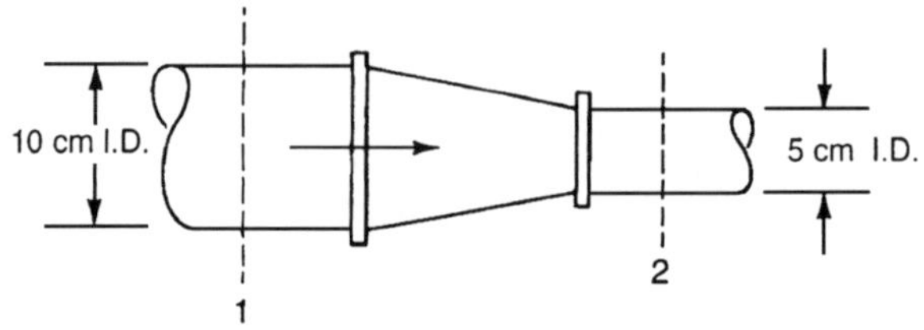

Exhibit 12.18

12.19 Refer to Exhibit 12.18. If the static pressure at Section 1 is 700 kPa and the 10-cm-diameter pipe is full of water undergoing steady flow at an average velocity of 10 m/s at *A*, the mass flow rate in kg/s at Section 2 is nearest to:

a. 10.0 c. 78.5
b. 19.5 d. 98.6

12.20 Air flows in a long length of 2.5-cm-diameter pipe. At one end the pressure is 200 kPa, the temperature is 150°C, and the velocity is 10 m/s. At the other end, the pressure has been reduced by friction and heat loss to 130 kPa. The mass flow rate in kg/s at any section along the pipe is nearest to:

a. 0.008 c. 0.126
b. 0.042 d. 0.5

12.21 Water flows through a long 1.0 cm I.D. hose at 10 liters per minute. The water velocity in m/s is nearest to:

a. 1 c. 4.24
b. 2.12 d. 21.2

12.22 Gasoline ($\rho = 800$ kg/m^3) enters and leaves a pump system with the energy in N • m/N of fluid that is shown in the following table:

	Entering	Leaving
Potential energy, *Z* meters above datum	1.5	4.5
Kinetic energy, $V^2/(2g_c)$	1.5	3.0
Flow energy, p/γ	9.0	45
Total energy	12.0	52.5

The pressure increase in kPa between the entering and leaving streams is nearest to:

a. 283 c. 722
b. 566 d. 803

12.23 Use the data of Problem 12.22. If the volume flow rate of the gasoline (800 kg/m^3) is 55 liters per minute, the theoretical pumping power, in kW, is nearest to:

a. 0.3
b. 0.5
c. 3.2
d. 300

12.24 Water flowing in a pipe enters a horizontal venturi meter whose throat area at *B* is 1/4 that of the original and final cross sections at *A* and *C*, as shown in Exhibit 12.24. Continuity and energy conservation demand that which one of the following be *TRUE*?

a. The pressure at *B* is increased.
b. The velocity at *B* is decreased.
c. The potential energy at *C* is decreased.
d. The flow energy at *B* is decreased.

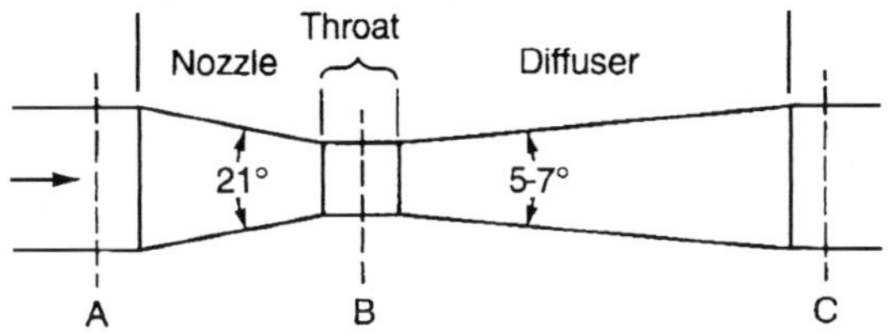

Exhibit 12.24

12.25 Given the energy data in N•m/N shown below existing at two sections across a pipe transporting water in steady flow, what frictional head loss in feet has occurred?

	Section A	Section B
Potential energy	20	40
Kinetic energy	15	15
Flow energy	100	75
Total	135	130

a. 0
b. 5
c. 130
d. 265

12.26 The power in kilowatts required in the absence of losses to pump water at 400 liters per minute from a large reservoir to another large reservoir 120 meters higher is nearest to:

a. 5.85
b. 7.85
c. 15.70
d. 30.00

12.27 The theoretical velocity generated by a 10-meter hydraulic head is:

a. 3 m/s
b. 10 m/s
c. 14 m/s
d. 16.4 m/s

12.28 What is the static head corresponding to a fluid velocity of 10 m/sec?

a. 5.1 m
b. 10.2 m
c. 16.4 m
d. 50 m

12.29 The elevation to which water will rise in a piezometer tube is termed the:
a. stagnation pressure
b. energy grade line
c. hydraulic grade line
d. friction head

12.30 A stream of fluid with a mass flow rate of 30 kg/s and a velocity of 6 m/s to the right has its direction reversed 180° in a "U" fitting. The net dynamic force in N exerted by the fluid on the fitting is nearest to:
a. 180
b. 360
c. 2030
d. 4300

12.31 The thrust in newtons generated by an aircraft jet engine on takeoff, for each 1 kg/s of exhaust products whose velocity has been increased from essentially 0 to 150 m/s, is nearest to:
a. 150
b. 1300
c. 3600
d. 7100

12.32 For the configuration in Exhibit 12.32, compute the velocity of the water in the 300-meter branch of the 15-cm-diameter pipe. Assume the friction factors in the two pipes are the same and that the incidental losses are equal in the two branches. The velocity in m/s is:
a. 10.0
b. 4.2
c. 1.8
d. 3.7

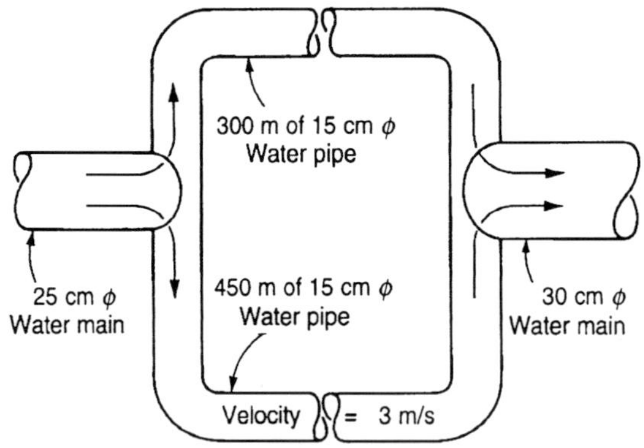

Exhibit 12.32

12.33 Which of the following statements most nearly approximates conditions in turbulent flow?
a. Fluid particles move along smooth, straight paths.
b. Energy loss varies linearly with velocity.
c. Energy loss varies as the square of the velocity.
d. Newton's law of viscosity governs the flow.

12.34 For turbulent flow of a fluid in a pipe, all of the following are true *EXCEPT*:

a. the average velocity will be nearly the same as at the pipe center
b. the energy lost to turbulence and friction varies with kinetic energy
c. pipe roughness affects the friction factor
d. the Reynolds number will be less than 2300

12.35 If the fluid flows in parallel, adjacent layers and the paths of individual particles do not cross, the flow is said to be:

a. laminar c. critical
b. turbulent d. dynamic

12.36 Which of the following constitutes a group of parameters with the dimensions of power?

a. ρAV c. $\dfrac{DV\rho}{\mu}$

b. pAV d. $\dfrac{\rho V^2}{P}$

12.37 At or below the critical velocity in small pipes or at very low velocities, the loss of head from friction:

a. varies linearly with the velocity
b. can be ignored
c. is infinitely large
d. varies as the velocity squared

12.38 The Moody diagram in Exhibit 12.38 is a log-log plot of friction factor vs. Reynolds number. Which of the lines *A*–*D* represents the friction factor to use for turbulent flow in a smooth pipe of low roughness ratio (ε/D)?

a. *A* c. *C*
b. *B* d. *D*

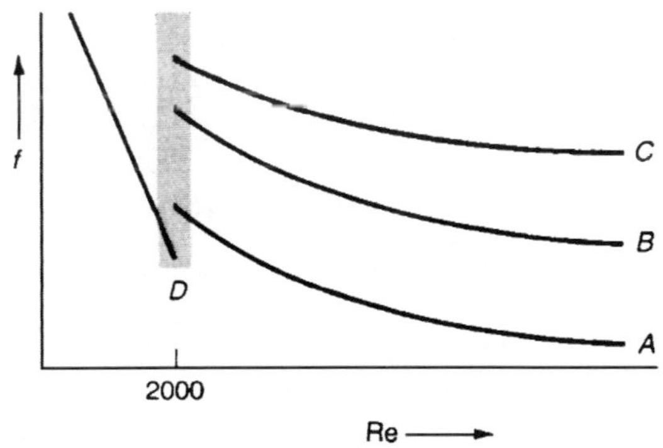

Exhibit 12.38

12.39 A 60-cm water pipe carries a flow of 0.1 m^3/s. At Point *A* the elevation is 50 meters and the pressure is 200 kPa. At Point *B*, 1200 meters downstream from *A*, the elevation is 40 meters and the pressure is 230 kPa. The head loss, in feet, between *A* and *B* is:

a. 6.94 c. 20.88
b. 15.08 d. 100.2

12.40 Entrance losses between tank and pipe, or losses through elbows, fittings, and valves are generally expressed as functions of:

a. kinetic energy
b. pipe diameter
c. friction factor
d. volume flow rate

12.41 A 5-cm-diameter orifice discharges fluid from a tank with a head of 5 meters. The discharge rate, Q, is measured at 0.015 m^3/s. The actual velocity at the *vena contracta* (v.c.) is 9 m/s. The coefficient of discharge, C_D, is nearest to:

a. 0.62
b. 0.77
c. 0.99
d. 0.86

12.42 At normal atmospheric pressure, the maximum height in meters to which a nonvolatile fluid of specific gravity 0.80 may be siphoned is nearest to:

a. 4.0
b. 6.4
c. 10.3
d. 12.9

12.43 The water flow rate in a 15-centimeter-diameter pipe is measured with a differential pressure gage connected between a static pressure tap in the pipe wall and a pitot tube located at the pipe centerline. Which volume flow rate Q in cubic meters per second results in a differential pressure of 7 kPa?

a. 0.005
b. 0.066
c. 0.50
d. 1.00

12.44 The hydraulic formula $CA\sqrt{2gh}$ is used to find the:

a. discharge through an orifice
b. velocity of flow in a closed conduit
c. length of pipe in a closed network
d. friction factor of a pipe

12.45 The hydraulic radius of an open-channel section is defined as:

a. the wetted perimeter divided by the cross-sectional area
b. the cross-sectional area divided by the total perimeter
c. the cross-sectional area divided by the wetted perimeter
d. one-fourth the radius of a circle with the same area

12.46 To calculate a Reynolds number for flow in open channels and in cross-sections, one must utilize hydraulic radius, R, and modify the usual expression for circular cross sections, which is

$$\text{Re} = \frac{DV\rho}{\mu} = \frac{VD}{\nu}$$

where D = diameter, V = velocity, ρ = density, μ = absolute viscosity, and ν = kinematic viscosity.

Which of the following modified expressions for Re is applicable to flow in open or noncircular cross sections?

a. $\dfrac{RD}{v}$　　c. $\dfrac{2RD}{v}$

b. $\dfrac{RV\rho}{\mu}$　　d. $\dfrac{4RV}{v}$

12.47 It is necessary to have a relationship regarding the shear stress of a fluid flowing in a pipe. Shear stress is thought to be a function of density, viscosity, and velocity of the fluid, as well as a function of the diameter and the roughness of the pipe containing the flowing fluid. The dimensionless equation relating these variables is

a. $C\rho V^2/\text{Re}^b$
b. $C\rho D/V^2$
c. $CD^2V/L\mu\rho$
d. $Cd\rho\mu/V$

Use the following information for Problems 12.48 & 12.49.

Given a 10-m × 10-m submerged gate used for flood control. The gate is hinged at the top and held in place by a force at the bottom. Consider that the top of the hinged gate is 10 meters below the surface of the water.

12.48 The center of pressure of the hinged gate is most nearly

a. 10 m
b. 12.67 m
c. 15.56 m
d. 18.43 m

12.49 The force, in kN, required to hold the gate in position is most nearly

a. 10,000
b. 12,500
c. 6,500
d. 8,200

Use the following information for Problems 12.50 through 12.53

A 40-cm-diameter pipeline has a flow rate of 1,000 m^3/hr of water at standard temperature. The pipeline is 5 km long, is made out of galvanized iron, and has a lift of 75 m. The water has a viscosity of 1.005×10^{-3} kg/m · s and the pump has an efficiency of 70%.

12.50 The friction factor for the above pipeline is most nearly
a. 0.014
b. 0.018
c. 0.022
d. 0.025

12.51 The pump work required, in meters of water added, is most nearly
a. 80 m
b. 90 m
c. 100 m
d. 120 m

12.52 The hp required for the pump is most nearly
a. 215
b. 330
c. 450
d. 620

12.53 If the flow rate were decreased to 300 m^3/hr how many hp is needed?
a. 75 hp
b. 125 hp
c. 105 hp
d. 135 hp

12.54 A heavy fuel oil with a specific gravity of 0.94 and a viscosity of 2.01 poise is to be pumped at the rate of 3000 m^3/day through an old 30 cm *ID* pipe that is badly corroded on the inside. What would be the friction factor?
a. 0.033
b. 0.055
c. 0.077
d. 0.093

12.55 An ore carrier 120 m long × 10 m wide displaces 8500 m^3 of fresh water. It is moved into a lock in fresh water that is 140 m long × 15 m wide and then is loaded with 3500 metric tons of ore. What will be the increase in the depth of the water in the lock after the ship is loaded?
a. 0.83 m
b. 1.89 m
c. 2.67 m
d. 3.89 m

12.56 A careless hunter shoots a 7.00-mm-diameter hole in the vertical side of a water tank one meter above the ground. The level of the water in the tank is 10.0 m above the ground. How far from the base of the tank will the water strike the ground?

Assume $C = 1.0$

a. 2.67 m
b. 4.75 m
c. 6.00 m
d. 7.88 m

12.57 The water level of a reservoir is 150 m above a power plant. A 30.0 cm *ID* pipe 200 m long connects the reservoir to the plant. If the friction factor for flow through the pipe is 0.02, how much water will flow to the plant?

a. 0.80 m^3/s
b. 1.0 m^3/s
c. 1.2 m^3/s
d. 1.4 m^3/s

12.58 A piece of glass weighs 125 g in air, 75 g in water, and 92 g in gasoline. What is the specific gravity of the gasoline?

a. 0.88
b. 0.80
c. 0.72
d. 0.66

12.59 A 30- × 30-cm timber 3.6 meters long floats level in fresh water with 11.0 cm above the water surface. One end is just touching a rock, which prevents that end from sinking any deeper. How far out from the supported end can a 68-kg man walk before the free end submerges? See Exhibit 12.59.

a. 1.0 m
b. 1.5 m
c. 2.1 m
d. 2.6 m

$\theta = 1.71°$ $\cos\theta = 0.9996$

Exhibit 12.59

12.60 An 20 cm *ID* pipeline, 1070 meters long, conveys water from a pump to a reservoir whose water surface is 450 ft above the pump which is pumping at the rate of 1.0 m^3/sec. Use constants for cast-iron pipe, disregard velocity head and minor losses, and determine the gauge pressure in pascals at the discharge end of the pump. The absolute viscosity of water is 1.0 centipoise.

a. 1.70 MPa
b. 2.04 MPa
c. 2.75 MPa
d. 3.01 MPa

12.61 Water is flowing, horizontally, from a reservoir through a circular orifice under a head of 50 ft. The coefficient of discharge is 0.7, and the diameter of the orifice is 10 cm. How many cubic meters per second are being discharged from the orifice?

a. 0.07 m^3/sec
b. 0.09 m^3/sec
c. 0.11 m^3/sec
d. 0.15 m^3/sec

12.62 A swimming pool is being cleaned with a brush fixture on the end of a rubber hose which is 5.0 cm inside diameter, through which the water is drawn by a vacuum pump. The hose is 15 meters long, leading to the pump 3.6 meters above the bottom of the pool. If the water in the pool is 2.4 meters deep, what suction (in Pascals) would be required at the pump to draw 2.25 m^3 per minute through the hose? (Assume a loss coefficient at the brush of 0.5 $v^2/2g$, where v is the velocity in the hose. Assume the absolute viscosity of water = 1.0 centipoise.)

a. −14 kPa
b. −17 kPa
c. −21 kPa
d. −24 kPa

12.63 Water flows from a supply tank through 25 meters of welded steel pipe 15 cm in diameter to a hydraulic mining nozzle 5.0 cm in diameter. The coefficient of the nozzle may be assumed to be 0.90, and the temperature of the water may be assumed to be 75F with a kinematic viscosity of 1.0 centipoise. If 28 l/s of water is flowing, how far must the nozzle lie below the elevation of the water in the tank?

a. 9 m
b. 12 m
c. 15 m
d. 18 m

12.64 Water falling from a height of 30 meters at the rate of 57 m^3/min drives a water turbine connected to an electric generator at 120 rpm. If the total resisting torque due to friction is 1.8 kN at 30 cm radius and the water leaves the turbine blades with a velocity of 3.70 m/sec find the power developed by the generator.

a. 308 kW
b. 289 kW
c. 275 kW
d. 266 kW

12.65 A venturi meter with a throat 5 cm in diameter is placed in a pipeline 10 cm in diameter carrying fuel oil. If the differential pressure between the upstream section and the throat is 25 kPa, how much oil is flowing in liters per minute? The discharge coefficient of the venturi meter is 0.97. The specific gravity of the oil is 0.94.

a. 790 *l*/min
b. 840 *l*/min
c. 890 *l*/min
d. 925 *l*/min

12.66 A rectangular sluice gate, 122 cm wide and 183 cm deep, hangs in a vertical plane. It is hinged along the top (122 cm) edge. If there are 6.4 meters of water above the top of the gate, what horizontal force applied at the bottom of the gate will be necessary to open it? Refer to Exhibit 12.66.

a. 75 kN
b. 83 kN
c. 92 kN
d. 98 kN

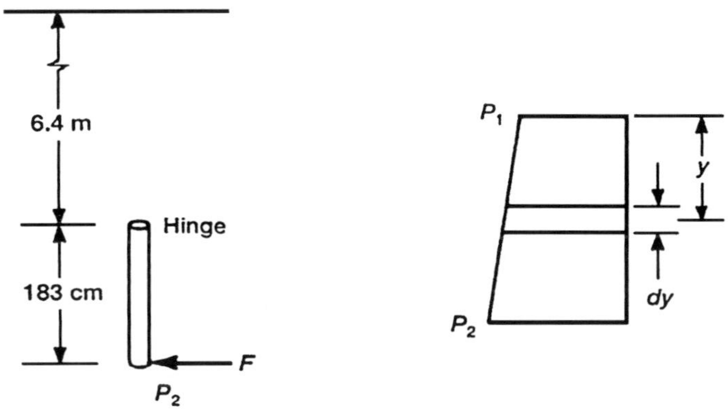

Exhibit 12.66

12.67 If a pipe is corroded on the *ID* the corrosion will affect the flow through it in the following manner.

a. For a constant pressure supply it will reduce the flow if the flow is laminar.
b. If the Reynolds number is above 10,000 the corrosion will not have any effect on the flow.
c. For a Reynolds number of 100,000 it will cause a higher friction factor than for an uncorroded pipe.
d. For a given flow rate the Reynolds number will be higher than for an un-corroded pipe.

12.68 A dam 15 meters high has water behind it to a depth of ten meters. What is the overturning-force per meter of width?

a. 1.6 MJ
b. 2.0 MJ
c. 2.3 MJ
d. 1.8 MJ

12.69 A two-meter *ID* pipe is flowing half full. What is the hydraulic radius?

a. 1.00 m
b. 0.50 m
c. 1.50 m
d. 1.75 m

12.70 Estimate the total time required to empty a tank which is a paraboloid of revolution with every horizontal section a circle whose radius equals the square root of the height above the bottom, through which is cut a 25-mm-round sharp-edged orifice. The depth of water at the start of time is 3.0 meters. Refer to Exhibit 12.70.

a. 2 hr 20 min
b. 2 hr 5 min
c. 1 hr 50 min
d. 1 hr 38 min

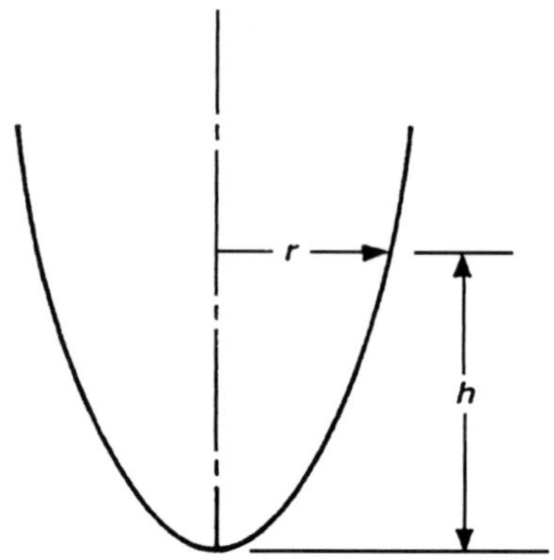

Exhibit 12.70

12.71 A 30-cm *ID* cast-iron pipe, 1220 meters in length, conveys water from a pump to a reservoir whose water surface is 76.0 meters above the pump, which is pumping at a rate of 85 liters per sec. Assuming the efficiency of the pump to be 89 percent, determine the power input. The viscosity of water is 1 centipoise.

a. 65 kW
b. 70 kW
c. 77 kW
d. 82 kW

12.72 A pitot tube having a coefficient of unity is inserted in the exact center of a long, smooth tube of 25.0 mm *ID* in which crude oil (sp. gr. = 0.9 and $\mu = 16.7$ centipoises) is flowing. Determine the average velocity in the tube if the velocity pressure is 100 mm of water.

a. 0.56 m/sec
b. 0.65 m/sec
c. 0.74 m/sec
d. 0.82 m/sec

12.73 Water flows through 915 linear meters of 90-cm-diameter pipe that branches into 610 linear meters of 45-cm-diameter pipe and 730 linear meters of 60-cm-diameter pipe. These rejoin, and the water continues through 460 meters of 75-cm-diameter pipe. All pipes are horizontal, and the friction factors are 0.016 for the 90-cm diameter pipe, 0.017 for the 60-cm and 75-cm diameter pipes, and 0.019 for the 45-cm pipe. Find the pressure drop in pascals between the beginning and the end of the system if the steady flow is 1.70 m^3/sec in the 90-cm pipe. Disregard minor losses. Refer to Exhibit 12.73.

a. 280 kPa
b. 300 kPa
c. 320 kPa
d. 335 kPa

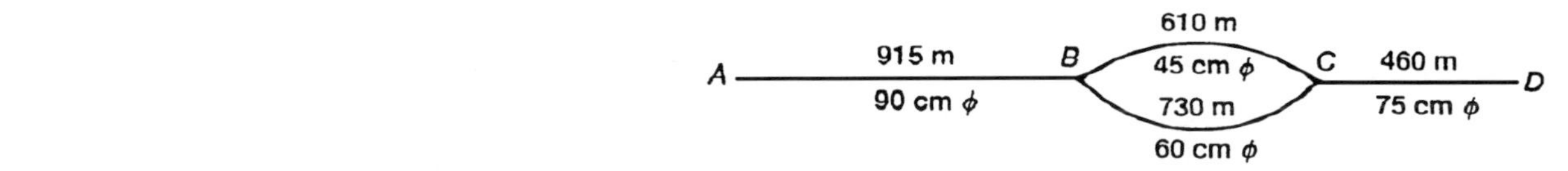

Exhibit 12.73

12.74 A piece of lead (sp. gr. 11.3) is attached to 40 cm^3 of cork (sp. gr. 0.25). When fully submerged the combination will just float in water. What is the mass of the lead?

a. 25 gm
b. 27 gm
c. 30 gm
d. 33 gm

12.75 A loaded timber 30 cm square and 1.83 meters long floats upright in fresh water with 60 cm of its length exposed. What will be the length of the timber projecting above the surface if the water is covered with a layer of oil 30 cm thick? The specific gravity of the oil is 0.8. Refer to Exhibit 12.75.

a. 54 cm
b. 56 cm
c. 59 cm
d. 61 cm

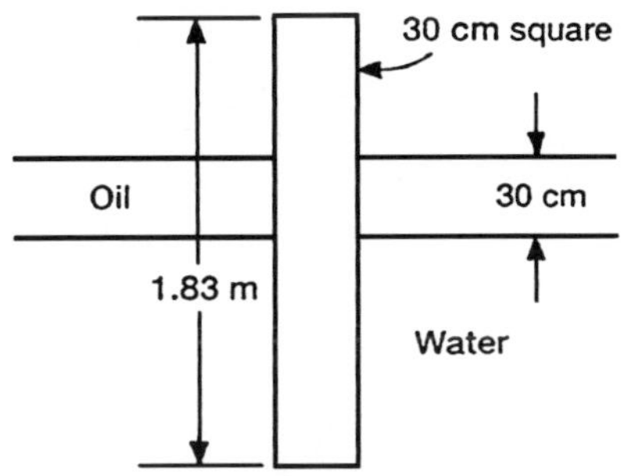

Exhibit 12.75

12.76 The outlet and inlet of a Venturi meter are each 10 cm in diameter, and the throat is 76 mm in diameter. The inlet velocity is 3.0 m/sec. The inlet has a static head of 3.05 meters of water. It there is no loss due to friction between the inlet and the throat of the meter, what will be the head in the throat?

a. 1.7 m
b. 2.1 m
c. 2.6 m
d. 2.9 m

12.77 A conduit having a cross section of an equilateral triangle of sides *b* has water flowing through it at a depth *b*/2. Find the "hydraulic radius."

a. 0.39*b*
b. 0.32*b*
c. 0.24*b*
d. 0.17*b*

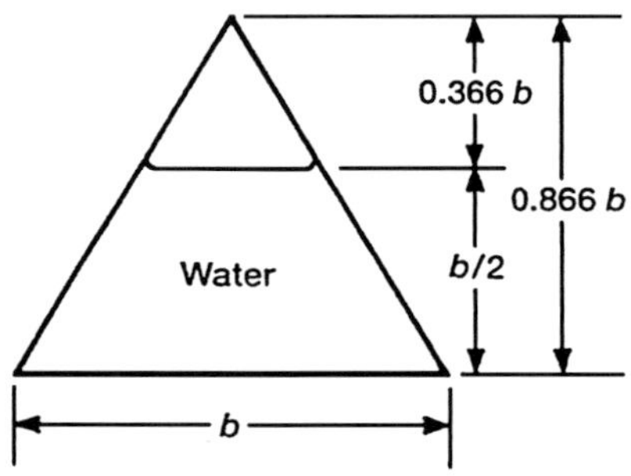

Exhibit 12.77

12.78 A mass of copper, suspected of being hollow, weighs 523 g in air and 447.5 g in water. If the specific gravity of copper is 8.92, what is the volume of the cavity, if any?

a. 11 cm^3
b. 14 cm^3
c. 17 cm^3
d. 21 cm^3

12.79 A 27-cm *OD* galvanized iron pipe 4.6 m long, having a mass of 47.6 kg/m, is closed at one end by a pipe cap with a mass attached so that the pipe will float upright. The cap and attached mass total 11.3 kg and displace 14.15l of water. How much of the pipe will show above the surface of seawater of 1.025 sp. gr.?

a. 0.69 m
b. 0.74 m
c. 0.79 m
d. 0.84 m

12.80 An oil pipeline with a 25-cm *ID* is 160 km long, and is made of welded steel. The discharge end is 150 meters above the intake. The rate of flow is 122 cm/sec. What is the total pressure in P pascals, gauge, at the intake when the whole length is full of gasoline at 16°C and 0.68 sp. gr.? The absolute viscosity of the gasoline is 0.30 centipoise.

a. 5.1 MPa
b. 5.4 MPa
c. 5.9 MPa
d. 6.2 MPa

The following information may be used for Problems 12.81 and 12.82.

A fan delivers 225 m^3/min of standard air against a back pressure of 10 cm of water while it is operating at 500 rpm. The efficiency of the fan is 85%.

12.81 The power, in kW, required to operate the fan is most nearly

a. 3.7
b. 4.3
c. 5.6
d. 6.8

12.82 If the speed is increased to 700 rpm how much power is required, and what is the new flow rate?

a. 7.6
b. 8.5
c. 10.1
d. 11.9

The following information may be used for problems 12.83 to 12.86

A pump uses a 1200 rpm motor to pump 100,000 kg/min of standard temperature water that has a density of 1000 kg/m^3 (which may be assumed constant through the pump), and has a pressure rise from 1 bar to 40 bar using a 10 cm diameter impeller. The inlet diameter is 35 cm, and the exit diameter is 25 cm. The outlet from the pump is at the same elevation as the inlet to the pump. The pump has an efficiency of 75%.

12.83 Power, in kW, required to operate the pump is most nearly

a. 7300
b. 5600
c. 8200
d. 9600

12.84 If a similar pump is used but has an impeller of 12 cm in diameter and an rpm of 1760, what is the power input in kW?

a. 75,500
b. 140,000
c. 160,000
d. 195,000

12.85 For the new pump, determine the new pressure rise in bar.

a. 120
b. 100
c. 85
d. 230

12.86 For the new pump, what is the new flow rate in kg/min?

a. 150,000
b. 193,000
c. 170,000
d. 211,000

The following information may be used for Problems 12.87 and 12.88.

A water-cooled air compressor operating on the polytropic process with $n = 1.2$ is used to compress 500 m^3/min of air at 1 bar and 20 °C to a pressure of 15 bar. The compressor has an isentropic efficiency of 75%.

12.87 Power input required, in kW, is most nearly

a. 2500
b. 3800
c. 3100
d. 4500

12.88 Heat transfer out of the compressor, in kJ/s, is most nearly

a. 1500
b. 1800
c. 2100
d. 2400

SOLUTIONS

12.1 **a.** Kinematic viscosity $v = \dfrac{\text{Absolute viscosity}}{\text{Density}} = \dfrac{\mu}{\rho}$

Units of absolute viscosity: N• s/m^2 or kg/m-s

Units of density: kg/m^3

The dimensions of kinematic viscosity would be

$$v = \frac{\mu}{\rho}\frac{(\text{kg/m}\bullet\text{s})}{(\text{kg/m}^3)} = \frac{\text{m}^2}{\text{s}}$$

12.2 **d.** The absolute viscosity is proportional to the shear stress (τ) divided by the angular deformation rate. Density is not involved in the definition. The rate of angular deformation $\cong \dfrac{dV}{dy}$.

$$\text{Thus, } \tau = \mu\frac{dV}{dy}.$$

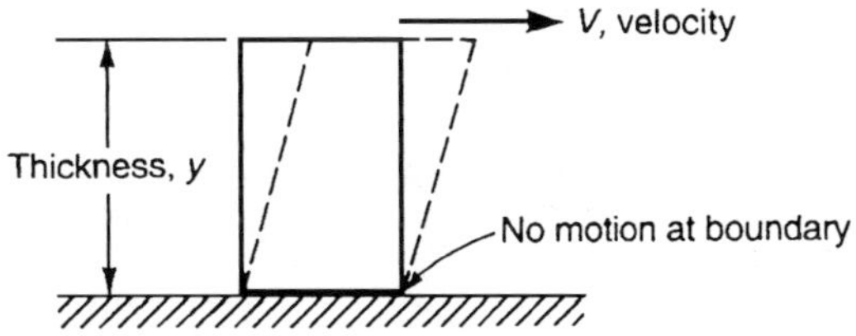

Exhibit 12.2

12.3 **c.** The internal pressure must equal the local external pressure. Externally, the pressure is

$$p = \gamma h = (\text{SG})(\gamma_{\text{water}})(h)$$

$$p = (1.03)\left(9.81\frac{\text{kN}}{\text{m}^3}\right)(50\text{ m}) = 505\text{ kPa (gage)}$$

12.4 **b.** Since the situation is static, gage pressure at the base is 1.5 m of water. In the air chamber it is 1.5 m of water, less 1 m of water less 1.2 m of oil.

$$p = \gamma h$$

$$p = 1.5\,(9810) - 1\,(9810) - 1.2\,(0.8)\,(9810) = -4513\text{ Pa}$$

12.5 **c.** Pressure $= \gamma h = 9810\,(100) = 981{,}000$ Pa $= 981$ kPa

12.6 **d.** The density of air can be calculated from the ideal gas law using

$$R = 0.286\frac{\text{kN}\bullet\text{m}}{\text{kg}\bullet\text{K}}$$

$$p = \frac{p}{RT} = \frac{100\text{ kN/m}^3}{0.286\dfrac{\text{kN}\bullet\text{m}}{\text{kg}\bullet\text{K}}(20+273)\text{K}} = 1.19\frac{\text{kg}}{\text{m}^3}$$

$$\text{The specific weight of air } \gamma = 1.19\frac{\text{kg}}{\text{m}^3} \bullet \left(\frac{9.81\frac{\text{m}}{\text{s}^2}}{1\frac{\text{kg} \bullet \text{m}}{\text{N} \bullet \text{s}^2}} \right) = 11.7\frac{\text{N}}{\text{m}^3}$$

$$p = \gamma h \text{ or } h = \frac{p}{\gamma} \text{ and } h = \frac{15{,}000\frac{\text{N}}{\text{m}^2}}{11.7\ \text{N/m}^3} = 1282\ \text{m}$$

12.7 **a.** Assuming atmospheric pressure at sea level at 101.3 kPa and a constant specific weight of air at 11.7 N/m^3 (as previously calculated):

$$p = p_{\text{SL}} - \gamma h = 101.3 - \frac{11.7(1200)}{1000} = 87.26\ \text{kPa}$$

12.8 **b.**

$$\text{Force} = PA = \gamma hA = 9.81\frac{\text{kN}}{\text{m}^3} \times 3\ \text{m} \times 500\ \text{cm}^2 \times \left(\frac{1\ \text{m}}{100\ \text{cm}} \right)^2 = 1.47\ \text{kN}$$

12.9 **d.** The pressure at the tank base $= p = \gamma_W h = (9810)(2.5) = 24{,}325\ \text{N/m}^3$

Area of tank base, $A = \frac{\pi}{4}(3)^2 = 7.07\,\text{m}^2$

Force on tank base $= pA = 24{,}325\ (7.07) = 171{,}978$ N

12.10 **c.** The average pressure exerted on one side is the pressure that exists at the centroid of the side times the area of the side.

$$F = \gamma h_c A$$

where h_c = the depth in meters from the fluid-free surface to the centroid of the area, and A = area. Since the sides are square, $h_c = 1.5/2 = 0.75$ m.

$$F = 9.81(0.75)(1.5 \times 1.5) = 16.55\ \text{kN}$$

12.11 **d.** The static force at $A = \left(700\frac{\text{kN}}{\text{m}^2}\right)\left[\frac{\pi}{4}(0.1\ \text{m})^2\right] = 5.50\ \text{kN}$ tension on the bolts at A.

The static force at $B = \left(700\frac{\text{kN}}{\text{m}^2}\right)\left[\frac{\pi}{4}(0.05\ \text{m})^2\right] = 1.37\ \text{kN}$ tension on the bolts at B.

The end restraint by the pipes opposes a net force of 5.50 – 1.37 = 4.13 kN to the right on the reducing section.

12.12 **d.** Area of port $= \frac{\pi}{4}D^2 = \frac{\pi}{4}(0.6)^2 = 0.283\,\text{m}^2$

Pressure $= 100\ \text{kPa} = 100{,}000\frac{\text{N}}{\text{m}^2}$

$F = pA = 100{,}000\ (0.283) = 28{,}300$ N

12.13 **b.** The volume of the bubble equals the volume of the displaced seawater, which equals

$$V_D = \frac{4}{3}\pi r^3 = \frac{4}{3}\pi(0.0125)^3 = 8.18\times10^{-6}\,\text{m}^3$$

Since the specific weight of seawater is

$$(\text{SG})(\gamma_W) = 1.03\left(9810\,\frac{\text{N}}{\text{m}^2}\right) = 10{,}104\,\frac{\text{N}}{\text{m}^2}$$

The buoyant force, B, is

$$B = \gamma V_D (10{,}104)(8.18\times10^{-6}) = 0.0826\ \text{N}$$

12.14 **d.** A buoyant force is equal to the weight of fluid displaced. At equilibrium, or floating, the weight downward is equal to the buoyant force.

Let V_1 = total volume of the iceberg in m^3. Its weight is $V_1(9810)(0.922) = 9045(V_1)$ N.

Let V_2 = immersed volume of the iceberg, which equals the volume of seawater displaced. The weight of seawater displaced is then $V_2(9810)(1.03) = 10{,}104(V_2)$ N.

Hence $\dfrac{V_2}{V_1} = \dfrac{9045}{10{,}104} = 0.895$ is the volume fraction of the iceberg immersed, and the volume fraction exposed is $1 - 0.895 = 0.105 = 10.5\%$.

12.15 **b.** Archimedes' principle applies equally well to gases. Thus a body located in any fluid, whether liquid or gaseous, is buoyed up by a force equal to the weight of the fluid displaced. A balloon filled with a gas lighter than air readily demonstrates the buoyant force.

Thus the weight of the cork is equal to the weight of water displaced plus the weight of air displaced. When the air within the vessel is removed, the cork is no longer provided a buoyant force equal to the weight of air displaced. For equilibrium, the cork will sink somewhat in the water.

12.16 **c.** $$V_D = \frac{W}{\gamma} = \frac{9.32\text{N}}{9810\dfrac{\text{N}}{\text{m}^2}} = 0.00095\ \text{m}^3 = 950\ \text{cm}^3$$

The change in total volume, ΔV, beneath the water surface equals the area of the cylindrical container, A, times the change in water level, dh, or $dV = A\ dh$. The depth of the water will increase

$$dh = \frac{dV}{A} = \frac{950\ \text{cm}^3}{\dfrac{\pi}{4}(20)^2} = 3.02\ \text{cm}$$

12.17 **d.** Let x = height of wood block, W = width of wood block, L = length of wood block, and γ = specific weight of water, N/m^3. The weight of the block is equal to the weight of the liquid displaced.

Weight of the block in water = (x – 15)WLγ(1.0)

Weight of the block in alcohol = (x – 10)WLγ(0.82)

Since the weight of the block is constant,

$$(x-15)WL\gamma = 0.82(x-10)WL\gamma$$
$$x-15 = 0.82x-8.2$$
$$x = \frac{6.8}{0.18} = 37.8\text{ cm}$$

The specific gravity of the wood block is, by definition,

$$\frac{\text{Volume of water displaced}}{\text{Total volume}} = \frac{(x-15)WL}{xWL} = \frac{37.8-15}{37.8} = 0.603$$

12.18 **d.** Continuity requires that

$$Q = A_1V_1 = A_2V_2 = A_1(3) = \frac{A_1}{4}V_2, \qquad V_2 = 12\text{ m/s}$$

12.19 **c.** Continuity requires that the mass flow rate be the same at all sections in steady flow. Calculate $\dot{m}$ at Section 1, where the velocity is given, using $\dot{m} = \rho AV$. This will also be the mass flow rate at Section 2.

Cross-sectional area at Section 1:

$$\frac{\pi}{4}(0.10)^2 = .00785\text{ m}^2$$

$$\dot{m} = \rho AV\left(1000\frac{\text{kg}}{\text{m}^3}\right)(.00785\,\text{m}^2)(10\,\text{m/s}) = 78.5\,\text{kg/s}$$

12.20 **a.** The mass flow rate $\dot{m} = \rho Q = \rho AV$. The density of air at 200 kPa and 150°C (423°K) is obtained from the ideal gas law:

$$\frac{p}{\rho} = RT$$

Use $R = 286\dfrac{\text{N}\bullet\text{m}}{\text{kg}\bullet{}^\circ\text{K}}$ for air.

$$\rho = \frac{P}{RT} = \frac{200,000\dfrac{\text{N}}{\text{m}^2}}{286\dfrac{\text{N}\bullet\text{m}}{\text{kg}\bullet{}^\circ\text{K}}\bullet 423^\circ\text{K}} = 1.65\frac{\text{kg}}{\text{m}^3}$$

The cross-sectional area $= A = \dfrac{\pi}{4}D^2 = 0.785(.025\text{ m})^2$ = 490 × 10^{-6} m^2, and

$$\dot{m} = \rho Av = \left(1.65\frac{\text{kg}}{\text{m}^3}\right)(490\times10^{-6}\text{ m}^2)\left(10\frac{\text{m}}{\text{s}}\right) = 0.00809\frac{\text{kg}}{\text{s}}$$

12.21 b.

$$Q = 10\frac{\text{L}}{\text{min}} \times \frac{\text{m}^3}{1000\ \text{L}} \times \frac{\text{min}}{60\ \text{s}} = 167 \times 10^{-6}\ \frac{\text{m}^3}{\text{s}}$$

The cross-sectional area $A = \frac{\pi}{4}D^2 = 0.785(0.01)^2 = 78.5 \times 10^{-6}\ \text{m}^2$

$$V = \frac{Q}{A} = \frac{167 \times 10^{-6}}{78.5 \times 10^{-6}} = 2.13\ \text{m/s}$$

12.22 a. The pressure (flow) energy change is 45 – 9 = 36 N•m/N, or

$$\gamma = \frac{g}{g_c}\rho = \frac{9.81}{1.0}(800) = 7848\ \text{N/m}^3$$

$$\frac{\Delta p}{\gamma} = 36, \qquad \Delta p = 36\gamma = 36(7848) = 282{,}500\ \text{Pa} = 282.5\ \text{kPa}$$

12.23 a. The volume flow rate is

$$55\frac{\text{L}}{\text{min}} \bullet \frac{\text{m}^3}{1000\ \text{L}} \bullet \frac{\text{min}}{60\ \text{s}} = 917 \times 10^{-6}\ \text{m}^3/\text{s}$$

Ignoring the head loss, h_L, from friction, the required energy input is $52.5 - 12 = 40.5\frac{\text{N}\bullet\text{m}}{\text{N}}$.

$$\text{Power} = \gamma Q h_A \left(7848\frac{\text{N}}{\text{m}^3}\right)(917 \times 10^{-6}\text{m}^3/\text{s})\left(40.5\frac{\text{N}-\text{m}}{\text{N}}\right)$$
$$= 291\ \text{W} = 0.291\ \text{kW}$$

12.24 d. In a venturi throat, the increased velocity required by continuity results in a *KE* (velocity) increase that occurs at the expense of pressure (flow) energy. Since the system is horizontal, no change in potential energy has occurred. At *B* the pressure (flow energy) decreases and *KE* increases. For a well-designed venturi, the conditions existent at *A* are essentially restored at *C*.

12.25 b. Apply an energy balance of the fluid flowing: Total energy in = Total energy out + Energy losses – Energy inputs. Thus, $135 = 130 + h_L - 0$. The head loss $h_L = 5$ N • m/m, or 5 meters.

12.26 b. Ignoring frictional losses, pump inefficiency, and noting that any changes in *KE* or pressure are essentially 0, pumping power is equal to the increase in potential energy between the reservoirs.

The potential energy increase per lb_m is *Z* or $h = 120$ meters or $\frac{\text{N}\bullet\text{m}}{\text{N}}$

The volume flow rate, *Q*, is

$$400\frac{\text{L}}{\text{min}} \bullet \frac{\text{m}^3}{1000\ \text{L}} \bullet \frac{\text{min}}{60\ \text{s}} = 6.67 \times 10^{-3}\ \text{m}^3/\text{s}$$

The power required is

$$P = \gamma Q h_a = 9.81\frac{\text{kN}}{\text{m}^3} \bullet 6.67 \times 10^{-3}/\text{s} \bullet 120\ \text{m} = 7.85\ \text{kW}$$

12.27 c.

$$h = \frac{V^2}{2g} \quad \text{or} \quad V = (2gh)^{1/2} = (2 \times 9.81 \times 10)^{1/2} = 14\,\text{m/s}$$

12.28 a. The head is

$$h = \frac{V^2}{2g} = \frac{10^2}{2(9.81)} = 5.10\ \text{m}$$

12.29 c. A **piezometer tube** indicates static pressure and is equivalent to a static pressure gage.

Stagnation pressure is an increased pressure developed at the entrance to a pitot tube when the velocity locally becomes zero.

The **hydraulic grade line** is a flow energy or pressure head in meters, which can be plotted vertically above the pipe centerline along the pipe.

The **energy grade line** is the total mechanical energy (flow energy or pressure head, plus kinetic energy or dynamic head, plus potential energy or height above datum) in meters, which may be plotted vertically above the datum along the pipe.

The **friction head** is the head loss h_f in meters caused by fluid friction.

The **critical depth** above the channel floor in open channels is the depth for minimum potential and kinetic energy for the given discharge. Tranquil flow (low *KE* and high *PE*) exists when the actual flow is above critical depth, and rapid flow (high *KE* and low *PE*) exists when the actual flow is below critical depth.

12.30 b. The steady impulse-momentum equation is $F_{\text{net}} = \dot{m}(\text{v}_2 - \text{v}_1)$ if the pressure in the fluid stream is zero at each end of the "U." Then

$$F = 30\ \text{kg/s}(-6-6)\,\text{m/s} = 360\,\frac{\text{kg}\bullet\text{m}}{\text{s}^2} = 360\ \text{N}$$

Since the original velocity was 6 m/s, the final reversed velocity is –6 m/s. This force from impulse-momentum is the force on the fluid to achieve the velocity change. In reaction, the fluid exerts an equal and opposite force, 360 N, to the right on the fitting.

12.31 a. The impulse-momentum equation is $F = pQ(V_2 - V_1)$. Here $\rho Q = 1$ kg/s, the final velocity of the exhaust is $V_2 = 150$ m/s, and $V_1 = 0$. Hence

$$F = \left(\frac{1\ \text{kg}}{\text{s}}\right)(150-0)\frac{\text{m}}{\text{s}} = 150\,\frac{\text{kg}\bullet\text{m}}{\text{s}^2} = 150\ \text{N}$$

12.32 **d.** There is a drop in the energy line from the 25-cm main to the 30-cm main. This head loss must be equal in both 15-cm lines, or

$$h_{f300} = h_{f450}$$

The Darcy equation is

$$h_f = f\frac{L}{d}\frac{V^2}{2g}$$

where h_f = head loss in meters, f = friction factor, L = length of pipeline in meters, d = diameter of pipe in meters, and $g = 9.81\ \text{m/s}^2$. Thus, in this situation,

$$f\frac{300}{0.15}\frac{V_{300}^2}{2(9.81)} = f\frac{450}{0.15}\frac{V_{450}^2}{2(9.81)}$$

which reduces to

$$300V_{300}^2 = 450(3)^2$$

$$V_{300} = \sqrt{\frac{4050}{300}} = 3.67\ \text{m/s}$$

12.33 **c.** Laminar (streamline, viscous) flow is compared with turbulent flow in the following table:

	Laminar Flow	Turbulent Flow
Motion of fluid particles	Parallel to stream velocity. Paths of particles do not cross.	Particle paths cross and move in all directions.
Energy loss, h_f	$h_f = f\left(\frac{L}{D}\right)\left(\frac{V^2}{2g}\right)$ f is independent of surface roughness and decreases with Re. $f = \frac{64}{\text{Re}}$	$h_f = f\left(\frac{L}{D}\right)\left(\frac{V^2}{2g}\right)$ f varies with surface roughness, decreases with Re to a constant value. See Moody diagram.
Velocity distribution in pipe	Average is $^1/_2$ of maximum at centerline. parabolic distribution. Zero at wall.	Essentially same throughout, except for thin boundary layer at wall. Follows 1/7 power law.
Reynolds number $\text{Re} = DV\rho/\mu$	Less than 2300	Greater than 2300

Newton's law of viscosity defines μ on the basis of shear stress and the rate of fluid angular deformation. The Reynolds number contains μ as a contributing parameter. Very viscous liquids usually move in laminar flow.

On the basis of the above data, select (c). Do not confuse the energy loss, h_f, with the friction factor, f.

12.34 **d.** In turbulent flow the Reynolds number is *greater* than 2300.

12.35 **a.** Turbulent flow is highly agitated flow with individual particles crossing paths and colliding; critical flow is a point at which some property of the fluid—or some parameter related to it—changes; dynamic flow is redundant; uniform flow is of constant rate.

12.36 **b.** Choice (a) is mass flow rate, $\dot{m}$, in kg/s.

Choice (b) has these dimensions:

$$pAV = \left(\frac{\text{N}}{\text{m}^2}\right)(\text{m}^2)\left(\frac{\text{m}}{\text{s}}\right) = \frac{\text{N}\bullet\text{m}}{\text{s}} = \text{W}$$

Choice (c) is the Reynolds number, Re; it is the dimensionless ratio of inertial force to viscous force.

Choice (d) is the Euler number, Eu; it is the dimensionless ratio of inertial force to pressure force.

12.37 **a.** Below the critical velocity (Re < 2300) flow is laminar, and $f = 64/\text{Re}$. Substitution of this term into the Darcy equation for friction loss in pipes, $h_f = f\frac{L}{D}\frac{V^2}{2g}$, yields the Hagen-Poiseuille equation,

$$h_f = \frac{32\mu LV}{\gamma\, d^2}.$$

12.38 **a.** Line *D* applies to all roughness ratios in laminar flow (Re < 2300) because the boundary layer at the wall makes the friction factor independent of roughness ratio:

$$f = \frac{64}{\text{Re}}$$

In turbulent flow (Re > 2300), increasing roughness is represented by *A* for a smooth pipe to *C* for a very rough pipe; moreover, only the thinnest boundary layer exists in a turbulent flow, so the friction factor is very dependent on surface roughness.

12.39 **a.** Use an energy balance to determine h_f. Upon substituting the given data, the resulting equation is

$$Z_A + \frac{V_A^2}{2g} + \frac{P_A}{\gamma} = Z_B + \frac{V_B^2}{2g} + \frac{P_B}{\gamma} + h_f$$

Since the pipe diameter is unchanged, continuity requires that *V* be the same at both points. Thus the kinetic energy terms can be deleted from both sides of the equation.

$$Z_A + \frac{P_A}{\gamma} = Z_B + \frac{P_B}{\gamma} + h_f$$

$$h_f = \frac{200 - 230}{9.81} + 50 - 40 = 6.94\,\text{m}$$

12.40 **a.** Typical head losses for the above items are expressed as an empirical average constant, K or C times the kinetic energy, $V^2/2g$:

$$h_f = K\frac{V^2}{2g}$$

12.41 **b.** The discharge coefficient is $C_D = C_c C_v$, where C_c = coefficient of contraction = (area of v.c.)/(area of orifice) and C_v = coefficient of velocity = (actual velocity at v.c.)/(theoretical velocity at v.c.) ignoring losses.

The theoretical velocity at the v.c. is

$$V = \sqrt{2gh} = \sqrt{2(9.81)(5)} = 9.9 \text{ m/s}$$

$$C_v = \frac{9.0}{9.9} = 0.909$$

The area of the v.c. is

$$A = \frac{Q}{V} = \frac{.015}{9} = 0.00167\,\text{m}^2$$

The area of the orifice $= \frac{\pi}{4}D^2 = \frac{\pi}{4}(.05)^2 = 0.00196\,\text{m}^2$. Thus,

$$C_c = \frac{0.00167}{0.00196} = 0.852 \quad \text{and} \quad C_D = C_c C_v = (0.852)(0.909) = 0.774$$

12.42 **d.** The maximum height to which a fluid may be siphoned is determined when the pressure of the fluid column plus its vapor pressure equals the external pressure. The minimum pressure at the highest point is 0 kPa plus vapor pressure.

Ignoring the vapor pressure (small),

$$p = \gamma h \quad \text{or} \quad h = \frac{P}{\gamma}$$

$$h = \frac{-101.3 \text{ kN/m}^2}{(0.8)9.81 \text{ kN/m}^3} = 12.91 \text{ m in depth (or height)}$$

12.43 **b.** A pitot tube generates a stagnation pressure as fluid kinetic energy is converted to pressure head. Hence

$$V = \sqrt{2g\left(\frac{\Delta p}{\gamma}\right)}$$

$$V = \sqrt{2(9.81)\frac{\text{m}}{\text{s}^2}\left(\frac{7000 \text{ N/m}^3}{9810 \text{ N/m}^3}\right)} = 3.74 \text{ m/s}$$

$$Q = AV = \frac{\pi}{4}D^2V = \frac{\pi}{4}(0.15)^2(3.74) = 0.0661 \text{ m}^2/\text{s}$$

12.44 **a.** For a static head orifice discharging freely into the atmosphere

$$Q = CA\sqrt{2gh}$$

12.45 **c.** Hydraulic radius, R, is defined as cross-sectional area, divided by wetted perimeter.

12.46 **d.** Choices (a) and (c) are not dimensionless, as required for a Reynolds number. Since the hydraulic radius R = (cross sectional area)/(wetted perimeter), for a circular cross section,

$$R = \frac{\frac{\pi D^2}{4}}{\pi D} = \frac{D}{4} \quad \text{or} \quad D = 4R$$

Therefore,

$$\text{Re} = \frac{4RV\rho}{\mu} = \frac{4RV}{\nu}$$

12.47 **a.** Starting with a generalized grouping of the possible variables in terms of generic equation

$$\tau = f(\rho, \mu, V, D, R)$$

Selecting the characteristic dimensions for the above generic equation based on Newton's 2nd law we obtain

τ = Shearing stress	F/L^2
ρ = Density	M/L^3
μ = Viscosity	M/LT
V = Velocity	L/T
D = Diameter	L
R = Roughness	L/L

Using Newton's second law relationship

$$\text{F} = \text{ML/T}^2$$

Our equation becomes

$$(\text{ML/T}^2)/\text{L}^2 = \text{M/LT}^2 = C(\text{M/L}^3)^{a}(\text{M/LT})^{b}(\text{L/T})^{c}(\text{L})^{d}(\text{L/L})^{e}$$

Since the roughness relation is non dimensional it will drop out of the relationship. The only way that it will again appear is through experimentation.

Set up table for like exponents

M: $1 = a + b$ $\qquad a = 1 - b$

L: $-1 = -3a - b + c + d + e - e$

T: $-2 = -b - c$ $\qquad c = 2 - b$

Solving for d

$$d = -1 + 3(1 - b) + b - (2 - b) = -1 + 3 - 3b + b - 2 + b = -b$$

Thus

$$\tau = C(\rho)^{1-b}(\mu)^{b}(V)^{2-b}(D)^{-b}(\text{R})^{0}$$

Grouping into terms of like exponents

$$\tau = C(\rho i^2)(\mu/\rho VD)^{b} = C(\text{Re})^{-b}\rho V^2$$

12.48 **c.** Selecting the equation from the *Handbook* for the coordinates of the center of pressure

$$Y^* = (\gamma I \sin\alpha)/(p_c A) + Y_c$$

For a vertical plate sin a = 1.0, and the equation may be rearranged to be

$$Y_p = Y_c + I/(Y_c A) = 15 \text{ m} + [(1/12)(bh^3)\text{m}^4/[(15)\text{m}(10 \times 10)\text{m}^2]$$

$$Y_p = 15 \text{ m} + [(1/12)(10)(10^3)]/1500 = 15.56 \text{ m}$$

12.49 **d.** Taking summation of moments about the hinge (A) we find

$$\Sigma M_A = Y'_p F - R(10) = (5.56 \text{ m})(\gamma A Y_c) - 10\,R = 0$$
$$(5.56 \text{ m})(1000 \text{ kg/m}^3)(9.81 \text{ m/s}^2)(10 \times 10 \text{ m}^2)(15\ Y_c m)(\text{N/kg} \cdot \text{m/s}^2) = (10\text{m})R$$
$$R = 8{,}182 \text{ kN}$$

12.50 **a.** To determine the friction factor of the pipe we need first to determine the velocity of flow using the continuity equation, calculate the Reynolds number, calculate the e/D ratio, and obtain the friction factor off the Moody diagram.

$$(AV)_1 = (AV)_2 = (1000 \text{ m}^3/\text{hr}) = (\pi D^2/4 \text{ m}^2)(V \text{ m/s})$$
$$V = [(1000 \text{ m}^3/\text{hr})/(3600 \text{ s/hr})]/[\pi(0.40)^2/4)\text{m}^2] = 2.21 \text{ m/s}$$

Therefore

$$\text{Re} = \rho DV/\mu = [(1000 \text{ kg/m}^3)(0.40 \text{ m})(2.21 \text{ m/s})]/1.005 \times 10^{-3} \text{kg/m} \cdot \text{s}$$
$$\text{Re} = 8.796 \times 10^5$$

Turning to the Moody diagram in the *Handbook* we find that a galvanized iron pipe has a roughness of 0.15 mm; therefore, e/D = 0.15 mm/400 mm = 0.000375.

The friction factor from the Moody diagram is:

$$f = 0.014$$

12.51 **d.** The work input needed to cause the above flow rate may be determined by using the Bernoulli equation simplified for this problem.

$$\text{Pump Work} = Z_2 - Z_1 + (f\,L/D)V^2/2g$$
$$= 75 \text{ m} + [(0.014)(5000 \text{ m})/0.4 \text{ m}][(2.21)^2 \text{ m}^2/\text{s}^2]/(2)(9.81 \text{ m/s}^2)$$
$$= 118.6 \text{ m}$$

12.52 **d.** The power required may be calculated using the pump power equation in the *Handbook*.

$$W = (Q\gamma h)/\eta = [(1000 \text{ m}^3/\text{hr})/(3600 \text{ s/hr})](1000 \text{ kg/m}^3) \times (9.81 \text{ m/s}^2)(118.6 \text{ m})/0.70$$
$$W = (461.7 \text{ kW})(1.341 \text{ hp/kW}) = 619 \text{ hp.}$$

12.53 b. It is simplest to make ratios using the previous calculations to determine the new hp needed.

$$V = (300/1000)(2.21\text{ m/s}) = 0.663\text{ m/s}$$

Therefore

$$\text{Re} = 0.3(8.796 \times 10^5) = 2.639 \times 10^5$$

Using the same e/D ratio and the new Re, a new friction factor can be selected

$$f = 0.0175$$

Calculating the new pump work in head of water added to the system

$$W = 75 + (0.0175)(5000/0.40)(0.663)^2/2(9.81) = 79.9\text{ m}$$
$$\text{Power} = [(300/3600)(1000)(9.81)(79.9)/0.70](1.341)/1000 = 125\text{ hp}$$

12.54 d. The friction will be a function of the Reynolds number

$$\text{Re} = \rho D v/\mu$$

where

$$\rho = 0.94 \times 1{,}000 = 940\text{ kg/m}^3$$
$$D = 0.300\text{ m}$$
$$A = 0.0707\text{ m}^2$$
$$Q = 3{,}000/(24 \times 3{,}600) = 0.0347\text{ m}^3\text{/sec}$$
$$v = 0.0347/0.0707 = 0.491\text{ m/s}$$
$$\mu = 201 \times 0.001 = 0.201\text{ Pa}\cdot\text{s}$$
$$\text{Re} = 940 \times 0.300 \times 0.0471/(0.201/g) = 648$$

This indicates that the flow will be well into the laminar-flow range so $f = 64/\text{Re} = 64/648 = 0.0988$

12.55 d. A floating body displaces a mass of fluid equal to its own mass, so the loaded ship would displace an additional mass of water equal to 3500 metric tons, which would equal—

$$(3500 \times 1000\text{ kg})/(1000\text{ kg/m}^3) = 3500\text{ m}^3$$

The area of the lock $= 140 \times 15 = 2100\text{ m}^2$

But the area of the ore carrier is equal to $120 \times 10 = 1200\text{ m}^2$, so the net open area of water is only equal to $2100 - 1200 = 900\text{ m}^2$, assuming the ore carrier's area remains constant as it sinks farther into the water. The water would thus rise $3500/900 = 3.889$ meters.

Another way of looking at this problem is that the ore carrier would sink an amount equal to $3500/1200 = 2.917$ meters, which would displace a volume of water in the lock equal to $2.917 \times 1200 = 3500\text{ m}^3$. The total volume of water in the lock would remain constant, so, as the ore carrier sank, the level of the water in the lock would rise. The total volume of water plus the submerged portion of the ore carrier would increase by 300 m^3. The water would then rise in the area outside the ore carrier but inside the walls of the lock. The water would rise $3500/(2100 - 1200) = 3.889$ meters.

12.56 c. The velocity of the water out of the tank will be equal to

$$v = C\sqrt{(2gh)} = \sqrt{(2)(9.8066)(9.00)} = 13.286 \text{ m/s}$$

The jet is 1.00 m above the ground.

$$h = v_0 t + \tfrac{1}{2}at^2$$

where a is the acceleration of gravity.

For this case, the vertical velocity at time zero is zero. The time for the water to fall 1.00 m is

$$t = (2h/g) = 0.452$$

The distance from the base of the tank the water will strike is

$$s = 0.452 \times 13.286 = 6.00 \text{ m}$$

12.57 b. Apply Bernoulli's equation

$$P_1/\rho g + v_1^2/2g + z_1 + h_A = P_2/\rho g + v_2^2/2g + z_2 + h_L$$

between the surface of the reservoir and the discharge from the pipe. There is no energy added, and the velocity of the surface of the water is zero. The pressures at the surface and at the pipe discharge are both atmospheric, so they cancel. The only term remaining on the left side of the equation is the elevation of the surface of the reservoir. The difference in elevation between the surface of the reservoir and the outlet of the pipe, is equal to $Z_1 - Z_2 = 150$ m. On the right hand side of the equation would be the velocity term, $v^2/2g$, and the work done in overcoming pipe friction

$$h_L = fL/D \times v^2/2g$$
$$fL/D = 0.02 \times 200/0.30 \times v^2/2g = 13.33\, v^2/2g$$

so,

$$150 = (1 + 13.33)v^2/2g$$
$$v^2/2g = 10.47 \text{ m, and}$$
$$v = 14.33 \text{ m/s}$$
$$Q = (14.33 \text{ m/s})(0.0707 \text{ m}^2) = 1.013 \text{ m}^3\text{/s}$$

12.58 d. The glass displaces 125 – 75 = 50 grams = 50 cm^3 of water. 50 cm^3 of gasoline has a mass of 125 – 92 = 33 g; therefore, the specific gravity of the gasoline is equal to 33/50 = 0.66.

12.59 c. See Exhibit 12.59.

$$\text{Volume of timber submerged} = (0.3 - 0.11)(0.3)(3.60) = 0.205 \text{ m}^3$$
$$\text{Mass of timber} = (0.205)(1000) = 205 \text{ kg}$$

Take moments about the point touching rock

$$\Sigma M_R = [(-205)(3.6/2) - 68D] + [(3.6/2)(205)] + [(0.667 \times 3.6)(\tfrac{1}{2})(0.30)(0.11)(3.6)(1000)] = 0$$
$$369 + 68D = 369 + 142.6$$
$$D = 142.6/68 = 2.10 \text{ meters}$$

The mass of the timber and the mass of the man would produce moments in the clockwise direction. These would be opposed by counterclockwise moments due to the buoyancy of the timber and the triangular section shown by the dotted line. The moment arm of the buoyancy of the triangular section would be equal to 2/3 of the distance from the apex to the base, 2.4 meters. The buoyancy would be equal to the volume of water displaced times the density of water, $0.0594 \times 1000 = 59.4$ kg. This buoyancy times the moment arm would be equal to 142.6 kg · m. The small amount of difference in the length of the moment arm introduced by the cosine of angle θ (0.9995) would not be enough to markedly affect the accuracy of the answer.

12.60 a. This problem can best be solved with the aid of Bernoulli's equation

$$P_1/\rho g + v_1^2/2g + z_1 + h_A = P_2/\rho g + v_2^2/2g + z_2 + h_L$$

where

P_1 and P_2 are zero
z_1 is equal to zero
z_2 is equal to 137 meters
v_1 is equal to zero
h_L is a function of the Reynolds number and the type of pipe.

$$\begin{aligned}
A &= 0.0314 \text{ m}^2 \\
v &= 0.085/0.0314 = 2.707 \text{ m/sec} \\
v^2/2g &= 0.3736 \text{ m} \\
\text{Re} &= \rho D v/\mu \\
\mu &= 0.01 \text{ poise, or } \mu = 0.001 \text{ kg/sec} \cdot \text{m} \\
\rho &= 1000 \text{ kg/m}^3 \\
D &= 0.20 \text{ m} \\
\text{Re} &= (1000)(0.20)(2.707/0.001) = 541{,}400 \\
\text{Head loss} &= f(L/d)(v^2/2g)
\end{aligned}$$

where f is the the friction factor for $\text{Re} = 5.4 \times 10^5$

From the *FE Handbook*, for cast iron pipe

$$\begin{aligned}
e/D &= 0.25/200 = 0.00125 \\
f &= 0.0205 \\
h_L &= 0.0205(1{,}070/0.200)(0.3736) = 40.97 \text{ m}
\end{aligned}$$

The head to be supplied by the pump (ignoring the velocity head and minor losses) is equal to

$$40.97 + 137 = 177.97, \text{ or } 178 \text{ m}$$

The required output pressure is thus equal to

$$178\rho g = 178(1000)(9.807) = 1{,}745{,}000 \text{ Pa, or } 1.75 \text{ MPa}$$

12.61 b.

Orifice area, $A = 0.10^2(\pi/4) = 0.007854 \text{ m}^2$ The head is 15 meters.

$$Q = 0.70(\sqrt{2gh})A$$
$$Q = 0.70A\sqrt{(2)(9.807)(15)} = 0.094 \text{ m}^3/\text{s}$$

12.62 d. Again, we should look at Bernoulli's equation

$$P_1/\rho g + v_1^2/2g + z_1 + h_A = P_2/\rho g + v_2^2/2g + z_2 + h_L$$

The differential head is equal to 3.6 − 2.4 = 1.2 meters, which must be supplied by the pump. The pump must also supply the frictional head loss in the hose, plus one-half of the velocity head at the entrance to the brush, plus one velocity head. Flow area is equal to

$$0.05^2(\pi/4) = 0.001964 \text{ m}^2$$
$$Q = (225/60) = 3.75 \text{ l/s} = 0.00375 \text{ m}^3\text{/s}$$
$$v = 1.909 \text{ m/s}$$
$$v^2/2g = 0.186 \text{ m}$$

To calculate the frictional head loss it is necessary to first calculate the Reynolds number

$$\text{Re} = \rho D v/\mu$$

where

$D = 0.05$ m
$\mu = 0.01$ poise $= 0.001$ kg/s · m

$$\text{Re} = 1000(0.05)(1.909/0.001) = 95{,}450$$

The rubber hose would have a smooth *ID* similar to drawn tubing, so e can be taken as 0.0015 mm. The relative roughness would be equal to 0.0015/50 = 0.00003 and $f = 0.018$ from the chart in the *Handbook* for this Re and e/D. The frictional head loss would be equal to

$$h_L = f(L/D)(v^2/2g) = 0.018(15/0.05)(0.186) = 5.40(0.186) = 1.004 \text{ m}$$

Adding 1.5 velocity heads, 0.279 m, and the increase in elevation of 1.2 m yields a total suction requirement of 1.00 + 0.28 + 1.20 = 2.48 m of water.

$$P_1/\rho g - P_2/\rho g = -2.48 \text{ m}$$
$$P_1 - P_2 = -1000(9.807)(2.48) = -24.3 \text{ kPa}$$

12.63 b. Again, check Bernoulli's equation

$$qP_1/\rho g + v_1^2/2g + z_1 + h_A = P_2/\rho g + v_2^2/2g + v_2^2/2g + z_2 + h_L$$

In this case there will be frictional line losses, one velocity head in the line, and a loss of 0.10 velocity head through the nozzle.

Area of the pipe $= 0.15^2(\pi/4) = 0.0177 \text{ m}^2$
Area of nozzle $= 0.00196 \text{ m}^2$

$$\text{Velocity through pipe, } v = 0.028/0.0177 = 1.582 \text{ m/s}$$
$$v^2/2g = 0.1276 \text{ m}$$
$$\text{Re} = \rho D v/\mu = 1000(0.15)(1.582/0.001) = 237{,}300$$
$$e/D = 0.046/150 = 0.0003$$

From the curve in the *Handbook*, $f = 0.017$ (welded steel pipe is the same as commercial steel pipe).

$$\text{Frictional head loss} = f(L/D)(v^2/2g) = 0.017(25/0.15)(0.1276) = 0.362 \text{ m}$$

For the nozzle

$$v = 0.028/0.00196 = 14.286 \text{ m/s}$$
$$v^2/2g = 10.405 \text{ m}$$

The nozzle has a discharge coefficient of 0.9, so the head required at the nozzle would be equal to

$$10.405/0.9 = 11.561 \text{ m}$$

The total head required to offset the total head loss would thus be equal to

$$11.561 + 0.362 + 0.1276 = 12.051 \text{ meters}$$

12.64 d. The potential energy of the water

$$(57/60)(1000)(30) = 28{,}500 \text{ kg}\cdot\text{m/s}$$
$$28{,}500(9.807) = 279{,}500 \text{ kg} \times \text{m}^2/\text{s}^3 = 279{,}500 \text{ N}\cdot\text{m/s, or } 279 \text{ kW}$$

Energy contained in discharge

$$KE = ½\, mv^2$$

For each kg

$$KE = ½(1)(3.70^2) = 6.845(\text{N}\cdot\text{sec}^2/\text{m}) \times \text{m}^2/\text{sec}^2 = 6.845 \text{ J energy leaving the turbine}$$

For (57/60)(1000) kg/s, this becomes 6.50 kW of energy lost in the discharged water.

Power to overcome friction = (1.8 kN)(0.30)(2π)(120/60)/s = 6.79 kN/s, or 6.79 kW

So, the power developed by the generator is

$$P = 279 - 6.50 - 6.8 = 265.7 \text{ kW}$$

12.65 c. Again, we resort to Bernoulli's equation

$$P_1/\rho g + v_1^2/2g + z_1 + h_A = P_2/\rho g + v_2^2/2g + z_2 + h_L$$

The pressure difference between the throat and the line is 25 kPa = 25 kN/m^2.

The density of the oil is 940 kg/m^3, so

$$\rho g = (940)(9.807) = 9219 \text{ kg/m}^3$$
$$25{,}000/9219 = 2.712 \text{ meters of oil pressure difference.}$$

So Bernoulli's equation reduces to—

$$2.712 + v_1^2/2g = v_2^2/2g$$
$$v_2/v_1 = (10/5)^2$$

or

$$v_2 = 4v_1$$
$$(16 - 1)(v_1^2/2g) = 2.712$$
$$v_1 = \sqrt{[(2)(9.807)(2.712/15)]} = 1.883 \text{ m/s}$$
$$Q = (1.883)(0.10^2)(\pi/4) = 0.0148 \text{ m}^3/\text{s}$$

or

$$Q = (14.8)(60) = 888 \text{ l/min}$$

12.72 **c.** The velocity at the center of the tube, $v = \sqrt{(2gh)}$

$$h = 0.10/0.9 = 0.111 \text{ m}$$

$$\text{Velocity at the center of the tube} = \sqrt{(2 \times 9.807 \times 0.111)} = 1.476 \text{ m/sec}$$

$$\text{Re} = \rho D v/\mu$$

where μ is the viscosity = (0.10)(0.167) = 0.0167 kg/s · m.

$$\text{Re} = (900)(0.025)(1.476/0.0167) = 1989$$

so the flow is in the laminar range. Flow in the laminar range is parabolic, so the average velocity is one-half the maximum velocity at the center of the tube. The average velocity is thus equal to 1.476/2 = 0.738 m/s.

12.73 **b.** The problem of split-flow, or flow through two branches, is discussed in the *Handbook*. The pressure drop, or head loss, must be the same through each branch, so the drop through the 45-cm-diameter pipe and that through the 60-cm-diameter pipe will be the same. The instructions say to disregard minor losses, so the losses due to the fittings will be ignored, and only the frictional losses will be calculated. $Q = 1.70 \text{ m}^3/\text{s}$

See Exhibit 12.73

The velocity from A to B, $v = 1.70/0.636 = 2.672$ m/s

$$L/D = 915/0.90 = 1017$$
$$v^2/2g = 0.364 \text{ m}$$
$$h_L = (0.016)(1017)(0.364) = 5.923 \text{ m}$$

From B to C, top branch

$$h_L = (0.019)(610/0.45)(v^2/2g) = 1.313\ v_T^2$$

From B to C, bottom branch

$$h_L = (0.017)(730/0.60)(v^2/2g) = 1.055 v_B^2$$

The flow splits between the top and bottom branches so

$$1.70 = v_T(0.159 \text{ m}^2) + v_B(0.283 \text{ m}^2)$$

which gives

$$v_T = 10.692 - 1.780\ v_B$$

The head loss through the top branch is equal to the head loss through the bottom branch, so

$$1.313\ v_T^2 = 1.055\ v_B^2$$

Taking square roots of both sides gives

$$v_T = 0.896 v_B$$

so,

$$0.896\ v_B = 10.692 - 1.780\ v_B$$
$$v_B = 3.996 \text{ m/s}$$

and

$$v_T = 3.580 \text{ m/s}$$
$$\text{Head loss, top branch} = (1.313)(12.816) = 16.83 \text{ m}$$
$$\text{Head loss, bottom branch} = (1.055)(15.968) = 16.85 \text{ m}$$
$$\text{Head loss from C to D} = (0.017)(460/0.75)(v^2/2g) = 0.532v^2$$
$$A = (0.563)(\pi/4) = 0.442 \text{ m}^2$$
$$v = 1.70/0.441 = 3.848 \text{ m}$$
$$v^2 = 14.807$$
$$h_L = 7.88 \text{ m}$$
$$\text{Total head loss} = 5.92 + 16.84 + 7.88 = 30.64 \text{ meters of water}$$
$$\text{Pressure drop} = (30.64)(1000)(9.807) = 300{,}486 \text{ Pa, or } 300 \text{ kPa}$$

12.74 d. Assume that the lead and the cork, together, displace $(40 + C)$ cm^3 of water. The cork has a mass of 10 g, so its mass will account for (40)(0.25) g = 10 g of the displacement.

The lead would add C cm^3 to the volume displaced

$$(C \text{ cm}^3)(11.3) = 11.3C \text{ cm}^3$$
$$(40 + C) \text{ cm}^3 = (10 + 11.3C) \text{ gram}$$

Since one cc of water has a mass of one gram,

$$40 - 10 = (11.3 - 1)C$$
$$C = 30/10.3 = 2.913 \text{ cm}^3$$
$$\text{Mass of lead} = (2.913)(11.3) = 32.92 \text{ g}$$

12.75 a. See Exhibit 12.75

$$1.83 - 0.60 = 1.23 \text{ m submerged}$$
$$\text{Volume submerged} = (0.30)(0.30)(1.23) = 0.111 \text{ m}^3$$
$$\text{Mass of timber} = (0.111)(1000) = 111 \text{ kg} = \text{mass of water displaced.}$$

The same mass of liquid would be displaced when the oil layer was added.

$$[(0.30)(800) + L(1000)](0.090) = 111 \text{ kg}$$
$$L = 0.993 \text{ m}$$

1.293 meters are immersed, and 1.83 − 1.293 = 0.537 m, or 54 cm, of timber projects above the surface of the oil.

12.76 b. Check Bernoulli's equation

$$P_1/\rho g + v_1^2/2g + z_1 + h_A = P_2/\rho g + v_2^2/2g + z_2 + h_L$$

Reviewing the factors in the equation we see that it reduces to

$$P_1/\rho g + v_1^2/2g = P_2/\rho g + v_2^2/2g$$

The difference in the pressures at the two points is equal to the difference in the velocity heads.

$$v_1/v_2 = (7.6/10.0)^2 = 0.578$$
$$v_2^2 = 2.993v_1^2$$
$$v_1 = 3.0 \text{ m/s}$$
$$v_2 = 3.0/0.578 = 5.190 \text{ m/s}$$

The difference in the velocity heads is then equal to

$$(26.936 - 9.00)/2g = 0.914 \text{ meters}$$

The head in the throat is equal to 3.05 – 0.914 = 2.14 meters
Or, the difference in the two heads is

$$\Delta h = v_2^2/2g - v_1^2/2g$$
$$= (2.993 - 1.00)(9.00)/2g = 0.914 \text{ m},$$

The head in the throat is equal to 2.14 m, as before.

12.77 d. The hydraulic radius is equal to the area of flow divided by the wetted perimeter.

Since this is an equilateral triangle, all of the internal angles are 60°. The height of the triangular duct would be

$$H = b(\sin 60°) = 0.866b$$

The width of the flow surface would be

$$W = (0.366/0.866)b = 0.423\text{b}$$

The area of the trapezoidal cross-section would be

$$A = [(0.423b + b)/2](b/2) = 0.356b^2$$

The wetted perimeter would be

$$P = b + 2(b/2)/\sin 60° = 2.155b$$
$$R_H = 0.356b^2/2.155b = 0.165b$$

12.78 c. The volume of water displaced is

$$V_W = 523 - 447.5 = 75.5 \text{ cm}^3$$

If it had been solid copper the volume of the copper would have been

$$V_C = 523/8.92 = 58.6 \text{ cm}^3$$

The volume of the void is, therefore

$$V_V = 75.5 - 58.6 = 16.9 \text{ cm}^3$$

12.79 a. The mass of the length of pipe is

$$M_P = (4.6)(47.6) = 218.96 \text{ kg}$$

Adding the mass of the cap and the attached mass gives a total mass of

$$M_T = 218.96 + 11.3 = 230.26 \text{ kg}$$

The pipe assembly would therefore displace a 230.26 kg mass of seawater, or 230,260/1.025 = 223,644 cm^3.

The area of the *OD* of the pipe is

$$A = \pi(27/2)^2 = 572.6 \text{ cm}^2$$

The pipe would therefore sink 223,644/572.6 = 390.6 cm into the seawater, leaving 4.6 – 3.91 = 0.69 m or 69 cm of the pipe extending above the surface.

12.80 **c.** Apply Bernoulli's equation

$$P_1/\rho g + v_1^2/2g + z_1 + h_A = P_2/\rho g + v_2^2/2g + z_2 + h_L$$

Simplifying the equation for the pertinent factors

$$P_1/\rho g = v^2/2g + z_2 + h_L$$

The head loss is

$$h_L = f(L/D)(v^2/2g)$$
$$v^2/2g = (1.22^2)/2g = 0.0759 \text{ m}$$
$$L/D = 160{,}000/0.25 = 640{,}000$$

The friction factor is a function of both the Reynolds number and the type of pipe.

$$\text{Re} = \rho D v/\mu$$

$\mu = 0.30$ centipoises or 0.0030 poise, which is equal to 0.00030 kg/s · m
$\rho = 680 \text{ kg/m}^3$
$\text{Re} = (680)(0.25)(1.22/0.00030) = 691{,}333$

For welded-steel pipe the roughness would be the same as that for commercial steel pipe, which is equal to 0.046 mm (from the *Handbook*).

$$e/D = 0.0046/25 = 0.000184$$

From the Moody Diagram in the *Handbook*, $f = 0.15$.

$$h_L = (0.015)(640{,}000)(0.0759) = 728.6 \text{ m}$$

Adding one velocity head gives a total head H of 729 + 150 = 879 m ($z_2 = 150$ m).

$$P_1 = \rho g H = (680)(9.807)(879) = 5{,}861{,}840 \text{ Pa, or } 5.86 \text{ MPa}$$

12.81 **b.** Knowing that the equation for fan power is P = Pressure × Volume rate of flow, we may write the equation as

$$\text{Power} = \rho gh \times \text{Volume rate of flow}$$
$$= [(1000 \text{ kg/m}^3)(9.81 \text{ m/s}^2)(0.10 \text{ m})(225 \text{ m}^3/\text{min})]$$
$$= 3678.8 \text{ kg} \cdot \text{m}^2/\text{s}^3$$

Or, power may be written in more conventional units as

$$\text{Air Power} = (3678.8 \text{ kg} \cdot \text{m}^2/\text{s}^3)(\text{N})/[(\text{kg} \cdot \text{m})/\text{s}^2]$$
$$= 3.678 \text{ kN} \cdot \text{m/s} = 3.678 \text{ kW}$$

The power to operate the fan is then

$$\text{Power} = 3.678/0.85 = 4.32 \text{ kW}$$

12.82 **d.** Since we are looking at a fan in which the basic conditions are unchanged, with the exception of the fan rpm we must make use of the fan laws.

$$(\text{Volumetric flow rate})_2 = (\text{rpm}_2/\text{rpm}_1)(\text{Volumetric flow rate})_1$$
$$(\text{Volumetric flow rate})_2 = (700/500)(225) = 315 \text{ m}^3/\text{min}$$
$$\text{Power}_2 = (700/500)^3(4.32 \text{ kW}) = 11.85 \text{ kW}$$

12.83 **d.** The Bernoulli equation must be written to determine the energy added to the water and thus the energy needed to power the pump.

$$(P_1/\rho gh) + Z_1 + V_1^2/2g + E = (P_2/\rho gh) + Z_2 + V_2^2/2g + F$$

Noting that ΔZ and the friction loss F are equal to zero, and solving first for the velocity using the continuity relationship

$$m = (\rho AV_1) = (\rho AV_2)$$

Assuming that ρ is constant

$$\begin{aligned} AV &= m/\rho = (100{,}000 \text{ kg/min})/(1000 \text{ kg/m}^3) = 100 \text{ m}^3\text{/min} \\ V_1 &= [(100 \text{ m}^3\text{/min})/(60 \text{ s/min})]/(\pi D_1^2/4 \text{ m}^2) \\ &= (1.67 \text{ m}^3\text{/s})/[(\pi)(0.35)^2/4 \text{ m}^2] = 17.36 \text{ m/s} \\ V_2 &= V_1(A_1/A_2) = V_1(D_1/D_2)^2 = (17.36 \text{ m/s})(35/25)^2 = 34.02 \text{ m/s} \end{aligned}$$

Therefore

$$\begin{aligned} E &= (V_2^2 - V_1^2)/2g + (P_2 - P_2)/\rho g \\ &= \{[(34.02)^2 - (17.36)^2] \text{ m}^2\text{/s}^2\}/[2(9.81 \text{ m/s}^2)] \\ &\quad + [(40 - 1) \text{ bar}]/(1000 \text{ kg/m}^3)(9.81 \text{ m/s}^2) \\ &= 43.63 \text{ m} + \{[(39 \text{ bar})(10^5 \text{ N}\cdot\text{m/bar})] \\ &\quad (\text{kg}\cdot\text{m/s}^2)/\text{N}\}/(9810 \text{ kg/m}^2\cdot\text{s}^2) = 441.2 \text{ m} \end{aligned}$$

Solving now for power

$$\begin{aligned} \text{Power} &= \text{Pressure} \times \text{Volume flow rate} \\ &= (441.2 \text{ m})(1000 \text{ kg/m}^3)(9.81 \text{ m/s}^2)\ (100 \text{ m}^3\text{/min})/(60 \text{ s/min}) \\ &= 7{,}213{,}600 \text{ kg}\cdot\text{m}^2\text{/s}^2 = 7{,}213 \text{ kJ/s} = 7{,}213.6 \text{ kW} \\ &\quad \text{power to the water} \end{aligned}$$

$$\text{Power to the pump} = 7{,}213.6/0.75 = 9618 \text{ kW}$$

12.84 **a.** Using the *scaling laws* in the *Handbook* we may calculate the new power required.

$$\begin{aligned} [W/(\rho N^3D^5)]_1 &= [W/(\rho N^3D^5)]_2 \\ W_2 &= W_1/(\rho N_2^3D_2^5)/(\rho N_1^3D_1^5) \\ &= (9{,}618)[(1760)^3(12)^5]/[(1200)^3(10)^5] \\ &= 75{,}500 \text{ kW} \quad \textit{New Power} \end{aligned}$$

12.85 **a.** To calculate the pressure rise, again use the scaling law in the *Handbook* for pressure.

$$\begin{aligned} [P/(\rho N^2D^2)]_1 &= [P/(\rho N^2D^2)]_2 \\ P_2 &= P_1(\rho N_2^2D_2^2)/(\rho N_1^2D_1^2) \\ &= 39[(1760)^2(12)^2]/[(1200)^2(10)^2] \\ &= 120.8 \text{ bar} \qquad \textit{New Pressure Rise} \end{aligned}$$

12.86 **d.** The scaling law for flow shall be used

$$\begin{aligned} [m/(\rho ND^3)]_1 &= [m/(\rho ND^3]_2 \\ m_2 &= m_1(\rho N_2D_2^3)/(\rho N_1D_1^3) \\ &= [100{,}000 \text{ kg/min } (1760)(12)^3]/[(1200)(10^3)] \\ &= 211{,}200 \text{ kg/min} \quad \textit{New Mass Flow Rate} \end{aligned}$$

12.87 **b.** Select the work equation from the *Handbook* for the steady-flow polytropic compression system.

$$\begin{aligned} w &= n(P_2V_2 - P_1V_1) \\ &= [n/(n-1)]RT_1[1 - (P_2/P_1)^{(n-1)/n}] \\ &= 1.2/(1.2-1)[(8.314/29)\ \text{kJ/kg}\cdot\text{K}]293\ \text{K}[1 - (15/1)^{(1.2-1)/1.2}] \\ &= -287\ \text{kJ/kg} \end{aligned}$$

To find the total work we must find the mass flow, which can be obtained using the perfect gas equation.

$$\begin{aligned} m &= PV/RT \\ &= [(1\ \text{bar})(10^5\ \text{N/m}^2/\text{bar})][(500\ \text{m}^3/\text{min})]/\{[(8.314/29)\ \text{kJ/kg}\cdot\text{K}] \\ &\quad \times (10^3\ \text{N}\cdot\text{m/kJ})(293\ \text{K})\} \\ &= 595.2\ \text{kg/min} = 9.92\ \text{kg/s} \end{aligned}$$

Therefore

$$W = (9.92\ \text{kg/s})(-287\ \text{kJ/kg}) = 2847\ \text{kJ/s Input to the air}$$

Total work supplied to the compressor

$$W_c = W/\eta = 2847/0.75 = 3796\ \text{kW}$$

12.88 **c.** Writing the first law of thermodynamics as a basis for the solution

$$Q/m + h_1 + V_1^2/2g + Z_1 = h_2 + V_2^2/2g + Z_2 + W/m$$

For a compressor the effect of the elevation is essentially equal to zero ($\Delta Z = 0$).

In addition, many compressors are sized such that $\Delta KE = 0$

Therefore

$$Q = m(h_2 - h_1) + W$$

To solve for heat transfer, the final temperature needs to be found using the polytropic relation found in the *Handbook*.

$$T_2 = T_1(P_1/P_2)^{(1-n)/n} = 293\ \text{K}\ (1/15)^{(1-1.2)/1.2} = 460\ \text{K}$$

Therefore, the heat transfer is

$$Q = mc_p\Delta T + W$$

where c_p is found in the *Handbook*.

$$\begin{aligned} Q &= (9.92\ \text{kg/s})(1\ \text{kJ/kg}\cdot\text{K})(460 - 293) - 3796\ \text{kJ/s} \\ Q &= 1656.6 - 3796 = 2139\ \text{kJ/s out of the compressor.} \end{aligned}$$

CHAPTER 13

Thermodynamics

Robert F. Michel and Jerry H. Hamelink

OUTLINE

This chapter reviews the basic knowledge and working tools required to pass the thermodynamics portion of the Fundamentals of Engineering (FE/EIT) exam. It excludes areas of thermodyamics not normally included on the exam.

Typical solved problems are grouped at the end of this chapter and are referred to in the text section of the review material.

THE FIRST LAW OF THERMODYNAMICS

Except for nuclear physics, which involves the conversion of mass into energy, mass is conserved in a process. In thermodynamics, the mass in a closed system (such as a piston-cylinder) is constant. In an open system, where mass is flowing in and out, the sum of the mass flowing into the system equals the mass flowing out if there is no accumulation of mass within the system. This is called conservation of mass.

Energy is also conserved and must be accounted for. The types of energy that are important in thermodynamics are the following:

- Internal energy
- Flow energy
- Kinetic energy
- Potential energy
- Heat
- Work

The first four energy types are a function of the state, or condition, of the substance. Heat and work are forms of energy that cross the boundary of systems and are not a function of the state.

The first law of thermodynamics is a bookkeeping system to keep track of these energies. As for notation, variables with capital letters generally refer to values with units standard for those variables. Lowercase letters are specific or *per mass*. For example, internal energy U is related to specific internal energy u by: $u = U/m$ where m is mass.

Also, refer to the Table of Selected Symbols and Abbreviations at the end of this chapter for variable definitions.

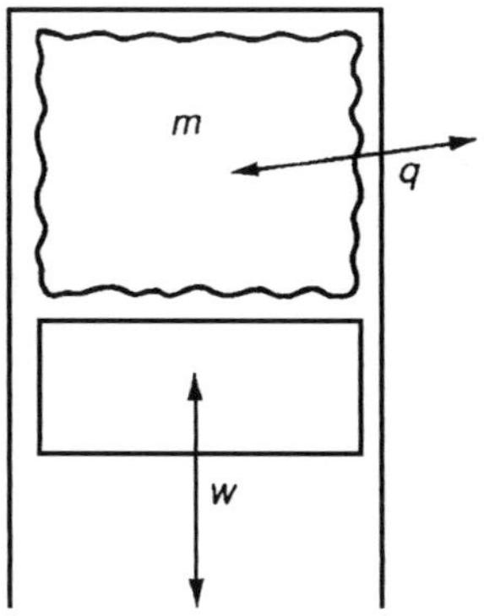

Figure 13.1

Closed System

For a typical closed thermodynamic system (see Figure 13.1), the kinetic and potential energy are not important, and since there is no flow, the bookkeeping is simple and reduces to

$$u_1 + q = u_2 + w$$

Open System

For an open system (Figure 13.2), the bookkeeping system yields

$$u_1 + p_1 v_1 + \frac{V_1^2}{2g_c} + \frac{Z_1 g}{g_c} + q = u_2 + p_2 v_2 + \frac{V_2^2}{2g_c} + \frac{Z_2 g}{g_c} + w$$

In many common open-system processes, the kinetic energy ($V^2/2g_c$) and the potential energy (Zg/g_c) are not important. For convenience, u and pv are combined into h (enthalpy), and the equation then reduces to

$$h_1 + q = h_z + w$$

The thermodynamic sign convention for heat and work is that q_{in} is positive and w_{out} is positive. This is the normal flow of heat and work in an engine or power plant.

Note also that a "change in" a property or state refers to the second point value minus the first point value.

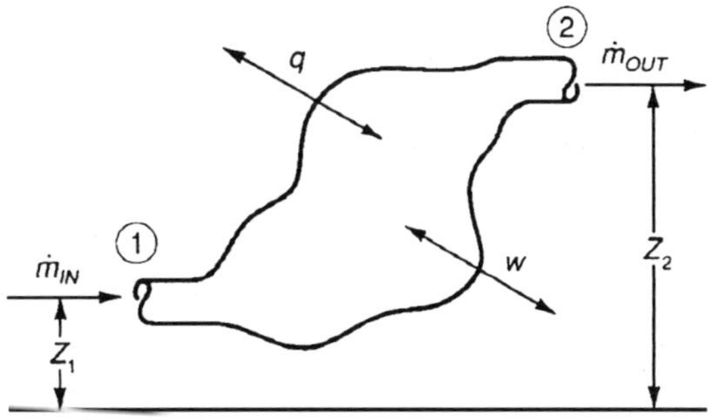

Figure 13.2

Care must be taken to keep the units consistent. For example, the normal units for internal energy, heat, and work are kJ/kg. In the SI system, $V^2/2g_c$ and Zg/g_c must be divided by 1000. Remember that g_c in the SI system has a value of 1 but is 32.2 in the U.S. system.

The first law is the most popular one by far in solving thermodynamics problems!

Example **13.1**

A piston-cylinder contains 5 kg of air. During a compression process 100 kJ of heat is removed while 250 kJ of work is done on the air. Find the change in internal energy of the air.

Solution

The system is a closed system since the mass is fixed. So

$$u_1 + q = u_2 + w$$

Here, $q = -100$ kJ, since heat leaving a system is negative, and $w = -250$ kJ, since work done *on* a system is negative. Thus

$$u_2 - u_1 = -100 - (-250) = 150 \text{ kJ}, \qquad \frac{150 \text{ kJ}}{5 \text{ kg}} = 30 \text{ kJ/kg}$$

Example 13.2

Heated air enters a turbine at a flow rate of 5 kg/s. The entering and leaving conditions are shown in the following table. The heat loss from the turbine is 50 kW. Find the power produced.

	Inlet	Exit
Pressure, kPa	1000	100
Temperature, °K	800	500
Velocity, m/s	100	200
Specific internal energy, kJ/kg	137	85
Specific volume, m^3/kg	0.23	1.44
Elevation, m	3	10

Solution

The system has mass flowing across the boundaries, so it is an open system.

$$u_1 + p_1 v_1 + \frac{Z_1 g}{g_c} + \frac{V_1^2}{2g_c} + q = u_2 + p_2 v_2 + \frac{Z_2 g}{g_c} + \frac{V_2^2}{2g_c} + w$$

$$w = (u_1 - u_2) + (p_1 v_1 - p_2 v_2) + \frac{(Z_1 - Z_2)g}{g_c} + \frac{V_1^2 - V_2^2}{2g_c} + q$$

$$= (137 - 85) + (1000)(0.23) - 100(1.44)$$

$$+ \frac{(3-10)\ 9.81}{1000 \quad 1.0} + \frac{100^2 - 200^2}{2(1)(1000)} + \left(-\frac{50}{5}\right)$$

$$= 52 + 86 - 0.069 - 15 - 10 = 112.9 \ \frac{\text{kJ}}{\text{kg}}$$

$$\dot{W} = w(\dot{m}) = 112.9 \times 5 = 564.5 \text{ kW}$$

PROPERTIES OF PURE SUBSTANCES

Use of Steam Tables

Property tables for substances that go through a phase change during normal thermodynamic processes, such as H_2O, R-12, and NH_3, are divided into three groups:

- Saturated Tables. These show the properties (v, u, h, s) of the saturated liquid R-134a at $P = 2$ and saturated vapor. For convenience, there are usually two sets: one using T as the entering argument and one using P. The highest temperature/pressure entry is usually the critical point. The quality x is needed to define properties in the mixture region.

- Superheated Tables. These show the properties (v, u, h, s) as a function of T and P in the superheated area to the right of the saturated vapor curve and are usually grouped by pressure. *Any two* properties (v, u, h, s, T, P) may be used to define the state. The saturated state is usually noted. At moderate pressures and temperatures well away from saturation, perfect gas relationships may be used as an approximation.
- Subcooled or Compressed Liquid Tables. These tables show properties (v, u, h, and s) as a function of T and P in the area to the left of the saturated liquid curve and are usually grouped by pressure. As with the Superheated Tables, any two properties may be used to define the state, and the saturated state is usually noted. For points in the region that are below the tabulated pressures, the properties are approximated as those of the saturated liquid at the same temperature (v, u, h, and s are weak functions of pressure).

The procedure for finding the state-point properties for given data where the condition of the substances is not defined is best done in a structured manner:

1. Always look at the saturation tables first to determine whether the state point is liquid, vapor, or "wet" (Figure 13.3).

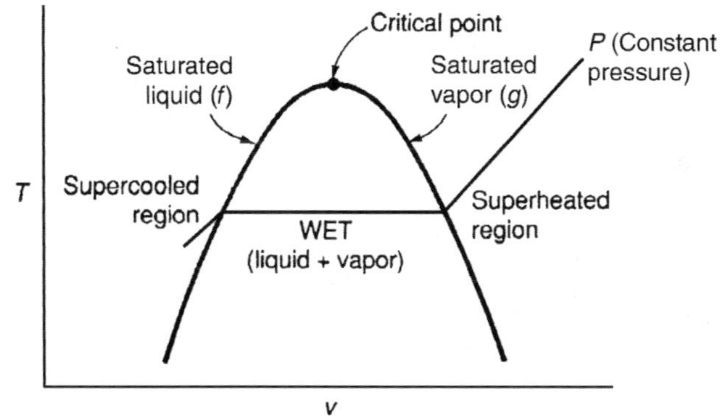

Figure 13.3

2. When one is given T or P and h, v, or u:

   a. If h, v, or u is between f and g, the point is in the wet region (inside the dome). Calculate the quality x using saturation properties and given properties. For example,

   $$x = \frac{u - u_f}{u_{fg}}$$

   b. If h, v, or u is greater than the saturated-vapor value, the point is in the superheated region. Locate the properties in the superheated tables.

   c. If h, v, or u is less than the saturated-liquid value, the point is in the subcooled liquid region. Locate the properties in the liquid tables and/or calculate from the saturated-liquid properties.

3. Given T or P and x:

   Go directly to the saturated tables, since an intermediate value of x implies it is in the "wet" region.

4. Given T and P:

   Compare the saturation temperature with the given temperature. The saturation temperature is read at the given pressure.

   If T is greater than T_{sat}, the point is superheated vapor.
   If T is less than T_{sat}, the point is subcooled liquid.

   If x is given, then v, u, h, and s can be calculated using values from the steam tables inserted into the following equations:

$$v = xv_g + (1 - x)v_f \qquad v = v_f + xv_{fg}$$
$$u = xu_g + (1 - x)u_f \qquad u = u_f + xu_{fg}$$
$$h = xh_g + (1 - x)h_f \qquad h = h_f + xh_{fg}$$
$$s = xs_g + (1 - x)s_f \qquad s = s_f + xs_{fg}$$

Example 13.3

Fill in the following table for steam (water) using the extracted tabulations.

	T, °C	P, kPa	x, %	h, kJ/kg	u, kJ/kg	v, m³/kg
a)	200	—	—	852.45	—	—
b)	—	143.3	—	—	1000	—
c)	300	800	—	—	—	—
d)	200	5000	—	—	—	—
e)	—	400	—	—	—	0.85
f)	300	—	80	—	—	—

Superheated Water Tables

T Temp. °C	v m³/kg	u kJ/kg	h kJ/kg	s kJ/(kg·K)	v m³/kg	u kJ/kg	h kJ/kg	s kJ/(kg·K)
	P = 0.01 MPa (45.81°C)				P = 0.05 MPa (81.33°C)			
Sat.	14.674	2437.9	2584.7	8.1502	3.240	2483.9	2645.9	7.5939
50	14.869	2443.9	2592.6	8.1749				
100	17.196	2515.5	2687.5	8.4479	3.418	2511.6	2682.5	7.6947
150	19.512	2587.9	2783.0	8.6882	3.889	2585.6	2780.1	7.9401
200	**21.825**	**2661.3**	**2879.5**	**8.9038**	**4.356**	**2659.9**	**2877.7**	**8.1580**
250	24.136	2736.0	2977.3	9.1002	4.820	2735.0	2976.0	8.3556
300	26.445	2812.1	3076.5	9.2813	5.284	2811.3	3075.5	8.5373
400	31.063	2968.9	3279.6	9.6077	6.209	2968.5	3278.9	8.8642
500	35.679	3132.3	3489.1	9.8978	7.134	3132.0	3488.7	9.1546
600	**40.295**	**3302.5**	**3705.4**	**10.1608**	**8.057**	**3302.2**	**3705.1**	**9.4178**
700	44.911	3479.6	3928.7	10.4028	8.981	3479.4	3928.5	9.6599
800	49.526	3663.8	4159.0	10.6281	9.904	3663.6	4158.9	9.8852
900	54.141	3855.0	4396.4	10.8396	10.828	3854.9	4396.3	10.0967
1000	58.757	4053.0	4640.6	11.0393	11.751	4052.9	4640.5	10.2964
1100	**63.372**	**4257.5**	**4891.2**	**11.2287**	**12.674**	**4257.4**	**4891.1**	**10.4859**
1200	67.987	4467.9	5147.8	11.4091	13.597	4467.8	5147.7	10.6662
1300	72.602	4683.7	5409.7	11.5811	14.521	4683.6	5409.6	10.8382
	P = 0.10 MPa (99.63°C)				P = 0.20 MPa (120.23°C)			
Sat.	1.6940	2506.1	2675.5	7.3594	0.8857	2529.5	2706.7	7.1272
100	1.6958	2506.7	2676.2	7.3614				
150	1.9364	2582.8	2776.4	7.6134	0.9596	2576.9	2768.8	7.2795
200	2.172	2658.1	2875.3	7.8343	1.0803	2654.4	2870.5	7.5066
250	**2.406**	**2733.7**	**2974.3**	**8.0333**	**1.1988**	**2731.2**	**2971.0**	**7.7086**
300	2.639	2810.4	3074.3	8.2158	1.3162	2808.6	3071.8	7.8926
400	3.103	2967.9	3278.2	8.5435	1.5493	2966.7	3276.6	8.2218
500	3.565	3131.6	3488.1	8.8342	1.7814	3130.8	3487.1	8.5133
600	4.028	3301.9	3704.4	9.0976	2.013	3301.4	3704.0	8.7770
700	**4.490**	**3479.2**	**3928.2**	**9.3398**	**2.244**	**3478.8**	**3927.6**	**9.0194**
800	4.952	3663.5	4158.6	9.5652	2.475	3663.1	4158.2	9.2449
900	5.414	3854.8	4396.1	9.7767	2.705	3854.5	4395.8	9.4566
1000	5.875	4052.8	4640.3	9.9764	2.937	4052.5	4640.0	9.6563
1100	6.337	4257.3	4891.0	10.1659	3.168	4257.0	4890.7	9.8458
1200	**6.799**	**4467.7**	**5147.6**	**10.3463**	**3.399**	**4467.5**	**5147.5**	**10.0262**
1300	7.260	4683.5	5409.5	10.5183	3.630	4683.2	5409.3	10.1982
	P = 0.40 MPa (143.63°C)				P = 0.60 MPa (158.85°C)			
Sat.	0.4625	2553.6	2738.6	6.8959	0.3157	2567.4	2756.8	6.7600
150	0.4708	2564.5	2752.8	6.9299				
200	0.5342	2646.8	2860.5	7.1706	0.3520	2638.9	2850.1	6.9665
250	0.5951	2726.1	2964.2	7.3789	0.3938	2720.9	2957.2	7.1816
300	**0.6548**	**2804.8**	**3066.8**	**7.5662**	**0.4344**	**2801.0**	**3061.6**	**7.3724**
350					0.4742	2881.2	3165.7	7.5464
400	0.7726	2964.4	3273.4	7.8985	0.5137	2962.1	3270.3	7.7079
500	0.8893	3129.2	3484.9	8.1913	0.5920	3127.6	3482.8	8.0021
600	1.0055	3300.2	3702.4	8.4558	0.6697	3299.1	3700.9	8.2674
700	**1.1215**	**3477.9**	**3926.5**	**8.6987**	**0.7472**	**3477.0**	**3925.3**	**8.5107**
800	1.2372	3662.4	4157.3	8.9244	0.8245	3661.8	4156.5	8.7367
900	1.3529	3853.9	4395.1	9.1362	0.9017	3853.4	4394.4	8.9486
1000	1.4685	4052.0	4639.4	9.3360	0.9788	4051.5	4638.8	9.1485
1100	1.5840	4256.5	4890.2	9.5256	1.0559	4256.1	4889.6	9.3381
1200	**1.6996**	**4467.0**	**5146.8**	**9.7060**	**1.1330**	**4466.5**	**5146.3**	**9.5185**
1300	1.8151	4682.8	5408.8	9.8780	1.2101	4682.3	5408.3	9.6906
	P = 0.80 MPa (170.43°C)				P = 1.00 MPa (179.91°C)			
Sat.	0.2404	2576.8	2769.1	6.6628	0.194 44	2583.6	2778.1	6.5865
200	0.2608	2630.6	2839.3	6.8158	0.2060	2621.9	2827.9	6.6940
250	0.2931	2715.5	2950.0	7.0384	0.2327	2709.9	2942.6	6.9247
300	0.3241	2797.2	3056.5	7.2328	0.2579	2793.2	3051.2	7.1229
350	**0.3544**	**2878.2**	**3161.7**	**7.4089**	**0.2825**	**2875.2**	**3157.7**	**7.3011**
400	0.3843	2959.7	3267.1	7.5716	0.3066	2957.3	3263.9	7.4651
500	0.4433	3126.0	3480.6	7.8673	0.3541	3124.4	3478.5	7.7622
600	0.5018	3297.9	3699.4	8.1333	0.4011	3296.8	3697.9	8.0290
700	0.5601	3476.2	3924.2	8.3770	0.4478	3475.3	3923.1	8.2731
800	**0.6181**	**3661.1**	**4155.6**	**8.6033**	**0.4943**	**3660.4**	**4154.7**	**8.4996**
900	0.6761	3852.8	4393.7	8.8153	0.5407	3852.2	4392.9	8.7118
1000	0.7340	4051.0	4638.2	9.0153	0.5871	4050.5	4637.6	8.9119
1100	0.7919	4255.6	4889.1	9.2050	0.6335	4255.1	4888.6	9.1017
1200	0.8497	4466.1	5145.9	9.3855	0.6798	4465.6	5145.4	9.2822
1300	**0.9076**	**4681.8**	**5407.9**	**9.5575**	**0.7261**	**4681.3**	**5407.4**	**9.4543**

Saturated Water - Temperature Table

Temp. °C T	Sat. Press. kPa P_{sat}	Specific Volume m³/kg Sat. liquid v_f	Sat. vapor v_g	Internal Energy kJ/kg Sat. liquid u_f	Evap. u_{fg}	Sat. vapor u_g	Enthalpy kJ/kg Sat. liquid h_f	Evap. h_{fg}	Sat. vapor h_g	Entropy kJ/(kg·K) Sat. liquid s_f	Evap. s_{fg}	Sat. vapor s_g
0.01	0.6113	0.001 000	206.14	0.00	2375.3	2375.3	0.01	2501.3	2501.4	0.0000	9.1562	9.1562
5	0.8721	0.001 000	147.12	20.97	2361.3	2382.3	20.98	2489.6	2510.6	0.0761	8.9496	9.0257
10	1.2276	0.001 000	106.38	42.00	2347.2	2389.2	42.01	2477.7	2519.8	0.1510	8.7498	8.9008
15	1.7051	0.001 001	77.93	62.99	2333.1	2396.1	62.99	2465.9	2528.9	0.2245	8.5569	8.7814
20	**2.339**	**0.001 002**	**57.79**	**83.95**	**2319.0**	**2402.9**	**83.96**	**2454.1**	**2538.1**	**0.2966**	**8.3706**	**8.6672**
25	3.169	0.001 003	43.36	104.88	2304.9	2409.8	104.89	2442.3	2547.2	0.3674	8.1905	8.5580
30	4.246	0.001 004	32.89	125.78	2290.8	2416.6	125.79	2430.5	2556.3	0.4369	8.0164	8.4533
35	5.628	0.001 006	25.22	146.67	2276.7	2423.4	146.68	2418.6	2565.3	0.5053	7.8478	8.3531
40	7.384	0.001 008	19.52	167.56	2262.6	2430.1	167.57	2406.7	2574.3	0.5725	7.6845	8.2570
45	**9.593**	**0.001 010**	**15.26**	**188.44**	**2248.4**	**2436.8**	**188.45**	**2394.8**	**2583.2**	**0.6387**	**7.5261**	**8.1648**
50	12.349	0.001 012	12.03	209.32	2234.2	2443.5	209.33	2382.7	2592.1	0.7038	7.3725	8.0763
55	15.758	0.001 015	9.568	230.21	2219.9	2450.1	230.23	2370.7	2600.9	0.7679	7.2234	7.9913
60	19.940	0.001 017	7.671	251.11	2205.5	2456.6	251.13	2358.5	2609.6	0.8312	7.0784	7.9096
65	25.03	0.001 020	6.197	272.02	2191.1	2463.1	272.06	2346.2	2618.3	0.8935	6.9375	7.8310
70	**31.19**	**0.001 023**	**5.042**	**292.95**	**2176.6**	**2569.6**	**292.98**	**2333.8**	**2626.8**	**0.9549**	**6.8004**	**7.7553**
75	38.58	0.001 026	4.131	313.90	2162.0	2475.9	313.93	2321.4	2635.3	1.0155	6.6669	7.6824
80	47.39	0.001 029	3.407	334.86	2147.4	2482.2	334.91	2308.8	2643.7	1.0753	6.5369	7.6122
85	57.83	0.001 033	2.828	355.84	2132.6	2488.4	355.90	2296.0	2651.9	1.1343	6.4102	7.5445
90	70.14	0.001 036	2.361	376.85	2117.7	2494.5	376.92	2283.2	2660.1	1.1925	6.2866	7.4791
95	**84.55**	**0.001 040**	**1.982**	**397.88**	**2102.7**	**2500.6**	**397.96**	**2270.2**	**2668.1**	**1.2500**	**6.1659**	**7.4159**
	MPa											
100	0.101 35	0.001 044	1.6729	418.94	2087.6	2506.5	419.04	2257.0	2676.1	1.3069	6.0480	7.3549
105	0.120 82	0.001 048	1.4194	440.02	2072.3	2512.4	440.15	2243.7	2683.8	1.3630	5.9328	7.2958
110	0.143 27	0.001 052	1.2102	461.14	2057.0	2518.1	461.30	2230.2	2691.5	1.4185	5.8202	7.2387
115	0.169 06	0.001 056	1.0366	482.30	2041.4	2523.7	482.48	2216.5	2699.0	1.4734	5.7100	7.1833
120	**0.198 53**	**0.001 060**	**0.8919**	**503.50**	**2025.8**	**2529.3**	**503.71**	**2202.6**	**2706.3**	**1.5276**	**5.6020**	**7.1296**
125	0.2321	0.001 065	0.7706	524.74	2009.9	2534.6	524.99	2188.5	2713.5	1.5813	5.4962	7.0775
130	0.2701	0.001 070	0.6685	546.02	1993.9	2539.9	546.31	2174.2	2720.5	1.6344	5.3925	7.0269
135	0.3130	0.001 075	0.5822	567.35	1977.7	2545.0	567.69	2159.6	2727.3	1.6870	5.2907	6.9777
140	0.3613	0.001 080	0.5089	588.74	1961.3	2550.0	589.13	2144.7	2733.9	1.7391	5.1908	6.9299
145	**0.4154**	**0.001 085**	**0.4463**	**610.18**	**1944.7**	**2554.9**	**610.63**	**2129.6**	**2740.3**	**1.7907**	**5.0926**	**6.8833**
150	0.4758	0.001 091	0.3928	631.68	1927.9	2559.5	632.20	2114.3	2746.5	1.8418	4.9960	6.8379
155	0.5431	0.001 096	0.3468	653.24	1910.8	2564.1	653.84	2098.6	2752.4	1.8925	4.9010	6.7935
160	0.6178	0.001 102	0.3071	674.87	1893.5	2568.4	675.55	2082.6	2758.1	1.9427	4.8075	6.7502
165	0.7005	0.001 108	0.2727	696.56	1876.0	2572.5	697.34	2066.2	2763.5	1.9925	4.7153	6.7078
170	**0.7917**	**0.001 114**	**0.2428**	**718.33**	**1858.1**	**2576.5**	**719.21**	**2049.5**	**2768.7**	**2.0419**	**4.6244**	**6.6663**
175	0.8920	0.001 121	0.2168	740.17	1840.0	2580.2	741.17	2032.4	2773.6	2.0909	4.5347	6.6256
180	1.0021	0.001 127	0.194 05	762.09	1821.6	2583.7	763.22	2015.0	2778.2	2.1396	4.4461	6.5857
185	1.1227	0.001 134	0.174 09	784.10	1802.9	2587.0	785.37	1997.1	2782.4	2.1879	4.3586	6.5465
190	1.2544	0.001 141	0.156 54	806.19	1783.8	2590.0	807.62	1978.8	2786.4	2.2359	4.2720	6.5079
195	**1.3978**	**0.001 149**	**0.141 05**	**828.37**	**1764.4**	**2592.8**	**829.98**	**1960.0**	**2790.0**	**2.2835**	**4.1863**	**6.4698**
200	1.5538	0.001 157	0.127 36	850.65	1744.7	2595.3	852.45	1940.7	2793.2	2.3309	4.1014	6.4323
205	1.7230	0.001 164	0.115 21	873.04	1724.5	2597.5	875.04	1921.0	2796.0	2.3780	4.0172	6.3952
210	1.9062	0.001 173	0.104 41	895.53	1703.9	2599.5	897.76	1900.7	2798.5	2.4248	3.9337	6.3585
215	2.104	0.001 181	0.094 79	918.14	1682.9	2601.1	920.62	1879.9	2800.5	2.4714	3.8507	6.3221
220	**2.318**	**0.001 190**	**0.086 19**	**940.87**	**1661.5**	**2602.4**	**943.62**	**1858.5**	**2802.1**	**2.5178**	**3.7683**	**6.2861**
225	2.548	0.001 199	0.078 49	963.73	1639.6	2603.3	966.78	1836.5	2803.3	2.5639	3.6863	6.2503
230	2.795	0.001 209	0.071 58	986.74	1617.2	2603.9	990.12	1813.8	2804.0	2.6099	3.6047	6.2146
235	3.060	0.001 219	0.065 37	1009.89	1594.2	2604.1	1013.62	1790.5	2804.2	2.6558	3.5233	6.1791
240	3.344	0.001 229	0.059 76	1033.21	1570.8	2604.0	1037.32	1766.5	2803.8	2.7015	3.4422	6.1437
245	**3.648**	**0.001 240**	**0.054 71**	**1056.71**	**1546.7**	**2603.4**	**1061.23**	**1741.7**	**2803.0**	**2.7472**	**3.3612**	**6.1083**
250	3.973	0.001 251	0.050 13	1080.39	1522.0	2602.4	1085.36	1716.2	2801.5	2.7927	3.2802	6.0730
255	4.319	0.001 263	0.045 98	1104.28	1596.7	2600.9	1109.73	1689.8	2799.5	2.8383	3.1992	6.0375
260	4.688	0.001 276	0.042 21	1128.39	1470.6	2599.0	1134.37	1662.5	2796.9	2.8838	3.1181	6.0019
265	5.081	0.001 289	0.038 77	1152.74	1443.9	2596.6	1159.28	1634.4	2793.6	2.9294	3.0368	5.9662
270	**5.499**	**0.001 302**	**0.035 64**	**1177.36**	**1416.3**	**2593.7**	**1184.51**	**1605.2**	**2789.7**	**2.9751**	**2.9551**	**5.9301**
275	5.942	0.001 317	0.032 79	1202.25	1387.9	2590.2	1210.07	1574.9	2785.0	3.0208	2.8730	5.8938
280	6.412	0.001 332	0.030 17	1227.46	1358.7	2586.1	1235.99	1543.6	2779.6	3.0668	2.7903	5.8571
285	6.909	0.001 348	0.027 77	1253.00	1328.4	2581.4	1262.31	1511.0	2773.3	3.1130	2.7070	5.8199
290	7.436	0.001 366	0.025 57	1278.92	1297.1	2576.0	1289.07	1477.1	2766.2	3.1594	2.6227	5.7821
295	**7.993**	**0.001 384**	**0.023 54**	**1305.2**	**1264.7**	**2569.9**	**1316.3**	**1441.8**	**2758.1**	**3.2062**	**2.5375**	**5.7437**
300	8.581	0.001 404	0.021 67	1332.0	1231.0	2563.0	1344.0	1404.9	2749.0	3.2534	2.4511	5.7045
305	9.202	0.001 425	0.019 948	1359.3	1195.9	2555.2	1372.4	1366.4	2738.7	3.3010	2.3633	5.6643
310	9.856	0.001 447	0.018 350	1387.1	1159.4	2546.4	1401.3	1326.0	2727.3	3.3493	2.2737	5.6230
315	10.547	0.001 472	0.016 867	1415.5	1121.1	2536.6	1431.0	1283.5	2714.5	3.3982	2.1821	5.5804
320	**11.274**	**0.001 499**	**0.015 488**	**1444.6**	**1080.9**	**2525.5**	**1461.5**	**1238.6**	**2700.1**	**3.4480**	**2.0882**	**5.5362**
330	12.845	0.001 561	0.012 996	1505.3	993.7	2498.9	1525.3	1140.6	2665.9	3.5507	1.8909	5.4417
340	14.586	0.001 638	0.010 797	1570.3	894.3	2464.6	1594.2	1027.9	2622.0	3.6594	1.6763	5.3357
350	16.513	0.001 740	0.008 813	1641.9	776.6	2418.4	1670.6	893.4	2563.9	3.7777	1.4335	5.2112
360	18.651	0.001 893	0.006 945	1725.2	626.3	2351.5	1760.5	720.3	2481.0	3.9147	1.1379	5.0526
370	**21.03**	**0.002 213**	**0.004 925**	**1844.0**	**384.5**	**2228.5**	**1890.5**	**441.6**	**2332.1**	**4.1106**	**0.6865**	**4.7971**
374.14	22.09	0.003 155	0.003 155	2029.6	0	2029.6	2099.3	0	2099.3	4.4298	0	4.4298

Solution

Fill in the following table for steam (water) using the extracted tabulations.

	T, °C	P, kPa	x, %	h, kJ/kg	u, kJ/kg	v, m³/kg
a)	**200**	1553.8	0	**852.45**	850.65	0.0012
b)	110	**143.3**	26.2	1046	**1000**	0.318
c)	**300**	**800**	—	3056.5	2797.2	0.3241
d)	**200**	**5000**	—	852	851	0.0012
e)	466	**400**	—	3413	3075	**0.85**
f)	**300**	8581	**80**	2467.9	2316.8	0.0176

(i) This state falls directly on the saturated liquid line.

(ii) The given value for internal energy falls between u_f and u_g, so the state is inside the vapor dome. Calculate x and use x to find h and v.

(iii) The temperature of 300°C is higher than T_{sat} = 170.4°C, so the state is superheated.

(iv) The temperature of 200°C is less than T_{sat} = 264°C, so the state is subcooled. Find the properties in the saturation tables at 200°C (good approximation).

(v) The specific volume is greater than v_g = 0.6058 at the saturation pressure of 300 kPa, so it is superheated (interpolation required).

(vi) A quality value (x) is specified, so the state is obviously in the vapor dome. Use x to find h, u, and v.

Use of R-134a Tables, NH_3 Tables, and *P-h* Diagrams

Typically, the four working fluids used in thermodynamic texts and on the FE/EIT exam are H_2O, R-134a, NH_3, and air. The refrigerant R-22 has characteristics similar to R-134a, so it is not customarily tabled. The refrigerants R-134a and NH_3 exhibit the same phase-diagram characteristics as water, so they have the same type of tabled data, that is, subcooled liquid, saturated mixture, and superheated vapor. The guidelines given for finding one's way around the steam tables apply equally well to R-134a and NH_3 tables.

By tradition and for convenience, refrigerant properties are shown in a *P-h* diagram; this is especially useful when analyzing a vapor compression refrigeration cycle, since it operates at two basic pressure levels. A skeleton *P-h* chart is shown in Figure 13.4. A larger *P-h* diagram on the next page will be used in the next example and in the section on the vapor compression refrigeration cycle.

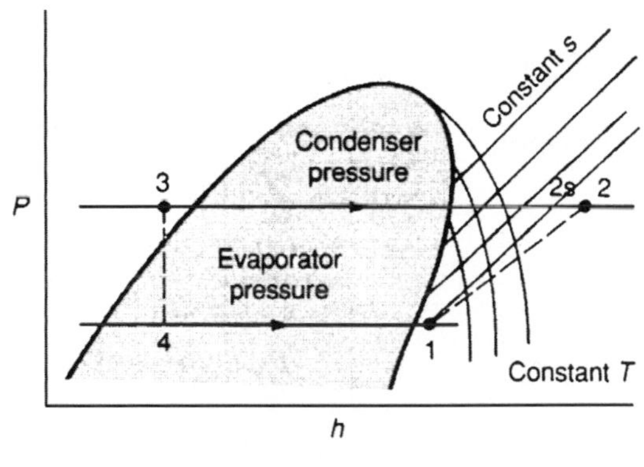

Figure 13.4

P-h DIAGRAM FOR REFRIGERANT HFC-134a
(metric units)
(Reproduced by permission of the DuPont Company)

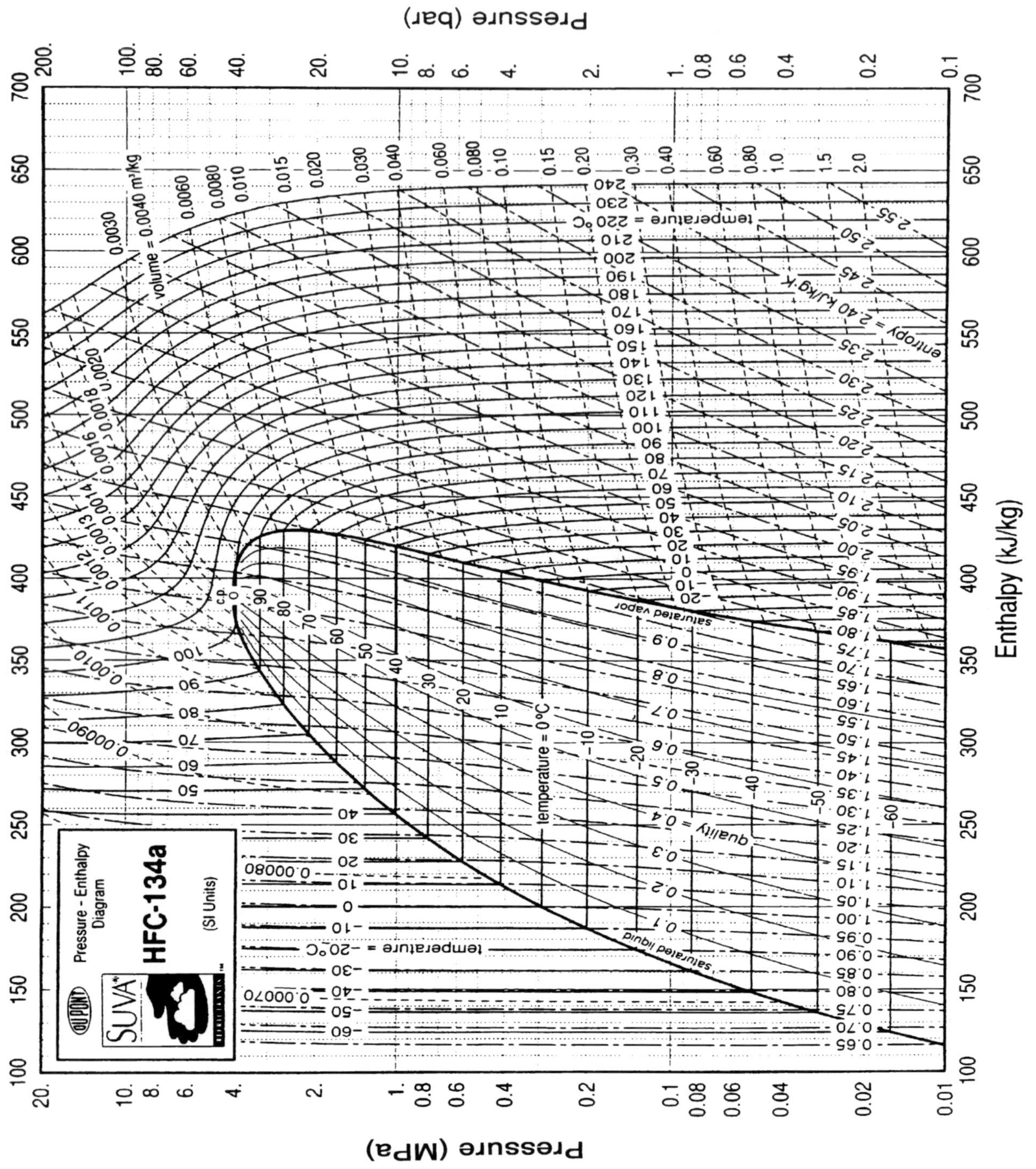

Example 13.4

Saturated liquid R-134a at $P = 2$ bars is heated to 100°C at constant pressure. Find the original and final state-point properties.

Solution

At $P = 2$ bars (.2 MPa) on the saturated liquid line,

$$v_f = .00075 \text{ m}^3/\text{kg}$$

$$h_f = 186 \text{ kJ/kg}$$

$$T_f = -10°\text{C}$$

$$s_f = 0.95 \text{ kJ/(kg} \bullet °\text{K)}$$

At $P = 2$ bars and 100°C, the R-134a is superheated.

$$v = .148 \text{ m}^3/\text{kg}$$
$$h = 491 \text{ kJ/kg}$$
$$s = 2.0484 \text{ kJ/(kg} \bullet \text{°K)}$$

IDEAL GASES

Equations of State

There are many relationships that relate the state-point properties of gases with varying degrees of accuracy. In order of increasing accuracy they are as follows:

1. Ideal gas
2. Van der Waals
3. Beattie-Bridgeman
4. Generalized compressibility
5. Property tables such as H_2O, R-134a, NH_3

Tables for these relationships are more conveniently simulated with software programs.

When one is dealing with a wide range of temperature and pressure for gases that are far from their saturated vapor state, the ideal gas relationship is accurate for most engineering work. Some working expressions are

$$Pv = \text{R}T$$

$$PV = m\text{R}T = n\bar{\text{R}}T, \qquad \text{R} = \frac{\bar{\text{R}}}{\text{m.w.}}$$

where

v = specific volume, m^3/kg
V = total volume, m^3
R = gas constant for the specific gas kJ/(kg • °K)
$\bar{\text{R}}$ = universal gas constant
n = number of moles
m = mass, kg
m.w. = molecular weight

Example 13.5

Calculate and compare the specific volumes of water (H_2O) at P = 10 MPa and T = 400°C. The gas (steam) table value is $v = 0.02641 \text{ m}^3/\text{kg}$. For an ideal gas,

$$\text{R} = \frac{8.31}{\text{m.w.}} = \frac{8.31}{18} = 0.462 \quad \frac{\text{kJ}}{\text{kJ/(kg} \bullet \text{°K)}}$$

$$v = \frac{\text{R}T}{P} = \frac{0.462(673)}{10{,}000} = 0.0311 \text{ m}^3/\text{kg}$$

The steam table value is the most accurate, and the ideal gas value is the least accurate.

Enthalpy and Internal Energy Changes

The other condition that is usually considered to be part of the definition of ideal gas is that the specific heats or heat capacities (c_p and c_v) are constant. This allows a number of working relationships to be developed:

$$h_2 - h_1 = c_p (T_2 - T_1) \qquad c_p = \frac{kR}{k-1}$$

$$u_2 - u_1 = cv (T_2 - T_1) \qquad c_v = \frac{R}{k-1}$$

$$c_p = c_v + R \quad \text{(always true)}$$

$$k = \frac{c_p}{c_v} = \text{constant}$$

PROCESSES

Thermodynamic processes usually involve a "working fluid" such as a pure substance (like water) or a gas (like air), so tables of properties or ideal gas relationships are used. For the process path to be known, the process must be reversible. If the process involves friction or turbulence and is irreversible, then only the first law of thermodynamics applies.

So, generally, the processes are considered to be reversible (no friction or turbulence) and are one of the following:

- Constant pressure (Isobaric)
- Constant volume (Isometric)
- Constant temperature (Isothermal)
- No heat flow (adiabatic) (Isentropic)

Table 13.1 shows the applicable relationships for the various processes.

Table 13.1 First and second thermodynamics law formulas for reversible processes of an ideal gas*

Process	Closed System (nonflow)	Open System (steady flow)
General	$q = c_p\ (T_2 - T_1) + w$	$q = c_p\ (T_2 - T_1) + w$
$(Pv = RT\ )$	$w = \int_1^2 P\ dv$	$w = \int_1^2 P\ dv$
$\frac{p_1 v_1}{T_1} = \frac{p_2 v_2}{T_2}$	$s_2 - s_1 = c_v \ln \frac{T_2}{T_1} - R \ln \frac{v_2}{v_1}$	$s_2 - s_1 = c_v \ln \frac{T_2}{T_1} - R \ln \frac{v_2}{v_1}$
	($s_2 - s_1$ for closed or open systems)	
Polytropic Pv^n = constant	$q = \frac{k-n}{1-n} c_v (T_2 - T_1)$	$q = \frac{k-n}{1-n} c_v (T_2 - T_1)$
$\frac{p_2}{p_1} = \left(\frac{T_2}{T_1}\right)^{n/(n-1)} = \left(\frac{v_1}{v_2}\right)^n$	$w = \frac{k-1}{1-n} c_v (T_2 - T_1)$	$w = n \frac{k-1}{1-n} c_v (T_2 - T_1) - \Delta KE - \Delta PE$
$\frac{T_2}{T_1} = \left(\frac{p_2}{p_1}\right)^{(n-1)/n} = \left(\frac{v_1}{v_2}\right)^{n-1}$	$s_2 - s_1 = c_p \ln \frac{T_2}{T_1} - R \ln \frac{p_2}{p_1}$	$s_2 - s_1 = c_v \ln \frac{T_2}{T_1} - R \ln \frac{v_2}{v_1}$
$\frac{v_2}{v_1} = \left(\frac{p_2}{p_1}\right)^{1/n} = \left(\frac{T_1}{T_2}\right)^{1/n}$	$s_2 - s_1 = \phi_2 - \phi_1 - R \ln \frac{p_2}{p_1} = \phi_2 - \phi_1 + R \ln \frac{v_2}{v_1}$	
Constant with volume (isometric)		$q = c_v (T_2 - T_1)$ $q = cv\ (T_2 - T_1)$
$v_2 = v_1,\quad n = \infty$	$w = 0$	$w = -v\ (p_2 - p_1) - \Delta KE - \Delta PE$
$\frac{p_2}{T_2} = \frac{p_1}{T_1}$	$s_2 - s_1 = c_v \ln(T_2/T_1)$	$s_2 - s_1 = c_v \ln(T_2/T_1)$
Constant pressure (isobaric)		$q = c_p (T_2 - T_1)$ $q = c_p (T_2 - T_1)$
$p_2 = p_1,\quad n = 0$	$w = p\ (v_2 - v_1)$	$w = -\Delta KE - \Delta PE$
$\frac{v_2}{T_2} = \frac{v_1}{T_1}$	$w = R(T_2 - T_1)$	$s_2 - s_1 = c_p \ln (T_2/T_1)$
	$s_2 - s_1 = c_p \ln(T_2/T_1)$	
Constant temperature (isothermal)	$q = w = T(s_2 - s_1)$	$q = T(s_2 - s_1) = RT \ln \frac{v_2}{v_1}$ or $\frac{p_1}{p_2}$
$T_2 = T_1,\quad n = 1$	$q = w = RT \ln \frac{v_2}{v_1}$ or $\frac{p_1}{p_2}$	$w = RT \ln \frac{v_2}{v_1} - \Delta KE - \Delta PE$ or $w = RT \ln \frac{p_1}{p_2} - \Delta KE - \Delta PE$
$p_2 v_2 = p_1 v_1$	$s_2 - s_1 = R \ln \frac{v_2}{v_1}$ or $\frac{p_1}{p_2}$	$s_2 - s_1 = R \ln \frac{v_2}{v_1} = R \ln \frac{p_1}{p_2}$
Adiabatic (isentropic)	$q = 0$	$q = 0$
$n = k$	$w = c_v\ (T_1 - T_2)$	$w = c_p\ (T_1 - T_2) - \Delta KE - \Delta PE$
	$w = \frac{p_1 v_1 - p_2 v_2}{k-1}$	$w = \frac{k(p_1 v_1 - p_2 v_2)}{k-1}\ \Delta KE - \Delta PE$
	$w = R(T_1 - T_2)/(k-1)$	$w = kR(T_1 - T_2)/(k-1) - \Delta KE - \Delta PE$
$s_2 = s_1$	$s_2 - s_1 = 0$	$s_2 - s_1 = 0$

*Per-unit mass basis and constant (average) specific heats (c_v, c_p) assumed. $R = c_p - c_v$, $k = c_p / c_v$, $c_p = kR/(k-1)$, $C_v = R/(k-1)$.

$\Delta u = u_2 - u_1 = c_v (T_2 - T_1)$, $\Delta h = h_2 = h_1 = c_p (T_2 - T_1)$

$\Delta KE\ \text{(S.I.)} = \frac{v_2^2 - v_1^2}{2000 \times g_c}$, $\Delta PE\ \text{(S.I.)} = \frac{g(z_2 - z_1)}{1000 \times g_c}$

ΔKE and ΔPE may be negligible for many open systems.

Example 13.6

Air at 27°C is heated to 927°C. Find the change in enthalpy and internal energy, treating air as a perfect gas (c_p and c_v constant).

Solution

For air at room temperature, $c_p = 1.00$ and $c_v = 0.718$ kJ/kg, so

$$h_2 - h_1 = (927 - 27) \times 1.00 = 900.0 \text{ kJ/kg}$$
$$u_2 - u_1 = (927 - 27) \times 0.718 = 646.2 \text{ kJ/kg}$$

THE SECOND LAW OF THERMODYNAMICS

In the morning section of the FE examination, there will typically be several problems that involve the application of one of the statements of the second law or the Carnot cycle. Useful statements of the second law are as follows:

1. Whenever energy is transferred, some energy is reduced to a lower level.
2. No heat cycle is possible without the rejection of some heat.
3. A Carnot cycle converts the maximum amount of heat into work; it has the highest thermal efficiency.
4. All Carnot cycles operating between two temperature reservoirs have the same efficiency.
5. A Carnot machine's efficiency, or coefficient of performance (COP), is a function of only the two reservoir temperatures.

A Carnot cycle consists of the following four processes:

4–1 Reversible adiabatic compression
1–2 Reversible adiabatic constant temperature heat addition
2–3 Reversible adiabatic expansion
3–4 Reversible constant temperature heat rejection

The normal property diagrams are shown in Figures 13.5 and 13.6. If the processes proceed in a clockwise direction, the Carnot engine operates as a power-producing engine; if in a counterclockwise direction, the engine is a refrigerator or a heat pump.

The efficiencies, or COPs, are

$$\eta = \frac{W_{\text{out}}}{Q_{\text{in}}} = 1 - \frac{T_C}{T_H}$$

$$\text{COP}_{\text{REFR}} = \frac{Q_{\text{in}}}{W_{\text{in}}} = \frac{T_C}{T_H - T_C}$$

$$\text{COP}_{\text{HEATPUMP}} = \frac{Q_{\text{out}}}{W_{\text{in}}} = \frac{T_H}{T_H - T_C}$$

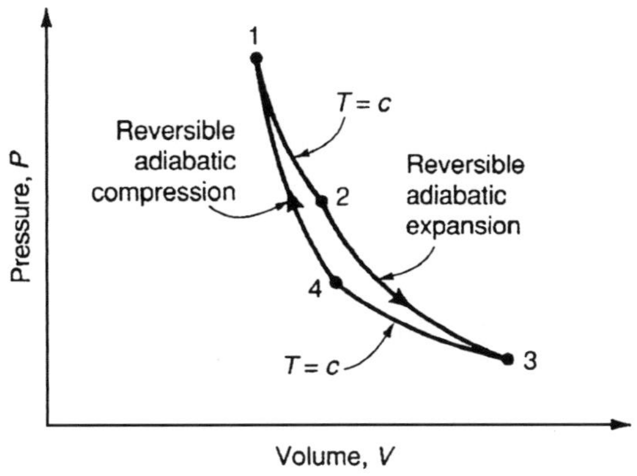

Figure 13.5

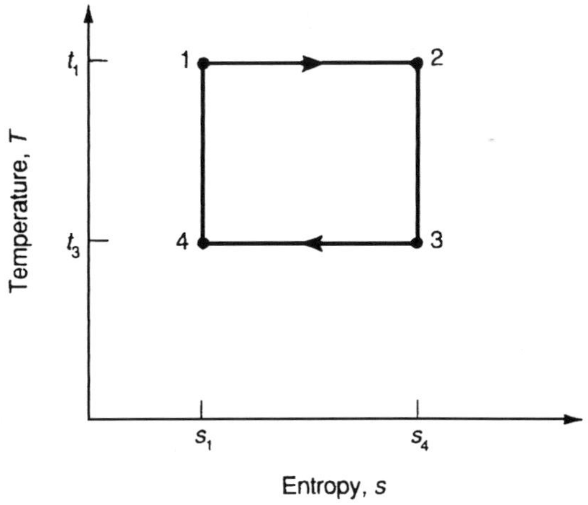

Figure 13.6

Example 13.7

A Carnot machine operates between a hot reservoir at 200°C and a cold reservoir at 20°C. (i) When operated as an engine, it receives 1000 kJ/kg; find the work output. (ii) Find the COP when the machine is operated as a refrigerator and as a heat pump.

Solution

(i)

$$\eta = \frac{T_H - T_C}{T_H} = \frac{200 - 20}{473} = 0.381$$
$$W = \eta\,(Q_{\text{in}}) = 0.381\,(1000) = 381 \text{ kJ/kg}$$

(ii)

$$\text{COP}_{\text{REFR}} = \frac{T_C}{T_H - T_C} = \frac{293}{200 - 20} = 1.63$$
$$\text{COP}_{\text{HEATPUMP}} = \frac{T_H}{T_H - T_C} = \text{COP}_{\text{REFR}} + 1.0 = 2.63$$

ENTROPY

Entropy is another thermodynamic property that is useful in the evaluation of thermodynamic systems and processes. The following statements are useful in solving problems:

1. Natural processes (which typically involve friction) result in an increase in entropy.
2. Entroy will always *increase* when heat is added.
3. Entropy will remain *constant* when processes are reversible and adiabatic.
4. Entropy can *decrease* only when heat is removed.

For reversible processes,

$$ds = \frac{dq}{T}$$
$$T\,ds = du + P\,dv = dh - v\,dP$$

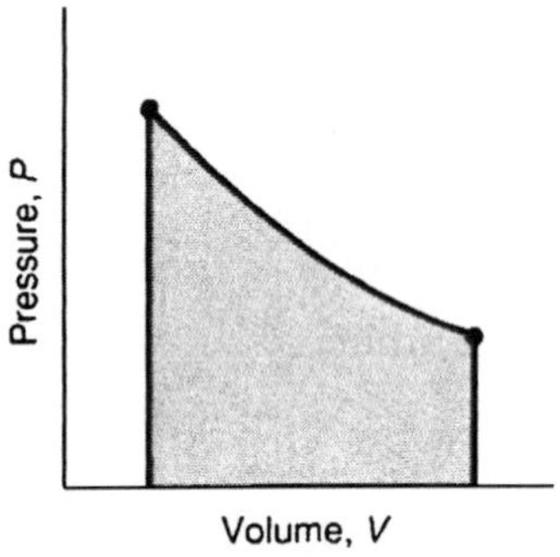

Figure 13.7

For an ideal gas,

$$s_2 - s_1 = c_v \ln\frac{T_2}{T_1} + \mathrm{R}\ln\frac{v_2}{v_1} = c_p \ln\frac{T_2}{T_1} - \mathrm{R}\ln\frac{p_2}{p_1}$$

Just as work for a closed system,

$$W = \int_1^2 P\,dv$$

can be shown as an area on a *P-V* diagram (Figure 13.7), so can heat,

$$Q = \int_1^2 T\,ds$$

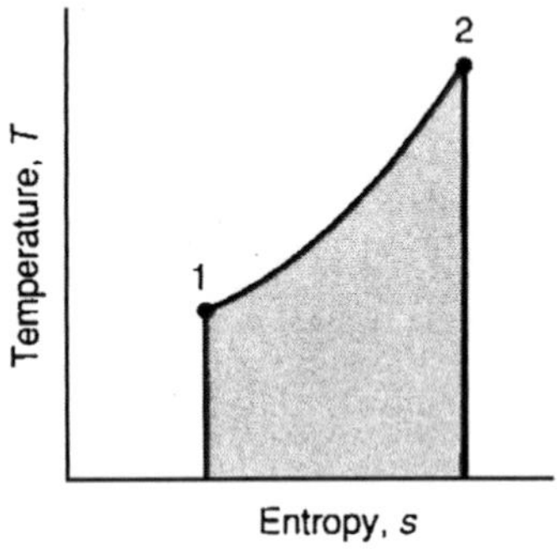

Figure 13.8

be shown as an area on a *T-s* diagram (Figure 13.8), and the area *enclosed* by the process lines on a *T-s* diagram shows the *net* heat flow in a cycle. This, of course, is equal to the *net* work.

An **isentropic process** is defined as one that is reversible and adiabatic. Of course, on a property diagram showing entropy (s), the process would appear as a straight line. For several important thermodynamic devices, the isentropic process is a standard of comparison and is used in the calculation of the component efficiency (turbine, compressor, pump, nozzle).

Since all natural processes produce an increase in entropy, the ideal (isentropic) and the actual processes can be compared, as shown in Figures 13.9 and 13.10.

$$\eta_{\text{turbine nozzle}} = \frac{h_1 - h_3}{h_1 - h_2}$$

$$\eta_{\text{compr., pump}} = \frac{h_2 - h_1}{h_3 - h_1}$$

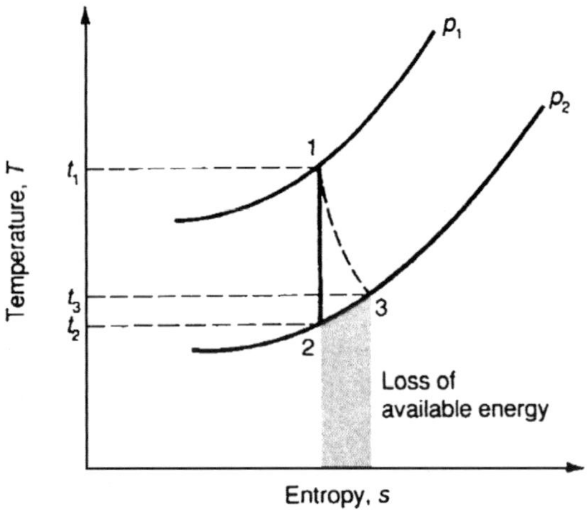

Figure 13.9

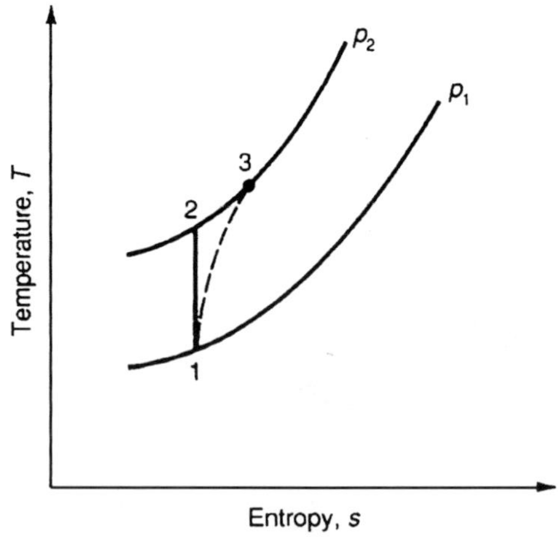

Figure 13.10

CYCLES

The FE afternoon problems will usually include a Rankine cycle problem and a vapor compression problem. Typically, there will be others involving a knowledge of the Brayton, Otto, or Diesel cycles.

Rankine Cycle (Steam)

An ideal Rankine cycle with superheated steam flowing into the turbine is shown in Figure 13.11. The four open-system components are analyzed as follows:

1. Boiler

$$q_{in} = h_2 - h_1 \left(\frac{\text{kJ}}{\text{kg}}\right)$$
$$\dot{Q}_{in} = \dot{m}_{stm}(h_2 - h_1)(\text{kW})$$

2. Turbine

$$w_T = h_3 - h_2 \text{ (kJ/kg)}$$
$$\dot{W}_T = \dot{m}_{stm}(h_3 - h_2) \text{ (kW)}$$

3. Condenser

$$q_{out} = h_3 - h_4 = c_p \Delta T \text{ (kJ/kg)}$$
$$\dot{Q}_{out} = \dot{m}_{stm}(h_3 - h_4) = \dot{m}_{cw} c_p \Delta T_{cw} \text{ (kW)}$$

4. Pump

$$w_p = h_1 - h_4 = v_4(p_1 - p_4) \text{ (kJ/kg)}$$
$$\dot{W}_p = \dot{m}_{stm}(h_1 - h_4) \text{ (kW)}$$

State points are found in property (steam) tables. State 2 is usually given by pressure and temperature and can be either saturated or (normally) superheated. State 2 can be found in the steam tables. State 3 is found by using an isentropic process, so that entropy (State 2) and pressure (condensing) are known. State 4 is saturated liquid found in the tables at the condensing pressure.

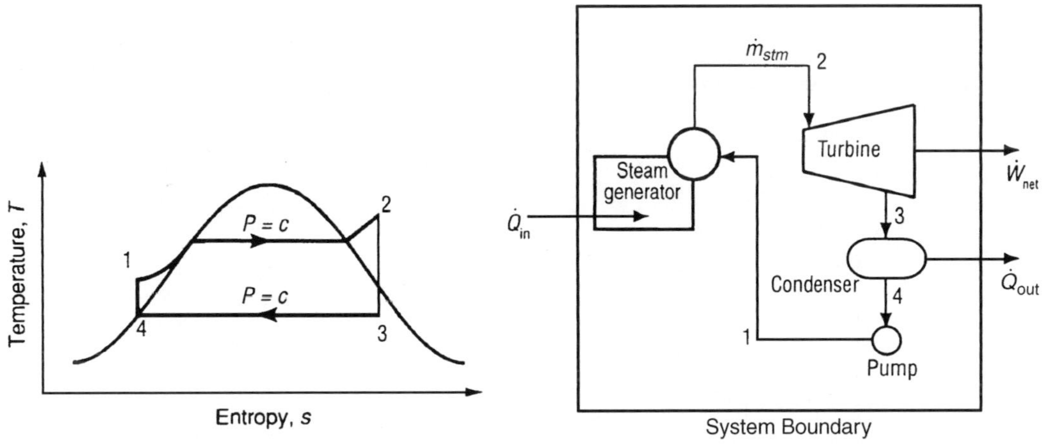

Figure 13.11

Also, State 1—using an isentropic process—can be found by the tables but is more easily found by

$$h_1 = h_4 + v_4(p_1 - p_4)$$

The thermal efficiency is

$$\eta = \frac{W_T - W_P}{Q_{in}} = \frac{h_2 - h_3 - (h_1 - h_4)}{h_2 - h_1}$$

Adding $h_1 - h_4$ (the pump work) to both the numerator and denominator results in

$$\eta_{approx.} = \frac{h_2 - h_3}{h_2 - h_4}$$

Example 13.8

A Rankine cycle using steam has turbine inlet conditions of $P = .4$ MPa, $T = 300°C$, and a condenser temperature of 70°C. The turbine efficiency is 90%, and the pump efficiency is 80%.

For both the ideal cycle and the cycle considering the component efficiencies, find (a) the thermal efficiency (η), (b) the turbine discharge quality (x), and (c) the steam flow rate ($\dot{m}$) for 1 MW of net power.

Properties of water (SI units): superheated-vapor table

Temp °C	v	u	h	s
	0.4 MPa (T_{sat} = 143.6 °C)			
300	.6548	2805	3067	7.566

v, m³/kg; u, kJ/kg; h, kJ/kg; s, kJ/(kg •°K)

Properties of saturated water (SI units): pressure table

		Specific volume		Internal energy		Enthalpy			Entropy	
Press. Bars kPa	Temp. °C T	Sat. Liquid vf	Sat. Vapor v_g	Sat. Liquid u_f	Sat. Vapor u_g	Sat. Liquid h_f	Evap. h_{fg}	Sat. Vapor h_g	Sat. Liquid s_f	Sat. Vapor s_g
31.2	70	.00102	5.04	293	2570	293	2334	2627	9549	7.7553

v, m³/kg; u and h, kJ/kg; s, kJ/(kg • °K)

Solution

Starting with State 2 (turbine inlet), the properties h and s can be found with the steam tables.

$$h_2 = 3067 \frac{\text{kJ}}{\text{kg}} \qquad s_2 = 7.566 \frac{\text{kJ}}{\text{kg} \bullet °\text{K}}$$

The ideal turbine discharge (State 3) is found at $T_3 = 70°C$ and $s_3 = 7.566$:

$$x = \frac{s - s_f}{s_{fg}} = \frac{7.566 - .9549}{7.7553 - .9549} = 0.972$$

$$h_3 = h_f + xh_{fg} = 293 + .972 \times 2334 = 2562 \text{ kJ/kg}$$

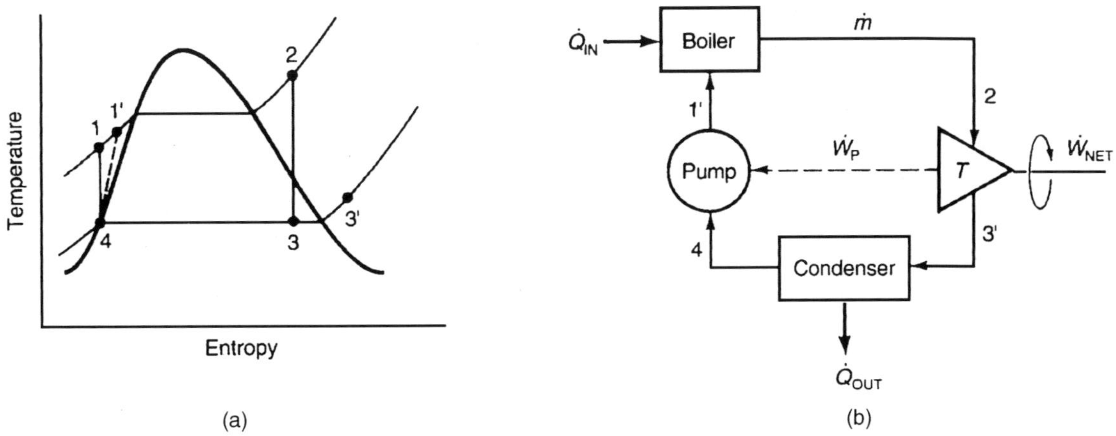

Figure 13.12

The saturated liquid state leaving the condenser (State 4) is read from the tables:

$$h_4 = 293\,\text{kJ/kg}$$
$$s_4 = .9549\,\frac{\text{kJ}}{\text{kg} \bullet {}^\circ\text{K}}$$
$$v_4 = 0.00102\,\text{m}^3/\text{kg}$$

The compressed (subcooled) liquid leaving the pump (State 1) can be found in the tables if values are available for the condition. The usual approximation is to calculate

$$\begin{aligned} h_1 &= h_4 + v_4(p_4 - p_1) \\ &= 293 + .00102(400 - 31.2) \\ &= 293 + .4 = 293.4\,\frac{\text{kJ}}{\text{kg}} \end{aligned}$$

For the ideal cycle,

(a) $$\eta = \frac{W_{\text{net}}}{Q_{\text{in}}} = \frac{W_T - |W_p|}{Q_{\text{in}}} = \frac{(3067 - 2562) - .4}{3067 - 293.4} = \frac{505 - .4}{2773.6} = 0.182$$

(b) $x = 0.972$

(c) $$\dot{m} = \frac{1000\ \text{kW}}{W_{\text{net}}} = \frac{1000}{505} = 2\ \text{kg/s}$$

For the cycle considering the component efficiencies (Figure 13.12),

$$\eta_{\text{turb.}} = \frac{h_2 - h_{3'}}{h_2 - h_3}$$

$h_{3'} = 3067 - .9(3067 - 2562) = 2613$ kJ/kg, and P = 31.2 kPa

$$h_{1'} = h_4 + \frac{h_1 - h_4}{\eta_p} = 293 + \frac{.4}{.8} = 293.5\frac{\text{kJ}}{\text{kg}}$$

$$x = \frac{h_{3'} - h_f}{h_{fg}} = \frac{2613 - 293}{2334} = 0.99$$

$$\eta = \frac{W_T - W_P}{Q_{in}} = \frac{h_2 - h_{3'} - h_{1'} - h_4}{h_2 - h_{1'}}$$

$$= \frac{(3067 - 2613) - \dfrac{.4}{.8}}{3067 - \left(293 + \dfrac{.4}{.8}\right)} = \frac{453.5}{2773.5} = .164$$

$$\dot{m} = \frac{1000}{453.5} = 2.2\frac{\text{kg}}{\text{s}}$$

The pump work makes little numerical difference and can usually be ignored in the calculation of thermal efficiency and steam flow rate.

Vapor Compression Cycle (Refrigeration)

An ideal vapor compression cycle is shown in Figure 13.13. The four open-system components are analyzed below:

1. Compressor

$$W_{in} = h_2 - h_1$$

$$\dot{W}_{in} = \dot{m}\,(h_2 - h_1)$$

2. Condenser

$$Q_{\text{out}} = h_2 - h_3$$
$$\dot{Q}_{\text{out}} = \dot{m}(h_2 - h_3)$$

3. Expansion valve

$$h_3 = h_4$$

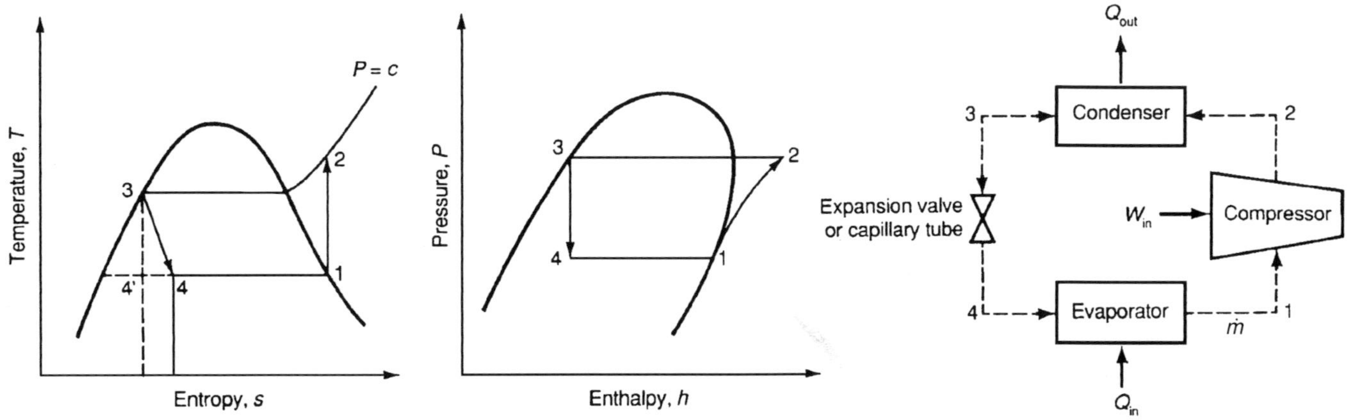

Figure 13.13

4. Evaporator

$$Q_{in} = h_1 - h_4$$
$$\dot{Q}_{in} = \dot{m}(h_1 - h_4)$$

The coefficient of performance (COP), if the cycle is used as a refrigerator, is

$$\text{COP}_{\text{REFR}} = \frac{Q_{in}}{W_{in}} = \frac{h_1 - h_4}{h_2 - h_1}$$

If the cycle is used as a heat pump,

$$\text{COP}_{\text{HEATPUMP}} = \frac{Q_{out}}{W_{in}} = \frac{h_2 - h_3}{h_2 - h_1} = \text{COP}_{\text{REFR}} + 1.0$$

The state points are found in property tables and/or property diagrams (*P*-*h*). State 1 is usually given as saturated vapor at a given pressure or temperature. State 2 is found by assuming an isentropic process so that the entropy and pressure (or corresponding condensing temperature) are known. This is best done on a *P*-*h* diagram. State 3 is saturated liquid at the given condensing pressure. State 4 is found at the same enthalpy as State 3 and at the evaporating pressure.

Example 13.9

An ideal vapor compression refrigeration cycle using R-134a operates between 100 kPa and 1000 kPa. Find the COP and the mass flow rate required for 100 kW of cooling.

Solution

Refer to the *P*-*h* diagram on page 732 and a schematic of the components (Exhibit 1). State 1 (saturated vapor at 100 kPa), from *P*-*h* diagram or from property tables:

$$h_1 = 383 \text{ kJ/kg}$$
$$s_1 = 1.746 \text{ kJ/(kg} \bullet \text{°K)}$$
$$T_1 = -26.31\text{°C}$$

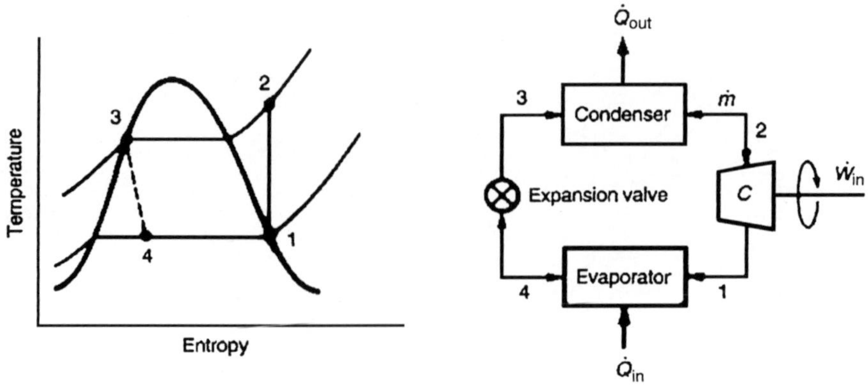

Exhibit 1

State 2 (P_2 = 1000 kPa)

$$s_1 = s_2 = 1.746 \text{ kJ/(kg} \bullet \text{°K)}$$
$$h_2 = 431 \text{ kJ/kg}$$
$$T_2 = 49.8\text{°C}$$

State 3 (saturated liquid)

$$h_3 = 255.6 \text{ kJ/kg}$$
$$T_3 = 39.3\text{°C}$$

State 4 (liquid + vapor)

$$h_3 = h_4 = 225.6 \text{ kJ/kg}$$
$$\text{COP}_\text{R} = \frac{\dot{q}_\text{evap}}{w_c} = \frac{h_1 - h_4}{h_2 - h_1} = \frac{383 - 255.6}{431 - 383} = \frac{127.4}{48} = 2.65$$
$$\dot{q}_\text{evap} = \dot{m}(h_1 - h_4), \qquad \dot{m} = \frac{100}{127.4} = 0.478\,\frac{\text{kg}}{\text{s}}$$

Psychrometrics

Our atmosphere is a mixture of noncombustible gases, namely air and water vapor. The state of this mixture at a particular pressure may be specified by two intensive properties. We may determine the conditions of the air-water mixture by using equations or by the use of the psychrometric chart. Note, however, that to use the psychrometric chart, the pressure of the air-water mixture must be 1 atm, or 14.7 lbf/in^2 in the English system, or 101.35 kPa in the SI system.

The psychrometric chart provides easily readable values of the properties of moist air at various conditions. Refer to the NCEES *Fundamentals of Engineering Supplied-Reference Handbook* for a complete psychrometric chart. Figure 13.14 shows a schematic diagram of the psychromatic chart.

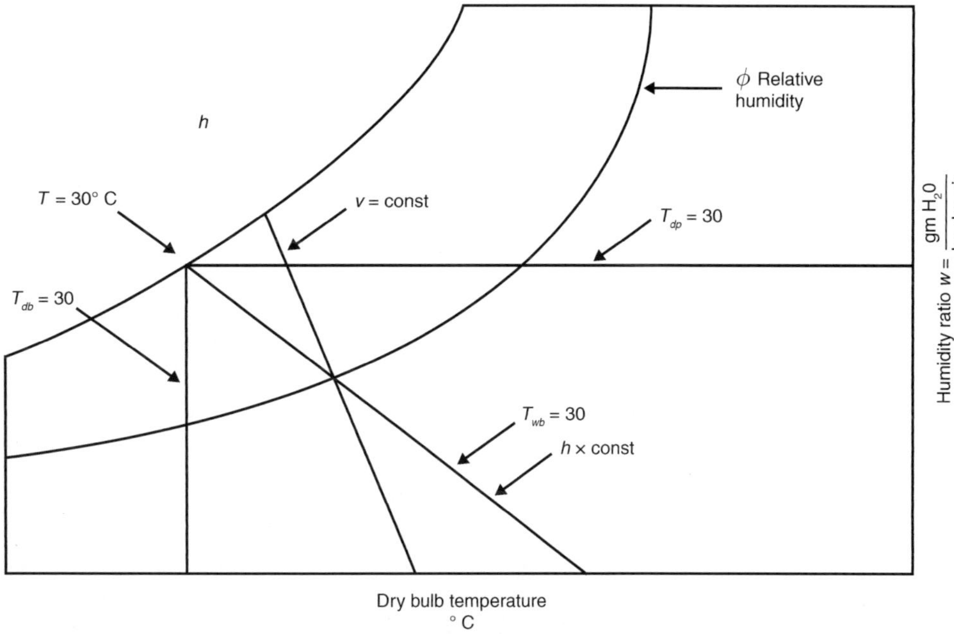

Figure 13.14

Example 13.10

A room at atmospheric pressure contains an air-water mixture at 25°C and relative humidity of 50%. Using the psychrometric chart, determine the humidity ratio, wet bulb temperature, dew point temperature, enthalpy, and specific volume.

Solution

Exhibit 2 is what the solution looks like on a psychrometric chart.

The solution to the problem is:

Humidity ratio (w)	=	10 grams water/kg dry air
Wet bulb temperature	=	17°C
Dew point temperature	=	14°C
Enthalpy (h)	=	51 kJ/kg
Specific volume (v)	=	0.8575 kg/m³

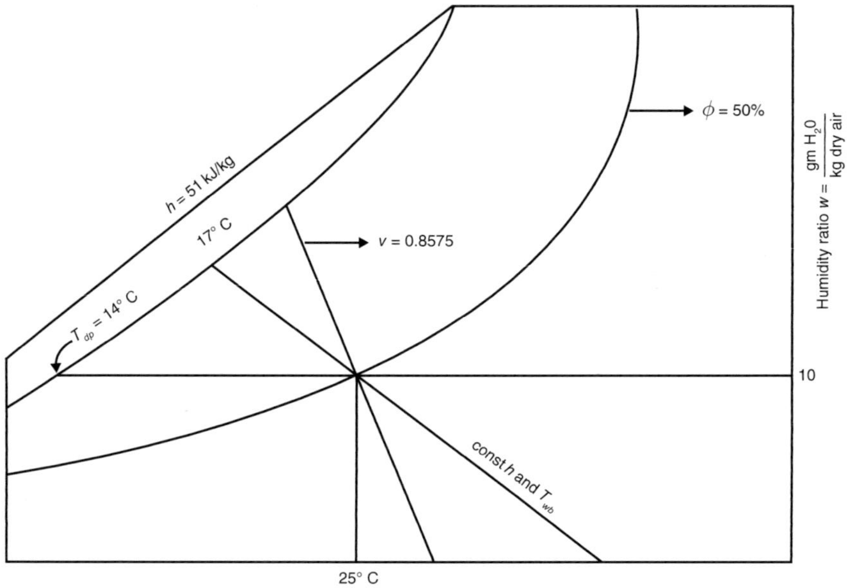

Exhibit 2

Example 13.11

Outside air at 10°C and relative humidity of 20% enters a heating duct. The air is heated to 25°C. Steam is then injected into the air stream, and the air leaves the unit at 30°C and 60% relative humidity.

Using the psychrometric chart, determine the amount of heat and water added, relative humidity, and wet bulb temperature at State 2.

Exhibit 3 shows the schematic diagram of the system.

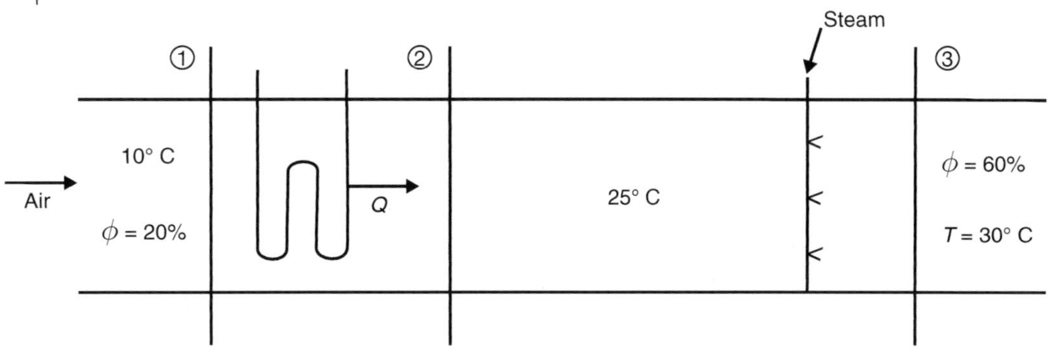

Exhibit 3

Solution

Using the psychrometric chart in Exhibit 4, obtain the values.

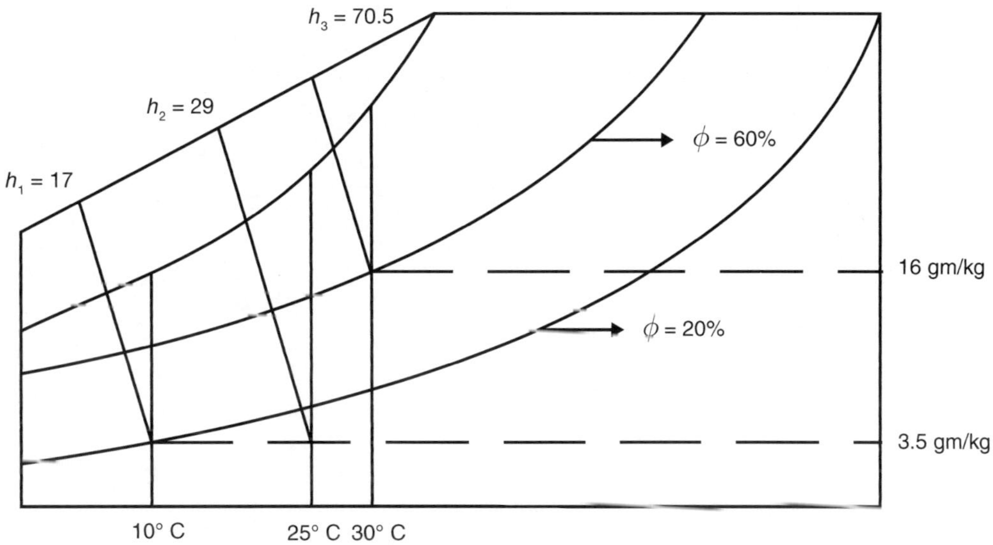

Exhibit 4

The heat added is

$$h_1 + ({}_1q_2) = h_2$$

$$({}_1q_2) = 29 - 17 = 12 \text{ kJ/kg}$$

The water added in the complete process:

$(\omega_3 - \omega_2) = 16$ gm water/kg dry air – 3.5 gm water/kg dry air = 12.5 gm/kg

Note that an additional amount of energy is added to the air in terms of steam. Determine, then, the output of enthalpy and the wet bulb temperature of the exiting air.

$$h_3 = 70.5 \text{ kJ/kg} \qquad T_{wb3} = 23.4°\text{C}$$

Otto Cycle (Gasoline Engine)

An ideal Otto cycle is shown in Figure 13.15. It consists of the following four processes:

1. An isentropic compression for 1 to 2
2. A constant volume heat addition from 2 to 3
3. An isentropic expansion from 3 to 4
4. A constant volume heat rejection from 4 to 1

The Otto cycle is an air standard cycle, that is, a cycle that uses ideal air as the working media and has ideal processes. An equipment sketch consists of only a piston and cylinder, since it is a closed-system cycle using a fixed quantity of mass. The four closed-system processes reduce to

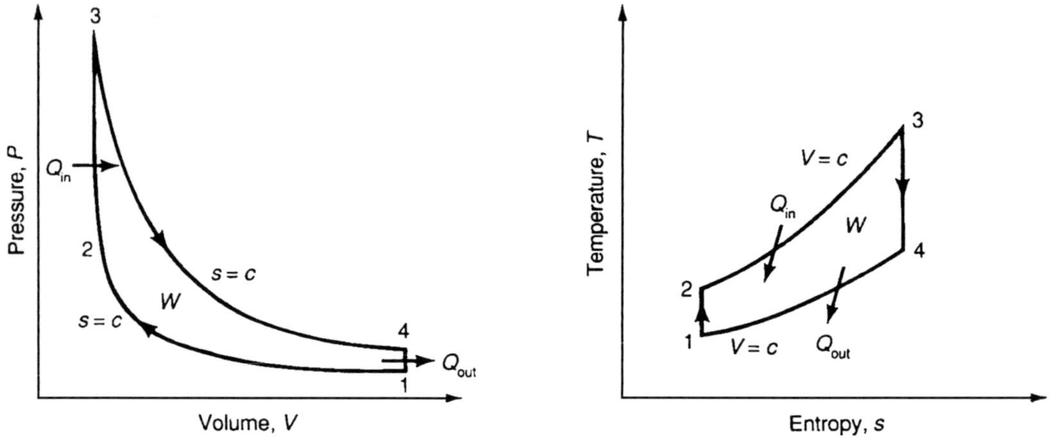

Figure 13.15

1. Isentropic compression

$$u_1 + q = u_2 - W$$
$$q = 0$$
$$w_{\text{comp}} = u_2 - u_1 = c_v(T_2 - T_1)$$

2. Heat addition

$$u_2 + q_{\text{in}} = u_3 + w$$
$$w = 0$$
$$q_{\text{in}} = u_3 - u_2 = c_v(T_3 - T_2)$$

3. Isentropic expansion

$$u_3 + q = u_4 + w$$
$$q = 0$$
$$w_{\text{exp}} = u_3 - u_4 = c_v(T_3 - T_4)$$

4. Heat rejection

$$u_4 + q_{\text{out}} = u_1 + w$$
$$w = 0$$
$$q_{\text{out}} = u_1 - u_4$$

The thermal efficiency is

$$\eta = \frac{W_{\text{net}}}{Q_{\text{in}}} = \frac{W_{\exp} - W_{\text{comp}}}{Q_{\text{in}}} = \frac{u_3 - u_4 - (u_2 - u_1)}{u_3 - u_2} = \frac{T_3 - T_4 - (T_2 - T_1)}{T_3 - T_2} = 1 - \frac{1}{r_c^{k-1}}$$

Note that r_c is the compression ratio, a ratio of the *volume* at the bottom of the piston stroke (bottom dead center) to the *volume* of the top of the stroke (top dead center). This is also equal to v_1/v_2.

The state points are found by ideal gas laws or air tables: State 1 is usually given by T and P. State 2 is found by using an isentropic process; for an ideal gas,

$$\frac{T_2}{T_1} = \left(\frac{v_1}{v_2}\right)^{k-1} = r_c^{k-1}$$

If air tables are used:

$$\frac{v_{r_2}}{v_{r_1}} = \frac{v_2}{v_1} = \frac{1}{r_c}$$

State 3 is usually found by knowing the heat addition:

$$Q_{\text{in}} = u_3 - u_2 = c_v(T_3 - T_2)$$

$$T_3 = \frac{q}{c_v} + T_2$$

State 4 is found by an isentropic process, for an ideal gas,

$$\frac{T_3}{T_4} = \left(\frac{V_4}{V_3}\right)^{k-1} = r_c^{k-1}$$

$$T_4 = \frac{T_3}{r_c^{k-1}}$$

Example 13.12

An engine operates on an air standard Otto cycle with a temperature and pressure of 27°C and 100 kPa at the beginning of compression. The compression ratio is 8.0, and the heat added is 1840 kJ/kg. Find the state-point properties and the thermal efficiency. The properties for ideal air are c_p = 1.008 kJ/(kg • °K), c_v = 0.718 kJ/(kg • °K), R = 0.287 kJ/(kg • °K), and k = 1.4.

Solution

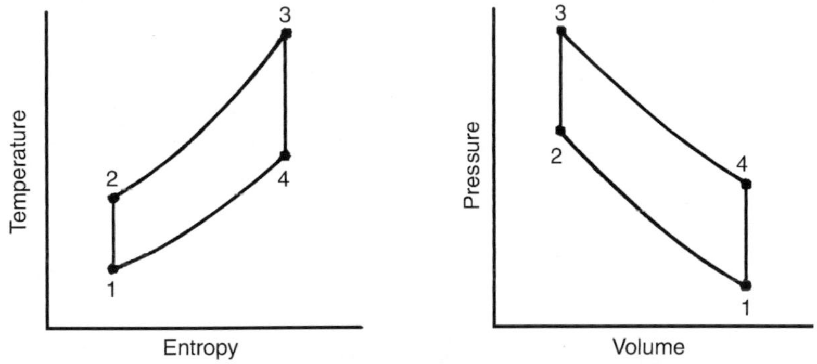

Exhibit 5

Refer to the *T-s* diagram and *P-V* diagram in Exhibit 5.

State 1

$$P_1 = 100 \text{ kPa}$$
$$T_1 = 300 \text{ °K}$$
$$v_1 = \frac{RT_1}{P_1} = 0.287\frac{300}{100} = 0.861 \text{ m}^3/\text{kg}$$

State 2

$$v_2 = \frac{v_1}{8} = 0.1076 \text{ m}^3/\text{kg}$$
$$T_2 = T_1\left(\frac{v_1}{v_2}\right)^{k-1} = 300\,(8.0)^{0.4} = 689\text{°K}$$
$$P_2 = P_1\left(\frac{v_1}{v_2}\right)^{k} = 100\,(8.0)^{1.4} = 1838 \text{ kPa}$$

State 3

$$v_3 = v_2 = 0.1076 \text{ m}^3/\text{kg}$$
$$u_3 = u_2 + q_{\text{in}}$$
$$q_{\text{in}} = u_3 - u_2 = c_v(T_3 - T_2)$$
$$T_3 = \frac{1840}{0.718} + 689 = 3255\text{°K}$$
$$P_3 = P_2\left(\frac{T_3}{T_2}\right) = 1838\frac{3255}{689} = 8683 \text{ kPa}$$

State 4

$$v_4 = v_1 = 0.861 \text{ m}^3/\text{kg}$$
$$T_4 = \frac{T_3}{r_c^{k-1}} = \frac{3255}{(8.0)^{0.4}} = 1417\text{°K}$$
$$P_4 = \frac{P_3}{r_c^{k}} = \frac{8676}{(8.0)^{1.4}} = 472 \text{ kPa}$$
$$\eta_{\text{TH}} = \frac{W_{\text{net}}}{Q_{\text{in}}} = \frac{W_{3-4} - W_{1-2}}{Q_{\text{in}}} = \frac{c_v(T_3 - T_4) - c_v(T_2 - T_1)}{1840}$$
$$= \frac{0.718(3255 - 1416) - 0.718(689 - 300)}{1840}$$
$$= \frac{1318 - 279}{1840} = \frac{1039}{1840} = 0.565$$

Check:

$$\eta_{\text{TH}} = \frac{Q_{\text{in}} - Q_{\text{out}}}{Q_{\text{in}}} = \frac{1840 - c_v(T_4 - T_1)}{1840} = \frac{1840 - 800.9}{1840} = 0.565$$
$$\eta_{\text{TH}} = 1 - \frac{1}{r_c^{k-1}} = 1 - \frac{1}{(8.0)^{0.4}} = 0.565$$

Combustion

All materials that can burn, and therefore produce heat, are called fuels. Most of these fuels are combinations of carbon and hydrogen atoms and are thus called hydrocarbons. Hydrocarbon fuels react with air and produce heat, carbon dioxide, and water.

Consider a simple hydrocarbon fuel, namely methane, and burn it in oxygen.

$$CH_4 + O_2 \rightarrow CO_2 + H_2O$$

Now, we must set up a balance across the equation so that we have equal numbers of C, O, and H on each side of the arrow.

C:	1	=	1
H:	4	=	2(2) = 4
O:	2(2)	=	2 + 2(1) = 4

Therefore, the equation becomes a stoichiometric equation of methane burning in oxygen:

$$CH_4 + 2O_2 \rightarrow CO_2 + 2H_2O$$

What about the case where the fuel is burned in air? We must then look at the major constituents of air. Air consists of 21% oxygen and 79% nitrogen. Therefore, for each mole of oxygen entering the combustion process, there are 3.76 moles of nitrogen entering the process (.79/.21).

Now, taking our methane and burning it in a complete combustion process in air, the equation becomes:

$$CH_4 + 2O_2 + 2(3.76)N_2 \rightarrow CO_2 + 2H_2O + 7.52N_2$$

A term often used in combustion processes is the *air/fuel ratio*. This term indicates the mass of air in the combustion process divided by the mass of fuel in the process. Consider the previous methane equation where one mole of methane is burned in air. The air/fuel ratio may be calculated as:

$$AF = \frac{\text{mass air}}{\text{mass fuel}} = \frac{NM_{\text{air}}}{NM_{\text{fuel}}}$$

where

N = number of moles

M = molar mass

Therefore:

$$AF = \frac{2(1+3.76)(28.92)}{(12+4)} = 17.2$$

Example 13.13

Octane is burned in air, and the combustion is complete. Determine (i) the stoichiometric mixture equation, and (ii) the air/fuel ratio.

Solution

$$C_8H_{18} + \alpha(O_2 + 3.76N_2) \rightarrow \beta CO_2 = \gamma H_2O + \alpha(3.76)N_2$$

$$C: 8 = \beta = 8$$

$$H: 18 = 2\gamma \quad \gamma = 9$$

$$O: 2\alpha = 2(\beta) + \gamma = 2(8) + 9 = 25 \qquad \alpha = 12.5$$

(i) Therefore, the stoichiometric equation is

$$C_8H_{18} + 12.5(O_2 + 3.76N_2) = 8CO_2 + 9H_2O + 47N_2$$

(ii) The air/fuel ratio is

$$AF = \frac{NM_{air}}{NM_{fuel}} = \frac{(12.5)(4.76)(28.92)}{(12)(8)+18} = 15.09$$

Example 13.14

Ethylene C_2H_4 is burned in 300% theoretical air. Assume complete combustion and a total pressure of combustion products is 75 kPa. Determine (i) the stoichiometric equation, (ii) the air/fuel ratio, and (iii) the dew point temperature.

Solution

$$C_2H_4 + \alpha(O_2 + 3.76N_2) \rightarrow \beta CO_2 + \gamma H_2O + \alpha 3.76N_2$$

$$C: 2 = \beta = 2$$

$$H: 4 = 2\gamma \quad \gamma = 2$$

$$O: 2\alpha = 2(\beta) + \gamma = 2(2) + 2 = 6 \qquad \alpha = 3$$

(i) Therefore, the stoichiometric equation is

$$C_2H_4 + 3(O_2 + 3.76N_2) \rightarrow 2CO_2 + 2H_2O + 11.28N_2$$

(ii) For the air/fuel ratio with 300% theoretical air, first write the actual combustion equation.

$$C_2H_4 + 9(O_2 + 3.76N_2) \rightarrow 2CO_2 + 2H_2O + 3O_2 + 33.84N_2$$

$$AF = \frac{9(4.76)(28.92)}{2(12)+4} = 44.2$$

(iii) The dew point requires that we calculate the vapor pressure of the combustion products.

$$P_v = \frac{N_v}{N_{products}}(P_{products}) = \frac{2}{44.2}(75) = 3.39 \text{ kPa}$$

Then, from the steam tables, find that the dew point temperature is 29.9°C.

ENERGY CONVERSION AND POWER PLANTS

The conversion of energy to useful power constitutes a major portion of the endeavors of the mechanical engineering profession. Energy is available in many forms of organic fuels, including coal, oil, natural gas, wood and peat moss. In addition, energy is available from other sources, such as ocean water in the form of tides and temperature differentials due to ocean currents, water flowing in rivers, the sun, nuclear reactors, and underground thermal sources, to mention a few. The major types of conversion systems include internal combustion engines, gas and vapor turbines, hydroelectric turbine systems, and reciprocating steam engines.

The largest amount of energy conversion is done in the many electric-generating facilities throughout the United States and the world. The energy is produced by electric generators that are powered by prime movers. Other things being equal, these systems are selected generally on the basis of their thermal efficiency. Probably the best definition of thermal efficiency is:

$$\eta_{th} = \text{(Energy wanted)/(Energy that costs \$)}$$

Normally this equation is then written in terms of output power in kW divided by the input energy in kJ/s.

Example 13.15

A power plant for a small company produces 54 kW of useful electric power. The generator has an efficiency of 96%, and the pump that is used to circulate the fluid takes 2.5 kW. The heat source supplies 10.2 MJ/min of energy to the system.

Determine

a). Power output of the heat engine

b). Thermal efficiency of the power plant

c). Amount of rejected heat from the plant

Solution

a). Since the power that is produced by the plant is equal to the useful (or net) power plus any other power need for electric generation, we must add the power that is used by the circulating pump and then divide this total output of the generator by the generator efficiency to obtain the power produced by the prime mover.

$$\text{Power produced} = (54\text{ kW} + 2.5\text{ kW})/(0.96) = 58.9\text{ kW}$$

b). Thermal efficiency is 54×10^3 J/s divided by $10.2 \times 10^6/60$ J/s

$$\eta = 31.8\%$$

c). Whatever energy that is not used is rejected; that is, 68.2% of the energy that was supplied to the system is lost.

$$\text{Heat lost} = 0.682 \times 10.2\text{ MJ/min} = 6.96\text{ MJ/min.}$$

Example 13.16

A windmill is directly coupled to a generator that is coupled to a battery. If the generator is able to produce 10 kW of power and the losses, in the form of heat losses due to line resistance and battery heat transfer, are 1.0 kW,

a). Determine the total amount of energy, in kJ, stored during a 10-hour period.

Solution

$$\begin{aligned}\text{Total energy produced} &= (\text{net production}) \times (\text{hours})\\ &= (10 - 1)\text{kW}(10 \text{ hrs})(3600 \text{ s/hr})(1 \text{ (kJ/s)/kW})\\ &= 3.24 \times 10^5 \text{ kJ}\end{aligned}$$

SOLAR ENERGY

Example 13.17

One of the least exploited forms of available energy is the energy from the sun. This is a renewable form of energy, with a fresh supply each day. Consider a flat plate solar collector used to supply energy to the working fluid (HFC 134a) in a 10-MW developmental power plant operating on a Rankine cycle. Saturated vapor enters the turbine at 20 bar and exits at 0.4 bar. Consider the turbine to be 85% efficient, the pump to be 70% efficient, and the rate of energy from the sun to the working fluid to be 0.3 kW/(m^2 of collector area). Cooling water at 20 °C is available for the condenser.

Determine

a). η_{th} of the cycle

b). Mass flow rate through the turbine

c). Mass flow rate of cooling water through condenser if $\Delta T = 10\,°\text{C}$

d). Size of collector in m^2

Solution

Property values are found on the HFC 134a chart in the *FE Handbook*

	1	2s	2	3	4s	4
P	20 bar	0.4bar	⇒	⇒	20 bar	⇒
T	Sat. Vap		Sat. Liq			
h	430	351	362.9	143	144.37	144.96

In order to find the enthalpy of the HFC 134a leaving the turbine, one must use the turbine efficiency equation:

$$\begin{aligned}\eta_t &= (\text{Work turbine actual})/(\text{Work turbine ideal})\\ &= (h_1 - h_2)/(h_1 - h_{2s}).\\ 0.85 &= (430 - h_2)/(430 - 351)\\ h_2 &= 362.9 \text{ kj/kg}\end{aligned}$$

To calculate the enthalpy of the HFC 134a at state 4, one must use the work of a pump equation.

$$h_{4s} = h_3 + W_p$$
$$= 143 + \int V\,dP$$

$$= 143 \text{ kJ/kg} + (0.0007)\text{m}^3/\text{kg}(20 - 0.4)\text{bar}[(10^5\text{N/m}^2)/(\text{bar})][1 \text{ kJ}/(10^3\text{Nm})]$$
$$= 143 + 1.37$$
$$= 144.37 \text{ kJ/kg}$$

To calculate the enthalpy at the actual state one must use the pump efficiency equation, which is:

$$\eta_p = (\text{Isentropic work of pump})/(\text{Actual work of the pump})$$
$$= 0.70 = (1.37)/W_p$$

Therefore,

$$W_p = 1.96 \text{ kJ/kg}$$
$$h_4 = h_3 + W_p = 143 + 1.96 = 144.96$$

a).
$$\eta_{th} = [\text{Energy wanted}]/[\text{Energy that costs}]$$
$$= [W_t - W_p]/[Q_{in}]$$
$$= [(h_1 - h_2) - 1.96]/[h_1 - h_4]$$
$$= [430 - 362.9 - 1.96]/[430 - 144.96]$$
$$= 22.8\%$$

b).
$$\text{Mass flow rate} = [\text{Power out}]/[W_{net}]$$
$$= (10\text{MW})(1000\text{kJ/kW})(3600 \text{ s/hr})/(65.14)\text{kJ/kg}$$
$$= 5.527 \times 10^5 \text{ kg/hr}$$

c). Mass flow rate of cooling water through the condenser requires an energy balance through the condenser.

$$m_t(h_2 - h_3) = m_{cond}(\Delta h_{cooling\ water})$$
$$[5.527 \times 10^5 \text{ kg/hr}][362.9 - 143]\text{kJ/kg} = [m_{cond}\text{kg/hr}][°\text{C kJ/kg}\cdot\text{K}][\Delta T]\text{K}$$
$$m_{cond} = [1.221 \times 10^8]/[4.179 \times 10] = 2.92 \times 10^6 \text{ kg/hr}$$

d). Collector size in square meters is obtained by dividing the heat into the working fluid by the rate into the collector per unit area.

$$\text{Area} = 10{,}000 \text{ kW}/0.3 \text{ kW/m}^2 = 33{,}333 \text{ m}^2$$

Example 13.18

Water is the working fluid in a power cycle that operates according to an ideal Rankine cycle. Superheated vapor enters the turbine at 10 MPa and 520 °C and is exhausted into the condenser at 8 kPa. The net output of the Rankine cycle is 120 MW.

Determine

a). Mass flow rate in kg/hr

b). Heat flow into the boiler in kJ/hr

c). Heat flow out of the condenser kJ/hr

Property values:

	1	2	3	4	
P	100 bar	8 kPa	8 kPa	100 bar	
T	520 C				
h	3425.1	2071.38		173.88	183.96
s	6.6622	6.6622			
v				1.0084×10^{-3}	

Obtaining quality at state 2:

$$s_2 = (x_2)(s_g) + (1 - x_2)(s_f) = 6.6622 = (x_2)(8.2287) + (1 - x_2)(0.5926)$$
$$x_2 = 0.7896$$

Therefore

$$\begin{aligned} h_2 &= (x_2)(h_g) + (1 - x_2)(h_f) \\ &= (0.7896)(2577.0) + (1 - 0.7896)(173.88) \\ &= 2071.38 \text{ kJ/kg} \end{aligned}$$

and

$$h_4 = h_3 + \int v\mathrm{d}P$$

$$\begin{aligned} &= 173.88 + (0.0010084)\text{m}^3/\text{kg}\ (100 - 0.08)\ \text{bar}(10^5\ \text{N/m}^2)/\text{bar}(\text{kJ}/10^3\ \text{Nm}) \\ &= 173.88 + 10.08 \\ &= 183.96 \text{ kJ/kg} \end{aligned}$$

a). Mass flow rate = $W_{net}/(W_t - W_p)$
$= 120{,}000 \text{ kW}/[(h_1 - h_2) - 10.08]$
$= (120{,}000 \text{ kJ/s})(3600 \text{ s/hr})/(3425.1 - 2071.38 - 10.08) \text{ kJ/kg}$
$= 4.837 \times 10^6 \text{ kg/hr}$

b). Heat flow into the boiler = (mass flow rate) $(h_1 - h_4)$
$= (4.837 \times 10^6) \text{kg/hr} (3425.1 - 183.96) \text{ kJ/kg}$
$= 1.568 \times 10^{10} \text{ kJ/hr}$

c). Heat flow out of the condenser = (mass flow rate)$(h_2 - h_3)$
$= (4.837 \times 10^6 \text{ kg/hr})(2071.38 - 173.88 \text{ kJ/kg})$
$= 9.178 \times 10^9 \text{ kJ/hr}$

HEAT ENGINES

In thermodynamics there are two basic, fundamental engine classifications, which are based on the working fluid. The first classification is an engine that operates on a fluid that is alternately vaporized and then condensed. This type of engine may be of the heat engine type in which the Rankine cycle is utilized, as discussed previously in this chapter, or may be of the heat pump/refrigeration type (see Chapter 12). The second classification of engine uses a working fluid that remains in the gaseous state throughout the operation of the engine. This second engine classification may be further divided into internal combustion engines and external combustion engines. Let us first consider the internal combustion engine.

INTERNAL COMBUSTION ENGINES

Internal combustion engines are engines in which the fuel air mixture that powers the engine ignites and burns in a closed, limited, and very dynamic region called the cylinder. These engines may be spark ignition engines (gasoline powered engines) or they may be compression ignition engines (diesel powered engines).

MEASUREMENT OF POWER

Power produced may be measured in various ways. Generally, the measurement methods may be divided into two classifications, external measurement and internal measurement. The external measurement variety uses a device on the outside of the engine, and it tends to brake the engine until it begins to lose power. This method therefore measures *brake horsepower*. The internal method uses a device mounted internally to the cylinder to measure the pressure and an external device to measure the corresponding volume. This method measures *indicated horsepower*.

Example 13.19

A prony brake is one of the common devices used to measure the actual output power of engines. It does this by means of a braking system mounted on the output shaft. A moment arm is connected to the brake shoe and the force required to hold the arm in place is measured. The prony brake equation from the *PE Handbook*:

$$W_b = 2\pi TN$$

here

W_b = Brake power
T = Torque, N · m
N = Rotational speed, rev/s

A prony brake system, as shown in Exhibit 6, is used to measure the output of an engine. For this setup $L = 700$ mm, and the brake is tightened until the rotational speed starts to drop. The engine operates at 1,150 rpm, and the force is measured to be $F = 2{,}200$ N.

Determine the output power of the engine.

Solution

Using the prony brake equation:

$$
\begin{aligned}
W &= 2\pi TN \\
&= 2\pi(2{,}200 \times 0.700)\text{N}\cdot\text{m}(1{,}150/60)\ \text{rev/s} \\
&= 185{,}460\ \text{J/s} \\
&= 185.5\ \text{kW}
\end{aligned}
$$

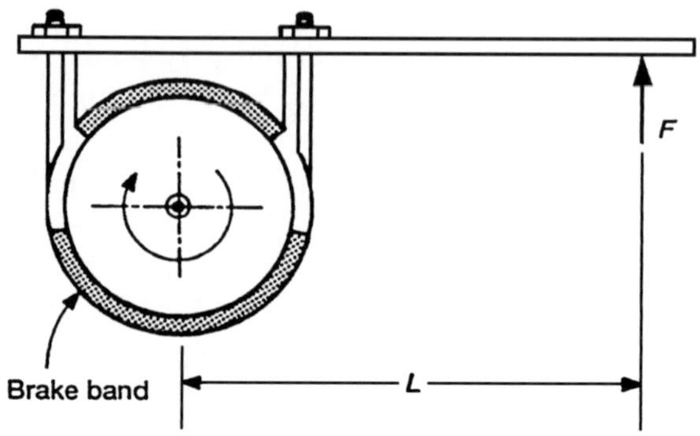

Exhibit 6

Example 13.20

A four-cylinder, four-cycle engine with 12.0-cm-diameter pistons and an 18-cm stroke operates at a speed of 500 rpm and yields an indicator diagram as shown in Exhibit 7. The area under the curve (PV diagram) is equal to 10.42 cm². The length of the diagram is 8.23 cm, and the spring constant of the indicator spring is 550 kPa/cm.

Determine

a). Mean Effective Pressure

b). Indicated power

Solution

The net area under the curve on the indicator card is proportional to the net work developed by the engine. The *Y*-axis represents a function of the pressure and the *X*-axis represents the stroke/volume of the cylinder.

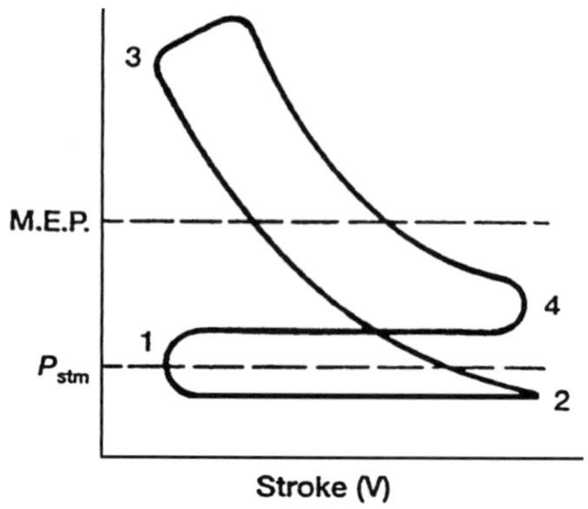

Exhibit 7

a). The *Mean Effective Pressure (M.E.P.)* is the average pressure felt by the piston over the range of the stroke.

$$\begin{aligned}\text{MEP} &= [(\text{Area under curve})/(\text{stroke})][\text{spring constant}]\\ &= [10.42\ \text{cm}^2/8.23\ \text{cm}][550\ \text{kPa/cm}]\\ &= 696.4\ \text{kPa}\end{aligned}$$

b). Work of a single stroke of a piston, $W/\text{cyl} = \text{M.E.P.} \times \text{Area} \times \text{Stroke}$

$$\begin{aligned}&= [696{,}400\ \text{Pa}][\pi D^2/4\ \text{cm}^2][\text{stroke cm}]\\ &= [696{,}400\ \text{Pa}][\pi(12.0)^2\ \text{cm}^2]/4\ [18\ \text{cm}]/10^6\ \text{cm}^3/\text{m}^3\\ &= [1418\ \text{Pa}\cdot\text{m}^3][(\text{N/m}^2)/\text{Pa}][\text{J/N}\cdot\text{m}]\\ &= 1{,}418\ \text{J}\end{aligned}$$

Since this is a four-cycle engine it makes a power stroke once every two revolutions for each piston. Therefore, the total power output is:

$$\begin{aligned}W &= [1{,}416\ \text{J}][500\ \text{revolutions/min}]/[2\ \text{revolutions/(power stroke)}(4\ \text{cylinders})]\\ &= 1416\ \text{kJ/min} = 23.6\ \text{kJ/s} = 23.6\ \text{kW}\end{aligned}$$

OTTO CYCLE

The gasoline engines that are so familiar to us operate on the Otto cycle. This is a spark ignition engine in which the fuel-air mixture is compressed until the piston is near top dead-center and the spark plug fires, igniting the fuel in the cylinder. The fuel combines with the oxygen in the air producing CO_2 and H_2O, along with a great amount of heat that expands the gasses and forces the piston down rapidly, thus producing power that may be used to our advantage.

Example 13.21

An air-standard Otto cycle with a compression ratio of 8.5 has air entering at 105 kPa and 20 °C. There is an input of heat of 1500 kJ/kg during the heat addition process.

Determine

a). Thermal efficiency

b). Net work in kJ/kg

c). Maximum temperature

Solution

a). Select the thermal efficiency equation from the *FE Handbook*:

$$\eta_{th} = 1 - r^{1-k} = 1 - 8.5^{1-1.4} = 1 - 0.42 = 58\%$$

b). Set up the equation for thermal efficiency in terms of the heat addition and heat rejection.

$$\eta_{th} = [Q_{in} - Q_{out}]/Q_{in} = W_{net}/Q_{in}$$

$$0.58 = W_{net}/1500 \quad \text{or} \quad W_{net} = 870\ \text{kJ/kg}$$

c). To find the maximum temperature we need to calculate T_2 using the isentropic relationships,then equate the heat addition to the internal energy rise, and, finally, obtain the temperature using $c_v \Delta T$.

$$T_2 = T_1[v_1/v_2]^{k-1} = 293[8.5]^{1.4-1} = 689.7 \text{ K}$$

$$Q_{\text{added}} = 0.718\ \Delta T$$

or

$$T_3 - T_2 = 1500/0.718 = 2089 \text{ K}$$

$$T_3 = T_2 + 2089 = 689.7 + 2089 = 2778.7 \text{ K}$$

DIESEL/DUAL CYCLE

The diesel cycle also represents an internal combustion engine; however, it is a compression ignition engine. Air, alone, is compressed up to top dead-center when the fuel is sprayed into the cylinder. Since the temperature is quite high, the fuel ignites spontaneously and propels the piston downward in its power stroke. The *PV* and *TS* diagrams are shown in the *FE Handbook*. The dual cycle is a closer representation of what really happens in a diesel engine. After the compression stroke, the fuel is ignited and the pressure increases rapidly before the downward motion of the piston can level off the pressure rise.

Example **13.22**

An air-standard dual cycle receives air at 27 °C and 1 atmosphere pressure. The compression ratio is 18, and the cutoff ratio is 3. Pressure doubles during the constant volume heat addition process and the maximum temperature is 2200 K.

Determine

a). Heat added in the constant volume process

b). Heat added in the constant pressure process

c). Thermal efficiency

Solution

The most effective way to attack this problem is to make a table of values. Notice the *PV* and *TS* diagrams (Exhibit 8 and 9).

	1	2	3	4	5
P	1.014 bar	58	116	116	9.4
T	300 K	953	1906	2200	1074

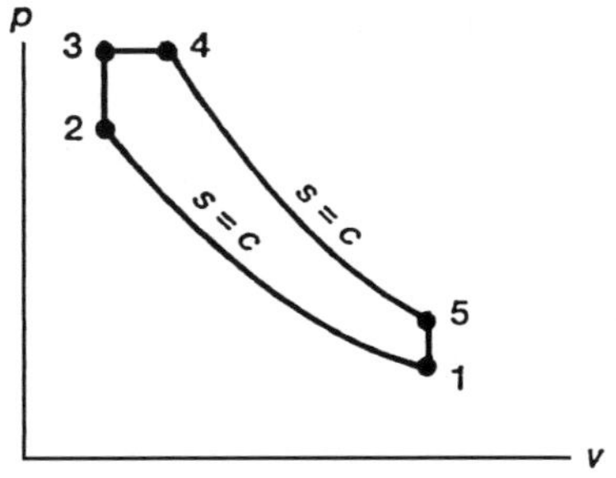

Exhibit 8 *PV* diagram

Since process 1 to 2 is isentropic, we use the isentropic relations listed in the *FE Handbook* to find T_2 and P_2.

$$P_2 = P_1[v_1/v_2]^k = 1.014[18]^{1.4} = 58 \text{ bar}$$
$$T_2 = T_1[v_1/v_2]^{k-1} = 300[18]^{1.4-1} = 953 \text{ K}$$

Process 2 to 3 is a constant volume ($V_2 = V_3$) process with $P_3 = 2P_2$, so Charles and Boyles law reduces to:

$$P_3/P_2 = T_3/T_2 = 2$$
$$T_3 = 2T_2 = 2(953) = 1906 \text{ K}$$

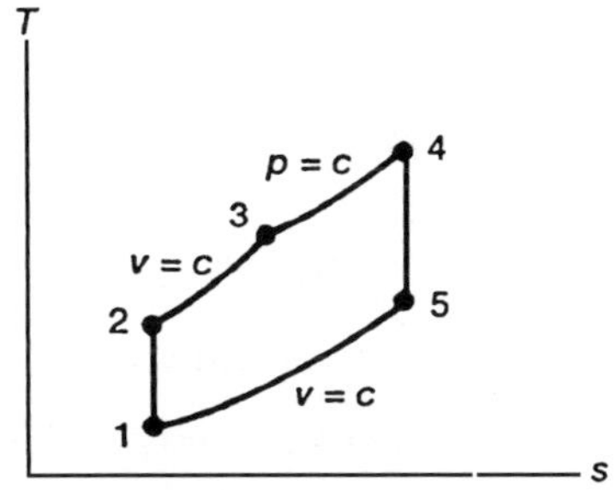

Exhibit 9 *T-s* diagram

Process 3 to 4 is a constant pressure process with $T_4 = 2200$ K. Process 4 to 5 is an isentropic expansion process. Note that $V_5 = V_1$; therefore:

$V_4/V_5 = [V_4/V_3][V_2/V_1] =$ (cut off ratio)/(compression ratio) $= 3/18 = 1/6$

Using the isentropic relations:

$$T_5/T_4 = [V_4/V_5]^{k-1} = (1/6)^{1.4-1} = 0.488$$
$$T_5 = 0.488 \times 2200 = 1074 \text{ K}$$
$$P_5/P_4 = [V_4/V_5]^k$$

Therefore

$$P_5 = [116][1/6]^{1.4} = 9.4 \text{ bar}$$

Now that all of the critical values have been obtained, it is a straightforward process to get the remaining answers.

a). The heat transfer during the constant volume process is:

$$Q_{2-3} = c_v[T_3 - T_2] = 0.718[1906 - 953] = 684 \text{ kJ/kg}$$

b). The heat transfer in the constant pressure process is:

$$Q_{3-4} = c_p[T_4 - T_3] = 1.00[2200 - 1906] = 294 \text{ kJ/kg}$$

c). The thermal efficiency is:

$$\eta_{th} = [Q_{in} - Q_{out}]/Q_{in} = [Q_{2-3} + Q_{3-4} - Q_{5-1}]/[Q_{2-3} + Q_{3-4}]$$
$$= [684 + 294 - c_v \Delta T]/[684 + 294] = [978 - 0.718(1074 - 300)]/978$$
$$= 43.2\%$$

BRAYTON CYCLE

The Brayton cycle is the basis for the operation of the jet engine. It is considered to be an external combustion engine because the fuel and air are mixed and burned in a combustion chamber and then the exhaust products are fed to the turbine. The turbine is directly coupled to the compressor and the excess power generated is used to power the aircraft. This excess power may be in the form of high velocity gasses that pass through a nozzle and thus propel the aircraft, or may be through extra rows of turbine blades that are used to power a propellor or a fan. The cycle is shown in the *FE Handbook* and consists of two isentropic processes and two constant pressure processes.

Example 13.23

Air at 3 bar and 500 K is extracted from a jet engine compressor and is to be used in cabin cooling. The extracted air is cooled at a constant pressure heat exchanger down to 400 K. It then expands isentropically through a turbine down to 1 bar. The power developed by the turbine is used for cabin lighting. The mass flow rate is 5 kg/min.

Determine

a). Temperature of air to cabin

b). Power developed by the turbine in kW

c). Rate of heat transfer in the heat exchanger in kW

Solution

Exhibit 10 illustrates the processes discussed above.

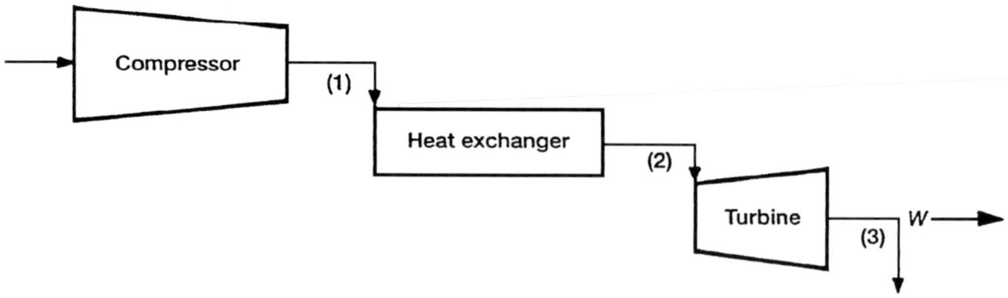

Exhibit 10

a). T_3 has to be found using the isentropic equation:

$$T_3 = T_2[P_2/P_3]^{(1-k)/k} = 400[3/1]^{(1-1.4)/1.4} = 292 \text{ K}$$

b). $W = m[c_p(T_2 - T_3)]$; therefore:
$$= 5\text{kg/min}[1.00(400 - 292)]$$
$$= [540 \text{ kJ/min}]/60$$
$$= 9 \text{ kW}$$

c). The heat transfer $Q = m[h_1 - h_2]$
$= 5 \text{ kg/min}[c_p \text{ kJ/kg} \cdot \text{K}(T_1 - T_2) \text{ K}]$
$= 5[1.00(500 - 400)]/60$
$= 8.33 \text{ kW}$

REFRIGERATION AND HEATING, VENTILATING AND AIR CONDITIONING

Refrigeration is a process whereby heat is extracted from a system. In the past it was done by means of ice stored in an ice box. The heat of fusion of the ice as it melts, came from the substance that was being cooled. This method has essentially been replaced by the mechanical refrigeration process, which operates on the Clausius statement of the *Second Law of Thermodynamics*. Mechanical refrigeration extracts heat from a cold temperature reservoir and, through the means of a work input, deposits this heat into the high temperature reservoir. It is interesting to note that the old concept of melting ice is still with us in terms of the rating of refrigeration units. A ton of refrigeration is the amount of heat that is absorbed by a ton of ice as it melts in a 24-hour period. This term was defined in the English system of units by:

Mass of ice × Heat of fusion = Heat required in a 24-hr day
(200 lb ice)(144 Btu/lb) = 288,000 Btu/day which is equal to 12,000 Btu/hr

The most common term is in Btu/min, which is 200 Btu/min. In the SI system this is 211 kJ/min or, dividing by 60 s/min, 3.517 kJ/sec.

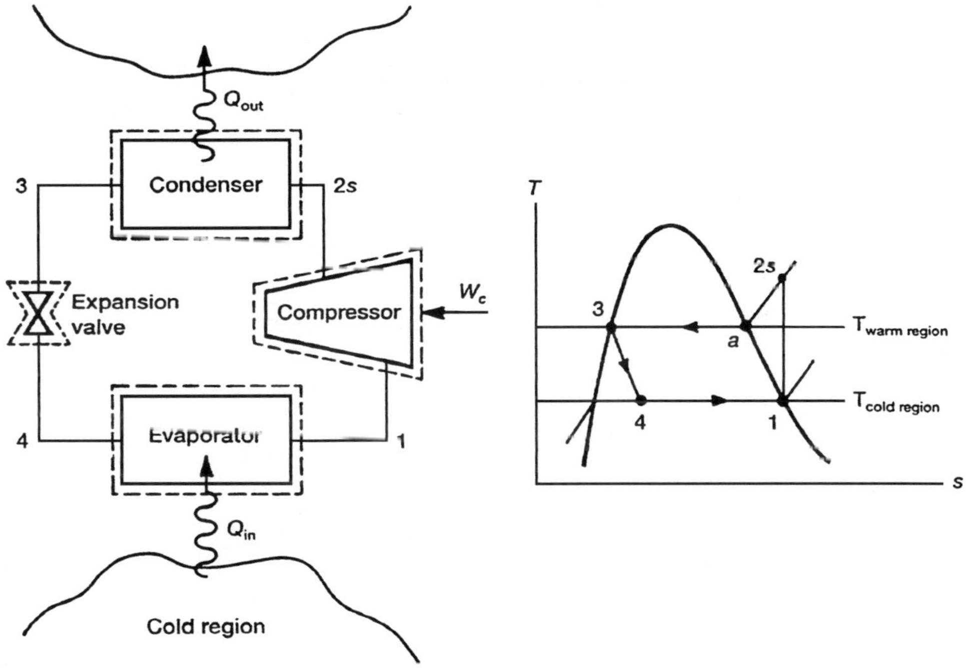

Figure 13.16

MECHANICAL REFRIGERATION

Mechanical refrigeration is based on the principle that, as a liquid vaporizes at a specific pressure, it will have a controlled absorption of the heat of vaporization. The principle of operation is patterned after the Carnot cycle with some changes due to physical constraints. The ideal refrigeration cycle is depicted in Figure 13.16 and also in the Thermodynamics section of the *Handbook*. The mechanical system consists of a compressor that compresses the cold vapor coming out of the evaporator to a high temperature and high pressure vapor. This vapor then enters the condenser where the vapor is cooled to a saturated or possibly slightly subcooled liquid. The liquid then expands through a throttling valve to a lower pressure and, consequently, a lower temperature liquid-vapor mixture and enters the evaporator where it continues to vaporize to a saturated or slightly superheated vapor for entrance into the compressor.

ABSORPTION REFRIGERATION

Another type of refrigeration system that is commonly used in facilities that have an abundant amount of heat energy available is the absorption system. This type of system uses a refrigerant such as lithium bromide or ammonia which, when evaporated, will absorb energy and, when cooled or condensed, will give off much energy. The advantage of this type of system is that the machinery operating costs are very low. All that is necessary are a few low-power pumps and a cheap source of energy and cooling water. This system is currently being researched for use in residential and commercial applications. The Coefficients of Performance (C.O.P.) of the absorption systems compare very favorably with the compression refrigeration systems.

EFFICIENCY OF REFRIGERATION SYSTEMS

Thermal efficiency has been defined as:

$$\eta_t = \text{(Energy wanted)/(Energy that costs \$)}$$

The same general philosophy can be used whether the system is an engine, pump, refrigeration system or almost any other system one can imagine. Considering an Air Conditioning system, the equation would be:

Efficiency = (Energy extracted from cold reservoir)/(Energy used in compression)

The difficulty with using the efficiency term is that efficiency can never be greater than one, or we would have a perpetual motion machine. Therefore, the term *Coefficient of Performance* has been coined to indicate relative efficiency terms. As can be determined from the equation, a higher C.O.P. is desired. These values typically range from about 2.0 to 5.0 or 6.0.

Example 13.24

A mechanical refrigeration system operates on a Carnot cycle. Saturated vapor leaves the compressor and enters the constant temperature condenser at 100 °C and exits from the constant temperature evaporator at –10 °C. If the unit provides 3 tons of air conditioning, determine

a). the C.O.P.

b). the amount of work needed to transfer this heat in kJ/min.

Solution

A Carnot cycle is the ideal cycle for heat engines and refrigeration systems. The schematic of the reverse cycle is in the Thermodynamics section of the *FE Handbook*. Using the refrigeration model of the efficiency equation:

a). C.O.P $= (T_L)/(T_H - T_L)$ where temperatures are in Kelvin.

$$\text{C.O.P.} = (-10 + 273)/(100 - [-10]) = 263/110 = 2.39$$

b). Now, using the efficiency equation in terms of heat transfer:

$$\text{C.O.P.} = (Q_L/(Q_H - Q_L)$$

where

Q_L = Heat transfer at low temperature
Q_H = Heat transfer at high temperature
$(Q_H - Q_L)$ = Work

Substituting in the known values:

$$2.39 = (211\text{kJ/min-ton})(3\text{ tons})/W$$

Solving for W:

$$W = 264.8\text{ kJ/min}$$

Example 13.25

An ideal vapor-compression refrigeration cycle operates with HFC 134a as the refrigerant. Saturated vapor enters the compressor at –10 °C. Saturated liquid leaves the condenser at 2 MPa. The mass flow rate through the unit is 10 kg/min. Determine the work of the compressor in kW, the refrigeration capacity in tons and the Coefficient of Performance.

Solution

An ideal vapor compression refrigeration system is patterned after the *TS* diagram that is shown in the Thermodynamics section of the *FE Handbook*. Note that the problem above is for just a one-stage cycle where, as in the Mechanical Engineering section, a diagram is included for a two-stage cycle. The schematic diagram is also listed on the same page. One should then turn to the *P-h* diagram for HFC 134a to obtain the enthalpy values. Note that Points 1 and 3 are saturated vapor and saturated liquid, respectively. Point 2 represents the gas after it has been isentropically compressed. The table of values shown below is taken from the HFC 134a diagram.

	1	2	3	4
P		2 Mpa	2 Mpa	
T	–10			–10
h	390	445	300	300
s	1.74	1.74		

a).
$$\begin{aligned} W_c &= m(h_2 - h_1) \\ &= (10\text{ kg/min})(445 - 390)\text{ kJ/kg}(1/60)(\text{min/sec}) \\ &= 4.3\text{ kW} \end{aligned}$$

b). $Q_{tons} = m(h_1 = h_2)$
$= (10 \text{ kg/min})(390 - 300)\text{kJ/kg}(1/211)(\text{tons/kJ})$
$= 4.3 \text{ tons}$

c). $\text{C.O.P.} = (Q_{in}/W)$
$= (h_1 - h_4)/(h_2 - h_1)$
$= (390 - 300)/(445 - 390)$
$= 90/55$
$= 164$

HVAC

HVAC stands for the words Heating, Ventilating and Air Conditioning, which is the science of creating the ideal environment for man and machine in a selected space. The creation of this ideal environment indicates that not only is temperature control required, but also humidity control. These controls are for comfort, for the dimensional stability of equipment and instrumentation, and for the stability of other characteristics.

BODY HEAT

The human body releases both sensible heat, since the body is usually higher in temperature than the surroundings, and latent heat due to perspiration. Perspiration tends to add water vapor to the air in the form of superheated steam. This water vapor, and the water vapor already existing in the air, can only be removed by means of condensation. The amount of heat that must be removed to maintain the room at a constant humidity is called the "latent heat gain" of the air-conditioned space. The rates of heat given off by the human body have been measured for various situations and conditions in order to determine the necessary size of air conditioners.

Example **13.26**

A theater holds 600 people for an evening performance. Measurements have indicated that an average person gives off 205.7 kJ of sensible heat per hour and 163.5 kJ of latent heat per hour, when sitting in a theater for an evening. What would the air-conditioning load be to maintain a comfortable temperature during this period of time? Give the answer in tons of air-conditioning.

Solution

Six hundred people will give off heat at the rate of:

$$\text{Total heat} = (\text{Number of people}) \times (\text{Sensible heat} + \text{Latent heat})$$

$$= (600)(205 \text{ kJ/hr} + 163.5\text{kJ/hr})$$

$$= \frac{(221.100)\text{kJ/hr}}{(211 \text{ kJ/min ton})(60 \text{ min/hr})}$$

$= 17.46$ tons, which is the amount of refrigeration required to counteract the heat supplied by the people.

GAS REFRIGERATION

The use of mechanical, vapor-compression air conditioning systems in aircraft is not a feasible alternative due to the added weight. Gas systems in aircraft utilize high temperature and high pressure air from the compressor. This air then passes through a heat exchanger, which exhausts the heat to the ambient air. The cooled, yet high-pressure, air then passes into an air turbine that acts as an auxiliary power source for the cabin of the aircraft. Upon exhausting from the turbine, the air is cooled considerably below the ambient temperature and, thus, may be exhausted into the cabin as cooling air. The Mechanical Engineering section of the *FE Handbook* gives a schematic diagram for the use of a Brayton cycle in the production of refrigeration.

Example **13.27**

Air at 3 bar and 500 K is extracted from a jet engine compressor to be used for cabin cooling and for the generation of auxiliary power for the cabin. The extracted air is cooled in a constant pressure heat exchanger down to 400 K. It then enters an isentropic turbine and expands to 1 bar before being rejected into the cabin. If the mass flow is 5 kg/min, determine:

a). The temperature of the gas as it leaves the turbine

b). The power developed by the turbine in kW

c). The rate of heat transfer out of the constant-pressure heat exchanger

Solution

Assuming that the air acts as an ideal gas, and using the ideal-gas relationships for each of the processes, a table of values can be constructed. Note that the schematic diagram for this process can be illustrated by reducing the air refrigeration cycle shown in the Mechanical Engineering section of the *FE Handbook* to include only the air from the compressor running through the heat exchanger and exhausting through the turbine into the air-conditioned space.

	1	**2**	**3**
P	3	3	1
T	500	400	257.8

a). Process 2–3 is isentropic, and T_3 must be obtained using the relationship:

$$T_3 = T_2(P_3/P_2)^{(k-1)/k} = 400(1/3)^{(1.4-1)/1} = 257.8 \text{ K}$$

b). Power developed by the turbine $= m(h_2 - h_3)$

$$= mc_p(T_2 - T_3)$$
$$= (5 \text{ kg/min})(1.00 \text{ kJ/(Kg}\cdot\text{K})(400 - 257.8)\text{K } (1 \text{ min/60 s})$$
$$= 11.85 \text{ kW}$$

c). Heat transfer from the heat exchanger $= m(h_1 - h_3)$

$$= (5 \text{ kg/min})(1.00 \text{ kJ/kg}\cdot\text{K})(500 - 400)\text{K } (1/60) \text{ min/s}$$
$$= 8.33 \text{ kW}$$

INFILTRATION

Living and working spaces that need to be conditioned, whether by heating in the winter or by cooling in the summer, need to be analyzed for infiltration. This infiltration can be intentional, such as in maintaining a positive pressure in *clean rooms*, etc., or unintentional, such as from cracks under doors, around windows, etc. The *FE Handbook* lists formulas in the Mechanical Engineering section that can be used in each of the above cases.

Example 13.28

A computer chip manufacturing plant of 5000 square meters has 4-meter-high ceilings and changes its inside air three times per hour by maintaining a slightly positive air pressure (1.01 bar). The room temperature is maintained at 21°C. To assist in the sizing of a heating system it is assumed in the design conditions that the outside temperature is –25 °C. Basing the heating requirements on the indoor conditions, determine the heating required to compensate for the infiltration.

Solution

Select the equation for the air change method listed under *Infiltration* in the *FE Handbook*. Determine the density of air using the perfect gas law:

$$PV = RT$$

$$\rho = \left[\frac{(1.01\ \text{B})(10^5\ \text{N/m}^2\ \text{B})}{\dfrac{(8.314\ \text{kJ/kmol}\cdot\text{K})(292\ \text{K})}{(28.97\ \text{kg/kmol})}}\right] \times \left[1\frac{\text{kJ}}{10^3\ \text{N}\cdot\text{m}}\right]$$

$$= 1.205\ \text{kg/m}^3$$

$$\begin{aligned}\text{Heat required, } Q &= (1.205\ \text{kg/m}^3)(c_p\ \text{kJ/kg K})(V\ \text{m}^3)(n_{AC}\ \text{changes/hr})\\ &\quad (T_i - T_o)\text{K}/(3600\ \text{sec/hr})\\ &= (1.205)(1.00)(5000 \times 4)(3)[21 - (-25)]/3600\\ &= 307.9\ \text{kW}\end{aligned}$$

PSYCHROMETRICS

Psychrometrics is the study of systems consisting of dry air and water. This combination of air and water is called moist air. The particular composition of the water and air mixture may be indicated by means of the *humidity ratio*, which is the ratio of the mass of water vapor to the mass of dry air. The make-up of moist air can also be described in terms of the relative humidity, which is the ratio of the partial pressure of the water vapor in the moist air to the saturation pressure of water vapor at the temperature of the air. These values may be readily obtained from the psychrometric chart in the *FE Handbook* merely by knowing the wet-bulb and dry-bulb temperatures of the moist air. Note the schematic diagrams in the Mechanical Engineering section of the *FE Handbook* that illustrate various HVAC systems and the representation of these processes on the schematics of the psychrometric chart.

Example 13.29

Air at 30 °C, 1 atm, and 70% relative humidity enters a dehumidifier operating at steady state. Saturated air and water leave both leave at 10 °C in two different streams.

Determine

a). The heat transfer from the incoming air stream to the coils of the dehumidifier in kJ/kg dry air

b). The amount of water condensed in kg per kg of dry air.

Solution

Turn to the Thermodynamic Section of the *FE Handbook* where the psychrometric chart is shown. The 30 °C is the dry bulb temperature of the incoming air stream. Locate the incoming condition by sketching a vertical line from the 30 °C position up to the 70% *Relative Humidity* line. Record the humidity ratio from the right-hand side of the chart (19 grams moisture/kg dry air) and the enthalpy from the left side of the chart (78.2 kJ/kg dry air). Now run the line directly to the left-hand side, which lists the wet-bulb temperature, or the saturation temperature, of the moist air. Run down the curve until you reach 10 °C. Again record the humidity ratio (7.7 grams $H_2 0$/kg dry air, and pick off the value of the enthalpy of 29.5 kJ/kg dry air). Now you have all of the values necessary for the solution of the problem.

a). Heat transfer = $(h_1 - h_2) = (78.2 - 29.5) = 48.7$ kJ/kg dry air

b). Water out = (19 – 7.7) gm water/kg dry air = 11.3 gm/kg = 0.0113kg/kg.

Example 13.30

Air enters a steam humidifier at a temperature of 21°C and a wet-bulb temperature of 10 °C. The mass flow rate of the dry air is 100 kg/min and saturated steam is injected at 100 °C at 1 kg/min. Consider that there is no heat transfer to or from the surroundings. Consider that the pressure is constant at 1 atm throughout.

Determine

a). The exit humidity ratio, in (kg moisture)/(kg air)

b). The exit temperature, in °C

	State 1	State 2
$T_{wb} = 10\,°C$	$T_1 = 21\,°C$ Moist air →	Moist air →
		↗ Injected steam $T = 100\,°C$, 1 kg/min

Solution

An energy balance equation should be written for the schematic diagram of the process shown above.

$$m_a h_{a1} + m_{w1} h_{g1} + m_{st} h_{st} = m_a h_{a2} + m_{w2} h_{g2}$$

Knowing that $m_{w1}/m_a = \omega_1$ and $m_{w2}/m_a = \omega_2$ the equation may be rewritten as:

$$(h_a + \omega_1 h_g)_1 + (\omega_2 - \omega_1) h_{g100} = (h_a + \omega_2 h_g)_2$$

The value for the enthalpy of the incoming water vapor—air mixture can be obtained from the psychrometric diagram in the *FE Handbook.*

$$h_1 = 30 \text{ kJ/kg dry air}$$
$$\omega_1 = 3.5 \text{ g water/kg dry air}$$
$$\text{The steam injected} = 1 \text{ kg/min steam}$$
$$\text{Therefore, } (\omega_2 - \omega_1) = (1000 \text{ g/min steam})/(100 \text{ kg/min air}) = 10 \text{ g moisture/kg air}$$

Therefore, $\omega_2 = 3.5 + 10 = 13.5$ g water/kg dry air, $h_{g100} = 2676.3$ kJ/kg

a). $\omega_2 = 13.5/1000 = 0.0135$ kg moisture/kg air

b). Substituting values into (i):

$$(30 \text{ kJ/kg}) + 0.0135 \text{ kg moisture/kg dry air } (2676.3)\text{kJ/kg} = h_2$$
$$h_2 = 66.13 \text{ kJ/kg; therefore, temperature} = 30\,°\text{C}$$

Example 13.31

A research laboratory building is to be heated and ventilated. Outside air is to be supplied to a room 12 m by 18 m at the rate of 1 m³/min of replacement air per square meter of floor space. The design conditions are 20 °C inside air temperature with a –20 °C outside air temperature. The air is to be pressurized in the room to a positive air pressure of 7.0 kPa to ensure that the room does not get contaminated from the outside. How much heat would be required to condition the air for this room?

Solution

The floor area is 12 m × 18 m = 216 m², so the air-flow rate required would be 216 m³/min at a positive pressure of (101,300 + 7000) Pa absolute pressure. The mass of air equals:

$$\begin{aligned} m &= PV/RT \\ &= (108{,}300 \text{ Pa})(216 \text{ m}^3/\text{min}) \\ &\quad (1 \text{ N/m}^2 \cdot \text{Pa})/(8314/28.97)\text{J/kg}\cdot\text{K } (293 \text{ K})\ 1 \text{ N}\cdot\text{m/J} \\ &= 278 \text{ kg/min} \end{aligned}$$

Therefore, calculating the heat transfer:

$$\begin{aligned} Q &= mc_p(\Delta T) \\ &= 278 \text{ kg/min } (1.00 \text{ kJ/kg}\cdot\text{K})(T_i - T_o) \text{ K} \\ &= (278)(1.00)(40) \\ &= 11{,}120 \text{ kJ/min} \\ &= (11{,}120 \text{ kJ/min})(1 \text{ min}/60 \text{ s}) \\ &= 185.3 \text{ kJ/s or } 185.3 \text{ kW heating} \end{aligned}$$

Example 13.32

An industrial air conditioning system mixes 70% conditioned air with 30% recirculated air before returning it to the plant. The system handles some 285 m^3/min of mixed air. The conditioned air leaves the cooling coils at 9 °C and 90% relative humidity and then is mixed with recirculated air at 25 °C and 50% relative humidity.

Determine

a). Dry bulb temperature of the air returning to the plant

b). Wet bulb temperatures of the air returning to the plant

Solution

Since the air is being mixed on a 70%:30% basis, the final temperature may be considered as a weighted average of these temperatures:

a). $T_{\text{final}} = (0.70)(9 + 273) + (0.30)(25 + 273) = 286.8 \text{ K or } 13.8\,^\circ\text{C}$

b). The wet bulb temperature may be found from the psychrometric chart by first locating the points indicated by the two streams of air. Then draw a line from point 9 °C dry bulb and 90% relative humidity to point 25 °C dry bulb and 50% relative humidity. Where this line intersects, the dry-bulb temperature of 13.8 °C, marks the condition of the exit air. One should merely follow the constant wet bulb line to the saturation line and pick off the wet-bulb temperature of 12.1 °C.

Example 13.33

A line carrying dry, saturated steam at 790 kPa gage pressure passes through an instrument room measuring 6m × 6m × 4m. The line develops a leak and the steam seeps into the room at a rate of 1 kg/hr and diffuses throughout the room. If the room is initially at 20 °C and has a relative humidity of 50%, and the pressure of the room remains at one standard atmosphere, estimate what the relative humidity will be after one hour. Assume there is no exchange of heat between the room and the surroundings and that the steam leak displaces some of the air in the room.

Solution

Assume that the steam and the air both act as perfect gasses. Calculation of the volume of the room is 6m × 6m × 4m = 144 m^3.

Determine the mass of the dry air in the room and the mass of the superheated steam at initial condition. According to Dalton's law of partial pressures we know that the total pressure in the room is equal to the partial pressure of the steam plus the partial pressure of the air. The partial pressure of the steam may be estimated as 50% of the saturation pressure of steam at 20 °C, which is obtained from the steam tables in the *FE Handbook* as 2.339 kPa.

Therefore

$$P_{\text{steam}} = (0.50)(2.339)\text{kPa} = 1{,}170 \text{ Pa}$$

Thus, the partial pressure of the air is:

$$P_{\text{air}} = 101{,}300 \text{ Pa} - 1170 \text{ Pa} = 100{,}130 \text{ Pa}.$$

The original masses of air and water vapor are calculated as:

$$m_{steam} = PV/RT$$
$$= (1{,}170\ Pa)(N/m^2 \cdot Pa)(144\ m^3)/$$
$$[8.314\ kJ/mol \cdot K)/(18\ kg/kmol)][293]K(10^3 Nm/kJ)$$
$$= 1.2449\ kg$$
$$m_{air} = (101{,}300 - 1{,}170)(144)/[8.314/28.97][293][10^3] = 171.5\ kg$$

Assuming that the steam displaces the air in the room, the amount of heat that is dissipated into the room by the saturated steam at 891 kPa (175 °C from the steam tables) is approximated by:

$$Q = mc_p\Delta T = (1\ kg)(1.87\ kJ/kg \cdot K)(175 - 20)K = 290\ kJ$$

The addition of the 290 kJ into the room will raise the temperature in the room by:

$$\Delta T = Q/mc_p$$
$$= (290\ kJ)/(171.5\ kg)(1.00\ kJ/kg)$$
$$= 1.69\ °C$$

This change in temperature is not significant when compared with the original temperature. It may be determined that the temperature rise does not materially affect the humidity of the room for a first approximation.

It may now be estimated that the relative humidity after 1 kg of steam has been injected into the room is:

$$\text{Relative Humidity} = [(1.2449 + 1.0)/1.2449][50\%] = 90\%$$

MISCELLANEOUS

Mixture of Gases

The composition of a closed mixture of gases may be expressed in terms of volume (mol) fractions or mass fractions. These are related through the component molecular weight (m.w.) and are best shown in tabular form. If volume fractions are given, convert to mass fraction:

Gas	Volume Fraction	m.w.	Volume Fraction × m.w.	Mass Fraction
O_2	0.2	32	6.4	$\frac{6.4}{38.4} = 0.167$
N_2	0.2	28	5.6	$\frac{5.6}{38.4} = 0.146$
CO_2	0.6	44	26.4	$\frac{26.4}{38.4} = 0.687$
			38.4	1.000

If mass fraction is given, convert to volume fraction:

Gas	Mass Fraction	m.w.	$\frac{\text{Mass Fraction}}{\text{m.w.}}$	Volume Fraction
O_2	0.1	32	0.00313	$\frac{0.00313}{0.03135} = 0.100$
N_2	0.6	28	0.0214	$\frac{0.0214}{0.03135} = 0.683$
CO_2	0.3	44	0.00682	$\frac{0.00682}{0.03135} = 0.217$
			0.03135	1.000

The mass fraction is sometimes called the gravimetric fraction. Component pressure and molecular weight are volume fraction functions; u, h, c_p, c_v, and R are mass fraction functions.

Heat Transfer

The three modes of heat transfer are conduction, convection, and radiation. The heat transfer "laws" are based on both empirical observations and theory but are consistent with the first and second laws of thermodynamics. That is, energy is conserved and heat flows from hot to cold.

Conduction

Conduction occurs in all phases of materials (Figure 13.17). The equation for one-dimensional, planar, steady-state conduction heat transfer is

$$\dot{Q} = kA\frac{T_H - T_C}{x} \text{ (watts)}$$

The **conductivity**, k, is a property of the material and is evaluated at the average temperature of the material. The **heat flow rate**, q, is sometimes expressed as a heat flux $\dot{Q}/A$.

For multiple layers of different materials (Figure 13.18), as in composite structures, it is usually best to use an electrical analogy:

$$\dot{Q}_1 = \frac{T_H - T_{x_1}}{R_1} \qquad R_1 = \frac{x_1}{A_1 k_1}$$

$$\dot{Q} = \dot{Q}_1 = \dot{Q}_2 = \dot{Q}_3$$

$$\dot{Q}_2 = \frac{T_{x_1} - T_{x_2}}{R_2} \qquad R_2 = \frac{x_2}{A_2 k_2}$$

$$\dot{Q}_3 = \frac{T_{x_2} - T_C}{R_3} \qquad R_3 = \frac{x_3}{A_3 k_3}$$

$$\dot{Q}_T = \frac{T_H \quad T_C}{R_1 + R_2 + R_3}$$

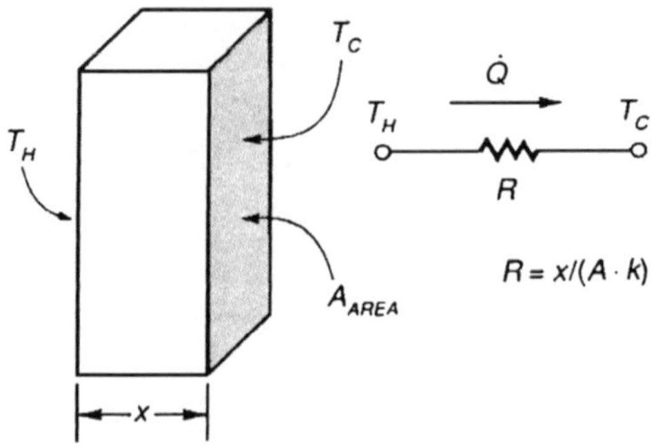

Figure 13.17

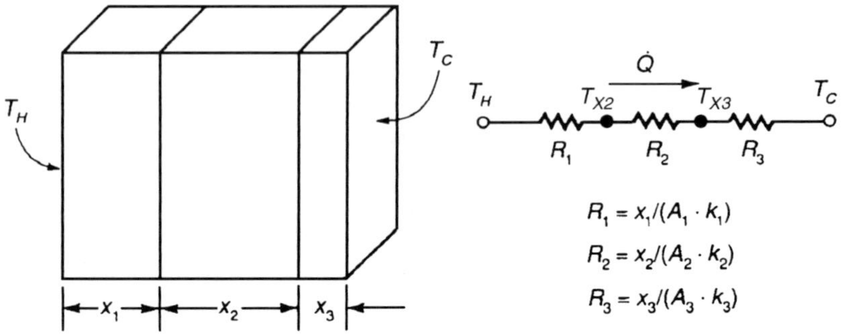

Figure 13.18

Example 13.34

A plane wall is 2 m high by 3 m wide and is 20 cm thick. It is made of material that has a thermal conductivity of 0.5 W/(m • °K). A temperature difference of 60°C is imposed on the two large faces. Find the heat flow, the heat flux, and the conductive resistance.

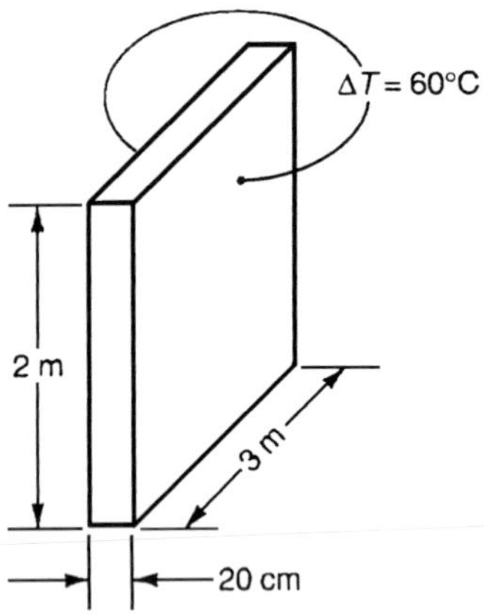

Exhibit 11

Solution

Refer to Exhibit 11.

$$\dot{Q} = \frac{kA(T_H - T_C)}{x} = \frac{0.5 \times 3 \times 2 \times 60}{0.20} = 900 \text{ W}$$

$$\frac{\dot{Q}}{A} = \frac{900}{3 \times 2} = 150 \frac{\text{W}}{\text{m}^2}$$

$$R = \frac{x}{kA} = \frac{0.2}{0.5 \times 3 \times 2} = 0.0667 \frac{°\text{K}}{\text{W}}$$

Convection

Convection occurs at the boundary of a solid and a fluid (liquid or gas) when there is a temperature difference. The mechanism is complex and can be evaluated analytically only for a few simple cases; most situations are evaluated empirically. The equation for convective heat transfer is

$$q = hA(T_{\text{surface}} - T_{\text{fluid}})$$

The evaluation of h, the heat transfer coefficient, normally involves use of data correlated in the form of dimensionless parameters, for example, Nussult number, Reynolds number, Prandtl number.

The conduction and convection mechanisms can be combined as shown in Figure 13.19. So the temperature of the surface, T_s, is dependent on the relative magnitude of the two resistances.

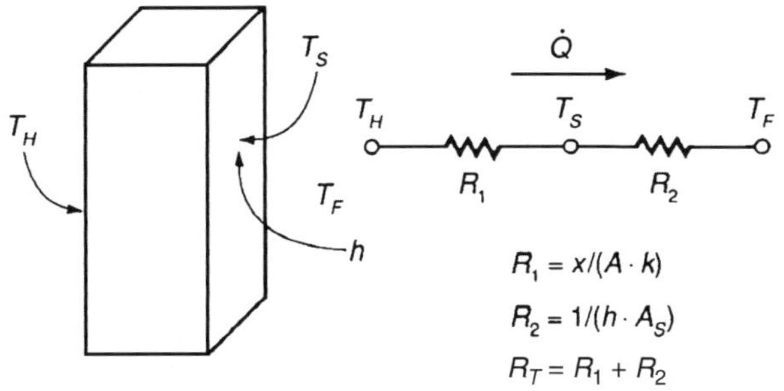

Figure 13.19

Example 13.35

Water at an average temperature of 20°C flows through a 5-cm-diameter pipe that is 2 m long. The pipe wall is heated by steam and is held at 100°C. The convective heat transfer coefficient is 2.2×10^4 W/(m^2 • °K). Find the heat flow, the heat flux, and the convective resistance.

Solution

Refer to Exhibit 12.

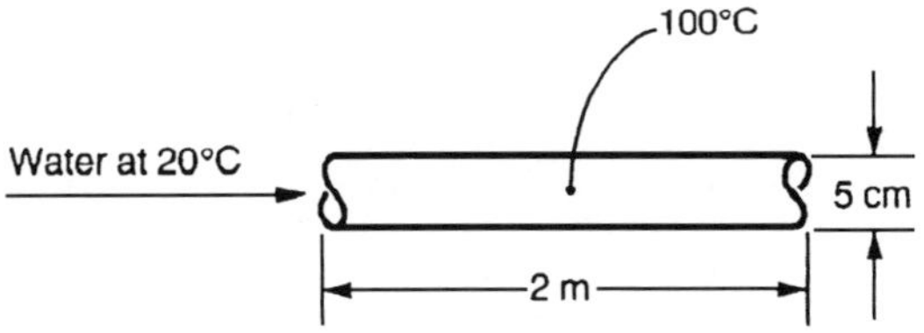

Exhibit 12

$$\dot{Q} = hA(T_H - T_C) = 2.2 \times 10^4 (\pi)(0.05)(2)(100 - 20) = 5.53 \times 10^5 \text{ W}$$

$$\frac{\dot{Q}}{A} = \frac{5.53 \times 10^5}{\pi(0.05)(2)} = 1.76 \ \frac{\text{MW}}{\text{m}^2}$$

$$R = \frac{1}{hA} = \frac{1}{2.2 \times 10^4 (\pi)(0.05)(2)} = 1.45 \times 10^{-4} \ \frac{°\text{K}}{\text{W}}$$

Radiation

Radiation heat transfer occurs between two surfaces via electromagnetic waves and *does not* require an intervening medium to permit the energy flow. In fact, it travels best through a vacuum as radiant energy does from the sun. The equation for radiation energy exchange between two surfaces is

$$q = \sigma A_1 F_e F_s \left(T_1^4 - T_2^4\right)$$

where the Stefan-Boltzmann constant is $\sigma = 5.67 \times 10^{-8}$ W/(m^2 • °K^4), F_e is a factor that is a function of the emissivity of the two surfaces with a value from 0 to 1.0, and F_s is a modulus that is a function of the relative geometries of the two surfaces with a value from 0 to 1.0.

Note that the heat flow is not proportional to the linear temperature difference but is a function of the temperature of the surfaces to the fourth power.

The simplest, and by far the most common, case of radiation energy exchange occurs in the case of a small surface radiating to large surroundings. In this case, the equation simplifies to

$$q = \sigma A_1 \varepsilon_1 \left(T_1^4 - T_2^4\right)$$

where ε_1 is the emissivity of the radiating surface.

Example 13.36

A steam pipe with a surface area of 5 m² and a surface temperature of 600°C radiates into a large room (which acts as a black body), the surfaces of which are at 25°C. The pipe gray-body surface emissivity is 0.6. Find the heat flow and heat flux from the surface to the room.

Solution

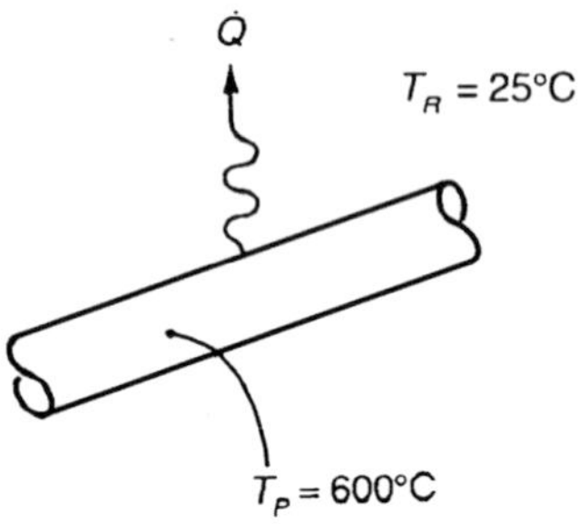

Exhibit 13

Refer to Exhibit 13.

$$\dot{Q}_{1-2} = \sigma A F_e F_s \left(T_1^4 - T_2^4\right)$$

For a gray body radiating to a black-body enclosure,

$$F_e F_s = \varepsilon_1$$

$$\dot{Q}_{1-2} = 5.67 \times 10^{-8} \times 5 \times 0.6 \times (873^4 - 298^4) = 9.75 \times 10^4 \text{ W}$$

$$\frac{\dot{Q}}{A} = \frac{9.75 \times 10^4}{5} = 1.95 \times 10^4 \ \frac{\text{W}}{\text{m}^2}$$

SELECTED SYMBOLS AND ABBREVIATIONS

Symbol or Abbreviation	Description
c_p	specific heat at constant pressure
c_v	specific heat at constant volume
H	enthalpy
h	specific enthalpy
h	heat transfer coefficient
k	thermal conductivity
m	mass
P	absolute pressure
p	partial pressure
P_r	relative pressure
Q	heat taken in or given off
q	heat per mass
r_c	compression ratio
R	gas constant
$\overline{\mathrm{R}}$	universal gas constant
S	entropy
s	specific entropy
T	absolute temperature (in Kelvin)
T_H	high or hot temperature
T_L, T_C	low temperature, cold temperature
U	internal energy
u	specific internal energy
V	total volume
v	specific volume
V_r	relative volume
W	work
w	specific work
Z	elevation
Z	compressibility factor

PROBLEMS

13.1 Equations of state for a single component can be any of the following, *EXCEPT*:

a. the ideal gas law, $Pv = RT$
b. the ideal gas law modified by insertion of a compressibility factor, $Pv = ZRT$
c. any relationship interrelating three or more state functions
d. a mathematical expression defining a path between states

13.2 Given the following data for a fluid, what is its state at 40°C and 3 kPa?

Saturated property table

T, °C	P, kPa	v_f, m³/kg	v_g, m³/kg	h_f, kJ/kg	h_g, kJ/kg	s_f, kJ/(kg • °K)	s_g, kJ/(kg • °K)
40	7.38	.001008	19.52	167.57	2574.3	.5725	8.257
80	47.39	.001029	3.407	334.9	2643.7	1.1343	7.5445
120	198.5	.001060	.8919	503.71	2706.3	1.5276	7.1296

a. Saturated liquid
b. Superheated vapor
c. Compressed liquid
d. Saturated vapor

13.3 Using the data table in Problem 13.2, what is the fluid's entropy in kJ/(kg • °K) at 120°C and 80% quality?

a. 1.53
b. 6.009
c. 7.13
d. 28.8

13.4 Using the refrigerant data table in Problem 13.2, what is its latent heat (heat of vaporization) in kJ/kg at 80°C?

a. 198.5
b. 2706
c. 1306
d. 2308.8

13.5 A nonflow (closed) system contains 1 kg of an ideal gas ($c_p = 1.0$, $c_v = .713$). The gas temperature is increased by 10°C while 5 kJ of work are done by the gas. What is the heat transfer in kJ?

a. –3.3
b. –2.6
c. +12.1
d. +7.4

13.6 Shaft work of –15 kJ/kg and heat transfer of –10 kJ/kg change the enthalpy of a system by:

a. –25 kJ/kg
b. –15 kJ/kg
c. –10 kJ/kg
d. +5 kJ/kg

13.7 A quantity of 55 cubic meters of water passes through a heat exchanger and absorbs 2,800,000 kJ. The exit temperature is 95°C. The entrance water temperature in °C is nearest to:

a. 49
b. 56
c. 68
d. 83

13.8 A fluid at 690 kPa has a specific volume of .25 m^3/kg and enters an apparatus with a velocity of 150 m/s. Heat radiation losses in the apparatus are equal to 25 kJ/kg of fluid supplied. The fluid leaves the apparatus at 135 kPa with a specific volume of .9 m^3/kg and a velocity of 300 m/s. In the apparatus, the shaft work done by the fluid is equal to 900 kJ/kg. Does the internal energy of the fluid increase or decrease, and how much is the change?

a. 858 kJ/kg (increase) c. 908 kJ/kg (increase)
b. 858 kJ/kg (decrease) d. 908 kJ/kg (decrease)

13.9 Exhaust steam from a turbine exhausts into a surface condenser at a mass flow rate of 4000 kJ/hr, 9.59 kPa, and 92% quality. Cooling water enters the condenser at 15°C and leaves at the steam inlet temperature.

Properties of saturated water (US units): Temperature table

		Specific volume		Internal energy		Enthalpy			Entropy	
Temp. °C T	Press. kPa P	Sat. Liquid vf	Sat. Vapor v_g	Sat. Liquid u_f	Sat. Vapor u_g	Sat. Liquid h_f	Evap. h_{fg}	Sat. Vapor h_g	Sat. Liquid s_f	Sat. Vapor s_g
15	1.705	.001	77.9	62.99	2396	62.99	2466	2529	.2245	8.781

v, m^3/kg; u and h, kJ/kg; s, kJ/(kg • °K)

The cooling water mass flow rate in kg/hr is closest to:

a. 157,200 c. 95,000
b. 70,200 d. 88,000

13.10 The mass flow rate of a Freon refrigerant through a heat exchanger is 5 kg/min. The enthalpy of entry Freon is 238 kJ/kg, and the enthalpy of exit Freon is 60.6 kJ/kg. Water coolant is allowed to rise 6°C. The water flow rate in kg/min is:

a. 24 c. 83
b. 35 d. 112

13.11 The maximum thermal efficiency that can be obtained in an ideal reversible heat engine operating between 833°C and 170°C is closest to:

a. 100% c. 78%
b. 60% d. 40%

13.12 A 2.2-kW refrigerator or heat pump operates between –17°C and 38°C. The maximum theoretical heat that can be transferred from the cold reservoir is nearest to:

a. 7.6 kW c. 15.6 kW
b. 4.7 kW d. 10.2 kW

13.13 In any nonquasistatic thermodynamic process, the overall entropy of an isolated system will:

a. increase and then decrease c. stay the same
b. decrease and then increase d. increase only

13.14 For spontaneously occurring natural processes in an isolated system, which expression best expresses ds?

a. $ds = \frac{dq}{T}$
b. $ds = 0$
c. $ds > 0$
d. $ds < 0$

13.15 Which of the following statements about entropy is *FALSE*?

a. The entropy of a mixture is greater than that of its components under the same conditions.
b. An irreversible process increases the entropy of the universe.
c. The entropy of a crystal at 0°C is zero.
d. The net entropy change in any closed cycle is zero.

13.16 A high-velocity flow of gas at 250 m/s possesses kinetic energy nearest to which of the following?

a. 3.13 kJ/kg
b. 313 kJ/kg
c. 31,300 kJ/kg
d. 31.3 kJ/kg

13.17 $(u + Pv)$ is a quantity called:

a. flow energy
b. shaft work
c. entropy
d. enthalpy

13.18 In flow process, neglecting *KE* and *PE* changes, $-\int v\, dP$ represents which item below?

a. Heat transfer
b. Shaft work
c. Closed system work
d. Flow energy

13.19 Power may be expressed in units of:

a. joules
b. watts
c. kJ
d. newtons

13.20 The temperature-entropy diagram in Exhibit 13.20 represents a:

a. Rankine cycle with superheated vapor
b. Carnot cycle
c. diesel cycle
d. refrigeration cycle

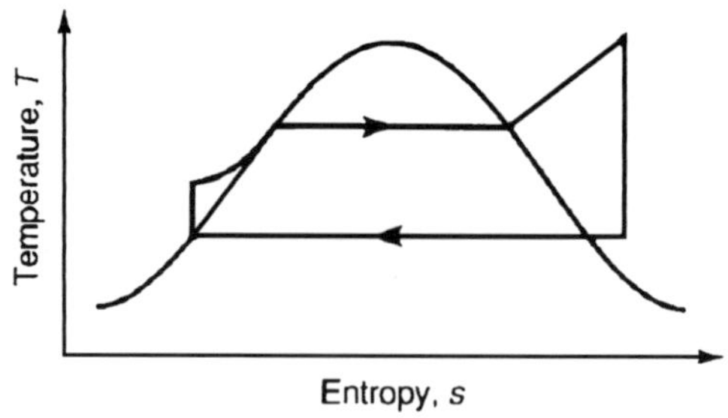

Exhibit 13.20

13.21 Entropy is the measure of:

a. the change in enthalpy of a system
b. the internal energy of a gas
c. the heat capacity of a substance
d. randomness or disorder

13.22 A Carnot heat engine cycle is represented on the T-s and P-V diagrams in Exhibit 13.22. Which of the several areas bounded by numbers or letters represents the amount of heat rejected by the fluid during one cycle?

a. Area 1–2–6–5
b. Area B–C–H–G
c. Area 3–4–5–6
d. Area D–A–E–F

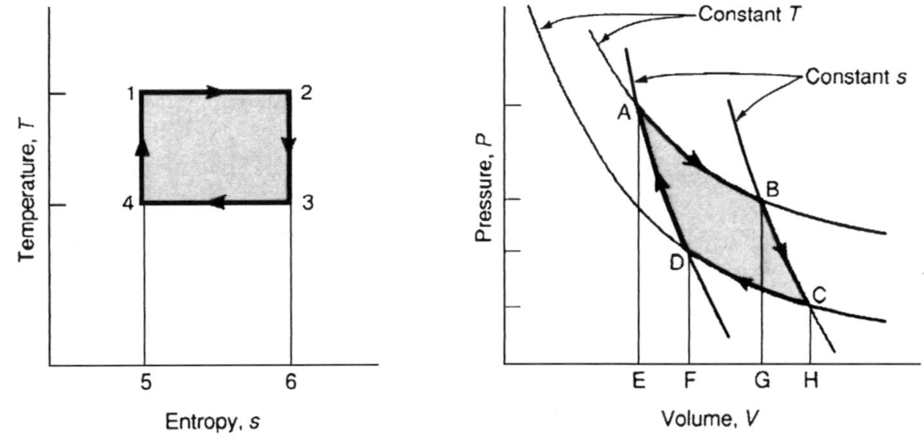

Exhibit 13.22

13.23 A Carnot engine operating between 70°C and 2000°C is modified solely by raising the high temperature by 150°C and raising the low temperature by 100°C. Which of the following statements is *FALSE*?

a. The thermodynamic efficiency is increased.
b. More work is done during the isothermal expansion.
c. More work is done during the isentropic compression.
d. More work is done during the reversible adiabatic expansion.

13.24 In the ideal heat pump system represented in Exhibit 13.24, the expansion valve 4–1 performs the process that is located on the T-s diagram between points:

a. A and B
b. B and C
c. C and D
d. E and A

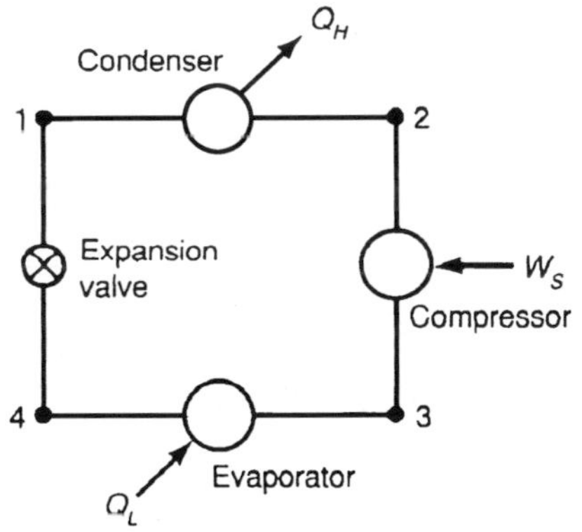

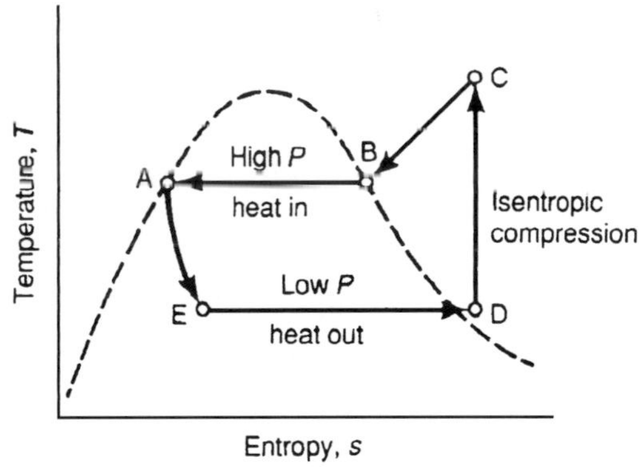

Exhibit 13.24

Use the psychrometric chart in solving problems 13.25 through 13.27.

13.25 Air is flowing through a duct and is heated using an electric coil (see Exhibit 13.25). The entering air is 20°C and $\phi = 40\%$. The exiting air is 35°C. The relative humidity of the exiting air is most nearly:

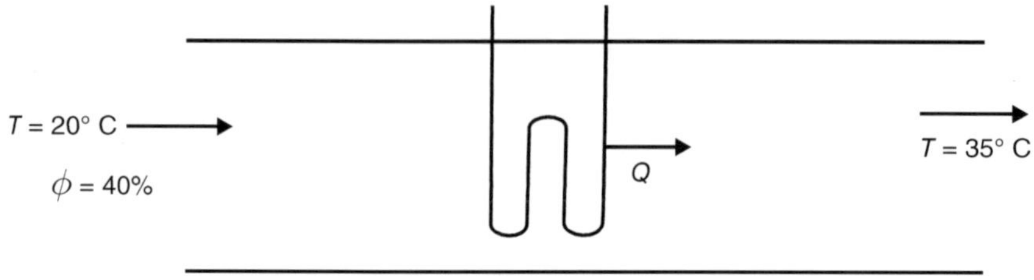

Exhibit 13.25

a. 60% c. 25%
b. 40% d. 17%

13.26 The humidity ratio of the exiting air in Problem 13.25 is most nearly:

a. 5.8 c. 4.5
b. 20 d. 16

13.27 An industrial air-conditioning system (see Exhibit 13.27) consists of 5 kg/min of outside air at 10°C and $\phi = 80\%$ mixing with 5 kg/min of inside air at T = 30°C and $\phi = 40\%$. The temperature at Position 3 is most nearly:

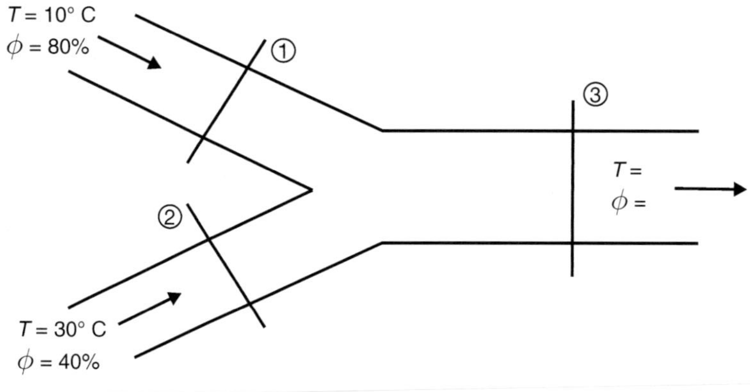

Exhibit 13.27

a. 30°C c. 0°C
b. 25°C d. 15°C

13.28 For the system in Problem 13.27 , the relative humidity at Position 3 is most nearly:

a. $\phi = 58\%$ c. $\phi = 45\%$
b. $\phi = 40\%$ d. $\phi = 64\%$

Use the following information for problems 13.29 through 13.31. Ethane is burned in 150% theoretical air. Assume complete combustion and that the process takes place in atmospheric air.

13.29 The air/fuel ratio in 150% theoretical air is most nearly:

a. 32 kg air/kg fuel
b. 28 kg air/kg fuel
c. 24 kg air/kg fuel
d. 16 kg air/kg fuel

13.30 The air/fuel ratio for stoichiometric conditions is:

a. 10 kg air/kg fuel
b. 16 kg air/kg fuel
c. 22 kg air/kg fuel
d. 24 kg air/kg fuel

13.31 The dew point temperature for 150% theoretical air is:

a. 62°C
b. 40°C
c. 48°C
d. 55°C

13.32 Data in the following table describe two states of a working fluid that exist at two locations in a piece of hardware:

	P, kPa	***v*, m³/kg**	***T*, °C**	***h*, kJ/kg**	***s*, kJ/(kg·K)**
State 1	25	0.011	20	19.2	0.0424
State 2	125	0.823	180	203.7	0.3649

Which of the following statements about the path from State 1 to 2 is *FALSE*?

a. The path results in an expansion.
b. The path determines the amount of work done.
c. The path is indeterminate from these data.
d. The path is reversible and adiabatic.

13.33 Name the process that has no heat transfer.

a. Isentropic
b. Isothermal
c. Quasistatic
d. Reversible

13.34 In a closed system with a moving boundary, which of the following represents work done during an isothermal process?

a. $W = P(V_2 - V_1)$

b. $W = 0$

c. $W = P_1V_1 \ln\left(\dfrac{P_1}{P_2}\right) = P_1V_1 \ln\left(\dfrac{V_2}{V_1}\right) = mRT \ln\left(\dfrac{P_1}{P_2}\right)$

d. $W = \dfrac{P_2V_2 - P_1V_1}{1-k} = \dfrac{mR(T_2 - T_1)}{1-k}$

13.35 The work of a polytropic ($n = 1.21$) compression of air ($c_p/c_v = 1.40$) in a system with moving boundary from $P_1 = 15$ kPa, $V_1 = 1.0$ m³ to $P_2 = 150$ kPa, $V_2 = 0.15$ m³ is:

a. –35.7 kJ
b. –324 kJ
c. 1080 kJ
d. 5150 kJ

13.36 The isentropic compression of 1 m³ of air, $c_p/c_v = 1.40$, from 20 kPa to a pressure of 100 kPa gives a final volume of:

a. 0.16 m³ c. 0.32 m³
b. 0.20 m³ d. 0.40 m³

13.37 An ideal gas at a pressure of 500 kPa and a temperature of 75°C is contained in a cylinder with a volume of 700 m³. Some of the gas is released so that the pressure in the cylinder drops to 250 kPa. The expansion of the gas is isentropic. The specific heat ratio is 1.40, and the gas constant is .287 kJ/(kg•K). The mass of the gas (in kg) remaining in the cylinder is nearest to:

a. 900 c. 1500
b. 1300 d. 2140

13.38 The theoretical power required for the isothermal compression of 800 m³/min of air from 100 to 900 kPa is closest to:

a. 70 c. 130
b. 90 d. 290

13.39 Which of the following statements is *FALSE* concerning the deviations of real gases from ideal gas behavior?

a. Molecular attraction interactions are compensated for in the ideal gas law.
b. Deviations from ideal gas behavior become large near the saturation curve.
c. Deviations from ideal gas behavior become significant at pressures above the critical point.
d. Molecular volume becomes significant as specific volume is decreased.

13.40 There are 3 kg of air in a rigid container at 250 kPa and 50°C. The gas constant for air is .287 kJ/(kg • °K). The volume of the container, in m³, is nearest to:

a. 2.2 c. 2.8
b. 1.1 d. 3.1

13.41 A mixture at 100 kPa and 20°C that is 30% by weight CO_2 (m.w. = 44) and 70% by weight N_2 (m.w. = 28) has a partial pressure of CO_2 in kPa that is nearest to:

a. 21.4 c. 68.3
b. 31.5 d. 78.6

13.42 Dry air has an average molecular weight of 28.79, consisting of 21 mole-percent O_2, 78 mole-percent N_2, and 1 mole-percent Ar (and traces of CO_2). The weight-percent of O_2 is nearest to:

a. 21.0 c. 23.2
b. 22.4 d. 24.6

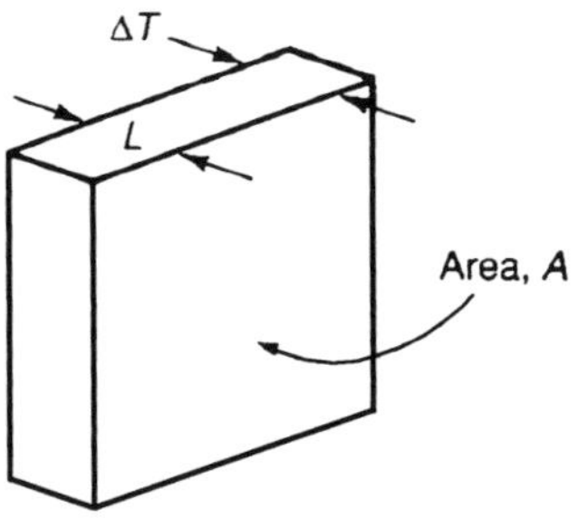

Exhibit 13.43

13.43 The temperature difference between the two sides of a solid rectangular slab of area A and thickness L, as shown in Exhibit 13.43, is ΔT. The heat transferred through the slab by conduction in time, t, is proportional to:

a. $AL\Delta Tt$
b. $AL\dfrac{\Delta T}{t}$
c. $AL\dfrac{t}{\Delta T}$
d. $\dfrac{A\Delta Tt}{L}$

13.44 The composite wall in Exhibit 13.44 has an outer temperature T1 = 20°C and an inner temperature T4 = 70°C. The temperature T3, in °C, is nearest to:

a. 38
b. 46
c. 58
d. 69

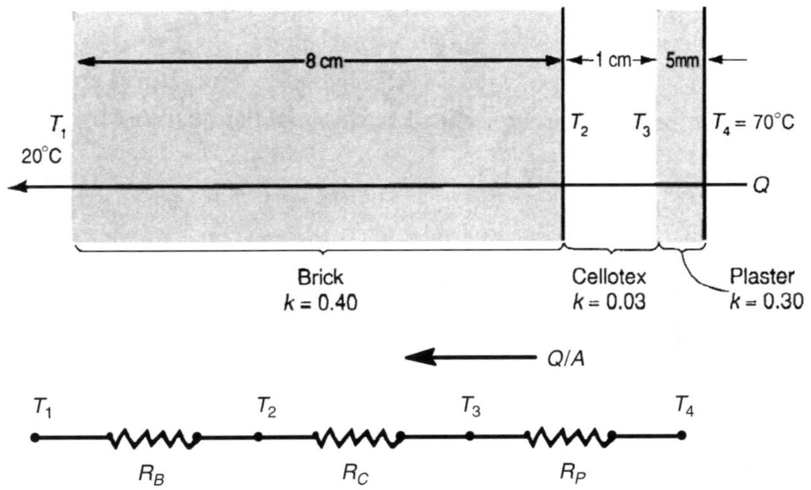

Exhibit 13.44

13.45 In Exhibit 13.45, the inner wall is at 30°C, and the outer wall is exposed to ambient wind and surroundings at 10°C. The film coefficient, h, for convective heat transfer in a 7-m/s wind is about 20 W/m² • °C. Ignoring any radiation losses, an overall coefficient (in the same units) for the conduction and convection losses is most nearly:

a. 1.4
b. 2.6
c. 12.5
d. 7.1

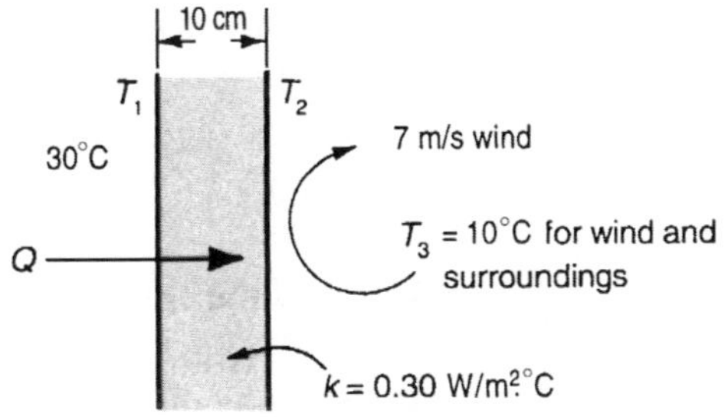

Exhibit 13.45

13.46 Heat is transferred by conduction from left to right through the composite wall shown in Exhibit 13.46. Assume the three materials are in good thermal contact and that no significant thermal resistance exists at any of the interfaces. The overall coefficient U in W/(m^2 • °C) is most nearly:

a. 0.04
b. 0.20
c. 0.35
d. 0.91

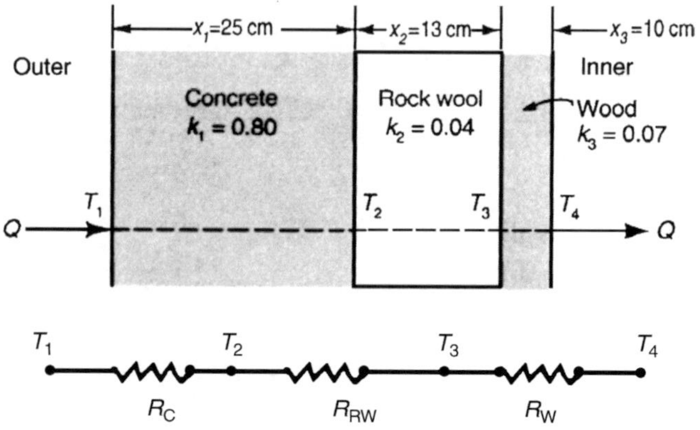

Exhibit 13.46

13.47 The heat loss per hour through 1 m^2 of furnace wall that is 40 cm thick is 520 W. The inside wall temperature is 1000°C, and its average thermal conductivity is 0.61 W/(m^2 • °C).

The outside surface temperature of the wall is nearest to:

a. 100°C
b. 300°C
c. 700°C
d. 1000°C

13.48 Which of the following is the usual expression for the power/unit-area Stefan-Boltzmann constant for black-body radiation?

a. 1.36×10^{-12} cal/(s • cm^2 • °K^4)
b. 5.67×10^{-5} ergs/(s • cm^2 • °K^4)
c. 5.67×10^{-8} watts/(m^2 • °K^4)
d. 5.67×10^{-8} coulombs/(s • m^2 • °K^4)

The following information may be used for questions 13.49 and 13.50.

A large municipal power plant produces 1,000 MW of power. The overall plant efficiency is 41%. Coal with a heating value of 24 MJ/kg is used for heating the water in the boilers.

13.49 How much coal in kg is burned each day?
a. 5,000
b. 8,778,000
c. 5,000,000
d. 85,000

13.50 If cooling water is allowed to raise 12 °C as it goes through the condenser, how much cooling water is necessary in kg/min?

a. 450,000
b. 850,000
c. 1,100,000
d. 1,380,000

The following information may be used for questions 13.51 to 13.55.

Water is used as the working fluid in a power plant that produces a net output of power of 100 MW. The constant pressure condenser operates at 8 kPa and the constant pressure boiler operates at 1 MPa. The steam enters the turbine, which operates at 85% efficiency at 600 °C, and the water leaving the condenser at 8 kPa enters the 75% efficient pump.

13.51 The mass flow rate kg/hr of the working fluid is most nearly:

a. 310,000
b. 256,000
c. 185,000
d. 375,000

13.52 The heat supplied to the boiler in kJ/hr is most nearly:

a. 2.1×10^9
b. 1.5×10^8
c. 8.5×10^6
d. 1.1×10^9

13.53 The heat rejected to the condenser in kJ/hr is most nearly:

a. 7.4×10^8
b. 6.0×10^7
c. 8.4×10^6
d. 8.6×10^9

13.54 The work of the pump in kW is most nearly:

a. 250
b. 400
c. 115
d. 525

13.55 The thermal efficiency is most nearly:

a. 51%
b. 45%
c. 38%
d. 33%

The following information is to be used for questions 13.56 and 13.57.

Air enters the diffuser of a ramjet engine at 25 kPa and 250 K with a velocity of 4000 km/hr and is decelerated to a low velocity at the exit of the diffuser of the engine. The heat addition in the combustion chamber is 1000 kJ/kg. The air exits the nozzle at 25 kPa.

13.56 The pressure at the diffuser exit, in kPa, is most nearly:
a. 850
b. 1200
c. 1450
d. 1717

13.57 The velocity at the nozzle exit is most nearly:
a. 1000 m/s
b. 1200 m/s
c. 1400 m/s
d. 1600 m/s

13.58 To maintain a comfortable temperature, what would be the cooling requirement to offset the emission of body heat in a dance hall that had an attendance of 1500 people if it has been determined that 75% of the people will be dancing and the others will be seated. Tests have shown that an adult will give off 71.8 W of sensible heat while engaging in moderate dancing, and 57.1 W of sensible heat while sitting. The latent heat emission for similar situations has been found to be equal to 177.3 W for dancers and 45.4 for sitters.
a. 200kW
b. 267 kW
c. 301 kW
d. 319 kW

13.59 An evaporative cooling device is used in Mesa, Arizona for the *conditioning* of the air. Outside air is brought in at 40 °C and a relative humidity of 10%. Water is then sprayed in at a temperature of 15 °C before it is introduced back into the room. The exit air is 20 °C.

Determine the exit relative humidity and the amount of water sprayed into the air (kg water/kg dry air).
a. RH = 75%, water added 0.1kg/kg dry air
b. RH = 85%, water added 0.0081 kg/kg dry air
c. RH = 65%, water added 0.0099 kg/kg dry air
d. RH = 55%, water added 0.011 kg/kg dry air

13.60 A residential heat pump provides some 3.2×10^6 kJ/day in order to maintain the house at 21 °C. The design condition is for the outside temperature to be –20 °C. If the electricity cost \$0.09/kW · h, determine (1) the minimum theoretical operating cost per day and (2) the ideal C.O.P. of the system.
a. Operating cost = \$5.48, C.O.P. = 12.2
b. Operating cost = \$10.50, C.O.P. = 8.5
c. Operating cost = \$7.56, C.O.P. = 6.8
d. Operating cost = \$9.23, C.O.P. = 9.48

SOLUTIONS

13.1 d. All *except* (d) are correct. The ideal gas law is the simplest equation of state; it is often applied to real gases by using a compressibility factor Z. Any relationships that interrelate thermodynamic state function data are equations of state. Answer (d) expresses the path of a process between states rather than a relationship between variables at a single point or state.

13.2 b. At 40°C, equilibrium between liquid and gas exists at 7.38 kPa. Below 7.38 kPa superheated vapor exists, and above 7.38 kPa only pressurized liquid exists.

13.3 b. At 120°C, s_f = 1.5276 and s_{fg} = 7.1296 – 1.5276 = 5.602. Here s_f is saturated liquid at 0% quality and s_g is saturated vapor of 100% quality. Thus s at 80% quality = s_f + (0.80) s_{fg} = 1.5276 + .8 × 5.602 = 6.009 kJ/(kg • °K).

13.4 d. Here, $h_{fg} = h_g - h_f$ = 2643.7 – 334.9 = 2308.8 kJ/kg.

13.5 c. The thermodynamic sign convention is + for heat in and + for work out of a system. Apply the first law for a closed system and an ideal gas working fluid:

$$\Delta U = mc_v \, \Delta T = q - w$$

$$.713(10) = q - (+5), \qquad 7.13 = q - 5, \qquad q = 12.13$$

13.6 d. The first law applied to a flow system is $h = q - w_s = -10 - (-15) = +5.$

13.7 d. For liquid water, c_p = 4.18 kJ/(kg • °C):

$$Q = mc_p \Delta T = mc_P(T_2 - T_1)$$

$$2{,}800{,}000 = (55\,\text{m}^3)\left(\frac{1000\,\text{kg}}{\text{m}^3}\right)(4.18)(95 - T_1)$$

$$12.2 = 95 - T_1 \qquad T_1 = 82.8°\text{C}$$

13.8 d. The basis of the calculation will be 1 kg. Use the thermodynamic sign convention that heat in and work out are positive. The first-law energy balance for the flow system is $h_2 + KE_2 - h_1 - KE_1 = Q - W_s$. Since the working fluid is unspecified and the internal energy change is desired, use the definition $h = u + Pv$. Then

$$u_2 + P_2v_2 + KE_2 - u_1 - P_1v_1 - KE_1 = Q - W_s \quad \text{or}$$

$$u_2 - u_1 = Q - W_s + P_1v_1 + KE_1 - P_2v_2 - KE_2$$

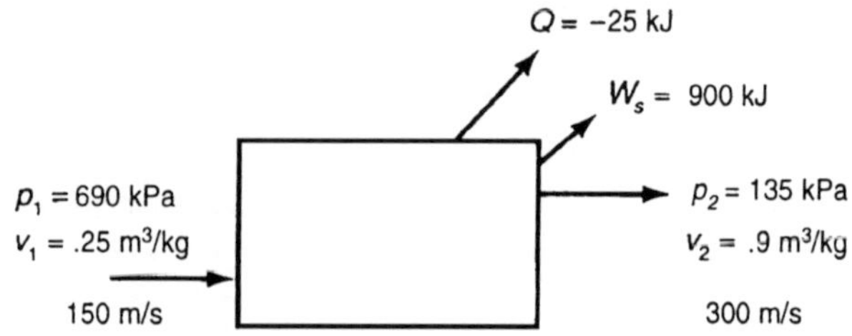

Exhibit 13.8

Now calculate numerical values for all terms except $u_2 - u_1$:

$$P_2 v_2 = 135 \times .9 = 121.5 \text{ kJ/kg} \qquad P_1 v_1 = 690(.25) = 172.5 \text{ kJ/kg}$$

$$KE_2 = \frac{V^2}{2gJ} = \frac{300^2}{2000} = 45 \text{ kJ/kg}$$

$$KE_1 = \frac{V^2}{2gJ} = \frac{(150)^2}{2000} = 11.3 \text{ kJ/kg}$$

$$W_s = +900 \text{ kJ/kg}$$

Therefore,

$$u_2 - u_1 = -25 - 900 + 172.5 + 11.3 - 121.5 - 45 = -907.7 \text{ kJ/kg}$$

13.9 Saturated steam table data at 9.59 kPA are

T, °C	h_f, kJ/kg	h_{fg}, kJ/kg	h_g, kJ/kg
45	188.45	2394.8	2583.2

The enthalpy of steam at 92% quality = h_1 = h_f + $0.92h_{fg}$ = 188.45 + 92 × 2394.8 = 2391.7. The enthalpy of liquid water at 45°C = h_2 = 188.45 kJ/kg. The enthalpy of liquid water at 15°C = h_3 = 62.99 kJ/kg above reference of 0°C.

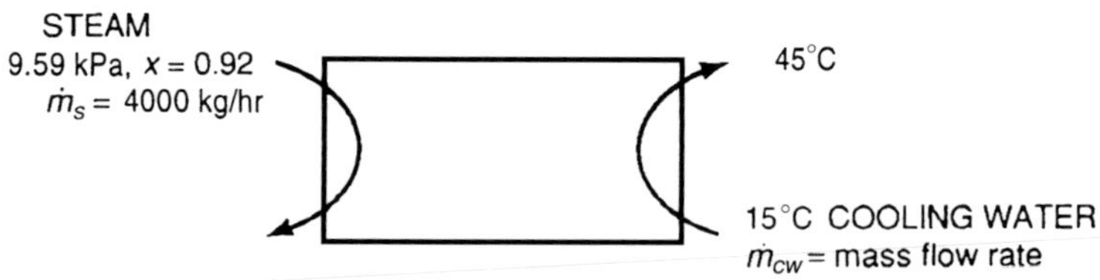

Exhibit 13.9

In the absence of data, assume that the steam condensate leaves at 45°C; if a heat balance is written over a 1-hour period, then the heat from steam = heat to cooling water, or

$$\dot{m}_s(h_1 - h_2) = \dot{m}_{cw}(h_2 - h_3)$$

$$4000(2391.7 - 188.45) = \dot{m}_{cw}(188.45 - 62.99)$$

$$\dot{m}_{cw} = 70,245 \text{ kg/hr}$$

13.10 **b.** Over a 1-minute period, the heat gain by water equals heat loss by Freon:

$$\dot{m}_{CW} c_p \Delta T = \dot{m}_F(h_1 - h_2)$$

$$\dot{m}_{CW} \times 4.2 \times 6 = 5(238 - 60.6)$$

$$\dot{m}_{CW} = \frac{887}{25.2} = 35.2 \text{kg/min}$$

13.11 **b.** Maximum efficiency is achieved with a Carnot engine.

$T_L = 170 + 273 = 443°K \qquad T_H = 833 + 273 = 1106°K$

$$\begin{aligned}\eta_{TH} &= \frac{w}{q_H} = \frac{q_H - q_L}{q_H}\\ &= 1 - \frac{Q_L}{Q_H} = 1 - \frac{T_L}{T_H}\\ &= 1 - \frac{443}{1106} = 1 - 0.40 = 0.60 = 60\%\end{aligned}$$

13.12 **d.** The coefficient of performance of a Carnot refrigerator or heat pump is

$$COP = \frac{T_L}{T_H - T_L} = \frac{256°K}{311°K - 256°K} = 4.65 = \frac{q_L}{w} = 4.65$$

$$COP = \frac{q_L}{q_H - q_L} = \frac{q_L}{w} = \frac{q_L}{2.2}; \qquad q_L = 4.65(2.2) = 10.2 \text{ kW}$$

13.13 **d.** Quasistatic means infinitely slow, lossless, hypothetical, by differential increments. The overall entropy will increase for an isolated system or for the system plus surroundings.

13.14 **c.** $ds > 0$ is the correct answer because all naturally occurring spontaneous processes are irreversible and result in an entropy increase.

(a) $ds = \frac{dq_{rev}}{T}$ only. The reversible requirement is necessary to generate the exact height vs. rectangular area equivalence on the Carnot cycle *T-s* diagram.

(b) Only a reversible adiabatic process is isentropic by definition.

(d) An energy input from the surroundings is required to reduce the entropy.

13.15 **c.** All are true *except* (c). The entropy of a perfect crystal at absolute zero (0°K or 0°R) is zero. This is the third law of thermodynamics. There is presumably no randomness at this temperature in a crystal without flaws, impurities, or dislocations.

13.16 **d.** Per 1 kg of flowing fluid,

$$KE = \frac{V^2}{2g_c} \qquad \text{where } V \text{ is in m/s, and } g_c = 1.0$$

Use 1000 to convert J to kJ:

$$KE = \frac{250^2}{2(1000)} = 31.3 \text{ kJ/kg}$$

13.17 d. Flow energy is Pv. Shaft work, W_s, is $-\int v\,dP$. Entropy is s. Internal energy is u. Enthalpy h is defined as $u + Pv$, the sum of internal energy plus flow energy.

13.18 b. Shaft work is work or mechanical energy crossing the fixed boundary (control volume) of a flow (open) system. Shaft work W_s is defined, in the absence of PE and KE changes, by $dh = T\,ds + v\,dP$, where $T\,ds = dq_{rev}$ and $-v\,dP$ is dW_s. In integrated form, $\Delta h = \int T\,ds + \int v\,dP = q_{rev} - W_s$, where W_s is represented by $-\int v\,dP$. Closed system work W is defined by $du = T\,ds - P\,dv$, or $\Delta u = \int T\,ds - \int P\,dv = q_{rev} - W$. Thus, closed system work is $+\int P\,dv$. Flow energy is the Pv term, and enthalpy change is ΔH.

13.19 b. Power is energy per unit time. The usual power units are watts.

13.20 a.

13.21 d.

13.22 c. The table below gives the significance of each area of the diagrams:

Process	*T-s* Diagram: Area Representing Heat	*P-V* Diagram: Area Representing Work
Isothermal expansion, 1–2 and A–B	1–2–6–5 = heat in from high-temp. reservoir	A–B–G–E = work done by fluid
Isentropic expansion, 2–3 and B–C	2–3–6 = 0 heat transfer	B–C–H–G = work done by fluid
Isothermal compression, 3–4 and C–D	3–4–5–6 = heat out to low-temp. reservoir	C–D–F–H = work done on fluid
Isentropic compression, 4–1 and D–A	4–1–5 = 0 heat transfer	D–A–E–F = work done on fluid
Net result of process	1–2–3–4 = net heat converted to work	A–B–C–D = net work done by process

13.23 a. The Carnot cycle efficiency is originally

$$\eta = \frac{T_H - T_L}{T_H} = \frac{2273 - 343}{2273} = 0.849$$

After the change,

$$\eta = \frac{2423 - 443}{2423} = .817 \qquad \text{(efficiency is reduced)}$$

On the *T*-*s* and *P*-*V* diagrams in Exhibit 13.23, the original cycle is shown as ABCD, and the modified cycle is shown as A′B′C′D′.

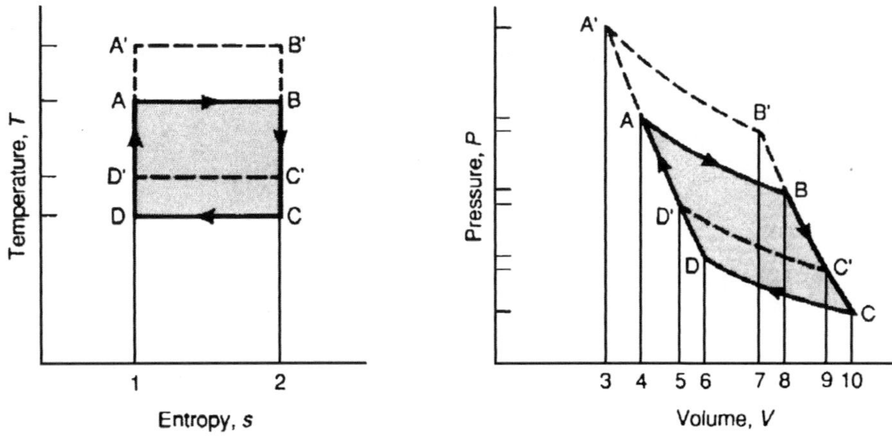

Exhibit 13.23

Compare the work done during the isothermal expansion (A to B vs. A′ to B′):

Original: area A–B–8–4

Modified: area A′ –B′ –7–3 is larger

Compare the work done during the isentropic compression (D to A vs. D′ to A′):

Original: area D–A–4–6

Modified: area D′–A′–3–5 is larger

Compare the work during the reversible (isentropic) expansion (B to C vs. B′ to C′):

Original: area B–C–10–8

Modified: area B′–C′–9–7 is larger

Compare the work during the isothermal compression (C to D vs. C′ to D′):

Original: area C–D–6–10

Modified: area C′–D′–5–9 is larger

Statements (b), (c), and (d) are correct.

13.24 **d.** The vapor compression-reversed Rankine cycle is conducted counterclockwise on both the schematic and the *T*-*s* diagram. Numbers on the schematic and letters on the *T*-*s* diagram are related: 1 = A, 2 = B, 3 = D, and 4 = E. Process C–B–A occurs in the condenser between 2 and 1. The expansion process A–E occurs between 1 and 4.

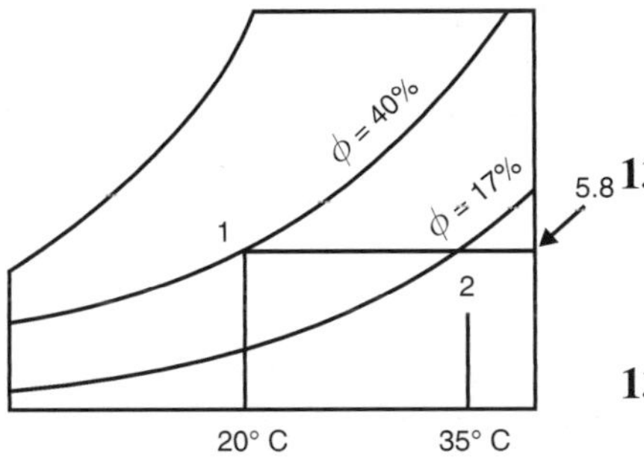

Exhibit 13.25a

13.25 d. On the psychrometric chart, locate point 1, which is $T = 20°C$ and $\phi = 40\%$ (see Exhibit 13.25a). Draw a horizontal line to point 2, which is at 35°C. Now, read ϕ from the chart to find $\phi = 17\%$.

13.26 **a.** From the psychrometric chart for the preceding solution, extend the line from point 1 to point 2 over to the humidity ratio to find that humidity ratio = 5.8 gm water/kg dry air.

13.27 c. On the psychrometric chart, locate the points at which $T = 10°C$, $\phi = 80\%$ and $T = 30°C$, $\phi = 40\%$. Draw a line between the two points and bisect it to find $T = 20°C$ (see Exhibit 13.27a).

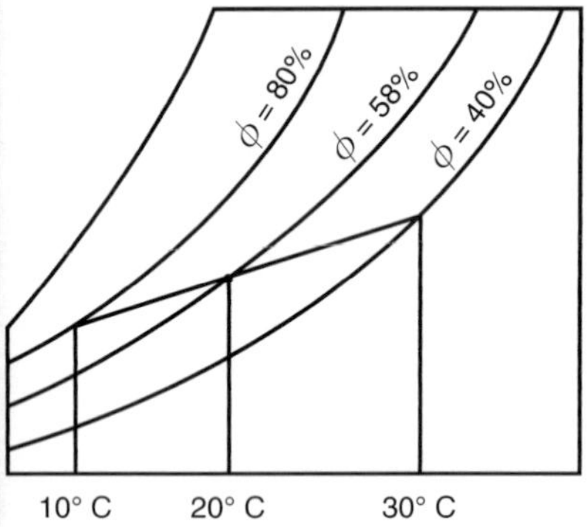

Exhibit 13.27a

13.28 a. From Exhibit 13.27a, $\phi = 58\%$.

13.29 c. Ethane $C_2H_6 + \alpha(O_2 + 3.76N_2) = 2CO_2 + 3H_2O + 3.76\alpha N_2$

C: $2 = 2$

H: $6 = 6$

O: $2\alpha = 4 + 3 \quad \alpha = 3.5$

The equation at 150% theoretical air is:

$$C_2H_6 + (1.5)(3.5)(O_2 + 3.76N_2) = 2CO_2 + 3H_2O + .5(3.5)O_2 + 1.5(13.16)N_2$$

Air/fuel ratio at 150% air = $(5.25)(28.92)(4.76)/(2 \times 12 + 6)$ = 24.06 kg air/kg fuel

13.30 b. The air/fuel ratio at stoichiometric conditions:

$$C_2H_6 + 3.5(O_2 + 3.76N_2) = 2CO_2 + 3H_2O + 13.16N_2$$

Air/fuel ratio = $(3.5)(28.92)(4.76)/(2 \times 12 + 6)$ = 16 kg air/kg fuel

13.31 c. The dew point temperature of the products of 150% theoretical air:

$$P_v = \frac{N_v}{N_{\text{products}}} P_{\text{products}} = \frac{3 \text{ k mol water}}{26.49} 101.35 \text{ kPa} = 11.48 \text{ kPa}$$

$$T_{dp} = 48.2°$$

13.32 d. The large volume and entropy changes indicate a change from a condensed phase to a vapor phase. Temperature, pressure, and enthalpy increases require an energy input. The path from 1 to 2 is indeterminate because no information on intermediate states is given. Work is always path dependent. The entropy increase means the process cannot be reversible and adiabatic (isentropic).

13.33 a. An *isentropic* process is reversible and adiabatic. An *adiabatic* process has no heat exchange with its surroundings. An *isothermal* process is conducted at constant temperature. A *quasistatic* (almost static) process departs only infinitesimally from an equilibrium state. A *reversible* process can have its initial state restored without any change (energy gain or loss) taking place in the surroundings.

13.34 c. For a closed system (piston-cylinder type, nonrepetitious) the work done is $W = \int P\,dV$. The equations in the problem are valid for ideal gases in the following processes, respectively:

a. constant pressure
b. constant volume
c. isothermal process
d. isentropic process

13.35 a. The work of a closed system (moving boundary) polytropic process for an ideal gas is

$$W = \frac{P_2V_2 - P_1V_1}{1-n} = \frac{[150(0.15) - 15(1.0)]}{1-1.21} = -35.7 \text{ kJ}$$

which is work done on the gas.

13.36 c. An isentropic process for an ideal gas follows the path

$$PV^k = P_1V_1^k = P_2V_2^k = \text{constant} \qquad \text{where } k = c_p/c_v$$

$$20(1)^{1.4} = 100(V_2)^{1.4}; \qquad V_2^{1.4} = 0.20; \qquad \text{hence, } V_2 = 0.317 \text{ m}^3$$

13.37 d. Given:

$k = c_p/c_v = 1.40$	$R = .287 \dfrac{\text{kJ}}{\text{kg} \bullet {}^\circ\text{K}}$
$P_1 = 500$ kPa	$P_2 = 250$ kPa
$V_1 = 700$ m^3	$V_2 = 700$ m^3
$T_1 = 75°\text{C} + 273 = 348\ °\text{K}$	$T_2 = ?$
	$w_2 = ?$

Basis: The ideal gas law may be written $PV = m\text{R}T$, and the basic equation for reversible adiabatic (isentropic) expansion is

$$\frac{T_2}{T_1} = \left(\frac{P_2}{P_1}\right)^{(k-1)/k}$$

The gas remaining in the tank cools as it expands; the new temperature is

$$T_2 = T_1\left(\frac{P_2}{P_1}\right)^{(k-1)/k} = 348\left(\frac{250}{500}\right)^{(1.4-1)/1.4} = 348\left(\frac{1}{2}\right)^{0.2857} = 285°\text{K}$$

Now apply the gas law at State 2, $P_2V_2 = m_2\text{R}T_2$:

$$250 \times 700 = m_2(.287)(285)$$

$$m_2 = \frac{(250)(700)}{(.287)(285)} = 2139 \text{ kg}$$

13.38 d. Since a volume flow rate is specified, the process is a flow process. The work of isothermal compression of an ideal gas is numerically the same in a steady flow process as in a closed system:

$$PV = \text{constant} = P_1V_1 = P_2V_2 = m\text{R}T$$

In a closed system,

$$W = \int_{V_1}^{V_2} P\,dV = P_1V_1 \ln\frac{V_2}{V_1} = P_1V_1 \ln\frac{P_1}{P_2} = m\text{R}T \ln\frac{V_2}{V_1} = m\text{R}T \ln\frac{P_1}{P_2}$$

In a flow system,

$$W_s = \int_{P_1}^{P_2} V\,dP = -P_1V_1 \ln\frac{P_2}{P_1} = P_1V_1 \ln\frac{P_1}{P_2} = P_1V_1 \ln\frac{V_2}{V_1} = m\text{R}T \ln\frac{V_2}{V_1} = m\text{R}T \ln\frac{P_1}{P_2}$$

Over a 1-minute interval,

$$W_s = P_1V_1 \ln\frac{P_1}{P_2} = (100)(800)\left(\frac{100}{900}\right) = -17,580 \text{ kJ/min}$$

$$W_s\left(\frac{-17,580}{60}\right) = -293 \text{ kW}$$

13.39 **a.** All statements except (a) are true. The ideal gas law does not consider the volume of the molecules or any interaction other than elastic collisions.

13.40 **b.** The ideal gas law is $PV = m\text{R}T$. Here $P = 250$ kPa and $T_1 = 50°\text{C} + 273 = 323°\text{K}$. Hence,

$$250\text{V} = 3 \times .287 \times 323$$

$$V = \frac{(3)(.287)(323)}{250} = 1.11 \text{ m}^3$$

13.41 **a.** The calculation is based on 1 kg of mixed gases. (1) Calculate the weight of each component and the number of moles of each that is present. (2) Compute the mole fraction of each, and apportion the total pressure in proportion to the mole fraction. The computations are in the following table:

Component	**Weight, kg**	**Number of kg-mol**	**Mole Fraction**	**Partial Pressure, kPa**
CO_2	0.30	$\frac{0.30}{44} = 0.00682$	$\frac{0.00682}{0.03182} = 0.214$	21.4
N_2	0.70	$\frac{0.70}{28} = 0.0250$	$\frac{0.0250}{0.03182} = 0.786$	78.6
Total	1.00	0.03182	1.000	100

Since the mole fraction of a gas is the same as the volume fraction, the composition of the mixture is 21.4% vol. CO_2 and 78.6% vol. N_2. From the table, the correct partial pressure of CO_2 is 21.4 kPa.

13.42 **c.** The calculation will be based on 1 kg-mol of dry air and arranged in the following table:

Component	**m.w.**	**Mole Fraction**	**Weight, kg**	**Weight, %**
O_2	32.0	0.21	6.72	23.2
N_2	28.0	0.78	21.80	75.4
Ar	40.0	0.01	0.40	1.4
Totals		1.00	28.92	100.0

13.43 d. The heat transfer rate through the slab by conduction is governed by the equation

$$Q = kA\Delta T/L$$

In time t the amount of heat transfer is proportional to

$$A\frac{\Delta T}{L}t$$

The symbol k is the coefficient of thermal conductivity of the material; hence, the heat transfer in a given material is proportional to the other variables.

13.44 d. At steady state the same Q flows across each material, and the temperatures descend in direct proportion to the thermal resistances (reciprocal of conductivity).

$$\text{Resistance of brick} = \frac{x}{k} = \frac{.08\text{m}}{.4\frac{w}{m \bullet °\text{C}}} = .2\frac{\text{m}^2}{\text{W} \bullet °\text{C}}$$

$$\text{Resistance of Cellotex} = \frac{x}{k} = \frac{.01}{.03} = .333\frac{\text{m}^2}{\text{W} \bullet °\text{C}}$$

$$\text{Resistance of plaster} = \frac{x}{k} = \frac{.005}{.3} = .017\frac{\text{m}^2}{\text{W} \bullet °\text{C}}$$

Total resistance = .2 +.333 + .017 = .55 $\text{m}^2/(\text{W} \bullet °\text{C})$

$$Q/A = \frac{\Delta T_{\text{total}}}{\text{total resistance}} = \left(\frac{\Delta T}{x/k}\right)_{\text{layer}}$$

Hence,

$$Q/A = \frac{50}{.55} = \frac{T_4 - T_3}{.017} = \frac{T_3 - T_2}{.333} = \frac{T_2 - T_1}{.2} = 90.9 \text{ W/m}^2$$

$T_4 - T_3 = 1.5°\text{C}$, since $T_4 = 70°\text{C}$, $T_3 = 68.5°\text{C}$
$T_3 - T_2 = 30.3°\text{C}$, since $T_3 = 68.5°\text{C}$, $T_2 = 38.2°\text{C}$
$T_2 - T_1 = 18.2°\text{C}$, since $T_2 = 38.2°\text{C}$, $T_1 = 20°\text{C}$
(in agreement with given data)

13.45 b. Since conduction and convection are based on ΔT, absolute temperatures are not required. For steady state, the heat conducted through a wall must equal the heat lost by convection:

$$Q = \frac{kA(T_1 - T_2)}{x} = hA(T_2 - T_3) \qquad \textbf{(13.1)}$$

In a similar way, Q can be expressed by an overall coefficient

$$Q = UA(T_1 - T_3) \qquad \textbf{(13.2)}$$

Here, U is calculated in a manner analogous to that used for thermal conductivities in series:

$$U = \frac{1}{\frac{1}{h_1} + \frac{x_1}{k_1}} \quad \textbf{(13.3)}$$

In this case,

$$U = \frac{1}{\frac{1}{20} + \frac{.1}{.3}} = \frac{1}{.05 + .333} = 2.61 \text{ W/(m}^2 \bullet {}^\circ C)$$

13.46 b. The overall coefficient U, the thermal conductivity k/x, and the film coefficient h are the reciprocals of their thermal resistances. Thermal resistances in series are handled analogously to series electrical resistances; hence

$$U = \left(\frac{1}{\sum_i \frac{x_i}{k_i}} \right)^{-1} = \frac{1}{R_T} = \frac{1}{R_1 + R_2 + R_3}$$

The overall coefficient U is then used in the simplified conduction equation $Q = UADT$.

In this problem

$$U = \frac{1}{\frac{.25}{.80} + \frac{.13}{.04} + \frac{.10}{.07}} = \frac{1}{.313 + 3.25 + 1.43} = 0.20 \text{ W/(m}^2 \bullet {}^\circ C)$$

13.47 c. The heat conduction equation is

$$Q = k\frac{A}{L}(T_1 - T_2)$$

where $T_1 = 1000°\text{C}$, T_2 = outside temperature, $k = 0.61$ W/(m^2 • °C), $Q/A = 520$ W/m^2, and $L = .4$ m.

Solving for T_2, one has

$$T_2 = -\frac{Q}{A}\frac{L}{k} + T_1 = -520\frac{.4}{.61} + 1000 = 659°C$$

13.48 c. All are numerically correct conversions of the constant in terms of power per unit area. The units of watts/(m^2 • °K^4), however, are normally used in heat transfer.

13.49 b. Thermal efficiency equation is *energy wanted/energy that costs\$*. It may then be written as:

$[1{,}000 \text{ MW}]/\eta_{th}$ = energy in the coal = 1,000/0.41 = 2439 MJ/s

kg coal needed = energy needed/energy/kg coal = [2439 MJ/s/24 MJ/kg]

kg coal needed = 101.6 kg/s [3600 s/hr × 24 hr/day] = 8.778 × 10^6 kg/day

13.50 d. To find the amount of water needed for cooling we go back to the energy in and subtract the work obtained. The difference between the two is equal to the heat rejected. This amount of heat is to be divided by the enthalpy difference due to the increase in temperature of the cooling water.

$$Q_{in} - W_{out} = Q_{out}$$

$$2439 \text{ MJ/s} - 1000\text{MW} = 1439 \text{ MJ/s}$$

$$\text{Water needed} = Q_{out}/\Delta h$$
$$= [1439 \text{ MJ/s}]/[(4.18 \text{ kJ/kg-K})(15 \text{ K})][1000\text{M/k}]$$
$$= [22.95][1000] \text{ kg/s}$$
$$= 22{,}950 \text{ kg/s} \times 60 \text{ s/min}$$
$$= 1{,}377{,}000 \text{ kg/min}$$

Problems 13.51–13.55

The most efficient method of solution is to set up a table in which to put the calculated values. Once the table is completed then the problem is very straightforward to solve. The values for position 1 which is entering the turbine, can be obtained directly from the steam tables listed in the *FE Handbook*. Position 2s is the isentropic drop to 8 kPa pressure.

Knowing:

$$s_1 = s_{2s} = 8.0290 = xs_g + (1 - x)s_f = x(8.2287) + (1 - x)(0.5926)$$
$$x_2 = 0.97$$

	1	2s	2	3	4s	4
P	1 MPa	8 kPa	—	—	1 Mpa	—
T	600 °C					
h	3697.9	2504.9	2683.9	173.88	174.88	175.21
s	8.0290	—				
x		0.97				

Solving for h_{2s} using the quality equation:

$$h_{2s} = xh_g + [1 - x]h_f = 0.97[2577] + 0.03[173.88] = 2504.9 \text{ kJ/kg}$$

Use the turbine efficiency equation to determine h_2:

$$\eta_T = [h_1 - h_2]/[h_1 - h_{2s}]$$
$$h_2 = h_1 - \eta_T[h_1 - h_{2s}] = 3697.9 - 0.85[3697.9 - 2504.9] = 2683.9 \text{ kJ/kg}$$
$$h_3 = \text{the saturated water value} = 173.88 \text{ kJ/kg}$$

Use the pump equation and add the result to the enthalpy at Point 3 to give h_{4s}.

$$h_{4s} = h_3 + \int v dP = 173.88 \text{ kJ/kg} + [1.0084/1000]\text{m}^3/\text{kg}$$
$$\times [1 \times 10^6 - 8{,}000]\text{Pa}/10^3 \text{ J/kJ}$$
$$h_{4s} = 173.88 + 1.0 = 174.88$$

To find h_4 we must use the efficiency equation for the pump:

$$\eta_p = [h_{4s} - h_3]/[h_4 - h_3]$$

$$h_4 = h_3 + [h_{4s} - h_3]/\eta_p = 173.88 + [174.88 - 173.88]/[0.75] = 175.21$$

Now that we have filled out the table, it is very straightforward to solve each of the questions.

13.51 a.

$$\text{Mass flow rate} = \text{Work output of system/Work output/kg}$$
$$\text{Mass} = 100{,}000 \text{ kW}/[(h_1 - h_2) - (h_4 - h_3)]\text{kJ/kg}$$
$$= 100{,}000/[(3697.9 - 2540.7) - (175.21 - 173.88)]$$
$$= 86.5 \text{ kg/s}$$
$$= 311{,}400 \text{ kg/hr}$$

13.52 d.

$$Q_{in} = m[h_1 - h_4] = 311{,}400 \text{ kg/hr } [3697.9 - 175.21] \text{ kJ/kg}$$
$$= 1.096 \times 10^9 \text{ kJ/hr}$$

13.53 a.

$$Q_{out} = m[h_2 - h_3] = 311{,}400[2540.7 - 173.88] = 7.37 \times 10^8 \text{ kJ/hr}$$

13.54 c.

$$W_p = m[h_4 - h_3]$$
$$= [311{,}400\text{kg/hr}]/3600 \text{ s/hr } [175.21 - 173.88] \text{ kJ/kg}$$
$$= 115 \text{ kW}$$

13.55 d.

$$\eta_{th} = W_{net}/Q_{in} = [(3697.9 - 2540.7) - 1.33]/[3697.9 - 175.21] = 32.8\%$$

Problems 13.56 and 13.57

13.56 d. Again, the best way to solve this problem is to set up a table of values, find the values, and then make the final calculations. Point 1 is entering the diffuser, Point 2 is exiting the diffuser, Point 3 is after combustion and just entering the nozzle and Point 4 is leaving the nozzle. It shall be assumed that the ramjet acts on an ideal Brayton cycle.

	1	2	3	4
P	25 kPa	1716.7	—	25
T	250 K	837	1837	548.7

Write a first law equation for a diffuser to get the condition at state 2:

$$h_1 + v_1^2/2 = h_2 + v_2^2/2$$

$$c_p[T_2 - T_1] = v_1^2/2$$

Noting that

[4000 km/hr/3600 s/hr][1000 m/km] = 1111 m/s

or

$$1.00\ \text{kJ/kg}\cdot\text{K}[T_2 - 220]\text{K} = [1111\ \text{m/s}]^2/2[\text{N/kg}\cdot\text{m/s}^2][\text{kJ}/10^3\text{N}\cdot\text{m}]$$
$$T_2 = 837\ \text{K}$$

Then, using the perfect gas relationship:

$$P_2 = P_1\,[T_1/T_2]^{k/(1-k)}$$
$$= 25\ \text{kPa}[250/837]^{1.4/(1-1.4)}$$
$$= 1716.7\ \text{kPa}$$

13.57 d. The heat addition into the engine occurs between states 2 and 3.

$$Q = c_p[T_3 - T_2];\ \text{therefore:}$$
$$1000\text{kJ/kg} = 1.00\ \text{kJ/kg}\cdot\text{K}[T_3 - 837]$$
$$T_3 = 1837\ \text{K}$$

To find T_4 one must sagain use the isentropic relationship, relating pressure and temperature.

$$T_4 = T_3[P_3/P_4]^{(1-k)/k} = 1837[1716.7/25]^{(1-1.4)/1.4} = 548.7\ \text{K}$$

Using the general energy equation for a nozzle we may find the exit velocity:

$$h_3 = v_3^2/2 = h_4 + v_4^2/2$$

The velocity entering the nozzle is taken as zero, thus:

$$v_4 = [2(h_3 - h_4)]^{1/2} = [2c_p(T_3 - T_4)]^{1/2} - [2(1)(1000)\,(1837 - 549)]^{1/2}$$
$$= 1604\ \text{m/s}$$

13.58 d. Total body heat $= 0.75(1500)(71.8 + 177.3) + 0.25(1500)(57.1 + 45.4)$
$= 318{,}675\ \text{W} = 319\ \text{kW}$

13.59 **b.** Write an energy balance of the system:

$$(m_a h_a + h_g)_{in} + m_f h_f = (m_a h_a + h_g)_{out}$$

Rearrange the terms and divide by the mass of the air:

$$c_p(T_1 - T_2) + \omega_1 h_{g1} + (\omega_2 - \omega_1)h_{f15} = \omega_{g2}$$

Obtain values from the steam tables and the psychrometric chart in the *FE Handbook*:

$$(1.00)(40 - 20) + (0.0045)(2574.3) + (\omega_2 - 0.0045)(62.99) = \omega_2(2538.1)$$

$$w_2 = 0.0126$$

Therefore, water added = (0.0126 – 0.0045) = 0.0081 kg water/kg dry air

Relative Humidity may be selected from the psychrometric chart as 85%

13.60 **d.** The ideal Coefficient of Performance is that of the Carnot cycle:

b) C.O.P. = $T_H/(T_H - T_L) = (21 + 273)/[(21) - (-10)] = 9.48$

a) C.O.P. = $Q_H/(Q_H - Q_L) = 9.48 = (3.5 \times 10^6)$/work

Work = 3.69×10^5 kJ/day

Cost = $(3.69 \times 10^5)(0.09)(1/3600)$ = \$9.23/day

CHAPTER 14

Heat Transfer

OUTLINE

Heat transfer is that branch of thermodynamics that is concerned with the transfer of energy from one point to another by virtue of temperature difference. There are three modes of heat transfer: conduction, convection and radiation. Conduction and radiation are true forms of heat transfer since these modes depend only on temperature differential and the characteristics of the materials involved in the temperature exchange. Convection, however, not only depends upon the materials and temperature differentials, but is also dependent upon mass transport.

CONDUCTION

Conduction is the molecule-to-molecule transfer of energy. The one-directional equation of energy flow for a steady-state condition is given by Fourier's Law of Conduction:

$$Q = [kA/L][T_1 - T_2]$$

where

Q = rate of heat flow in watts or J/s
k = thermal conductivity W/(m-K)
A = area perpendicular to heat flow
L = thickness in direction of heat flow
ΔT = $(T_1 - T_2)$ difference in temperature within the material in the direction of heat flow, may be given in terms of C or K

Example 14.1

What is the rate of heat flow through a wall constructed of brick and mortar, which is 25 cm thick and 3.0 m × 2.0 m in area. The temperature on one side is 165 °C and 55 °C on the other side. The average coefficient of thermal conductivity is equal to 0.750 W/m · K.

Solution

$$Q = [kA/L][T_1 - T_2] = [(0.750\text{ W/m}\cdot\text{K})(6\text{ m}^2)/(25/100)\text{m}][165 - 55]\text{ K or °C}$$
$$Q = 1980\text{ watts} = 1.98\text{ kW}$$

SERIES COMPOSITE WALL

A series composite wall is a wall that is made with layers of various materials sandwiched together. The concept is that the heat will transfer through the wall in a straight line perpendicular to the wall. The same amount of energy that goes into one side of the wall comes out the other side. Each of the various layers of material adds to the thermal resistance of the wall and thus reduces the amount of energy transferred through the wall.

Example 14.2

A series composite wall of a furnace consists of 22 cm of fire brick, 11 cm of a high temperature insulating material, 10 cm of ordinary brick, and, lastly, 6.5 mm of asbestos cement board. The inside of the furnace wall is 980 °C and the temperature of the outside wall is 90 °C. Determine the heat transfer through the wall per m^2.

Solution

Since a series composite wall is a wall that is layered with the heat transfer passing through each of the layers in series, the general equation for the solution of a series composite wall is:

$$Q = [\Delta T_{\text{overall}}]/\Sigma R_{\text{th}}$$

Since the furnace size was not given, it will be taken on a m^2 basis. It is also assumed that, since this is a wall, the area of conduction of every portion of the wall is the same. We may therefore write the equation:

$$Q = [980 - 90]/[L_1/k_1 + L_2/k_2 + L_3/k_3 + L_4/k_4]$$

where

Fire Brick	$k_1 = 1.4$ W/m-K
High Temperature Insulation	$k_2 = 0.22$ W/m-K
Ordinary Brick	$k_3 = 0.90$ W/m-K
Asbestos Cement board	$k_4 = 0.39$ W/m-K

$$Q = [890\text{ C}]/[(0.22/1.4) + (0.11/.22) + (0.10/0.90) + (0.065/0.39)]\text{W/m}^2\cdot\text{K}$$
$$Q = 832\text{ watts/m}^2$$

CIRCULAR PIPES

An insulated circular pipe behaves much as a series composite wall. In this case, however, the heat is transferred in a straight line radially out from the center of the pipe. Calculation of the heat transfer, however, is modified from the simple equation used for a plane wall because, as the heat travels out from the center, it goes through a larger and larger area. As a consequence, the equation becomes a function of a logarithmic mean radius. The generic equation is shown both in the *FE Handbook* and below.

$$Q = \frac{2\pi kL[T_1 - T_2]}{\ln(r_2/r_1)}$$

Example 14.3

A 10-cm *OD* and 9.5-cm *ID* cast iron pipe with conductivity of 80 W/m · K has 3.5 cm of polystyrene insulation with a conductivity of 0.027 W/m · K. Saturated steam at 400 °C is flowing through the pipe and the outside surface temperature is 40 °C.

Determine

a). Heat transfer per 100 meters of pipe

b). The temperature of the outside surface of the pipe

Solution

a). The specific equation for the determination of the heat transfer is:

$$Q = \frac{T_1 - T_3}{\dfrac{\ln(r_2/r_1)}{2\pi k_1 L_1} + \dfrac{\ln(r_3/r_2)}{2\pi k_2 L_2}}$$

where

$$[\ln(r_2/r_1)]/2\pi k_1 L_1 = \ln(10/9.5)/[2\pi\ 80\ \text{W/m} \cdot \text{K}][100\ \text{m}] = 1.02 \times 10^{-6}\ \text{K/W}$$

and $[\ln(r_3/r_2)]/2\pi k_2 L_2 = \ln(17/10)/[2\pi(0.027)][100] = 0.03128$ K/W;

therefore

$$Q = [400 - 40]/[1.02 \times 10^{-6} + 0.03128] = 11{,}509\ \text{W} = 10.5\ \text{kW}$$

b). The temperature of the surface of the pipe can be obtained by using the generic equation, knowing that the heat transfer does not change and by writing the equation across temperature difference from the surface of the pipe to the outside surface of the insulation. The thermal resistance includes only what is between the two temperatures.

$$Q = [T_2 - T_3]/[\ln(r_3/r_2)/2\pi k_2 L_2] = [T_2 - 40]/0.03128$$
$$T_2 = (0.03128)(11{,}509) + 40 = 399.99\ ^\circ\text{C}$$

Notice that the pipe does not give any significant thermal resistance.

RADIATION

Radiation is the transfer of energy from one body to another without the necessity of a intermediate medium. In fact, the transfer is done much more efficiently without a medium. The transfer takes place due to the difference in the fourth power of the absolute temperatures of the bodies. Examining first the energy emitted by a body:

$$Q = \varepsilon \sigma A T^4$$

where

Q = Energy transferred in watts
ε = Emissivity of the radiating body (varies from 0 to 1)
σ = Stefan Boltzmann Constant (5.67×10^{-8} W/m$^2 \cdot$ K^4)
A = Surface area of the body in m^2
T = Absolute temperature K

If we now consider the case where the body radiates to outer space and that none of the radiation comes back, the equation above changes to a slightly different form.

$$Q = \varepsilon \sigma F_{1-2} A [T_1^4 - T_2^4]$$

where

F_{1-2} = Shape factor between body 1 and body 2 (varies from 0 to 1)
T_1 = Absolute temperature of body sending the radiation
T_2 = Absolute temperature of body receiving the radiation

Example 14.4

A body at 20 °C is set out on a roof top during the night. The body "sees" nothing but the sky which has an effective temperature of 120 K. Determine the heat transfer rate from the body to the sky if the body temperature is maintained at 20 °C, the surface emissivity of the body is equal to 0.90, and none of the radiation going out of the body comes back.

Solution

Since the body sees nothing else than the sky, the shape factor is equal to one. Putting the temperatures into K, the heat transfer may be calculated using the equation above:

$$Q = (0.90)(5.67 \times 10^{-8})\text{W/m}^2 \cdot \text{k}^4 (1.0)(1\ \text{m}^2)(293^4 - 120^4)\ \text{K}^4$$
$$Q = 365\ \text{W}$$

Example 14.5

Consider the same body at the same conditions setting in a 20 m^2 room at 120 K and an emissivity of 0.4. Determine the heat transfer from body one to the room.

Solution

The situation may be best represented by an electric analog of a series of resistances as shown in Exhibit 1.

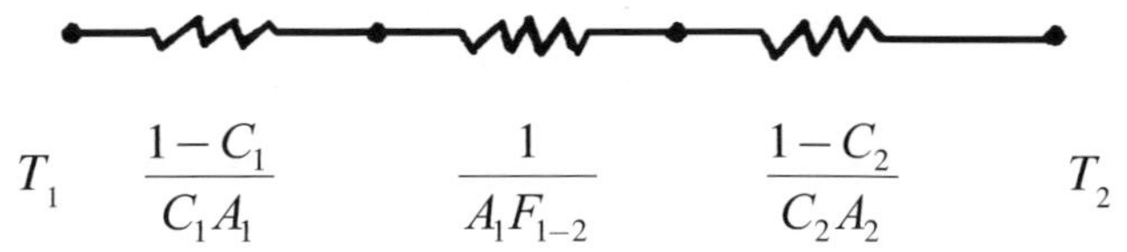

Exhibit 1

Therefore the equation for heat transfer becomes:

$$Q = \frac{\sigma\left(T_1^4 - T_2^4\right)}{\frac{1-\varepsilon}{\varepsilon_1 A_1} + \frac{1}{A_1 F_{1-2}} + \frac{1-\varepsilon_2}{\varepsilon_2 A_2}}$$

Substituting in the values:

$$Q = \frac{(5.67 \times 10^{-8})(293^4 - 120^4)}{\frac{1-.90}{(0.90)(1)} + \frac{1}{(1)(1)} + \frac{1-.40}{(0.40)(20)}}$$

$$Q = 342 \text{ W}$$

CONVECTION

Convection is a mode of heat transfer that combines molecule-to-molecule conduction with mass transport. There are two types of convection: natural and forced convection. Natural convection is circulation caused by differences of densities of fluids. These differences of densities are caused by expansion of the fluid due to elevated temperatures. Forced convection, on the other hand, is caused by mechanical circulation of the fluids. As the fluid moves faster over the heat transfer surface the film layer gets thinner and, as a consequence, the amount of heat transfer increases.

The generic heat transfer equation for convection may be written as:

$$Q = hA(T_s - T_\alpha)$$

where

Q = Heat transfer—watts
h = Convection coefficient, W/$m^2 \cdot$ K
A = Surface area, m^2
T_s = Temperature of the heated surface
T_a = Temperature of the flowing fluid

Example 14.6

Strawberry blossom time is critical in the upper Midwest. If the blossoms get a touch of frost on them, the harvest can be reduced significantly. Consider a very clear night with a slight breeze of about 10 km/hr making the convection coefficient equal to 20 W/m²·K. The outer space temperature may be assumed to be 120 K and the emissivity of the strawberry is 0.6. At what surrounding temperature will the strawberries freeze?

Solution

The approach to this problem is to write an energy balance equation.

$$\text{Heat in by convection} = \text{Heat out by radiation}$$
$$hA(T_{air} - T_{berry}) = \varepsilon\sigma F_{1-2}A(T_b^{\,4} - T_{o\,sp}^{\,4})$$

Notice that the area of convection to the strawberry is the same as the area of radiation away from the strawberry. Therefore, the areas may be eliminated.

$$20\ \text{W/m}^2\cdot\text{K}\ (T_{air} - 0\,°\text{C}) = (0.6)(5.67 \times 10^{-8})\text{W/m}^2\cdot\text{K}^4(1)(273^4 - 120^4)\text{K}^4$$
$$T_{air} = 9.1\,°\text{C}$$

Example 14.7

5 cm spheres at 300 °C are dropped on a conveyer belt traveling 5 m/min through an environment that is 10 °C. The convection coefficient of the spheres is 25 W/ m²·K and the other properties are: $\rho = 8933$ kg/m³, $c_p = 385$ J/kg·K and $k = 400$ W/m·K. How long must the belt be to have the spheres drop off the end at 35 °C?

Solution

This is a conduction/convection system. The object is relatively small and has a large thermal conductivity compared to the convection coefficient. Since it is possible that the system is a *lumped capacity* type of problem it must be checked. The *FE Handbook* shows that if Bi << 1 the lumped capacity method may be used.

$$Bi = hV/kA_s << 1$$

where

h = Convection coefficient = 25 W/m²·K
V = Volume = $4/3\pi r^3$ m³
k = Thermal conductivity = 400 W/m·K
A_s = Surface area = $4\pi r^2$ m²

$$Bi = [(25\ \text{W/m}^2\cdot\text{K})(4/3\ \pi r^3)]/[(400\ \text{W/m}\cdot\text{K})(4\pi r^2)]$$
$$= [(1/16)(0.025)]/3 = 0.00052$$

The problem fits the restrictions for the lumped-capacity type of problem.

Using the equation given in the *FE Handbook*:

$$Q = hA_s(T - T_\infty) = -\rho c_p V(dT/dt)$$

Integrating the equation and taking the natural log of both sides of the equation:

$$\ln\left(\frac{T - T_\infty}{T_o - T_\infty}\right) = -\left(\frac{hA}{\rho c_p V}\right)t = -\left(\frac{(h)(3)}{\rho c_p r}\right)t$$

$$\ln[(35 - 10)/(300 - 10)] = -[(25\ \text{W/m}^2 \cdot \text{K})(3)]/[(8933\ \text{kg/m}^3)(385\ \text{J/kg} \cdot \text{K})(0.025)\text{m}]t$$

$$t = 46.8\ \text{min}$$

Therefore, the belt must be 46.8 min × 5 m/min = 234 meters long.

HEAT EXCHANGERS

A heat exchanger is a device that is used to transfer heat. There are various types of heat exchangers, such as the counterflow heat exchanger, in which the cold fluid comes in one end and the hot fluid comes in the other end. There is also the parallel-flow heat exchanger where the cold fluid and the hot fluid come in on the same end and go out on the same end. In general, it has been found that the counterflow heat exchanger is significantly more efficient and more economical than the parallel-flow heat exchanger. In both cases the temperature difference between the two fluids is best approximated by the *Logrithimetic Mean Temperature Difference*, or *LMTD*. The equation is shown below:

$$LMTD = [\Delta T_{\text{in}} - \Delta T_{\text{out}}]/[\ln(\Delta T_{\text{in}}/\Delta T_{\text{out}})]$$

The nice thing about this form of the equation, it does not matter what end is called "in" and what end is called "out". One merely has to put the temperature difference of the fluids in the same order for the numerator as well as for the denominator.

Example **14.8**

A heat exchanger is needed to operate under the following conditions: the hot fluid ranges from 300 °C to 200 °C and the cold fluid ranges from 30 °C to 70 °C. If the other conditions such as the required heat transfer and the heat transfer coefficients regarding the heat exchangers are similar, determine the size ratio needed, A_c/A_p.

Solution

The general equation for the heat transfer in a heat exchanger is:

$$Q = UA(LMTD)$$

So for the above conditions one would write:

$$Q = UA_{\text{counterflow}}(LMTD) = UA_{\text{parallel flow}}(LMTD)$$

or

$$A_c/A_p = [U_p(LMTD)_p]/[U_c(LMTD)_c]$$

and

$$U_p = U_c$$

Therefore:

$$\frac{A_c}{A_p} = \frac{\left(\dfrac{(300-20)-(200-70)}{\ln\left(\dfrac{300-20}{200-70}\right)}\right)}{\left(\dfrac{(300-70)-(200-20)}{\ln\left(\dfrac{300-70}{200-20}\right)}\right)}$$

and

$$A_c/A_p = (195.5)/203.9 = 0.958$$

Therefore, the heat transfer area required of the counterflow heat exchanger is 95.8% that of the parallel-flow heat exchanger.

PROBLEMS

The following information may be used for questions 14.1 and 14.2.

A 2-cm-diameter power wire used in a transmission line transmits 14,000 volts and 600 amps. The resistance of the wire is 0.0005 ohms/m and is wrapped with 0.01 cm insulation of $k = 0.07$ W/m · K. The convection coefficient is an average of 25 W/m² · K and the outside temperature is a –10 °C.

14.1 The power in kW dissipated in the wire per km is most nearly:

a. 50 kW
b. 100 kW
c. 130 kW
d. 180 kW

14.2 The temperature in °C of the outside radius of the wire is most nearly:

a. 154
b. 175
c. 186
d. 121

The following information may be used for questions 14.3 and 14.4.

A liquid metal is flowing inside of a 3-cm-diameter pipe, 5 meters long with an entering temperature of 400 °C. The wall temperature is maintained at 500 °C, and the mass flow rate of the metal is 50 kg/min. $\rho = 5500$ kg/m³, $P_r = 0.0025$, $\mu = 0.0025$ N · s/m², $k = 6.2$ W/m · K, $c_p = 0.33$ kJ/kg · K

14.3 The convection coefficient of the liquid metal in W/m² · K is most nearly:

a. 560
b. 760
c. 895
d. 1,290

14.4 The exit temperature of the liquid metal is most nearly:

a. 430
b. 445
c. 465
d. 500

The following information may be used for questions 14.5 and 14.6.

Oil at 300 K is flowing over a flat plate maintained at 400 K. The plate is 0.5 m long and 2 m wide. The velocity of flow is 0.2 m/s. The properties are evaluated at the mean temperature of 350 °C. Properties of the oil are $v = 41.7 \times 10^{-6}\text{m}^2/\text{s}$, $k = 0.138$ W/m · K, $P_r = 546$, $\rho = 853$ kg/m³.

14.5 The Nusselt number is most nearly:
a. 260
b. 295
c. 230
d. 180

14.6 The heat transfer in kW from the plate to the oil is most nearly:
a. 4.5
b. 5.3
c. 6.4
d. 7.1

The following information may be used for questions 14.7 through 14.9.

A constant cross-section, rectangular fin 6 cm long, with a base of 1cm × 2mm is used to dissipate heat from a 400 °C surface. The convection coefficient is $h = 20$ W/m² · K and the conductivity is 40 W/m · K. The temperature of the air surrounding the fin is 35 °C.

14.7 The heat transfer in watts from the fin is most nearly:
a. 6.5
b. 5.8
c. 4.9
d. 7.1

14.8 The effectiveness of the fin is most nearly:
a. 21
b. 28
c. 32
d. 44

14.9 The efficiency of the fin is most nearly:
a. 32%
b. 61%
c. 45%
d. 58%

SOLUTIONS

14.1 d. The best approach to this problem is to write an energy balance equation.

$$\text{Heat generated} = \text{Heat convected away from the wire}$$
$$Q = I^2R = (600)^2(0.0005) = 180\ \text{W/m}$$
$$\text{Total } Q = (180)(1000)/1000 = 180\ \text{kW/km}$$

14.2 a. The temperature of the outside wall of the wire is obtained by using the composite wall equation for a pipe with insulation.

$$Q = \frac{T_w - T_\infty}{\dfrac{\ln(r_o/r_i)}{2\pi kL} + \dfrac{1}{hA}}$$

Rearranging and substituting in the values:

$$180\ \text{W}[\ln(2/1)]/[(2\pi)(0.07\ \text{W/m}\cdot\text{K})(1\ \text{m}) + 1/\{(25\ \text{W/m}^2\cdot\text{K})(2\pi)(0.02\ \text{m})(1\text{m})\}]$$
$$= (T_w - (-10))\,^\circ\text{C}$$
$$T_w = 154.6\,^\circ\text{C}$$

14.3 d. Writing an energy balance equation we get:

$$\text{Heat transfer} = \text{Convection into the fluid from the pipe}$$
$$= \text{Internal energy change of fluid}$$
$$Q = hA(T_w - T_{bave}) = mc_p(T_o - T_i)$$

To find the convection coefficient one has to obtain the Nusselt number, which has to be obtained using the liquid-metal equation for constant-wall temperature listed in the *FE Handbook*.

$$\text{Nu} = 7.0 + 0.025\ \text{Re}^{0.8}\,Pr^{0.8} = 7.0 + 0.025[(\rho DV)/\mu]^{0.8}\,Pr^{0.8}$$

where

$$V = m/\rho A = (50\ \text{kg/min})/(5500\ \text{kg/m}^3)(\pi)(0.03)^2/4\ \text{m}^2$$
$$V = 12.86\ \text{m/min} - 0.214\ \text{m/s}$$

$$\text{Nu} = 7.0 + 0.025[(5500\ \text{kg/m}^3)(0.03\ \text{m})$$
$$(0.214\ \text{m/s})/0.0025\ \text{N}\cdot\text{s/m}^2]^{0.8}(0.0025)^{0.8}$$
$$\text{Nu} = 7.433 - hD/k$$

Solving for h:

$$h = (\text{Nu}k)/D = [(7.433)(6.2\ \text{W/m}\cdot\text{K})]/0.03\ \text{m} = 1288\ \text{W/m}^2\cdot\text{K}$$

14.4 d. Go back to the energy balance equation:

$$Q = mc_p(T_o - T_i) = hA_s(T_w - T_{bave})$$

where

$$T_{bave} = (T_o - T_i)/2 \qquad \text{Assume mean temperature} = 450\,°\text{C}$$

$$(1000\text{ J/kJ})(50\text{ kg/min})(0.35\text{ kJ/kg}\cdot\text{K})(T_o - 400)\,°\text{C}$$
$$= (1288\text{ W/m}^2\cdot\text{K})(\pi)(0.03\text{m})(5\text{m})(500 - 450)\,°\text{C}$$
$$(17{,}500)\text{J/min}(T_o - 400) = 30{,}348\text{ J/s}$$
$$T_o = 504\,°\text{C}$$

14.5 a. In order to determine which equation must be used, the Re must first be calculated.

$$\text{Re} = (LV)/v = [(0.5\text{ m})(0.2\text{ m/s})]/[41.7 \times 10^{-6}\text{ m}^2\text{/s}] = 2398$$

The equation for the Nusselt number may now be selected from the *FE Handbook*:

$$\text{Nu} = 0.648\ \text{Re}^{0.5}Pr^{0.33} = (0.648)(2398)^{0.5}(546)^{0.33} = 258.8$$

14.6 d. Therefore the heat transfer is:

$$Q = hA(T_p - T_{oil})$$

and

$$\text{Nu} = (hX)/k$$

or

$$h = (\text{Nu}k)/X$$

$$h = [(258.8)(0.138\text{ W/m}\cdot\text{K})]/0.5\text{ m} = 71.4\text{ W/m}^2\cdot\text{K}$$

$$Q = (71.4\text{ W/m}^2\cdot\text{K})(0.5 \times 2)\text{m}^2(400 - 300)\,°\text{C} = 7140\text{ watts or }7.14\text{ kW}$$

Note that when k and/or h are used the temperature may be either °C or K because the problem involves a change in temperature.

14.7 a. Select the heat transfer equation for a fin from the *FE Handbook*.

$$Q = \sqrt{hpka_c}\,(T_b - T_\infty)\tanh mL_c$$

where

h = Convection coefficient $\text{W/m}^2\cdot\text{K}$
p = Perimeter m
k = Conductivity $\text{W/m}\cdot\text{K}$
A_c = Cross-sectional area
T_b = Temperature at the base of the fin
T_∞ = Temperature of the fluid
m = $[\text{hp/kA}_c]^{.5}$
L_c = Corrected length = $\text{L} + \text{A}_c/\text{p}$

$$Q = [(20\ \text{W/m}^2\cdot\text{K})(0.024\ \text{m})(40\ \text{W/m}\cdot\text{K})(0.01\times 0.002)\text{m}^2]^{0.5}$$
$$(400-35)\ ^\circ\text{C}\ \tanh mL_c$$

$$Q = (0.0196\ \text{W/K})(365)\text{K(or C)}\tanh[(20)(0.024)/(40)(2\times 10^{-5})]^{0.5}$$
$$[(0.06) + (2\times 10^{-5})/0.024]$$

$$Q = 7.154\ \tanh 1.49 = 6.47\ \text{W}$$

14.8 d. Effectiveness of the fin is equal to the heat transfer dissipated by the fin divided by the heat transfer from the area of the base of the fin if there were no fin.

$$Eff = Q_f/[(hA)(T_b - T_\infty)]$$
$$= (6.47\ \text{W})/[20\ \text{W/m}^2\cdot\text{K}(0.01\times 0.002)\text{m}^2(400-35)]$$
$$Eff = 44.3$$

14.9 b. Efficiency of the fin is equal to the heat transfer dissipated by the fin divided by the heat transfer that the fin would dissipate if the fin were at the base temperature.

$$\eta_f = Q_{act}/Q_{ideal} = (6.47)/[hpL_c(T_b - T_f)]$$
$$\eta_f = (6.47\text{W})/[(20\text{W/m}^2\cdot\text{K})(0.024\ \text{m})(0.0608)(400-35)] = 60.7\%$$

CHAPTER 15

Measurements, Instrumentation, and Controls

OUTLINE

This chapter reviews classical control systems equations in addition to an introduction to modern control systems (i.e., state variable analysis). In the past, the examination did not include questions on state variables; now, however, several questions may well be on this subject. While in the last few years many undergraduate control systems college courses have included an introduction to z-transforms, it is unlikely the examination will include any questions on this subject. One of the more popular texts widely used throughout the United States that does include both classical and modern control is *Automatic Control Systems* by Kuo (Prentice-Hall, 1991), which is written for the undergraduate level. There are many other books that cover both areas, but it's best to stick with your particular college text that you are familiar with if it includes state variables. (Caution: no need to go overboard on all of the details concerning state variables; know the notation and concepts and you should be ok.)

CLASSICAL CONTROL SYSTEMS

More than likely, the bulk of the questions will involve classical control theory, which includes: block diagram formulation and reduction, linear system stability, system error, second order system specifications (with higher ordered systems being approximated by second order ones), frequency response methods, and system compensation. The background for this material requires a good understanding of Laplace transforms, pole-zero maps, and transient response analysis.

BLOCK DIAGRAMS

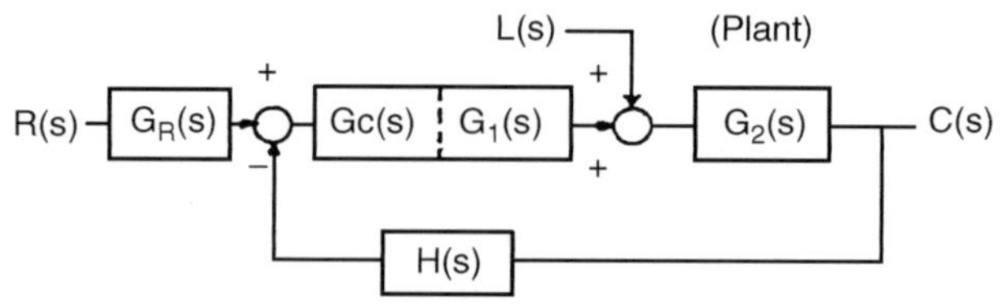

Figure 15.1 A system block diagram with negative feedback

The general form of the block diagrams will usually be some variation of Figure 15.1 where $R(s)$ is the referenced input, $C(s)$ is the controlled output, and $L(s)$ may be a disturbance (or loading) signal. For a simple single loop system, the closed loop system equation is easily found by realizing that the numerator(s) are nothing more than the products of the transfer functions in the forward path(s) of the signals from source to output. The denominator(s) are merely one plus the loop gains (the product of the transfer functions in the loop). If more than one loop exists, then somewhat more formal definitions and techniques are involved; the equations are derived from Mason's rules and flow graph methods. The equation relating the input(s) to the output, using Laplace transforms (rather than time domain relationships—where multiplying now becomes convolving in the time domain) are clearly shown to be (and here, since the complex frequency domain is implied, the "of(s)" notation will be dropped) as follows

$$C = \frac{G_c G_1 G_2}{1 + G_c G_1 G_2 H} G_R R + \frac{G_2}{1 + G_c G_1 G_2 H} L \qquad \textbf{(15.1)}$$

The denominator, $1 + G_c G_1 G_2 H$, if equal to zero defines the poles of the system (i.e., for values of s that make the denominator zero—or C equal to infinity) and, of course, is the characteristic equation of the system. It is clear that if these poles could be easily found, then system stability is known and there would be no need for the Routh-Hurwitz test or for going through the tests for root-locus paths. Also note that this characteristic equation is not a function of either R or L, which means that stability for linear systems is independent of the inputs.

ROUTH-HURWITZ CRITERION FOR STABILITY

The Routh-Hurwitz Criterion gives information as to whether any poles lie in the RHP but does not give their actual location. Another technique, the root locus method, will give the root location in both planes but is far lengthier.

However, for the Routh-Hurwitz method (frequency referred to as just the Routh Test), one needs only to set up an array from the characteristic polynomial and examine the first column of this array to determine stability. This test is used for any equation; but here it refers to the characteristic equation polynomial, which is simplified to

$$a_0 s^n + a_1 s^{n-1} + a_2 s^{n-2} + \dots a_n = 0 \qquad \textbf{(15.2a)}$$

The test then involves making an array out of the coefficients, a's, with additionally created new coefficients, b's. The first two rows, starting with the highest coefficient, a_0, are formed by making the first row of all even values of the a's, the second row is made up of the odd coefficients. The next row of b's are created as shown in equation 15.2b. If all of the resulting coefficients in the first column of a's and b's are the same sign and nonzero, then all of the roots will be in the left-hand portion in the s-plane (i.e., and will be positive),

s^n	a_0	a_2	a_4	a_6	…	…
s^{n-1}	a_1	a_3	a_5	a_7	…	
s^{n-2}	b_1	b_2	…			
…	…					
s^0	b_n					

where

$$b_1 = \frac{a_1 a_2 - a_0 a_3}{a_1}$$

Eq. 15.2b

$$b_2 = \frac{a_1 a_4 - a_0 a_5}{a_1}$$

Consider the following simple example (refer to Figure 15.1 with $G_R = 1$, $G = G_c G_1 G_2$)

$$G = 10(s + 2)/[s(s + 1)(s + 3)], \qquad H = (s + 4)$$

Then,

$$G_{\text{system}} = 10(s + 2)/[s^3 + 14s^2 + 63s + 80].$$

The characteristic polynomial, $s^3 + 14s^2 + 63s + 80$, may be arranged in a Routh array as follows

s^3	**1**	**63**
s^2	14	80
s^1	b_1	
s^0	b_2	

$b_1 = [(14)(63) - (1)(80)]/14 = 57.3$

$b_2 = [(b_1)(80) - (14)(0)]/b_1 = 80$

Since the first column is all positive, the system will be stable in closed loop. (Try this problem for H being unity and G being 3 times larger—you will get an unstable system!) Of course stability may be found by frequency response techniques (i.e., Nyquist criterion, etc.); but, if only the question of stability is to be answered, then use the easy Routh method.

STANDARDIZED SECOND-ORDER SYSTEMS

Although actual systems are usually higher than second-order, the standardized response of this kind of a system to either step or variable frequency input is customarily used as a standard. Their equations are

$$\omega p = \omega res = \omega n\sqrt{1-2\zeta^2}, \text{ the "driven" resonant frequency.} \qquad \textbf{(15.3a)}$$

$$\omega_r = \omega_n\sqrt{1-\zeta^2}, \text{ the "ringing" or damped frequency.} \qquad \textbf{(15.3b)}$$

$$t_p = \pi/\omega_r, \text{ the time-to-first-peak, step response.} \qquad \textbf{(15.3c)}$$

$$C_p = M_p = 1 + e^{-\pi\zeta/\sqrt{1-\zeta^2}}, \text{ the peak magnitude of the step response} \qquad \textbf{(15.3d)}$$

The quantities that describe this response are usually used to describe and approximate higher order ones. From Figure 15.1, where $H = G_R = G_C = G_1 = 1$

and, the plant, $G_2 = K/[s(s + a)]$, results in Figure 15.2. Standardized curves are well known and are published for normalized second-order system; these relationships and notation will simplify problem solutions.

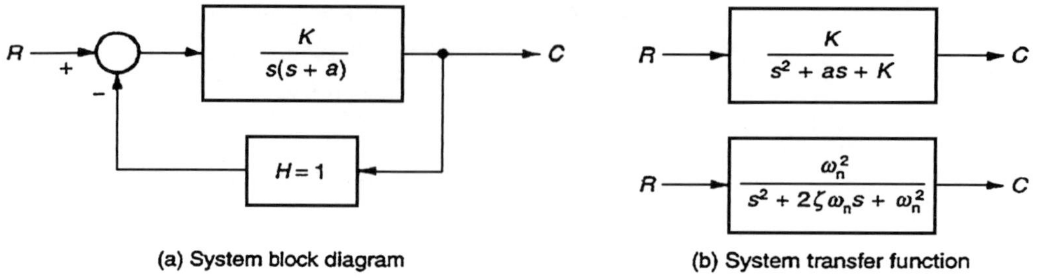

Figure 15.2 An ideal second-order system

As previously noted, the parameter zeta is related to whether the system is under-damped ($\zeta < 1$), critically damped ($\zeta = 1$), or over-damped ($\zeta > 1$). Furthermore, by defining the undamped natural frequency as ω_n (the frequency of a pure sinusoid), one may normalize a time axis with this value and generate the standardized curves for the ideal second-order system. For this simple system and its system transfer function (see Figure 15.2b), the solutions for step responses are as shown in Figure 15.3 with standardized notation. This notation normally becomes part of the system specifications.

Furthermore, the curves are given for the time solution (eliminating the need for finding any inverse Laplace transforms). As zeta approaches zero, the undamped curve becomes sinusoidal and the time-to-the-first peak is obviously π on the normalized axis. Clearly, for damped sinusoids, the peak will always be a number somewhat larger than 3.14; if one remembers this relationship, it will frequently give a check for an approximate answer on the examination! As a short example, assume for Figure 15.2a, $G = 25/[s(s + 2)]$, then the system transfer function is,

$$G_{sys} = 25/(s^2 + 2s + 25) = \omega_n^2/(s^2 + 2\zeta\omega_n s + \omega_n^2)$$

Here $\omega_n = 5$, $\zeta = 0.2$, from the normalized curves, one immediately determines that the percent overshoot is approximately 52% and the peak value occurs at about 3.2 on the normalized time scale; the 3.2 value is equal to $\omega_n t$, therefore $t_p = 3.2/\omega_n = 3.2/5 = 0.64$ seconds.

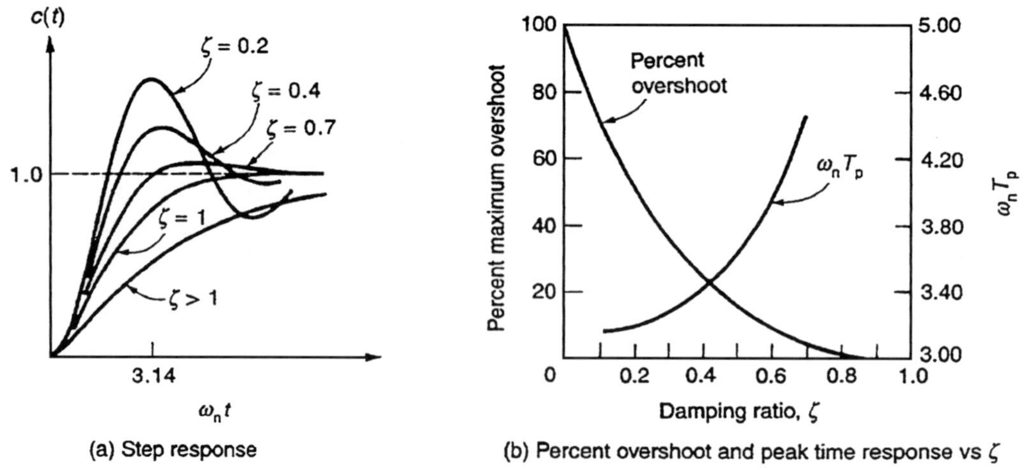

Figure 15.3 Standardized second-order curves

STEADY STATE GAIN AND SYSTEM ERROR

The steady state gain and the error of a system is found from the final value theorem (FVT); making use of the FVT requires that all poles of G are located in the LHP (i.e., negative real parts). In Figure 15.1 assume that $H = G_R = G_C = G_2 = 1$, then gain is given by equation 15.4a and for system error ($R - C$) by 15.4b,

$$\text{dc Gain Constant, } K_p = \lim_{s \to 0} G(s) \quad \textbf{(15.4a)}$$

$$\text{System Error } (R - C) = \lim_{s \to 0} s\left(\frac{R(s)}{1 + G(s)}\right) \quad \textbf{(15.4b)}$$

INSTRUMENTATION AND MEASUREMENT

Most engineering design requires data for use in the analysis phase of the design process. This data must be obtained through measurement and these measurements must be obtained by means of instrumentation. The generic instrumentation system consists of the primary sensing device (which is the transducer), the lead wires, and the secondary instrumentation.

TRANSDUCERS

A transducer is a device that changes one form of energy into another form. The final form of energy is usually an electric signal that is read and interpreted by secondary instrumentation (i.e., a readout device). There are many types of transducers, such as thermocouples, strain gages, piezoelectric crystals, linear variable displacement transformers (LVDT), and a host of others.

THERMOCOUPLES

A thermocouple is a device that operates on the principle of basic material property variations caused by temperature differences. There are three thermoelectric effects caused by these temperature differences. Fortunately, the most significant one is the Seebeck effect, which is the junction potential produced when two dissimilar are joined together. The emf that is produced is directly proportional to the temperature difference between the junction of the dissimilar metals and the junctions of the readout device or the reference temperature.

Example 15.1

A simple thermocouple is used to measure the temperature of boiling water at standard conditions of the environment. Determine the emf output of the thermocouple in terms of the thermoelectric coefficient a, and the temperatures.

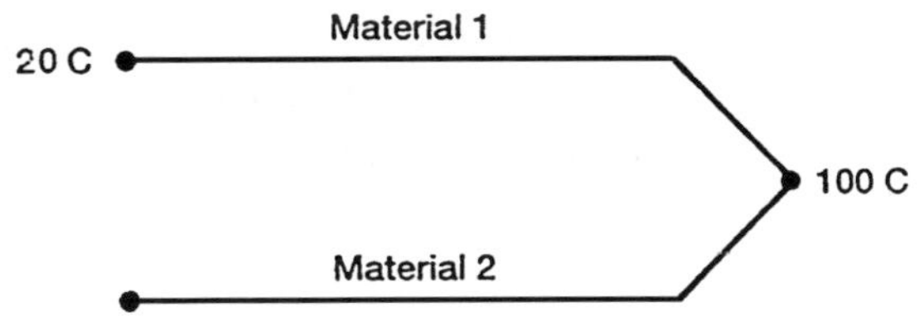

Exhibit 1

Solution

Because the electromotive force produced by the thermocouple is directly dependent upon the thermoelectric effect of the material times the temperature difference, or

$$\text{emf} = \alpha_{\text{mat 1}}(100 - 20) + \alpha_{\text{mat 2}}(20 - 100)$$

and because the thermoelectric effect of material one is opposite in sign to the thermoelectric effect of material two, the equation may be written as

$$\text{emf} = (\alpha_{\text{mat 1}} - \alpha_{\text{mat 2}})(100 - 20)$$

The above equation indicates, then, that the emf of any combination of materials is equal to the summation of the emf s produced by each of the constituents.

Example 15.2

A multijunction thermocouple (as shown hereafter) is used to determine the temperature of some unknown fluid. The measuring device is at a temperature T_1, the bath temperature is at T_2, and the unknown temperature is at T_3.

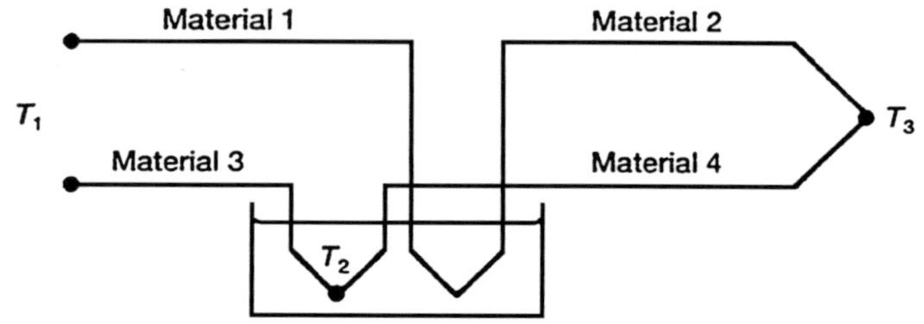

Exhibit 2

Determine whether the temperature of the measuring device's binding posts makes any difference in the output emf.

Solution

Write the equation for the summation of the emf's produced by the several combined materials.

$$\text{emf}_{\text{total}} = \alpha_{\text{mat 1}}(T_2 - T_1) + \alpha_{\text{mat 2}}(T_3 - T_2) + \alpha_{\text{mat 3}}(T_2 - T_3) + \alpha_{\text{mat 1}}(T_1 - T_2)$$

Notice that the first and fourth terms are equal and opposite each other. As a consequence the total emf obtained by the thermocouple is proportional to the difference in temperature of T_2 and T_3. We may therefore say that the binding post temperature has no effect upon the total emf.

STRAIN GAGES

A strain gage is a device that falls in the non–self-generating class of transducers, which fits the following model.

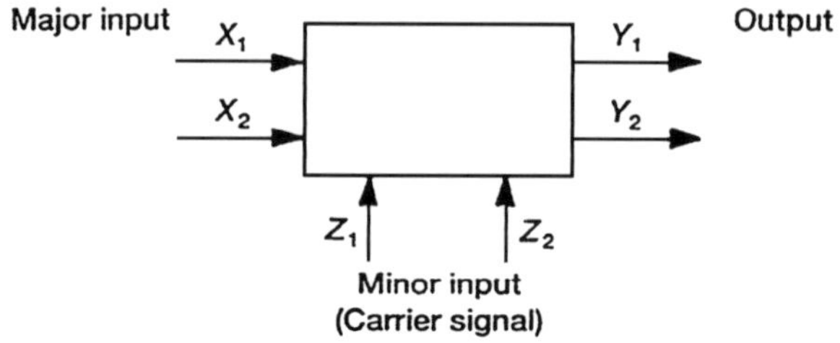

Figure 15.4

The major input is the effect caused by the environmental change that is being measured. This effect then is carried by the excitation/carrier signal and comes out as the output signal, which must then be translated into useful information. Typical strain gages operate on the principle of resistance change due to an applied force. The effect of the resistance change is measures by a bridge circuit, such as the Wheatstone bridge circuit shown in Figure 15.5.

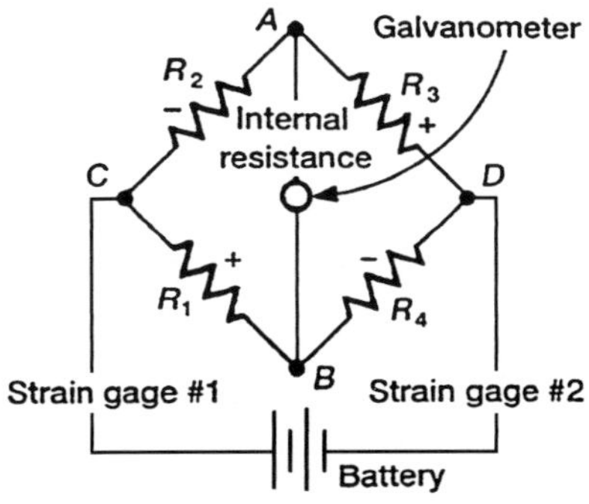

Figure 15.5

The signs of the values of each of the resistances are positive for R_1 and R_3 and negative for R_2 and R_4. These characteristics may be used to obtain various outputs of the bridge circuit.

Example 15.3

Determine the output of the cantilever beam in Exhibit 3, wired up to a simple two-arm bridge.

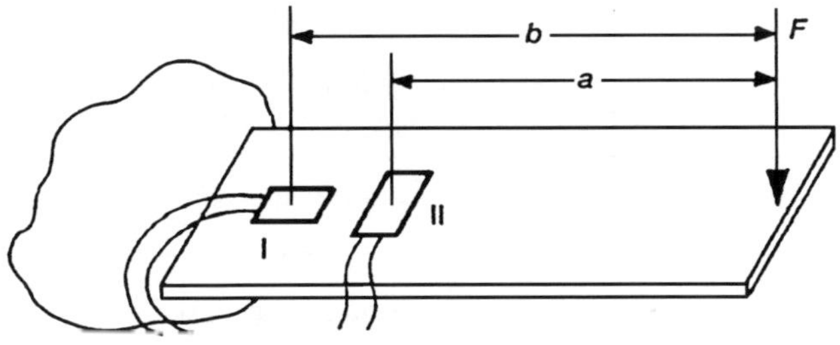

Exhibit 3

Solution

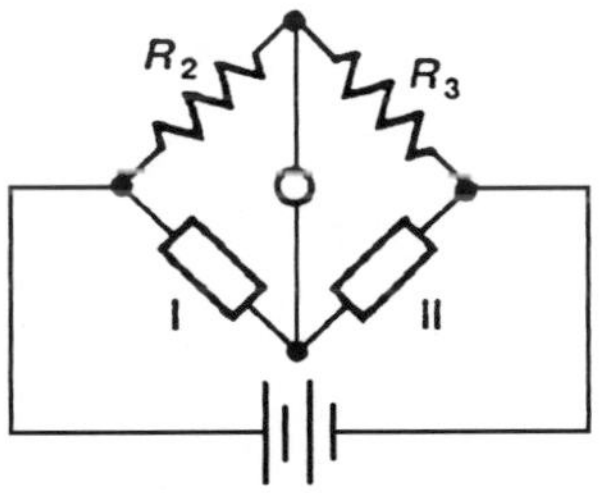

Exhibit 4

Both of the gauges are mounted on the top of the cantilever beam, thus the top (I) gauge is stretched directly proportional to the elongation of the upper fibers of the beam and is connected to the #1 position on the Wheatstone bridge. The second gauge (II) is perpendicular to the first gauge and thus reads a negative effect proportional to Poisson's ratio and is connected to the #4 position on the Wheatstone bridge. The output of the arrangement produces an effect equal to the output of one gauge plus the Poisson's ratio effect.

$$\text{output} = [1 + (a/b)\,\mu]$$

Note that the output is temperature compensated because the temperature will affect each gauge exactly the same and the values are subtracted from each other.

Example 15.4

Determine the output characteristics of the four-strain gauge cantilever beam shown in Exhibit 4. The numbers adjacent to the strain gauges indicate which resistance positions they are attached to.

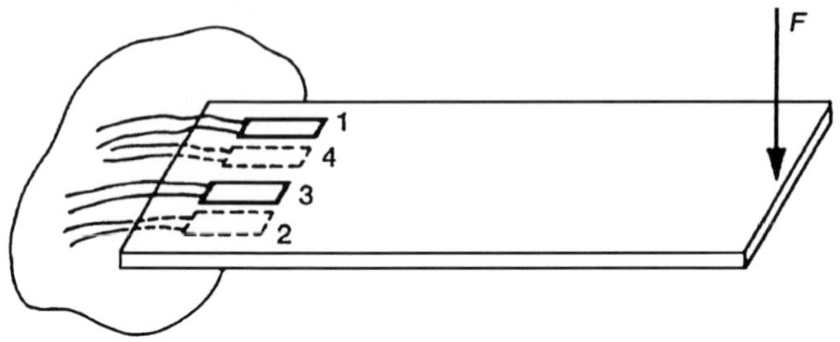

Exhibit 5

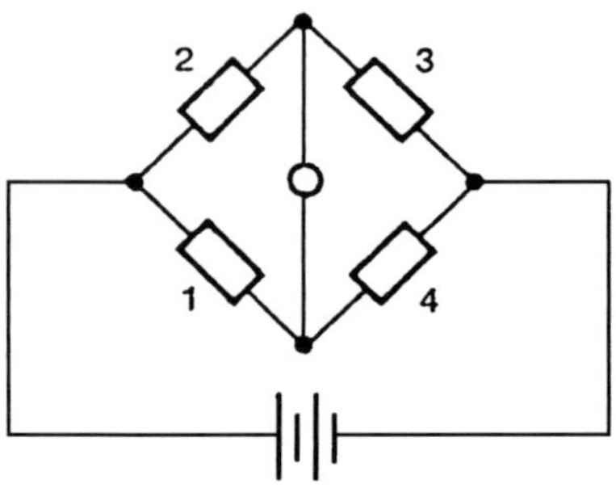

Exhibit 6

Solution

Note that gauges 1 and 3 are mounted directly over gauges 4 and 2, respectively. The characteristics are as follows:

a). Gauge 4 compensates for the temperature effect of 1, and gauge 2 compensates the temperature effect of 3.

b). For the load indicated, gauges 1 and 3 reflect a positive effect and read into positive arms, gauges 2 and 4 (the bottom gauges) reflect a negative effect and read into negative arms giving a net effect of 4 positive strain gauges.

c). Any axial load on the system would be compensated out giving a net effect of zero.

Example 15.5

Two resistance based strain gages are bonded upon two pieces of metal as shown. The first gage is mounted axially on a circular bar and attached to position #1 on the bridge circuit, and the second gauge is mounted on a different piece of metal and is wired into position #4 on the bridge circuit. A half bridge output will produce an output voltage in response to the following loads or environmental changes.

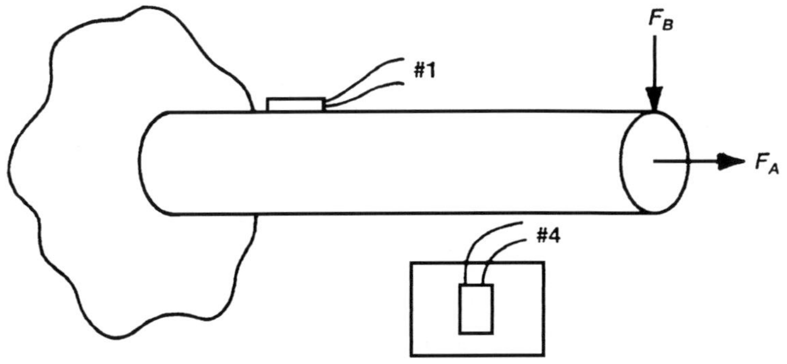

Exhibit 7

Solution

If it can be assumed that both of the gages are mounted in the same environment, there will be no temperature effect. The output then will be a summation of the effects caused by both the axial load F_A and the bending load F_B.

PROBLEMS

15.1 For a typical SISO system (refer to Figure 15.1), assume the transfer functions, $G_C G_1$, have a dc gain greater than unity (and $G_R = G_2 = 1$) and also assume a load disturbance, L. Which has the most effect on the output, C, the input, R, or the input, L?

15.2 If, for system stability, a Routh array is made for the system characteristic equation of 4th order is made, the first column indicates that all coefficients are positive except the 3rd entry of this first column. How many roots lie in the right-hand plane of the s-plane?

15.3 For a second-order control systems with $\zeta < 1$, will the damped natural resonant frequency (i.e., ringing frequency) to a step input be the same the driven resonant frequency?

15.4 A typical simple second-order system may be described by the following linear differential equation

$$\ddot{\theta}(t) + 2\dot{\theta}(t) + 25\theta(t) = 25u(t)$$

Find the values of the system matrix, A, the input distribution matrix, B, and the output matrix, C, for the following state block diagram (Exhibit 15.4). Hint: Define

$$x_1 = \theta(t) \quad \text{and} \quad x_2 = d\theta(t)/dt.$$

Also find the percentage of the maximum displacement of θ for a unit step input to the system (this is the point located on the $\theta(t)$ axis at t_2 in Figure 15.3b).

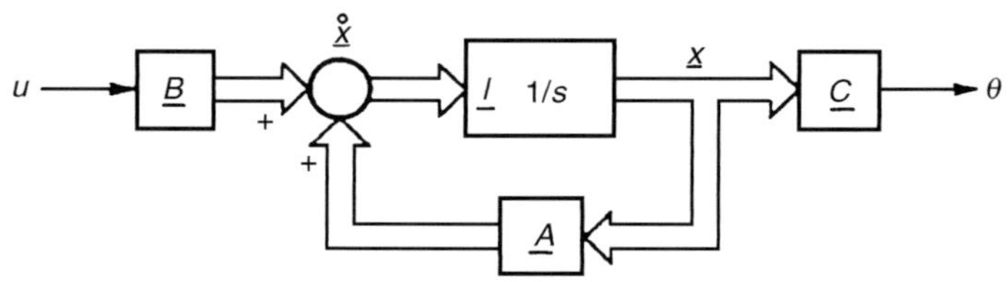

Exhibit 15.4 State block diagram

The following information may be used for problems 15.5 through 15.8.

Four resistance based strain gages are bonded to the metals as shown in Exhibit 15.5. Two are axially bonded to the circular bar and two to the additional piece of metal. All four of the strain gauges are subjected to the same environmental conditions. The gauges may be wired to a Wheatstone bridge in a one-quarter, one-half or a full bridge configuration.

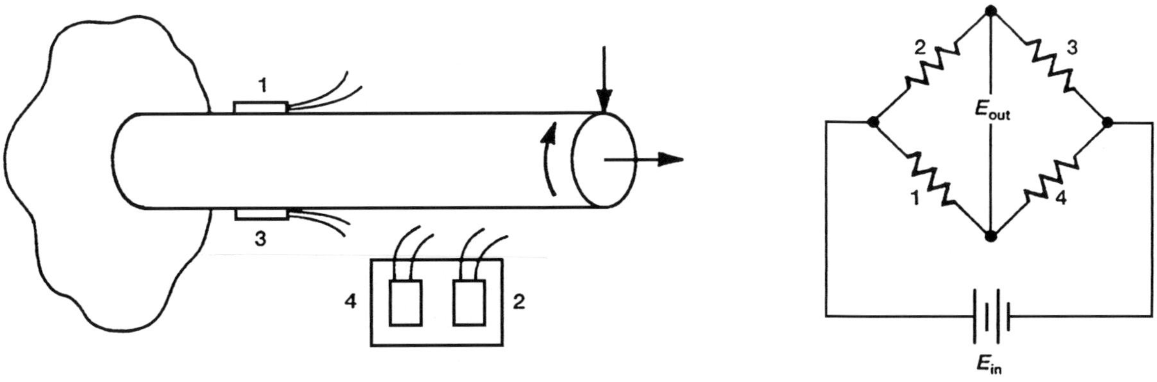

Exhibit 15.5

15.5 A quarter bridge utilizing only gauge 1 will produce an output voltage in response to the following conditions.
a. Torsional load (τ) and temperature effects
b. Axial load (F_A), bending load (F_B), and temperature effects
c. Axial load (F_A) only
d. Axial load (F_A), bending load (F_B), and temperature compensated

15.6 A one half bridge utilizing gauges 1 and 2 will produce an output voltage in response to the following conditions.
a. Axial load (F_A), bending load (F_B), and temperature compensated
b. Axial load (F_A), bending load (F_B), and temperature effects
c. Torsional load (τ) only
d. Torsional load (τ) and axial load (F_A) and temperature effects

15.7 A full bridge has gauges 1 and 3 attached to positions #1 and #3, and gauges 2 and 4 are attached to positions #2 and #4 on the Wheatstone bridge. This system will produce an output voltage that is equivalent to:
a. Axial load (F_A) with temperature compensation
b. Axial load (F_A) bending load (F_B), and temperature affects
c. Absolutely nothing
d. Bending load (F_B) and temperature effects

15.8 A one half bridge has gauges 1 and 3 attached to positions #1 and #4. If the voltage in to the circuit is equal to 10 volts, the strain gauges are 350 ohms, and the resistance change due to the loading is 1%, what is the output voltage?
a. 1.0 volts
b. 0.15 volts
c. 0.05 volts
d. 0.01 volts

SOLUTIONS

15.1 The output C for an input at L is

$$C = [G_2/(1 + G_C G_1 G_2 H)]L,$$

whereas that for R is

$$C = [(G_C G_1 G_2)/(1 + G_C G_1 G_2 H)]R.$$

Because the denominators are the same and the numerator term of $G_C G_1$ of the second equation is >1, obviously the answer is R.

15.2 As two sign changes appear (one from the second to third term, and another from the third to fourth term), it is clear that two roots appear in the RHP.

15.3 Refer to equations 15.3a,b. The term under the radical sign will be smaller for the driven resonant frequency. Therefore $\omega_p < \omega_r$.

15.4

$$dx_1/dt = 0x_1 + 1x_2$$
$$dx_2/dt = -25x_1 - 2x_2 + 25u(t)$$

Then

$$\mathbf{A} = \begin{bmatrix} 0 & 1 \\ -25 & -2 \end{bmatrix}, \quad \mathbf{B} = \begin{bmatrix} 0 \\ 25 \end{bmatrix}, \quad \mathbf{C} = \begin{bmatrix} 1 & 0 \end{bmatrix}$$

(Note: Mixed notation of time and frequency is common in the state variable world.)

Because this is a second-order system it is obvious that $2\zeta\omega_n = 2.0$, and $\omega_n^2 = 25$, $\omega_n = 5$, therefore $\zeta = 0.2$; and, from the standardized second-order curves (see Figure 15.3), the percent maximum overshoot is found to be approximately 52%.

15.5 **b.** Because this is a quarter bridge, only one strain gage is connected to the bridge circuit. The strain gauge in question is mounted at the top of the bar and thus will be stretched both when the bar is pulled in tension by the axial load and when the bar is bent due to the bending load. Notice also that there is no compensating gauge, thus any temperature effects will confound the output voltage.

15.6 **a.** The one half bridge uses the top gage and a temperature compensating gauge.

15.7 **d.** The full bridge has both strain gauges mounted in the positive positions of the bridge circuit. Therefore, the axial load will be measured but the bending will be subtracted out as well as any of the temperature effects.

This chapter was written by Lincoln D. Jones and also appears in *Electrical Engineering: FE Exam Preparation*.

15.8 **c.** The one half bridge sends the top gauge to the positive arm of the bridge and the bottom gauge to the negative arm of the bridge. The results are temperature compensation and bending shall be measured. The axial load effect is zero. Using the equation for

$$\text{G.F.} = (\Delta R/R)/(\Delta L/L) = (\Delta R/R),$$

and

$$E_{out} = E_{in} \times (\varepsilon_1 - \varepsilon_4)/2$$

we get from the *FE Handbook*.

$$E_{out} = (10 \text{ volts})(3.5/350)/2 = .05 \text{ volts}$$

CHAPTER 16

Mechanical Design

OUTLINE

Mechanical design, or machine design as it is often called, combines all of the disciplines in the study of mechanical engineering. A typical design problem requires first the development of a mechanism to perform the desired function. Next it has to be demonstrated that the device will perform without failing and without imperiling the user or the public. Attention should also be paid to the cost of the device including material, manufacturing, operating, and maintenance costs.

Example 16.1

Design a square key for a gear wheel attached to a 2.5 cm diameter shaft to transmit 10.0 kw of power at a speed of 1750 rpm. The gear wheel is 1.5 cm thick. If the key is to be made of steel with an allowable shear stress of 200 MPa, what size key should be used if safety factor of three is required assuming shear stress controls?

Solution

The allowable stress on the key material would equal

$$200/3 = 66.67 \text{ MPa.}$$
$$10.0 \text{ kW} = 10.0 \text{ kJ/s and a joule is a N} \times \text{m.}$$
$$\text{Power} = 2\pi \times (\text{rev/s}) \times \text{Torque} = \text{watts}$$

The shaft rotates at a speed of 1750/60 = 29.17 rev/s and transmits energy at the rate of

$$10{,}000 \text{ J/s} = F \times \text{m/s}$$
$$F \times (\pi \times 0.025 \times 29.17) = 2.2910 \times F = 10{,}000 \text{ N} \times \text{m/s}$$

where

F = force on key in newtons
F = 10,000 watts/2.2910 = 4364.9 N

The required shear area of key = 4365/66.67 × 10^6 = 65.47 × 10^{-6} m^2

The key would be 0.015 meters long (the thickness of the gear wheel) so the key would be 0.00006547/0.015 = 0.00436 meters, or the required key width would = 4.36 mm

INTERFERENCE FIT

Interference fits are some times desirable to provide a force fit. In such a case, the O.D. of the shaft will be larger than the I.D. of the mating part. For small interferences, this may be accomplished by forcing the shaft into the smaller diameter hole. For larger interferences, it will be necessary to expand the hole's diameter and/or reduce the shaft diameter by heating or cooling as, for example, shrinking rims onto railroad car wheels.

Example **16.2**

If a car wheel is 60 cm in diameter and a rim is 0.9 mm smaller in I.D., to what temperature would it be necessary to heat the rim to provide a 0.100 mm clearance to assemble the rim on the wheel? Both members are made of steel. Dry ice at −79 °C is available to cool the wheel. The coefficient of thermal expansion for steel is 11.7 × 10^{-6} m/(m × °C)

Solution

The total ΔD required is 1.00 mm. For an ambient temperature of 20 °C, the wheel diameter would reduce

$$\Delta T = 90\,°\text{C} \quad \Delta D = 600 \times 99 \times 11.7 \times 10^{-6} = 0.6950 \text{ mm}$$

The rim diameter would then have to expand 0.3050 mm, which means that the increase in the temperature of the rim is

$$\Delta T = 0.3050/(600 \times 11.7 \times 10^{-6}) = 43.4\,°\text{C}$$

The rim would have to be heated to 63.4 °C.

Example 16.3

To examine Example 16.2 a little further, assume all deformation took place in the rim after assembly; that is, the wheel was strong enough not to deform after the rim was installed. What would be the stress in the rim after it had been shrunk onto the wheel?

Solution

$$\Delta D = 1 \text{ mm} \quad \text{Circumference} = \pi D \quad \text{so} \quad \Delta C = \pi \times 1.0 \text{ mm}$$
$$\varepsilon_c = \pi/(600\ \pi) = 1/600 \quad \text{stress} = E \cdot \varepsilon_c = 2.1 \times 10^{11} \times 1/600 = 350 \text{ MPa}$$

Example 16.4

A 10-kW motor operates at speed of 1750 rpm. What diameter shaft should be used to transmit this power using a steel with a tensile strength of 572 MPa and a tensile yield strength of 496 MPa? Use a safety factor of 2.5, and assume the shear yield strength of the shaft steel is 0.60 times the yield strength in tension.

Solution

$$\text{Shear stress} = Tq \times c/J$$

where

$J = \pi D^4/32$, the polar moment of inertia and $c = D/2$

so

$$S_s = 16 \times Tq/(\pi \times D^3)$$
$$\text{Watt} = \text{joule/s} = \text{N} \times \text{m/s} \quad 1750 \text{ rpm} = 29.17 \text{ rev/s}$$
$$\text{Allowable shear stress} = 496 \times 0.60/2.5 = 119.04 \text{ MPa}$$
$$Tq = 10{,}000 \text{ watts}/(29.167 \text{ rev/s} \times 2 \times \pi) = 54.57 \text{ joules}$$
$$D = [16 \times Tq/(S_s \times \pi)]^{1/3} = 0.0133 \text{ m or } 1.33 \text{ cm}$$

Example 16.5

If it is desired to use a 2.00 cm O.D. hollow shaft for the conditions given in the above example, what would be the size of the bore to produce the same shear stress in the hollow shaft?

Solution

Since the stress is directly proportional to the polar moment of inertia, the J of the hollow shaft would have to be the same as for the solid shaft calculated previously.

$$J = \pi \times D^4/32 \quad \text{since } J_1 = J_2 \quad D^4 = D^4_{OD} - D^4_{ID}$$
$$1.33^4 = 2^4 - D^4_{ID} \quad D_{ID} = (16 - 3.129)^{1/4} = 1.894 \text{ cm},$$

which gives a wall thickness of 0.529 mm.

Example 16.6

In Example 16.5, a 1.500 cm solid shaft was selected. The power from the motor was delivered to the driven machine through a 60.00 cm diameter pulley. If the pulley were attached to the shaft at a distance of 15.00 cm from a bearing support, what would be the bending stress in the shaft (a), and what would be the total shear stress, τ, (b)?

Solution

The torque output of the motor is 54.57 J (N × m). The force acting on a belt at a distance of 60.00/2 cm would equal (1.00/0.300) × 54.57 = 181.9 N. The maximum bending moment would equal

$$0.15 \times 181.9 = 27.29 \text{ J}$$
$$S = M \times c/I,$$

where I is the transverse moment of inertia, and since $J = I_x + I_y$ it will equal one-half of the polar moment of inertia for a circular section.

$I_x = I_y$ so $I_x = \pi D^4/64 = 2.485 \times 10^{-9}$ m^4 $S = (27.29 \times 0.015/2)/(2.485 \times 10^{-9})$

S equals 82.36 MPa, bending stress (a) $J = 2 \times I_x = 4.970 \times 10^{-9}$

$$S_s = (54.57 \times 0.0150/2)/(4.970 \times 10^{-9}) = 82.35 \text{ MPa}$$

$$\text{Maximum shear stress } \tau = \left[s_s^2 + (S/2)^2\right]^{1/2} = 92.07 \text{ MPa}$$

The relationship for the maximum shear stress can be derived from the diagram of Mohr's Circle given in the *FE Handbook*, using only two tensile stresses (i.e., let $\sigma_2 = 0$). It would be a good idea for an examinee to study the section on Mohr's Circle before the exam if not already acquainted with it.

COMPOSITE BEAM

Example 16.7

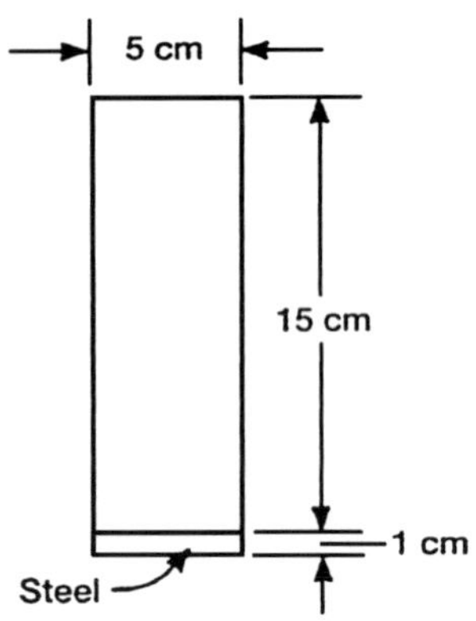

Exhibit 1

How much stiffer would an aluminum beam 5.00 cm wide by 16.00 cm deep be if the bottom 1.00 cm were replace by steel? See Exhibit 1.

$$E_S = 2.1 \times 10^{11} \text{ Pa} \qquad E_{Al} = 6.9 \times 10^{10} \text{ Pa}$$

Solution

$$n = E_S/E_{Al} = 210/69 = 3.0435,$$

so the equivalent all aluminum beam would be as shown in the figure, a top section 5.00 cm by 15.00 cm with a 1.00 cm by n × 5 = 15.218 cm flange section at the bottom. The stiffness of a beam is proportional to its moment of inertia. $I = b \times h^3/12$ for a rectangular section. For the initial all aluminum beam

$$I = 5 \times 16^3/12 = 1706.67 \text{ cm}^4$$

If the bottom 1.00 cm was replaced by a steel strip, the location of the neutral axis would shift. Taking area moments about the top of the beam

$$5 \times 15 \times 7.5 + 1 \times 15.218 \times 15.5 = (75 + 15.218) \times y$$
$$y = 798.079/90.218 = 8.846 \text{ cm}$$

from the top of the beam is the location of the neutral axis of the composite beam.

The new I, I_c, can be determined by means of the parallel axis theorem, with the new neutral axis 8.846 cm from the top of the composite beam.

$$I_c = 5 \times 15^3/12 + [75 \times (8.846 - 7.5)^2] + (15.218 \times 1^3/12) + (15.218)(15.5 - 8.846)^2 = 2217.19 \text{ cm}^4$$
$$I_c/I = 2217.19/1,706.67 = 1.30$$

So the composite beam would be 30% stiffer.

Example 16.8

A swing is to be designed for a public park. The support beam will be 3.00 m high, and the swing will be designed to hold a 100-kg person. What should be the strength of the lines holding a swing seat? Since this is a public park, it has to be assumed that the swing will be subjected to uses (or misuses) far beyond what any one might imagine, so a safety factor of five has been decide upon.

Solution

The maximum load on the swing seat will occur at the bottom of the arc and will equal the gravitational force exerted by the swinger and the centrifugal force, F_c.

$$Fc = m \cdot v^2/R$$

Assume the swinger goes as high as the support beam, then

$$v = /(2gh) \quad \text{or} \quad v^2 = 2 \times 9.8066 \times 3.0 = 58.84 \text{ m}^2/\text{s}^2$$

assuming the swing seat is at the level of the ground when at rest.

$$F_c = 100 \times 58.84/3 = 1961 \text{ N}$$

The total force down would equal

$$1961 + 100 \times 9.8066 = 2942 \text{ N}$$

Assume that a swinger would hold onto only one side of the swing. The strength of each support line should equal 5 × 2942 = 14,710 N.

Example 16.9

A punch press is to be designed to punch 2.50 cm diameter holes in 1.0-cm thick mild steel plate. The shear strength of the steel is 250 MPa. If 15% is added to the shearing force as a friction allowance, what force must the punch exert?

Solution

$$\text{Shear area} = 0.025 \times \pi \times 0.01 = 0.0007854 \text{ m}^2$$
$$\text{Force} = 250 \times 10^6 \times 785.4 \times 10^{-6} = 196.4 \text{ kN}$$
$$\text{Add allowance for friction} \quad F = 1.15 \times 196.4 = 225.9 \text{ kN}$$

Example 16.10

How much energy must be expended to punch one hole? The design calls for the punch press to be operated by a 400-watt motor running at a speed of 1750 rpm. The motor will drive a flywheel through a hydraulic clutch which is 85% efficient. Energy from the flywheel will be used to provide the required power to operate the press. The requirement is that the flywheel should not slow down by more than 20% to punch a hole. What should the moment of inertia of the flywheel be if rotates at a speed of 250 rpm before its energy is used to operate the punch press?

Solution

$$\text{Energy} = F \times \text{distance} = 225.9 \text{ kN} \times 0.01 \text{ m} = 2259 \text{ J}$$

KE of a rotating flywheel equals $\frac{1}{2} I\omega^2$

$$\omega_1 = (250/60) \times 2\pi = 26.180 \text{ rad/s}$$
$$\omega_2 = (200/60) \times 2\pi = 20.944 \text{ rad/s}$$
$$\Delta KE = \tfrac{1}{2} \times I \times (685.4 - 438.7) = 123.4 \times I = 2259 \text{ J}$$
$$I = 18.31 \text{ kg} \times \text{m}^2 \ (\text{N} \times \text{m} \times \text{s}^2)$$

Example 16.11

If a 65.0-cm diameter, steel flywheel of constant thickness is used, how thick should it be? The specific gravity of steel is 7.85.

Solution

$$I = \tfrac{1}{2}MR^2 = 18.31 \quad R = 0.325 \text{ m} \quad M = 2 \times I/R^2 = 2 \times 18.31/0.1056 = 346.8 \text{ kg}$$

The density of steel is 7850 kg/m^3, so the volume of the flywheel should be 346.8/ 7850 = 0.0442 m^3. The area = 0.3318 m^2 and required thickness = 13.33 cm.

MECHANICAL SPRINGS

The most common type of spring is the helical spring, made with round wire. Such springs are made to withstand either tensile or compressive loads. The spring rate is a function of the number of active coils, wire diameter, spring diameter, and the torsional modulus of the wire material. For steel wire the torsional or shear modules, G_s, equals 8.3×10^{10} Pa (see *Reference Handbook*). The apparent stress in the wire equals $S_t = 8 \times F \times D/(\pi d^3)$ equals torsional stress. However, this relationship does not include the effect of the curvature of the wire, which increases the stress in the wire by a factor K_s, which equals

$$(2C + 1)/(2C)$$

where

C = (mean coil diameter)/(wire diameter)

Example **16.12**

A coil spring 4.0 cm in outside diameter made of 5.0-mm wire is to support a mass of 45 kg. What is the stress in the wire?

Solution

The load would equal 45.0 × 9.807 = 441.3 N. D is the mean spring diameter

From the *Handbook*,

$$\tau = K_s \times (8FD)/(\pi d^3)$$

where

d = wire diameter = 0.005 m
F = load 45 × g = 441.3 N
D = mean spring diameter = 0.040 − 0.005 = 0.035 m
K_s and C are as given previously, C = 0.035/0.005 = 7

The apparent stress,

$$S_t = (8FD)/(\pi d^3) = 8 \times 441.3 \times 0.035/(\pi \times 0.125 \times 10^{-6}) = 314.6 \text{ MPa}$$
$$K_s = (2 \times 7 + 1)/(2 \times 7) = 1.071$$
$$\text{The actual maximum stress} = 314.6 \times 1.071 = 336.9 \text{ MPa}$$

DEFLECTION

The deflection, ΔH, of a constant diameter circular spring is given by the relationship

$$\Delta H = 8FD^3N/Gd^4$$

where

ΔH = deflection
N = number of active coils

This comes from the equations given in the *Handbook*, such as $k = d^4 \times G/(8D^3N)$ and $F = kx$, where the deflection $x = F/k$, which gives, for the deflection, the relationship for the change in height given previously.

Example **16.13**

For the spring in the previous example, a compression spring 4 cm in diameter made of 5-mm diameter wire, determine the number of active coils required for the spring to deflcct 6.0 cm under the load of 45 kg.

Solution

The calculation to determine the deflection of the spring equals

$$\Delta H = 8FD^3N/Gd^4$$

where

$F = 45 \times g = 441.3$ N
$\Delta H = 0.060 = 8 \times 441.3 \times 0.035^3 \times N\,(8.3 \times 10^{10} \times 0.005^4)\ 3.113 = 0.151 \times N$ or number of active coils = 20.6, say 21.

For a compression spring the ends would probably have squared and ground ends giving one dead coil at each end, so the total number of coils would equal 23.

SPRING LOAD

The load on a bolt supplied by a spring plus an additionally applied load will equal the load due to the force of the spring plus the additional load until the bolt has stretched (increased in length) enough to relieve the spring load. Another way of stating this is that the force acting on a bolt which is loaded by a spring plus an applied load equals the force applied by the spring before the additional load was applied plus the force of the applied load minus the reduction in the force applied by the spring due to the elongation of the bolt. When the assembly consists of an ordinary coiled spring the strain of the bolt due to the additionally applied load will be small and can usually be neglected. However, when the spring load is supplied by a composition gasket or a copper or aluminum washer or gasket, the reduction in the force applied by the spring can be appreciable.

Example **16.14**

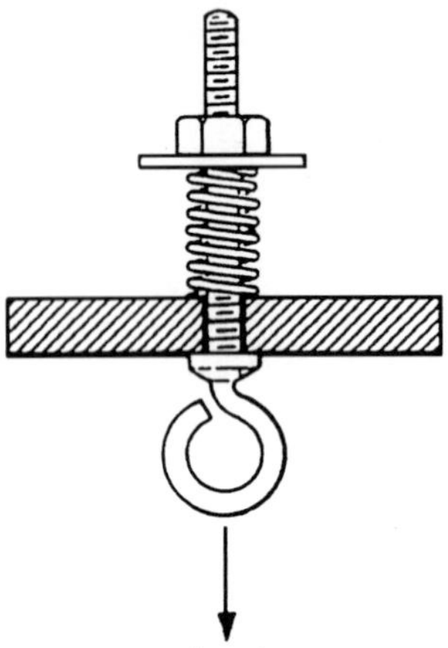

Exhibit 2

An eyebolt support is attached to a rigid steel beam as shown in Exhibit 2. A very stiff compression spring is used to preload the eyebolt. The spring is loaded by means of a nut screwed down on the end of the eyebolt until it exerts a force of 5000 N. What would be the force exerted on the shank of the bolt if a mass of 50 kg were attached to the eyebolt?

Solution

The force of gravity acting on the support by a mass of 500 kg would equal $(500 \times g)$ or

$$F = 500 \times 9.8066 = 4908.3 \text{ N}$$

This is less than the upward force exerted by the spring, so the load held by the bolt would equal 5.0 kN.

Example **16.15**

What would be the load held by the bolt if a mass of 800 kg were attached to the eye of the bolt?

Solution

$$F = 800 \times g = 7845 \text{ N}$$

This is greater than the initial load on the spring, so the spring would compress further, and the load on the bolt would be 7.85 kN.

When a bolt is tightened (preloaded) against a soft substanceb such as the condition shown in Exhibit 3, and a load P is applied, it will increase the load on the bolt, but not by the amount of the newly applied load. Stress and strain are proportional, the length of the bolt will increase, but the compression of the soft material will decrease by the same amount that the length of the bolt increases. The most common example of this is the case of a soft seal wherein the gasket material is compressed to seal the joint. The definition of a soft material is any material that has a lower coefficient of elasticity than that of the bolt, which is usually steel.

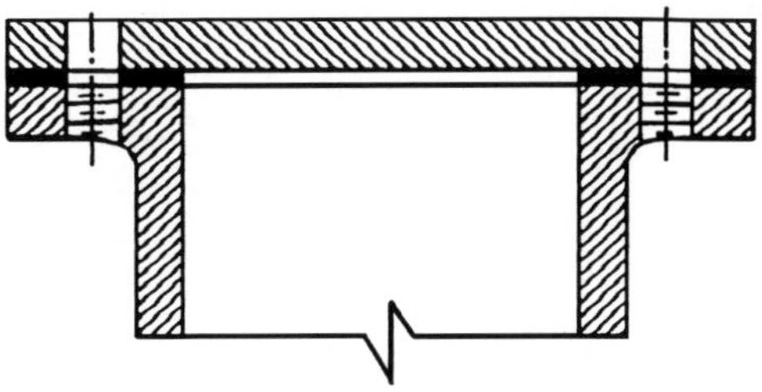

Exhibit 3

Example 16.16

Assume that a soft gasket which has a modulus of elasticity equal to 10.0×10^8 Pa is used to seal a pressure vessel as shown in the figure. The net area of the gasket (subtracting area of bolt holes) is 88 cm^2 and its thickness is 3.0 mm. The gasket is compressed by 12 steel bolts 15 cm long and 1.0 cm in diameter. Assume the joint has been carefully made up and that each bolt holds exactly 1/12 of the applied load. The preload on each bolt is 11.0 kN. If the force added to the joint due to pressurization of the vessel equals 54.0 kN, how much would the force acting on each bolt increase?

Solution

The bolt would strain an additional amount to withstand the added load, but the force exerted by the gasket would be reduced since the gasket would expand as the bolt lengthened. Both the bolt and the gasket can be treated as springs. The increase in the bolt load would equal the increase in bolt length, Δ, times the bolt-spring constant minus the gasket length decrease, Δ, times the gasket-spring constant. The magnitude of the added load would thus amount to

Added load $= \Delta \times (k_B + k_G)$ and the increase in the load held by the bold would equal $\Delta \times k_B$. The bolt-spring, k_B, constant equals $E_B A_B / L_B$, where E_B equals the modulus of elasticity of the bolt material, A_B, equals the cross-sectional area of the bolt, and L_B equals the free length of the bolt.

$$k_B = 2.1 \times 10^{11} \times 0.00007854/0.150 = 11.00 \times 10^7 \text{ N/m}$$

similar for the gasket, with $A_G = 0.0088/12 = 0.000733$ m^2

$$k_G = 10.0 \times 10^8 \times 0.000733/0.0030 = 24.43 \times 10^7 \text{ N/M}$$
$$4.5 \text{ kN} = \Delta \times (1.10 + 2.44) \times 10^8$$
$$\Delta = 4{,}500 \text{ N}/(35.43 \times 10^7 \text{ N/m}) = 127.0 \times 10^{-7} \text{ m}$$

The increase in the load on each head bolt would equal $127.0 \times 10^{-7} \times 11.0 \times 10^7 = 1397$ N, which is considerably less than the load increase on the head of 4500 N or 54,000/12 = 4500 N/bolt.

It might be easier for some engineers who are completely unacquainted with the metric system to convert all the quantities to U.S. units, work the problem, and then covert the answer to metric units. There is an excellent table of conversion factors in the *Handbook* supplied to all examinees. Using the foregoing problem as an example

Bolt: Diameter = 0.3937 in Area = 0.1217 in^2 $L = 5.9055$ in

$$E = 2.1 \times 10^{11}/6895 = 30.456 \times 10^6 \text{ lb/in}^2$$

Gasket: Area = 1.1367 in^2 per bolt $L = 0.1181$ in.

$$E = 10^9/6895 = 145{,}033 \text{ psi}$$

Then

$$k_B = 0.6276 \times 10^6 \text{ lb/in} \quad \text{and} \quad k_G = 1.395 \times 10^6 \text{ lb/in}$$

The added head load per bolt = 4500/4.448 = 1017 lb

$\Delta = 1017/(2.023 \times 10^6) = 503 \times 10^{-6}$ in or 0.0005 in the increase in the bolt load would equal $0.6267 \times 503 = 315.2$ lb, and converting to newtons, $315.2 \times 4.448 = 1402$ N, which is the same answer as obtained by the metric system calculation, considering slight errors in the conversion factors.

O-RING SEAL

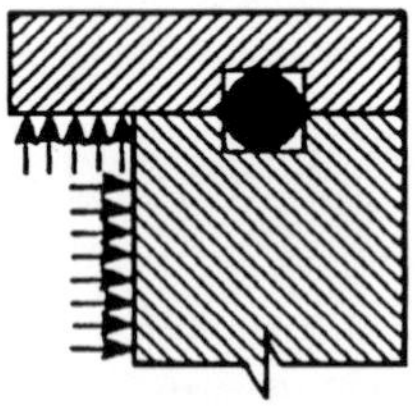

Figure 16.1

It is interesting to note that there would have been no increase in the load on the vessel-head bolts in Example 16.16 if an O-ring seal had been used; see Figure 16.1. This is a hard joint since the O-ring would not apply any load on the boltsb and thus the total load would equal only the original preload, at least until the pressure load exceeded the preload and the joint leaked.

FATIGUE AND ENDURANCE LIMITS

Metal parts that are subjected to alternating loads will often fail when the applied stress is well below the strength of the material as determined by a static tensile test. Such failure usually starts at a point of discontinuity such as a change in cross-section or a local surface blemish where the localized stress is greater than the endurance strength of the material. Large forgings such as, for example, a large high-pressure triplex pump, can fail because of a rough machining mark at a point that is subjected to a stress that is well below the static strength of the material. A fatigue failure is a progressive failure and occurs over a period of time. Allowance is made for such reduced operating stress through the application of a stress-concentration factor that is applied to the endurance limit of the metal (usually steel) of which the part is made. In an extreme case—a high strength, heat-treat steel operating in seawater—the endurance limit may drop to as low as 11% of the ultimate strength.

The stress-concentration factor depends upon a number of different factors and includes the condition of the surface (roughness, polishing the surface of parts subjected to repeated stress variations has resulted in a considerably longer life, even showing an endurance strength approaching the static strength of a steel for a mirror-polished part), a size factor (sudden changes in size, like a change in diameter or a hole in the part), load factor (whether the loading reverses from tensile to compressive or varies entirely in tension), temperature factor, the medium in which the part is to be used (air, water, sea water), and other factors which may affect a particular component.

The endurance limit of a given metal is usually obtained by a rotating test wherein a sample of the metal that has been machined to a specific size and shape (see Figure 16.2) has been rotated through many revolutions. Each revolution applies an equal tensile and compressive loading. When failure occurs, the stress and number of cycles to failure are recorded and plotted on a graph. The endurance limit is that stress at which the specimen will withstand an infinite number of such cycles. The stress and number of cycles to failure are plotted on an S-N diagram, usually a semi-log diagram. The curve of stress vs. failure usually flattens out for steels at about one to ten million cycles or more, but may not flatten out for aluminum even at many million cycles. The endurance limit may be defined as that stress at which a standard test specimen may withstand an infinite number of cycles of reversed stress. For ferritic carbon steels the endurance limit will usually average out to about 60% of the ultimate tensile strength, for pearlitic steels to about 40% of the UTS, and for plain carbon martensitic steels about 25%. For alloy martensitic alloy steels the ratio will average out to about 35%.

Figure 16.2

For design purposes this stress must be further modified (reduced) to take into account the stress-concentration effects previously noted. One design method is that of the modified Goodman relation

$$S_a/S_e + S_m/S_{ult} = 1/\mathrm{N}$$

where

N = safety factor
S_a = the alternating tensile stress
S_e = endurance limit
S_m = mean tensile stress, and
S_{ult} = ult. tensile strength

Another design method that is more conservative is with the Soderberg relation in which the value for the yield strength is used in place of the ultimate strength, giving

$$S_a/S_e + S_m/S_y = 1$$

where

S_y = yield strength

For shear or torsional stresses the Goodman relationship becomes

$$S_{vs}/S_{ns} + S_{ms}/S_{ults} = 1/N$$

where

S_{vs} is the variable shear stress
S_{ns} the endurance limit in shear
S_{ms} the mean shear stress
S_{ults} the ultimate shear strength
N the safety factor

The Soderberg relationship would be the same except the ultimate shear strength would be replaced by the shear yield stress S_{ys}.

Example 16.17

A medium-carbon steel with $S_u = 590$ MPa and $S_y = 380$ MPa is to be used for the design of a shaft that is to be subjected to varying torques ranging from +340 J to −115 J. What diameter should be specified if a safety factor of 2.0 is desired? The shaft will have an ordinary smooth surface, that is, it will not be a polished surface, nor will it have any rough machine marks.

Solution

Data for shearing stress for a specific steel may not be available, so in that case it must be estimated. The yield in shear is about 0.60 times yield in tension, so for this steel

$$S_{ys} = 0.60 \times 380 = 228 \text{ MPa}.$$

The endurance limit in tension is approximately half the ultimate or about 295 MPa. The endurance limit in torsion for steel has been found to equal approximately 60% in both tests and by distortion energy failure theory for polished specimens, so

$$S_{ns} = 0.60 \times 295 = 177 \text{ MPa}.$$

But this value must be further modified for surface effect and size effect. The surface effect for hot-rolled steel with a strength of 590 MPa is found from a graph to equal some 60%, and the size effect for a rod with a diameter between 13 and 50 mm equals some 85%. This gives a value for $S_{ns} = 0.60 \times 0.85 \times 177 = 90.3$ MPa.

The mean torsional loading equals

$$T_m = [340 + (-115)]/2 = 112.5 \text{ J}$$

and the variable torque equals

$$T_v = [340 - (-115)]/2 = 227.5 \text{ J}$$

The corresponding stresses are then, from

$$S = Tc/J' \text{ where } J'/c = \pi D^3/16, \text{ so } S = 16T/(\pi D^3)$$
$$S_{ms} = 112.5 \times 16/\pi D^3 = 572.96/D^3 \text{ Pa}$$
$$S_{vs} = 227.5 \times 16/\pi D^3 = 1159.65/D^3 \text{ Pa}$$

Using the Soderberg relationship for a more conservative design gives

$$1/N = S_{ms}/S_{ys} + S_{vs}/S_{ns}$$
$$1/2 = 572.96/(228 \times D^3) + 1159.65/(90.3 \times D^3)$$
$$D^3 = (5.03 + 26.68)/1{,}000{,}000$$
$$D = 0.0317 \text{ m or } 31.7 \text{ mm}$$

Example 16.18

Estimate the endurance of a 4.0 cm diameter cold-drawn steel bar for 99% reliability, if it is made of steel with an ultimate tensile strength of 400 MPa.

Solution

The term *endurance limit* is generally considered to mean the bending endurance limit unless otherwise specified. From a table of recorded data we find that the size factor for this diameter equals 0.869. Also, for this size bar, the surface finish factor will equal 0.84. From another source it is found that the reliability factor for 99% reliability equals 0.814. Since we do not have any endurance test results for this particular steel we can estimate its endurance limit as 50% of the ultimate strength, or

$$S'_e = 0.5 \times 400 = 200 \text{ MPa}$$

combining these factors gives—

$$S_e = 0.869 \times 0.84 \times 0.814 \times 200 = 0.594 \times 200 = 119 \text{ MPa}$$

LOW CYCLE FATIGUE

There is another type of fatigue termed *low cycle fatigue* in which failure occurs when a metal is stressed beyond its yield point in tension and then compressed beyond its yield point when the loading is reversed, thus causing the item to first to elongate plastically and then to shorten plastically, or vice versa.

Example 16.19

One place where such failures have occurred is in the exhaust system of a diesel engine. If the exhaust pipe of a diesel engine is rigidly held at a point one meter from its attachment to the exhaust manifold, what would happen to the stainless steel pipe when the diesel was operated at full load for a full shift?

Solution

The stainless steel exhaust pipe of the diesel near the exhaust manifold would probably become a bright red and its temperature would approach some 600 °C or more. The coefficient of thermal expansion for stainless steel is $16.92 \times 10^{-6}/°C$. The pipe would tend to expand in length

$$\varepsilon = 580 \times 16.92 \times 10^{-6} = 0.00981 \text{ m/m}$$

This would imply a compressive stress of

$$S = E\delta = 0.00981 \times 1.93 \times 10^{11} = 1894 \text{ MPa.}$$

However, the yield strength of the stainless steel at room temperature is only 275.8 MPa, and this would decrease to as low as 130 MPa at the higher temperature, so the fixed-end piece of pipe would compress. The time at temperature is long enough for the pipe to creep to an equilibrium condition at an applied stress of 130 MPa. At the higher temperature

$$\varepsilon = 130 \times 10^{6}/1.93 \times 10^{11} = 0.000674 \text{ m/m}$$

so the pipe would yield plastically— $0.00981 - 0.000674 = 0.00914$ m/m or 9.14 mm over the one meter fixed length. When the diesel engine was turned off, the pipe would cool and would attempt to shrink an amount equal to

$$\varepsilon = 580 \times 16.92 \times 10^{-6} = 0.00981 \text{ m/m,}$$

but this would indicate a stress of

$$S = E \times \varepsilon = 0.00981 \times 1.93 \times 10^{11} = 1894 \text{ Mpa.}$$

However, as previously noted, the yield strength of the stainless steel at room temperature is only 275.8 MPa, so the pipe would elongate plastically. This sequence of events would occur every time the diesel engine was operated, and after a number of such cycles (many less than the endurance limit) the pipe would rupture. It is interesting to note that such a design was actually constructed and such a failure did occur.

SCREW THREADS

Example 16.20

A power screw to raise a load is designed with a single square thread and a major diameter of 35.0 mm. The pitch is 5.0 mm. The bearing surface of the nut is 50 mm in diameter where it fits against the end of the structure to provide the resisting force. It is to be used to raise a load of 815 kg. It is assumed that the coefficient of frication for all surfaces equals 0.10. What would be the efficiency of the power screw, and what would be the stress in the screw?

Solution

The pitch is 5 mm, so the width and depth of a square thread would equal 2.5 mm. The mean diameter would then equal OD minus the thread height or, $d_m = 35.0 - 2.5 = 32.25$ mm. When the screw is rotated, each turn will raise the mass 5.0 mm. The distance of the screw wedge would equal the circumference at the circumference at the pitch diameter or $\pi \times d_m$ and the thread lead angle

$$\lambda = \tan^{-1}/(10/2\pi\, d_m) = 2.825° \qquad \lambda = 2.825°.$$

Probably the easiest way to analyze this problem is with the aid of the friction angle, μ, where the friction angle adds to, or subtracts from the mechanical angle to give an equivalent frictionless system. The friction angle equals the $\tan^{-1}$ of the friction factor. A friction factor of 0.10 gives a friction angle equal to $\tan^{-1} 0.10 = 5.711°$. The total equivalent frictionless wedge angle would equal $(\mu - \lambda)$ or 8.536° for raising the mass and $(\mu - \lambda) = 2.886°$ for lowering the mass. It might be noted here that when the friction angle is greater than the lead angle, the screw will be statistically stable, but if the lead angle is greater than the friction angle the force of the load will turn the screw and the load would lower unless the screw were held to keep it from turning. The torque required to turn the screw, considering only thread pitch and thread friction would equal

$$Tq_1 = (d_m/2) \times \tan(\mu - \lambda) \times F$$
$$\text{from } 2\pi Tq = \pi d_m \times \tan(\mu - \lambda) \times F$$

Torque would also be necessary to overcome the frictional resistance to turning due to the friction force acting between the bearing surface of the nut and the support structure.

$$Tq_2 = F \times \mu \times (D_{od} + D_{id})/2 = F \times 0.1 \times (50 + 35)/2 = 4.25F$$

the units in this case equal mm $\times$ N so to state it in joules or Nm $Tq_2 = 0.00425 \times \text{F}$ joules

The force, F, exerted by a load of 815 kg would equal

$$F = 815 \times 9.807 = 7993 \text{ N so } Tq_2 = 33.970 \text{ Nm}$$
$$Tq_1 = (32.25/2) \text{ mm} \times \tan 8.536° \times 7993 \text{ N}$$
$$= 19{,}345 \text{ Nmm, or } 19.345 \text{ Nm} = 19.345 \text{ J}$$

The total torque required to raise the load would equal

$$\text{Torque} = Tq_1 + Tq_2 = 19.345 + 33.970 = 53.315 \text{ Nm} = 53.315 \text{ J}$$

Similarly, torque required to lower the load would equal

$$3.970 + 257.774 \times \tan 2.866° = 46.875 \text{ J}$$

The efficiency of the power screw would equal $F \times$ pitch/2π Tq so

$$e = 7993 \times 0.005/(2\pi \times 53.315) = 11.93\%$$

If a ball bearings thrust bearing were placed under the head of the bolt, assuming the same dimensions and a friction factor of 0.01, the torque i_2 would be reduced to 3.397 J and the total torque would equal 22.742 J. The efficiency would increase to

$$7993 \times 0.005/(2\pi \times 22.742) = 27.97\%$$

The shear area per thread would equal π times the root diameter (major diameter -2 times thread height) times the width of the thread

$$\text{Shear area} = \pi \times (0.035 - 0.005) \times 0.0025 = 2.356 \times 10^{-4}\text{m}^2$$

Stress for one thread of engagement

$$S_s = 7993/(2.356 \times 10^{-4}) = 33.93 \text{ MPA}$$

Which is a relatively low shear stress for steel, so, since there would be more than one thread of engagement, the screw thread is more than strong enough for the job.

BEAM DEFLECTION

Example 16.21

A walkway is made from two steel tubes (pipes) 10 cm O.D. by 1.0 cm wall, spaced 60 cm apart and covered with wooden strips. It is used to span a 10 m ditch. If the mass of the wooden strips is ignored, what would be the maximum deflection if a man plus a wheelbarrow with a combined mass of 250 kg were to go across the bridge?

Solution

The maximum deflection would occur when the man was at the center of the span. The deflection

$$\delta_{max} = P \times \ell^3/(48 \times E \times I)$$

This can be derived from the relationship for deflection for a simply supported beam with $b = \ell/2$ in the *FE Handbook*. I for one tube equals $(\pi/64) \times (OD^4 - ID^4)$

$$I = 0.04909 \times (1.000 \times 10^{-4} - 0.4096 \times 10^{-4})$$
$$I = 2.898 \times 10^{-6} \text{ per tube so } \quad I_{total} = 5.7966 \times 10^{-6} \text{ m}^4$$
$$P = 250 \times 9.807 = 2452 \text{ N}$$
$$\delta_{max} = 2{,}452 \times 1000/(48 \times 2.1 \times 10^{11} \times 5.7966 \times 10^{-6}) = 4.20 \text{ cm}$$
$$\delta_{max} = 4.20 \text{ cm}$$

Example 16.22

What is the maximum stress in the span in Example 16.21?

Solution

This is a simply supported beam so the load on the support at each end would equal $P/2$ and the maximum moment would occur at the center of the span.

$$M_{max} = \ell/2 \times P/2 = S = M \times c/I$$
$$= (10 \times .0500 \times 2{,}452/4)/(5.7966 \times 10^{-6}) = 52.88 \text{ MPA}$$

PROBLEMS

16.1 A factor of safety is
a. yield stress divided by design stress.
b. ultimate stress divided by design stress.
c. ultimate stress divided by yield stress.
d. maximum expected load divided by average load.

16.2 A swing is to be designed for a public park. The support beam will be 3.00 m high, and the swing will be designed to hold a 100-kg person. What should be the strength of the lines holding a swing seat? Since this is a public park, it has to be assumed that the swing will be subjected to uses (or misuses) far beyond what any one might imagine, so a safety factor of five has been decided upon.
a. 12.2 kN
b. 14.7 kN
c. 13.9 kN
d. 15.6 kN

16.3 A design requirement for a pogo stick is a spring that will compress 18 cm when a 100-kg person drops one meter. What is the required spring constant?
a. 65.7 kN/m
b. 71.4 kN/m
c. 68.8 kN/m
d. 75.2 kN/m

16.4 It is proposed to use a standard gasoline engine with a compression ratio of 6:1 for an industrial operation. A test is conducted to determine the efficiency of the engine. Gasoline with a density of 0.70 and a heating value of 44.432 MJ/kg is used for the test. A torque device measures 200 N at a distance of one meter at an engine speed of 1500 r/min. The engine uses 3¾ liters of gasoline in 15 min. while developing that torque. Which of the following most nearly equals the thermal efficiency of the engine?
a. 19.8%
b. 28.3%
c. 24.2%
d. 32.1%

16.5 A transformer core is to be built up of sheet-steel laminations, using two different strip widths, x and y, so that the resultant symmetrical cross section will fit within a circle of diameter D. As plant engineer you are asked to determine the values of x and y in terms of D so that the cross section of the core will have maximum value. See Exhibit 16.5. The correct values for the X and Y dimensions are as follows:

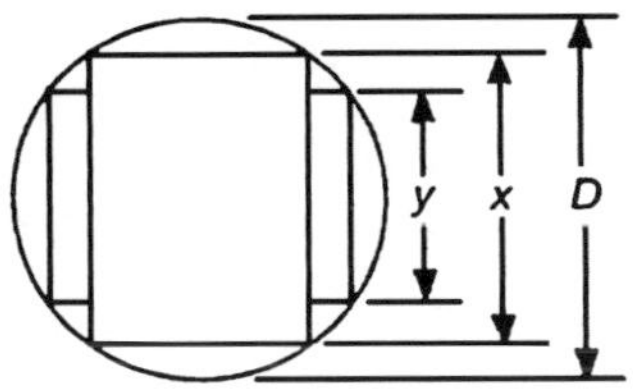

Exhibit 16.5

a. $y = 0.53D$ $x = 0.85D$
b. $y = 0.75D$ $x = 0.62D$
c. $y = 0.48D$ $x = 0.87D$
d. $y = 0.58D$ $x = 0.78D$

16.6 A jackscrew having threads spaced 3.2 mm apart, 312.5 threads/m, is to be operated manually by a handle. The point at which the handle is gripped to move is 60 cm from the axis of the screw. If friction is disregarded, what force must be applied at the grip on the handle to just raise a load of 1050 kg?

a. 7.6 N
b. 9.0 N
c. 8.4 N
d. 9.7 N

16.7 A proposed design contains three wires, each having a cross-sectional area of 1.30 cm^2 and the same unstressed length of 5.0 meters at 20 °C, which hang side by side in the same plane. The outer wires are copper. The middle wire, equidistant from each of the others, is steel. Given: (see *Handbook*)

$$E \text{ for steel} = 2.1 \times 10^{11} \text{ Pa} \quad \text{and} \quad E \text{ for copper} = 1.17 \times 10^{11} \text{ Pa.}$$

If a mass of 450 kg is gradually picked up by the three wires, the part of the load carried by the steel wire would most nearly equal which of the following?

a. 225 kg
b. 213 kg
c. 150 kg
d. 275 kg

16.8 The assembly described in the above problem will be located in a test bay in which the temperature can rise. Referring to the data in the above problem, to what temperature would the system have to rise for the steel wire to support the entire load?

The coefficient of linear thermal expansion for steel

$$\alpha_{\text{Steel}} = 11.7 \times 10^{-6}/°\text{C}$$

The coefficient of linear thermal expansion for copper $\alpha_{Cu} = 16.7 \times 10^{-6}/°C$

Other data are as given in problem 16.7.

a. 48 °C
b. 56 °C
c. 53 °C
d. 61 °C

16.9 A penstock 3.0 m in diameter carrying 7.0 cubic meters of water per second to a generating plant bends through a 45° angle just before it enters the plant. Which of the values below most nearly equals the force that the support post at the bend must be designed to withstand? Assume that the discharge is free (exhaust pressure equals zero gauge.) See Exhibit 16.9.

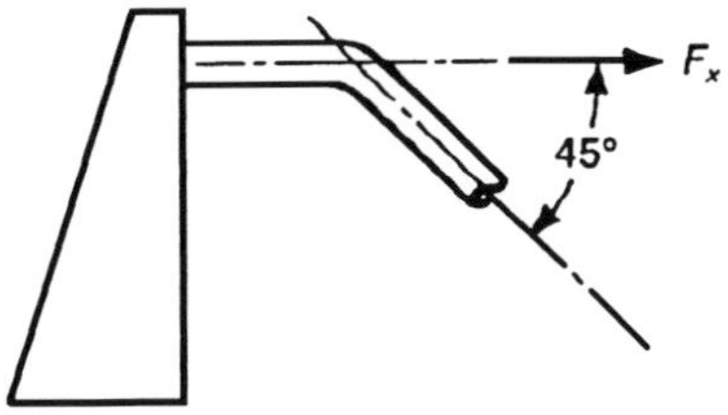

Exhibit 16.9

a. 5260 N
b. 6650 N
c. 5950 N
d. 6930 N

16.10 A mine elevator cage with a mass of 2.5 tons empty is descending at a rate of 32 km/hr when the hoist mechanism jams and stops suddenly. If the cable is steel with a cross-sectional area of 6.5 cm^2 and the cage is 1500 meters below the hoisting drum when the drum jams, which of the following most nearly equals the stress in the cable? (*E* for hoist cable equals 84,000 × 10^6 Pa, neglect the weight of the cable)

a. 165 MPa
b. 188 MPa
c. 173 MPa
d. 226 MPa

16.11 Shown in Exhibit 16.11 is a vertical tank, open to the atmosphere at the top. At the bottom of the tank is a valve that initially is closed, and the tank is filled to a level *L* with a liquid of zero viscosity. How long would it take the tank to empty if some one should mistakenly open the valve and fluid escaped through a round edged orifice ($C = 1.0$) which has an area 0.010 times the area of the tank ($A/a = 100$) and the height of the liquid in the tank is two meters?

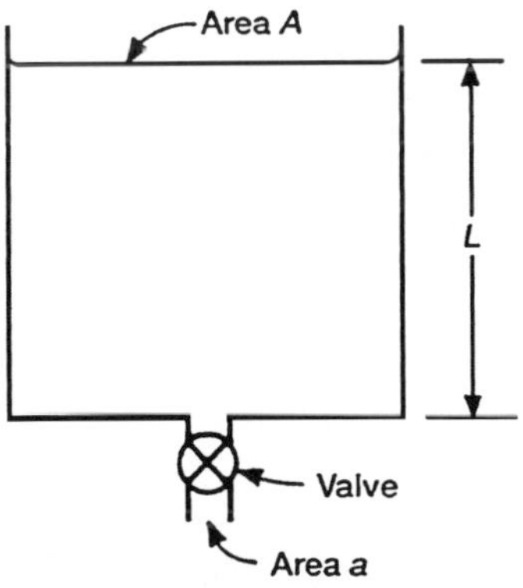

Exhibit 16.11

a. 97 s
b. 76 s
c. 82 s
d. 64 s

16.12 A coupling connecting a motor to a pump transmits 3.36 kW at 1750 rpm. Two 6.35-mm diameter pins hold the coupling together. What stress must the pins withstand when the motor transmitting the design power at the rated speed?

See Exhibit 16.12.

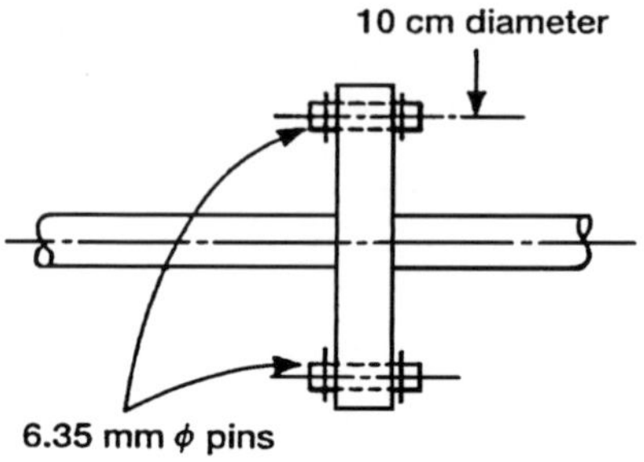

Exhibit 16.12

a. 6.89 Mpa
b. 5.79 Mpa
c. 6.21 Mpa
d. 4.89 MPa

16.13 A water jet is used on a mixing tank as shown in Exhibit 16.13. When the gate valve is full open the pressure gage reads 276 kPa. At what velocity is the water discharging through the jet when the water surface is 61 cm above the centerline of the jet? Assume no head loss in the line or jet.

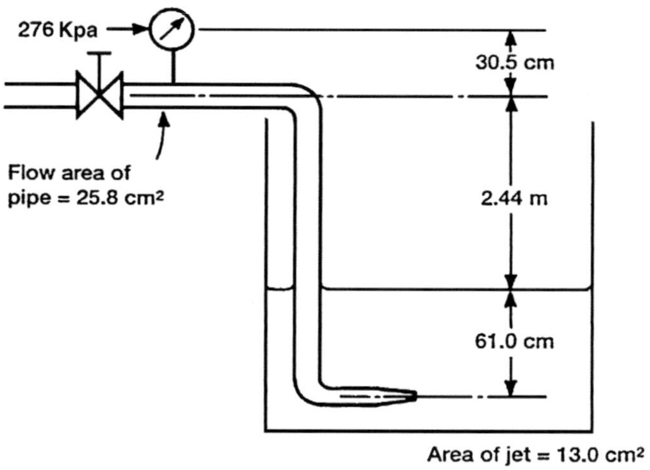

Exhibit 16.13

a. 28.5 m/s
b. 23.8 m/s
c. 25.2 m/s
d. 22.1 m/s

16.14 Determine which of the following is most nearly equal to the amount of heat which would be lost through a window which measured 60 cm by 75 cm if the inside temperature were 25 °C greater than the temperature outside. The window glass is 3.0 mm thick, and the surface coefficients, h, and glass conductivity, *k* are as follows

$$h_i = 8.52 \text{ W/m}^2 \times \text{K}$$
$$h_o = 34.07 \text{ W/m}^2 \times \text{K}$$
$$k = 1.04 \text{ W/m} \times \text{K}$$

a. Q = 58 W
b. Q = 71 W
c. Q = 64 W
d. Q = 75 W

16.15 A rigid bar is to be held by three wires of length *L*, with a modulus of elasticity, *E*, and a cross-sectional area, *A*. They are to be spaced a distance of a apart as shown in Exhibit 16.15a. All three wires are the same length. Which of the following is the maximum held by a wire if a force, *P*, is applied at a distance of *a*/2 from one wire as shown in the figure? Neglect the mass of the bar.

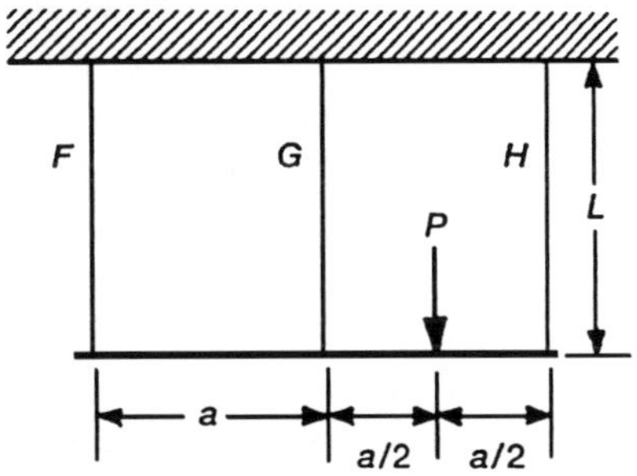

Exhibit 16.15a

a. *F* = P/2
b. *F* = 5P/7
c. *F* = 7P/12
d. *F* = 3P/10

16.16 The plant engineer in a steam power plant that burns a hydrocarbon fuel makes an Orsat analysis of the stack gas. The analysis shows the following:

$$11.3\%\ CO_2,\ 4.5\%\ O_2,\ \text{and}\ 84.2\%\ N_2$$

The discharge temperature of the stack gas is 315 °C and the pressure is 29.65 kPa gauge. The specific heats at constant pressure for the products of the combustion are: CO_2, 0.846 kJ(kg × K) O_2, 0.918 kJ/(kg × K), N_2, 1.04 kJ/(kg × K) and H_2O (steam), 1.989 kJ/(kg × K).

The fuel is a member of the methane series ($C_nH_{2n}+_2$)

Which of the following is the chemical formula of the fuel?

a. C_6H_{14}
b. C_4H_{10}
c. C_5H_{12}
d. C_2H_6

16.17 Using the data from Problem 16.16, how much excess air in percent, was used in the combustion process?

a. 10%
b. 25%
c. 17%
d. 32%

16.18 Determine what size load can be lifted with a hydraulic jack that has a 3.8 cm diameter ram if the diameter of the piston has a diameter of 1.6 cm and the force on the jack handle is limited to 135 N. The jack handle mechanism has a mechanical advantage of 15:1. Neglect any friction that may act.

a. 1162 kg
b. 1294 kg
c. 1239 kg
d. 1331 kg

16.19 An engineer is required to lay out a linear distance of 320 meters. The tape to be used is a 100 meter tape which is actually 99.992 meters long at 20 °C and is 99.999 meters long at 26 °C. Which of the following is the distance to be indicated by the tape to give a true distance of 320 meters if the ambient temperature on the day the measuring takes place is 33 °C?

a. 320.000 meters
b. 320.023 meters
c. 319.974 meters
d. 319.981 meters

16.20 A threaded bolt has a root diameter of 1.6 cm. The limiting stress for the bolt material is 344.8 MPa. Which of the following is the maximum load that can be safely supported by the bolt?

a. 7980 kg
b. 7450 kg
c. 7750 kg
d. 7070 kg

16.21 If a body has a mass of 100 kg in air, but tips a scale at 25 kg when suspended in fresh water, its specific gravity is closest to which of the following?

a. 1.52
b. 1.27
c. 1.33
d. 0.98

16.22 A torsion member is made of steel 15 cm in diameter. To save mass it is proposed to replace it with a hollow bar of the same type of steel with a 10 cm diameter hole through its center. What should the outside diameter of the hollow bar be to provide the same torsional stiffness?

a. 15.7 cm
b. 16.7 cm
c. 16.2 cm
d. 17.3 cm

16.23 Air is compressed rapidly in an insulated cylinder to 1/4 its initial volume. What will a gage show as the final pressure if the initial pressure is 13.79 kPa gage?

a. 575 kPa
b. 660 kPa
c. 620 kPa
d. 700 kPa

16.24 A solid steel flywheel 150 cm in diameter rotates at a speed of 600 rpm about an axis through its center. If it has a mass of 1460 kg, how much kinetic energy does it possess?

a. 692 kJ
b. 810 kJ
c. 775 kJ
d. 853 kJ

16.25 The surface temperatures of the faces of a 15 cm thick slab of material are 77 °C and 20 °C. What is the rate of heat loss, Q, through the slab if the thermal conductivity, k, of the slab equals 1.039 W/m × K?

a. 365 W/m^2
b. 432 W/m^2
c. 395 W/m^2
d. 451 W/m^2

SOLUTIONS

16.1 **b.** The factor of safety is defined as the ultimate stress divided by the design stress.

16.2 **b.** The maximum load on the swing seat will occur at the bottom of the arc and will equal the gravitational force exerted by the swinger and the centrifugal force, F_c.

$$F_c = m \times v^2/R$$

Assume the swinger goes as high as the support beam, then

$$v = /(2gh) \quad \text{or} \quad v^2 = 2 \times 9.8066 \times 3.0 = 58.84 \text{ m}^2/\text{s}^2$$

assuming the swing seat is at the level of the ground when at rest.

$$F_c = 100 \times 58.84/3 = 1961 \text{ N}$$

The total force down would equal

$$1961 + 100 \times 9.8066 = 2942 \text{ N}$$

Assume that a swinger would hold onto only one side of the swing.

The strength of each support line should equal 5 × 2942 = 14,710 N.

16.3 **b.** The energy to be absorbed equals

$$100 \times 9.8066 \times (1 + 0.18) = 1{,}57.18 \text{ J}$$
$$1{,}157.18 = ½ \times k \times 0.18^2 = 71.431 \text{ kN/m}$$

16.4 **c.** The ordinary gasoline engine operates on the Otto cycle, see *Handbook*.

The power output of the engine equals (1500/60) × 2π × 200 = 31,400 Nm/s = 31.4 kW or 31.4 kJ/s.

The heat supplied equals 3.75 × 0.70 × 44.432 × 10^6/(15 × 60) = 129,600 J/s.

$$\text{Efficiency} = 31.4/129.6 \times 100 = 24.2 \text{ percent.}$$

16.5 **a.** Refer to Exhibit 16.5. The system is symmetrical, so

$$x^2 + y^2 = D^2 \qquad A = xy + y(x - y)$$
$$x = /(D^2 - y^2)$$

Set $dA/dy = 0$ then

$$A = 2xy - y^2 = 2y/(D^2 - y^2) - y^2 = 2/(y^2D^2 - y^4) - y^2 \; dA/dy$$
$$= (2/2)(2yD^2 - 4y^3)/(y^2D^2 - y^4) - 2y$$

which reduces to $yD^2 - 2y^3 - 2y/(y^2D^2 - 4y^4) = 0$

$$yD^2 - 2y^3 = 2y/(y^2D^2 - y^4)$$

divide through by y and square both sides

$$D^4 - 4D^2y^2 + 4y^4 = y^2D^2 - y^4$$

which reduces to

$$D^4 - 5D^2y^2 + 5y^4 = 0$$

solve for y^2 using the general solution for a quadratic equation as given in the *Handbook*

$$y^2 = 0.5D^2 \pm 0.224D^2$$

which gives $y = 0.851D$ or $y = 0.525D$

Substitution in the equation for x gives that

$$x = 0.525D \quad \text{for } y = 0.851D \quad \text{and} \quad x = 0.851D \quad \text{for} \quad y = 0.525D$$

16.6 b. Work per revolution equals $F \times 2\pi \times 0.60 = 3.770$ N $\times$ m or 3.770 J

Work also equals $1050 \times g \times 0.0032 = 33.95$ J so $F = 33.95/3.77 = 9.01$ N

16.7 b. The load held by a wire will equal

$$\text{Stress} \times \text{area} = E \times \varepsilon \times A$$

The elongations of the three wires will be the same, so the strains will be equal. So

$$450 \times g = (2 \times E_{cu} + E_{\text{Steel}}) \times \varepsilon \times 0.00013$$
$$4413 = (2.34 + 2.1) \times 10^{11} \times \varepsilon \times 0.00013 = \varepsilon \times 577.2 \times 10^6$$

which gives $\varepsilon = 7.646 \times 10^{-6}$ m/m

Each copper wire will thus hold

$$1.17 \times 7.646 \times 10^{-6} \times 0.00013 = 1163 \text{ N or } 118.6 \text{ kg}$$

The steel wire will hold

$$2.1 \times 7.646 \times 10^{-6} \times 0.00013 = 2087 \text{ N} \quad \text{or} \quad 212.8 \text{ kg}$$

16.8 c. In order for all the load to be held by the single steel wire, the strain due to the load alone would have to equal

$$450 \times g/(E \times A) = 4413/(2.1 \times 10^{11} \times 0.00013)$$
$$= 4413/(27.3 \times 10^6) = 0.000162 \text{ m/m}$$

The total strain in the steel wire will equal $0.000162 + \alpha_{\text{Steel}}\,\Delta T$

Strain in copper wire due to thermal expansion alone will equal $\alpha_{\text{Cu}}\,\Delta T$

The two strains will be equal $0.000162 + 11.7 \times 10^{-6} \times \Delta T = 16.7 \times 10^{-6} \times \Delta T$

Which gives $\Delta T - 0.000162/(5.0 \times 10^{-6}) = 32.4$ C

The temperature would have to rise $20 + 32.4 = 53.4$ C

16.9 d. Flow area equals

$$\pi/4 \times 3^2 = 7.069 \text{ m}^2 \quad Q = 7.00 \text{ m}^3\text{/s} \quad V = 7.00/7.069 = 0.990 \text{ m/s}$$

Take the axis of the pipe into the bend as the X-axis. The force the water would exert on the bend would equal

$$\text{mass} \times v = (7.00 \times 1{,}000) \text{ kg/s} \times 0.990 \text{ m/s} = 6930 \text{ kg} \times \text{m/s}^2$$
$$= 6930 \text{ N into the bend}$$

The reaction forces on the discharge side of the bend would equal

$$6930 \cos 45° = 6930 \times 0.707 = 4900 \text{ N}$$

The vertical reaction force on the bend would equal

$$6930 \sin 45° = 4900 \text{ N upward, or in a positive direction}$$

The force that the elbow would exert on the support post, R, would equal

$$(4900^2 + 4900^2) = 6930 \text{ newtons; see Exhibit 16.9}$$

16.10 c. Treat the cable as a spring, k = N/meter. Energy absorbed by spring $= \frac{1}{2}k(\Delta L)^2$ where ΔL is the elongation of the spring due to the applied load. One metric ton equals 1000 kg.

Energy that must be absorbed by spring $= \frac{1}{2}mv^2 + (mg \times \Delta L)$

$$KE + \text{change in } PE$$

Next determine the spring constant of the cable

$$\text{Stress} = P/A \qquad A = 6.5 \text{ cm}^2 = 0.00065 \text{ m}^2$$

For one kg load,

$$\text{stress} = 1.00 \times \text{g}/0.00065 = 15.09 \text{ kPa}$$
$$\text{Stress} = E \times \varepsilon \quad \text{or} \quad \varepsilon = \text{S/E}$$

Change in length of the spring

$$1500 \times \varepsilon \text{ for a load of 1.0 kg or 9.807 N} \quad \varepsilon = 15{,}090/\text{E}$$
$$\Delta \text{L} = 1500\ \varepsilon$$

which equals $1500 \times 15{,}090/(84{,}000 \times 10^6) = 0.0002695\ m$

For mass of 1.0 kg, $k = 9.807 \times 1/0.0002695 = 36{,}390$ N/m, which equals the spring constant of the cable spring.

The KE of the hoist cage $= \frac{1}{2}mv^2 = \frac{1}{2} \times 2500 \times (32{,}000/3600)^2 = 98.76$ kJ or 98,760 Nm

$$98{,}760 + (2500 \times g \times \Delta L) = \tfrac{1}{2} \times 36{,}390 \times (\Delta L)^2 = 18{,}195\ (\Delta L)^2,$$

which reduces to

$$(\Delta L)^2 - 1.347\ \Delta L - 5.428 = 0$$

Using the general solution for a quadratic equation, see *Handbook*,

$$\Delta L = [1.347 \pm /(1.814 + 21{,}712)]/2 = (1.347 + 4.850)/2 = 3.099,$$

which gives that the elongation of the supporting cable would be 3.099 m. The strain, ε would equal $3.099/1500 = 0.002066$ m/m.

The stress would be $E \times \varepsilon = 84{,}000 \times 10^6 \times 0.002066 = 173.5 \times 10^6$ Pa or 173.5 MPa.

16.11 d. Call the height of the liquid above the discharge h. The velocity of efflux will then be $v = C/(2gh)$.

The rate at which the liquid drains from the tank would thus equal $a \times v$ or dQ/dt m³/s.

The rate at which the liquid drains from the tank would also equal the velocity of the surface of the liquid times the cross-sectional area of the tank or $dQ/dt = A \times dh/dt$ m³/s.

The height, h, is decreasing so dh/dt is negative. The two equivalent rates of flow can be set equal to one another

$$a \times v = -A \times dh/dt \quad \text{or} \quad a \times C/(2gh) = -A \times (dh/dt)$$

Rearranging terms gives

$$dt = -A/[a \times C \times /(2g)] \times h^{-1/2}dh$$

Integration gives

$$t = -2 \times h^{1/2} \times A/[a \times C \times /(2g)] + \text{constant of integration}$$

At $t = 0$, $H = L$, which gives constant $= 2AL^{1/2}/[a \times C\, /(2g)]$. The length of time to empty is the time required for h to reach zero.

For $A/a = 100$ and $L = 2.0$ meters,

$$C = 1.00 \quad t = 0 + \text{constant} \quad \text{or} \quad t = 2 \times 100 \times /2/(2g)$$
$$\text{or} \quad i = 200/g = 63.9 \text{ seconds}$$

16.12 b. A power of 3360 watts equals 3360 J/s or 3360 N×m/s. 1750 rpm = 29.17 rev/s.

Power $= 2\pi NT$ so these values correspond to a torque of 3360/(29.17 × 2π) = 18.33 N × m

The pins are at a distance of 10/2 = 5.0 cm from the axis of rotation (see Exhibit 16.12 or 0.050 meter).

The force acting on the pins equals 18.33/0.050 = 366.6 N, or 183.3 N/pin. The area of a pin equals $0.00635^2 \times \pi/4 = 31.67 \times 10^{-6}\ \text{m}^2$

$$\text{stress in pin} = 183.3/(31.67 \times 10^{-6}) = 5.79 \text{ MPa}$$

16.13 a. Using the data in Exhibit 16.13, Bernoulli's equation can be applied between the point at which the gauge is attached to the supply pipe and the discharge to solve for the discharge velocity.

$$P_1/\rho g + v_1^2/2g + z_1 + h_A = P_2/\rho g + v_2^2/2g + z_2 + h_L$$

At point (1) the pressure in the pipe will equal 276 kPa plus 0.305 m of water.

$$P_1/\rho g = 276{,}000/(1000 \times 9.807) + 0.305 = 28.45 \text{ meters of water}$$
$$P_2/\rho g = 0.61 \text{ meter of water}$$
$$P_1/\rho g - P_2/\rho g = 27.84 \text{ meters} \quad h_A \text{ and } h_L \text{ are zero} \quad z_1 \text{ minus } z_2 \text{ equals } 2.44 + 0.61 = 3.05 \text{ m}$$
$$v_1 = Q/0.00258 \text{ and } v_2 = Q/0.0013 \text{ so } v_1 = 0.5039\, v_2 \text{ and } v_1^2 = 0.254\, v_2^2$$

Combining values and substituting in Bernoulli's equation gives

$$27.84 \text{ m} + 3.05 \text{ m} = 0.746\, v_2^2/2g$$
$$27.84 \text{ m} + 3.05 \text{ m} = 0.746\, v_2^2/2g \quad V_1^2 = 41.41 \times 2 \times 9.807 = 812.22$$
$$V_1 = 28.49 \text{ m/s}$$

16.14 d. The overall heat transfer coefficient can be calculated from the relationship

$$1/U = 1/h_i + L/k + 1/h_o$$

So

$$1/U = 1/8.52 + 0.003/1.04 + 1/34.07 = 0.117 + 0.00288 + 0.0294 = 0.1493$$

$$\text{so } U = 6.698 \text{ W/m}^2 \times \text{K}$$

The area of the window equals

$$0.60 \times 0.75 = 0.45 \text{ m}^2$$

The heat loss, Q, would equal $6.698 \times 0.45 \times 25 = 75.35$ watts.

16.15 c. Label the three members F, G, and H as shown in the problem figure. It is assumed that all three wires are in tension. They will all stretch somewhat, and since they are all the same length, same cross-sectional area, and same material, i.e., are identical, the load held by each wire will be proportional to the amount it is stretched. Knowing that the bar is rigid, a deflection diagram can be drawn, see Exhibit 16.15b.

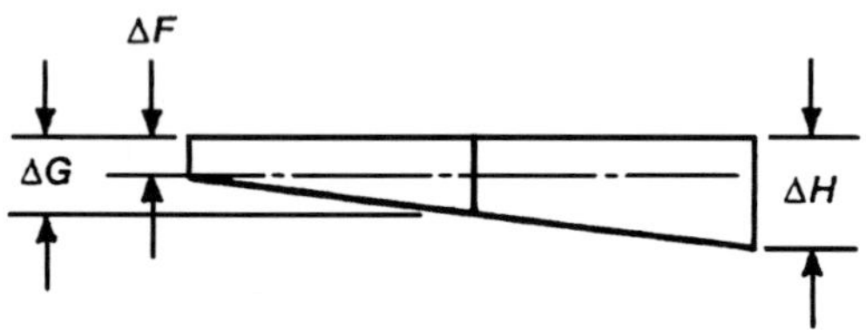

Exhibit 16.15b

Looking at the deflection diagram it is seen that—

$$\Delta G = \Delta F + D \quad \textbf{(16.1)}$$

and

$$\Delta H = \Delta F + 2D \quad \textbf{(16.2)}$$

In addition it is known that the force exerted by F,

$$F_F = \Delta F/L \times AE$$

also, similarly

$$F_G = \Delta G/L \times AE \quad \text{and} \quad F_H = \Delta H/L \times AE$$

summing the vertical forces and setting then equal to zero gives

$$P = AE/L(\Delta F + \Delta G + \Delta H) = AE/L(3\Delta F + 3D) \quad \textbf{(16.3)}$$

Taking moments about the point where wire F is connected to the bar gives

$$F_G \times a - 3/2a \times P \times 2a \times F_H = 0 \qquad F_G - 3/2P + 2 \times F_H = 0$$

Then

$$\Delta G/L \times AE + 2 \times \Delta H/L \times AE = 3/2 \times P \quad \text{and} \quad \Delta G + 2 \times \Delta H = 3/2 \times PL/AE \quad \textbf{(16.4)}$$

Taking moments about the point where G connects to the bar gives

$$F_H \times a - F_F \times a - P \times a/2 = 0 \quad \text{giving } \Delta H - \Delta F = PL/2AE \quad \textbf{(16.5)}$$

Solving Equations 16.1, 16.2, 16.3, 16.4, and 16.5 gives

$$3\Delta F = D$$

Substituting for D in Equation 16.3 gives

$$P = AE/L \times (12 \times \Delta F) \text{ since } F_F = (\Delta F/L) \times AE$$

Then

$$F_F = P/12 \; F_G = P/3 \quad \text{and} \quad F_H = 7P/12$$

16.16 a. The combustion equation would be of the following form

$$C_xH_y + O_2 + 84.2\ N_2® \ 11.3\ CO_2 + 4.50\ O_2 + 84.2\ N_2$$

The nitrogen does not burn, so the amount of pure nitrogen would be the same after combustion as before. Air is approximately 79% nitrogen and 21% oxygen by volume. The volumetric ratio of nitrogen to oxygen is thus 0.79/0.21 = 3.76 so that ratio would have existed in the air supplied. Thus there would have been 84.2/3.76 = 22.39 moles of oxygen supplied for each 100 moles of air. (For the purpose of calculation 100 moles of stack gas is considered.)

Since only 4.5 moles of uncombined oxygen is in the stack gas and 11.3 moles is accounted for in the carbon dioxide 22.39 − 4.5 − 11.3 = 6.59 moles of oxygen is unaccounted for and would have existed in the form of water. Water is H_2O so the molecular weight of water is 2 + 16 = 18 (see the *Handbook*) and since there would have been 2 × 6.59 = 13.18 moles of atomic oxygen supplied for the combustion, there would be 13.18 moles of water vapor in the stack gas. There would then have been 26.36 moles of hydrogen in the fuel that was burned, assuming complete combustion.

This means that 11.3 moles of carbon would have combined with 26.36 moles of hydrogen, or the ratio of hydrogen in the fuel to carbon, H/C = 26.36/11.3 = 2.33. So $(2n + 2)/n = 2.33$ (This is a form of a Diophantine equation; that is, an equation whose solution is in integers). The possibilities are few (only four) so it can be easily solved. For $n = 2$ we get 6/2 = 3.33. For $n = 4$ we get 10/4 = 2.5. For $n = 5$, 12/5 = 2.4. For $n = 6$ 14/6 = 2.33, so the answer is C_6H_{14}.

16.17 b. There were 84.2/3.76 = 22.39 moles of oxygen in the input air. The combustion equation can be written as $C_{11.3}H_{26.36} + O_2 \rightarrow 11.3\ CO_2 + 13.18\ H_2O$ so 11.3 + 13.18/2 = 17.89 moles of oxygen were required for combustion.

$$22.39/17.89 = 1.25 \quad 25\% \text{ excess air}$$

16.18 a. The area of piston equals $1.6^2 \times \pi/4 = 2.01\ cm^2 = 2.01 \times 10^{-4}\ m^2$. The force acting on the piston would equal 15 × 135 = 2025 N, so the pressure under the ram would equal

$$2025/(2.01 \times 10^{-4}) = 10.07\ MPa$$

The force exerted by the ram would thus equal

$$10{,}070{,}000 \times (3.8^2 \times \pi/4 \times 10^{-4}\ m^2) = 11.4\ kN = 1162\ kg$$

16.19 b. The tape increases 99.999 − 99.992 = 0.007 meters in length with an increase of 6 °C in temperature. It would, then, increase (7/6) ×

0.007 = 0.0082 meters for an additional increase in temperature of 7 °C. At the measuring conditions the tape would actually measure a distance of only 100 meters times 100/(99.999 + 0.0082) = 100/100.0072 = 0.999928 or 99.9928 meters when 100.000 meters was indicated. As a result, 0.0072 meters would have to be added to every tape length to obtain an accurate measurement. Thus 3.2 × 0.0072 = 0.023 meters would have to be added to the indicated length of the tape to obtain the desired dimension, meaning that the tape would have to indicate a distance of 320 + 0.023 = 320.023 meters.

16.20 d. The root area equals

$$1.6^2 \times \pi/4 = 2.011\ \text{cm}^2 = 2.011 \times 10^{-4}\ \text{m}^2$$

The permissible load thus equals

$$2.011 \times 10^{-4}\ \text{m}^2 \times 344.8 \times 10^6\ \text{Pa} = 69{,}339\ \text{kN}$$

which corresponds to 69,339/g = 7069 kg.

16.21 c. The body shows an apparent mass of 25 kg when weighed in fresh water, so it displaces 75 kg of fresh water. It thus has a volume of 75/1000 = 0.075 m^3 and has a density of

$$100\ \text{kg}/0.075 = 1333\ \text{kg/m}^3$$

so its specific gravity equals 1333/1000 = 1.33.

16.22 a. From the *Handbook* we see that $\phi = TL/GJ$ or torsional stiffness = $T/\phi = GJ/L$, where J equals the polar moment of inertia and G is the shear modulus. Since G and L will be the same, the torsional stiffness in this case will be proportional to the polar moment of inertia. From the table in the *Handbook*, taking the outer radius as a and the radius of the hole as b, we find that for the solid bar

$$J = \pi \times 7.5^4/2 = \pi \times 3164/2 = 4970\ \text{cm}^4$$

The J of the hollow bar must be the same for the torsional stiffness to be the same.

$$J = \pi(a^4 - b^4)/2 = 4970 \qquad 4970 \times 2/\pi = a^4 - 5^4$$
$$a = [3164 + 625]^{1/4} = 7.846$$

The outside diameter of the hollow torsion bar would thus equal 15.692 cm.

16.23 d. The air is compressed rapidly in an insulated cylinder so no heat is transferred during the process, thus the process is adiabatic and PV^k = constant

$$P_2 = P_1 \times (V_1/V_2)^{1.4} \text{ since } k \text{ for air equals } 1.4$$
$$V_1/V_2 = 4 \qquad P_1 = 13.79 + 101.3 = 115.09\ \text{kPa absolute}$$
$$P_2 = 115.09 \times 4^{1.4} = 115.09 \times 6.964$$
$$P_2 = 801.5\ \text{kPa absolute or } 700.2\ \text{kPa gauge}$$

16.24 b. The kinetic energy for a rotating mass equals $\frac{1}{2} \times I \times \omega^2$. From the *Handbook* the polar moment of inertia about an axis through its center for a rotating cylinder equals $MR^2/2$

$$I \text{ for the flywheel thus equals } 1460 \times 0.75^2/2 = 410.6$$
$$\omega = (600/60) \times 2\pi = 62.83 \text{ radians/s}$$
$$I = \tfrac{1}{2} \times I \times \omega^2 = 0.5 \times 410.6 \times 62.83^2 = 810{,}444 \text{ or } 810 \text{ kJ}$$

16.25 c. The rate of heat transfer through the slab will equal

$$k \times \Delta T/L = \text{watts/m}^2$$

k is given as 1.039 W/m × K

$$\Delta T = 77 - 20 = 57\,°\text{C} \quad L = 0.15 \text{ m} \quad Q = 1.039 \times 57/0.15 = 394.8 \text{ W/m}^2$$

Notes

Notes

Notes

Notes

Notes

Notes

Notes

Notes